4th International Conference on Mechanical Engineering Research (ICMER 2017)

IOP Conference Series: Materials Science and Engineering
Volume 257

Kuantan, Malaysia
1 – 2 August 2017

ISBN: 978-1-5108-5113-9
ISSN: 1757-8981

Printed from e-media with permission by:

Curran Associates, Inc.
57 Morehouse Lane
Red Hook, NY 12571

Printed by Curran Associates, Inc. (2017)

Institute of Physics
Dirac House, Temple Back
Bristol BS1 6BE UK

Phone: 44 1 17 929 7481
Fax: 44 1 17 920 0979

techtracking@iop.org

Additional copies of this publication are available from:

Curran Associates, Inc.
57 Morehouse Lane
Red Hook, NY 12571 USA
Phone: 845-758-0400
Fax: 845-758-2633
Email: curran@proceedings.com
Web: www.proceedings.com

Table of contents

Volume 257

4th International Conference on Mechanical Engineering Research (ICMER2017)
1–2 August 2017, Swiss Garden Beach Resort, Kuantan, Pahang, Malaysia

Accepted papers received: 11 October 2017
Published online: 1 November 2017

Preface

Papers

IOP Conf. Series: Materials Science and Engineering **257** (2017) 011001 doi:10.1088/1757-899X/257/1/011001

Preface The 4th ICMER2017

The 4th International Conference on Mechanical Engineering Research (ICMER2017) is the continuity of the National Conference on Mechanical Engineering Research (NCMER2010). The year 2010 represents a significant milestone in the history for Faculty of Mechanical Engineering, Universiti Malaysia Pahang (UMP) Malaysia with the organization of the first and second national level conferences (1st and 2nd NCMER) at UMP on May 26-27 and Dec 3-4 2010. The Faculty then has changed the name of National Conference on Mechanical Engineering Research (NCMER) to International Conference on Mechanical Engineering Research (ICMER) in 2011 and this year, 2017 is our 4th ICMER.

These proceedings contain the selected scientific manuscripts submitted to the conference. It is with great pleasure to welcome you to the "International Conference on Mechanical Engineering Research ICMER2017" that is held at Swiss Garden Hotel and Spa, Kuantan, Malaysia. The call for papers attracted submissions of over one hundred abstracts from 10 different countries such as Algeria, Indonesia, Germany, Japan, Australia, India, Qatar, Morocco, China and Iran and many more.

The scientific papers published in these proceedings have been revised and approved by the technical committee of the 4th ICMER2017. All of the papers exhibit clear, concise, and precise expositions that appeal to a broad international readership interested in mechanical engineering, combustion, metallurgy, materials science as well as in manufacturing and biomechanics. The reports present original ideas or results of general significance supported by clear reasoning and compelling evidence, and employ methods, theories and practices relevant to the research. The authors state clearly the questions and the significance of their research to theory and practice, describe how the research contributes to new knowledge, and provide tables and figures that meaningfully add to the narrative. In this edition of ICMER representatives coming are from academia, industry, governmental and private sectors. The plenary and invited speakers present, discuss, promote, and disseminate research in all fields of mechanical engineering in a constructive manner. Numerous social events that provide opportunities for the participants to renew old contacts and establish new ones were scheduled to accompany and follow the scientific activities.

IOP Conf. Series: Materials Science and Engineering **257** (2017) 011001 doi:10.1088/1757-899X/257/1/011001

They also aim to acquaint the participants with the capital of state Pahang, Kuantan. In addition, booths were also prepared for the industries to showcase their state-of-the-art products.

The organizing committee of the conference thanks all the participants for their fruitful work and personal contribution to the development of these conference proceedings.

Professor Dr. Rizalman Mamat
The Chairman of ICMER2017

IOP Conf. Series: Materials Science and Engineering **257** (2017) 011002 doi:10.1088/1757-899X/257/1/011002

Peer review statement

All papers published in this volume of *IOP Conference Series: Materials Science and Engineering* have been peer reviewed through processes administered by the proceedings Editors. Reviews were conducted by expert referees to the professional and scientific standards expected of a proceedings journal published by IOP Publishing.

The use of an improved technique to reduce the variability of output voltage in real-time Fibre Bragg Grating based monitoring system

E Vorathin[1*], Z M Hafizi[1], S A Che Ghani[2], K S Lim[3], A M Aizzuddin[1]

[1]Advanced Structural Integrity and Vibration Research (ASIVR), Faculty of Mechanical Engineering, University Malaysia Pahang (UMP), Pekan 26600 Pahang, Malaysia
[2]Human Engineering Group (HEG), Faculty of Mechanical Engineering, University Malaysia Pahang (UMP), Pekan 26600 Pahang, Malaysia
[3]Photonics Research Centre, Faculty of Science, University of Malaya, 50603 Kuala Lumpur, Malaysia

*Corresponding author: vora.91.11@hotmail.com

Abstract. Fibre Bragg Grating (FBG) sensors have been widely utilized in the structural health monitoring (SHM) of structures. However, one of the main challenges of FBGs is the existence of inconsistency in output voltage during wavelength intensity demodulation utilizing photodetector (PD) to convert the light signal into digital voltage readings. Thus, the designation of this experimental work is to develop a robust FBG real-time monitoring system with the benefit of MATLAB graphical user interface (GUI) and voltage normalization algorithm to scale down the voltage inconsistency. Low-cost edge filter interrogation system has been practiced in the experimentation and splitter optical component is make use to reduce the intensity of the high power light source that leads to the formation of noise due to unwanted reflected wavelengths. The results revealed that with the advancement of the proposed monitoring system, the sensitivity of the FBG has been increased from 2.4 mV/N to 3.8 mV/N across the range of 50 N. The redundancy in output voltage variation data points has been reduced from 26 data/minute to 17 data/minute. The accuracy of the FBG in detecting the load induced falls in the acceptable range of total average error which is 1.38 %.

1. Introduction

It is well apprehended that FBG sensors are far more superior in providing closer look upon defects compared to conventional Non-destructive testing (NDT) inspections. Most of the conventional NDT inspections are limited by the bulky size of the whole system making it impossible to be integrated with the structure as a whole. Furthermore, these inspections provide an off-line monitoring on the structures. Thus, specialized characteristics such as small physical size, immune to electromagnetic interference, light weight, relative signal stability and wavelength multiplexity [1-4] have turned researchers to make use of the FBGs in SHM of structures, especially in composite materials. However, much endeavour is still needed to bring the FBGs real-time monitoring system to a fully mature readiness level. One of the main challenges in converting the wavelength intensity demodulation utilizing PD is the existence of inconsistency in output voltage readings that could precede in error of desired readings and difficulty in value calibration. This complication has been encountered by the authors of this paper as well as other researchers [5-7] who make use of the photodetector to alter the intensity demodulation to digital voltage readings. For instance, work by Allwood et al [6] has shown a significant variation of the voltage output which leads to errors in the measurement as shown in figure 1. High power from the light source is one of the reasons which outcome in a high intensity of the reflected wavelength in which unwanted wavelengths will present in the spectrum that could lead to wrong mismatching condition and affect the output voltage consistency. Implementation of voltage normalization algorithm will be the proposed problem-solver to reduce the voltage inconsistency.

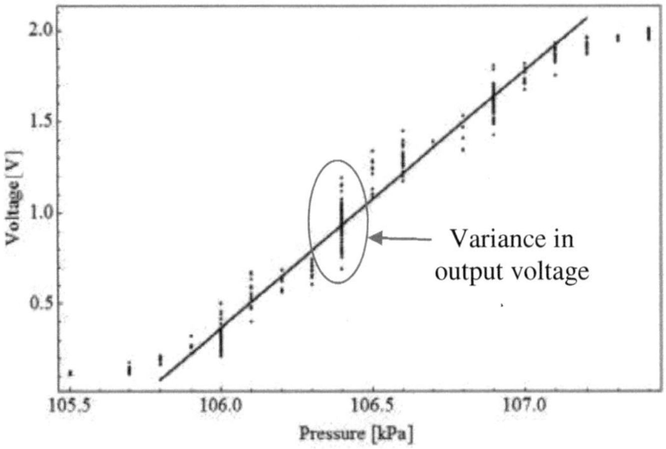

Figure 1. Variance in output voltage when photodetector was utilized for intensity to voltage conversion [6].

Normalization is a statistical analysis to scale down the redundancies of data in a database in which it is a technique of designing and redesigning a database [8]. Researches predominantly in social science [9-12] implement normalization in data analysis to reduce the variability of the desired output. Researchers in the field of FBGs [13, 14] have also employed this mathematical algorithm in normalized out the optical power or intensity values, however, the utilization of expensive signals precision devices such as optical interrogation unit and logging of data for post-processing are still the gap to be amended. Thus in this study, several configurations of interrogation system will be experimentally performed to demonstrate the root of the voltage inconsistency and at the end, a solution will be proposed.

2. Fibre Bragg Grating (FBG) sensor instrumentation and working principle

FBG is an optical fibre with laser inscribed in the core region known as Bragg grating. The grating reflects the wavelength as shown in figure 2 of the emitted light known as Bragg wavelength (λ_B) in which concerning with the grating period (Λ_o) formulate as [15]:

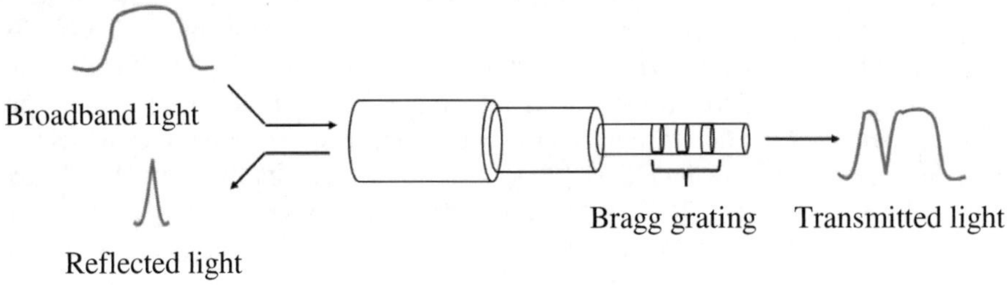

Figure 2. The working principle of FBG sensor.

$$\boldsymbol{\lambda_B = 2\eta_o\Lambda_o/\kappa} \tag{1}$$

where κ is the gratings order and η_o is the initial refractive index before subjected to strain. The Bragg wavelength also can be defined in term of refractive index of the grating (η_{eff}) as:

$$\lambda_B = 2\eta_{eff}\Lambda_o \tag{2}$$

Bragg wavelength shifts also can be modulated due to change of temperature, ΔT, and strain, ε can be denoted as [16, 17]:

$$\frac{\Delta\lambda}{\lambda_B} = (\hat{a} + \xi)\Delta T + (1 - p_e)\varepsilon \tag{3}$$

where $\Delta\lambda$ is the change in wavelength, â is the thermal expansion, ξ is thermo-optic coefficient and p_e is the effective photo-elastic constant of the fibre. However, for an isothermal condition the change of temperature can be neglected which simplified the expression as [17]:

$$\frac{\Delta\lambda}{\lambda_B} = (1 - p_e)\varepsilon \tag{4}$$

2.1. Experimental procedure and results

40 cm (length) x 40 cm (width) x 0.8 cm (thickness) of ten layers woven fibreglass composite plate was fabricated by the method of hand lay-up and cured by applying glycidyl (GL) epoxy and hardener as resin. A single 1544 nm FBG sensor with 99.9 % reflectivity was embedded at the centre between the ninth and tenth layers of the composites plate. The composite plate was clamped fixed at four edges and a weight support stand was positioned at the centre of the plate for equivalent and constant load distribution. The set-up was shown in figure 3a below. A 10 N and two pieces of 20 N known loads as in figure 3b were induced repeatedly on the composite plate starting from 0 N to 50 N with the increment of every 10 N. The experimental set-up undergoes three configurations and manipulation in interrogation system as detailed in figure 4.

Three different configurations of optical components namely optical circulator (15-PICIR- 3-SCL-1) - SET A, 50:50 intensity ratio optical coupler (CS-5250-S115-03) - SET B, and 1:8 intensity ratio optical splitter (OFC-CP1xN-034C) - SET C have been experimentally tested to determine the effect of high power ASE broadband light source on the reflected sensing FBG wavelength. Bayspec (FBGA-F-1525-1565-FP) optical spectrum analyzer (OSA) was used to view reflected spectrum.

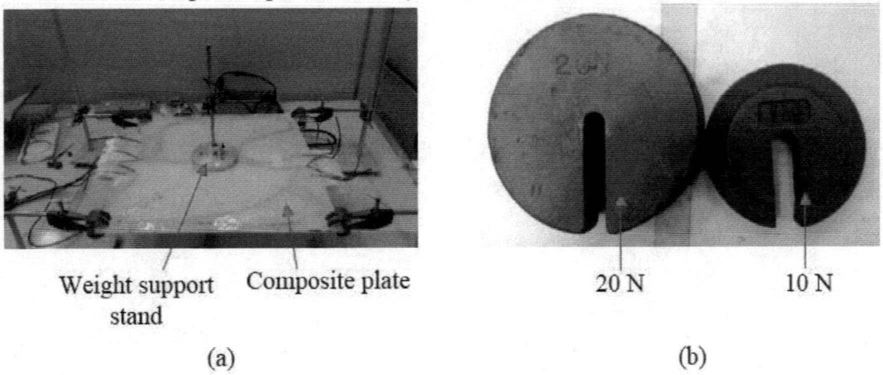

Figure 3. Experimental set-up: (a) Fixture of the specimen, (b) Loads used to induce on the plate.

Figure 5 shows the recorded and analyzed reflected wavelengths of the sensing FBG when no load induced from three different optical components namely SET A, SET B, SET C and also the overlaid graphs of the three components. Reflected wavelength for SET A shows the most severe wavelength distortions in which four unwanted wavelengths are present as shown in figure 5(a). Existent of peak splitting is also present in the spectrum and SET A shows the highest intensity recorded. 10.78% of intensity was reduced when SET B was adopted and the unwanted wavelengths were reduced to two as shown figure 5(b). SET C shows the most intensity reduction in which 66.47% of intensity has been reduced as shown in figure 5(c) compared to SET A and an ideal single Bragg wavelength was obtained. From the overlay of all the reflected spectrum as shown in figure 5(d), clearly shows that different utilization of optical components result in different power intensity.

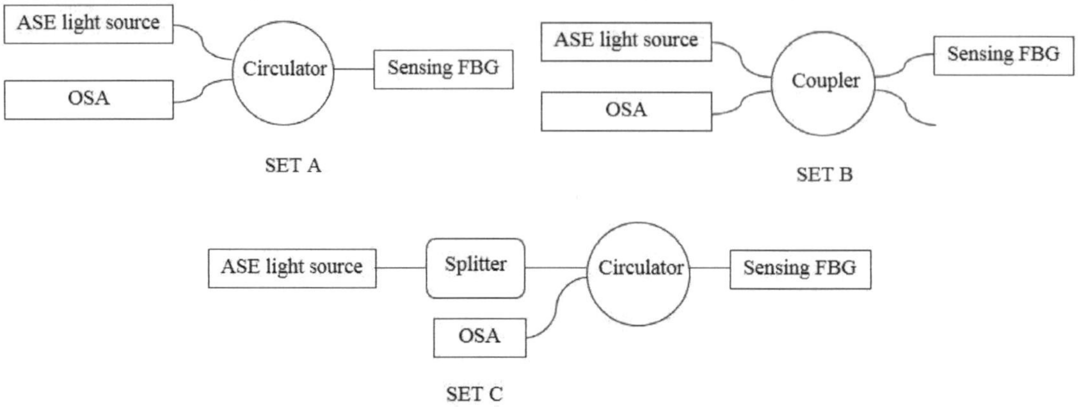

Figure 4. Configurations of the experimental set-up interrogation system.

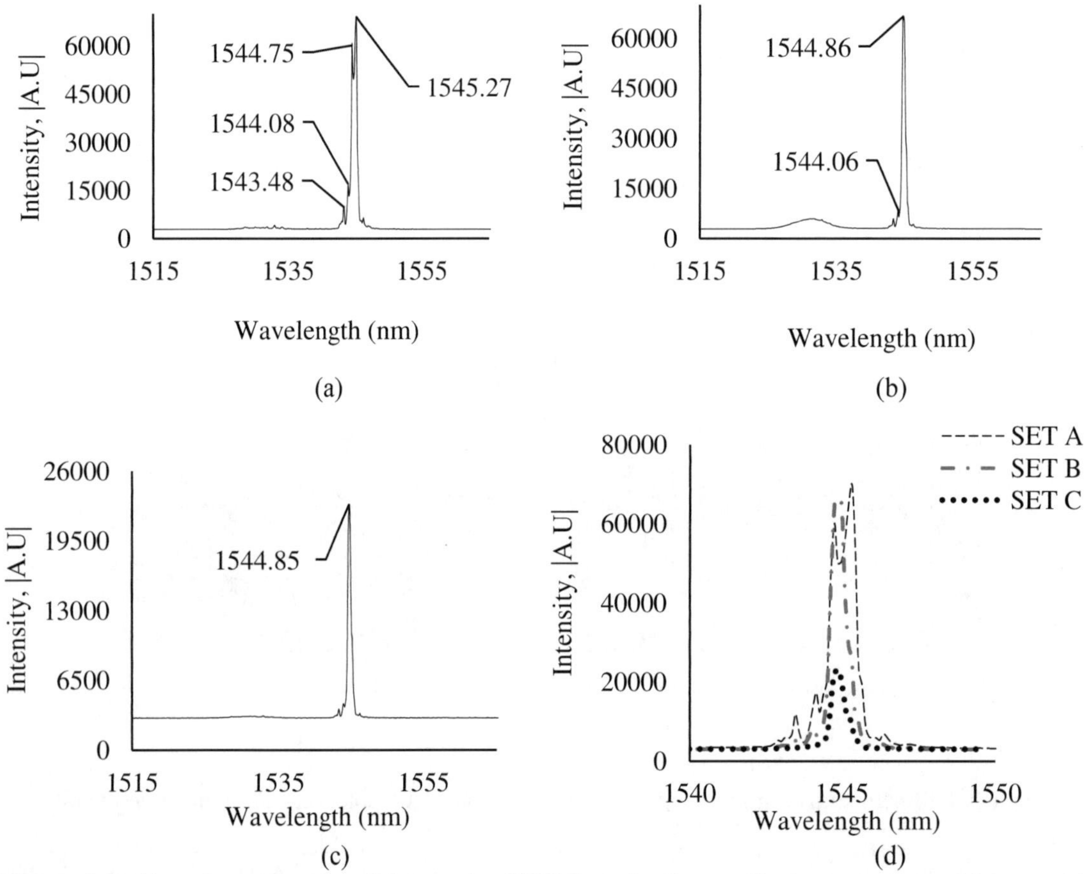

Figure 5. Reflected wavelengths of the sensing FBG from the three optical components which are: (a) SET A, (b) SET B, (c) SET C and (d) overlaid of the three.

3. Theoretical and experimental of edge filter detection due to static strain

Interrogation system is the working system used to illuminate the FBG sensors. There are two types of interrogation system namely power detection and edge filter detection [18]. In general, power detection interrogation benefits in dynamic strain detection due to the use of laser types light sources such as linear edge source [19] and narrow bandwidth source [20]. Edge filter detection is the most economical and simplest among the interrogation system [21]. Edge filter detection utilized the use of broadband light source and employed a reference FBG to mismatch with the sensing FBG. Edge filter detection interrogation system has been reported earlier than the year 2000s, however, direct output voltage without normalization is being measured utilizing photodetector making this low-cost interrogation system non-stable due to the fluctuation of output voltages.

A PD is commonly practiced to read the demodulation shift of the reflected intensity into an analog voltage signal. Figure 6 below shows the working principle of the photodetector. The wavelength shift of the sensing FBG as labelled A will result in intensity variation when sensing FBG slide across the slope of the reference FBG as labelled B. As a sequel, the photodetector will transform the intensity variation into a change of analog voltage signal as labelled C.

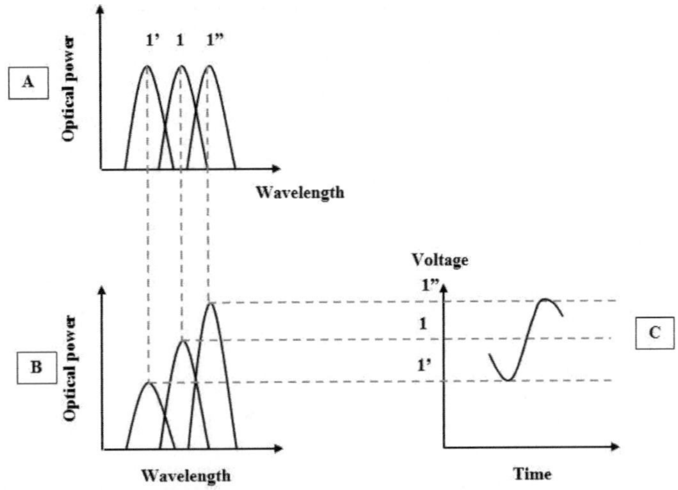

Figure 6. The working principle of photodetector [22].

3.1. Experimental procedure and results

The same experimental set-up as shown in figure 3 was maintained and adopted. Edge filter detection interrogation system configuration as shown in figure 7 was used for the experimentation.

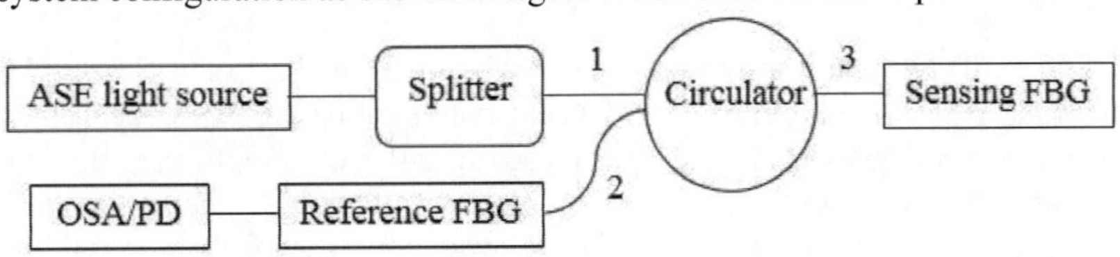

Figure 7. The configuration of edge filter detection interrogation system.

Figure 8 shows the reflected wavelength from sensing and reference FBG when both the FBGs were connected to port 3 and viewed from OSA connected to port 2. Reflected spectrum from the reference FBG is positioned to the left of the sensing FBG prior to measurement. It is also observed that the sensing FBG has slightly higher intensity than the reference FBG of about 3.76%. The reference FBG was then connected to port 2 and sensing FBG remained at port 3, figure 9 shows the recorded reflected wavelength and intensity from the mismatched between sensing and reference FBG when no load exerted.

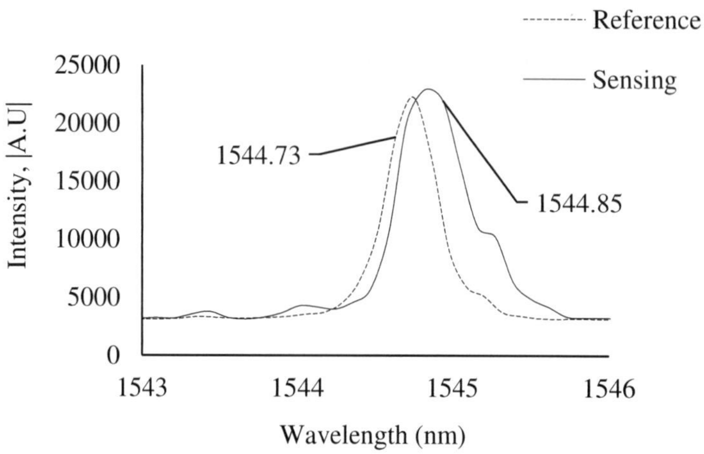

Figure 8. Reflected wavelength spectrum from sensing and reference FBG.

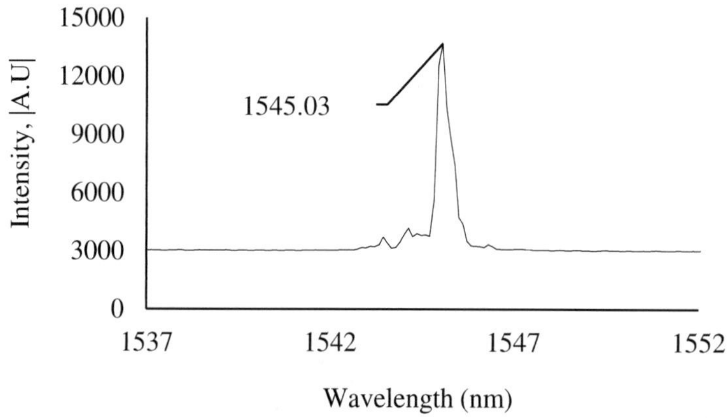

Figure 9. Reflected wavelength and intensity from the mismatched between sensing and reference FBG.

As load from 0 N to 50 N started to induce on the specimen, the reflected mismatched wavelength started to shift to the right with a constant and linear pattern. For a clear viewing in the shifting of reflected mismatched peak wavelength and increase of intensity amplitude, figure 10 shows the simplified Gaussian Bragg wavelength shift with an offset of 0.1 nm/N. The spectrum shift shows that when the load was induced from 0 N to 50 N, the peak wavelength shift from initial of 1545.03 nm to 1545.07 nm and an increasing intensity waveform was recorded. The maximum intensity recorded at 50 N of the load which is 17068 while the minimum intensity was at 0 N which is 13880.40, as expected.

Lastly, the intensity shift was converted to voltage readings with the replacement of OSA with PD. Five repetitions of voltage response against applied load were recorded and calibration to convert the voltage response to load induced were carried out. Initially, the calibration was formulated by averaging the linearity curves and without the voltage normalization algorithm as in figure 11. MATLAB real-time GUI with the substitution of linearity equation obtained was used to record the voltage response variation as the plate was induced with load as shown in figure 12. The data points correlate to the recorded voltage variations and the solid line is the line of best fit where the sensitivity is 2.4 mV/N. The dashed line is the expected voltage due to applied load calculated from the linearity equation.

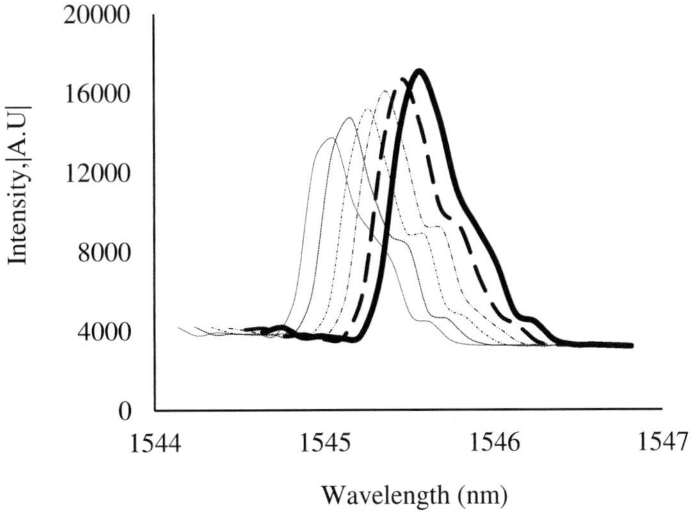

Figure 10. Reflected mismatched Bragg wavelength and intensity shift from 0 N to 50 N. The increase in line's weight correlate to the increase in loadings.

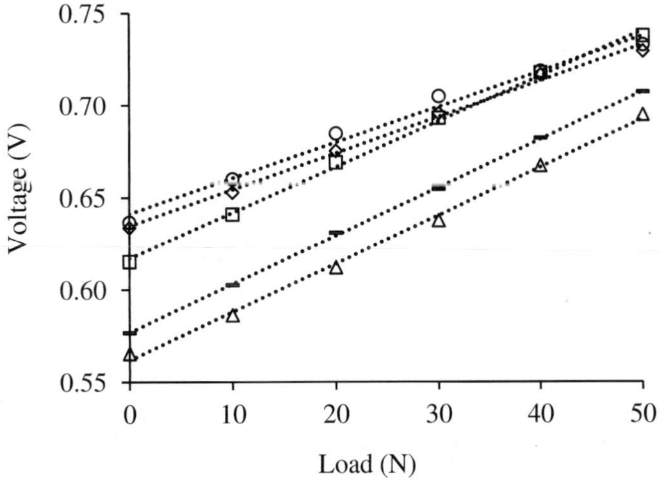

Figure 11. The linearity curves of voltage response against applied load without normalization.

Next, with the same voltage response the calibration was formulated by averaging the linearity curves and ratio scale voltage normalization algorithm was implemented as shown in figure 13. The ratio voltage normalization, $\boldsymbol{V_{norm}}$ can be expressed as in Equation 5.

$$\boldsymbol{V_{norm}} = \frac{\boldsymbol{V_s}}{\boldsymbol{V_r}} \tag{5}$$

where $\boldsymbol{V_s}$ is the voltage reading from the FBG sensor subjected to strain. Reference voltage, $\boldsymbol{V_r}$ is the reference voltage which is the initial voltage from the light source. Substituting Equation 5 into averaging linearity equation obtained from figure 13 can be expressed as in Equation 6.

$$\frac{\boldsymbol{V_s}}{\boldsymbol{V_r}} = \boldsymbol{m_{avg}\chi + C_{avg}} \tag{6}$$

where m_{avg} is the average gradient of the five voltage response, χ is the independent variable load induced and C_{avg} is the average y-intercept.

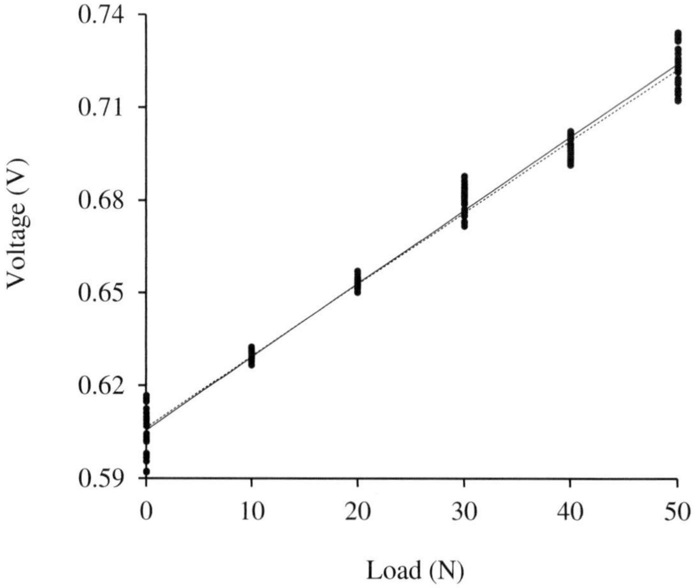

Figure 12. Output voltage variations as a function of applied load without normalization.

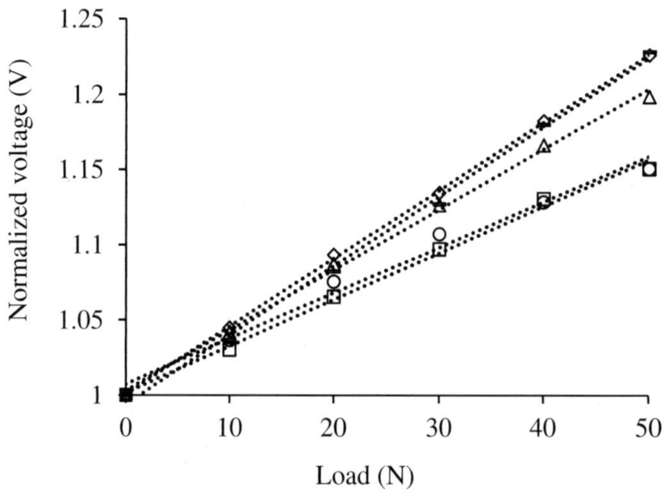

Figure 13. The linearity curves of voltage response against applied load with voltage normalization.

MATLAB real-time GUI with the substitution of voltage normalization linearity equation obtained was used to record the voltage response variation as shown in figure 14. A delay of 1 second has been programmed in the real-time GUI to reduce the data response. The data points correlate to the recorded voltage variations and the solid line is the line of best fit where the sensitivity is 3.8 mV/N. The dashed line is the expected voltage due to applied load calculated from the linearity equation. Table 1 summarizes the results of the range of load detected using non-normalization method obtained from the voltage variations in figure 12. The highest average percentage of error is 2.19% which is during 10 N load induction. The total average percentage of error is 5.35%. Table 2 summarizes the results of the range of load detected using normalization method obtained from the voltage variations in figure 14. The highest percentage of error is 1.65% which is during 10 N load induction. The total average percentage of error is 1.38%.

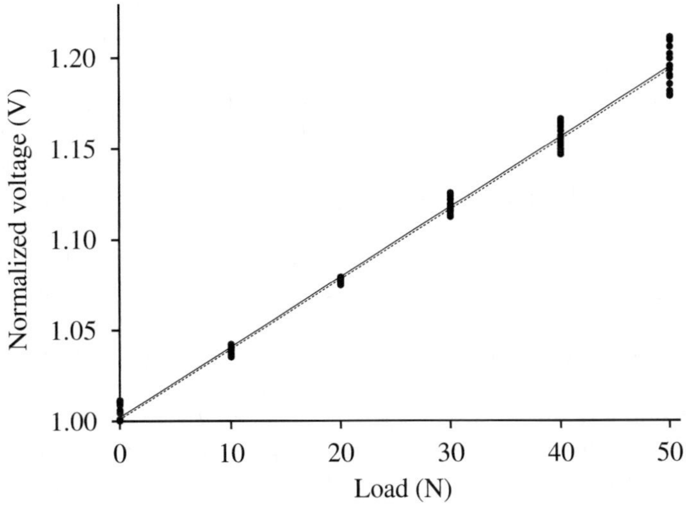

Figure 14. Output voltage variations as a function of applied load with voltage normalization.

Table 1. Range of load detected and percentage of error from the voltage variations obtained in figure 12 for non-normalization method.

Load (N)	Average load estimated (N)	Average percentage of error (%)
10	11.22	12.19
20	18.77	6.14
30	30.16	0.55
40	41.82	4.54
50	48.34	3.32
	Average percentage of error	5.35

Table 2. Range of load detected and percentage of error from the voltage variations obtained in figure 14 for normalization method.

Load (N)	Average load estimated (N)	Average percentage of error (%)
10	9.84	1.65
20	19.76	1.22
30	30.63	2.12
40	40.41	1.02
50	50.44	0.88
	Average percentage of error	1.38

4. Discussion

Optimization of optical circulator as a component in optical signal interrogation system resulted in high optical intensity. This is due to the working principle of the circulator itself in which its function is to circulate the full intensity of the light signal from the light source in one direction from port to port. In 50:50 optical coupler, the light entering as input is split into two 50% equally divided outputs. However in this experimentation, due to the high power of the ASE light source, only 10.78% of intensity was reduced. The high power from the light source resulted in the present of unwanted wavelengths that consequence in noise and a huge range of voltage variations during mismatching with reference FBG. 1:8 splitter has the same function as an optical coupler in which the input is split into eight equally divided ratio outputs. In the experimentation, 66.4% of intensity has been reduced and an ideal reflected Bragg wavelength was achieved.

Splitter configuration has been used throughout the experimentation since it is capable of reducing the intensity and eliminate possible noise. However, a well-built real-time monitoring system with the systematic algorithm is still necessary to attain the accurate or close to exact desired values. Replacement of OSA with PD and the capitalization of edge filter interrogation system authenticate that the low-cost interrogation system eradicates the use of expensive interrogation devices and at the same demonstrate that the interrogation system adequate with any low-cost electronic data acquisition devices such as NI-DAQ or programmable logic controller (PLC). From the results of the experimentation, it is noted that when the Bragg wavelength of the sensing FBG is adjusted to the right of the filter Bragg wavelength, the PD electrical voltage output will increase as the Bragg grating elongated.

Ratio normalization of voltage response decent in organizing a cluttered and scattered data as in figure 11 to a well-organized and correlate statistical data as in figure 13. The time delay coded in MATLAB GUI has reduced the voltage data points from 26 data/minute (figure 12) to 17 data/minute (figure 14) hence restraint the range of voltage variations from redundancy. The sensitivity of the plate has increased excessively to 3.8 mV/N after normalization. Normalization method also shows great accuracy against load induced in which the total average percentage of error is only 1.38% which falls in the acceptable range of errors. Although the total average percentage of error between non-normalization method and normalization method do not have much difference. However, the greatest dominance of voltage normalization is the cut-out of pre-calibration each time before operation since the voltage normalization is expressed in the ratio of dependent voltage due to load induced over initial no load voltage readings. Non-normalization linearity curve is only accurate at once and needs to calibrate each time before use since the output voltage is varied and not consistent.

5. Conclusions

Three optical components namely optical circulator, optical coupler, and splitter have been experimentally tested and results show that splitter which has the most intensity reduction obtained an ideal reflected Bragg wavelength. The practice of low-cost edge filter interrogation system with photodetector gives an upper-hand to the low-cost interrogation system to be compatible with electronic devices. However, the utilization of photodetector to convert the intensity shift to voltage readings resulted in the inconsistency of voltage readings. Thus, the implementation of robust MATLAB GUI real-time monitoring system with voltage normalization algorithm shows that the sensitivity has increased to 3.8 mV/N and the total percentage of average error is only 1.38%. The employment of the low-cost edge filter interrogation system has eliminated the use of expensive and bulky size interrogation devices while the implement of MATLAB GUI real-time monitoring system with voltage normalization has put the FBG monitoring system into a more mature readiness level.

Acknowledgments

The authors would like to thank the Faculty of Mechanical Engineering, University Malaysia Pahang for providing laboratory facilities and financial support. They would also like to thank the Photonics Research Centre University Malaya (PRCUM), for their support in fabricating the FBGs. Finally, special thanks to the Malaysian Ministry of Higher Education for providing the FRGS grant RDU160136.

References

[1] Pereira G, Frias C, Faria H, Frazao O and Marques A T 2013 On the improvement of strain measurements with FBG sensors embedded in unidirectional composites *Polymer Testing* **32** 99-105

[2] Dai Y, Li P, Liu Y, Asundi A and Leng J 2014 Integrated real-time monitoring system for strain/temperature distribution based on simultaneous wavelength and time division multiplexing technique *Optics and Lasers in Engineering* **59** 19-24
[3] Zou H, Liang D and Zeng J 2012 Dynamic strain measurement using two wavelength matched fiber Bragg grating sensors interrogated by a cascaded long period fiber grating *Optics and Lasers in Engineering* **50** 199-203
[4] Ling H Y, Lau K T, Cheng L and Jin W 2006 Viability of using an embedded FBG sensor in a composite structure for dynamic strain measurement *Measurement* **39** 328-34
[5] Ma, Wang X, Ma Y, Zhou P and Liu Z 2014 Analysis of multi-wavelength active coherent polarization beam combining system *Optics Express* **22** 16538-51
[6] Allwood G, Wild G, Lubansky A and Hinckley S 2015 A highly sensitive fiber Bragg grating diaphragm pressure transducer *Optical Fiber Technology* **25** 25-32
[7] Ma, Zhao M, Huang X, Bae H, Chen Y and Yu M 2016 Low cost, high performance white-light fiber-optic hydrophone system with a trackable working point *Optics Express* **24** 19008-19
[8] Stephens R, Plew R and Jones A D 2008 *Sams teach yourself SQL in 24 hours*: Pearson Education)
[9] Ioffe S and Szegedy C 2015 Batch normalization: Accelerating deep network training by reducing internal covariate shift *arXiv preprint arXiv:1502.03167*
[10] Pelz C R, Molly K M, Bagby G and Sears R C 2008 Global rank invariant set normalization (GRSN) to reduce systematic distortions in microarray data *BMC bioinformatics* **9** 1-
[11] Sboner A, Karpikov A, Chen G, Smith M, Dawn M, Freeman L C, Schweitzer B and Gerstein M B 2009 Robust linear model normalization to reduce technical variability in functional protein microarrays *Journal of proteome research* **8** 5451-64
[12] Wang L and Wei Y 2016 Revised normalized difference nitrogen index (NDNI) for estimating canopy nitrogen concentration in wetlands *Optik - International Journal for Light and Electron Optics* **127** 7676-88
[13] Fusiek G, Orr P and Niewczas P Temperature-independent high-speed distributed voltage measurement using intensiometric FBG interrogation. In: *2015 IEEE International Instrumentation and Measurement Technology Conference (I2MTC) Proceedings,* pp 1430-3
[14] Kanellos G T, Papaioannou G, Tsiokos D, Mitrogiannis C, Nianios G and Pleros N 2010 Two dimensional polymer-embedded quasi-distributed FBG pressure sensor for biomedical applications *Optics Express* **18** 179-86
[15] Kashyap R 1999 *Fiber bragg gratings*: Academic press)
[16] Park J M, Lee S I, Kwon O Y, Choi H S and Lee J H 2003 Comparison of nondestructive microfailure evaluation of fiber-optic Bragg grating and acoustic emission piezoelectric sensors using fragmentation test *Composites Part A: Applied Science and Manufacturing* **34** 203-16
[17] Majumder M, Gangopadhyay T K, Chakraborty A K, Dasgupta K and Bhattacharya D K 2008 Fibre Bragg gratings in structural health monitoring - Present status and applications *Sensors and Actuators A: Physical* **147** 150-64
[18] Wild G and Hinckley S 2010 Optical fibre Bragg gratings for acoustic sensors. In: *Proc. 20th International Congress on Acoustics,* pp 23-7
[19] Lee, Jeong Y, Yin S, Ruffin P B and Yu F T S 2008 Interrogation techniques for fiber grating sensors and the theory of fiber gratings *Fiber Optic Sensors. Taylor & Francis* 253-331
[20] Webb D J, Surowiec J, Sweeney M, Jackson D A, Gavrilov L R, Hand J W, Zhang L and Bennion I 1996 Miniature fiber optic ultrasonic probe. In: *SPIE's 1996 International Symposium on Optical Science, Engineering, and Instrumentation,* pp 76-80
[21] Tsuda H 2005 Ultrasound and damage detection in CFRP using fiber Bragg grating sensors *Composites Science and Technology* **66** 676-83
[22] Hafizi Z M 2014 The applications of near infra-red fibre Bragg grating sensors for wave propagation based structural health monitoring of thin laminated composite plates.

Double- diffusive convection in an anisotropic and inclined porous layer with cooperating horizontal gradients of temperature and concentration

S Safi[1] **and S Benissaad**[1]

[1] Laboratoire d'Energétique Appliquée et de Pollution
Faculté des Sciences de la technologie
Université des Frères Mentouri- Constantine- Algeria

*Corresponding author: s.safisafia@gmail.com

Abstract. Double-diffusive natural convection in porous layer saturated with a binary fluid is presented in this study. The layer is anisotropic with respect to both its permeability and diffusivity and is inclined with respect to the horizontal axis. The vertical walls of porous cavity are subjected to uniform temperature and concentration where the other surfaces are assumed to be adiabatic and impermeable. The governing equations are solved by the finite volume method. The study focuses on the effects of anisotropy in permeability and the inclination angle of the flow structures on heat and mass transfer. The results are presented and analyzed in terms of streamlines, isotherms, isoconcentrations lines and average Nusselt and Sherwood numbers. They indicate the existence of three regimes. A diffusive regime, a transition regime and an asymptotic regime.They are in good agreement with those found in the literature.

1. Introduction

The heat and mass transfer phenomena within porous configurations are particularly important to study because of the large number of industrial or environmental applications in which they are present. In most cases, natural convection of heat and/or solutal source plays a decisive role in the transfers involved.

Cheng and Chang [1] studied the natural convection heat transfer from impermeable horizontal surfaces in a saturated porous medium. Bejan and Khair [2] examined the natural convection boundary layer flow driven by both temperature and concentration gradients. Most studies [3-5] of thermosolutal convection induced in isotropic or anisotropic porous medium saturated by a fluid consider a single layer of porous medium in the case of rectangular cavities. Among these studies, it can be mentioned the numerical investigations of the simplified problem of square porous cavity where the vertical walls are held at constant temperatures and concentrations and the horizontal surfaces are considered adiabatic and impermeable. The scale analysis permits to treat this problem in the two extreme cases of heat-driven and solule-driven natural convection.

A special focus has been made by several authors on the case where a plane porous layer is inclined to the horizontal and bounded by impermeable isothermal walls. A temperature difference between the walls may lead to an unstable stratification that has a different from that of a horizontal layer, Darcy–Bénard [6]. In fact, the inclination to the horizontal results in a basic stationary state where the fluid is not at rest, but circulates along a single cell of infinite width. The basic velocity field is parallel, bidirectional, and with a vanishing mass flow rate. However, most of the porous materials are anisotropic due to preferential orientation for either enhancing or for deteriorating heat transfer. Hadadi et al. [7] studied numerically thermosolutal convection within inclined porous collector vertically. The effect of the inclination angle is taken into account and analyzed. The results are presented in terms of streamlines, isotherms, and isoconcentrations and are mainly analyzed in terms of the average heat and mass transfers. Dimensional analysis is applied to predict analytically the evolution of transfer and is compared to numerical results obtained from simulations of double-diffusive natural convection in such bi-layered porous medium for the different situations.

The problem of double-diffusive convection in an inclined horizontal bi-layered porous cavity was also considered by Hadadi and Bennacer [8]. The results were also presented and analyzed in terms of streamlines, isotherms, isoconcentrations lines and average Nusselt and Sherwood numbers. Numerical and a scale analysis are used to characterize the effect of the permeability ratio on the heat and mass transfer in vertical bi-layered porous cavity. Singh and al. [9] have studied the natural convection heat transfer in inclined porous square cavities. They concluded that the larger inclination angle may be optimal for the energy efficient processes involving inclined enclosures due to larger heat flow circulations with enhanced thermal mixing.

The aim of the present study is to emphasize on the natural convection and heat and mass transfer in inclined porous layer with reference to horizontal. The saturated porous medium is anisotropic in permeability. The vertical walls of the cavity are subjected to uniform temperature and concentration. The study focuses on the effects of anisotropy in permeability and the inclination angle of the flow structures on heat and mass transfer. The results are presented in terms of streamlines, isotherms, and isoconcentrations and are mainly analyzed in terms of the average heat and mass transfers.

2. Problem formulation

The physical model is presented in Figure 1. It is represented by an inclined porous layer saturated by a binary fluid. The inclination angle of the enclosure with respect to the horizontal plane is denoted by α. The porous medium is considered homogeneous and anisotropic in permeability and thermal conductivity. The vertical walls of the porous cavity are subjected to uniform conditions of temperature and concentration, whereas, the horizontal walls are assumed to be adiabatic and impermeable. A general Brinkman-Forchheimer extended Darcy model is used to account for the flow in the porous medium.

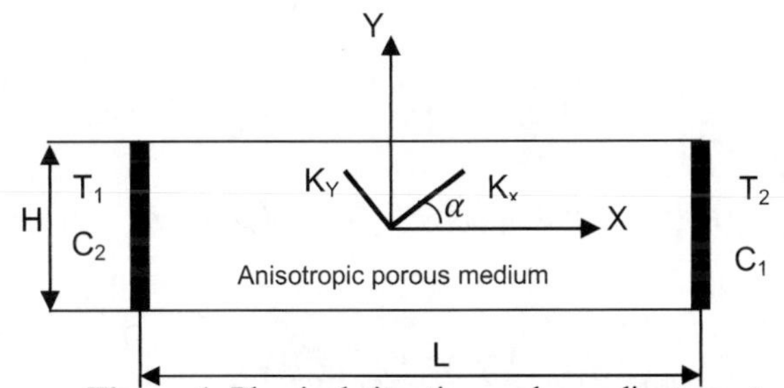

Figure 1. Physical situation and coordinate system.

Under the usual Boussinesq approximation, the governing equations for steady, laminar, two-dimensional boundary layer flow under above assumptions can be written as:

$$\vec{\nabla}\,.\,\vec{V} = 0 \tag{1}$$

$$\rho_f\left[\frac{1}{\varphi}\frac{\partial\vec{V}}{\partial t} + \frac{1}{\varphi^2}\,(\vec{V}\vec{\nabla}\,)\,\vec{V}\right] = -\vec{\nabla}P - \mu\frac{\vec{V}}{\overline{\overline{K}}} - \frac{C_f}{\overline{\overline{K}}^{1/2}}|\vec{V}|\vec{V} + \mu_{eff}\vec{\nabla}^2\vec{V} + \rho_f\vec{g} \tag{2}$$

$$\sigma\frac{\partial \mathrm{T}}{\partial \mathrm{t}} + \vec{\mathrm{V}}\vec{\nabla}\ \mathrm{T} = \vec{\nabla}\,.\,(\overline{\overline{\lambda}}\,\vec{\nabla}.\mathrm{T}) \tag{3}$$

$$\varphi\frac{\partial \mathrm{c}}{\partial \mathrm{t}} + \vec{\mathrm{V}}\vec{\nabla}\mathrm{C} = \vec{\nabla}\,.\,(\mathrm{D_{eq}}\vec{\nabla}\,\mathrm{C}) \tag{4}$$

where ρ_f is the fluid density(kg/m^3), g is the gravitational acceleration (m/s^2), μ_{eff} apparent dynamic viscosity for Brinkman's model(kg/m.s) ,D the equivalent mass diffusivity(m^2/s) and σ ratio of heat capacities.

$\overline{\overline{K}}$ and $\overline{\overline{\lambda}}$ are the second order flow permeability and thermal diffusivity tensors defined, respectively, by

$$\bar{\bar{K}} = \begin{bmatrix} K_1 cos\alpha^2 + K_2 sin\alpha^2 & (K_1 - K_2) sin\alpha \cos\alpha \\ (K_1 - K_2) sin\theta \cos\theta & K_2 cos\alpha^2 + K_1 sin\alpha^2 \end{bmatrix} \tag{5}$$

$$K = \frac{K_x}{K_y} \tag{6}$$

Substituting K (equation (6)) in equation (5) and by applying inverse, the permeability tensor can be reduced to

$$\bar{\bar{K}}^{-1} = \frac{1}{K_1} \begin{bmatrix} cos\alpha^2 + K sin\alpha^2 & (1-K) sin\alpha \cos\alpha \\ (1-K) sin\alpha \cos\alpha & K cos\alpha^2 + sin\alpha^2 \end{bmatrix} \tag{7}$$

$$\bar{\bar{\lambda}} = \begin{bmatrix} \lambda & 0 \\ 0 & 1 \end{bmatrix} \tag{8}$$

$$\lambda = \frac{\lambda_x}{\lambda_y} \tag{9}$$

Substituting λ (equation (9)') in equation (8)', the thermal conductivity tensor can be reduced to

$$\bar{\bar{\lambda}}^{-1} = \begin{bmatrix} \lambda & 0 \\ 0 & 1 \end{bmatrix} \tag{10}$$

The dimensionless governing equations are written as:

$$\frac{\partial U}{\partial X} + \frac{\partial V}{\partial Y} = 0 \tag{11}$$

$$\frac{1}{\varphi}\frac{\partial U}{\partial t} + \frac{1}{\varphi^2}\left(U\frac{\partial U}{\partial X} + V\frac{\partial U}{\partial Y}\right) = -\frac{\partial P}{\partial X} + \frac{Pr}{\varphi}\left(\frac{\partial^2 U}{\partial X^2} + \frac{\partial^2 U}{\partial X^2}\right) - \frac{Pr}{Da}[(U cos\alpha^2 + K sin\alpha^2) + V((1-K) sin\alpha \cos\alpha)] - \frac{C_f}{\sqrt{Da}}\left[U\left(cos\alpha^2 + \sqrt{K} sin\alpha^2\right) + V\left(\left(1-\sqrt{K}\right) sin\alpha \cos\alpha\right)\right] \times \sqrt{U^2+V^2} + PrRa(T + NC) \tag{12}$$

$$\frac{1}{\varphi}\frac{\partial U}{\partial \tau} + \frac{1}{\varphi^2}\left(U\frac{\partial V}{\partial X} + V\frac{\partial V}{\partial Y}\right) = -\frac{\partial P}{\partial X} + \frac{Pr}{\varphi}\left(\frac{\partial^2 V}{\partial X^2} + \frac{\partial^2 V}{\partial Y^2}\right) - \frac{Pr}{Da}[U((1-K) sin\alpha \cos\alpha) + V(K cos\alpha^2 + sin\alpha^2)] - \frac{C_f}{\sqrt{Da}}\left[U\left(\left(1-\sqrt{K}\right) sin\alpha \cos\alpha\right) + V\left(\sqrt{K}\, cos\alpha^2 + sin\alpha^2\right)\right] \times \sqrt{U^2+V^2} + PrRa(T + NC) \tag{13}$$

$$\frac{\partial T}{\partial t} + \left(U\frac{\partial T}{\partial X} + V\frac{\partial T}{\partial Y}\right) = \frac{1}{Pr}\bar{\bar{\lambda}}^{-1}\left(\frac{\partial^2 T}{\partial X^2} + \frac{\partial^2 T}{\partial Y^2}\right) \tag{14}$$

$$\frac{\partial C}{\partial t} + \frac{1}{\varphi}\left(U\frac{\partial C}{\partial X} + V\frac{\partial C}{\partial Y}\right) = \frac{1}{Le}\left(\frac{\partial^2 C}{\partial X^2} + \frac{\partial^2 C}{\partial Y^2}\right) \tag{15}$$

The non-dimensional boundary conditions can be expressed as follows:

$\mathrm{U = V = 0, T = 0\ et\ C = 0\ at\ t = 0}$

$\frac{\partial T}{\partial x} = 0, \frac{\partial C}{\partial x} = 0, \mathrm{U = V = 0\ at\ Y = 0}, \forall X$

$\frac{\partial T}{\partial x} = 0, \frac{\partial C}{\partial x} = 0, \mathrm{U = V = 0\ at\ Y = 1}, \forall X$

$\mathrm{T = 1, C = 0, U = V = 0\ at\ X = 0}, \forall Y$

$\mathrm{T = 0, C = 1, U = V = 0\ at\ X = 4}, \forall Y$

Where φ is the porosity, C_f: Forchheimer coefficient, U et V : dimensioneless horizontal and vertical velocities, X and Y are dimensioneless horizontal and vertical coordinates. T, C, P and t: are respectively the dimensioneless, temperature, concentration, pressure and time.

The dimensionless variables are defined as: Darcy number *(Da),* Prandtl number *(Pr)*, Rayleigh number *(Ra)*, Lewis number *(Le)* and Buoyancy ratio number *(N)*.

The average Nusselt and Sherwood numbers are defined by

$$Nu = \int_0^1 \left(\frac{\partial T}{\partial X}\right)_{X=0} dY \quad (16)$$

$$Sh = \int_0^1 \left(\frac{\partial C}{\partial X}\right)_{X=0} dY \quad (17)$$

3. Numerical method

The governing equations along with the boundary conditions are solved numerically employing finite volume method. For pressure and velocity coupling, the SIMPLER algorithm is used. The system of algebraic equations is solved iteratively by means of the Thomas algorithm. The iteration process is continued until the following convergence criterion is achieved.

$$\left(\left|\frac{\phi_{i,j}^{t+\Delta t} - \phi_{i,j}^{t}}{\phi_{i,j}^{t+\Delta t}}\right|\right) \leq 10^{-4} \quad (18)$$

where ϕ is a dependent variable U, V, T et C . The indices i, j indicate a mesh point and Δt being the increment of time.

For the accuracy of the numerical results, the present study is compared with the previous study Bennacer et al. [10] and Ni and Beckermann [11] for horizontal rectangular porous cavity as shown in table 1. It is observed that the present results are in good agreement with that of [10] and [11]. This favorable comparison leads confidence in the numerical results to be reported in the next section. The relative differences observed between the results being less than about 1.5%.

Table 1. Comparison the Nusselt number of present results with those of Bennacer et al. [8] and Ni and Beckerman [9].
$Da = 10^{-7}$, $A = 1$, $Ra = 10^3$, $N = 0$, $\alpha = 0°$.

$K = K_y/K_x$	10^3	10^2	10^1	10^0	10^{-1}	10^{-2}
Bennacer et al [10]	1.00	1.29	4.17	13.48	37.56	80.62
Ni and Beckermann [11]	1.01	1.30	4.17	13.41	37.37	80.34
Present study	1.04	1.26	4.19	13.09	37.31	80.33

4. Results and discussion

In the non-dimensional governing equations the Forchheimer constant (C_f) and porosity are taken respectively 0.55 and 0.8 according to Ergun relation [12]. The Prandtl number is taken to be 0.71 throughout the computations.

In 'figure 2' the effects of the inclination are illustrated for K= 10 and N = 10. The results are presented in terms of streamlines (on the left), isotherms (at the center) and isoconcentration (on the right) contours for different values of α. When the inclination decreased to $\alpha = 0°$ the pattern of streamlines of Figure 2 indicates that a large portion of the fluid in the center of the cavity is now stagnant due to the blocking effect of the vertical stratification of the density field in this area and the flow circulation is restricted to thin boundary layers of almost constant thickness, along the vertical walls. The isotherms are distorted and span in the diagonal direction of the entire cavity, the thermal boundary layer tends to be formed, located respectively on the lower and upper parts of the lateral walls. It is noted that the deformation of isoconcentrations increases as the tilt angle increases.

Figure 3 shows the average Nusselt and Sherwood numbers for different inclination of principal axes α with different permeability ratio K for $Da = 10^{-4}$, $Ra=10^6$ and k =10. The effects of *K* on the transfers are noticeable. The heat and mass transfers increase with the increase of α, passes through a peak, and then begins to decrease. The peak in Nusselt and Sherwood number occurs at about 30 ° for

IOP Conf. Series: Materials Science and Engineering **257** (2017) 012002 doi:10.1088/1757-899X/257/1/012002

different values of K. The increase of permeability in vertical direction with the variation of α from 0° to 90° leads to decrease in energy and mass transport, resulting in decrease of velocity.

The effect of the (buoyancy ratio) N on heat and mass transfer Nusselt (Nu) and Sherwood (Sh) are illustrated in 'figure 4' for α=30° and Da =10^{-4}. For a fixed value of N the heat and mass transfers increase with the increase of When K >10, the Nusselt and Sherwood numbers are found to be independent of K. Indeed, the resistance to the flow and then transfers decreases. For a fixed K and in a diffusive regime, one observes an increase of Nusselt and Sherwood with an increase of N, however it remains moderate. However over this regime, the transfers intensify. In fact, the buoyancy forces that induce the fluid motion are assumed to be cooperative. Nusselt reaches a maximum for K=10^2. However, for Sherwood the behavior is reversed, the maximum mass transfer occurs for K=10.

α=0°

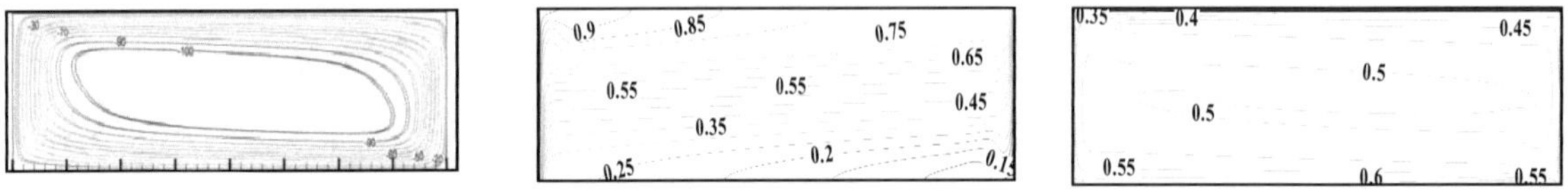

α=30°

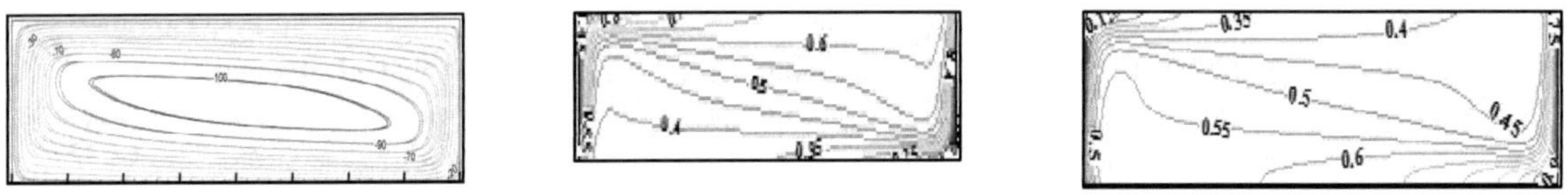

α =60°

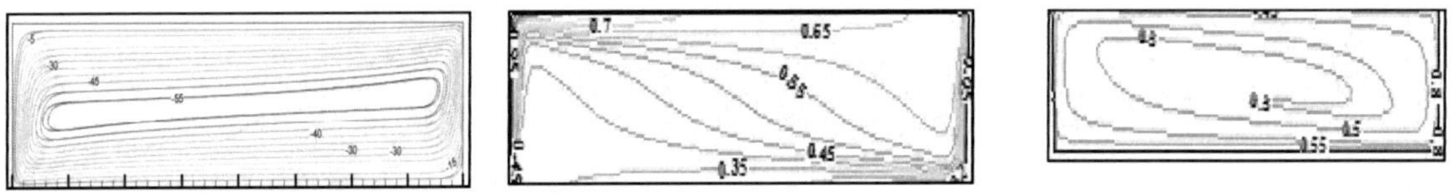

Figure 2. Streamlines (left), isotherms (center) and isoconcentration (right) for Ra = 10^6, Da = 10^{-4} , λ=10 and K=10 for different tilted angle α.

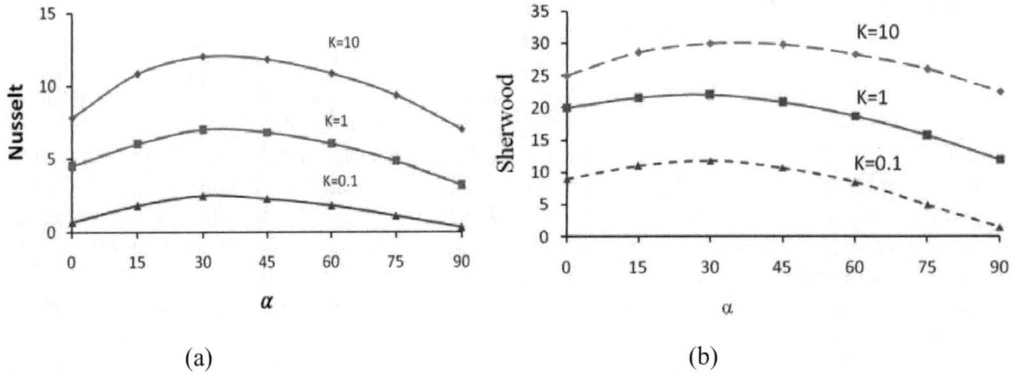

Figure 3. Effect of α on (a) average Nusselt and (b) Sherwood for various values of K for N = 10, Ra = 10^6 and Da = 10^{-4}, λ=10

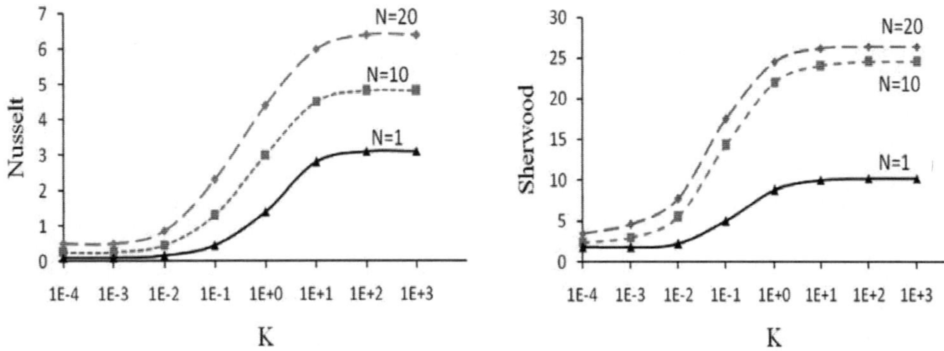

(a) (b)

Figure 4. Effect of K on (a) average Nusselt and (b) Sherwood for various values of N for α=0°, Ra=10^6 and Da=10^{-4}, λ=10.

5. Conclusions

A study is made of natural convection heat and mass transfer in a horizontal porous cavity. The porous medium is assumed to be both thermally and hydrodynamically anisotropic with the principal axes of anisotropic permeability inclined with respect to the gravity force. The major results obtained in the present investigation can be summarized as follows.

For large permeability ratios (K > 1) as the inclination angle varies from α=0° to 90°,the heat and mass transport is maximum for α = 30°. For small permeability ratios (K < 1) as the anisotropic orientation angle varies from α=0° to 90°, the convective strength of the flow field gradually diminishes.

The numerical results indicate the existence of three regimes, namely, a diffusive regime for low values of K, a transition regime where Nusselt and Sherwood numbers increase with an increase of K and an asymptotic regime where Nusselt and Sherwood become independent of K.

6. References

[1] Cheng P and Chang I D 1976 Buoyancy induced flows in a saturated porous medium adjacent to impermeable horizontal surfaces *Int. J. Heat and Mass Transfer* vol 19, pp.1267–1272

[2] Bejan A and Khair K R 1985 Heat and mass transfer by natural convection in a porous medium *Int. J. Heat and Mass Transfer*, vol.28, pp. 909–918

[3] Nishimura T, Wakamatsu M and Morega A M 1998 Oscillatory double-diffusive convection in a rectangular enclosure with combined horizontal temperature and concentration gradients *Int J Heat Mass Transfer*, vol.41, pp.1601–1611

[4] Amahmid A , Hasnaoui M and Vasseur P 1999 Etude Analytique et Numérique de la Convection Naturelle dans une Couche Poreuse de Brinkman Doublement Diffusive *Int. J. Heat and Mass Transfer*, vol. 42, pp.2991-3005

[5] Mohamad AA and Bennacer R 2001 Natural convection in a confined saturated porous medium with horizontal temperature and vertical solutal gradients *Int J Thermal Sci*, vol.40, pp.82-93

[6] Rees D A S 2000 The stability of Darcy–Bénard convection *Handbook of Porous Media, CRC Press,* New York, vol.12, pp.521–558

[7] Hadidi N and Y Ould-Amer 2013 Bi-layered and inclined porous collector: Optimum heat and mass transfer *Energy*, vol.51, pp. 422-430

[8] Hadidi N , Bennacer R and Ould-Amer Y 2015 Two-dimensional thermosolutal natural convective heat and mass transfer in a bi-layered and inclined porous enclosure *Energy*, vol.93, pp. 2582-2592

IOP Conf. Series: Materials Science and Engineering **257** (2017) 012002 doi:10.1088/1757-899X/257/1/012002

[9] Singh A, Basak T and Nag A 2015 Heatlines and thermal management analysis for natural convection within inclined porous square cavities *Int. J. Heat and Mass Transfer*, vol. 87, pp.583-597
[10] Bennacer R ,Tobbal A and Beji H 2001 Double Diffusive Convection in a Vertical Enclosure Filled with Anisotropic Porous Media Int. *J. Thermal Sciences*, vol.40, pp. 30-41
[11] NI J and Beckermann C 1991 Natural Convection in a Vertical Enclosure Filled With Anisotropic Porous Media *J. of Heat Transfer*, vol.113, pp. 1033-1037
[12] Ergun S 1952 Fluid flow through packed columns *Chemical Eng. Progress*, vol. 48, pp. 89–94, 1952.

A review of techniques to determine alternative selection in design for remanufacturing

A Z Mohamed Noor[*], M H F Md Fauadi, F A Jafar, N R Mohamad A S **Mohd Yunos**

Faculty of Manufacturing Engineering, Universiti Teknikal Malaysia,Hang Tuah Jaya, 76100, Durian Tunggal Malacca, Malaysia

*Corresponding author:ahamadzaki.mohamednoor@gmail.com

Abstract. This paper discusses the techniques used for optimization in manufacturing system. Although problem domain is focused on sustainable manufacturing, techniques used to optimize general manufacturing system were also discussed. Important aspects of Design for Remanufacturing (DFReM) considered include indexes, weighted average, grey decision making and Fuzzy TOPSIS. The limitation of existing techniques are most of them is highly based on decision maker's perspective. Different experts may have different understanding and eventually scale it differently. Therefore, the objective of this paper is to determine available techniques and identify the lacking feature in it. Once all the techniques have been reviewed, a decision will be made by create another technique which should counter the lacking of discussed techniques.In this paper, shows that the hybrid computation of Fuzzy Analytic Hierarchy Process (AHP) and Artificial Neural Network (ANN) is suitable and fill the gap of all discussed technique.

1. Introduction

Design for Remanufacturing (DFReM) is a concept of manufacturing environment which was developed for future generation of manufacturing and process technology. Remanufacturing is a process of returning used product by rework, reassemble, disassemble reprocess, inspect and testing in order to bring back current appearance to brand new part [1]. There are three indicators in Design for Remanufacturing. The indicators are economic, social and ecological [2]. The problem with current situation is that the techniques used are very difficult to apply in the assessment of the remanufacturing process [2]. Total of alternatives can be more than five, could also be twenty and the techniques used are not suitable to the situation. The techniques that are look into are narrowed down to a technique used for optimization in manufacturing field. Therefore, the objective of this paper is to determine the technique used for selection purpose of criterion in DFReM economy indicator. The next objective is to further identify the lacking in method and proposed another alternative for decision making purposes in DFReM.

2. Methodology

This paper starts with identifying the techniques used for cost optimization in economy indicator under DFReM. The techniques are index, weighted average, grey decision making method and fuzzy TOPSIS. All the techniques are presented with the steps to be carried out in order for determination of final answer. The steps are introduced with example from published experiment carried out by previous author. Next step is to propose personal argument regarding the steps especially the advantage and lacking. This paper further carried out by suggesting a new method that will be introduced in Design for Remanufacturing to be proven that this new method covers all the lacking from current introduced method. Figure 1 shows the overall flow chart of this paper.

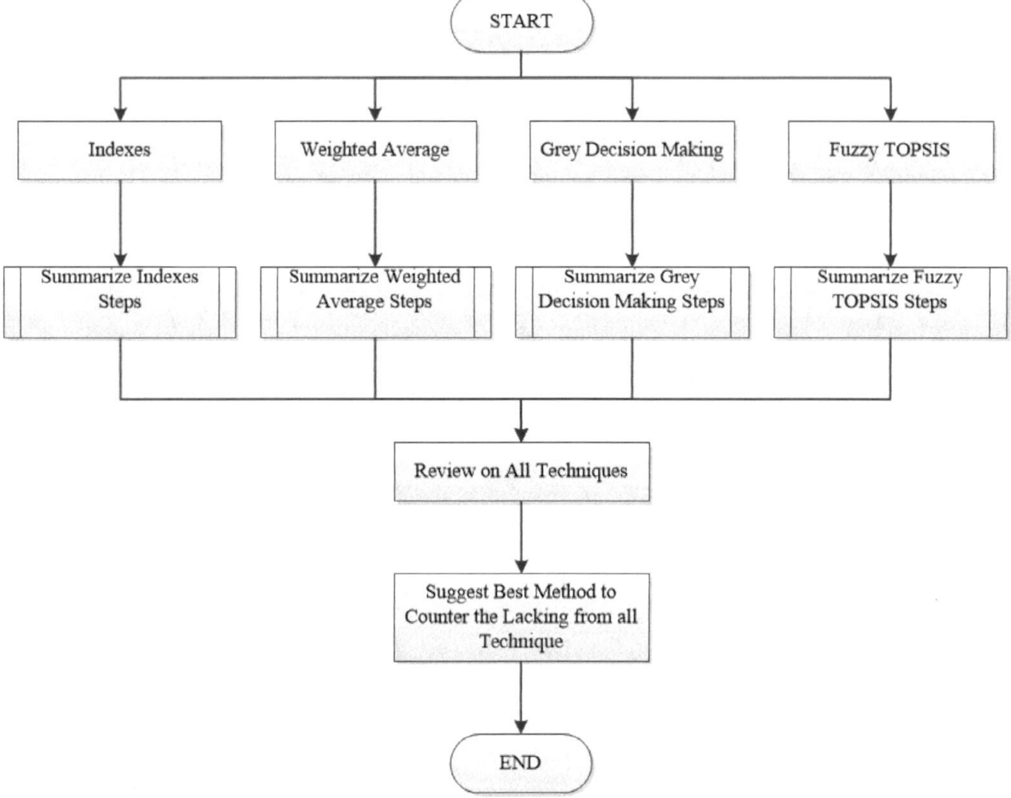

Figure 1. Overall Flow Chart of the Review

3. Techniques Used for Decision Making in DFReM

There are several techniques which being used for decision making in remanufacturing process. The techniques used are indexes, weighted average, grey decision making and Fuzzy TOPSIS the next section of this paper will show the step carried out in order to demonstrate these methods.

3.1 Indexes

The authors wants to evaluate product recyclability to be applied during the design phase, acting as a supporting technique for designer decision – making. Aguiar et al [3] have proposed indexes to be used with the product's Bill of Material (BOM) during conceptual and embodiment phase of product design. The indexes were used in basically two process which are recycling and disassembly. To be more specific, the criteria being evaluated under these processes are:

Table 1: Comprehensive Criterion Evaluation for Recycling and Disassembly Processes.

Recycling Process	Disassembly Process
Number of fastener	Existence of contaminants
Type of fastener	Material compatibility
Accessibility of fasteners	Presence of hazardous materials

These criteria have several options to be decided as the exact number of fastener or the best type of fasteners and other suitable criteria needed. The products that the author chooses to solve are portable cassette and CD player. The index scale is determined through colour scale for recyclability indexes. The green shows the ideal state and the red colour shows undesired state. Green will indicate that the selection is highly desired and should be selected in any situation. For this paper, author placed this colour for fasteners type in portable cassette and CD player. The presentation of ideal and undesired state is shown in figure 1.

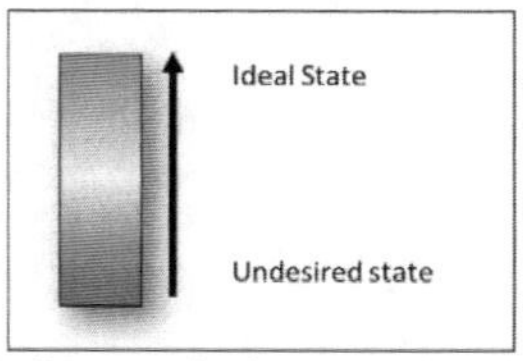

Figure 2. Colour scale for recyclability Index [3]

The colour have been placed its scale according to colour. The green is known as value "one" and the red shows value "four"[3]. In this paper, quantity of types of fasteners index (QTFI) is discussed in different types of fastener. The least type of fastener used is the ideal selection. However, the more types of fastener show code red. Therefore the indexes used are between values 1 to 4. Table shows value of indexes according to the colour code.

Table 2. Quantity of Types of Fastener Indexes[3]

QTFI Values	Description
1	1 Type
2	2 Types
3	3 Types
4	4 or more types

Meanwhile Ijomah et al [2] use indexes in the same field which is design for remanufacturing. The product used in this paper is engine block. They proposed fundamental steps required to build on past work. The authors have improved the robustness of DFReM methodologies.

Table 3. Design features affecting product remanufacturability[4]

Design Feature	Severity of Impact
A)Assembly Type	1 - 4
B)Product Complexity	3
Materials	4
Design Cycle	3

Table 3 (a). Problem identified from design feature affecting product remanufacturability[4]

Problem Identified	Severity of Impact
A)Screws	1
Rivets	2
Welding	3-4
Strong Adhesive	4
B)Numerous Component	2
Product Dimension	2
Internal Component Arrangement	2 - 3
Coatings	2

From the table 3, the index used is from 1 to 4. Value 1 shows low impact and value 4 shows very high impact[4]. The indexes used are still between ranges hardly to differentiate between criteria. Example of paper [5] uses index for the purpose of creating quantitative approach is assessing product design for remanufacturing. The authors created an assessment for remanufacturing of initial product design. The assessment consists of technical and economic feasibility. However, the author focuses usage of indexes in technical feasibilities. Figure 2 shows the exact operation where the index is used. To determine the index, authors use formulas. Each process has own calculation of index. The highest index shows high impact. Table 4 shows the index calculation for all process in technical feasibilities.

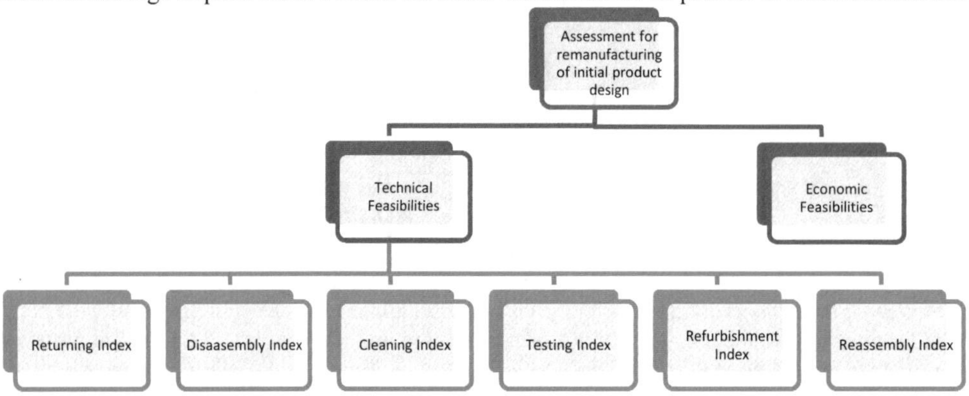

Figure 3. An assessment system for remanufacturing of initial product design returning[5]

Table 4. Computation of index with respective formula[5]

Type of Index	Formula
Returning Index	$\mu_H = \left(\frac{\text{The returning number of waste products}}{\text{The potential returning number of waste products}}\right)$
Disassembly Index	$\mu_D = \left[\frac{(\text{The number of disassembly parts})_L \times 1.5 \text{ seconds}}{\text{The total time of disassembly}}\right]$
Cleaning Index	$\mu_C = \left[\frac{(\text{The number of cleaning parts})_L \times 1(\text{ example: the score of minimum cleaning category})}{\sum \text{The score of cleaning category}_{L(j)}}\right]$
Testing Index	$\mu_I = \left[\frac{(\text{The number of testing parts})_L \times 10 \text{ seconds}}{\text{The total time of testing}}\right]$
Refurbishment Index	$\mu_R = \left(1 - \frac{\text{The number of testing parts}}{\text{The total number of parts}}\right) \times \text{Refurbishment index probability}$
Reassembly Index	$\mu_A = \left[\frac{(\text{The number of reassembly parts})_L \times 3 \text{ seconds}}{\text{The total time of reassembly}}\right]$

The authors [5] mentioned an evaluation model based on the remanufacturing process which comprehensive assess product design for remanufacturing need to be further tested and verified. Therefore, there are still room for improvement to improve quantitative approach using index.

3.2 Weighted Average

Weighted is another quantitative method that is used in design for remanufacturing. Basically this method uses decision maker's perception to determine weightage of criteria. When humans make decision, it may influence by mood, environment sometime time constraint making the weightage assigned sometimes gives incorrect answer. [1] create a metric for assessing remanufacturability by using weighted average method. The objective of the paper is to establish efficient and effective design metrics to measure the remanufacturability of product designs. This paper carried out experiment in remanufacturing automotive parts.

The method for weighted average uses for the same process as the previous paper which are assembly, disassembly, testing, repair, cleaning, inspection, refurbishment and replacement process.

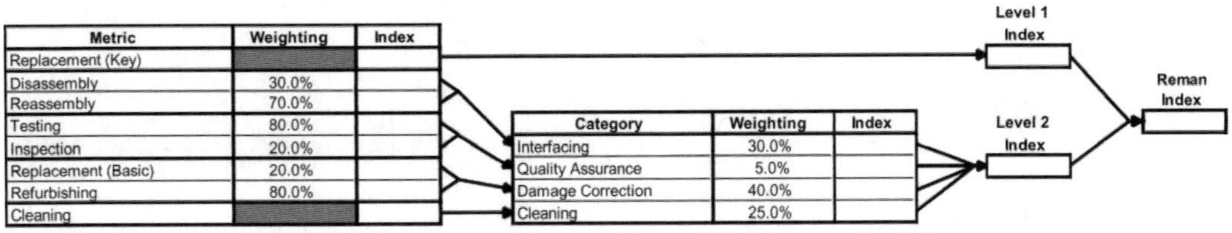

Figure 4. Structure of Remanufacturability Assessment[1]

Observed in figure 3, the weight placed need to achieve equivalent to 100%. The weight can be either 20% - 80%, or 40% - 60% can also be 30% - 70%. All depends on the decision maker's perception. Figure 4 shows the amount of investment needed for each method by pairwise comparison in order to generate their relative importance. Since the relative importance were based on incremental decisions, as shown in the legend, it is not appropriate to take the exact relative importance as being precisely correct. Therefore, the calculated importance is rounded to a set of approximations to the true relative importance.

Prioritization Matrix Legend

5	(row) requires much more investment than (column)
3	(row) requires more investment than (column)
1	(row) requires the same investment as (column)
1/3	(row) requires less investment than (column)
1/5	(row) requires much less investment than (column)

	Blown	Abraded	Baked	Washed	Score	Relative Importance	Approximate Cleaning Score	Usable Cleaning Score
Blown	1.0	0.3	0.2	0.2	1.7	7%	1.00	1
Abraded	3.0	1.0	0.3	0.3	4.7	18%	2.69	3
Baked	5.0	3.0	1.0	1.0	10.0	38%	5.77	6
Washed	5.0	3.0	1.0	1.0	10.0	38%	5.77	6
					26.4	100%	15.23	

Figure 5. Example Prioritization of Cleaning Process[1]

The authors comment that due to the linearity of this technique, a poor response (even a zero value) is compensated for by a higher response. Using this technique, it is possible to have undesirable scenarios such as the following: 30% of the indices are set to zero, and 70% are set to 100%, resulting in a total remanufacturability index of 70%.

3.3 Grey Decision Making

The advantage of grey theory is that able to deal flexibly with the fuzziness situation. Supplier selection can be viewed as a grey system process [6]. Grey Method of computing consist of 8 steps depends on its application. For this paper, the use of grey decision making is to determine best supplier selection.

Step 1 is to determine the attribute weight. K shows the total number of decision makers.

$$\otimes w_j = \frac{1}{K}[\otimes w_j^1 + \otimes w_j^2 + \cdots + \otimes w_j^K] \qquad (1)$$

The next step which is step 2 is to identify the rating value

$$\otimes G_{ij} = \frac{1}{K}[\otimes G_{ij}^1 + \otimes G_{ij}^2 + \cdots + \otimes G_{ij}^K] \qquad (2)$$

Once step 2 is done, the next procedure step 3 is to establish the grey decision matrix.

$$D = \begin{bmatrix} \otimes G_{11} & \otimes G_{12} & \cdots & \otimes G_{1n} \\ \otimes G_{21} & \otimes G_{22} & \cdots & \otimes G_{2n} \\ \vdots & \vdots & \ddots & \vdots \\ \otimes G_{m1} & \otimes G_{m2} & \dots & \otimes G_{mn} \end{bmatrix} \tag{3}$$

Since the scaling performed in matrices shows different sum and hardly to relate, therefore the fourth step is to normalize the grey decision matrix.

$$D^* = \begin{bmatrix} \otimes G_{11}^* & \otimes G_{12}^* & \cdots & \otimes G_{1n}^* \\ \otimes G_{21}^* & \otimes G_{22}^* & \cdots & \otimes G_{2n}^* \\ \vdots & \vdots & \ddots & \vdots \\ \otimes G_{m1}^* & \otimes G_{m2}^* & \dots & \otimes G_{mn}^* \end{bmatrix} \tag{4}$$

Whereby for benefit attribute,

$$\otimes G_{ij}^* = \left[\frac{\underline{G}_{ij}}{G_j^{max}}, \frac{\overline{G}_{ij}}{G_j^{max}} \right] \tag{5}$$

for cost attribute is expressed as

$$\otimes G_{ij}^* = \left[\frac{G_j^{min}}{\overline{G}_{ij}}, \frac{G_j^{min}}{\underline{G}_{ij}} \right] \tag{6}$$

Step 5: The weighted normalize grey decision matrix is expressed as

$$D^* = \begin{bmatrix} \otimes V_{11} & \otimes V_{12} & \cdots & \otimes V_{1n} \\ \otimes V_{21} & \otimes V_{22} & \cdots & \otimes V_{2n} \\ \vdots & \vdots & \ddots & \vdots \\ \otimes V_{m1} & \otimes V_{m2} & \dots & \otimes V_{mn} \end{bmatrix} \tag{7}$$

where $\otimes V_{ij} = \otimes G_{ij}^* \times \otimes w_j$.

Step 6: Supplier alternatives can be obtained by

$$S^{max} = \left\{ \begin{matrix} \left[\max_{1\le i\le m} \underline{V}_{i\,1}, \max_{1\le i\le m} \overline{V}_{i\,1}\right], \\ \left[\max_{1\le i\le m} \underline{V}_{i\,2}, \max_{1\le i\le m} \overline{V}_{i\,2}\right], \\ \left[\max_{1\le i\le m} \underline{V}_{i\,n}, \max_{1\le i\le m} \overline{V}_{i\,n}\right] \end{matrix} \right\} \tag{8}$$

Step 7: Calculate grey possibility degree between compared suppliers alternatives and ideal referential supplier alternatives S^{max}.

$$P\{S_i \le S^{max}\} = \frac{1}{n} \sum_{j=1}^{n} P\{\otimes V_{ij} \le \otimes G_j^{max}\} \tag{9}$$

Step 8: The order of supplier alternatives were ranked. When $P\{S_i \le S^{max}\}$ is smaller, the ranking order is better. According to [6], the ranking order of all suppliers alternatives can be determine and selection of the best among a set of feasible suppliers. Next paper which uses grey decision making as a technique for classification of sustainability level in remanufacturing companies. The objective of [2] are to provide a new technique for decision making.

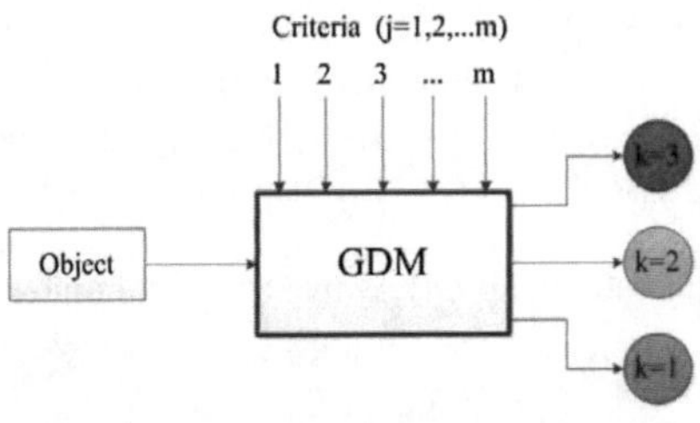

Figure 6. The classification using grey decision making[2]

In this paper, grey decision making is used to determine the final answer obtained. The answer will be further classified in three different categories. There are three values of k represent in three different colours. The values with respect to colours are $k = 1$(green), $k = 2$(yellow) and $k = 3$(red). When the colour show green, which means there is no immediate improvement action needed to remanufacture the product. If the colour show yellow, it means that the level of sustainability is conditional. There will be a little remanufacture needed to fix the product.

Lastly if the colour code show red after being computed using grey decision making, the level of sustainability is unacceptable. Major remanufacture or reject will take place. The classification will depends on the final computation from grey decision making method. However, the limitation of the presented approach is the fact that, it relies on the experts' knowledge of the decision makers. Furthermore, the prioritization scheme is elaborated based on experts' knowledge, so further research might be needed in order to verify the assigned criteria weights

3.4 Fuzzy Technique for Order Performance by Similarity to Ideal Solution (TOPSIS)

Fuzzy TOPSIS consists of eight steps. The first step is to determine the weightage of evaluation criteria. Most Fuzzy TOPSIS use triangular fuzzy AHP to find fuzzy preference weight[7], [8], [9].Therefore, only three values are used under Fuzzy TOPSIS for evaluation. It is expressed in $\widetilde{w}_i = (lw_i, mw_i, uw_i)$. Second step is to choose the appropriate linguistic judgement based from table 5 for choices or alternatives with respect to criteria. The table scaling is performed by[7].

Table 5. Linguistic Term for Fuzzy TOPSIS[8]

Linguistic Term	Scale
Very High	(9,10,10)
High	(7,9,10)
Fair	(5,7,9)
Average	(3,5,7)
Low	(1,3,5)
Very Low	(0,0,3)

Once all the terms have been identified and pairwise comparison made. The third step is to construct a fuzzy decision matrix. Fuzzy decision matrix is shown in equation (10).

$$D = \begin{matrix} & C_1 & C_2 & \cdots & C_n \\ A_1 \\ A_2 \\ \vdots \\ A_m \end{matrix} \begin{bmatrix} x_{11} & x_{21} & \cdots & x_{1n} \\ x_{21} & x_{22} & & x_{2n} \\ \vdots & \vdots & \ddots & \vdots \\ x_{m1} & x_{m2} & \cdots & x_{mn} \end{bmatrix} \quad i = 1,2, \ldots .m; j = 1,2, \ldots ., n \qquad (10)$$

The fourth step is to perform normalization. The normalized values are calculated using equation (11).

$$r_{ij} = \frac{x_{ij}}{\sqrt{\sum_{i=1}^{m} x_{ij}^2}} \quad i = 1,2, \ldots ., m; \; j = 1,2, \ldots ., n \qquad (11)$$

Previous steps were done for continuing in the fifth step. The fifth step is to identify the weighted normalized value (v_{ij}) which can be identified through $v_{ij} = \widetilde{w}_j . r_{ij}$. Weight is symbolized as ($\widetilde{w}_{ij}$) whereby it is the weight of jth attribute[9]. Next stage is to determine the ideal solution either to be positive or negative. This is basically the sixth step in Fuzzy TOPSIS.

For fuzzy positive ideal solution (FPIS):

$$A^+ = (v_1^+, v_2^+, \ldots, v_n^+) \text{ where } v_j^+ = (1,1,1) \qquad (12)$$

For fuzzy negative ideal solution (FNIS):

$$A^- = (v_1^-, v_2^-, \ldots, v_n^-) \text{ where } v_j^- = (0,0,0) \qquad (13)$$

After FPIS and FNIS are identified after using equations (12) and (13), the distance of each alternative from A^+ and A^-[10].The seventh step is to determine distance between alternative through equation (14) will be employed to calculate (d_i^+, d_i^-)[11]:

$$d(A_i, A^+) = d(A_i, A^-) = \sqrt{\frac{1}{3}[(l_1 - l_2)^2 + (m_1 - m_2)^2 + (u_1 - u_2)^2]} \quad (14)$$

Last step is to calculate the closeness coefficient for each alternative. The closest final answer to 1 will be rank and score as the best alternatives to carry out as a decision[12], [13]. Equation (15) shows how to calculate the closeness coefficient.

$$CC_i = \frac{d_i^-}{d_i^+ + d_i^-}, \ i = 1,2, \ldots., m \quad (15)$$

A paper written by [14] perform an experiment to carry out material selection. The objective of the paper is to evaluate the performance of candidate material using Fuzzy TOPSIS. Table 6 shows crisp value and table 7 shows representation in terms of fuzzy with respect to the crisp value.

Table 6. Crisp value for subjective importance rating

Rating of relative importance	Very low (VL)	Low (L)	Medium (M)	High (H)	Very high (VH)
Crisp value	1	3	5	7	9

Table 7. Linguistic rating and its triangular fuzzy number[14]

Rating of material performance	Poor (P)	Medium Poor (MP)	Fair (F)	Medium Good (MG)	Good (G)
Fuzzy Value	[0, 1, 3]	[1, 3, 5]	[3, 5, 7]	[5, 7, 9]	[7, 9, 10]

The authors discover that in this framework, the concept of entropy evaluates weight factor for each evaluation criterion. Fuzzy TOPSIS ranks the candidate materials in a fuzzy environment, are utilized to minimize personal bias and vagueness of the evaluation process. This research work has distilled a list of the material properties that are relevant to remanufacturing performance and formulated a systematic and rational methodology for evaluating the materials based on the proposed selection criteria.

To conclude, the techniques used in design for remanufacturing are indexes, weighted average, grey decision making and Fuzzy TOPSIS. Strength and weakness of this method with application will be reviewed and suggest a way to overcome the lacking feature of every technique.

4. Review of Techniques Used for Decision Making in DFReM

There are several methods that are being used for the field of Design for Remanufacturing. The methods are Indexes, Weighted Average, Grey Method and Fuzzy TOPSIS. Below table shows the limitation of all the method discuss in this paper.

4.1 Review on Indexes Method

The advantage of scaling is that easy to use and the index can also show ranking. The process is short will save more time. However the accuracy of decision making is very low because placing index is hardly can be control. The index given can be subjective depends on decision maker's perception. Index is created according to their application. Usually the indexes are between scales from 1 to 4 representing undesired states to ideal state. Other application of several options may need higher scale. Therefore changes in scale needed and takes longer time to create and to make sure its adaptability.

4.2 Review on Weighted Average Method

The advantage of this step is simple and the weightage is placed according to prioritization. The score will be multiplied to weighted average. However, the final decision can be manipulated depends on the weighted average. If alternative A is needed to be selected, therefore, the weight for alternative A should be highest or score placed higher than the rest. Weighted given are basically from decision maker. Weighted average can be manipulated by selecting alternatives which is more than average value. This will make the option which is correct theoretically rejected due to fail achieve the average value.

4.3 Review on Grey Decision Making Method
Advantage of this method is able to process uncertainty. The scale is presented in matrices form usually for every researcher who carried out Grey Decision Making. The scale is based on human's perspective. The accuracy is high because able to process uncertainty. However, the scaling can never be control and it can vary according to decision makers. Grey Decision Making is used before the condition was used determine what action should be taken. For this application either the product is good to go to customer (green), minor rework of product (yellow) or total reject (red). If the alternatives are more than three, therefore the model is no more relevant for decision making purpose.

4.4 Review on Fuzzy TOPSIS Method
Fuzzy TOPSIS can process uncertainty and make the final answer to be well interpreted and easy understanding regardless if the user has low knowledge regarding the technique. Fuzzy AHP is used to calculate relative weights of each coordination criterion and then partners are ranked based on closeness coefficient, calculated for each partner using Fuzzy TOPSIS. Authors failed to calculate relative weights and instantly assume the weight. The weakness is that, this method needed to perform in every changed environment.

4.5 Suggestion of Method
To counter all the lacking above, we suggest using Fuzzy Analytic Hierarchy Process (FAHP) hybrid with Artificial Neural Network in Design for Remanufacturing. This is because, to our knowledge, there is never been an implementation of hybrid computation of Fuzzy AHP and Artificial Neural Network in design for remanufacturing for cost optimization. The ANN feature has the ability to learn whereby it can counter the lacking of hardly adaptive in Fuzzy TOPSIS. The scale is hardly manipulated and must be kept in range due to consistency analysis. The scaling is performed once for all environments. The weights are determined several time, ANN will act as learning agent to adopt scale if any parameter needed to be placed as high importance.

5. Conclusions
There are four design aspects that are required to optimize Design for Remanufacturing. Those are indexes, weighted average, grey decision making and fuzzy TOPSIS. Indexes and Weighted average are easily manipulated if the desired alternative is determined before the aspects were selected. The weightage and the indexes scaling needed to be change according to every changes within the environment. For grey decision making, the approach is capable to process uncertainty however, the scaling placed in matrices form is not presented such as Fuzzy AHP. This is due to the scaling is hardly can be control and depends on decision maker's perception. The same goes for Fuzzy TOPSIS but the scaling is control due to the presence of consistency and sensitivity analysis. Therefore, future recommendation, to cover the lacking from all this four techniques, we suggest using hybrid computation of Fuzzy AHP and ANN for optimization purposes. Future work, we will use hybrid computation of Fuzzy AHP and ANN to determine the highest optimization in sustainability field especially in design for remanufacturing for economy indicator.

Acknowledgement
We would like to thank Mybrain15 by Malaysian Ministry of Higher Education for financial support given throughout this research.

References
[1] B. Bras and R. Hammond, "Towards Design for Remanufacturing – Metrics for Assessing Remanufacturability," *Proc. 1st Int. Work. Reuse*, pp. 5–22, 1996.
[2] P. Golinska, M. Kosacka, R. Mierzwiak, and K. Werner-Lewandowska, "Grey Decision Making as a tool for the classification of the sustainability level of remanufacturing companies," *J. Clean. Prod.*, vol. 105, pp. 28–40, 2014.

[3] J. de Aguiar, L. de Oliveira, J. O. da Silva, D. Bond, R. K. Scalice, and D. Becker, "A design tool to diagnose product recyclability during product design phase," *J. Clean. Prod.*, p. , 2016.

[4] W. L. Ijomah, C. a. McMahon, G. P. Hammond, and S. T. Newman, "Development of robust design-for-remanufacturing guidelines to further the aims of sustainable development," *Int. J. Prod. Res.*, vol. 45, no. 18–19, pp. 4513–4536, 2007.

[5] J. Li and G. Q. Liang, "A Quantitative Approach to Assessing Product Design for Remanufacturing," *Appl. Mech. Mater.*, vol. 110–116, pp. 4893–4898, 2011.

[6] G.-D. Li, D. Yamaguchi, and M. Nagai, "A grey-based decision-making approach to the supplier selection problem," *Math. Comput. Model.*, vol. 46, no. 3–4, pp. 573–581, 2007.

[7] Y. Beikkhakhian, M. Javanmardi, M. Karbasian, and B. Khayambashi, "The application of ISM model in evaluating agile suppliers selection criteria and ranking suppliers using fuzzy TOPSIS-AHP methods," *Expert Syst. Appl.*, vol. 42, no. 15–16, pp. 6224–6236, 2015.

[8] O. Taylan, A. O. Bafail, R. M. S. Abdulaal, and M. R. Kabli, "Construction projects selection and risk assessment by fuzzy AHP and fuzzy TOPSIS methodologies," *Appl. Soft Comput. J.*, vol. 17, pp. 105–116, 2014.

[9] R. Moghimi and A. Anvari, "An integrated fuzzy MCDM approach and analysis to evaluate the financial performance of Iranian cement companies," *Int. J. Adv. Manuf. Technol.*, vol. 71, no. 1–4, pp. 685–698, 2014.

[10] S. Avikal, R. Jain, and P. K. Mishra, "A Kano model, AHP and M-TOPSIS method-based technique for disassembly line balancing under fuzzy environment," *Appl. Soft Comput.*, vol. 25, pp. 519–529, 2014.

[11] S. K. Mangla, P. Kumar, and M. K. Barua, "Prioritizing the responses to manage risks in green supply chain: An Indian plastic manufacturer perspective," *Sustain. Prod. Consum.*, vol. 1, no. May, pp. 67–86, 2015.

[12] J. M. Sánchez-Lozano, M. S. García-Cascales, and M. T. Lamata, "Evaluation of suitable locations for the installation of solar thermoelectric power plants," *Comput. Ind. Eng.*, vol. 87, pp. 343–355, 2015.

[13] K. Shaw, R. Shankar, S. S. Yadav, and L. S. Thakur, "Supplier selection using fuzzy AHP and fuzzy multi-objective linear programming for developing low carbon supply chain," *Expert Syst. Appl.*, vol. 39, no. 9, pp. 8182–8192, 2012.

[14] S. S. Yang, N. Nasr, S. K. Ong, and A. Y. C. Nee, "Designing automotive products for remanufacturing from material selection perspective," *J. Clean. Prod.*, 2015.

Safety and health practice among laboratory staff in Malaysian education sector

Nurul Husna Che Hassan[1*], Ahmad Rasdan Ismail[2], Nor Kamilah Makhtar[4], Muhammad Azwadi Sulaiman[3], Noor Syuhadah Subki[1], Noor Adilah Hamzah[2]

[1]Faculty of Earth Science, University Malaysia Kelantan (UMK) Locked Bag 100, Jeli, Kelantan, Malaysia
[2]Faculty of Creative Technology and Heritage, University Malaysia Kelantan (UMK)
Locked Bag 01,16300 Bachok, Kelantan. Malaysia
[3]Faculty of Bioengineering and Technology, University Malaysia Kelantan, 17600 Jeli, Kelantan, Malaysia
[4]Institut Pendidikan Guru Kampus Kota Bharu, Jalan Maktab, Pengkalan Chepa, 16109 Kota Bharu Kelantan, Malaysia

*Corresponding author: nhusnahassan1390@gmail.com

Abstract. Safety is the most important issue in industrial sector such as construction and manufacturing. Recently, the increasing number of accident cases reported involving school environment shows the important of safety issues in education sector. Safety awareness among staff in this sector is crucial in order to find out the method to prevent the accident occurred in future. This study was conducted to analyze the knowledge of laboratory staff in term of safety and health practice in laboratory. Survey questionnaires were distributing among 255 of staff laboratory from ten District Education Offices in Kelantan. Descriptive analysis shows that the understanding of safety and health practice are low while doing some job activities in laboratory. Furthermore, some of the staff also did not implemented safety practice that may contribute to unplanned event occur in laboratory. Suggestion that the staff at laboratory need to undergo on Occupational Safety and Health training to maintain and create safe environment in workplaces.

1. Introduction

Safety and health has become important issue, which has been developing in the social context that increasing the concern for prevention of danger situations for people at surrounding [1]. Safety is defined as a method of preventing accidents or reducing personal injury or property damage that may be caused by an accident [2]. Recent years have showed increasing reports of accidents that occur in school area. The reported accidents that involved not only report on injuries but also to deaths. Parents normally feel that school is a safe place for their children to be in. Schools become one of the largest categories of the workplace because children, teachers and other members of staff spend a lot of their time in schools exposing them to a variety of risks and hazards through physical and social activities. Students spend their 180 days in a year, 6 hours daily in a school [3].This condition may negatively affect their life [1].

In year of 2016 till early 2007, according to report, there are 16 cases of accidents in school involving mercury spill in laboratories. Mercury is one of the chemical hazards that will harmful to human body. Table 1 shows the accident cases that involve within that year. Besides that, there are also accidents occurs in school involving death which are student hit by goalpost at the field, student fall into sewer and fall from a high building and so on. However, school accidents causing injuries have not been fully investigated as those occurring at home or in road traffic accidents rather than manufacturing, construction and other sector [4].

Table 1. Accident cases in school laboratory from 2016 till March 2017.

No	Accident cases	Date	Sources
1.	Anxious broken thermometer in the laboratory in Kota Tinggi	2 March 2017	[5]
2.	Thermometer breaks, the mercury spilled in the laboratory in Alor Gajah	23 Feb. 2017	[6]
3.	Liquid mercury spilled on the floor in a school laboratory in Kota Bharu	18 October 2016	[7]
4.	Reliable liquid mercury spilled in the school laboratory. In the scene at 2 pm, thermometer in the school laboratory believed to have fallen and broken in Machang	18 October 2016	[8]
5.	The thermometer used a student fell and caused the spill of mercury in Kota Bharu	13 October 2016	[9]
6.	5 students in Balik Pulau expose to mercury	5 October 2016	[10]
7.	Three girls are exposed to mercury in Labuan	29 Sept. 2016	[11]
8.	Thermometer breaks, the mercury spilled in the laboratory school in Kota Bharu	28 Sept. 2016	[12]
9.	Mercury spilled in the car students in Ipoh	28 Sept. 2016	[13]
10.	Sulphuric Acid in the laboratory spill in Sungai Petani	27 Sept. 2016	
11.	Nitric acid spilled in the laboratory school in Kuala Lumpur	1 Sept. 2016	[15]
12.	Students exposed to mercury in Seremban	25 August 2016	[16]
13.	Liquid mercury contained in thermometers broken, resulting in spillage of chemical in Kota Bharu	15 August 2016	[17]
14.	Chemical Spill at School in Port Dickson	12 August 2016	[18]
15.	Two Students horticulture Mercury Thermometer Bursts On Science Laboratory in Kepala Batas	28 July 2016	[19]
16.	56 students were quarantined due to mercury spill in Bukit Mertajam	26 May 2016	[20]

Accidents rate in the industry can be reduced if students at the school are exposed to issues of occupational safety and health [21]. This means that teachers and staff play an important role in exposing how students should believe and practice safety and health while they were in the school surroundings [22]. Thus, the school community needs to create a culture of safety in school in order to maintain a safer work environment. This suggests that the involvement of all community is the foundation of development safety culture in school [23].

The issue of safety is not only exposing in industries but also in educational institutions such as school laboratory which always involved with the use of chemical risk. Laboratory is one place used to create a practical work-related engineering. Safety practices are very important when they are working or exercising in the laboratory [40]. The students are constantly exposed to the danger of accidents at work if they are not handled properly and according to the regulations. Knowledge of the use of equipment, safe environment and safety regulations is important enough to change the attitude of the employees to carry out the task safely and effectively. Awareness among staff in this safety aspect is quite important and must be known and should be researched [24].

Apart from that, safety rules in laboratory should be practiced from time to time and staff should be responsible for avoiding accidents from happening [25]. Therefore, to prevent and reduce the risk of accidents, awareness of the importance of safety practices need to be improved. Normally accidents have particular relevance contributed by negligence, the lack of knowledge of any works to be carried out as well as damage or failure either on materials, equipment and chemical used [26]. According studied by some researcher, they found that 85% to 98% of injuries in the workplace is not attributed to the act safe practices of certain individuals who are negligent. Negligence can also associate with awareness [27]. This shows that safety awareness an important aspect in ensuring safety in the laboratory.

Accidents happen very easily if they do not adopt measures safety when working in laboratory [28]. In laboratory, there are rules that must be followed before doing practical work. Teachers, students and staff need to understand each of these rules. Good management will helped reduce accidents in laboratory. Before practical work students will be reminded about the safety aspect. Safety must be practiced while in the laboratory or during experiments conducted [24]. Safety is the most important aspect in the management of a laboratory science [29]. Security and Safety Unit states that the laboratory are expose with high risk for accidents and injuries occur and will cause disease as a result of the work conducted in the laboratory [30]. The use of chemicals can cause accidents if not handled in a safe manner.

Staff laboratory are responsible for providing a healthy and safe environment for students. They should ensure that health, safety and environmental aspects of the practical are considered in laboratory. Besides that, students will receive the appropriate information and good supervision for them to carry out their studies and work practices safely. Students also are warned about the particular hazards, and how to avoid, eliminate or minimize their exposure to them. Meanwhile, laboratory staff also need show good attitude in term of health and safety practices and transferred to the students so that the students can conduct work in safety condition [31]. Thus, this paper is aiming at safety and health practice among staff laboratory in school laboratory.

2. Research Methodology

This study used a quantitative approach to assess the safety and health practice among laboratory staff in the state of Kelantan, Malaysia. Kelantan is a state located in the north east of peninsular Malaysia with a total number of 592 schools spreading all over ten District Education Offices. The questionnaire was adopted from The University of Texas Health Science Centre which conduct survey on Laboratory Safety [38]. Thus, this questionnaire was distributed among 255 of staff laboratory from 111 schools that randomly selected from the ten District Education Offices. The purpose of the questionnaires is to measure awareness level of laboratory staff toward the importance of safety and health practices in school laboratory. The questionnaire was divided into two parts. Section A is for demographic background of the respondents. Section B is for safety and health elements. The question were using closed-ended question with 4 likert scale. The respondent need to answer the question whether they 1- Strongly Disagree, 2- Disagree, 3- Agree or 4- Strongly Agree. Thus, the data collected were analysed using Statistical Package for the Social Sciences (SPSS).

3. Results and Discussion

This section presents the result obtained from the questionnaire survey. The results were analyzed by using Statistical Package for the Social Sciences (SPSS). Figure 1 (a) shows the percentage of respondents by gender. It shows that 37% are males and 63% are females. While, Figure 1 (b) shows the percentage of safety knowledge among staff laboratory. From the result, it shows that, most of them have basic knowledge regarding safety and health aspect.

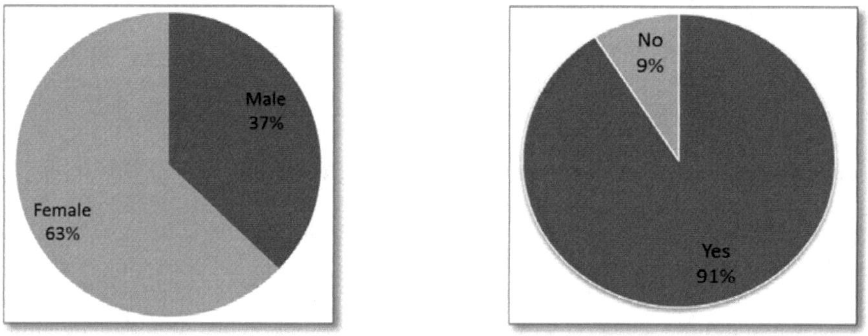

(a) **(b)**

Figure 1. (a) Percentage of respondent's gender (b) Percentage of respondent's safety knowledge.

Besides that, Figure 2 shows the percentage of working experience among laboratory staff. Based on the result, it shows that 48.2 % of laboratory staff are work more than 15 year, 18% work for 10 to 15 year, 11.3% of staff work for 7 to 9 year, 11% work for 4 to 6 year, 9% for 1 to 3 year and 2.35 % of staff work less than 1 year.

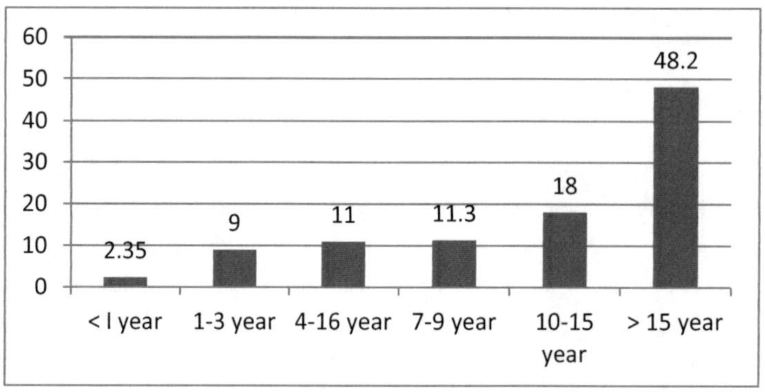

Figure 2. Percentage of working experience for respondents.

Table 2 show the reliability statistics for the questions used in this study. The result shows that Cronbach's Alpha is 0.97 and 0.91, which is considered reliable for the data [41]. Meanwhile, this study also measures the validity of each question. The alpha value for total correlation is lower than Cronbach's Alpha If Item Deleted, which means the questions used in this study are valid and accepted [32].

Table 2. Reliability statistic for question.

Question	Cronbach's Alpha	No. of Items
Safety awareness in Laboratory	0.969	12
Safety practice in Laboratory	0.910	9

Table 3 shows the data of safety awareness among staff laboratory. The mean score for each question is between 3 and 4, indicating that respondents agree with the questions in term of those aspects. According to the finding collected, most of them understand about the safety and health element that would be implementing in laboratory. The highest awareness was regarding the safety environment and surrounding in laboratory. They know and realize that, the surrounding in laboratory need to be free and safe from any obstacle or hazard that could contribute accidents such as electrical equipment. They also ensure that all electrical were switch off after laboratory activity. Cable and wires can be serious tripping hazards that pose electrical shock and fire hazards [39]. Besides that, most of the staff also agreed about the important of supervision that should give to students while doing some practical works in laboratory.

Appropriate supervisory practice helps to ensure safe environment in school for all academic activities that they had involved including activity in laboratory [33]. Other researcher found that some of students ignoring safety practices while doing practical work in the laboratory, which was disturbing their friend, and play around while doing some activity [34]. This behavior can contribute to accident happen in laboratory. Therefore, monitoring from staff or teachers is required when students do their practical work.

Table 4 shows the data regarding safety practice that implementing in laboratory. From the result, the mean score for each question is between 2 and 3. It is means that some of the staff agree and disagree with those aspects. Behavior also plays significant role in each organizational. According to researcher stated that, the factors that led to the accident is caused by the individual attitude on matters related to the safety and health [35]. Poor behavior regarding safety practice will effect smooth operation in school [33]. Based on the result, most of the staff was skilled to handle chemical in laboratory, know how to identify the hazard, danger and risk and know how to read Material Safety Data Sheet (MSDS) for chemical. This practice show good attitude for create safety environment in laboratory. They can establish a positive work culture and always concern about safety and health in any work done [24].

Besides that, most of staff not knows how to give cardiopulmonary resuscitation (CPR) training if emergency happen. The International Liaison Committee on Resuscitation published a consensus document on education in year 2003 which strongly recommended that CPR should be a standard part of the school curriculum. This skill is for their preparation for a response to medical emergencies in school [36]. Moreover, some studies also found and recommend that training in CPR should be required elements in school curricula and provides the rationale for implementation of CPR training for guidance and safety measure in overcoming barriers or accident that may occur in school [37]. Thus, staff should attend the training and know how to act while emergency happen.

Table 3. The mean score for safety and health element in laboratory.

No	Questions	Mean	Standard Deviation (SD)
Q1	Ensure that electrical wires are not tangled on the floor	3.69	0.54
Q2	Ensure switch off all the electrical equipment's in laboratory after use	3.6	0.48
Q3	Ensure the environment in laboratory are clean and tidy	3.67	0.49
Q4	Students need supervise by staff or teachers while doing practical work	3.66	0.47
Q5	Chemicals are labeled and storage in safe place	3.65	0.48
Q6	Laboratory have good ventilation	3.63	0.54
Q7	Fire extinguisher are put in suitable place and well-function	3.63	0.52
Q8	The exit way from laboratory should free from obstacles	3.62	0.48
Q9	There is a complete set of first-aid kit are put in laboratory and should always be checks to ensure there is no running out of medication.	3.60	0.53
Q10	Safety sign and symbol are display in laboratory	3.60	0.54
Q11	Safe operating procedure (SOP) of equipment should be displayed conspicuously	3.55	0.51
Q12	Teachers, staff or students can be exposed to chemicals that can be inhaled, ingested or absorbed by the body.	3.51	0.64

Table 4. The mean score for safety and health practice among staff laboratory.

No	Questions	Mean	Standard Deviation (SD)
Q1	Laboratory staff skilled in handling chemical in laboratory	2.97	0.66
Q2	I know the different between hazard, danger and risk that occur in laboratory	2.92	0.59
Q3	Laboratory staff know how to read Material Safety Data Sheet (MSDS) for chemical	2.92	0.67
Q4	I use Personal Protective Equipment while performing tasks in the laboratory	2.85	0.69
Q5	I know how to use fire extinguisher in the event of an emergency	2.87	0.71
Q6	I know the procedure of Emergency Response Plan	2.87	0.71
Q7	I can conduct hazard identification, risk assessment and risk control in schools	2.80	0.65
Q8	I know how to label and classify the chemical properly	2.73	0.40
Q9	I can give first aid and CPR training in case of emergency	2.41	0.80

4. Conclusions

The study shown that some of the staff in school laboratory understand and practice on safety and health matter in school. Meanwhile, some of them not develop safety and health practice while conducting activity in laboratory. This condition would contribute to unplanned event occur in laboratory. Knowledge and safety practice is important because it can create safety environment in workplace. For improvement, laboratory staff must be provided with information and occupational safety and health training that relevant to the hazards present in their laboratory. The training must be provided including new exposure situations. Knowledge regarding safety practice is very important while working in laboratory to prevent or minimize accidents. It is hoped that the number of accidents in school (including the loss of property whether directly or indirectly) also can be reduced with the good safety practice of staff laboratory. Thus, they can create safe environment in laboratory.

Acknowledgement

The authors would like to extend gratitude to Ministry of Higher Education in providing grant for this Knowledge Transfer Programme (R/KTP/A02.00/01163A/001/2015/000269). Special thanks goes to State Education Department, District Education Offices, school headmasters, school principal and teachers for willingness to participate in this study.

References

[1] Anna Diaz Vicario. Safety Management in Catalonia's school. *Social and Behavioral Sciences* **46**, 2012 : 3324 – 28

[2] Ngowi, A. B. & Mothibi, J. (1996). Culture and safety at work site. A case study of Botswana. In Alves Dias, L. M. & conle, R. J. Eds. Implementation of safety and Health and construction sites. Rotterdam. A A. Balkema. 417 – 27.

[3] Erkan T, T. B. (2009). Student Safety in Prmary Schools: A sample of Buyukcekmece county. *Procedia Social and Behavioral Sciences* **1**, 281-84.

[4] Scheps S.B, D. E. (1987). Epidemiology of School Injuries: A 2-year experience in a municipal health department. *Pediatrics* **79**, 96-75.

[5] Sinar Harian. (2017). Anxious broken thermometer in the laboratory in Kota Tinggi. *http://www.sinarharian.com.my/semasa/cemas-termometer-pecah-dalam-makmal-1.636999*

[6] Sinar Harian.(2017). Thermometer breaks, the mercury spilled in the laboratory in Alor Gajah. *http://www.sinarharian.com.my/edisi/melaka-ns/termometer-pecah-merkuri-tumpah-dalam-makmal-1.633485*

[7] GPS Bestari. (2016). Liquid mercury spilled on the floor in a school laboratory in Kota Bharu. *http://www.gpsbestari.com/berita/lima-kes-tumpahan-merkuri-di- makmal-sekolah-di-kelantan-1.582718*

[8] GPS Bestari. (2016). Reliable liquid mercury spilled in the school laboratory in Machang. *http://www.gpsbestari.com/berita/lima-kes-tumpahan-merkuri-di-makmal-sekolah-di-kelantan-1.582718*

[9] GPS Bestari. (2016). The thermometer used a student fell and caused the spill of mercury in Kota Bharu. *http://www.gpsbestari.com/berita/lima-kes-tumpahan-merkuri-di-makmal-sekolah-di-kelantan-1.582718*

[10] Sinar Harian. (2016). 5 students in Balik Pulau expose to mercury. *http://www.sinarharian.com.my/edisi/utara/lima-pelajar-mrsm-balik-pulau-terdedah-merkuri-1.569616*
[11] Metro. (2016). Three girls are exposed to mercury in Labuan. *http://www.hmetro.com.my/node/170245*
[12] Sinar Harian. (2016). Thermometer breaks, the mercury spilled in the laboratory school in Kota Bharu.*http://www.sinarharian.com.my/edisi/kelantan/termometer-pecah-merkuri-tumpah-di-makmal-sekolah-1.567188*
[13] Sinar Harian. (2016). Mercury spilled in the car students in Ipoh. *http://www.sinarharian.com.my/edisi/perak/hazmat-bantu-neutral-kesan-merkuri-dalam-kereta-1.586798*
[14] Sinar Harian. (2016). Sulphuric Acid in the laboratory spill in Sungai Petani. *https://translate.google.com/#ms/en/Asid%20dalam%20makmal%20tumpah*
[15] Metro. (2016). Nitric acid spilled in the laboratory school in Kuala Lumpur. *http://www.hmetro.com.my/node/163503*
[16] Metro. (2016). Students exposed to mercury in Seremban. *http://www.hmetro.com.my/node/161818*
[17] GPS Bestari. (2016). Liquid mercury contained in thermometers broken, resulting in spillage of chemical in Kota Bharu. *http://www.gpsbestari.com/berita/lima-kes-tumpahan-merkuri-di-makmal-sekolah-di-kelantan-1.582718*
[18] Berita Harian. (2016). Chemical Spill at School in Port Dickson. *http://www.bharian.com.my/node/182508*
[19] Mstars. (2016). Two Students horticulture Mercury Thermometer Bursts On Science Laboratory in Kepala Batas. *http://www.mstar.com.my/berita/berita-jenayah/2016/07/28/dua-murid-cemari-merkuri/*
[20] Utusan. (2016). 56 students were quarantined due to mercury spill in Bukit Mertajam. *http://www.utusan.com.my/berita/nasional/56-pelajar-dikuarantin-akibat- tumpahan-merkuri-1.335022*
[21] Fong, C. (2000). *Teks Ucapan Pelancaran Kempen Bulan Keselamatan dan Kesihatan Pekerjaan.* Kuala Lumpur: Kementerian Sumber Manusia.
[22] Brad, D. (2012, June 26). Perceptions of School Safety. *University of Nebraska at Omaha, EDAD 9550 Symposium on School Leadership.*
[23] Misnan, M. S. (2009). Modul Pembangunan Budaya Keselamatan Dalam Firma Pembinaan di Malaysia. *Universiti Teknologi Malaysia: Tesis PhD.*
[24] Fazreen, N.S. (2013). Kesedaran terhadap amalan keselamatan dalam kalangan pelajar di makmal kejuruteraan UTHM. Ijazah Sarjana Pendidikan Teknik dan Vokasional. Fakulti Pendidikan Teknikal dan Vokasional Universiti Tun Hussein Onn Malaysia
[25] Sohin, N.F. (2002). Pengurusan Keselamatan Bengkel di Kalangan Pelajar Tingkatan 4 Aliran Jentera di Sekolah Menengah Teknik Johor Bahru. Satu Tinjauan. Skudai: Universiti Teknologi Malaysia. Tesis Sarjana Muda Strasser,
[26] Jamaludin, J.Z. (2001). Amalan Keselamatan Bengkel Di Kuittho : Satu Kajian Kes Terhadap Sikap Pelajar Kursus Sarjana Muda Kejuruteraan Awam. Batu Pahat : Kolej Universiti Teknologi Tun Hussein Onn. Tesis Sarjana PendidikanTeknikal.
[27] Dilley, H. dan Kleiner, B.H. (1996). Creating a Culture of Safety. *Work Study* **45** (3): 5-8
[28] Jaafar, M.I. (2001) Amalan keselamatan di bengkel kerja kayu Kejuruteraan Awam (sat tinjauan di Politeknik Ungku Omar, Politeknik Port Dickson dan Politeknik Shah Alam. Batu Pahat: Kolej Universiti Teknologi Tun Hussein Onn. Tesis Sarjana.

[29] Kamaruddin, M. I., & Yazit, N. H. (2011). Tahap Pengetahuan Amalan Keselamatan Makmal Sains Dalam Kategori Guru Pelatih Sains, Fakulti Pendidikan, Universiti Teknologi Malaysia. *Jurnal of Science & Mathematic Education* **4** December 2011: p1.

[30] Unit Keselamatan dan Kesihatan Pekerjaan (UKKPK) (2005). Kursus Asas Keselamatan Makmal dan Bengkel. Dicapai Mac 20, 2012, From http://www.usm.my/ukku

[31] OSH guideline(2013). http://www.uq.edu.au/ohs/pdfs/OHYG-LabSafety-UG.pdf: 22 January 2013.

[32] Mohd, Y. (2010). Analisis Data Penyelidikan Untuk Pendidikan (Pendidikan & Sains Sosial). Kuantan: Bandar Ilmu.

[33] Chima, S. U. (2016). Improving school safety climate in public schools through supervision at 1st and 2nd tiers of Nigerian educational system. *IOSR Journal of Research & Method in Education (IOSR*-JRME) e-ISSN: 2320–7388,p-ISSN: 2320–737X. Volume **6**, Issue 5 Ver. II : 12-17 www.iosrjournals.org

[34] Boon, Y. & Kamarudin, S. (2010). Tahap Penguasaan Amalan Keselamatan Bengkel Kemahiran Hidup Dalam Kalangan Pelajar Tingkatan Empat Di Sekolah Menengah Kebangsaan Dato Yunus Sulaiman, Pekan Nanas, Pontian, Johor Darul Takzim. Skudai: Universiti Teknologi Malaysia

[35] Laney, J.C. (1987). Psychology : The science of People. Englewood Cliff: Prentice Hall Inc.

[36] Chamberlain DA, Hazinski MF. (2003). Education in resuscitation: an ILCOR symposium *Circulation* **108** : 2575–94

[37] Diana M.C, RN, MSN, CEN,Chair. Tom, P.A (2011). Importance and Implementation of Training in Cardiopulmonary Resuscitation and Automated External Defibrillation in Schools. *The American Heart Association Science Advisory, Circulation* **123**: 691-706.

[38] UTHealth -University of Texas Health Science Centre. (2008). Safety, Health, Environment and Risk Management. Laboratory Survey. https://www.uth.edu/safety/chemical-safety/laboratory-surveys.htm

[39] Department of Administrative and Financial Services, (2016). Codes, Wires and Tripping Hazards. http://maine.gov/bgs/riskmanage/tipofthemonth/tip45.htm

[40] Sabri. W. I, (2012). Kesedaran Keselamatan Dalam Kalangan Pelajar Kolej Kemahiran Tinggi Mara Di Lembah Kelang. Fakulti Kejuruteraan Teknikal dan Vokasional Universiti Tun Hussein Onn Malaysia

[41] Bahaman Abu Samah. 2012. Exploring statistics using SPSS. Nota Kuliah. Serdang, Selangor: Universiti Putra Malaysia

IOP Conf. Series: Materials Science and Engineering **257** (2017) 012005 doi:10.1088/1757-899X/257/1/012005

Increasing FSW join strength by optimizing feed rate, rotating speed and pin angle

Djarot B Darmadi[1], Anindito Purnowidodo[1] and Eko Siswanto[1]

[1]Mechanical Engineering Department, Brawijaya University, Malang, Indonesia

*Corresponding author: b_darmadi_djarot@ub.ac.id

Abstract. Principally the join in Friction Stir Welding (FSW) is formed due to mechanical bonding. At least there are two factors determines the quality of this join, first is the temperature in the area around the interface and secondly the intense of mixing forces in nugget zone to create the mechanical bonding. The adequate temperature creates good flowability of the nugget zone and an intensive mixing force produces homogeneous strong bonding. Based on those two factors in this research the effects of feed rate, rotating speed and pin angle of the FSW process to the tensile strength of resulted join are studied. The true experimental method was used. Feed rate was varied at 24, 42, 55 and 74 mm/minutes and from the experimental results, it can be concluded that the higher feed rate decreases the tensile strength of weld join and it is believed due to the lower heat embedded in the material. Inversely, the higher rotating speed increases the join's tensile strength as a result of higher heat embedded in base metal and higher mixing force in the nugget zone. The rotating speed were 1842, 2257 and 2904 RPMs. The pin angle determines the direction of mixing force. With variation of pin angle: 0°, 4°, 8° and 12° the higher pin angle generally increases the tensile strength because of more intensive mixing force. For 12° pin angle the lower tensile strength is found since the force tends to push out the nugget area from the joint gap.

1. Introduction

Recently, in the name of efficiency, metals with high strength to weight ratio are more preferable [1,2]. One of the prospective candidate is aluminum, which has strength to weight ratio higher than steel the most commonly used metals today. The drawback of the aluminum is its low weldability in term of fusion welding [3]. On the other hand, the welding institute (TWI) developed solid state welding, such as friction welding and FSW [4,5]. Basically, FSW mechanically joins two piece of metals in the same manner of unifying two lumps of clay. Hopefully this solid state welding fills the gap of aluminum low weldability.

Figure 1 shows how FSW is applied. The heat obtained from friction between tool and workpiece increases temperature, especially in the area close to the interface. Around 90% of heat is obtained from friction between shoulder and workpiece, while 5% is attained from the probe (pin) and the rest is collected by plastic strain developed in the base metal [6].

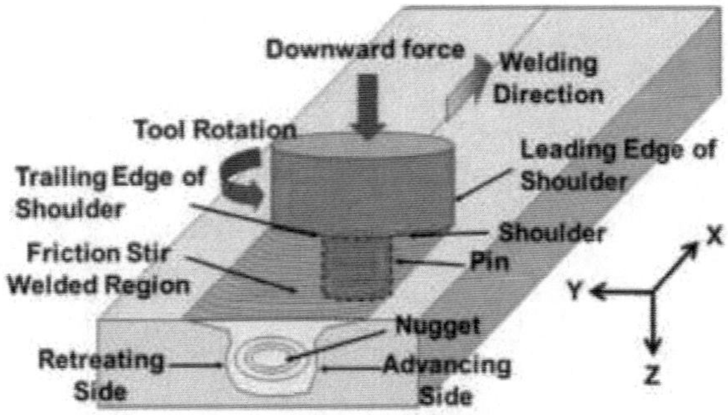

Figure 1. Schematic of FSW process [7].

In FSW process are developed five region types as shown in figure 2, those are: nugget zone, thermo-mechanically affected zone (TMZ), heat affected zone (HAZ), flow arm and base metal (BM) [8, 9]. The heat certainly increases the temperature in the base metal and this increased base metal temperature forms nugget zone, which can be analogized by a lump of clay. Nugget zone, also known as stirred zone or dynamically recrystallized zone (DRZ), can be identified by the presence of fine equiaxis grain when the FSW process has been completed. The stirred zone refers to the area where the pin movement has direct effects. The pin movement causes high temperature, strain and deformation which lead to grain recrystallization. It was found that this new grain is finer than initial base metal. Since the recrystallization is due to plastic deformation it is called as DRZ [7, 10]. The mixing forces due to friction, centrifugal and centripetal forces act as forces to mix two lumps of clay. With this mixing forces in the nugget zone, finally mechanical bonding is formed.

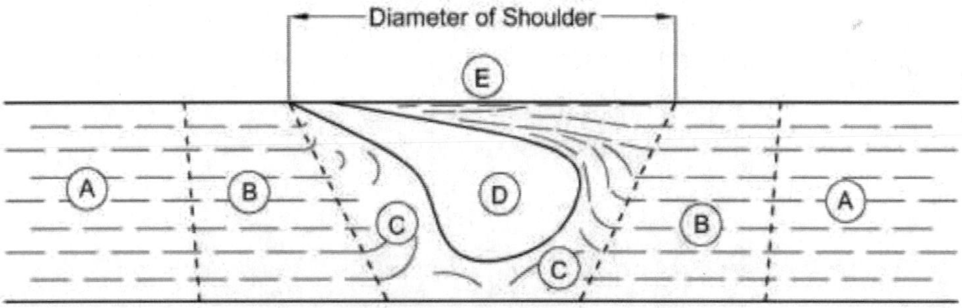

Figure 2. Schematic cross-section of a typical FSW weld: (A) base metal, (B) heat affected zone, (C) thermo mechanically affected zone, (D) nugget Zone and (E) flow arm [9]

The friction tool and its component clearly are the most important instrument in FSW process. In figure 3 is shown the friction tool and nomenclature of its element. Dimensions of tool's element which was used for this research are also attached and all measurements are in mm.

Considering the principle of how mechanical bonding is formed in FSW, it is plausible to consider feed rate, rotating speed and pin angle as important parameters that influence the quality of the joint. With a different feed rate while other parameters are constant varied heat is embedded in the base metal which creates different volume and temperature distribution of nugget zone. Variations on rotating speed determine the amount of heat and the value of forces. These forces mix the nugget area and creates mechanical bonding. Since forces as a vector are not only determined by its value, but also its direction, the pin angle influences the mechanical bonding by means of alternating the force's direction. In short, it can be said that, the objective of this research is observing the effect of feed rate, rotating speed and pin angle to the final resulted joint's strength (i.e. tensile strength).

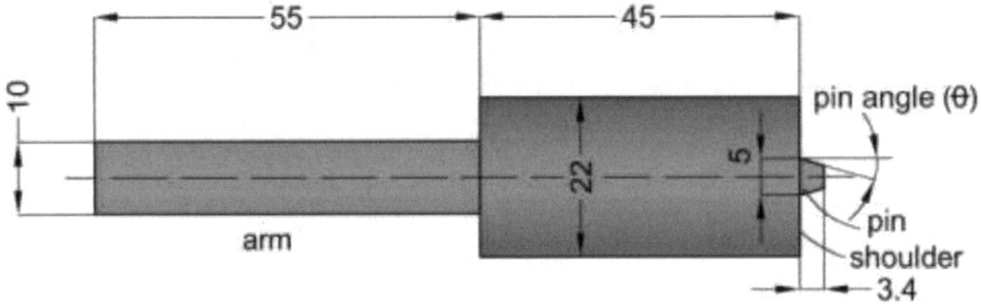

Figure 3. The nomenclature of friction tool and its dimensions

2. Experiment set up

There are three ways to understand a phenomenon, especially welding: analysis, modelling and experiment [11, 12]. Those methods enhance each other. This paper discusses true experiment method of FSW phenomenon. The experiment was carried out in Manufacturing Processes Laboratory of Mechanical Engineering, Brawijaya University. The FSW was performed on modified vertical milling machine. The feeding speed was varied using the available variation which is provided by the machine, those are: 24, 42, 55 and 74 mm/minute. The rotating speed was determined by belt and pulley installation and the provided rotating speed are: 1842, 2257 and 2904 rpm. Tools were made from EMS 45 hardenable tool steel. After pin, shoulder and arm have been formed from rod steel using turning machine the tool was hardened. For practical production reason the pin angle was varied at 0°, 4°, 8° and 12°. These angles were considered will give enough data to know the effect of pin angle on join's strength. In figure 4 is shown the tools before they are hardened. The hardening was done by heating the tools in furnace up to 1100°C and suddenly quenched in oil. After the tool cooled down to room temperature the tempering was applied by heating the tool in furnace up to 300°C and naturally cooled down to room temperature by exposing the tool in ambient temperature.

Figure 4. Friction tools with varied pin angle, from left: 0°, 4°, 8° and 12°

The base metal is a commercial A1090 Aluminum sheet with thickness equal to 3.5mm. First, the sheet is cut in a 70mm x 125mm rectangular, then friction stir welded on the 70mm side as shown in figure 5. How the FSW was carried out on milling machine is shown in figure 6.

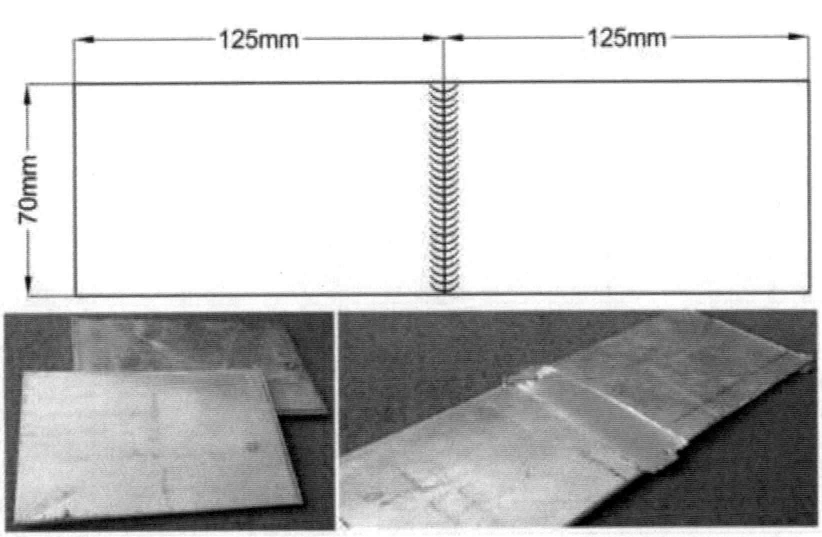

Figure 5. Aluminum plate as friction stir welded

Figure 6. Aluminum plate as friction stir welded

After the join has been formed the welded sheet was cut as a tensile specimen following AWS B4.0; 2007 as shown in figure 7. The tensile specimen, then is attached at Kai Wei universal testing machine to obtain ultimate tensile stress. These ultimate tensile stress for alternated parameters (feed rate, rotating speed and pin angle) which are considered to have significant influence on quality of resulted join, then tabulated and presented in graphics.

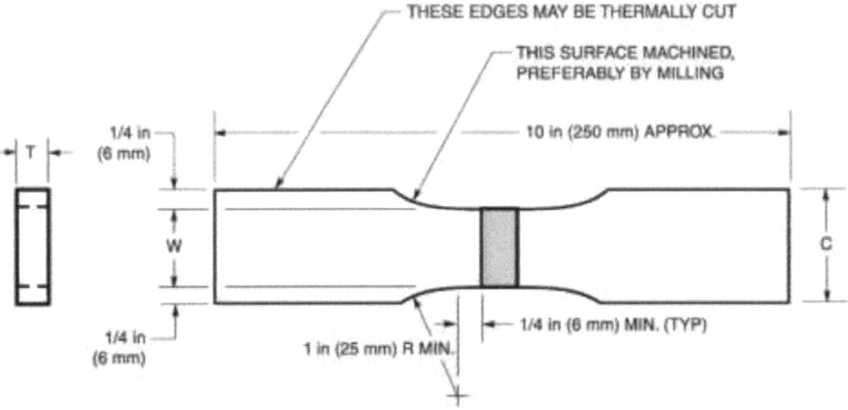

T	W
<1 in (25 mm)	1.50 in ± 0.01 in (38 mm ± 0.25 mm)
≥1 in (25 mm)	1.00 in ± 0.01 in (25 mm ± 0.25 mm)

Figure 7. Tensile test spesimen

3. Results and discussion

Table 1 shows the results for varied feed rate (24, 42, 55 and 75 mm/minute) with rotating speed equal to 1842, 2257 and 2904 rpms. The numbers in table 1 are presented in graphic as shown in figure 8. figure 8 shows normalized ultimate tensile strength for the miscellaneous parameters means the obtained tensile strength are compared to base metal ultimate tensile strength and presented in percent. The tensile strength of the FSW join for sundry parameters is below a hundred percent.

The maximum tensile strength is equal to 89.81% for specimen no. 3 which is obtained for feed rate equal to 24 mm/minute and rotating speed is 2904 rpm. As it is expected, the higher feed rate decreases the ultimate tensile stress. The lower feed rate means higher heat produced by friction force is embedded in the base metal which in turn increases ultimate tensile stress. In contrast, the embedded heat to the base metal is proportionate to the rotating speed means with higher rotational speed the bigger heat is embedded into base metal. On the other hand, higher rotating speed also means higher centrifugal and centripetal force. These forces analogous to mixing force of joining two lumps of clay and with higher centrifugal and centripetal forces better mechanical bonding will be produced. These two reasons, higher heat and forces, together contribute to higher ultimate tensile strength of the resulted FSW join. The second order polynomial for regression lines also attached in figure 8. For 1842 rpm the regression line follows the expression: $0.003f^2 - 0.978f + 101.3$ (%) with f is feed rate. The R^2 for the line is 0.952 which means the expression can very well represent the data. For 2257 rpm the mathematical expression for regression line and the value of R^2 are $0.006f^2 - 1.231f + 110.8$ (%) and 0.988 respectively again describes the equation very well represents the data. For 2904rpm the equation for normalized ultimate strength and R^2 are $0.007f^2 - 1.21f + 113.3$ (%) and 0.999 respectively. The R^2 describes the equation well match the data.

IOP Conf. Series: Materials Science and Engineering **257** (2017) 012005 doi:10.1088/1757-899X/257/1/012005

Table 1. The resulted ultimate tensile strength for varied feed rate.

No.	Feed rate (mm/min)	Rotating speed (rpm)	Spec	σ_{ult} (MPa)	Means (MPa)
1	24	1842	1	55.14	59.21
			2	58.36	
			3	64.14	
		2257	1	65.43	63.55
			2	59.86	
			3	65.36	
		2904	1	65.21	66.36
			2	66.50	
			3	67.36	
2	42	1842	1	51.32	52.54
			2	51.89	
			3	54.41	
		2257	1	54.84	54.43
			2	52.64	
			3	55.82	
		2904	1	55.14	56.83
			2	57.71	
			3	57.64	
3	55	1842	1	38.50	40.98
			2	46.14	
			3	38.29	
		2257	1	46.43	46.33
			2	44.64	
			3	47.93	
		2904	1	49.64	51.57
			2	52.07	
			3	53.00	
4	74	1842	1	38.50	36.76
			2	35.21	
			3	36.57	
		2257	1	42.14	42.57
			2	43.57	
			3	42.00	
		2904	1	47.14	48.10
			2	48.43	
			3	48.71	
Base metal					75MPa

Table 2 shows the result for various pin angles (0°, 4°, 8° and 12°) with rotating speed equal to 1842 and 2257 rpms. The maximum ultimate tensile strength is 70.52 MPa for specimen 1 with pin angle 8° and rotating speed 2257 rpm. For rotating speed equal to 1842 rpm the maximum ultimate tensile strength is 61.98 MPa for specimen 3 with pin angle again equal to 8°.

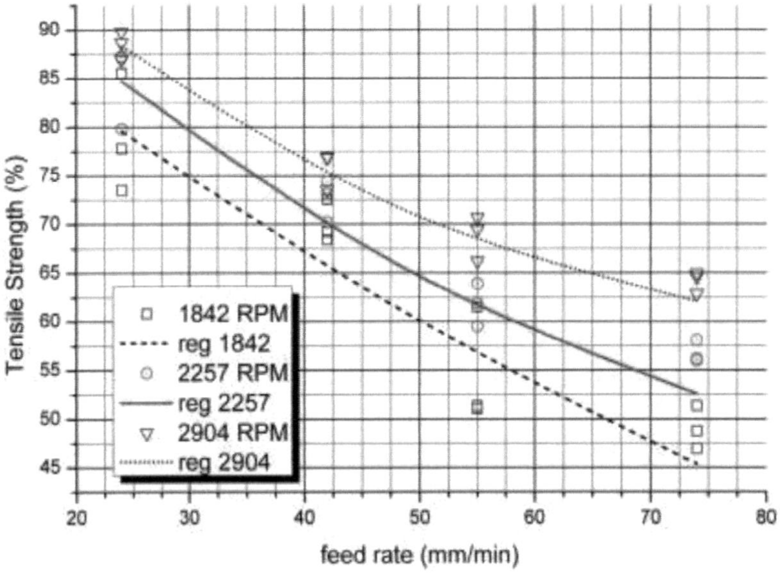

Figure 8. The effect of feeding speed on ultimate tensile strength

The numbers in table 2 are presented in a graphic as shown in figure 9 in which the ultimate strength is presented as normalized strength. The third order polynomial for regression line also included. For 1842 rpm the normalized tensile strength can be expressed as pin angle $-0.15\theta^3 + 1.984\theta^2 - 5.012\theta + 70.05$ and for 2257 rpm the normalized tensile strength is $-0.197\theta^3 + 2.731\theta^2 - 6.984\theta + 72.5$. The ultimate tensile strength for specimen 1 with pin angle 8° and rotating speed 2257 rpm in figure 9 is presented as normalized strength: 94.02% whilst for specimen 3 with rotating speed equal to 1842 rpm and the 8° pin angle is 82.64%. As previous results (figure 8), the higher the rotating speed also yields stronger FSW join. Pin without angle (0°) generates lower ultimate tensile strength of FSW join. Pin with angles exhibits more complex force's direction that mixes the nugget zone more intensive as shown schematically in figure 10. This intensive force generates more homogeneous and stronger mechanical bonding. Generally, the pin angle increases the resulted tensile strength.

However, the 12° pin angle generates lowest tensile stress. As shown in figure 10 the higher pin angle tends to push the nugget zone downward. This phenomenon may cause a part of nugget zone is pushed out of the bonding area, especially if the pin angle is too high (12°). The macro photos of 8° and the 12° cross section as shown in figure 11 proved this assumption. Since a part of the nugget zone for 12° is driven out of bonding area, small holes found in the joint area which in turn severely decrease the ultimate strength of the FSW join.

IOP Conf. Series: Materials Science and Engineering **257** (2017) 012005 doi:10.1088/1757-899X/257/1/012005

Table 2. The resulted ultimate tensile strength for varied pin angle.

No.	pin angle	Rotating speed (rpm)	Spec	σ (MPa)	means (MPa)
1	0°	1842 rpm	1	51.32	52.54
			2	51.89	
			3	54.41	
		2257 rpm	1	54.84	54.43
			2	52.64	
			3	55.82	
2	4°	1842 rpm	1	53.07	54.10
			2	53.80	
			3	55.41	
		2257 rpm	1	56.68	56.79
			2	55.74	
			3	57.94	
3	8°	1842 rpm	1	58.76	59.93
			2	59.04	
			3	61.98	
		2257 rpm	1	70.52	67.86
			2	64.75	
			3	68.31	
4	12°	1842 rpm	1	28.55	26.69
			2	25.31	
			3	26.21	
		2257 rpm	1	29.99	30.81
			2	31.35	
			3	31.09	
Base metal					75MPa

IOP Conf. Series: Materials Science and Engineering **257** (2017) 012005 doi:10.1088/1757-899X/257/1/012005

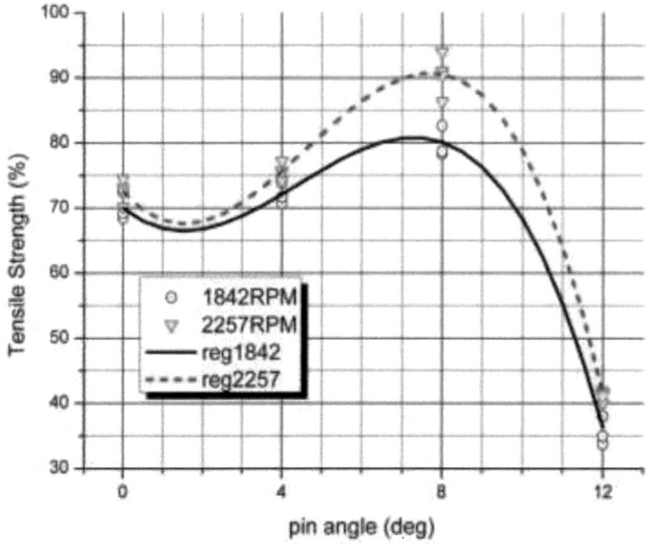

Figure 9. The effect of pin angle on ultimate tensile strength

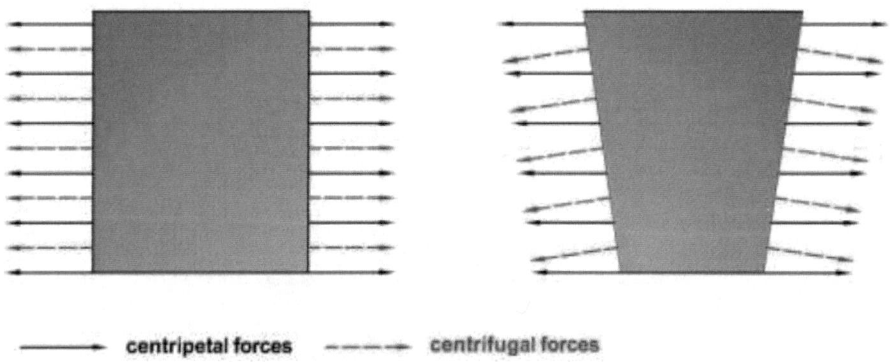

Figure 10. Force vectors for a pin without angle (left) and with angle (right)

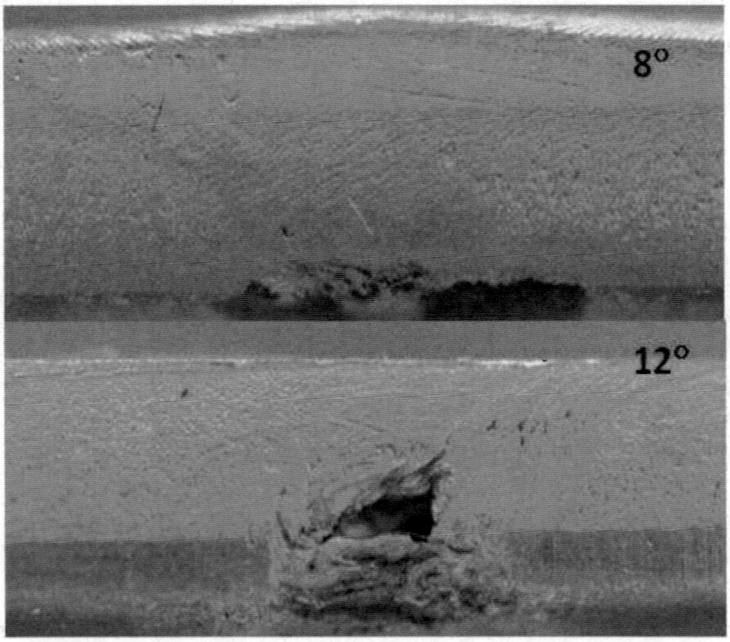

Figure 11. Macrophoto for 8° and 12° cross sections

4. Conclusions

Two factors determine the quality of FSW join are temperature and force. In terms of temperature, the feed rate and rotating speed have significant influence on resulted ultimate stress of FSW join. Lower feed rate and higher rotating speed increase the joint strength. The rotating speed also increases the value of centrifugal and centripetal forces whilst the pin angle determines the direction of centrifugal force. With the increasing value of forces the strength of the FSW will be improved. With higher pin angle, a better join strength is provided. This phenomenon is correct for pin angle up to 8° since 12° pin angle drives out nugget zone cause holes in join area which in turn decreases tensile strength of the FSW join.

References

[1] Djarot B. Darmadi 2016, *Evaluation of the effects of melting phenomenon on the residual stress formation in ferritic pipeline multi pass girth-weld joints,* International Journal of Engineering Systems Modelling and Simulation, vol. 8, no. 3, pp. 205-217.

[2] Djarot B. Darmadi 2015, John Norrish and Anh Kiet-Tieu *Residual stress analysis of pipeline girth weld joints* Lambert Academic Publishing, Deutschland / Germany.

[3] R. Hariharan dan RJG. Renjith Nimal 2014 *Friction stir weld of dissimilar aluminium alloys (6061 & 7075) by using computerized numerical control machine*, Middle-East Journal of Scientific Research, vol. 20 (5), pp. 601 – 605.

[4] W. M. Thomas, E. D. Nicholas, J. C. Needham, M. G. Murch, P. Temple-Smith and C. J. Dawes 1991: *Friction stir butt welding*, GB patent no. 9125978· 8, 1991.

[5] W. M. Thomas, E. D. Nicholas, J. C. Needham, M. G. Murch, P. Temple-Smith and C. J. Dawes 1991: *Improvements relating to friction welding*, US patent no. 5 460 317; EPS 0 616 490.

[6] Renju Mohan, N.R. Rajesh and Satheesh Kumar S. 2014 *Finite element modeling for maximum temperature in friction stir welding of AA 1100 and optimization of process parameter by Taguchi method* Int. J. of Research in Engineering and Technology, vol. 3, Issue 5, pp. 728 – 733.

[7] Rajiv Sharan Mishra, Parha Sarathi De & Nilesh Kumar 2014, *Friction stir welding and processing* Science and Engineering, Springer International Publishing, Switzerland.

[8] R. Nandan , T. DebRoy , H.K.D.H. Bhadeshia 2008 *Recent advances in friction-stir welding – Process,weldment structure and properties* Department of Materials Science and Engineering, Pennsylvania State University, University Park, PA 16802, USA Department of Materials Science and Metallurgy, University of Cambridge, Cambridge CB2 3QZ, UK.

[9] Rizki Prasetyo Utomo Toha, Achmad As'ad Sonief & Djarot B. Darmadi 2017, *Pengaruh kecepatan putar dan feed rate terhadap kekuatan tarik sambungan las friction stir welding aluminium A1090*, Jurnal Rekayasa Mesin, vol. 8, no. 2, *in-printed.*

[10] A. L. P Threadgill, H Shercliff P Wither 2009, *Friction stir welding of aluminium aloys* International Materials Reviews, vol. 54, pp. 44-93.

[11] Djarot B. Darmadi, John Norrish & Anh Kiet Tieu 2011, *Analytic and finite element solutions for temperature profiles in welding using varied heat source models*, World Academy of Science, Engineering and Technology, vol. 81, pp. 154-162.

[12] Djarot B. Darmadi, Anh Kiet Tieu & John Norrish 2014, *A validated thermo mechanical FEM model of bead-on-plate welding*, International Journal of Materials and Product Technology, vol. 48, nos. (1-4), pp. 146-166.

Parameter optimization and stretch enhancement of AISI 316 sheet using rapid prototyping technique

M Moayedfar[1, 2, 3, *], A M Rani[1, 2], H Hanaei[1], A Ahmad[3] and A Tale[4]

[1] Department of Mechanical Engineering, Universiti Teknologi PETRONAS, 32610 Bandar Seri Iskanadar, Perak Darul Ridzuan, Malaysia
[2] Centre of Intelligent Signal and Imaging Research, Universiti Teknologi PETRONAS, 32610 Bandar Seri Iskanadar, Perak Darul Ridzuan, Malaysia
[3] Faculty of Engineering and Technology, DRB-HICOM University of Automotive Malaysia, 26607 Pekan, Pahang Darul Makmur, Malaysia
[4] Wind Energy Technology Institute, Flensburg University of Applied Sciences, Flensburg, Germany

*Corresponding author: majid@dhu.edu.my

Abstract Incremental sheet forming is a flexible manufacturing process which uses the indenter point-to-point force to shape the sheet metal workpiece into manufactured parts in batch production series. However, the problem sometimes arising from this process is the low plastic point in the stress-strain diagram of the material which leads the low stretching amount before ultra-tensile strain point. Hence, a set of experiments is designed to find the optimum forming parameters in this process for optimum sheet thickness distribution while both sides of the sheet are considered for the surface quality improvement. A five-axis high-speed CNC milling machine is employed to deliver the proper motion based on the programming system while the clamping system for holding the sheet metal was a blank mould. Finally, an electron microscope and roughness machine are utilized to evaluate the surface structure of final parts, illustrate any defect may cause during the forming process and examine the roughness of the final part surface accordingly. The best interaction between parameters is obtained with the optimum values which lead the maximum sheet thickness distribution of 4.211e-01 logarithmic elongation when the depth was 24mm with respect to the design. This study demonstrates that this rapid forming method offers an alternative solution for surface quality improvement of 65% avoiding the low probability of cracks and low probability of crystal structure changes.

1. Introduction

There have been significant developments over the past decades regarding the availability of proper metals and metal alloys for sheet metal parts in low and batch production series. Different forming methods are often used by sheet metal industry which is mostly founded on the application of dies and punches with the correct dimensions of the final part to be planned. Generally, these manufacturing processes are employed for mass production series because the cost of the dies will be shared among the products. Nevertheless, the formal approaches based on dies, such as drawing or stamping are not useable if a small amount of product is needed and, consequently, due to the meeting of demands levied by the short and batch series production industries, new manufacturing techniques have to be established. The incremental sheet forming (ISF) process has encountered to be a trustworthy solution among the different possibilities. This technique is built on the shaping of a metal sheet via a computer numerical controlling (CNC) tool or indenter which deforms the blank sheet plastically with respect to the preferred design [1-3]. The major factors affecting localised deformation were found to contact stress and shear mechanisms [4].

The current forming process of sheet metal parts suffers from the mechanical defects such as cracks and non-homogenize sheet distribution which cause fatigue, sheet failure and inaccuracy [5, 6]. Moreover, based on the Micari et al. (2007) explanation, some basic issues related to ISF exist. Their study demonstrated that geometric accuracy can be thought as one of the major limitations of ISF, which can be recompensed through toolpath optimisation [7]. Hence, this study aims to find the minimum defect and maximum surface quality (SQ) of the sheet metal parts using optimum forming parameters in ISF process.

One of the important parameters is spindle speed, which affects friction and consequently surface quality [1]. Some other parameters, which are mentioned in Hussain et al. (2011), include sheet thickness, tool radius, width step, and wall angle. The effective combination of these parameters will produce a highly significant and efficient result. Their study showed that the small radius of the tool combined with a lower width step caused a reduction in profile error, especially when a large wall angle was considered [8]. However, Attanasio et al. (2008) believed that step down (that is a changed form of depth of cut in milling process) also is influential in this process [8, 9]. Both works were used a longstanding version of the design of experiment method which had not a respond method for the parameters interaction. In another study, Ambrogio et al. (2012) found that the quality of the bottom surface in a hot ISF (Ra 0.234 μm) is almost 20% higher when compared with the one in contact with the punch but more than Hussain et al. (2011) study. This is because of both the lower temperatures (around 200°C) reached on the bottom side, which decreased the oxidation phenomenon and caused the absence of mechanical actions on the sheet [10].

It should be mentioned that based on Duflou et al. (2007), in some cases, sheet strain is directly due to the influence of wall angle [11]. An increase in wall angle would result in a strong part being produced and would also prevent any failures and defects; however, it also limits the design variability in sheet metal parts. Fritzen et al. (2013) explained a different case study, which showed that it is difficult to produce a product with a long 90° wall angle, according to the sine law, as per equation (1) [12].

$$\boldsymbol{t} = \boldsymbol{t_0}\, sin\, \boldsymbol{\theta} \tag{1}$$

Where $\mathbf{t}$ is the final thickness, $\mathbf{t_0}$ is the initial thickness, and $\boldsymbol{\theta}$ is the wall angle. There are some other processes such as thinning and ironing that make inaccuracy in the final results of experiments compare to the prediction by the Sine low.

It is assumed that formability can be enhanced with forming speed because of heating effects. In Ham and Jeswiet (2006), an investigation related to the formability of aluminium AA3003 sheets in ISF was carried out through several experiments. The friction heat at the contact zone was high, as the rotational speeds for tool setups were high and thus, it was concluded that formability can be improved with a higher rotational speed. Additionally, the reduction in feed rate also improved blank formability [13]. Another study produced a dome geometry to describe the effect of spindle speed on the formability of St 37-2 steel. The results demonstrated that the tool rotational speed has an impact on formability and that when tool rotational speed is raised, formability will be reduced [14]. Besides that, the tool feed rate has a low impact on formability, as the results demonstrate reduced formability with an increase in feed rate. However, a year before this study, Hussain et al. (2008) conducted a cold incremental forming process to investigate the impacts of various features related to the formability (maximum wall angle) of a commercially pure titanium sheet. It was observed that, against the result of Rattanachan (2009), a raise in the feed rate reduced formability, and hence, the connection between feed rate and maximum wall angle followed a quadratic curve [15].

It should be mentioned that based on Nimbalkar et al. (2013), the forming process achieved higher feed rates compared to typical machining feeds. Due to the fact that the tool is hemispherical, no concern exists, related to the amount of material cut per tooth (flute) for each revolution; a critical factor in determining feed rates in conventional milling [16]. In addition, they noted that step-down regulates the z-axis increments. For some of the cases, an adjusted step-down for various wall angles was utilised for maintaining a constant traverse distance over the metal surface during the manufacture of a range of pyramids and cones with various wall angles. The case became very complex when complicated geometry was involved with various wall angles at diverse locations. Normally, in this case, a constant diameter forming tool with a standardised step-down would be used. The step-down also regulates the surface quality. Consequently, the optimum parameters can have a significant influence on the surface quality, and sheet thickness distribution value of the final part.

2. Methodology

In order to determine the optimum value for each of the forming parameters, an experimental study was designed. This set of experiments was combined with the numerical simulation to reduce the time, tool and sheet material waste. For this purpose, the Design-Expert® (State-Ease Corporation) software was employed to design the experiments for a better interaction between the parameters as the input data and the sheet thickness distribution (STD) value as the output factor. Hence, the parameters were inserted in the software environment and the Response Surface Method (RSM) with central composite design (CCD) technique was applied. There were some possibilities of using the DOE method for this study but CCD that finds the optimal levels of the design variables by adding a few more experiments to the Full Factorial design let to it being selected. Table 1 presents the range of input data for the software for creating the set of experiments in this study. Based on the Hussain et al. (2011) and Attanasio et al. (2008), the minimum and maximum values for each parameter were selected and inserted into the software.

Table 1. The range of forming parameters inserted to the DOE.

Parameter	Range of Variables
Feed Rate	10-200 mm/min
Spindle Speed	50-200 rpm
Step down	0.1-0.7 mm
Width Step	1-4 mm

The width step was based on a percentage of the tool diameter and was selected from 1 to 4 millimetres since the tool diameter is considered to be 10 mm for this study. In regard to some inapplicable values suggested by the software, for example, -0.1 mm for depth step (or DOC), the Alpha function was activated to bring in real values and reduce the negative values since all the applied variables should be more than zero in these experiments. Finally, the output data was created and then transferred to Microsoft Excel®.

The experiment for applying ISF on AISI 316L is adopted from previous work [17] which was clamping the metal sheet alloy from the edges and fix on the table of CNC milling machine using complete clamping system. This clamping includes a main part which is a cylindrical shape and a lid with 6 screws to fix the metal sheet. Toolpath also generated from computer-aided manufacturing (CAM) system and then modified with optimized parameters for ISF process. Tool trajectory applied in combination of manual and software programming as the "combined programming system" to increase the time efficiency and also reduce the amount of scratch [17, 18]. The frustum of a pyramid design include surfaces, edges and fillets used with a 24mm depth and the lubricant also flow on the surface of sheet reducing friction which causes the high temperature on the contact area, between the tool and sheet metal. Since the temperature is the main cause of the shear failure in this process, it will control sequent. After some cycles of the process, one sink will be created where the lubricant liquid flow inside it, hence, the heat will rise up to 80 °C when the tool is inside the pool of oil and stay steady until the end of the experiment as it is shown in Moayedfar (2014) work.

The pressure direction defined as Zhu et al. (2012) study, which explained that an optimized tool approach angle on the surface of sheet metal can effect on the depth of drawing [19]. In this case, 5 axes CNC milling machine is employed to make the motion similar to a robot actuator and then, shape the metal sheet alloy incrementally. Figure 1 shows the installed set of tool and clamp on the CNC machine. Toolpath also is directly taken from the CAM systems for the part to be formed.

Figure 1. Setup of ISF process on the CNC milling machine table [17].

In order to find the morphology of the sheet metal surface, the Field Emission Scanning Electron Microscope (FESEM) is employed to capture the surface structure of sheet material before and after the process and also the behind side of sheet metal after the process since that side will be the outer side of the implant (The inner side will contact directly with bone surface). Finally, roughness test machine was employed to magnifier the surface roughness of the sheet metal to detect a number of scratches introduced by the indenter during the process in all three mentioned surfaces.

3. Results and discussion

There were only 22 samples out of 30 samples suggested by DOE that provide the complete sheet stretching to the final depth design. In the other 8 samples, UTS were appeared before shaping the complete pyramid. Therefore, those samples considered as the failed tests and did not use in the characterization. However, 22 samples are examined in roughness test and surface morphology determination even those with low surface quality which could be detected visually.

The roughness test is employed to determine the surface quality of samples in three different sides of the raw material, the contact surface and backside surface. The results of tests illustrated the optimum set of parameters for the forming process. Then, respond for all 22 samples sent to the DOE for analysing and defining the parameters interaction. Figure 2 shows the best sample regarding the SQ of the contact surface in roughness test compare with the other surfaces and STD of 4.211e-01 (LE). As can be seen in figure 2 (a), the raw material SQ is Ra 0.359 μm while the contact surface as shown in figure 2 (b) has Ra 3.073 μm. Figure 2 (c) also illuminate the roughness of Ra 1.590 μm for the sheet backside of the part.

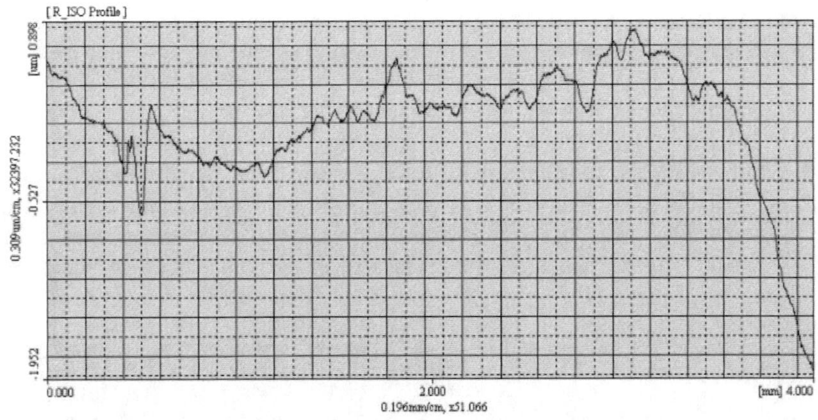

(a)

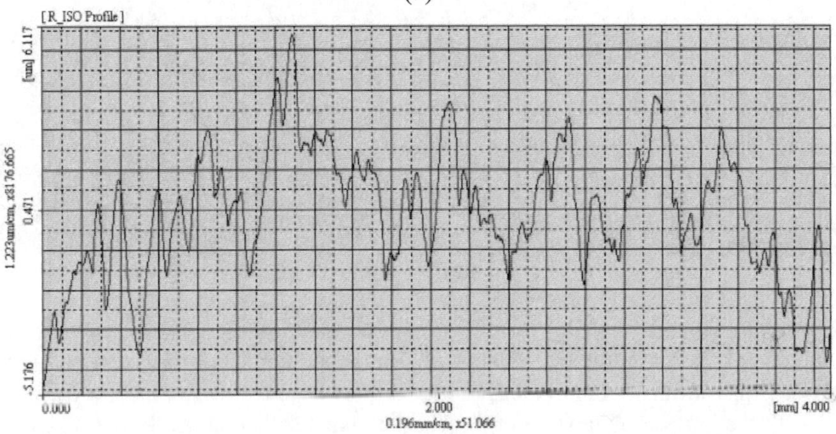

(b)

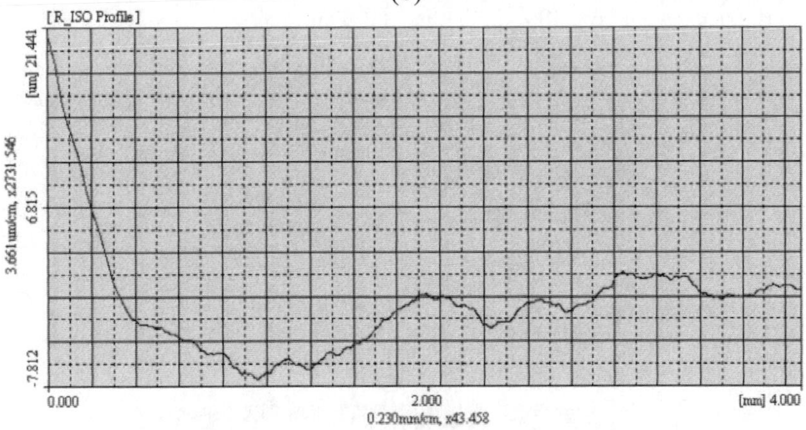

(c)

Figure 2. Roughness test of the best sample. (a) Raw material, (b) contact surface, (c) back-side surface.

As it shown, the surface quality of the contact surface 10 times is lower than the raw material but the back-side surface is around four times lower than the base material. Consequently, the amount of scratch in ISF is the big challenge in SQ of final parts that can be minimize via proper forming parameters. However, the back-side surface can be applied in this method as the final part. For instance, the outer side of the pyramid can be considered as the final part which is carrying a better SQ.

Parts are collected to characterise with FESEM resolution of 4 nm at 0.1 kV. The surface of parts is considered and presented in figure 3. As can be seen in figure 3 (a), the surface area of AISI 316L is for blank metal sheet alloy which is presented from steel company after tension reduction process. The final mechanical work on these sort of sheet alloys is cold rolling so the surface quality will be improved. Figure 3 (b) illustrates that there was some forming process on sheet metal that changes the surface morphology. There are some scratches on the surface of sheet metal made by the tool that reduce the surface quality of the final part. Hoverer, this amount is not considerable in sheet metal forming processes using stretching such deep drawing.

Figure 3. (a) Row material. (b) Sheet part after ISF process.

The back side of the sheet part after forming has more appropriate surface quality regarding the lower amount of scratches. Furthermore, the defects of ironing process and tool scratch that may effect on surface quality has few influences on the back-side surface. Figure 4 shows the similarity of the back side of sheet part with the surface of raw material. The amount of disorganization is lower, and also due to some work hardening on the sheet surface, the quality is increased. In general, the strength of part will be increased while the surface quality will remain almost the same.

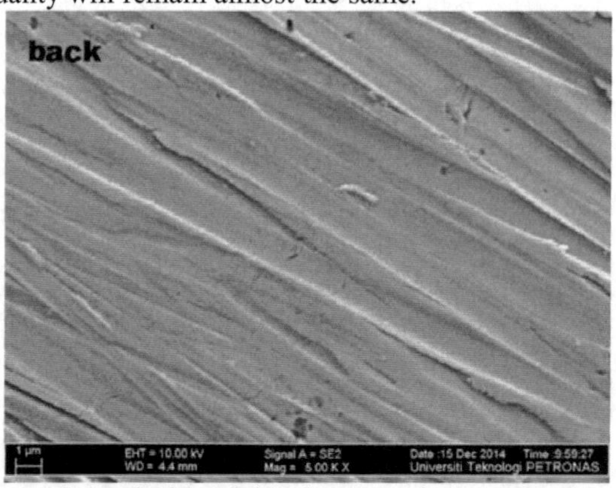

Figure 4. Back side of the sheet metal after ISF process.

The roughness test machine is employed to determine the roughness of the surfaces in this study. Among the all the samples. It should be mentioned that the temperature did not have a significant effect on the process since the cooling liquid is flowing on the surface of the metal sheet. Therefore, the effect of temperature is not considered in this study.

In this study, the influence of four parameters feed rate, spindle speed, step down, and width step versus the sheet stretching that is the representative of the sheet thickness distribution was examined via DOE. The Design of Experiment used in this respond method illustrates that some of the parameters such as feed rate and width step have a significant influence on STD compared to the other parameters. Figure 5 shows the behaviour of these four parameters versus STD during the experiment, where the (a) is the diagram of variation for feed rate, (b) spindle speed, (c) step down, and (d) width step.

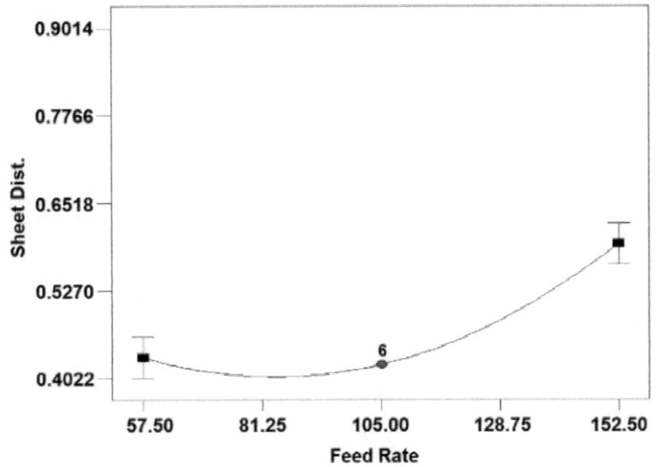

(a)

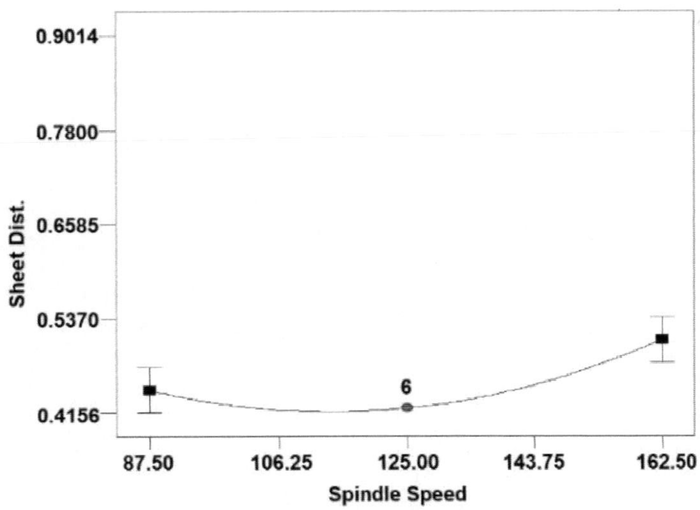

(b)

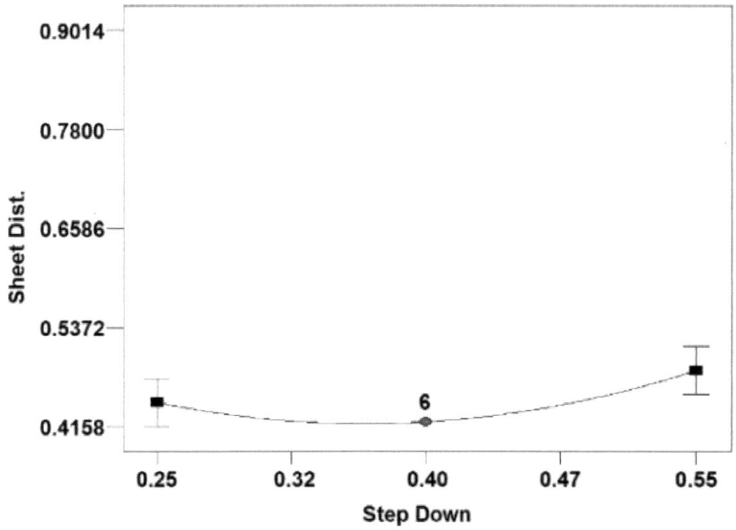

(c)

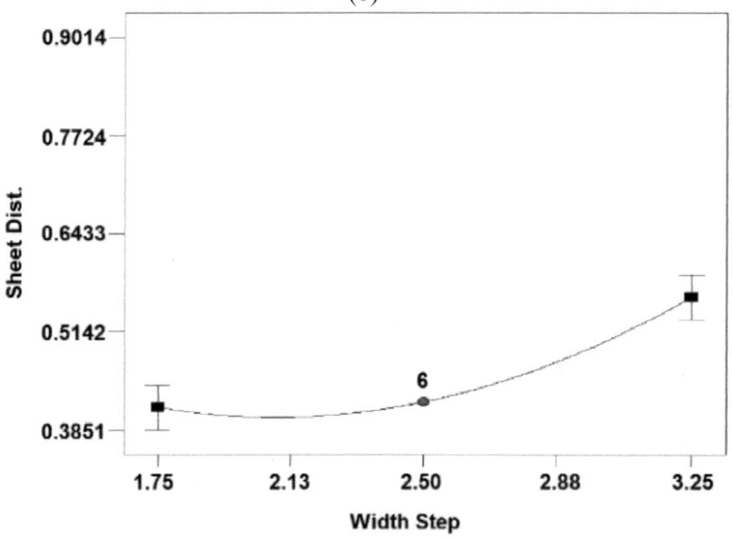

(d)

Figure 5. The interaction between effective parameters versus STD. (a) feed rate, (b) spindle speed, (c) step down, and (d) width step

As can be observed from the results, the feed rate is the most influential parameter in this study. Moreover, step width also showed a considerable effect on the amount of STD in this analysis, while the spindle speed and step down played a less important role in the forming process. The influence of step down just appeared once in a cycle, and the spindle speed effect could be minimised using a proper lubricant to reduce friction. However, high values of feed rate increased the friction and vibration of the process, which are the two major defects of ISF. In addition, a width step of more than 20% of the tool diameter increased the risk of sheet failure due to the huge pressure on the surface of the sheet metal,

IOP Conf. Series: Materials Science and Engineering **257** (2017) 012006 doi:10.1088/1757-899X/257/1/012006

which changes the process from stretching to chipping. This defect was accrued throughout the experimental investigation, as shown in figure 6.

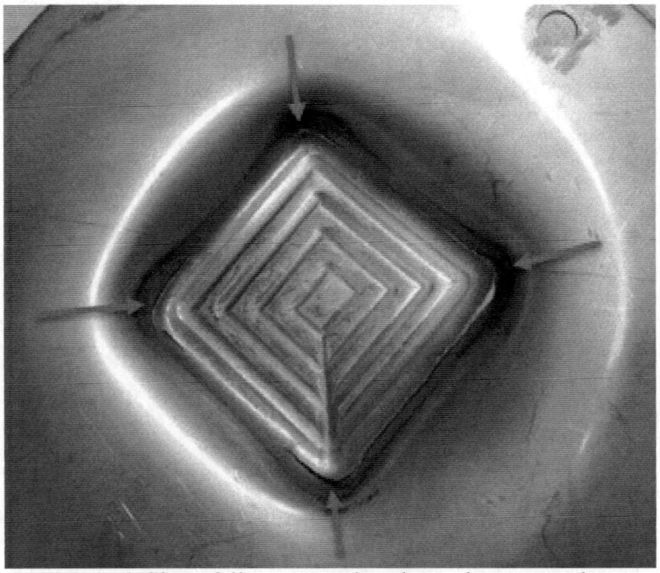

Figure 6. Sheet failure appearing due to large step size

The interaction of the two parameters on the STD was evaluated and shown in 3D diagrams to illustrate the influence of the combined parameters in a particular time. Figure 7 illustrates the influence of these significant parameters including feed rate and spindle speed, on the ISF process.

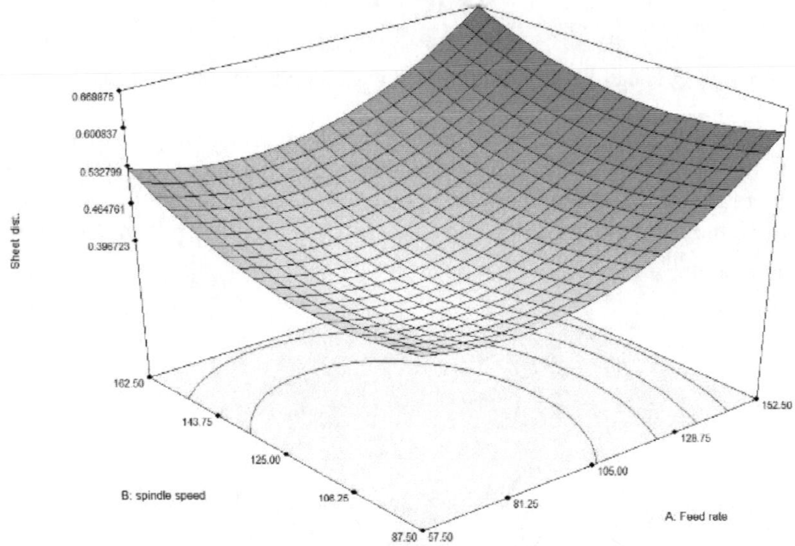

Figure 7. The effect of feed rate and spindle speed on sheet thickness distribution

Both feed rate and spindle speed have a linear effect on the response of the sheet thickness distribution while the combination of the two parameters also yielded a linear response. However, the combination of the parameters increased the amount of sheet stretch up to 0.668875 μm where every single parameter (~ 0532799 μm for spindle speed and 0.600837 μm for feed rate) was not able to increase the sheet stretch up to that point. Figure 7 shows that the combination of both effective parameters when the values are at maximum would cause sheet failure since the high amount of STD in this study is one reason for sheet fracture.

A high feed rate has a significant effect on sheet failure and a large width step increases the risk of sheet fracture. Figure 8 presents the considerable interaction between these parameters versus STD, where the thickness distribution shows a dramatic increase up to the highest point.

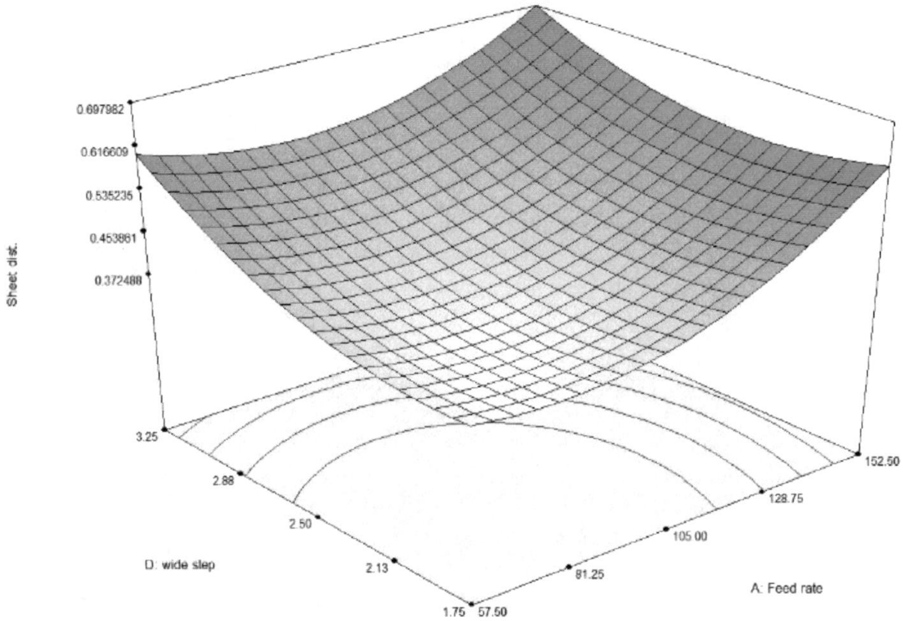

Figure 8. The effect of feed rate and wide step on sheet thickness distribution

As can be seen from figure 8, both factors sharply increased the STD up to ~ 0.697982 μm, whereas in figure 9, the combination of the two parameters of spindle speed and depth of cut increased the STD to ~ 0.585355 μm. Increasing the depth of cut to more than 0.47 together with an increase in spindle speed up to 143.75 would be the optimum condition for this response since the sheet distribution values increased sharply upward as a result.

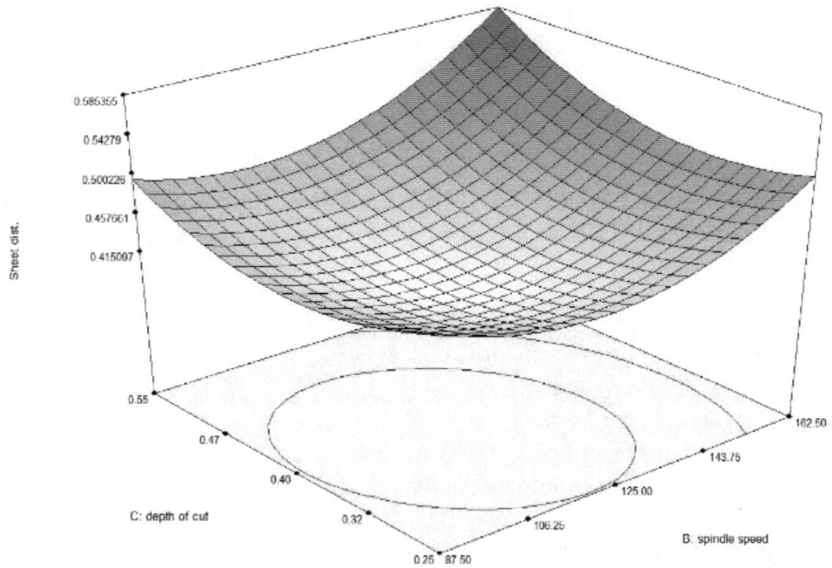

Figure 9. The effect of depth of cut and spindle speed on sheet distribution

However, using low parameters increased the time of production, hence, an optimum amount was obtained from equation (2). This equation was driven based on the parameter response in these experiments where the affected parameters are sheet thickness distribution (STD), feed rate (*A*), spindle speed (*B*), depth of cut (*C*), and (*D*) width step.

$$\boldsymbol{STD} = 0.42 + 0.081 * \boldsymbol{A} + 0.032 * \boldsymbol{B} + 0.020 * \boldsymbol{C} + 0.072 * \boldsymbol{D} + 0.091 * \boldsymbol{A}^2 + 0.056 * \boldsymbol{B}^2 + 0.044 * \boldsymbol{C}^2 + 0.066 * \boldsymbol{D}^2 - 0.012 * \boldsymbol{A} * \boldsymbol{B} + 4.919\boldsymbol{E}\text{-}003 * \boldsymbol{A} * \boldsymbol{C} - 0.033 * \boldsymbol{A} * \boldsymbol{D} + 0.013 * \boldsymbol{B} * \boldsymbol{C} + 4.069\boldsymbol{E}\text{-}003 * \boldsymbol{B} * \boldsymbol{D} - 4.244\boldsymbol{E}\text{-}003 * \boldsymbol{C} * \boldsymbol{D} \quad (2)$$

As stated in equation (2), the effect of feed rate and spindle speed on the STD is higher since the impact factors for these parameters are 0.081 and 0.072, respectively. Based on the analysis by the DOE software (DE®), the R-Squared ($\boldsymbol{R}^2$) value was 0.9521 and the Adjusting $\boldsymbol{R}^2$ value was calculated as 0.9073. Also, the Prediction R-Squared value of 0.7239 was found to be in reasonable agreement with the Adjusting R-Squared value.

4. Conclusions

A new prototyping method of production is in need to employ for batch and low production series in rapid forming techniques. Hence, incremental sheet forming as a potential process is presented to increase the variety of sheet metal parts. However, this process suffers from some drawback such as low surface quality and the high possibility of sheet failure. Hence, in this study, a set of experiment is designed and applied to determine the significant effective parameters in this sheet forming process and also find the optimum values of the parameters with respect to the maximum sheet thickness distribution and higher surface quality than current studies. For this purpose, the result of experiments is characterized via utilising of roughness test and the FESEM test. Then, the respond of the each set of parameters is applied in DOE to obtain the optimum forming parameters and their interaction. Consequently, the data presented in this study shows a range of optimized parameters in 24 mm pyramid design shape for ISF method with adequate surface quality (Ra 3.073 μm for contact surface) Ra 1.590 μm for back-side surface up to 35%

higher than the current research based on the graphs generated by surface roughness machine. Furthermore, the results illuminate that the ISF will not be the cause of crack and crystal structure defects on the final parts, and also presented as a low defect process for sheet metal accessories and parts production.

Acknowledgement

The authors are thankful to the Ministry of Higher Education, Malaysia (MOHE) for supporting this research via FRGS/1/2014/TK01/UTP/02/8 research grant.

References

[1] Jeswiet J, Micari F, Hirt G, Bramley A, Duflou J and Allwood J 2005 Asymmetric single point incremental forming of sheet metal *CIRP Annals-Manufacturing Technology* **54** 88-114

[2] Jackson K and Allwood J 2009 The mechanics of incremental sheet forming *Journal of materials processing technology* **209** 1158-74

[3] Allwood J M, Braun D and Music O 2010 The effect of partially cut-out blanks on geometric accuracy in incremental sheet forming *Journal of Materials Processing Technology* **210** 1501-10

[4] Emmens W and Van den Boogaard A 2009 Incremental forming by continuous bending under tension—an experimental investigation *Journal of Materials Processing Technology* **209** 5456-63

[5] Schaeffer L, Castelan J, Gruber V, Daleffe A and Marcelino R 2009 Development of customized products through the use of incremental sheet forming for medical orthopaedic applications. In: *3rd International Conference on Integrity, Reliability and Failure,* pp 65-6

[6] Shapiro F 2008 Bone development and its relation to fracture repair. The role of mesenchymal osteoblasts and surface osteoblasts *Eur Cell Mater* **15** 53-76

[7] Micari F, Ambrogio G and Filice L 2007 Shape and dimensional accuracy in single point incremental forming: state of the art and future trends *Journal of Materials Processing Technology* **191** 390-5

[8] Hussain G, Lin G and Hayat N 2011 Improving profile accuracy in SPIF process through statistical optimization of forming parameters *Journal of Mechanical Science and Technology* **25** 177-82

[9] Attanasio A, Ceretti E, Giardini C and Mazzoni L 2008 Asymmetric two points incremental forming: improving surface quality and geometric accuracy by tool path optimization *Journal of Materials Processing Technology* **197** 59-67

[10] Ambrogio G, Filice L and Gagliardi F 2012 Formability of lightweight alloys by hot incremental sheet forming *Materials & Design* **34** 501-8

[11] Duflou J, Callebaut B, Verbert J and De Baerdemaeker H 2007 Laser assisted incremental forming: formability and accuracy improvement *CIRP Annals-Manufacturing Technology* **56** 273-6

[12] Fritzen D, Daleffe A, Castelan J and Schaeffer L 2013 Brass 70/30 and Incremental Sheet Forming Process. In: *Key Engineering Materials*: Trans Tech Publ) pp 1419-31

[13] Ham M and Jeswiet J 2006 Single point incremental forming and the forming criteria for AA3003 *CIRP Annals-Manufacturing Technology* **55** 241-4

[14] Rattanachan K and Chungchoo C 2009 Formability in single point incremental forming of dome geometry *AIJSTPME* **2** 57-63

[15] Hussain G, Gao L and Zhang Z 2008 Formability evaluation of a pure titanium sheet in the cold incremental forming process *The International Journal of Advanced Manufacturing Technology* **37** 920-6

[16] Nimbalkar D and Nandedkar V 2013 Review of incremental forming of sheet metal components *Int J Eng Res Appl* **3** 39-51

IOP Conf. Series: Materials Science and Engineering **257** (2017) 012006 doi:10.1088/1757-899X/257/1/012006

[17] Moayedfar M, Leman Z and Baharudin B 2014 Incremental Sheet Forming (ISF) of AISI 316 Stainless Steel Sheet Using CNC Milling Machine. In: *Advanced Materials Research*: Trans Tech Publ) pp 322-7

[18] Moayedfar M, Rani A M B A, Salehi P and Tad V V L N R 2014 Comparative Study between Programming Systems for Incremental Sheet Forming Process. In: *MATEC Web of Conferences*: EDP Sciences) p 04009

[19] Zhu H, Li N and Liu Z 2012 The effect of pressing direction on the 5-axis CNC incremental forming quality *International journal of material forming* **5** 227-33

IOP Conf. Series: Materials Science and Engineering **257** (2017) 012007 doi:10.1088/1757-899X/257/1/012007

Efficient operation scheduling for adsorption chillers using predictive optimization-based control methods

Adrian Bürger[1,3], Parantapa Sawant[2], Markus Bohlayer[1], Angelika Altmann-Dieses[1], Marco Braun[1] and Moritz Diehl[3]

[1] Faculty of Management Science and Engineering, Karlsruhe University of Applied Sciences, Moltkestraße 30, 76133 Karlsruhe, Germany
[2] Institute of Energy Systems Technology, Offenburg University of Applied Sciences, Badstraße 24, 77652 Offenburg, Germany
[3] Systems Control and Optimization Laboratory, Department of Microsystems Engineering, University of Freiburg, Georges-Koehler-Allee 102, 79110 Freiburg, Germany

E-mail: `adrian.buerger@hs-karlsruhe.de`

Abstract. Within this work, the benefits of using predictive control methods for the operation of Adsorption Cooling Machines (ACMs) are shown on a simulation study. Since the internal control decisions of series-manufactured ACMs often cannot be influenced, the work focuses on optimized scheduling of an ACM considering its internal functioning as well as forecasts for load and driving energy occurrence. For illustration, an assumed solar thermal climate system is introduced and a system model suitable for use within gradient-based optimization methods is developed. The results of a system simulation using a conventional scheme for ACM scheduling are compared to the results of a predictive, optimization-based scheduling approach for the same exemplary scenario of load and driving energy occurrence. The benefits of the latter approach are shown and future actions for application of these methods for system control are addressed.

1. Introduction

Adsorption Cooling Machines (ACMs) provide possibilities for more ecological cooling energy production [1]. Therefor, multiple control aspects have been examined in the literature to increase the operation efficiency of ACMs, e. g. by (predictive) optimization of cycle times [2, 3] or by reducing the electrical energy consumption by situational control of pump and recooling tower fan speeds [4]. While series-manufactured machines typically are equipped with internal controllers whose strategies are determined by the manufacturer and cannot be altered or influenced externally, and/or are missing additional measurement devices (e. g. internal pressure sensors [2]) that allow for more elaborate control strategies, an aspect that can be optimized without changing a machine's internal control logic is to predictively schedule its operation times in a way to enable efficient operation, considering its internal functioning and control decisions.

The current temperatures of the mass flows entering an ACM are crucial for its operation efficiency [1]. Attractive providers of driving energy for ACMs are low-temperature waste heat of industrial processes or solar thermal power [5]. In both cases, the availability of driving energy is predictable in the form e. g. of production plans or weather forecasts, respectively. This favors the use of optimization-based predictive control methods, where in contrast to conventional set-point based controllers, predictions can be taken into account directly to determine at which times machine operation needs to be scheduled to be efficient, and how operating the ACM itself will affect these operation conditions. Such considerations can increase the operation efficiency, and with this, also reduce the operation time that is needed to provide a certain cooling demand.

Since predictive control methods are well-suited for control of systems with slow system response, such methods have been shown to be advantageous for control of Heating, Ventilation and Air Conditioning (HVAC) systems of buildings [6], which favors an inclusion of ACM scheduling. Further improvements to internal machine control strategies can be incorporated independently. However, compared to conventional controllers, these control methods need more information on the considered system in the form e. g. of suitable system models and forecasts.

Within this work, the benefits of using predictive control methods for operation scheduling of a silica-gel/water ACM are shown on a simulation study for a solar thermal HVAC system. First, the assumed system and the chosen modeling approach are described. Then, the model is simulated for a test scenario using a conventional control scheme. Afterwards, an optimization-based controller is applied to the system for the same test scenario, and the performance of both controllers is compared. The benefits of the optimization-based approach are shown, and future actions for application of these methods for system control are addressed.

2. Experimental setup

Within this simulation study, a solar thermal HVAC system as shown in Figure 1 is considered.

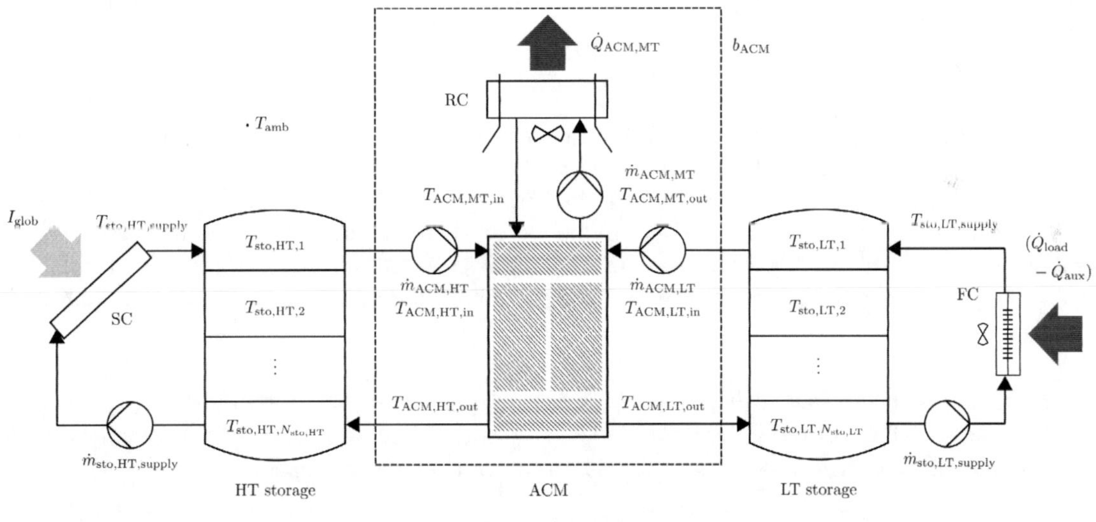

Figure 1. Schematic depiction of the solar thermal HVAC system considered within this study.

The system consists of solar collectors (SC) with total surface $A_{col} = 31.35\,\mathrm{m}^2$ and optical efficiency $\eta_{col} = 0.62$ that support a stratified high temperature water storage (HT storage) of volume $V_{sto,HT} = 2.0\,\mathrm{m}^3$.[1] If the ACM is in operation, it uses the heated water from the top of the HT storage to absorb heat from a stratified low temperature water storage (LT storage) of volume $V_{sto,LT} = 1.0\,\mathrm{m}^3$ and emits the combined heat through a recooling tower (RC).

The ACM operation conditions are determined by the current temperatures of the mass flows entering the machine. Favorable operation conditions are high values for the temperatures $T_{ACM,LT,in}(t)$ and $T_{ACM,HT,in}(t)$ and low values for the temperature $T_{ACM,MT,in}(t)$ [1] with limits shown in Table 1. At optimal conditions, the ACM has a cooling capacity of $\dot{Q}_{ACM,LT,max} = 12\,\mathrm{kW}$ and a Coefficient Of Performance (COP) of $\mathrm{COP}_{ACM,max} = 0.65$.

[1] For simplicity, usual system separations of frost-proof outdoor circuits from indoor water circuits are neglected.

Table 1. Minimum and maximum ACM inlet temperatures for operation.

Inlet temperature	Minimum value	Maximum value
$T_{\mathrm{ACM,LT,in}}$	10 °C	22 °C
$T_{\mathrm{ACM,HT,in}}$	55 °C	95 °C
$T_{\mathrm{ACM,MT,in}}$	15 °C	40 °C

The water from the bottom of the LT storage can be used to compensate a heat load $\dot{Q}_{\mathrm{load}}$ arising from the cooling demand of a building through fan coils (FC). If the cooling capacity available from the LT storage is not sufficient to compensate $\dot{Q}_{\mathrm{load}}$, an auxiliary cooling device (e. g. a compression chiller) is used to generate $\dot{Q}_{\mathrm{aux}}$ which decreases the load of the system.

Further information on the parametrization and control strategies of the relevant system components is given in the following section.

3. Methodology

In the following, the chosen modeling approach for the system introduced in Section 2 is described, as well as the design and implementation of the considered conventional and optimization-based controllers.

3.1. Modeling approach

For system modeling, physically motivated gray-box models based on mass and energy balances are used. Since the model is to be later used within gradient-based optimization methods (see Section 3.3), it must fulfill all necessary conditions e. g. regarding differentiability, cf. [7].

For simplicity, all heat losses of components to their surroundings are neglected. The system is assumed to be filled with incompressible water with constant specific heat capacity $c_{\mathrm{W}} = 4.128\,\mathrm{kJ/(kgK)}$ and density $\rho_{\mathrm{W}} = 1\,\mathrm{kg/l}$. The different component models and their internal control strategies are explained in the following.

3.1.1. Modeling of the solar collector cycle For the solar collectors, a simplified model variant of [8] is chosen that converts the solar radiation $I_{\mathrm{glob}}(t)$ available at a time t to a heat flow $\dot{Q}_{\mathrm{col}}(t)$, according to the collector surface A_{col} and optical efficiency η_{col} as in

$$\dot{Q}_{\mathrm{col}}(t) = \eta_{\mathrm{col}} A_{\mathrm{col}} I_{\mathrm{glob}}(t). \tag{1}$$

Further, it is assumed that the collector-side pump is controlled in a way that the temperature of the medium that enters the top of the HT storage is constant with $T_{\mathrm{sto,HT,supply}} = 80\,°\mathrm{C}$, and the mass flow $\dot{m}_{\mathrm{sto,HT,supply}}(t)$ is adjusted so that the medium extracted from the bottom of the HT storage $T_{\mathrm{sto,HT},N_{\mathrm{sto,HT}}}(t)$ can be heated up accordingly, i. e.,

$$\dot{m}_{\mathrm{sto,HT,supply}}(t) = \frac{\dot{Q}_{\mathrm{col}}(t)}{c_{\mathrm{W}}(T_{\mathrm{sto,HT,supply}} - T_{\mathrm{sto,HT},N_{\mathrm{sto,HT}}}(t))}. \tag{2}$$

3.1.2. Modeling of the load cycle On the load side, it is assumed that the temperature of the medium that enters the top of the LT storage is constant with $T_{\mathrm{sto,LT,supply}} = 20\,°\mathrm{C}$, and the mass flow $\dot{m}_{\mathrm{sto,LT,supply}}(t)$ is adjusted to compensate for the heat flow of the load $\dot{Q}_{\mathrm{load}}(t)$ that has not been covered by $\dot{Q}_{\text{aux}}(t)$, according to the temperature of the bottom of the LT storage $T_{\text{sto,LT},N_{\text{sto,LT}}}(t)$, so that

$$\dot{m}_{\text{sto,LT,supply}}(t) = \frac{\dot{Q}_{\text{load}}(t) - \dot{Q}_{\text{aux}}(t)}{c_{\text{W}}(T_{\text{sto,LT,supply}} - T_{\text{sto,LT},N_{\text{sto,LT}}}(t))}. \tag{3}$$

For this assumption to be applicable, it is ensured that the mass flow $\dot{m}_{\text{sto,LT,supply}}(t)$ at all times is less than some maximum value $\dot{m}_{\text{sto,LT,supply,max}}$, i. e.,

$$\dot{m}_{\text{sto,LT,supply}}(t) \leq \dot{m}_{\text{sto,LT,supply,max}}. \tag{4}$$

The auxiliary cooling device is only used to reduce the load for the system by $\dot{Q}_{\text{aux}}(t)$ if $\dot{Q}_{\text{load}}(t)$ at some time cannot be compensated from the LT storage alone.

3.1.3. Modeling of adsorption cooling machine and recooling system From the amount of heat exchanged by the machine's three circuits during an entire adsorption cycle [9] at certain operation conditions, energy balances can be formulated that allow to calculate the mean heat flows at a time during that cycle as in

$$\begin{aligned}
\dot{Q}_{\text{ACM,LT}}(t) &= f_{\dot{Q}_{\text{ACM,LT}}}(T_{\text{ACM,HT,in}}(t), T_{\text{ACM,MT,in}}(t), T_{\text{ACM,LT,in}}(t)) \\
\dot{Q}_{\text{ACM,HT}}(t) &= \frac{\dot{Q}_{\text{ACM,LT}}(t)}{f_{\text{COP}}(T_{\text{ACM,HT,in}}(t), T_{\text{ACM,MT,in}}(t), T_{\text{ACM,LT,in}}(t))} \\
\dot{Q}_{\text{ACM,MT}}(t) &= \dot{Q}_{\text{ACM,LT}}(t) + \dot{Q}_{\text{ACM,HT}}(t).
\end{aligned} \tag{5}$$

The functions $f_{\dot{Q}_{\text{ACM,LT}}}(\cdot)$ and $f_{\text{COP}}(\cdot)$ are fittings of quadratic polynomials to machine specific data for cooling power and COP at different operation conditions. Using the mass flows of the several circuits, here $\dot{m}_{\text{ACM,LT}} = 48\,\text{kg/min}$, $\dot{m}_{\text{ACM,HT}} = 41.8\,\text{kg/min}$ and $\dot{m}_{\text{ACM,MT}} = 85.0\,\text{kg/min}$, mean output temperatures can be obtained from

$$\begin{aligned}
T_{\text{ACM,LT,out}}(t) &= T_{\text{ACM,LT,in}}(t) - \frac{\dot{Q}_{\text{ACM,LT}}(t)}{c_{\text{W}}\dot{m}_{\text{ACM,LT}}} \\
T_{\text{ACM,HT,out}}(t) &= T_{\text{ACM,HT,in}}(t) - \frac{\dot{Q}_{\text{ACM,HT}}(t)}{c_{\text{W}}\dot{m}_{\text{ACM,HT}}} \\
T_{\text{ACM,MT,out}}(t) &= T_{\text{ACM,MT,in}}(t) + \frac{\dot{Q}_{\text{ACM,MT}}(t)}{c_{\text{W}}\dot{m}_{\text{ACM,MT}}}.
\end{aligned} \tag{6}$$

Though the model is neglecting the cyclic behavior of an ACM as it is shown e. g. in [9], the model becomes applicable since the HT and LT storages connected to the ACM as well as the mass flow of the MT circuit are of sufficient size to dampen the effects of the fluctuating output temperatures. The model can be adapted to a specific machine without additional measurement devices apart from temperature and flow sensors, and depicts the functioning of the machine including it's internal control decisions.

Since the booting- and shutdown-phases of the ACM are not considered within this model, it must be ensured that the machine is switched on and off only with a reasonable frequency. Within this work, this is realized either through switching hystereses or minimum operation times.

The recooling system of the ACM is controlled to always provide cooling medium with temperature $T_{\text{ACM,MT,in}}(t)$ that meets a constant temperature difference $\Delta T_{\text{ACM,MT}}$ to the current ambient temperature $T_{\text{amb}}(t)$, i. e.,

$$T_{\text{ACM,MT,in}}(t) = T_{\text{amb}}(t) + \Delta T_{\text{ACM,MT}}. \tag{7}$$

3.1.4. Modeling of the storage tanks For modeling of the stratified storage tanks, their volumes are discretized into several layers for which an energy balance is calculated. Variants of this common approach can be found e. g. in [10, 11]. The model of the HT storage contains $N_{\text{sto,HT}} = 4$ layers of mass $M_{\text{sto,HT}}$ and the LT storage model $N_{\text{sto,LT}} = 3$ layers of mass $M_{\text{sto,LT}}$ according to

$$M_{\text{sto,HT}} = \frac{\rho_{\text{W}} V_{\text{sto,HT}}}{N_{\text{sto,HT}}}, \quad M_{\text{sto,LT}} = \frac{\rho_{\text{W}} V_{\text{sto,LT}}}{N_{\text{sto,LT}}}. \tag{8}$$

The temperature within a stratum is influenced by the amount and temperature of the mass flows entering and leaving it. The current existence of a possible mass flow is determined by the current status $b_{\text{ACM}}(t)$ of the ACM, which is a binary variable that shows whether the machine is running ($b_{\text{ACM}} = 1$) or not ($b_{\text{ACM}} = 0$), and by the current values of $\dot{m}_{\text{sto,HT,supply}}(t)$ and $\dot{m}_{\text{sto,LT,supply}}(t)$.[2] The energy balance of the HT storage top layer with index 1 is shown in Eq. (9) and for the LT storage top layer with index 1 in Eq. (10), respectively:

$$\begin{aligned}\dot{T}_{\text{sto,HT,1}}(t) = M_{\text{sto,HT}}^{-1}[&\dot{m}_{\text{sto,HT,supply}}(t) T_{\text{sto,HT,supply}} \\ &- (1 - b_{\text{ACM}}(t))\dot{m}_{\text{sto,HT,supply}}(t) T_{\text{sto,HT,1}}(t) \\ &+ b_{\text{ACM}}(t)(\dot{m}_{\text{ACM,HT}} - \dot{m}_{\text{sto,HT,supply}}(t)) T_{\text{sto,HT,2}}(t) \\ &- b_{\text{ACM}}(t)\dot{m}_{\text{ACM,HT}} T_{\text{sto,HT,1}}(t)],\end{aligned} \tag{9}$$

$$\begin{aligned}\dot{T}_{\text{sto,LT,1}}(t) = M_{\text{sto,LT}}^{-1}[&\dot{m}_{\text{sto,LT,supply}}(t) T_{\text{sto,LT,supply}} \\ &- (1 - b_{\text{ACM}}(t))\dot{m}_{\text{sto,LT,supply}}(t) T_{\text{sto,LT,1}}(t) \\ &+ b_{\text{ACM}}(t)(\dot{m}_{\text{ACM,LT}} - \dot{m}_{\text{sto,LT,supply}}(t)) T_{\text{sto,LT,2}}(t) \\ &- b_{\text{ACM}}(t)\dot{m}_{\text{ACM,LT}} T_{\text{sto,LT,1}}(t)].\end{aligned} \tag{10}$$

The balance equations for the HT storage middle layers are shown in Eq. (11) and for the LT storage middle layers in Eq. (12):

$$\begin{aligned}\dot{T}_{\text{sto,HT,k}}(t) = M_{\text{sto,HT}}^{-1}[&(1 - b_{\text{ACM}}(t))\dot{m}_{\text{sto,HT,supply}}(t) T_{\text{sto,HT},k-1}(t) \\ &- b_{\text{ACM}}(t)(\dot{m}_{\text{ACM,HT}} - \dot{m}_{\text{sto,HT,supply}}(t)) T_{\text{sto,HT},k}(t) \\ &+ b_{\text{ACM}}(t)(\dot{m}_{\text{ACM,HT}} - \dot{m}_{\text{sto,HT,supply}}(t)) T_{\text{sto,HT},k+1}(t) \\ &- (1 - b_{\text{ACM}}(t))\dot{m}_{\text{sto,HT,supply}}(t) T_{\text{sto,HT},k}(t)], \qquad k = 2, \ldots, N_{\text{sto,HT}} - 1,\end{aligned} \tag{11}$$

$$\begin{aligned}\dot{T}_{\text{sto,LT,k}}(t) = M_{\text{sto,LT}}^{-1}[&(1 - b_{\text{ACM}}(t))\dot{m}_{\text{sto,LT,supply}}(t) T_{\text{sto,LT},l-1}(t) \\ &- b_{\text{ACM}}(t)(\dot{m}_{\text{ACM,LT}} - \dot{m}_{\text{sto,LT,supply}}(t)) T_{\text{sto,LT},l}(t) \\ &+ b_{\text{ACM}}(t)(\dot{m}_{\text{ACM,LT}} - \dot{m}_{\text{sto,LT,supply}}(t)) T_{\text{sto,LT},l+1}(t) \\ &- (1 - b_{\text{ACM}}(t))\dot{m}_{\text{sto,LT,supply}}(t) T_{\text{sto,LT},l}(t)], \qquad l = 2, \ldots, N_{\text{sto,LT}} - 1.\end{aligned} \tag{12}$$

The energy balance for the HT storage bottom layer with index $N_{\text{sto,HT}}$ is shown in Eq. (13) and for the LT storage bottom layer with index $N_{\text{sto,LT}}$ in Eq. (14):

[2] For the system considered here, it can be assumed that $\dot{m}_{\text{ACM,HT}}(t) \geq b_{\text{ACM}}(t)\dot{m}_{\text{sto,HT,supply}}(t)$ and $\dot{m}_{\text{ACM,LT}}(t) \geq b_{\text{ACM}}(t)\dot{m}_{\text{sto,LT,supply}}(t)$ at all times. If this is not the case, further expressions that take into account the additional mass flow directions need to be added to the models.

$$
\begin{aligned}
\dot{T}_{\text{sto,HT,N}_{\text{sto,HT}}}(t) = M^{-1}_{\text{sto,HT}}[&(1 - b_{\text{ACM}}(t))\dot{m}_{\text{sto,HT,supply}}(t)T_{\text{sto,HT,N}_{\text{sto,HT}}-1}(t) \\
&- \dot{m}_{\text{sto,HT,supply}}(t)T_{\text{sto,HT,N}_{\text{sto,HT}}}(t) \\
&+ b_{\text{ACM}}(t)\dot{m}_{\text{ACM,HT}}T_{\text{ACM,HT,OUT}}(t) \\
&- b_{\text{ACM}}(t)(\dot{m}_{\text{ACM,HT}} - \dot{m}_{\text{sto,HT,supply}}(t))T_{\text{sto,HT,N}_{\text{sto,HT}}}(t)],
\end{aligned} \tag{13}
$$

$$
\begin{aligned}
\dot{T}_{\text{sto,LT,N}_{\text{sto,LT}}}(t) = M^{-1}_{\text{sto,LT}}[&(1 - b_{\text{ACM}}(t))\dot{m}_{\text{sto,LT,supply}}(t)T_{\text{sto,LT,N}_{\text{sto,LT}}-1}(t) \\
&- \dot{m}_{\text{sto,LT,supply}}(t)T_{\text{sto,LT,N}_{\text{sto,LT}}}(t) \\
&+ b_{\text{ACM}}(t)\dot{m}_{\text{ACM,LT}}T_{\text{ACM,LT,OUT}}(t) \\
&- b_{\text{ACM}}(t)(\dot{m}_{\text{ACM,LT}} - \dot{m}_{\text{sto,LT,supply}}(t))T_{\text{sto,LT,N}_{\text{sto,LT}}}(t)].
\end{aligned} \tag{14}
$$

3.2. *Formulation of a conventional control scheme*

To simulate a conventional, set-point based control scheme, the model introduced in Section 3.1 is used within a system simulation in OpenModelica [12]. The ACM operation status $b_{\text{ACM}}(t)$ is determined using the control scheme

$$
b_{\text{ACM}}(t) = \begin{cases} 1, & \text{if } T_{\text{sto,HT,1}}(t) \geq 60\,°\text{C} \wedge T_{\text{sto,LT,N}_{\text{sto,LT}}}(t) \geq 12\,°\text{C}, \\ 0, & \text{if } T_{\text{sto,HT,1}}(t) < 55\,°\text{C} \vee T_{\text{sto,LT,N}_{\text{sto,LT}}}(t) < 10\,°\text{C}. \end{cases} \tag{15}
$$

If $T_{\text{sto,HT,1}}$ is sufficient for operating the ACM, and $T_{\text{sto,LT,N}_{\text{sto,LT}}}$ is not too low, the ACM is switched on. If either $T_{\text{sto,HT,1}}$ or $T_{\text{sto,LT,N}_{\text{sto,LT}}}$ are too low, the ACM is switched off. A hysteresis is introduced for both temperature constraints to avoid frequent switching (or in simulation, chattering). Predictions for driving energy and load are not considered.

3.3. *Formulation of an optimization-based scheduling method*

The aim of the optimization-based control approach is to run the system in a way that minimizes the total use of auxiliary cooling over the considered period, taking into account system dynamics and operation conditions and boundaries as well as cooling demand, solar radiation and ambient temperature profiles. This results in the Mixed-Integer Nonlinear Optimal Control Problem (MINOCP)

$$
\begin{aligned}
\underset{x(\cdot),b_{\text{ACM}}(\cdot)}{\text{minimize}} \quad & \int_{t_0}^{t_f} \dot{Q}_{\text{aux}}(t)\,\mathrm{d}t && \text{(16a)} \\
\text{subject to} \quad & \text{(1)-(14)} && \text{(16b)} \\
& x(t) = \begin{pmatrix} T_{\text{sto,HT},k}(t), & k = 1,\ldots,N_{\text{sto,HT}} \\ T_{\text{sto,LT},l}(t), & l = 1,\ldots,N_{\text{sto,LT}} \end{pmatrix}, \quad u(t) = \begin{pmatrix} T_{\text{amb}}(t) \\ \dot{Q}_{\text{Load}}(t) \\ I_{\text{glob}}(t) \end{pmatrix} && \text{(16c)} \\
& T_{\text{ACM,LT,IN}}(t) + M_{\text{ACM,LT,IN,min}}(1 - b_{\text{ACM}}(t)) \geq T_{\text{ACM,LT,IN,min}} && \text{(16d)} \\
& T_{\text{ACM,HT,IN}}(t) + M_{\text{ACM,HT,IN,min}}(1 - b_{\text{ACM}}(t)) \geq T_{\text{ACM,HT,IN,min}} && \text{(16e)} \\
& b_{\text{ACM}}(t) \in \{0,1\}, \quad t \in [t_0, t_f] && \text{(16f)} \\
& x_{\min} \leq x(t) \leq x_{\max}, \quad t \in [t_0, t_f] && \text{(16g)} \\
& x(0) = x_0. && \text{(16h)}
\end{aligned}
$$

The aim of minimizing the total auxiliary cooling is stated in the objective in Eq. (16a), the system dynamics are considered in Eq. (16b) for all states x and profiles u in Eq. (16c). The linear inequalities in Eq. (16d) and (16e) ensure that the ACM is operated only within valid conditions, see e. g. [13].[3] In Eq. (16f) it is assured that the ACM can only be switched on or off. Maximum and minimum system temperatures are specified in Eq. (16g) to prevent the water from freezing or boiling. Finally in Eq. (16h), it is ensured that the initial values of x match the initial state values of the simulation.

For solving, the MINOCP in Eq. (16) is transformed into a Mixed-Integer Non-Linear Program (MINLP) using direct collocation [7] with Lagrange polynomials on an equidistant time grid of step size $\Delta t = 120\,\mathrm{s}$ using the symbolic optimization framework CasADi [14, 15]. Power-cycling the ACM is made possible only every 1800 s to keep the amount of integer variables of the MINLP reasonable, as well as to incorporate the minimum operation time of the ACM (cf. Section 3.1). The MINLP is solved from CasADi using the MINLP solver BONMIN [16] that incorporates IPOPT [17] as NLP solver, which here uses MA57 [18] as linear solver.

4. Results and discussion

A comparison of the controller performances is carried out for a typical scenario of energy availability and load appearance where the load appears delayed from the main availability of the solar irradiation, for a time horizon of 24 hours. The load is calculated from the current ambient temperature $T_{\mathrm{amb}}(t)$ and the target building temperature $T_{\mathrm{target}} = 22\,°\mathrm{C}$ as

$$\dot{Q}_{\mathrm{load}}(t) = \begin{cases} 0.0\,\mathrm{kW}, & \text{if } T_{\mathrm{amb}}(t) \leq T_{\mathrm{target}}, \\ 0.9\,\frac{\mathrm{kW}}{\mathrm{K}}(T_{\mathrm{amb}}(t) - T_{\mathrm{target}}), & \text{if } T_{\mathrm{amb}}(t) > T_{\mathrm{target}}. \end{cases} \tag{17}$$

4.1. Application of the conventional controller

A depiction of the simulation results for the conventional controller, computed in less than 1 s on an Intel Core i5-4570 3.20 GHz CPU, is shown in Figure 2. It can be seen that the ACM is switched on relatively early in the day and working with high values for $\dot{Q}_{\mathrm{ACM,LT}}$ and COP due to the low values of T_{amb} that facilitate low values for $T_{\mathrm{ACM,MT,in}}$. However, this is merely coincidental, since the controller does not actively consider T_{amb}. During that time, the ACM is power-cycled several times, since the driving energy supplied by the collectors to the HT storage is not yet sufficient for permanent operation. Since no load occurs before 10:00, the operation of the ACM causes low LT storage temperatures, and with this, inefficient operation conditions.

The occurring load cannot be compensated completely by the system alone, and auxiliary cooling power becomes necessary.[4] It is shown that the ACM is running with a cooling power lower than the occurring load for a long time. One reason for this is that the driving energy from the early day has already been utilized, often within inefficient operation conditions.

These results reveal a systematic problem, since this situation is likely to happen if the load occurs later than the availability of the driving energy, and the controller is neither aware of the time when the load occurs nor of the amount of driving energy available throughout the day. Changing the temperature boundaries for switching the ACM cannot finally eliminate this problem, since the specific occurrence of the problem is dependent on the daily load and driving energy situation. However, this problem can be countered by applying predictive, optimization-based control strategies, as shown in the next section.

4.2. Application of the optimization-based controller

The solution of the MINLP for the optimization-based controller, which takes about 415 s on an Intel Core i5-4570 3.20 GHz CPU, is depicted in Figure 3, where multiple aspects show that the

[3] Only these operation limits have to be actively considered within this study. Other limits for ACM operation as shown in Table 1 cannot be reached due to the problem setup.

[4] In case such an auxiliary cooling device would not be available, insufficient cooling power would result in the building temperature to rise above the target temperature.

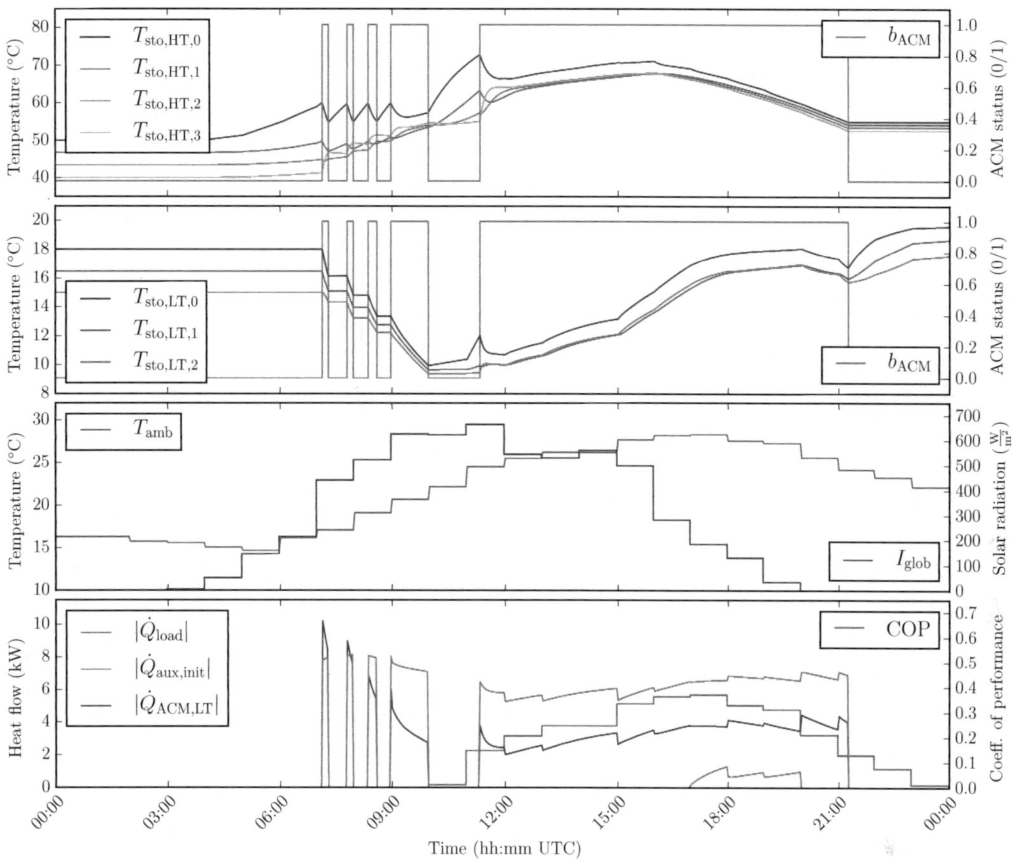

Figure 2. Simulation results using the control scheme in Eq. (15).

controller actively aims towards running the ACM more efficiently.

In the early day, it first increases the HT storage temperature more than the conventional controller, but still operates the ACM for some period before load occurrence to profit from the still comparatively low ambient temperatures. Afterwards, it heats up the HT storage, and also heats up the LT storage to compensate the now occurring load, which again creates favorable ACM operation conditions. Then, it runs the ACM for a longer period with comparatively high LT storage temperatures to again favor efficiency. At the end of the day, it runs the ACM one more time to profit once more from lower ambient temperatures to compensate for the final cooling demands of the day. At all times, the ACM is operated only within valid conditions. In that way, the auxiliary cooling demand has been reduced widely, as shown in the next section.

4.3. Comparison of controller performances

A comparison of the controller performances is given by a comparison of the integrated auxiliary cooling power as well as the integrated ACM operation time for the conventional and optimization-based control strategy in Figure 4.

In comparison to the conventional control strategy, the total demand for auxiliary cooling is reduced for this example by approximately a quarter using the optimization-based control strategy. This is a direct result of the objective formulation in Eq. (16a), and is obtained via a more efficient operation of the ACM.

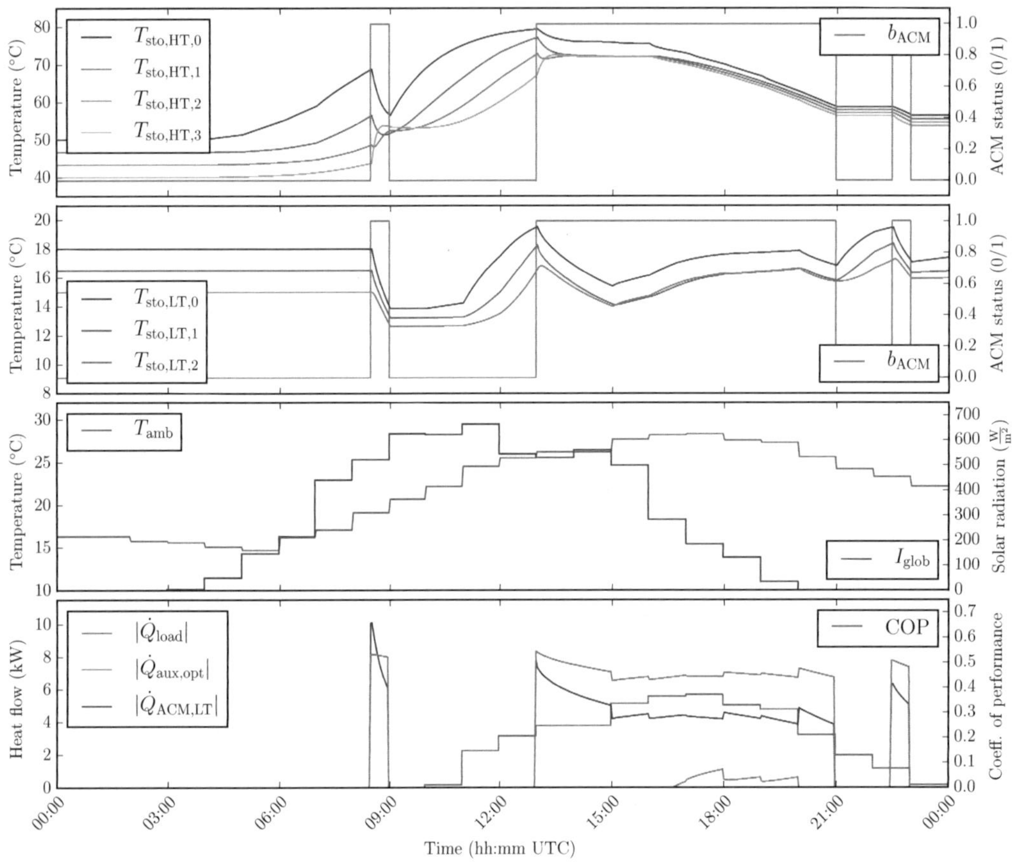

Figure 3. Solution of the MINLP arising from the MINOCP in Eq. (16).

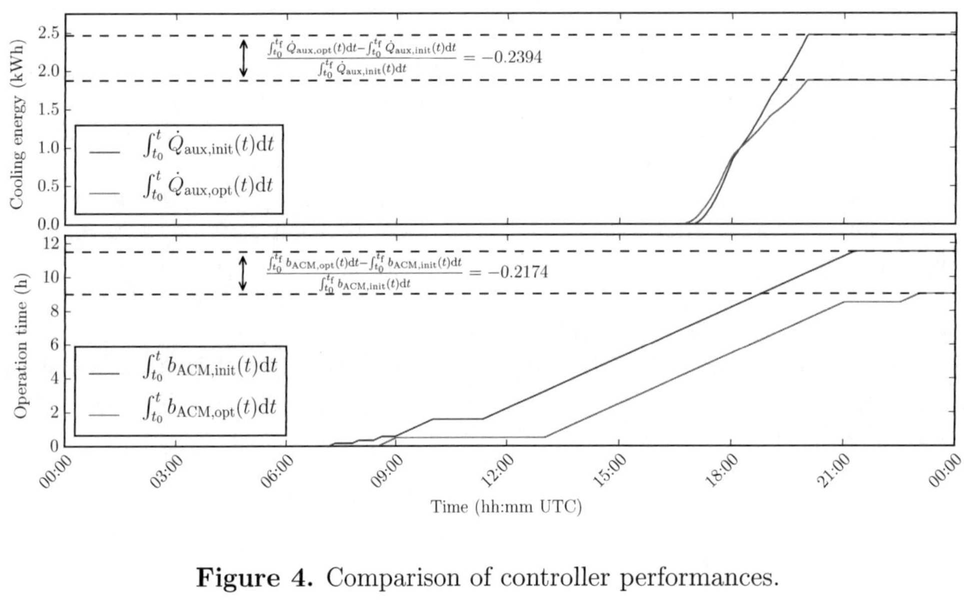

Figure 4. Comparison of controller performances.

Implicitly, a more efficient operation of the ACM, and with this, a more efficient use of the available driving energy, results in a shorter integrated operation time of the ACM. Thus, in addition to the reduced auxiliary cooling demand, the total ACM runtime is reduced by more than 20 % using the optimization-based control strategy, saving electricity cost for ACM, pump and especially recooling tower operation.

5. Conclusions and future work

Within this work, advantages of using predictive control methods for operation scheduling of a silica-gel/water ACM have been exemplified on a simulation study. Using the optimization-based control approach, the use of auxiliary cooling power supply could be reduced, which simultaneously resulted in a reduction of the total operation time of the ACM.

The examinations should now be carried out for other scenarios of driving energy and load occurrence. Due to its predictive nature, it can be expected that the optimization-based approach is well suited also for scenarios of fluctuating availability of driving energy.

For practical application of this approach for system control, an extended, closed-loop model-predictive control approach should be implemented, i. e., the MINLP should be solved repeatedly during operation of the system based on the current system state to compensate for model mismatches, prediction errors and other operation disturbances. This creates need for an optimization routine that can solve the problem almost in real-time and requires further investigation.

Acknowledgments

Support by the EU via ITN-TEMPO (607957) and ITN-AWESCO (642 682), by DFG via Research Unit FOR 2401, by BMWi via the project eco4wind and by Ministerium fuer Wissenschaft, Forschung und Kunst Baden-Wuerttemberg (Az: 22-7533.-30-20/9/3) is gratefully acknowledged.

References

[1] Chang W S, Wang C C and Shieh C C 2007 *Applied Thermal Engineering* **27** 2195 – 2199 ISSN 1359-4311
[2] Gräber M 2013 *Energieoptimale Regelung von Kälteprozessen* Ph.D. thesis
[3] Bau U, Braatz A L, Lanzerath F, Herty M and Bardow A 2015 *International Journal of Refrigeration* **56** 52 – 64 ISSN 0140-7007
[4] Nienborg B, Dalibard A, Schnabel L and Eicker U 2017 *Applied Energy* **185, Part 1** 732 – 744 ISSN 0306-2619
[5] Choudhury B, Saha B B, Chatterjee P K and Sarkar J P 2013 *Applied Energy* **104** 554 – 567 ISSN 0306-2619
[6] Afram A and Janabi-Sharifi F 2014 *Building and Environment* **72** 343 – 355 ISSN 0360-1323
[7] Biegler L 2010 *Nonlinear Programming* (Society for Industrial and Applied Mathematics)
[8] Ferhatbegovic T, Zucker G and Palensky P 2011 *Proceedings of 10th IEEE AFRICON*
[9] Núñez T 2010 *Journal of Sustainable Energy* **10**
[10] Streckiene G, Martinaitis V and Vaitiekunas P 2011 *8th International Conference on Environmental Engineering*
[11] Ferhatbegovic T, Palensky P, Fontanella G and Basciotti D 2012 *21st IEEE International Symposium on Industrial Electronics - ISIE2012*
[12] The Open Source Modelica Consortium 2017 OpenModelica URL `https://www.openmodelica.org`
[13] Glover F 1975 *Management Science* **22**
[14] Andersson J 2013 *A General-Purpose Software Framework for Dynamic Optimization* Ph.D. thesis Department of Electrical Engineering (ESAT/SCD) and Optimization in Engineering Center, KU Leuven
[15] Andersson J, Gillis J and Diehl M 2017 *User Documentation for CasADi* URL `http://casadi.sourceforge.net/users_guide/casadi-users_guide.pdf`
[16] Bonami P, Biegler L T, Conn A R, Cornuéjols G, Grossmann I E, Laird C D, Lee J, Lodi A, Margot F, Sawaya N and Wächter A 2008 *Discrete Optimization* **5** 186 – 204 ISSN 1572-5286
[17] Wächter A and Biegler L T 2006 *Mathematical Programming* **106** 25–57 ISSN 1436-4646
[18] HSL 2017 A collection of fortran codes for large scale scientific computation URL `http://www.hsl.rl.ac.uk/`

IOP Conf. Series: Materials Science and Engineering **257** (2017) 012008 doi:10.1088/1757-899X/257/1/012008

6 DOF articulated-arm robot and mobile platform: Dynamic modelling as Multibody System and its validation via Experimental Modal Analysis.

A Toledo Fuentes[1,*], M Kipfmueller[1], M A José Prieto[2]

[1] Institute of Materials and Processes, University of Applied Science, Karlsruhe 76131, Germany

[2] Department of Electrical Engineering, University of Oviedo, Gijón 33203, Spain

[*]Corresponding author: aishe.toledo-fuentes@hs-karlsruhe.de

Abstract. Mobile manipulators are becoming a key instrument to increase the flexibility in industrial processes. Some of their requirements include handling of objects with different weights and sizes and their "fast" transportation, without jeopardizing production workers and machines. The compensation of forces affecting the system dynamic is therefore needed to avoid unwanted oscillations and tilting by sudden accelerations and decelerations. One general solution may be the implementation of external positioning elements to active stabilize the system. To accomplish the approach, the dynamic behavior of a robotic arm and a mobile platform was investigated to develop the stabilization mechanism using multibody simulations. The methodology used was divided into two phases for each subsystem: their natural frequencies and modal shapes were obtained using experimental modal analyses. Then, based on these experimental results, multibody simulation models (MBS) were set up and its dynamical parameters adjusted. Their modal shapes together with their obtained natural frequencies allowed a quantitative and qualitative analysis. In summary, the MBS models were successfully validated with the real subsystems, with a maximal percentage error of 15%. These models will serve as the basis for future steps in the design of the external actuators and its control strategy using a co-simulation tool.

1. Introduction

Recent developments in the field of the fourth industrial revolution have emphasized in the ambitious trends and achievement criteria in automate production [1, 2]. Most studies in this area have highly focused on the development of the so-called Internet of Things, underestimating all the technology that has to carry out the tasks in an autonomous way without critical limitations. In recent years, mobile manipulators based on costly light-weight robot arms mounted on the top of relatively bulky mobile autonomous platforms have been introduced in production lines to achieve a higher degree of flexibility in transport and handling tasks. However, this form of robotic solutions is limited by their restricted payload (nowadays, 14 kg maximal payload [3, 4]), avoiding the potential risk of losing their stability while the mobile platform moves through the plant. The demand for handling higher payloads implies the use of bigger manipulators (with higher center of gravity) mounted on large-footprint and heavier mobile platforms to prevent the robot arms from tipping over [5]. Likewise, large-footprint platforms are directly associated with more required workspace, and thus higher costs for the plant surface area. There is, therefore, a definite need to develop a robot system which comprises a high payload robot arm on a small-footprint autonomous mobile platform.

The first serious discussions and analyses aimed to stabilize mobile manipulators with a high center of gravity emerged in 1997 with Graf and Dillmann [6], implementing a parallel manipulator using a Stewart-platform to actively compensate any accelerations. Further examples of research include the optimization of this approach [7, 8, 9]. While it is an efficient way to ensure the stability during the transport of fragile objects on its platform base, the system is not designed to handle them. Besides that, some researchers (e.g. Zhao et al. [10]; Acar and Murakami [11]) have attempted to implement a three DOF manipulator on a two-wheeled mobile platform. They applied the inverted pendulum principle to model the robot system dynamic as one body: its center of gravity, an important parameter to keep the system stabile, varies depending on displacements, velocities and accelerations of the mobile platform, as well as on the weight and height of the robot arm. The design approaches developed at the Institute for Cognitive Systems at the Technical University of Munich [12] as well as at the Korea Institute of Science and Technology [13] and at the Fraunhofer Institute of Optronics in Karlsruhe [14] consist of a two-arm humanoid upper body. The robot body is mounted on a "Compact Omni-directional Mobile Platform" and on Segway devices, respectively, to allow it to be moved.

Such approaches, however, have tended to focus on the development of completely new robotic systems rather than the design of an active external mechanism to ensure the system stability at standstill o during normal plant operation, regardless of the robot arm and mobile platform model. In order to realize this concept, an upgradeable system is proposed: With in the fabric currently available robot arms and an active external actuator system, an affordable mobile manipulator could be built, simply by integrating them with a small footprint platform. The particular importance of this approach lies in the possible use of low-cost robot arms or the re-use of already existents in the plant.

The main challenge to face in the design of the actuator system is to guarantee the stability of the manipulator during its operation, including when the mobile platform abruptly starts or stops. In other words, the system should react to any external influences and, consequently not to affect the human safety in the working areas [15]. Their adequate control strategy to compensate the accelerations should ensure the equilibrium of dynamical forces affecting the system balance in every moment.

As an essential phase of the development process, the performance of the control strategies to be implemented should be evaluate under critical conditions using multibody simulation models. These developed dynamical models should resemble as accurate as possible the real testing system [16], consisting of a mobile robotic platform (see figure 1) and an articulated-arm robot (see figure 2).

This paper focused on the modelling process of both mentioned robots as multibody systems by adapting the procedure used in [17]: As a first step, the dynamic properties (modal shapes and natural frequencies) of each real robot were separately identified by means of experimental modal analyses. Subsequently, CAD-Models of each robotic system were prepared in order to build them as multibody models into a MBS-Software (MSC.Adams/View). The obtained experimental findings (from experimental modal analyses) were compared with the results produced by the multibody simulation. By adjusting the modal parameters of the simulative models, the dynamic behaviors of the real system were replicated. Thus, the expected result will be the development of close to reality MBS models of the testing systems. Further information about this method is presented in [18, 19].

Figure 1. Robot I to be modelled: Autonomous mobile platform (MetraLabs X3).

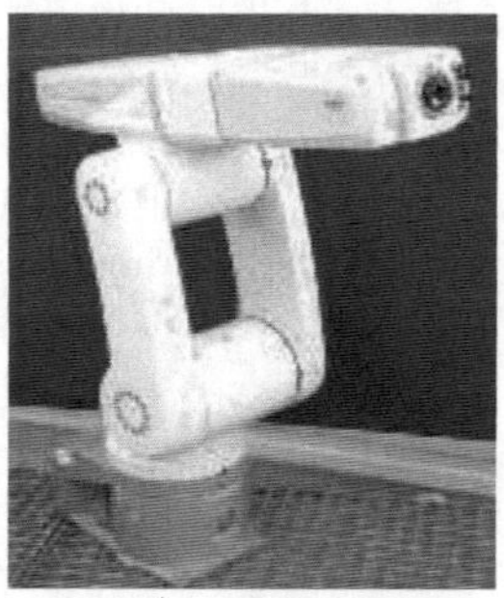

Figure 2. Robot II to be modelled: 6 DOF articulated-arm robot (Mitsubishi RV-3AL).

2. Experiment / Simulation set up

2.1. Identification of modal parameters via experimental modal analysis

The dynamic behavior of any system can be estimated by means of modal analysis. The so-called experimental modal analysis consists in exciting externally a real mechanical system (input forces) and in measuring its dynamic behavior in response to the applied stimulus (output accelerations). This Response Function describes the relation force/acceleration and allows to validate simulation models of multibody systems.

The presented study seeks to obtain experimental data which will help to validate the dynamical models of robotic systems.

The correct election of the excitation and measurement form were the initial aspects to consider in order to perform a proper experimental modal analysis (the abbreviation EMA will be used hereafter to refer to it). For this particular set up, the relevant parameters for the impulse hammer (including tip), 3D accelerometers and the measurement frequency range are shown in table 1.

Table 1. Configuration parameters for the experimental modal analyses.

Parameter	Value (for both robotic systems)
Measured Response Frequency range	800 Hz
Impulse hammer – Tip	Soft impact tip
Impulse hammer – Weight / sensitivity	1.0 pound / 1.0 mV/LB
Accelerometer – Voltage Sensitivity	1.0 +12/-8% mV/ms^{-2} 10 ±10% mV/g

The election of the suitable excitation and measurement points was another important object of analysis. The challenge lay in choosing the proper points without having previous knowledge about the dynamic behavior of the real systems. In our case, the risk of ignoring some natural frequencies and modal shapes during the excitation or measuring was low, as long as the excitation points were consciously collocated non orthogonal to the measurement points (all orthogonal shapes concerned to the impact force direction may not be measured by the accelerometers [20]). The excitation and measuring conditions, the used impulse hammer, tip and hammer head as well as the sensors, excitation/measurement points and their directions were the same for all tests.

Although the analysis was carried out in a frequency range from 0 Hz to 800 Hz, the most important frequencies were expected within a low frequency range, under 100 Hz.

2.1.1. EMA of mobile platform. First, the structure of the mobile platform was analyzed. The red plastic covering (see figure 1) was removed in order to identify all the surfaces where the sensors had to be placed and the force impacts had to be applied. The 3 excitation points (see green dots in figure 3) were selected to guarantee the application of stimulus in all coordinates of the Cartesian space. Each experimental series consisted in an excitation of 5 stimuli with delays between each other using the same impulse hammer. Additionally, the 25 chosen measuring points were mainly collocated on the rigid aluminum profiles and near to the screw joints to ensure a good reproduction of the system oscillation but do not emulate the vibration of each single element. These points were the basis for the geometry built into the processing software OROS Modal 2 (see figure 4), which would graphically display the platform modal shapes as result of the EMAs.

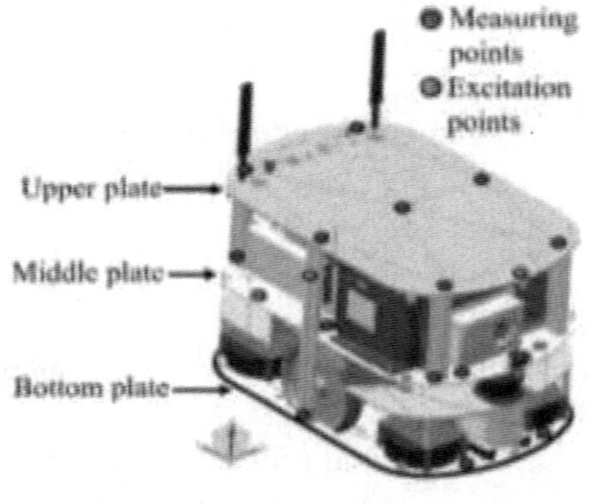

Figure 3. CAD model of the mobile platform and its measurement points.

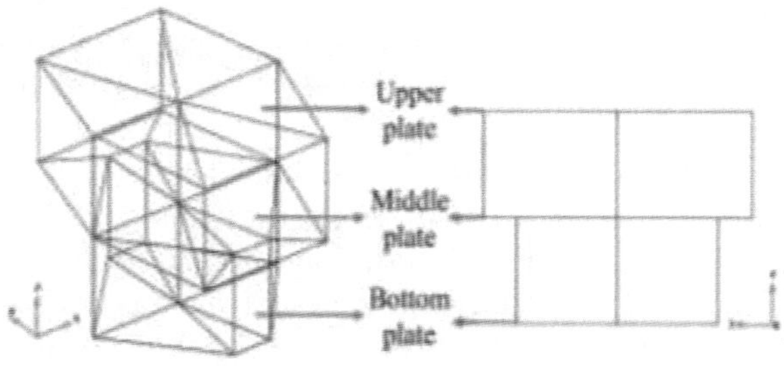

Figure 4. Representation of the mobile platform in OROS Modal 2 (based on the points in figure 3).

2.1.2. EMA of articulated-arm robot. The EMA of the articulated-arm robot was carried out in Home configuration, as presented in the figure 5.

Similar to the mobile platform, the system was analyzed, and the excitation and measurement points that guaranteed an appropriate excitation and detection of the whole modal parameters were identified (see red dots in figure 5): The simplification of the robot geometry in OROS Modal 2, shown in figure 6, resulted in 8 excitation points and 23 measurement points.

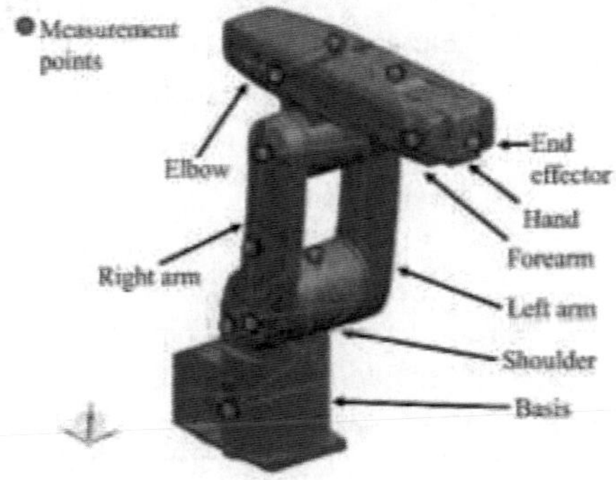

Figure 5. CAD model of the articulated-arm robot and its measurement points.

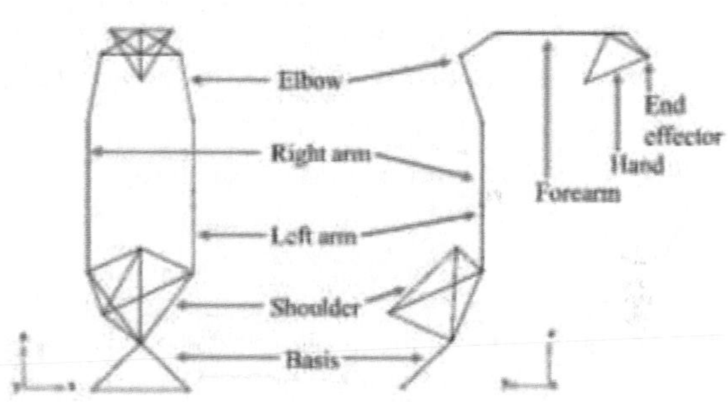

Figure 6. Representation of the main parts of the articulated-arm robot in OROS Modal 2.

2.2. Modelling of multibody systems

The attainment of the experimental analyses, in terms of natural frequencies and modal shapes, allowed the reliable modelling of the mobile platform and the articulated-arm robot as multibody systems. The modelling process started with the analysis of the system dynamic behavior according to the information obtained from the EMAs. Hence, the systems were modelled with a low level of detail to get good computing performance, but as complex as the reproduction of the real dynamic behavior may not be affected by excluding some elements in the model (see "Model validation"). In other words, the multibody systems had to contain justly all necessary information to describe the dynamic phenomenon of the real mobile platform and articulated-arm robot.

During this analysis, the potential to model the structure of both systems as rigid bodies in the MBS was detected: No relevant deformations appears in the main bodies of the structures in comparison to the elastic joint elements, which were responsible to generate the detected model shapes (see 4.1- 4.2).

2.2.1. MBS of mobile platform. In order to build the MBS model of the mobile platform, the estimation of the inertial parameters was carried out based on its CAD model, which contained all the information about its geometry and materials. This CAD model with inertial properties was exported into the used multibody simulation software, MSC.Adams/View.

According to [21] there are different variants to model bodies as MBSystems: rigid, elastic and their combination (so-called hybrid). In the cases presented in this paper, all bodies were implemented as rigid (except for the joints elements), since the purpose was not to represent the deformation of each particular element of the system when a force is acting on them, but rather to reproduce the oscillation of the whole system into a fixed space. This implied that each element tends to suffer an insignificant deformation in comparison to the oscillation of the whole system exhibited from EMA.

In order to simplify the system, elements with the same material were merged into one. Without considering the red outer covering, each element of the structure was examined to determine its possible influence in the oscillation phenomenon: screws, screw nuts, washers and slot nuts were removed and replaced by fix joints.

The next step was to connect each of the rigid bodies using massless elastic joint elements. They defined the degree of freedom of the whole mobile platform, and thus, restrict the number of the modal shapes within the low frequency range. All existing joint elements of the real mobile platform were classified based on their kinematic (geometrical relation) and their physical force properties, acting between two or more links [22]. The most influential components were recognized between the drive wheel and the aluminum support: wheel suspensions, swing arms, dampers, spacer sleeves and rigid struts. To define the systems degrees of freedom and thus to limit the constraining forces affecting their bodies, so-called kinetic virtually joints with floating characteristics had to be implemented in the model. These massless elastic joint elements are shown in table 2. The drive wheels were fixed to the ground.

Table 2. Joint elements implemented between the main rigid bodies of the mobile platform.

Joint elements found in the real mobile platform	Massless joint elements used in the MBS
Cylinder-piston system / spacer sleeves	Damper and linear spring / Bushing
Drive shaft and bearings	Bushing
Coupling and motor shaft	Torsion spring
Rigid strut and structure	Bushing
Swing arm and structure	Linear and torsion spring

The modelling of those elements, together with their inherent stiffness and damping parameters, enabled the dynamic oscillation of the system. A first estimation of these parameters for each joint element was performed via Finite Element Analysis. The multibody model was dynamically simulated with these first approximations: The results obtained in terms of natural frequencies and their corresponding modal shapes were analyzed and compared with the experimental findings from the EMA. If there was no coincidence between the experimental and the simulative procedure, the stiffness and damping coefficients of specific joints had to be adjusted until the results were equivalent. The visualization of the modal shapes in the simulations presented a helpful qualitative analysis to identify which specific joint element had to be modified and how much.

2.2.2. MBS of articulated-arm robot. The modelling of the articulated-arm robot in MSC.Adams/View was very similar to the procedure followed for the mobile platform, but with a main difference: It available CAD model just contained information merely related to sizes and geometries, but not aspects about materials, weight, center of mass and moment of inertia, which played a crucial role in the modelling process. The bodies with bigger and heavier properties may mainly affect the dynamic behavior of the system because of their inertial properties. A material identification was carried out, in particular, to estimate the density of the materials in the outer structure by performing some simple non-destructive tests (Archimedean principle as well as magnetic proof). Then, the material specifications were introduced into the shell CAD model and were adjusted so as to match up the weight specified by the robot manufacturer. This detailed model with centers of mass and moments of inertia was exported into MSC.Adams/View.

The model simplification of the robot was achieved by identifying all the mechanical joints existing in the real robot (based on [23]). Analogous to the mobile platform, the robot components with the higher mass and inertial moment were abstracted and merged into basic elements: Basis, shoulder, right arm, left arm, elbow, forearm, hand and end effector (see figure 6). They were modelled as rigid bodies for the same reason as the mobile platform.

These 8 basic elements were joined to each other using the so-called bushings, massless elastic joint elements with up to 6 DOF, which can replicate both translational and rotational constraints by introducing certain stiffness and damping values for each available degree of freedom; in every joint, they reproduced kinematic effects and, during a dynamic simulation, applied forces and torques to each junction point. For the articulated-arm robot, each bushing had to allow only one rotational DOF, emulating the motors in the robot joints. The correlation between the rigid bodies and the entire set of bushings in the system defined the robot oscillation behavior. Detailed information about MBS methodologies could be found in [23, 24].

3. Model validation

3.1. EMA and MBS: Mobile platform

The results of the EMA for the mobile platform are shown in the table 3. An overlapping method [25] allowed to identify the first eight natural frequencies (ω) under 100 Hz: The first, third and fourth natural frequencies (7 Hz, 21 Hz and 28 Hz) appeared in all tests. Following the overlapping method, the remaining values were obtained at 10 Hz, 54 Hz, 62 Hz, 75 Hz and 81 Hz, which were visible just in no more than two different directions of excitation. The corresponding first five modal shapes are graphically represented in the table 3.

Table 3. Modal shapes and natural frequencies of the mobile platform obtained from the EMAs.

Natural freq. and modal shapes	ω1= 7.75 Hz	ω2= 10.3 Hz	ω3= 21.4 Hz	ω4= 28.0 Hz	ω5= 54.0 Hz
EMA Mobile platform	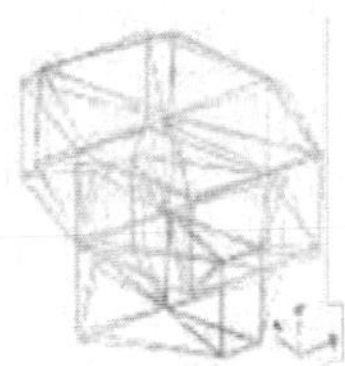	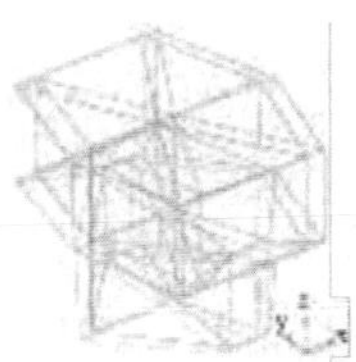	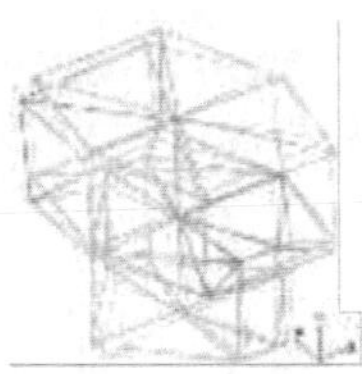	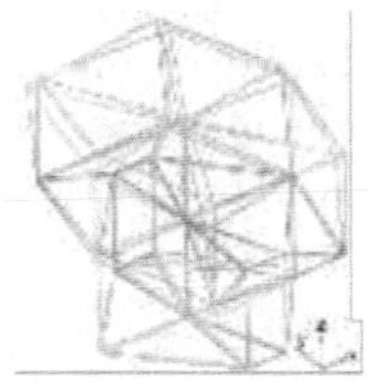	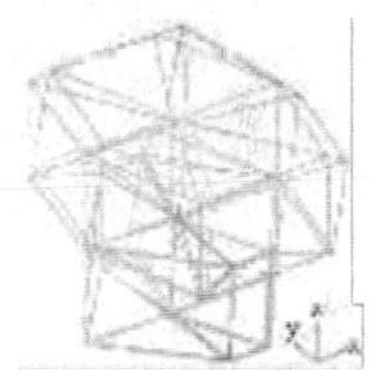

The grey lines in the modal forms represent the initial state, the red lines the maximum amplitude attained, the yellow lines the proportionately smaller amplitudes and the blue lines the smallest one. Taking figure 4 as reference for the geometry, the first modal shape from table 3 consisted in a rotation of the platform about the Z-axis (yaw). The second modal shape showed a translational movement in X-direction. The third one exhibited a petty rotation around the Y-axis (roll) and another around the X-axis (pitch). The fourth and fifth modal shapes displayed a slightly tilting around the Y-axis (roll) and the X-axis (pitch), respectively. All axes are related to the light blue coordinate system.

After completing the multibody modelling of both systems with rigid bodies and massless joint elements, proceed the validation of these simulation models: Their stiffness and damping parameters had to be iteratively adjusted as long as the simulation reproduced similar natural frequencies and modal forms related to the experimental results.

For both mobile platform and articulated-arm robot the modal shapes from EMA were the key to validate the calculated natural frequencies of the MBS: The modal shape of the desired natural frequency had to exhibit the same oscillation as in the real system (from EMA). Thus, the natural frequency could be validated and the iterative process was concluded.

The results for the mobile platform after the adjustment of their natural frequencies and modal shapes are shown in the table 4. The red traces in the illustrations represent the modal shapes.

Table 4. Modal parameters of the mobile platform obtained from multibody simulations.

Natural freq. and modal shapes	ω1= 8.92 Hz	ω2= 12.96 Hz	ω3= 21.54 Hz
MBS Mobile platform			

3.2. *EMA and MBS: 6DOF articulated-arm robot*

Analogous to the mobile platform, the table 5 gives the first natural frequencies and modal shapes from EMA acquired by the excitation of the system in different directions.

Table 5. Modal shapes and natural frequencies of the articulated-arm robot obtained from the EMAs.

Natural frequencies and modal shapes	ω1= 11.0 Hz	ω2= 21.0 Hz	ω3= 27.0 Hz	ω4= 72 Hz	ω5= 89 Hz
EMA Articulated-arm robot		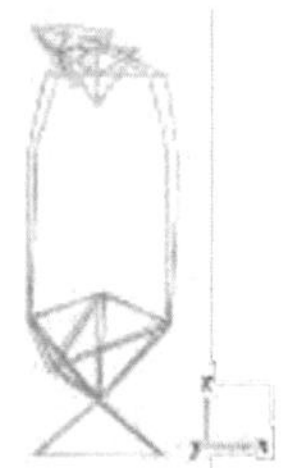		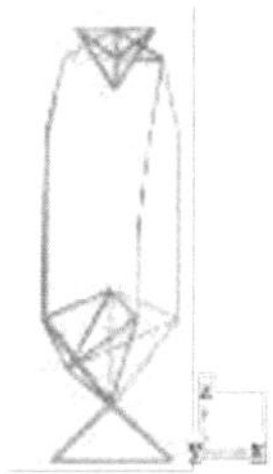	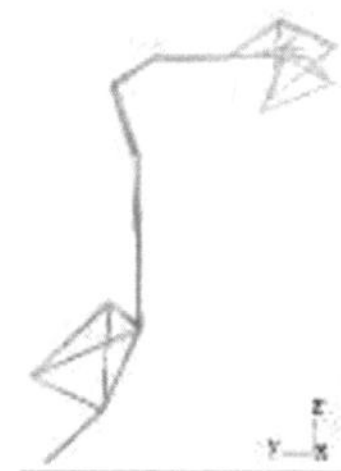

It was observed that the first three natural frequencies (11 Hz, 21 Hz and 27 Hz) were found in all directions of excitation. However, the natural frequencies at 72 Hz and 89 Hz appeared only in two separately excitation points. All frequencies of interest were found within the first 100 Hz.

Each modal shape, and thereby the natural frequencies, depended on the current position of the system. For the illustrated home position, the first modal shape (at 11 Hz) was a pitching movement (around the X-axis) generated by the elbow, the forearm and the end effector; the second and third modal shapes (21 Hz and 27 Hz) affected only the lightest four components, causing a slightly rolling movement around the Z-axis and a rotation around the X-axis (pitching) in the elbow, forearm, hand and end effector; the fourth modal shape (at 89 Hz) presented a deflection on the left arm; and finally, the fifth modal shape also showed a slightly pitching movement (rotation around X-axis) but just starting in the hand and continuing with the end effector.

Likewise, in the same way as for the mobile platform, these experimental outcomes were the reference for the successive adjustment of the stiffness and damping parameters in the multibody simulation model. These simulation results are presented in table 6. The red traces in the illustrations represent the modal shapes.

Table 6. Modal parameters of the articulated-arm robot obtained from the multibody simulations.

Natural frequencies and modal shapes	ω1= 13.0 Hz	ω2= 22.5 Hz	ω3= 27.2 Hz	ω4= 68.3 Hz	ω5= 88.9 Hz
MBS Articulated-arm robot			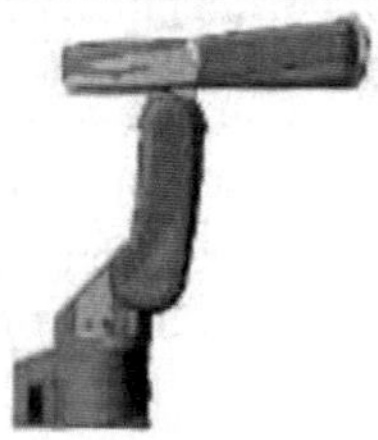		

4. Results and discussion

The graphic in figure 7 presents the calculation of the MBS reliability related to the EMA, using the "three-sigma limit" method [26]. The natural frequencies from EMA are represented by the black solid line. It can be seen that the first modal frequency of the MBS (red solid line) was not kept within the calculated limits (grey and green dashed lines) in contrast to the second and third values, which are in the tolerated area. The deviation of this first natural frequency is about 15%, which could have been caused by the model simplification procedure. Despite this and according to [27, 28], this value could be considered as a good approximation.

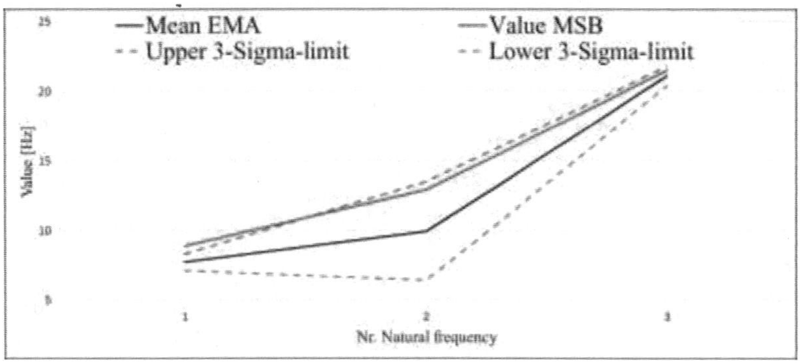

Figure 7. Correlation between EMA- and MSB-mobile platform using 3-sigma limits.

Meanwhile, the MBS of the articulated-arm robot presented a better fitting (see figure 8). However, the second natural frequency was slightly outside of the tolerance margin either (red solid line). Although this deviation between the MBS and the EMA (black solid line) resulted about 10.6%, it fulfilled the condition mentioned above [27, 28]. As can be seen from the graphic, the remaining values were within the "three-sigma limits".

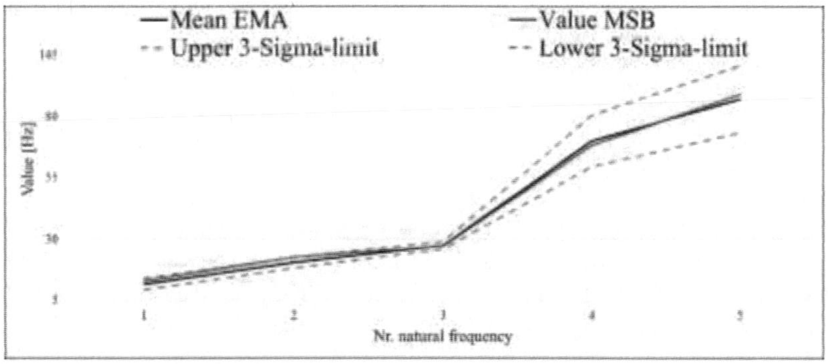

Figure 8. Correlation between EMA- and MSB-articulated arm robot using 3-sigma limits.

Hence, some variations between the MBS- and the EMA- results could have been caused, on the one hand, by the model simplification carried out to decrease the complexity of the systems (e.g. friction coefficients of the components were not experimentally determined, rigid bodies were used to avoid large computational calculations, etc.). On the other hand, random errors could have emerged due to parametric uncertainties as lack of material homogeneity, tolerances in the geometry, unknown clearance between the robot elements, etc., having non-ideal real conditions as in MBS models.

5. Mechatronic simulation

Once both systems were modelled and validated, it was possible to join them into a whole mobile manipulator model. This new MBS system had to be integrated with external actuators, whose task should be the dynamic compensation of external forces to assure its equilibrium, and thus to prevent the manipulator from tilting by abrupt acceleration and deceleration events. The whole system is shown in figure 9. In order to speed up the design and testing phase of the required control strategy for the integrated actuators, a co-simulation between MatLab/Simulink and MSC.Adams/View must be performed: the closed loop control of the external actuators communicates with the dynamic MBS and sends the control signals to each of them; the MBS, as feedback loop, returns measurements from virtual sensors as force, angle, speed, acceleration, etc., at certain time intervals. The figure 10 illustrates the coupled simulation. Currently, the integrated model is used to develop control algorithms for the system's active stabilization. A full discussion about the selection process, the design and testing phase of the actuators lies beyond the scope of this article.

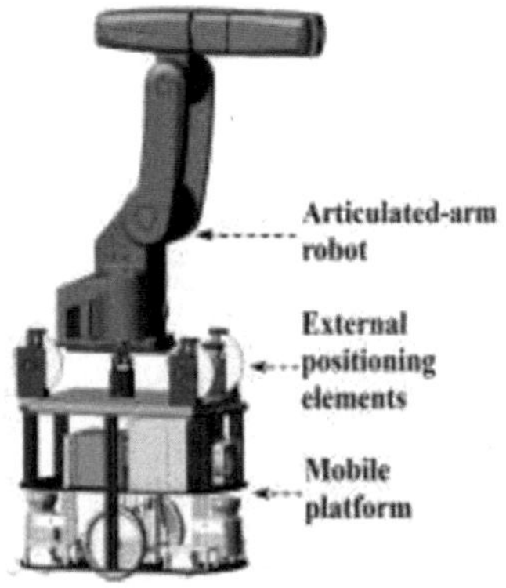

Figure 9. MBS-mobile manipulator.

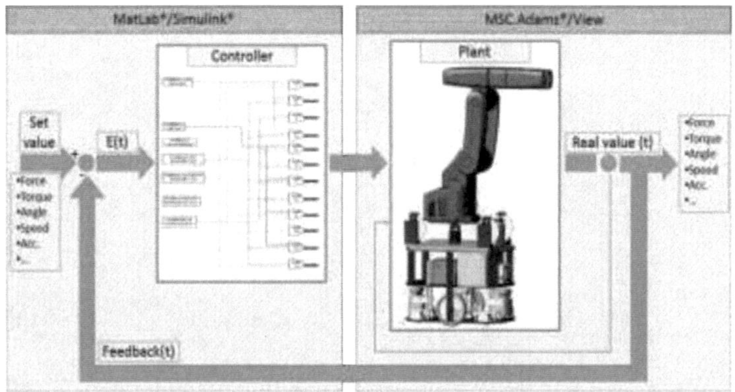

Figure 10. Schema of the co-simulation loop.

6. Conclusions

The aim of the present article was to determine experimentally the dynamic behavior of two robotic systems and, based on the results achieved, to model them as multibody simulations. At the first stage, experimental modal analyses have identified the natural frequencies and modal shapes of both real systems. These experimental results supported the implementation of rigid bodies in the multibody simulation to model all the components of both systems (except for their elastic joint elements). A key strength of their modelling as MBS was to define the stiffness and damping values for all joint elements of the systems. Each dynamical simulation generated magnitudes of natural frequencies depending on those stiffness and damping values. Together with their corresponding modal shapes, they were compared with the empirical values until they matched suitably to perform the development of the control strategy for the active stabilization.

The findings verified the good reproduction of the dynamical behavior of the real systems in the multibody simulation models. Although the current study is based on particular robots, it offers a helpful insight into the methodology for further complex systems, especially with regards to mobile platforms, whose dynamic behavior still little known.

A limitation of the implemented MBS modeling technique, particularly for the mobile platform, was the iterative adjustment process of the stiffness and damping values to obtain the desired natural frequencies. This suggested moving on to discuss the demand to improve this method. What is now needed is the development of an automated optimization algorithm able to mathematically determine the approximate stiffness and damping values for each joint element of the system. All these resulting parameters together would be reflected in their proper natural frequencies and modal shapes (see 3.2.), thereby avoiding the time-consuming manual iterative process. Thus, the articulated-arm robot could be easily modeled in different poses (e.g. in vertical and horizontal configuration) for further investigations. The authors are currently working on the implementation of those optimization strategies in a second investigation project.

The described MBS model is currently used to complete the development of the control strategy based on mechatronic co-simulations. The control system should ensure the stability of the whole mobile manipulator under accelerations and decelerations. Alongside, the optimization of its functionality should be carried out, taking in account sensitivity studies for the simulation model. The intention is to implement the control algorithm in a prototype developed and already set up by the authors without making relevant changes.

Acknowledgement
The authors would like to thank the state Baden-Württemberg for the kind support within the funding program "Innovative Projekte".

References

[1] Qin J, Liu Y and Grosvenor R 2016 A categorical framework of manufacturing for industry 4.0 and beyond *Procedia 6th CIRP Int. Conf. CARV (September 4-6, Bath)* vol 52 ed A Nassehi and S Newman (Amsterdam: Elsevier) pp 173-8

[2] Schleipen M, Gilani S S, Bischoff T and Pfrommer J 2016 Industrie 4.0 - enabling technology with high diversity and variability *Proc.49th CIRP Conf. Manufacturing Systems (May 25-27, Stuttgart)* vol 57 ed E Westkämper and T Bauernhansl (Amsterdam: Elsevier) pp 315-20

[3] Fang H, Lijie G and Shaoping B 2015 A light weight arm designed with modular joints *Proc. 3rd IFToMM Symp. on Mechanism Design for Robotics (Mechanisms and machine science* vol 33*) (June 2-4, Aalborg)* ed B Shaoping and M Ceccarelli (Switzerland: Springer) p 49

[4] SCHUNK Mobile Greifsysteme 2017 Powerball Lightweight Arm / Dextrous Lightweight Arm Available from: http://www.schunk-mechatronik.com/startseite/schunk-mobile-greifsysteme.html [Accessed 26th July 2017]

[5] Acar C and Murakami T 2008 Underactuated two-wheeled mobile manipulator control using nonlinear backstepping method *Proc. 34th Annual Conf. IEEE IECON (November 10-13, Orlando)* (Piscataway, NJ: IEEE) pp 1680-5

[6] Graf R and Dillmann R 1997 Active acceleration compensation using a Stewart-platform on a mobile robot *Proc. 2nd EUROMICRO Workshop on Advanced Mobile Robots (October 22-24, Brescia)* (Los Alamitos, California; Piscataway, NJ: IEEE) pp 59-64

[7] Dang A X H and Ebert-Uphoff I 2004 Active acceleration compensation for transport vehicles carrying delicate objects *IEEE Trans. Robot.* **20** 830-9

[8] Bai X, Davis J, Doebbler J, Turner J D and Junkins L J 2007 Dynamics, control and simulation of a mobile robotic system for 6-DOF motion emulation *Proc. WCECS (October 24-26, San Francisco)* ed S I Ao et al (Newswood Limited) pp 753-8

[9] Danko T W, Chaney K P and Oh P Y 2015 A parallel manipulator for mobile manipulating UAVs *Proc. 7th IEEE Int. Conf. TePRA (May 11-12, Woburn)* (Piscataway, NJ: IEEE) pp 1-6

[10] Zhao Y, Woo C and Lee J 2015 Balancing control of mobile manipulator with sliding mode controller *Proc. 15th ICCAS (October 13-16, Busan)* (S.I.: IEEE) pp 802-5

[11] Acar C and Murakami T 2011 Multi-task control for dynamically balanced two-wheeled mobile manipulator through task-priority *Proc. 20th IEEE ISIE (June 27-30, Gdansk)* (Piscataway, NJ: IEEE) pp 2195-200

[12] Leboutet Q, Dean-León E and Cheng G 2016 Tactile-based compliance with hierarchical force propagation for omnidirectional mobile manipulators *Proc. 16th IEEE RAS Int. Conf. Humanoid Robots (November 15-17, Cancun)* ed T Asfour (Piscataway, NJ: IEEE) pp 926-31

[13] Cha Y S, Kim K G, Lee J Y, Lee J, Choi M, Jeong M H, Kim C H, You B J and Oh S R 2011 MAHRU-M: A mobile humanoid robot platform based on a dual-network control system and coordinated task execution *Rob. Auton. Syst.* **59** 354-66

[14] Milighetti G, Petereit J and Kuntze H B 2010 Mobile experimental platform for the development of environmentally interactive control algorithms towards the implementation on a walking humanoid robot *Proc. 41st ISR and 6th ROBOTIK (June 7-9, Munich)* (S.I.: VDE) pp 682-8

[15] Brettel M, Fischer F G, Bendig D, Weber A R and Wolff B 2016 Enablers for self-optimizing production systems in the context of industrie 4.0 *Proc. 48th CIRP Conf. Manufacturing Systems (June 24-26, Ischia)* vol 41 ed R Teti (Amsterdam: Elsevier) pp 93-8

[16] Karnopp D C, Margolis D L and Rosenberg R C 2012 *System dynamics: Modeling, simulation, and control of mechatronic systems* (New Jersey: John Wiley & Sons) p 4

[17] Webhofer M 2000 *Modellierung eines Mehrkörpersystems zur Simulation der Querpendelbewegung von Einseilumlaufbahnen bei der Stationseinfahrt* (Student research project TU Munich, Institute of Materials Handling, Material Flow, Logistics)

[18] Rahnejat H and Rothberg S 2004 *Multi-body dynamics: Monitoring and simulation techniques III* (London and Bury St Edmunds: John Wiley & Sons) pp 373-90

[19] Bauchau O A 2010 *Flexible multibody dynamics* (Dordrecht – Heidelberg – London - New York: Springer) pp 571-77

[20] Karim A, Verl A and Höhne R 2014 Schwingungsanalyse an Bearbeitungsrobotern: Durchführung einer experimentellen Modalanalyse an Industrierobotern *wt online* **1/2-2014** 49

[21] Kreuzer E, Lugtenburg J B, Meißner H G and Truckenbrodt A 2012 *Industrieroboter: Technik, Berechnung und anwendungsorientierte Auslegung* (Berlin - Heidelberg: Springer)

[22] Gross D, Hauger W, Schröder J, Wall W and Govindjee 2014 *Engineering mechanics 3: Dynamics* (Berlin - Heidelberg: Springer)

[23] Blundell M and Harty D 2014 *The multibody systems approach to vehicle dynamics* (Oxford - Massachusetts: Butterworth-Heinemann)

[24] Bremer H 2008 *Elastic multibody dynamics: A direct Ritz approach* (Netherlands: Springer)

[25] Neugebauer R, Kolouch M, Richter M and Schulten M 2009 Fehlerquellen bei einer Modalanalyse: Untersuchung von Einflussfaktoren während der praktischen Durchführung *wt online* **11/12-2009** 889-94

[26] Devore J L, Farnum N R and Doi J A 2013 *Applied statistics for engineers and scientists* (Stamford: Cengage Learning) p 257

[27] Kipfmüller M and Munzinger C 2009 Efficient simulation of parallel kinematic machine tools *Proc. ASME IDETC/CIE 33rd Mechanisms and Robotics Conf. (August 30- September 2, San Diego)* vol 7 (ASME) pp 523-30

[28] Vinayak H and Singh R 1998 Multi-body dynamics and modal analysis of compliant gear bodies *J. Sound Vib.* **210** 171-214

Development of Six Sigma methodology for CNC milling process improvements

M N Ismail[1, 2], A N M Rose, [1], N Z Mohammed [1], M F F Ab Rashid [1]

[1]Manufacturing Focus Group, Faculty of Mechanical Engineering, University Malaysia Pahang, 26600 Pekan Pahang, Malaysia
[2]Department of Quality and Productivity, Kolej Kemahiran Tinggi MARA Kuantan, Km 8 Jln Gambang, 25150 Kuantan Pahang, Malaysia

[*]Corresponding author: mohdnasrul.ismail@gmail.com

Abstract. Quality and productivity have been identified as an important role in any organization, especially for manufacturing sectors to gain more profit that leads to success of a company. This paper reports a work improvement project in Kolej Kemahiran Tinggi MARA Kuantan. It involves problem identification in production of "Khufi" product and proposing an effective framework to improve the current situation effectively. Based on the observation and data collection on the work in progress (WIP) product, the major problem has been identified related to function of the product which is the parts can't assemble properly due to dimension of the product is out of specification. The six sigma has been used as a methodology to study and improve of the problems identified. Six Sigma is a highly statistical and data driven approach to solving complex business problems. It uses a methodical five phase approach define, measure, analysis, improve and control (DMAIC) to help understand the process and the variables that affect it so that can be optimized the processes. Finally, the root cause and solution for the production of "Khufi" problem has been identified and implemented then the result for this product was successfully followed the specification of fitting.

1. Introduction

In 1987, Motorola developed and organized the Six Sigma process improvement methodology to achieve "world-class" performance, quality, and total customer satisfaction. Since that time, at least 25% of the Fortune 200, including Motorola, General Electric, Ford, Boeing, Allied Signal, Toyota, Honeywell, Kodak, Raytheon, and Bank of America, to name a few, have implemented a Six Sigma program [1] . Six Sigma (SS) is a quality tool for process improvement program in any organisation. It is striving to eliminate the defects up 3.4 parts per million. This research will present the step-by-step application of the Define–Measure–Analyse–Improve–Control (DMAIC) approach as to eliminate the defects caused by CNC Milling process. This study will help the organisation to identify and reduce defects in the process. Hence, this effort could improve productivity and quality of the product. This study is conducted at Kolej Kemahiran Tinggi MARA Kuantan (KKTM), based on facing problem due to fitting on "Khufi". This product is produced by KKTM as a souvenir product for the visitors and also as a present in any event conducted in KKTM. This souvenir is processed by CNC milling machines by using the aluminium material. It is comprised of three parts which called base, part 1 and part 2. A complete set of Khufi should has part 1 and part 2 assembled on the base. The responsible lecturer has assigned this project to the Semester 3 student, Session July- December 2016 to produce that product for the trial phase. There were 25 units of semi-finished product (based part) produced by the students. The assembled result show the "part 1 and part 2" cannot be slotted properly to the base part. It was due to the dimension of the based, part 1 and part 2 are not accurate as the required specification.

The main objectives of this study are to investigate the opportunities for quality and process improvement based on six sigma methodology, to eliminate potential wastes and variability and the last objective is to develop a model for effective implementation of Six Sigma for CNC milling process. End of the study, the researchers could understand the fundamental elements or factors that contribute to the unfit khufi due to CNC milling manufacturing process and able to determine best practices six sigma implementation during milling process.

1.1 Literature Review
Normally the studies conducted on six sigma methodologies were particularly designed for large organizations but this research is carried out to study the application of six sigma in Computer Numerical Control (CNC) machine operation. In recent years, the interest from the academic community towards six sigma has increased dramatically. However, to date only few papers can be identified for literature review on Six Sigma focusing on the basic concept, implementation and future of Six Sigma Six Sigma has been defined as the statistical unit of measurement, a Sigma that measures the capability of the process to achieve a defect free performance. Six Sigma has the ability to produce products and services with only 3.4 defects per million, which is a world-class performance. Six Sigma has also been described as a high performance data driven approach in analyzing the root causes of business problems and solving them [2] [3].

1.2 Process Sigma
Process Sigma is defined by numeric levels that are related to a process's output of defects per million opportunities. Defects are defined as any failure to meet the customer's specifications. Process yield is used to look up the Process Sigma level from a Table. Yield is based on Defects (D), Units Processed (N), and the number of Opportunities (O) for a defect to occur. Once the yield is calculated the Process Sigma can be found in the process sigma level Table.

$$Process\ Yield = \left(1 - \frac{D}{N*O}\right)*100 \tag{1}$$

1.3 Define, Measure, Analyze, Improve and Control (DMAIC)
Six Sigma implementation uses five step DMAIC (Define, Measure, Analyze, Improve and Control) methodology, somewhat similar to Plan-Do-Check-Act problem solving methodology defined by Deming. DMADV (Define, Measure, Analyze, Design and Verify) methodology is adopted for new product developments [4].

DMAIC is a closed-loop process that eliminates unproductive steps, often focuses on new measurements, and applies technology for continuous improvement. Some papers focus on explaining the DMAIC contents, with some authors discussing each phase of DMAIC in detail [5]. For example, (Wang 2008) present self-learning training material for DMAIC, using a fictitious application. This paper helps the readers to learn how to carry out a small-scale Six Sigma project, including guidance on the application of tools. It indicates a perceived need for training material and suggests that an avenue for further research is to develop training material to cover a wider range of applications and larger scale projects [2].

Other papers concentrate on specific aspects of DMAIC, such as the project selection process in the Define phase or process control in the Control phase, explaining some key measures in Six Sigma, such as project metrics and Roll Throughput Yield (RTY). For example, (Bożek and Hamrol 2012) emphasizes the importance of the project selection process in the Define phase for the successful implementation while Mason suggests using multivariate statistical process control in the Control phase [6].

2. Methodology

This section explains the methodology adopted for this case study. Scientific investigation on innovating a system or improvement to the existing one needs to begin with good plan. The plan of investigation was conceived so as to obtain answers to research questions in the research design. This research will pursue a systematic approach for empirical study. It consists of six steps as in Figure 1. Step 1 literature review is intended to understand the implementation of Lean Six Sigma in the manufacturing process. The main focus of this reviewing process is to identify the most important lean six sigma implementing techniques. Step 2 data collection will be carried out through sampling from the student project in advance machining subject. The sampling part will be take and measure for the dimension and parameter based on specification given. For Step 3 the data gathered will be analysed using the statistical process analysis. The data analysis will go through the DMAIC phase in six sigma methodology. Then, step 4 model will be used in developing the integration model for lean six sigma implementation in CNC milling manufacturing process. Step 5 is the validation of the model developed will be carried the case study to implementing the solution. The data from the case study will be used in this validation process.

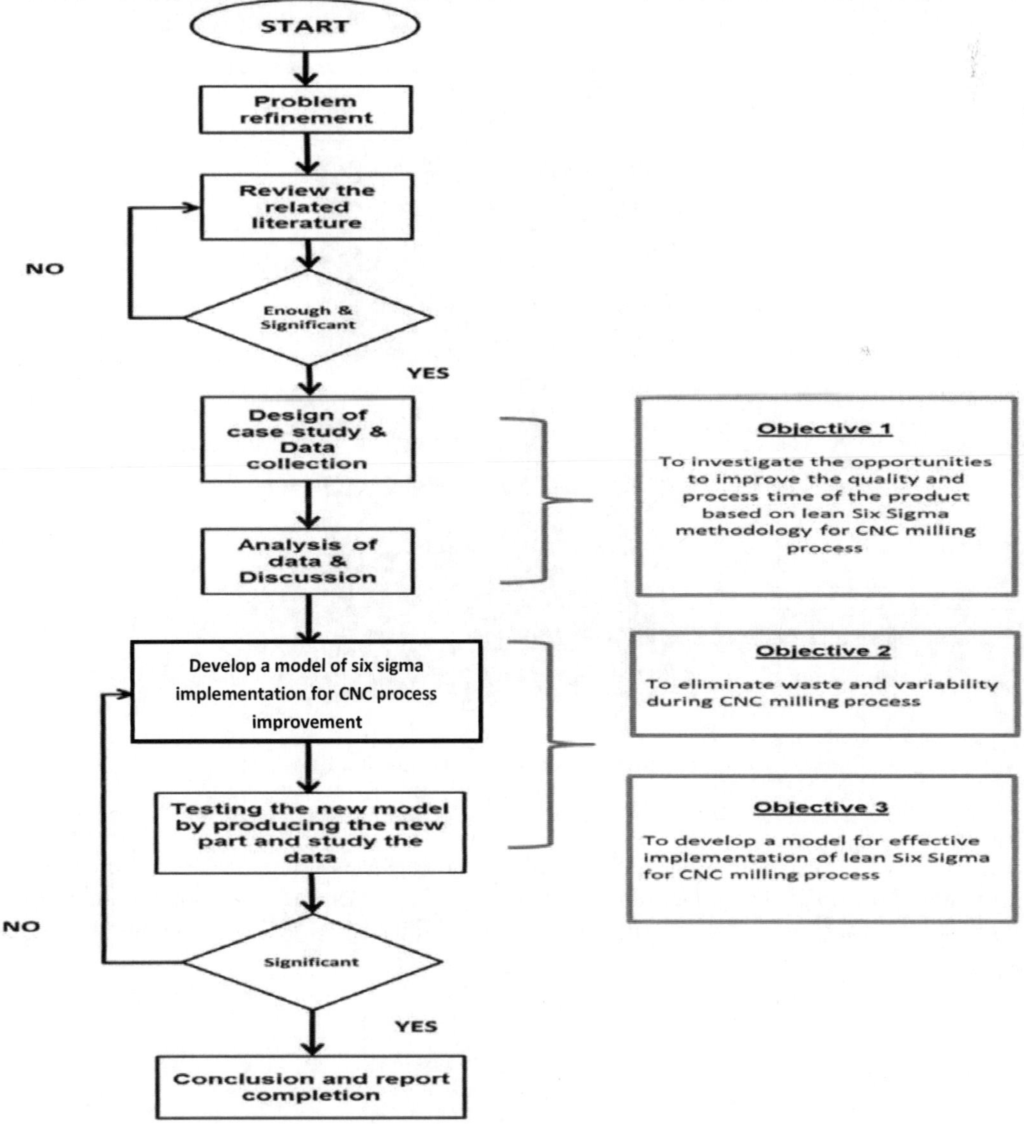

Figure 1. Process flow for the project methodology

IOP Conf. Series: Materials Science and Engineering **257** (2017) 012009 doi:10.1088/1757-899X/257/1/012009

3. Results and Discussion

This case study conducted based on problem of part 1 and 2 could not be slotted into the base part. Based on the initial design of Khufi, the base part is used as the female part which there are two slots on the top. While part 1 and 2 as a male part which its design have a rib at the bottom of the part for assembly purpose. Some samples were taken and conducted the fitting test. The results discover all parts cannot be assembled correctly because some of them are loose and too tight. Therefore, a study based on the DMAIC methodology was conducted to identify the cause of this problem.

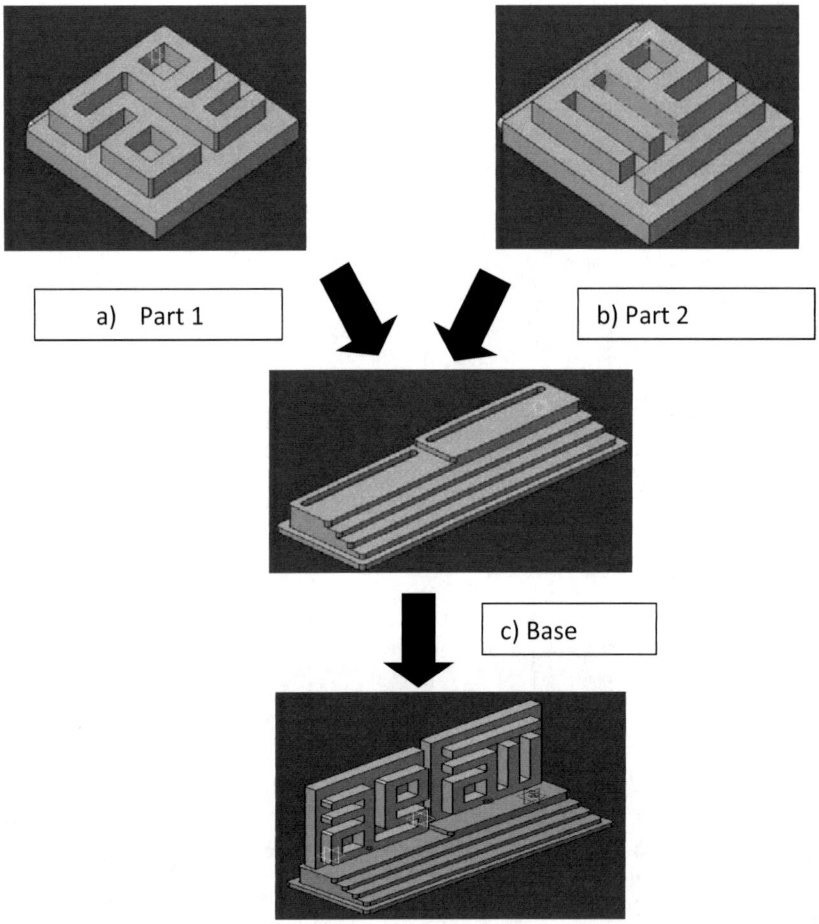

Figure 2. Assembled part of Khufi

3.1 Define

25 units of base part had been measured using the Coordinate Measuring Machine (CMM) .Before the measurements are made, two dimensions (A & B) have been identified as a critical dimension to be control during machining process. Therefore, this two dimension has been measured and verified based on the required specification.

IOP Conf. Series: Materials Science and Engineering **257** (2017) 012009 doi:10.1088/1757-899X/257/1/012009

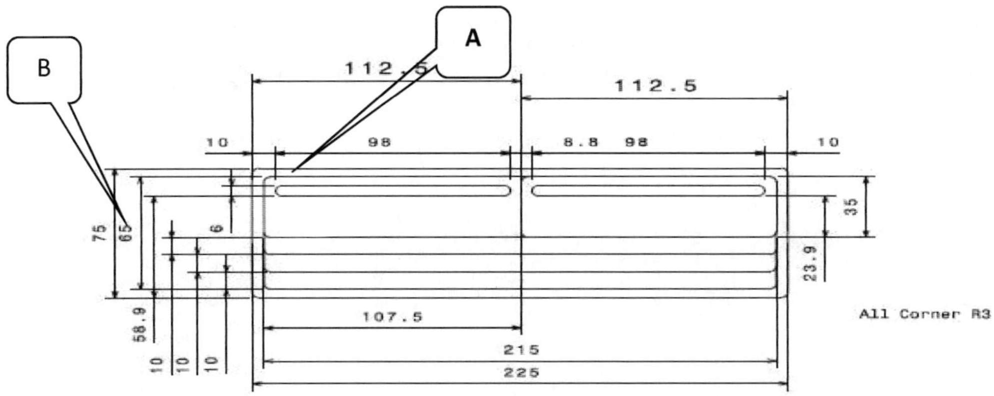

Figure 3. Technical drawing for base part

3.2 Measure

In this stage, a check sheet has been developed in order to record all the data from the measured part. The data has been separated into two categories, dimension A and dimension B. All measured data for dimension of A & B are highlighted in Table 1

Table 1 Measurement for dimension A and B

Parts	Dimension A (mm)	Dimension B (mm)
1	97.99656	6.04060
2	98.03733	6.10923
3	98.0542	6.08104
4	98.07508	5.98632
5	98.01746	6.09286
6	97.95087	6.09393
7	98.03882	6.10794
8	98.06297	6.10600
9	98.00606	5.99135
10	98.06782	6.01106
11	98.01281	5.99395
12	97.98396	6.05732
13	98.09409	6.05423
14	98.07179	6.02820
15	98.00491	6.05689
16	98.04085	6.02283
17	97.97587	6.00241
18	97.94062	6.09148
19	98.04602	6.09875
20	98.07852	6.08838
21	98.07089	6.07921
22	97.99526	6.00217
23	98.01653	6.10740
24	98.17584	6.12667
25	98.05817	6.09221

The I-MR control chart as in Figure 4 and 5 have been developed based on collected data as in Table 2.

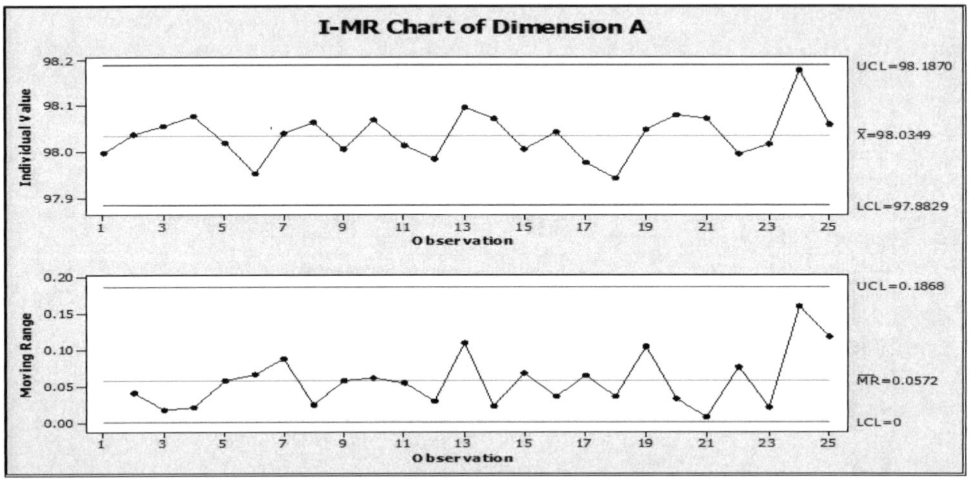

Figure 4. Control chart for dimension A

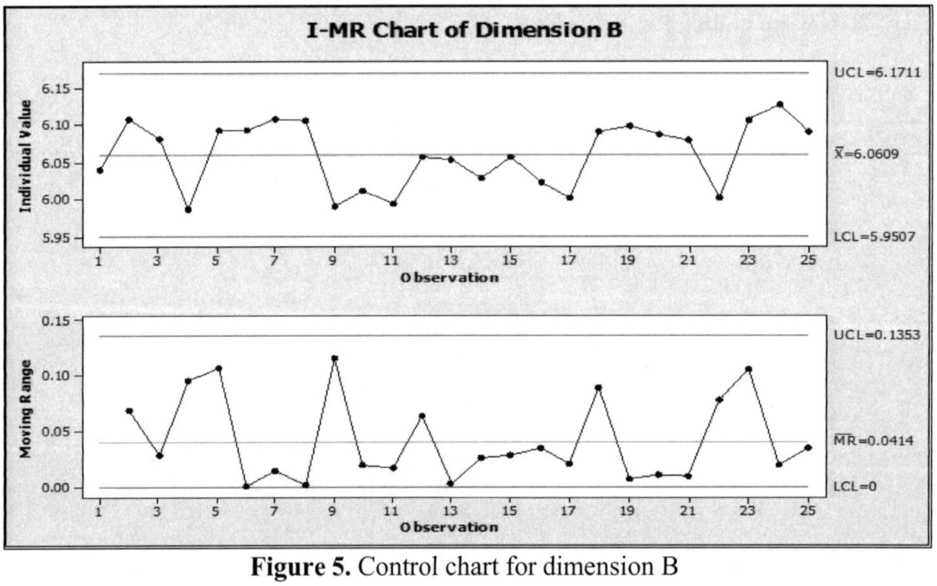

Figure 5. Control chart for dimension B

Figure 4 and 5 show the I-MR chart for the dimension A and B. All data obtain from both chart are still within the UCL and LCL value. The variation of the data from the mean value is not too high as shown in individual value chart but for the moving range is quite large compared to dimension A. The UCL and LCL value for this dimension are almost same with dimension A which the data obtained are over the specification limit. Therefore, this dimension also needs to be analysed in order to identify the cause of dimension out of specification.

Further analysis is to calculate the standard deviation and also process capability (Cp) value based on process capability chart. The Cp is a measurable property of a process to the specification, expressed as a process capability index. Two parts of process capability are measure the variability of the output of a process, and compare that variability with a proposed specification or product tolerance.

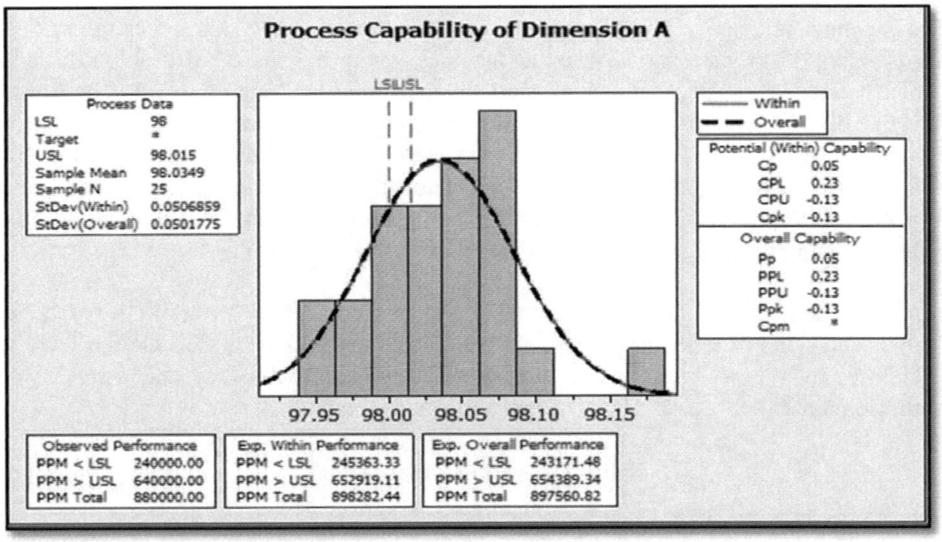

Figure 6. Process capability chart for dimension A

From the process capability chart in Figure 6, it shows almost all the dimensions from 25 unit parts are out of specification. The standard deviation for this data is not too large (0.05), meaning it does not deviate much from the average. While for the Cp we get the value is 0.05. Based on literature review, the best Cp must be above than 1.2. If the value is less than 0.5 it means that the variability of the process is more than the specification limits.

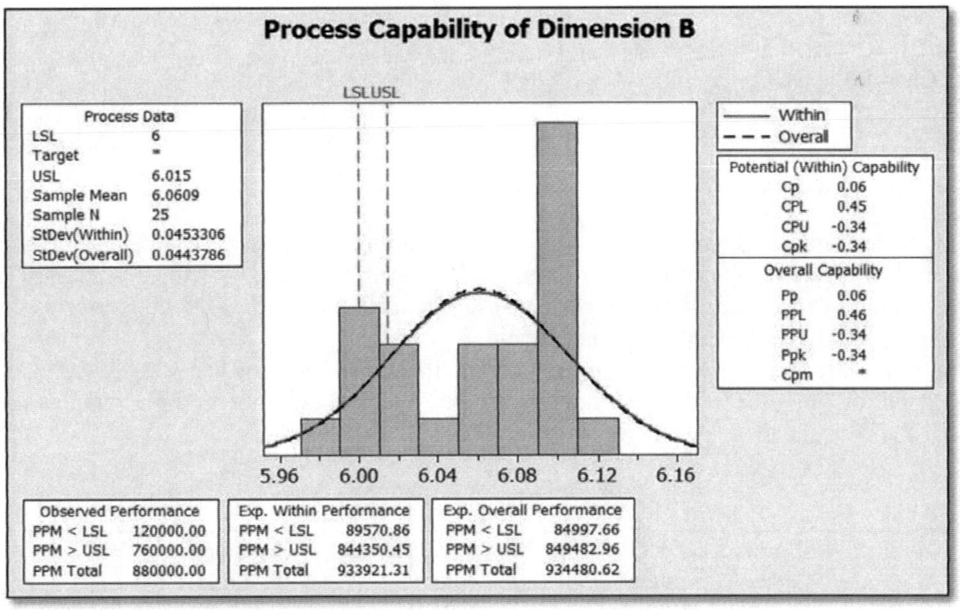

Figure 7. Process capability chart for dimension B

Base on Figure 7, it shows the dimension B is same results as dimension A. There were 22 over 25 units are not according to specifications. The standard deviation and Cp value also almost the same, 0.04 and 0.06. Conclusion for this issue shows both of dimensions have a same problem and root cause. A solution is necessary to improve this process in order to get results according to the required specifications.

3.3 Analysis

Data analysis were conducted to identify the root cause and the possible solution as to improve out of dimension problem. Fishbone diagram used as a tool to study the causes of problem based on five criteria which is man, machine, method, measurement, material and environment. The first cause is probably came from measurement factor which is because of inaccurate cutting tool size. Second cause is from material which is maybe due to cutting tool worn out. Man also one of the factor that cause to this problem which is CNC machine was conducted by low skill operator. The third cause is came from method which is inaccurate parameter setting will also contribute to this problem. All causes were verified through experiment for improve the existing process.

Experiment 1 has been conducted to study the effect of feedrate and depth of cut parameter to the product dimension. This experiment is divided into two which is first to increase the feed rate with constant depth of cut and second is increase depth of cut with constant feed rate. Table 2 shows the result obtain from the experiment.

Table 2. Data for Experiment 1

No	Comment	Spindle speed (RPM)	Feed rate (mm/m)	DOC	Measurement (6mm + 0.015)	Production time (min)	Result
1	Change the Feed rate value	4000	500	0.1	5.982	38m 49s	All data is out of spec and trend is decreased
2		4000	800	0.1	5.981	24m 19s	
3		4000	1000	0.1	5.979	19m 29s	
1	Change the DOC value	4000	1000	0.2	5.978	9m 47s	All data is out of spec and trend is decreased
2		4000	1000	0.3	5.977	6m 41s	
3		4000	1000	0.4	5.976	5m 08s	

In this experiment, 6 samples were proceed for machining process based on parameter setting defined in Table 2. The dimension of the sample has been measured and the result show all the dimension is out of specification. Based on that result, it show the trend of the dimension is decrease when value of feed rate and depth of cut is increase.

Based on finding on experiment 1, the second experiment has been conducted by reducing the feed rate and depth of cut value to the minimum parameter. For this experiment, three samples has been used for machining and the result shows in Table 3

Table 3. Data for Experiment 2

No	Comment	Spindle speed (RPM)	Feed rate (mm/m)	DOC	Measurement (6mm + 0.015)	Production time (mnt)	Result
1	Reduce to minimum value of Feed Rate & DOC	4000	300	0.1	5.984	1h 4m 37s	All data is out of spec and trend is increased
2		4000	200	0.1	5.984	1h 36m 51s	
3		4000	100	0.1	5.985	3h 13m 35s	

IOP Conf. Series: Materials Science and Engineering **257** (2017) 012009 doi:10.1088/1757-899X/257/1/012009

In this experiment, the feed rate has been set to 100 mm/m while the depth of cut set to 0.1mm. The result obtain is increase but still didn't archive to the required specification. In order to get the exactly value for this two parameter, a regression equation is an option to get the right formula to define the actual value of parameter.

3.4 Regression analysis

Linear regression attempts to model the relationship between two variables by fitting a linear equation to observed data. One variable is considered to be an explanatory variable, and the other is considered to be a dependent variable. Multiple linear regression attempts to model the relationship between two or more explanatory variables and a response variable by fitting a linear equation to observed data. Every value of the independent variable x is associated with a value of the dependent variable y. For this experiment, the dependent variable is the dimension of B while the independent variables are feed rate and DOC. I have use the formula below in order to generate the regression equation for my case study.

$$Y = a + b_1 X_1 + b_2 X_2 \tag{2}$$

Y = Dependent variable/Output (Required dimension = 6 +0.0015)
a = intercept
b = slope
X1= Independent variable 1 (Feed rate)
X2 = Independent variable 2 (Depth of Cut)

All the value has been calculated using the Microsoft Excel software based on regression formula. The result from calculation is obtain in Table 4 below

Table 4. Regression analysis

	Y	X1	X2	X1*Y	X2*Y	X1*X2	X1^2	X2^2
	5.982	500	0.1	2991	0.5982	50	250000	0.01
	5.981	800	0.1	4784.8	0.5981	80	640000	0.01
	5.979	1000	0.1	5979	0.5979	100	1000000	0.01
	5.977	1200	0.1	7172.4	0.5977	120	1440000	0.01
	5.976	1500	0.1	8964	0.5976	150	2250000	0.01
	5.978	1000	0.2	5978	1.1956	200	1000000	0.04
	5.977	1000	0.3	5977	1.7931	300	1000000	0.09
	5.976	1000	0.4	5976	2.3904	400	1000000	0.16
	5.974	1000	0.5	5974	2.987	500	1000000	0.25
	5.984	300	0.1	1795.2	0.5984	30	90000	0.01
	5.984	200	0.1	1196.8	0.5984	20	40000	0.01
	5.985	100	0.1	598.5	0.5985	10	10000	0.01
SUM	**71.8**	**9600**	**2.2**	**57386.7**	**13.2**	**1960**	**9720000**	**0.62**
AVG	**5.9794**	**800**	**0.183**	**4782.225**	**1.0959**	**163.3333**	**810000**	**0.05167**
^2		92160000	4.84	3293233337	172.95	3841600	9.44784E+13	0.3844

Table 4. Regression analysis (continued)

x1y	-15.7
x2y	-0.003816667
x1x2	200
x1^2	2040000
x2^2	0.216666667
B1	-0.00000656302
B2	-0.011557214
a	5.986785904

From the Table 4, the result shows the value of b1, b2 and a. This value will substitute into the regression equation of Y and become a complete equation.

Y= 5.99 – 0.000007 X1 – 0.0116 X2

Then, using this equation to get the X1(feed rate) and X2 (DOC) value based on required Y (dimension). For example, the required Y (dimension) is 6 then the X1 and X2 get from the calculation below:

Y= 5.99 – 0.000007 X1 – 0.0116 X2

Rearrange the equation to make X1 & X2 as subject

X1 = 6 + 0.0116X2 – 5.99/(-0.000007)-------------1

X2 = 6-5.99+0.000007X1/(-0.0116) ----------------2

Sub 2----->1

X1 = 6+0.0116[-517.24+516.38-0.0006X1]-5.99/ (-0.00007)

X1 = -42.25

Sub -42.25----->2

X2 = 6-5.99 + (0.000007 x -42.25)/-0.0116

X2 = -0.837

The results from these calculations can be concluded, in order to get the dimension Y for 6mm, the required parameter for feed rate and DOC are -42.25 and -0.837. The value of feed rate and DOC that obtain from the calculation is impossible to be implemented at the actual process because of value is too small and negative. Therefore, the parameter setting problem is not the root cause of this problem and no possible solution from this parameter and need to proceed for the next experiment.

The next experiment (experiment 3) is to change the tool radius. In the CNC programme, there is a setting namely tool wear offset for length and radius. The function of this setting is to change the cutting tool setting (length and radius) if that was worn out. For this experiment, only the radius value is changed and monitor whether affect the product dimension or not

The tool radius for the first sample is change to +0.018, while the second sample reduce to -0.018. This change value is based on the best result obtain from the experiment 2.

Table 5. Data for Experiment 3

No	Comment	Spindle speed (RPM)	Feed rate (mm/m)	DOC	Measurement (6mm + 0.015)	Production time (min)	Result
13	Change tool radius ware (+0.018)	4000	500	0.1	5.951	38m 45s	This parameter is not the original setting
14	Change tool radius ware (-0.018)	4000	500	0.1	6.081	38m 48s	

From the Table 5, the result shows the measurement for No 13 is reduce and far from the specification. While, for No 14, result shows the dimension is increase and closer to the specification. Therefore, this experiment shows the change of tool dimeter is drastically effect the dimension of product change. This situation gives a clue that, this new tool size is not exactly as specified by the manufacturer. To prove this matter, next experiment has been conducted to measure the actual size of the cutting tool.

The experiment 4 has been conducted as to verify the dimension of the cutting tool as indicated by the manufacturer. The dimension of diameter claimed by the manufacturer is 4mm as shown in Figure 8.

However, the measured diameter by vernier calliper was 3.86mm, which is different as claimed by the manufacturer. By using this new measured diameter, two samples were machined for the result verification. The result shows these samples were over than specification. This phenomenon might due to the measurement by vernier calliper is not accurate on cutting tool which has three tooth. Second measurement has been conducted by using the Coordinate Measuring Machine (CMM) to measure the cutting tool. The new measurement was 3.95mm which is a bit different compared to previous measurement. Then, the next verification on new diameter was performed on two samples. The positive results obtained when the dimension between the required specifications but at maximum area. Therefore, the next experiment has been conducted in order to get the optimum dimension for this product.

This final experiment (experiment 5) is intended to get the optimum dimension of the product. The result which obtained in the experiment 4 shows the dimension is already in specification but almost at the maximum limit. Therefore, experiment 5 will combine the theory that get from the experiment 2 which is the dimension will slightly decrease when the value of Feed rate and DOC is increase. In this experiment, only DOC is increase, while the feed rate value was remain at 1000mm/m. The final result obtained as in Table 7.

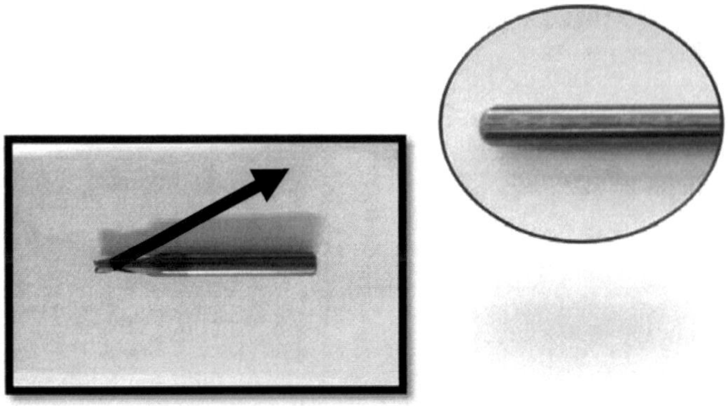

Figure 8. cutting tool diameter

Table 6. Data from experiment 4

No	Comment	Spindle speed (RPM)	Feed rate (mm/m)	DOC	Measurement (6mm + 0.015)	Production time (min)	Result
1	Measure the cutting tool size using Vernier calliper, Dimension =3.86mm	4000	1000	0.2	6.196	9m 45s	Measurement using calliper is not accurate
2		4000	1000	0.2	6.194	9m 44s	
1	Measure the cutting tool size using CMM Dimension =3.95mm	4000	1000	0.2	6.015	9m 44s	Data in spec but hit the maximum limit
2		4000	1000	0.2	6.015	9m 46s	

Table 7. Data for Experiment 5

No	Comment	Spindle speed (RPM)	Feed rate (mm/m)	DOC	Measurement (6mm + 0.015)	Production time (mnt)	Result
1	Increase the DOC value for reduce the processing time	4000	1000	0.5	6.013	3m 58s	All data is accepted with faster processing time
2		4000	1000	0.5	6.012	3m 58s	
3		4000	1000	0.5	6.013	3m 56s	

Experiment 5 was conducted to verify the new cutting tool diameter with the best parameters setting get from previous experiment whether effect the final product dimension. Based on Table 7, it show all dimensions are in the specification and accepTable. Beside that the production time also was improved a lot and become more faster than the initial process. The finding and solution from this experiment can be accepted for improving product quality on dimensions and also become more productive to manufacture in large quantity.

3.5 Improve

In the improve phase, it is about the implementation of true solution that obtained from the analysis phase. There are lot of inputs and findings were gained from the experiment conducted during the analysis phase. For this case study, all inputs were considered for developing a model for effective lean six sigma implementation for CNC milling process. This model will used as a guideline for machinist in order to get the best quality with high productivity.

In statistics, a mediation model is used to identify and explain the mechanism or process that underlies an observed relationship between an independent variable and a dependent variable via the inclusion of a third hypothetical variable, known as a mediator variable. In this project, the mediation model has been developed based on three independent variables to achieve the two dependent variables.

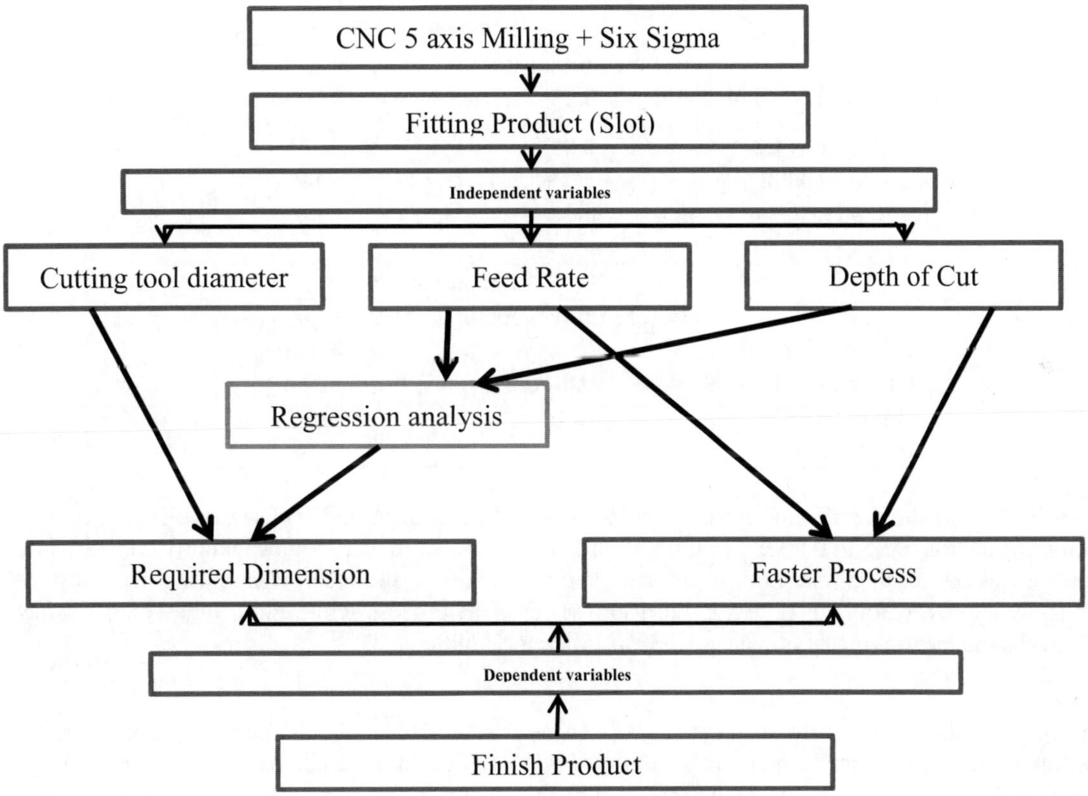

Figure 9. Mediation Model

This model was established by examining the inter relationship between cutting tool diameter, feed rate, depth of cut (DOC) and quality & productivity of the product. In this model, cutting tool, feed rate and DOC were made as a predictor variable. While the product quality & productivity (dimension and process time) as the outcome variable as illustrated in Figure 9.

From the five experiment conducted, the result shows that the feed rate and DOC have a direct positive influence on process time and the tool diameter is influence to the dimension Most interesting findings from this research is the combination of this three independent variables will produce the best product quality and productivity in term of product dimension and also the faster process time.

Therefore, by implementing this model, it will eliminate the all the wastes that contribute to the process and directly increase the production productivity.

3.6 Validate the model
The validation model was established as to ensure the final result obtained is according to the specification target. A case study has been designed and developed for producing 25 samples of machining parts based on procedure and parameters as suggested in the developed model. This case study used the new cutting tool diameter 3.97mm with the suggested parameters of feed rate 1000mm/m and depth of cut 0.5mm.

In the improvement phase, the investigation was carried out by doing the comparison between the improvements result and the initial process. Four factors have been selected as the benchmarking values to prove the improvement process is better than the initial process. These factors are standard deviation, process capability (Cp), Yield and sigma level.

Table 8. Comparison for dimension A and B

	Dimension A	
	Before improvement	**After improvement**
Standard deviation	**0.05**	**0.0008**
CP value > 1.2	**0.05**	**2.75**
Yield =100%	**12%**	**100%**
Sigma level=6σ	**0.3σ**	**6σ**

	Dimension B	
	Before improvement	**After improvement**
Standard deviation	**0.04**	**0.0008**
CP value > 1.2	**0.06**	**2.86**
Yield =100%	**12%**	**100%**
Sigma level=6	**0.3σ**	**6σ**

Table 8 shows the significant result obtained for both dimension A and B. The dimension variation has improved from 0.05 to 0.0008. In term of Cp, the process show very high capability compared to initial process, from 0.05 and 0.06 to 275 and 2.86. In addition, the result also show the process is drastically improves from the 12 % yield to 100% yields. The process achieve the highest sigma value of 6 means the measurement is the nearest of specification limits.

3.7 Control
Control is the last step of the Six Sigma five step process DMAIC. The objective of Control is to develop and implement the best controls to maintain the gains and to celebrate, share and reward the successes. In this project, the control has been set by documenting and standardizes the procedures and parameter as carried out in the analysis phase. The machinist who will manufacture khufi must understand the model and how it is work. A proper training and explanation from the expert is necessary to ensure the process is running as per plan in order to produce the good quality product.

4. Conclusions
The implementation of six sigma in the production of kuhfi has been considered successful because the process performance was improved tremendously. This research has been successfully eliminate the process variability and unfit product through identifying actual cutting tool diameter. In order to ensure the process is sustained, the improvement process was transformed into model development as per objective no 3. This model has been positively validated through four selected factors such as standard deviation, process capability, yield and level of sigma. As a conclusion, Lean Six Sigma

implementation can be helpful in eliminating the nonconforming product or improving the organization quality and cost reduction.

Acknowledgement
The authors gratefully acknowledge to Kolej Kemahiran Tinggi MARA Kuantan and all staff that involved in this project for their support.

References

[1] Ansari, a et al. 2011. "Application of Six-Sigma in Finance : A Case Study." *Journal of Case research in business and economics* 3: 1–14. http://0-search.proquest.com.wam.leeds.ac.uk/docview/902798758/fulltextPDF/1364932AF561623ECEE/6?accountid=14664.

[2] Wang, Hongbo. 2008. "A Review of Six Sigma Approach: Methodology, Implementation and Future Research." *2008 International Conference on Wireless Communications, Networking and Mobile Computing, WiCOM 2008*: 2–5.

[3] Vest, Joshua R, and Larry D Gamm. 2009. "A Critical Review of the Research Literature on Six Sigma, Lean and StuderGroup's Hardwiring Excellence in the United States: The Need to Demonstrate and Communicate the Effectiveness of Transformation Strategies in Healthcare." *Implementation Science* 4(1): 35. http://www.implementationscience.com/content/4/1/35.

[4] Raghunath, a, and R V Jayathirtha. 2012. "Lean and Six Sigma Approach for Manufacturing SMEs." : 1–10.

[5] Snee, "*Lean Six Sigma—Getting Better All the Time*," International Journal of Lean SixSigma, Vol. 1, No. 1, 2010, pp. 9-29

[6] Bożek, Mariusz, and Adam Hamrol. 2012. "Analysis of Efficiency of Lean Manufacturing and Six Sigma in a Production Enterprise." *Management and Engineeing Review* 3(4): 14–25.

IOP Conf. Series: Materials Science and Engineering **257** (2017) 012010 doi:10.1088/1757-899X/257/1/012010

Effect of sintering temperature on physical properties & hardness of CoCrMo alloys fabricated by metal injection moulding process

Ahmad Ridhwan Abdullah[1], Nur Aidah Nabihah Dandang[2], Nur Zalikha Khalil3, Wan Sharuzi Wan Harun[3]

[1]Mechanical Engineering Department, Politeknik Sultan Haji Ahmad Shah (POLISAS), Jalan Semambu, Pahang, 25350 Kuantan, Malaysia.
[2]Institute of Postgraduate Studies, Universiti Malaysia Pahang, Lebuhraya Tun Razak, 26300 Gambang, Kuantan,Pahang Malaysia.
[3]Green Research for Advanced Materials Laboratory, Human Engineering Group, Faculty of Mechanical Engineering, Universiti Malaysia Pahang, 26600 Pekan, Pahang, Malaysia.

*Corresponding author: ridhwan@polisas.edu.my

Abstract. Metal Injection Moulding (MIM) process is one of the Powder Metallurgy manufacturing techniques utilised to produce Cobalt Chromium Molybdenum (CoCrMo) compacts. The objective of this study is to determine physical properties and hardness of CoCrMo alloy compact sintered at three different sintering temperature at the similar soaking time. At the beginning, sample were fabricated by using Injection Moulding machine. Cobalt Chrome Molybdenum (CoCrMo) metal powder was selected for this study. A morphological study was conducted using optical microscope (OM) and micro-Vickers hardness testing. From the result obtained, it shows upward trend either on the hardness or physical properties of the samples. CoCrMo sintered compact become harder and volume of pores on surface become less due to the increase on sintering temperature. However, effect of increasing sintering temperature shows significant shrinkage of the sample, beginning losses in dimensional accuracy. It is discovered that a little change in sintering temperature gives significant impact on the microstructure, physical, mechanical of the alloy.

1. Introduction

The CoCrMo alloys that containing 28% weight of Chromium, 6% weight of Molybdenum (balance Co) [1] is one of the important materials in the biomedical field, particularly in the manufacture of body implant and dental surgery. The advantages of this material are durable, high mechanical strength and resistant to rust. This alloy has a resistance to corrosion due to the presence of chromium in the material. Molybdenum also serves to increase the strength of the alloy [2-4]. But, in any cases, unfavourable side effect like allergic cause by phenomenon called 'metallosis' commonly happen in biomedical field[5]. This situation makes the material less used in industry. Therefore, there is opportunities for researchers to study this material further in order to improve their mechanical properties.

Compacts from CoCrMo could be processed by using at least three techniques; powder metallurgy, casting, and hot forging. Powder metallurgy provide attractive resolution in order to explore and study new material [6]. Well organized open cavity or porosity and probability of fabricating any materials are the advantages of Powder Metallurgy methods. Some examples of techniques available in Powder Metallurgy are Metal Injection Moulding (MIM) and Selective Laser Melting (SLM).

Metal injection moulding (MIM) is an innovation technique, savvy financial process for large scale manufacturing for produce complex, accurate, net shape parts or component by using either metal powder or clay powder. Capability of MIM depends on capacity to adapt complexity of design of plastic injection moulding and infinite selection of material offered, make it potential to merge with several parts to become solitary significant component [7, 8]. Internal crack in casting, defect cause by tolerance error, capital gain in casting manufacturing, and shape limitation are common problems that been faced by industries that easily can be solved by implementing MIM [9, 10]. MIM production is conquered by automotive sectors, particularly in Europe. Meanwhile, in the USA, focus for MIM is in biomedical/ healthcare. Instead, in Asia region, application of MIM is more to electronics and information technology field. Also, another growing factor is the development in the manufacturing of medical tools in Asia region, known as a greater population gains access to enhanced health care [11].

Basically, to improve mechanical properties, there are several methods that been used by researchers such as Severe Plastic Deformation (SPD). This method had been used to induce high strain into metallic materials. CoCrMo that been induced with high strain can cause the development of grain refinement. It is conceivable to refine grain size of cobalt in range 10-20 nm, heading to sharply increment up to 5 times in hardness and tensile strength [12]. Pure Cobalt (Co) recorded ultimate tensile strength and hardness normally around 210 MPa and 1700 MPa, and can be reached up to 750 MPa and 3530 MPa after grain refinement achieved [13]. In the other words, because of grain refinement, it improves mechanical properties of the material as well. Bolton and Becker [14] in their research have studied the effects sintering atmosphere on physical properties of CoCrMo alloys for surgical implants. It shown that by changing or manipulated certain parameters, level of porosity on materials could be changed either high or low. Compared sintering of CoCrMo alloy in argon with vacuum atmosphere shows that it produces high corrosion resistance. Densification and microstructure growth of CoCrMo during sintering were studied by many researchers to evaluate the compatibility of dissimilar grain sizes for co-sintering. Two components must have the same sintering behaviour to decrease gap stress during heating and cooling cycles. A study had been conducted by Heaney et al [15] to determine effect of sintering condition to final product of CoCrMo by using MIM. They stated that in the case of liquid phase sintering, the materials must have approximately 5% or less difference in shrinkage at or near the prescribed sintering temperature.

The objective of this present work is to study the effect of different sintering temperature and constant soaking time on the microstructure, physical properties, and hardness of CoCrMo alloy compact fabricated by metal injection moulding process.

2. Experimental Procedure

Table 1 representing the composition of raw material of CoCrMo. The CoCrMo alloy powder supplied from industry was fabricated through water atomization. Figure 2 shows Particle's profile is in spherical shape. Table 2 shows sample conditions involved in the study. Metal powder were mixed with polymer binder that consisted of Paraffin Wax, Polypropylene, and Stearic Acid to produce feedstock. Mixing ratio selected for feedstock is 65 volume % metal powder and 35 volume % binder system.

Table 1. CoCrMo alloy powder composition

Composition	Co	C	Cr	Mo	Ni	Mn	Si	Fe
Volume (%)	Balance	0.4	28.5	6.0	1.0	1.0	1.0	0.8

Table 2. The sample conditions

Sintering			Type of Metal Powder	Type of Binder	Mixing Ratio
Temperature	Time	Atmosphere			
1250 °C	2 hours	High purity Argon gas flow	Cobalt Chrome Molybdenum (CoCrMo)	Paraffin Wax, Polypropylene, and Stearic Acid	65 % metal powder + 35 % binder
1300 °C					
1350 °C					

Cylindrical samples with 20 mm diameter and 10 mm thickness were produced by using Metal Injection Moulding (MIM), equipped by standard moulding equipment under high pressure unit. Moulded part is referred to as a 'green compact'. Meanwhile, solvent de-binding is done by submerging green part in gaseous ethanol. This process had been done in closed container. Heat supplied by heater and temperature of the container was set at 60 °C. Next, compacts were put in a tube furnace in atmosphere of argon. Thermal de-binding and sintering was implemented at temperature of 1250, 1300 and 1300 °C separately. Figure 1 shows the temperature pattern for de-binding and sintering process. De-binding process is initiated by heating green compacts from room temperature to 500 °C with heating rate at 5 °C/min, takes at least one and a half hours. Then, compacts left exposed at temperature 500 °C for another one hour. This process is called 'Soaking'. Soaking process is to remove residual binder remaining in the compact. Then, furnace temperature was raised at rate 10 °C/min until it reaches the desired sintering temperature (1250, 1300, and 1350°C) respectively. Sintering process is carried out at constant temperature for one hour. Finally, compact is allowed to cool slowly in over six hours in furnace after shutdown. Sample that been burned in furnace at sintering temperature called sintered compacts.

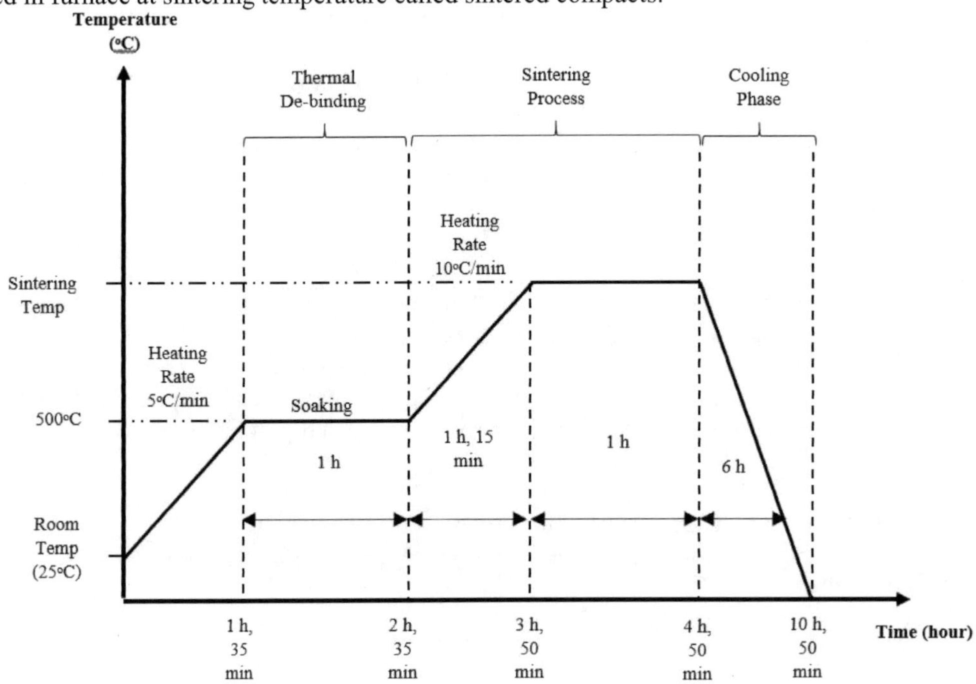

Figure 1. Temperature pattern for de-binding and sintering

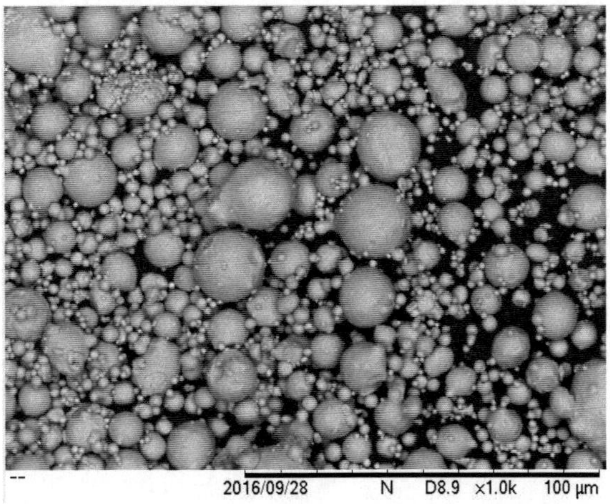

Figure 2. Spherical shape of CoCrMo particles

Sintered compacts were analysed by using optical microscope (OM) to determine porosity level via microstructure image. Correlation between microstructure result and mechanical properties of the sample were investigated via hardness test by using Vickers Micro Harness Machine.

3. Results and discussion

3.1 Microstructure

Figure 3 describe the sintered and polished compacts without etching with chemical reagent to expose the level of porosity. Analysis of the porosity level reveal that CoCrMo sintered compact which burned in 1350°C showed the lowest amount of pore. Meanwhile, sintered compact at sintering temperature 1250°C showed huge volume of porosity or open cavity as well. Forming of porosity on compact began in solvent de-binding process. Pores on the compacts surface actually becomes passage way for remaining binder left out from compact during thermal de-binding. During sintering process, exposure to high temperature causing amount of pores shrink. High temperature when performing the sintering procedure diminished pores on the surface of the parts. It obviously as a result of the recrystallization procedure of the metal particles, where the molecule of metal bonded each other together as inter-particle necks grow during sintering and porosity level is reduce[4, 16, 17]. When sintering process remains at higher temperature, the residual porosity in the compact parts reduces because pores are eliminated by bulk diffusion to grain boundaries, which contributes to an increase in relative density of the powder compact. Besides, high melting point temperature of CoCrMo powder lead to a diffusion, which porosity removed by increasing the sintering temperature [2, 3, 10].

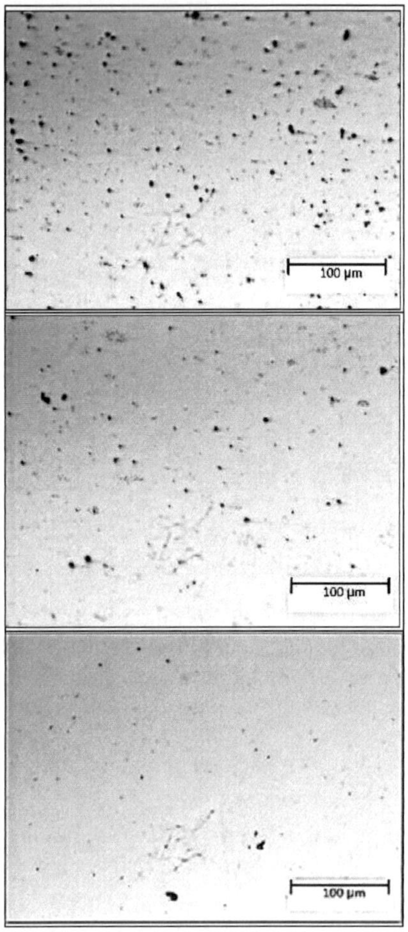

b.

c.

Figure 3. Amount of porosity for sintered compacts; sintering temperature; a) 1250 °C; b) 1300 °C; and c) 1350 °C

Shrinkage during solidification is the amount transformation of iron or metal in transitory from the melted state to the hard state at the freezing temperature, or the transition in volume of an alloy in passing from the liquid state at the beginning of the freezing range to the solid state at the end of the freezing range. The shrinkage undergone during solidification of an inter-metallic compound, freezing at a constant temperature, is similar to that of a pure metal. In figure 4, sintering temperature 1250 °C recorded the lowest shrinkage rate. With the increasing of sintering temperature, shrinkage rate also increased. So, at the highest temperature 1350°C, sintered compacts recorded the highest shrinkage rate. Figure 5 shown the decreasing or beginning loss in dimensional accuracy. At high temperature 1350°C, the sintered part showed lower than 73% similarity with eventual desired product specification. Component become increasingly smaller and relatively far away from net shape.

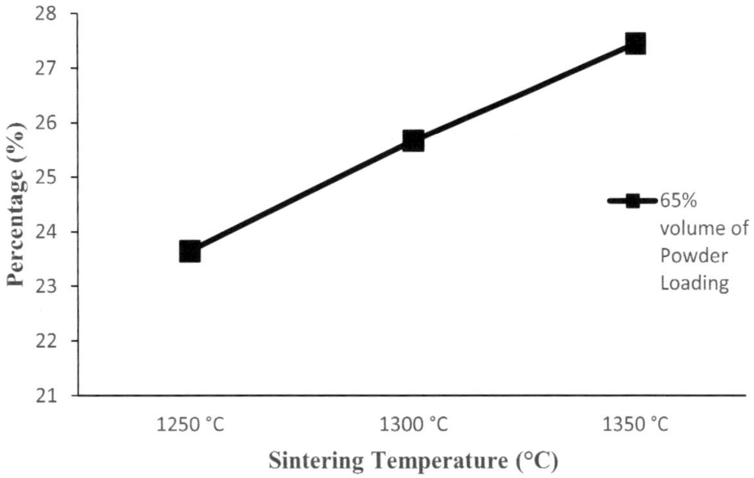

Figure 4. Shrinkage amount at different sintering temperature

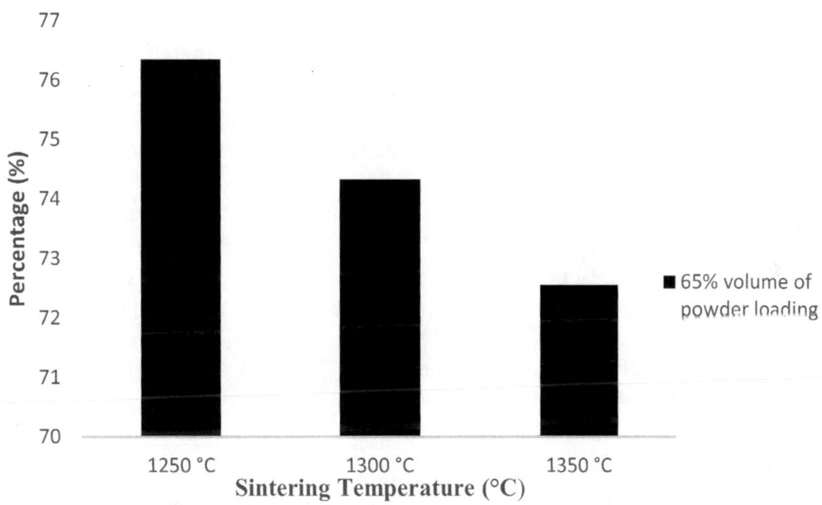

Figure 5: Percentage of dimensional accuracy for sintered compact at different sintering temperature

Figure 6 shows the hardness value of the sintered samples, showing sharp increase trend in hardness value from sintered temperature 1250 to 1350 °C. Sintered compact at sintering temperature 1250 °C recorded the lowest value of hardness, meanwhile compact at 1350 °C is the hardest one. By increasing the temperature, metal powder tends to melt and particle bonded together, after that penetrate porosity and fill large portion on it, helps to increase hardness [18]. Low level of porosity of CoCrMo (1350°C) sample can be assumed as the main factor of the high hardness value for that sample [19, 20].

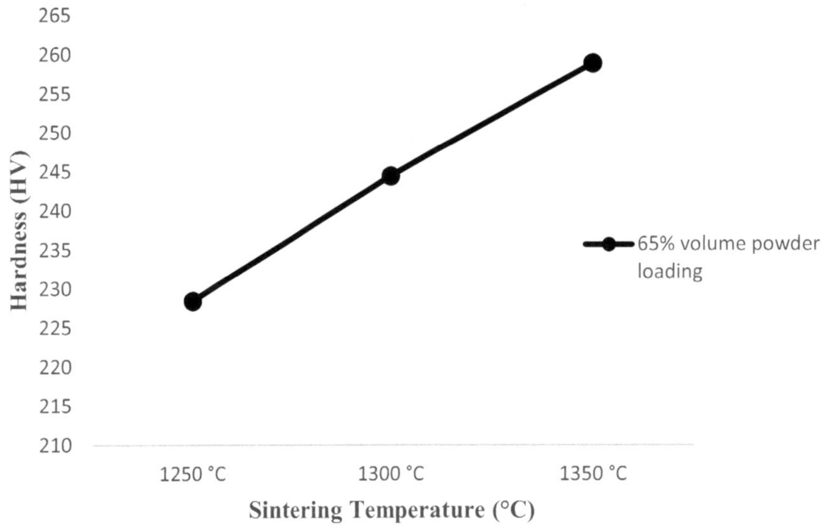

Figure 6. Hardness Value at different sintering temperature

4. Conclusions

From these three sintering temperatures tested, 1250 °C provided poor result in terms of microstructure of the sintered sample, leading to poor quality of mechanical properties as well, and it recorded lowest amount shrinkage. Sintering at 1350 °C recorded the highest hardness result showed that mechanical properties improved by the increasing the sintering temperature. Less amount of pores were detected on the sample sintered at 1350 °C, leading to good result for microstructure. But, it also led to the beginning loss of dimensional accuracy, where it recorded the highest amount of shrinkage. All sintering temperatures recorded below 80 % of dimensional accuracy where it does not meet the objectives of Powder Metallurgy; to produce near net shape components and below 5 % shrinkage that suggested by Heaney et al[15]. For next study, result will be improved by reduce sintering temperature lower than 1200 °C in order to achieve main goal in powder metallurgy method; to produce near net shape product. Sintering temperature was the parameter that strongly give impact on microstructure, physical and mechanical properties.

Acknowledgement

It is a great pleasure to acknowledge my deepest thanks and gratitude to my Superviser and Post-graduate students at *Green Research for Advanced Materials (GRAMs) Laboratory*, *Human Engineering Focus Group, Faculty of Mechanical Engineering, Universiti Malaysia Pahang*, for their kind endless help and support in completing the project.

References

[1] Y. Chen Y L, S. Kurosu, K. Yamanaka, N. Tang, A. Chiba 2014 Effect of Microstructures on the Sliding Behaviour of Hot-Pressed CoCrMo alloys *Wear* **319** 200-10

[2] Z. Doni A C A, F. Toptan, F.R. Gomes, A. Ramalho, M. Buciumeanu, L. Palaghian, F.S. Silva, 2014 Dry sliding and tribocorrosion behaviour of hot pressed CoCrMo biomedical alloy as compared with the cast CoCrMo and Ti6Al4V alloys, *Mater. Des* **52** 47-57

[3] Z. Doni A C A, F. Toptan, L.A. Rocha, M. Buciumeanu, L. Palaghian, F.S. Silva 2015 Tribocorrosion behaviour of hot pressed CoCrMo-HAP biocomposites *Tribology Int*

[4] K. Yamanaka M M, A. Chiba 2014 Effects of nitrogen on microstructural evolution of biomedical Co-Cr-W alloys during hot deformation and subsequent cooling *Mater. Des* **57** 421-5

IOP Conf. Series: Materials Science and Engineering **257** (2017) 012010 doi:10.1088/1757-899X/257/1/012010

[5] Koizumi Y, Suzuki,S.,Yamanaka,K.,Lee,B.S.,Sato,K.,Li,Y.,Kurosu,S.,Matsumoto,H.,Chiba,A., 2013 Strain Induced Martensitic Transformation Near Twin Boundaries In A Biomedical Co-Cr-Mo Alloy with Negative Stacking Fault Energy *Acta Mater* **61** 1648-61

[6] Muhammad Aslam F A, Puteri Sri Melor Binti Megat Yusoff, Khurram Altaf, Mohd Afian Omar, Randall M. German 2016 Powder Injection Moulding of Biocompatible Stainless Steel Biodevices *Powder Technology*

[7] M. Javanbakht E S, M.J. Hadianfard 2016 The effect of sintering temperature on the structure and mechanical properties of medical-grade powder metallurgy stainless steels *Powder Technology* **Volume 289**

[8] Hausnerová B 2011 Powder injection moulding- An alternative processing method for automotive items. *New Trends and Developments in Automotive System Engineering, Chiaberge M. (Ed.), InTech* pp.129-45

[9] A. Arifin e a 2014 Material processing of hydroxyapatite and titanium alloy (HA/Ti) composite as implant materials using powder metallurgy: a review *Mater. Des* **55** 165-75

[10] Jue Liu L C, Hairong Liu, Yongsheng Li , Hailin Yang, Jianming Ruan 2016 Microstructure, mechanical behavior and biocompatibility of powder metallurgy Nb-Ti-Ta alloys as biomedical material *Materials Science and Engineering C* **C17** 512-9

[11] German R M 2008 PIM breaks the $1 bn barrier *Metal Powder Report* **Vol.63** pp.8-10

[12] Karimpoor A A, Erb,U.,Aust,K.T.,Palumbo,G. 2003 High Strength Nanocrystalline Cobalt with High Tensile Ductility *Scr. Mater* **49** 651-6

[13] Edalati K, Toh,S.,Arita,M.,Watanabe,M.,Horita,Z. 2013 High Pressure Torsion of Pure Cobalt: HCP-FCC phase Transformations and Twinning during Severe Plastic Deformation *Appl. Phys. Lett.* **102** 1-5

[14] B.S. Becker J D B 1999 Cobalt-Base Alloys for Biomedical Applications *ASTM STP 1365* 47–61

[15] Heany. D 2012 Handbook of Metal Injection Moulding

[16] Boljanovic V 2010 In Metal Shaping Processes: Casting and Molding, Particulate Processing, Deformation Processes, Metal Removal *Powder metallurgy* pp. 75-106

[17] Koizumi Y, Suzuki. S, Yamanaka K, Lee. B.S. et al 2013 Strain Induced Martensitic Transformation Near Twin Boundaries in A Biomedical Co-Cr-Mo Alloy with Negative Stacking Fault Energy, *Acta Mater* **61** 1648-61

[18] Wilson Correa Rodriguez L R B, Lirio Schaeffer, Gerhard Knorschild, Fidel Romel Mallqui Espinoza 2010 Powder Metallurgical Processing of Co-28% Cr-6% Mo for dental implants: Physical, Mechanical and Electrochemical Properties *Powder Technology* **Vol. 206** 233-8

[19] Sh. Zangeneh H R L, A. Roshani 2012 Microstructure and Tribological Characteristics of Aged Co-28Cr-5Mo-0.3C alloy *Mater. Des.* **37** 292-303

[20] M. Dourandish D G, A. Simchi, V. Firouzdor 2008 Sintering of biocompatible P/M Co–Cr–Mo alloy (F-75) for fabrication of porosity-graded composite structures *Materials Science and Engineering: A* **Volume 472** Pages 338-46

IOP Conf. Series: Materials Science and Engineering **257** (2017) 012011 doi:10.1088/1757-899X/257/1/012011

Improvement of mechanical strength of sintered Mo alloyed steel by optimization of sintering and cold-forging processes with densification

Y Kamakoshi[1,2*], I Shohji[2], Y Inoue[3] and S Fukuda[4]

[1] Dept. Applicational-Mechanical Engineering, Gunma Industrial Technology Center, Maebashi, Gunma 379-2147, Japan
[2] Division of Mechanical Science and Technology, Graduate School of Gunma University, m
Kiryu, Gunma 376-8515, Japan
[3] Inoue Heat-treatment Industrial Co.,Ltd, Maebashi, Gunma 379-2101, Japan
[4] Gunma Industrial Technology Center, Maebashi, Gunma 379-2147, Japan

*Corresponding author: kamakoshi-yu@pref.gunma.lg.jp

Abstract. Powder metallurgy (P/M) materials have been expected to be spread in automotive industry. Generally, since sintered materials using P/M ones contain many pores and voids, mechanical properties of them are inferior to those of conventional wrought materials. To improve mechanical properties of the sintered materials, densification is effective. The aim of this study is to improve mechanical strength of sintered Mo-alloyed steel by optimizing conditions in sintering and cold-forging processes. Mo-alloyed steel powder was compacted. Then, pre-sintering (PS) using a vacuum sintering furnace was conducted. Subsequently, cold-forging (CF) by a backward extrusion method was conducted to the pre-sintered specimen. Moreover, the cold-forged specimen was heat treated by carburizing, tempering and quenching (CQT). Afterwards, mechanical properties were investigated. As a result, it was found that the density of the PS specimen is required to be more than 7.4 Mg/m^3 to strengthen the specimen by heat treatment after CF. Furthermore, density and the microstructure of the PS specimen are most important factors to make the high density and strength material by CF. At the CF load of 1200 kN, the maximum density ratio reached approximately 99% by the use of the PS specimen with proper density and microstructure. At the CF load of 900 kN, although density ratio was high like more than 97.8%, transverse rupture strength decreased sharply. Since densification caused high shear stress and stress concentration in the surface layer, micro-cracks occurred by the damages of inter-particle sintered connection of the surface layer. On the contrary, in case of the CF load of 1200 kN, ultra-densification of the surface layer occurred by a sufficient plastic flow. Such sufficient compressed specimens regenerated the sintered connections by high temperature heat treatment and thus the high strength densified material was obtained. These processes can be applicable to near net shape manufacturing without surface machining.

1. Introduction

In recent years, automotive industry has promoted not only cost reduction but also solution of global environment problems. Those are enabled by saving resources and energy. To save them, strengthening materials and reducing used materials are effective. Sintered materials of iron series are able to be obtained with comparative ease. They can perform mass-production of small complicated shape parts, save resources and do not have to throw away garbage parts nearly. Furthermore, they enable the abbreviation of secondary machining in near net shape manufacturing and cost saving. For many years, sintered materials have not been used much in Japan as compared with Europe and USA. However, the use of sintered materials has increased year by year in Japan. Low-alloy steel powder including Cr, Ni, Mo, Cu and so on have been developed for high-strengthening of sintered materials of iron series [1-2]. Lowering mass of such added elements is expected for low costs. Since sintered materials contain many pores and voids within itself, they have a limit of strengthening. Especially, fatigue strength and wear resistance are decreased by stress concentration in pores of the surface layer. As has often been pointed out, one of effective strengthening processes is reduction of pores i.e. high densification. For this reason, sintering industry has used high temperature sintering and compression molding by high pressure. However, the general processes such as forging, sizing press and coining press cause small dimensional changes and can not strengthen the materials. As the post-compacting processes for densification, warm forming with a warming die and powder materials [3] and sinter-hot forging at high speed and high temperature have been also used [4-5]. These processes need a large scale facilities and high costs. Thus they cause the problems of lubricants shortage, die scoring and adhering oxidized scale. These problems spoil above advantages of sintering and its high cost performance. Conventionally, although wrought steel by machining and forging has been used for high strength automotive parts, such materials need high costs. Sintering and cold-forging (CF) have been used to manufacture densified powder metallurgy materials with near net shape [6]. These processes can produce the good exterior products and deform the materials at a low load to produce near net shape parts. Also, compared to other steel powder with many expensive elements, low cost iron powder is useful. Densification of those improves the mechanical properties (static strength, fracture toughness, fatigue strength, wear resistance and so on) of the powder material. However, there are few reports about sintering and cold-forging process for densified sintered materials [6]. Usually, the mechanical properties and macro deformation behaviours of the material have been investigated using the specimen that is cut out from the core area of it. The research on the surface layer of the near net shape sintered material with high density is little. In this study, the densified sintered material was produced with Mo diffusion alloyed atomized iron by pre-sintering and cold-forging. In particular, the optimum conditions of sintering and cold-forging were investigated to obtain superior mechanical properties by the use of the near net shape specimen with the natural forged surface.

2. Experimental procedure

2.1. Specimens preparation

In this study, Mo-alloyed steel powder (JFE steel, JIP SGM10MO-CMX) for cold-forging was prepared. This is atomized iron powder with diffusionally adhered 1 mass% Mo and additional 0.35 mass% C (graphite) to the surfaces of powder particles. The chemical composition of this iron powder and the particle distribution investigated by the maker are shown in table 1 and table 2, respectively. Figure 1 shows schematic of the fabrication process of the specimen. In this study, green compacting, pre-sintering (PS), cold-forging (CF) and heat treatment (CQT: gas carburizing, quenching and tempering) were conducted sequentially. Firstly, powder was compacted under pressure of 380-980 MPa using a floating die method. Density (ρ_c) of green compacts were approximately 6.8 Mg/m^3, 7.2 Mg/m^3 and 7.4 Mg/m^3 at pressure of 380 MPa, 620 MPa and 980 MPa, respectively. The width and the length of the

Table 1. Chemical composition of iron powder [mass%] (Lubricant: 0.5 mass%)

Element	C	Si	Mn	P	S	Mo	O	Additional graphite
	0.02	0.01	0.15	0.09	0.006	0.99	0.115	0.35

IOP Conf. Series: Materials Science and Engineering **257** (2017) 012011 doi:10.1088/1757-899X/257/1/012011

Table 2. Particle size distribution of iron powder

Particle size [μm]	over 180	150≦ <180	106≦ <150	75≦ <106	63≦ <75	45≦ <63	below 45	Total
Ratio [%]	0.1	7.4	18.0	24.7	7.3	20.2	22.3	100

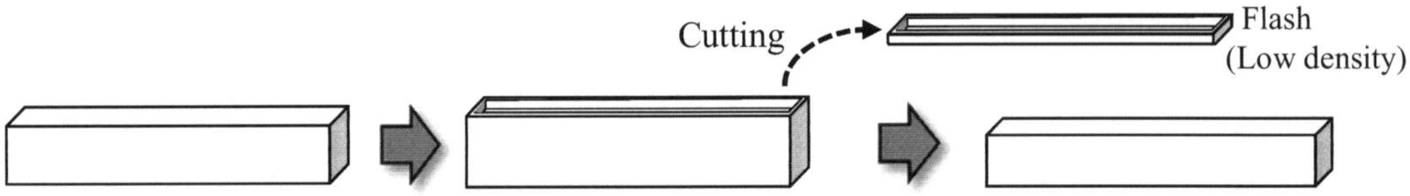

Figure 1. Schematic of fabrication process of specimen.

compacted specimen were 10 mm and 54 mm, respectively. The height of them were 13, 12.5 and 12 mm for ρ_c of 6.8, 7.2 and 7.4 Mg/m^3, respectively. PS was conducted by the use of a vacuum furnace at sintering temperature from 775°C to 1125°C and sintering time of 20, 50 and 80 min. The degree of vacuum was approximately 30 Pa. Then PS density (ρ_{PS}) was almost equal to ρ_c. Generally, decarburization in steel powder occurs and it causes to decrease the strength of the material after pre-sintering. Moreover, in case of the N_2-H_2 conveyer furnace for mass-production, surface decarburization occurs by discharging CO_2 derived from H_2O in the furnace. When the graphite on the surface of the particle is exposed to high temperature, it easily diffuses into the material and reacts with dissolved oxygen in the material. Thus, decarburization occurs and the strength of the pre-sintered material decreases. Usually, the green products are sunk in a tray filled with carbon powder and are sintered so that those are carburized little. However, it is not proper for sinter-forging because hardening and reducing of the deformability occur at PS. Therefore, the vacuum furnace should be used to avoid the above-mentioned problems. After PS, CF was conducted using backward extrusion with an oil hydraulic press machine (Tanakakame, TMS-200X100). The setting load range was from 400 kN to 1300 kN. Pressing time was 1 sec. After CF, heat treatment was conducted by the use of a vacuum carburizing furnace (Nakanihonrokogyo, NVF-15). Gas carburizing was conducted at hold temperature of 900°C for 180 min. Quenching was conducted by oil quenching (quenching conditions: 840°C, 60 min). Tempering was conducted at 180°C for 90 min. Before heat treatment, the flash formed by cold-forging was cut as shown in figure 1.

2.2. Experiment and analysis

Dimensions, density ratios and microstructures of specimens were investigated after each process. Density was measured as apparent density by the Archimedes method. The Charpy impact test was conducted using non-notched specimens with a 500 kJ Charpy tester and an instrumented 300 kJ Charpy tester (Tokyo testing machine, IC-50 and CIEM-300D). Transverse rupture strength was measured using a fatigue testing machine (Shimadzu, EHF-EM100kN/1kNm). Fracture surfaces were observed using a field-emission scanning electron microscope (FE-SEM) after tests. Hardness measurement was conducted with a rockwell hardness tester and a micro vickers hardness tester. Microstructural observation was conducted with an optical microscope and a SEM. Mapping analysis and quantitative analysis were conducted using an electron probe X-ray microanalyzer (EPMA). Quantitative analysis of carbon in the pre-sintered specimen was conducted by combustion and infrared absorption spectroscopy. For analysis, test pieces of 0.5 g sliced from the core of the pre-sintered specimen was prepared. The cross section polisher by Ar-ion beam was used when it is difficult to observe a slight change of microstructures by conventional mechanical polishing.

3. Results and Discussion

3.1. Properties of pre-sintered specimens

3.1.1 Impact properties and fracture surface. Figure 2 shows results of the Charpy impact test. In the figure, average Charpy impact energies of three times measurement are plotted. As has often been pointed out, the Charpy impact value increases exponentially with densification [7]. The similar trend was observed in this study. With higher density of the green compact and higher PS temperature, Charpy impact energy is higher too. In the upper optical images in the figure, two typical fracture surfaces are shown. Right side and left side images show the fracture surfaces of specimens with high and low impact energy, respectively. In case of high impact energy, light silver grains were observed in the fracture surface. It corresponds to cleavage fracture confirmed by SEM. Exactly, cleavage fracture area is caused by hardening with advanced sintering and transgranular fracture. On the other hand, grain boundary fracture occurs due to insufficient sintering in the specimen with low impact energy and the frosted gray area was observed as an uneven fracture area with many pores.

3.1.2 Morphology of side surfaces of specimens. Figure 3 shows back-scattered electron images of the side surfaces of pre-sintered specimens at various PS temperatures. Since the side surface of the specimen did not contact with the upper punch directly, superficial depression of boundaries occurred by growth of sintering connection on the side surface. To increase toughness and density of the specimen by CF, proper morphology of the side surface of the specimen after PS is required. From figure 3, the sufficient sintering surface was confirmed in the image at PS temperature above 975 °C because PS temperature is approximately over A_3 transformation point of steel used. On the basis of these results shown in figure 2 and 3, we selected 7.4 Mg/m^3 as proper density of green compact and 1000 - 1100°C as proper PS temperature, respectively.

3.1.3 Hardness and microstructure. Table 3 shows contents of carbon and hardness of core areas of pre-sintered specimens. When PS temperature increased, although the carbon content in the core area of the specimen decreased, the Rockwell hardness (H_{RA}) increased. H_{RA} saturated at PS temperature above 1050°C. This means that

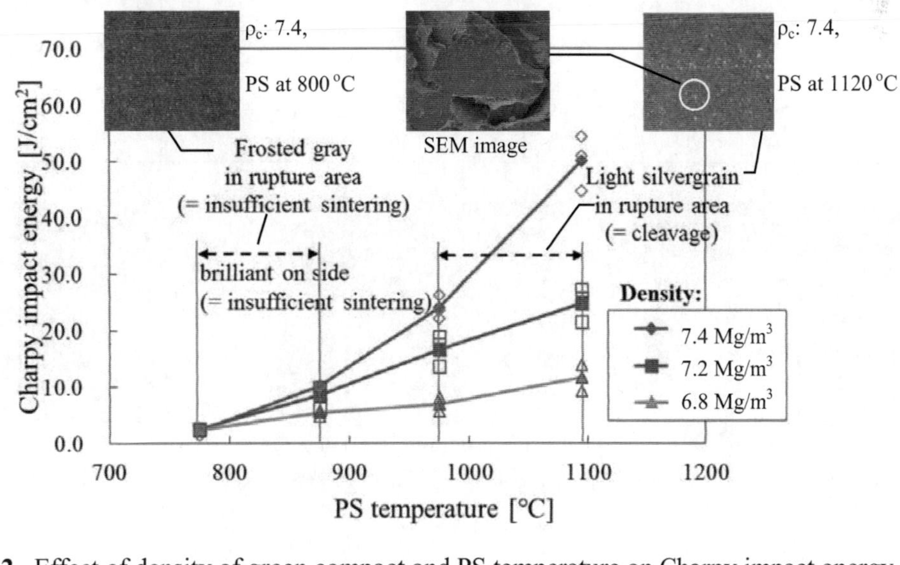

Figure 2. Effect of density of green compact and PS temperature on Charpy impact energy (PS time: 20 min).

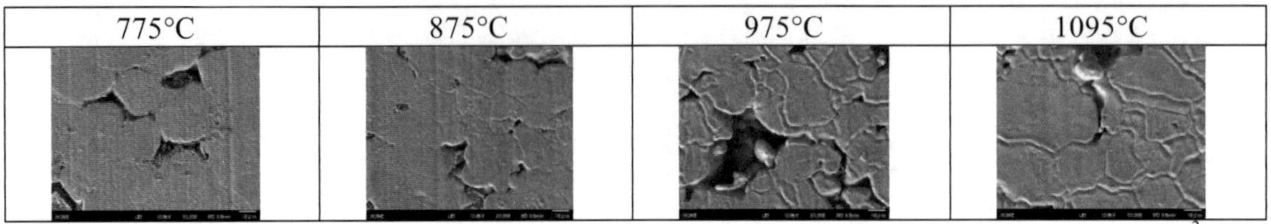

Figure 3. Back-scattered electron images of side surfaces of specimens after PS (ρ_c: 7.4 Mg/m^3, PS time: 20 min).

sufficient sintering was conducted at those temperatures and that hardening by sintering exceeded the weakening by decarburization. Figure 4 shows microstructures of core sections of pre-sintered specimens. White phases and gray phases in the figures are ferrite and pearlite, respectively. It seems that the pearlite area increases with increasing PS temperature. Also, as shown in table 1, the powder material contains 0.115 mass% O and a little lubricant. As shown in tables 1 and 3, in case of vacuum sintering, it was found that 0.04-0.07 mass% graphite is decarburized by oxidizing of graphite contained in the material, instead of oxidizing carbon of lubricant. Therefore, it was clarified that the weight of added carbon is desirable to be a half of the content of dissolved O in addition to the weight of original carbon.

3.1.4 Transverse rupture strength (TRS). Figure 5 shows TRS of pre-sintered specimens under various PS conditions. TRS is relatively stable at PS temperature above 1050°C, irrespective of sintering time. Generally, the possible maximum temperature in mass production is approximately 1100°C. Therefore, recommended PS temperature and time seem to be 1050°C-1075°C and 20 min, respectively. The pre-sintered material under such conditions has both high strength by pearlite and deformability by ferrite, and it is economical. The previous studies reported that Mo promotes sintering [8] but Mo alloying in ferrite is few [9]. Similar tendency is confimed in the hardness shown in table 3.

Table 3. Results of quantitative analysis of carbon and average hardness for core area of PS specimen (ρ_{PS}: 7.4 Mg/m^3, PS time: 20 min).

PS temperature	C [mass%]	H_{RA}	HV0.2kgf (Pearlite)	HV0.2kgf (Ferrite)
975°C	0.31	44.4	201	130
1000°C	0.30	45.0	198	128
1025°C	0.30	46.0	199	123
1050°C	0.29	47.3	197	129
1075°C	0.28	47.6	197	123
1100°C	0.28	47.4	205	134

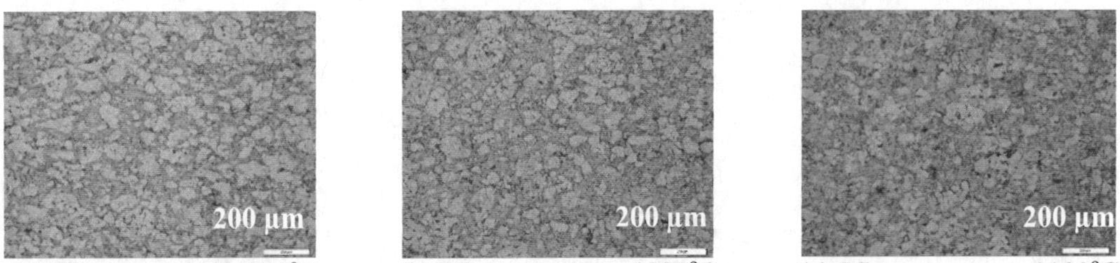

(a) PS temperature: 975°C (b) PS temperature: 1025°C (c) PS temperature: 1100°C

Figure 4. Microstructures of core sections of PS specimens (ρ_{PS}: 7.4 Mg/m^3, PS time: 20 min).

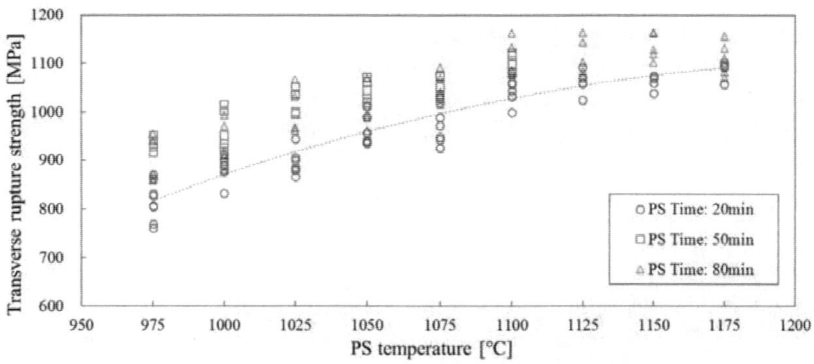

Figure 5. Effect of PS temperature and time in transverse rupture strength (ρ_{PS}: 7.4 Mg/m^3).

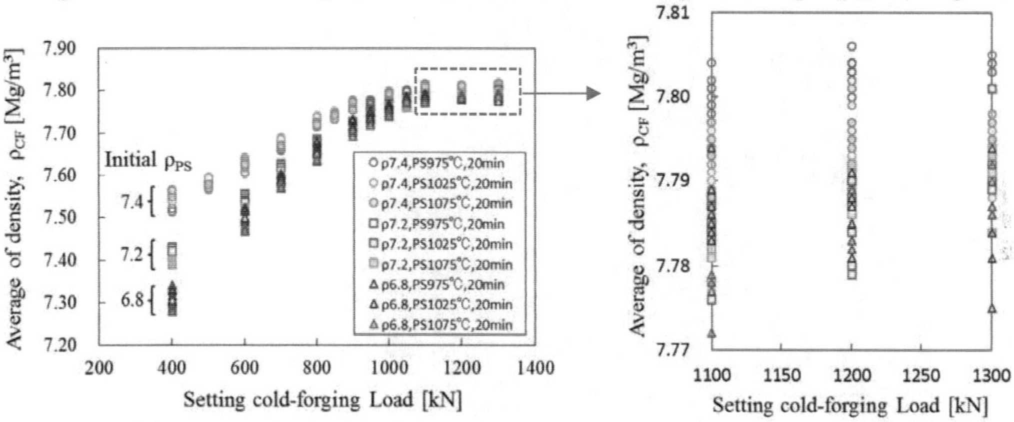

(a) Average density at CF load range of 400-1300 kN (b) Magnified graph of (a)

Figure 6. Average density of cold-forged specimens (ρ_{PS}: 6.8, 7.2, 7.4 Mg/m^3).

3.2. Properties of cold-forged specimens

3.2.1 Effect of PS conditions on cold-forged (CF) specimens.

Figure 6(a) and 6(b) show densification properties of PS specimens by CF. Density (ρ_{CF}) of the specimen saturated at the setting CF load above 1100 kN. At such CF load, with increasing initial PS density (ρ_{PS}), an upper limit of ρ_{CF} increases. This modification of density (ρ_{CF}) scarcely depend on PS temperature.

Also, maximum CF density (ρ_{CF}) is from 7.78 to 7.81 Mg/m^3 for each PS condition. These density are higher than that of the previous study by using free cold upsetting [10]. Thus, the proper density of the specimen after PS (ρ_{PS}) seems to be 7.4 Mg/m^3. Figure 7 shows microstructures of cross sections of cold-forged specimens at various CF loads. In case of setting CF load of 1100 kN, it was found that ultra-densification occurs on the surface layer by the plastic flow of ferrite phases. The similar densified microstructures were observed in cold-forged specimens at the setting CF load above 1100 kN, and the number of pores decreased with increasing the CF load. Since the CF load of 1100 kN with the upper limit of ρ_{CF} in figure 6 equals to the load for the flat ferrite grains elongation in figure 7, these microstructural deformations of the surface layer will be related to average density of the CF specimen (ρ_{CF}).

IOP Conf. Series: Materials Science and Engineering **257** (2017) 012011 doi:10.1088/1757-899X/257/1/012011

Optical microscope images	Setting CF Load
	600 kN
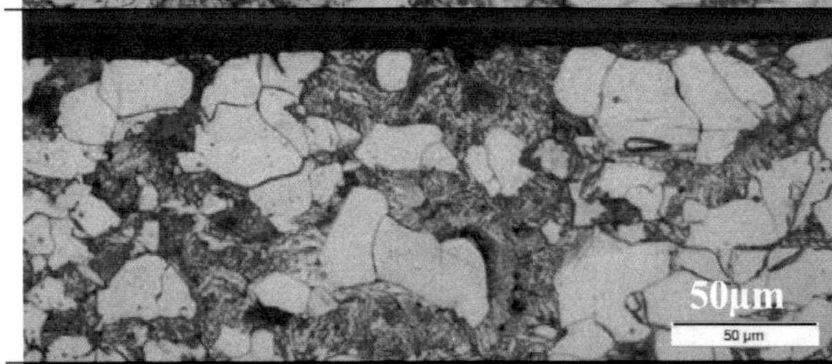	900 kN
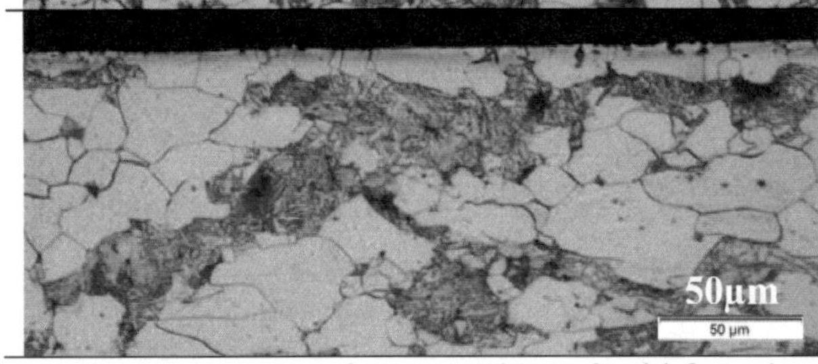	1100 kN

Figure 7. Microstructures of cross sections of cold-forged specimens at various CF loads (ρ_{PS}: 7.4 Mg/m^3, PS conditions: 1050°C, 20 min).

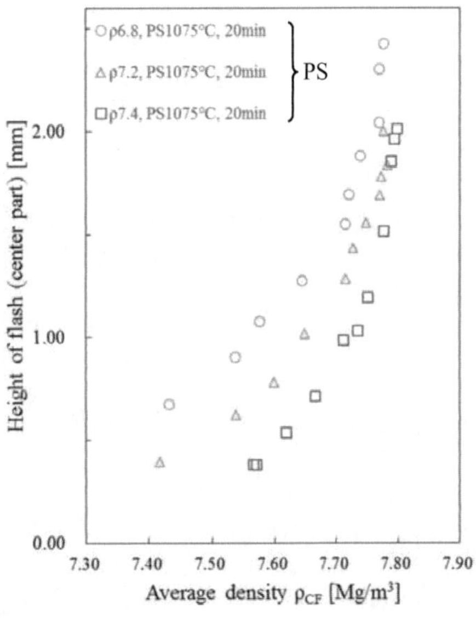

(a) Flash height and average density of cold-forged specimen (ρ_{CF}) in case of PS temperature of 1075°C.

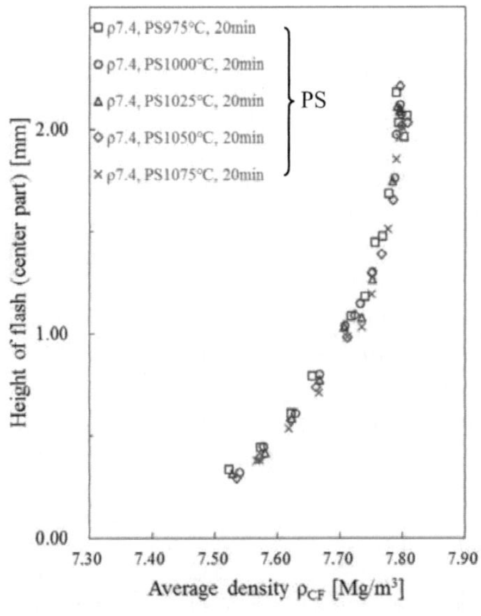

(b) Flash height and average density of cold-forged specimen (ρ_{CF}) in case of ρ_{PS} of 7.4 Mg/m^3.

Figure 8. Relationship between flash height and average density after cold-forging.

3.2.2 Flash height and average density. Figure 8 shows a relationship between the flash height and average density (ρ_{CF}). From figure 6(a) and 8, it was found that the flash increases exponentially with the higher CF load. Moreover, at the CF load range investigated in this study, in case of a low initial PS density (ρ_{PS}), final flash heights were higher. Regardless of ρ_{PS}, ρ_{CF} approached to almost the same value. Also, as shown in figure 8(b), in case of the same initial PS density (ρ_{PS}), the effect of PS temperature of the flash height is negligible. This result means that the initial PS density (ρ_{PS}) is an important factor to obtain higher average density (ρ_{CF}).

3.3. TRS and microstructure after carburized quenching and tempering (CQT)

Figure 9 shows TRS of PS-CF specimens after CQT heat treatment. In case of the range of the CF load is from 400 kN to 800 kN, TRS gradually increases with increasing the CF load. As the previous studies indicated, generally, TRS is directly proportional to a sintered density [11]. However, TRS decreases sharply at the CF load range from 800 kN to 900 kN. With increasing the CF load above 900 kN, TRS also increases. Such change in TRS with the change of the CF load is a "death valley" type. Similar phenomena were reported as "abnormal brittleness" in results of decreasing Charpy impact values at using specimens with the densities of 7.2-7.4 Mg/m^3 [12-14]. However, results of this study in higher

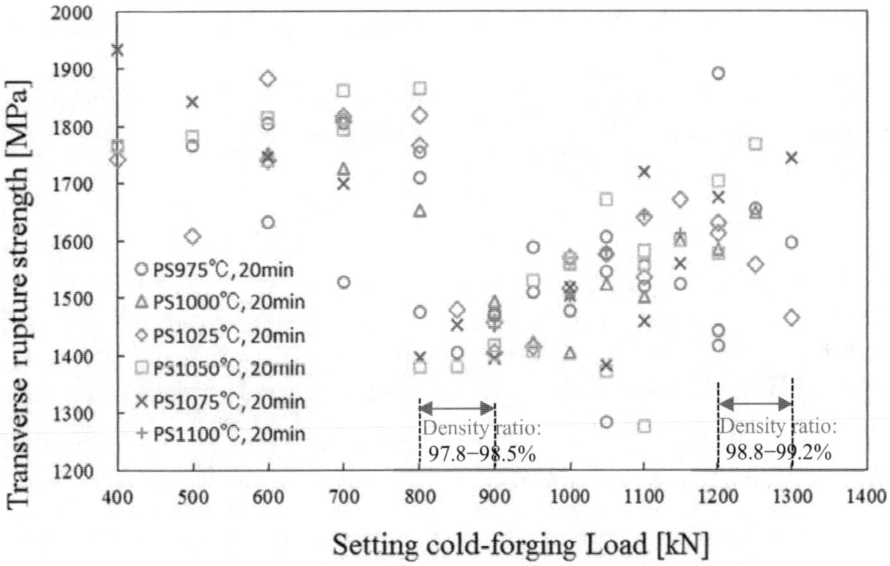

Figure 9. TRS evaluation results of heat treated PS-CF specimens (ρ_{PS}: 7.4 Mg/m^3).

Density range could indicate a different trend of the transverse rupture strength. At the CF load above 900 kN, irrespective of PS temperature, TRS increased gradually again. In case of the CF density (ρ_{CF}) of 7.70-7.76Mg/m^3 (density ratio: 97.8-98.5%), TRS decreases and not suitable for production of near net shape parts.

Figure 10 shows back-scattered electron images of cross sections of surface layers after cold-forging. Figure 10 (a) and (b) show the residual pores existed on the surface layer. In figure 10 (c) and (d), channelling contrasts indicating a large scale of a plastic flow are observed. In case of image (c) at the CF load of 950 kN, although pores were compressed, micro-cracks formed at triple or quadruple points of grain boundaries in the surface layer. Such micro-cracks cause TRS decreasing. When a high pressure is loaded on the surface layer for ultra-densification, inter-particle sintered connection is damaged. Then, the micro-cracks along grain boundaries are introduced by stress concentration and shearing stress derived from a compression stress and a plastic flow.

As shown in figure 6(a) and figure 7, the CF load above 1100 kN is needed to achieve the maximum density. Furthermore, for ultra-densification of the surface layer and disappearance the micro-cracks, a sufficient plastic flow is required as shown in figure 10(d). The setting CF load of 1200 kN can compress the specimen sufficiently and heat treatment at high temperature could regenerate the sintered connection. These processes could give high strength densified materials stably and be applicable to not only the improvement of spalling fracture on the surface of densified sintered iron [15] but also the near net shape manufacturing by sinter forging.

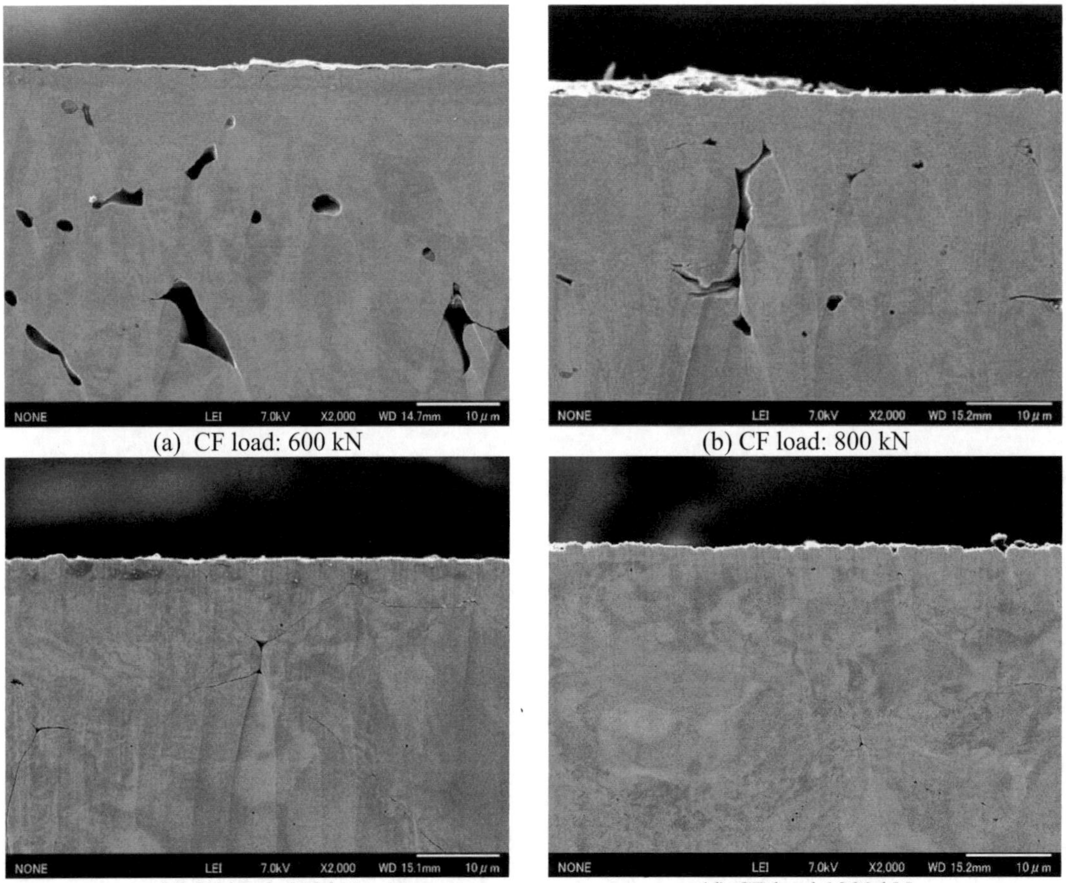

(a) CF load: 600 kN (b) CF load: 800 kN

(c) CF load: 950 kN (d) CF load:1200 kN

Figure 10. Back-scattered electron images of cross sections of surface layers after cold-forging (ρ_{PS}: 7.4 Mg/m^3, PS temperature:1050°C, PS time: 20 min).

4. Conclusions

In this study, we made the high density sintered material with Mo diffusion alloyed atomized iron by sintering and cold-forging. Then the optimum conditions of pre-sintering (PS) and cold-forging (CF) to obtain the good mechanical properties were investigated. The obtained results are as follows.

- Since the specimen is decarburized a little in vacuum sintering, the weight of added carbon is desirable to be a half of the content of dissolved O in addition to the weight of original carbon.

- Density of PS specimen is necessary more than 7.4 Mg/m^3 to obtain good mechanical properties. Furthermore, in case of PS process time of 20 minutes, PS temperature is necessary more than 1050°C. The microstructure and density of the PS specimen are most important for high densification by CF. The maximum density became approximately 7.8 Mg/m^3 (density ratio: 99%) at the setting CF load of 1200 kN in case of the backward extrusion.

- At the setting CF load range from 800 kN to 900 kN, TRS decreased sharply. The safety threshold level of the density ratio was approximately under 97.8% (7.70 Mg/m^3) or over 98.8% (7.78 Mg/m^3) to obtain high TRS. When the densification of the core of the sintered specimen progresses, shearing stress generates in the surface layer and stress concentration occurs in the pores of the surface layer. Thus inter-particle sintered connections are damaged and micro-cracks initiate in the surface layer. Moreover, in case of the backward extrusion, ultra-densification of the surface layer occurs by the sufficient plastic flow at the setting CF load of 1200 kN. Such sufficient compressed specimens could regenerate the sintered connections by heat treatment at high temperature and could obtain the high strength densified materials stably.

Acknowledgements
The authors express their sincere gratitude to Noriko Inoue (Inoue Heat-treatment Industorial Co., Ltd.), Ryuichi Abe (Fuji Sintering Alloy Co., Ltd.), Sakae Fujiki (Fujiki engineering office), Tomoaki Suganuma (Chubureikan Co., Ltd.), Shigeyoshi Kabe, Masahiro Gotoh and Naohiko Ogino (Gunma Industrial Technology Center).

References
[1] S Oki, T Akiyama, K Shoji 1983 Effects of Molybdenum Content and Carbon Content on the Transformation Features of Sintered Cr-Mo Steels, *Japan Society of Powder and Powder Metallurgy* **30** 223-228
[2] T K Kandavel, R Chandramouli, D Shanmugasundaram 2009 Experimental study of the plastic deformation and densification behaviour of some sintered low alloy P/M steels, *Materials & Design* **30** 1768-1776
[3] A Fujiki, H Kojima, Y Maekawa, M Yamaguchi, T Murata, Y Sugaya, M Iwakiri, T Shibano 2001 Development of Warm Compacted P/M Sprockets for Automotive Engines, *Japan Society of Powder and Powder Metallurgy* **48** 440-444
[4] K Takada, R Kogure, T Kobayashi 2007 Research of High Fatigue Strength for Powder-Forged Connecting Rods, *Honda R&D Technical Review* **19** 129-137
[5] S Ishihara 1999 Development of High strength Connecting Rod -Powder Forging and Forging-, *Japan Society of Powder and Powder Metallurgy* **46** 505-509
[6] N Nakamura, S Uenosono, M Fujinaga, S Koizumi, H Anma, T Yoshimura 2005 Sintering and Cold-Forging Process for High Density Sintered Materials, *JFE GIHO* **7** 19-23
[7] A Squire 1947 Density Relationships of Iron Powder Compacts, *AIME Transactions* **171** 485-505
[8] S Unami, T Goushi, K Kudo, F Tsumori, H-G Kang, H Miura 2010 Improvement of Surface Densification and Fatigue Strength by Tooth Root Rolling of Sintered Ferrous Alloy Gear Added Mo,*Japan Society of Powder and Powder Metallurgy* **6** 430-434
[9] T Kouno, K Obara 1979 Effects of Starting Raw Materials, Admixed and Alloyed Powder on the Mechanical Properties of 4600 Sinter-forged Steels, *Japan Society of Powder and Powder Metallurgy* **1** 27-33
[10] Narayan S, Rajeshkannan A 2011 Densification behaviour in forming of sintered iron-0.35% carbon powder metallurgy preform during cold upsetting, *Mater. Des.* **32** 1006-1013
[11] T Iwaoka, A Fujiki 2014 Microstructure and Strength of Fe-2%Cu-0.8%C Sintered Steel using Fe-Cu Alloy Powder, *Japan Society of Powder and Powder Metallurgy* **61** 290-297
[12] H Kuroki, Y Tokunaga 1969 Abnormal Brittleness of High Density Sintered Iron, *Japan Society of Powder and Powder Metallurgy* **16** 266-273
[13] H Kuroki, Y Tokunaga 1969 Charpy Impact Tests on Sintered Iron, *Japan Society of Powder and Powder Metallurgy* **16** 259-265
[14] H Kuroki, T Wakabayashi, Y Tokunaga 1971 Investigation on Abnormal Brittleness of High Density Sintered Iron, *Japan Society of Powder and Powder Metallurgy* **17** 307-313

[15] H Miura, F Tsumori, S Hamamoto, K Komoto, S Unami 2010 Improvement of the Contact Fatigue Strength by Rolling of Sintered Ferrous Alloy Roll -2nd Report-, *Japan Society of Powder and Powder Metallurgy* **6** 430-434

IOP Conf. Series: Materials Science and Engineering **257** (2017) 012012 doi:10.1088/1757-899X/257/1/012012

Finite element method analysis of cold forging for deformation and densification of Mo alloyed sintered steel

Y Kamakoshi[1, 2*], S Nishida[3], K Kanbe[2] and I Shohji[2]

[1]Dept. Applicational-Mechanical Engineering, Gunma Industrial Technology Center, Maebashi, Gunma 379-2147, Japan
[2]Division of Mechanical Science and Technology, Graduate School of Gunma University, Kiryu, Gunma 376-8515, Japan
[3]Division of Mechanical Science and Technology, Graduate School of Gunma University, Ota, Gunma 373-0057, Japan

*Corresponding author: kamakoshi-yu@pref.gunma.lg.jp

Abstract. In recent years, powder metallurgy (P/M) materials have been expected to be applied to automobile products. Then, not only high cost performance but also more strength, wear resistance, long-life and so on are required for P/M materials. As an improvement method of mechanical properties of P/M materials, a densification is expected to be one of effective processes. In this study, to examine behaviours of the densification of Mo-alloyed sintered steel in a cold-forging process, finite element method (FEM) analysis was performed. Firstly, a columnar specimen was cut out from the inner part of a sintered specimen and a load-stroke diagram was obtained by the compression test. 2D FEM analysis was performed using the obtained load-stroke diagram. To correct the errors of stress between the porous mode and the rigid-elastic mode of analysis software, the analysis of a polynominal approximation was performed. As a result, the modified true stress-true strain diagram was obtained for the sintered steel with the densification. Afterwards, 3D FEM analysis of backward extrusion was carried out using the modified true stress-true strain diagram. It was confirmed that both the shape and density of the sintered steel analyzed by new FEM analysis that we suggest correspond well with experimental ones.

1. Introduction

Mechanical parts need a stable quality, a reliable supply, a high mass-productivity and low costs. Furthermore, superior mechanical, chemical, thermal, anti-wear properties are also required for them. Wrought steel has been mainly used for automobile components. Sintered steel has been also adopted in some parts of the components. Not only requirements of more cost down for the components but also various demands for sintered steel components has been increasing in many countries year by year [1]. Compared with wrought steel, sintered steel is brittle and weak because of inner pores and microstructural defects [2]. However, it can be used as oil-less bearing utilizing porosity and as a composite material which is mixed completely different materials. Moreover, the sintered steel has superior yield rate and enables near-net-shaping, furthermore it has high form accuracy and mass-productivity of small complex shape parts. On the contrary, although the wrought steel has good mechanical properties, it is not resource saving due to high cost reprocessing. Many researchers have reported that the mechanical properties of the sintered iron materials such as tensile strength, toughness, fatigue strength, wear resistance and so on, are improved by the densification [3]. For examples, those reports are about additional elements, cold-forging and hot-forging [4], sintered cold-forging [5], rolling of sintered ferrous alloy gear [6], powder-forged connecting rods [7] and so on. Although many studies about powder-forging with non-plastic deformation particular by methods of numerical analysis were reported [8], the numerical analysis of the sintered cold-forging with the densification and the plastic deformation was few [9]. The aim of this study is to research the behaviour of the densification of Mo-alloyed sintered steel in the cold-forging process by finite element method (FEM) analysis. Moreover, the densification and the plastic deformation of sintered steel were investigated by 3D FEM analysis.

2. Experimental Procedure

2.1. Specimen

In this study, partial diffusion alloyed steel powder (JFE steel, JIP SGM10MO-CMX) for cold-forging was prepared. It is atomized iron powder with diffusional adhered 1 mass% Mo and 0.35 mass% C (graphite powder) to the surface of powder. Then, cold-forging analysis and experiment were performed by use of sintered steel. Figure 1 shows the schematic flow of compacting and cold-forging. Steel powder was compacted to a cuboid specimen by pressure of 980 MPa (floating die method). Density (ρ_c) of the green compact was approximately 7.4 Mg/m^3. In this study, average density was measured as apparent density by the Archimedes method. Width, height and length of the cuboid specimens were 10 mm, 12 mm and 54 mm, respectively. The compacts were sintered in a vacuum furnace at vacuum degree of approximately 30 Pa. Sintering temperatures were approximately 975°C and 1050°C, and sintered time was 20 min. After sintering, density (ρ_{PS}) of the specimens were approximately 7.4 Mg/m^3. Cold-forging (CF) was performed by backward extrusion method at setting load of 1200 kN and setting time of 1 sec. In this study, only cold-forging process of the sintered specimen was simulated by FEM.

2.2. Experiment and Analysis

2.2.1. Compression test of sintered specimen. To investigate a compression load-stroke (cross head stroke of testing machine) diagram of sintered materials, the compression test was performed using an electromechanical universal testing machine (TENSILON RTF-2430 (max load: 300 kN, head speed: 1000 mm/min)). A smooth die set was used for the compression test. To reduce friction force, graphitic lubricant was used and teflon sheet was set between the die set and the specimen. Figure 2 shows the sintered specimens for the compression test. Columnar specimens (φ: 5 mm, length: 5 mm) were cut out from the inner part of the specimen. The upper columnar specimen was more densified than the lower one. The result of the compression test about the upper specimen was used as basic data for FEM analysis. The cross sectional microstructure of cold-forged specimen was observed with an inverted metallurgical microscope and the side surface of the specimen was observed with a field emission-scanning electron microscope (FE-SEM).

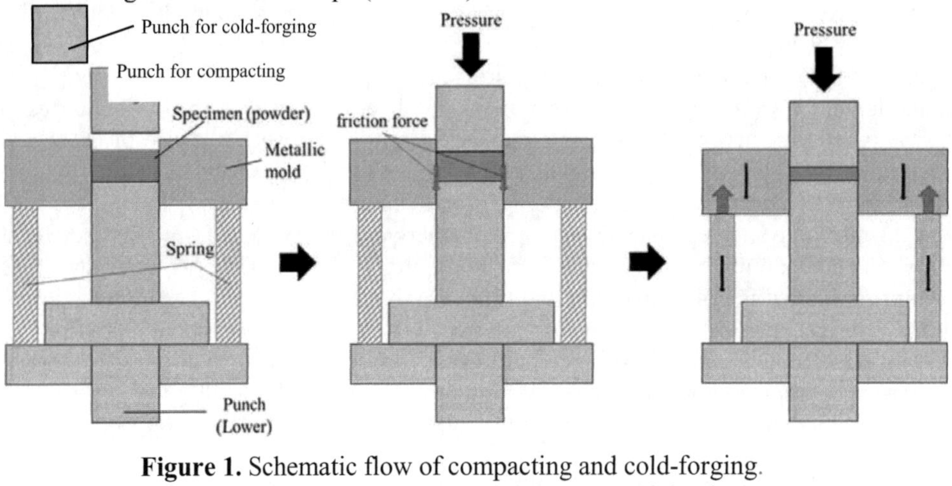

Figure 1. Schematic flow of compacting and cold-forging. (Spring is changed to rigid-bar at cold-forging).

2.2.2. FEM analysis model. Using software "DEFORM (ver.11.0)", FEM analysis was performed to reappear a plastic flow by deformation during the cold-forging. At first, the load-stroke diagram obtained by the compression test was used for FEM analysis. Figure 3 shows the 2D analysis model for the compression test. The model was axisymmetrical one-half model and both the upper punch and the lower die were rigid material. The columnar specimen was defined as an elasto-plastic material. The 2D FEM analysis conditions are shown in table 1(1). Table 1(2)-(4) also shows the 3D FEM analysis conditions. The object type of the specimen was defined as rigid-plastic or porous. Young's modulus was 210 GPa. In case of the porous mode, initial sintered density ratio was 0.94. The mesh window was set to reduce element size of upper part of the specimen which is deformed especially. Remeshing was performed for greatly deformed area by a method of local remeshing. A normalized Cockroft & Latham model was used as the fracture model. In the model, a damage value (critical value) and the fracture element were 100 and 1, respectively. Then the load-stroke diagram obtained by 2D FEM analysis was used for 3D FEM.

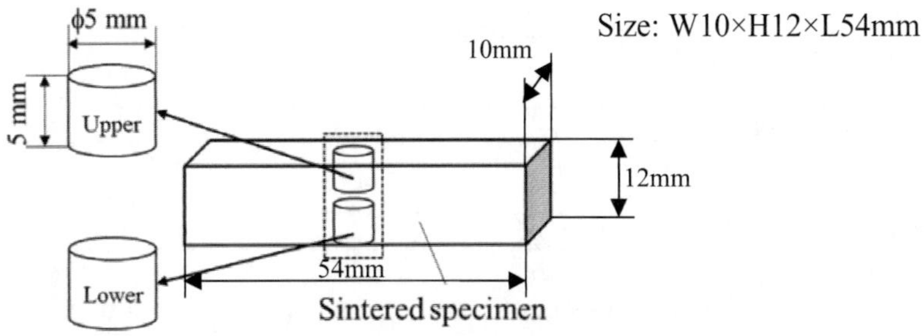

Figure 2. Specimens for compression test.

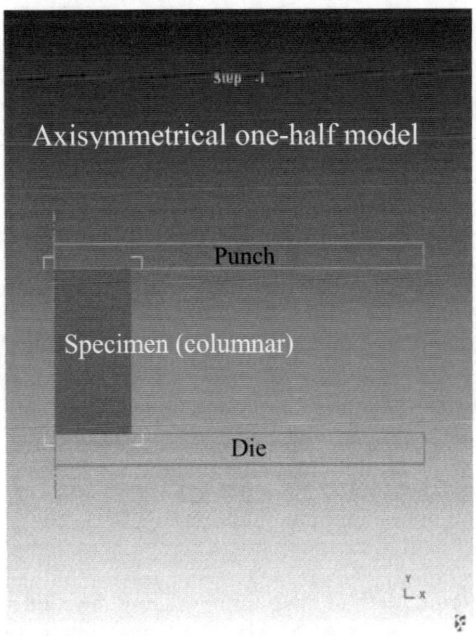

(a) Layout of punch, specimen and die before compression test.

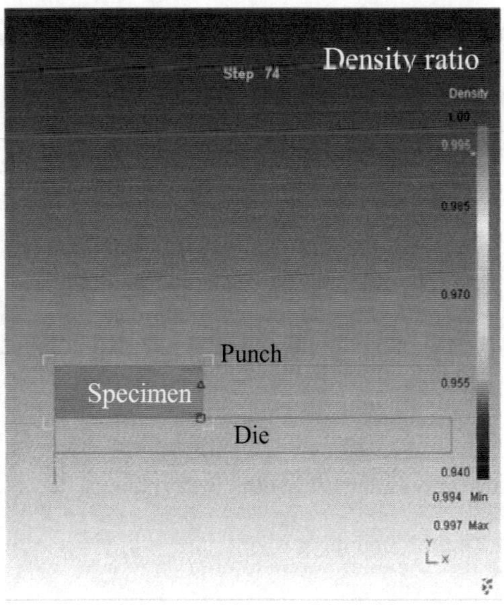

(b) Example of analysis result after compression test.

Figure 3. 2D analysis model for compression test.

Table 1. FEM analysis conditions for cold-forging of sintered material.

Object Type/ Model/ Method	(1) Elasto-plastic material/ Axisymmetric 2D model	(2) Rigid-plastic material/ 3D model	(3) Porous material/ 3D model/ remeshing	(4) Porous material/ 3D model/ remeshing and element elimination method
Density (relative density)	–	–	0.94	0.94
Element count [count]	2,103	158,389	148,148	148,148
Calculation step [step]	90	500	2000	2000
Punch speed [mm/sec]	16.667	1	1	1
Step increment [mm/step] (die displacement)	0.05	0.01	0.001	0.001
Shear friction coefficient between a specimen and a punch	0.01	0.4	0.4	0.4
Shear friction coefficient between a specimen and a die	0.01	0.08	0.08	0.08
Remeshing [mm] (Maximum stroke increment)	–	–	0.01	0.01
Target volume (for meshing)	–	–	Valid	Valid
Cockroft & Latham marginal damage value	–	–	–	100
Fracture element	–	–	–	1

Table 2. Result of density measurement after compression test.

Specimen	Sintering temperature [°C]	Cut position	Density before compression test [Mg/m^3]	Density after compression test [Mg/m^3]	Density ratio [%]*
A	975	Upper	7.4	7.76	98.5
B	1050			7.80	99.1
A	975	Lower	7.4	7.70	97.8
B	1050			7.72	98.0

* Density of wrought steel is defined as 7.874 Mg/m^3.

Table 3. Comparison in size of specimens before and after compression test

Specimen	Sintering temperature [°C]	Cut position	Radius [mm] (before)	Radius [mm] (after)	Height [mm] (before)	Height [mm] (after)	Compression ratio [%] (using height)
A	975°C	Upper	5.0	9.6	4.90	1.27	74.1
B	1050°C		5.0	9.0	5.22	1.54	70.6
A	975°C	Lower	5.0	8.3	5.22	1.83	65.0
B	1050°C		5.0	8.8	5.49	1.72	68.7

3. Results and Discussion

3.1. Properties and microstructure of specimen after compression test

Table 2 shows density change of the columnar specimen by the compression test. Density ratios were above 98.5% or equals at the upper part of the specimen. These results mean that the sintered specimen has high-deformational properties. Table 3 shows size change of the specimens before and after the compression test. Compression ratios were above 70% at the upper part of the specimens. The upper part was slightly more densified than the lower part because of its contacting the upper punch and one-way press at the cold-forging. Figure 4 shows a side surface area and a cross section at the upper part of cut specimen-B after the compression test. No crack was observed in both the cross section and the side surface area. Figure 4(c) shows that the upper part of the cut specimen-B was successfully compressed with the plastic flow. These results mean that the sintered specimen has a compressibility which is nearly equivalent to the wrought steel.

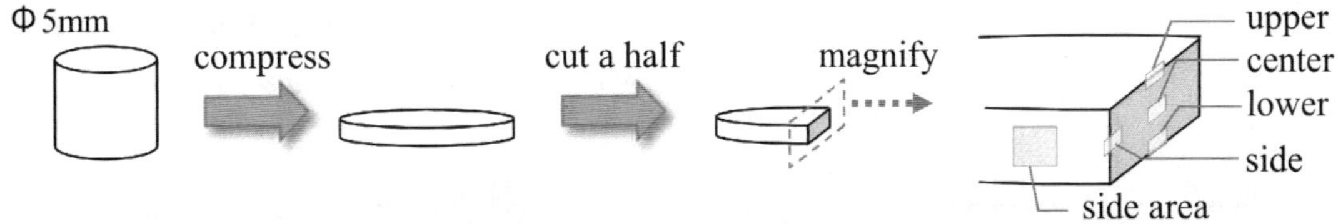

(a) Observed area of specimen

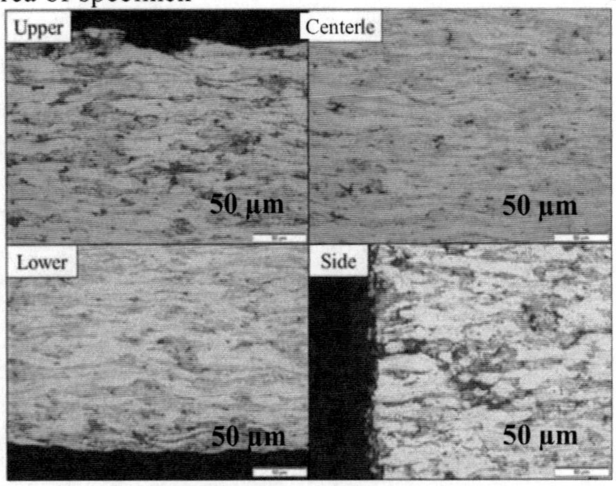

(b) SEM image of surface of side area (c) OM images of cross section

Figure 4. Surface of side area and cross section of specimen-B after compression test.

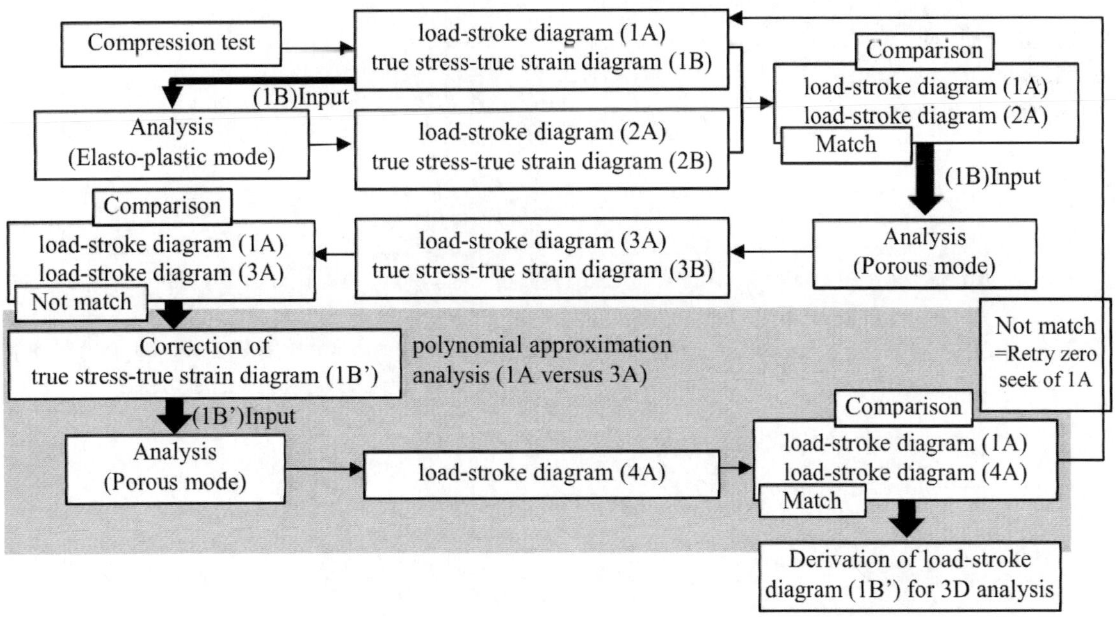

Figure 5. 2D deformation analysis procedure for sintered material.

3.2. Derivation of load-stroke diagram for sintered material by 2D deformation analysis

Since this subsection, the data of the specimen-B are used. Figure 5 shows process of 2D analysis. Suffixes A and B of diagram number shown in the figure mean load-stroke and true stress-true strain, respectively. At first, load-stroke diagram (1A) was obtained by the compression test of the columnar specimen. The diagram (1A) was corrected by zero seek and changed a nominal stress-nominal strain diagram.

Table 4. Comparison in compression amount and density ratio for experiment and analysis values.

	Compression amount [mm]	Density ratio [-]	Error in density ratio [%]
Experiment	3.68	0.991	-
Analysis	3.70	0.995	0.4

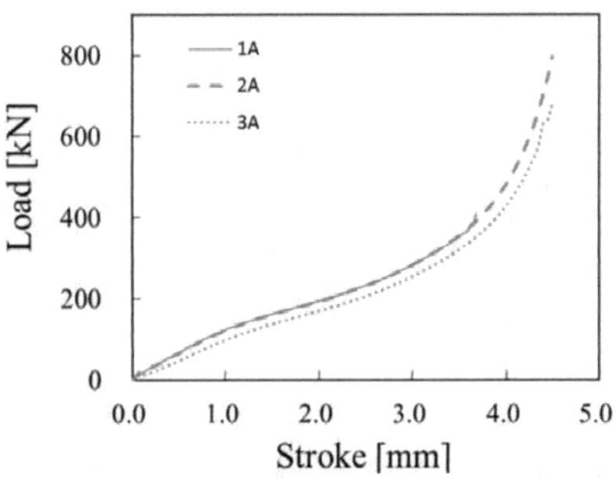

(a) Load-stroke diagrams (1A, 2A, 3A).

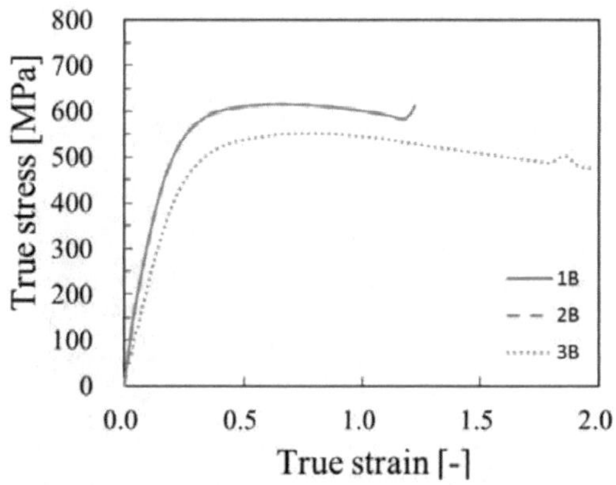

(b) True stress-true strain diagrams (1B, 2B, 3B).

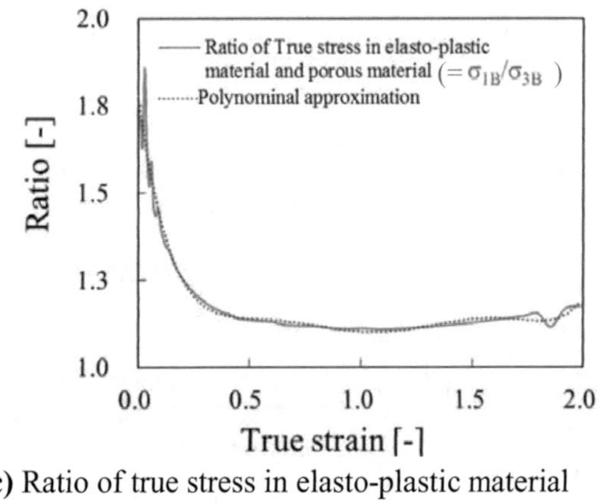

(c) Ratio of true stress in elasto-plastic material and porous material.

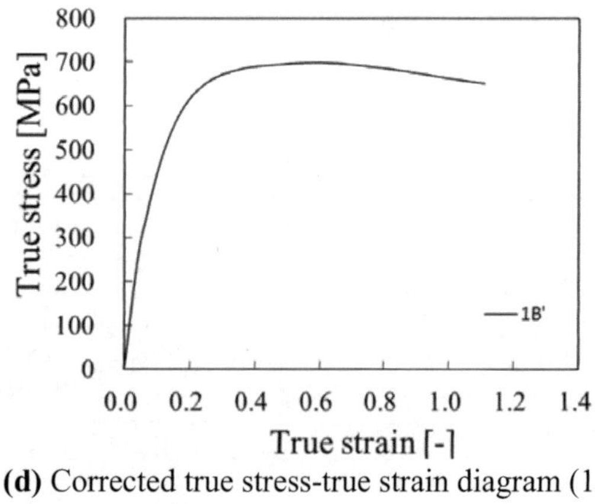

(d) Corrected true stress-true strain diagram (1B').
(1B'=1B × ratio of true stress).

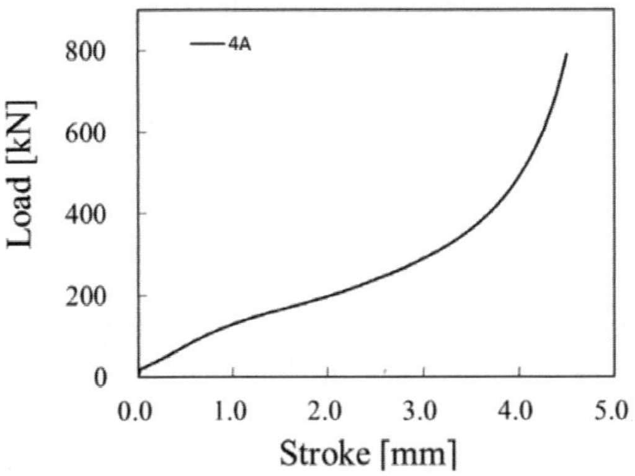

(e) Load-stroke diagram (4A).

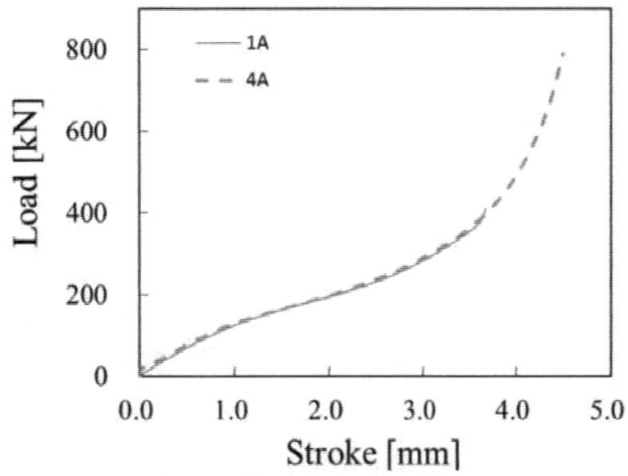

(f) Comparison of load-stroke diagrams (1A, 4A)

Figure 6. Results of 2D deformation analysis used axisymmetrical columnar model.

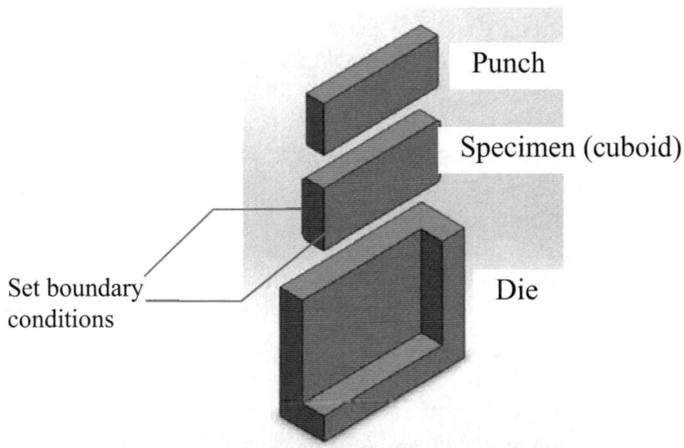

Figure 7. 1/4 model for 3D FEM analysis in cold-forging.

It was converted to a true stress-true strain diagram (1B). Using the true stress-true strain diagram (1B), the compression test was analyzed by the elasto-plastic material mode. Then a load-stroke diagram (2A) and a true stress-true strain diagram (2B) were obtained. When compared with the load-stroke diagrams (1A) and (2A), match of both diagrams was confirmed. Next, the compression test was analyzed with the porous material mode using the true stress-true strain diagram (1B). As a result, a load-stroke diagram (3A) was obtained. When compared with the load-stroke diagrams (1A) and (3A), both diagrams did not correspond constantly. The reason seems that the analysis result was performed with the porous material mode although the data of the diagram (1A) already involved compression property of the porous. Therefore, to correct difference of both diagrams (1A) and (3A), a polynomial approximation analysis was used anew. Processes are as follows.

(i) Check what times load of diagram (1A) is that of diagram (3A) at the same stroke value.

(ii) To obtain amplification coefficient in the true stress-true strain diagram, the polynominal approximation is performed. In this study, the polynominal of sixth degree (shown in figure 6(c)) is used.

(iii) Multiply the obtained amplification coefficients by the true stress in the true stress-true strain diagram (1B). Then the modified true stresses of diagram (1B') is obtained.

(iv) Analyze the compression test with the porous material mode using the diagram (1B') and obtain diagram (4A).

As a result, the load-stroke diagram (4A) matched load-stroke diagram (1A). Therefore, the modified diagram (1B') was used for 3D deformation analysis after this. Figure 6 shows all of diagrams which were obtained according to above mentioned procedure. Table 4 shows results of comparing in compression amount and density ratio. The results indicated that analysis values were quite close to experimental values for the compression amount and the density ratio. Also the results indicated that 2D analysis model in above procedure is proper in this study.

3.3. 3D deformation analysis of porous material

3.3.1. Analysis model. Figure 7 shows the 3D FEM model for the compression test. Considering the symmetry, the 1/4 model was used. The cuboid specimen was inserted into the die and compressed by the upper punch. The area of the die inserted the specimen and that of the upper punch were 10×55 mm and 8×53 mm, respectively. Then, after cold-forging, the flash (equals to rim) width by deformation was almost 1mm.

3.3.2. Results of analysis and improvement of analysis method. Figure 8 and 9 show comparison of equivalent strain distribution in rigid-plastic material model and porous material model analyzed under conditions of table 1(2) and (3), respectively. In case of rigid-plastic material model, the punch load was saturated at approximately 1000 kN. On the other hand, in case of porous material model, equivalent strain of the corner part increased to 28.7 and the punch load increased to above 1800 kN. Since the real load of the upper punch which was measured by a strain gauge was approximately 800 kN, the result is invalid. To improve the analysis method, the forced remeshing mode was selected. In this mode, when maximum stroke increment reaches to 0.01 mm in the specimen, remeshing is performed to the elements in the upper parts at 4 mm height. The target volume mode which hold volume constantly by interpolation is also used to avoid volume reduction at calculation and remeshing. The target volume mode is only valid in remeshing. As shown in figure 8 and 9, the height of flash in the porous material model is lower than that of the rigid-plastic material model. This phenomenon indicates that volume reduction occurs by not only microstructure deformation but also reducing pores at the compression test. Thus, to avoid increasing strain locally, analysis with the element elimination method is effective. In particular, this method is valid for the porous material model that strain concentration occurs easily by volume reduction. With increasing density of the sintered material, the properties of plastic deformation approaches those of the rigid-plastic material. In this case, local strain concentration is unnatural. In this study, when the damage value of the element greatly exceeds those of surrounding elements, the element is vanished. The damage value was determined using the normalized Cockroft and Latham equation. The marginal damage value was selected 100. Figure 10 shows the analysis result using the porous material model, the modified true stress-true strain diagram (1B'), and the remeshing and element elimination method under conditions of table 1(4). In figure 10 (a), a example photo of the cold-forged specimen is also shown. The shape of the analyzed specimen almost corresponds to that of the cold-forged specimen.

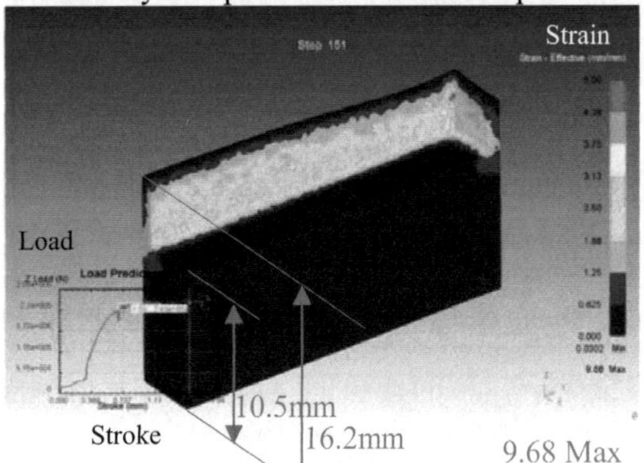

Figure 8 (a). Equivalent strain distribution in rigid-plastic material model analyzed under conditions of table 1(2).

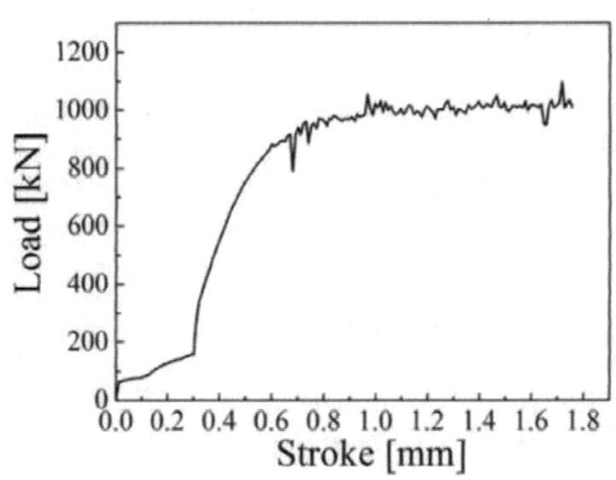

Figure 8 (b). Magnified graph of load-stroke diagram in figure 8 (a).

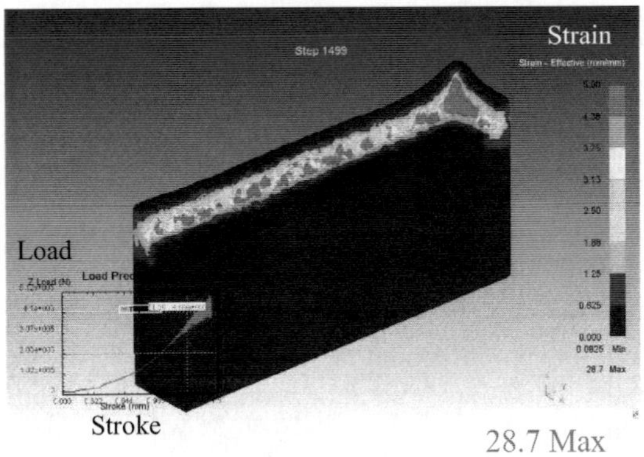

Figure 9 (a). Equivalent strain distribution in porous material model analyzed under conditions of table 1(3).

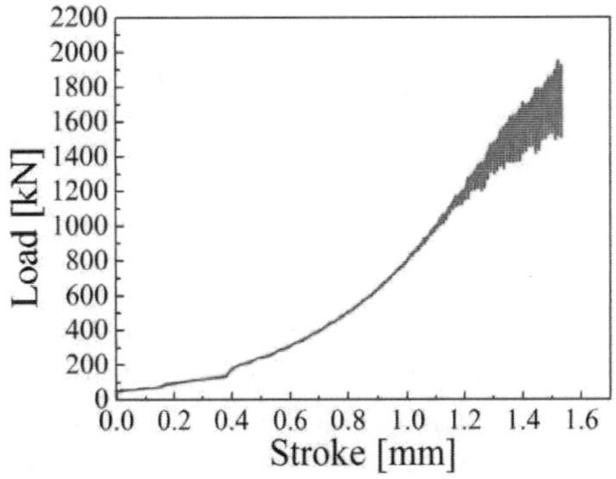

Figure 9 (b). Magnified graph of load-stroke diagram in Fig. 9a.

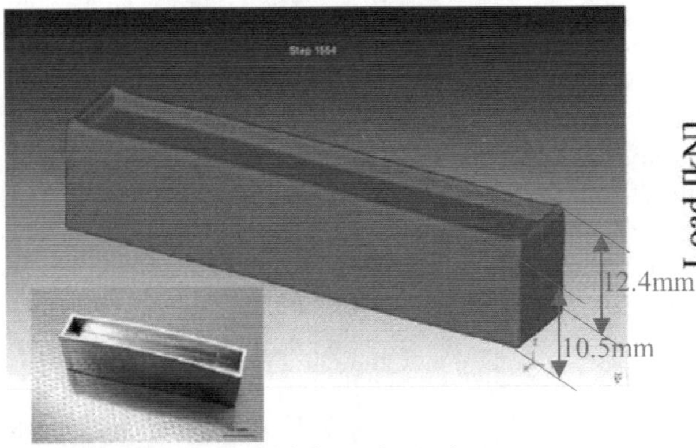

Figure 10 (a). Analysis result using remeshing and element elimination method for cold-forged specimen under conditions of table 1(4).

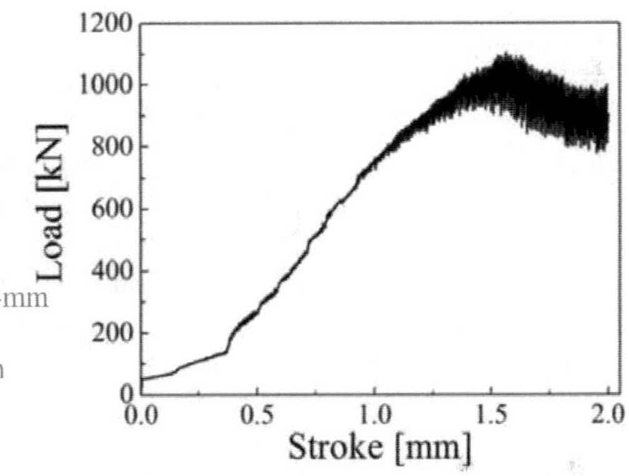

Figure 10 (b). Load-stroke diagram obtained by FEM analysis using remeshing and element elimination method in cold-forging.

Figure 11 (a). Density distribution by cold forging analyzed using element elimination method (specimen-B, sintering temperature: 1050°C).

Figure 11 (b). Density distribution by cold forging analyzed using element elimination method (specimen-A, sintering temperature: 975°C).

Table 5. Comparison of experiment and 3D analysis values. ( ) : Error for experimental value[%]

	Specimen-B (sintering at 1050°C)		Specimen-A (sintering at 975°C)	
	Density ratio [-]	Flash height [mm]	Density ratio [-]	Flash height [mm]
Experiment	0.991	2.03	0.990	2.05
Analysis	0.978 (1.3%)	1.95 (3.9%)	0.978 (1.2%)	1.98 (3.4%)

As shown in Fig.10b, the load-stroke diagram in the cold-forging process can be obtained by FEM analysis and the load greatly fluctuates at the stroke above 1.0 mm. The fluctuation is caused by the remeshing. Fig. 8 shows the analysis result in case of the punch stroke 1.55 mm by rigid-plastic material model. The height from the bottom surface to the upper surface in the center was 10.5 mm and the maximum height of flash was 16.2 mm. In case of porous material model, both heights were 10.5 mm and 12.4 mm, respectively (Fig. 10a). These values approximately corresponded to sizes of real sintered cold-forged specimens (ρ_{CF}: 7.8 Mg/m^3, PS: 1000°C, 20 min). The shape of flash was similar to that of the real sintered cold-forged specimen and gradually increased from the edge to the center part. The fluctuation in the load in Fig. 10b is smaller than that in Fig. 9b and saturates at approximately 1000 kN. The results mean that the remeshing and element elimination method is effective to simulate cold-forging process. The load-stroke diagram in Fig. 10 is relatively similar to that in Fig. 8 and more reliable than the result in Fig. 9. Fig. 11 shows density distribution of the specimen which was analyzed by using the element elimination method. Table 5 shows comparison of experimental and 3D analysis values for the density ratio and the flash height. As shown in table 5, regardless of sintering temperature (1050°C and 975°C), the averages of the density ratios of the real sintered cold-forged specimens were approximately 0.99 and those evaluated by FEM analysis were approximately 0.978. As for the flash height, the error for the experimental value of the specimen-B and A are 3.9% and 3.4%, respectively. These analysis results almost corresponded with those of the real sintered cold-forged specimens. Therefore, the analysis model suggested in this study would be applicable to simulate the densification of the sintered iron products.

4. Conclusions

To examine behaviours of the densification of Mo-alloyed sintered steel in the cold-forging process, the plastic deformation properties of the sintered steel were investigated by 3D FEM analysis. The suggested 3D analysis method enables to express the properties of the densification and the plastic deformation of the sintered cold-forged specimen. At first, the true stress-true strain diagram was obtained from the compression test. In the next, to determine the modified true stress-true strain diagram for 3D FEM analysis, the analysis of polynomial approximation was performed by use of analytical and experimental load-stroke diagrams. Using the modified true stress-true strain diagram, the analysis of backward extrusion was carried out. As a result, it was confirmed that the shape and the density of the sintered steel analyzed by 3D FEM correspond well with the actual values. Thus, validity of the analysis model for the sintered cold-forged steel was indicated. The analysis model suggested in this study would be applicable to simulate the local densification for the sintered steel parts.

Acknowledgements

The authors express their sincere gratitude to Y Morokuma (Gunma University), R Abe (Fuji Sintering Alloy Co., Ltd.), and T Suganuma (Chubureikan Co., Ltd.).

References

[1] K S Narasimhan 2007 Technology advances for the growth of powder metal in automotive applications *Japan Society of Powder and Powder Metallurgy* **54** 499-505

[2] O Yukiko, O Tomoshige, T Tsuguyuki, U Satoshi 2011 Effects of Apparent Density on Compacting Behavior of Atomized Iron Powder, *Japan Society of Powder and Powder Metallurgy* **58** 83-90

[3] M Kondoh 2007 Development of High Accuracy, High Density Powder Compaction Technology, *Japan Society of Powder and Powder Metallurgy* **54** 506-512

[4] Kandavel T K, Chandramouli R, Shanmugasundaram D 2009 Experimental study of the plastic deformation and densification behaviour of some sintered low alloy P/M steels, *Materials & Design* **30** 1768-1776

[5] N Nakamura, S Uenosono, M Fujinaga, S Koizumi, H Anma, T Yoshimura 2005 Sintering and Cold-Forging Process for High Density Sintered Materials, *JFE GIHO* **7** 19-23

[6] S Unami, T Goushi, K Kudo, F Tsumori, H Kang and H Miura 2011 Implovement of Surface Densification and Fatigue Strength by Tooth Root Rolling of Sintered Ferrous Alloy Gear Added Mo, *J.Jpn.Soc.Powder and Powder Metallurgy* **58** 692-696

[7] K Takada, R Kogure, T Kobayashi 2007 Research of High Fatigue Strength for Powder-Forged Connecting Rods, *Honda R&D Technical Review* **19** 129-137

[8] S Yi, L Yuan-yuan, Z Zhao-yao, Z Zhen-xing, C Pu-quing 2010 Improved model and 3D simulation of densification process for iron powder, *Transactions of Nonferrous Metals Society of China* **20** 1470-1475

[9] T Nagata, T Takemasu, T Koide and S Nishida 2016 Bending Durability of Ni-Mo pre-alloyed Sintered Steel Case-Carburizing Gears with Different Densities and Tooth Root Bending Stress Analysis Using FEM Model Considering Voids, *Japan Society of Powder and Powder Metallurgy* **63** 17-22

IOP Conf. Series: Materials Science and Engineering **257** (2017) 012013 doi:10.1088/1757-899X/257/1/012013

Comparison of three approaches to determine the projected area in contact from finite element Berkovich nanoindentation simulations in tungsten

T Volz[1,3], R Schwaiger[3], J Wang[3] and S M Weygand[2]

[1] Institute of Applied Research, Karlsruhe University of Applied Sciences, Moltkestraße 30,
76133 Karlsruhe, Germany
[2] Faculty of Mechanical Engineering and Mechatronics, Karlsruhe University of Applied Sciences, Moltkestraße 30, 76133 Karlsruhe, Germany
[3] Institute for Applied Materials (IAM), Karlsruhe Institute of Technology (KIT), Hermann-von-Helmholtz-Platz 1, 76344 Eggenstein-Leopoldshafen, Germany

*Corresponding author: tillmann.volz@hs-karlsruhe.de

Abstract. Nanoindentation nowadays is a standard method for the mechanical characterization of thin films and small volumes of material. One of the most meaningful parameters determined in nanoindentation experiments and simulations is the hardness of the tested material. For its determination, the knowledge of the exact value of the projected area in contact between the indenter and the specimen is essential. Inaccurate results for the projected area will result in noticeable errors in hardness. The determination of this area in finite element (FE) nanoindentation simulations is challenging because it cannot be determined directly and phenomena like pile-up and geometric imperfections of the indenter have to be considered. Hence, a new method, namely the triangulation method has been developed. It provides a reliable way to determine the projected area in FE-simulations, even under the occurrence of material pile-up. It is based on the nodes in contact between the indenter and the specimen as well as on the coordinates of the nodes. With this information, a Delaunay triangulation and Alpha shapes can be used to calculate the projected area. The triangulation method is compared to two established methods, one following the Oliver-Pharr analysis and the other one based on the computed true area in contact between specimen and indenter. The three methods are applied to results from an elastic-plastic FE simulation. Bland-Altman plots are used to compare the results of the three methods and to validate the triangulation method.

1. Introduction

Tungsten is becoming more and more important as a structural material e.g. in nuclear fusion reactors [1]. However, its brittleness and its brittle-to-ductile transition far above room temperature represent major difficulties to an application, even more so as these properties are strongly influenced by the microstructure of the material [2]. One of the most popular parameters to characterize a material is the hardness determined by nanoindentation. Nowadays nanoindentation is a standard method to investigate the mechanical properties of thin films and small volumes of materials [3-5]. The basic idea of nanoindentation is to measure an elastic-plastic load-displacement curve of a hard spherical, conical or pyramidal diamond indenter with a known geometry penetrating into a softer specimen.

Although this idea is simple in theory, the accurate recording of material-relevant data at sub-micron range, e.g. the load-displacement curve, requires very thorough preparation of the specimen and little influence from the experimental environment. Due to great progress in instrumentation development over the last few decades, nanoindentation is a reliable method to determine a variety of mechanical parameters. These include elastic modulus, hardness, strain hardening exponent, viscoelastic properties and fracture toughness [4], all of them determined from the load-displacement curves that are the "fingerprints" of the tested materials. One meaningful parameter for material scientists and engineers determined in nanoindentation experiments is the hardness H defined by the following equation:

$$H = \frac{P}{A_\mathrm{p}}, \tag{1}$$

where P is the load on the indenter and A_p the projected area in contact. The applied load can be recorded accurately by the nanoindenter, whereas, A_p is the critical parameter for an exact calculation of the hardness. It has to be noted that the given definition of hardness differs from the macroscopic hardness which is defined as the applied force divided by the area of the residual impression, measured optically directly after unloading [4]. Furthermore, the projected area cannot be measured directly like the load P and indentation depth h but has to be calculated indirectly. The direct measurement of the imprints is technically difficult at the length scale relevant to nanoindentation experiments. In case of a perfectly sharp Berkovich indenter, the projected area A_p and the plastic indentation depth h_c are related to each other purely geometrically via the equation:

$$A_\mathrm{P} = 24.5\, h_\mathrm{c}^{\,2}. \tag{2}$$

Following the Oliver-Pharr analysis [5] the plastic indentation depth h_c can be obtained by the equation:

$$h_\mathrm{c} = h - \varepsilon \frac{P}{S}. \tag{3}$$

Here, S refers to the contact stiffness and ε is a geometry parameter that is 0.75 in case of a Berkovich indenter. To determine the contact stiffness in nanoindentation experiments, continuous stiffness measurement [6] is frequently used. For this dynamic measurement method, a load oscillation with a small amplitude is superimposed on the static load to determine the contact stiffness continuously over the loading portion of the indentation experiment. Alternatively, the contact stiffness can be calculated by determining the slope of an unloading curve after the loading sequence following the Oliver-Pharr analysis [5]:

$$S = \frac{\mathrm{d}P}{\mathrm{d}h}. \tag{4}$$

Beside nanoindentation experiments, numerical simulations of nanoindentation are used to investigate the mechanical properties and deformation behavior of small scale specimens. Simulations of nanoindentation processes can support experiments and help to understand the deformation behavior and stress distributions. FE simulations along with advanced plasticity models like crystal plasticity or strain gradient plasticity have turned out to be a convenient approach for a detailed description of nanoindentation processes. A variety of methods for the determination of A_p in experiments were developed [5, 7- 10]. In nanoindentation FE simulations, additional information regarding the contact are available and therefore simulations offer the possibility to determine the projected area in contact directly for various geometries of indenters.

In this paper, we compare three approaches to determine the projected area in contact in FE nanoindentation simulations. The first presented method follows the Oliver-Pharr analysis and is based on the simulation of multiple unloading and reloading sequences that allow the determination of the contact stiffness at various indentation depths and thus, the calculation of h_c and A_p. The second method uses the true area in contact which is calculated in the FE software directly together with the geometry of the Berkovich indenter to convert the true area into the projected area in contact. The third method, called triangulation method, has been newly developed in our research and is based on

IOP Conf. Series: Materials Science and Engineering **257** (2017) 012013 doi:10.1088/1757-899X/257/1/012013

the contact status of the surface nodes. This method is universally, independent of the occurrence of pile-up (see figure 5) and the shape of the indenter.

2. Finite Element (FE) Modeling

A FE model for indentation was set up to compare the thee proposed methods for the determination of A_p. The FE simulation of the Berkovich indentation was performed with the commercial finite element software Abaqus. The presented three-dimensional model consists of two parts, namely the three-sided Berkovich indenter and the cylindrical specimen with a diameter of 80 μm and a height of 35 μm. The chosen dimensions of the specimen are much larger than the indentation depth to make sure that the stress field does not reach the specimens boundaries. The geometry of the indenter shown in figure 1 is defined by the two half angles of 77.05° and 65.3° which are the values for a Berkovich indenter. As the indenter's deformation is negligible compared to the deformation of the specimen, it is modelled as a non-deformable rigid surface. Like in experiments, the modelled indenter is not perfectly sharp but exhibits a tip radius of 150 nm. The indenter geometry is discretized with 720 rigid four-node elements (R3D4) and 25 rigid three-node elements (R3D3). The specimen shown in figure 1 consists of 33.240 cubic eight-node elements (C3D8R) with reduced integration. The mesh is refined underneath the indenter tip and in the regions where plasticity is expected. Different element sizes were implemented and tested to ensure the convergence of the load-displacement curve. The dimensions of the smallest elements under the tip in the used mesh are about 0.075 μm × 0.075 μm × 0.32 μm. To keep computational time reasonable, the mesh is coarser in regions further away from the contact region. The bottom of the specimen is fixed in all directions while the lateral surface is not restricted. The rotation and lateral movement of the indenter is fixed. Liu et al. [11] have shown, that load and displacement are independent of the friction coefficient. Therefore, the contact between indenter and specimen is treated frictionless.

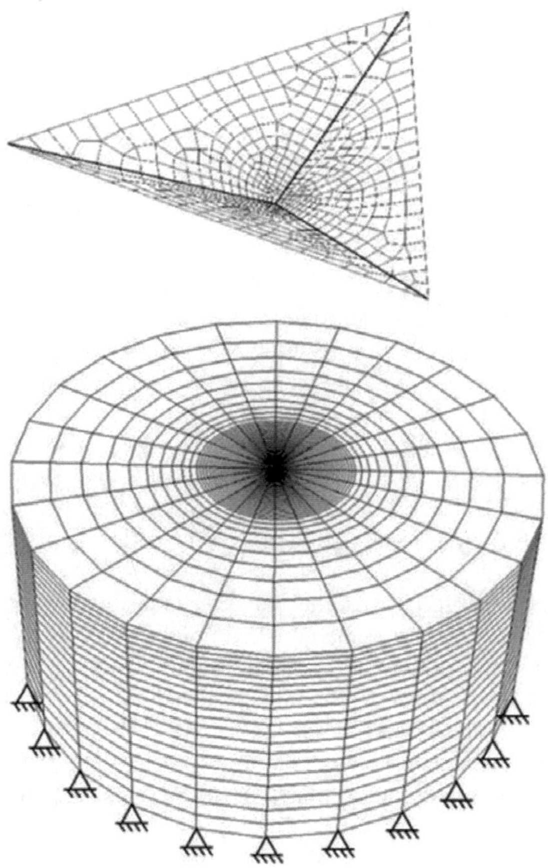

Figure 1. Illustration of the fe-model for the indentation simulation. The rigid Berkovich indenter exhibits the tip radius $r = 150$ nm.

The mesh of the deformable specimen is refined in the indented region and the bottom is fixed in all directions. The material behavior of tungsten is modelled as elastic-plastic with an elastic modulus $E = 408$ GPa, Poisson's ratio $v = 0.28$ and an initial yield strength of 320 MPa using J_2-plasticity with isotropic hardening. The implemented hardening behavior is given in figure 2. This stress-strain curve is based on the experimental work on tungsten single crystals of Argon and Maloof [12]. The described model is applied to simulate displacement-controlled indentation. Based on the results of the simulation, the three proposed methods for the determination of the projected area can be performed.

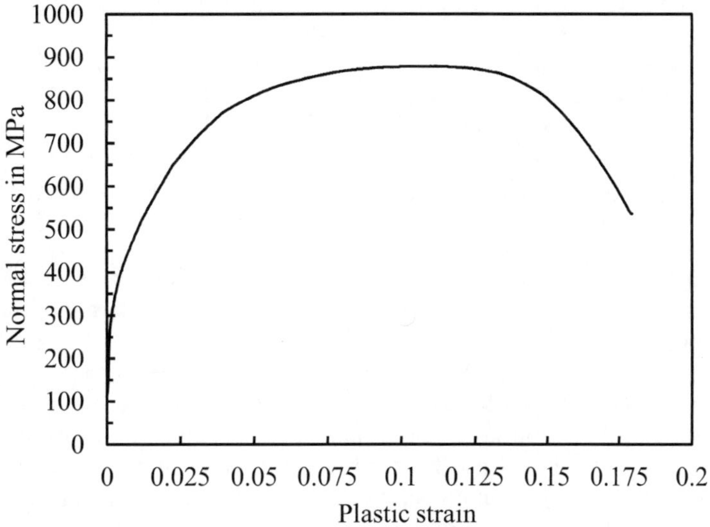

Figure 2. Stress-strain curve implemented in the plastic material model for tungsten.

3. Proposed methods to determine the projected area A_p

In the following section, three methods are presented to determine the projected area of the intender in contact with the specimen in the performed nanoindentation simulations. The aim is to compare these methods among themselves. Thereby the newly developed triangulation method should be validated by comparing its results regarding the projected area in contact A_p and the plastic indentation depth h_c with the established methods based on the Oliver-Pharr analysis and the true area in contact A_C that is determined in Abaqus directly.

3.1. Determination of A_p following the Oliver-Pharr analysis

This method follows the Oliver-Pharr analysis (see equation (3) and (4)) and relies on the assumption that A_p can be calculated from the plastic indentation depth h_c via the purely geometrical relation given in equation (2) for a Berkovich indenter. However, deviations are most likely as the modelled indenter exhibits a tip radius of 150 nm which is not taken into account in equation (2).

In order to obtain the contact stiffness S at different depths according to equation (4), additional unloading sequences were simulated evenly distributed over total depth. Figure 3 shows the computed load-displacement curve with ten unloading and reloading sequences. Based on this load-displacement curve, the slopes of the unloading curves $\mathrm{d}P/\mathrm{d}h$ are determined via a linear fit to the first four data points of every unloading sequence. Furthermore, the simulations give the corresponding load on the indenter at maximum indentation depth before the unloading sequence. These values together with the geometry parameter ε of 0.75 for a Berkovich indenter are applied in equation (3) to calculate h_c and via equation (2) the corresponding A_p can be calculated. The resulting values are given in table 1.

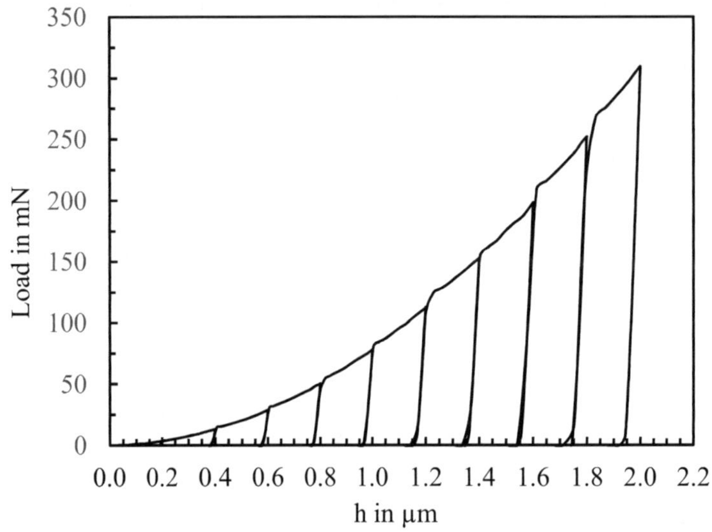

Figure 3. Computed load-displacement curve with ten unloading and reloading sequences during indentation to determine the contact stiffness *S* from the unloading slope.

3.2. Determination of A_p *through the true area in contact* A_c (A_p-A_c method)

The second method is based on the true area in contact A_c. In contrast to the experimens this value is accessible in the FE simulations and can be determined directly in every increment, i.e. for every indentation depth *h*. The geometric relation between A_c and A_p is used in this method. In the case of a perfectly sharp Berkovich indenter, A_c equals the lateral surface of the three sided pyramide shown in figure 4 and is related to h_c by:

$$A_c = 27.5\, h_c^{\,2}. \tag{5}$$

The projected area in contact A_p equals the base surface of the pyramid shown in figure 4 and is calculated following equation (2). Eliminating h_c in equation (2) and (5) leads to the following geometric relation between A_p and A_c:

$$A_p = 0.89\, A_c. \tag{6}$$

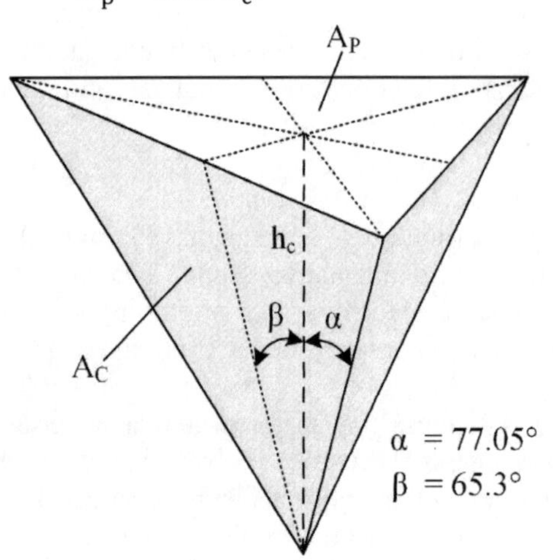

Figure 4. Geometry of the Berkovich indenter.

Applying this procedure to the nanoindentation simulation gives the values for h_c and A_p that are listed in table 1.

3.3. Triangulation method

As mentioned above, the presented two methods require a relation between A_p and h_c or A_c. The given geometric relations (equation (2) and (5)) are only accurate for a Berkovich indenter with a perfectly sharp tip. Furthermore, in soft materials that show high strain-hardening rates, the indentation causes far-field plasticity so that the specimen exhibits plastic deformation further away from the contact impression. This phenomenon is called sink-in and leads to a smaller projected area in contact compared to the original triangle area (see figure 5). Whereas materials with low strain-hardening rates and a large ratio for E_{eff}/σ_y exhibit locally deformation that results in piled up material at the three faces of the indentation (see figure 5). This leads to an underestimation of the real plastic indentation depth h_c and causes a projected area in contact A_c that is larger than the original triangle area and as a result, the hardness calculated from equation (1) is overestimated [7, 13]. Pile-up is not taken into account in the Oliver-Pharr analysis because the plastic indentation depth cannot be larger than the maximum indentation depth following equation (3). However, in the simulation of nanoindentation in tungsten, there is almost no pile-up occurring but one has to be aware of this downside applying the Oliver-Pharr analysis to materials with low strain-hardening rates.

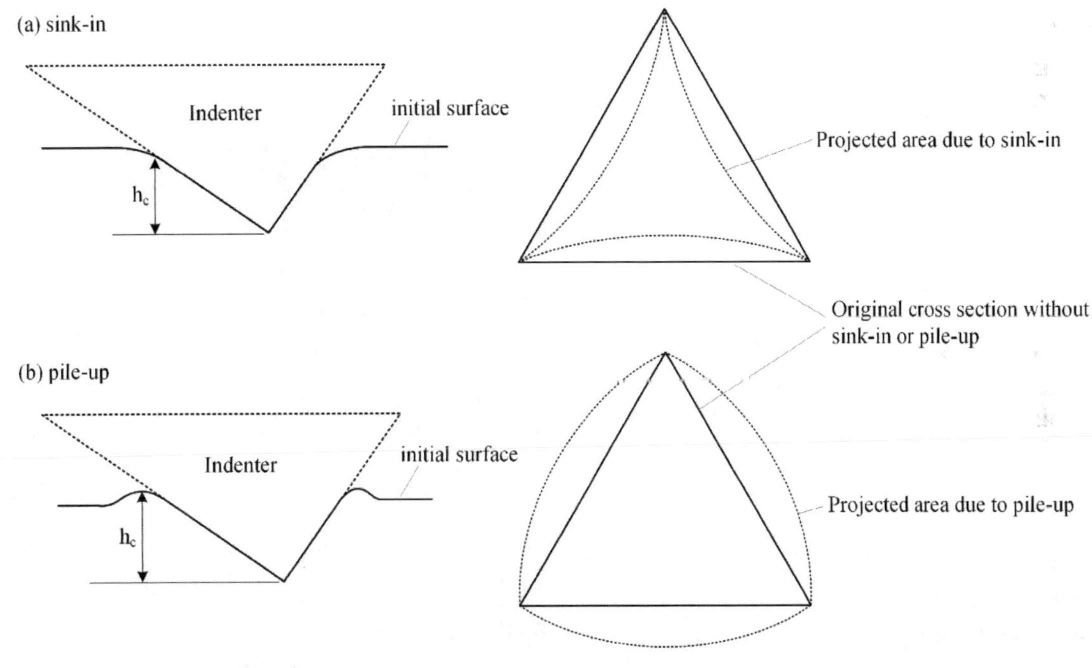

Figure 5. Illustration of sink-in (a) and pile-up (b).

In contrast to experiments, the projected area is directly accessible in the nanoindentation simulations as the positions of all nodes at the surface are known and updated for each indentation depth. A so-called triangulation method is presented which makes use of this information. This method offers a procedure for the determination of the projected area in contact A_p which does not require any geometric relation. It is not restricted to Berkovich indenters, and sink-in and pile-up effects are automatically taken into account. The idea of the newly developed method is based on the contact information of single nodes that are determined in the simulation in every increment. In post-processing, the nodes in contact can be determined and the internal node labels in Abaqus are stored in every increment, i.e. at every indentation depth. With the known node label of all nodes in contact, the node coordinates can be obtained and stored as well. For the determination of the projected area, the node coordinate in the direction of indentation is irrelevant since the projected area is desired. However, the curve that defines the projected area is not a planar curve but a three-dimensional closed curve that is defined by the connection of all the outer nodes in contact in the simulation. Furthermore, there is not just one plastic indentation depth h_c but a variation along the edge of the projected area [14]. This effect occurs because of the shape of the indenter, sink-in and pile-up and is observed in both, namely simulations as well as in experiments. The projected area of contact between the indenter and the specimen is simply the projection of this three-dimensional curve on a plane normal to the direction of indentation. Figure 6 shows exemplarily the plot of all nodes in contact at an indentation depth of 1400 nm and the projected three-dimensional curve based on the connection of the outer nodes. For the calculation of the desired projected area, a convex hull that includes all nodes in contact could be applied. However, if sink-in occurs the shape of the projected area becomes a concave hull. A more general way is the application of alpha shapes based on the Delaunay triangulation, where the alpha parameter has to be adjusted until the desired shape is found. The result of the triangulation method for the indentation depth of 1400 nm is shown in figure 7. The grey highlighted area as the sum over all triangles represents the desired projected area in contact and can easily be calculated. In summary, the effort for this method is reasonable. It is the most versatile method to determine the projected area in contact since it is independent of the indenter shape and sink-in and pile-up. The method also works for all other standard indenter shapes, independent of blunting and geometric imperfections.

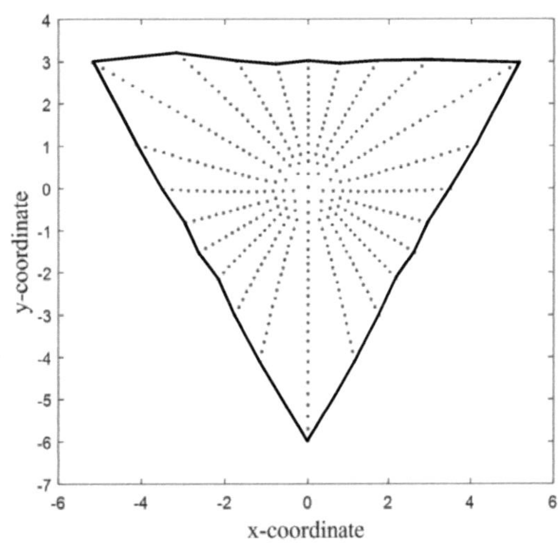

Figure 6. Plot of all nodes in contact at an indentation depth of 1400 nm.

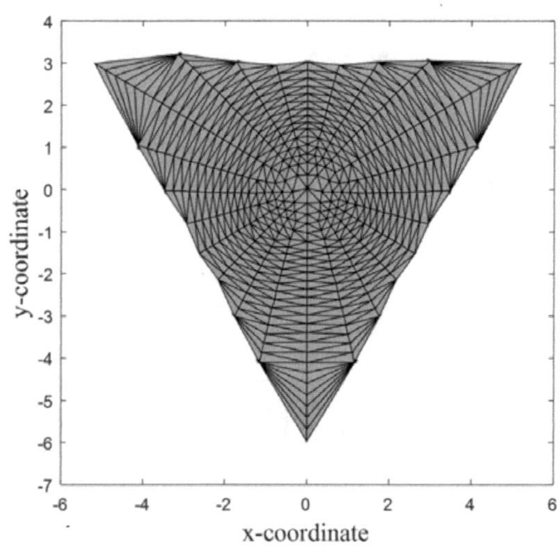

Figure 7. Triangulation to determine A_p based on the nodes in contact.

4. Results and discussion

In the following, the results for the plastic indentation depth h_c and the projected area in contact A_p that are determined in the three presented methods are compared. Table 1 lists the results for h_c and A_p for indentation depths ranging 0.2 μm to 2 μm. The same results are plotted in figure 8 and figure 9 as a function of the indentation depth h. In the Oliver-Pharr analysis, the plotted h_c is calculated with equation (3) and in the A_p-A_c method, h_c is determined with equation (5). In both cases, just one value for the plastic indentation depth is obtained. In the triangulation method, each outer surface node has its own h_c. For a comparison, all these plastic indentation depths determined in the triangulation method are taken to calculate a mean value that is comparable to the two other methods. The values of h_c in the A_p-A_c method are larger than the values of the overall indentation depth h. This is reasonable because a perfectly sharp tip would lead to a smaller true area at the same indentation depth than the true indenter shape with a radius at the tip. Vice versa, the plastic indentation depth in the A_p-A_c method is overestimated since it is determined with the geometric description of the perfect indenter geometry. The occurrence of material pile-up can be another possible explanation for high values of

h_c. However, no material pile-up is observed in the simulation. Moreover, this is confirmed by the results for h_c in the Oliver-Pharr method and the triangulation method. However, it has to be noted that the occurrence of material pile-ups leads to an underestimation of h_c (and of A_p) in the Oliver-Pharr analysis while the triangulation method can handle this properly.

Table 1. Comparison of the results for h_c and A_c determined in the three methods.

Indentation depth in µm	0.2	0.4	0.6	0.8	1.0	1.2	1.4	1.6	1.8	2.0
Oliver-Pharr analysis										
dP/dh in N/mm	560	1186	1766	2358	2965	3722	4172	4973	5616	6262
h_c in µm	0.17	0.39	0.59	0.79	0.98	1.18	1.37	1.57	1.77	1.96
A_p in µm^2	0.95	3.81	8.58	15.26	23.86	34.41	46.78	61.21	77.45	95.66
A_p-A_c method										
h_c in µm	0.22	0.43	0.64	0.84	1.04	1.26	1.42	1.65	1.85	2.02
A_p in µm^2	1.20	4.54	10.23	17.37	27.04	39.75	49.91	67.78	84.60	101.58
Triangulation method										
h_c in µm	0.19	0.39	0.57	0.79	0.98	1.18	1.38	1.60	1.75	1.96
A_p in µm^2	0.99	4.41	9.44	16.18	25.35	34.74	47.52	64.40	80.95	97.36

Figure 9 shows the projected area in contact for all methods. Since no pile-ups occur, only the tip radius could affect the area. We assume that the influence of the radius is negligible and state that all three methods should determine the same values for A_p in the elastic-plastic nanoindentation simulation, the results should be exactly the same in all three methods. This hypothesis is checked with the help of Bland-Altman plots given in the appendix. They prove that the triangulation method is a versatile method for determining the projected area in contact.

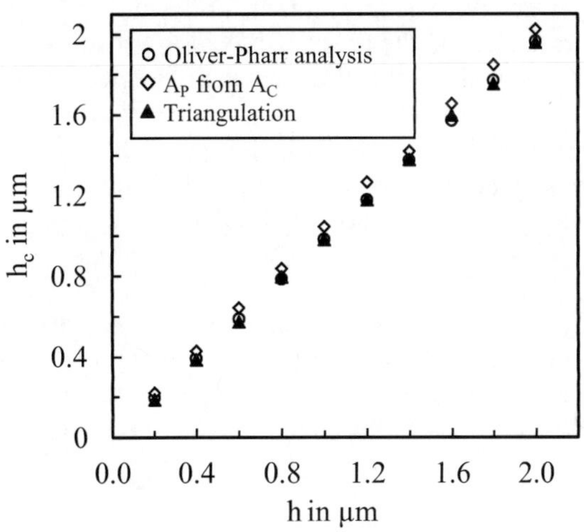

Figure 8. Comparison of the plastic indentation depth determined by the three methods.

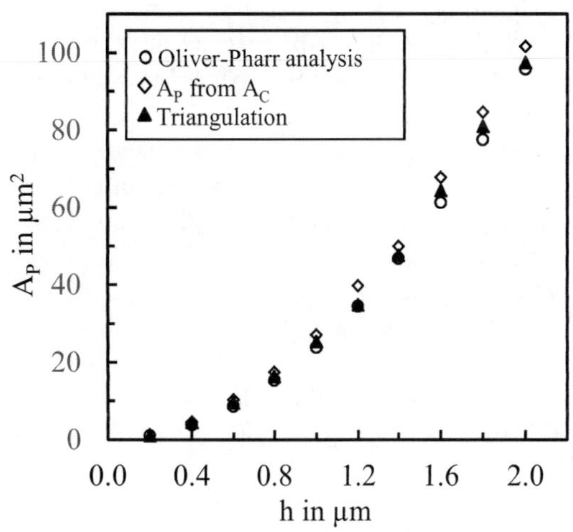

Figure 9. Comparison of the projected area in contact determined by the three methods.

The values of A_p obtained with the triangulation method are used together with the calculated load-displacement curve shown in figure 10 to calculate the nanoindentation hardness according to equation (1). The resulting hardness is given in figure 11. As expected, the hardness is constant over the indentation depth. Nanoindentation experiments in tungsten single crystals indicate an increase of the hardness with decreasing indentation depth at small scale, the so-called indentation size effect (ISE). Nix and Gao established a relation between hardness and indentation depth [15]. As the implemented elastic-plastic material model is a standard local version, it cannot describe size dependent behavior. An approach for the simulation of the ISE is the implementation of strain gradient plasticity (SGP) into the simulation to describe the strain gradients underneath the tip of the indenter.

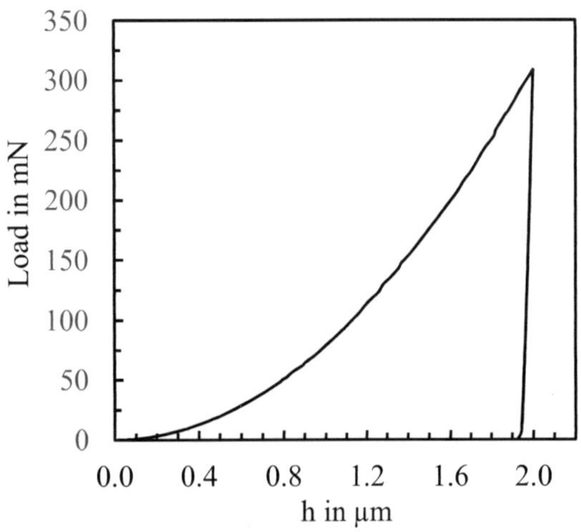

Figure 10. Computed load-displacement curve.

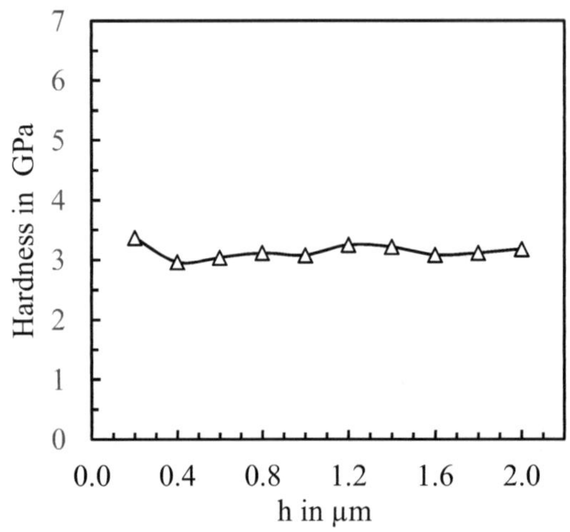

Figure 11. Hardness calculated based on the triangulation method.

5. Conclusions

In this paper three methods are presented to determine the projected area in nanoindentation simulation, namely the Oliver-Pharr method, the A_p-A_c method, and the triangulation method. A Bland-Altman plot shows the agreement of the newly developed triangulation method and the Oliver-Pharr and the A_p-A_c method. However, the Oliver-Pharr method in the presented form is not suited for materials where pile-up is expected to occur. The A_p-A_c method has the disadvantage of neglecting the actual radius of the indenter. It can be concluded that the triangulation method is validated and is the best choice as it takes into account the geometry of the indenter and can handle the appearance of pile-ups. Moreover, it is expected to be suitable for other indenter tip shapes as well. Therefore, it is a versatile method for determining the projected contact area in nanoindentation simulation.

Acknowledgement

The authors would like to thank the Karlsruhe University of Applied Sciences and the Karlsruhe Institute of Technology (KIT) for providing the necessary software and equipment. This research was financially supported by the fellowship of the "Kooperatives Promotionskolleg: Gefügestrukturanalyse und Prozessbewertung" funded by the Postgraduate Research Grants Program of Baden-Württemberg (Germany).

Appendix

A Bland-Altman plot [16] is a suited method to assess the agreement between the three methods. In the Bland-Altman plot, the average of the differences of A_p between two compared methods are plotted against the difference between the particular values of A_p in percent. Furthermore, the mean (correlates to the bias) and the standard deviation (SD) of the difference between two methods are determined. The mean of the differences in the comparison of the triangulation method and the Oliver-Pharr method is at about -5.4%. So, the values for A_p in the triangulation method are on average 5,4% larger than in the Oliver-Pharr method. The mean of the differences in the comparison between the triangulation method and the A_p-A_c method is at 7.6%, so A_p is on average 7.6% larger in the A_p-A_c method than it is in the triangulation method. The standard deviation of the differences in the comparison of the triangulation and the Oliver-Pharr method is about 3.9%, so about 95% of the determined values for A_p are likely to be in the range of the bias ± 7.8%. In the comparison of the triangulation and the A_p-A_c method, 95% of the values are likely to be in the range of the bias ± 9.6% as the standard deviation is about 4.8%. Beside the comparison of the new method and the established methods, a Bland-Altman plot for the two established methods is prepared. Here, the bias is 12,9% and the standard deviation is 5.1%. The fact, that the bias and the standard deviation in the comparison of the established methods is even larger than in the comparison of the triangulation method with the established methods indicates that the determination of the projected area is a challenging exercise. However, it also proves that the triangulation method is suited for the determination of the projected contact area. The Bland-Altman plots are shown in figure A1.

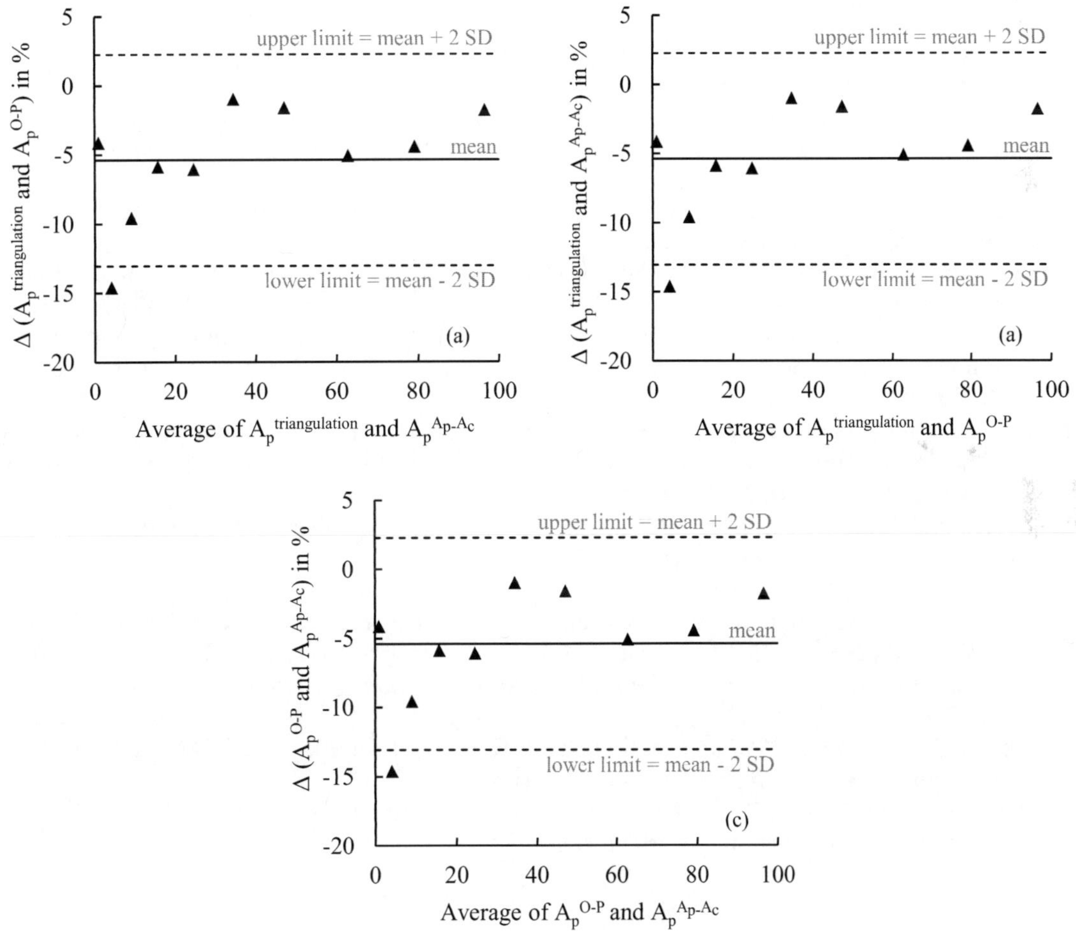

Figure A1. Plot of the difference between the triangulation method and Oliver-Pharr method (a), the triangulation method and the A_c-A_p method (b) and between the Oliver-Pharr method and A_c-A_p method (c) vs. the average of the determined values.

References

[1] Linsmeier C et al. 2017 Development of advanced high heat flux and plasma-facing material *Nuclear Fusion* **57**
[2] Bohnert C, Schmitt N J, Weygand S M, Kraft O and Schwaiger R 2016 Fracture toughness characterization of single-crystalline tungsten using notched micro-cantilever specimens *International Journal of Plasticity* **81** 1-17
[3] Pharr G M, Oliver W C and Brotzen F R 1992 On the generality of the relationship among contact stiffness, contact area, and elastic modulus during indentation *J. Mater.Res.* **7** 613-17
[4] Fischer-Cripps A C 2011 *Nanoindetation* vol 3 (New York: Springer)
[5] Oliver W C and Pharr G M 1992 An improved technique for determining hardness and elastic modulus using load and displacement sensing indentation experiments *J. Mater. Res.* **7** 1564-83
[6] Pethica J B and Oliver W C 1988 Mechanical Properties of Nanometre Volumes of Material: Use of the Elastic Response of Small Area Indentations *MRS Proceedings* **130** 13-23
[7] Oliver W C and Pharr G M 2003 Measurement of hardness and elastic modulus by instrumented indentation: advances in understanding and refinements to methodology *J. Mater.Res.* **19** 3-20
[8] Zhao Z X, Jiang Z D, Li X X and Gao Z K 2006 Calculation of indent's real contact projectionarea in nanoindentation *J. Phys.: Conf. Ser.* **48** 1121-26
[9] Guillonneau G, Kermouche G, Bergheau J-M and Loubet J-L 2015 A new method to determine the true projected contact area using nanoindentation testing *C.R. Mecanique* **343** 410-18
[10] Guillonneau G, Kermouche G, Bec S and Loubet J-L 2012 Determination of mechanical properties by nanoindentation independently of indentation depth measurement *J. Mater. Res.* **27** 2551-58
[11] Liu Y, Wang B, Yoshino M, Roy S, Lu H and Komanduri R 2005 Combined numerical and nanoindentation for determining mechanical properties of single crystal copper at mesoscale *J. Mech. Phys. Solids* **53** 2718-41
[12] Argon A S and Maloof S R 1966 Plastic deformation of tungsten single crystals at low temperatures *Acta Metallurgica* **14** 1449-62
[13] McElhaney K W, Vlassak J J and Nix W D 1997 Determination of indenter tip geometry and indentation contact area for depth-sensing indentation experiments *J. Mater. Res.* **13** 1300-06
[14] Lee W B and Chen Y P 2010 Simulation of micro-indentation hardness of FCC single crystals by mechanism-based strain gradient crystal plasticity *International Journal of Plasticity* **26** 1527-40
[15] Nix W D and Gao H 1997 Indentation size effects in crystalline materials: a law for strain gradient plasticity *J. Mech. Phys. Solids* **46** 411-25
[16] Bland J M and Altman D G 1986 Statistical methods for assessing agreement between two methods of clinical measurement *The Lancet* **327** 307-10

IOP Conf. Series: Materials Science and Engineering **257** (2017) 012014 doi:10.1088/1757-899X/257/1/012014

Formation mechanism of pillar-shaped intermetallic compounds dispersed lead-free solder joint

Y Nakata[1, 2], T Hashimoto[1], M Kurasawa[1], Y Hayashi[2], and I Shohji [2]

[1]Calsonic Kansei, Co., Ltd., Global Technology Division, Sano, Tochigi 327-0816, Japan
[2]Faculty of Science and Technology, Gunma University, Kiryu, Gunma 376-8515, Japan

*corresponding author: yusuke_nakata@ck-mail.com

Abstract. Large area joining between a substrate and a heat sink is desirable for high performance power modules. An intermetallic compounds (IMCs) pillar dispersed solder joint has been developed as a highly durable joint to achieve large area joining. The aim of this study is to clarify the generation and growth mechanism of the IMC pillar during soldering process. The structural characteristic of the IMC pillar was also examined by cross-sectional observation. The area ratio of the IMC pillars in the cross section of the joint increased with increasing the joining temperature. The shape of the IMC pillar became finer when the cooling rate was fast. In addition, the IMC pillar grew along the cooling direction.

1. Introduction

The demand for electric vehicles is expected to increase, and inverters are required to be had high power density for cost reduction and downsizing of them [1]. In a power module that dominates the cost and the size of the inverter, it is possible to further increase the current density by increasing per unit area (cooling performance) to further miniaturize the module [2]. Moreover, to improve heat radiation performance, it is effective to change a direct cooling structure joined with solder from a separate structure filled with heat dissipating grease between the power module and the heat sink [3]. The structure enables miniaturization, weight reduction, and cost reduction of the module. However, in the case of the direct cooling structure, large thermal distortion occurs at the joint portion between the insulating substrate with a small coefficient of thermal expansion (CTE) and the aluminum or copper plate with a large CTE. The authors have developed a pillar shaped intermetallic compounds (IMCs) dispersed lead-free solder joint that can suppress the crack generation in the joint under thermal fatigue conditions and improve durability of the joint. Figure 1 shows cross sectional views of as-soldered joint with Sn-3.0Ag-0.5Cu-5.0In and the joint after thermal cycle test. Cu dissolved from the substrate forms a pillar shaped intermetallic compound (hereinafter referred to as an IMC pillar) in the soldered joint and the IMC pillar can suppress crack growth. On the basis of the quantitative analysis with an electron probe X-ray microanalyzer (EPMA), it was found that the IMC pillar consists of Cu and Sn [4]. The aim of this study is to investigate the formation mechanism of the IMC pillar.

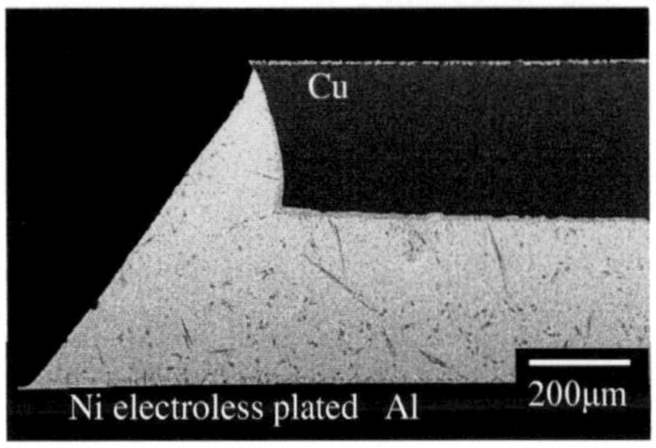

(a)

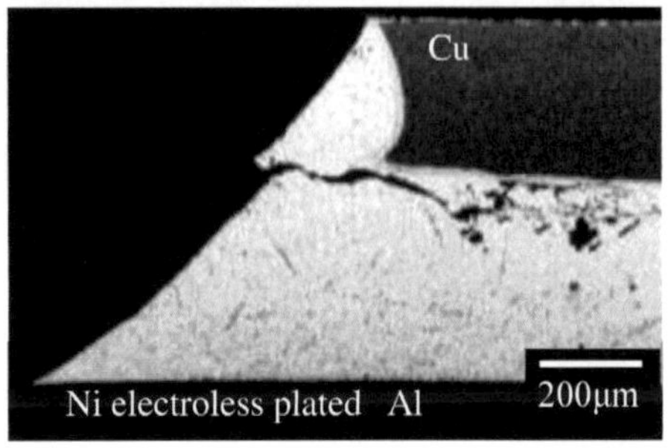

(b)

Figure 1. Cross sectional secondary electron images of pillar-shaped IMC dispersed lead-free solder joints.
(a) As-soldered, (b) Crack formed in solder layer after thermal cycle test (500 cycles, -40 - 120°C).

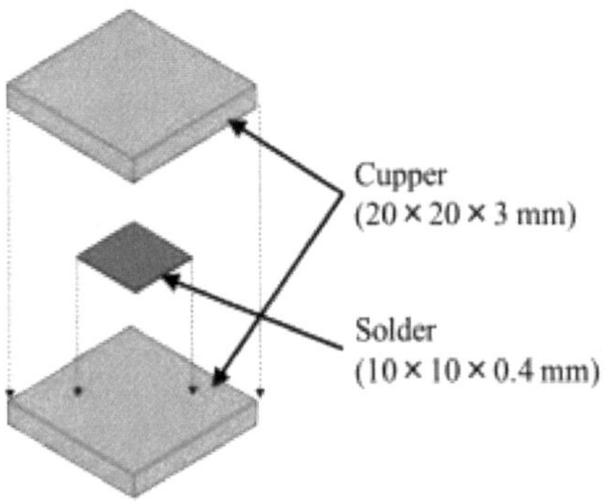

Figure 2. Shape and dimensions of specimen to investigate effect of joining conditions. conditions.

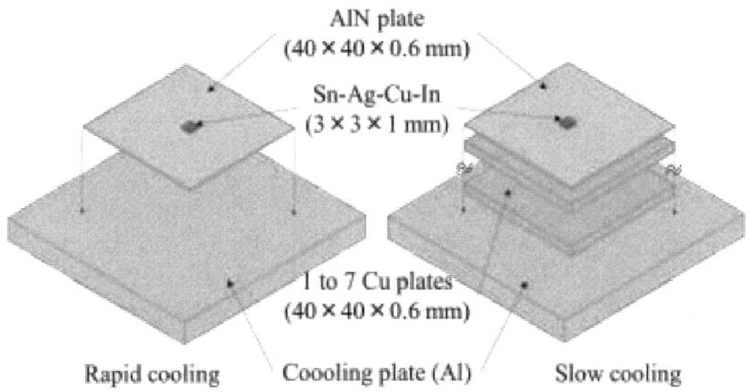

Figure 3. Schematic of specimen preparation to investigate effect of unidirectional cooling

2. Experimental Procedure

To elucidate the formation mechanism of the IMC pillar, the effects of joining conditions and cooling conditions in solidification on the shape of the IMC pillar were investigated.

2.1. Investigation of the effect of joining conditions on the shape of the IMC pillar

Figure 2 shows a specimen to investigate the effect of joining conditions on the shape of the IMC pillar. The specimen was prepared by sandwiching a Sn-3.0Ag-0.5Cu-5.0In (mass%) alloy with 10×10×0.4 mm size between two Cu plates with 20×20×3 mm size. Joining was conducted in a vacuum below 1 kPa. In this study, heating and cooling system are located under the specimen. Joining temperature and joining time were 270, 300 and 330 °C and 10, 20 and 30 min, respectively. The cooling rate in solidification was 5 °C/sec. After joining, each specimen was filled with an epoxy resin and the cross section of it was cut out with a cutter. The cross section was polished with emery papers from #320 to #1000, and buffing was carried out with diamond suspension of 15 to 1 μm. Thereafter, the cross-sectional observation was carried out using a scanning electron microscope (SEM). In SEM observation, the working distance and the acceleration voltage were 8 mm and 20 kV, respectively. Back-scattered electron (BSE) images were observed. For the obtained image of the microstructure of the joint, binarization processing was performed with an image processing software (WinROOF), and the area ratio of the IMC pillar in the joint was measured. In order to confirm the effect of the joining conditions on the dissolution amount of Cu, the abundance ratio of Cu in the cross section of the solder layer was semi-quantitatively analyzed using energy dispersive X-ray spectroscopy (EDS). In EDS analysis, the working distance and the take-off angle of X-ray were 10 mm and 35 degrees, respectively.

2.2. Investigation of the effect of cooling conditions in solidification on the shape of the IMC pillar

2.2.1. Investigation of the effect of cooling rate. In order to eliminate the effect of dissolution of Cu in the molten solder and confirm the effect of only the cooling rate in solidification, an alloy in which Cu was added in advance to the Sn-3.0Ag-0.5Cu-5.0In alloy was prepared. Figure 3 shows the schematic of the specimens. The Sn-3.0Ag-0.5Cu-5.0In alloy of 3×3×1 mm in size was completely melted at 500 °C that is higher than the melting point of Cu_6Sn_5 on a ceramic (AlN) plate and then cooled. The specimen with the Sn-3.0Ag-0.5Cu-5.0In alloy and the ceramic plate was heated to 500 °C and the ceramic plate with the molten alloy was placed on a cooling plate. The cooling rate was varied by stacking 1 to 7 Cu plates between the solder material and the cooling plate. In this study, the average cooling rate from 500 °C to 200 °C in the vicinity of the solidus temperature of solder was defined as cooling rate.

2.2.2. Investigation of the effect of cooling direction. As shown in figure 1, it was suggested that the IMC pillar is formed along the cooling direction. Therefore, the structure of the IMC pillar was also investigated when giving directionality to the direction of heat flow during solidification. The experiment was conducted in the same manner as shown in figure 3. The heating temperature was set to 500 °C, and for the cooling rate, a rapid cooling and a slow cooling were examined. In the rapid cooling, the solder material was melted on the ceramic plate and then cooled on a cooling plate. In the slow cooling, seven Cu plates were added between the ceramics plate and the cooling plate. The cooling of the specimen was conducted approximately uniformly. In the rapid cooling, a billet of Sn-3.0Ag-0.5Cu-5.0In alloy with 91×15.5×24 mm size was melted and then cooled in water, cooling rate was 50 °C/s. In the slow cooling, Sn-3.0Ag-0.5Cu-5.0In alloy of 3×3×1 mm in size was melted in a muffle furnace and then furnace cooled, cooling rate was 10 °C/s.

3. Results and Discussion

3.1. Effect of joining conditions on the shape of the IMC pillar

Figure 4 shows the measurement results of the area ratio of IMC pillars in the solder joints under several conditions. Figure 5 shows back-scattered electron images of the cross sections of the joints at 270, 300 and 330 °C for 30 min. In the case of the same joining temperature, the area ratio of IMC pillars does not change even if the joining time increases. On the other hand, the effect of the boning temperature on the area ratio of IMC pillars is remarkable. Figure 6 shows the semi-quantitative analysis result of Cu in the solder layer by EDS plotted on the Sn-Cu phase diagram with the liquidus line. The abundance ratio of Cu in the solder layer increases with increasing the liquidus temperature and changes along the liquidus line. When the joining temperature rises, the solubility limit of Cu dissolved in the molten solder also increases. Since the dissolved Cu crystallizes as Cu_6Sn_5 during cooling from the peak temperature of reflow soldering, the area ratio of IMC pillars increases with increasing the joining temperature.

IOP Conf. Series: Materials Science and Engineering **257** (2017) 012014 doi:10.1088/1757-899X/257/1/012014

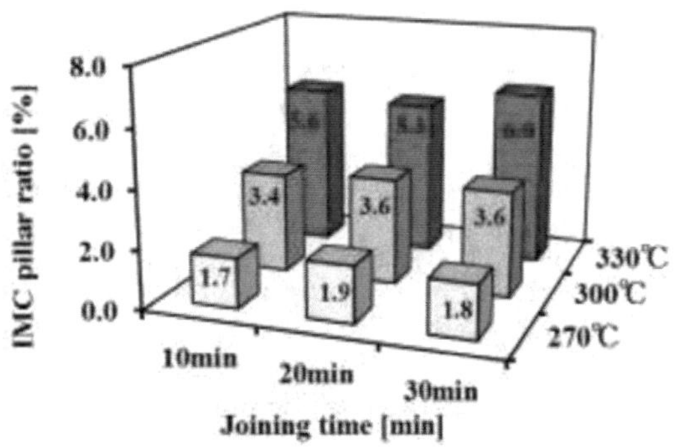

Figure 4. Effect of joining conditions on rate of IMC pillar.

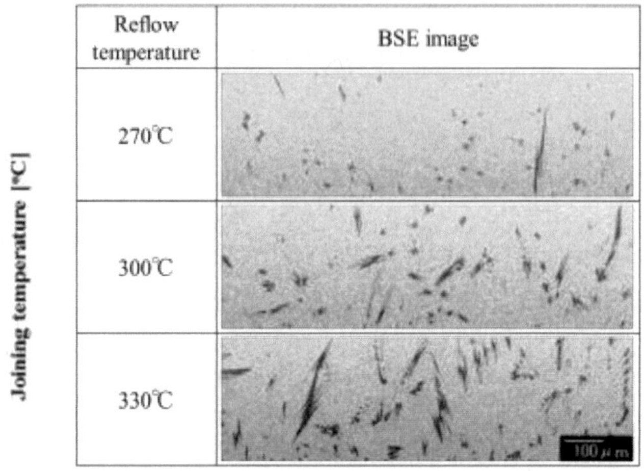

Figure 5. Back-scattered electron (BSE) images of sections of joints (joining time: 30 min).

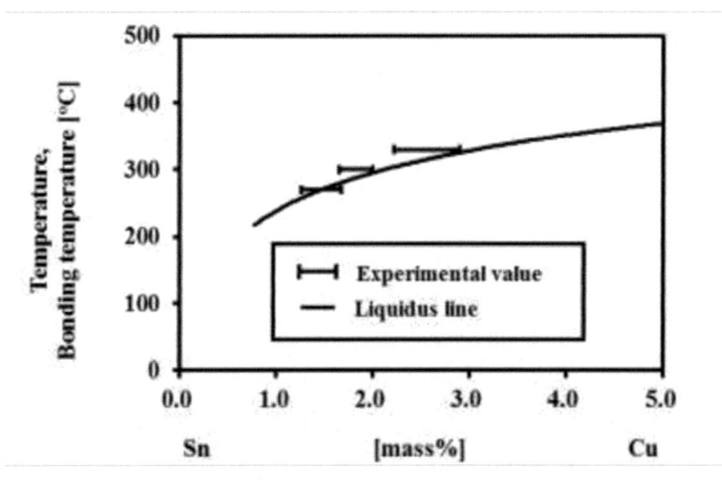

Figure 6. Cu dissolution rate in joint layer to solubility limit of Cu.

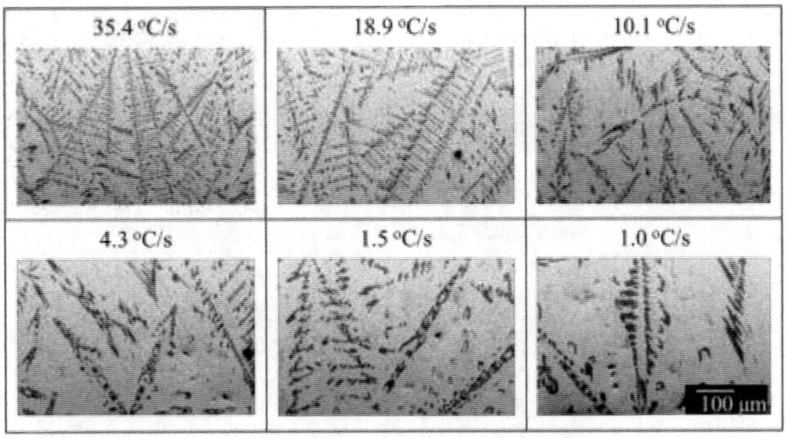

Figure 7. Effect of cooling rate on morphology of IMC (BSE images).

3.2. Effect of cooling conditions on the shape of the IMC pillar

3.2.1. Effect of cooling rate. Figure 7 shows morphology of IMCs in solder at each cooling rate. The IMC becomes fine when the cooling rate increases and it becomes coarse when the cooling rate is slow. The morphology of IMCs is dendritic. This means that the formation mechanism of the IMC pillar is the same as the crystallization mechanism of dendrite. When the cooling rate is high, extrusion of Cu from molten solder in the cooling process becomes fast so that the IMC grows fast. Since the cooling time from the peak temperature to the solidus temperature is short, the joint is cooled without sufficient growth of IMC pillars. Thus the IMC has a fine shape. On the contrary, IMC grows in a thick columnar shape when the cooling rate is slow [5]. In the case of the slow cooling rate, the dendritic growth of the IMC is slow and it coarsens.

3.2.2. Effect of cooling direction. Figure 8 shows BSE images of the cross sections of solder which were cooled under various cooling conditions. In the case of uniform cooling, IMCs grew randomly. On the other hand, directionality was seen in the growth direction of IMC in the case of unidirectional cooling. As shown in figure 3, the cooling plate was placed under the specimen and solder was cooled from below. In figure 8, the heat flow direction is vertical direction. Therefore, it was confirmed that most IMCs form along the cooling direction. This tendency was observed in both rapid and slow cooling conditions. Cu_6Sn_5 crystallizes and it grows toward the liquid phase because the phase transformation from the cooling side to the solid-liquid coexisting phase easily occurs when solder is cooled from one direction. In the specimen shown in figure 3, the phase transformation occurred from the portion below the liquidus line of solder to the solid-liquid coexisting phase by cooling molten solder from the bottom. Extrusion of Cu from molten solder in the cooling process can nucleate Cu_6Sn_5 IMC. Therefore, IMC easily forms along the cooling direction because IMC grows to a high-temperature side where the liquid phase amount is high [6].

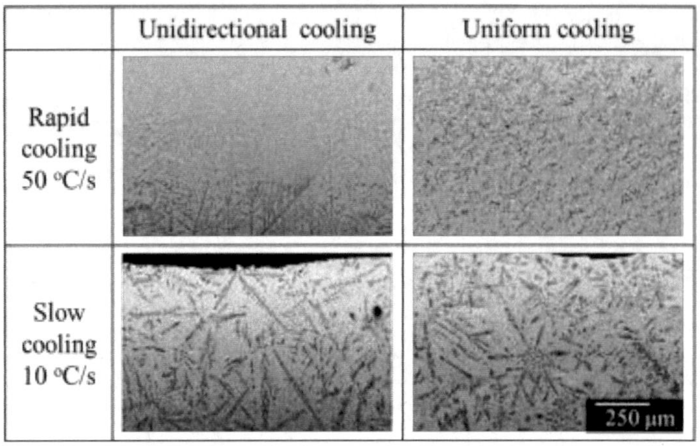

Figure 8. Effect of heat flow direction on morphology of IMC (BSE images).

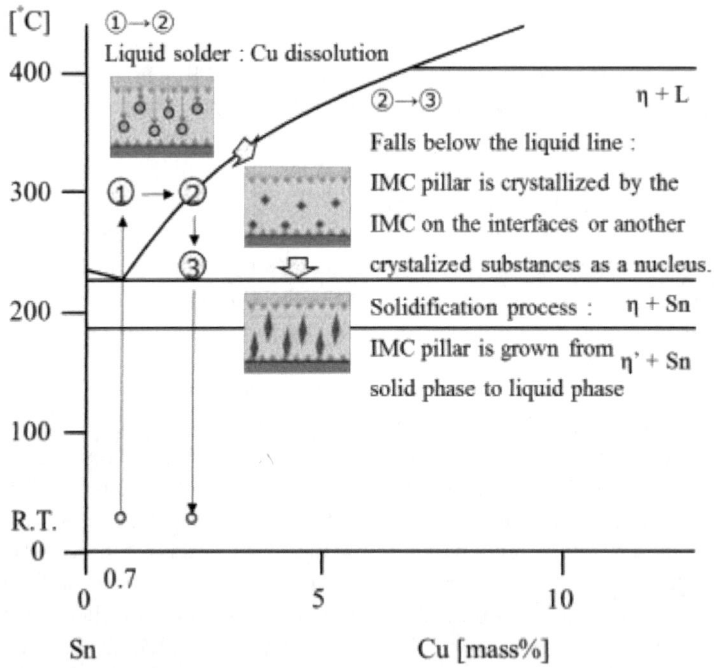

Figure 9. Formation mechanism of IMC pillar shown in Sn-Cu phase diagram.

3.3. IMC pillar formation mechanism

Figure 9 shows the formation mechanism of the IMC pillar with the Sn-Cu phase diagram. Formation of the IMC pillar is considered as follows.

① to ②: Cu dissolves in molten solder until the solubility limit of Cu in solder at the joining temperature [7, 8].

② to ③: When the temperature of solder reduces below the liquidus line, crystallization of Cu_6Sn_5 IMCs starts. In the nucleation stage, the joint interface, additive metals other than Cu and oxides originally present can be nucleation sites. IMCs grow during the cooling process from liquidus temperature to solidus temperature and thus IMC pillars form in solder. IMC pillars easily grow along the cooling direction [9, 10].

4. Conclusions

In this study, the formation mechanism of the IMC pillar which remarkably improves thermal fatigue life was investigated. The following results were obtained from the experimental results under various joining and cooling conditions.

(1) The area ratio of the IMC pillars increases with increasing the joining temperature. This is caused that the solid solution limit of Cu dissolved in molten solder increases with increasing the joining temperature.
(2) The shape of the IMC pillar becomes fine when the cooling rate is fast. The morphology of IMCs is dendritic. The formation mechanism of the IMC pillar seems to be the same as the crystallization mechanism of dendrite.
(3) The growth direction of the IMC pillar becomes from the low temperature side toward the high temperature side in the heat flow direction during cooling.

References

[1] C. C. Chan 2007 *Proceedings of the IEEE* 705

[2] T. Kimura, R. Saito, K. Kubo, K. Nakatsu, H. Ishikawa, K. Sasaki 2014 *Hitachi Review* 63 96

[3] H Arai, K Higuchi and T Koyama 2015 *Fuji Electric Review* 61 252

[4] Y. Hayashi, I. Shohji, Y. Nakata and T. Hashimoto 2015 *Electronics Packaging Technology Conference* 139

[5] Y. Natsume and H Ishiba 2010 *International Conference on Aluminium Alloys* 1746

[6] M. A. A. Mohd Salleh, S. D. McDonald, H. Yasuda, A. Sugiyama and K. Nogita 2014 *Scripta Materialia* 100 17

[7] C.E. Ho ,W.Z. Hsieh, C.S. Liu, C.H. Yang 2014 *Thin Solid Films* 238

[8] C.S. Liu, C.E. Ho, C.S. Peng, C.R. Kao 2011 *Journal of Electronic Materials* 1912

[9] K.L. Lin, Y.W. Lin, C.H. Yu, 2012 *The Journal of The Minerals, Metals & Materials Society* 1184

[10] M. A. A. M. Salleh, S. D. McDonald, C. M. Gourlay, S. A. Belyakov, H. Yasuda, K. Nogita 2016 Journal of Electronic Materials 154

IOP Conf. Series: Materials Science and Engineering **257** (2017) 012015 doi:10.1088/1757-899X/257/1/012015

Development open source microcontroller based temperature data logger

M H Abdullah[1,3], S A Che Ghani[1], Z Zaulkafilai[1,2], S N Tajuddin[3]

[1]Faculty of Mechanical Engineering, Universiti Malaysia Pahang, Pekan, Pahang, Malaysia.
[2]Faculty of Engineering, DRB Hicom University, Pekan, Pahang, Malaysia.
[3]Bioaromatic Centre, Universiti Malaysia Pahang, Gambang, Pahang, Malaysia.

*Corresponding Author: ahaslizam@gmail.com

Abstract. This article discusses the development stages in designing, prototyping, testing and deploying a portable open source microcontroller based temperature data logger for use in rough industrial environment. The 5V powered prototype of data logger is equipped with open source Arduino microcontroller for integrating multiple thermocouple sensors with their module, secure digital (SD) card storage, liquid crystal display (LCD), real time clock and electronic enclosure made of acrylic. The program for the function of the datalogger is programmed so that 8 readings from the thermocouples can be acquired within 3 s interval and displayed on the LCD simultaneously. The recorded temperature readings at four different points on both hydrodistillation show similar profile pattern and highest yield of extracted oil was achieved on hydrodistillation 2 at 0.004%. From the obtained results, this study achieved the objective of developing an inexpensive, portable and robust eight channels temperature measuring module with capabilities to monitor and store real time data.

1. Introduction

Agarwood is a resinous, fragrant and highly valuable heartwood. Healthy agarwood is relatively light and pale colored. Occasionally the heartwood gets contaminated by a parasitic ascomycetous mold, haeoacremonium parasitical. As a reaction, the tree produces a resin high in volatile organic compounds that aids in suppressing or retarding the fungal growth [1]. There are numerous grades of agarwood, and the most astounding quality wood is extremely expensive [2]. Indeed, the first grade agarwood is one of the costly natural product on the planet, with costs up to $13,000 per pound. However, the finest grade of agarwood is produce from naturally occurring fungal infection which happens gradually and rarely. There are also several technique to produce the resin such as inoculation of grown agarwood with the fungus as well as purposeful harming of the tree to urge fungal infection [3-5]. However, resin harvested in this way is considered to be optional quality and has inferior market value contrasted with naturally harvested agarwood resin.

Agarwood resin and its oil are valuable for their use in medicine, perfumery and other aromatic products [6]. They are volatile oils, generously odorous, which occur in certain plants or specified parts of plants, and are recovered by accepted procedures, such that the nature and composition of the product is, as nearly as practicable, unchanged by such procedures [7, 8]. It specifies clearly that the nature and composition of the oil must be unchanged by the process of extraction. Currently, there are a few conventional and modern methods of extracting essential oils such as by hydro-distillation, supercritical fluid extraction and microwave extraction. Hydro-distillation is the oldest and most common method of extracting essential oil since it is economically viable and safe. During hydro-distillation the essential oil components form an azeotropic mixture with water. The extraction period influences not only the yield but also the extract composition. Hydro-distillation can be achieved by one of the two methods which are Clevenger or steam distillation.

In both methods the vapours of the volatile components are carried by the steam to a condenser. On condensation oil-rich and water-rich layers are formed. Clevenger distillation generates steam continuously and on a large scale. Controlling the boiling process is extremely difficult, it is a highly nonlinear process, its dynamics vary with load and it is strongly multivariable. It is also inherently unstable due to the integrator effect of the drum. Boiler for distillation process is different to boiler for energy generation, where the boiler in energy generation is commonly operated at maximum capacity [9], whereas the boiler for essential oil distillation process must consider the critical temperature that may potentially degrade the thermolabile bioactive components [10]. However, both boiler systems have process parameters that are strongly dependent on load. The process parameters drift over time for a number of reasons such as the build-up of soot on heating surfaces, actuator wear and variations in raw materials quality. It may be necessary to continuously update the controller parameters [11, 12].

The yield, taste, flavour and chemical composition (amount and ratio of components) of essential oil depends on a number of parameters, such as plant variety, season, soil, environmental conditions, drying procedure, storage conditions, method of distillation, and the analytics used for identification of the compounds [13, 14]. Despite of the generally successful practical hydro-distillation technology used to extract essential oils, there is still a need to consider a procedure or method in detail that would enable the production of essential oils at an optimum output [15]. Up to now many investigators have studied the thermolability and thermostability of the fragrant components when the plant material is under distillation process [10, 16]. They conclude that optimum temperature should be determined to avoid losses of volatile and thermolabile substances during extraction of the essential oils.

Automatic data acquisition systems are currently used in wide variety of applications which include environmental monitoring, renewable energy, medical and many significant fields. With numerous benefit of automated measurements, data are collected continuously throughout the practice with minimal human intervention. Systems based on microcontrollers have been widely used for the monitoring of general physical parameters [17] and, in particular, for solar energy applications research and weather data acquisition [18-22]. Such systems enable the use of complex control strategies and can improve significantly the energy efficiency of solar thermal facilities [23-27]. Fuentes et al. [28], following a slightly different approach, have developed a data logger for a photovoltaic system in compliance with the standard of the International Electrotechnical Commission (IEC) IEC 61724:1998 [29].

The aims of this work is to develop stand alone, inexpensive, reliable, portable and easily-programmable control systems and monitoring, oriented to temperature data acquisition research in general and temperature management of agarwood extraction process in particular. From the acquired temperature variation data, recommended temperature variations have been outlined for the purpose of managing optimum mixture temperature of hydrodistillation and improving the cooling system effectiveness. In-situ temperature control of traditional hydrodistillation is highly needed to ensure the high yield and consistent quality of producing agarwood essential oil.

2. Materials and Methods

The proposed control system is based on the Arduino electronic platform, which is an open source electronic prototyping platform based on ATMEL microcontroller[30]. It offers several products such as boards, modules, shields and also Integrated Development Environment (IDE) software to program them. Because of its uncomplicated programming and economical features, this platform has already taken a significant role in recent data acquisition research application [31-35].

2.1. Data logger hardware

The presented control system of data logger has been designed and built to measure and record real time temperature readings as an improvement of previous simpler prototypes (Figure 1)[36], and it is based on the Arduino Mega 2560 model due to the enhanced features offered. This second version of the temperature data logger (Figure 2) is able to register and monitor up to eight channel temperature variable in real time format. Then it records the acquired data in Secure Digital (SD) card attachedi to the system, in order to retrieve the temperature variation data on the targeted application. The acrylic exterior case also has been fabricated using laser cutter machine to meet the robust and portable design criteria.

Figure 1. Previous prototype version of temperature data logger system

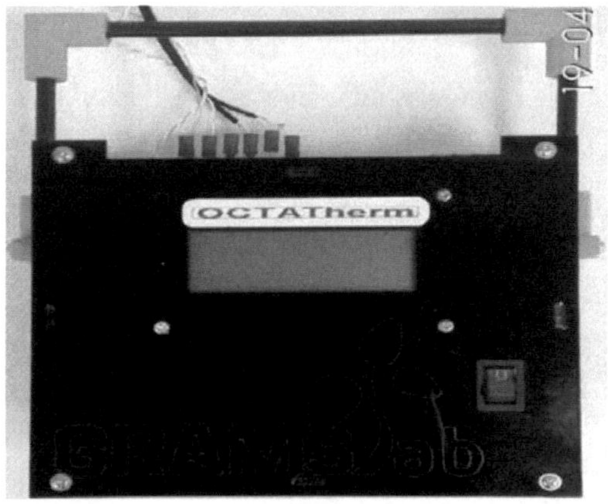

Figure 2. Latest version of temperature data logger system

The latest version of the temperature data logger system that has been develop for this study consists of the following parts.

2.1.1. Microcontroller board. An Arduino Mega 2560 board equipped with a powerful and inexpensive ATMega 2560 microcontroller [37] is used as the system mainboard. This board works with a 16 MHz clock and has 256 KB of flash memory, 54 digital Inputs and Outputs, of which 15 provide Pulse Width Modulation (PWM) [38], and 16 analog inputs. This board is the main core of the system managing all sensors, actuators and shields. It can be programmed, from the Arduino IDE, with the desired control algorithm from a PC by means of a USB port. All the other module boards must be connected to it, since it controls the whole system.

2.1.2. Sensor shield. The Arduino MEGA has many more Input/output (I/O) pins (54 Digital pins and 16 Analog pins) than a standard Arduino. This shield has connectors specifically for some external devices like an SD Card, Radio communications (Bluetooth and others). It also has an option to power the voltage (V) pins on all the Digital I/O 3-pin groups from an external power supply instead of the Arduino +5V. Using an external supply allows more current than the Arduino or Mega can supply, such as current for control of many sensors and other attached modules.

2.1.3. SD Card module. This module is in charge of saving the experimental data in a secure digital (SD) memory card to be analyzed later. The SD card module makes use of the Serial Peripheral Interface (SPI) bus [39] in order to communicate with the microcontroller board.

2.1.4. Real time clock (RTC) module. The RTC chip [40], employed in this module to put data time stamp in the data log file, it uses the Inter-Integrated Circuit bus (I2C) [41] for communication with the Arduino board. The RTC module is also equipped with a battery backup device to keep the time and date going even when the Arduino is disconnected from the power supply.

2.1.5. LCD module. An Adafruit Character LCD board with I2C module is used as a display interface. It makes use of a 20×4 Liquid Crystal Display (LCD) [42] which shows information about the monitoring temperatures. Since this shield uses the I2C bus to communicate with the Arduino board, it requires only a two digital pins connection.

2.1.6. Temperature sensor module. Maxim Integrated MAX6675 digital thermocouple amplifier [43] are used. The MAX6675 performs cold-junction compensation and digitizes the signal from a type-K thermocouple. The data is output in a 12-bit resolution, these amplifier carry out the analog to digital conversion directly and make use of the Serial Peripheral Interface (SPI™)-compatible [39], read-only format via digital pin of the Arduino board. This converter is able to measure temperatures values in range of 0°C to 1024°C with a maximum resolution of 0.25°C, and since their power supply range goes from 3.0 V to 5.5 V, they can be powered directly from the Arduino board data line. The selected k-type thermocouple sensor are most applicable to avoid moisture and water-related problems [34].

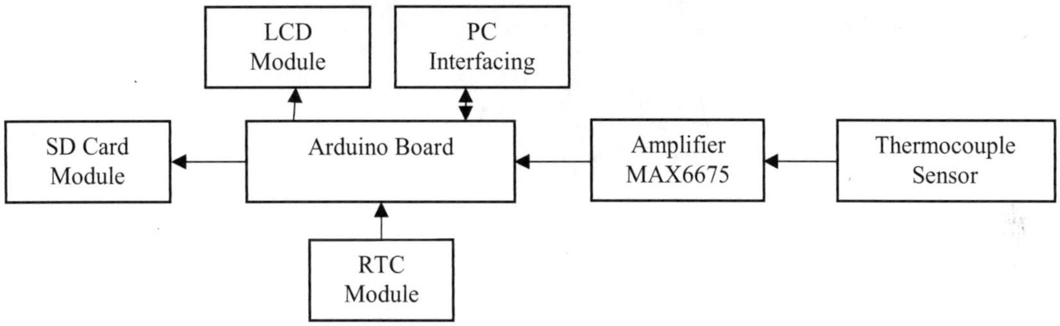

Figure 3. Block diagram of the control system

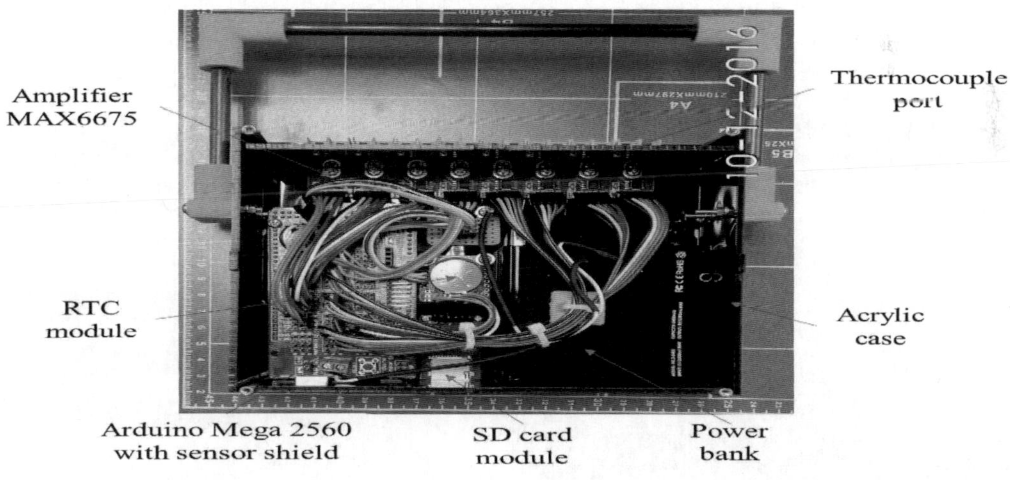

Figure 4. Latest version prototype-detail view

A block diagram of the control system is shown in Figure 3, while detail view of the latest prototype version and a wiring diagram of the temperature data logger are shown in Figure 4 and Figure 5 respectively.

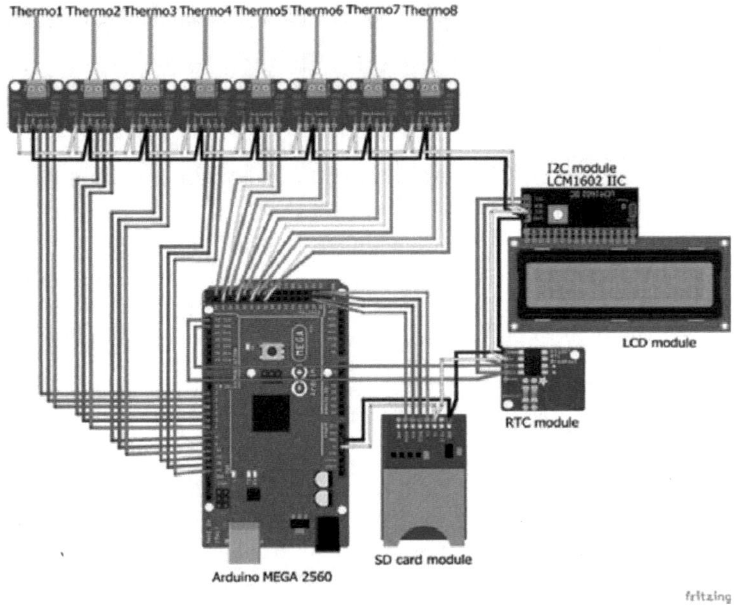

Figure 5. Latest version prototype-wiring diagram

2.2. Software programming

The control algorithms are coded in the Arduino C/C++ subset language combined with the AVR Libc library [44] using the Arduino IDE in a PC (Figure 6). Then they are uploaded to the Arduino board by means of a USB cable. The serial interface employed allows also to receive data from the board to the PC. Thus, the monitored data can be optionally shown on the Arduino IDE Serial Monitor on a PC at the same time the control system is working. The sequence of steps followed in a typical programming code to perform the desired actions. After this common structure, the selected control algorithm included in the loop function runs indefinitely for continuous data acquisition and monitoring function.

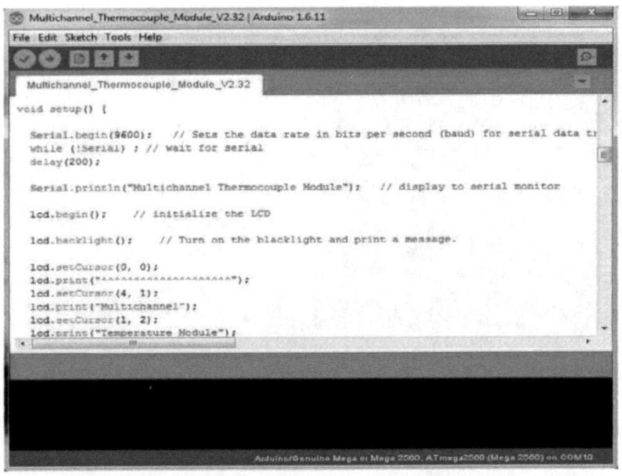

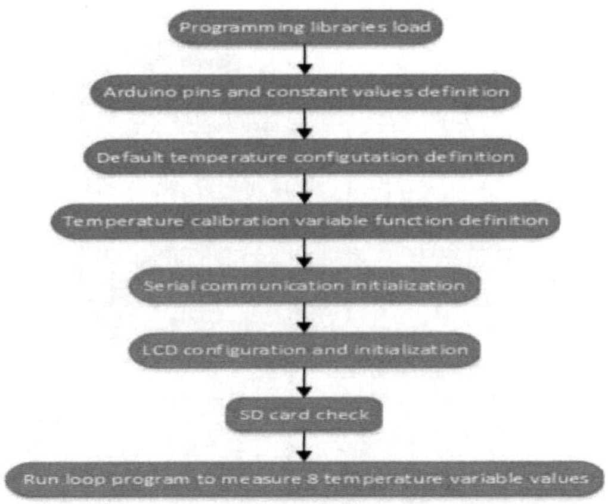

Figure 6. Arduino IDE software

Figure 7. Arduino code step sequence

Data acquisition structure. The control system is in charge of managing the temperature data measurement and storage. The flowchart presented in Figure 8 illustrates the working structure of temperature data gathering on this data logger. The program is divided into three procedure; the initialization, measurement, recording.

IOP Conf. Series: Materials Science and Engineering **257** (2017) 012015 doi:10.1088/1757-899X/257/1/012015

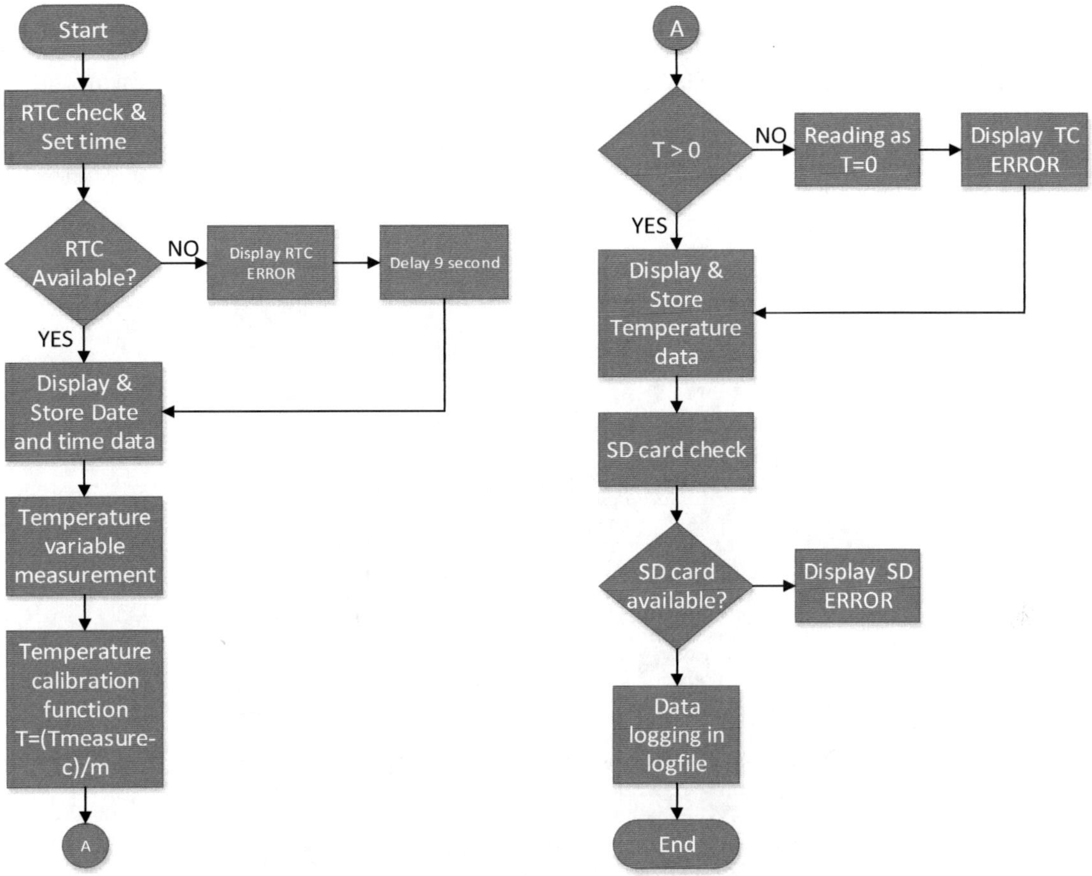

Figure 8. Data acquisition structure flowchart

2.3. Experimental setup

The developed data logger has been used in the monitoring process of agarwood oil extraction using conventional hydrodistillation method. The process started with immersion of agarwood chips in distilled water for 14 days at room temperature (30°C) and then hydrodistillation was conducted with variations of heating temperature between 150°C and 250°C. Temperature measurement at five different points of two hydrodistillations were taken in real time for 24 hours using k-type thermocouples that were connected to in-house built data logger and the produced oil yields from both distillation systems were compared.

2.3.1. Raw material. Inoculated agarwood (Aquailaria agallocha) plant sample for this research was cultivated by artificial method to stimulate the resin production. This agarwood was purchased from R&Z Agarwood's plantation in Sungai Udang, Melaka, Malaysia. Observation of sample composition ratio between resinous wood and non-resinous is about 80:20. The sample then ground into approximately 2 mm chips after drying. 30 kg of grinded agarwood chips were then immersed in distilled water for 14 days at ambient temperature (about 30°C). After 14 days, soaked chips were filtered manually to remove the excessive water and weighed before hydrodistillation process. Hypothetically, soaking process caused the degradation of wood cell wall which enlarges the agarwood pore size to improve yield of extracted oil [45].

2.3.2. Hydrodistillation system. Conventional hydrodistillation technique was used as extraction method in this study. The hydrodistillation apparatus consists of boiler, evaporator, condenser and glass container with overflow. The boiler is a custom design stainless steel with the capacity of 250 L. The condenser of Clevenger apparatus was also fabricated using stainless steel. Recycled water from coolant tank was used as heat exchanging medium of the condenser. The hydrodistillation and its apparatus were shown in Figure 9. Using distilled water as a solvent, the ratio of agarwood chips to distilled water was 1:10 (kg/L). Extraction process was conducted continuously at boiling temperature of water for 24 hours on two separate hydrodistillation systems with 15 kg chips in each boiler. Liquid propane gas (LPG) stove was used to supply heat. The gas flow rate for heating was set instinctively by experienced operator for entire experiment to ensure constant heating of 100°C inside the boiler.

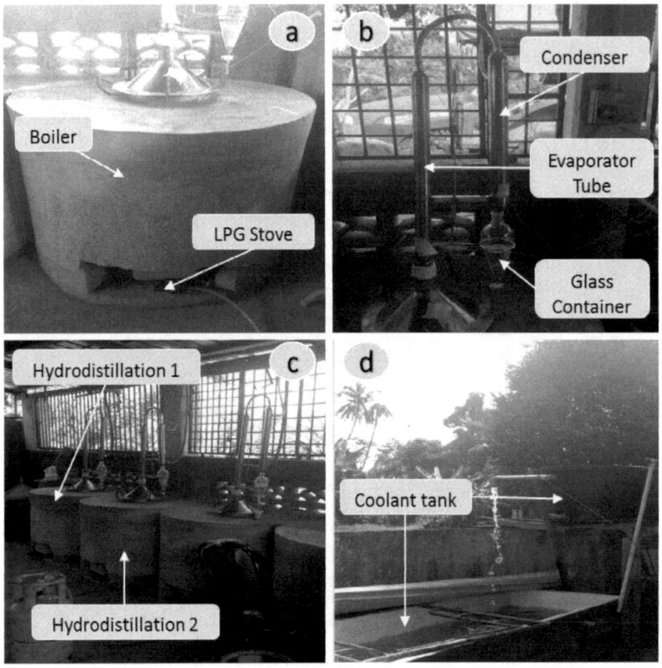

Figure 9. (a) Hydrodistillation boiler inside concrete design is heated using LPG stove. (b) Custom stainless steel apparatus. (c) Extraction facilities setup. (d) Reservoir tank to supply coolant water to condenser of hydrodistillation.

2.3.3. Temperature measurement. The temperature measurement of hydrodistillation setup was monitored by attaching thermocouples at several points and signals from the thermocouples were acquired by in-house built eight channel thermocouple data logger controlled by Arduino microprocessor. The thermocouples were aimed to measure four locations in the hydrodistillation setup for 24 hours. The locations include mixture temperature, vaporize temperature as well as condenser's inlet and outlet temperature (Figures 10 and 11). Then, data will be logged in real-time format and can be retrieved from an 8 Gigabyte memory card although the size of the log file is just in kilobyte. Instant temperatures were also showed on a LCD display of the data logger as depicted in Figure 10(d). The sampling rate was set at 0.3Hz which recorded one sample every 3 seconds.

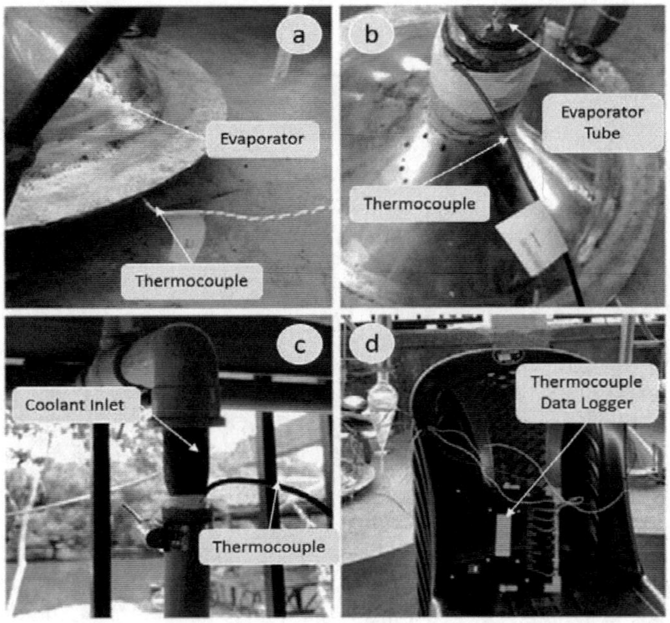

Figure 10. (a) Thermocouple arrangement to measure mixture temperature inside the boiler. (b) Thermocouple arrangement to measure vapour temperature at evaporator. (c) Thermocouple arrangement to measure coolant temperature at condenser. (d) All thermocouples are attached to multichannel temperature data logger.

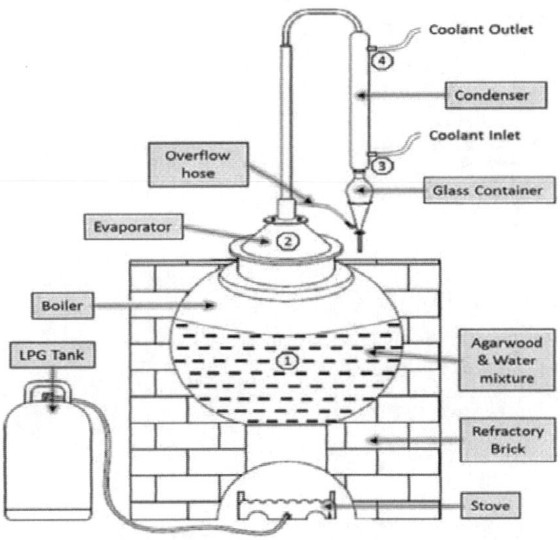

Figure 11. Schematic diagram of custom made traditional hydrodistillation system (1) Mixture temperature measuring point (2) Vapor temperature measuring point. (3) Coolant inlet temperature measuring point. (4) Coolant outlet temperature measuring point

Yield of extracted oil. After 24 hours of hydrodistillation process, mixture of agarwood oil and water collected in container flask were separated into 3 mL bottle. Then, the bottle was weighed to determine the yield of oil. The yield of oil extracted was calculated using following formula.

$$\text{Yield of Oil } (\%) = \frac{\text{agarwood oil (g)}}{\text{agarwood chips (g)}} \times 100\% \quad (1)$$

3. Results and discussion

The result presented the current progress from DAQ module deployment test at industrial facilities. The main objective is to emphasizes on the temperature distribution as a process control for standard operation procedure by the industry for the extraction process towards better control of the yield, and quality of the extracted essential oil.

3.1. Temperature profile of hydrodistillation system

The extraction process was executed for 24 hours to monitor the temperature profile on four (4) points of two hydrodistillation system respectively. It is observed from Figure 12 that both hydrodistillation systems were able to achieve higher mixture temperature (more than 100°C) at evaporator point. Boiling point of water at normal room temperature and atmospheric pressure is 100°C. The higher boiling temperature than the water's suggests that addition of agarwood into the system increases the mixture boiling points. The addition of the agarwood chips became impurities in the water and elevates the boiling points. According to Pornpunyapat, J. (2011) [46], the temperature of distillation should not reach beyond 120°C because it may burn out the plant material. As observed in Figure 12, both hydrodistillation system does not exceeding 120°C and hence met the condition criteria.

Figure 12 also shows that both hydrodistillation system have similar temperature profile patterns. The graphs show extraction process has archived its temperature stability after approximately 4.5 hours and continue to maintain for the rest of the time. The average of recorded temperature was calculated with the standard deviation and summaries in table 1. The heat supplied for the extraction process was less utilised for breaking the cell wall but consumed more energy for vaporizing the volatile compounds. The amount of energy required for the both processes depend on the hydration process of the agarwood chips [47]. The compounds were then carried by the steam along the condenser and finally collected as distillate.

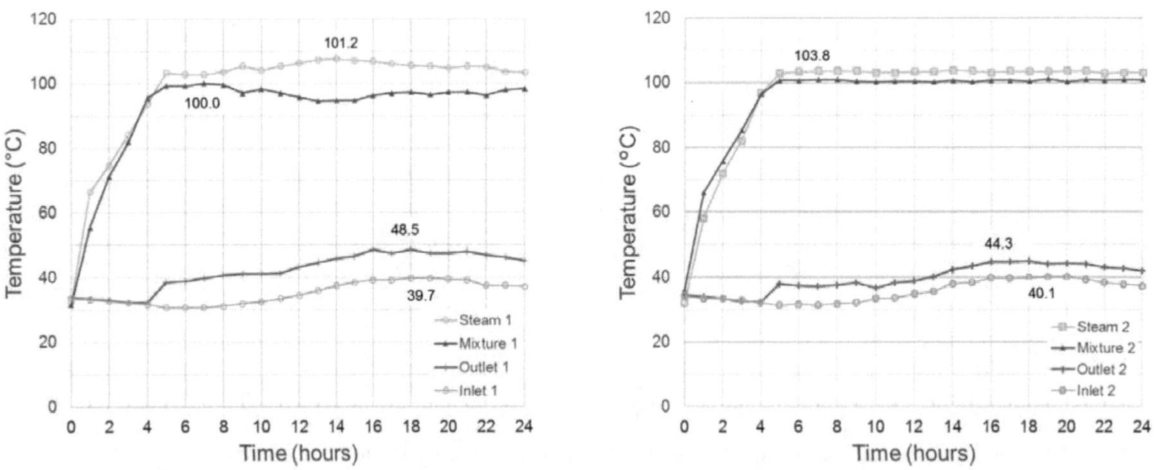

Figure 12. Temperature profile on **(a)** hydrodistillation 1 and **(b)** hydrodistillation 2. The numeral shows the highest temperature reading for each point.

Table 1. Comparison of average temperature on both hydrodistillation system after 4.5 hours.

	Hydrodistillation 1		Hydrodistillation 2	
	Steam	Mixture	Steam	Mixture
Average Temperature (°C)	104.889	97.256	103.206	100.665
Standard Deviation (°C)	1.580	1.606	0.355	0.313

3.2. Yield of extracted oil

Both hydrodistillation setup shows similar outcome based on yield of agarwood oil extracted using conventional hydrodistillation as shown in Figure 13. Higher yield was obtained from hydrodistillation 2 which is 0.004% compared with hydrodistillation 1 samples as summaries in table 2. Our findings seemed contradicts with other researchers who found that the yield of essential oil derived from agarwood soaked in water was between 0.03 and 4.43% [48, 49]. However, the difference results presented in Wetwitayaklung, Thavanapong and Charoenteeraboon [48] and Tajuddin and Yusoff [49] due to the laboratory setup might be the reason of the contradiction with the yield obtained in this study.

Table 2. Yield of extracted agarwood oil.

	Hydrodistillation 1	Hydrodistillation 2
Yield of oil (%)	0.0036	0.0040

Figure 13. Yield observation of extracted agarwood oil on (a) hydrodistillation 1 and (b) hydrodistillation 2.

4. Conclusions

As a vital part of any systematic studies nowadays, data logger has to be developed with minimum power consumption but yet able to acquire data accurately from multisensors. Taking open source Arduino microcontroller-based data logger for monitoring temperature distribution in hydrodistillation systems, this article established an architecture of software and hardware to provide datalogging functionalities for real time process monitoring that requires durability for long term operation in rough environment. Furthermore, the deployment in the real industry application can also correlate the measurement taken to the quality and quantity of extracted essential oil from the process. Distinction research findings from this research that should be paid attention are that both hydrodistillation showed similar temperature profile pattern with stable readings after 4.5 hours, mixture temperature for both hydrodistillation achieved the average stable temperature of 97.26 °C to 100.66 °C. Whereas, steam temperature shows small difference between both hydrodistillation setup. Higher yield of extracted agarwood oil achieved 0.004% on hydrodistillation 2 compare to hydrodistillation 1 with 0.0036%. From the obtained results, this study achieved the objective of developing an inexpensive, portable and robust eight channels temperature measuring module with capabilities to monitor and store real time data.

Acknowledgement
This work was supported under the KTP Grant number RDU151007 by the Universiti Malaysia Pahang, Malaysia. The authors would like to thank Faculty of Engineering and Bio Aromatic Centre of Universiti Malaysia Pahang. Also to the industrial partner which is IGawa Resourses for the strong support and priceless contribution during this research.

References

[1] Chang I, Ng L and Kadir A 1997 A review on agur (gaharu) producing Aquilaria species *Journal Tropical Forest Product*

[2] Ahmad Fadzli Z 2006 Extraction of gaharu essential oil using spinning band distillation. Universiti Malaysia Pahang)

[3] Bhore S J, Preveena J and Kandasamy K I 2013 Isolation and identification of bacterial endophytes from pharmaceutical agarwood-producing Aquilaria species *Pharmacognosy research* **5** 134

[4] Xu Y, Zhang Z, Wang M, Wei J, Chen H, Gao Z, Sui C, Luo H, Zhang X and Yang Y 2013 Identification of genes related to agarwood formation: transcriptome analysis of healthy and wounded tissues of Aquilaria sinensis *BMC genomics* **14** 227

[5] Mohamed R, Jong P and Zali M 2010 Fungal diversity in wounded stems of Aquilaria malaccensis *Fungal Diversity* **43** 67-74

[6] Buchbauer G and Wallner I M 2016 *Encyclopedia of Food and Health,* ed B Caballero, *et al.* (Oxford: Academic Press) pp 558-62

[7] Bakkali F, Averbeck S, Averbeck D and Idaomar M 2008 Biological effects of essential oils – A review *Food and Chemical Toxicology* **46** 446-75

[8] Joy P, Thomas J, Mathew S, Jose G and Joseph J Aromatic Plants.

[9] Molloy B 1997 Modelling and predictive control of a drum-type boiler. In: *School of Electronic Engineering,* (Dublin, Ireleand: Dublin City University)

[10] Cheah L C E 2009 Study of extraction processes and their impact on bioactivity of botanicals. In: *Department of Pharmacy,* (Singapore: National University Singapore)

[11] Unbehauen H and Kocaarslan I 1990 Experimental modelling and adaptive power control of a 750 MW once-through boiler. In: *Proceedings of 11th IFAC world congress, Tallin, SU,* pp 226-31

[12] Unbehauen H 1969 The load dependent multivariable steam temperature control system in a boiler *Automatica* **5** 421-32

[13] Stanković M Z, Nikolić N Č, Stanojević L and Cakić M D 2004 The effect of hydrodistillation technique on the yield and composition of essential oil from the seed of Petroselinum crispum (Mill.) Nym. ex. AW Hill *Hemijska Industrija* **58** 409-12

[14] Stanojević L, Stanković M, Cakić M, Nikolić V, Nikolić L, Ilić D and Radulović N 2011 The effect of hydrodistillation techniques on yield, kinetics, composition and antimicrobial activity of essential oils from flowers of Lavandula officinalis L *Hemijska industrija* **65** 455-63

[15] Milojević S Ž, Stojanović T D, Palić R, Lazić M L and Veljković V B 2008 Kinetics of distillation of essential oil from comminuted ripe juniper (Juniperus communis L.) berries *Biochemical engineering journal* **39** 547-53

[16] Fisher J A, Scarlett M J and Stott A D 1997 Accelerated solvent extraction: an evaluation for screening of soils for selected US EPA semivolatile organic priority pollutants *Environmental science & technology* **31** 1120-7

[17] Davies A J 1978 Microprocessors and their use in Physics *Advances in Electronics and Electron Physics* **47** 51-121

[18] Mukaro R and Carelse X F 1999 A microcontroller-based data acquisition system for solar radiation and environmental monitoring *IEEE Transactions on Instrumentation and Measurement* **48** 1232-8

[19] Benghanem M, Arab A H and Mukadam K 1999 Data acquisition system for photovoltaic water pumps *Renewable Energy* **17** 385-96

[20] Koutroulis E and Kalaitzakis K 2003 Development of an integrated data-acquisition system for renewable energy sources systems monitoring *Renewable Energy* **28** 139-52

[21] Rosiek S and Batlles F J 2008 A microcontroller-based data-acquisition system for meteorological station monitoring *Energy Conversion and Management* **49** 3746-54

[22] Benghanem M 2009 Measurement of meteorological data based on wireless data acquisition system monitoring *Applied Energy* **86** 2651-60

[23] Johnson G R 1977 A microprocessor based solar controller. In: *1977 IEEE Conference on Decision and Control including the 16th Symposium on Adaptive Processes and A Special Symposium on Fuzzy Set Theory and Applications,* pp 336-40

[24] Moen R L 1979 Solar energy management system. In: *1979 18th IEEE Conference on Decision and Control including the Symposium on Adaptive Processes,* pp 917-9

[25] Kotob S, Salman M, Ayyash S and Farhat M 1983 Low cost microprocessor based optimal controller for solar cooling systems. In: *The 22nd IEEE Conference on Decision and Control,* pp 958-63

[26] Muralidhar G K, Nagaraju J and Mohan S 1988 *Advances In Solar Energy Technology,* (Oxford: Pergamon) pp 2824-8

[27] Nhut L M and Park Y C 2013 A study on automatic optimal operation of a pump for solar domestic hot water system *Solar Energy* **98, Part C** 448-57

[28] Fuentes M, Vivar M, Burgos J M, Aguilera J and Vacas J A 2014 Design of an accurate, low-cost autonomous data logger for PV system monitoring using Arduino™ that complies with IEC standards *Solar Energy Materials and Solar Cells* **130** 529-43

[29] Commission I E 1998 Photovoltaic system performance monitoring-guidelines for measurement, data exchange and analysis *International Standard IEC* **61724** 1998-11

[30] Arduino website. pp Arduino open-source electronics platform

[31] Ali A S, Zanzinger Z, Debose D and Stephens B 2016 Open Source Building Science Sensors (OSBSS): A low-cost Arduino-based platform for long-term indoor environmental data collection *Building and Environment* **100** 114-26

[32] Fatehnia M, Paran S, Kish S and Tawfiq K 2016 Automating double ring infiltrometer with an Arduino microcontroller *Geoderma* **262** 133-9

[33] Claros-Marfil L J, Padial J F and Lauret B 2016 A new and inexpensive open source data acquisition and controller for solar research: Application to a water-flow glazing *Renewable Energy* **92** 450-61

[34] Gad H E and Gad H E 2015 Development of a new temperature data acquisition system for solar energy applications *Renewable Energy* **74** 337-43

[35] Lin Y and Sriyudthsak M 2016 Design and Development of Standard 12-Lead ECG Data Acquisition and Monitoring System *Procedia Computer Science* **86** 136-9

[36] Leik T T 2015 Experimental study on cutting temperature in sustainable machining of alluminium alloy and mild steel. In: *Fakulti Kejuruteraan Mekanikal,* (Pahang: Universiti Malaysia Pahang)

[37] 2014 Atmel ATmega640/V-1280/V-1281/V-2560/V-2561/V Datasheet. ed A Corporation (San Jose, USA
[38] Webpage A Pulse Width Modulation (PWM).
[39] Maxim Integrated Products I 2007 SPI/I²C Bus Lines Control Multiple Peripherals. In: *Application Note 4024*: Maxim Integrated Website)
[40] 2015 DS1307 64 x 8, Serial, I2C Real-Time Clock Datasheet. ed M I Products
[41] Maxim Integrated Products I 2000 Comparing the I²C Bus to the SMBus. In: *Application Note 476*: Maxim Integrated website)
[42] 2009 LCM Module (TC2004A-01) Datasheet. ed Adafruit
[43] 2002 MAX6675 Datasheet. ed M I Products (USA
[44] Corporation A 2016 AVR Libc Reference Manual.
[45] Fazila K N and Halim K K 2012 Effects of soaking on Yield and Quality of Agarwood Oil *Journal of Tropical Forest Science* 557-64
[46] Pornpunyapat J, Chetpattananondh P and Tongurai C 2011 Mathematical modeling for extraction of essential oil from Aquilaria crassna by hydrodistillation and quality of agarwood oil *Bangladesh Journal of Pharmacology* **6** 18-24
[47] Jok V A, Radzi N C and Hamid K H K 2015 Agarwood Oil Yield As A Result of Changes in Cell Morphology Due To Soaking Process *Procedia-Social and Behavioral Sciences* **195** 2443-50
[48] Wetwitayaklung P, Thavanapong N and Charoenteeraboon J 2009 Chemical constituents and antimicrobial activity of essential oil and extracts of heartwood of Aquilaria crassna obtained from water distillation and supercritical fluid carbon dioxide extraction *Silpakorn University Science and Technology Journal* **3** 25-33
[49] Tajuddin S N and Yusoff M M 2010 Chemical composition of volatile oils of Aquilaria malaccensis (Thymelaeaceae) from Malaysia *Natural product communications* **5** 1965-8

Comparison of hydro-pneumatic accumulator's charging performance under different thermal process for dual hybrid driveline

F Wasbari[1, 2*], R A Bakar[1], L M Gan[1], A A Yusof[2], N M Daud[2], T Ali[3]

[1]Faculty of Mechanical Engineering, Universiti Malaysia Pahang (UMP), 26600 Pekan, Pahang, Malaysia
[2]Faculty of Mechanical Engineering, Universiti Teknikal Malaysia Melaka (UTeM), Hang Tuah Jaya, 76100 Durian Tunggal, Melaka, Malaysia
[3]Concentrated Solar Power (M) Sdn. Bhd., Taman Subang Permai, 47600 Subang Jaya, Selangor, Malaysia

*Corresponding Author: faizilwasbari13@gmail.com

Abstract. The hydro-pneumatic accumulator is broadly used in heavy industry as leaking compensator. Lately, much interest has been developed to study the component as energy storage. The important element in the selection of accumulator is performance. Hence, research was conducted to examine the use of thermal process condition and its impingement on the accumulator as an energy store. For charging process, the performance is influenced by the thermal process because it involves temperature change and heat transfer. Both processes are producing a different performance. The storage system is planned to be adapted in the dual hybrid driveline. A simulation study has been conducted by using Automation Studio software which focusses on two different processes called isothermal and adiabatic. The process has involved a schematic design, functional testing, parameter setting, and pretense. The result has shown that the thermal process affected the fluid power parameters such as power, effective storage capacity, and temperature differentials. The isothermal process produced higher effective volume compared to adiabatic process, stored higher power and had lower temperature differentials. Regarding charging speed, the adiabatic was faster. However, it was a lack of storage capacity.

1. Introduction

The hydro-pneumatic accumulator is one of a critical component in the heavy industry sector. They are employed in the hydraulic system for processing high workload processes such as stamping and pressing. This is imputable to the ability of this component to run as a leaking compensator, backup energy system, and surge control. In the sectors of automobiles, the hydro-pneumatic accumulator is used as shock dampening component and also energy storage [1][2]. As a shock dampening, the component reduces fatigue of hydraulic and mechanical components. As for the energy storage, it is used to store the energy obtained from the active or passive way to be used as a backup system, and sometimes it functions as a secondary energy source. In this project, the focus is the application of hydro-pneumatic accumulator to the hybrid driveline of the hybrid vehicle. It will be installed to the dual hybrid hydro-pneumatic driveline. This component acts as secondary energy storage. In this hybrid system, accumulator conducts three (3) main functions, which is charging, storage and discharge process. Charging is the process of compression to allow conversion of the kinetic energy transformed into potential energy. It will cut down the bulk of the compressed gas and increase the force per unit area. Storage function, on the other hand, is the process of storing energy and serves as a cooling system. While the discharge is the process of converting the stored potential energy, transmitted to the propulsion system to move the vehicles when needed. The primary focal point of this report is the charging process which is tender to the thermal reaction. Thermal change can affect the performance of components in terms of power, charging time and the capacity of effective volume.

There are three (3) types of accumulators used in the industry, namely, piston, diaphragm and the bladder type. In terms of output power, the bladder type is the best. While about compression ratio, piston accumulator would be preferred. For the application of dual hybrid driveline, the bladder-type accumulator is most suitable based on the characteristics of high power output, average effective volume, and high operating pressure. Bladder-type accumulator consists of various components, namely shell, bladder, nitrogen gas and safety valve. The shell is the main casing and usually made of steel or composite material. It is designed based on the standards of the pressure vessel. The bladder is elastomer materials having characteristics of high elastic to keep the pre-charge pressure and also to stand the compression of hydraulic oil. Nitrogen gas is an ideal gas to absorb the compression energy because it delivers excellent properties to withstand the temperature change. In this matter, the nitrogen is better than the air [3]. The charging process detail described through the figure 1. The process of compression results in an increase of temperature of the system. It can cause some unwanted behavior such as the volume change due to the thermal expansion. This behavior at once affects the system's maximum pressure, power and shorten the lifespan of the parts. Thus, the temperature of the compression process needs to be controlled to obtain the optimal operation of the energy storage. In this case, the most common process condition that has been applied to thermal storage are isothermal and adiabatic. In isothermal condition, the fluid temperature remains constant. The heat exchange between gas and atmosphere is instantaneous. Meanwhile, the adiabatic occurs without transfer of heat or matter between a thermodynamic system and its surroundings. In an adiabatic process, energy is transferred only as work. It also defined as zero heat transfer, and all the work that has been done on the system raises the temperature of the system. So, the internal energy between these two systems are different and affected the performance of the application system.

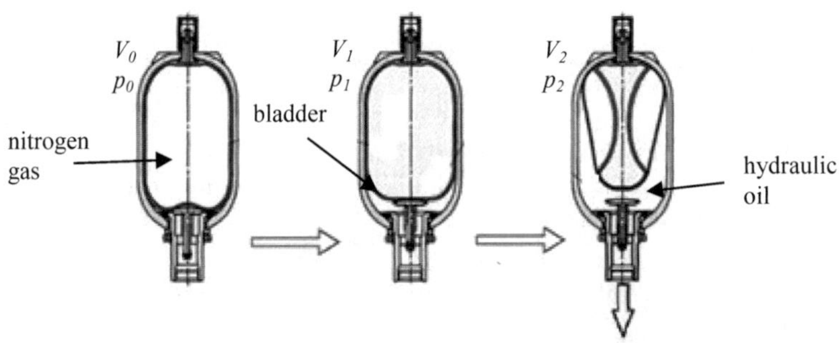

Figure 1. Bladder condition during the charging process [4]

One of the important element that affects the performance of compression is the thermal process. The thermal process commonly studied by researchers are isothermal and adiabatic. The energy storage divided into two because each process has their unique behavior. The concept of thermal energy storage is mostly applied to electric energy generation system, hydraulic system and also vehicle system [3]. Isothermal energy storage is a storage system that maintains the inside temperature during the process of compression. As there is no change in internal temperature, $\Delta T = 0$, the temperature increase due to pressure change converted in the form of heat transfer. Since the internal temperature does not alter, then higher pressure could be kept in the storage. While adiabatic energy storage has some sort of insulation resulting occurrence of heat transfer is zero. The temperature in the system as a whole is transformed into internal energy. Therefore, each of the process has unique features and contribute to the application of the storage system. However, application compatibility is dependent on the process used [5]. In wind-tidal electric energy generation [6], energy storage is used to store pressure energy produced by wind blade. This pressure energy stored and cooled. When needed, it will be channeled to Pelton turbine to turn the generator. The hydraulic medium used is seawater. The analysis indicated that the energy efficiency could exceed 100% as the small input speeds are seen, and the exergy efficiency has the consistent change trends with demand power. At the same time, the study also said that there is an inversely proportional relationship between the accumulator and hydraulic equipment including the pump and nozzle in terms of proportions. Among the energy storages that have been introduced, two well-known energy storages have been proposed called isothermal compressed air energy storage (Isothermal CAES) and adiabatic compressed air energy storage (Adiabatic CAES) [7]. Each has its unique solutions. Based on the study, isothermal has a potential round-trip efficiency of 70-80%, meanwhile adiabatic efficiencies more than 70%. Although the two systems have advantages, however, several challenges must be solved to optimize the efficiency of the compression process. Then it can be manipulated to suits for expansion process [8]. In an ideal solution, the isothermal process is used to maintain the compression so that a higher pressure is obtained and expand the energy in adiabatic[9]. Means, there are no losses occurred during the expansion process. However, to acquire a combination of two processes in one system is very challenging.

Due to the above statement, simulations are carried out to study the effect of isothermal, and adiabatic compression (charging process) on the performance of hydro-pneumatic accumulator for dual hybrid driveline system. By knowing the characteristics of each charging condition, all aspects can be considered in the process of designing energy storage as well as facilitate optimization process.

2. Methodology

2.1. Schematic diagram

The schematic diagram consists of various components. Pump system was employed to supply hydraulic oil in hydrostatic approach. Accumulator works as energy storage, a pressure gauge is used to measure the pressure change and pressure relief valve to secure the circuit. The pump runs to create a flow that will go to the accumulator. The accumulator consists of the bladder that filled with nitrogen gas. Once the bladder being compressed, it stored energy in terms of potential energy. The pressure will keep increased, and when it reaches the setting limit, the pressure relief valve will open and secure the system as illustrated in Figure 2.

2.2. Equation derivation

Based on figure 2, derivation has been developed. The first approach to study the effect of the input parameter to the capacity of the storage and the effect of the different thermal process during compression. Charging process starts with pumping fluid from pump to accumulator unit. The pressure different through the pump can be expressed as follows [10],

$$\Delta p = p_{out} - p_{in} \tag{1}$$

Where Δp is the pressure different (N/m^2), p_{out} is pump output pressure (N/m^2) and p_{in} is pump input pressure (N/m^2). Pressure entering the accumulator is defined as

$$p_{acc} = p_{out} - h_m - h_f \tag{2}$$

Where p_3 is the input pressure (N/m^2), p_2 is pump output pressure (N/m^2), h_m is minor losses and h_f is friction losses (N/m^2). Since the main function of the pump is to create flow, the actual output flows produced by the pump can be expressed as

$$Q_{out} = C \cdot \omega \cdot \eta_V \tag{3}$$

Where Q_{out} is pump output flow rate (m^3/s), C is displacement (m^3/rad), ω is nominal speed (rad/s) and η_V is volumetric efficiency. The pressure relief valve (PRV) works as a safety valve. The preset value is a maximum pressure allowed to be working in the system. If the charging pressure were exceeded, the PRV opened to stabilize the pressure. So, the pressure in the system can be controlled.

For the internal compression effects on nitrogen bladder, the analysis made by using equations derived from Boyle-Mariotte's law for gasses [10][4], which takes into account the compressibility of the gas.

$$\Delta V_{eff} \leq V_1 - V_2 \tag{4}$$

Where ΔV_{ideal} is the effective volume, V_1 is volume before compression and V_2 is volume after compression.

$$p_0 \times V_0 = p_1 \times V_1 = p_2 \times V_2 \tag{5}$$

Where p_0, p_1, p_2 is pressure at a certain stage of compression and V_0, V_1, V_2 is the volume at a certain stage of compression. Refer to figure 1 for detail stage of charging. In terms of storage capacity, in adiabatic condition, the total volume represented as;

$$V_{total} = \frac{p_{\min} \cdot Z_0 \cdot \Delta V_{eff;T\min}}{p_{0;T\min} \cdot Z_1} \cdot \frac{1}{1 - \left(\frac{p_{\min} \cdot Z_2}{Z_1 \cdot p_{\max}}\right)^{\frac{1}{n_c}}} \tag{6}$$

However, if the condition is not in isothermal then, the equation is as follows,

$$V_{total} = \frac{p_{\min} \cdot Z_0 \cdot \Delta V_{eff;T\min}}{p_{0;T\min} \cdot Z_1} \cdot \frac{\left(\frac{Z_1 \cdot p_{\max}}{p_{\min} \cdot Z_2}\right)^{\frac{1}{n_d}} \cdot \left(p_{\min} \cdot Z_2\right)^{\frac{1}{n_d}}}{\left(Z_1 \cdot p_{\max}\right)^{\frac{1}{n_d}} - \left(p_{\min} \cdot Z_2\right)^{\frac{1}{n_d}}} \tag{7}$$

Where V_{total} is the total volume of the accumulator. ΔV_{eff} is effective volume. $p_{\min}$ is minimum working pressure, $p_{0;T\min}$ is pre-charge pressure at minimum temperature, $p_{\max}$ is maximum working pressure, $z_{0,1,2}$ are compressibility factor, n_c is charge coefficient and n_d is discharge coefficient. This is the simplified equation to calculate the average time required for the filling process.

$$V_{eff} = \Delta V_{total} \cdot K_i \tag{8}$$

$$V_{eff} = Q_i \cdot t_i \tag{9}$$

Where V_{eff} is effective volume, and K_i is correction factor, Q_i is the flow rate and t_i is the filling time. The equation (12) is then rearranged,

$$t_i = \frac{\Delta V_{effective}}{Q_i} \tag{10}$$

2.3. Simulation

Automation Studio is a tool for design, functional simulation of complex automation and documentation. The software includes hydraulic, pneumatic and electrical operative devices as well as a command part diagram as shown in figure 2(a). It also provides technical and commercial data for simulation. Undoubtedly, software such as Matlab has high flexibility to run a simulation related to the case study, but the Automation Studio has the advantage in terms of circuit design, functional simulation, fluid power component sizing, system design, validation and virtual simulation [6]. These advantages had resulted the selected software was more suited for this project. Currently, Automation Studio has improved their functional by connecting some applications with Simulink, Matlab [11]. All specifications for components, fitting, and measurement tool can be inserted into the provided dialogue box as illustrated in figure 2(b). Since the study involves bulk modulus and heat transfer, so the two functions were enabled. Bulk modulus effect occurs because of the process of nitrogen compression in the bladder [12]. By enabling the bulk modulus setting, the simulation considers the effects of compression in charging and discharging process. The thermal setting considers the fluid heating in contact with the pipe component inner surface. Pressure elevation is set as an independent variable. Temperature, effective volume and charging time as dependence variable. As for pre-charge pressure and volume displacement, the parameters are set as a control variable as shown in table 1.

The system powered by motor and pump which turns at 2000 RPM and 90% thermal efficiency. The accumulator is 225 mm in diameter set zero as initial volume with 25°C nitrogen gas in the bladder. The simulation system secured by pressure relief valve that set as 400 bar cracking pressure.

Table 1. Simulation parameters

Motor rotational speed	2000 RPM	Initial liquid volume	0 liter
Pump thermal efficiency	90%	Accumulator internal diameter	225 mm
Pump heat transfer coefficient	10 W/m^2K	Type of process	Isothermal/Adiabatic
Hydraulic oil	Hydraulic AW-32	Gas type	Nitrogen
Ambient temperature	25°C	Gas temperature	25°C
Cracking pressure	400 bar	Hydraulic line	Steel
Port 1 (C_v)	12	Line type	NPS ¼-DN8

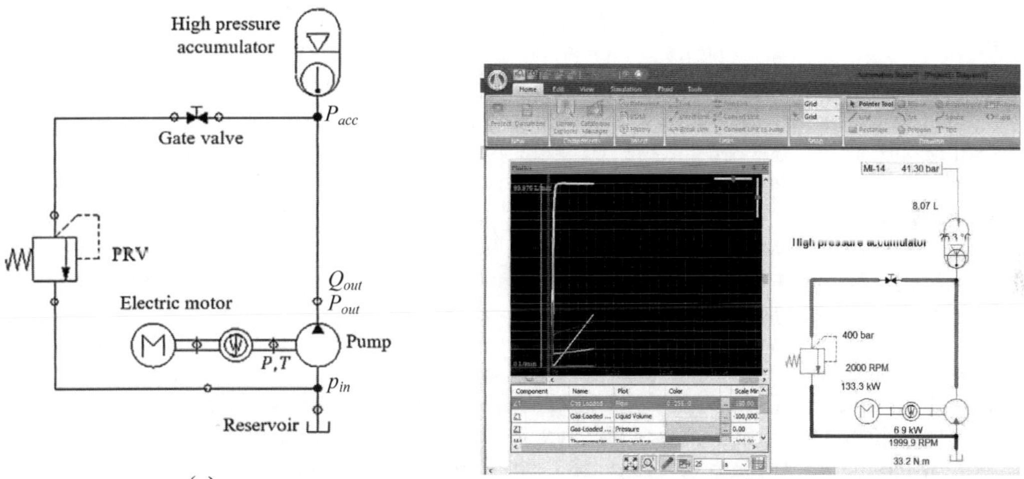

(a) (b)

Figure 2. Schematic diagram (a) and Automation Studio GUI (b)

3. Results and discussion

3.1. Effect of pre-charge pressure to effective volume

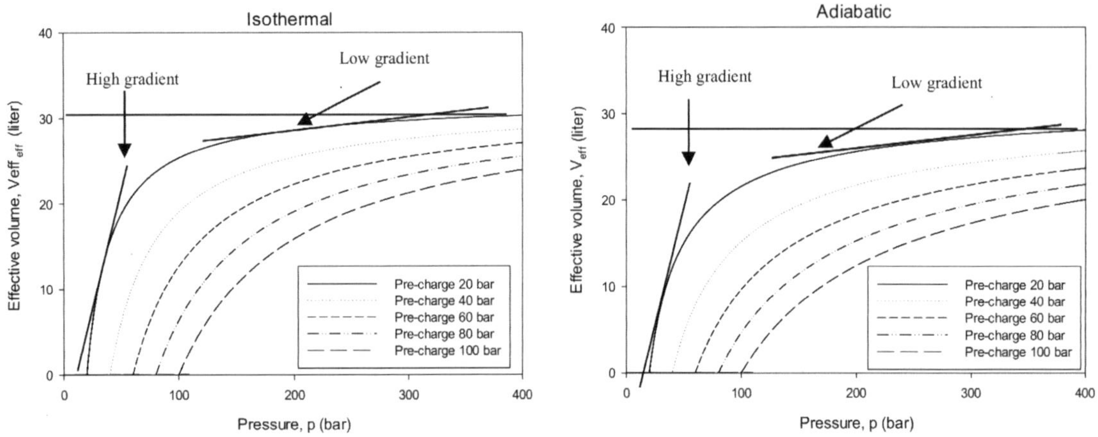

(a) (b)

Figure 3. Effects on effective volume

Figure 3(a) Isothermal process produced better output in term of effective volume compare to adiabatic. The highest Isothermal output is 30.35 liter at pre-charge 20 bar and the lowest is 23.8 liter at 100 bar pre-charge. Both occurred at 400 bar system pressure as illustrated in figure 3(b). These are due to the resistive force created by 100 bar pre-charge are more superior compared to the 20 bar. The highest adiabatic output is 28 liter at pre-charge 20 bar and the lowest is 20 liter at 100 bar pre-charge. In term of graph gradient for the pre-charge 20 bar, rapid pressure changes occurred below 80 bars. For the pre-charge 40 bar, rapid pressure changes occurred below 160 bars. At pre-charge 60 bar, rapid pressure changes occurred below 230-240 bars. At pre-charge 80 bar, rapid pressure changes occurred below 80 bars. At pre-charge 100 bar, rapid pressure changes occurred below 380-400 bars. The value of high gradient showed that the filling proses (compression) happen efficiently. However, the lower gradient compression consuming time and it is representing over-pressure condition. Over pressure is not good for the accumulator since it will shorten the lifespan of the bladder and service time. Based on the graph, it might be the reason why a hydraulic company such as Bosch Rexroth, Parker and Hydac decided to limit the compression ratio of the accumulator. It is to prevent overpressure [13][4]. In term of efficiency, at 400 bar, the isothermal process at pre-charge pressure of 20 bar shown 3% higher volume than the adiabatic process at the same pre-charge pressure. Meanwhile for the pre-charge pressure 100 bar, the isothermal shown about 10% higher volume than the adiabatic process at the same pre-charge pressure.

To ensure that the simulation data and result are reliable and correct, the pre-charge pressure effect on storage pressure and effective volume has been compared with Bosch type industrial accumulator. Since all the specifications are different, then the comparison cannot be made a point to point. However, it is just a reference to proceed with the next simulation. In this case, the resulting profile is the same. It started at the pre-charge pressure, increased in quadratic, once it achieved maximum pressure, the gradient start fall which translated to inefficient compression. It called inefficient because the input given was high, but the volumetric output is small. It also called over-pressure condition.

Figure 4 has shown the comparison of effective volume based on pre-charge pressure. It was found that the higher the pre-charge pressure, the wider the difference of effective volume between isothermal and adiabatic. The lowest difference is at 20 bar pre-charge, and the highest was 100 bar pre-charge pressure. It was noted that the volume changes from 20 bar to 100 bar pre-charge was 6.32

liter in isothermal and 8.04 liter in adiabatic. These differentials are due to the effect of temperature change and heat exchange.

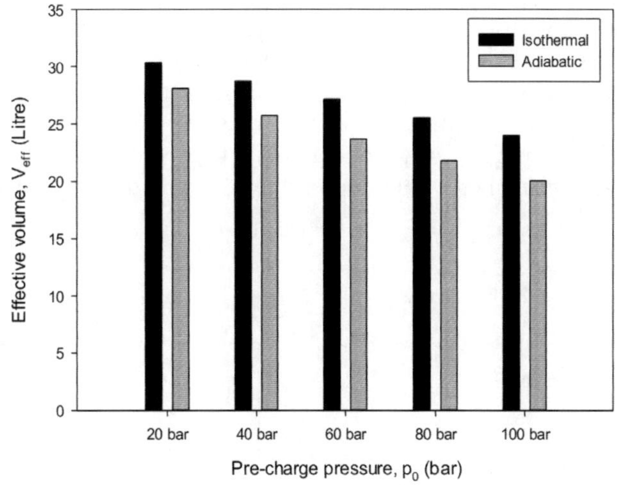

Figure 4. Comparison between isothermal and adiabatic effective volume capacity

Based on figure 3 and figure 4, it was found that the safe, reasonable effective volume and comes with lower inefficient volume is 80 bar pre-charge pressure. The pre-charge pressure does not experience significant over-pressure condition and maintains adequate effective volume. However, this decision does not count the input power given to the system. More detail selection of input type and performance characteristic is required to further the study. Therefore, the next simulation will focus on 80 bar pre-charge pressure.

3.2. Flowrate and effective volume

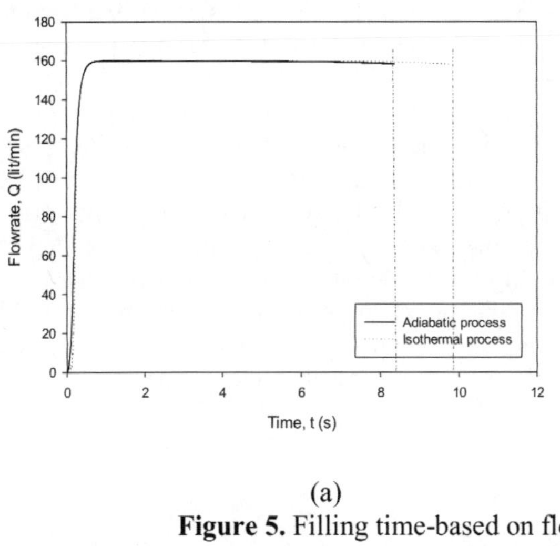

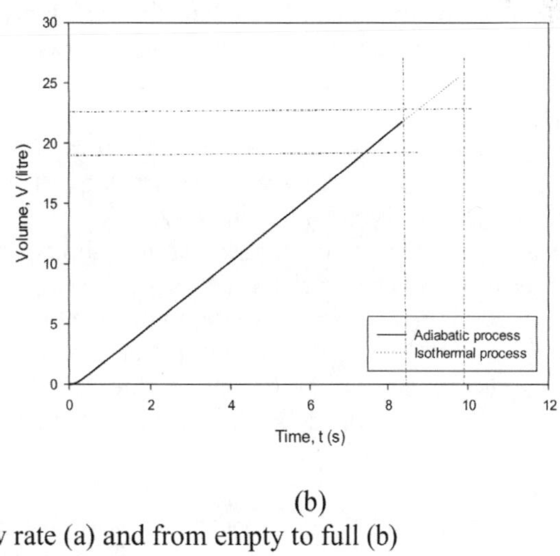

(a) (b)

Figure 5. Filling time-based on flow rate (a) and from empty to full (b)

Based on the Bosch catalog [14], the pump speed has been set to the average speed of 2000 rpm and 80 cm^3/rev and it took only 1 to 2 seconds for the pump to raise 0 – 160 liter/min for both process condition as illustrated in figure 5(a). However, from the beginning, the flow rate for the adiabatic process is higher compared to the isothermal until it reached consistency flow rate of 160 liters/min. The flow rate for both (isothermal and adiabatic conditions) had no significant difference when it achieved the stable flow rate. It remains stable at 160 liters/min until about 2 seconds before it reached its effective volume. The isothermal process requires more time about 9.8 seconds to achieve effective volume meanwhile adiabatic process took about 8.2 seconds which is 1.8 seconds faster. This statement has proven that the thermal process condition has affected the flow rate of the filling process and directly influence the filling time. Figure 5(b) shows the effective volume are related to the flow rate. The isothermal process requires more time about 9.8 s to achieve an effective volume of 25.5 liters. Meanwhile, the adiabatic process took about 8.2 seconds to achieve 22 liter which is 3.5 liters less than the isothermal process. The different volume between both processes consider significant (13.7%) and will affect the running time of the hydraulic motor. It can be seen clearly that adiabatic compression able to fill the accumulator much faster but as a consequence, the volume was less than the isothermal.

3.3. Pressure elevation

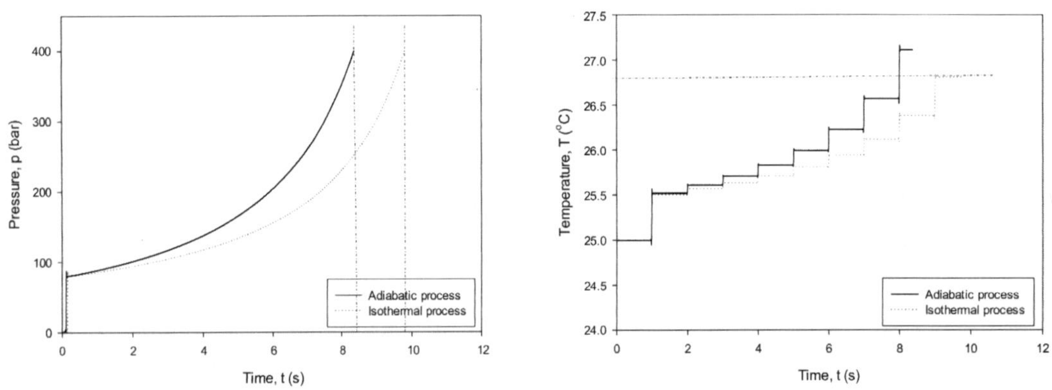

(a) (b)

Figure 6. Pressure profile during filling process (a) and temperature elevation (b)

Figure 6(a) has shown that isothermal process requires more time about 9.8 seconds to achieve effective volume meanwhile adiabatic process took about 8.2 s which is 1.6 s faster. The adiabatic process elevates pressure quicker than the isothermal from the beginning to the cracking pre-charge pressure and finally to the final system pressure of 400 bar. It clearly showed that the adiabatic was faster in term of the filling process compared to isothermal. The flow rate difference and the non-existence of heat exchange are the reasons to justify this behavior. The setting temperature of Nitrogen is 25°C. Maximum temperature recorded by the adiabatic process is 27.2 °C from 25°C. Meanwhile, the isothermal recorded maximum 26.7°C. Less about 0.5 °C as illustrated in figure 6(b). This value seems small, but it contributes to the ability to store the energy. Adiabatic process kept higher temperature, but as consequences, it stores less capacity. Isothermal produces less heat, but it can store more volume capacity. One thing to remember that the temperature value extract at the fluid temperature. Since the hydraulic oil is cooling medium, so it neutralized the heat produced by the compression process inside and of the bladder. There is a tendency that the temperature at the compressed nitrogen and bladder is higher than the recorded.

3.4. *Performance: Hydraulic power and correction factor*

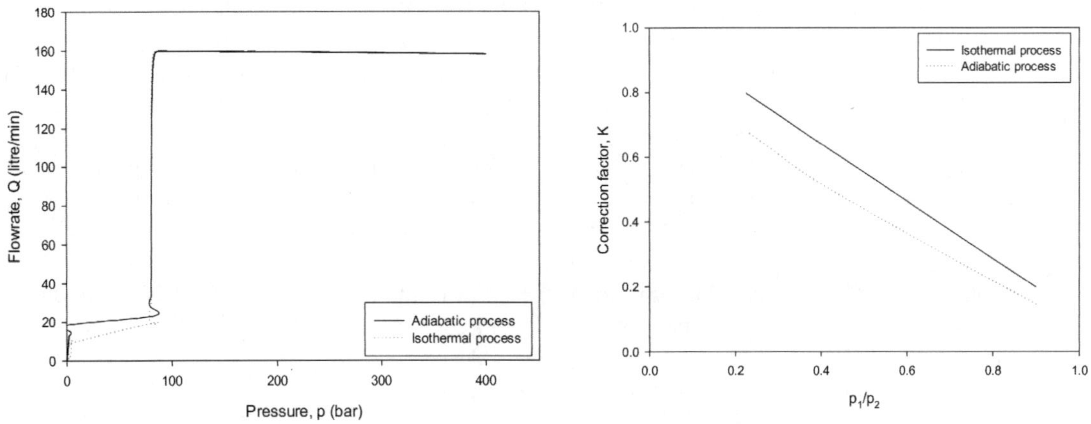

(a) (b)

Figure 7. Flowrate profile based on pressure storage increase (a) and correction factor for isothermal and adiabatic condition (b)

Power is a correlation between pressure and flow rate as represented in figure 7(a). At the beginning of the compression, the graph shows fluctuation until the pressure exceeds pre-charge pressure. This behavior happens because of the resistive force produced by the gas and bladder. When the flow rate increase, the pressure remains constant. Moreover, when the flow stabilized, the pressure increased from 80 bar to 400 bar. This represents that the hydraulic power keeps increase when the pressure increase. Validation has been done by comparing the graph pattern to the existing industrial Bosch Rexroth accumulator that available in the market. It was found that the profile of effective volume is similar, but the value is different because of the differences in parameter used. The ratio of maximum and minimum operating pressure also within the same range as illustrated in figure 7(b). The isothermal correction factor is higher compared to the adiabatic. The maximum correction factor for an isothermal process is 0.8 while the lowest is 0.22. Adiabatic correction factor started lower than isothermal from 0.17 to 0.7.

4. Conclusions

For the pre-charge pressure, the higher the pre-charge pressure, the lower the effective volume capacity. Due to the resistive produced by the nitrogen gas and bladder. The isothermal process produced higher effective volume compared to adiabatic process. In term of charging speed, the adiabatic was faster however it is lack of storage capacity. This is due to its higher flow rate. No significant thermal increase recorded since the hydraulic oil itself acts as a coolant to the system. The results of this simulation clearly showed that if the capacity and temperature change is decisive to the selection process, then the isothermal process is the best. While if the filling time is decisive, then the adiabatic is the best results.

Acknowledgement

This paper was made possible by a scholarship from the Ministry of Higher Education and Universiti Teknikal Malaysia Melaka. The authors also would like to be obliged to Universiti Malaysia Pahang for giving the opportunities and providing facilities to complete this study. This research was supported by UMP FRGS RDU130135.

References

[1] Wasbari F, Bakar R A, Gan L M, Tahir M M and Yusof A A 2017 A review of compressed-air hybrid technology in vehicle system *Renew. Sustain. Energy Rev.* **67** 935–53
[2] Wu B, Lin C C, Filipi Z, Peng H and Assanis D 2004 Optimal Power Management for a Hydraulic Hybrid Delivery Truck *Veh. Syst. Dyn.* **42** 23–40
[3] Puddu P and Paderi M 2013 Hydro-pneumatic accumulators for vehicles kinetic energy storage: Influence of gas compressibility and thermal losses on storage capability *Energy* **57** 326–35
[4] Boschrexroth 2016 Accumulators *boschrexroth.com*
[5] Tavares F 2011 *Thermally Boosted Concept for Improved Energy Storage Capacity of a Hydro - Pneumatic Accumulator*
[6] Fan Y J, Mu A Le and Ma T 2016 Modeling and control of a hybrid wind-tidal turbine with hydraulic accumulator *Energy* **112** 188–99
[7] Mahlia T M I, Saktisahdan T J, Jannifar A, Hasan M H and Matseelar H S C 2014 A review of available methods and development on energy storage; technology update *Renew. Sustain. Energy Rev.* **33** 532–45
[8] Energystorage.org 2016 Isothermal CAES *Energy Storage Assoc.*
[9] Wasbari F, Abu Bakar R, Gan L M, Yusof A A and Amat Jafar Z 2017 Simulation of storage performance on hydropneumatic driveline in dual hybrid hydraulic passenger car *MATEC Web Conf.* **90** 1052
[10] Esposito A 2013 *Fluid Power with Applications* (Pearson Education Limited)
[11] Famic 2016 Automation Studio Target for Simulink *br-automation.com*
[12] Man Z, Ding F, Ding C and Liu S 2015 Study of an energy regeneration system with accumulator for hydraulic impulse testing equipment *Stroj. Vestnik/Journal Mech. Eng.* **61** 196–206
[13] Parker 2016 Bladder Accumulators *parker.com*
[14] Boschrexroth 2017 Pumps *boschrexroth.com*

IOP Conf. Series: Materials Science and Engineering **257** (2017) 012017 doi:10.1088/1757-899X/257/1/012017

Design and simulation of integration system between automated material handling system and manufacturing layout in the automotive assembly line

S Seha[1*], J Zamberi[2] and A J Fairul[2]

[1] Centre of Graduate Studies, Universiti Teknikal Malaysia Melaka, Hang Tuah Jaya, 76100 Durian Tunggal, Melaka, Malaysia
[2] Faculty of Manufacturing Engineering, Universiti Teknikal Malaysia Melaka, Hang Tuah Jaya, 76100 Durian Tunggal, Melaka, Malaysia

*Corresponding author: M051420004@student.utem.edu.my

Abstract. Material handling system (MHS) is an important part for the productivity plant and has recognized as an integral part of today's manufacturing system. Currently, MHS has growth tremendously with its technology and equipment type. Based on the case study observation, the issue involving material handling system contribute to the reduction of production efficiency. This paper aims to propose a new design of integration between material handling and manufacturing layout by investigating the influences of layout and material handling system. A method approach tool using Delmia Quest software is introduced and the simulation result is used to assess the influences of the integration between material handling system and manufacturing layout in the performance of automotive assembly line. The result show, the production of assembly line output increases more than 31% from the current system. The source throughput rate average value went up to 252 units per working hour in model 3 and show the effectiveness of the pick-to-light system as efficient storage equipment. Thus, overall result shows, the application of AGV and the pick-to-light system gave a large significant effect in the automotive assembly line. Moreover, the change of layout also shows a large significant improvement to the performance.

1. Introduction

In a typical factory, material handling consists of 87% production time and an average of 50% total cost for company operation in manufacturing product [1, 2]. Hereby, material handling system is important in facilitating production layout. It has been several years that material handling has become a new and rapidly evolving science, hence a proper design of material handling system is important to facilitate worker safety and make the production and process flow become more efficient. A problem gap can be seen through an observation of case study at a selected real automotive assembly line, mainly in the material handling equipment or the layout facility that integrate material supply in the production line with inventory system in the warehouse.

Based on the case study observation which is done in an actual factory, the production background is start with logistic warehouse, body shop, paint shop and lastly assembly shop where the painted body is assembled in three main stations. The issues involving material handling can contribute to the reduction of production efficiency where the time required for an operator to do work is wasted. This research work focused on one main station only where one type of car is assembled. The semi-auto process is done by the operator during the assembly process. Studies are focused at the assembly shop and warehouse where it is noticed that the process flow of material from warehouse to the assembly line is done fully manually. Hence, transferring the material took a longer time and the risk of parts damage is high (either due to human or material) when transfering the material.

Integration between production system and material supply system is modeled by using a proper design of material handling system. By investigating the above problem, it is recommended to apply automated material handling equipment in order to see the effect on the production. But, to apply it in a real situation is high cost and time-consuming. At the same time, it holds high risk in wasting resource during investigating. Thus, simulation comes in mind. Simulation modeling solutions and tools are effectively used in addressing and challenging new issues to become more flexible, feasible and provide infinite integration of various simulation methods [3].

This paper intends to show the influences of material handling system and layout of material flow in an automotive assembly line by simulating the system in Delmia Quest software. The equipment focused on this research work is transport and storage equipment, where the new proposed system is compared with the simulation of the current system in a real situation of automotive assembly line factory. The improvement is done involving changes of transport equipment from manual tugger train into Automated Guided Vehicles (AGVs), and the changes of current storage equipment using manual order from the assembly line into semi-automatic order of the pick-to-light system in the warehouse. Therefore, a simulation of the automotive assembly line that is comparable to the case study on the real factory is done through simulation and the process performance is analyzed.

2. Case study

This section explained the data gathering needed for the simulation. All the data is taken from a case study done in one of the automotive factories in Malaysia and the data is gathered from observation, parts borrowing and reading on existing documentation. The process flow in designing the system, as shown in Figure 1 is started with planning, where all the data gathered in arrangement based on the elements needed in Delmia Quest software [4]. It consists of plant layout from warehouse to assembly line, distance traveled, distance between each station, including all parts involves in the process and their handling flow, and lastly is the transportation used to transfer part from warehouse to assembly line.

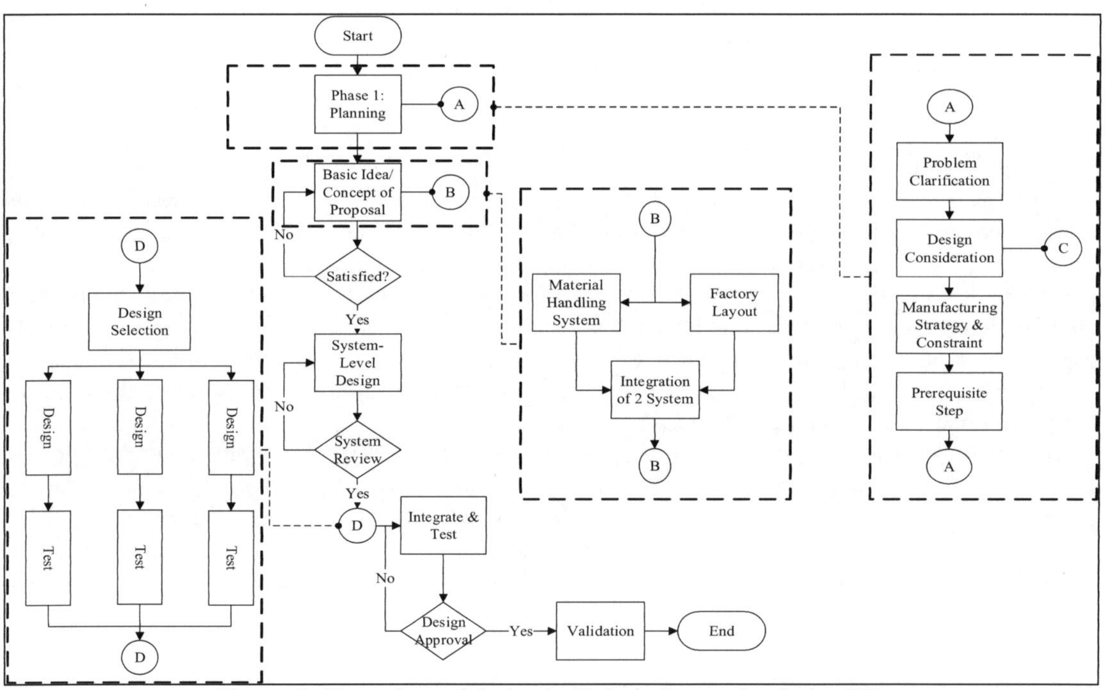

Figure 1. Flow chart of design in Delmia Quest simulation [4].

Starting from warehouse system, the error in parts arrangement cause improper part supply system. There is a possibility that operator might mistakenly sent parts to the carrier on the assembly line. Downline on the assembly line is caused by parts provision. This is because the manual system that used the list on paper can cause a human error in ordering parts. Thus, a pick-to-light system is recommended in this research work. Another issue is that, after in-depth observation and interview with operator and line captain, it is found that the operator in assembly line needs to walk and take the cart that supply from the supermarket and transfer the parts into assembly line. In the system level design, all the data planned is transfer into simulation elements. The model 1 is done similar to the case study situation, where the result of the model 1 is referred as the base performance of the current system to compare with other models. There are 3 models overall that are proposed for this research, All the design undergo evaluation test by comparing their production status and cycle time value with the current system values. Later on, the validation process is done by observing that all the components used are working properly and satisfied the current system requirement.

The factors considered in the experiment are type of layout, transport equipment, and storage equipment. Table 1 illustrates the variation of factors used for the experiments. Noted, that model 1 is the similar process used in the case study factory. The purpose of model 2 creation is to study the influence of material handling in the system. Meanwhile, model 3 brings the purposes of studying the new layout effect together with the improved material handling to the system performance. The comparison between all models can be used to analyze the significant effect of improving material handling and layout to the system. The comparisons between all models also act as a validation of the design capability to function similarly like the real situation of the factory.

Table 1. Variation of factors and list of models.

Models	MHS Configuration		
	Layout	Transport equipment	Storage equipment
Model 1	Old layout	Tugger train	Picking list system
Model 2		AGV	Pick-to-light system
Model 3	New layout	AGV	Pick-to-light system

In order to familiarize with the assembly line of the case study, observations and studies on all related documentations are done. The current layout of the assembly line is called old layout as shown in Figure 2 which state the transportation track from warehouse to assembly line including the travel distance. This research work only focuses on chassis line which consists of 1 transport pick-up station at the warehouse with 7 sub-pick-up stations where the labor is assigned to collect all the parts and load it into transportation. Also, the old layout has 10 drop off stations with different processes and part supplies. In the assembly line, each process has their own stations and each station is not fixed with only one process. It involves assembly of 3 to 4 types of different parts for each station.

The differences between current layout and proposed layout are the stations and the distance travel. Figure 3 shows the layout of the new system. It is based on the cellular and product layout concept, in which the concept is easier to monitor, easy to automate and most important it can increase the flexibility of the system [5, 6]. The new layout is designed in concerned of the design of the rack used as temporary storage while parts are transferred, and the allocation of each process is suited with the chassis line. As mention previously, the new layout consists of only 7 pick-up stations and 8 drop-off stations. Even though the stations have been reduced, the processes of the new system are the same with the current assembly processes due to a combination of 2 stations into 1 station.

Basically, the travel distance is according to movement of transportation from pick-up point at the warehouse and return back to its starting point in order to have a total distance traveled in 1 cycle of working hour. Apart from familiarizing with the assembly line, all parts involve have also been studied in order to know the requirement for parts distribution for every stations. All 31 parts are required for the assembly process of one type of car model. In order to avoid confusion and keep focusing on answering the research question, only one model of assembled car is focused.

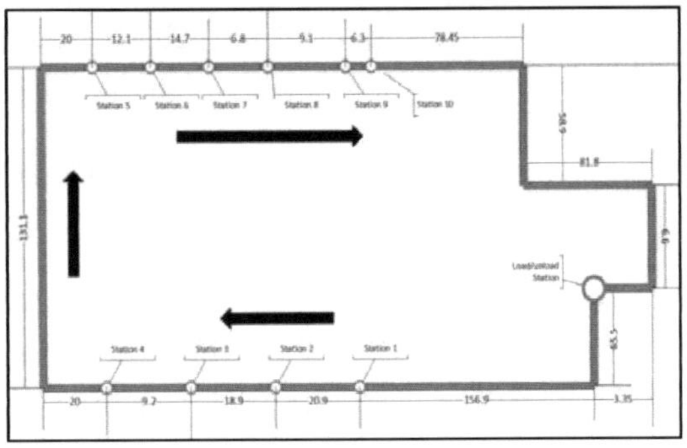

Figure 2. The drawing of old layout

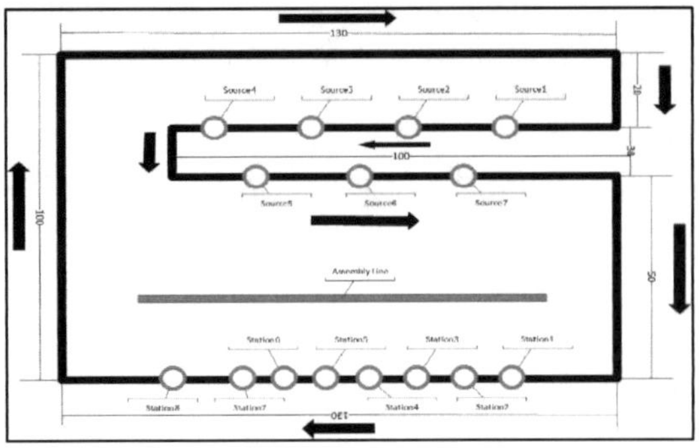

Figure 3. The drawing of new layout

In the case study, a fixed space for location of production and storage has been decided. Thus, on-floor transport equipment is the best recommendation to improve the system. AGV is introduced to replace the current tugger train transportation. Meanwhile, multiple parts are stored in different sections where the warehouse has already divided the parts into racks. The current warehouse system used in the case study is manual order picking list coming from the assembly line. A pick-to-light system is introduced to change the system into semi-automatic in the warehouse, and the order has been already programmed into the rack itself.

3. Simulation using Delmia Quest software

Delmia Quest includes various elements consist of physical model, parts, modeling element, logic and connections that justify parts flow through the model. This section discusses all elements needed to design the simulation. Delmia Quest is different from simple CAD approach in laying facility, where various parts and elements have interaction and logic between them. Even though the development of simulation is in sequence, but the order is not rigid as each step must be completed before simulating the model.

3.1. Elements in Delmia Quest

The layout arrangement consists of few elements in Delmia Quest software, which are parts as the main element, sources as the parts creation element, buffer function as the parts storage element, and lastly, the material handling elements (Labour, AGV, tugger train, picking list and pick-to-light).

Creating parts in the model is compulsory, in order to represent the models. No matter what kind of parts, they are the entities in Delmia Quest model that flow through the system, occupying resources and undergoing processing. Thus, when parts are created, source is made to produce the parts to allow other elements for processing and consumption. The source is very flexible elements that represent the entry-point of parts into the model. All models for this research work are using the same character of sources, but with a different arrangement of the layout based on the previous Figure 2 and 3. The label of the source also describe for parts arrangement for bins in the warehouse, the destination of the parts is set using simulation control language to assign a different quantity of parts for every sources. All models have 7 sources for storage elements in the warehouse. Meanwhile, 1 additional source is created at assembly line to resemble the arrival of car's frame into the conveyor to be assembled by labor.

Based on case study situation, transport equipment transfer parts using bins containing different parts based on workstations in the assembly line. The bins later are loaded into transportation and being transferred to the assembly line. This is where the function of buffers element is used, to represent temporary storage locations for parts before labor or transportation pick up the parts for next process. The function of the buffer at the warehouse is as hopper where the labor pick up a group of different parts that coming from the source and collect them into bins. Same as buffer at the assembly line, parts are stored in the buffer until labor pick up parts to do assembly work. The list of sources and buffers are described in Table 2 according to each model.

Table 2. List of sources and buffers for every model

Models	Sources	Warehouse's Buffers	Assembly's Buffers
Model 1 Model 2	Main_Car Source_OPT1 Source_OPT2 Source_OPT3 Source_OPT4 Source_RC_Bin Source_RR_Centre_Bracket	OPT1_Buffer_1 OPT1_Buffer_2 OPT2_Buffer_1 OPT2_Buffer_2 OPT3_Buffer_1 OPT3_Buffer_2 OPT4_Buffer_1 OPT4_Buffer_2 Dash_Buffer_1 Dash_Buffer_2 RC_Buffer_1 RC_Buffer_1	Assembly_Line_Buffer,α (α: 1 to 10)
Model 3			Assembly_Line_Buffer,α (α: 1 to 8)

Queuing discipline applied in all models is First-In-First-Out (FIFO), based on the concept where every time the buffer at assembly line is free, a new part will enter. Thus, a system where the first part to enter the buffer will be the first to be routed out is created. In Quest, transport and labor elements, loading and unloading parts at a decision point where the parts from other elements may enter and exit the path system only through the decision point.

In addition, traffic management of AGVs' and trains' stopping point is also at the decision point. In this research work, the decision point for AGV and train is divided into 2 sections that cover the 3 models used in the experiment. Section 1 consist of model 1 where their path system and decision point is created similar with case study's transportation path system. For model 2 and 3, the path system is a bit different due to the application of the pick-to-light system in the warehouse. The number of decision point for model 1 is 10 points at assembly line and functions as the parts drop off stations and 1 decision point for AGV pick-up station. Not forgetting, 7 decision points for labor inside the warehouse. As for model 2, the quantity of decision point is the same as model 1 at assembly line and has 7 decision points for tugger train at the warehouse.

In Quest, the user can set minimum requirements that need to be satisfied for the AGV, train or labor to leave the decision point. This setting is defined in depart requirement button. Until the requirement is satisfied, the AGV, train or labor will wait at the decision point, and unable to move. But it is restricted that the departure requirement will not be set to a value greater than the capacity of the AGV, if not the AGV will never move from the decision point.

The entry of parts is coming from the 7 sources as shown in the previous topic. In order to make the model realistic and as close to the actual process as possible, every time source detects zero queues (empty) at buffer, a part will be released from the source. It shows that the source monitors the buffer that is the first station in the production sequence of each part. Also, in order to show the uniqueness of the designed and determination of the outline structure in a fabrication of manufacturing layout, a Simulation Control Language (SCL) is applied in part entry of the model. SCL is used only in source and buffer process in order to have a unique sequence of part allocations in warehouse and assembly line.

3.2. Running simulation

The simulation can run after all elements are in place and connected with each other. The finished layout is illustrated in Figure 4 (model 1), Figure 5 (model 2) and Figure 6 (model 3) respectively. Each source for all models has 2 buffers together colored as purple in the Figures and the pickup station is at the center of the warehouse, allocate together with labor. The buffers at the assembly line are shown as cyan in the Figure. The characteristic of all models for warehouse buffers and assembly line buffers are using the default setting from Quest software.

The simulation is run with working day starting from 8.00 a.m. until 5.00 p.m. (8 hours). The 1 hour time is given as a rest time for labor. The result taken is only for 1 day of working time. The amount of time the labors take to transfer part between each assigned stations is established at 10 seconds of constant on the contribution.

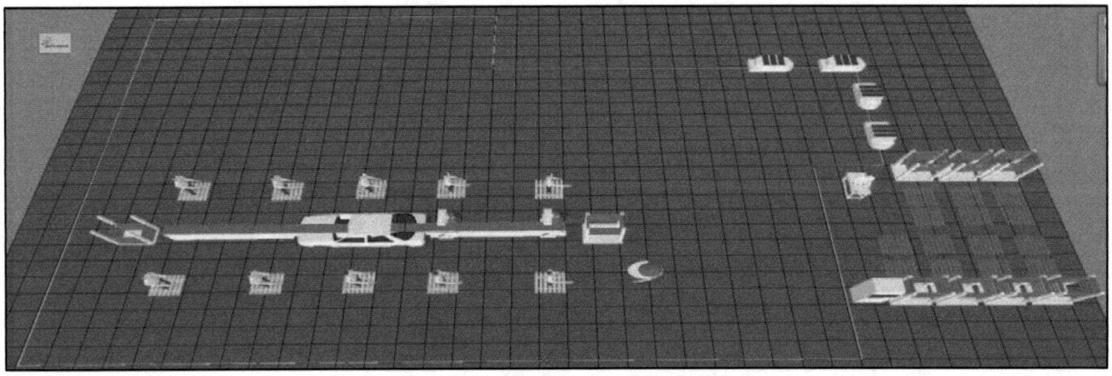

Figure 4. The layout for model 1

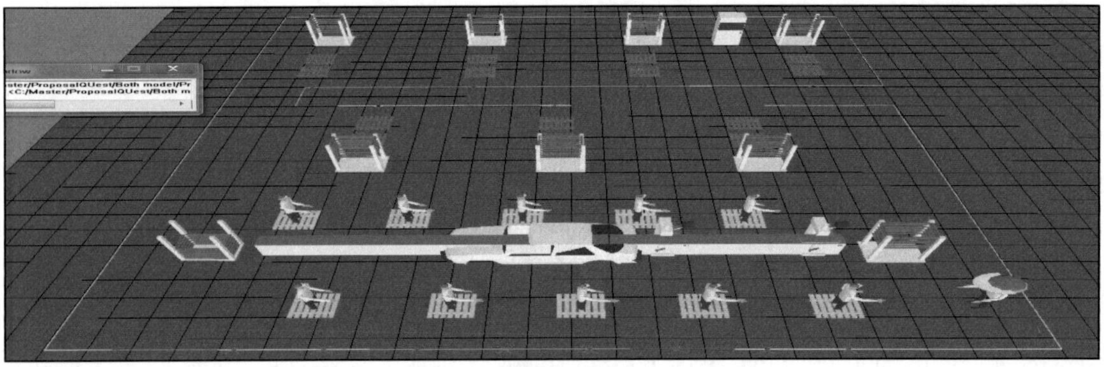

Figure 5. The layout for model 2

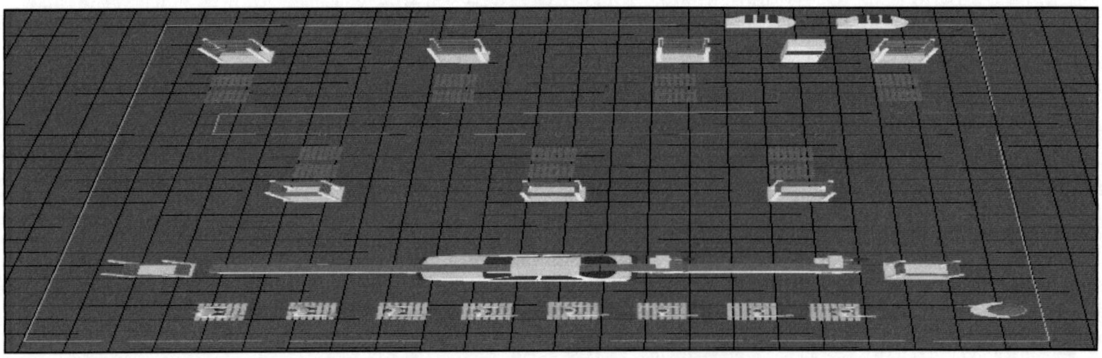

Figure 6. The layout for model 3

4. Improvement result of automotive assembly line performance

The result is compiled and presented in the statistical chart that consists of a few elements such as element class statistic (the total part created and the number of part supply), throughput rate in source statistic, the utilization of transportation statistic, the idle time and part entries of drop off buffer statistic and lastly a comparison of work in process for every models.

Based on the simulation result, the output increased sharply over the period where the output for a complete assembled car for the current system in model 1 is 14 units per working hour. The pie chart in Figure 7 illustrates the output (total part created), nearly half of the chart conquered by model 3's output. After changes of transport and storage equipment, the output increase as much as 36 units per working hour. When the layout is improved in model 3 the output increase 31% from the output of model 2, make the unit production is 47 units per working hour. It is obvious that the increment of output from the pie chart of Figure 7 is due to the effective material handling using AGV and pick-to-light system in the factory and the changes of the layout give more significant effect in the system.

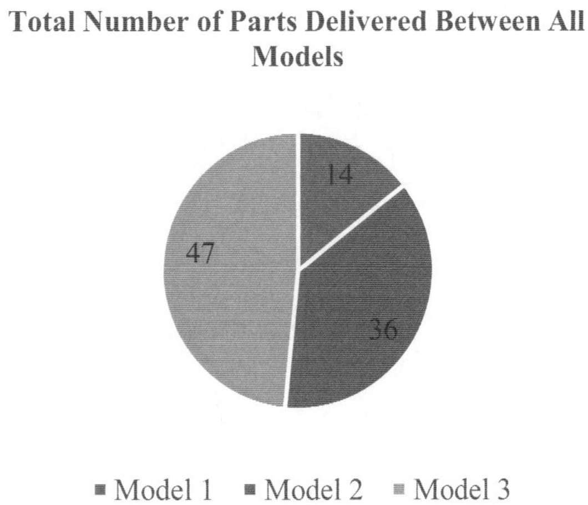

Figure 7. The pie chart of total number of parts delivered between all models

Figure 8 is the chart area that illustrates the throughput rate of the source in the warehouse. From model 1 to model 3, the chart shows remarkable changes of the area. That is most likely due to the total throughput rate data rose from only 53 units per working hour in model 1 status into 180 units per working hour in model 2. The source throughput rate average value went up into 252 units per working hour in model 3 and shows the effectiveness of the pick-to-light system as efficient storage equipment. Noted that in Figure 7, a few sources are able to give a large value of throughput rate, especially at Source_OPT1_1 where the creation rate of model 3 achieved 60 units per working hour compared with model 1 and 2, where the correlation of the value is not high. Same as throughput rate at Source_RC_Bin_1, even though it has the mode value, the correlation between model 2 is not strong. But, model 2 and 3 throughput rate give a very strong correlation than model 1. That is most likely because of the changes in material handling and layout of the system.

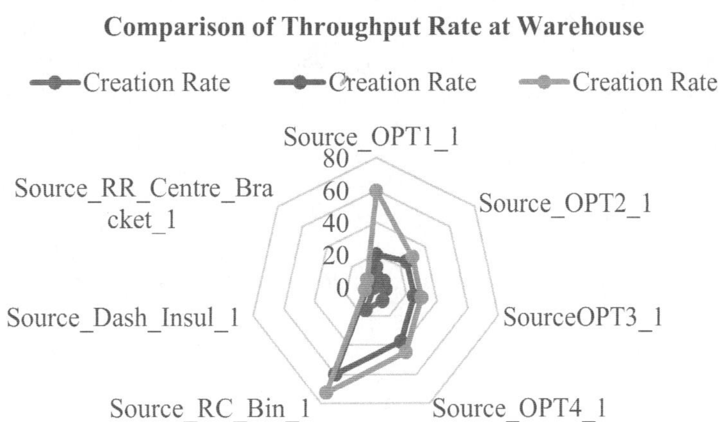

Figure 8. The chart area for comparison of throughput rate at warehouse

Figure 9 shows a linear graph for comparison of idle time in assembly line supply buffer. The data of model 1 shows that starting at Assem_Line_Buffer_1, the supply time reduce sharply into Assem_Line_Buffer_2. Based on layout configuration, the distance between both buffers is not too far from each other. Therefore, transportation is able to transfer parts faster, thus avoid from wasting time. But, as part transferring goes on, the idle time increasing along the other buffers. Differ with model 2 and 3, where the idle time reduces dramatically. The buffer that has highest idle time in model 1 is at Assem_Line_Buffer_10 (the last drop off stations), which is 2.877 hr. It is believe that the value is high due to the effect of waiting for transport to transfer parts first at 9 buffers before arrived at the last buffer. Noted that the idle time of model 2 and 3 does not show a strong correlation due to the application of transport equipment is same. The idle time still shows improvement where it reduces in average from 0.347 hour in model 2 into 0.292 hour in model 3. These averages shows, a 16% of idle time decrement in model 3. The buffer's idle time for model 2 and 3 shows a steady line and no major difference of idle time between every buffer. It's proved that AGVs are able to cope with the supply demand and distribute parts faster.

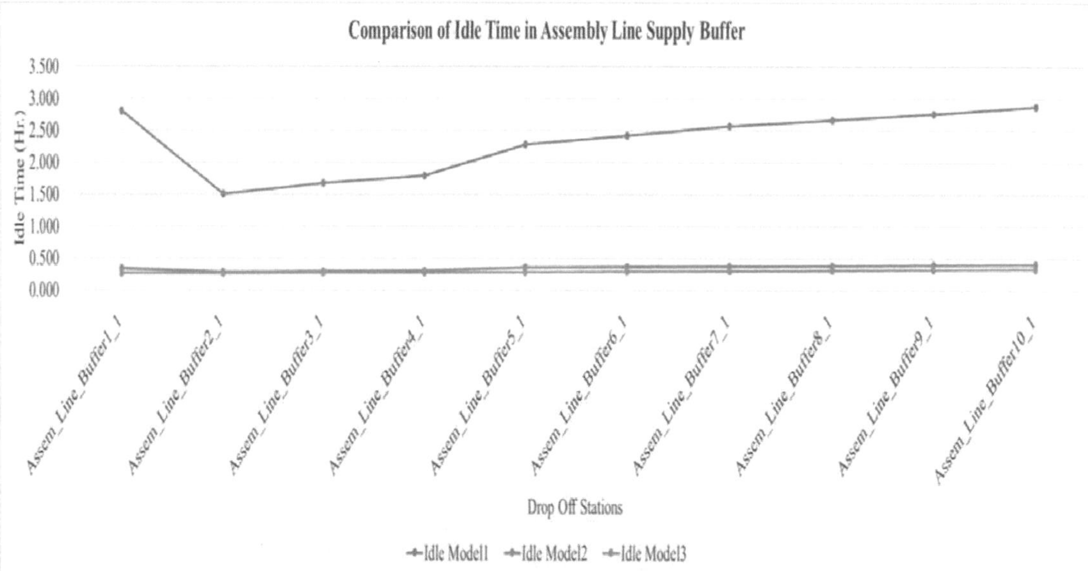

Figure 9. The line graph for comparison of idle time in assembly line supply buffer

A linear graph also is used to illustrate the comparison of parts entries in assembly line supply buffer as shown in Figure 10. The comparison is made to show how many parts are supplied at assembly line by transportation during working hour. The number of part entries in assembly line supply buffer maintain steadily in model 1. But, the average part supplies are only 32 units. Meanwhile, the value increases tremendously, in model 2 as much as 178 units. The increment percentage achieved more than 4 fold from the current system. Not to mention, the value of part entries at model 3 increase rapidly as much as 290 units in average. For an example, the upper extreme of part entries for every model holds by Assem_Line_Buffer6. This is due to higher demands from the assembly line at station 6. It also shows the part entries on model 2 and 3 give up and down values compared to model 1. The parts in every station have a different quantity to be transferred. Thus, the part entries are different from every station based on the demands along the period.

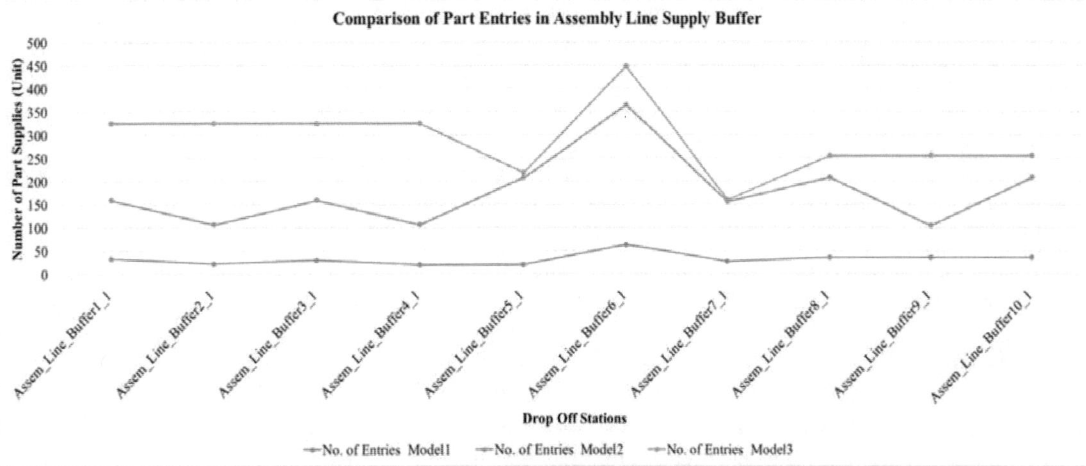

Figure 10. The line graph for comparison of part entries in assembly line supply buffer

In addition, the average waiting time of transportation reduced where the application of AGV allows transferring parts efficiently and saving time. As describe in Figure 11, the pie chart shows a comparison of average waiting time at warehouse pick up point. Model 2 and 3 shows a decrement value in warehouse buffer waiting time compared to model 1. Especially for model 3, the average waiting time improved better than model 2 average waiting times. In general, the current waiting time of parts in warehouse buffer from 0.716 hour (model 1) reduced to only 0.107 hour (model 3).

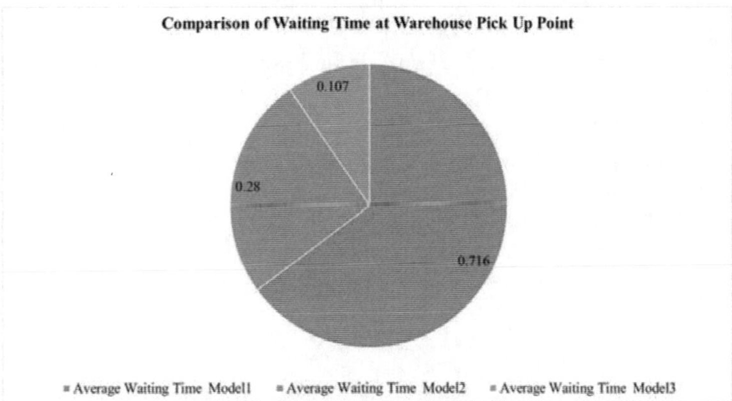

Figure 11. Pie chart for comparison of waiting time at warehouse pick up point

5. Conclusions

The purpose of this study is to propose a new design of integration between material handling and layout by investigating the influences of layout and material handling system in order to improve the automotive assembly line performance. For manufacturing strategy and constraint, this research work field is focused on the improvement of integration between material handling system and facility layout in current case study situation.

Thus, overall result shows, the application of AGV and the pick-to-light system gave a large significant effect in the automotive assembly line. Not to mention, the changes of layout show a bonus improvement to the performance. Hence, it can be concluded that the result obtained shows the influences of material handling system and facility layout in increasing the performance of the automotive assembly factory as well as other manufacturing fields.

A lot of aspects in this research study could not be covered at the same time, such as flexibility and industrial engineering scope. So, the constraint for this research work will be likely the manufacturing towards flexibilities and leans. Flexible Manufacturing is defined as to push the process ability to quickly adapt to changes in the volume of goods produced. The objective is to remain competitiveness ability to customize their products to offer an advantage. But the constraint for this research work is that it will give high complexity for automated material handling system and in-depth study in the simulation using Delmia Quest software.

Meanwhile, as for lean manufacturing which also called Just In Time (JIT) manufacturing, it aims to make the manufacturing process as efficient as possible by eliminating waste. Such aspect will make the company employ workers with multiple skills, sets and develop an efficient handling system that produces a low percentage of waiting/delay material. However, all those constraints are under this research consideration for future work.

Acknowledgment

First of all, thanks to Allah for allowing us to finish this paper. We would like to thank Universiti Teknikal Malaysia Melaka for giving us the opportunity to develop this study. Special thanks to research grant FRGS/1/2015/TK03/FKP/02/F00277 for funding the research. We also want to thank engineers and officers from Mawea Industries Sdn. Bhd. for their support and guidance on conducting a simulation in Delmia Quest software.

References

[1] Meyers F E 1993 *Plant Layout and Material Handling* (NJ: Regents/Prentice Hall-Englewood Cliffs)
[2] Kirkavak N 2010 *Introduction to Material Handling Systems* Cankaya University Ankara.
[3] Boteanu E and Zapciu M 2014 Improving Layout and Workload of Manufacturing System Using Delmia Quest Simulation and Inventory Approach *Int. J. of Innovative Research in Advanced Engineering* **6** 52 – 61.
[4] Seha S, Fairul A J and Zamberi J 2016 Methodology on Investigating the Influences of Automated Material Handling System in Automotive Assembly Process *IOP Conf. Series: Materials Science and Engineering* p 1 – 10
[5] Spur G, Seliger G and Viehweger B 1986 Cell Concepts for Flexible Automated Manufacturing. *J. of Manufacturing Systems* **3** 171 – 179.
[6] Weber A 2004 The Pros and Cons of Cell *assemblymag*/83136

"All talk no torque"– A novel set of metrics to quantify muscle fatigue through isometric dynamometry in Functional Electrical Stimulation (FES) muscle studies

M J Taylor[1,2*], C Fornusek[3], P de Chazal[1,2], A J Ruys[1]

[1] Faculty of Engineering and IT, The University of Sydney, Darlington, Australia.
[2]Charles Perkins Centre, The University of Sydney, Camperdown, Australia.
[3] Faculty of Health Sciences, The University of Sydney, Lidcombe, Australia.

*Corresponding author: matthew.j.taylor@sydney.edu.au

Abstract. Functional Electrical Stimulation (FES) activates nerves and muscles that have been ravished and rendered paralysed by disease. As such, it is advantageous to study joint torques that arise due to electrical stimulation of muscle, to measure fatigue in an indirect, minimally-invasive way. Dynamometry is one way in which this can be achieved. In this paper, torque data is presented from an FES experiment on quadriceps, using isometric dynamometry to measure torque. A library of fatigue metrics to quantify these data are put forward. These metrics include; start and end torque peaks, percentage changes in torque over time, and maximum and minimum torque period algorithms (MTPA 1 and 2), and associated torque-time plots. It is illustrated, by example, how this novel library of metrics can model fatigue over time. Furthermore, these methods are critiqued by a qualitative assessment and compared against one another for their utility in modelling fatigue. Linear trendlines with coefficients of correlation (R^2) and qualitative descriptions of data are used to achieve this. We find that although arduous, individual peak plots yield the most relevant values upon which fatigue can be assessed. Methods to calculate peaks in data have less of a utility, offset by an order of magnitude of ~10^1 in comparison with theoretically expected peak numbers. In light of this, we suggest that future methods would be well-inclined to investigate optimized form of peak analysis.

1. Introduction

The isokinetic dynamometer has been around since the 1800s [1] and has been used extensively in the area of muscle strengthening research. The field of Functional Electrical Stimulation (FES) has benefited greatly from the usage of this apparatus in a research context. Isokinetic dynamometers such as the Biodex [2,3] are used to assess the effects of various electrical stimulation protocols on fatigue as measured through quantification of force or torque. For example, Gorgey & colleagues [3] used a Biodex device to measure torque of knee extensor muscles in a study looking at the differential effects of pulse width on fatigue in FES cycling. While muscle biopsies can be used to measure fatigue in a physiological sense, such as in Martin et al. [4], this is a highly invasive procedure.

Torque-measuring devices such as the Biodex isokinetic dynamometer are less invasive, user-friendly and require minimal setup time. Therefore, they may be used to measure fatigue in a more user friendly fashion, albeit by measuring fatigue indirectly. The participant is seated on a chair, which is aligned to a moment arm connected to the dynamometer. This device is connected to a computer, with software that measures and displays torque produced as a muscle contracts. In isometric testing of quadriceps, the leg moves against a resistance and knee joint torque is measured. To perform isometric testing, the velocity of the device is set at a rate of zero degrees per second [5]. Therefore, dynamometry may be used to measure how various stimulation sequences can affect muscle responses. In addition, the Biodex is a "gold standard" against which other dynamometry methods such as hand-held systems may be compared against in the context of measuring strength [6], demonstrating its' utility in the field of muscle biomechanics.

Dynamometry may be used to measure how changing electrical stimulation parameters affects torque produced due to muscle contraction. One such electrical parameter is the duty cycle, which is the parameter describing ON time in relation to OFF time, or total stimulation time [7,8]. Changing the duty cycle for example, has been shown to produce differential decreases in force over time in biceps brachii [9].The aim of the larger study from which this data set was taken evolved around adapting the work of earlier authors Gentz & Moore [10] in a test subject to look at the differential effects of permuting ON, OFF or ON+OFF time of a 1:3 duty cycle on quadriceps fatigue.

The purpose of this paper is to demonstrate how a novel library of metrics developed can be used to describe fatigue from torque-time data, using a test case from an experiment looking at the effects of duty cycle on fatigue.

2. Methods

2.1. Electrical Stimulation Experiment and Biomechanical Setup

2.1.1. Biomechanical setup. A healthy subject (male, 25-years old, author MT) was used as the test subject for electrical stimulation experiments. He provided informed consent for all experiments. He had no history of neuromuscular pathology that excluded him from participating. The subject sat on an isokinetic dynamometer configured to setup mode where isometric torques could be measured. This machine has a moment arm that attaches to a limb. This allows torques to be measured around joints when muscles contract. Therefore, this experiment sought to measure knee joint torque when the quadriceps (thigh) contracted due to electrical stimulation. Upon calibration, the moment arm of the Biodex was set at an angle of 60 degrees knee joint flexion, relative to the horizontal. The subject was set up on the dynamometer such the lateral condyle of the right femur was directly in line with the pivot point (i.e. axis of rotation) of the moment arm, similar to other dynamometry studies [11-13]. The leg was strapped into the machine so the muscle contracted isometrically, without changing length.

2.1.2. Electrical stimulation protocol. There were 22 total sessions of electrical stimulation completed, (3 × 1:3 (baseline), 3 × 2:3, 3 × 3:3, 3 × 1:6, 3 × 1:9, 1 × 1:3 (test), 3 × 2:6, 3 × 3:9). Electrical stimulation was delivered using a hand-held Med4 (OttoBock Healthcare Products Austria GmbH, Vienna, Austria, TGA approved) stimulator, and applying electrodes on the quadriceps femoris muscle. Manual programming of the device was performed to make programs for different duty cycles. All programs were identical except for the duty cycle used to deliver stimulation. Each program was set to a protocol of frequency 30 Hz, pulse width 300 μs, with a 1-s ramp-up and 1-s ramp-down, at the voltage defined by factory settings of the stimulator. This protocol was an adapted form of Gentz & Moore [10]. Current amplitude was manually set at the beginning of each exercise, and was set at 45 mA on a basis of preliminary findings. For most experiments (except 1:3), stimulation was delivered at 1:3 for a 5-min warmup, then the test duty cycle for 20-min. Some trial runs were also performed prior to the actual experiment, for the purposes of optimisation. All permutations were performed on three separate occasions (runs 1,2,3). This with done with at least two days of rest in between trials, similar to Kesar & colleagues [14]. For the purposes of this paper, torque data from the 2:3 [2-s ON, 3-s OFF] duty cycle runs are discussed only. Further, only uncorrected data are presented for the purposes of discussing the data analysis procedures.

2.2. Data Analysis Procedures and Metrics

2.2.1. Metric conceptualization. The Biodex dynamometer produced datasets of time, torque and angle throughout the duration of stimulation. A list of metrics were developed from the literature, supplemented with other metrics to quantify these data (table 1). During preliminary analysis, there were several metrics added to this library in collaboration with members of the research team. However, only a subset of this comprehensive list were used in the final analysis, and these are presented. In addition, some were conceptualised *post-hoc.* In the context of this experiment, torque waveforms may be considered to be divided up into two distinct time periods: warm-up (*Wp*), and exercise (*Ex*), as generated from the two periods of stimulation. The overall exercise is a combination of these two periods. The term "run" (table 1) refers to repeats of the same duty cycle, however in this paper only run 1 of the 2:3 condition is discussed.

Table 1. Torque peak metrics and plots used in this study and computational procedures.

A. Fatigue Metrics: Metrics taken or derived from literature.

- **Strength decrement indices modified [SDIM]:** The decrease in torque over time during warm-up, exercise and overall stimulation. SDIM (Wp-Wp), SDIM (Ex-Ex) & SDIM (Wp-Ex) are computed by calculating FST and APL averages and comparing percentage decreases. Index adapted from Matsunaga et al. [15], who cite Clarke & colleagues [16]; and also from Hartkopp et al. [17].

B. Peak Metrics: Metrics to describe torque peaks.

- **Individual peaks [Z, F, S, T, A, P & L]:** Individual peak values. Z – zeroth peak, F – first peak, S – second peak, T – third peak, A – antepenultimate peak, P – penultimate peak, L – last peak. Z peak is that stimulation peak caused when the current is increased manually prior to the commencement of automated stimulation as manually set. F, S, T appear at start, A, P, L appear at end of a given region of the waveform.
- **Compound peak values [FST and APL]:** Averages of first and last three peaks [excluding Z peak].
- **Expected peak number [EPN]:** The expected number of peaks in a torque-time dataset based on the number of repeated units of stimulation [period] provided. The period is the sum of the ON, OFF and ramp-up (RU) and ramp-down (RD) times. Calculated by dividing the stimulation time by the period. The EPN may be compared with other maxima calculations, to test how accurate these are in peak detection.
- **Overall torque metrics [Maximum torque, minimum torque, range, average, median]:** Basic features of the torque-time waveform once it has been adjusted about the *x*- and *y*- axes.

C. Angle Metrics: Metrics to describe angle time data.

- **Overall angle metrics [Maximum angle, minimum angle, angle variation (AV), average, median]:** Similar computation to overall torque metrics.

D. Derived Torque Plots: Plots generated by applying different manipulations to torque-time data.

- **Raw plot:** Plot of torque-time waveform as generated by machine.
- **Processed plot:** Waveform shifted such that time zero is when the stimulation begins. Removal of noise either side.
- **Peak plots:** Individual torque peaks.
- **Maxima and minima plots:** Plots derived by applying maxima or minima algorithms to torque data. An **exclusion value (EV)** was chosen such that peaks above were used for maxima, and peaks below for minima.
- **Averaged plots:** Plots generating by averaging data over 1-s, one-half-period, one-period and two-periods respectively.
- **Maximum and minimum torque period algorithms [MTPA 1 and MTPA 2]:** These plots show maximum and minimum torques occurring in the period surrounding a torque value. The algorithm looks one half-period in front and behind a torque point, then the maximum is displayed.
- **Peak torque every 5 min and 1 min:** The maximum torque occurring every 5- or 1- min [relative to the half-period in front or behind a data point at $t = 5$ and $t = 1$ min, respectively]. Generated from MTPA 1 and MTPA 2 plots. Adapted from Fornusek [18].

2.2.2 Data processing using metric library. The pertinent metrics applied on the torque time data are listed in table 1. These metrics and plots were produced once the data is imported into MS Excel. The fatigue metric used in this study was the strength decrement index modified (SDIM) (section A, table 1). This calculated a percentage decrease in torque by comparing averages of first and last three peaks (section B, table 2). In addition, an expected peak number was calculated that was the number of peaks expected based on the stimulation applied to the muscle. This value was compared to the number of maxima calculated (section D, table 1) to describe differences in peak counts and those expected. Further, overall torque (section B) and angle (section C) metrics were also computed to describe basic characteristics of the waveform. Following computation of metrics, a global correction factor (GCF) was applied to account for the torque of gravity. These data are not discussed in this paper, as uncorrected data are used for illustrative purposes of how the library can be used in application.

In addition to metrics, plots were generated to give a visual representation of torque data over time (section D, table 1). Raw and processed plots showed the basic features of the waveform over time, whereas peak plots were used to see peaks at the start and end of regions of interest. Maxima and minima, averaged, MTPA and peak torques were also generated in an attempt to describe how torque peaks decrease over time. Computational plots were generated in MS Excel, and methods described in section D of table 1.

3. Results

3.1. *Waveform processing*

Raw data [time, torque, angle] were obtained in the .LVM format from the Biodex, and imported into MS Excel as .XLSX files. Data were truncated to remove "noise" either side caused by setting up of the leg in the dynamometer. The full waveform for the 2:3 (run 1) is shown in figure 1, with raw waveform (top) and processed waveform (bottom). There are two distinct time periods of the exercise, the 5-minute *warm-up* (*Wp*) [1:3], and 20-minute *exercise* (*Ex*) [2:3]. Either side of the raw waveform (figure 1, top) are two types of noise. Prior to the beginning of Wp, there is a lower baseline, corresponding to the torque reading before the leg was loaded onto the machine. Then, the baseline shifts when the leg is loaded (indicated by the second patch of varying signal, diamond in figure 1 top). The baseline changed due to the leg exerting a torque on the moment arm of the dynamometer, positioned at an angle of 60 degrees flexion.As can be seen on the far right of the waveform (square, figure 1), there is a return to a baseline that is similar to before the experiment was commenced (before diamond, figure 1).

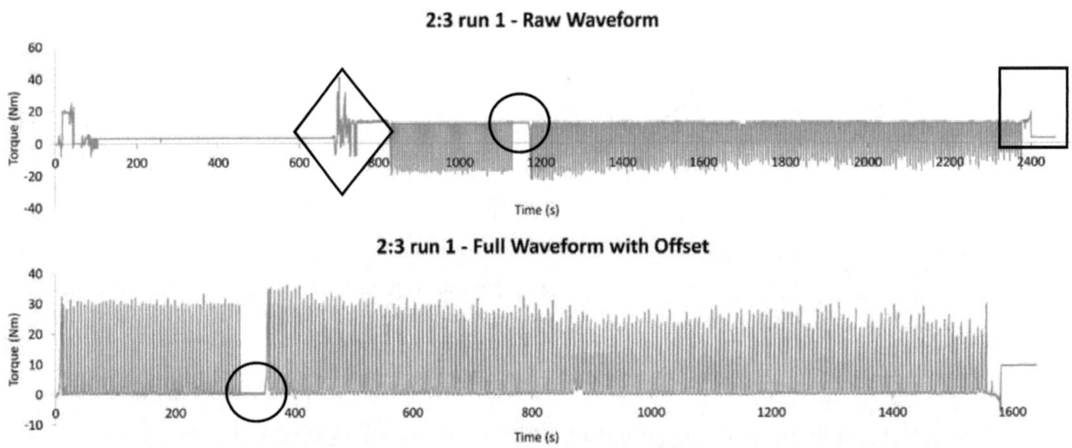

Figure 1. The full waveform for 2:3 (run 1). The torque due to gravity may be seen after the exercise period (square). There are two periods of torque production, warm-up (*Wp*) and exercise (*Ex*). These are separated by a short time gap corresponding to when stimulation was changed (circles).

The raw signal was then processed by performing baseline adjustment and shifting such that the zero time point corresponded to start of stimulation (figure 1, bottom).This provided a more clear representation of the generated torque waveform during the period of electrical stimulation. Pre-experimental noise was eliminated, and the torque waveform during the defined periods of *Wp* and *Ex* may be seen more explicitly.The direction of torque peaks were initially downward, with a negative sign convention (figure 1, top). They were changed to positive by multiplying the entire torque series by -1, and are easier to manually inspect (figure 1, bottom). Also, initially the baseline of the torque waveform during stimulation was raised above zero (figure 1, top). The processed waveform was adjusted such that peaks are zeroed about the baseline (by adding 13 Nm to every torque value). Accurate values of torque peaks could then be read from the signal (figure 1, bottom).

3.2. Fatigue and peak metrics

Adjustment of waveforms permitted calculation of fatigue metrics over time. The maximum torque for the 2:3 torque waveform was found to be 36.14 Nm, while the minimum torque was a value of -6.16 Nm [torque range, 42.30 Nm]. The average and median torques were 6.99 Nm & 0.42 Nm, respectively. Peak plots (figure 2) were used to derive fatigue metrics over time by using the magnitude of first and last peaks. Individual peak values for the warm-up period were found to be: 31.77, 30.02, 27.98 & 30.89 Nm [Z,F,S,T peaks], and: 29.55, 28.87 & 29.23 Nm [A,P,L peaks]. The compound peak values for Wp were found to be 29.63 Nm [FST(Wp)], and 29.22 Nm [APL(Wp)]. For the exercise period, peak metrics were: 34.30, 34.78, 34.10, 34.19 Nm [Z,F,S,T peaks], and: 21.54, 20.37, 29.09 Nm [A,P,L peaks]. The compound peak values for Ex were found to be 34.36 Nm [FST(Ex)], and 23.67 Nm [APL(Ex)]. These compound peak values were then used to calculate percentage decreases in torque over the periods of warm-up, exercise and total stimulation (warm-up+exercise). Corresponding strength decrement indices modified were: 1.39 Nm [SDIM(Wp-Wp)], 31.11 Nm [SDIM(Ex-Ex)], and 20.13 Nm [SDIM(Wp-Ex)].

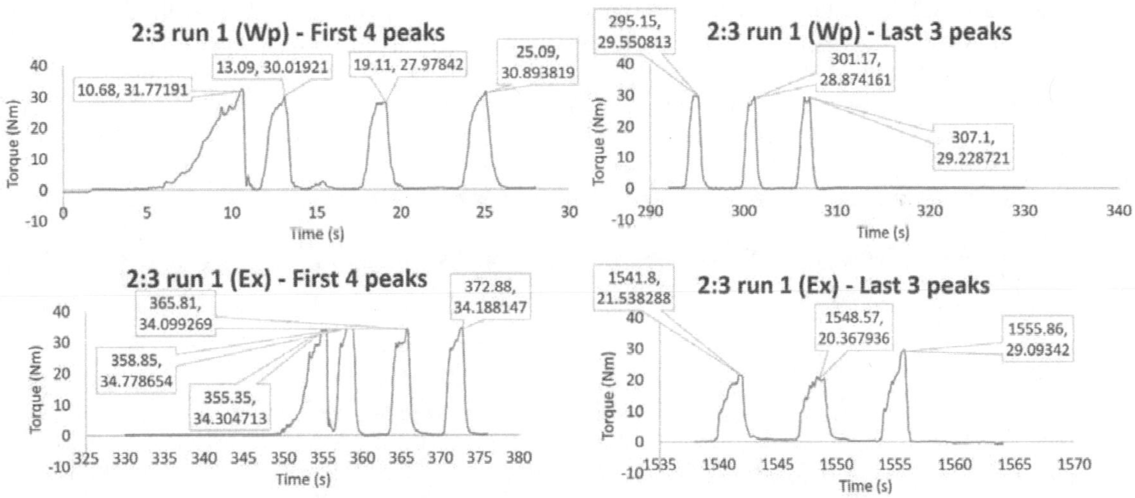

Figure 2. The first and last peaks for 2:3 (run 1).

3.3. Peak approximation – extrema plots

The maxima and minima (*extrema*) plots were used generated to approximate torque peaks and the baseline respectively. Linear trendlines were superimposed across all plots to assess the degree of appropriateness of a linear fit. The maxima and minima plots (figure 3) had coefficient of correlations of R^2 = 0.414, and R^2 = 0.115. Maxima plots were used to compare peaks with those expected by theory. The expected peak number (EPN) for the 2:3 condition was found by dividing the total time of stimulation by the period [2-s ON + 3-s OFF + 1-s ramp up + 1-s ramp down = 7-s], each repetitive unit of stimulation. This was found for the warm-up at 1:3 and exercise at 2:3. The EPN(Wp) was 50 peaks, while the EPN(Ex) was 171.43 peaks. Hence, the total EPN of the entire stimulation period was found to be 50 + 171.43 = 221.43 peaks. The number of maxima computed from the maxima algorithm was found to be 2080, by use of the function COUNT in Excel. Therefore, there was an approximate difference of one order of magnitude between peaks expected and peaks calculated by the maxima analysis in Excel (figure 3).

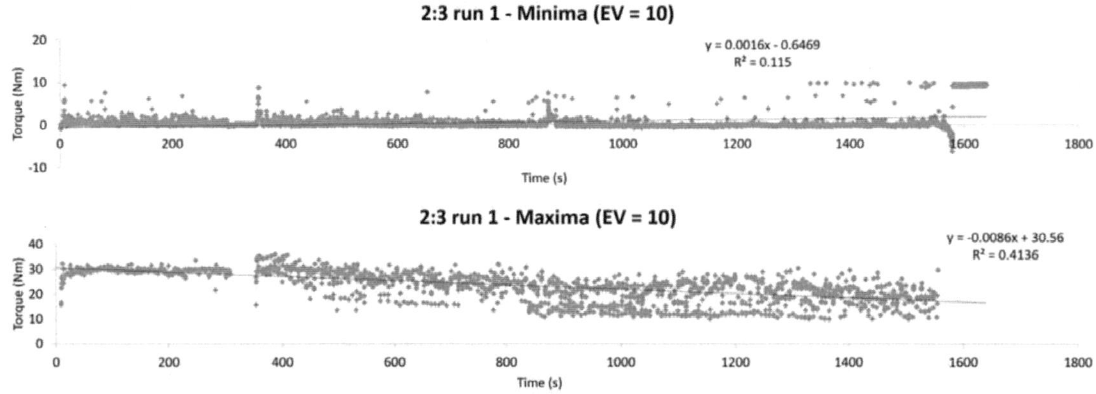

Figure 3. Maxima and minima plots. For maxima, Wp and Ex may be seen as two distinct data sets separated by a gap in stimulation.

3.4. Further peak approximation – MTPAs and peak torque every 5- and 1-minute

Following maxima and minima analysis of torque data, maximum and minimum torque period algorithms (MTPA 1 and MTPA 2) were generated (figure 4). Figure 4 depicts the MTPA 1 and MTPA 2 for all 3 runs of the 2:3 duty cycle experiment. For the run 1 data, MTPA 1 had an $R^2 = 0.216$ (figure 4, top), and MTPA 2 had an $R^2 = 0.049$ (figure 4, bottom). Also computed from these algorithms were maximum torque every 5-min [data not presented] and 1-min. Similar to figure 4, runs 2 and 3 are plotted alongside run 1. The peak torque for the 1-min data had $R^2 = 0.507$ (figure 5).

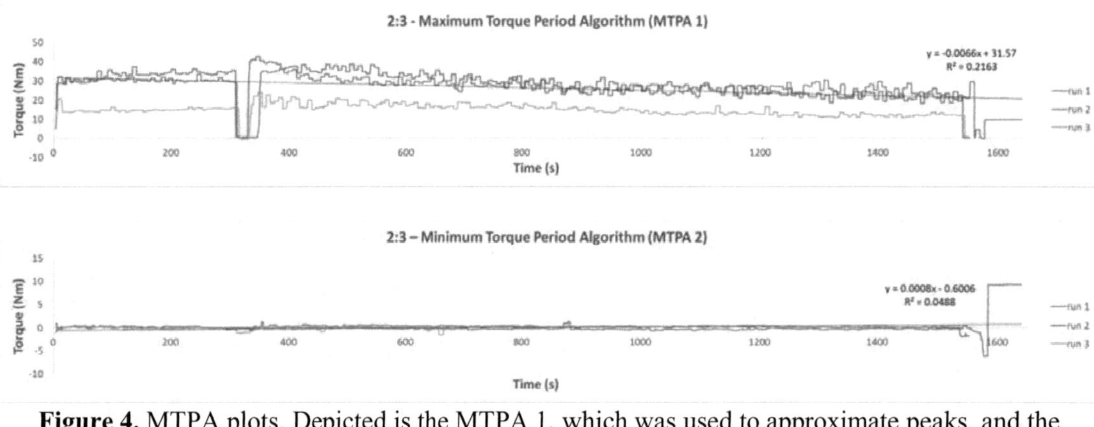

Figure 4. MTPA plots. Depicted is the MTPA 1, which was used to approximate peaks, and the MTPA 2 which was used to approximate the baseline. R^2 and trendline shown for run 1 only.

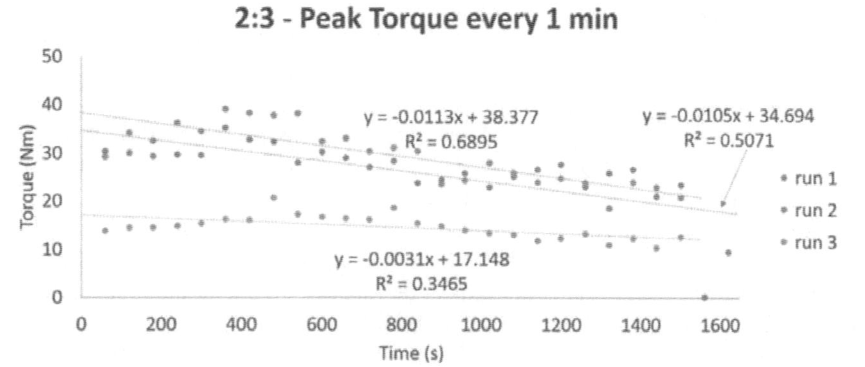

Figure 5. Peak torque plots. Shown is the peak torque in the period of the data point every 1-min. 5-min was also computed, 1-min shown only for illustrative purposes.

3.5. *Averaged plots*

Averaged plots were also found in attempt to reduce torque peak data to plots that could be compared between and across experimental conditions. For the averaged plots, the coefficients of linear fit were: $R^2 = 0.002$ (1-s), $R^2 = 0.005$ (3.5-s), $R^2 = 0.034$ (7-s), and $R^2 = 0.045$ (14-s). Averages over 1-s and two periods (14-s) are shown in figure 6. As may be seen, the 1-s averaging (left) has less of a linear approximation to peaks than the two-period average (right).

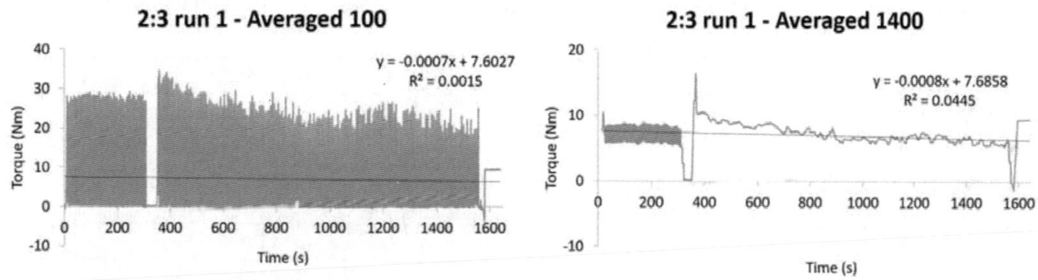

Figure 6. Averaged plots. Shown are plots with data averaged over two time periods: 1-s (100), and two periods 14-s (1400). Plots were also generated for one half-period 3.5-s (350), one period 7-s (700) [data not shown].

3.6. *Angle variation.*

For the variation in dynamometer angle over time (figure 6), the metrics similar to torque metrics (section C, table 1) were also computed. The maximum angle was 60.348 degrees, minimum angle 57.041 degrees, with an angle variation (i.e., range) of 3.307 degrees. The average angle was 58.737 degrees, and median angle 58.740 degrees. Hence, the "accuracy" (defined as the range as a percentage of 60) was a value of [3.307/60.000]*100 = 5.512%.

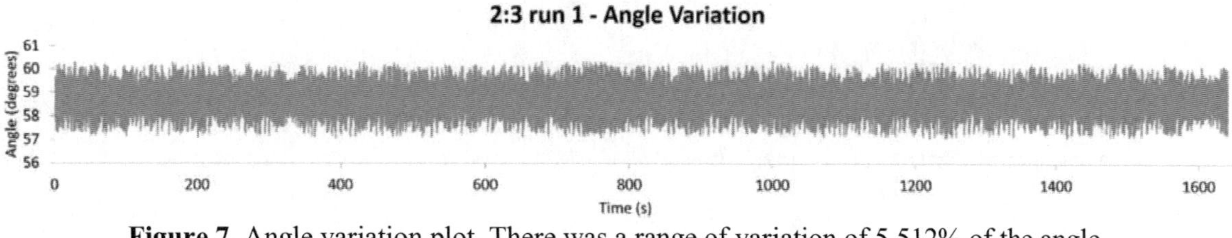

Figure 7. Angle variation plot. There was a range of variation of 5.512% of the angle.

3.7 Critique of data plot procedures.

In addition to the quantitative metrics derived from plots, a qualitative description of data analysis procedures was made by the author MT (table 1). A qualitative description of the relative ease of computation of the various plots (table 2) shows how the use of Excel is associated with different degrees of ease, or tedium in calculating various representations of torque (or angle)/time data. The raw waveform had the quickest relative time required to generate a graph, as simple importation and plotting of the data was required to achieve the desired result. The full angle plot was similar, but required an exact awareness of the time zero of Wp in which to commence the recording (and hence plot the waveform). Extrema and MTPA plots took longer to derive as the full waveform had to be plotted first (i.e., start of Wp and end of Ex had to be identified). Then, the data were operated on by relevant formulae to yield these representations of the torque waveform. The individual (first four, last three) plots were by far the most time consuming as iterations of manually selecting the torque/time data where these had to be performed, sometimes multiple times, to obtain the data in the zone of interest. Similar results are also suggested for the "manipulation" parameter (column 2, table 2).

In terms of usefulness, the individual peak plots were the most useful as values could be read from them and fatigue metrics could be derived as a result (e.g., SDIMs). Perhaps the least useful were the minima and MTPA2 plots, as due to the high sampling rate several local minima not representative of of the baseline were encompassed in the plot with the exclusion value of 10 Nm. The low R^2 values of 0.115 (minima), and 0.049 (MTPA2) are suggestive that these plots are not accurate representations of the baseline. Comparability trends across plots follow a similar trend (column 4, table 2).

Table 2. Qualitative description of the computational complexity and relevance of various plots (+ poor, +++ average, +++++ outstanding).

Waveform plot	Comp. time[a]	Manipulation[b]	Usefulness[c]	Comparability[d]
Raw waveform	++++	+++	++	+
Full waveform	+++	+++	+++	++
Individual peak plots	++	+	+++++	+++++
Maxima plots	+++	++	++	+++
Minima plots	+++	++	+	+++
MTPA 1 plot	+++	++	++	++
MTPA 2 plot	+++	++	+	++
Peak torque 5-min 1-min plots	+++	+.5*	+++**	++++
Averaged plots	+++	++	N/A***	++
Angle variation plot	+++.5	+++	+++	++

[a]*Computational time to generate plot. Some of these build off the full waveform, so the times shown are relative to once this data set has been obtained (e.g., maxima, MTPA's).*

[b]*How easy or difficult it is to manipulate torque/time (or angle/time) data to produce the desired plot. A similar argument also applies to a).*

[c]*Is it a meaningful metric in the context of examining torque responses for fatigue?*

[d]*Is is easy to compare plot with plots of other runs?*

Have to choose every 5 or 1 min and exclude other data which requires more complex analyses.

**Averaging over a few periods is perhaps a more meaningful analysis.*

***Depends on period (consistent with runs from this duty cycle – other runs, and other duty cycle). The averaged dataset changes greatly depending on time periods over which data is averaged.*

4. Discussion

This study has successfully shown how a list of derived fatigue metrics may be employed in an experiment looking at torque/time data from isometric dynamometry. Through creation of a library of metrics, a practical example has been provided of how the library can be applied to analyse a quadriceps' fatigue when stimulated at a duty cycle of 2:3 under isometric conditions.

Use of this library of fatigue metrics was able to demonstrate some profound torque changes over the 25-minute period chosen. The plots showed clearly fatigue arising, reflected as decreasing torque values from the muscle. Further, individual peak plots were able to show "snapshots" of this trend over time, and supplemented with SDIM and individual peak values provide a basis upon which this data could be compared with other experimental conditions (i.e., duty cycles) for the sake of comparison.

The applications of the proposed metric library are not limited to muscle experiments involving electrical stimulation protocol investigations. Isokinetic dynamometry has been used in other studies, such as that of Thorstensson et al. [19] in their work on the force-velocity correlation in isokinetic contractions of the knee extensor group, using a Cybex II system. Furthermore, isokinetic dynamometry has also been used in sports' contexts to measure isokinetic muscle properties [20,21], for example, to assess if there is added benefit of kinesio taping in muscle torque production [21].The metrics we have presented here may be of assistance to these fields and hence is not limited to FES. While we have utilised an isometric dynamometry methodology, the library canbe adapted to include variables relevant to isokinetic contractions. For example, in isokinetic experiments, contractions are elicited such that fixed rotational velocity are traversed by a muscle [22].Furthermore, work performed by Drouin et al. [23], which aimed to characterise the "reliability and validity" of a Biodex System 3 machine examined the variables ofposition and velocity in addition to torque. Therefore, a future refinement of our metric library could incorporate more isokinetic-specific metrics, to increase its usefulness across a range of experimental designs.

The issue of peak analysis is one problem, which has been studied in other fields such as computational cardiology. As such, future work would be well-inclined to apply similar methods to torque-time data. In our study, identification of peaks was attempted by the derivation of maxima and MTPA plots, in order to characterise torque peaks over time. There have been various attempts to quantify electrocardiogram waveforms for example, through identification of "fiducial points" [24]and differentiating the QRS complex [25]. Perhaps insights from the algorithms in this field may assist in identification of torque peaks of Biodex data. Furthermore, use of computational software such as MATLAB which has been used in similar contexts, such as that of ECG analysis to characterise R peaks [26] may solve the issue of slow computation time experienced in the derivation of some of our metrics (table 2).

The methodology of this work is not without limitations. The qualitative critique of various metric methods demonstrated for example, that although individual peak plots are indeed quite meaningful, their computation time is poor in relation to the other plots. Further, the extrema plots, with an average computational time, yield at best average meaningful data.In terms of experimental design, there were several avenues for optimisation, such as quantification of the exact site to provide optimal stimulation through use of a motor pen [27] and assessing whether to stimulate the femoral nerve or quadriceps directly as they differ in terms of aspects such as comfort [28]. Moreover, the lower limb and torso could be held still during the contraction process [29] to minimise the input of residual torques from other movements of the body. In comparison with other runs, there were marked differences in some metrics for example [data not presented]. Future refinements of both data analysis procedures, precise electrical stimulation and dynamometry set-up would assist in yielding meaningful, comparable results. However, this paper has successfully shown, in light of these limitations, how the library may be implemented in practice on torque-time data series.

5. Conclusions

Here it has been successfully demonstrated how a novel list of parameters can be adopted to model fatigue from a muscle undergoing isometric contractions due to FES. Through using data from the thigh, metrics were applied to torque-time waveforms over a 25-minute period. Individual peak plots were most useful in showing fatigue over time, but were computationally tedious. Future work would utilize advanced computational methods to increase computation time, and the utility of this proposed library in modelling muscle fatigue. Upon these refinements, this library can be adopted by researchers in the wider field of muscle biomechanics, namely within the scope of rehabilitation engineering and exercise and sports science applications. This will hopefully facilitate in making comparisons of muscle fatigue across different applied stimulus (mechanical or electrical) conditions.

References

[1] Gandevia S C 2001 *Physiol Rev* **81**(4): 1725-1789.
[2] Gregory C M, Dixon W and Bickel C S 2007 *Muscle Nerve* **35:** 504-509.
[3] Gorgey A S, Poarch H J, Dolbow D R, Castillo T and Gater D R 2014 *JRRD***51**(9): 1455-1468.
[4] Martin T P, Stein R B, Hoeppner P H and Reid D C 1992 *J. Appl. Physiol.* **72**(4): 1401-1406.
[5] Binder-Macleod S A, Halden E E and Jungles K A 1995 *Med. Sci. Sports Exerc.* **27**(4): 556-565.
[6] Martin H J, Yule V, Syddall H E, Dennison E M, Cooper C, Sayer A A 2006 *Gerontology* **52:** 154-159.
[7] Baker L L, Wederich C L, McNeal D R, Newsam C and Waters R L 2000 *NeuroMuscular Electrical Stimulation – A Practical Guide (4th ed)*(United States of America: Rancho Los Amigos National Rehabilitation Center).
[8] Carmick J 1997 *Pediatr Phys Ther* **9**(3): 128-136.
[9] Naeem J, Azman A W, Khan S, Mustafah Y M *IOP Conf. Series: Materials Science and Engineering, **53:** 5th Int. Conf. on Mechatronics, 2-4 July, 2013, Kuala Lumpur, Malaysia.*
[10] Gentz L and Moore C 1988 *Phys Ther.* **68**(5): 834.
[11] O Brien T D, Reeves N D, Baltzopoulos V, Jones D A and Maganaris C N 2009 *Eur J Appl Physiol* **106:** 849-856.
[12] Kubo K, Tsunoda N, Kanehisa H and Fukunaga T 2004 *Eur J Appl Physiol* **91:** 349-352.
[13] Gerrits K H, Maganaris C N, Reeves N D, Sargeant A J, Jones D A and De Haan A 2005 *Muscle Nerve* **32:** 73-80.
[14] Kesar T, Chou L-W and Binder-Macleod S A 2008 *J Electromyogr Kinesiol* **18:** 662-671.
[15] Matsunaga T, Shimada Y and Sato K 1999 *Arch Phys Med Rehabil* **80:** 48-53.
[16] Clarke H H, Shay C T and Mathews D K 1954 *Arch Phys Med Rehabil* **35:** 560-567.
[17] Hartkopp A, Murphy R J L, Mohr T, Kjaer M, Biering-Sorensen F 1998 *Arch Phys Med Rehabil* **79:** 1133-1136.
[18] Fornusek C 2005 *An Isokinetic Functional Electrical Stimulation Leg Cycle Ergometer for Individuals with Spinal Cord Injury. PhD Thesis, The University of Sydney, Australia.*
[19] Thorstensson A, Grimby G and Karlsson J 1976 *J Appl Physiol* **40**(1): 12-16.
[20] Gleeson N P and Mercer T H 1996 *Sports Med.* **21**(1): 18-34.
[21] dos Santos Gloria I P *et al.* 2017*J Back Musculoskelet Rehabil,* Uncorrected proof.
[22] Marchant D C and Greig M 2017 *Hum Mov Sci* **52:** 67-73.
[23] Drouin J M, Valovich-mcLeod T C, Shultz S J, Gansneder B M and Perrin D H 2004 *Eur J Appl Physiol* **91:** 22-29.
[24] Plesnik E, Malgina O, Tasic J F and Zajc M 2012 *Med Eng Phys* **34:** 524-529.
[25] Arzeno N M, Deng Z-D and Soon C-S 2008 *IEEE Trans Biomed Eng* **55**(2): 478-484.
[26] Thulasi Prasad S and Varadarajan S 2013 *IJRET* **2**(8): 508-513.
[27] Gobbo M, Maffiuletti N A, Orizio C and Minetto M A 2014 *J Neuroeng Rehabil* **11:** 17.
[28] Bergquist A J, Clair J M, Lagerquist O, Mang C S, Okuma Y and Collins D F 2011 *Eur J Appl Physiol,* **111:** 2409-2426.
[29] Laufer Y, Ries J D, Leininger P M and Alon G 2001 *Phys Ther.* **81**(7): 1307-1316.

Finite element analysis of residual stress in cold expanded plate with different thickness and expansion ratio

Kamarul Arifin Shariffudin[1], Saravanan Karuppanan[1] and Santosh S Patil[1, 2]

[1]Mechanical Engineering Department, Universiti Teknologi PETRONAS, 32610 Bandar Seri Iskandar, Perak, Malaysia.
[2]Mechanical Engineering Department, Manipal University Jaipur, 303007, Dehmi Kalan, Jaipur, India

*Corresponding author: santosh045@gmail.com

Abstract. Cold expansion of fastener/rivet holes is a common way to generate beneficial compressive residual stress around the fastener hole. In this study, cold expansion process was simulated by finite-element method in order to determine the residual stress field around two cold expanded holes by varying the plate thickness and expansion ratio of the hole. The model was developed in ANSYS and assigned to aluminium alloy 7475-T61 material model. The results showed that the residual stress become more compressive as the plate thickness is increased up to t/d = 2.6 and decreased for further level of thickness. In addition, the residual stress at the edge of the hole become more compressive as the expansion ratio is increased up to 4.5% and decreased for further level of expansion. This study also found that the residual stresses near the entrance and the exit face of the plate are less compressive than the residual stresses on the mid-thickness of the plate.

1. Introduction

Variety of techniques has been developed to enhance the fatigue life of fastener or riveted structural joints used in aircraft industry. These type of joints are widely used since most of the joints are connected by rivets or bolts through the fastener holes that exist on the aircraft structure. This type of joint will help the parts to be assembled and disassembled easily. However, these holes on the structure experience tensile stresses on the surface which subsequently will lead to fatigue failure. In order to increase the fatigue life of the material, cold expansion of rivet/fastener holes has been used for over four decades in the aircraft industry and it is very efficient approach to extend the fatigue life of the aircraft structure [1-3]. Today, two common cold expansion methods employed are roller burnishing technique and split-sleeve method.

Roller burnishing technique consists of a tapered mandrel and pin roller bearing held in a cage. The tapered mandrel is inserted into the hole and it is pulled through the roller bearing cage which will induce the compressive residual stress around the hole [4]. However, the split sleeve cold expansion method, in which a tapered mandrel fitted with lubricated split sleeve is pulled through a hole, is considered as the most effective way in creating beneficial compressive residual stress around expanded holes since compressive residual stress remains tangential to the hole edge as the mandrel is being removed from the hole [4]. The development of compressive residual stress around the holes due to the plastic deformation of the material enhances the fatigue life of fastener joints since the initiation of fatigue cracks at the stress concentration region is delayed. The role of split sleeve is to ensure the correct radial expansion of the hole and prevent damage to the hole by avoiding the direct contact between the sliding mandrel and the hole. Figure 1 shows the schematic diagram of the split sleeve cold expansion method.

Residual stress is the stress that exists in a body due to many conditions such as material properties, expansion ratio, thickness of the material, size of the hole and others. Odzemir and Herman [4] indicated that the compressive residual hoop stresses became increasingly compressive as the plate thickness increases for a given degree of expansion. The same goes to the research done by Nigrelli and Pasta [6] where the results showed that the hoop residual stress increased as the plate thickness was increased from 3 mm to 10 mm. In this research, the residual stress induced by split-sleeve cold-expansion process was simulated using finite element analysis. Gopalakrishna et al. [7] have proved that residual stress increases with the increase of expansion ratio until 5% and then decreased for 6% expansion. Rahman Seifi [8] applied cold expansion process to a plate made from elastic-perfectly plastic material. Residual stresses distributions were obtained by analytical and numerical method (using FEA method). The results showed that the stresses due to external loading were influenced by residual stresses. The maximum stresses found in the plate were observed to have been reduced by about 46%.

Most of the research done previously only focused on one factor at a time that can affect the residual stress distribution. However, in this work two factors that affect residual stress distribution, namely the expansion ratio and plate thickness were considered. This study is related to the previous work done by Karuppanan et al. [9] where the thickness and the centre distance between the holes were varied. They proved that the compressive residual stresses increased as the plate thickness increased. They also found that the residual stresses at the entrance face and the exit face were less compressive than the residual stresses at the mid-thickness of the plate.

In this study, the expansion ratio of the plate were varied from 1% to 6% and the diameter of the hole was kept constant with the centre distance between holes (c/d) ratio was fixed at 2. In addition, the thickness was varied according to the ratio between the thickness and the hole diameter (t/d) of 1, 2 and 3. Aluminium alloy 7475-T61 was used as the material for the model developed using ANSYS.

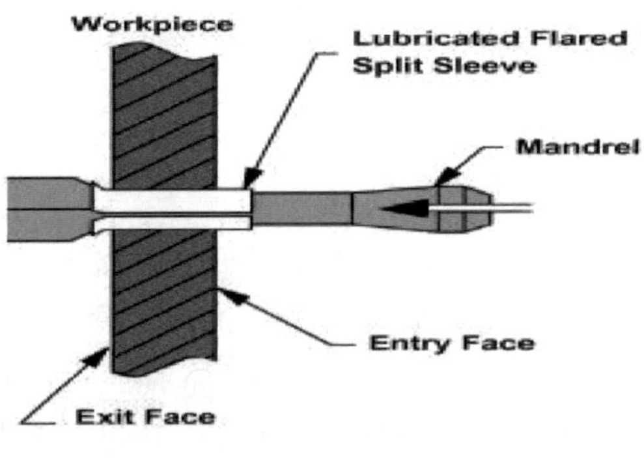

Figure 1. The schematic of the split sleeve cold expansion process [5].

2. Methodology

Figure 2 shows the plate model considered in this study. It consists of a plate containing two holes each with diameter, $d = 5$ mm with a centre distance between the holes, $c = 10$ mm. The dimensions of the plate were: width $w = 50$ mm, height $h = 50$ mm, and thickness over diameter ratio $t/d = 1$, 2 and 3. The expansion ratio of 1% to 6% was considered.

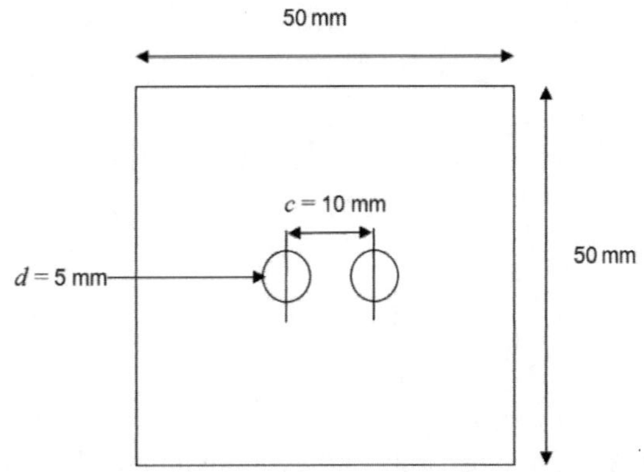

Figure 2. Dimension of the model.

However, only one quarter of the plate was modelled for the analysis since simultaneous hole expansion was considered. Figure 3 shows one quarter of the plate which was modelled in ANSYS where the stresses were measured from point B to point A. The dimensions used were: width w = 25 mm, height h = 25 mm and hole diameter d = 5 mm, as shown in Figure 3. Based on the research done by Papanikos and Meguid [1], the simultaneous expansion will result in higher compressive residual stresses compared with sequential expansion where half of the plate is modelled.

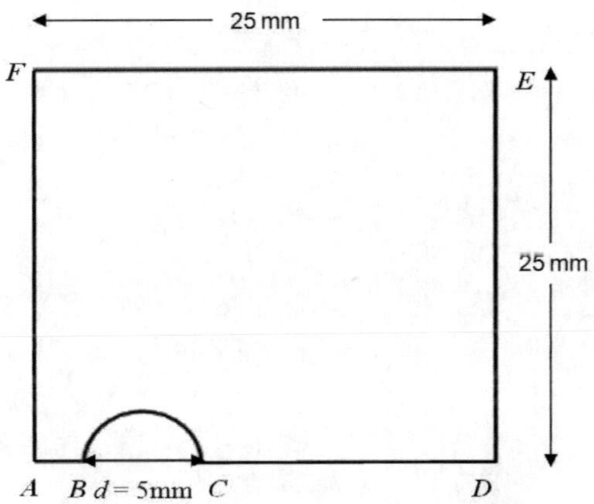

Figure 3. One quarter of the plate

Table 1 shows the material properties used in the model. These data were required to define the material in ANSYS.

Table 1. Mechanical properties of 7475-T61 Aluminium alloy [10].

Mechanical Properties	Unit
Yield Strength, σ_y	490 MPa
Ultimate Tensile Strength, σ_{UTS}	565 MPa
Modulus of Elasticity, E	70.3 GPa
Poisson's Ratio, v	0.33

Figure 4 shows the stress versus strain curve for Aluminium 7475-T61 alloy. This curve is required to define the elastic and inelastic properties of this material.

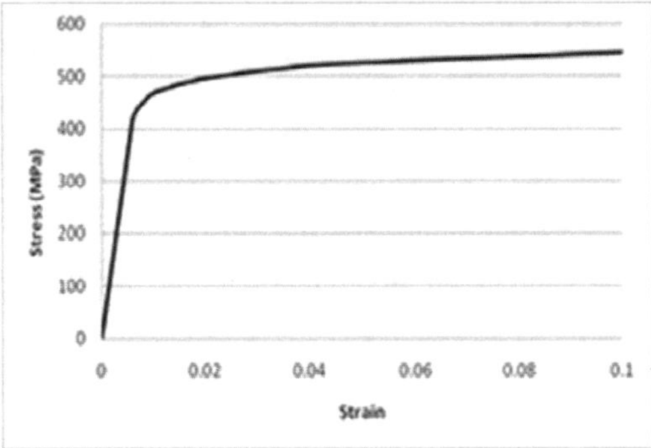

Figure 4. Stress versus strain curve for Aluminium 7475 alloy [5].

The element type considered for the model created in ANSYS was SOLID185. This type of element has plasticity, creep, stress stiffening, large deflection and large strain capabilities. Figure 5 shows the meshed model with fixed and symmetric boundary conditions. The symmetric boundary conditions were imposed on the side surfaces along the edges AB, CD, and AF as shown in Figure 5. Side surfaces along edges DE and EF were considered as the fixed boundary condition. The inner surface of the hole was then subjected to incremental pressure loading since the tapered mandrel was considered in this study.

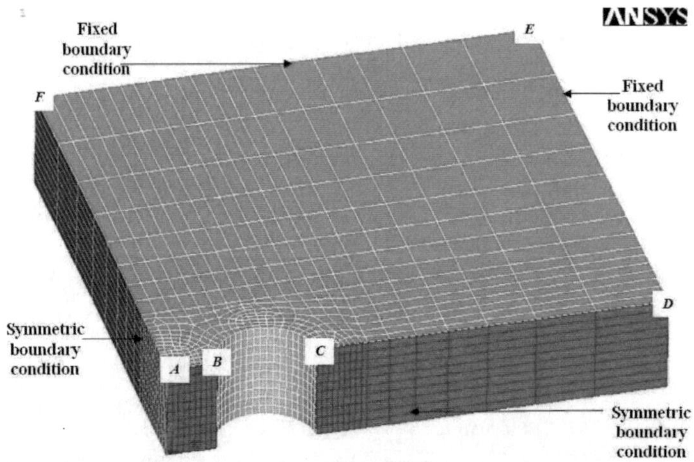

Figure 5. Meshed model with fixed and symmetric boundary conditions.

3. Results and Discussion

3.1. Validation

The experimental result from the study by Odzemir and Hermann [4] was used for the validation. Odzemir and Hermann [4] used 5 mm thick Aluminium 7050-T76 alloy plate with the size of 300 x 40 mm in their experiment. A hole was drilled at the centre of the plate and the hole was expanded using Fatigue Technology Incorporation (FTI) method with the different percentage of expansion, 2%, 4% and 6%. For validation, a quarter of the model was considered for modelling in ANSYS. Using this model, the stress distribution in x-axis and y-axis can be analysed at the edge of the model by plotting the results of the nodes. Figure 6 shows the model of the plate for validation which was developed in ANSYS.

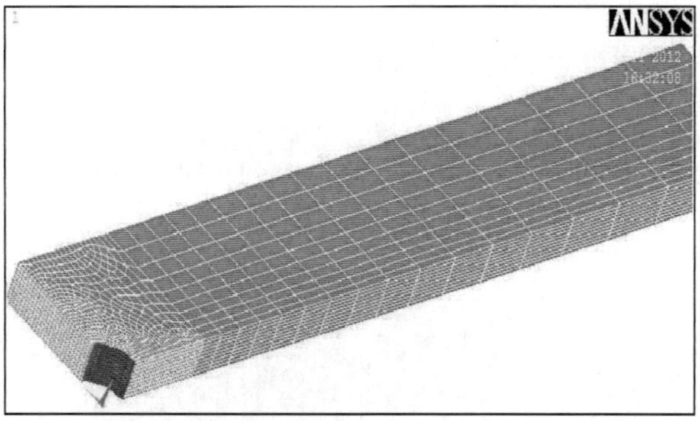

Figure 6. The 3D model of the plate for validation which was developed in ANSYS.

Experimental study done by Odzemir and Hermann showed that, 4% FTI expansion results the most compressive residual hoop stress compared with 2% and 6 % FTI expansion. The same goes to the finite element analysis done by the authors, where 4% expansion showed the most compressive residual stress compared with 2% and 6% expansion.

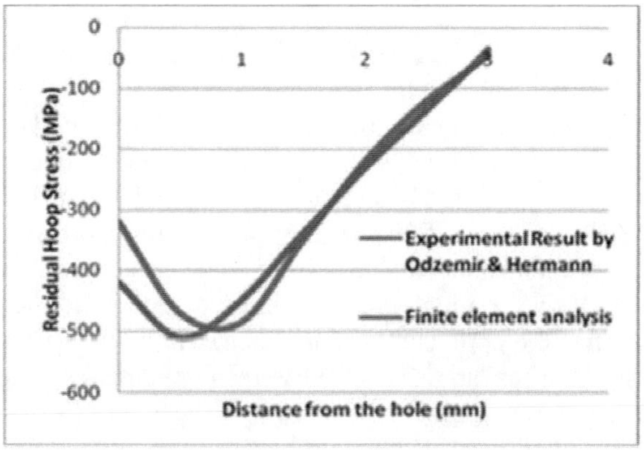

Figure 7. Comparison of residual stress developed in Aluminium 7050 T76 at 4% expansion using finite element analysis and experiment done by Odzemir & Hermann.

Figure 7 shows the comparison of residual stress results in 5 mm thickness Aluminium 7050-T76 at 4% expansion between finite element analysis and experiment conducted by Odzemir and Hermann. Based on the graphs, the residual stress developed at the edge of the hole using finite element analysis was slightly higher compared with the residual stress at the edge of the hole developed in experiment conducted by Odzemir and Hermann. In experiment done by Odzemir and Hermann, the lower value of compressive residual stress at the edge of the hole in that experiment was caused by the shear stress and frictional force due to the axial movement of the mandrel. In finite element analysis, the value of shear stress and frictional force were neglected since the lubricated split sleeve was considered in this study. Nigrelli and Pasta [6] stated that the low residual stress may be caused by the shear stress from the axial movement of the mandrel. This is the reason why the compressive residual stress at the edge of the hole in experiment done by Odzemir and Hermann was lower compared with the residual stress at the edge of the hole developed in finite element analysis done by the authors. However, the highest values of compressive residual stress developed by these two methods were almost the same which was about -500 MPa and the shape of the graphs was almost the same after this value. Therefore, the finite element analysis done was considered valid and this method of analysis can be used for this study.

3.2. Finite Element Analysis

In order to investigate the effect of thickness and expansion ratio on the residual stress distribution, finite element modelling was executed. The t/d ratios of 1, 2 and 3 were considered. Figure 8 shows the model of the 3-D finite element modelling using ANSYS with 4% cold expansion ratio for t/d = 2. The expansion ratios were varied from 1% up to 6% for each plate thickness.

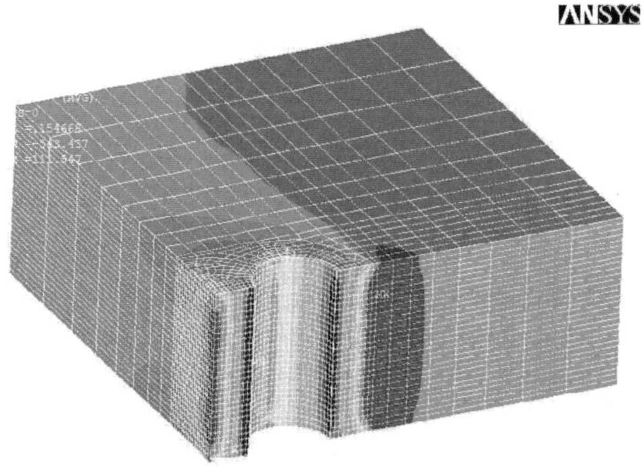

Figure 8. Model for t/d = 2 with expansion ratio = 4%.

3.2.1. Effect of plate thickness on residual stress. Odzemir and Hermann [4] noticed that the change in plate thickness can result in different values of residual stress around the cold expanded hole. Therefore, this effect has been investigated by evaluating the residual stress near the expanded hole for three different values of plate thickness. Residual stress in Y-direction was considered in this study and this stress is known as the tangential residual stress. The tangential residual stress is relevant in this study because the compressive residual stresses remain tangential to the hole edge as the mandrel is removed from the hole after the application of the load. The value of tangential residual stress is also higher compared with radial residual stress. These residual stresses were analysed on the surface of the plate starting from point B to point A as shown in the Figure 3. It is important to know the residual stress behaviour on the surface of the plate because the fatigue process initiates the crack at the surface of the plate due to excessive load. Figure 9 shows the stress behaviour around the expanded hole due to the different t/d for a fixed expansion ratio.

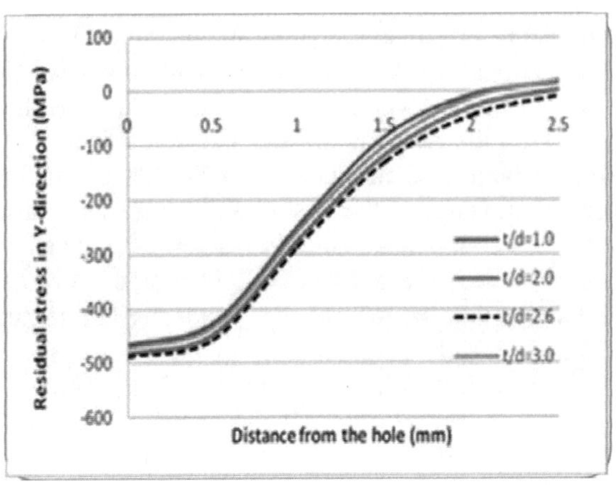

Figure 9. Residual stresses graph versus distance from the hole for 4% expansion ratio.

Based on the graphs shown in Figure 9, we can observe that the residual stress becoming more compressive for t/d = 2 as compared with t/d =1 and become less compressive for t/d = 3. From here, different plate thickness between t/d = 2 and t/d = 3 was used to find the optimum thickness to induce high compressive residual stress. t/d = 2.6 showed the highest compressive residual stress as the plate thickness was varied between t/d = 2 and t/d = 3. From the graphs, the maximum compressive stress was found close to the hole edge and then the compressive residual stress values decreased rapidly away from the edge of the hole. The residual stress become more compressive as the thickness is increased from 5 mm to 13 mm due to the change in the state of stress from the plane-stress to the plane strain condition. 13 mm plate thickness or t/d = 2.6 is considered as the optimum plate thickness to induce high compressive residual stress in this project and further thickness increment will cause the residual stress to become less compressive. The less compressive residual stress is due to the blocking action of plastic zone. Plastic zone is important in creating the residual compressive stress and due to blocking action of plastic zone, further thickness increment after t/d = 2.6 will reduce the value of residual compressive stress.

3.2.2 Effect of expansion ratio on residual stress. Figure 10 shows the stress behaviour around the expanded hole due to the different expansion ratio.

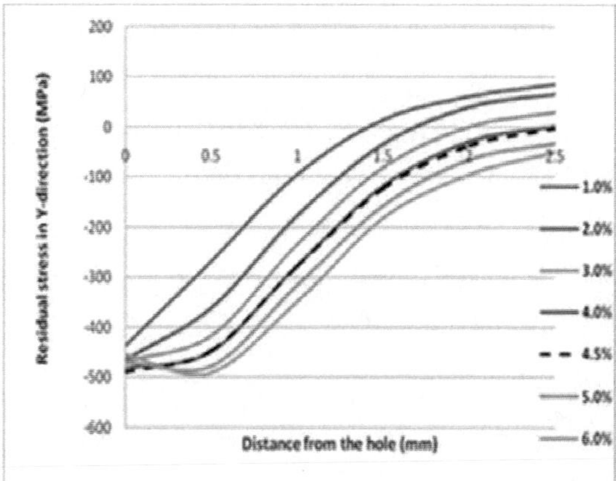

Figure 10. Residual stresses graph versus distance from the hole for t/d = 2.

Based on the graphs shown in Figure 10, the stress at edge of the hole becomes more compressive as the expansion ratio is increased up to 4 % and it become less compressive for further level of expansion for *t/d* = 1, *t/d* = 2 and *t/d* = 3. From here, different expansion ratio between 4% and 5% was applied at the hole to find the optimum expansion ratio to induce high compressive residual stress at the edge of the hole. 4.5 % expansion ratio showed the most compressive residual stress at the edge of the hole as the expansion ratio of the hole was varied between 4% and 5%. However, the area of the compressive part increases as the degree of expansion increases. This is due to the increase of applied load at the hole that will deform the material beyond the yield point and extends the plastic zone to a larger area. Plastic zone size is a critical factor in creating compressive stress. 4.5% of expansion ratio was considered as the optimum level of cold expansion and due to reverse yielding, residual compressive stresses at the edge of the hole become relaxed. Therefore, as the degree of expansion increases beyond 4.5%, the higher reverse yielding resulted in less compressive residual stresses at the edge of the hole. Reverse yielding is a condition that occurs when the stress in a material is decreased or relieved when the material is subjected to prolonged constant strain [11]. Expanding the hole beyond the limit only intensifies the stress relaxation at the edge of the hole which will reduce the compressive residual stress at the edge of the hole.

Expansion ratio of 4.5% with t/d = 2.6 are considered as the best condition to induce the high compressive residual stress which will reduce the propagation of fatigue cracks on the surface of fastened structure. Figure 11 shows the residual stress developed in optimum condition.

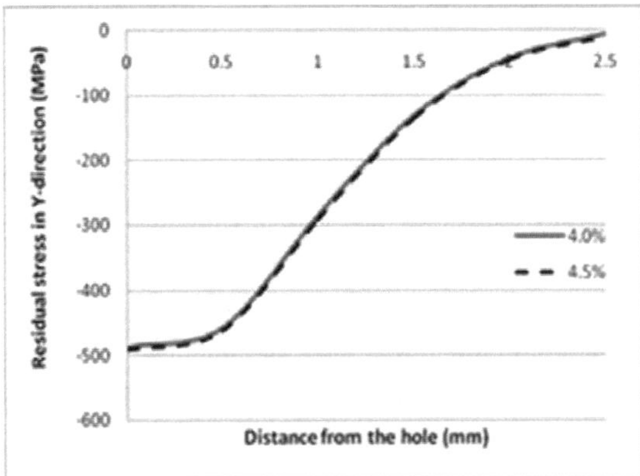

Figure 11. Residual stresses graph versus distance from the hole for t/d = 2.6 at 4% and 4.5% expansion ratio.

Based on the graph shown in Figure 11, we can observe that the residual stress becoming more compressive at 4.5% expansion ratio as compared with 4% expansion ratio for t/d = 2.6. This is the reason why t/d = 2.6 with 4.5% expansion ratio are considered as the best condition to induce the high compressive residual stress.

3.2.3 Stress analysis at different through-thickness position. Figure 12 shows the residual stresses through the thickness at the edge of the hole from entrance face (z = 0 mm) up to the exit (z = 10 mm) with different expansion ratios.

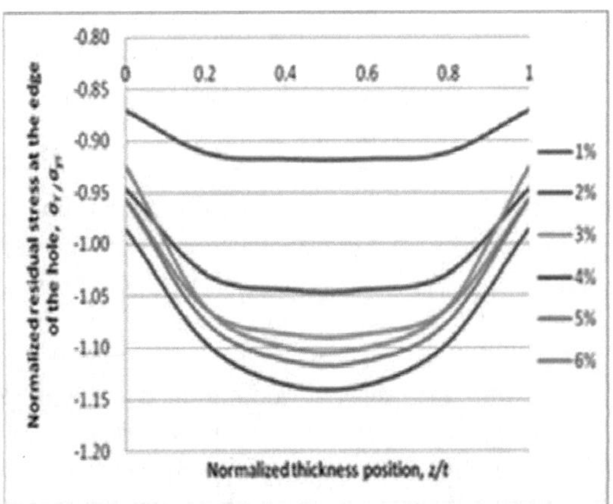

Figure 12. Residual stress at different through-thickness position with different expansion ratios for $t/d = 2$.

Based on the graphs, the compressive residual stresses near the entrance and the exit face are lower than the stresses on the mid-thickness for all expansion ratios. The presence of this less compressive residual stress near the entrance and exit face may be caused by the shear stress due to the axial movement of the mandrel. The graphs also show that the residual stress becomes more compressive as the expansion ratio is increased up to 4% expansion and it becomes less compressive for further level of expansion. As discussed before, the residual stress become more compressive due to the increment of plastic zone and it started to decrease beyond 4% expansion due to reverse yielding. The same behaviour of residual stress distribution was observed for other plate thickness where the compressive residual stresses start to decrease beyond 4% expansion ratio.

4. Conclusions

The effect of plate thickness and expansion ratio toward residual stress distribution around the cold expanded hole has been analysed using ANSYS on 7475-T61 plates. The failure of a component due to excessive tensile residual stress can be avoided by applying the findings in this study.

From the results, we conclude that the residual stress become more compressive as the plate thickness is increased up to t/d = 2.6 due to the change of the state of stress from plane-stress to plane strain and it starts to decrease for further level of thickness due to blocking action of plastic zone. Furthermore, t/d = 2.6 was considered as the optimum plate thickness to induce high compressive residual stress in this study.

The residual stress at edge of the hole becomes more compressive as the expansion ratio is increased up to 4.5% and decreased for further level of expansion. However, as the degree of expansion increases, the area of the compressive part also increases. The residual stresses distributions at the edge of the hole are controlled by the degree of reverse yielding when it reaches the optimum level of cold expansion.

Therefore, 4.5 % expansion ratio with t/d = 2.6 were considered as the best relation to induce high compressive residual stress which will reduce the propagation of fatigue cracks on the surface of fastened structure. This study also found that the residual stresses near the entrance and the exit face are less compressive than the residual stresses on the mid-thickness of the plate.

Acknowledgment

The authors would like to thank Universiti Teknologi PETRONAS for providing the facilities for the research work.

References

[1] Papanikos, P. and Meguid, S.A., 1998. Three-dimensional finite element analysis of cold expansion of adjacent holes, International Journal of Mechanical Science, **40(10)**, 1019-1028.
[2] Yucan F., Ende G., Honghua S., Jiuhua X. and Renzheng L., 2015. Cold expansion technology of connection holes in aircraft structures: A review and prospect, Chinese Journal of Aeronautics, **28 (4)**, 961-973.
[3] Moreira P. M. G. P., De Matos P. F. P., Pinho S. T., Pastrama S. D., Camanho P. P., De Castro P. M. S. T., 2004. Fatigue & Fracture of Engineering Materials & Structures, **27 (9)**, 879-886.
[4] Odzemir, T. and Hermann, R., 1999. Effect of expansion technique and plate thickness on near-hole residual stresses and fatigue life of cold expanded holes, Journal of Materials Science, **34**, 1243-1252.
[5] Burlat, M., Jullien, D., Levesque, M., Bui-Quoc, T. and Bernard, M., 2008. Effect of local cold working on the fatigue life of 7475-T7351 aluminium alloy hole specimens, Engineering Fracture Mechanics, **75 (8)**, 2042-2061.
[6] Nigrelli, V. and Pasta, S., 2008. Finite-element simulation of residual stress induced by split-sleeve cold expansion process of holes, Journal of Materials Processing Technology, **205**, 290-296.
[7] Gopalakrishna, H.D., Narasimha Murthy, H.N., Krishna, M., Vinod, M. S. and Suresh, A.V., 2009. Cold expansion of holes and resulting fatigue life enhancement and residual stress in Al 2024 T3 alloy, Engineering Failure Analysis, **17**, 361-368.

[8] Gomez, J. A., Lek, D., Song, I-H. and You, H. B., 2014. Study on stress evolution in the cooling process of micro hot embossing, International Journal of Mechanical and Materials Engineering, **9(20)**, 6 pages.

[9] Karuppanan, S., Hashim, M. H. and Wahab, A.A., 2013. Finite Element Simulation of Residual Stresses in Cold-Expanded Plate, Asian Journal of Scientific Research, **6(3)**, 518–527.

[10] Holt, J. M., Mindlin, H. and Ho, C. Y., 1996. Structural Alloys Handbook, West Lafayette, Indiana: CINDAS/Purdue University.

[11] Rees, D.W.A., 2007. Descriptions of Reversed Yielding in Bending, Proceedings of the Institution of Mechanical Engineers, Part C: Journal of Mechanical Engineering Science, **221(9)**, 981-991.

Paraffin wax removal from metal injection moulded cocrmo alloy compact by solvent debinding process

N A N Dandang[1*], W S W Harun[2], N Z Khalil[2], A H Ahmad[3], F R M Romlay[3], N A Johari [1]

[1] Institute of Postgraduate Studies, Universiti Malaysia Pahang, Lebuhraya Tun Razak, 26300 Gambang, Kuantan, Pahang, Malaysia
[2] Human Engineering Group, Faculty of Mechanical Engineering, Universiti Malaysia Pahang, 26600 Pekan, Pahang, Malaysia
[3] Manufacturing Focus group, Faculty of Mechanical and Materials Engineering, Universiti Malaysia Pahang, Pekan Pahang, 26600 Pekan, Pahang, Malaysia

*Corresponding author: nuraidahnabihahdandang@gmail.com

Abstract. One of the most crucial and time consuming phase in metal injection moulding (MIM) process is "debinding". These days, in metal injection moulding process , they had recounted that first debinding practice was depend on thermal binder degradation, which demanding more than 200 hours for complete removal of binder. Fortunately, these days world had introduced multi-stage debinding techniques to simplified the debinding time process. This research study variables for solvent debinding which are temperature and soaking time for samples made by MIM CoCrMo powder. Since wax as the key principal in the binder origination, paraffin wax will be removed together with stearic acid from the green bodies. Then, debinding process is conducted at 50, 60 and 70°C for 30-240 minutes. It is carried out in n-heptane solution. Percentage weight loss of the binder were measured. Lastly, scanning electron microscope (SEM) analysis and visual inspection were observed for the surface of brown compact. From the results, samples debound at 70°C exhibited a significant amount of binder loss; nevertheless, sample collapse, brittle surface and cracks were detected. But, at 60°C temperature and time of 4 hours proven finest results as it shows sufficient binder loss, nonappearance of surface cracks and easy to handle. Overall, binder loss is directly related to solvent debinding temperature and time.

1. Introduction

Debinding is a process that enhance the elimination of the binder component either by solvent or thermal or combination of both from the green compacts without disturbing the original form of the compact's body [1-3]. After mixing of feedstock and injection moulding, the green compact is then introduces to debinding process where the binder need to be fully removed from the feedstock [4, 5]. Besides sintering, there are one more critical phase in MIM which is call debinding [6-10]. Unsuccessful removing the maximum amount of the binder earlier can effect in compact's failings, such as cracking, collapsing and compacts will become very brittle. During sintering, inadequate removal of binder can cause infection from polymer's disintegration. Eliminating the binder without crumpling the metal particles is an elusive method and can be reached via few stages. Thermal, solvent debinding and catalytic are the most common debinding processes exist besides others method [11-13]. The aim for debinding process is to eliminate the binder's component in the possible shortest minutes besides the minimum effect on the compact. Premature debinding process is used only thermal debinding, which mandatory more than 200 h required time to accomplishment. These days, the complete process for debinding minimised to 120 minutes depend on binder used as thermal debinding is united with solvent debinding. The benefit of solvent debinding is based on high solubility in the organic solvent of small molecular weight binder as main constituent [8, 14]. It forms networks and pore channels in the green compact thus prepare the brown compact for thermal debinding [15, 16]. When the number of pores in the compacts are huge, they lets the degraded binder which is slight constituent to simply absorb to the external of the compact's body through thermal debinding. Hence, time for thermal debinding is shorten expressively [14, 17].

Evolving a systematic considerate in process of debinding shortens debinding time and temperature. For quantity manufacture, extensive of times for debinding are impracticable for mass production. Thomas-Vielma et. al. [14] and Liu et. al. [18] exposed when solvent and thermal debinding were combine they effectively cuts time for binder removal and avoids failings after establishing in green compacts.

In this work, a paraffin wax of a low molecular weight (MW) assortment, and polypropylene (PP) and stearic acid were used as the primary, backbone binder and surfactant agent correspondingly. Among the all the binder's component, major amount of paraffin wax must be detached from a green compact in the solvent debinding process, but after debinding, PP must endure in the compact, with the purpose of holding metal particles by one and other and sustain the green compact' shape. This research concentrate on the study of solvent debinding parameters and resolve an ideal conditions for this process.

2. Experimental method

2.1. Material

Metal powder used is CoCrMo. Table 1 indicates the characteristics of metal powder. Volume of powder Loading used: 67%. Binder's components comprised of 70 vol. % paraffin wax (PW), 25 vol. % Polypropylene (PP) and 5 vol. % stearic acid (SA). The morphology of the cobalt chromium molybdenum alloy powder is shown in Figure 1. While table 1 shows the characterization results for the binders.

Table 1. Characteristics of metal powder.

Powder	Size distribution	D_{50} (μm)	D_{90}	D_{10}	S_w	Density (g/cm^3)	Shape
CoCrMo	22	11.5	23.7	4.1	3.3	8.4	Spherical

[a] Sw (slope parameter distribution = 2.56/log(D90/D10),

[b] 22 represents a particle size distribution

[c] D50 is shows the mean diameter of particle size for distribution of particle size

[c] D90 and D10 indicate the distribution at 90% and 10%,(cumulative) for size distribution curve

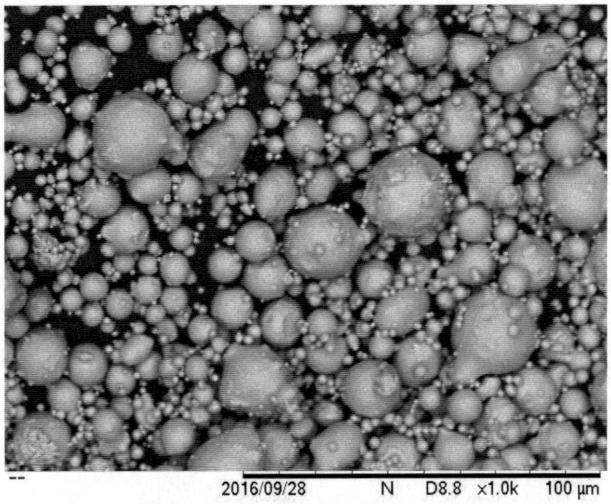

Figure 1. Morphology image of CoCrMo Alloy powder.

Table 2. A Characterization of the binders.

Binder	Chemical structure	Density (g/cm^3)	Melting temperature (˚C)	Decomposition temperature (˚C)
PW	$C_{31}H_{64}$	0.91	47-65	200-400
PP	(C3H6)n	0.90	>95	328-410
SA	$C_{18}H_{36}O_2$	0.94	69.4	180-380

2.2. Mixing of feedstock and injection moulding
Double planetary mixer at 160 °C for 90 minutes for 70 rpm rotational speed was used to prepare CoCrMo feedstock. Then, to produce a dog-bone shaped sample, injection moulding was conducted by using 200 tonne NISSEI Model NS20-2A injection moulding machine. The green compacts were successfully fabricated by injection moulding at 145 ˚ C. Based on the physical inspection, there were no defects observed on the green compacts.

2.3 Analysis of the feedstock
By using NETZSCH DSC 214 Polyma DSC21400A-01717-L machine, differential scanning calorimetry (DSC) analysis was conducted under nitrogen atmosphere on the feedstock. The start temperature for DSC test was at 0 °C until 550 °C as end temperature. Gas rate was 40ml/min and 10 °C/min was the heating rate used. While for starting and end temperatures, 10 minutes holding time was used throughout this process.

2.4 Solvent debinding
Figure 2 shows schematic diagram of solvent debinding process by wicking technique. Here, samples been placed in a condenser after been compacted. Then, the samples were debound at 50, 60 until 70 ˚C temperatures. The debinding times were 30 to 240 minutes (4 hours). From Figure 3, the illustrations of graph for solvent debinding process been utilised. About 40 ml n-heptane was used as a solvent in this debinding process by wicking technique using Al_2O_3 powders. This powder bed helps to avoid distortion and it offers more uniform heating through this process [1]. Then, as the brown compacts were obtained from this process, the amount of weight loss for each compacts for each conditions as stated will be measured.

From equation 1 below, Mpw indicates the mass loss of paraffin wax where the mass of PW and SA were been calculated based on the equation below:

$$Mpw\ (\%) = \frac{Mb - M}{Mb} x\ 100 \tag{1}$$

Where Mb is the mass before solvent debinding (green compact) and Ma is the mass after solvent debinding of the (brown compact)

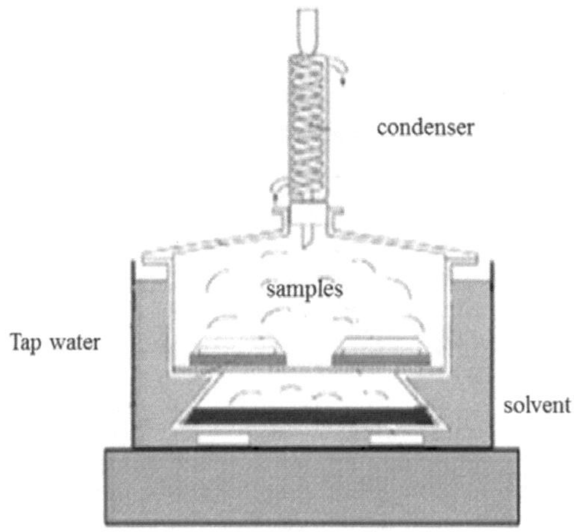

Figure 2. Schematic diagram for solvent debinding by wicking technique.

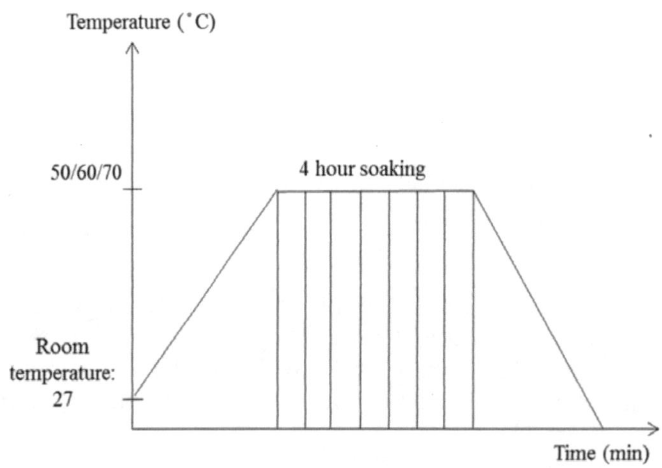

Figure 3. Graph illustrates Time vs Temperature plot of solvent debinding.

3. Results and discussion

Feedstock was successfully moulded using a Nissei NS20-2A injection moulding machine and debound by wicking technique. Defect free green and brown compacts are produced as shown in Figure 4.

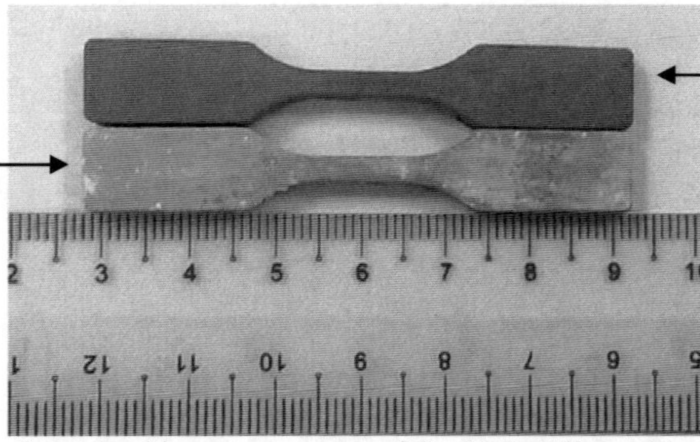

Figure 4. Green and brown compact of CoCrMo Alloy without defects.

3.1. DSC of the feedstock

The injection moulding and solvent debinding temperatures were predicted by thermal properties of feedstock as it offers an excellent guide for future experiment in this research. From Figure 5, it can be observed 57.5 – 167.5 ° C is the uttermost feedstock's melting temperatures. As shown in the Figure, first peak of the curve shows the melting point of paraffin wax and stearic acid. While, for polypropylene, the highest peak, which is the second peak specified the melting of polypropylene. From Hwang et. al. [12] it was stated that, when the melting point of the maximum melting binder component is surpassed, the binder becomes more flowable.

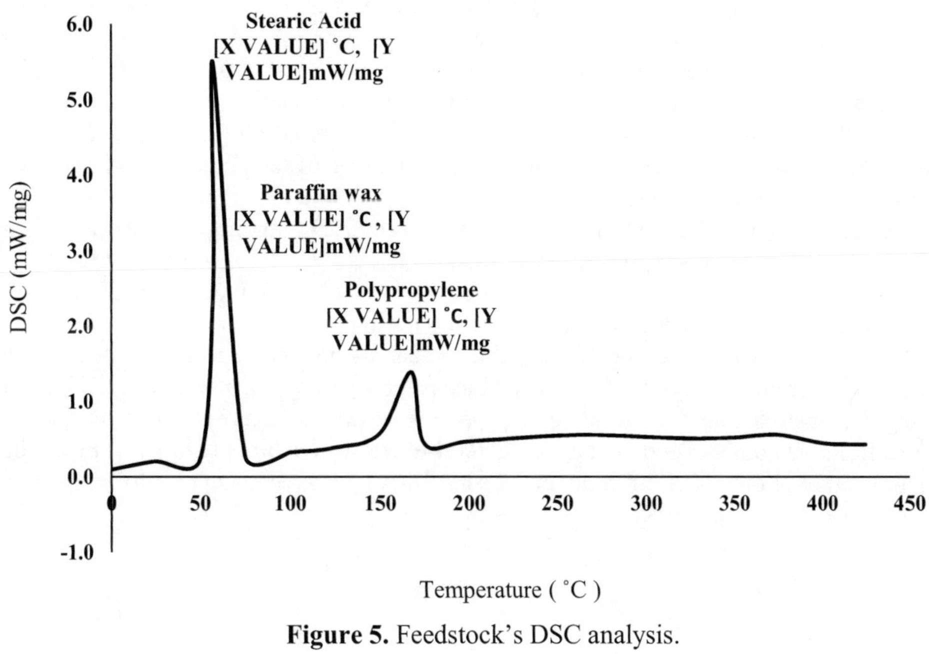

Figure 5. Feedstock's DSC analysis.

3.2. Solvent debinding

To reduce time for debinding, thermal debinding and solvent debinding techniques had been introduced. Then, the parameters for this process were enhanced using results from DSC test. The function of solvent solution used, which is n-heptane was to remove paraffin wax and stearic acid. Then, a porous brown compact was formed once they had been removed. The existence of pores in the compacts enables backbone binder which is polypropylene (PP) to diffuse out during thermal debinding process.

Figure 6 shows the mass loss of paraffin wax at numerous temperatures. In the first hour of process, it can be presumed that as the debinding temperature increase, the loss of the binder also increase. Hence, when temperatures increase, it boost binder diffusivity and solubility. Besides, more binder loss with the extended of debinding time. Correspondingly, it can be illustrated from Figure 7 that 4 phases occur throughout solvent debinding. Based on Md Ani et. al. work [19], the phases of process are discussed.

Phase I [0-60 minutes]: During this phase, loss of binder's component were quite fast as they are directy contact with the solvent on the surface.

Phase II [60-120 minutes]: The rate of debinding decelerates. It is due to the solvent that enter intensely into the compact and the binder was extracted and liquefy into the surface. Capillary forces helped the dissolved binder to diffuse to the exterior surface of compact by liquid extraction [3, 20].

Phase III [120-180 minutes]: The debinding rates start to rise due to an adequate channels formed in the compact that permits the dispersion of the binder dissolved rapidly to the exterior surface of compact. From the graph, the incline is the highest at 70 ˚C which can be proved that the debinding rate achieves it peaks at this phase. At 50 °C and 60 °C temperatures, debinding rates on graph seem to be going at the similar leap. While, on curve 70 °C, the debinding rate is reasonably high.

Phase IV [180-240 minutes]: From the graph, the mass loss of binder remains constant when temperature increase and cause the debinding rate reduces. From this phase, the saturation point of solvent and the binder had been achieved. But, it should be alert that the stability point among the latter is reliant on the solvent-binder component volume ratios.

The compassion of binder diffusivity and solubility to changes of temperature can be seen from the graph of debinding rate at 50 and 60˚C. For phases II-IV which is at 50°C, the debinding rate seems to be consistently reduces, where inclines are nearly constant. In contrast, at 60 °C, phases II – IV, the debinding rate seems to increase until 180 minutes. Figure 8 illustrates the overview of solvent debinding mechanism, besides the conditions of the binder and diffusion of solvent during debinding. Debinding rates throughout debinding process are reliant on movement of molecules from dissolved binder, where they commonly move more rapidly with increasing of temperatures and smaller solvent molecules as reported by German and Bose.

Researchers had stated that, when the temperature of solvent debinding exceeds highest melting point of the soluble binder components, MIM compacts will start to collapse. Figure 9 shows compacts are collapse and cracks after solvent debinding. High solvent temperatures, longer debinding times, and thickness variation of the component are the reasons behind all those conditions from the Figure 9. The compact will crack due to insoluble binder component due to longer soaking time. Besides, higher temperature makes binder soften which effects green compact to crack. Therefore, it is why cracking demonstrates on the surface of compact as shown in Figure 9. These external defects were detected from the compact debound at 70 °C for 240 minutes debinding time. Compact debound at 60 °C temperature shows efficient amount of paraffin wax mass loss after 240 minutes with no defects on the surface existing on the compact.

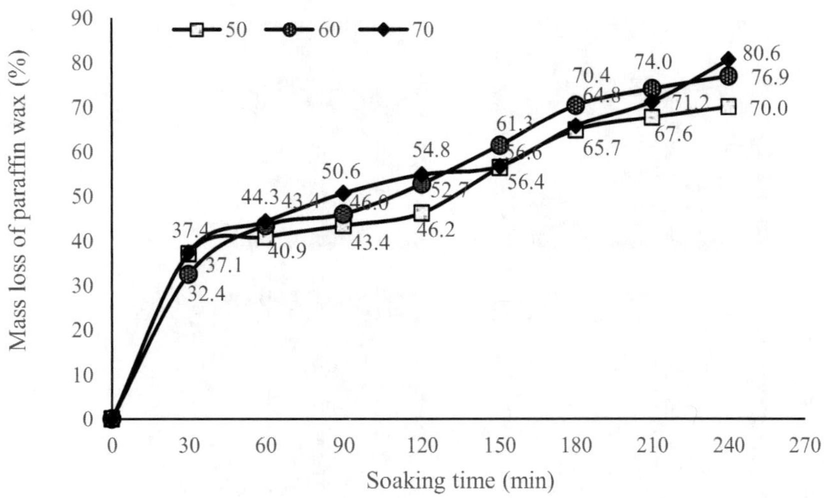

Figure 6. Time vs Temperature plot for paraffin wax's mass loss during solvent debinding.

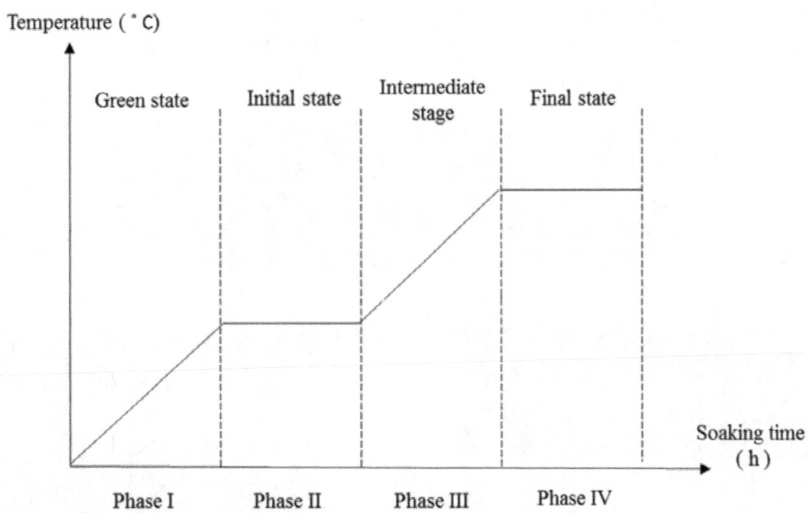

Figure 7. A schematic graph of paraffin wax loss for each phases.

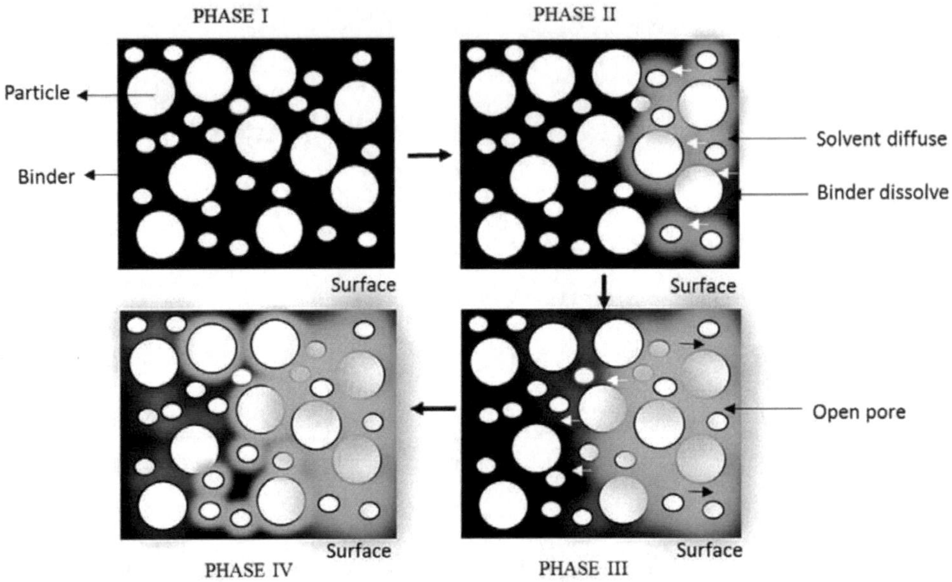

Figure 8. A schematic Figure presenting the interaction between solvent and binder diffusion occur during wicking debinding process.

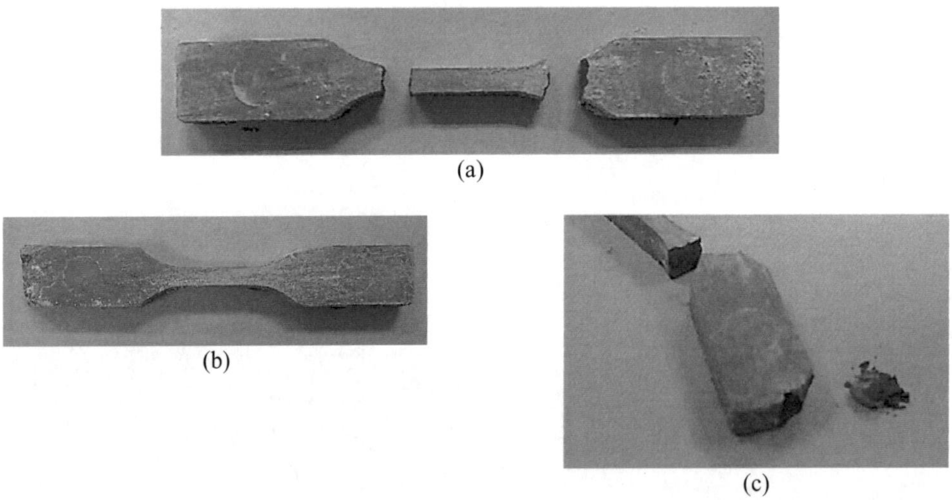

Figure 9. Defects observed on brown compact a) Fractures b) collapse c) cracks at conditions 70 ˚C and 4 hour.

From Figure 10, SEM micrographs demonstrate the microstructure image of compacts before and after solvent debinding. From the Figure, it can be realised that the metal particles was squarely scattered in the binder matrix. Then, uniform binders distribution can also been detected in the compact. After solvent debinding, it is clear that, open pore channels had formed besides PW and SA binders had been removed but there are still remaining binder to retain the shape of the compact before thermal debinding which is PP.

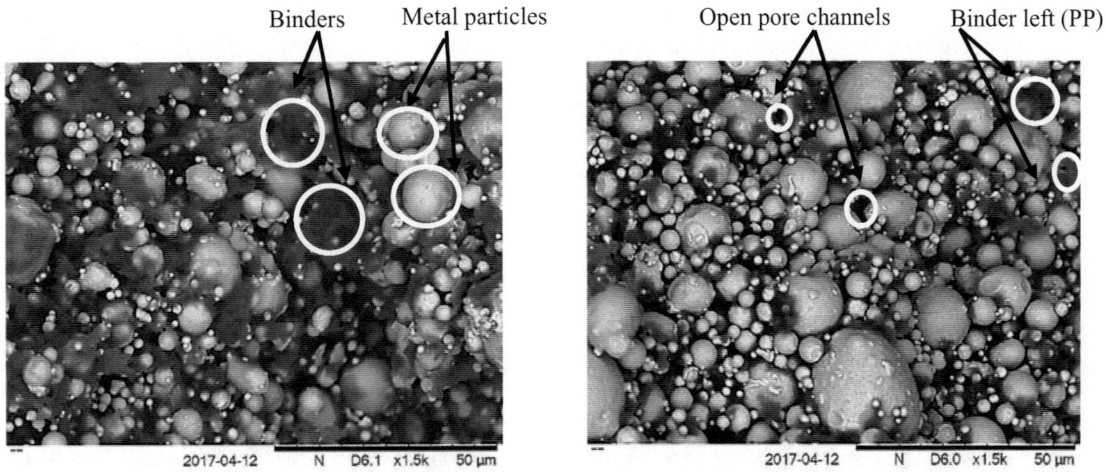

(a) (b)

Figure 10. SEM micrographs of green compact before solvent debinding (a) and brown compact after solvent debinding (b).

4. Conclusions

From the above research work, we can conclude following:

- Based on DSC graph result from Figure 5, it was proven that the melting point of each binder's component which is Paraffin Wax (PW), Polypropylene (PP) and Stearic Acid (SA) had been achieved. Therefore, it makes them becomes more flowable and appropriate to be used as selected binder component in this research study.
- To achieve an ideal conditions of solvent debinding, the process had been completed at different parameters. From the results, debinding temperatures and soaking time significantly influences the mass loss of paraffin wax of compacts. The significant amount of mass loss of paraffin wax whics 76.9% without no defects on the brown compacts shows an ideal condition for solvent debinding of CoCrMo alloy compact.
- The ideal and an optimum condition of solvent debinding for CoCrMo alloy is suitable at the condition of 60 °C for 4 hours, using n-heptane solution with 40 ml volume.
- This can be proved from SEM image results from Figure 10, where at conditions 60℃ and 4 hours soaking time, Paraffin Wax had been removed and open pore channels are formed which shows that, the binder is successfully been removed at this condition.

Acknowledgements

Universiti Malaysia Pahang fully supports the facilities and resources for this research. The author W.S. W. Harun would like to acknowledge the support of the internal grant of Universiti Malaysia Pahang RDU141101, RDU140354, RDU150337 and the support of Research Acculturation Collaborative Effort (RACE) provided by the Ministry of Higher Education, Malaysia RDU151314.

References

[1] German R M and Bose A 1997 *Injection molding of metals and ceramics*: Metal Powder Industries Federation)

[2] German R M, Bose A and Mani S S 1992 Sintering time and atmosphere influences on the microstructure and mechanical properties of tungsten heavy alloys *Metallurgical Transactions A* **23** 211-9

[3] BECKER F H 2006 Debinding processes: Physical and chemical conclusions and their practical realisations. In: *CFI. Ceramic forum international*: Göller)

[4] Ye H, Liu X Y and Hong H 2008 Sintering of 17-4PH stainless steel feedstock for metal injection molding *Materials Letters* **62** 3334-6

[5] Onbattuvelli V P, Chinn R, Enneti R K, Park S-J and Atre S V 2014 The effects of nanoparticle addition on binder removal from injection molded silicon carbide *Ceramics International* **40** 13861-8

[6] Zhu B, Qu X and Tao Y 2003 Mathematical model for condensed-solvent debinding process of PIM *Journal of materials processing technology* **142** 487-92

[7] OMAR M, SAUTI R and ABDULLAH N DEBINDING AND SINTERING CHARACTERISTIC OF INJECTION MOULDING CoCrMo ALLOY POWDER FOR BIOMEDICAL APPLICATIONS

[8] Yang W-W, Yang K-Y, Wang M-C and Hon M-H 2003 Solvent debinding mechanism for alumina injection molded compacts with water-soluble binders *Ceramics International* **29** 745-56

[9] Chen G, Cao P, Wen G and Edmonds N 2013 Debinding behaviour of a water soluble PEG/PMMA binder for Ti metal injection moulding *Materials Chemistry and Physics* **139** 557-65

[10] Wongpanit P, Khanthsri S, Puengboonsri S and Manonukul A 2014 Effects of acrylic acid-grafted HDPE in HDPE-based binder on properties after injection and debinding in metal injection molding *Materials Chemistry and Physics* **147** 238-46

[11] Sidambe A T, Figueroa I A, Hamilton H G C and Todd I 2012 Metal injection moulding of CP-Ti components for biomedical applications *Journal of Materials Processing Technology* **212** 1591-7

[12] Hwang K, Lin H and Lee S 1997 Thermal, solvent, and vacuum debinding mechanisms of PIM compacts *Material and Manufacturing Process* **12** 593-608

[13] Hayat M D, Wen G, Zulkifli M F and Cao P 2015 Effect of PEG molecular weight on rheological properties of Ti-MIM feedstocks and water debinding behaviour *Powder Technology* **270, Part A** 296-301

[14] Thomas-Vielma P, Cervera A, Levenfeld B and Várez A 2008 Production of alumina parts by powder injection molding with a binder system based on high density polyethylene *Journal of the European Ceramic Society* **28** 763-71

[15] Li B Q, Wang C Y and Lu X 2013 Effect of pore structure on the compressive property of porous Ti produced by powder metallurgy technique *Materials & Design* **50** 613-9

[16] Páez-Pavón A, Jiménez-Morales A, Santos T G, Quintino L and Torralba J M 2016 Influence of thermal debinding on the final properties of Fe–Si soft magnetic alloys for metal injection molding (MIM) *Journal of Magnetism and Magnetic Materials* **416** 342-7

[17] Zaky M 2004 Effect of solvent debinding variables on the shape maintenance of green molded bodies *Journal of materials science* **39** 3397-402

[18] Liu Z, Loh N, Tor S, Khor K, Murakoshi Y and Maeda R 2001 Binder system for micropowder injection molding *Materials Letters* **48** 31-8

[19] Md Ani S, Muchtar A, Muhamad N and Ghani J A 2014 Binder removal via a two-stage debinding process for ceramic injection molding parts *Ceramics International* **40** 2819-24

[20] Chikwanda H and Machaka R 2014 A study of solvent debinding variables on Ti6Al4V green bodies

Effect of heat treatment on mechanical properties and microstructure of selective laser melting 316L stainless steel

M S I N Kamariah[1*], W S W Harun[2], N Z Khalil[2], F Ahmad[3], M H Ismail[4], S Sharif[5]

[1]Institute of Postgraduate Studies, Universiti Malaysia Pahang, Lebuhraya Tun Razak, 26300 Gambang, Kuantan, Pahang, Malaysia.
[2]Green Research for Advanced Materials (Gramslab), Human Engineering Group, Faculty of Mechanical Engineering, Universiti Malaysia Pahang, 26600 Pekan, Pahang, Malaysia.
[3]Department of Mechanical Engineering, Universiti Teknologi PETRONAS, Malaysia.
[4]Faculty of Mechanical Engineering Universiti Teknologi MARA, Shah Alam, Selangor, Malaysia.
[5]Department of Materials, Manufacturing & Industrial Engineering, Faculty of Mechanical Engineering, Universiti Teknologi Malaysia.

*Corresponding author: nurul.kamariah01@gmail.com

Abstract. Selective Laser Melting (SLM) has been one of the preferred Additive Manufacturing process to fabricate parts due to its merits in terms of design freedom, lower material waste and faster production when compare to the conventional manufacturing processes. However, due to the thermal gradient experienced during the process, the parts are exposed to the residual stress that leads to parts distortion. This work presents the effect of heat treatments on the micro-hardness of 316L stainless steel parts. In current study, SLM has been employed to fabricate 316L stainless steel compacts. Different heat treatments of 650°C, 950°C, and 1100°C for 2 hours were applied on the compacts. Hardness test were performed on the as-built and heat-treated compacts. The relationship between the microstructures and micro-hardness were- discussed in this paper. The results revealed that the micro-hardness of the as-built compacts is between 209.0 and 212.2 HV, which is much higher than the heat-treated compacts.

1. Introduction

Selective Laser Melting (SLM) is the frequently adopted additive manufacturing technology that use powdered raw material that is melted by a high energy focused laser prior to its consolidation. The low material waste, high geometrical freedom in design, and faster production of parts are its main strength when compared to the conventional manufacturing methods [1]. The excellent mechanical properties which are comparable or better than those of the casted counterparts are another credit of this method.

The effectiveness of SLM fabrication method also rely on the design parameters, for examples, the building orientation of parts, and the scanning patterns or strategy [1, 2]. Nevertheless, the design parameters can be restricted by the SLM machine such as the size of the parts. Building orientation will alter the build schedule of the SLM process. Therefore, different building orientations of parts will cause the parts to experience different thermal history [3, 4]. However, the rapid heating and cooling experienced by a part during its fabrication lead to high temperature gradient that generates residual stresses which cause part distortion [5]. This will detrimentally give impact on the mechanical behaviour of fabricated parts [6].

316L stainless steel has been broadly study for SLM [7]. This is due to the widespread application of this material in various industry, as it exhibits outstanding corrosion resistance and excellent ductility [8]. Therefore, its applications can be found in the fields of biomedical, aerospace, automotive, and marine. Some of the research done has shown the ability of SLM to improve the mechanical properties of parts when compare to conventional methods. This includes the study done by Yadroitsev et al where they observed nearly 50% increase of yield stresses of fabricated 316L SLM parts compare to forging and 20-30% of elongation of the forging [9].

The mechanical properties of the SLM fabricated parts are strongly influenced by the grain structure of molten pool [10]. Besides that, cellular dendrites are exhibited at the microstructure after SLM process due to high thermal gradient and high cooling rate [10]. The cell orientations are likely towards the gradient inside the melting pool [11, 12]. In terms of mechanical properties, Sun et. al. work managed to achieve high micro-hardness value compare to the annealed 316L stainless steel counterpart [7]. It is believed that this result is influenced by the nano-scale amorphous inclusions quantity and the dislocation density.

It will be beneficial if the 316L stainless steel parts could be used directly after SLM fabrication or heat-treated if the mechanical properties are satisfied. Therefore, it is necessary to understand the relationship between microstructure and the mechanical performance of the as-built and the heat treated samples. This study aims to understand heat treatment effect on micro-hardness of 316L stainless steel compact fabricated by SLM.

2. Experimental Procedure

Gas-atomised 316L stainless steel (SS) powder provided by SLM GmbH with mean powder particle size of 30 μm was utilised in this research. The powder particle morphology is as shown in Figure 1. Mainly, spherical-shaped particles with some smaller satellite particles attached can be observed. The chemical composition of the as-received powder is shown in Table 1. SLM of dog-bone compacts were accomplished via an 125HL SLM with 400W YLR Faber laser. All compacts were fabricated under Argon atmosphere to preclude oxidation. The dimension of the dog-bone compacts is shown in Figure 2. The compacts were fabricated with different building orientation depicted in Figure 3. Table 2 summarised the processing parameters utilised in this research.

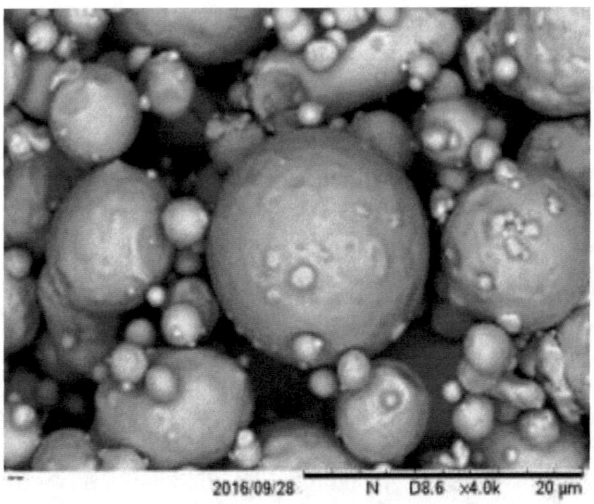

Figure 1. SEM image of as-received 316L SS particle powder morphology.

Table 1. Chemical compositions (weight %) of the as-received_316L stainless steel powder.

Element (wt.%)	Fe	Cr	Ni	Mo	Mn	Si	P	S	C
316L SS	Bal.	16.8	10.4	2.1	1.11	0.56	0.03	0.011	0.01

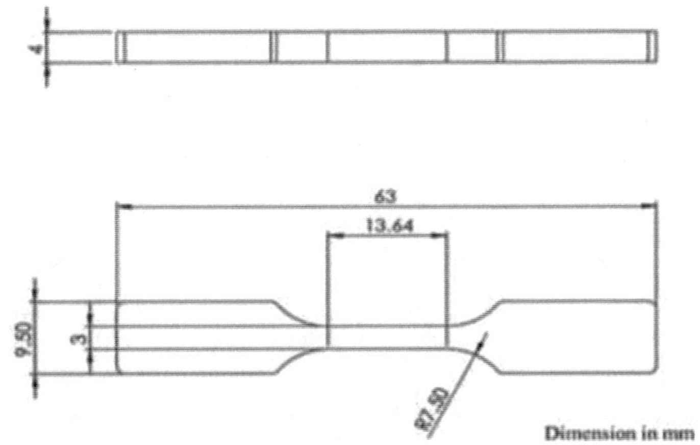

Figure 2. Detailed dimensions of tensile compacts based on ASTM E8 standard.

Identified process parameters were used to fabricate compacts at different angles of building orientation of compacts from building platform at 0°, 45°, and 90°, as shown in Figure 3. Each compact experienced same build schedule which can be consider as one single job as they were processed simultaneously, and cooling-heating rates during fabrication. Referring to Figure 3, three different axis are illustrated which represent building direction (BD), scanning direction (SD), and transverse direction (TD). For the microstructure study, the compacts were cross-sectioned at the BD and SD planes at the dotted-red box region.

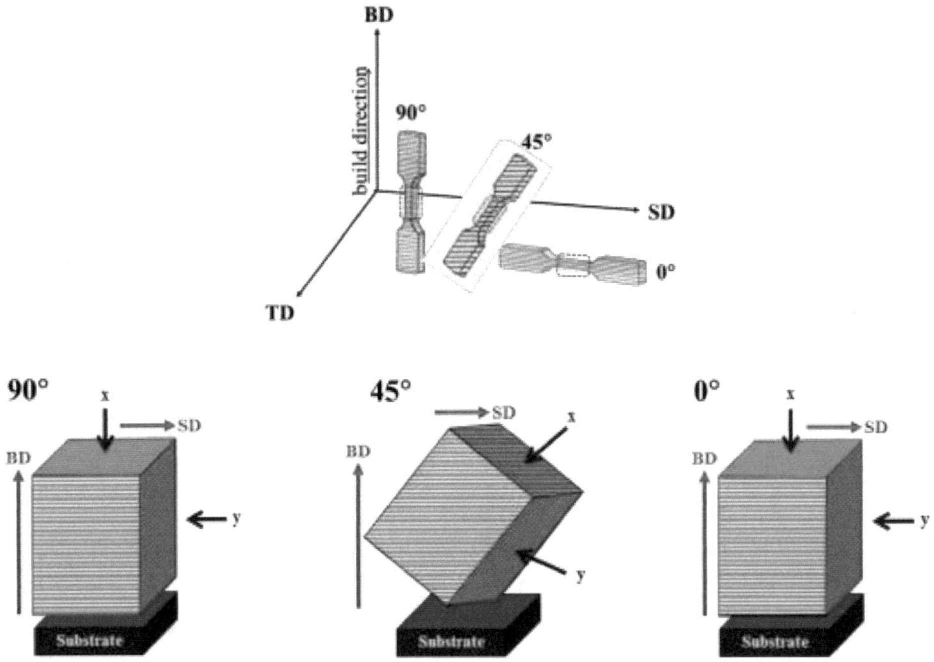

Figure 3. Schematic of compact positioning during SLM fabrication process.

Table 2. Process parameters utilised for 316L SS SLM compacts fabrication.

Laser Power (W)	Scanning speed (mm/min)	Layer thickness (μm)	Platform temperature (°C)	Argon gas consumption in operation (mL/min)
200	800	30	150	1.2

After fabrication, the compacts were detached from the build substrate. A total of twelve different conditions of 316L SS were taken into account. The "as-built" (AB) condition is referred to the untreated SLM samples. Some of the compacts from each building orientation were subjected to heat treatment (HT) using a horizontal tube furnace under argon gas with 10°C min^{-1} of heating rate. The heat treatment process were conducted at three difference temperature which are 650°C, 950°C, and 1100°C for 2 hours and were furnace cooled. Table 3 summarized the heat treatment conditions employed.

Table 3. Heat treatment condition to SLMed 316L stainless steel compacts.

Compact	**Heat treatment cycle**
As-Built	No heat treatment
HT1	650°C for 2 hours, furnace cooling
HT2	950°C for 2 hours, furnace cooling
HT3	1100°C for 2 hours, furnace cooling

The compacts were grinded, polished, and chemically etched for 5 minutes using solution of 7.5 ml nitric acid, 5.0 ml hydrofluoric acid, and 37.5 mL distilled water. The microstructures were analysed using optical microscopy (OM) (OLYMPUS BX51M).

Vickers micro-hardness tester (MATSUZAWA Type MMT X7) were utilised to conduct hardness measurements on the surface of mounted compacts under a 0.5 kg load for 10 s of dwell time. Diamond indenter was used with five repetitions per compacts.

3. Results and Discussion

The results of the Vickers hardness (HV) testing were evaluate through: (i) comparing SLM compacts hardness values for different heat treatment conditions; and (ii) comparing SLM compacts values for different building orientations.

Table 4. Results of average hardness values (HV) for SLMed 316L stainless steel compacts based on different conditions.

Sample	Building orientation (°)	Heat treatment cycle	Hardness (HV)
0-AB	0	-	209.0
45-AB	45	-	212.2
90-AB	90	-	212.0
0-HT 1	0	650°C for 2 hours, furnace cooling	201.8
45-HT 1	45	650°C for 2 hours, furnace cooling	215.5
90-HT 1	90	650°C for 2 hours, furnace cooling	204.0
0-HT 2	0	950°C for 2 hours, furnace cooling	181.7
45-HT 2	45	950°C for 2 hours, furnace cooling	186.0
90-HT 2	90	950°C for 2 hours, furnace cooling	191.5
0-HT 2	0	1100°C for 2 hours, furnace cooling	173.3
45-HT 2	45	1100°C for 2 hours, furnace cooling	185.7
90-HT 2	90	1100°C for 2 hours, furnace cooling	170.8

In Figure 4, the values of hardness are plotted against the building orientation angles (0°, 45°, 90°) from SLM build substrate plate for all four heat treatment conditions of compacts. Note that each data point that poses the hardness value is the average values of six micro-hardness measurements taken at the cross-sectioned parts at x and y, i.e. BD and SD planes, of the compacts, significantly. Thus, represents the average hardness values of the specific compact conditions.

It can be seen in Figure 4, the as-built and each heat treated conditions shows different trend of hardness values. Referring to 0° and 45° building orientations, the as-built and HT1 compact conditions shows similar variation of the hardness values. Although there are differences between the average hardness values of 90° building orientation for as-built compact, i.e. 212.0 HV, and HT1 compact, i.e. 204.0 HV, the range of values distribution shows that the difference is small. Thus, it can be said that hardness values gain for as-built condition is similar to HT1 compact condition. Nevertheless, the HT2 and HT3 compact conditions exhibit significant lower hardness values than as-built and HT1 compact conditions. This finding is aligned with the results gain by Sistiaga et. al. at similar HT3 compact condition that marked the beginning of the lower hardness values of the SLMed heat treated compact condition compare to the as-built and other lower heat treatment temperature [13]. HT2 and HT3 compact conditions illustrates dissimilar pattern of hardness values. 0° building orientation for HT2 shows a slightly higher than HT3 compact condition. However, a more significant difference can be seen for 90° building orientation, where HT2 shows an average value of 191.5 HV which is higher compare to the lower average value of 170.8 HV for HT3. However, the values gained for 45° building orientation for HT2 and HT3 are similar.

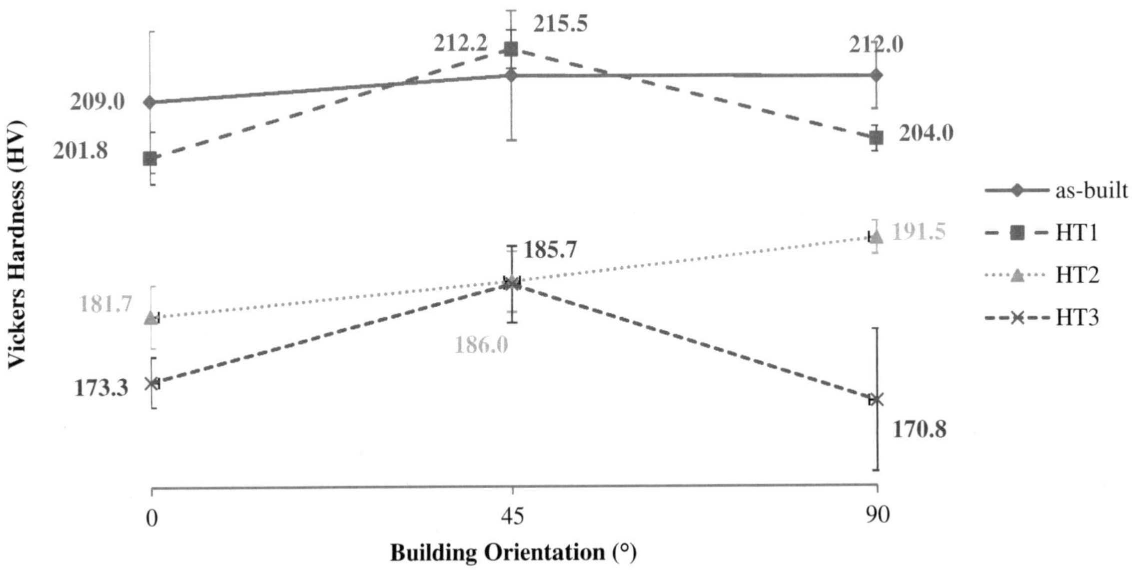

Figure 4. Average micro-hardness (HV) values for SLM compacts.

The microstructure was observed using optical microscope (OM) to identify the cellular dendrites structures, and the phase formation of austenite and ferrite on the compacts from each compact condition, i.e. at different heat treatment conditions.

Figure 5 (a) and (b) shows the structure of melt pools and fine cellular dendrites microstructure in the SLM-ed compacts for as-built and HT1 conditions. This microstructure is a general characteristics poses by metal materials manufactured through AM process as the consequences of the rapid solidification rates of a local melted region, precisely laser-scanned areas. This is typical for area that exposed to short laser-material interaction time during build process [14]. The high hardness value gained for as-built and HT1 conditions compare to HT2 and HT3 conditions are believed cause by the fine-grain microstructures, hence results in higher dislocation density of austenite cells [15]. This makes it difficult for slip motion along the grain boundaries, hence increases its strength and resistance towards deformation. Interestingly, as the compacts were heat treated up to a higher temperature of 950°C and 1100°C, the cellular dendrites microstructure was no longer can be observed, and the melt pool boundaries seems to be dissolved. This is the reason for low hardness value gained for HT2 and HT3 compact conditions.

The dispersion of two different iron based phases comprising of dominant face-centred cubic (FCC) structure of austenitic (γ) phase, and less prevalent body-centred cubic (BCC) structure of ferrite (δ) phase are reveal in Figure 5 (c) and (d). Material with full austenite can be gained at equilibrium according to the chemical composition of 316L stainless steel. However, some δ-ferrite formation is found due to the fast solidification and the existence of chemical elements conducive for ferrite formations which are Si, Cr, and Mo [4, 16]. Comparing Figure 5 (c) and (d), the δ-ferrite phase's volume fraction (darker phase) is slightly bigger in Figure 5 (c) than in Figure 5 (d). Hence, demonstrates that the employment of heat treatment greatly effect in the decrease of volume fraction

IOP Conf. Series: Materials Science and Engineering **257** (2017) 012021 doi:10.1088/1757-899X/257/1/012021

of δ-ferrite phase. It is known that the mechanical properties of materials are significantly affected by the volume fraction of δ-ferrite phase.

Referring to Figure 4, the micro-hardness values of HT3 for 0° and 90° building orientations are the lowest follows by HT2, and HT1 significantly. These are the result of the enforcement of microstructural coarsening enforced through the process of heat treatment. Note that a room temperature, the δ-ferrite phase is harder compare to the austenitic phase. Therefore, reducing the δ-ferrite phase during heat treatment will lead for the lower hardness values gained. This is supported by the research done by Yadollahi et. al [4, 17, 18].

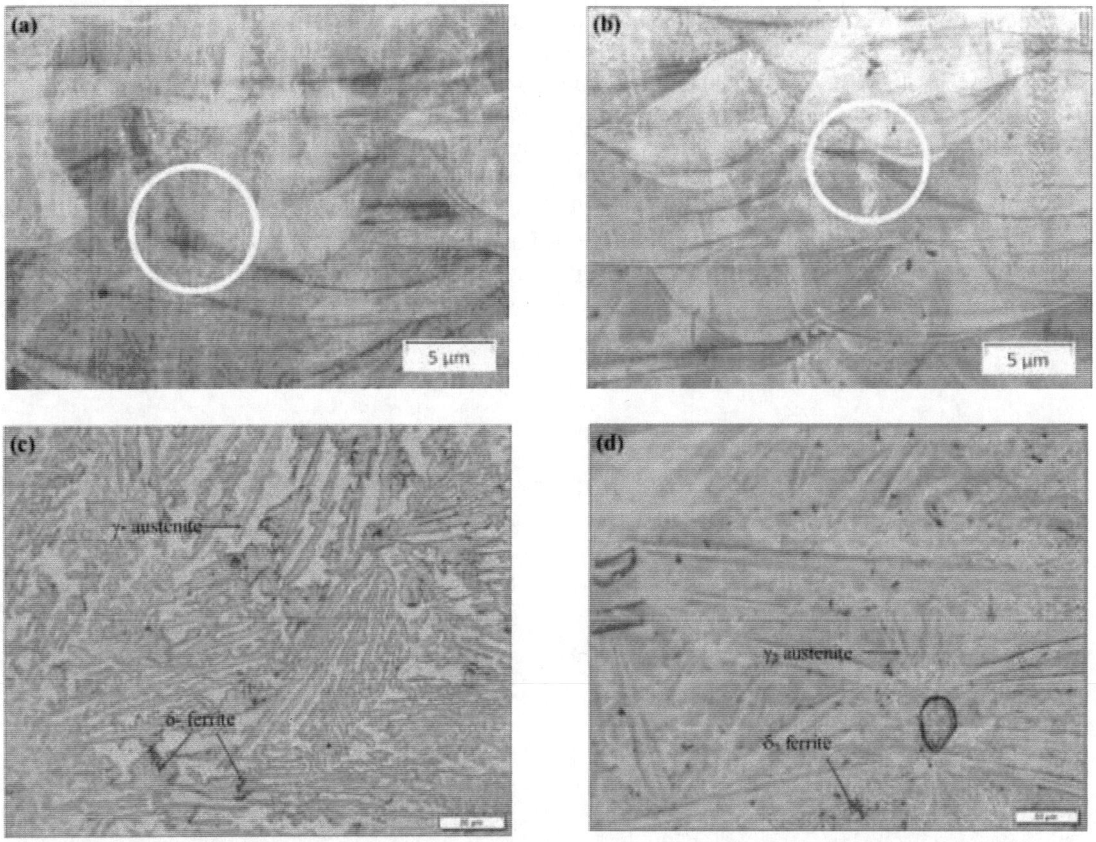

Figure 5. Optical microscope image of BD plane cross-sectioned 316L stainless steel compacts produced by SLM. (a) As-built, (b) HT1, (c) HT2, and (d) HT3 conditions.

Noted that Figure 6 represents the micro-hardness values were taken only at the 'y' cross-sectioned plane of the SD plane of the compacts plotted against the building orientation. Figure 6 reveals the variation of the micro-hardness vales for different building orientation. For as-built, HT1, and HT3 conditions, the lowest micro-hardness values are at the 0° building orientation, follows by the 90° building orientation. Both 0° building orientation and 90° building orientation show variation in the values. 45° building orientation on the other hand exhibits the highest micro-hardness values for those three conditions, i.e. as-built, HT1, and HT3, but with a much stable values. A different trend is shown by HT2 where the micro-hardness measurements increase with the increase of degree of building orientation. However, the average value gains for 0° building orientation, i.e. 181 HV, and 45° building orientation, i.e. 184 HV, do not show a significant different.

The micro-hardness variation of each orientation demonstrates anisotropy in SLM as reported in other literatures [14, 19]. This is a usual quality of additive manufacturing processes for metal parts [20]. This is believed due to the localised melting of powder particles that leads to non-homogeneous morphologies [14]. Besides that, it is also caused by the layer-by-layer approaches and the formation of grain structures and textures due to the thermal gradient [21]. The variations of micro-hardness values of 0° building orientation is probably due to the existence of high residual stresses. This phenomenon might be due to inhomogeneity of delivered powder layer which causes differences in layers' thicknesses caused by the roughness of the substrate and non-parallel between coater and the substrate. This leads to strains during cooling. Deformations and loss of the metallurgical contact with the substrate during manufacturing as observed during the fabrication 0° building orientation resulted in accumulation of heat as well as occurrence of redistribution of stresses. As for 45° building orientation, the micro-hardness values measured are more stable. During the fabrication, the compact experienced more layer-by-layer formation as compared to 0° building orientation's compact. During the process, powders are melted and solidification occurs under high laser energy. Thus, heat is conducted layer-by-layer in the compact. For each layer fabrication, the compact is contacted to support structures, and hence, helps as heat conductor by transferring the heat away from the working compact. As a result of the heat dissipation, the influence from residual stresses on the process is fairly stable. In contrast to 45° building orientation's compacts, the 90° building orientation's compacts have shown a variation in their micro-hardness values probably due to the existence potential of high residual stresses that mainly oriented along the building direction. This phenomenon is also aligning with other reported literatures [22-24]. As for real building application, parts with more consistence microhardness property and low residual stress potential is more preferable as shown by 45° building orientation in this study.

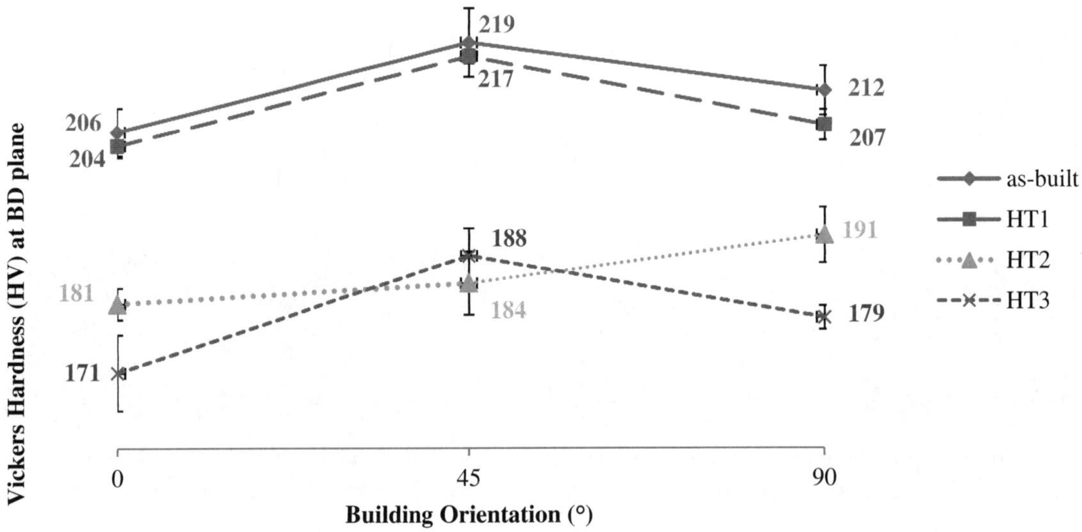

Figure 6. Average micro-hardness (HV) values for at BD plane cross-sectioned of SLM compacts.

4. Conclusions

This research concludes the following points,

1) The results reveal that the micro-hardness of the as-built compacts is between 209.0 and 212.2 HV, which is much higher than the heat-treated compacts. Heat treatment causes the hardness values of the SLM -fabricated 316L stainless steel compacts to decrease. The lowest hardness value gain was for the highest heat treatment temperature tested is 171 HV for 1100 °C with 2 hours soaking time.
2) Mechanical properties of SLM-produced 316L stainless steel are affected by the formation of fine cellular dendrites microstructures and heat treatment. The fine cellular dendrites of completed parts due to localized melting of powder and rapid heating-and-cooling rates during SLM fabrication.

Thus, leads to high micro-hardness values gained of the SLMed parts. While heat treatment reduced the volume fraction of δ-ferrite and become the reason for the lower micro-hardness values gained in the study.

3) Variation of micro-hardness values for different building orientations demonstrates anisotropy properties of the SLM fabricated compacts. This is due to the non-homogeneous morphologies of melted powder particles, and thermal gradient of the built compacts. The inconsistency hardness values of compacts for 0° and 90° building orientations are probably due to the high residual stresses existed in the compacts.

Future studies would be done to excess the existence of residual stress for building orientations conditions and their effects on other mechanical properties, and deeper microstructure study.

Acknowledgments
Authors would like to recognise the contributions of Universiti Malaysia Pahang (Grant no. RDU151314, RDU160354, GRS1503121), the Human Engineering Group (HEG), and Universiti Malaysia Pahang for its comprehensive facilities support in this research.

References

[1] Zhang B, Dembinski L and Coddet C 2013 The study of the laser parameters and environment variables effect on mechanical properties of high compact parts elaborated by selective laser melting 316L powder *Materials Science and Engineering: A* **584** 21-31

[2] Ahmadi A, Mirzaeifar R, Moghaddam N S, Turabi A S, Karaca H E and Elahinia M 2016 Effect of manufacturing parameters on mechanical properties of 316L stainless steel parts fabricated by selective laser melting: A computational framework *Materials & Design* **112** 328-38

[3] Ma M, Wang Z and Zeng X 2017 A comparison on metallurgical behaviors of 316L stainless steel by selective laser melting and laser cladding deposition *Materials Science and Engineering: A* **685** 265-73

[4] Yadollahi A, Shamsaei N, Thompson S M and Seely D W 2015 Effects of process time interval and heat treatment on the mechanical and microstructural properties of direct laser deposited 316L stainless steel *Materials Science and Engineering: A* **644** 171-83

[5] Wang D, Yang Y, Liu R, Xiao D and Sun J 2013 Study on the designing rules and processability of porous structure based on selective laser melting (SLM) *Journal of Materials Processing Technology* **213** 1734-42

[6] Krakhmalev P, Fredriksson G, Yadroitsava I, Kazantseva N, Plessis A d and Yadroitsev I 2016 Deformation Behavior and Microstructure of Ti6Al4V Manufactured by SLM *Physics Procedia* **83** 778-88

[7] Sun Z, Tan X, Tor S B and Yeong W Y 2016 Selective laser melting of stainless steel 316L with low porosity and high build rates *Materials & Design* **104** 197-204

[8] AlMangour B, Grzesiak D and Yang J-M 2017 In-situ formation of novel TiC-particle-reinforced 316L stainless steel bulk-form composites by selective laser melting *Journal of Alloys and Compounds* **706** 409-18

[9] Yadroitsev I and Smurov I 2011 Surface Morphology in Selective Laser Melting of Metal Powders *Physics Procedia* **12** 264-70

[10] Wang D, Song C, Yang Y and Bai Y 2016 Investigation of crystal growth mechanism during selective laser melting and mechanical property characterization of 316L stainless steel parts *Materials & Design* **100** 291-9

[11] Yasa E and Kruth J P 2011 Microstructural investigation of Selective Laser Melting 316L stainless steel parts exposed to laser re-melting *Procedia Engineering* **19** 389-95

[12] Zhong Y, Liu L, Wikman S, Cui D and Shen Z 2016 Intragranular cellular segregation network structure strengthening 316L stainless steel prepared by selective laser melting *Journal of Nuclear Materials* **470** 170-8

[13] Montero Sistiaga M, Nardone S, Hautfenne C and Van Humbeeck J 2016 Effect of heat treatment of 316L stainless steel produced by selective laser melting (SLM)

[14] Yusuf S M, Chen Y, Boardman R, Yang S and Gao N 2017 Investigation on Porosity and Microhardness of 316L Stainless Steel Fabricated by Selective Laser Melting *Metals* **7** 64

[15] Saeidi K, Gao X, Zhong Y and Shen Z J 2015 Hardened austenite steel with columnar sub-grain structure formed by laser melting *Materials Science and Engineering: A* **625** 221-9

[16] Bouche G, Bechade J, Mathon M, Allais L, Gourgues A and Naze L 2000 Texture of welded joints of 316L stainless steel, multi-scale orientation analysis of a weld metal deposit *Journal of nuclear materials* **277** 91-8

[17] Imandoust A, Zarei-Hanzaki A, Heshmati-Manesh S, Moemeni S and Changizian P 2014 Effects of ferrite volume fraction on the tensile deformation characteristics of dual phase twinning induced plasticity steel *Materials & Design* **53** 99-105

[18] Kocijan A, Merl D K and Jenko M 2011 The corrosion behaviour of austenitic and duplex stainless steels in artificial saliva with the addition of fluoride *Corrosion Science* **53** 776-83

[19] Miranda G, Faria S, Bartolomeu F, Pinto E, Madeira S, Mateus A, Carreira P, Alves N, Silva F S and Carvalho O 2016 Predictive models for physical and mechanical properties of 316L stainless steel produced by selective laser melting *Materials Science and Engineering: A* **657** 43-56

[20] Chen L Y, Huang J C, Lin C H, Pan C T, Chen S Y, Yang T L, Lin D Y, Lin H K and Jang J S C 2017 Anisotropic response of Ti-6Al-4V alloy fabricated by 3D printing selective laser melting *Materials Science and Engineering: A* **682** 389-95

[21] Aboulkhair N T, Stephens A, Maskery I, Tuck C, Ashcroft I and Everitt N M 2015 Mechanical properties of selective laser melted AlSi10Mg: nano, micro, and macro properties. In: *Proc. of Solid Freeform Fabrication Symposium,* pp 1-30

[22] Vrancken B, Cain V, Knutsen R and Van Humbeeck J 2014 Residual stress via the contour method in compact tension specimens produced via selective laser melting *Scripta Materialia* **87** 29-32

[23] Rangaswamy P, Griffith M L, Prime M B, Holden T M, Rogge R B, Edwards J M and Sebring R J 2005 Residual stresses in LENS® components using neutron diffraction and contour method *Materials Science and Engineering: A* **399** 72-83

[24] Chlebus E, Kuźnicka B, Kurzynowski T and Dybała B 2011 Microstructure and mechanical behaviour of Ti—6Al—7Nb alloy produced by selective laser melting *Materials Characterization* **62** 488-95

IOP Conf. Series: Materials Science and Engineering **257** (2017) 012021 doi:10.1088/1757-899X/257/1/012021

Nomenclature

γ	Austenite phase
δ	Ferrite phase
0-AB	Compact condition with 0° building orientation and no heat treatment
0-HT 1	Compact condition with 0° building orientation and 650°C heat treated
0-HT 2	Compact condition with 0° building orientation and 950°C heat treated
45-AB	Compact condition with 45° building orientation and no heat treatment
45-HT 1	Compact condition with 45° building orientation and 650°C heat treated
45-HT 2	Compact condition with 45° building orientation and 950°C heat treated
90-AB	Compact condition with 90° building orientation and no heat treatment
90-HT 1	Compact condition with 90° building orientation and 650°C heat treated
90-HT 2	Compact condition with 90° building orientation and 950°C heat treated
AB	As-built
BD	Building direction
BCC	Body-centred cubic structure
C	Carbon (chemical element)
Cr	Chromium (chemical element)
FCC	face-centred cubic structure
Fe	Iron (chemical element)
HT1	Heat treatment cycle at 650°C for 2 hours with furnace cooling
HT2	Heat treatment cycle at 950°C for 2 hours with furnace cooling
HT3	Heat treatment cycle at 1100°C for 2 hours with furnace cooling
Ni	Nickel (chemical element)
Mn	Manganese (chemical element)
Mo	Molybdenum (chemical element)
P	Phosphorus (chemical element)
S	Sulphur (chemical element)
SD	Scanning direction
Si	Silicon (chemical element)
SLM	Selective Laser Melting
SS	Stainless Steel
TD	Transverse direction

IOP Conf. Series: Materials Science and Engineering **257** (2017) 012022 doi:10.1088/1757-899X/257/1/012022

Effect of pin tool design on the material flow of dissimilar AA7075-AA6061 friction stir welds

Mohammed M Hasan[1,2*], M Ishak[1] and M R M Rejab[1]

[1]Faculty of Mechanical Engineering, Universiti Malaysia Pahang, Pahang, Malaysia
[2]University of Technology, Baghdad, Iraq

[*]Corresponding author: modmhasan@gmail.com

Abstract. Tool design is the most influential aspect in the friction stir welding (FSW) technology. Influence of pin tool geometry on material flow pattern are studied in this work during the FSW of dissimilar AA7075 and AA6061 aluminium alloys. Three truncated pin tool profiles (threaded, threaded with single flat, and unthreaded with single flat) were used to prepare the weldments. The workpieces were joined using a custom-made clamping system under 1100 rpm of spindle speed, 300 mm/min of traverse rate and 3° of tilt angle. The metallographic analysis showed that defect-free welds can be produced using the three pin tools with significant changes in the mixing stir zone structure. The results declared that the introducing of the flat on the cone of the probe deviates the pattern of the onion rings without changing the chemical composition of the created layers. This in turn improves the hardness distribution and tensile strength of the welded joint. It was also noted that both heat affected zone (HAZ) and thermal-mechanical affected zone (TMAZ) are similar in composition to their corresponding base materials (BM).

1. Introduction

Friction stir welding has become one of the widely-used technologies in modern industries, such as automotive and aerospace [1, 2]. This novel green welding method was invented in 1991 by The Welding Institute (TWI) of the UK to solve the problem of joining the hard-to-weld aluminium alloys [3, 4]. It also offered the ability of welding dissimilar materials from the same or different families, such as aluminium alloy to another aluminium alloy, aluminium to magnesium and aluminium to steel [5, 6]. This advantage permits to take the benefits of both adjoined materials [7, 8]. As seen in Figure 1, the materials to be weld are connected through this solid-state method by stirring a specially designed rotating tool, which is the main source of the process heat that not exceeds the melting point of the workpieces [9].

The success of the material mixing is mainly depending on the geometry of the tool pin or probe [10, 11] and the welding rotation speed (ω) and tool traverse rate (υ) [12]. Improper design of the pin tool may result in an internal defects or incomplete material mixing within the stir zone (SZ). For defect-free weld, the characterization of the probe design determines the flow pattern, which has not been widely considered in the previous research studies as the mechanical properties of the welding joints. Accordingly, the inspection of material flow in conjunction with tool design during the friction stir processing is vital for putting on the right track of the weld quality.

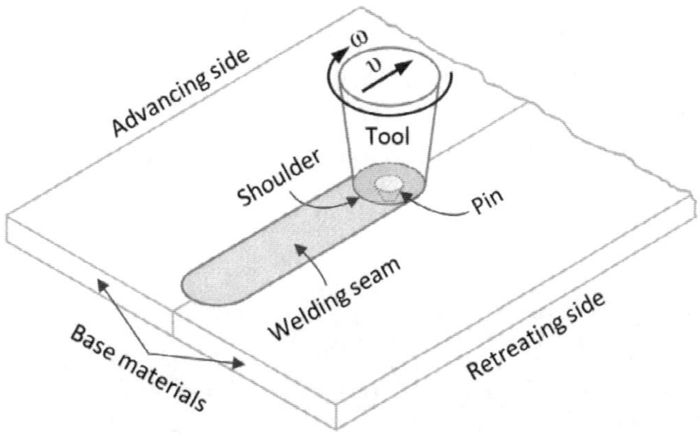

Figure 1. Schematic drawing of the FSW process.

Recently, Tongne et al. [13] studied the formation and evolution of banded structures observed in similar weld of AA6082-T6 aluminium alloy using trigonal tool. It was demonstrated that the formation of the banded structures is mainly related to the geometry of the pin whereas the friction conditions have a much smaller effect. Patel et al. [14] investigated the influence of polygonal pin profiles on friction stir processed super plasticity of AA7075-T651 aluminium alloy. They noticed that all polygonal pin profiles indicated sticking of the workpiece material around tool pin and resulted in non-uniform microstructure in the SZ. On the other hand, uniform microstructure without cavitation in the SZ was observed in the sample produced by the square pin tool. Schneider and his co-authors [15] examined the material flow modification in the FSW of AA2219-T87 aluminium alloy through introduction of flats into the threaded cylindrical pin tools. Significant changes in the FSW structure with varying numbers of flats were observed, but without significant changes in tensile strength. Changes in the textural banding shape, the addition of sub-bands, and a new set of bands from coalescence of band kinks constitute the structural changes observed. Based on the material flow characteristics, a mechanism of the friction stir AA7020-T6 weld formation was presented by Kumar and Kailas [16]. They stated that there are two different modes of material flow regimes involved in the friction stir weld formation; namely "pin-driven flow" and "shoulder-driven flow". These material flow regimes merge together to form a defect-free weld. It was also illustrated that the etching contrast in these regimes gives rise to the onion ring pattern in friction stir welds. Formation mechanism of these onion rings was rather difficult to be explained. Krishnan [17] attributed the creation of the onion rings to the process of friction heating due to the tool rotation and forward movement, which extrudes the metal around to the retreating side of the tool.

In the FSW of dissimilar alloys or materials, limited articles focused on the material flow and mixing, especially with the effect of tool design. Consequently, this paper aims to investigate the influence of the pin tool profile on the microstructure of the dissimilar AA7075-AA6061 friction stir welds. This study contributes to the advanced industrial technology.

2. Experimental set up

Welding coupons were prepared from 3-mm-thick AA7075-T6 and AA6061-T6 aluminium alloys, whose mechanical properties are presented in Table 1. The specimens were cut and ground by a precision grinding machine to the final dimensions of 125×50 mm and cleaned with acetone. According to our previous work [18], a specially designed clamping system was used to produce the welds, as shown in Figure 2. Butt-joints were formed in a route parallel to the rolling direction of the base materials.

Table 1. Mechanical properties of the base materials (normal to the rolling direction).

Alloy	Yield strength (MPa)	Tensile strength (MPa)	Vickers hardness	Elongation (%)
AA7075-T6	503	571	175	11
AA6061-T6	276	307	107	12

Figure 2. The custom-made clamping system and a sample photograph of the welding joint.

Three different welding pin tools, whose geometry are schematically viewed in Figure 3, were fabricated from H13 steel. Each tool consisted of concave shoulder of 12-mm-diameter and 8° of concave angle. The base diameter of the probes was 4 mm, which truncated through 10° along the length of 2.7 mm. The cone of the probe was threaded with M4×1 left-hand threads in the first and second tools, while it left smooth in the third one. A single flat was formed in the second and third tapered pin tools. Based on the results of our previous research studies [19, 20], the softer AA6061 alloy was placed on the advancing side (AS) of the weld and the tool rotation speed, traverse rate and tilt angle were fixed at 1100 rpm, 300 mm/min and 3°, respectively.

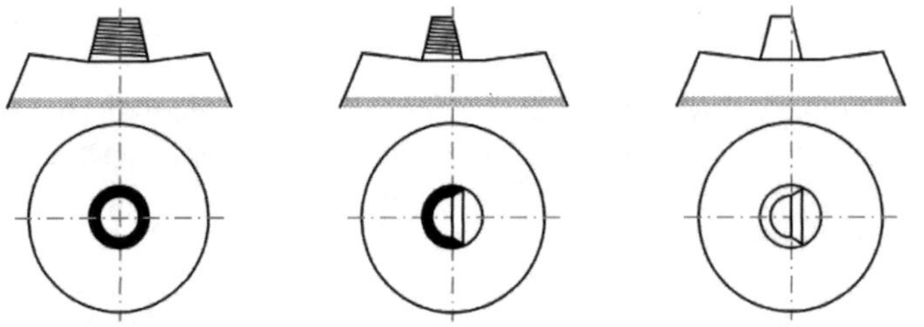

Figure 3. Schematic views of the three welding pin tools.

The produced welds were left two months for natural aging [21]. After that, tensile and metallographic specimens were cut by a wire cutting machine. The specimens were prepared as per the standard of American Society for Testing of Materials (ASTM: E8-11). Tests were then conducted at room temperature using a 50 kN universal testing machine with a speed of 1 mm/min. Three tension tests were conducted for each case study, and the average ultimate tensile strength and elongation were considered. The standard guide for preparation of metallographic specimens (ASTM: E3-11) was followed to prepare the metallographic specimens, which were then etched by a modified killer's reagent to reveal the grain structure of the weld zone. Hardness distribution were then observed across the weld centreline, in a direction normal to the weld seam, by a Vickers micro-hardness tester using HV0.5 test method with an indent time of 10 seconds. The metallographic analysis was performed using an optical microscope and scanning electron microscope (SEM) equipped with an energy dispersive spectrometry (EDS).

3. Results and discussion

The obtained micrographs showed that defect-free joints were produced using the three tools. The weld in the stir zone has exhibited the well-recognized onion rings, as seen in Figure 4. The nugget zone resulted from the threaded tools T_1 (without flat) and T_2 (with single flat) were compared to the that produced by T_3, which consisted of a non-threaded probe with additional flat.

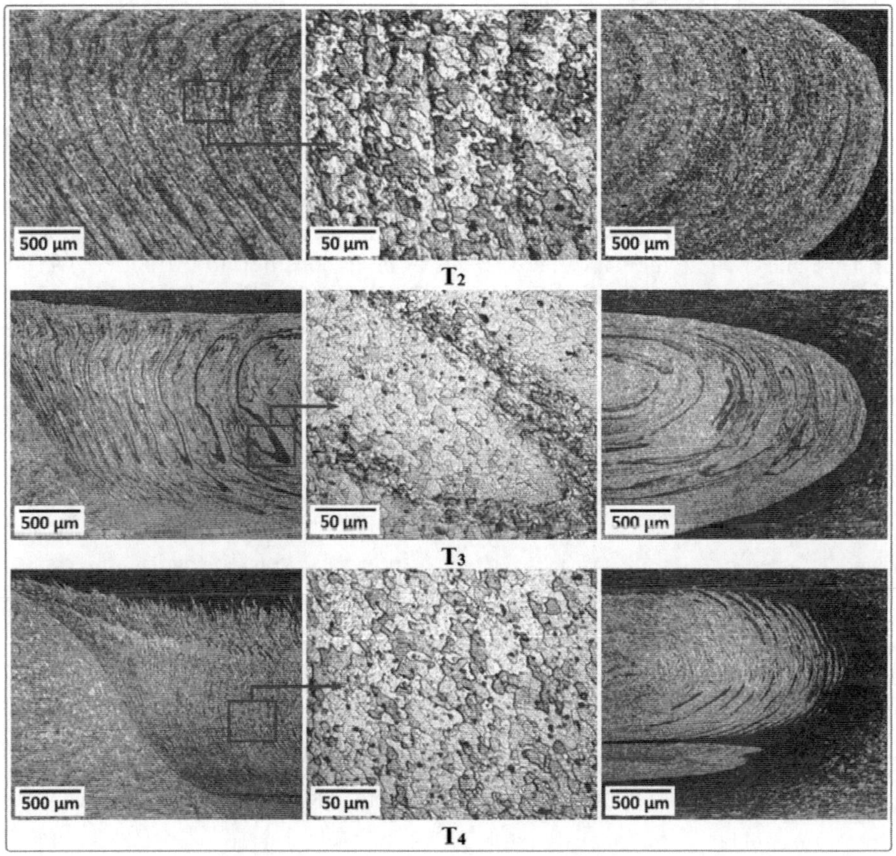

Figure 4. Microstructure of the welding stir zone related to the pin tool design.

The onion rings appeared more uniform when the threaded pin tool T_1 was used, while they nearly vanished when the unthreaded tool T_3 was used. This is due to that when using a threaded probe in similar and dissimilar FSW, the material moves downward and a helical vertical rotational flow within the intermixed region generates beside the periphery of the rotating pin and hence forming the sub-layers or onion rings [22]. The shape of the rings was significantly changed when a flat was added to the cone of the threaded pin tool T_2. The flat pushed the abutting materials away from the probe and reduced the effect of the threads.

Chemical composition of the weld was analysed in the onion rings, TMAZ, HAZ, and BM. The etching response of the dissimilar parent materials to the Keller's reagent are not the same due to the difference in their chemical composition [23]. For this reason, the micrographs revealed in dark and bright colours, as shown in Figure 5, which represents the SEM image of the onion rings and the corresponding EDS spectrum.

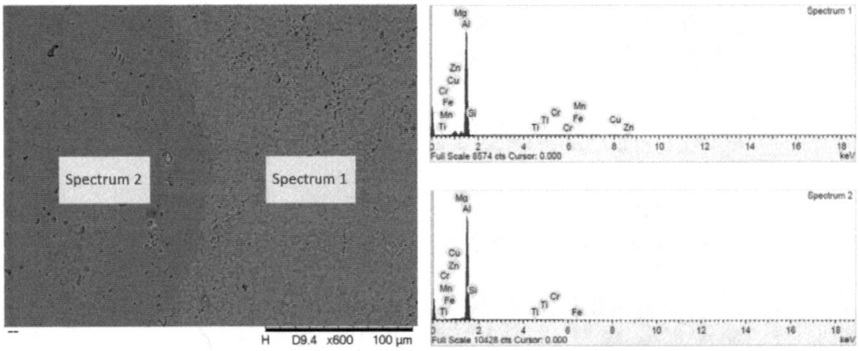

SEM image EDS spectrums

Figure 5. The SEM image of the onion ring layers and the corresponding EDS spectrums.

The results presented in Table 2 clarify that the bright region is similar in composition to that of the AA7075-T6 base alloy, while the dark one is identical in composition to the softer AA6061-T6 base material. Same observation was found in the works of İpekoğlu and Çam [24] and Guo et al. [25]. On the other hand, the composition of the TMAZ and HAZ of both welding sides are close to their corresponding base materials.

Table 2. The acquired chemical composition (wt.%) of the base materials, HAZ, TMAZ and the onion ring layers related to the spectrums shown in Figure 5.

Spectrum	Si	Fe	Cu	Mn	Mg	Cr	Zn	Ti	Al
1	0.543	0.183	2.261	0.031	2.610	0.224	5.203	0.136	Balance
2	0.720	0.322	1.046	0.018	1.077	0.090	0.393	0.072	Balance
TMAZ-6061	0.716	0.046	0.185	0.040	1.036	0.320	0.227	0.039	Balance
TMAZ-7075	0.097	0.241	1.943	0.061	2.704	0.196	5.853	0.115	Balance
HAZ-6061	0.716	0.046	0.185	0.040	1.036	0.320	0.227	0.039	Balance
HAZ-7075	0.097	0.241	1.943	0.061	2.704	0.196	5.853	0.115	Balance
BM-6061	0.701	0.090	0.161	0.023	0.801	0.258	0.215	0.047	Balance
BM-7075	0.168	0.212	1.801	0.040	2.512	0.251	5.525	0.132	Balance

As shown in Figure 6, the joint strength was not highly affected by the creation of the different layers or onion rings. The ultimate tensile strength of the weld ranged from 229.5 MPa for tool T_3, passing through 237.6 MPa for T_1 to reach the maximum value of 252.1 MPa for T_2. Since the joint strength of dissimilar friction stir weld is typically compared to the softer material [26], tensile strength of the harder AA7075 base material was not shown in the graph. These outcomes are coincide with the observation of Schneider and his research group [27]. They found that the existence of flat/flute in the pin tool design changes the flow pattern of the mixed materials without significant change in the weld strength. Figure 7 presents the micro-hardness distribution along the transvers centreline of the weld related to the welding pin tools. Overall, the welding joints have exhibited a noticeable decrease in the level of the Vickers hardness number (VHN) compared to the base materials. This could be attributed to the grain refinement or recrystallization within the stir zone and the over-aging in the heat affected zone [2, 28, 29].

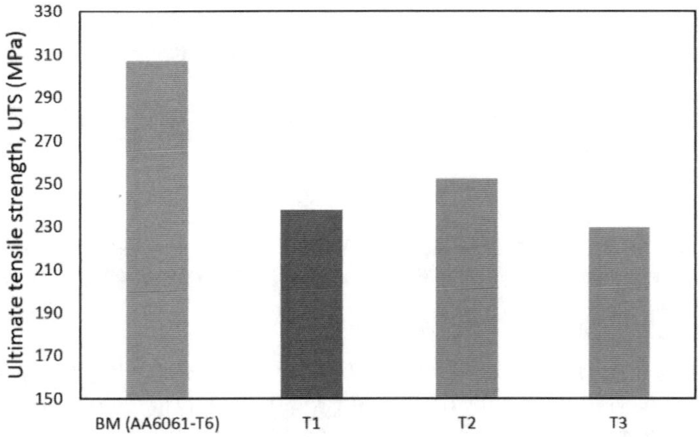

Figure 6. Tensile strength of the weld related to pin tool design.

Maximum reduction of the weld hardness was observed in the HAZ of the AA6061-T6 alloy, which was fixed on the AS. The lowest hardness values of 66, 71 and 63 were measured in the welding joints produced by the welding tools T_1, T_2 and T_3, respectively. The fracture of the tension test specimens was initiated from the corresponding locations of these minimum hardness values at the HAZ. It has been reported that the fracture takes place in the weakest region during the tension test of the similar and dissimilar friction stir welds of aluminium alloys. In case of defect-free dissimilar friction stir welds, the HAZ of the softer alloy represents the weakest region, where the fracture commonly occurs [24, 30]. It is obvious that the results of tension tests are proportional with the hardness level at the HAZ of the softer material. In other word, the minimum joint strength was measured in the specimen that exhibited the lowest VHN and vice versa.

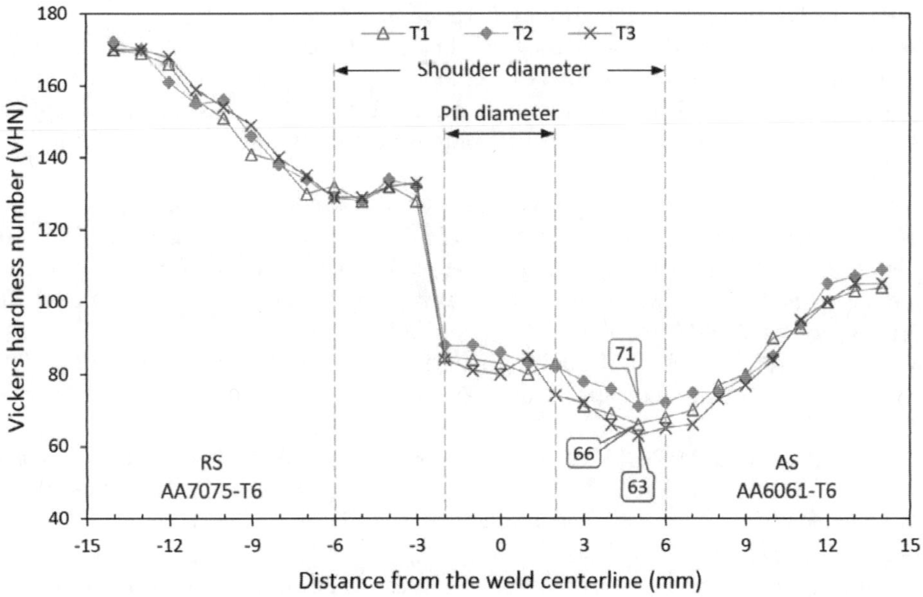

Figure 7. Hardness distribution of the weld related to the pin tool design.

4. Conclusions

The outcomes of this research showed that there is a clear relation between the design of the welding pin tool and the material flow pattern of dissimilar friction stir welds. Formation of the well-known onion rings is found to be sensitive to the profile of the probe. The introducing of a flat on the cone of the pin tool changes the flow pattern within the mixing stir zone and affect the hardness distribution, and hence the tensile strength of the welding joint. The etching response showed different layers in dark and bright colours inside the weld nugget, which are similar to their corresponding base materials. The chemical compositions of the TMAZ and HAZ regions were not change during the FSW process.

Acknowledgements
The authors would like to be obliged to the Malaysian Ministry of Education for providing research grant (RDU1403114). Thanks to Universiti Malaysia Pahang for laboratory facilities.

References

[1] Sathari N, Razali A, Ishak M and Shah L 2015 Mechanical strength of dissimilar AA7075 and AA6061 aluminum alloys using friction stir welding *International Journal of Automotive & Mechanical Engineering* **11** 2713-21

[2] Ahmed M, Ataya S, Seleman M E-S, Ammar H and Ahmed E 2017 Friction stir welding of similar and dissimilar AA7075 and AA5083 *J. Mater. Process. Technol.* **242** 77-91

[3] Selamat N, Baghdadi A, Sajuri Z and Kokabi A 2016 Friction stir welding of similar and dissimilar aluminium alloys for automotive applications *International Journal of Automotive & Mechanical Engineering* **13**

[4] Zhou Z, Yue Y, Ji S, Li Z and Zhang L 2016 Effect of rotating speed on joint morphology and lap shear properties of stationary shoulder friction stir lap welded 6061-T6 aluminum alloy *Int. J. Adv. Manuf. Technol.* 1-7

[5] Sahu P K, Pal S, Pal S K and Jain R 2016 Influence of plate position, tool offset and tool rotational speed on mechanical properties and microstructures of dissimilar Al/Cu friction stir welding joints *J. Mater. Process. Technol.* **235** 55-67

[6] Shi H, Chen K, Liang Z, Dong F, Yu T, Dong X, Zhang L and Shan A 2016 Intermetallic Compounds in the Banded Structure and Their Effect on Mechanical Properties of Al/Mg Dissimilar Friction Stir Welding Joints *J. Mater. Sci. Technol.*

[7] Shah L H and Ishak M 2014 Review of Research Progress on Aluminum–Steel Dissimilar Welding *Mater. Manuf. Processes* **29** 928-33

[8] Islam M R, Ishak M, Shah L H, Idris S R A and Meriç C 2016 Dissimilar welding of A7075-T651 and AZ31B alloys by gas metal arc plug welding method *Int. J. Adv. Manuf. Technol.* **88** 2773-83

[9] NAA S and AR R 2014 Investigation of Single-pass Double-pass Techniques on Friction Stir Welding of Aluminium *Journal of Mechanical Engineering and Sciences (JMES)* **7** 1053-61

[10] García-Bernal M, Mishra R, Verma R and Hernández-Silva D 2016 Influence of friction stir processing tool design on microstructure and superplastic behavior of Al-Mg alloys *Mat. Sci. Eng. A-Struct.* **670** 9-16

[11] Mehta K P and Badheka V J 2017 Influence of tool pin design on properties of dissimilar copper to aluminum friction stir welding *Transactions of Nonferrous Metals Society of China* **27** 36-54

[12] Pan Y and Lados D A 2017 Friction Stir Welding in Wrought and Cast Aluminum Alloys: Weld Quality Evaluation and Effects of Processing Parameters on Microstructure and Mechanical Properties *Metallurgical and Materials Transactions a-Physical Metallurgy and Materials Science* **48a** 1708-26

[13] Tongne A, Desrayaud C, Jahazi M and Feulvarch E 2017 On material flow in Friction Stir Welded Al alloys *J. Mater. Process. Technol.* **239** 284-96

[14] Patel V V, Badheka V and Kumar A 2017 Effect of polygonal pin profiles on friction stir processed superplasticity of AA7075 alloy *J. Mater. Process. Technol.* **240** 68-76

[15] Schneider J, Brooke S and Nunes A C 2015 Material Flow Modification in a FSW Through Introduction of Flats *Metall. Mater. Trans. B* **47** 720-30

[16] Kumar K and Kailas S V 2008 The role of friction stir welding tool on material flow and weld formation *Mat. Sci. Eng. A-Struct.* **485** 367-74

[17] Krishnan K 2002 On the formation of onion rings in friction stir welds *Mat. Sci. Eng. A-Struct.* **327** 246-51

[18] Hasan M M, Ishak M and Rejab M R M 2015 A simplified design of clamping system and fixtures for friction stir welding of aluminium alloys *JMES* **9** 1628-39

[19] Hasan M M, Ishak M and Rejab M R M 2017 Effect of backing material and clamping system on the tensile strength of dissimilar AA7075-AA2024 friction stir welds *Int. J. Adv. Manuf. Technol.* 1-17

[20] Hasan M M, Ishak M and Rejab M R M 2017 Influence of machine variables and tool profile on the tensile strength of dissimilar AA7075-AA6061 friction stir welds *Int. J. Adv. Manuf. Technol.* **90** 2605-15

[21] Mishra R S and Mahoney M W 2007 *Friction stir welding and processing*: ASM International)

[22] Teimurnezhad J, Pashazadeh H and Masumi A 2016 Effect of shoulder plunge depth on the weld morphology, macrograph and microstructure of copper FSW joints *J. Manuf. Processes* **22** 254-9

[23] Firouzdor V and Kou S 2010 Formation of liquid and intermetallics in Al-to-Mg friction stir welding *Metall. Mater. Trans. A* **41** 3238-51

[24] İpekoğlu G and Çam G 2014 Effects of initial temper condition and postweld heat treatment on the properties of dissimilar friction-stir-welded joints between AA7075 and AA6061 aluminum alloys *Metall. Mater. Trans. A* **45** 3074-87

[25] Guo J F, Chen H C, Sun C N, Bi G, Sun Z and Wei J 2014 Friction stir welding of dissimilar materials between AA6061 and AA7075 Al alloys effects of process parameters *Mater. Des.* **56** 185-92

[26] Giraud L, Robe H, Claudin C, Desrayaud C, Bocher P and Feulvarch E 2016 Investigation into the dissimilar friction stir welding of AA7020-T651 and AA6060-T6 *J. Mater. Process. Technol.* **235** 220-30

[27] Schneider J, Brooke S and Nunes Jr A C 2016 Material flow modification in a FSW through introduction of flats *Metall. Mater. Trans. B* **47** 720-30

[28] Sun Y, Tsuji N and Fujii H 2016 Microstructure and mechanical properties of dissimilar friction stir welding between ultrafine grained 1050 and 6061-t6 aluminum alloys *Metals* **6** 249

[29] Khan N Z, Siddiquee A N, Khan Z A and Mukhopadhyay A K 2017 Mechanical and microstructural behavior of friction stir welded similar and dissimilar sheets of AA2219 and AA7475 aluminium alloys *J. Alloys Compd.* **695** 2902-8

[30] Rodriguez R I, Jordon J B, Allison P G, Rushing T and Garcia L 2015 Microstructure and mechanical properties of dissimilar friction stir welding of 6061-to-7050 aluminum alloys *Mater. Des.* **83** 60-5

The two-stroke poppet valve engine. Part 1: Intake and exhaust ports flow experimental assessments

M Kamili Zahidi[1]*, M Razali Hanipah[1, 2], D Ramasamy[1, 2], M M Noor[1, 2], K Kadirgama[1, 2], M M Rahman[1, 2]

[1]Automotive Engineering Centre (AEC), Universiti Malaysia Pahang, 26600, Pekan, Pahang, Malaysia.
[2]Automotive Engineering Research Group (AERG), Faculty of Mechanical Engineering, Universiti Malaysia Pahang, 26600, Pekan, Pahang, Malaysia.

*Corresponding author: kamili.zahidi@gmail.com

Abstract. A two-stroke poppet valve engine is developed to overcome the common problems in conventional two-stroke engine designs. However, replacing piston control port with poppet valve will resulted different flow behaviour. This paper is looking at experimental assessment on a two-stroke poppet valve engine configuration to investigate the port flow performance. The aims are to evaluate the intake and exhaust coefficient of discharge and assess the two-stroke capability of the cylinder head. The results has shown comparable coefficient of discharge values as production engine for the intake while the exhaust has higher values which is favourable for the two-stroke cycle operation.

1. Introduction

Engine downsizing concept has been adopted by most automotive manufacturers to increase the power to weight ratio [1]. Engine downsizing are done by reducing the weight of the engine using smaller capacity and intake air boosting to increase the power.

Two-stroke engine cycle has been known to have higher power to weight ratio. However, the conventional two-stroke engines use piston control port which poses several problems. The first problem is bore distortion caused by an asymmetric temperature of the cylinder liner. Bore distortion is increasing the wear rate or worst case it will lock-up the piston[2]. The second problem is lubricant oil easily enter combustion cylinder via intake port resulting in higher pollutants in the exhaust emission[3].

Two-stroke poppet valve engine has been proposed by several researchers to eliminate the problem and at the same time increase the engine performance [4-7]. Each researcher has experiment different specification and design of two-stroke poppet valve engine. Nakano, et al reported the converted four-stroke engine into two-stroke poppet valve engine produce better power than original engine when compare at the same engine speed [8]. While, Sato, et al has design shrouded intake valve to overcome two-stroke poppet valve problem. The problem is short-circuit phenomenon during scavenging process [2, 3, 9].

In this paper, a production four-stroke cycle engine is converted into a two-stroke cycle engine. Then, the port flow performance is investigated by using flow bench.

2. Engine specifications

The investigation is conducted on a 65cc four-stroke gasoline engine. The bore of the engine is 50mm and stroke is 33mm while the compression ratio is 9.5:1. In terms of valves specifications, it has maximum valve lift of 4 mm, intake valve diameter of 20mm and for exhaust is 18mm. The engine specification is as shown in Table 1.

Table 1. The engine specification.

Parameter	Value
Displacement volume	65 cm^3
Bore	50 mm
Stroke	33 mm
Compression ratio	9.5:1
Maximum valve lift	4 mm
Intake valve diameter	20 mm
Exhaust valve diameter	18 mm

The head design is shown in Figure 1. The valve design is straight at 90° from the horizontal axis. It has one valve is for intake and another valve is for exhaust. The left valve in Figure 1 is exhaust valve and the right valve is intake valve.

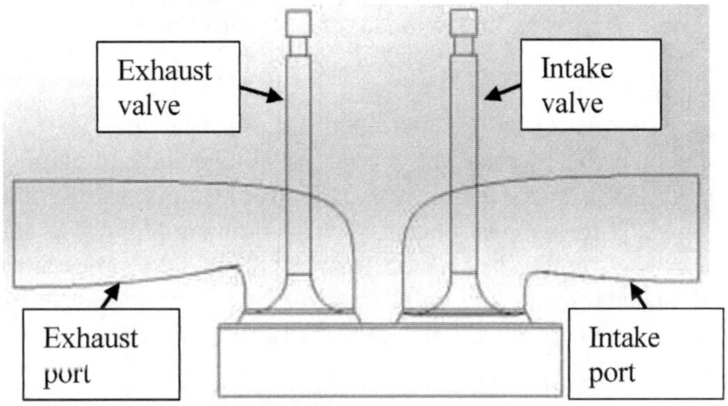

Figure 1. The engine design.

3. Experimental investigations

The experimental investigation is conducted into two phases. First phase involves experimental setup preparation and second is the experimental procedure. The preparation involved is the design and fabrication of adapter to place an engine head on a flow bench. Then, the experiment was run and the data is being calculated to find the coefficient of discharge. There are two types of testing mode; the intake test by setting the flow bench to draw the air from surrounding into flow bench through intake port. Exhaust test is for exhaust by setting the flow bench to blow the air through the exhaust port. Flow bench is an equipment used to measure air flow rate through orifice plate by means of differential pressure. The flow bench is capable of measuring velocity, temperature, and volume flow rate [10] with relevant sensors installed on the equipment during testing.

3.1. Experimental setup preparation

There two components need to be fabricated. First is engine the head adapter to mount engine head on SuperFlow SF-1020 flow bench. The flow bench can be used to test a wide range of cylinder bore diameters. Thus, the adapter is necessary to fit engine head with the flow bench standard sized bore. The adapter also to ensure there is no air leakage. The second part is valve pusher to vary the valve lift during the experiment. The experiment setup is shown in Figure 2. In the setup, dial gauge was used to measure the valve lift.

Figure 2. Engine head setup on flow bench.

3.2. *Experimental procedure*

In this paper, the engine head was tested without manifold or surge tank. The experiment started with intake port test during which the exhaust valve remains closed and the pump was set into suction mode. The data was recorded at every 0.5mm valve lift from 0mm until the maximum lift of 5mm. Next, the experiment resumed for exhaust port test where blowing mode was selected and the intake valve closed. Data were taken at similar valve lift intervals and the procedures were repeated five times for consistencies. The results are presented in terms of L/D ratio as it is an universal ratio used elsewhere [10]. The formula for L/D ratio is given in equation (1) and the results are shown in Table 2.

$$L/D\,ratio = \frac{Valve\ lift\ (L)}{Valve\ diameter\ (D)} \tag{1}$$

Table 2. Intake and Exhaust L/D ratio.

Valve Lift (mm)	Intake L/D ratio	Exhaust L/D ratio
0.5	0.025	0.028
1.0	0.050	0.056
1.5	0.075	0.083
2.0	0.100	0.111
2.5	0.125	0.139
3.0	0.150	0.167
3.5	0.175	0.194
4.0	0.200	0.222
4.5	0.225	0.250
5.0.	0.250	0.278

From the experimental data, the coefficient is calculated in order to assess the efficiency of air flow [11]. The higher the coefficient value results better air flow into the engine. Thus, the coefficients give practical advices to engine designers and tuner on sizing and location of ports and valves[12]. The coefficient of discharge, C_D can be calculated by using equation (2).

$$C_D = \frac{\dot{m}_a}{\dot{m}_t} \tag{2}$$

Where $\dot{m}_a$ is actual mass flow rate and $\dot{m}_t$ is theoretical mass flow rate. The actual mass flow rate can be calculated by using equation (3).

$$\dot{m}_a = Q\frac{P}{RT} \tag{3}$$

Where Q is volume flow rate, P is the local pressure, R is a gas constant of air and T is air temperature. While the theoretical mass flow rate can be calculated by using equation (4).

$$\dot{m}_t = \rho_s A_k V_s \tag{4}$$

Where ρ_s is the air density, A_k is the valve seat area and V_s is flow velocity. The flow velocity can be calculated by using equation (5).

$$v_s = \sqrt{\frac{2\gamma}{\gamma-1}RT[1-(\frac{P_2}{P_1})^{\frac{\gamma-1}{\gamma}}]} \tag{5}$$

Where P_1 is upstream pressure, P_2 is downstream pressure and γ is an index of isentropic expansion. As long as the flows remain in the subsonic region no choked flow occurs in the design.

The flow coefficient, C_f has same equation with C_D. But, it has different area calculation for $\dot{m}_t$ in equation (4). For flow coefficient, the area is at the valve throat area.

4. Results and discussion

The result from flow bench experiment is volume flow rate data of intake and exhaust port. Then the experimental result is calculated with theoretical equation of coefficient. The result from coefficient calculation is compared against small block Chevrolet (SB Chevy) engine coefficient.

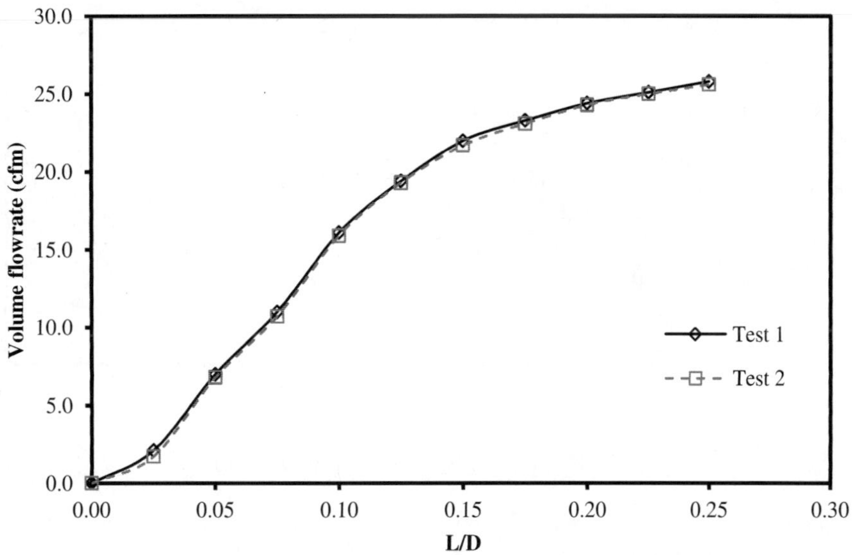

Figure 3. Volume flow rate through intake port.

Figure 3 shows the two result of the air volume flow rate through the intake port. Both results show a small different value of air volume flow rate. The volume flow rate is 0 cubic feet per minute, CFM at 0 L/D ratio. Then it increases gradually until 25.8 CFM at 0.25 L/D ratio.

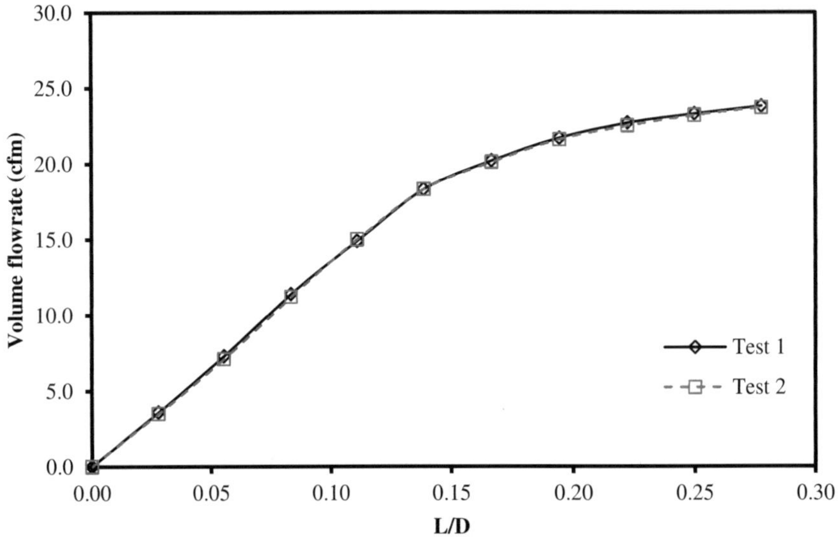

Figure 4. Volume flow rate exhaust port.

Figure 4 present the two result of the air volume flow rate through the exhaust port. The both results also have small different value. The volume flow rate at 0 L/D ratio is 0 CFM and increase gradually to 23.8 CFM at 0.278 L/D ratio.

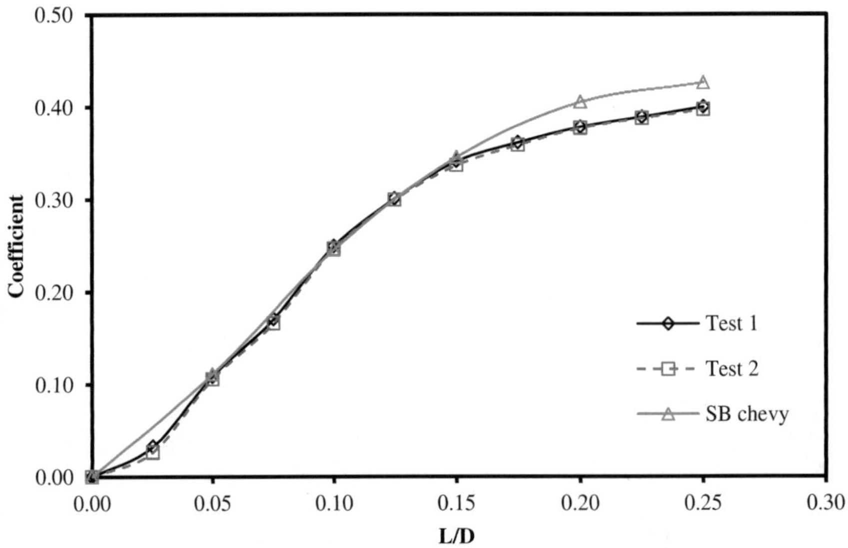

Figure 5. Flow coefficient of intake port.

Figure 5 shows the flow coefficient of intake port two-stroke poppet valve engine and compare with SB Chevy. The figure shows the intake port flow coefficient did not have much different with SB Chevy. The coefficient is 0 at 0 L/D ratio and increases gradually to 0.4 at 0.25 L/D for two-stroke poppet valve engine. While SB Chevy coefficient is 0 at 0 L/D and increases gradually to 0.42 at 0.25 L/D ratio.

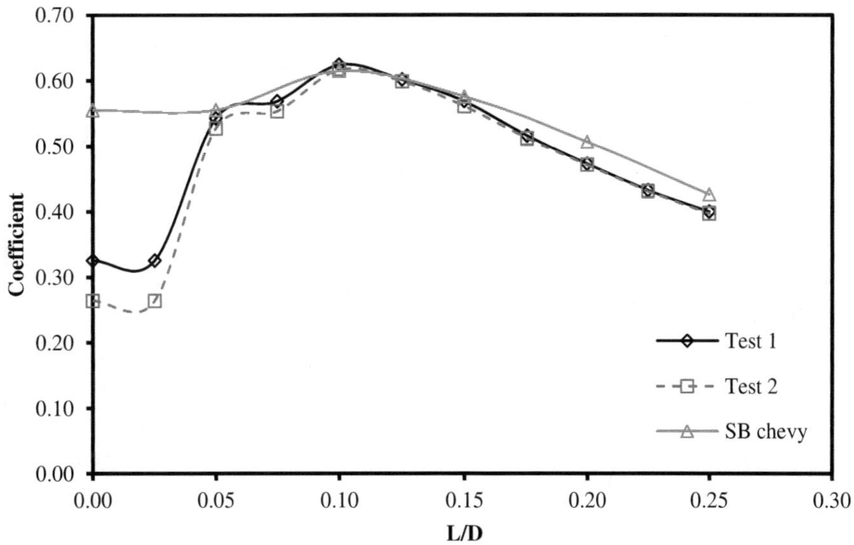

Figure 6. Discharge coefficient of intake port.

Figure 6 present discharge coefficient of intake port for both head engine. The result shows SB Chevy has higher coefficient at 0 until 0.05 L/D ratio. At 0 L/D ratio, SB Chevy has 0.55 coefficient and two-stroke poppet valve engine is 0.32 and 0.26 coefficient. Then the coefficient for both engine head have similar values at 0.05, 0.1 and 0.15 L/D ratio with 0.54, 0.61 and 0.57 coefficient respectively. The highest coefficient value is at 0.1 L/D ratio. Then, the values decrease gradually until 0.25 L/D ratio with 0.4 coefficients for two-stroke poppet valve and 0.42 coefficients for SB Chevy. The result show the intake port coefficient of two-stroke poppet valve engine is almost same value with SB Chevy. It is show the two-stroke poppet valve intake port is comparable with SB Chevy.

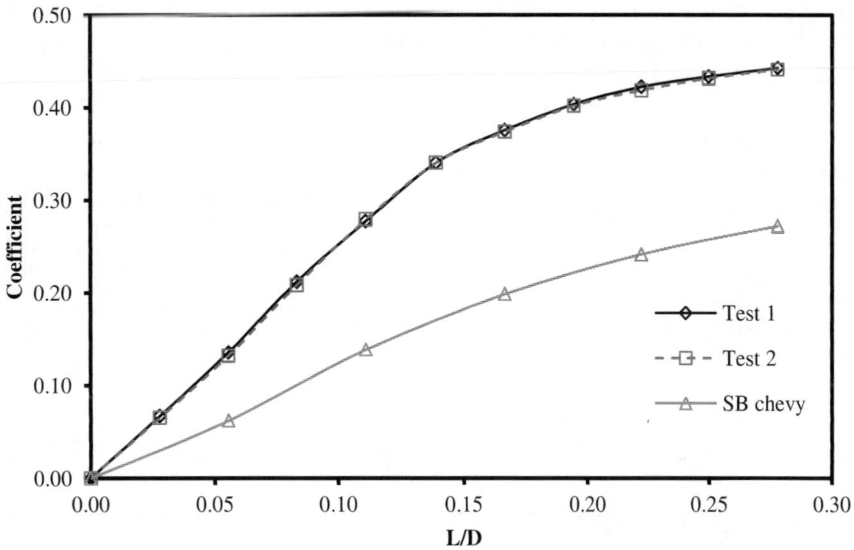

Figure 7. Flow coefficient of exhaust port.

Figure 7 shows flow coefficient result of the exhaust port. It shows two-stroke poppet valve has high coefficient value compare to SB Chevy. Both heads start with 0 coefficients at 0 L/D ratio. But at last or 0.278 L/D ratio, the two-stroke poppet valve has 0.44 coefficients and SB Chevy has 0.27 coefficients.

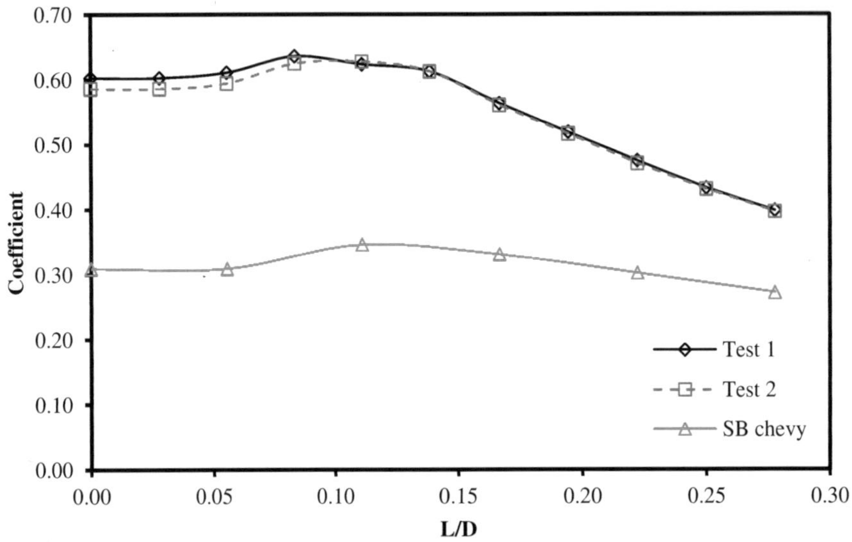

Figure 8. Discharge coefficient of exhaust port.

Figure 8 present discharge coefficient of the exhaust port. It also shows the two-stroke poppet valve has higher discharge coefficient compare to SB Chevy. The two-stroke poppet valve engine has 90% higher coefficient between 0 until 0.05 L/D ratio. Then the percentage of coefficient different is reducing gradually until 46% at 0.279 L/D ratio. The result show the coefficient of two-stroke poppet valve engine exhaust port has better coefficient compare to SB Chevy. The higher coefficient show the exhaust gas can release faster at limited time in two-stroke cycle duration.

5. Conclusions

The port flow experimental investigation of the two-stroke poppet valve has highlighted the following results: The flow and discharge coefficient of the intake port has similar values for two-stroke poppet valve and SB Chevy. Except at zero L/D ratio of intake C_D, the SB Chevy has 89% higher value than the two-stroke poppet valve. The flow and discharge coefficient of the exhaust port for the two-stroke poppet valve has higher values than SB Chevy. Thus, it is possible to operate two-stroke cycle on the head as the flow performance through the exhaust valve is higher at lower lift. Further, the flow is capable of maintaining the subsonic flow without choking as illustrated by the low lift performance trend.

Acknowledgement

The authors would like to thank Universiti Malaysia Pahang (www.ump.edu.my) for providing laboratory facilities and financial assistance under project no. RDU160313.

References

[1] Golloch R and Merker G P 2005 Internal combustion engine downsizing *MTZ worldwide* **66** 20-2

[2] Zhang Y 2014 Experimental investigation of CAI combustion in a two-stroke poppet valve di engine. Brunel University)

[3] Nora M D 2015 Experimental and numerical study of a two-stroke poppet valve engine fuelled with gasoline and ethanol. In: *Department of Mechanical, Aerospace and Civil Engineering*: Brunel University)

[4] Stokes J, Hundleby G E, Lake T H and Christie M J 1992 Development Experience of a Poppet-Valved Two-Stroke Flagship Engine. SAE International)

[5] Nomura K and Nakamura N 1993 *Development of a new two-stroke engine with poppet-valves Toyota S-2 engine*: Editions OPHRYS)

[6] Dunn P F N F, AU), Rutherford, Robert Matthew (Nundah, AU) 2003 Two-stroke engine. (United States: Rotec Design Ltd. (Queensland, AU))

[7] Rebhan M and Stokes J 2009 Two-stroke/four-stroke multicylinder gasoline engine for downsizing applications *MTZ worldwide* **70** 40-5

[8] Nakano M, Sato K and Ukawa H 1990 A Two-Stroke Cycle Gasoline Engine with Poppet Valves on the Cylinder Head. SAE International)

[9] Sato K, Ukawa H and Nakano M 1992 A Two-Stroke Cycle Gasoline Engine with Poppet Valves in the Cylinder Head - Part II. SAE International)

[10] Abdul R. Ismail R A B, Semin 2008 An Investigation of Valve Lift Effect on Air Flow and Coefficient of Discharge of Four Stroke Engines Based on Experiment *American Journal of Applied Sciences* **5** 9

[11] S.A.Sulaiman, S.H.M.Murad, Ibrahim I and Karim Z A A 2010 Study of flow in air-intake system for a single-cylinder go-kart engine *International Journal of Automotive and Mechanical Engineering (IJAME)* **1** 14

[12] Algieri A 2011 An Experimental Analysis of the Fluid Dynamic Efficiency of a Production Spark-Ignition Engine during the Intake and Exhaust Phase *ISRN Mechanical Engineering* **2011**

The two-stroke poppet valve engine. Part 2: Numerical investigations of intake and exhaust flow behaviour

M Kamili Zahidi[1]*, M Razali Hanipah[1, 2]

[1]Automotive Engineering Centre (AEC), Universiti Malaysia Pahang, 26600, Pekan, Pahang, Malaysia.
[2]Automotive Engineering Research Group (AERG), Faculty of Mechanical Engineering, Universiti Malaysia Pahang, 26600, Pekan, Pahang, Malaysia.

*Corresponding author: kamili.zahidi@gmail.com

Abstract. A two-stroke poppet valve engine is developed to overcome the common problems in conventional two-stroke engine designs. However, replacing piston control port with poppet valve will resulted different flow behaviour. This paper presents the model and simulation result of three-dimensional (3D) port flow investigation of a two-stroke poppet valve engine. The objective of the investigation is to conduct a numerical investigation on port flow performance of two-stroke poppet valve engine and compare the results obtained from the experimental investigation. The model is to be used for the future numerical study of the engine. The volume flow rate results have been compared with the results obtained experimentally as presented in first part of this paper. The model has shown good agreement in terms of the flow rate at initial and final valve lifts but reduced by about 50% during half-lift region.

1. Introduction

Engine downsizing concept has been adopted by most automotive manufacturers to increase the power to weight ratio[1]. The engine downsizing is done by reducing the weight of the engine using smaller capacity and intake air boosting to increase the power.

The two-stroke engine is known to have a higher power to weight ratio. However, the conventional two-stroke engines use piston control port which poses several problems. The first problem is bore distortion caused by an asymmetric temperature of the cylinder liner[2]. Bore distortion is increasing the wear rate or worst case it will lock-up the piston. The second problem is lubricant oil easily enter combustion cylinder via intake port resulting in higher pollutants in the exhaust emission[3].

Two-stroke poppet valve engine has been proposed by several researchers to eliminate the problem and at the same time increase the engine performance [4-7]. Each researcher has experimented different specification and design of two-stroke poppet valve engine. Nakano, et al reported the converted four-stroke engine into two-stroke poppet valve engine produce better power than original engine when compare at the same engine speed [8]. While, Sato, et al has design shrouded intake valve to overcome two-stroke poppet valve problem. The problem is the short-circuited phenomenon during scavenging duration[9].

The first part of this research has covered the experimental work of the intake and exhaust ports assessments. In this paper, the flow through intake and exhaust valves is visualised through three-dimensional (3D), computational flow dynamics (CFD) tool. Both qualitative and quantitative results are evaluated.

2. Engine specifications

The investigation is based on a 65cc four-stroke gasoline engine. The base engine is running on a four-stroke cycle utilising two-stroke lubrication and intake crankcase compression system [10]. The engine stroke is 33mm and the bore is 50mm. Then the compression ratio is 9.5:1. While the maximum valve lift is 4 mm. The valve diameter for the intake is 20mm and for exhaust is 18mm. The engine specification is as shown in Table 1.

Table 1. The engine specification.

Parameter	Value
Capacity	65cm^3
Bore	50mm
Stroke	33mm
Compression ratio	9.5:1
Maximum valve lift	04mm
Intake valve diameter	20mm
Exhaust valve diameter	18mm

The engine model is as shown in Figure 1. The valve is designed to be at the vertical position. One valve is for intake and another valve is for exhaust. The left valve in Figure 1 is exhaust valve and the right valve is intake valve.

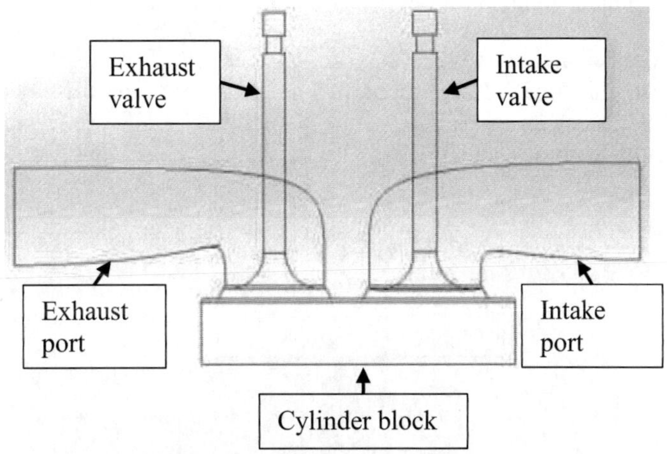

Figure 1. The engine model.

3. Method

The numerical investigation begins with the 3D model construction. The model for port flow simulation is as Figure 2 and Figure 1. Then, the model is imported into Ansys port flow IC engine simulation. The engine model is added with plenum box at bottom of cylinder block.

Figure 2. The engine geometry model.

Figure 3. The mesh for port flow investigation.

The complete model has undergone meshing process. The model meshed with hybrid type, which is the combination of tetrahedron and hexahedron. The hexahedron mesh is used at plenum box and three-quarter of cylinder block as present in . Figure 3. While the tetrahedron is used at port and top quarter of cylinder block as shown in Figure 4. The mesh total nodes are 1,430,193 and elements are 3,383,208.

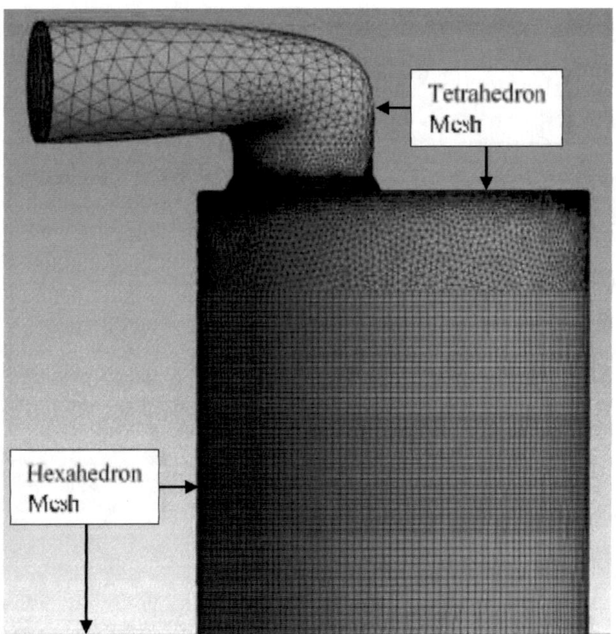

Figure 4. The mesh at engine head and block.

After meshing, the investigation continues with flow simulation. The flow simulations were run based on Navier-Stokes equations and k-ε two equation turbulent model at steady state condition.

4. Results and discussion

The successful simulation run produces various data regarding the flow through the ports. However, the main relevant outcome of the numerical investigation is the volume flow rate data which is necessary for comparing the flow obtained from the experiment.

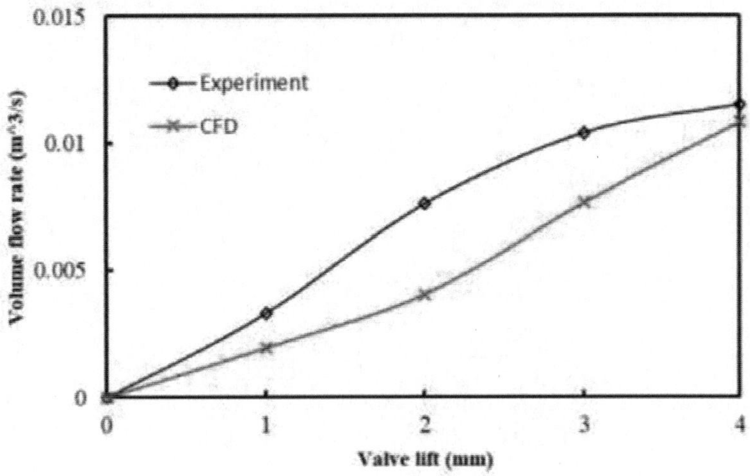

Figure 5. The comparison of numerical and experimental result.

Figure 5 shows the comparison between numerical simulation and flow bench experimental results of the volume flow rate through the intake port. The results show the experimental result is higher than CFD simulation result. Similar trends are observed but the CFD shows 50% lower flow rate during half-lift. It shows the model necessary to be tuning for a comparable result with flow bench experiment.

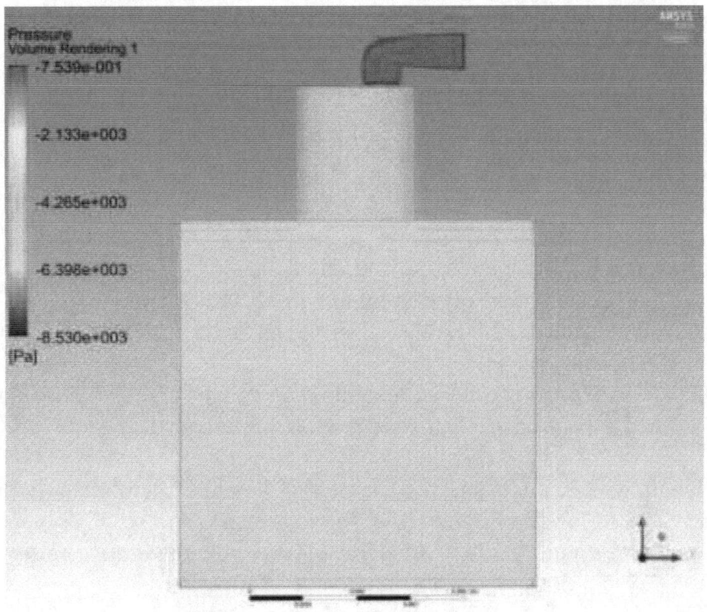

Figure 6. The Pressure contour.

Figure 6 shows the pressure rendering. It is present the pressure is highest at intake port compare to the cylinder block and plenum box. The pressure is -75.3 Pa. The value is negative because of the air suction simulation and the flow of air from upstream to downstream.

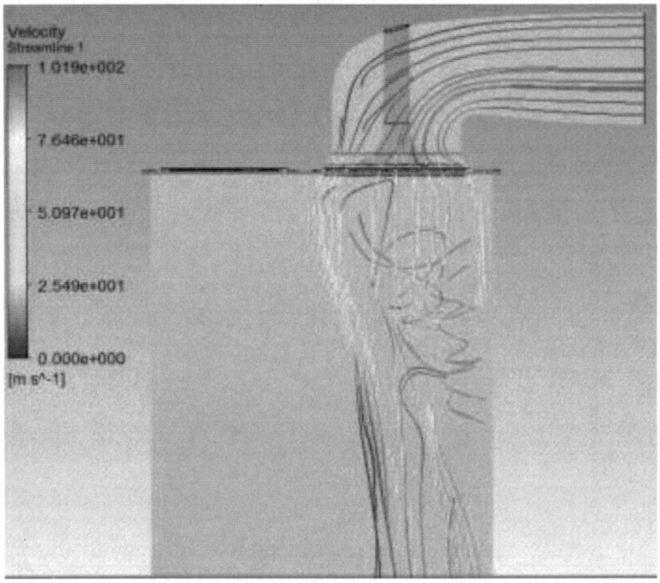

Figure 7. The velocity streamline.

As can be seen in Figure 7, the line is representing the velocity streamline. The streamline shows the air is swirling below the valve during port flow simulation.

5. Conclusions

This paper has presented the numerical model construction and flow investigation of the two-stroke poppet valve engine. The volume flow rate results have been compared with the results obtained experimentally as presented in part 1 of this paper. The model has shown good agreement in terms of the flow rate maximum lift but reduced by 50% at half-lift. The experimental results have higher value than numerical simulation result on several points but apparent trend is observed. Based on the findings from part 1 of the paper, the flow rate is sufficient for the two-stroke engine operation.

Acknowledgement
The authors would like to thank Universiti Malaysia Pahang (www.ump.edu.my) for providing laboratory facilities and financial assistance under project no. RDU160313.

References

[1] Wada Y, Nakano K, Mochizuki K and Hata R 2016 Development of a New 1.5L I4 Turbocharged Gasoline Direct Injection Engine. SAE International)
[2] Zhang Y 2014 Experimental investigation of CAI combustion in a two-stroke poppet valve di engine. Brunel University)
[3] Nora M D 2015 Experimental and numerical study of a two-stroke poppet valve engine fuelled with gasoline and ethanol. In: *Department of Mechanical, Aerospace and Civil Engineering*: Brunel University)
[4] Stokes J, Hundleby G E, Lake T H and Christie M J 1992 Development Experience of a Poppet-Valved Two-Stroke Flagship Engine. SAE International)
[5] Nomura K and Nakamura N 1993 *Development of a new two-stroke engine with poppet-valves Toyota S-2 engine*: Editions OPHRYS)
[6] Dunn P F N F, AU), Rutherford, Robert Matthew (Nundah, AU) 2003 Two-stroke engine. (United States: Rotec Design Ltd. (Queensland, AU))

IOP Conf. Series: Materials Science and Engineering **257** (2017) 012024 doi:10.1088/1757-899X/257/1/012024

[7] Rebhan M and Stokes J 2009 Two-stroke/four-stroke multicylinder gasoline engine for downsizing applications *MTZ worldwide* **70** 40-5
[8] Nakano M, Sato K and Ukawa H 1990 A Two-Stroke Cycle Gasoline Engine with Poppet Valves on the Cylinder Head. SAE International)
[9] Sato K, Ukawa H and Nakano M 1992 A Two-Stroke Cycle Gasoline Engine with Poppet Valves in the Cylinder Head - Part II. SAE International)
[10] Mohd Razali H and Zamri M 2016 Numerical Investigations on the Port Flow Field of a Two-Stroke Poppet Valve Engine

A study on the effect of feed rate and cutting speed on surface roughness and material removal rate of mild steel

A A Latif, M R Ibrahim*, A Z Amran, and E A Rahim

Precision Machining Research Center (PREMACH), Faculty of Mechanical and Manufacturing Engineering, Universiti Tun Hussein Onn Malaysia, Parit Raja, Batu Pahat, Johor, Malaysia

*Corresponding author: rasidi@uthm.edu.my

Abstract. In any machining process, achieving a good surface quality and maximization of material removal rate are absolute importance. This paper provides an insight to the effect of feed rate and cutting speed on NOVIANO and conventional cutting tool in term of surface roughness and Material Removal Rate (MRR). The output of the machining during slot milling of Mild Steel with 17 HRC were measured. Mitutoyo surface tester were used to measure average mean surface roughness (Ra) of the slot milling surface. For the MRR measurement, mathematical formula was used to calculate and evaluate the different weight of workpiece before and after milling. The result shows that, when the feed rate and cutting speed increase, better surface roughness and MRR will be occur. However, the comparison between both cutting tools found that the NOVIANO cutting tool produce much better result than conventional cutting tool where surface roughness and MRR is 13.97% and 51.62%.

1. Introduction

Mild steel or plain-carbon steel is currently the most well-known type of steel, since its price is moderately low compare to other material while it gives material properties that are adequate for some applications. Mild steel is one of the most versatile and resilient materials, and it forms the backbone of wide varieties of heavy machines as well as for various industries due to low cost and good mechanical strength [1]. Low-carbon steel contains around 0.05–0.25% carbon making it ductile and malleable. Generally, mild steel has low elasticity. However, it is easy to form when machining. There are many applications of mild steel were used in daily life such as mounting plates, tool and die sets. It also can provide a soft core and high surface hardness to parts that include machinery parts, special bolts and many more.

Milling is one the machine that widely used to machining this type of materials. This type of cutting operation is capable of producing different types of shapes with the use of multi-tooth cutting tools. The cutting action of the many teeth around the milling cutter provides a fast method of machining. In the milling process, a multi-tooth cutter rotates along various axes with respect to the workpiece. Among various sorts of milling processes, end milling is a standout amongst the most indispensable and regular metal cutting operations utilized for machining parts as a result of its ability to remove materials at faster rate with a sensibly good surface quality. Different application of end milling process can be found in almost every industry ranging from small tool maker units to large production industries. Components and parts made out of steel have to experience wide assortments of manufacturing operations including machining before they are used in auxiliary and also non-structural applications. This last stage of fabrication (lathing, milling, turning, grinding, etc.) tends to cause severe plastic deformation and extensive shearing effect, which removes the top-most layer of work-piece in the form of thin chips [2].

The machining process of metal work-piece is affected enormously by the relative velocity between the workpiece and the edge of the cutting tool. The integration of rotating and translator movement both of the workpiece or the cutting tool or both created the relative movement in the machining operations. The translator displacement of the cutting edge of the tool along the work surface during a given period of time is called 'feed', while the rate of traverse of the work surface past the cutting edge is designated as 'cutting speed' [3]. The most important interactions, that effect surface roughness & MRR of machined surfaces, were between the cutting feed and depth of cut, and between cutting feed and cutting speed [4]. Surface roughness is one of the significant characteristics in metal cutting that is being detected in machining processes. Surface roughness is defined as a group of irregular waves in the surface, measured in micrometres. It is produced by the fluctuations of short wavelengths characterized by asperities (local maxima) and valleys (local minima) of varying amplitudes and spacing [5]. Material Removal Rate is the main output factor for the productivity of machining. MRR is a measurement of productivity & it can be expressed by analytical derivation as the product of the width of cut, the feed velocity of milling cutter and depth of cut [6]. So, in this paper the effect of feed rate and cutting speed on NOVIANO and conventional cutting tool in term of surface roughness and Material Removal Rate (MRR) will be discussed.

2. Methodology

All the tests were carried out by using CNC Milling Machine Sodick MC 430L as shown in figure 1. This CNC machine can improve the productivity and increase the profit with it capability to machine the workpiece such as it can reach precision up to ± 1 μm. It also the only CNC milling machine that can rotate spindle up to 40000 rpm. In this experiment, two types of cutting tool had been chosen and apply throughout the experiments which is conventional and NOVIANO cutting tool manufactured by HPMT industries as shown in figure 2. The meaning of NOVIANO is "No Vibration and Noise" which apply at it particular geometry of the cutting tool makes it anti-vibration. The NOVIANO cutting tool composes a four-flute tungsten carbide tool with TiSi (Titanium-Silicon) multilayer coating, helix angle of 40°, and with maximum application of temperature below than 1200 °C. The conventional cutting tool composes a four-flute micro grain carbide and with helix angle of 35°. Both cutting tool were designed to machine carbon steels, tool steels, alloy steels and stainless steels. Table 1 show a detail terminology of two different cutting tool.

Figure 1. Sodick MC430L.

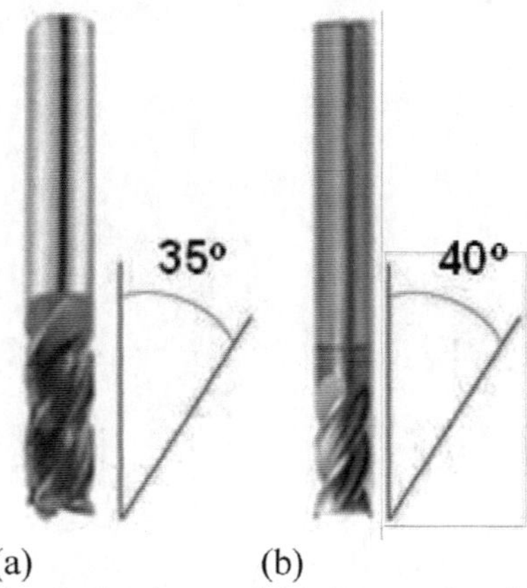

Figure 2. Cutting tool (a) Conventional (b) NOVIANO.

Table 1. Cutting tool terminology.

Terminology	Conventional	NOVIANO
Tool Diameter (mm)	6	6
Tool Length (mm)	50	57
Length of Cut (mm)	16	13
Flute	4	4
Helix Angle (^{0})	35	40
Tool Material	Micro Grain Carbide	Tungsten Carbide
Coating	Coated	Coated with TiSi Based-Multilayer

Table 2. Machining Parameter.

Type of Cutting Tool	Feed Rate (mm/min)	Cutting Speed (m/min)
Conventional	140	2.24
	280	4.48
	560	8.95
NOVIANO	179	2.74
	359	5.51
	718	11.00

In this research, Mild Steel was used as a material for the workpiece where it is categorized as world-wide usage material in structural component. Mild steel has an element of C, Si, Mn, S, P, Cr, Mo, V and Fe. One of the material advantage is contain high ductility and low carbon. On the other hand, mild steels are also used in applications such as boiler and pressure vessel. It also can withstand and experience high temperature up to 600-700 °C. The mild steel was cut into small rectangular plate with length, width and thickness dimension of 90 x 50 x 10 mm for slot milling process.

The slot milling was conducted using conventional and NOVIANO cutting tool on mild steel of 17 HRC rectangular block with the dimension of 90 x 50 x 10 mm. Workpiece was clamped on the machine worktable as demonstrated in figure 3 and the same chuck (tool holder) will be applied in this experiment. The inspection of the flatness on the workpiece were done before cutting operation started using mitutoyo 513-404T Dial Indicator as shown in figure 4. The 513-404T Dial Indicator series from Mitutoyo are horizontal type. These indicators were easy to use in a narrow or cramped area and still give an accurate measurement.

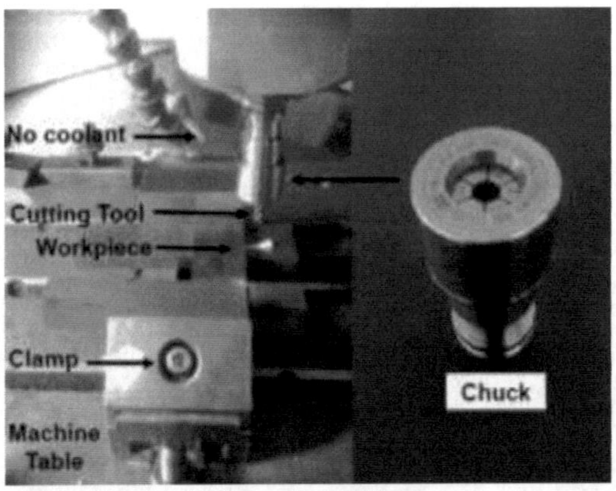

Figure 3. Experimental set up.

Figure 4. Mitutoyo dial indicator.

The measurement of surface roughness was done using Mitutoyo SJ-400 surface roughness tester as shown in figure 5. The measurement will be based on Japan International Standard (JIS 1994) with cut off length of 0.8 mm x 5. Measurement will be repeated at three different spots for each sample which is first, middle and last surface as shown in figure 6. Every test of the measurement will be printed on printed paper which showing the graph value for surface roughness and the average value of mean surface roughness was calculated and plotted in the graph for analysis.

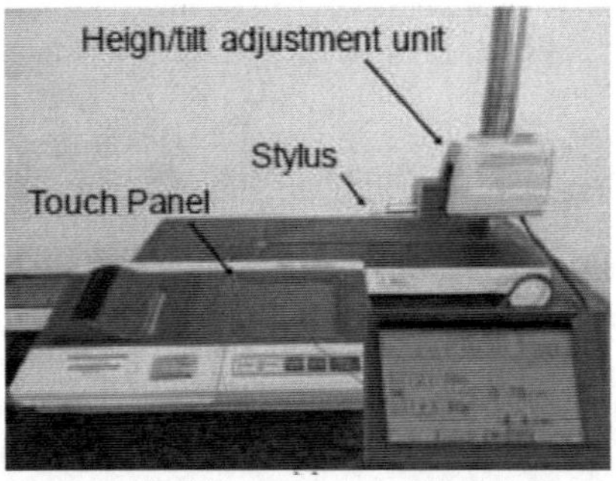

Figure. 5. Mitutoyo SJ-400 Surftest.

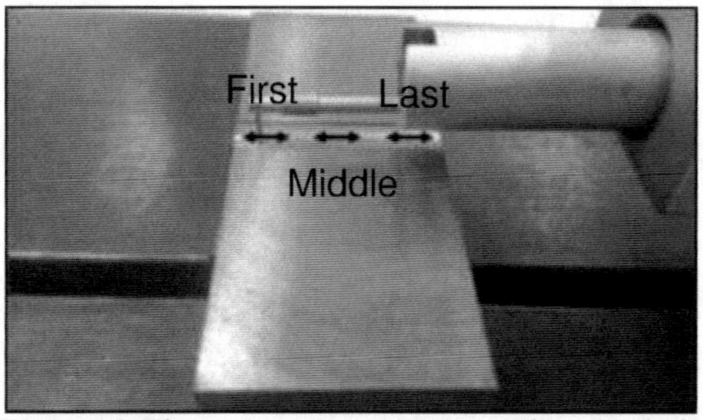

Figure 6. Three measurements on surface workpiece.

Before cutting process started, the mass for each of the workpiece was measured using Mettler Toledo Weighting Balance as shown in figure 7(a). It has sensitivity of three decimal places. The MRR measurement will be conducted by referring to Chahal et al., (2013). The equation can be presented as in equation (1):

$$\text{Mass}_{\text{remove}} = \text{Mass}_{\text{before cut}} - \text{Mass}_{\text{after cut}} \quad (1)$$

Where, the unit of mass remove in kg. The result gain from equation (1) will be represent to the volume remove model as in equation (2). That is,

$$\text{Volume}_{\text{remove}} = \text{Density x Mass}_{\text{remove}} \quad (2)$$

Where, the unit of volume remove in mm^3. The density of mild steel of 17 HRC had been analyze using Mettler Toledo Density kit as shown figure 7(b) which result 7.8360 g/cm^3. The relationship between MRR and machining independent variables (length of tool traveled and feed rate) can be written as:

$$\text{MRR} = \text{Volume}_{\text{remove}} / (\text{Length}_{\text{tool traveled}} / \text{feed rate}) \quad (3)$$

Where, the length of tool traveled had been set constant as 70 mm on the machine.

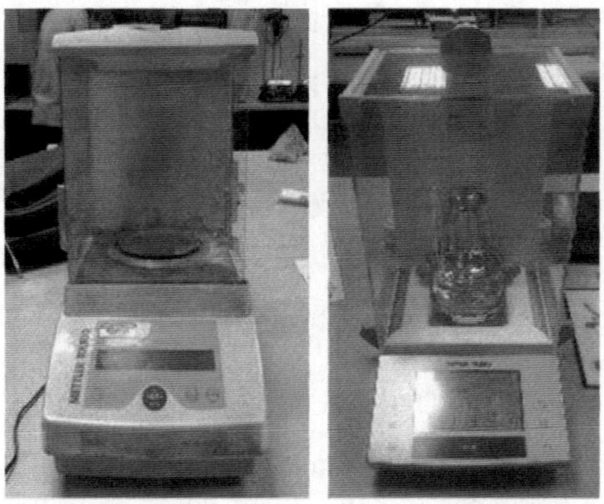

(a) (b)

Figure 7. (a)Weighting balance (b) Density kit.

3. Results and Discussion

3.1 Surface Roughness

Figure 8 shows the graph for surface roughness under various of feed rate using conventional and NITICO cutting tool. It can be observed that, with higher feed rate will produce lower surface roughness value for both cutting tool. Furthermore, at the depth of cut of 0.3 mm for both type of cutting tool show that the surface roughness slightly decreased when the feed rate increased. Figure 8(a) show the graph for surface roughness under various of feed rate using NITICO cutting tool. Based on figure 8(a), the minimum level of feed rate and cutting speed for NOVIANO cutting tool at 179 mm/min and 2.74 m/min will produce 0.520 μm average surface roughness. Apart from that, the medium level of feed rate and cutting speed at 359 mm/min and 5.51 m/min produce the output of 0.430 μm for average surface roughness. For the maximum level of feed rate and cutting speed at 718 mm/min and 11.00 m/min, it will produce 0.333 μm for average surface roughness. Furthermore, comparison from minimum level to medium level shows decrement of 18.95% for average surface roughness produce. Apart from that, the medium level to maximum level show decrement of 25.43% for average surface roughness occur. For minimum level to maximum level, occur large differences which have 43.85% of decrement for average of surface roughness value.

Figure 8(b) show the graph for surface roughness under various of feed rate using conventional cutting tool. Based on Figure 8(b), the minimum level of feed rate and cutting speed for conventional cutting tool is 140 mm/min and 2.24 m/min which give the value of 0.620 μm for average surface roughness. The medium level of feed rate and cutting speed at 280 mm/min and 4.48 m/min for conventional cutting tool produce 0.467 μm for average surface roughness. The maximum level of feed rate and cutting speed for conventional cutting tool at 560 mm/min and 8.95 m/min which give the value of 0.383 μm for average surface roughness.

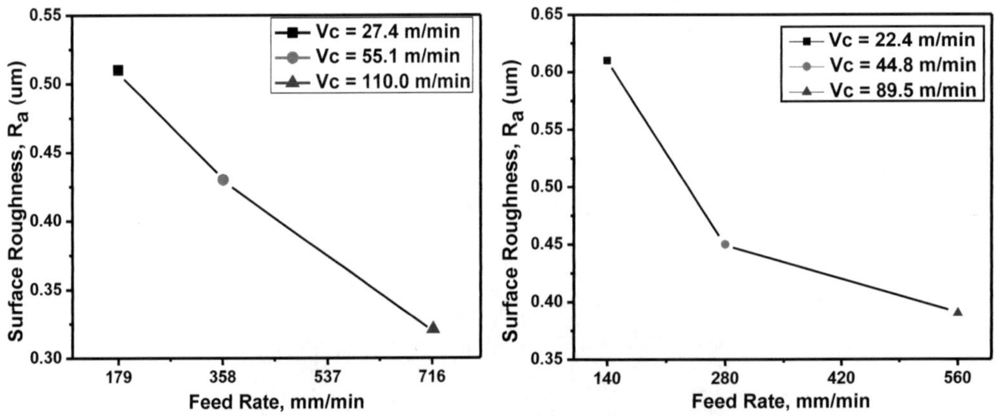

(a) (b)

Figure 8. Average surface roughness, R_a (μm) versus feed rate (mm/min)
(a) NOVIANO (b) Conventional.

From the comparison between both of the cutting tool, it shows that NOVIANO cutting tool produce better surface finish of end slot milling compare to conventional cutting tool within the feed rate and cutting speed suggest by each tool manufacturer. With the special geometry of NOVIANO cutting tool with 40° of helix angle compare to 30° of helix angle for conventional cutting tool, it assisted in improvement of surface finish. This supported by previous researcher that the increase in helix angle of the end mill cutter resulted in better surface finish [7]. Surface roughness will deteriorate if there any disruption occurs during machining process such as BUE and chip formation which stuck of the surface. But, a large helix angle can reduce tool deflection by transferring stress vertically which allows the end mill by produce vertical chip ejection. A high helix angle can effectively help the chip extraction away from the cutting zone whilst reducing the stuck-chip disruption [8].

3.2 Material Removal Rate

Figure 9 shows the graph for material removal rate under various of feed rate using conventional and NITICO cutting tool. It can be observed that, the MRR slightly increase when feed rate increases for both cutting tool. Furthermore, at the depth of cut of 0.3 mm for both type of cutting tool show that the MRR slightly increase when the feed rate increased. Figure 9(a) show the graph for MRR under various of feed rate using NITICO cutting tool. Based on figure 9(a), the minimum level of feed rate and cutting speed for NOVIANO cutting tool at 179 mm/min and 2.74 m/min will produce 74.413 mm^3/min MMR. Apart from that, the medium level of feed rate and cutting speed at 359 mm/min and 5.51 m/min produce the output of 223.917 mm^3/min MMR. For the maximum level of feed rate and cutting speed at 718 mm/min and 11.00 m/min, it will produce 634.917 mm^3/min MRR. Furthermore, comparison from minimum level to medium level shows decrement of 100.12% for MMR produce. Apart from that, the medium level to maximum level show decrement of 95.82% for MRR occur. For minimum level to maximum level, occur large differences which have 43.85% of decrement for MRR value.

Figure 9(b) show the graph for material removal rate under various of feed rate using conventional cutting tool. Based on Figure 9(b), the minimum level of feed rate and cutting speed for conventional cutting tool is 140 mm/min and 2.24 m/min which give the value of 49.6 mm^3/min for MRR. The medium level of feed rate and cutting speed at 280 mm/min and 4.48 m/min for conventional cutting tool produce 123.2 mm^3/min for MRR. The maximum level of feed rate and cutting speed for conventional cutting tool at 560 mm/min and 8.95 m/min which give the value of 374.4 mm^3/min for MRR.

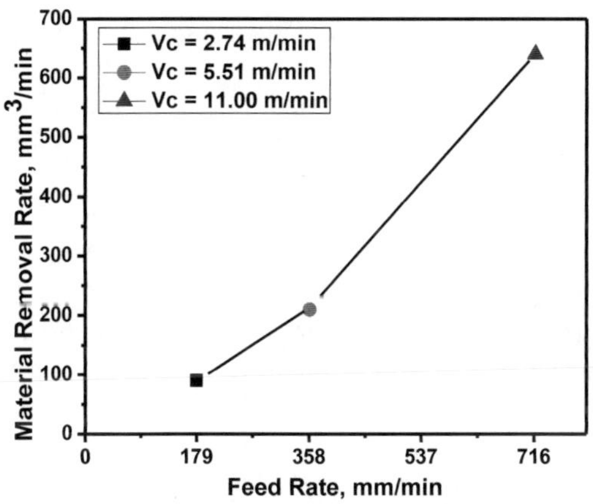

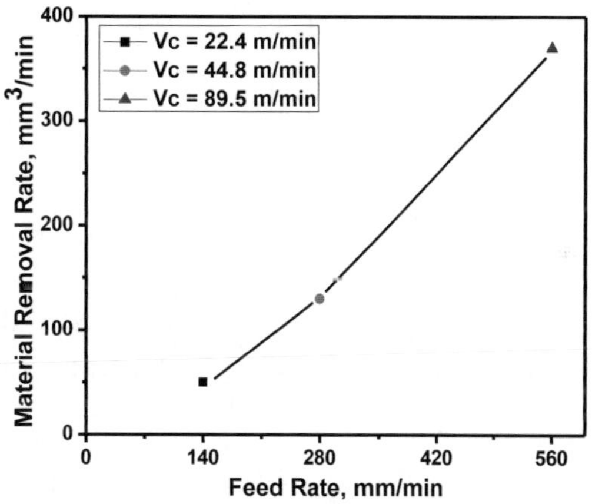

Figure 9. Material removal rate versus feed rate (mm/min)
(a) NOVIANO (b) Conventional.

It can be observed from the comparison between for both cutting tool that NOVIANO cutting tool remove better value of MRR compare to conventional cutting tool. NOVIANO cutting tool highest recommendation was 718 mm/min compare to conventional which is 560 mm/min in term of feed rate. Higher feed rate will produce higher MRR [9]. Also, at higher feed rate, tool traverses the work piece too fast, resulting high MRR [10]. The fact that must be considered is that, as feed increases, the maximum chip thickness also increased, thus MRR will also increase. Moreover, the difference of MRR produce due to the geometry and type of material of tool applied on NOVIANO cutting tool. As NOVIANO cutting tool had larger helix angle, it will minimize better formation of burr compare to conventional cutting tool. Other factor that expect to differentiate the result of MRR between NOVIANO and conventional cutting tool was type of material used as NOVIANO used Tungsten Carbide (WC). In fact, blunt tool with produce more heat that contribute the form of burr. It was proved that wear pattern in the tungsten carbide tool had lower effect to the sharpness of cutting edge compare to other material tool. The tool remains sharp as it will reduce the possibility of tool softening which reduce the formation of burr compare to conventional cutting tool. Furthermore, reduction of burr also will increase the MRR.

4. Conclusions

The main objective of this research was successfully achieved. The effect of feed rate and cutting speed on surface roughness and Material Removal Rate (MRR) of mild steel of 17 HRC using slot milling for conventional and NITICO cutting tool were evaluated. In term of different cutting tool, it showed the improvement in surface roughness and MRR using NOVIANO cutting tool. The following conclusion can be drawn:

- Surface roughness improved due to the increasing for both feed rate and cutting speed by taking maximum consideration tool manufacturer's suggestion.
- By increasing feed rate and cutting speed, higher Material Removal Rate (MRR) can be achieved by taking maximum consideration tool manufacturer's suggestion.
- The performance of NOVIANO cutting tool give better surface roughness and higher MRR compare to conventional cutting tool.
- The influence of special geometry and material used applied in NOVIANO cutting tool give better result compare to conventional cutting tool.
- The higher helix angle and Tungsten Carbide used on NOVIANO were the main factor it have better performance than conventional cutting tool.

References

[1] Prakash, M., Shekhar, S., Moon, A. P., & Mondal, K. 2015 Effect of machining configuration on the corrosion of mild steel. *Journal of Materials Processing Technology*, 219, 70-83.

[2] Majumdar, P., Shekhar, S., & Mondal, K. 2015 Effect of Machining Parameters on Oxidation Behavior of Mild Steel. *Journal of Materials Engineering and Performance*, 24(1), 484-498.

[3] Aslaliya, K., Patel, A. P., Doshi, H. P., Aslaliya, K., Patel, A. P., & Doshi, H. P. A Review on Effect of Process Parameters on MRR & Surface Roughness in Milling Machine. *International Journal*, 1, 135-138.

[4] Chahal, M., Singh, V., Garg, R., & Kumar, S. 2013 To estimate the range of process parameters for optimization of surface roughness & material removal rate in CNC milling. *Int. J. Eng. Trends Technol.(IJETT),* 4, 4556-4563.

[5] Rasidi, Ibrahim, E. A. Rahim, A. A. Ibrahim, N. A. Maskam, and S. C. Ghani 2014. "The effect on the application of coolant and Ultrasonic Vibration Assisted Micro Milling on machining performance." *In-Applied Mechanics and Materials*, vol. 660, pp. 65-69. Trans Tech Publications.

[6] Joshi, A., Kothiyal, P., & Pant, R 2012. Experimental investigation of machining parameters of CNC milling on MRR by taguchi method. *International Journal of Applied Engineering Research*, 7(11).

[7] Boujelbene, M., Bayraktar, E., Tebni, W., & Salem, S. B. 2009 Influence of machining parameters on the surface integrity in electrical discharge machining. Archives of Materials Science and Engineering, 37(2), 110-116.

[8] Izamshah, R., Husna, N., Hadzley, M., Amran, M., Shahir, M., & Amri, M. 2014 Effects of cutter geometrical features on machining polyetheretherketones (PEEK) engineering plastic. Journal of Mechanical Engineering and Sciences, 6, 863-72.

[9] Rasidi, I. I., Noor Hakim Rafai, E. A. Rahim, S. A. Kamaruddin, H. Ding, and K. Cheng. 2015. An investigation of cutting mechanics in 2-dimensional ultrasonic vibration assisted milling toward chip thickness and chip formation. In IOP Conference Series: Materials Science and Engineering, 100(1)1, 012057.

[10] Garg, S., Goyal, R. K., & Chabra, A. Effect of Material Removal Rate (MRR) on AISI H11 Die Steel in Drilling using Face Centered Design. IJSTE-International Journal of Science Technology & Engineering, ISSN (online), 400-41.

Application of displacement monitoring system on high temperature steam pipe

M H A Ghaffar[1*], S Husin[1] and J E Baek[2]

[1] Materials Engineering Group, TNB Research Sdn. Bhd., No.1, Lorong Ayer Itam, Kawasan Institusi Penyelidikan dan Latihan Bangi, 43000 Kajang, Selangor
[2] Korea Leading Engineering System, 40, Techno 3-ro, Yuseong-gu, Daejeon, Korea

*Corresponding author: hafiz.ghaffar@tnb.com.my

Abstract. High-energy piping systems of power plants such as Main Steam (MS) pipe or Hot Reheat (HR) pipe are operating at high temperature and high pressure at base and cyclic loads. In the event of transient condition, a pipe can be deflected dramatically and caused high stress in the pipe, yielding to failure of the piping system. Periodic monitoring and walk down can identify abnormalities but limitations exist in the standard walk down practice. This paper provides a study of pipe displacement monitoring on MS pipe of coal-fired power plant to continuously capture the pipe movement behaviour at different load using 3-Dimensional Displacement Measuring System (3DDMS). The displacement trending at Location 5 and 6 (north and south) demonstrated pipes displace less than 25% to that of design movement. It was determined from synchronisation analysis that Location 7 (north) and Location 8 (south) pipe actual movement difference has exceeded the design movement difference. Visual survey at specified locations with significant displacement trending reveals issues of hydraulic snubber and piping interferences. The study demonstrated that the displacement monitoring is able to capture pipe movement at all time and allows engineer to monitor pipe movement behaviour, aids in identifying issue early for remedy action.

1. Introduction

Piping system of a power plant is subjected to high pressure and high temperature during operation. As the pipe temperature changes from the installation condition to the operating condition, it expands or contracts. Thermal expansion of the pipe due to temperature changes generates enormous force and stress in the system [1]. Piping support systems are critical component in maintaining the forces and stresses within the allowable limit. Considerable stresses can be created in the piping and in the terminal point connections (boiler header and turbine) if the support system is not working properly as designed [2]. Support system malfunction can be caused by transient conditions such as start-up, shut down and water hammer where a pipe can be deflected dramatically and caused high stress in the system. Ignoring this can yield to failure of the system. ASME B31.1, "Power Piping" prescribes requirements for inspection of piping systems in Section VII- Operation and Maintenance, stated "the effects of unexpected piping position changes, and malfunctioning supports on the piping system's integrity and safety shall be evaluated, significant displacement variations from the expected design displacement shall be considered to assess the piping system's integrity" [3].

Appendix V, while listed in the code as nonmandatory, contains important information on piping displacement recording, documenting, marking and visual survey [3]. Conventional periodic monitoring and walk down of a pipe system has limitation in capturing pipe position. Very often, pipe position is recorded only at the time of inspection through hanger reading. Topped or bottomed out hanger will restricted the movement of pipe and the reading remains unknown until the hanger is adjusted to designed hot or cold position. The goal of preventing failure to the piping system can be attained by deploying adequate effective supervision and continuous frequent condition monitoring [4].

This paper provides a pipe displacement monitoring study on the MS pipe at coal-fired power plant to continuously capture pipe movement behaviour at different load by using real-time displacement monitoring using the 3-dimensional displacement measuring system (3DDMS). The deployment of the 3DDMS allows continuous real-time monitoring of the pipe movement. The system also makes capturing pipe movements during hot, transient and cold conditions possible for record and reference data.

2. 3-D displacement measuring system (3DDMS)

The on-line monitoring and measurement system known as 3DDMS was customized by Korea Electric Power Research Institute (KEPRI) and has given the right to Korea Leading Engineering System (KLES) as the sub-supplier to design and manufacture. The system consists of the measurement device (3DDMS transducer, see Figure 1) and the control unit or server. The displacement measurement device is the two-link type, which has the ability to measure 3-dimensional displacement of the pipe. Captured pipe displacement data will be sent to the system server known as PLUMBER to be stored. The PLUMBER displays the pipe movement trending on screen for monitoring purpose [5].

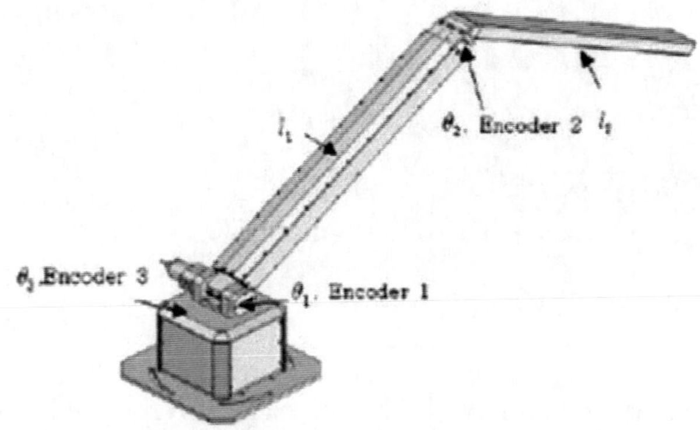

Figure 1. Two-link type displacement measurement device.

3. Displacement measurement transducer set-up

3DDMS transducers were installed on MSP at selected locations where pipe hangers and snubbers are installed. The device link l_2 end is attached to pipe hanger using a special clamp to prevent any harm on the pipe material. The other end of the device is connected by using jigs to a fix structure to secure its position. Figure 2 shows the location of 8 units of 3DDMS transducer on MS pipe of a power plant. Figure 3 shows the example of 3DDMS transducers installed on MS pipe. The pipe displacement data captured by the 3DDMS devices is transferred to the system server to be processed and visual display.

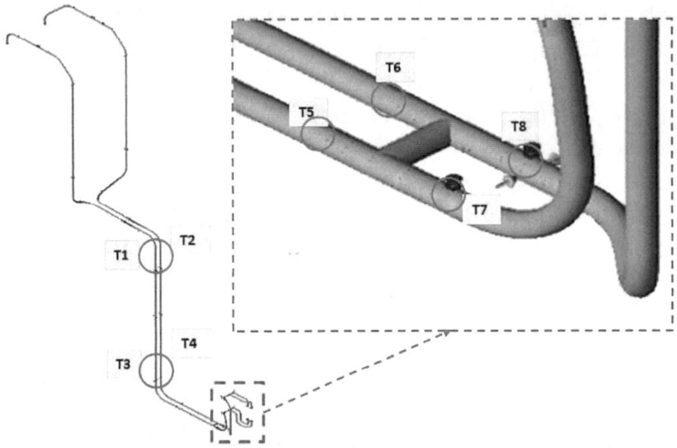

Figure 2. Location of 3DDMS transducers on MS pipe.

Figure 3. Example of 3DDMS transducers installed on MS pipe.

4. Pipe monitoring using 3DDMS

The system is essentially based on measuring pipe displacement. The 3DDMS transducer supplies information about the pipe behavior in x-, y- and z-directions and the magnitude of pipe movement. This is vital when it comes to non-ideal conditions such as over scaled or malfunction hanger where the movement of pipe is restricted. Malfunction hangers can cause high stress in piping system and cause damage development and cracking at base and weldments.

The captured displacement data is compared to pipe hanger design movement where upper and lower movement limits are introduced in the tabulated data trending to monitor the pipe movement in terms of movement direction and magnitude. Displacement measurement device and monitoring system can be directly used for evaluating the design and durability of the piping system [5].

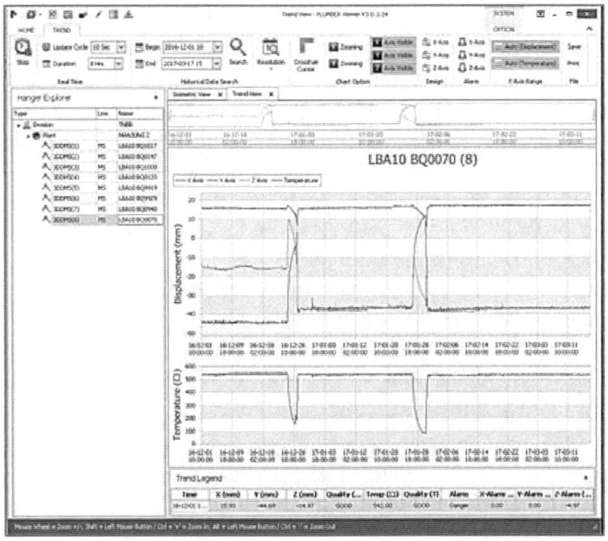

Figure 4. Pipe displacement data monitoring.

Figure 4 shows a window (screen shot) visualizing the pipe movement trending at specified location of the MS pipe. For the pipe isometric view, displacement measurement and history trend may be opened in other tabs as significant information on the hangers' location and number are shown. In total, 8 3DDMS transducers were installed on the MS pipe.

Compared to the conventional common practice without a monitoring system, the pipe system is significantly improved since

- Direction of the pipe movement is known.
- Magnitude of hanger movement is known.
- The pipe behavior check is continuous, compared to the common practice of pipe hanger survey and walk down.

5. Results: Pipe displacement trending

8 units of two-link type 3DDMS transducer were installed at MS pipe. Data of actual pipe movement has been obtained in real time and tabulated to compare to the design movement. For the purpose of monitoring the pipe movement, movement upper and lower limits are embedded in the trending graphs.

Based on the pipe displacement trending, it was found that the pipe displacement at all locations are within the control limits and consistent except at locations 7 and 8. Figure 5 shows pipe displacement trending in x-, y- and z-axis at Location 5 near hydraulic snubber location. As it can be seen, the pipe displaced within the upper and lower limits. The z-axis movement however demonstrates that the movement is 25% less compared to that of the design (Figure 6). This signifies that the movement in this particular axis is unusual and not according to the designed. Visual inspection during pipe walk down recorded evidence of hydraulic snubber fluid leaked (Figure 9). As a hydraulic snubber leaks fluid, the dampening capacity of the unit is reduced to the point where it freely slides from full compression to full extension. At this point the hydraulic snubber does not counteract the shock forces generated in the system and translate them back to the equipment nozzle such as turbine [6]. Other pipe displacement data with good condition of snubber such as at Location 1 exhibits movement more than 50% to that of the designed. This shows that the unusual trending of the pipe movement could be due to the performance of pipe snubber.

Figure 7 exhibits the pipe displacement trending at Locations 8 where the pipe movement is inconsistent and have exceeded the limits. Figure 8 shows the pipe movement actual difference between both Location 7 and 8 (north and south pipe) in z-axis are increasing from 18mm in December 2016 to 41mm in March 2017 after the unit re-start-up in December 2016. This demonstrates that both north and south sides of the pipe actual movement difference is more than the design movement difference which is 2.3mm. Visual inspection during pipe walk down found that

IOP Conf. Series: Materials Science and Engineering **257** (2017) 012026 doi:10.1088/1757-899X/257/1/012026

there is anomalous condition of the piping system where a beam structure is seen penetrated into pipe cladding (Figure 10). This signify that this condition in anyway could affect or disturb the pipe movement behavior. The increasing movement difference between these 2 pipes indicates that the pipes have distorted in the z-axis and this could amplify the risk of high stress on the connection pipe weldment in between (Figure 8). From the displacement trending, it clearly shows that the pipes did not return to its exact original position during cold condition. This is an example of the system degradation as a result of relaxation in the material, the piping rarely returns to its exact position after years in service at high temperature operation. Also, as high energy piping ages, creep deformation results in permanent shifting throughout the system [2].

With the pipe movement monitoring, one can knows pipe actual movement behavior in terms of magnitude and direction and also confirm on pipe hanger and support performance particularly for malfunction hanger or snubber so that remedy action can be taken to secure the piping system safety and reliability. Pipe displacement trending can greatly use in supporting the identification of inspection locations on steam pipe for future condition assessment plans.

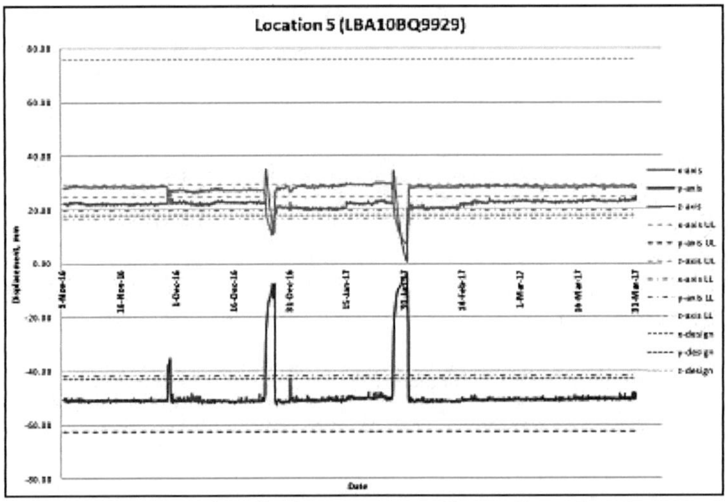

Figure 5. Pipe displacement trending at Location 5.

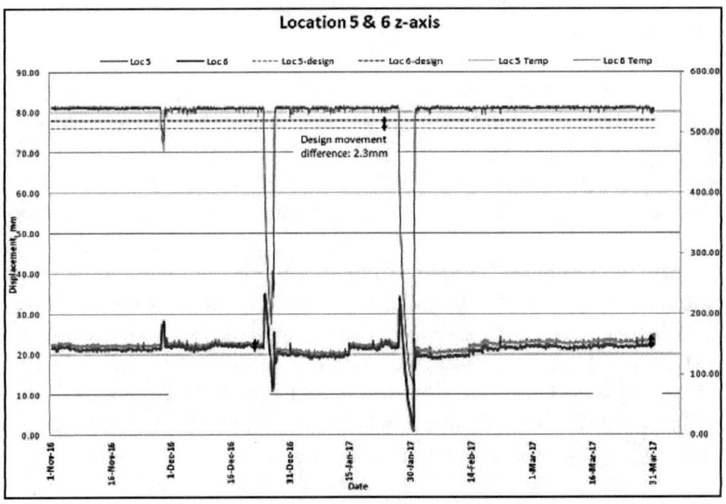

Figure 6. Location 5 and 6 z-axis displacement trending.

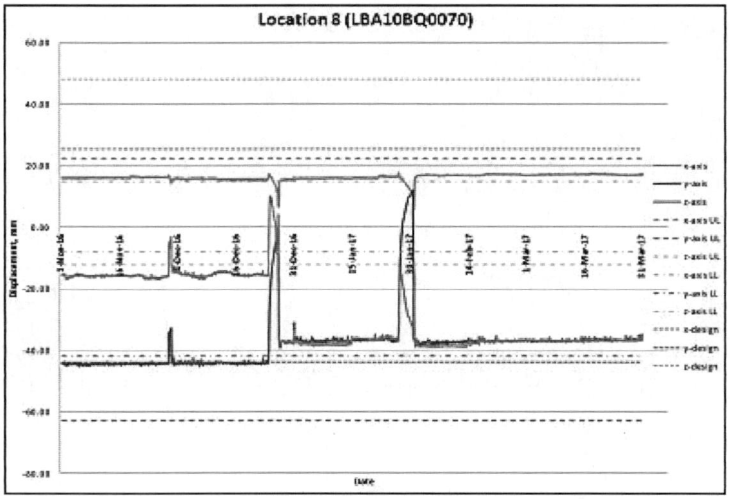

Figure 7. Pipe displacement trending at Location 8.

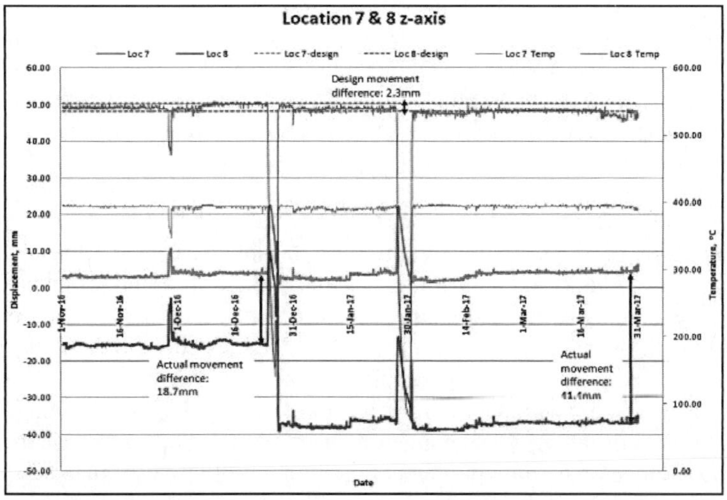

Figure 8. North and south pipe actual movement difference.

(a) (b)

Figure 9. (a) Snubber fluid leaked (b) Magnified view shows hydraulic fluid droplet.

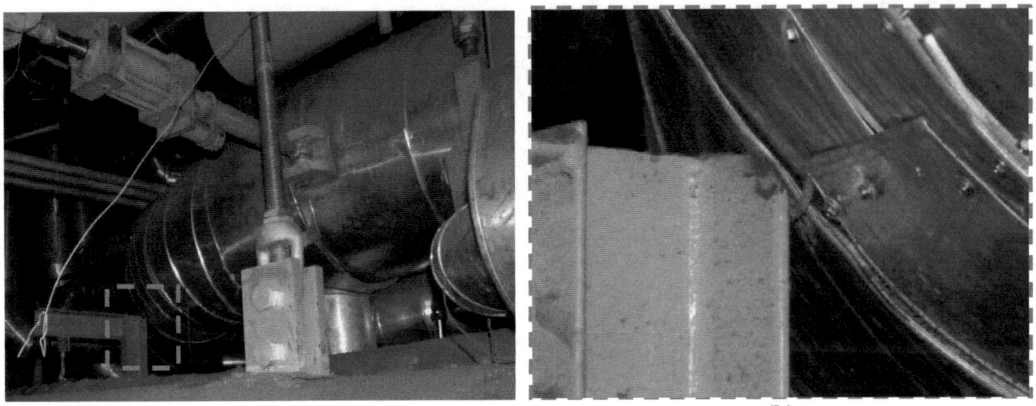

(a) (b)

Figure 10. (a) Beam structure penetrated into pipe cladding (b) Magnified view.

6. Conclusions

This paper provides a real-time pipe displacement monitoring study on the MS pipe at coal-fired power plant. The study demonstrated that the real-time pipe displacement measurement system is able to continuously capture pipe movement and allows engineer to monitor pipe displacement behavior. This aids in identifying issue of trouble pipe hanger/support early so that remedy action can be taken to ensure the piping system safe operation as pipe hanger/support malfunction is the first sign of trouble. This is significantly beneficial for inaccessible and hard-to-reach pipe hangers and snubbers where the condition can be known based on the pipe actual displacement trending captured in the system.

The field inspection at the specified location of abnormal pipe displacement trending reveals the abnormal condition of the piping support system. The pipe snubber at Location 6 found the evidence of hydraulic fluid leaking. This could contribute to the displacement trending less than 25% to that of the design displacement. At Location 8, a beam structure was found penetrated into pipe cladding. It signifies that this condition in any way could affect or disturb the pipe movement behavior and amplify the risk of high stress on the connection pipe weldment in between.

The significant displacement variations from the expected design displacement captured can be considered in piping system integrity assessment as a means to support the identification of critical locations along with the engineering judgement and flexibility stress analysis. This will greatly contribute to the piping system safety and reliability.

Acknowledgements

The cooperation of Sultan Azlan Shah Power Plant is acknowledged for their support and assistance during this project. This work is also supported by the Korea Leading Engineering System (KLES).

References

[1] Peng L C 2009 Quick check on piping flexibility *Peng Engineering*, Houston, Texas
[2] Coleman K 2007 Fossil Plant High-Energy Piping Damage: Theory and Practice", **1**: Piping Fundamentals *EPRI Interim Report* pp. 273-289
[3] ASME B31.1- 2014 Power Piping Code for Pressure Piping
[4] Hahn B, Buhl G, Nerger D 2002 In-service condition monitoring of piping systems in power plants requirements and advanced techniques *OMMI* Vol. **1** Issue 1 pp. 1-13
[5] Hamzah M Z, Zaihan, M H M 2011 TNBR/KEPRI Collaborative Development of High Temperature Steam Piping Monitoring and Assessment System." *Technical Report*, TNB 772/2006, TNB Research Sdn. Bhd., pp.1-64
[6] Huff S A, Leach J P, Vail D S, 2015 The Importance of High Energy Piping Support Maintenance To Enhance System Useful Life *Proceedings of the ASME 2015 Power Conference,* San Diego, California.

[7] Hyun J S, Song G W, Cho S Y, Lee Y S 2005 Real-time displacement monitoring system for high temperature steam pipe of fossil power plant *Key Engineering Materials*, Vols. **297-300**, pp. 2164-2168

[8] Ghaffar M H A, Seon L D, Husin S, Yusof W Y W, 2014 High temperature steam pipe real-time displacement monitoring system of gas-fired power plant", 20th *Conference of The Electric Power Supply Industry (CEPSI)* Jeju, South Korea.

IOP Conf. Series: Materials Science and Engineering **257** (2017) 012027 doi:10.1088/1757-899X/257/1/012027

Energy absorption capabilities of complex thin walled structures

F Tarlochan*, Sami AlKhatib

Department of Mechanical and Industrial Engineering, Qatar Transportation and Traffic Safety Center (QTTSC), Qatar University, Doha, Qatar

*corresponding author: faris.tarlochan@qu.edu.qa

Abstract. Thin walled structures have been used in the area of energy absorption during an event of a crash. A lot of work has been done on tubular structures. Due to limitation of manufacturing process, complex geometries were dismissed as potential solutions. With the advancement in metal additive manufacturing, complex geometries can be realized. As a motivation, the objective of this study is to investigate computationally the crash performance of complex tubular structures. Five designs were considered. In was found that complex geometries have better crashworthiness performance than standard tubular structures used currently.

1. Introduction

The demand for new automotive vehicles is increasing yearly with the increase of world's population. According to the International Energy Agency number of automobile vehicles is expected to reach 1.7 billion by 2035 [1]. With the increase in number of vehicles on road, an increase in the number of car accidents is observed. According to the World Health Organization (WHO) the total number of road traffic deaths remains unacceptably high at 1.2 million per year. From this value, nearly 50% are vehicle occupants [2]. Automotive engineers play a major role in tackling this issue by ensuring the maximum level of safety for occupants when designing the automobile vehicles. This is achievable through passive safety measures where the vehicle's structure is designed to provide higher level of protection and safety in any crash event.

The main vehicle's structures responsible for crash energy absorption in frontal impacts are the frontal longitudinal members, alternatively called side members (See Figure 1). The side member is a thin-walled hollow tubular structure of rectangular or square cross section that dissipates crash energy through plastic deformation. Thin-walled structures are most widely used for crashworthiness and energy absorption applications, due to their light weight and high energy absorption efficiency [3]. Stringent automotive safety standards have driven engineers to enhance the energy absorption capacity without compromising on the overall vehicle weight. Numerous studies have been published on investigating energy absorption of metal tubular thin-walled structures. These studies can be categorized as follows: (1) studying the effect of different geometries and configurations and (2) investigating the effect of foam fillers.

A lot of work looking into different geometrical modifications and configurations of tube thickness designs have been carried out to enhance the energy absorption of thin-walled tubes. Most of the published work in this area looked at simple tubular designs which can be achieved with conventional manufacturing techniques. One of these modifications is dividing the thin-walled column into simple extruded multi-cells profiles that can fill the tube partially or fully. Tang et al [4] studied the characteristics of multi-cells cylindrical tubes on energy absorption and found that multi-cell geometries enhance the energy absorption. Other work on multi-cells geometries studied the effect of different thin-walled tubes cross sections on energy absorption [5]. Varying the tubular structure thickness was another investigated geometrical alteration by researchers to enhance energy absorption and reduce weight [6, 7]. As well, the effect of varying the thickness of multi-cells tubular structures was studied [8]. Tailor made tubular structures were also studied as potential energy absorptions with varying thickness [9]. Expanded metal tubes was another approach taken by researchers to enhance the energy absorption and reduce weight of thin-walled structures [10 – 11]. Other researchers studied the effect of multi-cornered tubular structures on energy absorption [12, 13, and 14], while others investigated the energy characteristics of bi-tubular tubes [15] and nested tubular systems [16].

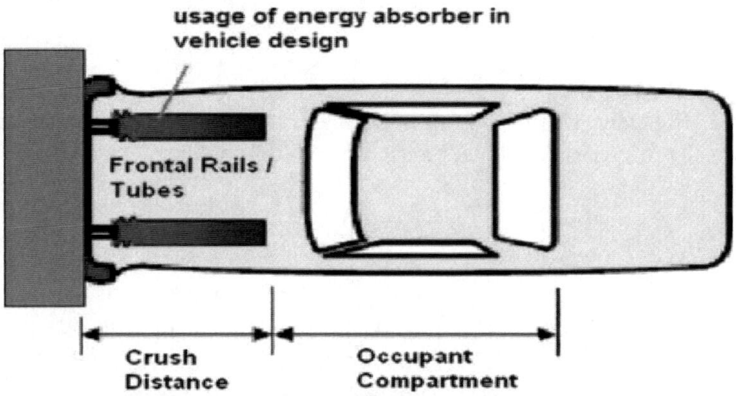

Figure 1. Frontal longitudinal thin-walled structures of the vehicle [3].

Other investigators looked into the effect of metal foam fillers into tubes. Because of its low initial stress peaks and extended load plateaus, foam is considered as an attractive energy absorber. Arriving to the perfect combination of foam and tubular structure requires the consideration of foam density, column width, thickness and cross-section profile and material properties. Some researchers looked into uniform foam density [17-25] fillers, while others looked into functionally graded foams [26-30]. In principle it was found that foams enhance crashworthiness performances, in particular functionally graded foams, if designed correctly.

The aforementioned solutions are limited to the available manufacturing techniques commonly used in producing thin-walled structures. Even the most recent studies in the area of crashworthiness applications are still focused on simple geometries. With the advancement in metal additive manufacturing (AM) it is now possible to produce complex geometries that can be integrated to the thin-walled structures. With this in mind, the main objective of the present work is to study numerically the energy absorption performance of complex thin walled structure geometries. This numerical study is the preliminary study to assess the capabilities of complex geometries in absorbing energy. The subsequent future step in this research framework is additively manufacture the best design selected numerically and to conduct dynamic crush test to actually assess the potential of such complex thin walled structures for energy absorption applications.

2. Performance indicators

The performance of the structures is mainly evaluated based on the overall crash response and based on the ease of manufacturing secondary. To assess the crash response the following parameters were consider.

2.1. Peak load

The peak load is the maximum load needed to start deforming the structure permanently. This parameter is of importance as it is the highest load felt by the occupant during the crash.

2.2. Energy absorption

The total energy absorption E in a crash equals the area under the load-displacement curve. It is defined as:

$$E = \int_0^{dmax} Fds \tag{1}$$

To relate how much energy a structure can absorb to it mass, specific energy absorption is obtained. It is defined as:

$$SEA = \frac{E}{m} \tag{2}$$

where *m* is the crushed mass of the structure. The SEA is very important to automotive industry where weight reduction is crucial.

2.3. Crush Force Efficiency (CFE)

The crushing force efficiency (CFE) is the ratio between the mean and peak crushing forces, the two parameters that are directly related to the deceleration level felt by the occupants. It is defined as:

$$CFE = \frac{P_{mean}}{P_{max}} \tag{3}$$

An ideal energy absorber would have a CFE value close to 100%. That is because an ideal absorber would preserve a peak load for its entire crushed length. If the CFE value is far from 100%, this indicates a large deceleration shift which will be dangerous for the passenger. To summarize, a good design would have a reasonable peak load, a high specific energy absorption, and a CFE value close to 100%

3. Design methodology

This work aims to investigate the effect of complex geometries and structures on the crashworthiness characteristics of thin-walled tubular structures. To achieve this aim, five different complex designs were generated for numerical crash tests as shown in Figures 2 - 6. Each design has two different parameters that were altered to result in 4 profiles for each design. This was done to investigate the effect of each parameter on the crashworthiness performance indicators. Two materials were used to model the generated profiles, steel and aluminum. This was done to examine the effect of the material on the crash response of each profile. Each design's parameters and profiles are listed in Table 1. The numerical simulation includes an axial impact using a striker of 275 kg mass with an initial velocity of 15.6 m/s. The numerical modeling values are discussed in details in the next section. All of the design's profiles are thin-walled tubes with a cross-sectional area of 80x80 mm and a thickness of 2 mm. These dimensions were selected based on the average sedan car side member's dimensions. From the generated profiles, one profile was chosen via a multi-criteria decision making process.

For the multi-criteria decision making process, the first step is choosing the performance criteria. Two categories of performance criteria were considered. The main category includes the performance indicators discussed previously, while the other includes the level of ease of manufacturing. After choosing the performance criteria, the weightage of each criterion is formulated using the Digital Logic (DL) method [31]. In this method, comparison is made between only two performance criterions at a time. An example of determining the relative importance of each performance criterion is shown in Table 2, where a numerical value of (3) is given to the important criterion, and a numerical value of (1) is given for the less important criterion. If the two criterions are equally important, they are both given the numerical value of (2). Total number of possible comparisons can be determined by the following equation:

$$N = \frac{n(n-1)}{2} \tag{4}$$

where *n* is the number of performance criteria.

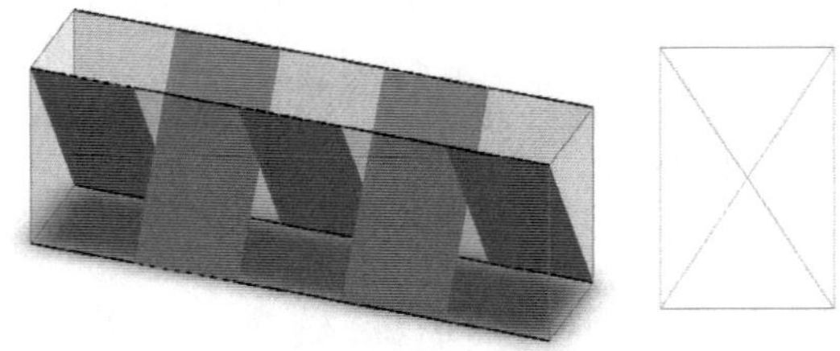

Figure 2. X-shaped internal plates.

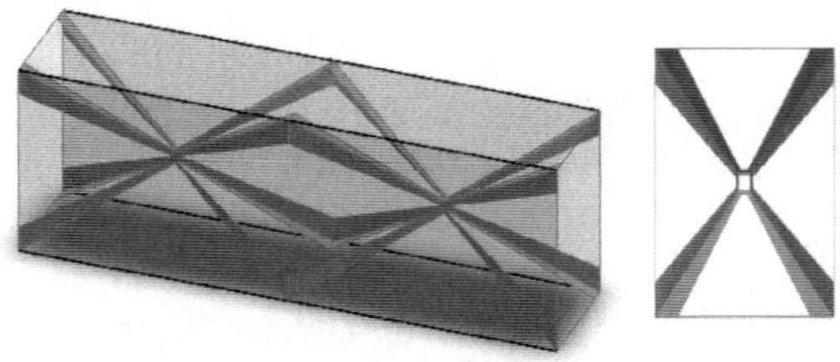

Figure 3. Pyramid's internal shapes.

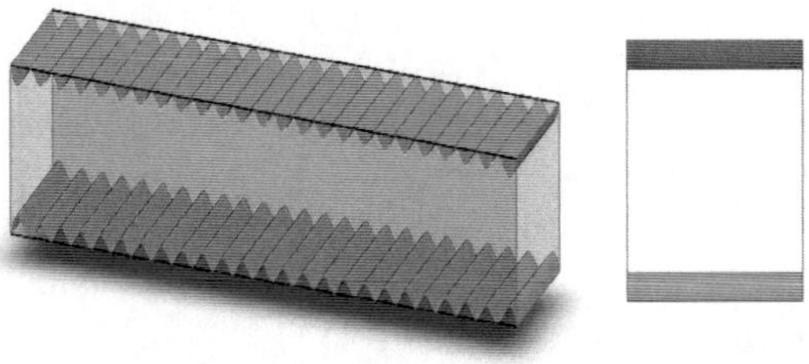

Figure 4. Wavy internal sheets.

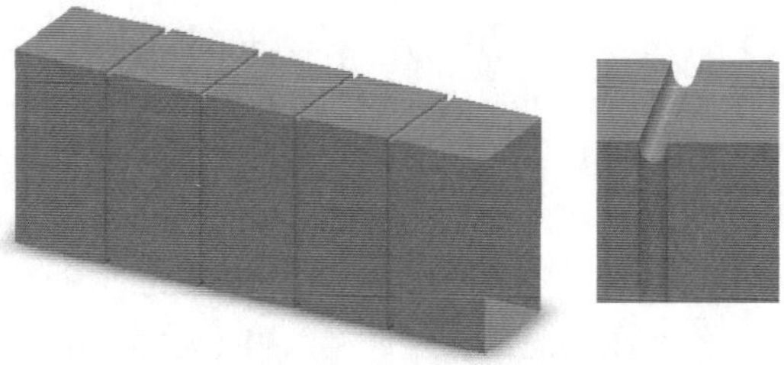

Figure 5. Circular surface grooves.

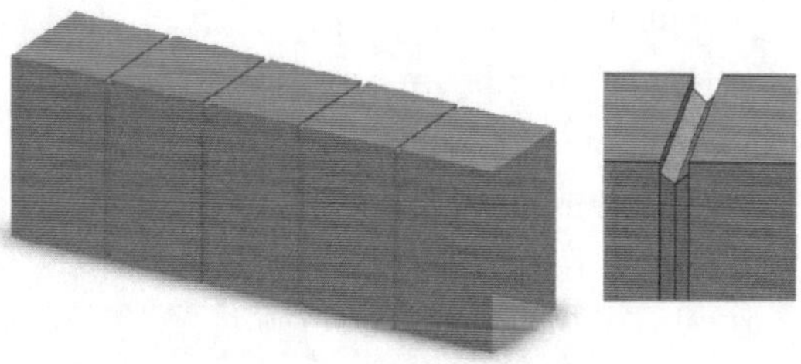

Figure 6. Hexagonal surface grooves.

Table 1. Design's parameters and abbreviations.

Design	Parameters	Abbreviation
Circular Grooves	Grooves Number=4, depth=5mm	CG4D5
	Number=6, D=5mm	CG6D5
	Number=4, D=7mm	CG4D7
	Number=6, D=7mm	CG6D7
Hexagonal Grooves	Grooves Number=4, depth=5mm	HG4D5
	Number=6, D=5mm	HG6D5
	Number=4, D=7mm	HG4D7
	Number=6, D=7mm	HG6D7
Pyramids	Pyramids Number=4, Thickness=1mm	P4T1
	Number=6, t=1mm	P6T1
	Number=4, t=2mm	P4T2
	Number=6, t=2mm	P6T2
X-plates	Plates Number=5, Thickness=1mm	XP5T1
	Number=7, t=1mm	XP7T1
	Number=5, t=2mm	XP5T2
	Number=7, t=2mm	XP7T2
Waves	Wave's width=5mm, Length=5mm	WW5L5
	W=5mm, L=7mm	WW5L7
	W=7mm, L=5mm	WW7L5
	W=7mm, L=7mm	WW7L7

Table 2. Example of weightage setting.

Criteria	*Number of possible comparisons* $N = n(n-1)/2 = 5(5-1)/2 = 10$										*Score*	*Relative Weightage,*
	1	**2**	**3**	**4**	**5**	**6**	**7**	**8**	**9**	**10**		
A	3	3	1	2							9	→ 9/40 = 0.225
B	1				2	3	1				7	→ 7/40 = 0.175
C		1			2			3	2		8	→ 8/40 = 0.200
D			3			1		1		1	6	→ 6/40 = 0.150
E				2			3		2	3	10	→ 10/40 = 0.250

4. Finite element model

In this work the finite element model for the thin-walled tubular profiles were modelled using ABAQUS. The non-linear finite element code ABAQUS explicit was used to predict the response of the generated profiles after they were impacted by a striker. To model the profiles, a square cross section tube of 80x80 mm, 350 mm length and 2 mm thickness is considered. The thin walled tube was modelled using 4 node shell continuum (S4R) element of 2.5 mm size with reduced integration points. The top plate (striker) was modelled as a rigid body with a reference point mass of 275 kg and a predefined velocity field of 15.6 m/s. The striker's mass was selected to be 25% of an average sedan vehicle's mass (1100 kg). This was based on the fact that less than 50% of the vehicle's mass kinetic energy can be absorbed by two longitudinal tubes [32].

The striker's initial velocity was selected as 15.6 m/s (56 km/h) as the value used in National Car Assessment Program (NCAP) crash tests by the National Highway Traffic Safety Administration (NHTSA).The motion of the top plate was restricted to only translational motion along the tube's axes.

The thin-walled tube is fixed on the lower rigid plate using a constraint between their two surfaces. To account for the contact between the two plates and the tube, a general contact algorithm was chosen with a penalty coulomb friction coefficient of 0.2 [33]. This is to avoid the interpenetration between the tube walls. To model the design's profiles materials, the A-36 mild-steel model from [3] and AA6060T4 aluminum alloy model from [4] models were adopted. The adopted steel model is characterized by Johnson-Cook constitutive hardening model. The adopted steel model parameters are listed in Table 3. The AA6060T4 aluminum is characterized as an isotropic plastic material with the plastic behavior shown in Figure 7. The adopted aluminum model has a young modulus E=68.21 GPa, Poisson's ratio of 0.3 and yield strength Sy = 80 MPa. text.

Table 3. Johnson-Cook Parameters [3].

Parameters	Values
A	146.7 MPa
B	896.9 MPa
Strain Power Coefficient, N	0.32
C	0.033
Temperature Power Coefficient, M	0.323
Reference Strain Rate	1.0 s^{-1}
Density	7850 kg/m^3
Melting temperature, Tm	1773 K
Specific Heat, Cp	486 J/kg-K

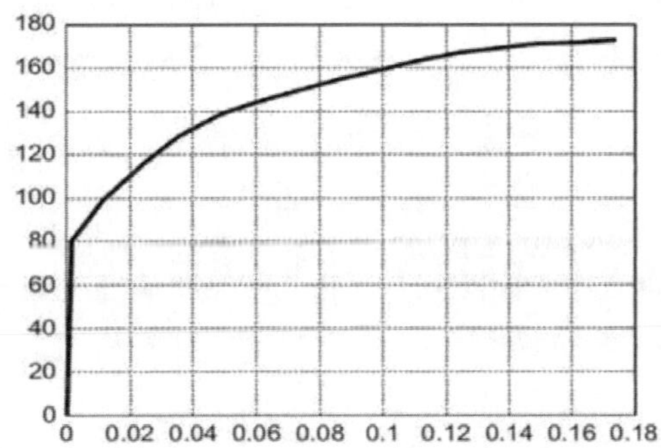

Figure 7. Tensile test graph of AA6060T4 aluminum alloy [4].

5. Results and discussion

The energy absorption is plotted against the deformation length (Figure 8 (a) and (b)) for the highest energy absorbing profiles for steel and aluminum respectively. This is done to facilitate the comparison between the different profiles. The force-displacement diagrams for all of the tested profiles are shown in Figures 9–13. The following sub-sections includes a detailed explanation on the obtained crashworthiness parameters for the different generated concepts.

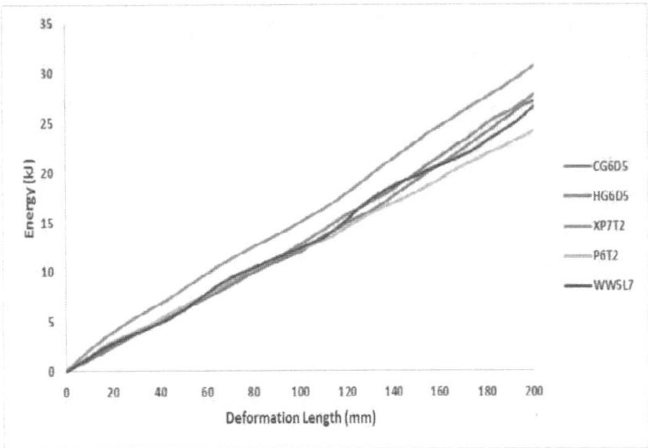

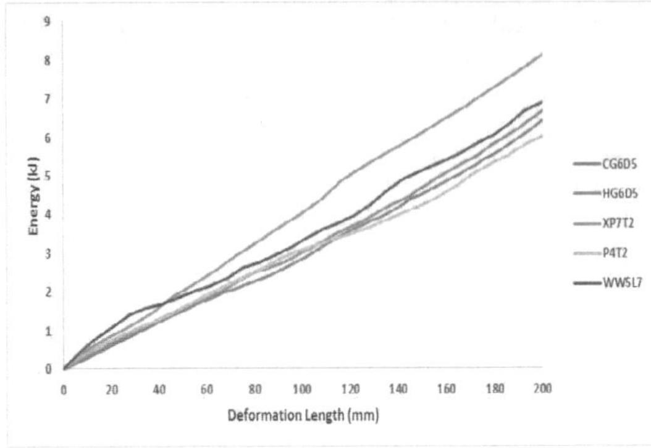

(a) (b)

Figure 8. Energy against deformation length for best energy absorbing (a) steel (b) aluminium profiles

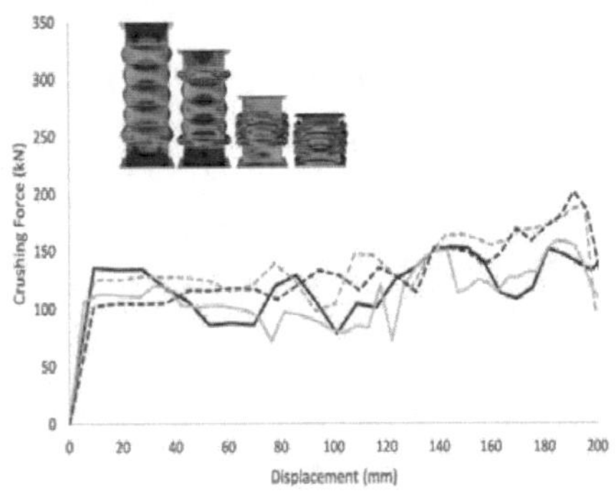

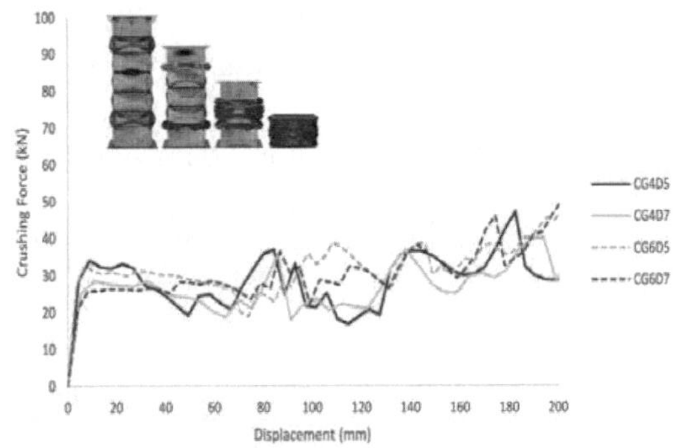

(a) (b)

Figure 9. Force-displacement diagram for circular grooves profiles (a) steel (b) aluminium

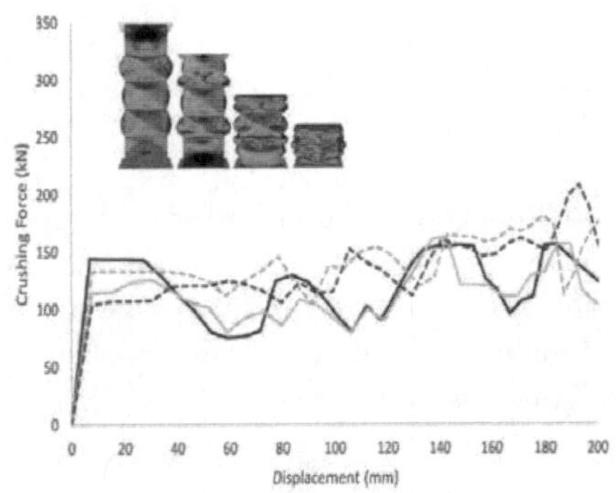

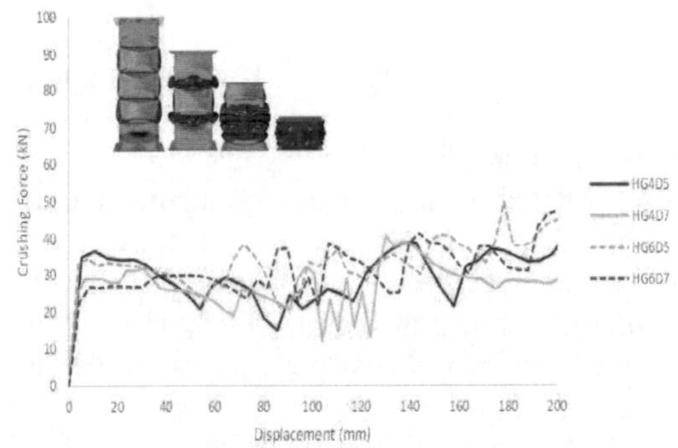

Figure 10. Force-displacement diagram for hexagonal grooves profiles (a) steel (b) aluminium

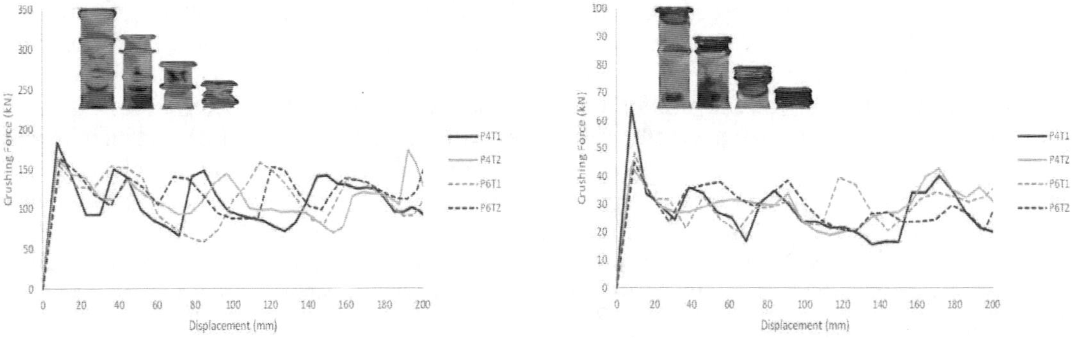

Figure 11. Force-displacement diagram for pyramid grooves profiles (a) steel (b) aluminium

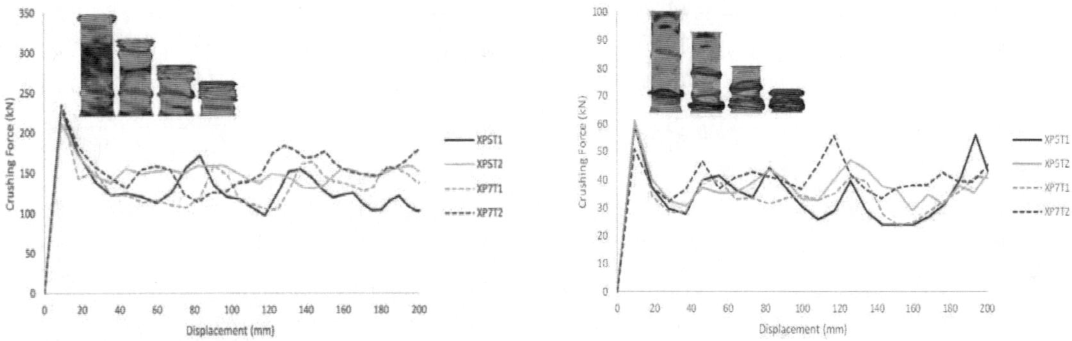

Figure 12. Force-displacement diagram for x-internal grooves profiles (a) steel (b) aluminium

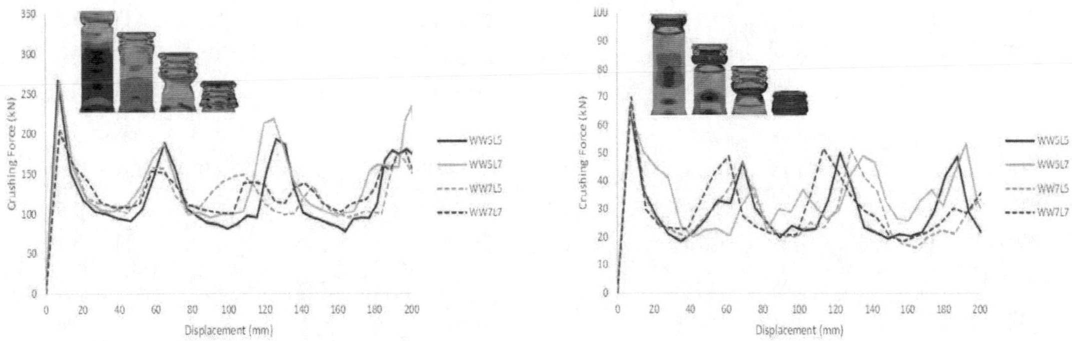

Figure 13. Force-displacement diagram for internal wavy sheets (a) steel (b) aluminium

The first step of the selection process is to choose the selection criteria. The main general principles that has to be achieved in an ideal energy absorber are: (1) restricted and constant reactive force, (2) light weight and high Specific Energy Absorption (SEA) and (3) low cost and easy installation [34]. To satisfy the first two principals, the following measured indicators were selected as evaluation criteria: the maximum crushing force (Pmax), Crushing Force Efficiency (CFE) and the Specific Energy Absorption (SEA). Similarly, the production criteria which assesses the complexity of the profile was selected to satisfy the last principle. The second step of the selection process is to determine the weight for each of the performance criteria. This is done using the Digital Logic method, in which two criteria are compared at a time. The details of this method have been explained previously. Tables 4 and 5 show the total score for each performance criteria for the steel and aluminum profiles respectively.

Table 4. Summary of the performance criteria scores for the steel-assigned profiles.

Model	CFE [%]	score	rating	SEA [J/kg]	score	rating	Pmax [kN]	score	rating	Production	rating	Total
CG4D5	84.7	9	2.97	24120.7	9	1.89	167.806	9	2.97	10	1.3	9.13
CG6D5	69.7	5	1.65	25658	9	1.89	189.419	7	2.31	8	1.04	6.89
CG4D7	85.1	9	2.97	23300.6	7	1.47	165.854	10	3.3	10	1.3	9.04
CG6D7	65	3	0.99	24273.2	9	1.89	201.188	6	1.98	7	0.91	5.77
HG4D5	88.6	10	3.3	24068.8	8	1.68	156.017	10	3.3	9	1.17	9.45
HG6D5	85	9	2.97	27154.2	10	2.1	181.138	8	2.64	7	0.91	8.62
HG4D7	87.3	10	3.3	23299.1	8	1.68	165.148	10	3.3	9	1.17	9.45
HG6D7	66.8	4	1.32	24831.1	8	1.68	214.405	5	1.65	8	1.04	5.69
P4T1	68.5	5	1.65	19863.1	5	1.05	183.717	8	2.64	2	0.26	5.6
P6T1	78.8	8	2.64	21278.1	6	1.26	159.834	10	3.3	1	0.13	7.33
P4T2	74.6	6	1.98	19773.5	5	1.05	173.887	9	2.97	2	0.26	6.26
P6T2	74	6	1.98	20985.6	6	1.26	182.951	8	2.64	1	0.13	6.01
XP5T1	58.3	1	0.33	21220.7	6	1.26	230.149	4	1.32	4	0.52	3.43
XP7T1	57.6	2	0.66	21783.9	6	1.26	231.409	4	1.32	3	0.39	3.63
XP5T2	68.9	5	1.65	21074.1	6	1.26	211.654	5	1.65	4	0.52	5.08
XP7T2	65.7	4	1.32	21633.2	6	1.26	235.354	3	0.99	3	0.39	3.96
WW5L5	58.4	2	0.66	15115.2	1	0.21	267.076	1	0.33	6	0.78	1.98
WW5L7	55	1	0.33	14041	1	0.21	265.562	1	0.33	6	0.78	1.65
WW7L5	73	6	1.98	15638.9	2	0.42	207.173	6	1.98	5	0.65	5.03
WW7L7	74.4	6	1.98	14941.6	1	0.21	202.397	6	1.98	5	0.65	4.82

It was found that the best steel profile is the hexagonal grooves (HG4D7), while the best aluminum profile is the circular grooves (CG4D7). Figure 14 illustrates the enhancements obtained for the maximum crushing force (Pmax), the crushing force efficiency (CFE) and the specific energy absorption (SEA) of the best selected profiles compared to the conventional empty steel and aluminum tubes. The selected steel profile (HG4D7) shows a reduction in (Pmax) of around 44 kN, an increase in (CFE) of around 30% and an increase in (SEA) of around 9 kJ/kg relative to the conventional empty steel tube tested in [3]. Similarly, the selected aluminum profile (CG4D7) shows a reduction in (Pmax) of around 85 kN, an increase in (CFE) of around 42% and an increase in (SEA) of around 10 kJ/kg compared to the conventional empty aluminum tube tested in [35].

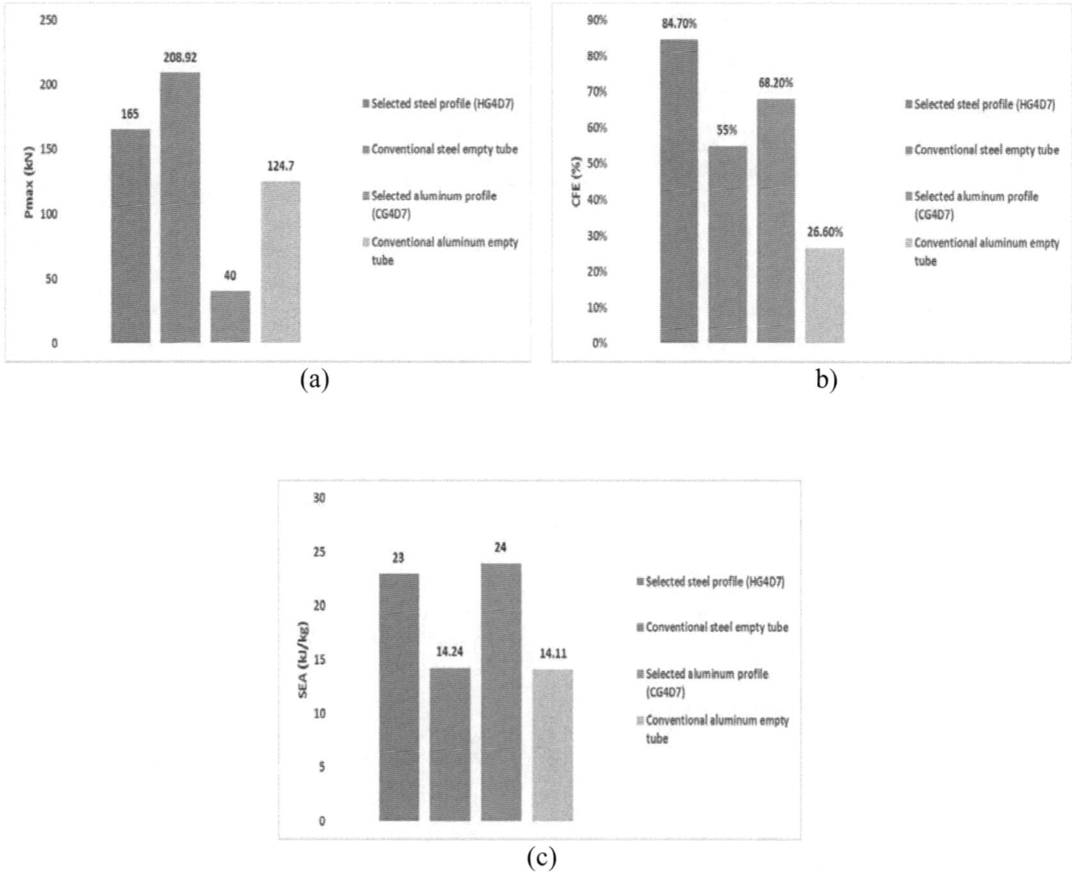

Figure 14. Comparison values for (a) peak force (b) crush force efficiency (c) specific energy

6. Conclusions

The crash response of different complex geometrical designs of steel and aluminum were examined numerically using ABAQUS software. The different complex-geometries designs parameters were altered which resulted in 20 profiles that were tested numerically for both steel and aluminum. All numerical tests were dynamics with speed of 15.6 m/s and an impact mass of 275 kg. Different crashworthiness indicators were obtained to assess the performance of the tested profiles. The indicators were then weighted according to their significance to select the most efficient crashworthy profile using a multi-criteria decision making process. It was found that the hexagonal grooves (HG4D7) profile was better for crashworthiness application than the other steel profiles, while the circular grooves (CG4D7) profile was better than the other aluminum profiles. The steel (HG4D7) profile had a maximum crushing force value of 165 kN, specific energy absorption value of 23 kJ/kg and crushing force efficiency of 87%. Similarly, the aluminum (CG4D7) profile had a maximum crushing force value of 40 kN, specific energy absorption value of 24 kJ/kg and crushing force efficiency of 68%. In conclusion, it can be stated that the best selected profiles (HG4D7) and (CG4D5) had enhanced the crashworthiness of the thin walled structure, showing the potential of using additive manufacturing techniques in automotive safety applications in the future. However, further experimental investigation is required to assure the accuracy of the numerical material's models.

Table 5. Summary of the performance criteria scores for the aluminum-assigned profiles.

Model	CFE [%]	score	rating	SEA [J/kg]	score	rating	Pmax [kN]	score	rating	Production	rating	Total
CG4D5	59.1	7	2.31	28463.7	8	1.68	47.135	8	2.64	10	1.3	7.93
CG6D5	67.3	10	3.3	28938.5	9	1.89	47.27	8	2.64	8	1.04	8.87
CG4D7	68.2	10	3.3	23928.3	6	1.26	40.066	10	3.3	10	1.3	9.16
CG6D7	61.9	8	2.64	32353.9	10	2.1	49.99	7	2.31	7	0.91	7.96
HG4D5	54.5	5	1.65	27165.7	8	1.68	54.254	6	1.98	9	1.17	6.48
HG6D5	60.5	8	2.64	28061	8	1.68	54.868	6	1.98	7	0.91	7.21
HG4D7	56.5	6	1.98	24809.1	7	1.47	50.378	7	2.31	9	1.17	6.93
HG6D7	65.5	9	2.97	25138.2	7	1.47	47.863	8	2.64	8	1.04	8.12
P4T1	42.2	1	0.33	15779.7	2	0.42	64.767	2	0.66	2	0.26	1.67
P6T1	51.2	4	1.32	23634.2	6	1.26	61.704	3	0.99	1	0.13	3.7
P4T2	66.5	9	2.97	15854.8	2	0.42	43.821	9	2.97	2	0.26	6.62
P6T2	66.2	9	2.97	20547.8	5	1.05	44.864	9	2.97	1	0.13	7.12
XP5T1	55.4	6	1.98	16905.8	3	0.63	59.249	4	1.32	4	0.52	4.45
XP7T1	54.6	5	1.65	16608.6	3	0.63	59.887	4	1.32	3	0.39	3.99
XP5T2	64.9	9	2.97	16955.7	3	0.63	61.315	3	0.99	4	0.52	5.11
XP7T2	69.8	9	2.97	20335	4	0.84	55.845	5	1.65	3	0.39	5.85
WW5L5	44.1	2	0.66	19349.5	4	0.84	66.338	2	0.66	6	0.78	2.94
WW5L7	49.9	4	1.32	18546.5	4	0.84	65.529	2	0.66	6	0.78	3.6
WW7L5	40.7	1	0.33	19688.6	4	0.84	68.01	1	0.33	5	0.65	2.15
WW7L7	43.6	1	0.33	12625.9	1	0.21	69.946	1	0.33	5	0.65	1.52

References

[1] *International Energy Agency* (Retrieved on 11th April, 2016) Available from: http://www.iea.org/

[2] *The Global Status Report on Road Safety 2013* World Health Organization (WHO) Available from: www.who.int

[3] Tarlochan F, Samer F, Hamouda A, Ramesh S and Khalid K 2013 *Thin-Walled Structures* **71** 7-17

[4] Tang Z, Liu S and Zhang Z A 2013 *Thin-Walled Structures* **62** 75-84

[5] Nia A and Parsapour M 2014 *Thin-Walled Structures* **74** 155-65

[6] Zhang X, Wen Z and Zhang H 2014 *Thin-Walled* Structures **84** 263-74

[7] Guangyao L, Fengxiang X, Guangyong S and Qing L 2015 *International Journal of Impact Engineering* **77** 68–83

[8] Jianguang F, Yunkai G, Guangyong S, Gang Z and Qing L 2015 *International Journal of Mechanical Sciences* **103** 63–73

[9] Shahi V and Marzbanrad J 2012 *Thin-Walled Structures* **60** 24-37

[10] Smith D, Graciano C and Martínez G 2014 *Thin-Walled Structures* **84** 170-6

[11] Smith D, Graciano C, Martínez G and Teixeira P 2014 *Thin-Walled Structures* **85** 42-9

[12] Liu S, Tong Z, Tang Z, Liu Y and Zhang Z 2015 *Thin-Walled Structures* **88** 70-81

[13] Reddy S, Abbasi M and Fard M 2015 *Thin-Walled Structures* **94** 56-66

[14] Abbasi M, Reddy S, Ghafari-Nazari A and Fard M 2015 *Thin-Walled Structures* **89**31-41

[15] Sharifi S, Shakeri M, Fakhari H and Bodaghi M 2015 *Thin-Walled Structures* **89** 42-53

[16] Abbas N, and Pourya H O 2016 *Thin-Walled Structures* **100** 113-23

[17] Yin H, Wen G, Liu Z and Qing Q 2014 *Thin-Walled Structures* **75** 8-17

[18] Ahmad Z, Thambiratnam D and Tan A 2010 *International Journal of Impact Engineering* **49** 475-88

[19] Hanssen A, Langseth M and Hopperstad O 2001 *International Journal of Mechanical Sciences* **43** 153-76

[20] Hanssen A, Langseth M and Hopperstad O 2000 *International Journal of Mechanical Sciences* **24** 347-83
[21] Zhang X and Cheng G 2007 *International Journal of Impact Engineering* **34** 1739-52
[22] Yin H, Wen G, Liu Z and Qing Q 2014 *Thin-Walled Structures* **75** 8-17
[23] Djamaluddin F, Abdullah S, Ariffin A and Nopiah Z 2015 *Thin-Walled Structures* **87** 1-11
[24] Goel M 2015 *Thin-Walled Structures* **90** 1-11
[25] Qiang G, Liangmo W, Yuanlong W and Chenzhi W 2016 *Thin-Walled Structures* **100** 105-12
[26] Sun G, Li G, Hou, S, Zhou S, Li W and Li Q 2010 *Materials Science and Engineering A* **527** 1911-9
[27] Omid M and Hashem G 2016 *Thin-Walled Structures* **98** 627–39
[28] Guangyao L, Zheshuo Z, Guangyong S, Fengxiang X and Xiaodong H 2014 *International Journal of Mechanical Sciences* **89** 439–52
[29] Jianguang F, Yunkai G, Guangyong S, Yuting Z, and Qing L 2014 *Computational Materials Science* **90** 265–75
[30] Hanfeng Y, Guilin W, Hongbing F, Qixiang Q, Xiangzheng K, Juru X and Zhibo L 2014 *Materials and Design* **55** 747–57
[31] Farag M M 2008 *Materials and Process Selection for Engineering Design* (Boca Raton: CRC Press)
[32] Witteman W J 1999 *Improved vehicle crashworthiness design by control of the energy absorption for different collisions situation* (PhD thesis) Netherlands: Eindhoven University of Technology
[33] Dehghan-Manshadi B, Mahmudi H, Abedian A and Mahmudi R 2007*Materials & Design* **28** 8-15
[34] Lu G and Yu T X 2003 *Energy absorption of structures and materials* (Boca Raton: CRC Press)
[35] Cheng Q, Altenhof W and Li L 2006 *Thin-Walled Structures* **44** 441-54

A review on the effect of welding on the corrosion of magnesium alloys

N S Mohamed and J Alias*

Structural Materials and Degradation, Faculty of Mechanical Engineering, Universiti Malaysia Pahang (UMP), 26600 Pekan, Pahang.

Corresponding Author: juliawati@ump.edu.my

Abstract. Welding is an important joining technique for lightweight alloys with their increasing applications in aerospace, aircraft, automotive, electronics and other industries. The applications of lightweight alloys particularly magnesium alloys increased rapidly due to their beneficial properties such as low density, high strength-to-mass ratio, good dimensional stability, electromagnetic shielding and good recyclability. The effect of welding on the corrosion of magnesium alloys are reviewed in this paper, which closely related to the developed microstructure by the welding process. The paper focuses particularly on friction stir and laser welding. The basic principles of friction stir and laser welding are discussed, to present the likelihood of defects which significantly affect the corrosion of magnesium alloy. The finding in corrosion demonstrated the morphology of corrosion occurrence on each welded region, and observation on the potential and current values are also included.

1. Introduction

Magnesium is extremely light metal with excellent specific strength [1], excellent sound damping capabilities [2], good cast-ability [3], hot formability [4], excellent machinability, good electromagnetic interference shielding and recyclability [5]. However, their poor resistance to corrosion and wear, limits their wider applications [6]. Mg is generally an active metal and its corrosion behavior is significantly influenced by the microstructure [6].

The alloying elements are normally added to increase the strength. This consequently led to the formation of large intermetallic precipitates during casting. These precipitates are too large to be greatly affected by sub-sequent thermo mechanical processing. However the presence of these precipitates in the microstructure could reduce the material resistance to corrosion [7].

Work by Cao [5] shows magnesium alloy components can be joined using mechanical fasteners as well as a variety of welding methods including tungsten-arc inert gas (TIG), metal-arc inert gas (MIG), plasma arc, electron, laser, friction, adhesive, explosion, stud, ultrasonic, and spot welding. TIG and MIG processes are the main common welding method methods for magnesium alloys. However, low welding speeds, large heat affected zone (HAZ) and fusion zone (FZ), high shrinkages, variations in microstructures and properties, evaporative loss of alloying elements, high residual stress and distortion of arc-welded joints have caused attention to be drawn towards laser welding [1, 5] . Besides laser welding, friction stir welding (FSW) is capable of joining magnesium alloys without melting and thus it can eliminate problems related to the solidification [2]. As FSW does not require any filler material, the metallurgical problems associated with it can also be eliminated and good quality weld can be obtained.

2. Welding process

Friction stir weld (FSW) is solid state material joining process that was invented in 1991 at The Welding Institute (TWI) of the United Kingdom [8]. It was initially applied to aluminum alloys and resulting large strain of material at high temperature without melting the base material [9]. The mechanism of friction stir welding working is by applying a non-consumable rotating tool with a particularly designed pin and shoulder that are inserted into the abutting edges of the joining plates. The tools then subsequently pass through the joint line definition of the tool and work piece as shown in Figure 1 [3].

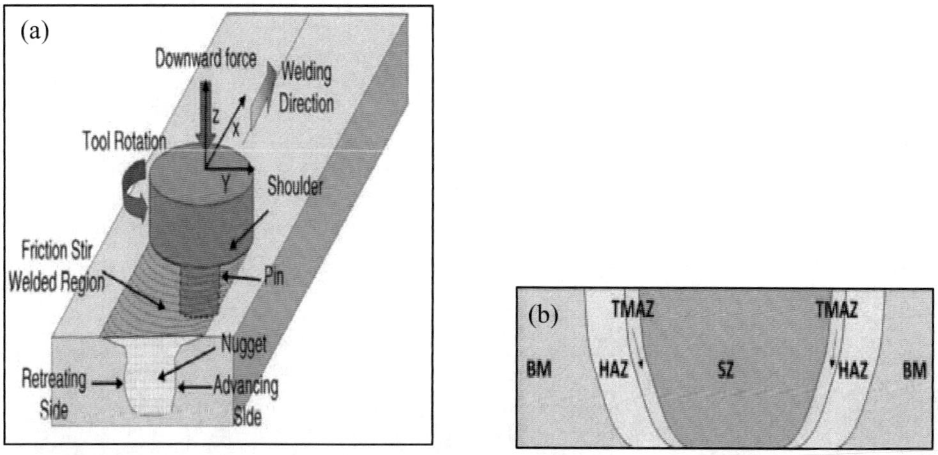

Figure 1. (a) Frictions stir welding process [8], (b) a typical micrograph showing various microstructural zones in FSW, stir zone (SZ), thermomechanical heat affected zone (TMAZ), heat affected zone (HAZ) and base metal (BM) [10].

Table 1. FSW parameters for mechanized welding [8].

No.	Parameter	Effect of parameter
a	Rotational speed	Frictional heat, stirring, oxide layer breaking and mixing of material
b	Welding speed	Appearance, heat control
c	Pressure on tool (down force)	Frictional heat, maintaining contact conditions
d	Tilting angle	Appearance of weld, thinning

The tool produces the thermo mechanical deformation and work piece frictional heating necessary for friction stirring. When the down force is applied on tool then tool is inserting in the base materials. The friction stirring tool consists of a pin or probe, and shoulder. Contact of the pin with the work piece creates frictional and deformational heating and softens the work piece material contacting the shoulder to the work piece increases the work piece heating, expands the zone of softened material, and constrains the deformed material [3, 8] The correct tool material determines material characteristics that are important for each friction stir application including wear resistance, coefficient of thermal expansion, elevated temperature strength, tool reactivity, fracture toughness, machinability and microstructure uniformity [8].

The shape of the tool pin (or probe) influences the flow of plasticized material and affects weld properties. Optimizing tool geometry to produce more heat or achieve more efficient "stirring" offers improvement in breaking and mixing of the oxide layer and more efficient heat generation, yielding higher welding speeds and, enhanced weld quality [8].

Most complex motion tools require specialized machinery or specially machined tools, making these tools unsuitable for basic applications [11].

Table 2. Complex motion tool [8, 11].

No.	Type of tools	Description of the type of tools
a	Skew-stir Tool	Increase the volume of material swept by pin to pin volume ratio by offsetting the axis of the pin from the axis of spindle
b	Com-stir Tool	Combine rotary motion (tool shoulder) with orbital motion (tool pin) to maximize the volume of material swept by pin to pin volume ratio
c	Dual-rotation Tool	The pin and shoulder rotate separately at different speeds and/or in different directions
d	Re-stir tool	Avoid the inherent asymmetry produced during friction stirring by alternating the tool rotation, either by angular reciprocation (direction reversal during one revolution) or rotary reversal (direction reversal every one or more revolutions) Alternating the tool rotation produces alternating regions of advancing and retreating side material through the length of the weld, thus eliminating the asymmetry issues

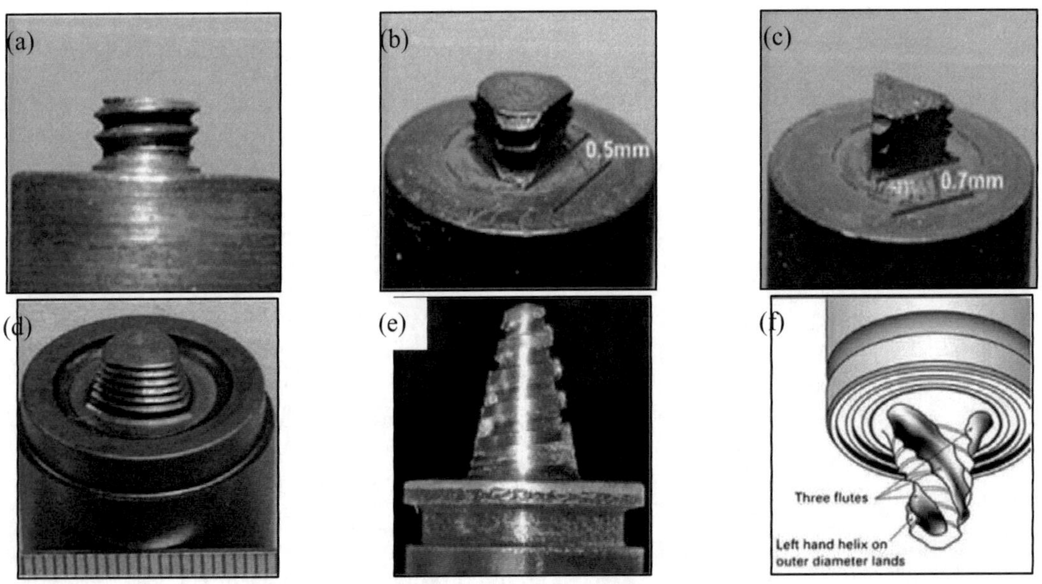

Figure 2. Commonly used pin geometries (a) cylinder threaded, (b) three flat threaded, (c) triangular, (d) trivex, (e) threaded conical, and (f) chematic of a triflute [8].

3. Laser welding

The effectiveness of laser welding depends on the physical properties of the material to be welded. Cao [5] state that magnesium alloys possess certain natural characteristics such as low absorptivity of laser beams, strong tendency to oxidize, high thermal conductivity, high coefficient of thermal expansion, low melting and boiling temperatures, wide solidification temperature range, high solidification shrinkage, a tendency to form low melting-point constituents, low viscosity, low surface tensions, high solubility for hydrogen in the liquid state, and absence of color change at the melting point temperature [1, 5, 12]. Therefore during laser welding of magnesium alloys, some processing problems and weld defects can be encountered. The following discussion focuses on some important processing variables and their influence on the weld quality of magnesium alloy.

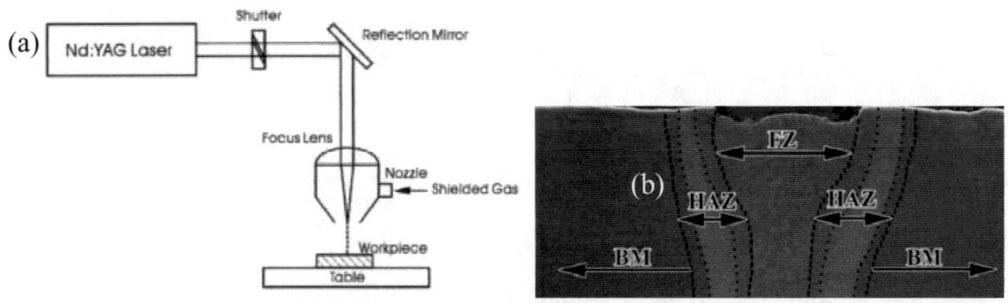

Figure 3. (a) The Nd:YAG laser welding system [1] (b) a typical micrograph showing various microstructural zones in laser welding, fusion zone (FZ), heat affected zone (HAZ) and base metal (BM) [13].

3.1. Laser power

High power density at the work piece is crucial to achieve keyhole welding and to control the formation of welds. Figure 4 shows the effect of laser power on the penetration depth (Figure 4(a)) and weld width (Figure 4(b)) for WE43 alloy welded at a speed of 33 mm/s and a focused diameter of 0.25mm [5, 14]. High beam powers led to deep and wide beads, and reduce both ripples and crowning [5]. Lehner [15] reported that a lower power level and a slower speed lead to better weld quality.

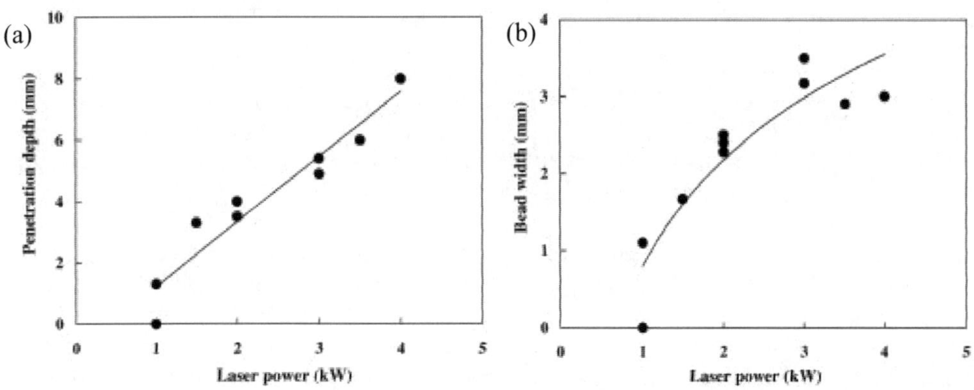

Figure 4. Effect of CO_2 laser power on (a) penetration depth (b) bead width of cast WE43 alloy joints [5].

3.2. Welding speed

Figures 5 and 6 show the effects of welding speed on penetration depth and weld width at different levels of power for CO_2 and Nd:YAG lasers, respectively. The penetration depth and weld width both decrease linearly with increasing welding speed.

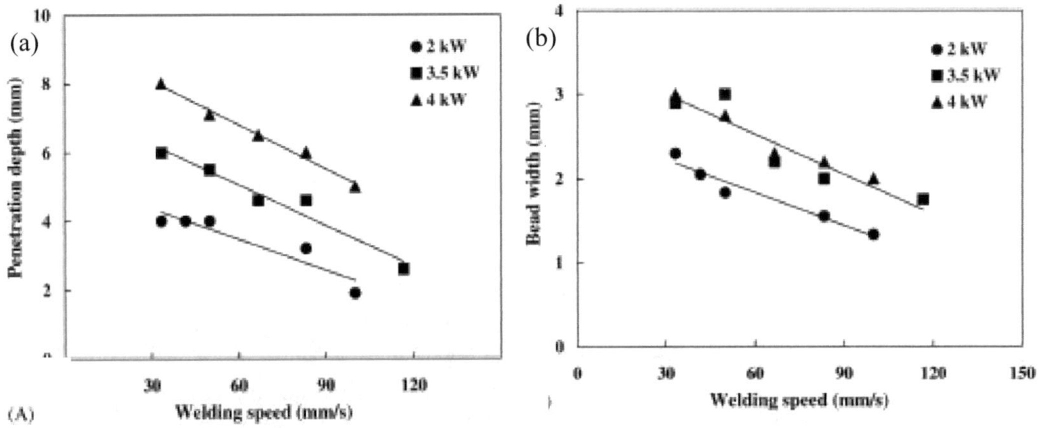

Figure 5. Effect of welding speed on (a) penetration depth, and (b) bead width of cast WE43 alloy joints welded using CO_2 laser [5], [14].

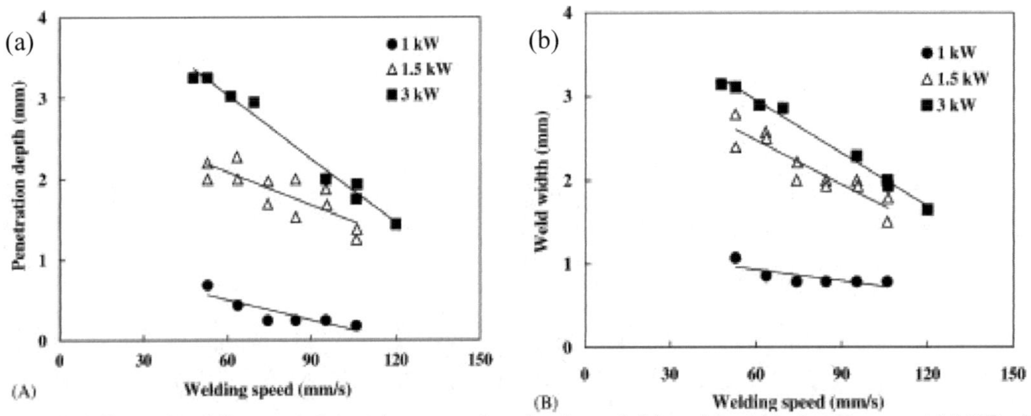

Figure 6. Effect of welding speed on (a) penetration depth, and (b) weld width for die-cast AM60B alloy joints welded using a ND:YAG [5], [16].

Too high welding speed was reported to reduce ripples but greatly increase crowning, or even to increase tendency to brittleness in the fusion zone [5]. Clearly the welding speed should be adjusted to provide the penetration depth required at a given power level. The research into laser welding processes for magnesium alloys is still in its infancy. Much more work, therefore, is needed to systematically investigate the laser welding characteristics of different magnesium alloys because of the difference in their thermal properties.

4. Corrosion of welded magnesium alloy

The corrosion behavior of magnesium alloy is depending on variety factor such as pH value [17], and if the pH value is above 12, a stable and self-healing passive layer developed, and is responsible for high corrosion resistance [18]. It is confirmed that the friction stir welded metal could enhance the corrosion rate with a decrease in pH value, remain constant in neutral pH value and a comparatively very low corrosion rate in alkaline solution.

Figure 7 shows the corrosion and pit morphology of the welded magnesium alloy after immersion tests at pH 3, 7 and 11 with constant chloride ion concentration of 0.60 mol/L. It is found that at a lower concentration of chloride ion the surface of the welded alloy was slightly corroded in neutral solution while it was severely corroded at a higher concentration of chloride ion [18]. From the morphology of welded metal after immersion test, it is observed that the base metal shows the pitting marks and pitting corrosion take place at the welded microstructure [19]. The number of pits are highly seen in the welded metal region when it is immersed in the solution of low pH [20]. Moreover, the increase of grain boundary in the weld metal region acts cathode causing micro-galvanic effect [17]. Corrosion tends to be concentrated in the area adjacent to the grain boundary until the grain may be undercut and fall out [18].

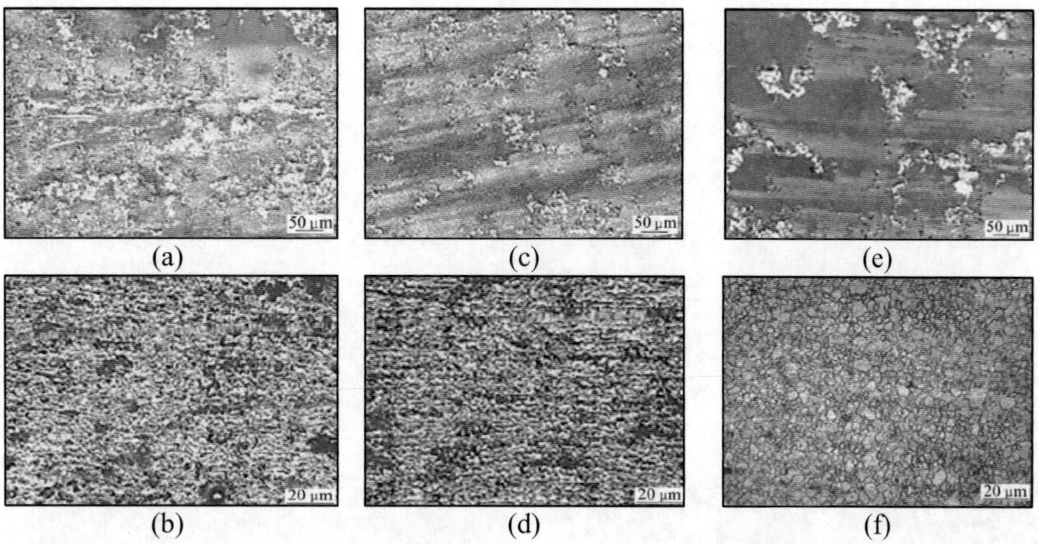

Figure 7. Effect of pH on corrosion morphology (a), (c), (e) and pit morphology (b), (d), (f), pH=3; (a), (b) pH=7; (c), (d) pH=11; (e), (f) [18].

A study by Ghali [21] examined that corrosion potential of Mg is slightly negative than -1.5 V in dilute chloride solution with respect to the standard electrode potential at 25°C is -2.37 V. Recent study by Argade [22] revealed that the corrosion properties of a welded Mg-Y-RE alloy in four different conditions, (i) A: as-received, (ii) B: as-received samples subjected to peak age hardening treatment of 210°C for 48 h, (iii) C: friction stir weld (FSW) of B-type material, and (iv) D: FSW of A-type material subjected to subsequent peak age hardening treatment of 210°C for 48 h. The potentiodynamic polarization plots for all four conditions of Mg-Y-RE alloy in 3.5 wt.% NaCl solution are shown in Figure 8.

For type-A and type-B, the anodic branch showed a linear region in the potential range of -1800 to -1700 mV vs. SCE as can be seen in Figure 8. In case of type-C and type-D a small linear region in the anodic branch was observed as compared to type-A and type-B. Among all the microstructural conditions, type-D showed the highest positive value of E_{corr} and the lowest value of i_{corr}. The comparative E_{corr} and i_{corr} values of different conditions were observed in the following order: (a) E_{corr}: type-D > type-C > type-B > type-A, and (b) i_{corr}: type-D < type-B < type-A < type-C. The details of the polarization parameter are summarized in Table 3. The SEM micrographs (Figures 9(a) and (b)) indicated the localized corrosion attack for type D material with deeper pit attack in comparison to the type B material.

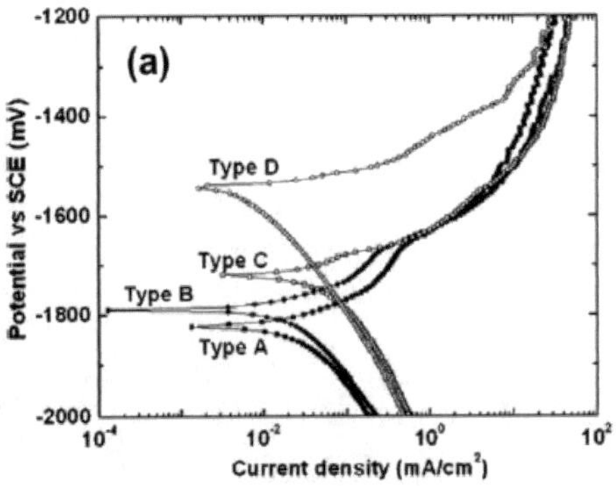

Figure 8. Polarization plots of Mg-Y-RE alloy conditions [22].

Table 3. Electrochemical parameters from LPP for Mg-Y-RE alloy conditions [22].

Sample Conditions	Corrosion Potential E_{corr} (mV vs. SCE)	Current Density i_{corr} (A/cm^2)
Type A	-1823 ± 22	5.9 ± 0.4 x 10^{-5}
Type B	-1788 ± 15	2.7 ± 0.2 x 10^{-5}
Type C	-1719 ± 25	8.7 ± 0.4 x 10^{-5}
Type D	-1544 ± 28	1.3 ± 0.1 x 10^{-5}

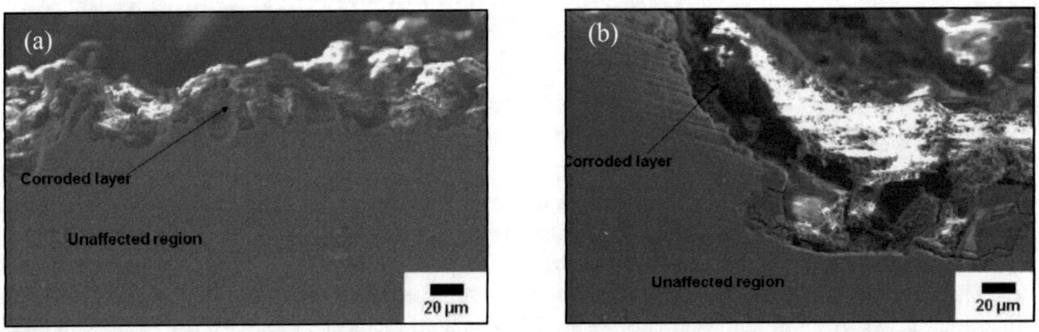

Figure 9. SEM micrographs of cross-section of LPP tested (a) type B material and (b) type D material [23].

5. Conclusions

The friction stir and laser welding of magnesium alloys have been reviewed. The welding parameters such as welding speed, tool material and shape, tilting angle could affect the welding quality and induces microstructure variation and surface character of welded magnesium alloy. It is consequently affect the pit attack morphology and corrosion behavior, as observed from the potentiodynamic polarization plot and cross-section images.

References

[1] L. Kwang, C. Chung, Y. Ching, and K. Chyn, “Optimization of Nd : YAG laser welding onto magnesium alloy via Taguchi analysis,” vol. 37, pp. 33–42, 2004.

[2] S. Mironov, T. Onuma, Y. S. Sato, and H. Kokawa, “Microstructure evolution during friction-stir welding of AZ31 magnesium alloy,” *Acta Mater.*, vol. 100, pp. 301–312, 2015.

[3] R. S. Mishra and Z. Y. Ma, “Friction stir welding and processing,” *Mater. Sci. Eng. R Reports*, vol. 50, no. 1–2, pp. 1–78, 2005.

[4] G. L. Song and Z. Xu, “The surface, microstructure and corrosion of magnesium alloy AZ31 sheet,” *Electrochim. Acta*, vol. 55, no. 13, pp. 4148–4161, 2010.

[5] X. Cao, M. Jahazi, J. P. Immarigeon, and W. Wallace, “A review of laser welding techniques for magnesium alloys,” *J. Mater. Process. Technol.*, vol. 171, no. 2, pp. 188–204, 2006.

[6] B. A. James, “Localized Corrosion of Friction Stir Spot Welds in Magnesium AZ31 Alloy Localized Corrosion of Friction Stir Spot Welds,” 2012.

[7] N. N. Aung and W. Zhou, “Effect of grain size and twins on corrosion behaviour of AZ31B magnesium alloy,” *Corros. Sci.*, vol. 52, no. 2, pp. 589–594, 2010.

[8] K. R. Pagar and P. A. D. Wable, “Review Paper : Friction Stir Welding (FSW),” vol. 6, no. 5, pp. 123–129, 2016.

[9] R. C. Zeng, J. Chen, W. Dietzel, R. Zettler, J. F. dos Santos, M. Lucia Nascimento, and K. U. Kainer, “Corrosion of friction stir welded magnesium alloy AM50,” *Corros. Sci.*, vol. 51, no. 8, pp. 1738–1746, 2009.

[10] R. Sharma and O. P. Singh, “Effect of FSW Process Parameters on Mechanical Properties of Polypropylene : An Experimental Study,” vol. 2, no. 12, pp. 7792–7798, 2013.

[11] R. Rai, A. De, H. K. D. H. Bhadeshia, and T. DebRoy, “Review: friction stir welding tools,” *Sci. Technol. Weld. Join.*, vol. 16, no. 4, pp. 325–342, 2011.

[12] C. Taltavull, B. Torres, A. J. Lopez, P. Rodrigo, E. Otero, A. Atrens, and J. Rams, “Corrosion behaviour of laser surface melted magnesium alloy AZ91D,” *Mater. Des.*, vol. 57, pp. 40–50, 2014.

[13] X. N. Wang, C. J. Chen, H. S. Wang, S. H. Zhang, M. Zhang, and X. Luo, “Microstructure formation and precipitation in laser welding of microalloyed C-Mn steel,” *J. Mater. Process. Technol.*, vol. 226, pp. 106–114, 2015.

[14] S. Ignat, P. Sallamand, D. Grevey, and M. Lambertin, “Magnesium alloys (WE43 and ZE41) characterisation for laser applications,” *Appl. Surf. Sci.*, vol. 233, no. 1–4, pp. 382–391, 2004.

[15] C. Lehner, G. Reinhart, and L. Schaller, “Welding of die-casted magnesium alloys for production,” *J. Laser Appl.*, vol. 11, no. 5, p. 206, 1999.

[16] E. Aghion and B. Bronfin, “Magnesium Alloys Development towards the 21st Century,” *Mater. Sci. Forum*, vol. 350–351, pp. 19–30, 2000.

[17] S. Bender, J. Goellner, A. Heyn, and E. Boese, “Corrosion and corrosion testing of magnesium alloys,” *Mater. Corros.*, vol. 58, no. 12, pp. 977–982, 2007.

[18] A. Dhanapal, S. Rajendra Boopathy, and V. Balasubramanian, “Corrosion behaviour of friction stir welded AZ61A magnesium alloy welds immersed in NaCl solutions,” *Trans. Nonferrous Met. Soc. China (English Ed.*, vol. 22, no. 4, pp. 793–802, 2012.

[19] J. Verma and R. V. Taiwade, "Effect of welding processes and conditions on the microstructure , mechanical properties and corrosion resistance of duplex stainless steel weldments — A review," vol. 25, pp. 134–152, 2017.
[20] W. Yuan, R. S. Mishra, B. Carlson, R. Verma, and R. K. Mishra, "Material flow and microstructural evolution during friction stir spot welding of AZ31 magnesium alloy," *Mater. Sci. Eng. A*, vol. 543, pp. 200–209, 2012.
[21] E. Ghali, W. Dietzel, and K. U. Kainer, "General and localized corrosion of magnesium alloys: A critical review," *J. Mater. Eng. Perform.*, vol. 22, no. 10, pp. 2875–2891, 2013.
[22] G. R. Argade, K. Kandasamy, S. K. Panigrahi, and R. S. Mishra, "Corrosion behavior of a friction stir processed rare-earth added magnesium alloy," *Corros. Sci.*, vol. 58, pp. 321–326, 2012.
[23] F. Gharavi, K. A. Matori, R. Yunus, N. K. Othman, and F. Fadaeifard, "Corrosion behavior of Al6061 alloy weldment produced by friction stir welding process," *J. Mater. Res. Technol.*, vol. 4, no. 3, pp. 314–322, 2015.

Effect of intake manifold water injection on a natural gas spark ignition engine: an experimental study

H Arruga[1], **F Scholl**[1], **M Kettner**[1*], **O I Amad**[2], **M Klaissle**[3], **B Giménez**[4]

[1] Karlsruhe University of Applied Sciences, Germany
[2] Universiti Malaysia Pahang
[3] SenerTec Kraft-Wärme-Energiesysteme GmbH, Germany
[4] University of Valladolid, Spain

[*]Corresponding author: maurice.kettner@hs-karlsruhe.de

Abstract. Design and development of gas CHP (combined heat and power) engines are strongly influenced by the progressively more severe European NO_x emissions normative. Water injection represents a promising approach to reduce these emissions while attaining high engine efficiency. In this work, the effect of intake manifold water injection on combustion parameters and performance of a single-cylinder naturally aspirated natural gas spark ignition engine is presented. First, the most appropriate injector was selected, using a spray test bed. Subsequently, engine experiments at constant indicated mean effective pressure (IMEP) and engine speed were conducted with water-fuel ratios of 0.1 to 0.3. IMEP was kept constant at about 6.3 bar by adjusting both air-fuel ratio and spark timing. A NO_x reduction of 0.2 g/kWh$_i$ (15 %) for a constant ISFC of about 204 g/kWh$_i$ was achieved. In the low NO_x regime, water injection allows for an improvement of the NOx-ISFC trade-off, while leading to poor fuel consumption at same NO_x in the high efficiency regime. Furthermore, water injection implies a reduction of intake mixture temperature, lengthened burning delay and combustion duration and a moderate increase of combustion instability.

1. Introduction

Combined heat and power systems have gained increasing relevance in the last years in the field of decentralised energy supply. By recovering waste heat while generating electric power, a high overall efficiency of the unit is achieved. Natural gas CHP units often operate with lean air-fuel mixtures ($\lambda > 1$) in order to meet the NO_x emissions limits imposed by legislation at high engine efficiency. These operating conditions further reduce the knocking tendency of the engine, allowing for high compression ratios. The reduction of HC and CO emissions is less problematic than that of NO_x, as oxidation catalysts still function under excess of oxygen and can be used for after treatment as long as exhaust gas temperature are above a certain level. Figure 1a shows the NO_x emissions of a gas engine as a function of λ (relative air-fuel ratio) qualitatively. As it can be seen, there is a limit determined by engine knock and a lean-burn limit due to misfires.

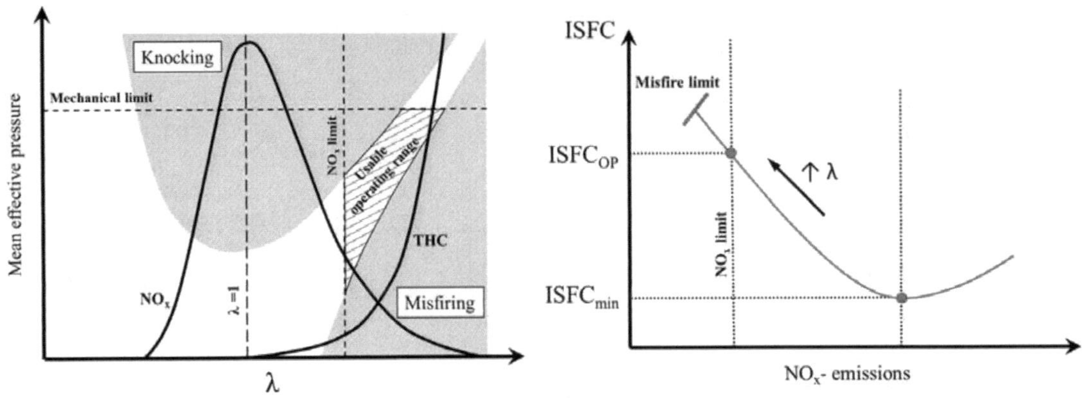

(a) (b)

Figure 1. Lean-burn combustion: Mean effective pressure operating map of a gas engine as a function of relative air-fuel ratio [9] (a), NO_x-ISFC trade-off for gas engines (b).

As λ is increased in order to reduce NO_x emissions, combustion temperature and flame speed decrease and combustion instability increases. The combustion then diverges from the optimal isochoric combustion at TDC, which is reflected in a worsened ISFC (and a decrease of indicated efficiency). This is shown in the NO_x-ISFC trade-off diagram in figure 1b. The engine works in an operating point with higher ISFC but with a NO_x emissions level that fulfils the normative.

In Germany, the emissions of gas engines with a fuel input power above 1 MW are regulated according to TA-Luft standards, which set the limits for NO_x emissions to 500 mg/Nm^3. Smaller gas engines, typical for micro-CHP units, however, are currently not subject to specific emission requirements. In 2018, a new European norm will be introduced. It restricts NO_x to 240 mg/kWh fuel input in terms of GCV [6]. This implies that new measures need to be found to meet this stringent emission requirement without losses in engine efficiency and power.

Among the internal measures to reduce NO_x emissions and knock in spark ignition engines as, for example, exhaust gas recirculation (EGR), water injection offers a high potential. The main advantage lies in the higher heat capacity of water compared to N_2, O_2 and CO_2 (remaining main components of exhaust gas), as it is illustrated in figure 2a.

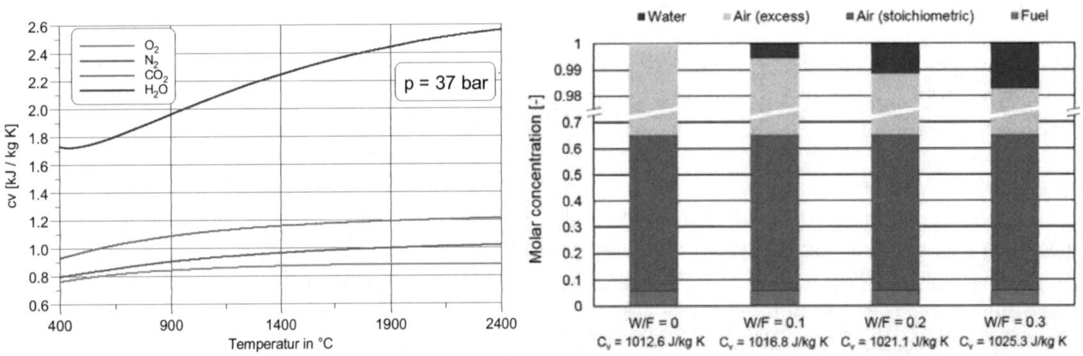

(a) (b)

Figure 2. Constant volume heat capacity of several gases for p = 37 bar as a function of temperature (calculated with EES v.10) (a) and variation of the heat capacity of the mixture with water addition (b)

Figure 2b shows the variation in constant volume heat capacity of the mixture for p = 37 bar and T = 890 °C when water is added to the mixture, supposing that the amount of fuel does not vary. This pressure and temperature levels match approximately with the conditions at the end of compression stroke for the engine configuration used for water injection in this study. As it can be seen, water

IOP Conf. Series: Materials Science and Engineering **257** (2017) 012029 doi:10.1088/1757-899X/257/1/012029

replaces a part of the excess air and, due to the higher capacity of water respect to air, the total capacity of the mixture increases. The injected water is normally quantified using the variable water-fuel ratio (W/F). It is defined as the ratio between the mass flows of water and fuel in a particular operating point:

$$W / F = \frac{m_{fuel}}{m_{water,inj}} \tag{1}$$

According to equation (2), the increase of the heat capacity of the mixture leads to a lower temperature rise during combustion:

$$Q = m \cdot c_V \cdot \Delta T \tag{2}$$

with Q = const., m = const., $c_V \uparrow \Rightarrow \Delta T \downarrow$

Q is the heat release in J

m is the mass of in-cylinder charge in kg

c_v is the specific heat capacity of the in-cylinder charge in J/(kg · K)

ΔT is the increase in temperature due to combustion

In internal combustion engines, NO is usually present in higher concentration in exhaust gas compared to NO_2. Besides, the "thermal" NO formation has a dominant role over other mechanisms of NO formation. This reaction takes place for temperature values over approximately 1800 K and is accelerated exponentially with the increasing temperature [15]. Therefore, when the combustion temperature levels are lower due to the increase of the heat capacity, less NO_x is formed.

Another positive effect of water injection is the active cooling of the aspirated charge due to the endothermal process of evaporating the water in the intake manifold. Because of the lowered intake temperature, mixture temperature at the end of compression is also reduced. This reduces the knock tendency of the engine in middle and high-load operating points, thus allowing for an increase of the compression ratio with an increment of the efficiency and a fuel reduction [5][11][20]. The mixture temperature reduction arising from the evaporation enthalpy of water can be deduced from the following equation:

$$m_w \cdot \Delta h_{ev} = m_{charge} \cdot c_P \cdot \Delta T_{charge} \tag{3}$$

where $\dot{m}_W$ is the mass of injected water in kg/s

Δh_{ev}; is the enthalpy of water evaporation in J/kg

$\dot{m}_{charge}$ is the mass of intake air in kg/s

c_P is the specific heat capacity in J/(kg · K)

ΔT_{charge} is the reduction of charge air temperature in K

The higher viscosity and the higher surface tension of water to petrol imply a worse atomization and breakup of the water spray. Furthermore, the lower vapour pressure of water compared to petrol causes worse evaporation of water droplets in the cylinder charge [11]. These factors need to be accounted for by choosing an optimal injector design and optimal injection parameters. The pre-warming of water before the injector could favour its evaporation by flash boiling of the droplets. It could be combined, for example, with exhaust gas heat recovery, using the energy of exhaust gas to heat up the water. The possibility of injecting water steam, as it was made by Kökkülünk et al. [14], is another strategy. However, the effect of charge cooling would be lost.

Water injection was introduced in 2015 by BMW in their M4 GTS model. Water is injected after the intercooler before the intake manifold of the turbocharged engine, using a water injector specifically designed for this purpose [4]. The water is taken from a tank that needs to be refilled periodically [2]. The second generation of water injection was investigated by BMW combining water injection downstream the throttle valve and high-pressure direct injection of a mixture of petrol and water by means of a specially designed injector [3]. In this case, condensed water from the air conditioning system of the vehicle was used. As a result, an increase of the compression ratio and a reduction of 23 % in fuel consumption in g/kWh were achieved. Additionally, with water injection, operating with stoichiometric amount of fuel is also allowed in operation points where normally fuel enrichment is necessary. Therefore, the three-way catalyst is operating even at high engine load.

Water injection has been investigated in several gas engine research works. Subramanian et al. [19] used intake manifold water injection in a naturally aspirated single cylinder hydrogen-fuelled engine. In this case, water injection reduced NO emissions by 32 % with no loss in brake thermal efficiency (BTE). Pucher et al. [17] investigated the influence of intake manifold water injection in a town gas driven internal combustion engine. The water was brought in after the throttle valve using the carburettor injection principle. This setup enabled a maximal decrement of 60 % in NO_x emissions without a loss in engine power. More recently, Tschalamoff et al. [21] investigated the influence of direct water injection for a turbocharged 13.6 litre natural gas cogeneration engine. The influence of water injection on NO_x reduction, engine efficiency, knocking and possible increase of mean effective pressure was reported. With a water-fuel ratio of approximately 20 %, it was possible to reduce engine-out NO_x by 50 %. At the same time, however, an increment of HC and CO emissions together with losses in engine efficiency due to the slower combustion and lower combustion temperature was observed. Furthermore, the authors determined the optimal injection timing as well as the shift of the knock limit for different W/F ratios. Water injection during intake stroke showed the higher reduction of NO_x emissions. Under the effect of water injection, the knock limit moved towards earlier ignition timings. Tschalamoff et al. [22] also conducted experiments on intake manifold water injection on a natural gas 4.8 litre research engine. They found that intake manifold water injection has a stronger effect on NO_x decrease compared with direct water injection. Analogously, a higher burning delay and a slower combustion was observed, together with higher HC and CO emissions and reduced engine efficiency.

In this paper, the influence of intake manifold water injection on a small naturally aspirated spark ignition engine fuelled with natural gas will be studied. The experiments were carried out in the framework of a research project where the aim is to recirculate condensate from the exhaust gas. The objective of this work was to determine the feasibility of the developed water injection system and study the effect of increasing water injection ratios on emissions, combustion stability and engine efficiency.

2. Experimental equipment

2.1. Spray test bed

A spray test bed was built up to analyse and simulate water injection into the intake manifold of the gas engine, as shown in figure 3a. To this end, the cross section of the test bed channel was adjusted geometrically to that of the intake manifold. The duct consists of a square geometry confined by four slabs made of acrylic glass.

An adjustable aluminium profile is placed in the middle of the duct. Its bottom is heated by two contact heaters (type Omega KH-103/10-P and KH-104/10-P) with a heating power of 1.55 W/cm^2. The total surface of the heaters is 45 cm^2 and, therefore, the maximum provided power by the heaters is around 70 W. Their function is to vary the temperature of the middle profile to simulate different conditions of intake manifold wall temperature. Temperature is controlled by an Omega CN7823 PID controller from the reading of a self-adhesive thermocouple. The PWM (pulse width modulation) output of the controller is connected to a solid-state relay, which powers the heaters intermittently. The PWM duty cycle is closed-loop controlled to maintain the rated temperature constant.

A pivoting structure was designed to test different injection angles and study their influence on wall wetting. Water is fed to the injector from a tank, pressurised with compressed air. The water is led to the injector using flexible polyethylene tubes and a special adapter for the injector. Tank pressure is controlled using a Festo LRP-1/4-10 high-precision pressure control valve.

An air conditioning unit provides the air mass flow through the test bed. It consists of a motor-driven compressor in series with a heat exchanger that allows adjusting both mass flow and air temperature at which injection testing is carried out. A Bosch HFM5-4.7 hot film air flow meter measures the intake air mass flow.

Spray images were captured employing a monochromatic CCD camera with a resolution of 9 megapixels and a maximum frame rate of 7 fps (figure 3b). The instant at which the image will be taken can be determined by either internal (software) or external triggers. The camera is fixed on a special mount, allowing for an adjustment of its horizontal and vertical position, as well as its distance to the spray test bed.

An infrared light source (Büchner HI-LIGHT-120) was used to light the water sprays evenly. This type of light source was selected because it provides higher contrast images. Moreover, the infrared light has a lower surface reflection as visible light on acryl glass and the water droplets can be better captured. An infrared filter was additionally coupled to the objective to reduce the distortion caused by other wavelengths from ambient light.

(a) (b)

Figure 3. Spray test bed used for the injector tests and simulation of the intake manifold flow conditions (a), CCD-Camera SVS-Vistek eco815MTLGEC and Myutron HF5018V-2 objective used for taking the spray images (b).

An ADwin Gold-II control unit and LabVIEW software were employed for data acquisition and for triggering both the injector and the camera. Software SVCapture of the company SVS-Vistek was used to adjust the camera settings.

Three different petrol injectors were tested with water: one port fuel injector (PFI) and two gasoline direct injectors (GDI). All of them will be used with low injection pressure. Their main characteristics are summarised in table 1.

Table 1. Main characteristics of the tested (gasoline) injectors.

	Siemens 98MF-9F593-EXP10	Denso E7T05072	Bosch 0261500020
Injector type	Multi-hole injector	Swirl injector	Swirl injector
Injection type	PFI	GDI	GDI
Rated injection pressure	< 4 bar	≈ 80 bar	≈ 120 bar
Number of holes	2	1	1
Spray type	Full cone	Hollow cone	Hollow cone
Spray angle (α)	≈ 20°	≈ 57°	≈ 65°
Inclination angle (β) spray / injector axis	0°	0°	10°

Although swirl injectors are designed for high pressure, they provide relative small droplet sizes even at low injection pressures [1]. The hollow cone of the spray is due to the tangentially rotating flow inducted by the tangential fluid inlets inside the injector's swirl chamber. The Denso swirl injector was previously used by Kettner et al. [12] for injecting water at the intake port of a diesel engine.

While injectors with lower spray angle cause less wall wetting, the spray covers less volume of the air mass flow and an even dispersion of the injected water becomes more challenging. Therefore, a compromise between these two characteristics must be found. In addition, the injector must provide an adequate quantity of water according to the engine working conditions. For this purpose, the flow characteristic of each injector, which gives the injected mass of water for a particular injection pressure as a function of the excitation time, must be obtained.

The injector is activated and controlled by the ADwin Gold-II control unit by means of a pulsed voltage signal. This signal is sent to a LM1949 injector driver controller. At high level, the driver lets current pass through the solenoid of the injector and the injector needle opens. At low level (0 V), the current through the injector will be cut off and the needle closes. In this way, the injection instant and the injection duration over the engine cycle can be easily controlled.

2.2. Engine test bed

The used CHP unit is a SenerTec Dachs G 5.5. It is driven by a naturally aspirated 4-stroke single-cylinder engine running on natural gas. The main characteristics and configuration of the engine used at the tests are listed in Table 2 and the CHP unit is shown in figure 4. For the experiments, the gas mass flow was regulated using a Maxitrol EXA Star modulating valve coupled to the gas multibloc. Trials were run with a modified piston to increase the compression ratio from 13.2:1 to 15.25:1.

Table 2. Engine specifications.

Engine type	SenerTec Dachs G 5.5 / swirl port
Nominal speed	2450 rpm
Compression ratio	15.25:1
Stroke	91 mm
Bore	90 mm
Displacement volume	580 ccm
Number of valves	2
Ignition system	prechamber spark plug

Figure 4. Dachs G 5.5 CHP unit [18].

Since water was to be injected before the intake valve of the engine, a special adapter was designed and coupled to the intake manifold as shown in figure 5a and figure 5b. Further, a thermocouple was mounted into the intake manifold and bent in order to measure gas temperature right before the intake valve in order to examine the effect of water injection on mixture cooling. The inclination angle of the injector was 45° with respect to the intake flow axis.

The water supply to the injector was made using flexible PTFE pipes, which resist the temperature of the engine housing and coupled to a connection situated at the upper part of the adapter.

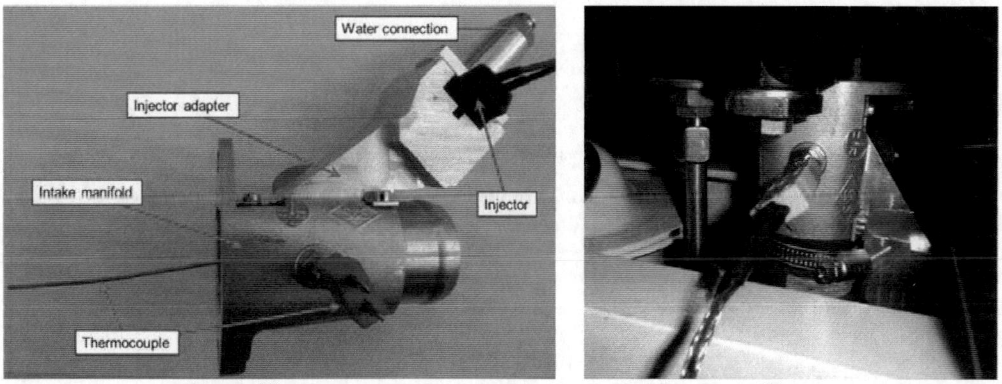

(a) (b)

Figure 5. Injector adapter coupled to the intake manifold and thermocouple to measure the temperature near the intake valve (a), installation of the adapter at the gas engine (b).

A proprietary LabVIEW graphic user interface was employed to communicate with an ADwin Pro-II system responsible for controlling all engine actuators and slow data acquisition. Pressure at the different measuring points of engine were obtained using SITRANS P200 pressure sensors; the air mass flow was measured with a HFM5-4.7 hot film air flow meter calibrated for low-range mass flows; fuel mass flow was determined using a Bronkhorst F-103-low-Δp gas flow meter; λ was measured using a Bosch LSU 4.9 wideband lambda sensor. K-Type thermocouples were used to measure temperatures, connected to PICO TC-08 data loggers. A DEWE-800-CA combustion analyser was used for recording the in-cylinder pressure against crank angle and calculating combustion parameters.

Constant ambient conditions in terms of pressure and temperature at the air intake of the CHP unit were supplied using the same air conditioning unit as for the spray test bed.

Exhaust gas analysis was made using an ABB 6-component exhaust gas analyser with heated sample lines. A Multi FID 14 unit with a flame ionisation detector was used to measure wet total hydrocarbons (THC). A Limas11 HW unit with a photometer after the DUV-RAS/NDUV (Differential Ultra Violet Resonance Absorption Spectroscopy / Non Dispersive Ultra Violet) principle was used to measure wet NO and NO_2 concentration. Dry CO_2 and CO were measured with an Uras14 unit equipped with an IR photometer. Dry O_2 concentration was measured using a Magnos 206 based on the paramagnetic properties of O_2.

3. Results and discussion

3.1. Injectors flow characteristics

The flow characteristics were obtained for the three injectors presented. They must be obtained to determine the relation between injector activation time and the water mass provided by the injector. Injector excitation time was set as described in section 2.1 and water was injected into a vessel, positioned on a Kern PLJ 6200-2AM precision balance. An average value over 3000 injection events was taken for each injector activation time. The maximum tank pressure that can be provided by the compressed air system and the given pressure reducer is 9 bar (absolute). Since the pressure regulation of the pressure reducer was slightly unstable at maximum pressure, the tested pressure level for the Denso and Bosch injector was reduced to 8.5 bar. For the Siemens injector, a pressure level of 4 bar was tested. Figure 6 shows the mass of water injected per injection event and the calculated equivalent water-fuel ratio.

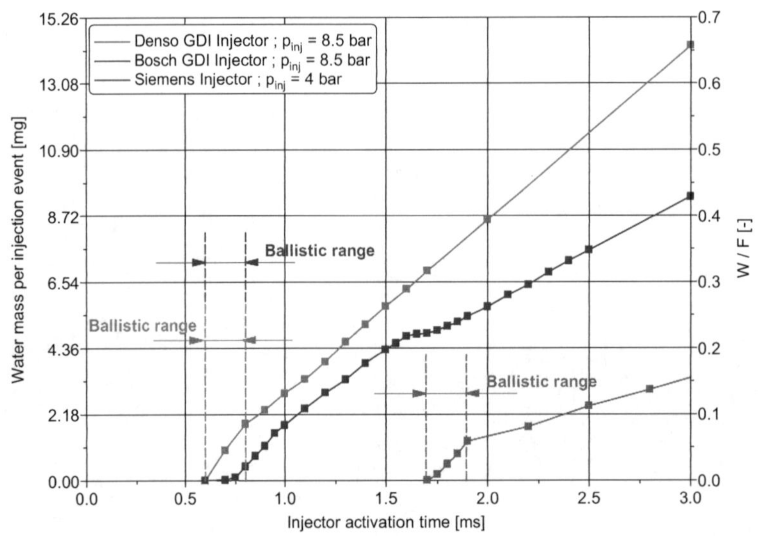

Figure 6. Flow characteristics of the three tested injectors.

As shown in the figure 6, the injector activation time, represented in x-axis, is not the same as the injector opening time. The injector needs a particular time to start opening, which is approximately 0.6 ms for the Denso and Bosch injectors, and 1.2 ms for the Siemens injector. Moreover, the injector needs a minimum opening duration to open the needle completely. This is known as the "ballistic" range of the injector and must be avoided for water injection operation. During this time, the water mass flow shows high cyclic fluctuations due to different needle lifts from cycle to cycle. Above the ballistic range, the injector shows a stable injection behaviour.

When comparing all injectors, it can be confirmed that the Siemens injector has a considerably lower water mass flow rate. The ballistic range for the three injectors is similar. The operating point of W/F = 0.05 is inside the ballistic range for all three injectors. For this reason, the lowest W/F ratio tested was set to 0.1. The Bosch injector has a lower mass flow rate as the Denso injector for the same ballistic range, allowing the injector to operate farther away from the ballistic range at W/F = 0.1, which can be taken as an advantage.

3.2. Tests at spray test bed

To study the spray propagation, images were taken at the spray test bed. For the GDI injectors, several injection pressure levels were tested. It was found that the angle of the cone was wider and the atomisation better as the pressures increased. Therefore, it was decided to operate the GDI injectors at the maximal possible injection pressure (8.5 bar). Figure 7 shows the images of the three sprays injected in an air mass flow of 40.8 kg/h, what corresponds approximately to the reading of the mean air mass flow of the engine at the series operating point. A binarization process of the images using Matlab was applied to further increase the contrast of the spray outline and the droplets.

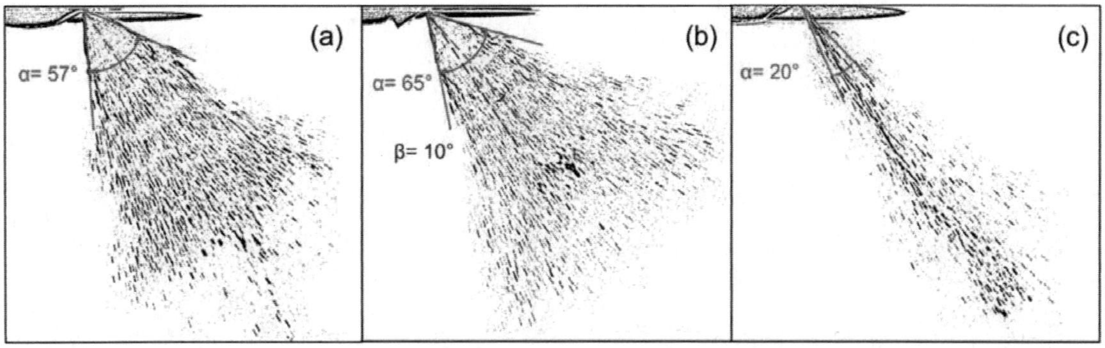

Figure 7. Water spray for Denso injector (a) and Bosch injector (b) after 3 ms injector excitation time; for Siemens injector (c) after 4.5 ms injector excitation time.

As can be seen from the images, the Siemens PFI injector has a smaller spray angle compared to the GDI injectors tested. This can cause a worsened dispersion of water, due to the lower volume of air mass flow covered by the spray. From the two swirl injectors, the Bosch injector has a larger cone angle and an inclination of the spray of $\beta = 10°$ with respect to the injector axis, which allows orienting the spray more parallel to the air mass flow. These two factors improve the entrainment of droplets, reducing wall wetting compared to the Denso injector. Moreover, the Bosch injector achieves a slightly better atomisation.

In due consideration of the spray test findings and having to operate it farther away from the ballistic range, the Bosch injector was chosen for conducting the experimental trials presented in the following section.

3.3. Experimental Results

Engine trials were carried out at a constant IMEP of 6.35 bar. Intake air was conditioned at a constant intake pressure of 995 mbar and a constant temperature of 35 °C. The tested water-fuel ratios during the experiments were 0, 0.1, 0.2 and 0.3.

For each water-fuel ratio series, combustion phasing (CA50, defined as 50 % fuel mass fraction burnt) was varied by adjusting spark timing and λ. Each operating point was measured three times and the average value calculated. Combustion parameters were averaged over 200 engine cycles.

Figure 8a shows NO_x emissions against CA50) for each water-fuel ratio. NO_x emissions are expressed as specific emission output referenced g/kWh to indicated engine power. As it can be observed, NO_x emissions decrease for a constant CA50 when the injected water mass flow increases, due to the increase of the heat capacity of the mixture and the reduction of the combustion temperature. The absolute decrease of NO_x emissions is greater for advanced CA50. The relative decrease for W/F = 0.3 compared to W/F = 0 varies between 20 % and 25 %, depending on combustion phasing.

Figure 8b shows the NO_x-ISFC trade-off for different water-fuel ratios. In the low NO_x regime of the diagram and at constant NO_x, water injection allows decreasing ISFC. For a constant ISFC of 204.1 g/kWh$_i$, which approximately matches with series combustion phasing (CA50 = 19°CA ATDC), the reduction in NO_x for W/F = 0.3 is approximately 0.2 g/kWh$_i$ (15 %).

On the contrary, in the high efficiency regime region of the diagram, increasing fractions of water appear to be detrimental to ISFC.

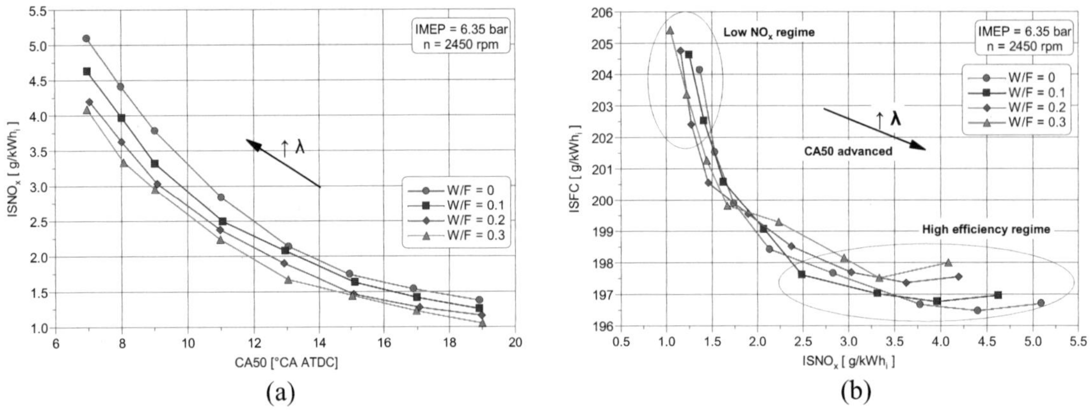

Figure 8. NO_x emissions against combustion phasing CA50 and water-fuel ratio (a), $ISNO_x$-ISFC trade-off for different water-fuel ratios (b).

Figure 9 shows NO_x emissions calculated according the 2018 new European normative. It can be seen that even a water-fuel ratio of 0.3 is not yet sufficient to meet the new limit.

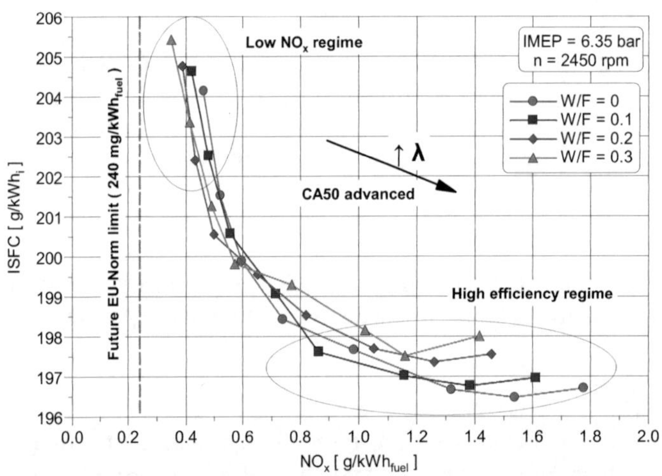

Figure 9. NO_x-ISFC trade-off calculated according to new 2018 EU normative.

Figure 10 shows the dependence of mixture temperature at the intake valve on the water-fuel ratio. Mixture temperature is fairly constant for each water-fuel ratio series and shows no dependence on combustion parameters. For W/F = 0.3 a cooling of the mixture at the intake valve of approximately 24 K is achieved.

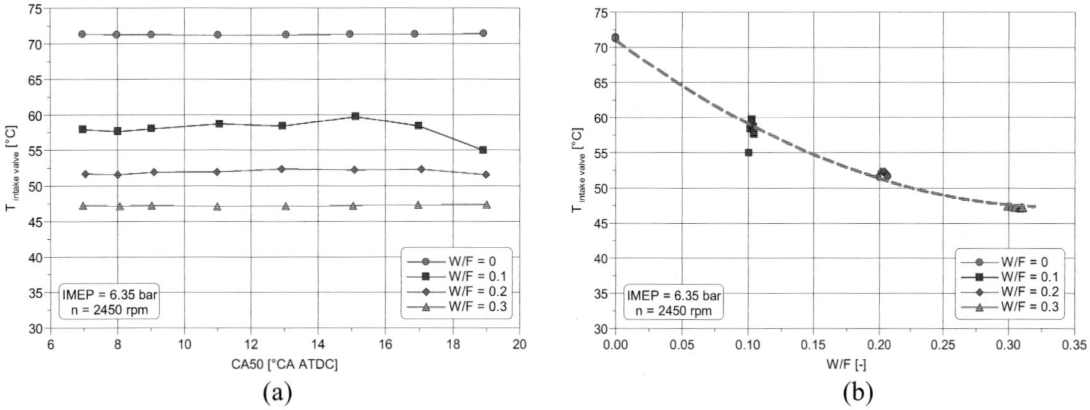

Figure 10. Mixture temperature at the intake valve depending as a function of CA50 (a) and the water-fuel ratio (b).

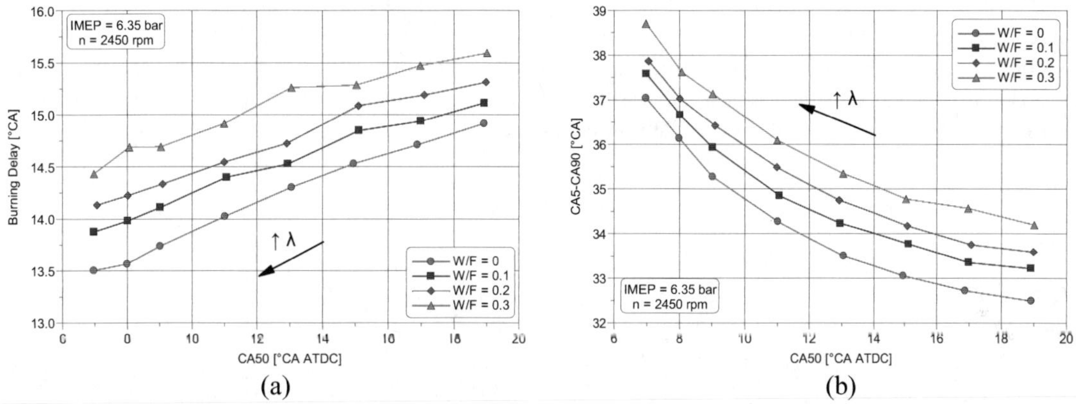

Figure 11. Burning delay (a) and combustion duration (b) as a function of combustion phasing CA50.

According to Halter et al. [7], the dilution of an air-fuel mixture with nitrogen and carbon dioxide causes a decrease of the laminar burning velocity, which leads to a slower combustion. In the same way, the addition of EGR (with the main constituents H_2O, CO_2, N_2 and O_2) to the intake mixture implies a decrement of the burning velocity, as determined by Neher et al. [16] for different dilution ratios. For water injection, a higher dilution of the charge also increases the burning delay periods and the combustion duration, as it was exposed by several authors [8][13][23]. Figure 11a and figure 11b show that both burning delay and combustion duration increase with water injection. Burning delay is defined as the time in °CA between spark timing and the start of combustion. CA5 is taken as start of combustion and it is defined as 5 % mass fraction burnt. Combustion duration is defined as the time in CA between start of combustion (CA5) and end of combustion (CA90, defined as 90 % mass fraction burnt). As expected, burning delay and the duration of the combustion are affected by water injection and the tendency agrees well with the data found in the previously mentioned research work.

HC and CO emissions are presented in figure 12. As it can be observed, water injection enhances the formation of HC. This is caused by the lower combustion velocity, which leads to incomplete combustion phenomena. The same effect can be found for a constant W/F. When λ increases, the fraction of air in the mixture increases; it is more diluted and leads to incomplete combustion and a rise of HC emissions [10]. A clear tendency of CO emissions with injected water mass flow (figure 12b) cannot be found. For a constant W/F, the CO emissions increase when λ increases, because the reactions of CO formation are favoured by the presence of O_2 [15].

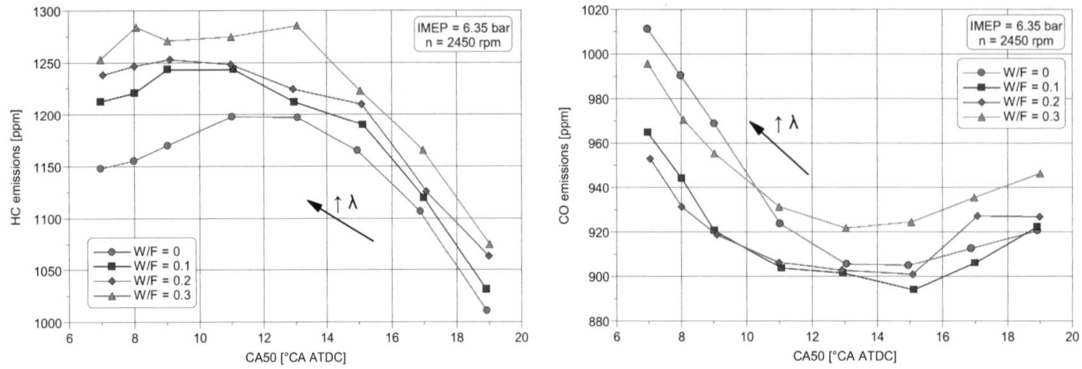

(a) (b)

Figure 12. HC emissions (a) and CO emissions (b) depending on the water-fuel ratio.

Figure 13 depicts the coefficient of variance of IMEP. It is defined as the standard deviation of the IMEP divided by the mean value of IMEP, calculated from the ensemble of 200 cycles recorded. Engine stability and combustion variability normally deteriorates when the mixture becomes more dilute, as it occurs when leaning or by addition of EGR [10]. In a similar way, charge dilution by water injection, increases combustion instability, as it can be deduced from figure 13. However, the increase in COV is marginal and the obtained instability levels are acceptable. Much larger fractions of water could be injected until the stability of the combustion would be significantly affected.

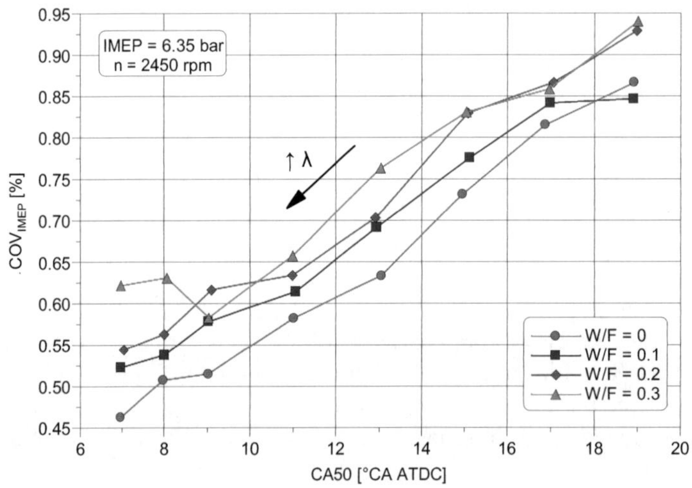

Figure 13. Calculated COV_{IMEP} of the combustion depending on the CA50.

4. Conclusions

In the present work, a system of intake manifold water injection for a small natural gas spark ignited engine was developed and tested. A Bosch swirl injector was selected in a spray test bed due to the better atomization, the better orientation of the spray and the most appropriate water mass flow of the injector for the engine operating conditions.

Subsequently, the selected injector was tested on the engine test bed. During the experiments, IMEP was kept constant at 6.35 bar. The following findings were obtained:

- In the low NO_x regime (late combustion phasing) at a constant ISFC of about 204 g/kWh$_i$, a reduction of $ISNO_x$ of approximately 0.2 g/kWh$_i$ (15 %) at W/F = 0.3 was achieved.
- For high efficiency regime (advanced combustion phasing), increasing the injected water mass causes a higher ISFC for constant NO_x.

- The evaporation enthalpy of the injected water allows decreasing intake mixture temperature by 24 K at W/F = 0.3.
- Increasing the water-fuel ratio increases burning delay, combustion duration and combustion instability.
- Injecting water increases HC emissions, while a tendency for CO emissions could not be found.

Intake manifold water injection has shown potential to reduce the NO_x output of the engine without losses in engine efficiency. Future research will focus on increasing the water-fuel ratio for further reductions in NO_x and analysing the potential of water injection for different compression ratios.

Acknowledgments

The authors would like to thank the German Federal Ministry of Education and Research, which supported this project through the grant programme “IngenieurNachwuchs”, as well as all persons that were involved in the project.

References

[1] Ashgriz N 2011 Handbook of Atomization and Sprays. Theory and Applications (Springer)
[2] Backhaus R Gemischbildung und Verbrennung 2016 *Motortechnische Zeitschrift* **77(4)** 14-15
[3] Böhm M, Mährle W, Bartelt C and Rubbert S Funktionale Integration einer Wassereinspritzung in den Ottomotor 2016 *Motortechnische Zeitschrift* **77(1)** 38-43
[4] Bosch GmbH H_2O statt CO_2. Wassereinspritzung für Benzinmotoren URL: http://www.bosch-waterboost.de
[5] Bozza F, De Bellis V and Teodosio L Potentials of cooled EGR and water injection for knock resistance and fuel consumption improvements of gasoline engines 2016 *Applied Energy* **169** 112-125
[6] European Commision Ecodesign requirements for space heaters and combination heaters, Regulation Nr. 813/2013 for execution of Norm 2009/125/EG
[7] Halter F, Foucher F, Landry L and Mounaim-Rousselle Effect of Dilution by Nitrogen and/or Carbon Dioxide on Methane and Iso-Octane Air Flames 2009 *Combustion Science and Technology* **181** 813-827
[8] Harrington J A Water Addition to Gasoline-Effect on Combustion, Emissions, Performance, and Knock 1982 *SAE Technical Paper* 820314
[9] Herdin G Skriptum Grundlagen Gasmotoren 2012
[10] Heywood J B 1988 *Internal combustion engine fundamentals* (McGraw-Hill)
[11] Hoppe F, Thewes M, Baumgarten H and Dohmen J Water injection for gasoline engines: Potentials, challenges, and solutions 2016 *International Journal of Engine Research* **17** 86-96
[12] Kettner M, Dechent S, Hofmann M, Huber E, Arruga H, Mamat R and Najafi G Investigating the influence of water injection on the emissions of a diesel engine 2016 *Journal of Mechanical Engineering and Sciences* **10(1)** 1863-1881
[13] Kim J, Park H, Bae C, Choi M and Kwak Y Effects of water direct injection on the torque enhancement and fuel consumption reduction of a gasoline engine under high-load conditions 2016 *International Journal of Engine Research* **17(7)** 795-808
[14] Kökkülünk G, Gonca G, Ayhan V, Cesur I and Parlak A Theoretical and experimental investigation of diesel engine with steam injection system on performance and emission parameters 2013 *Applied Thermal Engineering* **54** 161-170
[15] Merker G P, Schwarz C and Teichmann R 2012 *Combustion Engines Development. Mixture Formation, Combustion, Emissions and Simulation* (Springer)
[16] Neher D, Scholl F, Kettner M, Schwarz D, Klaissle M and Giménez Olavarría B The Effect of Cooled Exhaust Gas Recirculation for a Naturally Aspirated Stationary Gas Engine 2016 *SAE International Journal of Engines* **9(4)** 2477-2492

[17] Pucher H and Netzel H NO_x-Senkung bei Gasmotoren durch Saugrohr-Wassereinspritzung 1984 *Motortechnische Zeitschrift* **45(1)** 33-35

[18] SenerTec Kraft-Wärme-Energiesysteme GmbH

[19] Subramanian V, Mallikarjuna J M and Ramesh A Effect of water injection and spark timing on the nitric oxide emission and combustion parameters of a hydrogen fuelled spark ignition engine 2007 *International Journal of Hydrogen Energy* **32** 1159-1173

[20] Thewes M, Hoppe F, Baumgarten H and Seibel J Wassereinspritzung für ottomotorische Brennverfahren 2015 *Motortechnische Zeitschrift* **76(2)** 26-31

[21] Tschalamoff T, Laaß U and Janicke D Direkte Wassereinspritzung im mittelschnelllaufenden Gasmotor 2007 *Motortechnische Zeitschrift* **68(11)** 954-962

[22] Tschalamoff T and Sahl C Saugrohr-Wassereinspritzung an einem Ottogasmotor 2009 *Proceedings of 6th Dessau Gas Engine Conference* 186-196

[23] Wei M, Nguyen T S, Turkson R F, Guo G and Liu J The Effect of Water Injection on the Control of In-Cylinder Pressure and Enhanced Power Output in a Four-Stroke Spark-Ignition Engine 2016 *Sustainability* **8(10)** 993

IOP Conf. Series: Materials Science and Engineering **257** (2017) 012030 doi:10.1088/1757-899X/257/1/012030

Improvement of productivity in low volume production industry layout by using witness simulation software

V Jaffrey*, N M Z N Mohamed and A N M Rose

Faculty of Mechanical Engineering, Universiti Malaysia Pahang, 26600 Pekan, Pahang, Malaysia.

*Corresponding author: veyolla@gmail.com

Abstract. In almost all manufacturing industry, increased productivity and better efficiency of the production line are the most important goals. Most factories especially small scale factory has less awareness of manufacturing system optimization and lack of knowledge about it and uses the traditional way of management. Problems that are commonly identified in the factory are a high idle time of labour and also small production. This study is done in a Small and Medium Enterprises (SME) low volume production company. Data collection and problems affecting productivity and efficiency are identified. In this study, Witness simulation software is being used to simulate the layout and the output is focusing on the improvement of layout in terms of productivity and efficiency. In this study, the layout is rearranged by reducing the travel time from a workstation to another workstation. Then, the improved layout is modelled and the machine and labour statistic of both, original and improved layout is taken. Productivity and efficiency are calculated for both layout and then being compared.

1. Introduction

In a competitive market nowadays, the survival of any industry greatly depends on response time, production costs and flexibility in manufacturing. Food processing industry is a subset of manufacturing with unique challenges [1, 2]. A manufacturing company should have the efficient production system to accomplish the operation. The system should consist of people, equipment, and procedure designed for the combination of materials and processes that steer the company's operations [3]. The number of workstations makes the most of a strong impact on the performance of the manufacturing system in terms of production capacity, productivity, cost per unit, and maintainability. The amount of work that a system could accomplish will increase with the rise in the number of the workstation. It will cause the complexity of the system and thus cause the system hard to manage and sustain. The layout of the plants is an essential factor in determining the most appropriate material handling system. The layout is improved by minimizing the inventory of products and consequently enhances the total production [4]. The layout of a facility determines the structure and material flows of a production system, and thereby influences its operational performance over the long term [5].

Productivity is as the application of the numerous resources or inputs of an organization, industry or country, to achieve certain planned and desired results or outputs. In other words, productivity more broadly means production rate per unit of input, especially per unit of labour, for goods or services [6]. Productivity improvement as a process to achieves higher levels of output while consuming same or lesser amount of input resources, and can be defines as ratio of product being produced to what is required to produce it [7]. Well-designed facilities will give effect on material handling in term of efficiency and productivity, decreased production cycle time and resource transportation times [8, 9].

IOP Conf. Series: Materials Science and Engineering **257** (2017) 012030 doi:10.1088/1757-899X/257/1/012030

Witness simulation is an alternative to show how improvement will gives advantages as it is made at the lowest cost and less disruption to the existing production operations [10, 11].

Witness simulation environment is the product of the British Lanner Group [12] and it is one of the most successful world-class environments in manufacturing simulation. Good plant layout will have less distance of material flow which will raise production [13] [14]. Through simulation, significant advantages over real life experimentation are provided in term of cost, time and repeatability when variability, disruption, and complexity exist. It is necessary to conduct facility and layout planning before any factory set up to ensure sustainable process and reduce losses [15-17].

2. Methodology

This study is conducted in a low volume industry in Malaysia, and for this research, data is collected by observing the layout, identify types and number of machines in the production line .The data collection also includes the number of employees, the layout of the production; the total time to produce the products. Time study is defined as a work measurement technique for recording the times and rates of working for the elements of a specified job carried out under specified conditions, and for analysing the data so as to obtain the time necessary for carrying out the job at a defined level of performance[18].

Figure 1 shows the specific study area of the company to be studied and improved. The arrow represent on the process flow of the production line, as the focus of this study started from the filling and stamping line, and then the product is placed in an area labelled as buffer before sent to the expiry stamping line. The expiry date stamping line is the last workstation before it is placed in another buffer before packaging process.

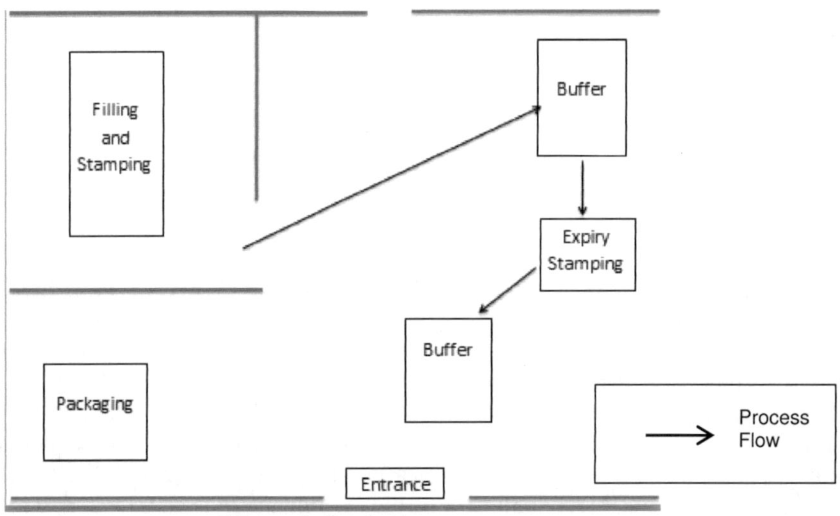

Figure 1. Study Area.

By using the data obtained from observation and through interview session with workers, the layout is designed by using Witness software and simulation is done to obtain the estimation of performance [19]. The efficiency and productivity are obtained by using the simulation result, which are from the machine statistic, labour statistic, and the output.

By using trial and error method, the efficiency and productivity of machine and labour of the existing layout are compared with the improved layout. The results are being analysed by referring to the previous study related to layout improvement. The new layout is being proposed when the result shows improvement in efficiency and productivity [17, 20]. The first step in building model is by observing the real layout, identifying how the real system works and the interaction among various components. It is critical to understand the manufacturing processes before attempting to mimic them by using simulation. The salient processes were studied separately in depth and understanding the process operational characteristic, timings, and products flow before developing models. The crucial part is on the problems while building the model.

Figure 2 below shows the existing layout that been modelled by using the simulation software, Witness. The study area is modelled, which consist of filling and stamping machine, buffer, and expiry date printing machine. Workers are assigned to the workstation, where three labours are assigned on filling and stamping line, while, at the expiry date printing line, two labours are assigned.

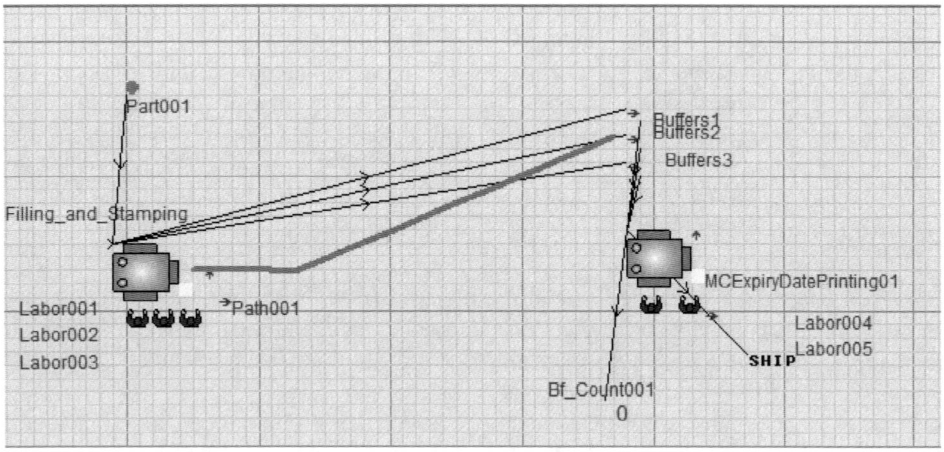

Figure 2. Layout modeled using Witness Software.

3. Results and Discussion

This study area selection is based on the problems identified through the observation. The person in charge and workers are also being interviewed to collect regarding the line to determine error. The problem detected in the layout is the high idle time of workers at the expiry date printing line, which causes the production to be lower than the target. Based on the simulation using Witness software, the idle and busy percentage of machine of the layout can be identified. To obtain an improved layout, try and error method had been applied. By reducing the distance, the time used to transfer the product can be reduced thus reducing the cycle time [21-23].

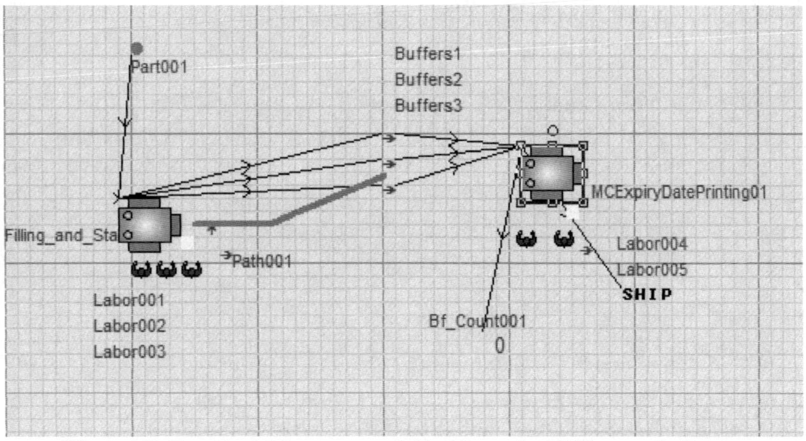

Figure 3. Improved layout modelled by using Witness Software.

Figure 3 shows the improved layout where the buffers are rearranged in order to decrease the distance from the filling and stamping workstation. Table 1 below shows the percentage of idle and busy of machines, filling and stamping machine and expiry date printing machine which are obtained from the original and improved layout. Based on the calculated percentage of improvement, it shows slight changes in the idle rate, as it reduced 2.84 on the filling and stamping machine that is from

14.4% to 12.00%. Meanwhile, for the idle percentage of the expiry date printing machine, the percentage of idle reduces from 47.61% to 46.53% which show a reduction of 1.08%. When the idle percentage decrease, busy time increase. For the filling and stamping machine, the busy percentage increases from 85.16% to 88.00%, while for expiry date printing machine, the busy percentage increases 1.98%, which is from 52.39% to 53.47%.

Table 1. Comparisons on machine utilization of original and improved layout.

Parameters		Original Layout		Improved Layout		Improvement	
		Filling & Stamping Machine	Expiry date printing machine	Filling & stamping machine	Expiry date printing machine	Filling & stamping machine	Expiry date printing machine
Machine utilization	Idle (%)	14.84	47.61	12.00	46.53	-2.84	-1.08
	Busy (%)	85.16	52.39	88.00	53.47	2.84	1.98

The busy percentage shows changes when the distance from the filling and stamping machine to the buffer are reduced. This will give effects on the transfer time and thus reducing cycle time. Since the scope of this study is limited to no cost addition, it still shows that the layout still have the chance to improve.

Table 2 below shows the improvement in labour efficiency and has been mentioned earlier, the efficiency of labour is calculated based on the busy time of labour. For labours that are assigned in the filling and stamping machine, the efficiency increases 2.80%, which is from 85.16% to 88.00%.

Table 2. Improvement of the labour efficiency.

	(Busy Time) Efficiency (%)		Percentage of Efficiency Change (%)
	Original Layout	Improved Layout	
Labour 1	85.16	88.00	2.84
Labour 2	85.16	88.00	2.84
Labour 3	85.16	88.00	2.84
Labour 4	52.39	53.47	1.08
Labour 5	52.39	53.47	1.08
Total Efficiency Change			1.07

For labour 4 and 5 who are placed in the expiry date printing line, their efficiency increases from 52.39% to 53.47%, which gives changes of 1.08% improvement. The overall change in the improved layout is 10.68% and the efficiency changes of the original and improved layout is presented in Figure 4 below. The increased in the percentage of efficiency affect the total production that lead to the increase in productivity. The increased efficiency is most probably caused by the reduced cycle time, which consequently reduced the idle percentage of labour.

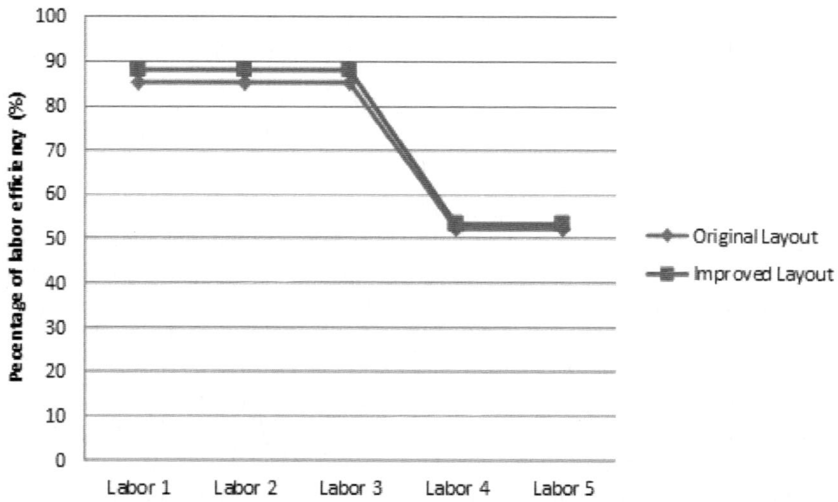

Figure 4. Efficiency changes of original and improved layout.

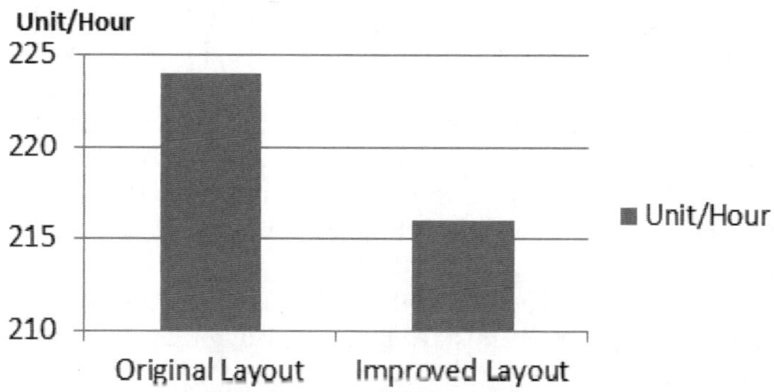

Figure 5. Productivity Comparisons of the Original Layout and the Improved Layout.

The original layout produces 216 units per hour, and after being improved, through simulation it can be seen that the productivity increases to 224 units per hour as shown in Figure 5. The study is limited to no cost addition, and it is recommended to the company to invest on addition of another expiry date printing machine so that efficiency and productivity can be increased. It is known that the printing machine will only able to print for one type of jelly drink flavour at one time as changing it frequently will take some times to change the code. Each box requires all flavour and thus, by having only one expiry date printing machine will cause the high idle time on the workstation. This will reduce the work in progress in the station. This recommendation is suggested for long-term consideration as adding machine requires a high cost.

Secondly, through evaluation using WITNESS software, there are high idle times in the expiry date printing machine, and it is recommended for the employees to be rotated to other stations to increase their efficiency [24, 25]. The company also should look into training their operators so they can perform work on other stations instead of hiring new labours.

4. Conclusions

A new layout has been designated by rearranging the layout, which is by moving the buffer closer to the filling and stamping machine. By simulating the layout by using the software, the efficiency of the machine is increased, productivity is also increased. The result obtained shows improvement, which is a reduction of 2.84% and 1.08% on idle time of the filling and stamping machine and expiry date printing machine respectively. Labour efficiency increase and this is done by reducing the distance between machines. Total of 10.68% changes in labour efficiency is calculated. Last but not least, in term of productivity, the original layout produces 216 units per hour while the newly improved layout could produce 224 units per hour. The knowledge of manufacturing system can be used to improve the system of a factory to get a better performance and consequently increase productivity and reduces waste to get more profit.

References

[1] Wanniarachchi, W.N.C., R.A.R.C. Gopura, and H.K.G. Punchihewa, Development of a Layout Model Suitable for the Food Processing Industry. *Journal of Industrial Engineering*, 2016: p. 8.

[2] ElMaraghy, H., et al., A model for assessing the layout structural complexity of manufacturing systems. *Journal of Manufacturing Systems*, 2014. **33(1)**: p. 51-64.

[3] K., N.M.Z.N.M.a.M., Decomposition Of Manufacturing Processes: A Review. *International Journal of Automotive and Mechanical Engineering (IJAME)*, 2012: p. 545-560.

[4]Seyed Ali Mirzapourrezaei, K.Y.W., Ahmad Dargi, Morteza Lalmazloumian, Simulation of a Manufacturing Assembly Line Based on WITNESS, in *Computational Intelligence, Communication Systems and Networks, International Conference*. 2011. p. 132-137.

[5] Bock, S. and K. Hoberg, Detailed layout planning for irregularly-shaped machines with transportation path design. *European Journal of Operational Research*, 2007. **177(2)**: p. 693-718.

[6] Mitchell, P. Productivity in Mining. 2014; Available from: http://www.ey.com.

[7] Helvoigt, T.L., An Analysis of Technical Efficiency and Productivity Growth in the Pacific Northwest Sawmill Industry. 2006, Oregon State University.

[8] SUBY, R.J., Facility Layout and Location—An Analytical Approach.By R. L. Francis and J. A. White. . **Vol. 13**. 1975.

[9] Gnanavel, S.S., V. Balasubramanian, and T.T. Narendran, Suzhal – An Alternative Layout to Improve Productivity and Worker Well-being in Labor Demanded Lean Environment. *Procedia Manufacturing*, 2015. **3**: p. 574-580.

[10] O'Kane, J.F., J.R. Spenceley, and R. Taylor, Simulation as an essential tool for advanced manufacturing technology problems. *Journal of Materials Processing Technology*, 2000. **107(1–3)**: p. 412-424.

[11] Pam Laney Markt, M.H.M. Witness Simulation Software A Flexible Suite Of Simulation Tools. in Proceedings of the 1997 Winter Simulation Conference. 1997.

[12] Waller, A. WITNESS SIMULATION SOFTWARE in Proceedings of the 2012 *Winter Simulation Conference* 2012.

[13] Azadivar, F. and J. Wang, Facility layout optimization using simulation and genetic algorithms. *International Journal of Production Research,* 2000. **38(17)**: p. 4369-4383.

[14] Anucha Watanapa, P.K., Patcharee Duangpitakwong , and Wisitsree Wiyaratn Analysis Plant Layout Design for Effective Production in *Proceedings of the International MultiConference of Engineers and Computer Scientists.* 2011. Hong Kong: IMECS.

[15] Ojaghi, Y., et al., Production Layout Optimization for Small and Medium Scale Food Industry. *Procedia CIRP*, 2015. **26**: p. 247-251.

[16] BRONISLAV CHRAMCOV, P.V., Use of computer simulation with the aim of achieving more efficient production in manufacturing systems.

[17] K.Hemanand, D.A., S.Chidambara Raja,G.Sundararaja, Improving Productivity Of Manufacturing Division Using Lean Concepts And Development Of Material Gravity Feeder –A Case Study. International Journal of Lean Thinking, 2012. **3(2)**.

[18] Kanawaty, G., Introduction to Work Study. 1992.

[19] Chramcov, B., The Optimization of Production System Using Simulation Optimization Tools in Witness. *International Journal Of Mathematics And Computers In Simulation*, 2013. **7(2)**.

[20] Helber, S., et al., A hierarchical facility layout planning approach for large and complex hospitals. *Flexible Services and Manufacturing Journal*, 2016. **28(1)**: p. 5-29.

[21] Amit, N., et al. Using simulation to solve facility layout for food industry at XYZ Company. in 2012 *IEEE Symposium on Humanities, Science and Engineering Research, SHUSER* 2012. 2012. Kuala Lumpur.

[22] Hiten Patel, S.C.S., Review on Cycle Time Reduction in Manufacturing Industries. *Journal of Emerging Technologies and Innovative Research (JETIR)*, 2014. **1(7)**: p. 955-956.
[23] Ali Naqvi, S.A., et al., Productivity improvement of a manufacturing facility using systematic layout planning. *Cogent Engineering*, 2016. **3(1)**: p. 1207296.
[24] Tidke, A.J., Designing Facilities Layout for Small and Medium Enterprises. *International Journal of Engineering Research and General Science*., 2013.
[25] F.G.Miller, T.S.D., and L.J. Mogas, Job Rotation Raises Productivity. *Industrial Engineering*, 1973. **5(6)**: p. 24-29

IOP Conf. Series: Materials Science and Engineering **257** (2017) 012031 doi:10.1088/1757-899X/257/1/012031

Uric acid detection using uv-vis spectrometer

N Norazmi, Z R Abdul Rasad, M Mohamad* and H Manap

Faculty of Engineering Technology, Universiti Malaysia Pahang, Lebuhraya Tun Razak, 26300 Gambang, Kuantan, Pahang

*Corresponding author: mazlee.greenfinite@gmail.com

Abstract. The aim of this research is to detect uric acid (UA) concentration using Ultraviolet-Visible (UV-Vis) spectrometer in the Ultraviolet (UV) region. Absorption technique was proposed to detect different uric acid concentrations and its UV absorption wavelength. Current practices commonly take a lot of times or require complicated structures for the detection process. By this proposed spectroscopic technique, every concentration can be detected and interpreted into an absorbance value at a constant wavelength peak in the UV region. This is due to the chemical characteristics belong to the uric acid since it has a particular absorption cross-section, σ which can be calculated using Beer's Lambert law formula. The detection performance was displayed using Spectrasuite sofware. It showed fast time response about 3 seconds. The experiment proved that the concentrations of uric acid were successfully detected using UV-Vis spectrometer at a constant absorption UV wavelength, 294.46 nm in a low time response. Even by an artificial sample of uric acid, it successfully displayed a close value as the ones reported with the use of the medical sample. It is applicable in the medical field and can be implemented in the future for earlier detection of abnormal concentration of uric acid.

1. Introduction

Gout is a common disease affecting the human joints. Men over the age of 40 years is the common population which suffers from this disease [8]. It is influenced by an excessive amount of uric acid in the human body due to the daily intake issue containing more purines than the optimal daily amount. In other words, high purine food may cause Gout disease to occur [3]. Uric acid is caused by the abnormal amount of purine which is broken down into uric acid and poorly dissolved in the human blood. Therefore, it gets concentrated and crystallized forming numerous crystallites which normally inhibit in the human joints [5]. Additionally, it also may riskily cause other diseases such as chronic renal failure due to the existence of uric acid stone. It is also known as a risk factor of Hyperuricemia (high level of uric acid) [9]. Sudden intense pain and swelling in the joints are classified as the common symptoms due to the condition [4]. Particular research studies has been carried out by researchers till today for the technology and diagnosis advancement in term of reducing the risk of this disease [3]. Gout diagnosis is necessary for the inspection of this disease even from the earlier stage. The uric acid level or concentration can be assumed as a 'fingerprinting' of the risk of this disease to happen [6,7]. There are various techniques used to detect uric acid and an abnormal concentration of uric acid in human blood or serum sample. However, the techniques take a long time for the analysis phase and even require complicated structures. An effective sensor has to be proposed by considering the simplicity, cost effectiveness and fast response time

[1].Optical fiber sensing system or UV-Vis spectrometer is well known as a steadfast technology in the medical, biomedical and industrial sector due to its high elasticity and effective characteristics [10].The characteristics afford high sensitivity, high selectivity, high accuracy, fast time response, low detection limit and it even comes in a small size [11,12]. In this research study, a UV-Vis spectrometer is used to detect the uric acid by examining its current concentration and wavelength absorption in the UV region. Absorption cross-section of each uric acid concentration is defined by the derivation of Beer's Lambert formula. It used to describe the ability of the sample area to absorb photons of UV light wavelength. That is the reason why uric acid only absorbs at a constant UV wavelength. Basically, different concentrations have a constant absorption cross-section which is due to the chemical characteristics of uric acid to absorb the UV light radiation.

The relationship of absorbance, A and absorption cross-section, σ can be interpreted by the following formula of Beer's Lambert law or equations.

$$A = -\ln (I/I_0) = \ln (I_0/I) = \sigma N b = \varepsilon c b \quad (1)$$

$$A = \log_{10} (I_0/ I) = \frac{\sigma N b}{2.303} = \varepsilon c b \quad (2)$$

$$\sigma = \frac{2.303 \log 10 (I_0/ I)}{Nb} = \varepsilon c b \quad (3)$$

2. Material and methods

2.1 Instrumentations

2.1.1 Instrumentation. A UV-VIS Ocean Optics Maya 2000 Pro spectrometer is used with 10mm cuvette (cell) for the detection and measurement of uric acid.

2.1.2 Reagent. Uric acid stock standard (1 mg/ml) is prepared by dissolving 100mg of uric acid powder in 15ml of distilled water. Heat this mixture up to 60°C on a hot plate. It should be stirred using a magnetic stirrer on the hot plate. Cool it down to a comfort room temperature, 20-26°C. Add it up with 85 of distilled water to make it 100ml of uric acid stock. This uric acid stock can be refrigerated for one week and can be repetitiously used for scientific research.

Uric acid working samples (0.8 mg/ml, 0.6 mg/ml, 0.4 mg/ml). Molarity formula, $M_1V_1 = M_2V_2$ is used to make calculations for the dilutions. Start it by preparing the first uric acid working sample, 0.8 mg/ml. Dilute 16ml of the uric acid stock standard with 4ml of distilled water to make it to 20ml. For the second uric acid working sample (0.6 mg/ml), dilute 15 ml of the first working sample with 5ml of distilled water. Afterward, dilute 13.33 ml of the second working sample with 6.67ml of distilled water to produce the third working sample, 0.4mg/ml. Thus, three diluted working samples are resulted with three different uric acid concentrations by the serial dilution.
*Note → Dr., I didn't use any term 'downsampled' here . "The temperature readings were then downsampled to 30 minutes per sample per channel. Check the term downsampled and please provide sampling rate instead e.g. 1 S/sec per channel etc. (standard representation)."

2.2 Methods

2.2.1 Experimental Procedure. The uric acid stock and working samples are tested using a UV-Vis spectrometer in the UV wavelength region. Connect it to the CPU using a USB cable to integrate with

Spectrasuite software installed on the PC. Before running the experiment, warm the Deuterium light source up to 20 minutes. Place the sample of uric acid stock or uric acid working sample into 10-mm path-length cuvette before placing it into the cuvette holder. Every sample has to be tested separately. Set up the integration time till there is no saturation displayed in the graph of intensity versus intended wavelength, 200-400nm UV wavelength. Transmission rate is considered by choosing UV absorption peak below 1.0 abs since the light transmission rates 1% at 2.0 abs and 100% at 0 abs. Fig. 2 shows a complete experimental setup for the spectroscopic research.

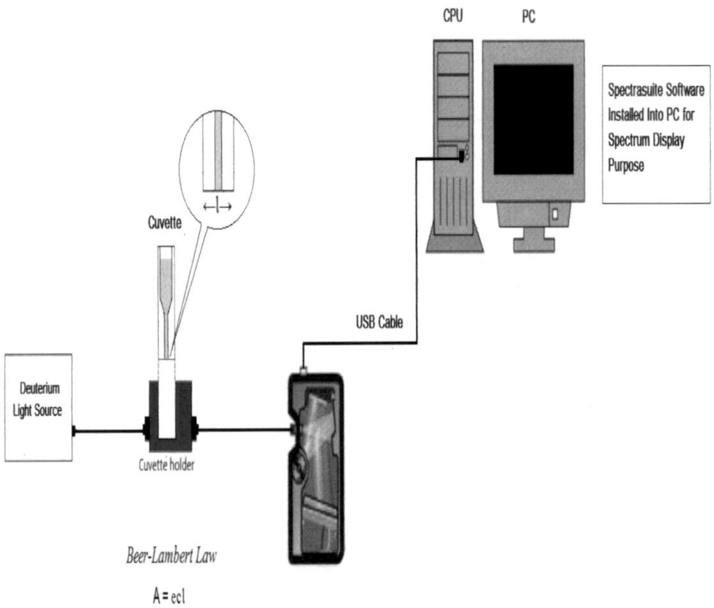

Figure 1. A complete experimental setup for spectroscopic study.

3. Results and discussion

The four different concentrations of uric acid showed four different absorbance values at a same UV wavelength peak in the UV region, 200nm to 400nm using Spectrasuite. The absorption cross-section value was calculated using Beer's Lambert law formula by substituting the absorbance and concentration value of uric acid. Each concentration of uric acid resulted a different absorbance value. However, uric acid constantly absorbs UV light radiation at a same wavelength peak. The UV-Vis spectrometer which was set to UV wavelength region successfully operated to observe the current concentration (each sample) and the UV absorption wavelength of uric acid. Fortunately, the samples were prepared with the artificial powder of uric acid which was mixed up with distilled water. The result shows that the UV absorption wavelength or absorption cross section was acquired at a very close value to the ones reported in the previous research studies. The previous research studies were carried out either by artificial sample (powder) or medical sample of uric acid which was extracted from the human serum.

Fig. 1 illustrates four different concentrations (absorbances) of uric acid acquired at maximum absorption wavelength peak, 294.46 nm using Spectrasuite software. There are some specific criteria had to be considered in the concentration measurement. The absorbance value should be below 1 absorbance (abs). This is because that it is related to the transmission rate which is one of the criteria used to indicate the light penetration and absorption. The light transmission rates 0% at 3.0 abs and 1% at 2.0 abs. 0 abs occurs when the light transmission rates 100%. It means that the absorption value below 1.0 abs rates multiple series of transmission light percentage. The curves of absorbance indicate the values of four uric acid concentrations against the intended UV wavelength region, 200nm-400nm. The highest absorbance spectrum was obtained by the uric acid stock, 1 mg/ml following by the other three absorbance spectrums obtained by the concentrations; 0.8 mg/ml, 0.6 mg/ml and 0.4 mg/ml. These three concentrations were resulted by the serial dilution which was based on the calculation with the molarity formula. The differences of the absorbance spectrum describe the current concentration of uric acid proportionally. Since every experimental sample is uric acid, it constantly absorbs the UV light radiation at a same wavelength peak, 294.46nm. This proved that there is a good potential of the spectrometer to acquire different uric acid concentrations at a constant UV wavelength peak in a real-time monitoring process.

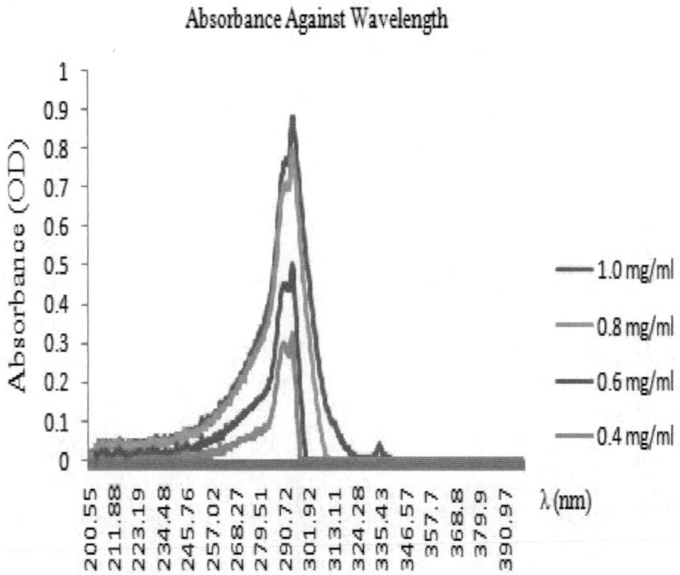

Figure 2. Absorption spectrums of four concentrations of uric acid.

Fig.3 illustrates absorption cross-section of uric acid concentration with four different samples or concentrations against the intended UV wavelength region. The absorption cross-section, $\sigma = 8.0 \times 10^{-19}$ molecules/cm^2 is selected as a maximum absorption value and it is relative to the molar absorptivity of uric acid. It is a constant variable that different uric acid concentrations should indicate the same or very close values of absorption cross-section.

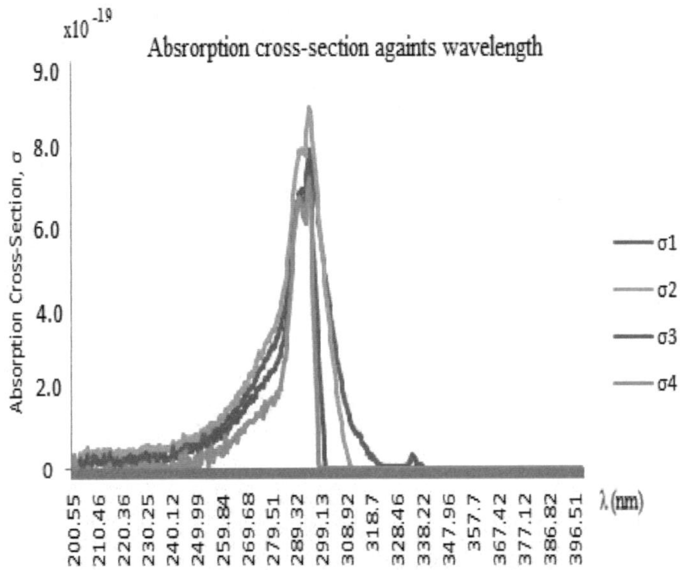

Figure 3. Absorption cross-section of uric acid.

Table 1 shows the values of absorption cross-section of each uric acid concentration.

Table 1. Uric Acid Concentration and Absorption Cross-Section.

ple Concentration	ross-Section, σ
g/ml	206e-19
ng/ml	300e-19
ng/ml	727e-19
ng/ml	221e-19

These four concentrations including the uric acid stock indicate a very close or discrete value of absorption cross-section. The variable differences are considerable since they are discrete values. This proves that uric acid has a particular molar absorptivity which is relative to its special chemical characteristics towards UV light radiation. The spectrometer was successfully functioned to measure the current concentration and observe the UV absorption wavelength of uric acid.

4. Conclusions

The research study was carried out to analyze the different concentrations of uric acid by respecting to the absorption technique. The maximum absorption wavelength peak was obtained at 294.46 nm. It is relative to the absorption cross-section of uric acid that it specifically absorbs the UV light radiation at that UV wavelength peak. The absorption cross-section was calculated using Beer's Lambert law formula by considering the current concentration and absorbance value of uric acid. Through the analysis, the spectrometer was able to observe the current concentration of uric acid and its UV absorption wavelength. Even by four different concentrations of uric acid, it was still able to indicate that all of them absorbed the UV light at a constant wavelength peak. This is due to the chemical characteristics or attributes belong to uric acid itself. Optical fiber sensor or UV-Vis spectrometer can be assumed as an effective sensor which may give a fast time response. Additionally, the experimental setup is simple and it doesn't require complicated structures for generating the detection process. It is applicable technology in medical or biomedical field in the future since it may give a solution for a fast detection system.

Acknowledgment
I would like to thank my supervisor, Dr. Hadi bin Manap for his full support and guidance. I would also like to thank all the laboratory staffs and science officers to provide me all the tools and instruments for the research use. I really appreciate all the knowledge and guidance as the directions for me complete this research.

References

[1] Ali S M U, Alvi N H, Ibupoto Z, Nur O, Willander M and Danielsson B 2011 Selective potentiometric determination of uric acid with uricase immobilized on ZnO nanowires. Sensors and Actuators B: Chemical, 152(2), 241-247.
[2] Choi H K, Atkinson K, Karlson E W, Willett W and Curhan G 2004 Purine-rich foods, dairy and protein intake, and the risk of gout in men. *N Engl J Med*, *2004*(350), 1093-1103.
[3] Choi H K, Mount D B and Reginato A M 2005 Pathogenesis of gout.*Annals of internal medicine*, *143*(7), 499-516.
[4] Dadig B and Wallace A E 2011 CE/CME. *Clinician Reviews*, *21*(7), 29.
[5] Hainer B L, Matheson E and Travis Wilkes R 2014 Diagnosis, treatment, and prevention of gout. *American family physician*, *90*(12).
[6] Idborg H, Zamani L, Edlund P O, Schuppe-Koistinen I and Jacobsson S P 2005 Metabolic fingerprinting of rat urine by LC/MS: Part 1. Analysis by hydrophilic interaction liquid chromatography–electrospray ionization mass pectrometry. *Journal of Chromatography B*, *828*(1), 9-13.
[7] Kutzing M K and Firestein B L 2008 Altered uric acid levels and disease states. Journal of Pharmacology and Experimental Therapeutics, 324(1), 1-7.
[8] Schlesinger N 2005 Diagnosis of gout: clinical, laboratory, and radiologic findings. *Am J Manag Care*, 11(15 Suppl)
[9] Spieker L E, Ruschitzka F T, Lüscher T F and Noll G 2002 The management of hyperuricemia and gout in patients with heart failure. *European journal of heart failure*, *4*(4), 403-410.
[10] Taffoni F, Formica D, Saccomandi P, Pino G D and Schena E 2013 Optical fiber-based MR-compatible sensors for medical applications: An overview. *Sensors*, *13*(10), 14105-14120.
[11] Lee B H, Kim Y H, Park K S, Eom, J B, Kim M J, Rho B S and Choi H Y 2012 Interferometric fiber optic sensors *Sensors*, *12*(3), 2467-2486.
[12] Lou J, Wang Y and Tong L 2014 Microfiber optical sensors: A review. *Sensors*, 14(4), 5823-5844.

Electrochemical deposited nickel nanowires: influence of deposition bath temperature on the morphology and physical properties

A G N Sofiah, J Kananathan, M Samykano*, S Ulakanathan, N A C Lah, W S W Harun, K Sudhakar, K Kadirgama, W K Ngui and J P Siregar

Faculty of Mechanical Engineering, Universiti Malaysia Pahang, Pahang, Malaysia.

Corresponding Author E-mail: mahendran@ump.edu.my

Abstract. This paper investigates the influence of the electrolytic bath temperature on the morphology and physical properties of nickel (Ni) nanowires electrochemically deposited into the anodic alumina oxide porous membrane (AAO). The synthesis was performed using nickel sulfate hexahydrate ($NiSO_4.6H_2O$) and boric acid (H_3BO_3) as an electrolytic bath for the electrochemical deposition of Ni nanowires. During the experiment, the electrolyte bath temperature varied from 40°C, 80°C, and 120°C. After the electrochemical deposition process, AAO templates cleaned with distilled water preceding to dissolution in sodium hydroxide (NaOH) solution to obtain free-standing Ni nanowires. Field Emission Scanning Electron Microscopy (FESEM), Energy Dispersive Spectroscopy (EDX) and X-ray Diffraction (XRD) analysis were employed to characterize the morphology and physical properties of the synthesized Ni nanowires. Finding reveals the electrodeposition bath temperature significantly influences the morphology and physical properties of the synthesized Ni nanowires. Rougher surface texture, larger crystal size, and longer Ni nanowires obtained as the deposition bath temperature increased. From the physical properties properties analysis, it can be concluded that deposition bath temperature influence the physical properties of Ni nanowires.

1. Introduction

Template-assisted electrochemical deposition is one of the dependable technique for synthesizing metallic nanomaterials with controlled shape and size [1]. This method is known for its simplicity, high-throughput and cost-effectiveness which allow the duplication of complex topology present on the surface a template in a single step [2-4]. The method typically operates at ambient temperatures and pressure as well as can be used for mass production of nanowires with controlled geometry and morphology [5-8]. For example, the diameters of nanowires can be well determined and maintained by the diameter of template pores, and the lengths determined by the thickness of template [9]. The process is relatively straightforward as the template serves as a scaffold within or around which nanowires can be shaped with the morphology complementary to the template used.

Miguel García et al. [10] describes that during the electrochemical deposition process, the template pores were initially filled with liquid precursor from the chemical reaction of the electrochemical deposition process. Later, solidification of the filled liquid precursor takes place. These are a very simple route, but it is critical to ensure that the pores are filled with the fluid. If a solution has an excellent wettability for the template, it can diffuse through the membrane producing an enrichment of the solid component in the interior of the pores. Once the pores filled, the template is removed from the solution, dried and further processed.

It is a well-established fact that synthesis process during the electrochemical deposition in porous templates initiates from the cathode at the bottom surface of the pore. The high surface area and the presence of sites with low coordination number in the porous part of the alumina provide energetically favorable sites for initiating the metal adsorption during the electrochemical deposition [11]. The metal ion M^{n+} is transferred from the electrolyte into the ionic metal lattice. Where else, the electrons are provided from the external electron source (power supply) to the electron gas of the metal M. Ohgai [12] in their study of Ni nanowires and Cobalt (Co) nanowires synthesis has explained that during the electrochemical deposition process, Ni^{2+} ions and Co^{2+} ions in the electrolyte of $NiSO_4.6H_2O$ and $CoSO_4.7H_2O$ respectively are surrounded by a hydration shell. This hydration layer will reduce the metal ion mobility, and the component of metal ion velocity become almost equal. The degree of reduction of metal ions thus becomes equal in both directions, resulting in the formation of nanowires.

Among the available synthesis method, Ni nanowires can be consistently synthesized using electrochemical deposition method [13]. The properties of synthesized nanowires by this approach significantly influenced by the processing parameters, such as solution pH, deposition bath temperature, magnetic field [14] and current density [15]. In this present study, the influence of the electrodeposition bath temperature on morphology and physical properties of the Ni nanowires synthesized is investigated. The synthesis performed using Ni Sulfate Hexahydrate ($NiSO_4.6H_2O$) as an electrolyte and boric acid (H_3BO_3) as a stabilizer. The deposition bath temperature was varied from 40 °C, 80 °C, and 120 °C, while other synthesis parameters were kept constant.

2. Experimental Details

Commercial Whatman, AAO membrane with a pore size of 200 nm was used as a template for electrochemical deposition of Ni nanowires. These templates have the advantage in controlling the diameter of nanowires precisely, as the diameter dictated by the pore size of the AAO membranes. In this study, the commercial template is employed to avoid the template fabrication step and emphasize more towards achieving the main objective the research.

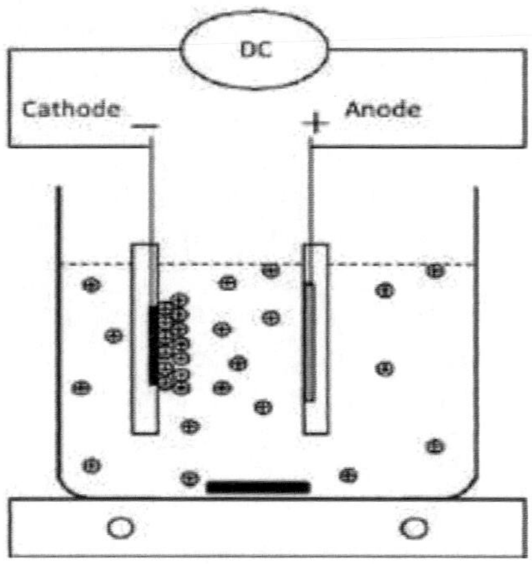

Figure 1. Electrochemical cell setup.

Figure 1 show the schematic illustration of the electrochemical cell setup. In cathode, the Copper tape attached to AAO template, where else, in the anode, the Copper tape is attached to Ni plate for the electrical connection in an electrolytic cell. The anode is placed into the electrolyte, parallel to the cathode. When an electric field is applied, cations diffuse through the channels and deposit on the cathode, resulting in the growth of nanowires inside the template.

The electrochemical cell placed on top of the heater and magnetic stirrer from Heidolph Instruments GmbH & Co. The whole system placed inside the fume hood to limit exposure to hazardous gasses released during the synthesis. The magnetic stirrer used to stir the electrolyte to keep the reaction active. As the deposition bath temperature act as manipulated parameter in this study, three different heating temperatures exploited at 40 °C, 80 °C, and 120 °C.

The electrolyte prepared by mixing the 176.8 g/L of nickel sulfate hexahydrate ($NiSO_4.6H_2O$) solution with 60 g/L boric acid (H_3BO_3) solution. The two solutions were combined and stirred for 30 minutes at 25 °C till all the chemical completely dissolved. The Keithley 2400-C systems voltage/current source meter used as the power supply and as an effective equipment that could be utilized both as potentiostat and amperostat. The applied current was set at 3.0 mA for all the cases. During the electrochemical deposition process, we closely monitor the current for continuous deposition. The sudden rise or decline in current gives an indication that the growth of nanowires has increased or decrease significantly, and the process needs to be terminated to prevent the excessive and non-growth of nanowires and also to avoid the collapse of the pore walls due to excessive growth pressure.

The electrochemical deposition duration is set at 1 hour for all the cases. The duration of electrochemical deposition determines the growth length of the nanowires. As a precaution, during the synthesis process, it is important to ensure copper tape on both cathode and anode not immersed in the electrolyte to avoid reduction and plating of copper in the template. The deposited template finally dissolved in NAOH solution and further cleaned using methanol to obtain freestanding nanowires for characterization. JOEL JSM-7800F FESEM equipped with EDX system was used to investigate the morphology characteristics and the elemental composition of the synthesized Ni nanowires. Where else, the crystallographic information of the synthesized Ni nanowires obtained by X-ray diffraction analyses conducted using Bruker D8Advance diffractometer.

3. Results and discussion

3.1. Effect of deposition temperature on Crystal Orientation

Figure 2 shows the XRD spectrum of Ni nanowires array grown at 3.0 mA at three different deposition bath temperatures: 40°C, 80°C and 120°C. For all cases, XRD spectrum shows the presence of multiple XRD peaks which suggest that the obtained Ni nanowires are polycrystalline in nature. Also, the XRD spectrum shows there was no a significant influence of deposition bath temperature on the orientation of the crystals. The preferred orientation of synthesized Ni nanowires in all cases was found to be (1 1 1), followed by (2 0 0) and (2 2 0) planes. However, obtained Ni nanowires with a deposition bath temperature of 120°C present the highest intensity of diffraction peaks demonstrating a better crystalline structure. From the obtained XRD spectrum pattern, the increase of bath temperature gives a strong advantage to the crystalline structure of the Ni nanowires. These are due to the effect of high deposition temperature on the crystalline structure enhancement of electrochemical deposition growth.

From the spectrum, the obtained 2-theta peak angle used to calculate the spacing between atoms for the nanowires synthesized at a deposition temperature of 40 °C, 80 °C, and 120 °C. Table 1 presents the calculated d-spacing for Ni nanowires synthesized at this process parameters as compared with bulk Ni literature data. Bragg's formulation was used to determine the d-spacing for Ni nanowires.

$$\lambda = 2d \sin \theta$$

where, d is the space between atomic planes in the crystalline phase, λ wavelength of incident X-ray and θ is the angle between incident X-ray and diffraction plane.

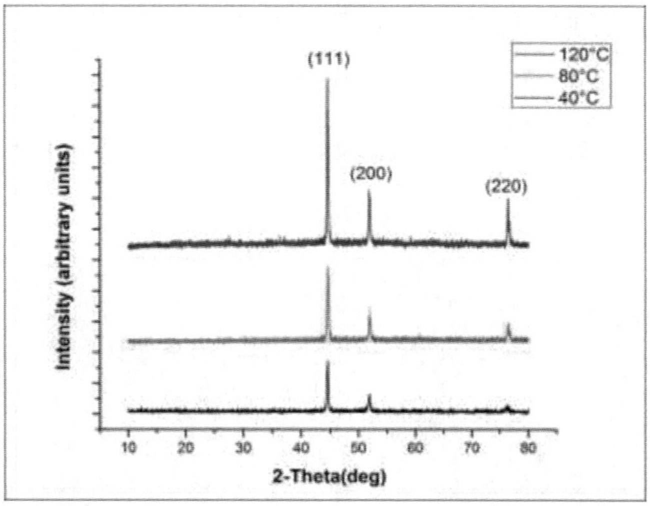

Figure 2. X-ray diffraction spectrum of Ni nanowires at different deposition temperature.

Table 1. 2θ angle and d-spacing for Ni nanowires synthesized at different deposition temperature compared with bulk nickel literature data.

Sample	(1 1 1)		(2 0 0)		(2 2 0)	
	2θ (deg)	d (Å)	2θ (deg)	d (Å)	2θ (deg)	d (Å)
Ni bulk (literature)	44.37	2.04159	51.596	1.77	76.0884	1.25098
			Ni nanowires			
40°C	44.620	2.02836	51.979	1.75716	76.533	1.24329
80°C	44.639	2.02754	52.003	1.75641	76.487	1.24393
120°C	44.627	2.02805	51.976	1.75725	76.683	1.24124

3.2. Effect of deposition temperature on Crystal Size

The obtained XRD spectrums were further analyzed to study the effect of different deposition bath temperature on the formed crystallite size. For a specific material, if the individual crystallite size of the material is less than 1000 Å or 100 nm, then the crystallite size can be estimated using the Scherrer equation.

$$D = \frac{k\lambda}{\beta cos\theta}$$

where, D is the diameter of crystallites, λ is the X-ray wavelength, K is crystallites shape factors (for spherical crystallites, K = 0.9) and β is the full width at half maximum (radians).

Table 2. Crystal size of synthesized nickel nanowires.

Sample	Deposition Temperature		
	40°C	80°C	120°C
Crystal Size Sample 1(nm)	28.550	32.277	35.745
Crystal Size Sample 2(nm)	29.875	34.311	39.967
Crystal Size Sample 3(nm)	24.195	25.740	28.612
Average	**27.540**	**30.776**	**34.775**

To obtain the crystal size, three XRD experiments performed for each deposition temperature. The full width half maximum (FHWM) for each experiment is obtained using Origin Lab Software. Table 2 lists the average crystallite sizes obtained for all the deposition temperature. The results clearly show a significant increment in crystallite size when the deposition bath temperature increased.

From the crystal size obtained as shown in table 2, further investigation needed to understand the physics behind the increasing of the crystal size when the deposition temperature increased. According to the past results, the increase in bath temperature gives a strong benefit on the crystalline structure of the Ni nanowires [16]. The higher bath temperatures promote the surface diffusion of atoms, favoring the growth of preexisting nuclei established during the first stages of the electrochemical process [17].

3.3. Effect of Deposition Bath Temperature on Surface Morphology

(a) (b) (c) (d) (e) (f)

Figure 3. FESEM surface texture images of grown Ni nanowires at a deposition temperature of 40 °C, 80 °C, and 120 °C.

FESEM (JOEL JSM-7800F) used to study the structure and morphology of the grown Ni nanowires. The study was performed through qualitative observation from a series of images taken using FESEM. The morphology imaging carried out at a beam source of 5 kV, and a working distance of 10 mm. Figure 3 (a-b) show the FESEM surface morphology images of Ni nanowires electrodeposited at a temperature of 40 °C. Figure 3 (c-d) and Figure 3 (e-f) show the FESEM surface morphology images of Ni nanowires deposited at a temperature of 80 °C and 120 °C respectively.

From the obtained FESEM images, the surface texture and roughness found to increase as the deposition temperature increase. The surface texture change to be very rough at the deposition temperature of 120 °C. The possible reasons for the formation of the rough surface are possibly due to the formation of larger grain size as the deposition temperature increased. Thus, the large Ni grains could not uniformly fill the AAO template pores during the deposition process. Typically, smooth surface textured nanowires are preferred since it is suitable for many applications, including sensing of electrical or optical signals. However, in some circumstances, an increased surface roughness is of interest [18]. Hochbaum et al. and Boukai et al. recently reported that rough textured Si nanowires exhibit a thermal conductivity up to 100 times smaller than their smooth counterparts, becoming promising objects to be implemented in thermoelectric devices [19, 20].

3.4. Effect of Deposition Bath Temperature on Growth Length

The growth length of Ni nanowires at different deposition temperature was measured. The measurement performed after dissolving the AAO templates in NaOH solution after the deposition to obtain freestanding nanowires. A total of 10 length measurement was taken for each deposition temperature. The measured growth length of deposited Ni nanowires for all deposition temperature listed in table 3. These are used to examine the influence of deposition temperature on the growth length of the nanowires.

Table 3. Measured length of grown Ni nanowire at different deposition bath temperature.

Heating Temperature (°C)	**Ni nanowires Length (μm)**
40	11.321
80	12.600
120	16.860

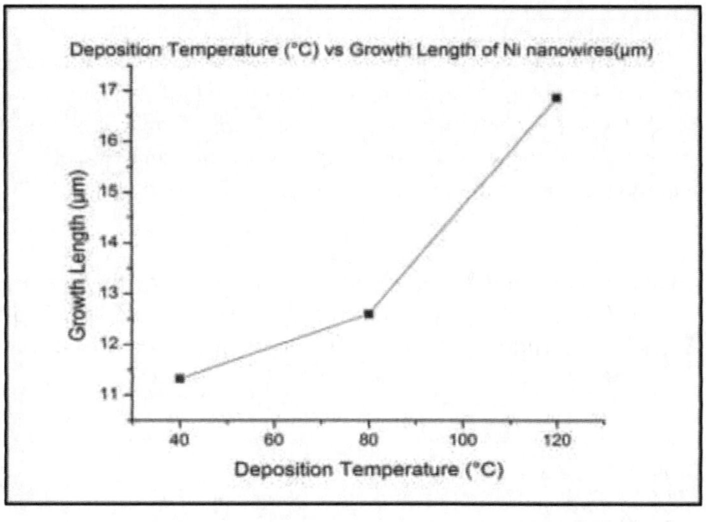

Figure 4. Deposition temperature versus Growth Length curve of synthesized nickel nanowires.

The grown length of the Ni nanowires is found to increase significantly as the deposition bath temperature increased. These are due to the higher rate of nanoparticles diffusion when the temperature increased [16]. Higher temperature promotes atomic diffusion on the growing face of the nanowires and thus accelerates growth relative to the nucleation of new grains [21]. The higher rate of nanoparticles diffusion improves the degree of growth of Ni nanowires. Apart from that, the length improvement with an increase in the temperature is due to the lowered energy barriers [22]. However, when the nanowire length increases, creates a higher center of gravity which makes their position unbalanced and easy to angle during the washing and processing stage [23].

3.5. Energy Dispersive X-Ray (EDX) Analysis

A quantitative EDX analysis is performed on all the samples to determine the elemental composition of the Ni nanowires deposited on the AAO templates at different bath temperatures and the results shown in Table 4. The EDX analysis is performed using Oxford Instruments EDX attached to JOEL JSM-7800F FESEM at a beam source of 15kV. The EDX analysis demonstrates that the synthesized Ni nanowires were consistently composed of 97.97 % of Ni and 2.04 % of oxygen. The small amount of oxygen observed in the entire spectrum indicates potential absorption from the air on the surface of Ni nanowires.

Table 4. EDX elemental composition of synthesized Ni nanowires

Elements	Value (%)	Standard deviation (%)
Nickel (Ni)	97.97	0.4
Oxygen (O_2)	2.04	0.4

4. Conclusions

The present research has focused on the effect of deposition bath temperature on the morphology and physical properties of electrodeposited Ni nanowires. Ni nanowires were synthesized using template-assisted electrochemical deposition method at a deposition bath temperature of 40 °C, 80 °C, and 120 °C. From the obtained trend of XRD spectrum, the increase of temperature gives a strong benefit on the crystalline structure of the Ni nanowires. These are due to the effect of high deposition temperature on the crystalline structure enhancement of electrochemical deposition growth. The crystal was also found to increase as the electrodeposition bath temperature increased. Qualitative FESEM image analysis has shown that the surface roughness increased as the deposition bath temperature increase. Overall growth length found to be significantly better as the deposition bath temperature increased. EDX analysis showed the synthesized structure was composed of 97.97 % of Ni and 2.04 % of Oxygen. The observed oxygen is primarily due to absorption from the air on the surface of the nanowires after synthesis. From the physical properties properties analysis, it can be concluded that deposition bath temperature influence the physical properties of Ni nanowires.

Acknowledgements

Universiti Malaysia Pahang fully supports the facilities and resources for this research. The authors would like to acknowledge the support from the Universiti Malaysia Pahang internal grant RDU160337, RDU170320, and Ministry of Higher Education, Malaysia (FRGS Grant) RDU160119.

References

[1] Bera, D., S.C. Kuiry, and S. Seal, *Synthesis of nanostructured materials using template-assisted electrodeposition.* JOM, 2004. **56**(1): p. 49-53.

[2] Xia, Y., et al., *One-Dimensional Nanostructures: Synthesis, Characterization, and Applications.* Advanced Materials, 2003. **15**(5): p. 353-389.

[3] Ghahremaninezhad, A. and A. Dolati, *A study on electrochemical growth behavior of the Co–Ni alloy nanowires in anodic aluminum oxide template.* Journal of Alloys and Compounds, 2009. **480**(2): p. 275-278.

[4] Samykano, M., R. Mohan, and S. Aravamudhan, *Effect of current density and magnetic field on the growth and morphology of nickel nanowires*, in *MEMS and Nanotechnology, Volume 8*. 2015, Springer. p. 75-83.

[5] Irshad, M., et al., *Preparation and structural characterization of template assisted electrodeposited copper nanowires.* Int. J. Electrochem. Sci, 2014. **9**: p. 2548-2555.

[6] Şişman, İ., *Template-assisted electrochemical synthesis of semiconductor nanowires*, in *Nanowires–Implementations and Applications*. 2011, InTech. p. 41-58.

[7] Zheng, M., et al., *Fabrication and optical properties of large-scale uniform zinc oxide nanowire arrays by one-step electrochemical deposition technique.* Chemical Physics Letters, 2002. **363**(1): p. 123-128.

[8] Al-Salman, R., et al., *Template-Free Electrochemical Synthesis of High Aspect Ratio Sn Nanowires in Ionic Liquids: A General Route to Large-Area Metal and Semimetal Nanowire Arrays?* Chemistry of Materials, 2015. **27**(11): p. 3830-3837.

[9] Wang, Z.-L., et al., *Controllable Template-Assisted Electrodeposition of Single- and Multi-Walled Nanotube Arrays for Electrochemical Energy Storage.* Scientific Reports, 2013. **3**: p. 1204.

[10] García, M., P. Batalla, and A. Escarpa, *Metallic and polymeric nanowires for electrochemical sensing and biosensing.* TrAC Trends in Analytical Chemistry, 2014. **57**: p. 6-22.

[11] Narayanan, T.N., et al., *On the growth mechanism of nickel and cobalt nanowires and comparison of their magnetic properties.* Nano Research, 2008. **1**(6): p. 465-473.

[12] Ohgai, T., *Magnetoresistance of nanowires electrodeposited into anodized aluminum oxide nanochannels*. 2012: INTECH Open Access Publisher.

[13] Samykano, M., *Physical and Mechanical Characterization of Electrodeposited Nickel Nanowires--Influence of Current Density and External Magnetic Field.* 2015.

[14] Samykano, M., R. Mohan, and S. Aravamudhan, *Structure and Morphology of Electrodeposited Nickel Nanowires at an Electrode Distance of 20mm.* Structure, 2015. **1**: p. 23704.

[15] Samykano, M., R. Mohan, and S. Aravamudhan, *Morphology and Crystallographic Characterization of Nickel Nanowires—Influence of Magnetic Field and Current Density During Synthesis.* Journal of Nanotechnology in Engineering and Medicine, 2014. **5**(2): p. 021005.

[16] Cortés, A., et al., Single-crystal growth of nickel nanowires: influence of deposition conditions on structural and magnetic properties. Journal of nanoscience and nanotechnology, 2009. 9(3): p. 1992-2000.

[17] Tian, M., et al., Electrochemical growth of single-crystal metal nanowires via a two-dimensional nucleation and growth mechanism. Nano Letters, 2003. 3(7): p. 919-923.

[18] Toimil-Molares, M.E., Characterization and properties of micro- and nanowires of controlled size, composition, and geometry fabricated by electrodeposition and ion-track technology. Beilstein Journal of Nanotechnology, 2012. 3: p. 860-883.

[19] Hochbaum, A.I., et al., Enhanced thermoelectric performance of rough silicon nanowires. Nature, 2008. 451(7175): p. 163-167.

[20] Boukai, A.I., et al., Silicon nanowires as efficient thermoelectric materials. Nature, 2008. 451(7175): p. 168-171.

[21] Wang, J.-G., et al., Controllable template synthesis of superconducting Zn nanowires with different microstructures by electrochemical deposition. Nano letters, 2005. 5(7): p. 1247-1253.

[22] Dayeh, S.A., E.T. Yu, and D. Wang, III– V nanowire growth mechanism: V/III ratio and temperature effects. Nano letters, 2007. 7(8): p. 2486-2490.

[23] Li, Z., et al., Effects of nanowire length and surface roughness on the electrochemical sensor properties of Nafion-free, vertically aligned Pt nanowire array electrodes. Sensors, 2015. 15(9): p. 22473-22489.

IOP Conf. Series: Materials Science and Engineering **257** (2017) 012033 doi:10.1088/1757-899X/257/1/012033

Influence of boric acid (H_3BO_3) concentration on the physical properties of electrochemical deposited nickel (Ni) nanowires

J Kananathan, A G N Sofiah, M Samykano*, S Ulakanathan, N A C Lah, W S W Harun, K Sudhakar, K Kadirgama, W K Ngui and J P Siregar

Faculty of Mechanical Engineering, Universiti Malaysia Pahang, Pahang, Malaysia.

E-mail: mahendran@ump.edu.my

Abstract. Authors have investigated the influence of the stabilizer (Boric Acid) concentration during the template-assisted electrochemical deposition of Nickel (Ni) nanowires in Anodic Alumina Oxide (AAO) templates. The synthesis was performed using Ni Sulfate Hexahydrate ($NiSO_4.6H_2O$) as metal salts and Boric Acid (H_3BO_3) as a stabilizer. The mixture of both solutions creates electrolyte and utilized for the electrochemical deposition of Ni nanowires. During the experiment, the boric acid concentration varied between 5 g/L, 37.5 g/L and 60 g/L with a deposition temperature of 80 °C (constant). After the electrochemical deposition process, AAO templates were cleaned with distilled water before dissolution in Sodium Hydroxide (NaOH) solution to obtain the freestanding Ni nanowires. Physical properties of the synthesized Ni nanowires were analyzed using Field Emission Scanning Electron Microscopy (FESEM), Energy Dispersive Spectroscopy (EDX) and X-ray Diffraction (XRD). The physical properties of obtained Ni nanowires has eloborated by taking into account the effect of boric acid concentration on the surface morphology, growth length, elemental composition and crystal orientation crystal of the synthesized nickel nanowires. The finding exposes that the boric acid concentration does not influence all aspects in the physicals properties of the synthesized Ni nanowires. The boric acid concentration did not affect the surface texture and crystal orientation. However, shorter Ni nanowires obtained as the concentration of boric acid increased.

1. Introduction
Synthesis of nanostructures with defined morphology and properties has attracted serious attention among nanotechnology researchers due to the importance of the dimensionality which classifies their properties [1, 2]. Among various available method, template-assisted electrochemical deposition is one of the popular and well-utilized approaches [3]. This method is relatively straightforward [4] with controlled diameter, cost efficient and allows the duplication of complex topology present on the surface a template in a single step [5, 6]. The method typically operates at ambient temperatures and pressure as well as can be used for mass production of nanostructures with controlled geometry and morphology [2, 7-9]. Additionally, among templates, Anodic Alumina Oxide (AAO) templates are the most widely used due to its effective medium for synthesis of metallic nanowires. These templates have an advantage over controlling the diameter of nanowires precisely and producing relatively straight nanowires [7, 10-15].

Figure 1. Synthesized Ni nanowires.

Ni nanowires with their ferromagnetic properties are of interest in numerous applications due to their lower cost, and they can be consistently synthesized via electrochemical deposition [16]. The properties of synthesized nanowires are significantly dependent on the synthesis parameters, such as solution pH, deposition bath temperature, magnetic field [17] and current density [18]. During the experiment, the boric acid concentration was varied between 5g/L, 37.5g/L, and 60 g/L to study the effect of different boric acid concentration during the synthesis on their physical properties. Boric acid is one of the important chemical compounds in the electrolyte for Ni deposition since they act as buffering agents in an electrolyte [19]. The nanowires then characterized by using Field Emission Scanning Electron Microscopy (FESEM), Energy Dispersive Spectroscopy (EDX) and X-ray Diffraction (XRD). The influence of boric acid concentration on surface texture, growth length, elementary composition and crystal orientation on synthesized Ni nanowires discussed in greater detail.

2. Experimental details

Alumina oxide membranes (Whatman, 99.9% pure) was used in this template-assisted electrochemical deposition. The synthesis procedure was carried out using a two-electrode electrochemical cell system. One side of AAO templates coated with an aluminum coating which acts as a cathode and Ni plate which serves as an anode, placed into electrodeposition bath, parallel to each other. When an electric field is applied, cations diffuse through the channels and deposit on the cathode, resulting in the growth of nanowires inside the pores of the template. For the electrolyte solution preparation, the procedure adopted by Ertan et al. [20] was used. It is by far the simplest method and has demonstrated success in synthesizing Ni nanowires using the template-assisted electrochemical deposition method. 71 g of Ni metal salt, (Ni Sulfate Hexahydrate, $NiSO_4.6H_2O$) dissolved in 400 mL distilled water. In a separate beaker, boric acid (H_3BO_3) dissolved in room temperature in distilled water. The two solutions were mixed and stirred for 30 minutes at 25 °C until the solutions well mixed. The correlation between the stabilizer agents concentration and the physical properties of Ni nanowires, the amount of boric acid used for synthesis had been manipulated at 5 g/L, 37.5 g/L, and 60 g/L respectively while the amount of Ni sulfate hexahydrate kept constant. The electrochemical bath cell was placed on top of a magnetic stirrer and heater from Heidolph Instruments GmbH & Co. The whole system was then placed inside the fume hood to limit the exposure of hazardous gas. The Keithley systems source meter was used to supply the required electrical current for electrodeposition. The applied current was set at 3 mA, and during the electrochemical deposition process, the current monitored for continuous and constant deposition. The sudden fluctuation (increase or decrease) in current gives an indication of the inconsistent growth of nanowires, and the process needs to be ended to prevent the disproportionate growing of the nanowires and also to avoid the collapse of the pore walls due to extreme growth pressure. The electrochemical deposition time duration is set to be 60 minutes (constant for all sample). As a deterrent, during the electrodeposition process, it is vital to ensure the copper tape used as the electrical connector on both cathode and anode are not soaked into the electrolyte to avoid contamination to the nanowires. Table 1 shows the simplified synthesis condition used for the experiment. After the electrodeposition of Ni nanowires, 10 g of sodium hydroxide (NaOH) dissolved in 15 ml distilled water. Then, etching process takes place by placing the electrodeposited AAO template inside NaOH solution. The whole mixture rapidly shook until the AAO template completely dissolved and a dark grey product (Ni nanowires) formed at the bottom part of the glass bottle. The dark grey products then cleaned using ethanol solution several times by magnetic decantation and ready for characterization.

Table 1. Synthesis condition of electrochemically deposited Ni nanowires.

Experiment	Boric Acid Concentration (g/L)	Electrochemical Deposition Time (min)	Applied Current (mA)
A	5	60	3
B	37.5	60	3
C	60	60	3

3. Results and discussion

Ni nanowires have been successfully deposited on the AAO templates. The influence of boric acid concentration on physical properties of obtained Ni nanowires investigated. In the conventional Ni plating, the electrolyte solution is known as Watts bath, which composed of Ni sulfate and Ni chloride as metal salts and boric acid act as buffering agents. Holm and O'Keefe demonstrated Ni plating using electrolyte solution without boric acid and produced brittle deposition of Ni due to the precipitation of Ni hydroxide on the substrate surface [21]. The surface appearance improvement can be performed by increasing the activity of Ni either by increasing the metal salt concentration or temperature or decreasing the solution pH. In the next section, the influence of boric acid concentration on surface morphology of Ni nanowires is discussed.

3.1. Influence of boric acid concentration on surface morphology

FESEM (JOEL JSM-7800F) used to investigate the morphology of the synthesized Ni nanowires. The analysis performed by qualitative observation on the surface synthesized Ni nanowires from a sequence of images taken using FESEM. The surface morphology imaging conducted at a beam source of 5kV, and a working distance of 10 mm. Fig. 2 (a-b) show the FESEM images of surface morphology Ni nanowires electrodeposited using boric acid concentration of 5g/L. Fig. 2 (c-d) and Fig. 2 (e-f) shows the FESEM images of surface morphology of Ni nanowires using boric acid concentration of 37.5 g/L and 60 g/L respectively.

From the FESEM images, it is observed that there are no significant differences in surface texture of the synthesized Ni nanowires. All the surface texture found to be rough and flaky. These needs further examination to precisely conclude the finding. However, from the study, the appearance of the Ni deposit was greatly improved with the presence and increased the concentration of boric acid. Instead of a brittle, irregular surface, a smooth, glassy nickel surface formed with the presence of the boric acid. The improved surface structure reported being attributed to the buffering ability of boric acid [22, 23]. In this present study, the observation obtained found in contrast with the previous finding reported and appears that the influence of boric acid on the surface texture of electrodeposited Ni nanowires remains complicated unclear.

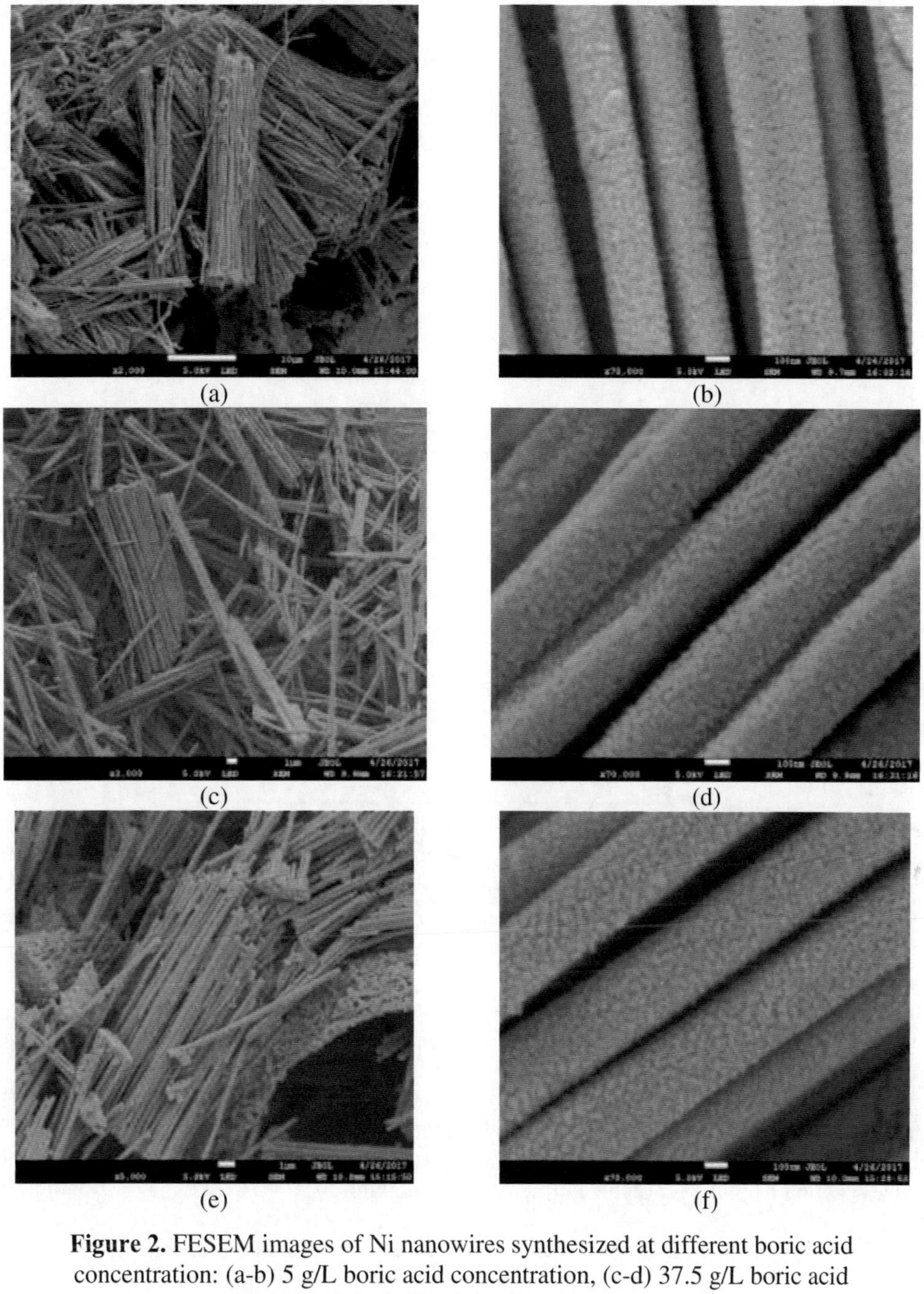

Figure 2. FESEM images of Ni nanowires synthesized at different boric acid concentration: (a-b) 5 g/L boric acid concentration, (c-d) 37.5 g/L boric acid concentration, (e-f) 60 g/L boric acid concentration.

3.2. Influence of boric acid concentration on growth length

The growth length of Ni nanowires at different boric acid concentration was measured using FESEM. Average ten (10) length measurements taken for each synthesis condition. The measured growth length of electrodeposited Ni nanowires at all boric acid concentration used for synthesis tabulate in Table 2. The finding used to investigate the influence of boric acid concentration on the growth length of the Ni nanowires.

The growth length of Ni nanowires found to decrease significantly as the concentration of boric acid increased. These are due to the presence of a high concentration boric acid which inhibits the Ni ion deposition [23].

Table 2. Measured length of grown Ni nanowire at different synthesis condition.

Boric Acid concentration	Ni nanowires Length (μm)
5g/L	33.125
37.5g/L	15.850
60g/L	12.600

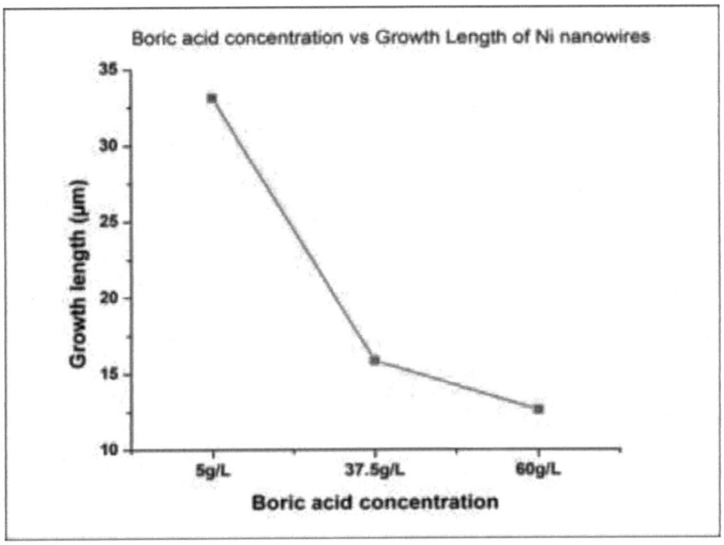

Figure 3. Nickel Nanowire grown length at different boric acid concentration.

3.3. Energy Dispersive X-Ray (EDX) analysis

A quantitative EDX analysis is performed on all the samples to determine the elemental composition of the grown Ni nanowires deposited on the AAO templates at different boric acid concentration and the results are shown in Table 3. The EDX analysis performed using Oxford Instruments EDX which attached to JOEL JSM-7800F FESEM. The experiments carried out at a beam source of 15kV. The EDX analysis demonstrates that the synthesized Ni nanowires were consistently composed of 97.83 % of Ni and 2.17 % of Oxygen. The presence of small amount of oxygen in the entire spectrum indicates potential absorption from the air on the surface of Ni nanowires.

Table 3. EDX elemental composition of synthesized Ni nanowires.

Elements	Value (%)	Standard deviation (%)
Nickel (Ni)	97.83	0.4
Oxygen (O_2)	2.17	0.4

3.4. Influence of boric acid concentration on crystal orientation

Figure 3.2 shows the XRD spectrum of Ni nanowires synthesized at different concentration of boric acid. The XRD spectrum shows there was no significant influence of boric acid concentration on the orientation of the crystals. The predominant orientation is still dictated by the concentration of the metal salts. The preferred orientation of synthesized Ni nanowires in all cases was found to be (1 1 1) plane, followed by (2 0 0) and (2 2 0) planes.

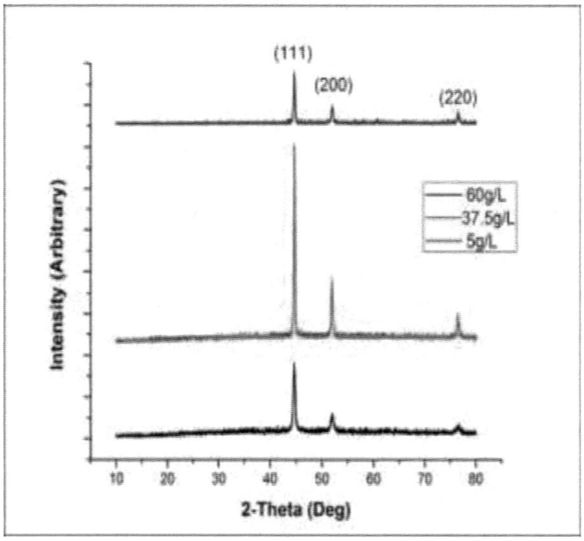

Figure 4. XRD spectrum of Ni nanowires synthesized at a different boric acid concentration.

However, the spectrum of Ni nanowires synthesized at a boric acid concentration of 37.5 g/L presents the highest intensity of diffraction peaks demonstrating a better crystalline structure. From the obtained trend of XRD spectrum, the optimum concentration of boric acid to synthesize the best crystallite structure could be suggested at 37.5 g/L. The obtained spectrum further used to derive the 2-theta peak angle which is used to calculate the spacing between atoms for the nanowires synthesized at a boric acid concentration of 5 g/L, 37.5g/L, and 60 g/L. Table 4 presents the calculated d-spacing for Ni nanowires synthesized at this process parameters as compared with bulk Ni literature data. Bragg's formulation was used to determine d-spacing for the deposited Ni nanowires. From the table, we can clearly see that there are no significant differences between the bulk nickels d-spacing from the literature with the present study.

Table 4. 2θ angle and d-spacing for nickel nanowires synthesized at different deposition temperature compared with bulk nickel literature data.

Sample	(1 1 1)		(2 0 0)		(2 2 0)	
	2θ (deg)	d (Å)	2θ (deg)	d (Å)	2θ (deg)	d (Å)
Ni bulk (literature)	44.37	2.04159	51.596	1.77	76.0884	1.25098
Ni nanowires						
5g/L	44.620	2.02836	51.979	1.75716	76.533	1.24329
37.5g/L	44.639	2.02754	52.003	1.75641	76.487	1.24393
60g/L	44.627	2.02805	51.976	1.75725	76.683	1.24124

4. Conclusions

As a conclusion, Ni nanowires has been successfully synthesized using the template-assisted electrodeposited technique. The properties analysis reveals that boric acid concentration does not influence all the aspects. From the FESEM analysis, for all the investigated cases, the surface texture found to be rough and flaky. The observation obtained was in contrast to the previous finding and appears that the influence of boric acid on Ni deposition is complicated and remains unclear. Where else, in the growth length, from the ten (10) length measurements taken for each boric acid condition, the growth length of Ni nanowires is found to decrease significantly as the concentration of boric acid increased. These are due to the presence of a high concentration boric acid which inhibits the Ni ion deposition. The EDX analysis demonstrates that the synthesized Ni nanowires were consistently composed of 97.87 % of Ni and 2.17 % of Oxygen. The small amount of oxygen present in the entire spectrum indicates the possible absorption from the air on the surface of Ni nanowires. The XRD spectrum also shows there was not a significant influence of boric acid concentration on the preferred orientation of the crystals. The predominant orientation still primarily dictated by the concentration of metal salts. The preferred orientation of synthesized Ni nanowires in all cases was found to be (1 1 1) plane, followed by (2 0 0) and (2 2 0) planes

Acknowledgements
Universiti Malaysia Pahang fully supports the facilities and resources for this research. The authors would like to acknowledge the support from the Universiti Malaysia Pahang internal grant RDU160337, RDU170320, and Ministry of Higher Education, Malaysia (FRGS Grant) RDU160119.

References

[1] Cao G 2004 *Nanostructures and nanomaterials: synthesis, properties and applications* (World Scientific).
[2] Al-Salman R et al 2015 *Chemistry of Materials* **27** 3830-3837.
[3] Samykano M, Mohan R and Aravamudhan S 2015 *Effect of current density and magnetic field on the growth and morphology of nickel nanowires*, in *MEMS and Nanotechnology* (Springer) **8** 75-83.
[4] Xia Y et al 2003 *Advanced Materials* **15** 353-389.
[5] Xia Y et al 2003 *Advanced Materials* **15** 353-389.
[6] Ghahremaninezhad A and Dolati 2009 *Journal of Alloys and Compounds* **480** 275-278.
[7] Irshad M et al 2014 *Int. J. Electrochem.* Sci. **9** 2548-2555.
[8] Şişman İ 2011 *Template-assisted electrochemical synthesis of semiconductor nanowires*, in *Nanowires–Implementations and Applications* (InTech) 41-58.
[9] Zheng M et al 2002 *Chemical Physics Letters* **363** 123-128.
[10] Guo Y Y et al 2010 *Acta Physico-Chimica Sinica* **26** 2037-2043.
[11] Mu C and He J H 2010 *Journal of Nanoscience and Nanotechnology* **10** 8191-8198.
[12] Bograchev D A, Volgin V M and Davydov A D 2013 *Electrochimica Acta* **96** 1-7.
[13] Choi J S et al 2003 *Journal of Materials Chemistry* **13** 1100-1103.
[14] Li X D et al 2012 *Acs Nano* **6** 831-836.
[15] Sharma G, Pishko M V and Grimes C A 2007 *Journal of Materials Science* **42** 4738-4744.
[16] Samykano M 2015 *Physical and Mechanical Characterization of Electrodeposited Nickel Nanowires--Influence of Current Density and External Magnetic Field.* 2015.
[17] Samykano M, Mohan R and Aravamudhan S 2015 *Structure* **1** 23704.
[18] Samykano M, Mohan R and Aravamudhan S 2014 *Journal of Nanotechnology in Engineering and Medicine* **5** 021005.
[19] Ji J et al 1995 *Journal of Applied Electrochemistry* **25** 642-650.
[20] Ertan A, Tewari S N and Talu O 2008 *Journal of Experimental Nanoscience* **3** 287-295.
[21] Holm M and O'keefe T 2000 *Journal of applied electrochemistry* **30** 1125-1132.
[22] Davalos C et al 2013 *Int J Electrochem Sci* 2013 **8** 9785.
[23] Santos J et al 2007 *Electrochimica Acta* **53** 644-649.

Engine performance and exhaust emission analysis of a single cylinder diesel engine fuelled with water-diesel emulsion fuel blended with manganese metal additives

Ahmad Muhsin Ithnin*, Wira Jazair Yahya, Jasmine Baun Fletcher and Hasannuddin Abd Kadir

Vehicle Systems Engineering, Malaysia-Japan International Institute of Technology, Universiti Teknologi Malaysia, Jalan Sultan Yahya Petra, 54000 Kuala Lumpur, Malaysia

*Corresponding author: ahmadmuhsin@utm.my

Abstract. Water-in-diesel emulsion fuel (W/D) is one of the alternative fuels that capable to reduce the exhaust emission of diesel engine significantly especially the nitrogen oxides (NOx) and particulate matter (PM). However, the usage of W/D emulsion fuels contributed to higher CO emissions. Supplementing metal additive into the fuel is the alternate way to reduce the CO emissions and improve performance. The present paper investigates the effect of using W/D blended with organic based manganese metal additives on the diesel engine performance and exhaust emission. The test were carried out by preparing and analysing the results observed from five different tested fuel which were D2, emulsion fuel (E10: 89% D2, 10% - water, 1% - surfactant), E10Mn100, E10Mn150, E10Mn200. Organic based Manganese (100ppm, 150ppm, 200ppm) used as the additive in the three samples of the experiments. E10Mn200 achieved the maximum reduction of BSFC up to 13.66% and has the highest exhaust gas temperature. Whereas, E10Mn150 achieved the highest reduction of CO by 14.67%, and slightly increased of NO_x emissions as compared to other emulsion fuels. Organic based manganese which act as catalyst promotes improvement of the emulsion fuel performance and reduced the harmful emissions discharged.

1. Introduction

Water-in-Diesel emulsion fuel (W/D) is an alternative fuel which progressively intrigues the world's attention due to its great impact to the environment as well as energy consumption. W/D is able to simultaneously reduce the formation of Nitrogen oxides (NOx) and Particulate Matter (PM) in a large extent and at the same time improving the combustion efficiency of the engine [1]. There is a special occurrence in W/D ignition called the micro-explosion phenomena that attract researchers worldwide, as it is non-existent in other normal diesel combustion. It is a secondary atomization of the primary spray as a result of the rapid evaporation process of water that is initially contained in the oil drop, making the combustion more efficient [2]. Water-in-Diesel emulsion fuel is one of example of oil phased emulsion which can be used as an alternative fuel. Surfactant was added to improve the composition and provide a stable emulsion fuel [3].

Numerous studies have been conducted concerning the W/D emulsion fuel involving the performance and emissions of an engine running on the said fuel, in return, explaining the combustion characteristics and micro-explosion process. According to Abu-Zaid [4] brake thermal efficiency is increased by 3.5% from that of the neat diesel fuel. Basha et al. [5] found that the measurement increased to 26.9% compared to diesel, which is 25.2%. A.K Hasannudin et al [6] conducted the durability study of W/D, fuelled the engine for 200 hour and study its effect to the lubricant oil. They reported that the water content in the fuel did not cause any excessive amount of metals or degraded the additive. However, these alternative fuels emitted higher emission of CO[7], [8]. This issue can be solved through the addition of metal additive to the W/D. Very few researchers worked, on the use of nano-additives to emulsion fuel for the enhancement of the performance and emission characteristics but many on biodiesel. Sajith et al. [9] carried out an extensive investigation on a diesel engine fuelled with and without cerium oxide additives, to study the performance and emission characteristics. They found that the viscosity and volatility hold direct relations with the dosing level of 20-80 ppm. The emission levels of HC and NOx are appreciably reduced with the inclusion of CeO2. El-Seesy et al [10] added the Multi-Walled Carbon nanotubes into the Jojoba methyl ester- diesel blended fuel (JB20D) and evaluate its combustion characteristic. The authors found that the combustion characteristics were improved compared to JB20D fuel where the peak cylinder pressure, the maximum rate of pressure rise and the peak heat release rate were increased by 7%, 4%, and 4%, respectively. In order to reduce the NOx and PM emission from the CI engines, emulsification techniques are adopted. Basha and Anand [11] investigated the performance and emissions of a diesel engine fuelled with biodiesel emulsion fuel incorporation of alumina nanoparticles in the mass fractions of 25, 50 and 100 ppm, with a higher concentration of water. They observed the improvement of NOx, and smoke emission.

From this motivation, the present paper investigates the effect of using W/D blended with organic based manganese metal additives on the diesel engine performance and exhaust emission. Three different volume of Manganese (100ppm, 150ppm, 200ppm) were blended into the W/D (E10: 89% D2, 10% - water, 1% - surfactant) and labelled as E10Mn100, E10Mn150, E10Mn200 respectively. The results are used for a comparison with neat diesel D2 and conventional emulsion fuel, E10. A single cylinder, direct injection diesel engine is tested under 5 different load conditions (0 to 5 KW) and with a constant engine speed of 3500 rpm. In the engine performance the specific fuel consumption is discussed in detail. In respect of the emission characteristics, NOx, HC, and CO are analysed before being further discussed.

2. Methodology

2.1 Fuel Preparation

In this study, W/D is made by mixing 10% of tap water, 1 % of surfactant and 89 % of diesel and labelled as E10. All the immiscible liquids were emulsified through the mechanical homogenizer at 2500rpm for 10 minutes. The surfactant used to form the emulsion is the Polyoxyethylenenonylphenyl ether, or so called span 80 with a HLB (Hydrophilic–Lipophilic Balance) of 4.3. The Malaysian Grade 2 diesel, D2 was selected as the base fuel; the detail specification is tabulated in Table 1. As for the metal additives blended W/D, three different volume of organic-based Manganese oxide were added to the E10; 100 ppm, 150ppm, and 200ppm and labelled as E10Mn100, E10Mn150, and E10Mn200 respectively. The blending process of the metal additives with the E10 where conducted by using ultrasonic mixer for 10 minutes.

2.2 *Engine Performance and Emission Test*

The schematic diagram of the engine testing setup is shown in Figure. 1. The type of engine used in the experiment is a 0.406 L single cylinder, four stroke, air-cooled, direct injection diesel engine. The combustion system of the engine is toroidal crown and the intake port type is helical. The other basic specifications of the engine are shown in Table 2. The aforementioned engine is coupled with an eddy current dynamometer for loading purposes. The eddy current used in the experiment is a 10kW KLAM RETARDER T10 dynamometer (dyno) which is able to sustain torque up to 25 Nm at 3700 rpm. The fuel flow into the engine is measured using the OMEGA FLR1007 flow meter sensor. The sensor is attached to the diesel fuel line, at the location just before the fuel enters the engine fuel pump, to measure the fuel consumption of the engine. As for the measurement of emissions, E – instrument E4500 emission analyser is used to measure the NOx, CO and HC contents in the exhaust gas emitted by the engine. The measurement is recorded for every 5 second interval during the 2 min duration of engine running and the average data are calculated for consideration and discussion. The engine tests are performed for different engine load conditions, 0 to 5kW, with a constant engine speed of 3500 rpm. The engine is initially run for 10-12 minutes and the corresponding exhaust temperature and speed are monitored. Once these parameters became steady, data are recorded.

Table 1. D2 fuel characteristic.

Properties	Unit	D2
Calorific value	MJ/kg	45.28
Cloud point	°C	18
Density @ 15 °C	kg/L	0.854
Total sulphur	mass%	0.28
Viscosity @ 40 °C	cSt	4.64
Distillation temperature, 90% recovery	°C	368
Flash point	°C	93
Pour point	°C	12
Cetane number	-	54.6
Carbon	wt%	84
Hydrogen	wt%	12.8
Sulphur	wt%	0.2
Nitrogen	wt%	<0.1
Oxygen	wt%	3.9

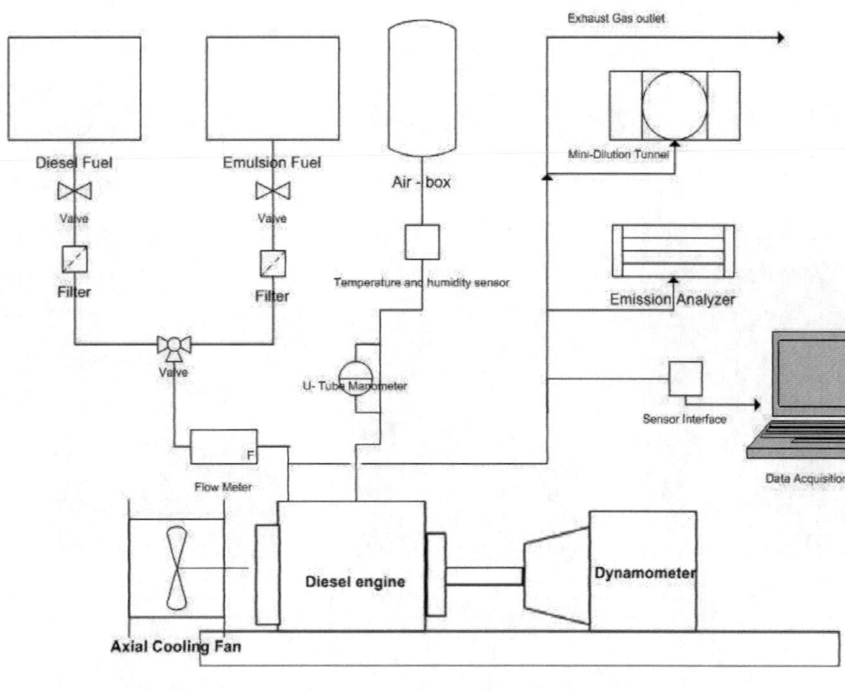

Figure 1. Schematic diagram of the engine emission test set up.

Table 2. Test engine specifications.

Parameter	Specification
Engine Type	4-Stroke, single cylinder, DI, air cooled,
Bore/ Stroke (mm)	86/70
Rated revolution (rpm)	3100
Compression ratio	19.3
Displacement (L)	0.406
Fuel injection pressure (Mpa)	19.6

3 Results and discussion

The brake specific fuel consumption (BSFC) of the engine fuelled with D2, E10, E10Mn100, E10Mn150, and E10Mn200 under varying engine load conditions (0-5kW) with constant speed of 3500 rpm is illustrated Figure 2. All three types of fuel show the same trend in the graph; the BSFC is significantly reduced when the load is increased indicating that the engine burns the fuel efficiently at high engine load. Maximum reduction of BSFC was obtained from the fuel E10Mn200. At 5kW load, E10Mn200 achieved maximum decreased of BSFC by 13.66% in comparison with E10. While, for E10Mn150 and E10Mn100 they reduced to 2.96%, and 0.49% of the brake specific fuel consumption. It was observed that the BSFC reduced when the load increased. The more quantity of additive being added increased reduction of BSFC. This proved that, supplementing additive which was organic-based manganese promotes better combustion of the injected fuel due to the additive that acts as an oxidation catalyser. Other than that, through introducing the organic-based manganese additive, the fuel properties were improved positively in which affect the atomization and penetration of the fuel, hence improving the fuel mixture.

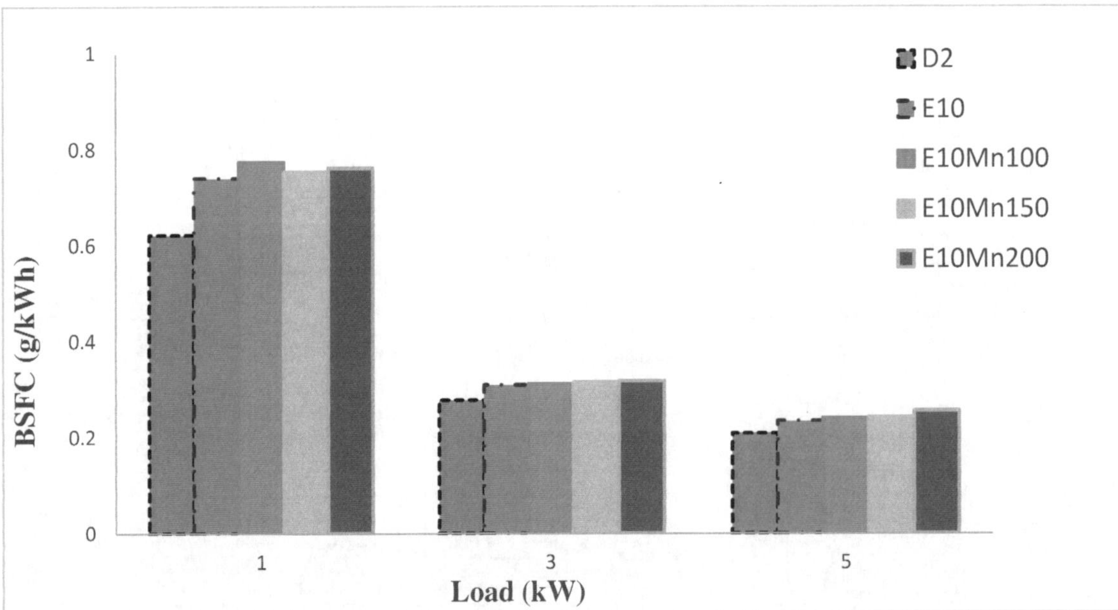

Figure 2. BSFC of tested fuel.

Figure 3 presents the CO emissions of tested fuel. As shown in the graph, overall trend in every fuels show that the formation of CO is drastically reduced as the load is increased, in which again this proves that the engine combust efficiently as higher load is applied. All the manganese blended emulsion fuels show significant reduction of CO compared to the neat E10. However, the said fuels still significantly higher than D2. The maximum reduction of CO was obtained from E10Mn150 as compared to other metal additive emulsions. At full load, E10Mn150 reduced up to 14.76% of CO emissions in comparison of E10. E10Mn100 reduced to 11.5% of CO emissions while E10Mn200 increased to 19.04% of CO emissions discharged at the full load. It was observed that the improvement of the CO emission start to decrease after 3 kW.

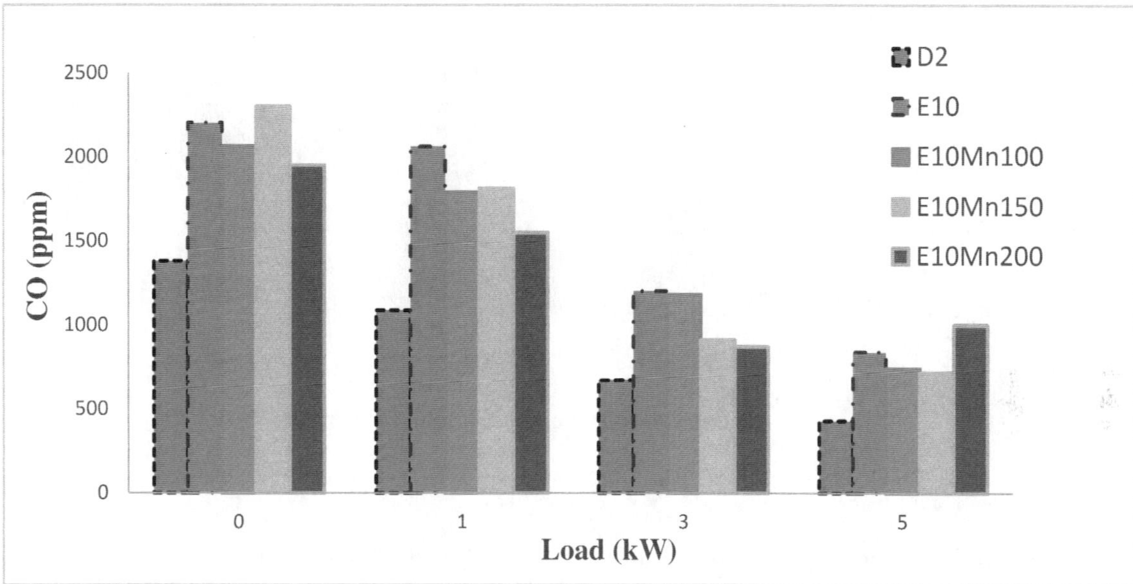

Figure 3. CO emission graph of tested fuel.

Figure 4 shows the plot of the formation of NO_x for D2, E10, E10Mn100, E10Mn150, and E10Mn200 under varying engine load conditions (0-5kW) with constant speed of 3500 rpm. As shown in the graph, overall trend for all the fuels plotted higher NO_x as the load is increased. Apart from that, all the manganese blended emulsion fuels show higher NO_x as compared with neat E10, however still significantly lower than D2. It was observed that adding of the manganese additive up to 150 ppm into the emulsion fuel indirectly increased the NO_x emissions. At maximum load, E10Mn150 emits the highest NO_x as compared to all emulsions 227ppm. However, the NO_x emission still much lower compared to D2 which is 302ppm. This is due to the to the water content in the emulsion that lowered the peak temperature of the flame during the combustion.

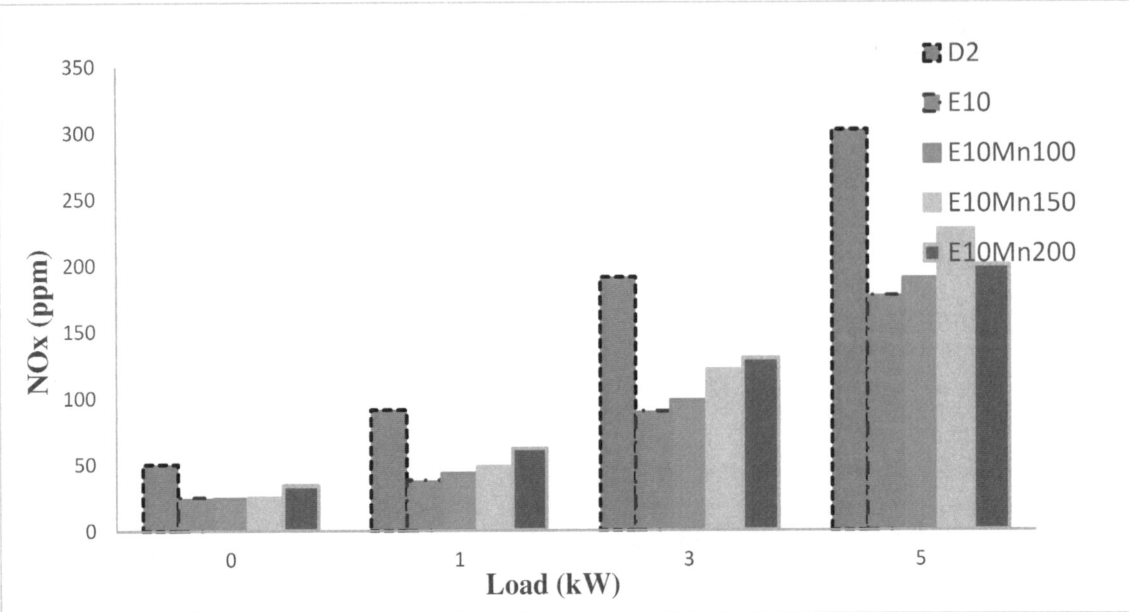

Figure 4. NOx emission graph of tested fuel.

Figure 5 presents the HC emissions of tested fuel. The HC emissions decrease in line as the load increased. The highest HC is reported when fuelled E10200 as compared to other fuels. At full load, the lowest HC was found at E10Mn100, followed by E10200, E10 and E10Mn150. The highest formation of HC was found at D2. This indicates that, by supplementing organic based manganese which acts as additive successfully decreased the HC emissions discharged.

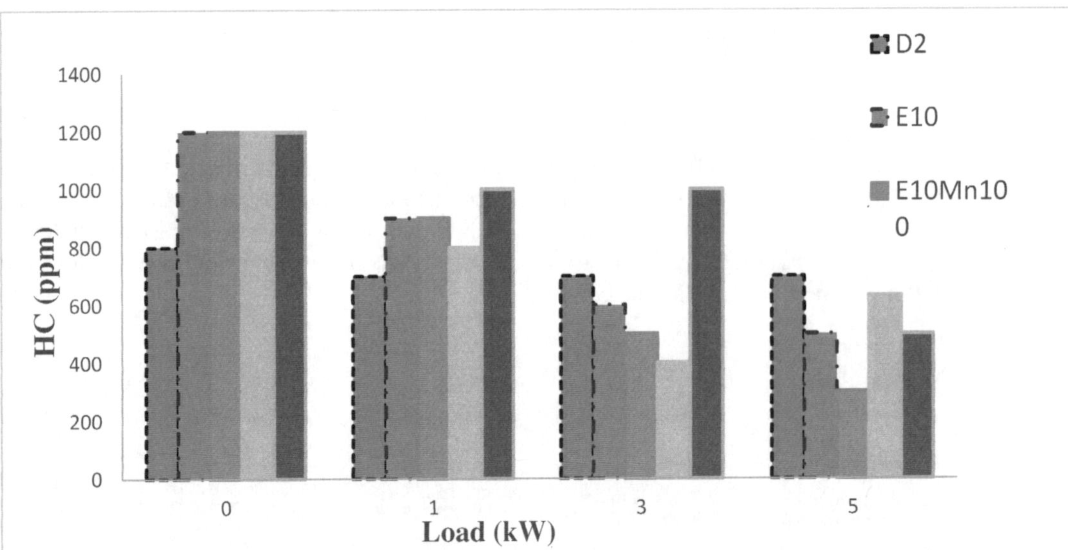

Figure 5. HC emission graph of tested fuel.

Figure 6 presents the exhaust gas temperature of tested fuel. Ignition delay has strong correlation with the increase of the exhaust gas temperature. From that graph shown, it is clearly shows that as the volume of manganese additive are increases, the exhaust gas temperature increased significantly. Maximum of exhaust gas temperature of E10, E10Mn100, E10Mn150, and E10Mn200 were 245 ^{0}C, 293 ^{0}C, 375 ^{0}C, and 399 ^{0}C respectively at full load. The highest exhaust gas temperature was observed to be obtained from E10Mn200 which is 62.86% followed by E10Mn150, E10Mn100, D2, and E10 respectively.

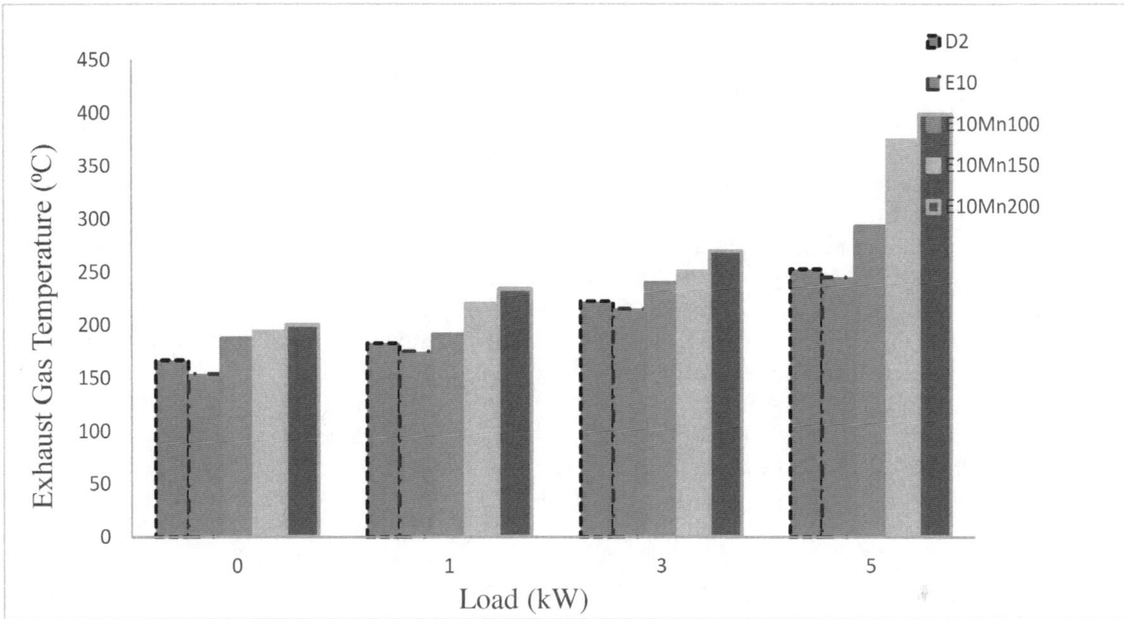

Figure 6. Exhaust gas temperature of tested fuel.

4. Conclusions

The detailed analysis of engine performance and exhaust emission analysis have been conducted using 3 different volume of organic manganese oxide (E10Mn100, E10Mn150, and E10Mn200) into E10 emulsion fuels. Based on the experimental investigations, E10Mn200 achieved the maximum decreased in BSFC. It reduced up to 13.66% of the BSFC in comparison with E10Mn100, and E10Mn150 which only reduced to 2.96%, and 0.49% of the BSFC respectively. The CO emission was reduced to 14.67% when fuelled with E10Mn150 compared to E10. However, E10Mn150 increased the NO_x emissions compared to other emulsions. E10Mn150 reduced the highest percentage of HC emissions at full load condition. Whereas, the highest exhaust gas temperature were achieved from E10Mn200 which it is 62.86% in comparison with E10.Supplementing organic based manganese additive into emulsion fuel promote the reduction of the fuel consumption, CO emissions, and HC emissions discharged from the engine. On the other hand, the manganese blended emulsion fuels slightly increased the NO_x emissions compared to the E10. Organic based manganese proved that able to improve the emulsion fuel performance and reduced the harmful emissions discharged.

Acknowledgments

The authors would like to express their gratitude to the Malaysia-Japan International Institute of Technology, Universiti Teknologi Malaysia who provided the research with related facilities and equipment. Highest appreciation goes to the Universiti Teknologi Malaysia for their financial support (Q.K.130000.2743.02K53).

References

[1] A. M. Ithnin, H. Noge, H. Abdul Kadir, and W. Jazair, "An overview of utilizing water-in-diesel emulsion fuel in diesel engine and its potential research study," *J. Energy Inst.*, vol. 87, no. 4, pp. 273–288, Nov. 2014.

[2] J. M. Ballester, N. Fueyo, and C. Dopazo, "Combustion characteristics of heavy oil-water emulsions," *Fuel*, vol. 75, no. 6, pp. 695–705, May 1996.

[3] a. K. Hasannuddin, J. Y. Wira, R. Srithar, S. Sarah, M. I. Ahmad, S. a. Aizam, M. a. B. Aiman, M. Zahari, S. Watanabe, M. a. Azrin, and S. S. Mohd, "Effect of emulsion fuel on engine emissions–A review," *Clean Technol. Environ. Policy*, 2015.

[4] M. Abu-Zaid, "Performance of single cylinder, direct injection Diesel engine using water fuel emulsions," *Energy Convers. Manag.*, vol. 45, no. 5, pp. 697–705, Mar. 2004.

[5] J. S. Basha and R. B. Anand, "An Experimental Study in a CI Engine Using Nanoadditive Blended Water–Diesel Emulsion Fuel," *Int. J. Green Energy*, vol. 8, no. 3, pp. 332–348, Apr. 2011.

[6] A. K. Hasannuddin, J. Y. Wira, S. Sarah, W. M. N. Wan Syaidatul Aqma, A. R. Abdul Hadi, N. Hirofumi, S. A. Aizam, M. A. B. Aiman, S. Watanabe, M. I. Ahmad, and M. A. Azrin, "Performance, emissions and lubricant oil analysis of diesel engine running on emulsion fuel," *Energy Convers. Manag.*, vol. 117, pp. 548–557, 2016.

[7] A. K. Hasannuddin, J. Y. Wira, S. Sarah, M. I. Ahmad, S. A. Aizam, M. A. B. Aiman, S. Watanabe, N. Hirofumi, and M. A. Azrin, "Durability studies of single cylinder diesel engine running on emulsion fuel," *Energy*, vol. 94, no. x, pp. 557–568, 2016.

[8] A. M. Ithnin, M. A. Ahmad, M. A. A. Bakar, S. Rajoo, and W. J. Yahya, "Combustion performance and emission analysis of diesel engine fuelled with water-in-diesel emulsion fuel made from low-grade diesel fuel," *Energy Convers. Manag.*, vol. 90, pp. 375–382, Jan. 2015.

[9] V. Sajith, C. B. Sobhan, and G. P. Peterson, "Experimental investigations on the effects of cerium oxide nanoparticle fuel additives on biodiesel," *Adv. Mech. Eng.*, vol. 2010, no. March, 2010.

[10] A. I. El-Seesy, A. K. Abdel-Rahman, M. Bady, and S. Ookawara, "Performance, combustion, and emission characteristics of a diesel engine fueled by biodiesel-diesel mixtures with multi-walled carbon nanotubes additives," *Energy Convers. Manag.*, vol. 135, pp. 373–393, 2017.

[11] J. S. Basha and R. B. Anand, "Role of nanoadditive blended biodiesel emulsion fuel on the working characteristics of a diesel engine," *J. Renew. Sustain. Energy*, vol. 3, no. 2, 2011.

IOP Conf. Series: Materials Science and Engineering **257** (2017) 012035 doi:10.1088/1757-899X/257/1/012035

Study of solvent debinding parameters for metal injection moulded 316L stainless steel

M F F A Hamidi[1*], W S W Harun[2], N Z Khalil[2], S A C Ghani[2], M Z Azir[3]

[1] Institute of Postgraduate Studies, Universiti Malaysia Pahang, Lebuhraya Tun Razak, 26300 Gambang, Kuantan, Pahang, Malaysia
[2] Green Research for Advanced Materials (GramsLab), Human Engineering Group, Faculty of Mechanical Engineering, Universiti Malaysia Pahang, 26600 Pekan, Pahang, Malaysia
[3] DRB-HICOM University of Automotive Malaysia, Kawasan Perindustrian Peramu Jaya, 26600 Pekan, Pahang, Malaysia

*Corresponding author: faizulfarhan06@gmail.com

Abstract. Solvent debinding is one of a crucial stage in Metal Injection Moulding (MIM) process. This process begins with the removal of the soluble binder components by using solvents such as heptane or hexane. In solvent debinding process, unsuccessful to achieve maximum binder removal will cause a defect to the compact such as crack and swelling. So to have an optimum solvent debinding parameters are very important to improve the quality of the compact. Optimisation of solvent debinding process parameters for MIM of Stainless Steel 316L has been testified in this study. Gas atomised stainless steel 316L powder was mixed with a multicomponent binder in a twin blade mixer at a temperature of 150 °C for 90 minutes. The feedstock was successfully injected at the temperature of 150 °C. The green compacts were kept in n-heptane for eight different debinding times ranging between 30 to 240 minutes at temperatures of 40, 50, 60 and 70 °C to remove the primary binder components. From the result, the optimum temperature and time for solvent debinding were recorded at 60 °C and 240 minutes. Solvent debinding temperature and time give a significant effect on the rate of paraffin wax removal.

1. Introduction

The fabrication of biocompatible metals as implant devices is restricted because of the rather high costs of raw materials, complex design geometry, and limitations of the current fabrication process. Metal injection moulding (MIM) could serve as an alternative means to overcome these problems. One of the reasons is that MIM process could reduce production costs due to its net-shape fabrication advantages, befitting for manufacturing of small parts, and combine high part complexity with large production quantities [1-4]. This technique was procured and adapted from the plastic injection moulding process, of which small metal particles replaced a significant volume fraction of plastic [5-8].

In MIM technology, there are four processing stages which are, mixing of powders and binders to produce feedstock, injection moulding, debinding, and sintering.The third step was the debinding process where all the binders will be extracted from the compacts. This process is crucial as it can influence MIM processes and the ultimate quality of the products. Long debinding times combined with relatively high tendency of compact distortion are the main challenge for MIM process [9-12]. To increase the rate of binder removal while preventing the compacts from defects, debinding is typically conducted in multiple stages, namely solvent and thermal debinding. For solvent debinding process, it has been studied first to perforate the binder structure before proceeded to thermal debinding stage [13-15]. The solvent debinding process was popularised to the metal injection moulding industries during the 1980s. In this process, it began with the removal of the soluble binder components by using solvents such as heptane or hexane [16, 17].

In this study, an experiment was conducted to optimise the solvent debinding parameters of 316L stainless steel. 62 vol.% powder loading of 316L stainless steel were mixed with the multi-component binders by using twin blade mixer. The green compacts were kept in vaporised n-heptane solvent at various time and temperatures. The microstructure of green and debound compacts were observed using SEM. It is expected that higher solvent debinding temperature and time will increase the removal rate of the primary binders, particularly paraffin wax.

2. Experimental Method

2.1 Material

The gas atomized 316L Stainless Steel (SS) alloy powder provided by Osprey Co, the UK with the mean particle size of 11.4 μm was utilised in this study. Chemical composition and particle morphology of the powder are shown in Table 1 and figure 1 respectively.

Table 1. Chemical composition of gas atomized 316L SS.

Element	Wt.%
Cr	16.7
Ni	10.3
Mo	2.2
Mn	0.99
Si	0.69
P	0.02
C	0.01
S	0.05

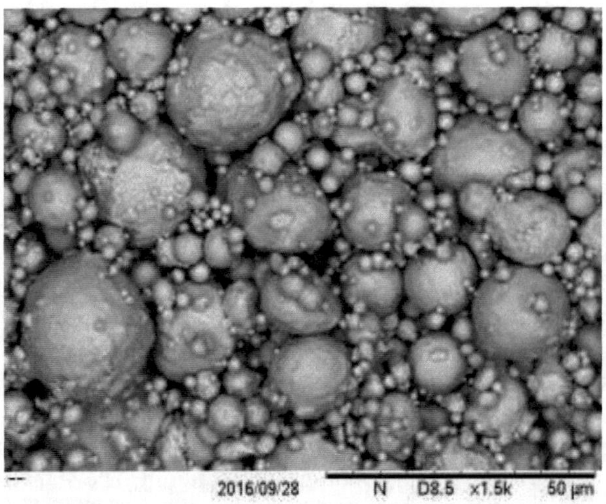

Figure 1. Particle morphology of 316L SS powder.

2.2 Feedstock Preparation

A formulation 62 vol % powder loading of stainless steel powders was prepared. 62 vol % powder loading was obtained from the preliminary investigations performed to estimate the optimum powder loading. A multicomponent binder system used to formulate the feedstock consists of Paraffin Wax (PW), Polypropylene (PP), and Stearic Acid (SA). Table 2 shows the composition of the multicomponent binder used in this work. Twin blade type mixer at a rotational speed of 70 rpm at 150 °C for 90 min has been utilised to mix the feedstock.

Table 2. Binder system for SS 316L.

Binder Components	Composition (%)
Paraffin wax (PW)	70
Polypropylene (PP)	25
Stearic acid (SA)	5

2.3 Thermal analysis of the feedstock
To identify the melting temperature for each binder component, Differential Scanning Calorimetry (DSC) has been used. The suitable temperature for mixing, injection moulding and solvent debinding also can be determined from this analysis. DSC analysis has been performed on an NETZSCH DSC 214 Polyma DSC21400A-01717-L equipment. The heating rate was set at 10°C/min and conducted under nitrogen atmosphere.

2.4 Injection Moulding of Feedstock
Injection process for feedstock was performed on a Nissei NS20-2A injection moulding machine to fabricate tensile shape compacts. The green compacts were produced by injection moulding at 150 °C. There is no defects were observed on the green compact after it have been checked physically.

2.5 Solvent Debinding
For solvent debinding process, wicking debinding technique was applied by using fine Al_2O_3 powders [13]. At this stage, the primary binder which is paraffin wax will be extracted from the green compacts. Compacts were kept in a solvent bath of the vaporised heptane at 40, 50, 60 and 70 °C for 30 to 240 minutes. The schematic diagram for solvent debinding process is shown in figure 2.

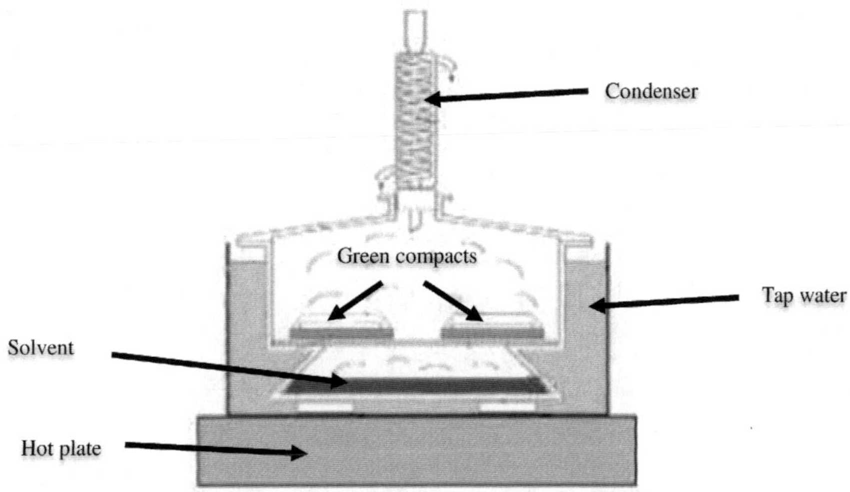

Figure 2. Schematic diagram for solvent debinding process.

2.6 SEM Analysis
The micrograph for green and solvent debound compacts was analysed using Scanning Electron Microscopy (SEM). This analysis also used to verify the complete leaching and homogenous distribution of primary binder from the solvent debound compacts correspondingly.

3. Results and Discussion

The green compacts were well moulded via Nissei NS20-2A injection moulding machine. Figure 3 shows a defect-free green compacts. The mould and injection temperatures were 40°C and 150°C respectively.

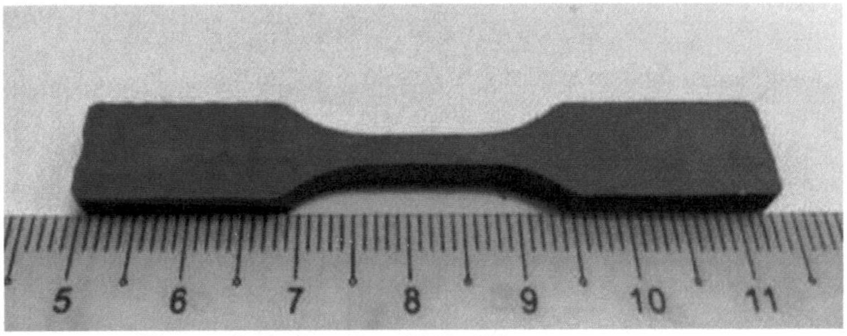

Figure 3. Defect-free 316L SS green compacts.

3.1 Solvent Debinding

To extract the soluble binders, solvent debinding was carried out by keeping the green compacts in the solvent bath of vaporised n-heptane solution at different temperatures and times. This technique has been chosed because paraffin wax can be dissolved in vaporised n-heptane. During solvent debinding process, open pore channels will be produced on the debound compacts, which allow the diffusion process of the remaining binder in the second debinding process. The schematic diagram for the solvent debinding process is shown in figure 4. This schematic diagram suggests how the binders was extracted from the green compacts due to the capillary forces that happened when these compacts start to dissolved by the vaporised solvent [18].

The indicator to determine the suitable temperature during solvent debinding which is the melting point of the binders was verified by using DSC analysis. Figure 5 depicts the DSC analysis result for the feedstock .From Figure 5, there was three peak melting temperature of the multicomponent binders have been observed which are 58 and 168 °C. Each peak shows an endothermic reaction. It is envisaged that peaks at lower temperature (58°C) correspond to the the melting point of Paraffin Wax and Stearic Acid whereas the peak at a higher temperature correspond to the melting point of Polypropylene. The solvent debinding temperature directly affected the diffusivity and solubility of the binders in the solvent. Thus, the selection of solvent debinding temperature should be guided by the melting temperatures of the binders. The melting temperature of Paraffin wax and Stearic acid were used as a reference at this stage.

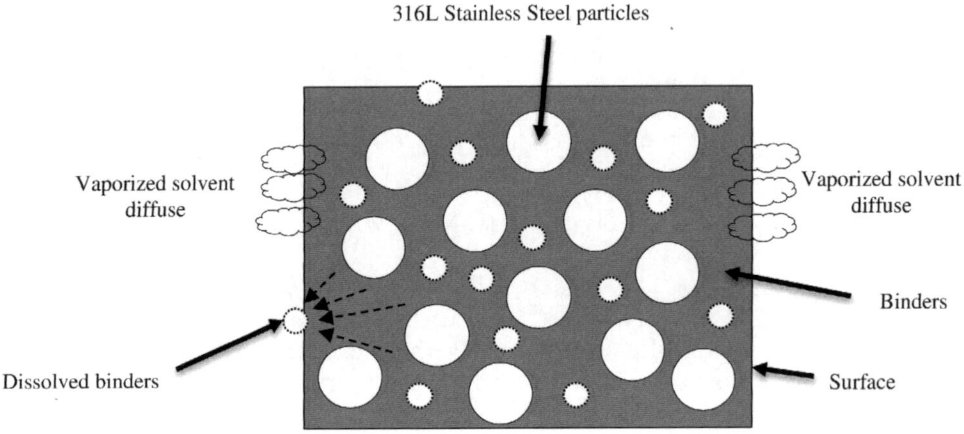

Figure 4. The schematic of solvent debinding stage.

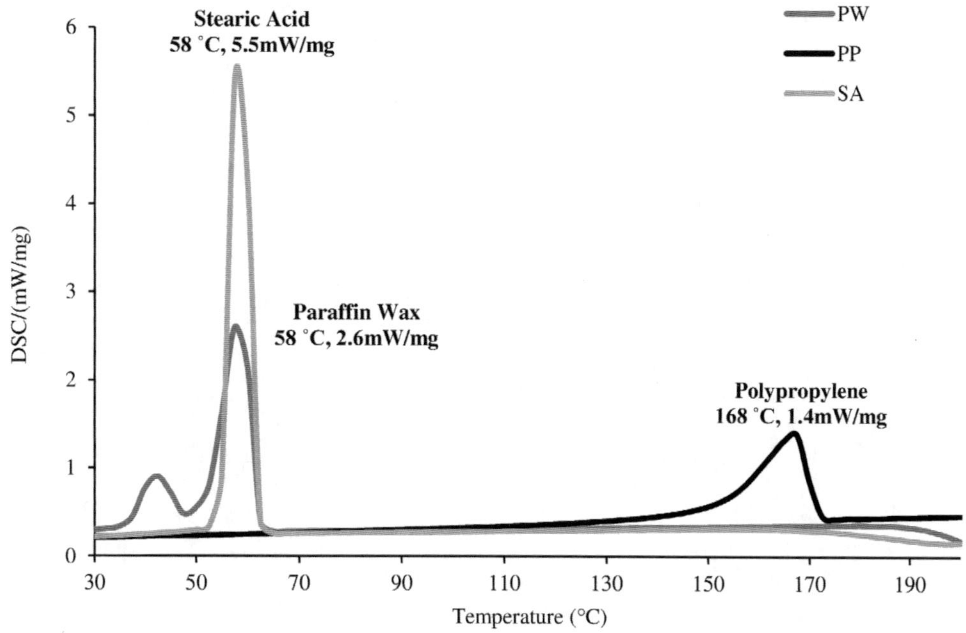

Figure 5. Differential Scanning Calorimetry analysis.

In this work, wicking debinding technique was applied by using Al_2O_3 powder. From the observation, at the end of the solvent debinding process, removed paraffin wax changed the Al_2O_3 powder from white to yellowish colour as shown in figure 6. It was suggested by [19] that this observation is due to the capillary suction that causes the soluble binder move toward the wicking powder as it is in contact with the powder.

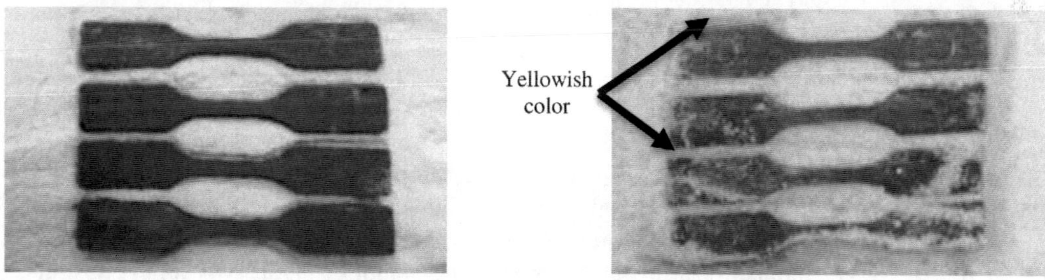

(a) (b)

Figure 6. Solvent debinding a) before solvent debind, b) After solvent debind.

To identify the optimum solvent extracted condition, four (4) debinding temperatures were chosen which were 40, 50, 60 and 70°C. The debinding time was varied between 30 to 240 minutes to observe the optimum extraction time. Figure 7 presents the influence of solvent debinding temperature and time on the rate of Paraffin Wax removal. Mass loss of the binder (M_{loss}), which is the paraffin wax, was calculated based on the following equation:

$$M_{loss} = \frac{Mi - Mf}{Mi} \text{ x } 100 \quad (1)$$

Where Mi is the mass of green compact and Mf is the mass of debound compact. It can be determined that the percent of paraffin wax loss from the green compacts is increased with the increases in debinding temperature, particularly in the first 30 minutes of the process. At this period, the rate of binder loss is extremely fast as the vaporised heptane were directly in contact with the binder on the surface of the compacts. Then, before the vaporised n-heptane dissolved the binder, it firstly diffused deeply into the green compacts. Thus, the rate of binders' loss significantly slowed till it achieved a plateau region after 210 to 240 minutes. This phenomenon occurred when a dynamic equilibrium has been reached between the solvent and binders as the percentage of paraffin wax removal maintained unchanged. This phenomenon also has been explained by the previous researcher [20].

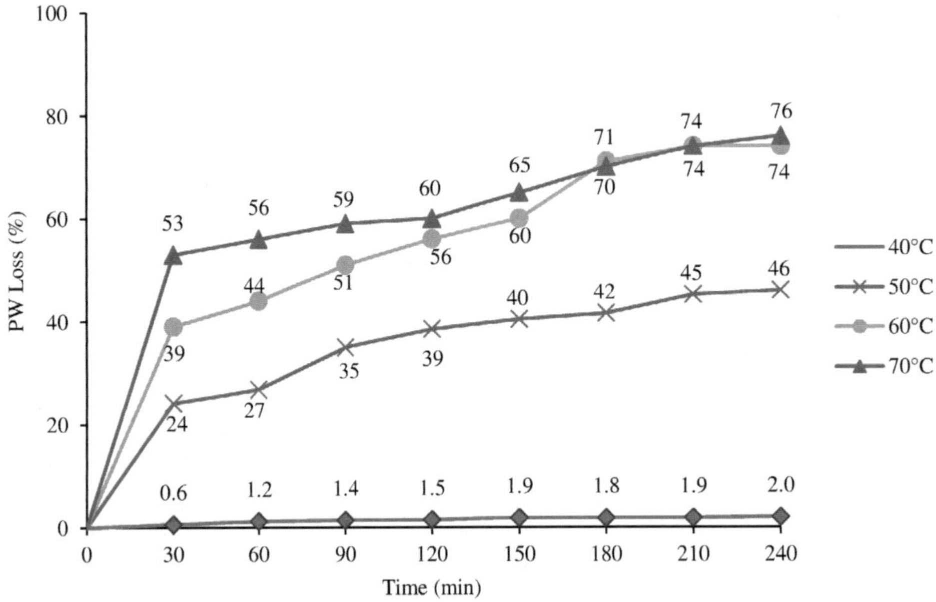

Figure 7. Effect of solvent debinding temperature and time on the rate of paraffin wax removal.

3.2 Defects

From Figure 7, 74 % and 76 % wax was removed at temperature 60 °C and 70 °C after 240 minutes. Both temperatures can be considered as the optimum temperature for solvent debinding process as they recorded the highest percent of paraffin wax loss from the green compacts. The amount of Paraffin Wax loss for both temperature also was acceptable because too much binder loss will cause the compact to become fragile and difficult to handle. However, for the first 30 minutes, 53 % of Paraffin Wax has been removed at temperature 70 °C, and it is higher compared to temperature 60 °C which recorded 39 % of Paraffin Was loss. The maximum amount of binder loss for 70 °C at the early stage of the debinding process inappropriate for the green compacts as some of the compacts were found to be broken as shown in figure 8.

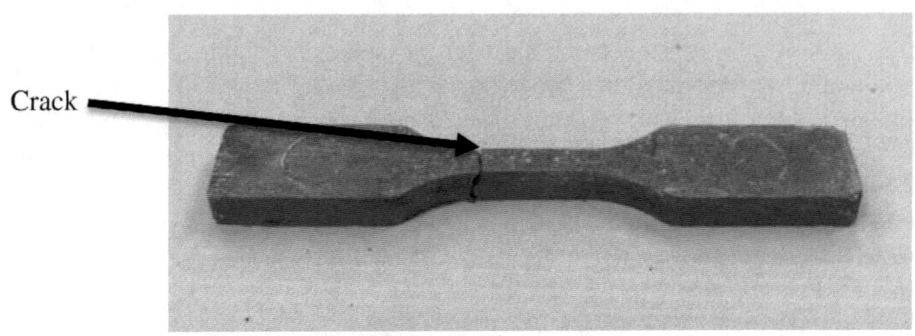

Figure 8. Damaged compacts.

Too high extracting temperature was not appropriate for solvent debinding process as it will affect the quality and can cause defects on the debound compacts. This result also align with previous work by [21] and [18] where they found that, after the debinding temperature had been increased higher than 60 °C, swelling and crack were generated on the surface of the compacts. They also explained that more binders are diffused and melted speedily out from the compacts when the debinding temperature was higher than the melting temperature of a binder hence leading to the initiation of defects on the debound compacts. The defects that experienced by the compacts also might be because of the thermal expansion of the binders because of the reaction at higher temperature among the binders and solvent. In this work, most of the compacts were cracked at 70 °C. Therefore, 60°C has been set as the optimum solvent debinding temperature where high percent rate of paraffin wax removal was recorded. After 240 minutes, 74 % of paraffin wax loss was achieved, which demonstrated in the creation of defect-free compacts.

3.3 SEM Analysis

SEM micrograph in figure 9 shows a cross-sectional view of the green and debound compact. The SEM micrograph for the debound compact was taken from the green compact that was debound in the optimum condition which is at 60°C for 240 minutes. It is observed that binders were uniformly distributed within the 316L stainless steel powders to prevent it from oxidation and maintain the shape of the compacts as shown in Figure 9(a). Uniform binders' distribution also can be observed in the green compact. Figure 9(b) presents the open pores channels that were produced after the removal of paraffin wax. Some were inter-particle pores, and several were in the interior of the binder, indicating that paraffin wax, polypropylene and stearic acid were interacted and mixed to a degree and that a few soluble binders were removed. This result has proved the effectiveness and success of n-heptane as a solvent in extracting the paraffin wax from the green compacts. After 240 minutes of solvent debinding process, a significant amount of paraffin wax has been extracted, leaving the remaining polypropylene either in the contact areas or as whiskers holding particles together. From the SEM observation, the open pore channels also have been developed at the centre of the compact. This shows that the penetration of vaporised was reached at the core of the compact. With these open pore channels, the remaining binders especially polypropylene can be removed easily without affecting the compacts in the thermal debinding stage. If there are no open pores were produced in this stage, the backbone binders especially polymer cannot be removed easily in the thermal debinding stage. The binders will keep pushing the compact surface outward and cause the surface of the compacts become rough and also have a potential to produce a crack on the compact [22]. Therefore, this process can help in shortening the thermal debinding process and maintained the integrity of the compact [17].

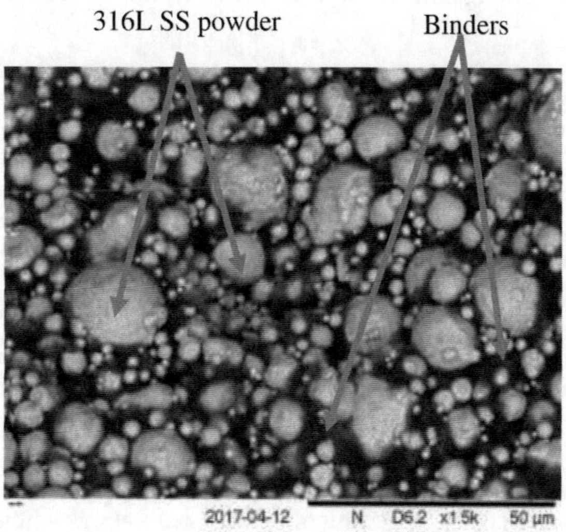

(a)

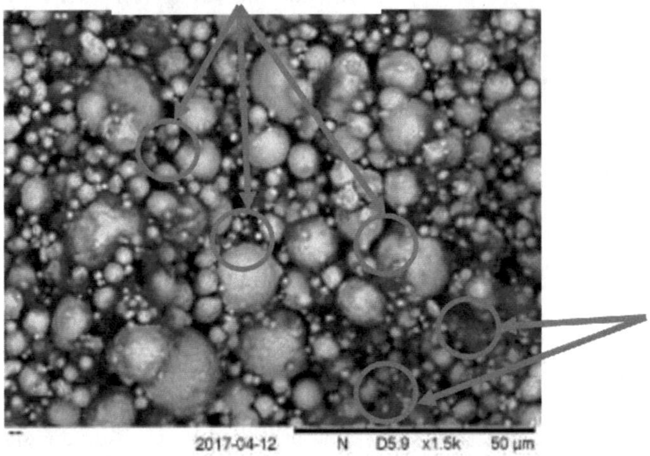

(b)

Figure 9. SEM micrograph (a) green compact (b) solvent debound compact.

4. Conclusions

This work concludes the following:

1) To attain the optimum parameters for solvent debinding process, the green compacts were solvent debound at different temperature and times. From the experiment, temperature and time proved to have a significant effect on the mass loss of paraffin wax as the removal rate increased with the increase of temperature and times particularly in the first 30 minutes.
2) The optimum solvent debinding parameters have been verified. Solvent debinding at 60°C for 240 minutes were considered as the optimum parameters for solvent debinding process because of the adequate quantity of paraffin wax loss. The total percentage of paraffin wax removal for this condition was recorded at 74 %.
3) Solvent debinding temperature at 70°C showed a rapid removal of paraffin wax and generated a defect which is crack on the compact. This might be due to thermal expansion of the binders as a result of the reaction at higher temperature among the binders and solvent.
4) From SEM analysis for green compact, the binders were very well mixed with the 316L stainless steel powder. For a debound compact, a lot of open pore channels were produced where they can help on removing the remaining binders in the thermal debinding stage easily.

Acknowledgments

Universiti Malaysia Pahang fully supports the resources and facilities for this work. The author W.S. W. Harun would like to acknowledge the funding of the internal grant of Universiti Malaysia Pahang RDU141101, RDU140354, RDU150337, PGRS160383, PGRS170388 and the support of Research Acculturation Collaborative Effort (RACE) RDU151314 and Research Acculturation Grant Scheme (RAGS) RDU151404 provided by the Ministry of Higher Education, Malaysia. Besides that, the project also has been supported by Qatar National Research Fund NPRP88762375 (UIC161504).

References

[[1] Santos P F, Niinomi M, Liu H, Cho K, Nakai M, Itoh Y, Narushima T and Ikeda M 2016 Fabrication of low-cost beta-type Ti–Mn alloys for biomedical applications by metal injection molding process and their mechanical properties *Journal of the Mechanical Behavior of Biomedical Materials* **59** 497-507

[2] Hayat M D, Goswami A, Matthews S, Li T, Yuan X and Cao P 2017 Modification of PEG/PMMA binder by PVP for titanium metal injection moulding *Powder Technology* **315** 243-9

[3] Hausnerova B, Mukund B N and Sanetrnik D 2017 Rheological properties of gas and water atomized 17-4PH stainless steel MIM feedstocks: Effect of powder shape and size *Powder Technology* **312** 152-8
[4] Marçal R L S B 2016 *Reference Module in Materials Science and Materials Engineering*: Elsevier)
[5] Aggarwal G, Smid I, Park S J and German R M 2007 Development of niobium powder injection molding. Part II: Debinding and sintering *International Journal of Refractory Metals and Hard Materials* **25** 226-36
[6] Raza M R, Ahmad F, Muhamad N, Sulong A B, Omar M A, Akhtar M N and Aslam M 2016 Effects of solid loading and cooling rate on the mechanical properties and corrosion behavior of powder injection molded 316 L stainless steel *Powder Technology* **289** 135-42
[7] Zhao D, Chang K, Ebel T, Nie H, Willumeit R and Pyczak F 2015 Sintering behavior and mechanical properties of a metal injection molded Ti–Nb binary alloy as biomaterial *Journal of Alloys and Compounds* **640** 393-400
[8] Mariot P, Leeflang M A, Schaeffer L and Zhou J 2016 An investigation on the properties of injection-molded pure iron potentially for biodegradable stent application *Powder Technology* **294** 226-35
[9] Li Y, Liu S, Qu X and Huang B 2003 Thermal debinding processing of 316L stainless steel powder injection molding compacts *Journal of Materials Processing Technology* **137** 65-9
[10] Omar M A, Ibrahim R, Sidik M I, Mustapha M and Mohamad M 2003 Rapid debinding of 316L stainless steel injection moulded component *Journal of Materials Processing Technology* **140** 397-400
[11] Enneti R K, Shivashankar T S, Park S-J, German R M and Atre S V 2012 Master debinding curves for solvent extraction of binders in powder injection molding *Powder Technology* **228** 14-7
[12] Setasuwon P, Bunchavimonchet A and Danchaivijit S 2008 The effects of binder components in wax/oil systems for metal injection molding *Journal of Materials Processing Technology* **196** 94-100
[13] Gorjan L, Kosmač T and Dakskobler A 2014 Single-step wick-debinding and sintering for powder injection molding *Ceramics International* **40** 887-91
[14] Hayat M D, Wen G, Zulkifli M F and Cao P 2015 Effect of PEG molecular weight on rheological properties of Ti-MIM feedstocks and water debinding behaviour *Powder Technology* **270** 296-301
[15] Shbeh M M and Goodall R 2015 Design of water debinding and dissolution stages of metal injection moulded porous Ti foam production *Materials & Design* **87** 295-302
[16] Aslam M, Ahmad F, Yusoff P S M B M, Altaf K, Omar M A and M.German R 2016 Powder injection molding of biocompatible stainless steel biodevices *Powder Technology* **295** 84-95
[17] Zaky M T, Soliman F S and Farag A S 2009 Influence of paraffin wax characteristics on the formulation of wax-based binders and their debinding from green molded parts using two comparative techniques *Journal of Materials Processing Technology* **209** 5981-9
[18] Md Ani S, Muchtar A, Muhamad N and Ghani J A 2014 Binder removal via a two-stage debinding process for ceramic injection molding parts *Ceramics International* **40** 2819-24
[19] Gorjan L, Dakskobler A and kosmač T 2010 Partial wick-debinding of low-pressure powder injection-moulded ceramic parts *Journal of the European Ceramic Society* **30** 3013-21
[20] Cheng J, Wan L, Cai Y, Zhu J, Song P and Dong J 2010 Fabrication of W–20 wt.%Cu alloys by powder injection molding *Journal of Materials Processing Technology* **210** 137-42
[21] Raza M R, Ahmad F, Omar M A and German R M 2012 Effects of cooling rate on mechanical properties and corrosion resistance of vacuum sintered powder injection molded 316L stainless steel *Journal of Materials Processing Technology* **212** 164-70
[22] Farhan M F, Hamidi A, Sharuzi W, Harun W, Faiz A and Abu Bakar S 2016 Effect of Binders on Physical and Mechanical Properties of Stainless Steel 316L Alloy Fabricated by Metal Injection Moulding Process

IOP Conf. Series: Materials Science and Engineering **257** (2017) 012036 doi:10.1088/1757-899X/257/1/012036

Diesel engine fuel consumption and emission analysis using steam generated non-surfactant water-in-diesel emulsion fuel

Dhani Avianto Sugeng[1,2], Mohamad Fathur Hafeezat Mohd Zahari[1], Ahmad Muhsin Ithnin[1], Wira Jazair Yahya[1*]

[1]Vehicle System Engineering, Malaysia-Japan International Institute of Technology, Universiti Teknologi Malaysia, Jalan Sultan Yahya Petra, 54100 Kuala Lumpur, Wilayah Persekutuan Kuala Lumpur, Malaysia
[2]Badan Pengkajian dan Penerapan Teknologi (BPPT - Agency for the Assessment and Application of Technology), Jalan MH Thamrin 8, Jakarta 10340, Indonesia

*Corresponding email: wira@utm.my

Abstract. Efforts in making water in diesel emulsion (W/D) with the absence of surfactant have been developed to address the issues of long-term stability and the dependence on surfactants. This paper discusses an alternative formation method of a non-surfactant W/D, e.g. by steam condensation. By injecting steam into a batch of colder diesel fuel, fine water droplets are formed and suspended in the fuel forming an emulsion. The droplets are confirmed to be in the size range of hundreds of nanometers. The emissions of NOx is reduced by a maximum of 71%, whereas the CO and UHC emissions are increased by maximum respectively 180% and a surprising 517%. Not less interesting is the lower BSFC which was measured at a maximum reduction of 18.4%. These results on emission analysis together with the brake specific fuel consumption confirm this method to resemble the combustion behaviour of a conventional emulsion fuel of lower NOx and BSFC, yet higher CO and UHC

1. Introduction

Interest in the research of water-in-diesel emulsion (W/D) is growing since while it is an alternative fuel that can reduce the emission of nitrogen oxides (NOx), black smoke and particulate matter (PM) simultaneously, it also improves the combustion efficiency. Moreover, W/D can be used in a regular diesel engine without major modification[1, 2].

Conventional W/D is defined as an immiscible blend of water droplets (dispersed/internal phase) in diesel (continuous/external phase) [3]. A surfactant in low concentrations is added to prolong emulsions stability and lower the energy requirement [4], which is often supplied by mechanical agitation [5].

In 2015 at the VSE Lab in MJIIT-UTM Malaysia, Muhsin developed a system called the RTES (Real Time non-surfactant Emulsion Fuel Supply System). This system produces non-surfactant W/D by combining an ultrasonic transducer with a high shear mixer and was proven to be able to feed a 5 kW stationary engine [6] and a light-duty lorry [7]. This method of fuel preparation utilises mechanically rotating and electronic components such as additional pumps, high-shear mixer, an ultrasonic transducer, which need extra investment, operating and maintenance costs. To reduce these other costs, finding alternatives on reducing these components makes a good case.

The alternative is found in steam, where water in vapour state will condense in contact with sub-cooled diesel. In this concept, steam would change phase and lose its sensible heat in the sub-cooled diesel [8]. Because of the nucleation phenomena [9] and the turbulent condition, very small water droplets may be produced, dispersed and suspended in the diesel.

Therefore, this articles main objective is to evaluate the feasibility of the concept of making non-surfactant W/D through steam condensation method. The evaluation is performed by characterising the water droplet size of the produced fuel and analysing its emission and the brake- specific fuel consumption (BSFC) on a single cylinder diesel engine.

2. Experimental

2.1. Materials

In this study, the main components are diesel and water. For diesel, Euro 2 standard diesel (D2) with physical characteristics tabulated in Table 1 was used as the continuous emulsion phase. As for water, Malaysian tap water was used with no pre-treatment, and its characteristics are tabulated in Table 2 [10].

2.2. Emulsion preparation

The preparation of W/D was set up with the help of a glass column with multiple nozzles as illustrated in Figure 1.The column is filled with diesel entering the column through nozzle 2. Steam is then supplied from a steam generator through nozzle 5 generating bubbles in the diesel pool. When emulsified fuel overflows nozzle 4, it will exit the flash and go to the fuel injector into the combustion chamber. Some condensate will enter the flash due to the saturation of the steam and will eventually discharge through nozzle
6. Through nozzle 1 some uncondensed steam will exit the column.

2.3. Measurements

After the emulsion is formed, a sample from the sampling point is taken for further analyses, which are water percentage and droplet size distribution.

2.3.1. Water percentages The amount of water forming the emulsion droplets cannot be directly measured since the vapour condensation rate is unknown as some steam leaves the system uncondensed. Instead, to find the water percentages in samples, distillation is performed at a temperature of 140C.

2.3.2. Droplet size and its distribution Water droplet size plays a significant role in altering the combustion behaviour, due to its role as an agent that causes the micro-explosion phenomenon [1]. To determine the droplet size distribution, a laserdiffraction particle-size-analyzer (Shimadzu SALD 2300) was used to analyse the droplet size distribution. The SALD claims to be able to examine across a broad particle concentration range of 0.1 ppm to 20%, and its measurement range covers particle sizes from 17 nm to 2,500 m. The refractive index of water as a dispersant is required for accurate determination of small particle sizes and should be at 1.3325 [11]. However, because such value could not be selected with the SALD, a value of 1.35 with an imaginary index of 0 was used.

2.3.3. Emission analysis The produced emulsion from subsection 2.2 is supplied to a 5.5 kW naturally aspirated indirect injection diesel engine with a configuration as illustrated in Figure 2. The setup followed the SAE Engine Test Code J816b and focused on monitoring the NOx, CO and unburned hydrocarbons (UHC) emission. Exhaust emission is sampled and monitored every 10 seconds for 5 minutes with E-Instruments E4500 portable emission analyser; an average was then considered for the discussion. The engine is coupled to an engine dyno with variable torque to exert variable loads between 0 to 5 kW to measure the emission. The BSFC of D2 and emulsion fuel is measured by using a burette installed after the mixer.

IOP Conf. Series: Materials Science and Engineering **257** (2017) 012036 doi:10.1088/1757-899X/257/1/012036

Table 1. Malaysian Euro 2 Diesel (D2) characteristics

Properties	Unit	Value
Calorific Value	MJ/kg	45.28
Cloud Point	C	18
Density @15C	kg/L	0.8538
Total Sulphur	mass %	0.28
Viscosity	cSt	7.6 @20C
5.1 @40C		
4.0 @60C		
Distillation Temperature, 90% recovery	C	367.9
Flashpoint	C	93.0
Pour Point	C	12
Cetane Number		54.6
Carbon	wt %	84.1
Hydrogen	wt %	12.8
Sulphur	wt %	0.2
Nitrogen	wt %	<0.1
Oxygen	wt %	3.9
Dielectric constant		2.1

Table 2. Characteristics of Malaysian tap water

Specification	
Density, g/cm, 25 C	1.02412
Specific conductivity, S/cm , 25 C	0.0532
Viscosity, milipoise , 25 C	9.02
Vapor pressure, mm Hg, 20 C	17.4
Isothermal compressibility, vol/atm, 0 C	46.4 E-6
Surface tension, dyne/cm , 0 C	72.74
Specific heat, J/g C , 17.5 C	3.898
Temperature of Maximum density, C	-3.25
Freezing point, C	-1.91

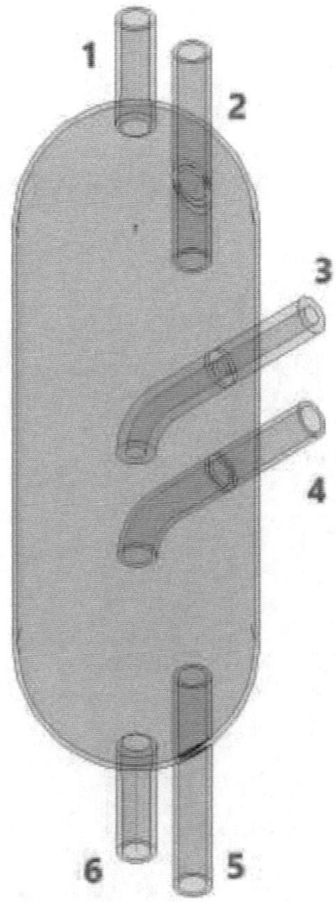

Figure 1. Mixer design

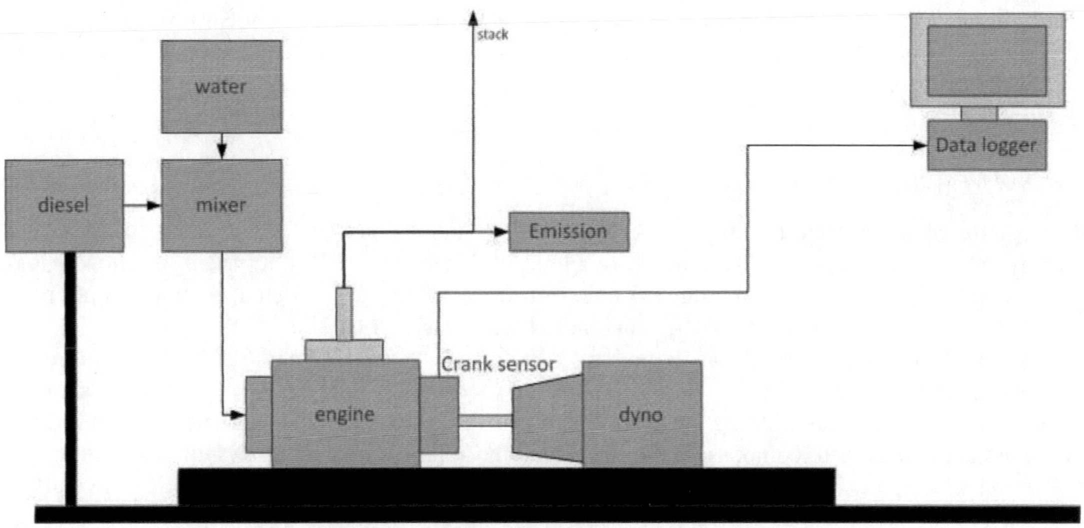

Figure 2. Engine test schematics

3. Results and discussion

The experiment confirms that an emulsion is obtained by the procedure described in 2.2. In Figure 3, it can be observed that the emulsions are yellowish but milky. Water droplets suspended in the diesel cause the milky appearance of the emulsions.

Figure 3. Comparison of D2 and steam-emulsion

When steam enters the mixer, it formed bubbles that rose upward and came into contact with sub-cooled diesel. Wu, et al. stated that the ascending bubbles form nuclei of condensates on its surface [9]. The nuclei will form a layer when the bubble stream is in laminar condition, but turbulence caused by its wake propels and mixes the condensate nuclei into the liquid bulk forming the milky looking substance.

3.1. Droplet size distribution

The presence of the submicron droplets is confirmed by the droplet size distribution measurement using the Shimadzu SALD 2300 as shown in Figure 4. There, the distribution curves show a log-normal distribution with series 1 through 3 belong to a sample measured after 60 minutes after the formation procedure with an interval of 15 minutes between each series.

The graph shows a shift in the average droplet size from around 350 m to 600 m. This shift infers the growth of the droplet size due to the coalescence of the small droplets into bigger ones. This trend can be attributed to the unstable nature of W/D. When the droplets condensed and the emulsion is formed, potential energy to compensate the interfacial tension is stored in the system. Emulsion with small droplets contains bigger potential energy than the one with the larger droplets, so the shift to bigger droplets is inevitable. Eventually, the separation will be complete, and a distinct layer of diesel and water can then be observed, as such layer of their respective indigenous phases is closer to equilibrium.

The fact that the observed sizes after 60 minutes are averaging 350 m also infers that the average droplet size after the formation would be much smaller. However, it could not be confirmed with the sampling method due to the particular preparation time needed before the sample is ready for measurement.

3.2. Water content

During condensation, the bulk temperature of the emulsion rose quickly because of the transfer of latent heat of condensation and conductive heat. With the rising temperature the condensation rate of the steam decreases, so the temperature of the bulk limits the water content of the resulting emulsion.

Abu Zaid [12] observed that the BSFC of W/D decreases by the water content which he evaluated up to 20%. In contrast with that result, the distillation procedure for the steam generated emulsion as described in subsection 2.3.1 yielded smallish and inconsistent average water content ranging from 2.13 to 6.22% as tabulated in Table 3.

3.3. Emission test results

A comparison of the Nitrogen Oxides (NOx), Carbon monoxide (CO) and unburned hydrocarbon (UHC) emission are shown in Figure 5 through Figure 7. It is noticeable that with steam generated emulsion a reduction of NOx at 5kW engine load around 20%

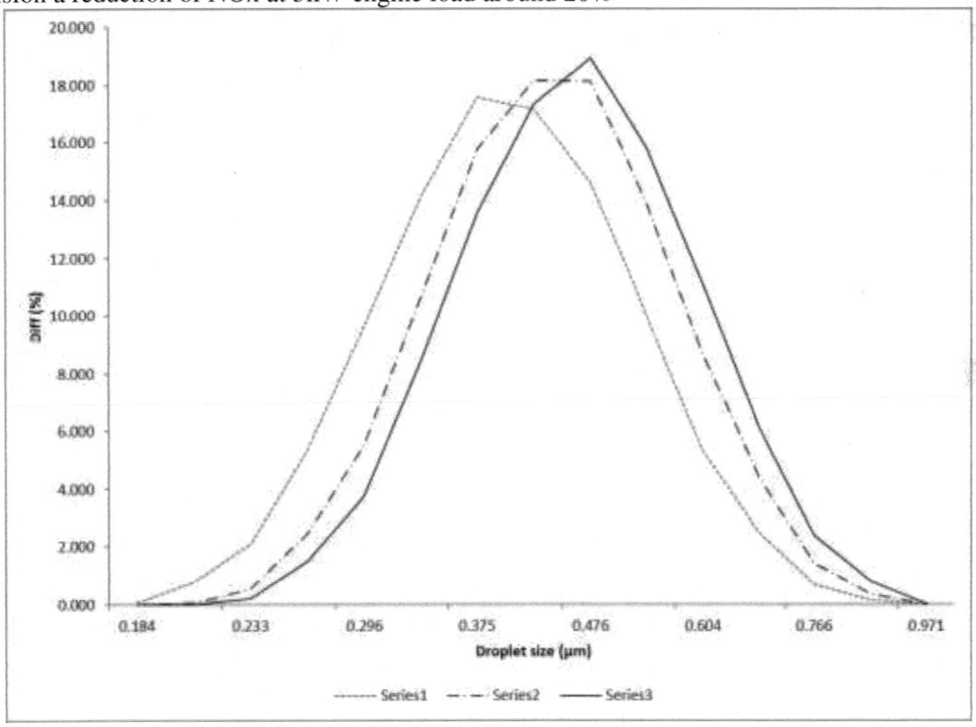

Figure 4. Droplet size distribution and its growth in average size

Table 3. The water content of samples

Load [kW]	Sample water content [mass %]
0	6.22
1	2.13
4	2.14
5	4.38

is achieved compared to D2. However, the CO emission is 50% higher, and the UHC also rises to almost 300%. With emulsion fuel, NOx production is suppressed due to the lower temperature of combustion [13]. CO and UHC increase is also attributed to this decrease in the combustion temperature, causing less complete carbon conversion. These results resemble the problems faced of using W/D as concluded by Murayama et.al.[14] and Subramanian et.al.[15].

3.4. Brake specific fuel consumption

Measurements of the BSFC resulting from both fuels are plotted in Figure 8. Here, the water content is extracted from the calculation. Steam emulsion resulted in overall lower fuel consumption which tally to general findings in W/D research [16, 17, 18]. Murayama et.al. concluded that the BSFC reduction is mainly caused by the microexplosion [14].

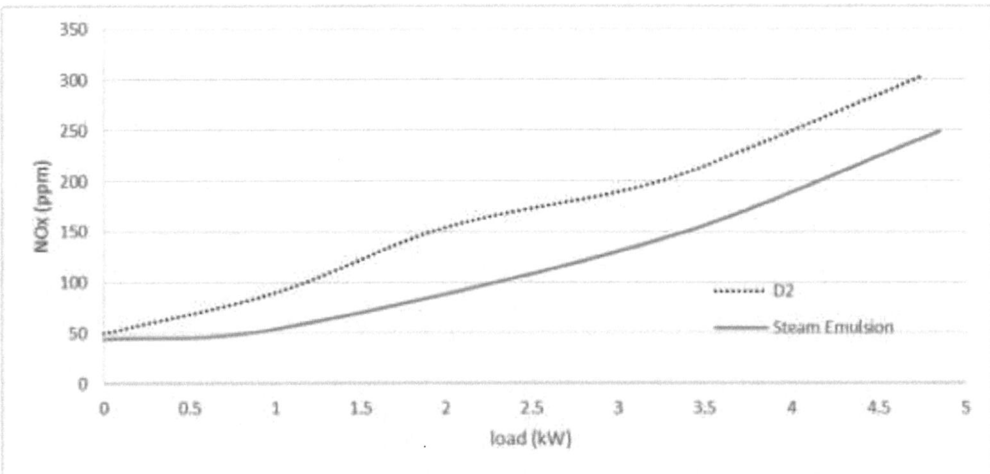

Figure 5. NOx emission

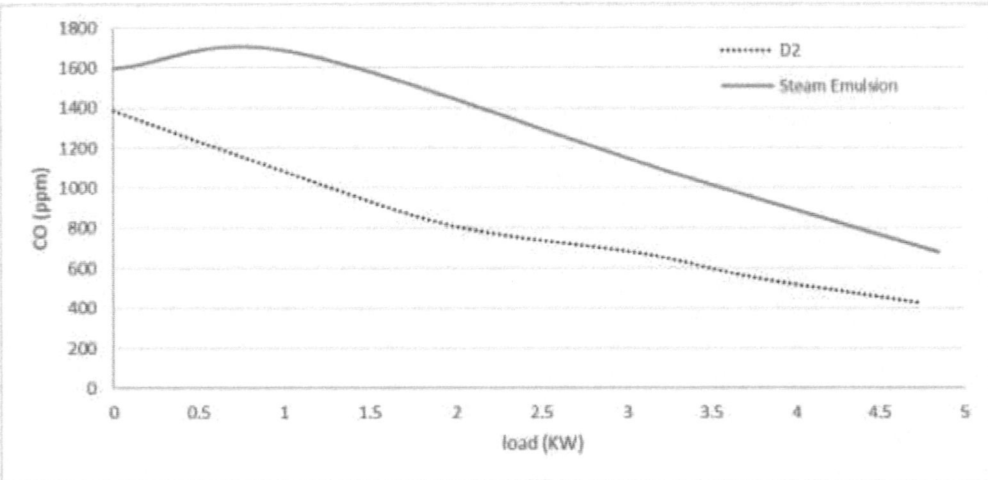

Figure 6. CO emission

The small and inconsistent water content does not hinder the performance of the steam generated emulsion. This boost in BSFC is caused mainly by micro-explosion phenomena. Marrone et al. implied that finer droplet size does not cause better microexplosion [19]. They found in an emulsion fuel with average water droplet size ranging from 2.1 to 4.5 m that the bigger droplets deliver more substantial explosion. The fact that larger droplets perform better micro-explosion confirmed the model by Fu et al. [20] that predicted that the optimum average droplet size is 4.7 m. However, Attia et al.

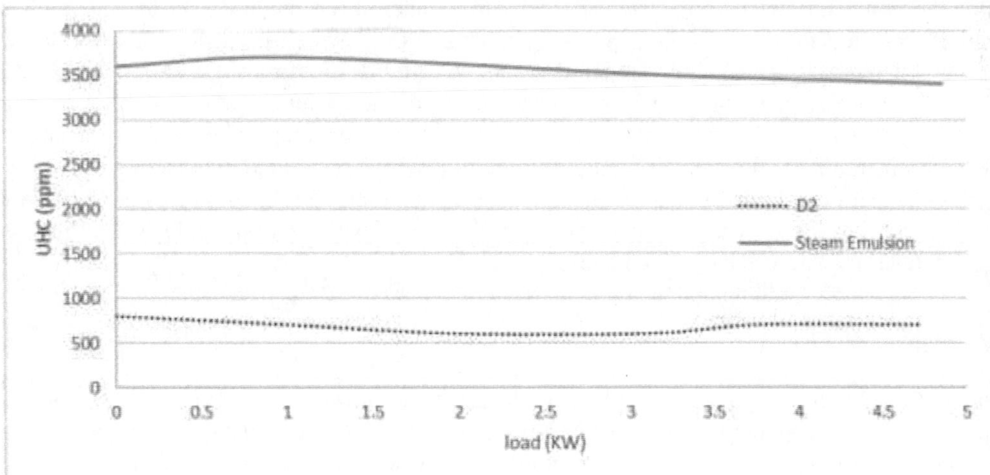

Figure 7. UHC emission

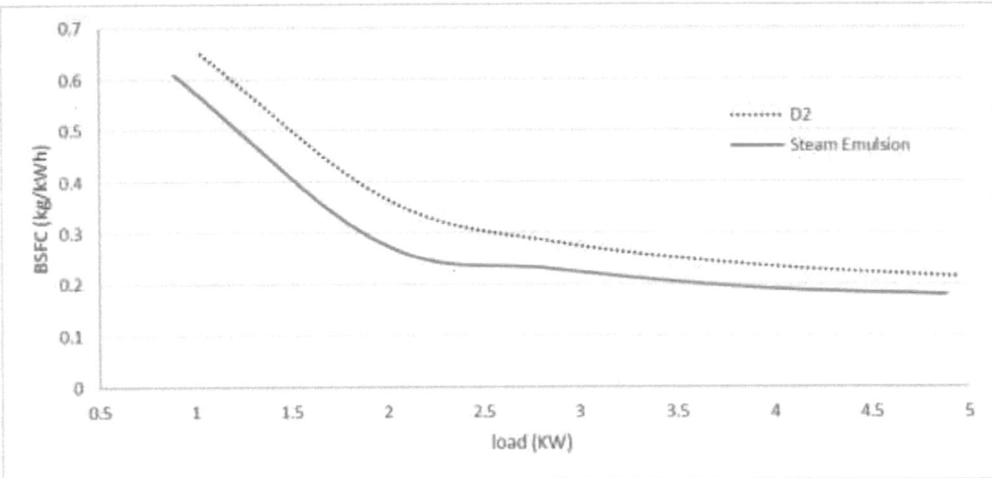

Figure 8. Break specific fuel consumption

reported that smaller droplets cause a notable reduction in the hydrocarbon and smoke level emission in addition to the decrease in engine brake specific fuel consumption. These results were caused by the improvement of the homogeneity of the size distribution and the increase in surface contact between fuel and water droplets which is found in the fine droplets of the steam generated emulsion [21].

4. Conclusions

Reproducible emulsions were formed using the steam condensation method. The mean droplet size observed 60 minutes after preparation process is around 350 m, which is far less than the modelled ideal droplet size of 4.7 m. After the preparation process, the mean droplet size shifts to bigger sizes with the release of the potential formation energy stored in the system. Even though the maximum water content is limited by the bulk emulsion temperature, the droplet population and uniformity of the emulsion compensate the low water content found in steam generated emulsion. The emulsion prepared with steam condensation is proven to be a potential alternative method to yield W/D since it has the characteristics of conventional W/D which are lower NOx (71% max) and BSFC (18.4% max) but higher CO (180%) and UHC (517%) emissions.

Acknowledgement

The authors would like to express their gratitude to the Malaysia-Japan International Institute of Technology (MJIIT), Universiti Teknologi Malaysia (UTM) for the provision of the state of the art instruments. We are also thankful to the Agency for the Assessment and Application of Technology (BPPT, Indonesia) for lending us the E4500 gas analyser for the duration of the research. Funding: Highest appreciation goes to the Ministry of Higher Education (Malaysia) for their financial support through PRGS no.
4L648.

References

[1] Ithnin A M, Noge H, Kadir H A and Jazair W 2014 *Journal of the Energy Institute* **87** 273–288

ISSN 17460220 URL http://dx.doi.org/10.1016/j.joei.2014.04.002

[2] Lif A and Holmberg K 2006 *Advances in Colloid and Interface Science* **123-126** 231–9 ISSN 00018686 URL http://www.sciencedirect.com/science/article/pii/S0001868606000546

[3] Tadros T 2013 *Emulsion Formation and Stability* 1–76

[4] Kunjappu J T and Rosen M J 2013 *Surfactants and interfacial phenomena* 4th ed (Hoboken, New Jersey: Wiley) ISBN 1118229029

[5] Walstra P 1993 *Chemical Engineering Science* **48** 333–349 ISSN 00092509 URL http://www.sciencedirect.com/science/article/pii/000925099380021H

[6] Ithnin A M 2015 *Real-time water-in-diesel emulsion fuel production system for diesel electric generator* Ph.D. thesis Malaysia-Japan International Institute of Technology, Universiti Teknologi Malaysia

[7] Atiqah Ramlan N, Jazair Yahya W, Muhsin Ithnin A, Amiliyana Norazni S, Aiyshah Mazlan N, Avianto Sugeng D, Dayana Bahar N and Koga T 2016 *Energy Conversion and Management* **130** 192–199

[8] Kalman H 2006 *International Journal of Heat and Mass Transfer* **49** 2391–2395 ISSN 00179310 URL http://www.sciencedirect.com/science/article/pii/S0017931006001220

[9] Wu W H and Maa J R 1976 *Journal of Colloid and Interface Science* **56** 365–369 ISSN 00219797 URL http://www.sciencedirect.com/science/article/pii/0021979776902630

[10] bin Ahmad M A 2016 *Performance of Real Time Water-in-Diesel Emulsion Fuel Supply System on Diesel Engine* Ph.D. thesis Universiti Teknologi Malaysia

[11] Hale G M and Querry M R 1973 *Applied Optics* **12** 555 ISSN 0003-6935

[12] Abu-Zaid M 2004 *Energy Conversion and Management* **45** 697–705 ISSN 01968904 URL http://www.sciencedirect.com/science/article/pii/S0196890403001791

[13] Alahmer A, Yamin J, Sakhrieh A and Hamdan M 2010 *Energy Conversion and Management* **51** 1708–1713 ISSN 01968904 URL http://www.sciencedirect.com/science/article/pii/S0196890409004944

[14] Murayama T, Tsukahara M, Morishima Y and Miyamoto N 1978 *SAE Transactions* **87**

[15] Subramanian K A 2011 *Energy Conversion and Management* **52** 849–857 ISSN 01968904 URL http://dx.doi.org/10.1016/j.enconman.2010.08.010

[16] Hountalas D, Mavropoulos G, Zannis T and Mamalis S 2006 *SAE Technical Papers* **2006**

[17] Ithnin A M, Ahmad M A, Bakar M A A, Rajoo S and Yahya W J 2015 *Energy Conversion and Management* **90** 375–382 ISSN 01968904 URL http://www.sciencedirect.com/science/article/pii/S0196890414009820

[18] Alahmer A 2013 *Energy Conversion and Management* **73** 361–369 ISSN 01968904 URL http://dx.doi.org/10.1016/j.enconman.2013.05.012

[19] Marrone N J, Kennedy I M and Dryer F L 1983 *Combustion Science and Technology* **33** 299–307 ISSN 00102202 URL http://dx.doi.org/10.1080/00102208308923682http://www.tandfonline.com/doi/abs/10.1080/00102208308923682

[20] Biao Fu W, Yun Hou L, Wang L and Hua Ma F 2002 *Fuel Processing Technology* **79** 107–119

[21] Attia A M and Kulchitskiy A 2014 *Fuel* **116** 703–708 ISSN 00162361

IOP Conf. Series: Materials Science and Engineering **257** (2017) 012037 doi:10.1088/1757-899X/257/1/012037

Comparative study on the pyrolysis behaviour and kinetics of two macroalgae biomass (Gracilaria changii and Gelidium pusillum) by thermogravimetric analysis

A N Roslee* and N F Munajat

School of Ocean Engineering, Universiti Malaysia Terengganu, 21030 Kuala Terengganu, Terengganu, Malaysia

*Corresponding author: nabylaamyra26@gmail.com

Abstract. Macroalgae are often referred as seaweed and could be significant biomass resource for the production of numerous energy carriers including biofuels. In this study, the chemical composition of *Gracilaria changii* (*G. changii*) and *Gelidium pusillum* (*G. pusillum*)were determined through proximate and ultimate analysis and the thermal degradation behaviour of *G. changii* and *G. pusillum* were investigated via thermogravimetric analysis (TGA) in determining the important main composition to be considered as biomass fuels. It has found the pyrolysis of *G. changii* and *G. pusillum* consists of three stages and stage II is the main decomposition stage with major mass loss of around 52.16% and 44.42%, respectively. The TGA data were then used for determination of kinetic parameters of the pyrolysis process using three model-free methods: Kissinger, Kissinger-Akahira-Sunose (KAS) and Flynn-Wall-Ozawa (FWO). The apparent activation energy calculated by using Kissinger method for *G. changii* was lower than *G. Pusillum*, i.e.173.12 kJ/mol and 193.22 kJ/mol, respectively. The activation energies calculated from KAS and FWO methods were increased with increasing the pyrolysis conversion with average activation energies of 172.32 kJ/mol and 181.19 kJ/mol for *G. changii* while for *G. pusillum* (177.42 kJ/mol and 187.4 kJ/mol). *G. pusillum* has lower and wider distribution of activation energy and revealed that the pyrolysis process for *G. changii* was easier than *G. pusillum*. These data provide information for further application for designing and modelling in thermochemical conversion system of macroalgae biomass.

1. Introduction

With continuously increasing the demand for fossil fuels, the utilization of renewable energy sources such as wind, solar, fuel cell, geothermal ocean and biomass also increased. Biomass energy is becoming more attractive among many researchers due to its ability to be transformed into various energy carriers including biofuel, clean and renewable, abundantly available, low contamination and wide distribution [1]. However, the utilization of land-based biomass resources needs large arable land and cause problems to develop it as a feedstock.

Compared with the existing energy crops, macroalgae have many outstanding advantages, such as high biomass production, short growth period, do not occupy arable land, and can be grown in open water (sea or pond) [2]. Macroalgae are seen as another potential fuel and can effectively reduce the environmental problems, greenhouse effect and have numerous species worldwide. Two species considered in this study are *changii* (*G. changii*) and *Gelidium pusillum* (*G. pusillum*). *G. changii* is commonly growing at sandy mudflats or on the substrate such as rock, shells or corals and commonly used in applications such as cosmetics, biomedical, biotechnology, food and pharmaceutical. In Malaysia, *G. changii* is potentially used in the agar industry due to its good agar properties as reported in literature and eaten as a food [3]. In addition, these macroalgae that have high growth rate may be used as biomass resource and also be converted to useful energy carriers in similar ways of land-based biomass via thermochemical process where the macroalgae may be converted into fuel and/or chemicals through a pyrolysis, gasification and combustion technology [4]. Pyrolysis is a thermal decomposition process of biomass materials in the absence of oxygen or any other reagent contains oxygen such as air, water or carbon dioxide. Therefore, this work is focus on the pyrolysis to understand the pyrolysis characteristic of macroalgae biomass.

Thermogravimetric analysis (TGA) is an useful method and frequently used for thermal characterization in which weight changes occur. The thermogravimetric data can be presented either as the mass against time or temperature (TGA curve) or alternatively as a plot of the rate of change of mass with respect of time or temperature, known as derivative thermogravimetry (DTG) curve. Both TGA and DTG curves can be used to elucidate the kinetic parameters such as activation energy and pre-exponential factors for the pyrolysis process of macroalgae biomass [5]. Kissinger, Kissinger-Akahira-Sunose (KAS) and Flynn-Wall-Ozawa (FWO) are among methods that can be used for this purpose [1, 6, 7]. In recent year, studies on the pyrolysis of macroalgae of different macrolage species and its kinetics were reported. For example *Tetraselmis suecica* and *Chlorella* sp. [8], *Sagarssum* sp. [9], and *Polysiphonia elongata* [10], however, no studies found on the pyrolysis-kinetic of two species considered in this study. On the whole, this present study mainly focused on determination of potential use as energy carriers of two macroalgae species, *G. changii* and *G. pusillum* based on its chemical composition, thermolytic characteristic and kinetic parameters in determining the important main composition to be considered as biomass fuels. These analyses are crucial for further application for designing and modelling in thermochemical conversion system of macroalgae biomass in a large-scale production.

2. Methodology

2.1. Materials and sample preparation

The macroalgae biomass used in this study, *G. changii* and *G. pusillum* collected from different places in Southern part of Malaysia on August and October 2016. The samples were identified and catalogued at Marine Algal Reference Collection (MARC) in Central Laboratory, Universiti Malaysia Terengganu (UMT). Prior to analysis, the samples were washed, dried in an oven at 60°C for 24 hours, milled and sieved to a particle with average size between 100-120μm.

(a) (b)

Figure 1. *Gelidium pusillum* and *Gracilaria changii.*

2.2. Material characterizations

The elemental analysis for both macro algae species were done using Elemental AnalyzerVarioMicroCube to identify the carbon (C), hydrogen (H), nitrogen (N) and sulphur (S) content. 10mg of the samples were used to examine its C, H, N and S content. The oxygen (O) was identifying by subtracting the value of C, H, N and S. The proximate analysis to determine ash, moisture content, volatile matter and fixed carbon were conducted. The moisture content are determined from TGA curve by the weight loss of the samples at 110°C while the volatile content are determined by the weight loss of the samples between 150 to 600°C. The ash content was determined by combustion of the samples. 1g of the samples was combusted in the furnace at 950°C and hold for 10 min. The fixed carbon was determined by subtracting the amount of M, VM and Ash from the dry mass of the samples.

2.3. Thermal analysis by using thermogravimetric analyzer (TGA)
The pyrolysis characteristic was performed by using thermal analyzer (TGA Q500 V20.13 Build 39). In each experiment, 10mg of the sample mass was heated at heating rates of 10, 20 and 30°C/min under a nitrogen flow of 100 ml min^{-1}. This experiment was conducted from the room temperature to 800°C.

2.4. Kinetic analysis
Kinetic reactions on solid-states are described by following equation [11-14]:

$$\frac{d\alpha}{dt} = k(T)f(\alpha) \tag{1}$$

The solid-state first order reaction is described from the derivative of f(α) where f´(α)=1. The function f(α) is expressed as:

$$f(\alpha)=(1-\alpha)_n \tag{2}$$

The weight loss of the decomposed sample was calculated by using the equations below:

$$\alpha = \frac{m_i - m_a}{m_i - mf} \tag{3}$$

wherem$_i$ is the initial mass of the sample, m_α is the actual mass and m_f is the final mass after the pyrolysis.
The reaction rate constant, k according to Arrhenius equation is defined as:

$$k=A\exp\left(-\frac{E_a}{RT}\right) \tag{4}$$

whereT is the absolute temperature, A is the pre-exponential factor (min^{-1}), E_α is the activation energy and R is the gas constant (8,314 J K^{-1} mol^{-1}). The equation for kinetic parameters obtained from the TGA shown as below:

$$\frac{d\alpha}{dt} = A\exp\left(-\frac{E_a}{RT}\right).f(\alpha) \tag{5}$$

By substituting equation (2) into equation (5) gives the following equations:

$$\frac{d\alpha}{dt} = A(1-\alpha)^n.\exp\left(-\frac{E_a}{RT}\right) \tag{6}$$

If the initial temperature of the sample is changed by a constant heating rate, $\beta = \frac{dT}{dt}$, equation (6) can be written as:

$$\frac{d\alpha}{dT} = \frac{A}{\beta} \exp\left(-\frac{E_a}{RT}\right).f(\alpha) \tag{7}$$

Model-free methods were used to calculate its activation energy and kinetic parameter which is Kissinger, Kissinger-Akahira-Sunose (KAS) and Flynn-Wall-Ozawa (FWO) method were applied in this study.

2.4.1. Kissinger method. Kissinger method was used to calculate the activation energy from the slope of $\ln(\beta/T_m^2)$ against $1000/T_m$ at different heating rates, βwhich is equal to$-\frac{E}{R}$. The equation can be written as equation (8) where T_m is the peak temperature at which the maximum reaction rate occurs:

$$\ln\left(\frac{\beta_i}{T_m^2}\right) = \ln\left(\frac{AR}{E}\right) - \frac{E}{RT_m} \tag{8}$$

2.4.2. Kissinger-Akahira-Sunose (KAS) method. The KAS method is based on the equation (9):

$$\ln\left(\frac{\beta_i}{T^2{\alpha_i}}\right) = \ln\left(\frac{A_\alpha R}{E_\alpha}\right) - \ln g(\alpha) - \frac{E_\alpha}{RT_{\alpha i}} \tag{9}$$

At every given value of conversion,α,$\ln(\beta_i/T_{\alpha i}^2)$ is linear with $1000/T_{\alpha i}$ and thus the activation energy is determined from the slope $-\frac{E\alpha}{R}$.

2.4.3. Flynn-Wall-Ozawa (FWO) method. The FWO method calculates the activation energy from a plot of $\ln(\beta_i)$ versus $1000/T_{\alpha i}$ at given conversion value based on the equation (10).

$$\ln(\beta_i) = \ln\left(\frac{A_\alpha E_\alpha}{R}\right) - \ln g(\alpha) - 5.331 - 1.502\frac{E_\alpha}{RT_{\alpha i}} \tag{10}$$

where$g(\alpha)$ is constant at a given value of conversion. The subscripts i and α denotes given value of heating rate and conversion, respectively. The activation energy is obtained from the slope $-1.502E_\alpha/R$.

3. Results and discussion

3.1. Characterization of macroalgae biomass

The ultimate and proximate analysis for *G. changii*and *G. pusillum* in comparison to other species of macroalgae and some terrestrial biomass feedstock were ascribed in table 1. The ultimate analysis of macroalgae biomass including *G. changii* and *G. pusillum* indicates that they are lower in carbon content than terrestrial biomass and higher in H, N, S and O. The low content of C can reduce the calorific value of the macroalgae but could significantly reduce the carbon dioxide (CO_2) emissions during the conversion process of macroalgae biomass [15]. Higher oxygen and hydrogen content showed that macroalgae biomass could have higher thermal reactivity compared to terrestrial biomasses. Therefore, macroalgae could release more bio-oil and bio-gas than terrestrial biomass. This is because H content of the macroalgae could aid the role of combustible hydrogen gas (H_2), hydrogen sulphide (H_2S), carbohydrates, hydrocarbons and H-radicals. Meanwhile, O content is related to the presence of oxygen containing elements in macroalgae biomass such as alcohol and/or carboxylic acids that can promote the conversion process [16].

The higher N and S content of both *G. changii* and *G. pusillum* than the terrestrial biomass can be disadvantages during thermochemical conversion of macroalgae. Slightly higher nitrogen content for both macroalgae species indicates that macroalgae has higher protein content. In contrast, the nitrogen values of these species are slightly lower than the *U. prolifera* from literature (table 1). The S content could produces and releases sulphur dioxide (SO_2) and generates fine particulate, while nitrogen could yields and frees nitric oxide (NO_x) emissions during the conversion process of macroalgae. All mentioned gasses have negative impacts on environment. Furthermore, S content could cause deposit formation, agglomeration, slagging and corrosion to the devices during conversion process. Therefore, high N and S content in macroalgae is undesired during macroalgae conversion. An additional treatment and/or reduction technique prior or during the conversion process are desired to minimize S and N content as well as the emission production.

The proximate analysis (table 1) indicated that the volatile content for macroalgae is lower than the terrestrial biomass. In contrast, the volatile content of *G. pusillum* is slightly higher than coal from the literature in Table 1. The higher content of volatile matter of lignocellulosic biomass could be assigned to the presence of lignocellulosic components such as cellulose, hemicellulose or lignin. In contrast, no lignin component present in macroalgae and the volatile content only depends on the cellulose component from its cell wall and some of hemicellulose component [17]. The corresponding volatile content of *G. changii* and *G. pusillum* are 50.01% and 44.81%, respectively. This shows that *G. changii* could produce more bio-oil and gas during the pyrolysis process due to the high proportion of condensable products. It also shows that *G. changii* is more reactive than *G. pusillum.*

The ash content of these macroalgae species is slightly lower than *Sargassum* sp. and *U. prolifera* but higher than other biomasses. With regard to the macroalgae biomass, the ash content for these macroalgae was higher, which in a good agreement those macroalgae biomasses have higher salinity thus led to higher ash content. Difference habitats will also give significant difference of ash content. *G. changii* is commonly growing and can be found abudantly at sandy mudflats while *G. pusillum* can be found on the substrate such as rocks, shells, or corals in upper intertidal sea water. The salinity of sea water and rocks could give significant effect. Usually, sea water has lower salinity (approximately 3.5%) compared to the salinity on the rocks. This is because the remaining sea water leaves the dissolved minerals when the sea water was evaporate thus makes the rock saltier than the sea water. The macroalgae species are very sensitive to the salinity of its surrounding [15]. Higher ash content would affect the designing and operation process where it cause slagging, fouling and other ash-related problems during conversion process but can be used as a catalyst material for bio char formation [18]. Therefore, an appropriate process of removing ash prior the conversion devices is desired for large-scale production of fuel or chemical feedstock.

The moisture content for *G. pusillum* was higher than terrestrial biomasses but lower than *Sargassum* sp. This is related to its natural habitats and environments surrounded by the sea water. High moisture content is unfavourable for thermochemical process and would give problems to the conversion devices such as flame stability, deposition of fouling agent and many more. Thus, further drying time is necessary to remove as much as possible the moisture content before the conversion process[19].The fixed carbons for macroalgae biomass were higher than terrestrial biomass but lower than coal. Fixed carbon is required for bio-char formation and could be possibly used as carbonaceous materials by pyrolysis process. Thus, *G. changii* could produce more bio-char as well as other biomass at the end of pyrolysis process compared to *G. pusillum*.

Table 1. Characterization of *Gracilariachangii* and *Gelidiumpusillum.*

Sample	Ultimate analysis					Proximate analysis				Ref.
	C	H	N	S	O^a	Ash	Volatile matter	Moisture	Fixed carbon	
Gracilaria changii	26.5	5.86	1.68	1.92	64.04	14.41	50.01	9.42	26.16	This work
Gelidium pusillum	23.58	5.97	1.83	2.44	66.18	17.83	44.81	11.49	25.87	This work
Ulva prolifera	37.44	7.01	1.87	2.88	50.8	24.46	57.87	9.92	7.77	[20]
***Sargassum* sp.**	26.7	4.23	1.35	0.19	67.53	27.45	48.2	11.55	12.8	[9]
Poplar wood	45.5	6.26	1.04	-	47.2	3.7	75.54	9.6	11.15	[21]
Coal	72.09	4.77	1.44	1.15	8.13	11.52	31.23	7.16	50.09	[22]

[a] Calculated by difference C, H, N and S

3.2. Thermal behavior of macroalgae biomass

The TGA and DTG curves obtained during the pyrolysis of *G. changii* and *G. pusillum* under inert atmosphere at heating rate of 10, 20 and 30°C/min are shown in figure 2 and figure 3. The dissimilar shaped curves of both samples suggesting that the pyrolysis process of both macroalgae species are different. However, three main stages of thermal degradation could be identified, similarly to other published works [1, 9, 10, 23]. The thermal degradation characteristics detected during the heating process of the samples are presented in table 2.

The initial mass loss was occurred at the first stage (I) as the temperature increased from room temperature to temperature around T_1. During this stage, the weight loss of 5.4% and 11.5% were observed for *G. changii*and *G. pusillum*, respectively. The decreased in weight of samples revealed that the external water bound attached on the sample's surface was evaporate or dehydrate continuously surface and some volatile compounds in macroalgae with the increasing temperature. There is one peak appeared in the DTG curve for *G. Changii* at around 71°C, while for *G. pusillum* one peak appeared in the DTG curve at temperature around 66°C, which simultaneously with the higher moisture content of *G. pusillum* than that *G. changii* as shown in table 1.

The main decomposition step took place in the second stage (II) between T_1 and T_2 (Figure 2, Figure 3 and Table 2) with *G. changii* showed wider decomposition range than that *G. pusillum.* A weight loss of 52.16wt% was observed for *G. changii* while a lower value was observed for *G. pusillum* (44.42wt%). This weight loss during this stage is commonly owing to the main devolatilization reactions in which most of the volatile matter released during the pyrolysis process. In addition, *G. changii* showed two peaks where the main decomposition took place at 259.5 and 322.2°C. However, only one peak at 274.2°C was observed for *G. pusillum.* The difference in thermal decomposition behaviour that presented by these macroalgae in TGA and DTG curves could be ascribed to the differences in the natural structural and chemical characteristics that make they went through different rates of decomposition. Furthermore, the two peaks shown in DTG curve of *G. changii* indicated that multi-zone pyrolysis existed during the main decomposition, which related to the component complexity of this species. Similar trend of thermal degradation during pyrolysis also has been reported for other biomass in previous study [9, 14, 24-26]. The stepwise of thermal degradation of macroalgae is occurred between 180 and 270°C; correspond to the decomposition of carbohydrates followed by protein within temperature range of 320 to 450°C as well as reported by Kim et al. [9]. In the case of *G. pusillum*, only one and larger peak was observed. This means that *G.*

pusillum has larger carbohydrates content than *G. changii*. This is in agreement with the fact mentioned by Vassilev et al., brown algae contains more carbohydrates than red algae. Red algae are unicellular or multicellular macroalgae which its cell wall consists of cellulose, carrageenan, or agarose [27]. Meanwhile, brown algae are multicellular macroalgae which contains cellulose and alginate. According to Rubén et al., [28], cellulose is not easily decomposed compared to alginate and thus may explained multi-zone decomposition of *G. changii*.

The volatiles produced during stage II may be further decomposed with remaining protein and char in the third stage (III) as the temperature increased. The decomposition of *G. changii* and *G. pusillum* progressed more slowly and attributed to the slow decomposition of solid residue. The weight loss of ca. 8.17 wt% was accounted for *G. changii* and 3.66wt% for *G. pusillum* at temperature ranging from T_2 to temperature around 800°C. There is one peak appeared in the DTG curve for *G. changii* at around 683.6°C, while for *G. pusillum* one peak appeared in the DTG curve at temperature around 680.1°C. The decomposition peak at high temperature results from the devolatilization of inorganic compounds, such as K, Cl, N and S. The char residue at 800°C was higher for *G. pusillum* (37.7 wt%) than *G. changii* (23.5 wt%). This high char yield in *G. pusillum* could be attributed to the high ash content of *G. pusillum* (table 1). Figure 2 and 3 ascribes the effect of the heating rates on the weight loss of *G. changii* and *G. pusillum*. The corresponding thermochemical characteristics detected during pyrolysis process of both samples at different heating rates are presented in table 2. In the case of *G. changii* and *G. pusillum*, the maximum rate of decomposition (DTG_{max}) increased slightly with increasing heating rate (figure 3 and table 2). The increase in the DTG_{max} and the maximum mass loss with the increasing heating rate is due to the increase in thermal energy which leads to higher rates of the thermal energy transfer between the surroundings and the samples [8]. In addition, the increase in heating rate not only increases the thermal energy thus promote the heat transfer, but could also change the pyrolysis reaction process inside the particles. Furthermore, the increase in heating rate leads to the slight wider temperature ranges (T_1 to T_2) for both samples with Tp_2 and Tp_3 for *G. changii* as well as Tp_1 and Tp_2 for *G. pusillum* departed to higher temperature. This indicates that higher temperature is required to set off the decomposition process at greater heating rate. On the other hand, the increase in heating rate only transferred the peak temperature to higher value, without altering the thermal profile of decomposition. Simultaneously, the mass loss of second stage of this sample increase as the heating rate changed from 10 to 30°C/min.

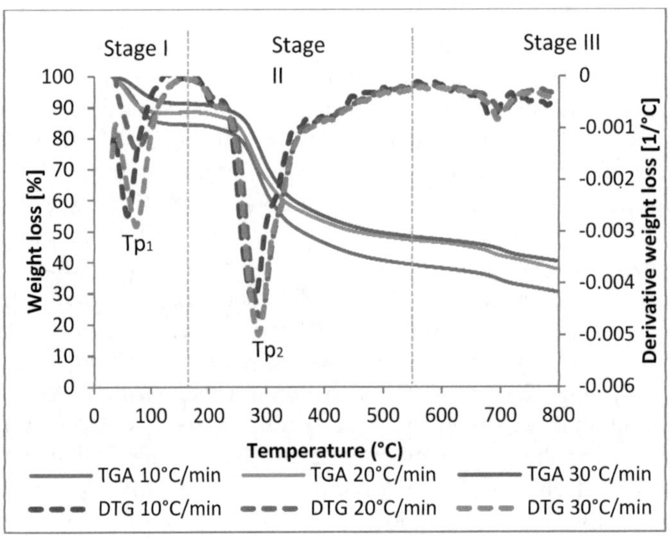

Figure 2. TGA and DTG curves of *Gelidium pusillum.*

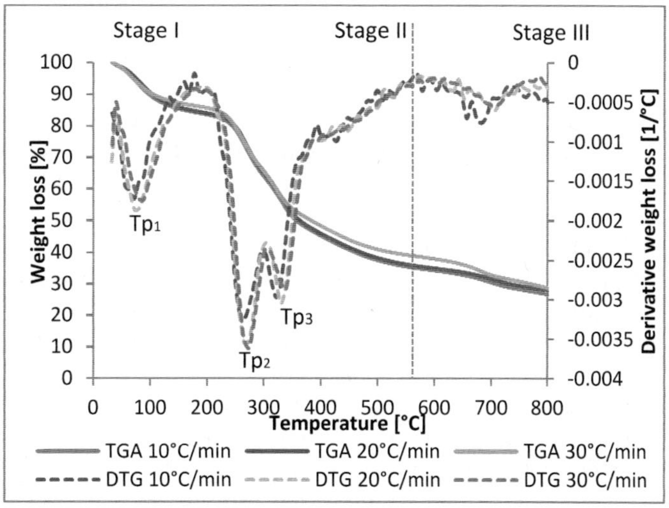

Figure 3. TGA and DTG curves of *Gracilaria changii.*

Table 2. Thermal degradation characteristics of *Gracilaria changii* and *Gelidium pusillum* at different heating rates.

	Heating rate [°C/min]	Temperature (°C) Tp1	T1	Tp2	Tp3	T2	DTGmax[1/°C]	Mass loss[a] [wt.%]	Mass [mg]
G. changii	10	71.97	161.68	259.58	322.25	580.79	-0.00323	50.17	6.03
	20	80.39	165.09	267.06	329.69	586.92	-0.00354	52.45	6.11
	30	89.22	170.3	274.24	339.27	590.69	-0.00362	53.84	6.23
G. pusillum	10	66.77	163.17	280.67		554.83	-0.04318	41.34	4.28
	20	73.66	170.37	286.19		558.62	-0.04614	45.35	4.71
	30	81.64	178.16	294.32		566.92	-0.04998	46.56	4.93

[a]Mass loss at main degradation stage

3.3. Pyrolysis kinetic analysis by using isoconversional methods: Kissinger, Kissinger-Akahira-Sunose (KAS) and Flynn-Wall-Ozawa (FWO)

The TGA data is use to study the kinetic the main stage (stage II) in the pyrolysis process of *G. Changii* and *G. pusillum*. The model free methods: Kissinger, KAS and FWO methods were used to calculate the apparent kinetic energy and pre-exponential factor for the pyrolysis of both macroalgae based on the differential method of equation (7) mentioned earlier.

The activation energy, E_α and pre-exponential factor, A, of Kissinger method, were determined based on eq. (8) from a plot of $\ln(\beta/T_m^2)$ against $1000/T_m$ at different heating rates, βas described in figure 4 (a). The activation energy of 173.12 kJ/mol with pre-exponential factor of 6.07×10^{12} min^{-1} was obtained for *G. changii* and 193.22 kJ/mol for *G. pusillum* with pre-exponential factor of 1.0×10^{12} $min^{-1}$for this method. For KAS and FWO methods, the kinetic parameters were defined based on eq. (9) and (10). The plots of $\ln(\beta_i/T_{\alpha i}^2)$ and $\ln(\beta)$ against $1000/T_{\alpha i}$ for KAS and FWO methods at different level of fractional conversion (0.1 to 0.9) of stage II and their linear-fitting line are shown in figure 4 (b) and (c), respectively. The activation energies obtained for *G. changii* for KAS method ranging from 94.05 to 252.43 kJ/mol and from 101.41to 264.83 kJ/mol for FWO method with average activation energy of 172.32 kJ/mol and 181.19 kJ/mol. While for *G. pusillum,* the activation energies obtained from KAS method ranging from 101.88 to 261.34kJ/mol and for FWO method was from 109.33 to 273.86 kJ/mol with average activation energy of 177.42and 187.4 kJ/mol.

The activation energy, E_α and pre-exponential factor, *A*, are listed in table 3 and the linear correlation coefficient (R^2) are in the range of 0.9362 to 1. From table 3, it can be seen that the activation energies defined by KAS and FWO method were increased with the increasing conversion values for *G. changii* and *G. pusillum.* This indicates that the complex reaction mechanism of pyrolysis is not in the same path during the whole decomposition process of stage II and the activation energy obtained is dependences on the conversion [29]. In the case of *G. pusillum,* the activation energies calculated by FWO and KAS methods were lower than Kissinger method while for *G. changii*the activation energies calculated using FWO methods was higher than KAS and Kissinger methods. This can be due to the fact that the Kissinger method only generates a single set of kinetic energy for the whole pyrolysis process in comparison to different degree of conversions for FWO and KAS methods. Thus, the Kissinger method might unable to reveal the complexity of the whole decomposition process. Comparing both macroalgae, the activation energy of *G. changii* was lower and has wider distribution range compared to *G. pusillum*. This revealed that the *G. changii* is more reactive compared to *G. pusillum*. In addition, the differences in the composition of macroalgae biomass will gives slightly significant to the differences in their kinetic parameters. Other than that, the kinetic energies calculated for both macroalgae in this study are comparable to the values calculated for poplar wood [29], *Chlorella vulgaris* [30], and *Polysiphonia elongata*[10].

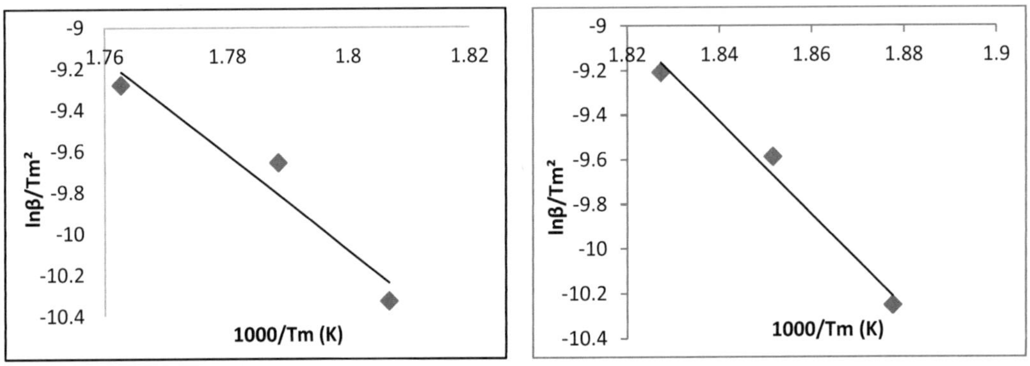

(a)

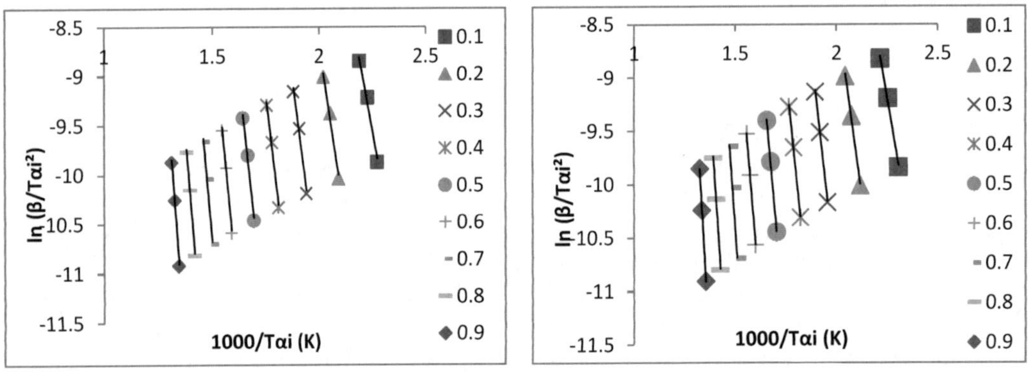

(b)

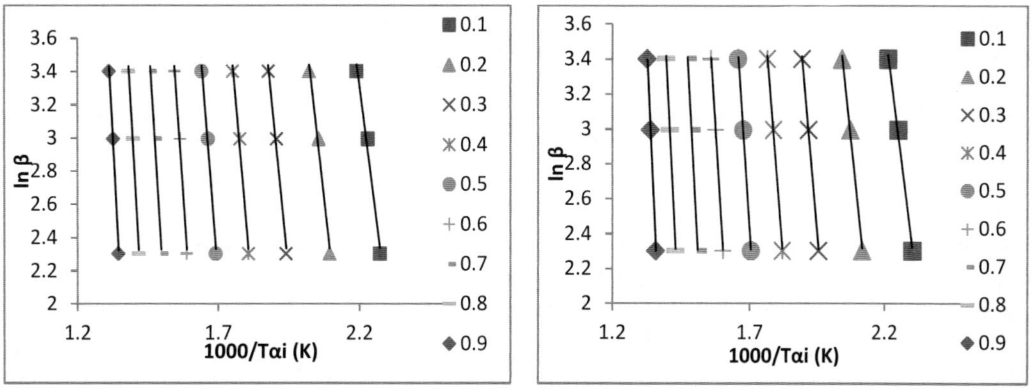

(c)

Figure 4. Plots of model free methods (a) Kissinger (b) KAS and (C) FWO for *Gracilaria changii* and *Gelidium pusillum.*

Table 3. Activation energies obtained from KAS, FWO and Kissinger methods for *Gracilaria changii* and *Gelidium pusillum.*

		KAS				
	Gracilaria changii			*Gelidium pusillum*		
α	E [kJ/mol]	A [min^{-1}]	R^2	E [kJ/mol]	A [min^{-1}]	R^2
0.1	101.41	$2.04X10^{12}$	0.9953	109.33	$1.32x10^{12}$	0.9855
0.2	123.79	$7.48X10^{12}$	0.9971	126.65	$1.11x10^{13}$	0.9870
0.3	142.31	$4.36X10^{13}$	0.9982	145.02	$8.84x10^{13}$	0.9882
0.4	162.90	$9.69X10^{14}$	0.9974	164.35	$6.57x10^{14}$	0.9894
0.5	177.49	$4.22X10^{15}$	0.9991	184.48	$4.47x10^{15}$	0.9905
0.6	199.98	$5.61X10^{16}$	0.9996	205.44	$2.85x10^{16}$	0.9916
0.7	222.53	$2.59X10^{17}$	0.9995	227.33	$1.75x10^{17}$	0.9925
0.8	235.44	$7.61X10^{17}$	0.9979	250.13	$1.04x10^{18}$	0.9934
0.9	264.83	$3.99X10^{18}$	1	273.86	$5.96x10^{18}$	0.9940
Av.	181.17			187.4		
		Kissinger				
	173.12	$6.07x10^{14}$	0.9812	193.22	$1.43X10^{12}$	0.9362

4. Conclusions

The TGA and DTG curves showed the differences of degradation pathways for *G. changii* and *G. pusillum.* The difference in thermal decomposition behaviour that presented by these macroalgae in DTG curves could be ascribed to the differences in the natural structural and chemical characteristics that make they went through different rates of decomposition. The increase in heating rate leads to the slight wider temperature ranges without altering the thermal profile of decomposition compositions, increases the maximum rate of decomposition (DTG_{max}), increases the thermal energy and also changes the pyrolysis reaction process inside the particles. In the terms of kinetic parameters, *G. changii* has lower and wider distribution of activation energy, suggesting that the pyrolysis process of *G. changii* could be easier than *G. pusillum* thus makes *G. changii* a better raw material than *G. pusillum.* Overall, *G. changii* and *G. pusillum* may well be a potential source of biomass based renewable fuels.

Acknowledgments
The authors gratefully thanked to Universiti Malaysia Terengganu for providing laboratory facilities and financial support by the Fundamental Research Grant Scheme (FRGS) under project no: 59352.

References

[1] Zhao H, Song Q, Huaxiao Y and Congwang Z 2012 Pyrolysis Characteristics and Kinetics of Macroalgae Biomass Using Thermogravimetric Analyzer *International Journal of Environmental, Chemical, Ecological, Geological and Geophysical Engineering* **4** 433-40

[2] Sanchez-Silva L, López-González D, Garcia-Minguillan A M and Valverde J L 2013 Pyrolysis, combustion and gasification characteristics of Nannochloropsis gaditana microalgae *Bioresource Technology* **130** 321-31

[3] Lee W-K, Lim P E, Phang S-M and Chai L 2016 Agar properties of Gracilaria species (Gracilariaceae, Rhodophyta) collected from different natural habitats in Malaysia. In: *Regional Studies in Marine Science,* pp 123-8

[4] Filippis P D, Caprariis B D, Scarsella M and Verdone N 2015 Double Distribution Activation Energy Model as Suitable Tool in Explaining Biomass and Coal Pyrolysis Behavior *Energies* **8** 1730-44

[5] Qiu H-W, Zhou Q-C and Geng J 2015 Pyrolytic and kinetic characteristics of Platycodon Grandiflorum peel and its cellulose extract *Carbohydrate Polymers* **117** 644-9

[6] Agrawal A and Chakraborty S 2013 A kinetic study of pyrolysis and combustion of microalgae Chlorella vulgaris using thermo-gravimetric analysis *Bioresour Technol* **128** 72-80

[7] Álvarez A, Pizarro C, García R, Bueno J L and Lavín A G 2016 Determination of kinetic parameters for biomass combustion *Bioresource Technology* **216** 36-43

[8] Kassim M A, Kirtania K, Cruz D D L and Bhattacharya S P 2014 Thermogravimetric analysis and kinetic characterization of lipid-extracted *Tetraselmis suecica* and *Chlorella* sp. *Algal Research* **6** 39-45

[9] Kim S-S, Ly H V, Kim J, Choi J H and Woo H C 2013 Thermogravimetric characteristics and pyrolysis kinetics of Alga Sagarssum sp. biomass *Bioresource Technology* **139** 242-8

[10] Ceylan S, Topcu Y and Ceylan Z 2014 Thermal behaviour and kinetics of alga Polysiphonia elongata biomass during pyrolysis *Bioresource Technology* **171** 193-8

[11] Guo W, Xiao H, Yasuda E and Cheng Y 2006 Oxidation kinetics and mechanisms of a 2D-C/C composite *Carbon* **44** 3269-76

[12] Matečić Mušanić, S. Fiamengo Houra and I. Sućeska M 2010 Applicability of Non-isothermal DSC and Ozawa Method for Studying Kinetics of Double Base Propellant Decomposition *Central European Journal of Energetic Materials* **7** 233-51

[13] Frączyk A 2011 The Activation Energy of Primary Crystallization of Fe95Si5 Metallic Glass *Technical Science* **14** 93-100

[14] Naihao Ye, Demao Li , Limei Chen, Xiaowen Zhang and Xu D 2010 Comparative Studies of the Pyrolytic and Kinetic Characteristics of Maize Straw and the Seaweed Ulva pertusa *PLOs One* **5** 1-6

[15] Cole A J, Mata L, Paul N A and Nys R D 2014 Using CO_2 to enhance carbon capture and biomass applications of freshwater macroalgae *Global Change Biology (GCB) Bioenergy* **6** 637-45

[16] Munir S, Daood S S, Nimmo W, Cunliffe A M and Gibbs B M 2009 Thermal analysis and devolatilization kinetics of cotton stalk, sugar cane bagasse and shea meal under nitrogen and air atmospheres *Bioresources Technology* **100** 1413-8

[17] Lam S S, Liew R K, Lim X Y, Ani F N and Jusoh A 2016 Fruit waste as feedstock for recovery by pyrolysis technique *International Biodeterioration & Biodegradation* **113** 325-33

[18] Blanco P H, Wu C, Onwudili J A and Williams P T 2013 Characterization and evaluation of Ni/SiO2 catalysts for hydrogen production and tar reduction from catalytic steam pyrolysis-reforming of refuse derived fuel *Applied Catalysis B: Environmental* **134–135** 238-50

[19] Li D, Chen L, Yi X, Zhang X and Ye N 2010 Pyrolytic characteristics and kinetics of two brown algae and sodium alginate *Bioresource Technology* **101** 7131-6

[20] Ceylan S and Goldfarb J L 2015 Green tide to green fuels: TG–FTIR analysis and kinetic study of Ulva prolifera pyrolysis *Energy Conversion and Management* **101** 263-70

[21] Slopiecka K, Bartocci P and Fantozzi F 2011 Thermogravimetric analysis and Kinetic study of poplar wood pyrolysis In: *Third International Conference on Applied Energy,* pp 1687-98

[22] Tumuluru J S, Hess J R, Boardman R D and Kenney K L 2012 Formulation, pretreatment,and densification options to improve biomass specifications for co-firing highpercentages with coal *Industrial Biotechnology* **8** 113-32

[23] Hui Z, Huaxiao Y, Mengmeng Z and Song Q 2010 Pyrolysis Characteristics and Kinetics of Macroalgae Biomass Using Thermogravimetric Analyzer *International Scholarly and Scientific Research & Innovation* **4** 148-53

[24] Shuping Z, Yulong W, Mingde Y, Chun L and Junmao T 2010 Pyrolysis characteristics and kinetics of the marine microalgae Dunaliella tertiolecta using thermogravimetric analyzer *Bioresource Technology* **101** 359-65

[25] Ceylan S, Topcu Y and Ceylan Z 2014 Thermal behaviour and kinetics of alga Polysiphonia elongata biomass during pyrolysis *Bioresour Technol* **171** 193-8

[26] Wang J, Wang G, Zhang M, Chen M, Li D, Min F, Chen M, Zhang S, Ren Z and Yan Y 2006 A comparative study of thermolysis characteristics and kinetics of seaweeds and fir wood *Process Biochemistry* **41** 1883-6

[27] Abbasi S M, Rashidi A, Nemati A and Arzani K 2013 The effect of functionalisation method on the stability and the thermal conductivity of nanofluid hybrids of carbon nanotubes/gamma alumina *Ceramics International* **39** 3885-91

[28] López-Mondéjar R, Zühlke D, Becher D, Riedel K and Baldrian P 2016 Cellulose and hemicellulose decomposition by forest soil bacteria proceeds by the action of structurally variable enzymatic systems *Scientific Reports* **6** 25279

[29] Slopiecka K, Bartocci P and Fantozzi F 2012 Thermogravimetric analysis and kinetic study of poplar wood pyrolysis *Applied Energy* **97** 491-7

[30] Agrawal A and Chakraborty S 2013 A kinetic study of pyrolysis and combustion of microalgae *Chlorella vulgaris* using thermo-gravimetric analysis *Bioresource Technology* **128** 72-80

Modal properties investigation of car body-in-white with attached windscreen and rear screen

Z M Hafizi[1*], A M Aizzuddin[1], N I A Halim[2] and M F Jamaludin[2]

[1]Advanced Structural Integrity and Vibration Research Group, Faculty of Mechanical Engineering, Universiti Malaysia Pahang, 26600 Pekan, Pahang, Malaysia
[2]PROTON Sdn. Bhd., HICOM Industrial Estate (Batu Tiga, PO Box 7100), 40000 Shah Alam, Selangor, Malaysia

*Corresponding author: hafizi@ump.edu.my

Abstract. Vibration analysis of car body in white (BIW) is crucial during design stage. Car BIW holds all the essential components; therefore, analysing its dynamic behaviour is necessary to understand how vibrations are transferred to the end rows from any vibration sources. This paper presents the modal properties of car BIW with attached windscreen and rear screen calculated by means of finite element analysis (FEA) and experimental modal analysis (EMA). The aim for the analysis is to observe the effect of windscreen and rear screen to the dynamic properties of car BIW. Detailed CAD models of BIW for both schemes were used for normal modes calculation in FEA, therefore providing the finest prediction of its modal properties. Actual car BIW was hanged on a metal frame to imitate free-free boundary condition as in FEA; prior to EMA. The EMA was done for both schemes, whereby the results for FEA were confirmed with EMA, at least for the first five modes with errors below 15%. It can be concluded from the analysis that the attached windscreen and rear screen is significantly affects the first five modes of the BIW. However, other higher frequencies remain unchanged.

1. Introduction

Vibration analysis of a car body in white (BIW) is important to understand the Noise, Vibration and Harshness (NVH) of a car during earlier design stage [1-3]. A car BIW must be able to meet the NVH design criteria as well as the stiffness, strength and fatigue life [4]. In most car manufacturing development cycle, this procedure take place during engineering design where computational aided engineering (CAE) is involved [5].

A car BIW can be referred to a car body's sheet metal components which have been welded together, before attachment of the moving parts (doors, hoods, and deck lids as well as fenders) the motor, chassis sub-assemblies, or trim (glass, seats, upholstery, electronics, etc.); and before painting [6]. The main parameters to be concerned in a vibration analysis of a car BIW are the natural frequencies and the vibration mode shapes [7-9]. Whenever these natural frequencies match with vibration sources, resonance will occur and causing serious effects to some components. The lower frequency range of vibration mainly contributes to the stress and critical loading on the individual component of the BIW, especially if it is subjected to vibration source from the rough road surface. On the other hand, the higher frequency range of vibration contributes to the interior and exterior noise to the vehicle. The main source can be from the second order vibration of engine, the wheel-road contact during cruising on highways or subcomponents such as air-conditioning system and cooling fans.

In most research works, the vibration analysis of a car BIW was done without the consideration of windscreen and rear screen attached to the BIW. If the car BIW definition is concerned, the vibration analysis with windscreen and rear screen attached to the BIW may contributes to the dynamic properties of the whole BIW as they are actually bind to the structure.

In the meantime, the analytical study of modal parameters using finite element analysis (FEA) is well establish and widely used in the engineering design stage of any structures. It significantly helps to save the development cost of car manufacturers. However, due to several assumptions made during the finite element modeling, there are some discrepancies to the actual measurement which may affect the credibility of the test result in the next car development stages. Therefore, the experimental modal analysis (EMA) is necessary steps to evaluate the integrity of the FEA [1, 10, 11]. Although EMA is traditional, yet is considered as the most accurate representation to the actual dynamic behavior of a structure, with the use of frequency response function (FRF) data [12, 13]. The combination of EMA with CAE makes the results more impressive and can provide theoretical simulations.

In this research work, normal modes investigation of a car BIW with attached windscreen and rear screen has been done, via FEA and EMA. The results were compared with the BIW only case.

2. Finite Element Analysis

The car model chosen for the analysis is a 4-door sedan car. A full meshing has been done by using *Hypermesh* software. The free-free boundary condition, with no constraint was chosen to match the experimental modal analysis setup. Figure 1 shows the FE model of the car BIW.

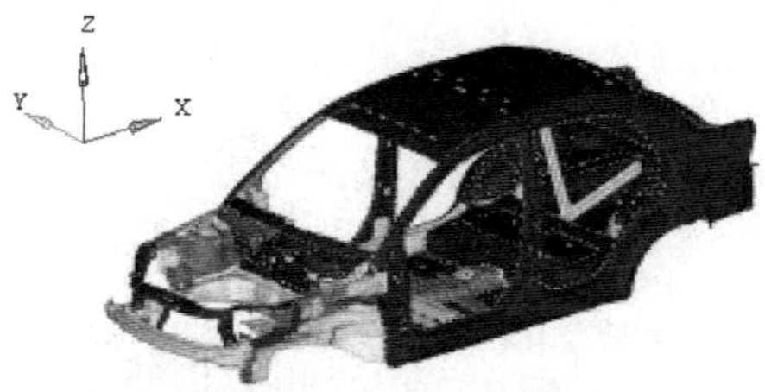

Figure 1. FEA model of the BIW.

The results have been calculated for the first 20 frequency modes, exclusive the rigid modes and frequency below 100 Hz was targeted. Figures 2 to 6 show the first five modes of car BIW only. It was predicted that the BIW experienced the first natural frequency at 30.11 Hz, in torsion mode about x-axis, with the maximum deformation appears to be localized at rear part of the roof, close to the rear screen.

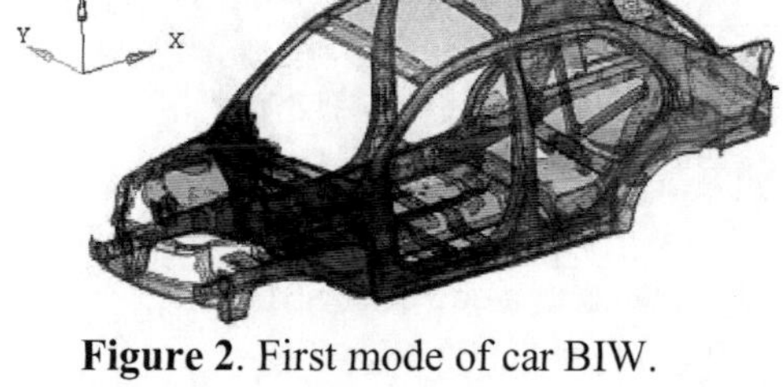

Figure 2. First mode of car BIW.

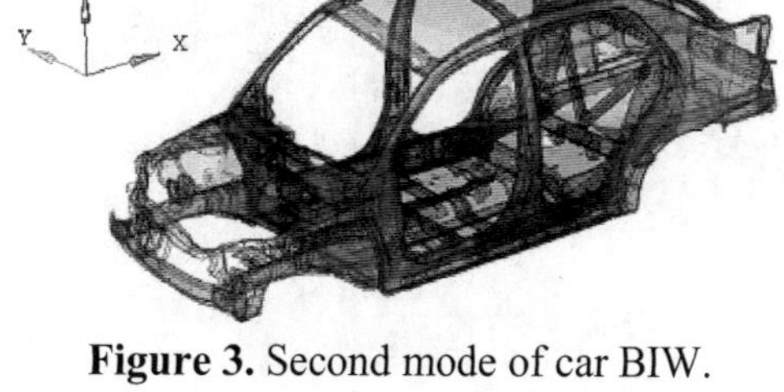

Figure 3. Second mode of car BIW.

Meanwhile the second mode was 34.31 Hz, also twisting about x-axis; however, the largest deflection was at the front engine room area. In this mode, the front roof also displays a significant deformation. The third mode was the mix mode, with a large bending in y-axis at the front end of engine area at frequency of 41.08 Hz. Large twisting also can be seen at the front roof area. In the meantime, the fourth and fifth mode shapes indicate the pure bending mode with frequency 45.06 Hz and 46.04 Hz, respectively.

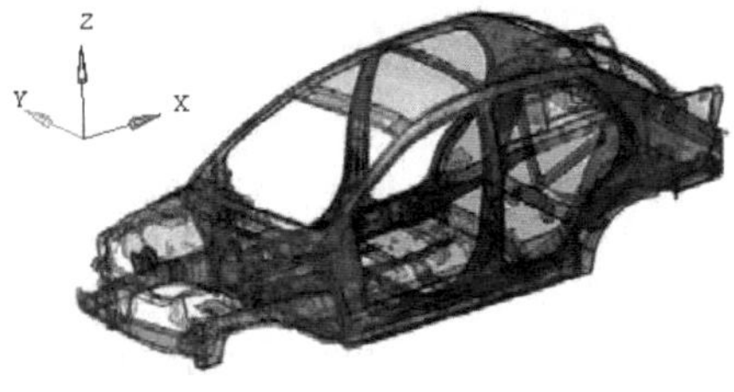

Figure 4. Third mode of car BIW.

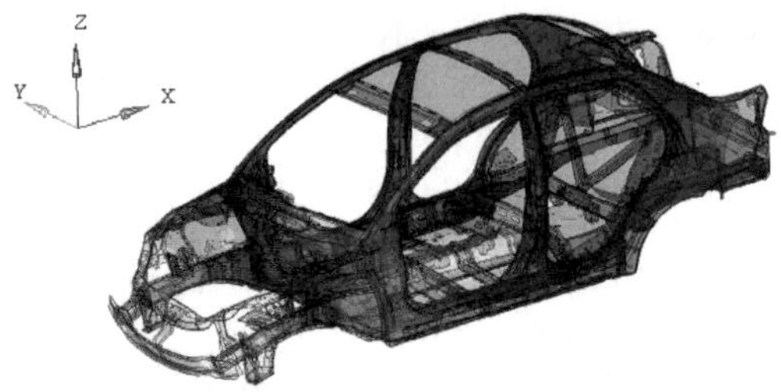

Figure 5. Fourth mode of car BIW.

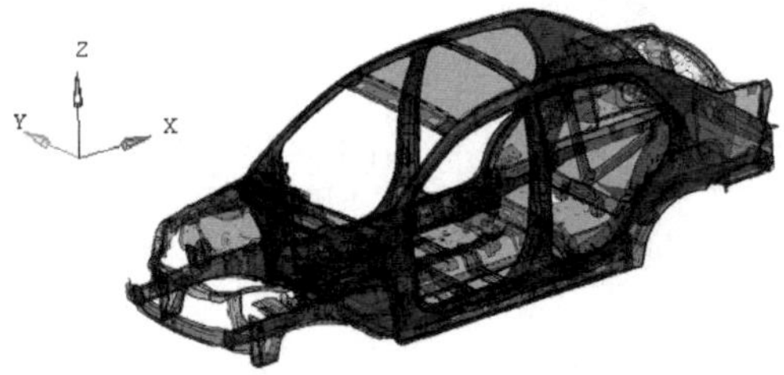

Figure 6. Fifth mode of car BIW.

Figures 7 to 11 present the result of normal modes analysis for car BIW with attached windscreen and rear screen. The first natural frequency appears to be in bending mode at y-axis, with the most deflected area is at the front end of the BIW. Its frequency increase 24% compared to the first mode of car BIW only; 37.43 Hz. However, the first mode of the BIW only is in twisting mode. In the meantime, the second and third mode shape also bending mode, but in z-axis, with frequency are 44.70 Hz and 46.91 Hz, respectively. The significant deformation occurs similarly for both modes, which are at the front roof and front end area, however with vary magnitude. The second mode displays maximum deflection at front end while the third mode indicates maximum deformation localized at front roof attached to the windscreen.

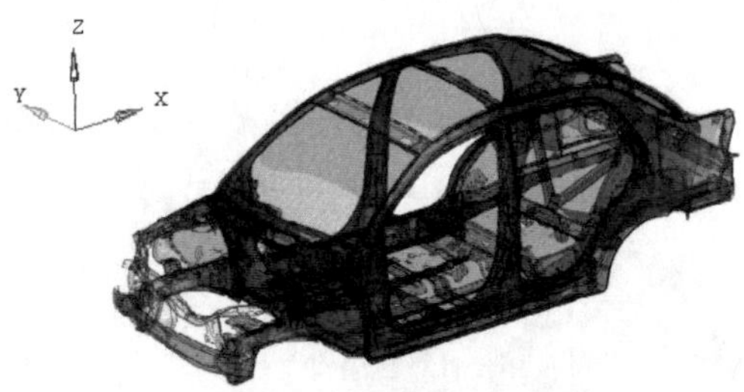

Figure 7. First mode of car BIW with windscreen and rear screen.

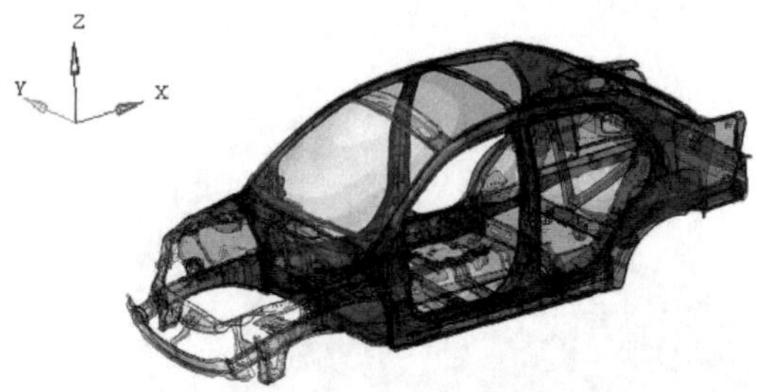

Figure 8. Second mode of car BIW with windscreen and rear screen.

The first torsion mode experienced by this BIW scheme appears at the fourth mode; twisting about x-axis with frequency 51.59 Hz. It is 70% higher than the first torsion in the BIW only result. However, the first torsion in BIW with windscreen and rear screen experienced more global vibration compared to BIW only. Meanwhile, the fifth mode indicates the bending in z-axis with frequency at 54.09 Hz and maximum translation at the mid of rear screen along x-axis.

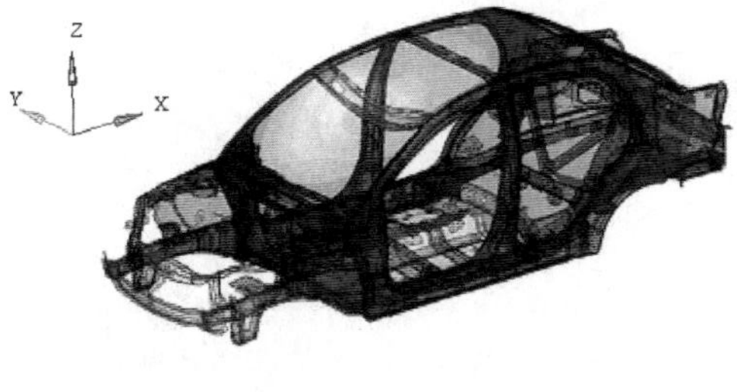

Figure 9. Third mode of car BIW with windscreen and rear screen.

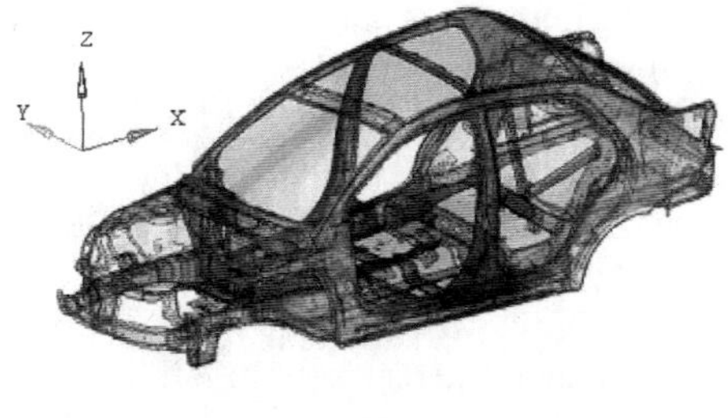

Figure 10. Fourth mode of car BIW with windscreen and rear screen.

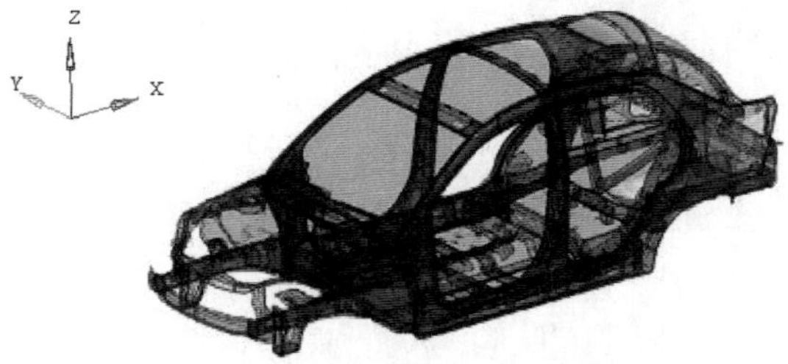

Figure 11. Fifth mode of car BIW with windscreen and rear screen.

The dissimilarity of the first 10 natural frequency values for the car BIW with windscreen and rear screen, over the car BIW only can be summarized as in Table 1. The frequency variances are significant only for the first five modes, while the rest fall below 5% discrepancies. Further comparisons for the mode shapes from mode 6 to mode 20 for both models also reveal no disagreement at all.

Table 1. Comparison between BIW only and BIW with windscreen & rear screen.

Mode	Mode BIW (Hz)	BIW with windscreen & rear screen (Hz)	Different (%)
1	30.11	37.43	24.29
2	34.31	44.7	30.31
3	41.08	46.91	14.21
4	45.64	51.59	13.03
5	46.04	54.09	17.49
6	54.47	56.15	3.08
7	56.18	57.61	2.54
8	58.57	58.97	0.68
9	60.78	60.94	0.27
10	62.39	63.52	1.82

3. EMA Set-up

The experimental modal analysis was performed for car BIW with attached windscreen and rear screen; and BIW only scheme. The structure was hanged up with bungee cord, which represent the free-free condition as setup in FEA. Total 213 points have been marked all over the BIW for modal testing. These points were selected based on clear deformation observed through the normal modes analysis in FEA and appropriate location for transducer. They were plotted in *MEscopeVES* software with actual *x, y, z* coordinates obtained from the FE model.

Figure 12 presents the model setup in *MEscopeVES.* For better presentation, all points were combined with lines, and surfaces were added at the appropriate locations. Roving accelerometer scheme has been chosen for the experiment; whereby a tri-axial accelerometer was used to measure the *x, y, z* responses of the BIW at all 213 points. Note that, the numbering marked in EMA weren't follow the number of points. The numbering for EMA start with 10 (10, 11, 12, 13, ...223). Meanwhile, a shaker was used for the input excitation. Excitation at positive z direction was applied at point no 223, located at the rear bottom left of the BIW, as shown in Figure 13. A chirp signal ranging from 20 Hz to 100 Hz was selected as shaker input, which is expected to excite more than 10 modes.

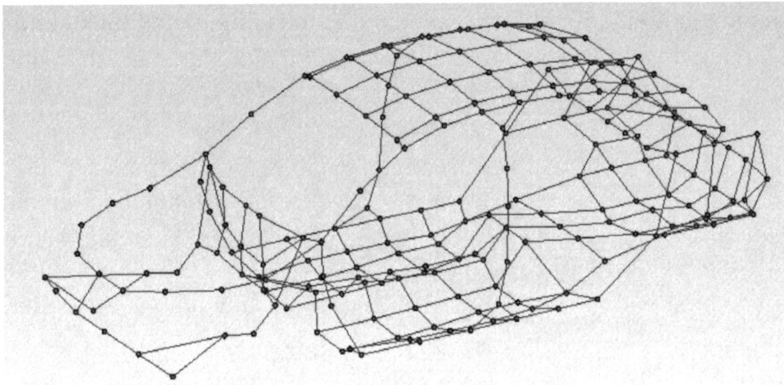

Figure 12. BIW modeling in MEscopeVES.

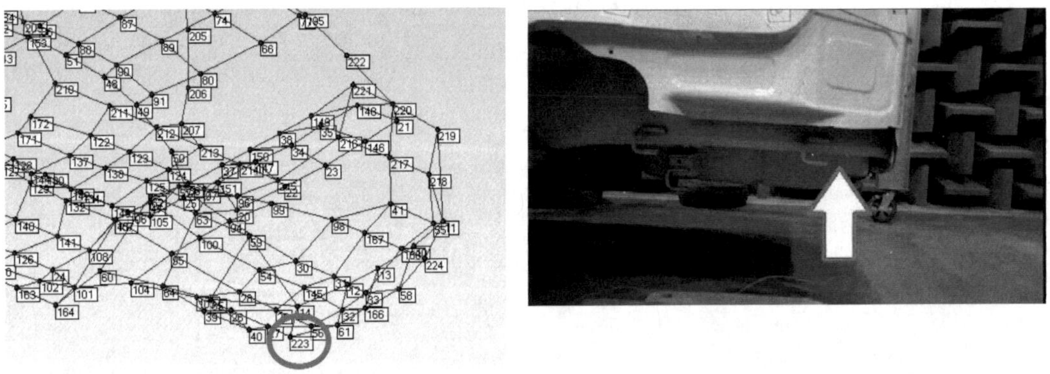

Figure 13. Excitation point, with the red circle and red arrow indicates the point number 223.

Data acquisition system used for the experiment was the 4-channel National Instruments NI9234. The time domain data; the ratio of output responses due to an input, were converted into frequency domain representation; the frequency response function (FRF). These data will be included in *MEscopeVES* for mode shapes calculation of the BIW structures.

4. EMA Results and Discussion

Figure 14 and 15 show the FRF data generated for 213 points during the shaker test, for the first five modes. Total 639 of FRF traces inclusive *x*, *y* and *z* axes responses, were plotted. All 5 modes were visible for both BIW schemes. To estimate the frequency for each mode, complex mode indicator function (CMIF) method was used.

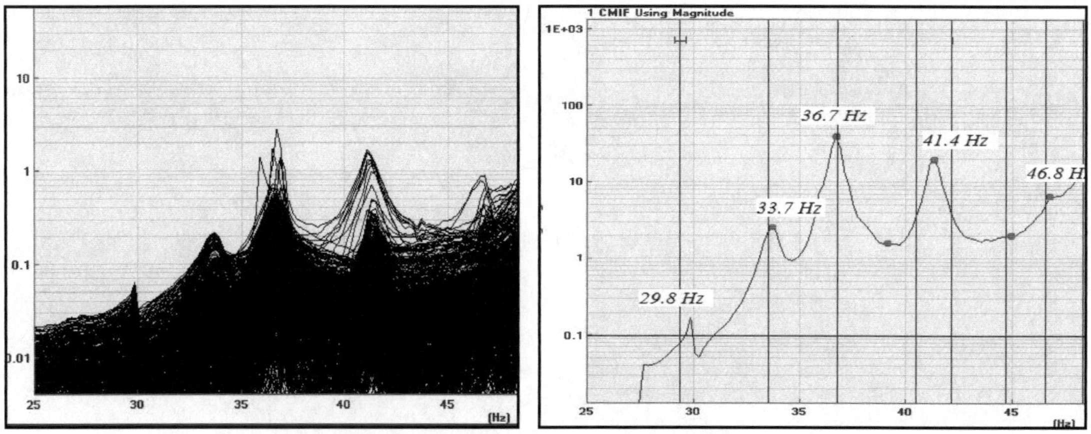

Figure 14. FRF's data (left) and mode indicator using CMIF (right); for BIW only.

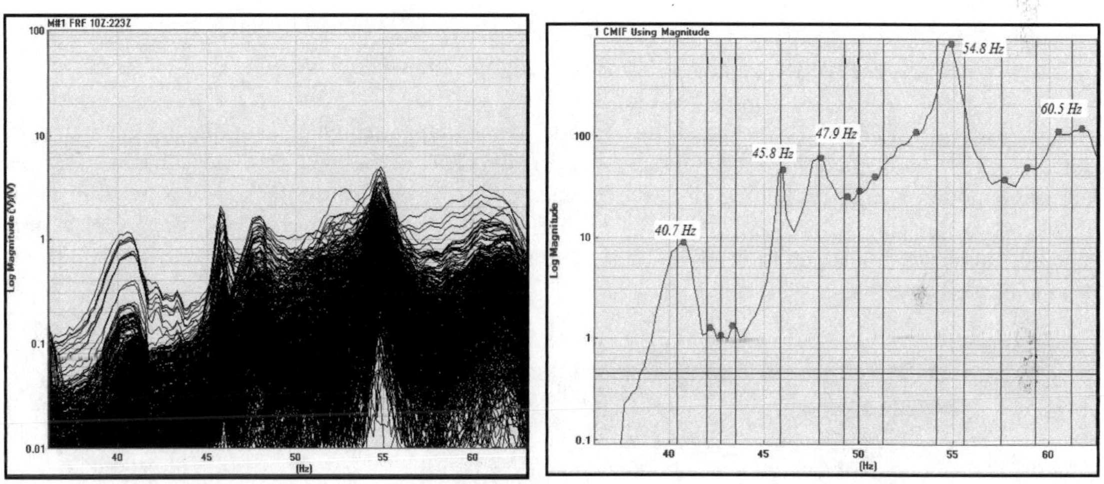

Figure 15. FRF's data (left) and mode indicator using CMIF (right); for BIW with screen.

In the meantime, Figures 16 to 20 reveal the mode shapes of BIW only structure, for mode 1 to 5, calculated based on FRF data. The results indicate similar mode shapes as were determined by FEA. The first 3 modes display a global twisting and torsion with several critical deformation areas. Meanwhile, mode 4 and mode 5 show huge deformation at the front end and rear roof area, respectively. Despite of the well correlation of FEA and EMA mode shapes, there are discrepancies in the natural frequency values; especially for mode 3 and 4, with 10.65% and 9.01% error respectively.

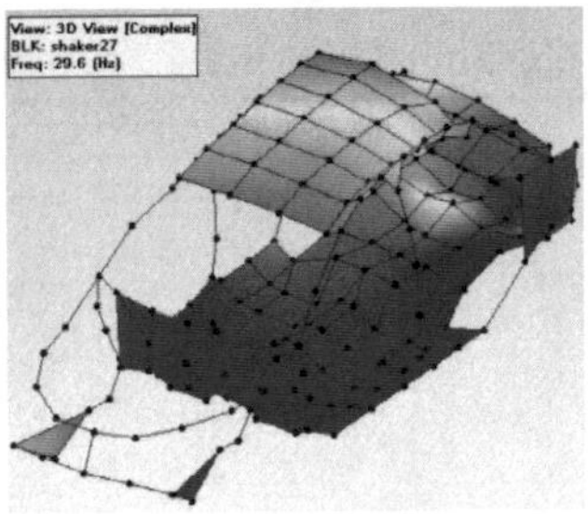

Figure 16. Mode 1 (29.8 Hz).

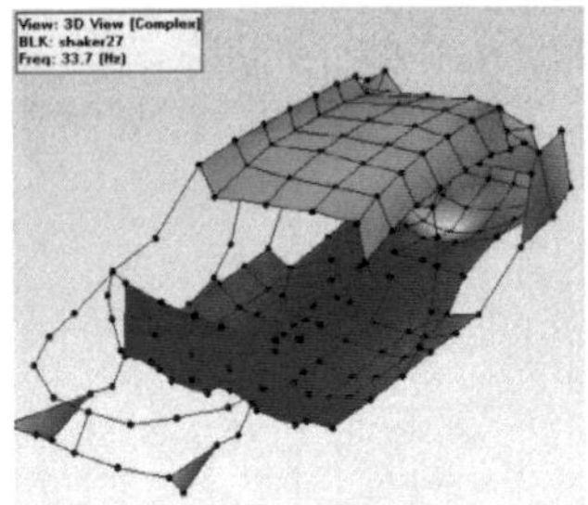

Figure 17. Mode 2 (33.7 Hz).

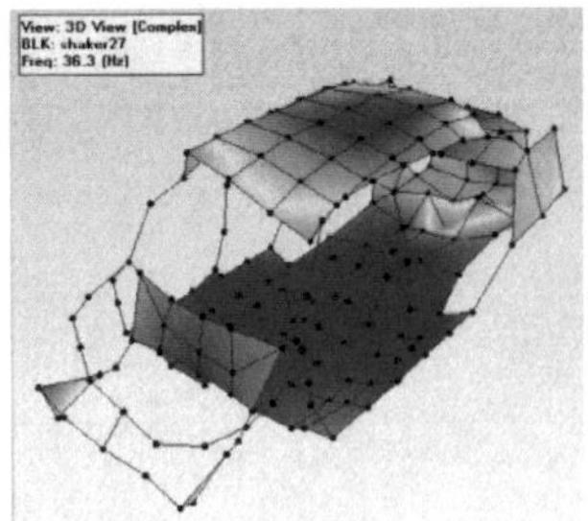

Figure 18. Mode 3 (36.7 Hz).

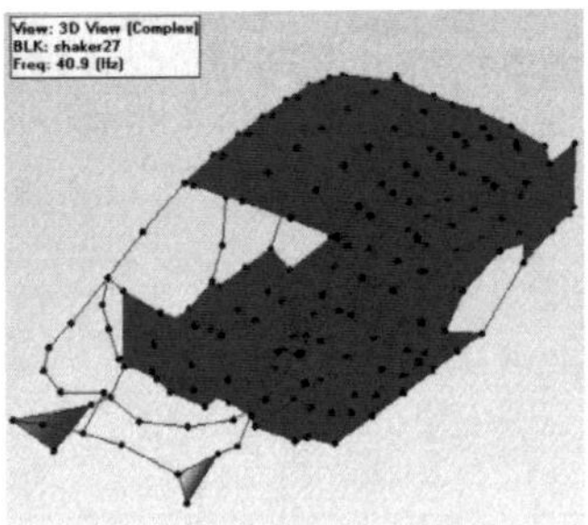

Figure 19. Mode 4 (41.4 Hz).

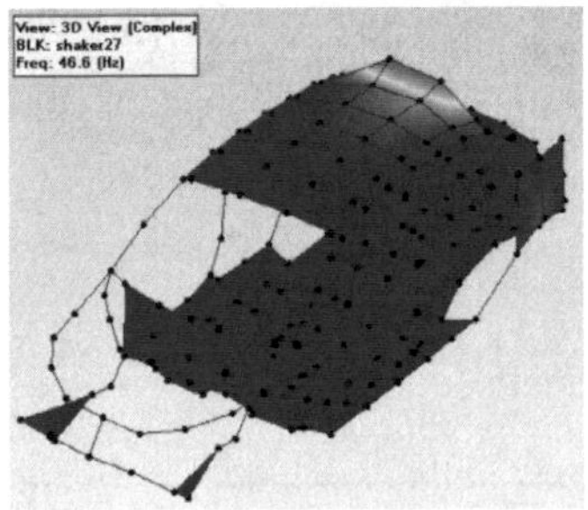

Figure 20. Mode 5 (46.8 Hz).

Figures 21 to 25, on the other hand, display the mode shapes for BIW with attached screen. The outcomes have confirmed the normal modes analysis by FEA, earlier. As predicted, the first, second and third mode experienced the bending mode. Meanwhile, the fourth mode shows a twisting mode about x-axis; and the fifth mode displays a big deformation at the rear roof. As for the BIW only scheme, there are small variation of the natural frequency values between FEA and EMA results, despite of well agreement of the mode shapes.

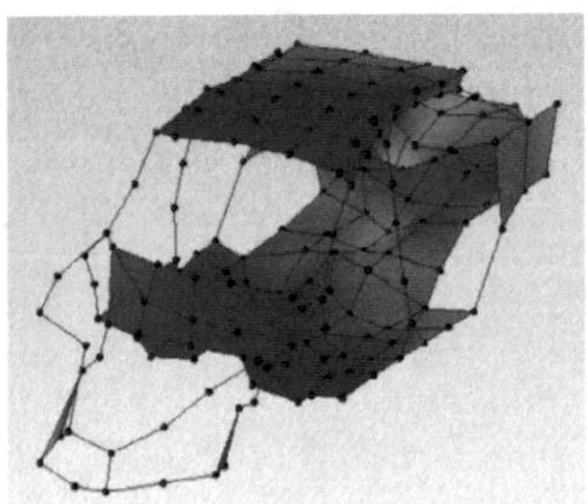

Figure 21. Mode 1 (40.7 Hz).

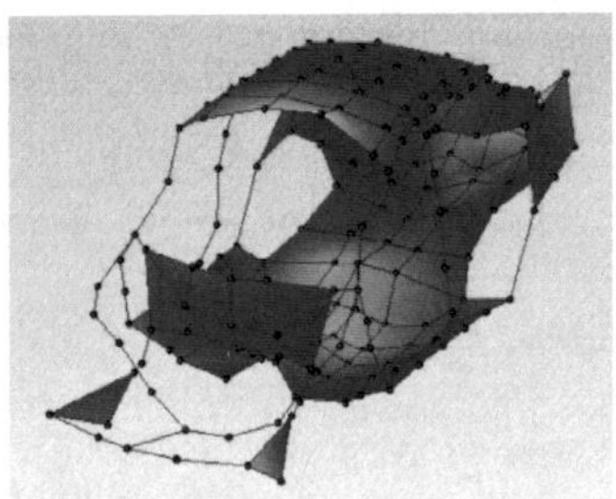

Figure 22. Mode 2 (45.8 Hz).

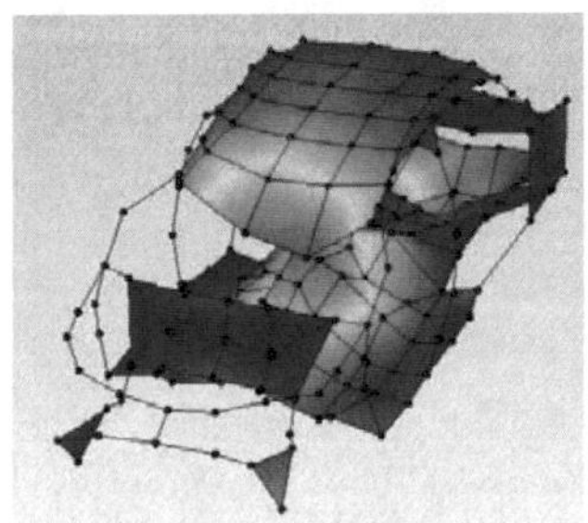

Figure 23. Mode 3 (47.9 Hz).

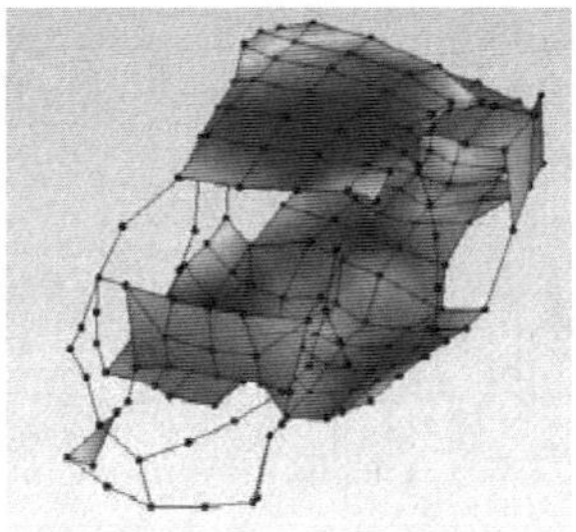

Figure 24. Mode 4 (54.8 Hz).

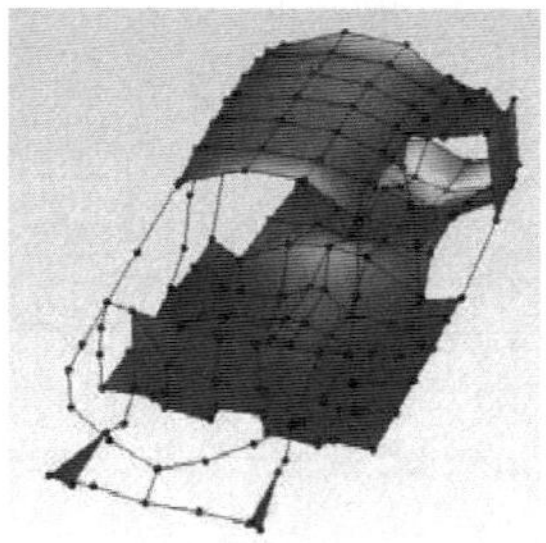

Figure 25. Mode 5 (60.5 Hz).

Table 2 presents the natural frequencies correlation of car BIW with attached windscreen and rear screen extracted from FEA and EMA, for the first five modes. Meanwhile, Figure 26 displays frequency pairing between both FEA and EMA. The comparison for both analyses is necessary to determine any inconsistencies. Mode 1 and mode 5 show excellent correlation, while mode 2, 3 and 4 have 10% errors. These discrepancies can be reduced by applying model updating techniques; however, beyond the scope of this study.

Table 2. Natural frequencies comparison from EMA and FEA.

	Mode	FEA Frequency (Hz)	EMA Frequency (Hz)	Error (%)
BIW only	1	29.8	29.79	0.0
	2	33.7	33.93	0.7
	3	36.7	40.61	10.7
	4	41.4	45.13	9.0
	5	46.8	45.70	2.4
BIW with screen	1	40.7	37.4	8.1
	2	45.8	40.5	11.6
	3	47.9	42.7	10.9
	4	54.8	47.2	13.9
	5	60.5	54.2	10.4

Figure 26. Frequency pair between FEA and EMA for the first five modes, for BIW only (left) and BIW with screen (right).

5. Conclusions

This research work has presented the modal analysis of a car body in white by means of finite element analysis and experimental modal analysis. Comparison between the modal parameter; natural frequencies and mode shapes of the BIW only and BIW with attached windscreen and rear screen reveals that there were significant variances, especially for the first five modes. As the increase of the first natural frequency is expected, the mode shapes also indicate significant dissimilarity; whereby the first torsion and the first bending appear at different mode. Thus, the examination of BIW with attached windscreen and rear shows indispensable information to be considered during development stage.

Acknowledgements

The authors would like to thank the Faculty of Mechanical Engineering, Universiti Malaysia Pahang (UMP) for providing laboratory facilities; and financial support under internal research grant RDU160136 from UMP Research and Innovation Department. Also, special thanks to Proton Sdn. Bhd. under Local Expert Attachment Program (LEAP) for the financial support.

References

[1] Rajesh H, Patwardhan M A, Karanth N V, Bhatkhande M, Ramkumar R, Pachhapurkar N, Saraf M and Reddy Y 2017 LCV Chassis Frame Optimization Using Combined Simulation and Experimental Approach. (SAE Technical Paper)

[2] Hanouf Z, Faris W F and Ahmad K 2017 An investigation into NVC characteristics of vehicle behaviour using modal analysis. In: *IOP Conference Series: Materials Science and Engineering*: (IOP Publishing) p 012027

[3] Sun L, Chen N and Zhao Z 2016 Experimental Modal and Dynamic Performances Analysis of Car's Body-in-white *Journal of Residuals Science & Technology* **13**

[4] Yahaya Rashid A S, Ramli R, Mohamed Haris S and Alias A 2014 Improving the dynamic characteristics of body-in-white structure using structural optimization *The Scientific World Journal* **2014**

[5] Hirz M, Dietrich W, Gfrerrer A and Lang J 2013 *Integrated Computer-Aided Design in Automotive Development*: (Springer Berlin Heidelberg)

[6] Genta G and Morello L 2009 *The automotive chassis* vol 1: (Springer Netherlands)

[7] Van der Auweraer H 2001 Structural dynamics modeling using modal analysis: applications, trends and challenges. In: *Instrumentation and Measurement Technology Conference, 2001. IMTC 2001. Proceedings of the 18th IEEE*: (IEEE) pp 1502-9

[8] Sani M, Rahman M, Noor M, Kadirgama K and Izham M 2011 Identification of dynamics modal parameter for car chassis. In: *IOP Conference Series: Materials Science and Engineering*: (IOP Publishing) p 012038

[9] Zhang H and Tong M 2014 The results of experiment analysis on body-in-white research *Applied Mechanics & Materials* 6358-61

[10] Rotondella V, Merulla A, Baldini A and Mantovani S 2017 Dynamic Modal Correlation of an Automotive Rear Subframe, with Particular Reference to the Modelling of Welded Joints *Advances in Acoustics and Vibration* **2017**

[11] Lu J, Zhan Z, Song H, Liu X, Yang X and Yang J 2017 Design Optimization of Vehicle Body NVH Performance Based on Dynamic Response Analysis. (SAE Technical Paper)

[12] Kranjc T, Slavič J and Boltežar M 2013 The mass normalization of the displacement and strain mode shapes in a strain experimental modal analysis using the mass-change strategy *J. Sound Vib.* **332** 6968-81

[13] Wang B-T and Cheng D-K 2011 Modal analysis by free vibration response only for discrete and continuous systems *J. Sound Vib.* **330** 3913-29

Design and optimize of 3-axis filament winding machine

Ma Quanjin*, M R M Rejab, M S Idris, D Bachtiar, J P Siregar, M N Harith

Structural Materials & Degradation Focus Group, Faculty of Mechanical Engineering, Universiti Malaysia Pahang, 26600 Pekan, Pahang, Malaysia

*Corresponding author: neromaquanjin@gmail.com

Abstract. Filament winding technique is developed as the primary process for composite cylindrical structures fabrication at low cost. Fibres are wound on a rotating mandrel by a filament winding machine where resin impregnated fibres pass through a pay-out eye. This paper aims to develop and optimize a 3-axis, lightweight, practical, efficient, portable filament winding machine to satisfy the customer demand, which can fabricate pipes and round shape cylinders with resins. There are 3 main units on the 3-axis filament winding machine, which are the rotary unit, the delivery unit and control system unit. Comparison with previous existing filament winding machines in the factory, it has 3 degrees of freedom and can fabricate more complex shape specimens based on the mandrel shape and particular control system. The machine has been designed and fabricated on 3 axes movements with control system. The x-axis is for movement of the carriage, the y-axis is the rotation of mandrel and the z-axis is the movement of the pay-out eye. Cylindrical specimens with different dimensions and winding angles were produced. 3-axis automated filament winding machine has been successfully designed with simple control system.

1. Introduction

Composite materials are increasingly used in the civil infrastructure and piping system to transport sewage, service and potable waters [1]. Because of light weight, high specific strength, stiffness, fatigue characteristics and better corrosion resistance. In the last decade, the use of filament wound composite materials has drastically increased. Filament wound composites have been replaced metal alloys in many applications, for example, the high-pressure vessel in chemical plant or aeronautical/space vehicles, or transportation tubes in oil, gas and nuclear industries [2, 3]. Until now more researchers have researched the experimental and investigations on filament wound GRP pipes, filament wound GERP and CFRP composite tubes. Different kinds of composite materials have processed specific mechanical properties by previous researchers [4].

The filament winding process is a low cost, efficient and relative maturity manufacturing process, which generally used in the mass production of fibre reinforced composite components [5]. Filament winding method is the most broadly used fabrication process to produce rotational symmetry [6]. This process can be accomplished in two different ways as reciprocal or continuous method [7]. Continuous filament winding is the simple and sufficient method to fabricate a fibre reinforcement structures, a continuous band of reinforcement is impregnated with resin and then wound over a rotating mandrel [8, 9]. The reinforced fibres are usually made of glass, kevlar or carbon. Owing to a simplicity of manufacturing process, the hardware involves two main sub-systems, which is the rotary assembly and the delivery system[10].

2. Review on the filament winding process

2.1 Winding Methods

2.1.1 Wet winding. There are two different winding methods, which are wet and prepreg winding. Wet winding, in which the fibres are passed through a resin bath and wound onto a rotating mandrel. This type of low-cost system is generally used in commercial applications with polyester and epoxy resins. The resin field is affected by several parameters: resin viscosity, the squeeze rollers, winding tensioner, the number of layers, etc [11]. Figure 1 shows the wet winding method general view. Compared to prepreg winding, wet winding has several main advantages: short processing time; low costs about the composite material, simple resin formulation model which can be easily selected to meet the specific needs in different severe working condition, convertibility of resin system which can be transformed.

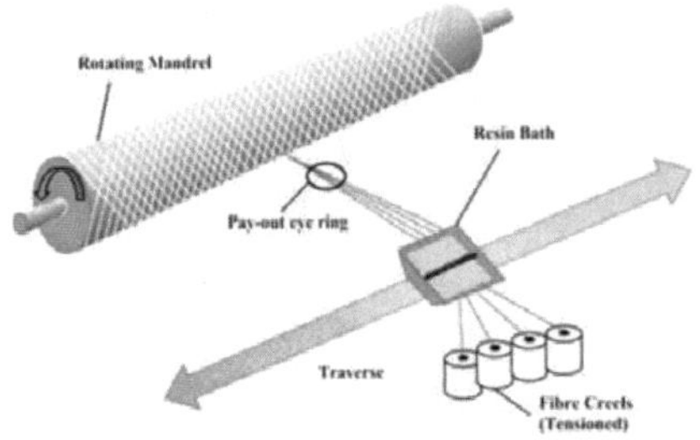

Figure 1. Wet winding method.

2.1.2 Prepreg winding. This winding method is shown in Figure 2. Prepreg winding, in which the pre-impregnated fibre tows are placed on the rotating mandrel [12]. Pre-impregnated rovings offer excellent quality control and reproducibility in contents, bandwidth and uniformity. These parameters can be well satisfied with the filament winding process. On the factory experiment laboratory, many quality control tests can be done for prepreg winding. The obvious advantage of the prepreg is better resin content control, due to the protection supplied by the resin and more consistent composite properties reduce the fibre tows damage. The general shortcomings of prepreg are poor surface layer availability, higher costs and without room temperature as possible.

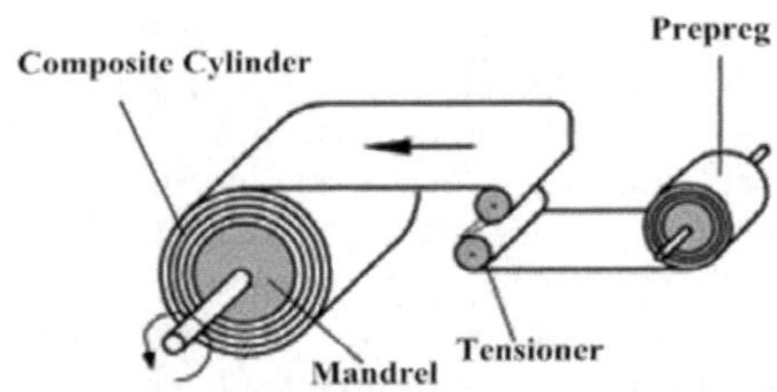

Figure 2. Prepreg winding method [11].

2.2 Winding pattern

There are basically three types of winding patterns: circumferential, helical, and polar winding.

2.2.1 Circumferential winding. It is known as the hoop or circumferential winding, Generally speaking, circumferential winding is a special winding angle which attains 90 degrees of winding angle. Each full rotation of the mandrel advances the carriage by full bandwidth as shown in Figure 3.

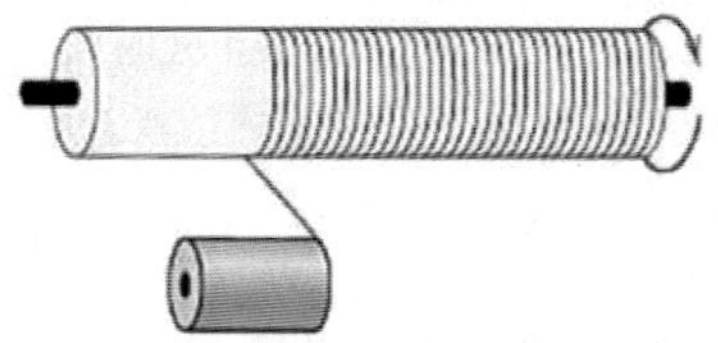

Figure 3. Circumferential or hoop winding [12].

2.2.2 Helical winding. In the helical winding, mandrel rotates at a constant speed while the carriage unit transverses back and forth at a regular speed, the composite fibre is taken out by mandrel rotation force from ring pay-out eye or delivery system. It has been shown in Figure 4. To control the rotation speed and carriage speed, the desired winding angle can be fabricated.

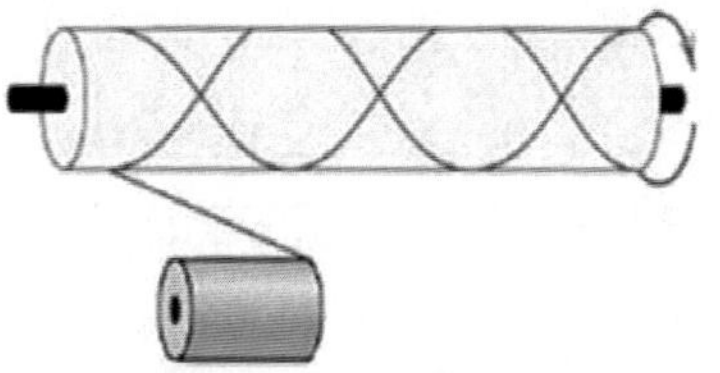

Figure 4. Helical winding [14].

2.2.3 Polar winding. In polar winding, which is shown in Figure 5. Composite fibres pass tangentially to the polar position and cover the fibre along the polar path, reverses direction, and passes tangentially to the opposite polar position at the other end. In one word, composite fibres are wound from one pole to pole, while the mandrel arm rotates around the longitudinal axis. Polar winding needs a high degree of freedom machine to execute.

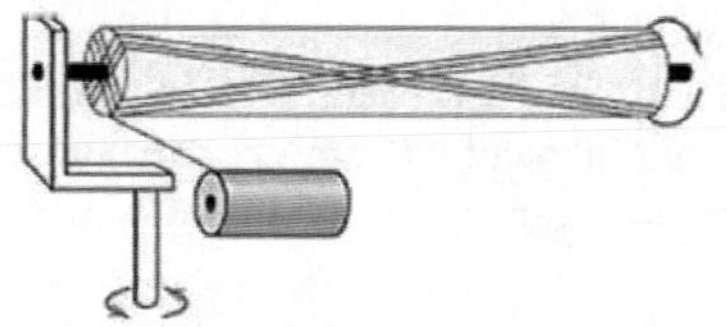

Figure 5. Polar winding [14].

2.3 Materials of Fabrication

Filament winding requires continuous fibre reinforcement and a resin system to stick together and fabricate desired composite material. There are various types of fibre reinforcement that can be impregnated with a resin matrix.

Reinforcing fibres for polymer-matrix composites are commonly glass, graphite and aramids. The fibre characteristics have affected the stiffness and the strength of the composite material performances. Table 1 shows a brief comparison of carbon reinforcements. The main fibres are summarised below:

a) Glass: glass fibres are the most commonly used and the least expensive of all fibres. The composite material is called glass-fibre reinforced plastic (GFRP) and it may contain between 30% and 60% glass fibre by volume[15];

1) E type: a calcium alumni borosilicate glass-good tensile strength, low tensile modulus, lowest cost fibre, generally used in commercial and industrial products, commonly used in filament winding;

2) S type: magnesia-aluminosilicate glass-applying high tensile strength and stiffness but at greater cost, used in aerospace and high-performance pressure vessel;

b) Aramids: higher tensile strength, higher tensile modulus, high cost, low density, excellent impact and damage tolerance properties.

c) Carbon fibre: The properties of carbon fibres, such as high stiffness, high tensile strength, low weight, high chemical resistance, high-temperature tolerance and low thermal expansion. When combined with a plastic resin and wound or moulded it forms the carbon-fibre-reinforced polymer (often referred to as carbon fibre) which has a very high strength-to-weight ratio, and is extremely rigid although somewhat brittle.

Table 1. Comparison of fibre reinforcements[13].

Reinforcement	*Density/Kg/m*^{3}	*Tensile strength/Mpa*	*Tensile modulus/GPa*
E-glass	2547	3447[a]	72.4
		2758[b]	
S-glass	2491	4585[a]	85.5
		3792[b]	
Carbon(AS4)	1799	3998	231
Kevlar 49(Aramids)	1449	2758	131

[a]Monofilament [b]Strand

3. Methodology

Research methodology in this project can be simply divided into three phases:

Phase 1: Literature review and purchase necessary components

i. It is essential that we should figure out identification of objectives, methods, potential problems and sufficient knowledge about filament winding manufacturing technique.

ii. It is a effective method to purchase the material and equipment structure, parameters and applications.

Phase 2: Design a 3-axis filament winding machine and optimize processing capability

i. Compared to previously related machine structure, start the preliminary design of the 3-axis lathe-type filament winding machine.

ii. Essential process is that a simple, stable, efficient control system should be designed, which includes different kinds of microcontrollers are more suitable for target machine structure. It spends much time on comparing the strengths and weaknesses.

Phase 3: Fabricate filament wound composite cylinders

i. Based on the test and check the control system of the machine, send the specific codes to fabricate cylindrical structures in different winding angles, which range from 0° to 90°.

3.1 Problem identification and solution

All steps in the engineering design process, so it is most important to give a problem definition. Figure 6 shows the basic progress in this project. This design project is to design a 3-axis filament winding machine and optimise it, then enhance the machine performance that can fabricate the desired product, which on account of costs and customer requirements. Based on the previous design of filament winding machine [16], 3-axis filament winding machine design concepts and methods are clear-cut in principle. Design concepts are the methods for machine function, which include the rotation unit, carriage unit, ring pay-out eye unit, control system unit, resin bath, machine structure, tensioner, comb device, mandrel and etc.

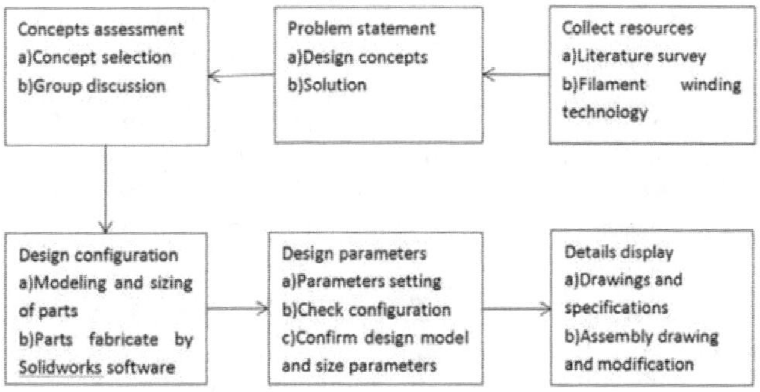

Figure 6. Basic progress in the design process.

4. Design and optimize 3 axes machine

Schematic layout of proposed 3-axis filament winding machine, which can be divided into four main units: the rotary unit, the carriage unit, the ring pay-out eye unit and the control system unit. Based on those main units, the machine prototype can gradually be designed. Other necessary components should be completed after the machine prototype is accomplished step by step. The 3-axis filament winding machine optimises design process includes the lead screw, resin bath, comb device and other components. The optimization process focuses on improvements about 3-axis filament winding machine processing capability.

4.1 The rotary unit

Figure 7 shows the machine rotary unit front view and left view. Like other common filament winders, the rotary unit consists of two basic blocks. Of the two blocks, one is fixed and can simply be removed, the other is mandrel rotation, which can be under controlled. It has many kinds of mandrels, the mandrel in this project is chosen as a 38mm diameter, 690mm length hollow mandrel, which is made of aluminium alloy material. It can support the burden on stepper motor y loading. The mandrel is held on the horizontal line by two mandrel holders, which are printed via 3D Printer Machine. Inside the mandrel, a stainless steel rod is designed to hold the mandrel and mandrel holders, which can prevent the mandrel holders from bending. Between the mandrel and mandrel holders, the rubber ring is put on the contact area, which will effectively enhance the mandrel cylindricity and horizontal extent, rubber ring can reduce vibration impact for composite cylinders when mandrel rotates. Figure 7 shows the rotary unit design drawing. The rotation process is optimized by modification on process stabilization, stepper motor position and proper belt pully parameters.

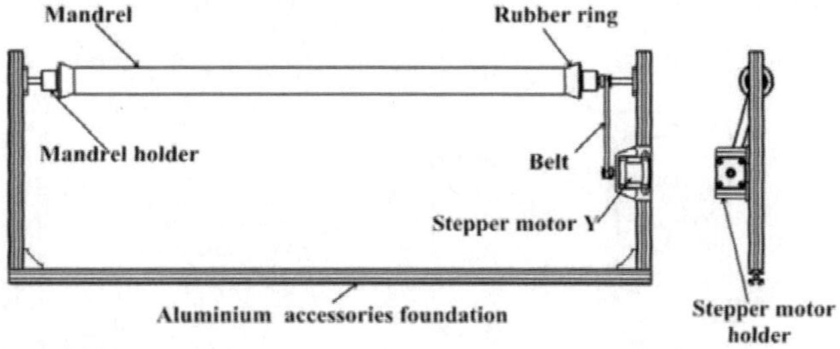

Figure 7. The rotary unit front view and left view.

4.2 The carriage unit

The carriage unit mainly consists of the tensioner, composite fibres creels, resin bath, two guide bars, stepper motor x, and comb device. The carriage unit is driven by a reversible stepper motor x, which transfers constant speed to carriage frame via belt holder fixed in carriage base. The tensioner has put beside the composite fibres creels, the tension force can be controlled by turning the screw to reduce the interval between the tensioner and lever. The comb device is designed to fabricate more than one kind of composite fibre. And two carbon fibre rods can remove redundant resins, the distance can be adjusted by twisting the screw. At last, it still has another carbon fibre rod, which belongs to the transmission device, can guide composite fibre to the ring pay-out eye unit. Figure 8 shows the carriage unit final design drawing. The carriage process is optimized by modification on movement stabilization, x belt pully levelness, appropriate limit switch position, comb device structure and resin bath configuration.

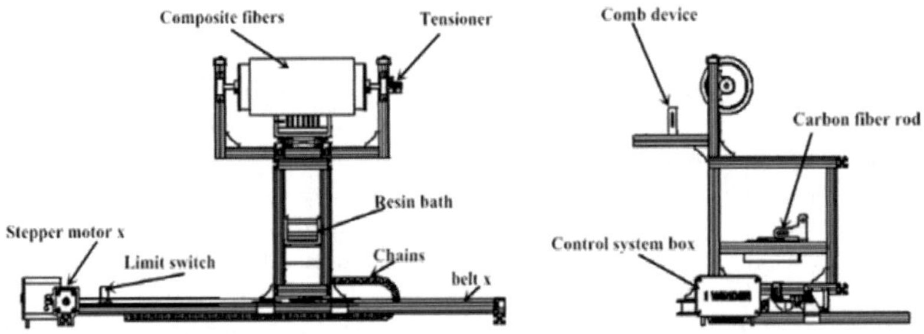

Figure 8. The carriage unit back view and right view.

4.3 The ring pay-out eye unit

The ring pay-out eye unit is one portion of the delivery system, which controlled by stepper motor z. The shape of pay-out eye unit is a ring or circular, a straight bar is put to guide fibres closely to the mandrel. The ring pay-out eye is fixed on the carriage unit by four square holders, which put four bearings into ring inside a slot. It can make the ring pay-out eye and the straight bar rotation in the meantime when the stepper motor z rotates. Based on the design concept, the straight bar rotates a certain angle clockwise or counterclockwise. Figure 9 shows the ring pay-out eye unit front view and left view structure. It is a highlighting unit compared to traditional delivery system design ideas. The ring pay-out eye unit has some advantages, such as maximum precision in control of bandwidth and density, good for the high angle or circular windings and good for wide bandwidth. During the carriage unit reaches a point of return, the ring pay-out eye unit just needs to rotate a certain angle clockwise or counter-clockwise. The ring pay-out eye unit process is optimized by modification on rotation stabilization, z belt pully framework, pay-out eye rotation angle, comb device structure and resin bath configuration.

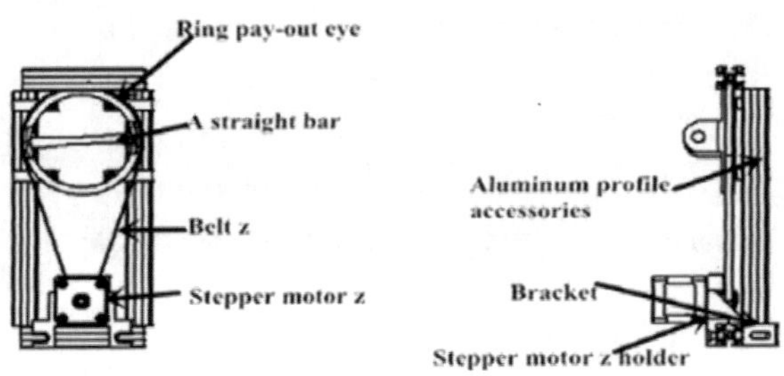

Figure 9. The ring pay-out eye unit front view and left view.

4.4 The control system unit

The control unit consists of Arduino Uno, CNC v1 shield, voltage regulator, Bluetooth module, power plug, DC brushless fan and limit switch. The main modules in control system are Arduino Uno and CNC v1 shield, which are connected with pins. CNC v1 shield pins are put on the back panel, which exactly connected with Arduino Uno pins. The CNC v1 shield is set above the Arduino Uno module, which cannot be drawn in the schematic diagram. Figure 10 shows other modules connection.The Bluetooth module receives the signal and sends commands to Arduino Uno to control the stepper motor x, y, and z through CNC shield. The voltage regulator transforms 220V to 35V. The power plug is a device that allows electrically operated equipment to be connected to the primary alternating current (AC) power supply in a building. DC brushless fan cools the temperature when control system works. The control system unit is optimized by modification on wire connection, stable signal transmission, the control system box position and resin protection.

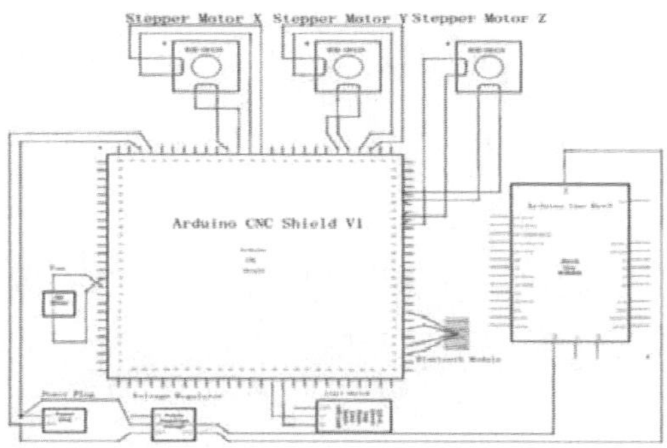

Figure 10. Control system schematic diagram for the 3-axis filament winding machine.

5. Final design scheme

Based on the above main concepts units, 3-axis filament winding machine has been mainly completed. It is even necessary to optimise some subsystem and devices, some components need to change the size and position, such as limit switch holder position, the control system box, comb device, tensioner and so on. Figure 11 shows the schematic diagram for proposed winding machine. The contrivable machine approach to our initial idea by means of the optimisation process, 3-axis filament winding machine only has few improvements for diversified devices to meet the needs in the future. Figure 12 shows the filament winding machine with 3-axis of winding system.

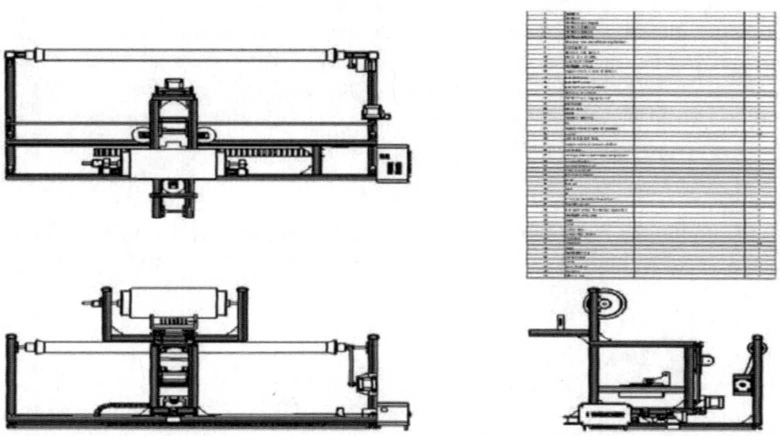

Figure 11. Schematic diagram for proposed a 3-axis filament winding machine.

Figure 12. The 3-axis filament winding machine with inexpensive control system.

5.1 Specimen fabrication

When we tested the machine movement capability, it had some problems with uploading commands to Arduino Uno control system. After we solved the matters, the 3-axis filament winding machine can fabricate the filament wound composite cylinders. Figure 13 shows specimen by this machine.

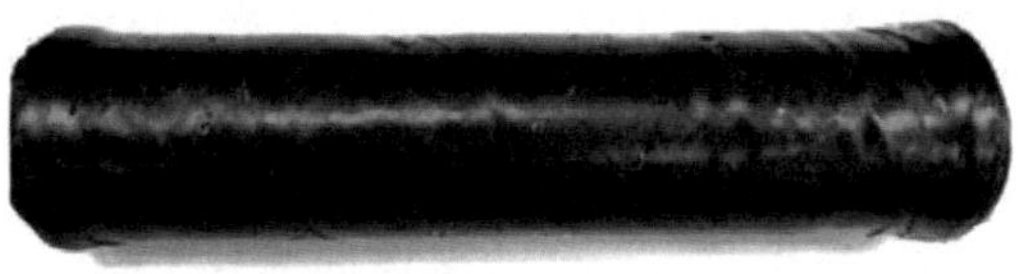

Figure 13. Specimen cylinder with carbon fibre and epoxy.

6. Conclusions

This research project was successfully designed, a 3-axis automated winding machine system was designed via Arduino Uno and CNC shield module. By means of controlling the spindle speed and carriage separately, the helical winding pattern with adjustable winding angle has been basically produced. Some specimens have been made to test this new winding system. The results obtained were relatively satisfactory and improved compared to the previous 2-axis or 4-axis machine.

From the specimens fabricated, it concludes that the defect in surface finish is due to several main factors. Firstly, surface finish is directly proportional to steering gear from a distance. By increasing steering gear the distance, the surface finish becomes poor. Secondly, winding angle also affects the surface finish. The higher winding angle provides a better surface finish. Lastly, movement vibration also affects the quality extent of surface finish, especially for the carriage movement. Due to many limitations of this research project, some problems in the main structure have not been solved, Therefore, the specimen fabricated does not perform some experiments to measure the mechanical performance of the cylindrical structure under static compression loading.

Acknowledgements

The authors are grateful to the Ministry of Higher Education and Universiti Malaysia Pahang (www.ump.edu.my) for funding this research with FRGS/1/2014/TK01/UMP/02/3 and RDU1603108.

References

[1] R. Rafiee 2013 "Experimental and theoretical investigations on the failure of filament wound GRP pipes," *Composites Part B: Engineering,* vol. 45, pp. 257-267.

[2] G. Perillo, R. Vacher, F. Grytten, S. Sørbø, and V. Delhaye 2014 "Material characterisation and failure envelope evaluation of filament wound GFRP and CFRP composite tubes," *Polymer Testing,* vol. 40, pp. 54-62.

[3] M. Gascons i Tarrés 2011 "Assessment of virtual design and manufacturing techniques for fibre reinforced composite materials."

[4] E. V. Rojas, D. Chapelle, D. Perreux, B. Delobelle, and F. Thiebaud 2014 "Unified approach of filament winding applied to complex shape mandrels," *Composite Structures,* vol. 116, pp. 805-813.

[5] X. Jia, G. Chen, Y. Yu, G. Li, J. Zhu, X. Luo, *et al.* 2013 "Effect of geometric factor, winding angle and pre-crack angle on quasi-static crushing behavior of filament wound CFRP cylinder," *Composites Part B: Engineering,* vol. 45, pp. 1336-43.

[6] S. Koussios, O. Bergsma, and A. Beukers 2006 "Filament winding: kinematics, collision control and process optimisation through application of dynamic programming," *Composites Part A: Applied Science and Manufacturing,* vol. 37, pp. 2088-2104.

[7] M. S. Babu, G. Srikanth, and S. Biswas, 2007 "Composite fabrication by filament winding-An insight," *Technology Information, Forecasting, and Assessment Council. Available: http://www. tifac. org. in/news/pub. htm.*

[8] M. Lossie and H. Van Brussel 1994 "Design principles in filament winding," *Composites manufacturing,* vol. 5, pp. 5-13.

[9] S. Mutasher, N. Mir-Nasiri, and L. C. Lin 2012 "Small-scale filament winding machine for producing fiber composite products," *Journal of Engineering Science and Technology,* vol. 7, pp. 156-168.

[10] F. Abdalla, S. Mutasher, Y. Khalid, S. Sapuan, A. Hamouda, B. Sahari, *et al.* 2007 "Design and fabrication of low cost filament winding machine," *Materials & design,* vol. 28, pp. 234-239.

[11] D. B. Miracle, S. L. Donaldson, S. D. Henry, C. Moosbrugger, G. J. Anton, B. R. Sanders, *et al.* 2001 *ASM handbook* vol. 21: ASM international Materials Park, OH, USA.

[12] M. Rejab, K. Kadirgama, M. Noor, M. Sani, and R. Daud 2008 "Modification and Testing of Four Axes Filament Winding Machine," in *International Conference on Science & Technology: Application in Industry & Education.*

[13] J. Kim, T. J. Moon, and J. R. Howell 2003"Transient thermal modeling of in-situ curing during tape winding of composite cylinders," *TRANSACTIONS-AMERICAN SOCIETY OF MECHANICAL ENGINEERS JOURNAL OF HEAT TRANSFER,* vol. 125, pp. 137-146.

[14] N. Hancox 1981 *Fibre composite hybrid materials*: *Applied Science.*

[15] S. Kalpakjian and S. R. Schmid 1996 "Manufacturing engineering and technology, 2001," *φ13mm W,* vol. 7, pp. 340-368.

[16] S. T. Peters, W. D. Humphrey, and R. F. Foral 1991 *Filament winding composite structure fabrication*: SAMPE International Business Office.

Investigation on musculoskeletal discomfort and ergonomics risk factors among production team members at an automotive component assembly plant

Fazilah Abdul Aziz[1*], Zakri Ghazalli[2], Nik Mohd Zuki Mohamed[3] and Amri Isfar[4]

[1,2,3]Faculty of Mechanical Engineering, University Malaysia Pahang, 26600 Pekan, Pahang, Malaysia
[4] Safety, Health and Environment Department, Ingress Technology Sdn Bhd, 48300 Rawang, Selangor, Malaysia

*Corresponding author: fazilahaa@ump.edu.my

Abstract. Musculoskeletal discomfort (MSD) is very common condition in automotive industry. MSD is affecting the worker's health, well-being and lower down the productivity. Therefore, the main objective of this study was to identify the prevalence of MSD and ergonomics risk factors among the production team members at a selected automotive component manufacturer in Malaysia. MSD data were collected by conducting structure interview with all participants by referring to the Cornell Musculoskeletal Disorder Questionnaire (CMDQ). Those production team members who achieved a total discomfort score for all body regions more than 100 was selected for job task assessment. The physical exposure risk factors of work related musculoskeletal disorders (WMSD) has evaluated by using Quick Exposure Check (QEC) techniques. The results of the study identified the severe MSD associated with production assembly team members. It is expected that the prevalence of MSD for those production assembly team members was lower back (75.4%), upper back (63.2%), right shoulder (61.4%), and right wrist (60%). The QEC analysis discovered that about 70% of job tasks had very high risks for neck posture and 60% had high risks for the back (in moving condition) and shoulder/arm postures. There were 80% of respondents have produced a high score for exposure risk to vibration. As a conclusion, the main implication of the current study is that special attention should be paid to the physical and psychosocial aspects in production team members with musculoskeletal discomfort to improve their safety, health, and well-being, maintain work ability and productivity.

1. Introduction

1.1 Research background

Nowadays, to increase the competitive market in the manufacturing industry that requires speedier time-to-market, higher quality of products [1], variety of product model, assembly processes grow into more globalized [2], and employees have good job satisfaction and well-being [3]. The manual assembly and flexibility processes have created for workers to make the assembly production survive in dynamic and uncertain market demand [4]. Manual assembly process is often required on physicality, human dexterity and cognitive capabilities of the production workers [5]. Assembly task in automotive production plant involve with repetitive motions, forceful exertions, standing postures and awkward postures. All these risk factors have exposed the workers to the risk of musculoskeletal disorders (MSDs) [6–8]. The major health problem among industry workers is work related musculoskeletal disorders (WMSDs) [9,10]. The WMSDs can caused the prolonged pain state regularly turns into an explanation behind inability and causes weakened work capacity, nonetheless many workers with pain continue to work [11]. Moreover, a high production volume prompts high physical workload [5]. This state is potential to develop more MSDs among employees.

1.2 Increasing trend of production rate, employees turn over and accident case in an automotive component manufacturer

Figure 1 shows that the selected automotive component manufacturer in this study faces a steady deficit in production output from 2011 to 2014. The production output dramatically bounces in 2015. We extrapolate that the hazardous workload and environment such as repetitive tasks, exposure to WMSD, injuries, ill-behaviour, morbidity, mortality and similar incidents will increase. Additionally, the analysis also disclosures that poor organisation and operation management such as working overtime leads to poor individual health. This is in agreement with Nur *et al.* [12],Mossa *et al.* [5], and Niu [13].

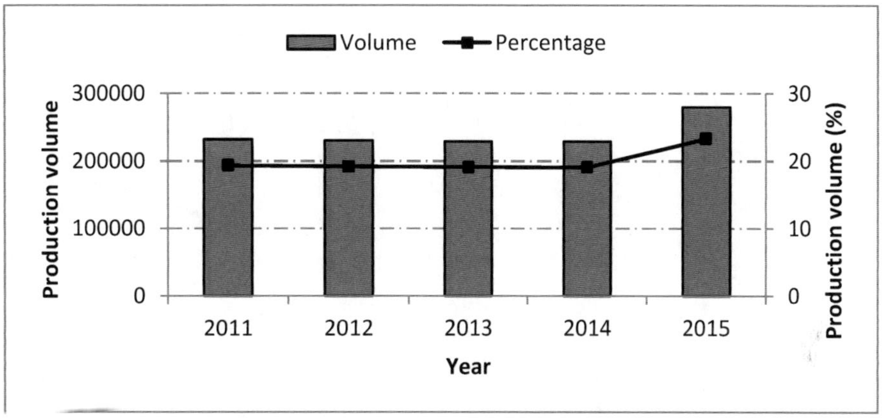

Figure 1. Assembly plant production volumes (2011~2015).

Figure 2 displays the average number of production team members who resigned per year is 20% of the total workforce of the production within five years. The turnover rate affected the productivity. One possible reason is that the workers are exposed to the ergonomically risky job task. Particularly, in 2015, the turnover is the highest as the production volume intensely increased (refer to Figure 1). During this period, the workers need to perform more repetitive and physically demanding tasks. These tasks might expose them to risks such as WSMD, low back injuries, for example. The results of this study is in agreement with the finding of Ferguson et al. [14] which found that most turnovers are due to physical demands of the job. It also supported Lavender and Marras [15] who suggested that high turnover rates could be used as supplemental incidence rates to recognize high-risk jobs for low back injury.

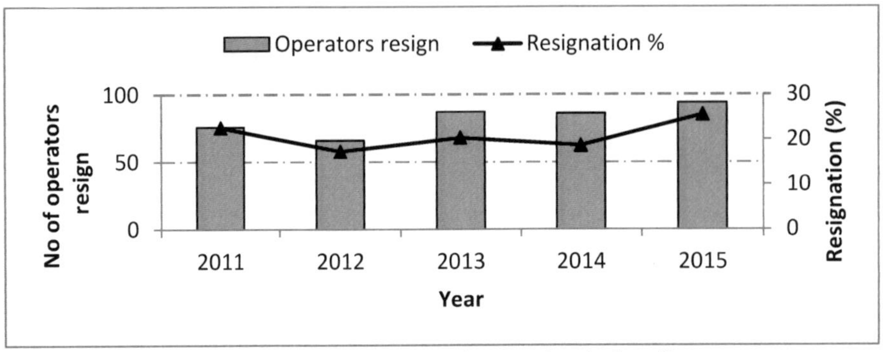

Figure 2. Production team members resign in last five years.

Figure 3 demonstrations that two or more accident cases are reported every year for the last five years. One possible explanation is that, as the output augmented the chances of workers to be involved in accidents due to stress, fatigue and tasks overloaded. A research conducted by Allahyari et al. [16] discovered that accidents are caused by lack of attention, interruption, and mental errors. The previous injury history were important risk factors for work-related injury and MSD.[17].

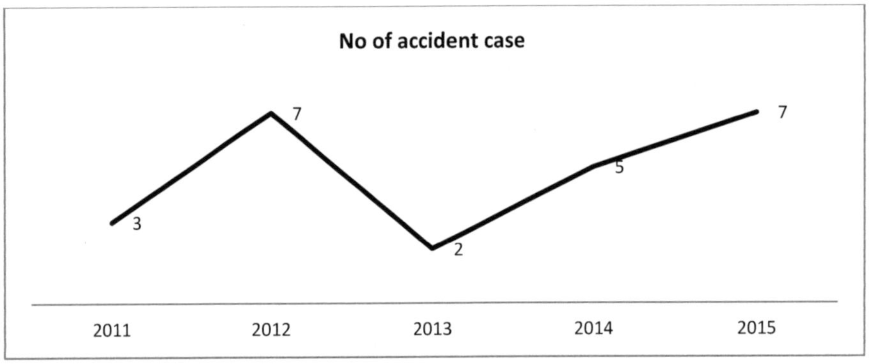

Figure 3. No of accident case reported in assembly plant.

1.3 Research aims and objectives

Although established studies have recognized MSD in several occupational, but there is still less information for an exact determination the prevalence of MSD and exposure risk level among production team members who performed high physical demand job task in challenging working environment. Furthermore various intervention studies encourage the adequacy of participatory ergonomics (PE) intervention as a method for decreasing exposure to biomechanical and psychosocial well-being risk factors [18–20]. Surprisingly PE method is seldom applied in assessing the ergonomics risk factors in workplace. In this study the PE has practiced by the involvement of production employee from different position and department as observer to conduct the job task assessment, and the selected production team members who has direct experience of the job task. The aim of this study was to identify the prevalence of MSD and ergonomic risks factors in an automotive component assembly plant. The specific objectives of this research were to recognize the prevalence musculoskeletal discomforts among production team members, to identify the job task exposure risk level, and to determine the ergonomics physical and psychosocial risk factors.

IOP Conf. Series: Materials Science and Engineering **257** (2017) 012040 doi:10.1088/1757-899X/257/1/012040

2. Method

2.1 Identification of prevalence musculoskeletal discomforts

2.1.1 Respondents. The study was conducted in an automotive component assembly plant in Malaysia. Respondents were production assembly team members from various assembly lines in the assembly plant including Line Model A (32%), Line Model B (27%), Line Model C (20%), Line Model D (5%) and Line Model E (15%). The sample size for this study was 40% from the total production assembly team members who worked in the different shift including normal shift (13%), day shift (41%) and night shift (46%).

2.1.2 Use of Cornell Musculoskeletal Discomforts Questionnaire. Data were collected by conducting structure interview with all participants by referring to the Cornell Musculoskeletal Disorder Questionnaire (CMDQ) [21]. The interview session has conducted in group, and each group consists of 7 to 8 respondents. There are 54 questionnaire items related to the prevalence of musculoskeletal pain or discomfort in 18 body regions (neck, shoulders, upper back, upper arm, lower back, forearm, wrists, hip/buttocks, thighs, knees, and lower legs). During the interview the respondents were required to indicate the frequency of discomfort on an ordinal scale from 0(never) to 4 (several times every day) and the severity of discomfort from 1(slightly uncomfortable) to 3(very uncomfortable). A pain level of at least "moderately uncomfortable" was selected as a severity threshold for determining prevalence and frequency. Moreover, the level at which the discomfort interfered with assembly task was scored from 0 (no interference) to 2 (substantial interference). Total discomfort score was calculated by using the following formula.

$$\text{Discomfort score} = \text{Frequency of discomfort} \times \text{Severity of discomfort} \times \text{Interference} \quad (1)$$

2.2 Job task assessment

2.2.1 Subjects. The participants in this study were recruited through a screening process using CMDQ. Those production team members who achieved a total discomfort score for all body regions more than 100 was selected for job task assessment. Production team members worked on a different shift, attached to a different workstation and assembly line, vary in numbers of the task and work output per hour (refer Table 1). There were six observers appointed for job task assessment including 2 persons each from safety, health and environment department and engineering department, and 1 person each from production assembly department and university researcher. This group of the observer has mixed background and experiences. Nine workstations were selected based on workstation's worker total discomfort score in exposure WMSD assessment using CMDQ.

2.2.2 Use of Quick Exposure Check techniques. The QEC was one of the observational methods and developed for ergonomists, health and safety practitioners in order to investigate musculoskeletal risk factors in workers [22–24]. In this study, the job task assessment was executed by applying the QEC technique. QEC has been previously applied to production team members and includes assessment of four body regions, namely the back, shoulder/arm, wrist/hand, and neck, with regards to postures and repetitive movements. This technique also assesses several psychosocial risk factors including driving, vibration, work pace and stress. The exposure levels for body regions and psychosocial factors are categorized into four exposure categories; low, moderate, high, and very high (refer Table 2).

Table 1. Summary of workstation and job task for exposure to musculoskeletal risk factors.

Production Team Member (PTM)	Total discomfort score	Assembly Line	Workstation	No of task	No of spot/ length welding	Target output (pcs/hour)
PTM 1	163	Model A	Workstation 1	1	4	45
PTM 2	144	Model A	Workstation 2	3	6	30
PTM 3	110	Model B	Workstation 3	4	15cm	40
PTM 4	117	Model B	Workstation 4	3	8	17
PTM 5	120	Model C	Workstation 5	2	7	35
PTM 6	110	Model D	Workstation 6	4	9	20
PTM 7	186	Model D	Workstation 7	4	6	25
PTM 8	118	Model D	Workstation 8	4	9	20
PTM 9	111	Model E	Workstation 9	2	-	40

Table 2. Priority levels for Quick Exposure Check scores [25].

Exposure factor	Exposure level			
	Low	Moderate	High	Very high
Back (static)	8 - 14	16 - 22	24 - 28	30 – 40
Back (moving)	10 - 20	21 - 30	32 - 40	42 – 56
Shoulder/arm	10 - 20	21 - 30	32 - 40	42 – 56
Wrist/hand	10 - 20	21 - 30	32 - 40	42 – 56
Neck	4 - 6	8 - 10	12 - 14	16 – 18
Driving	1	4	9	-
Vibration	1	4	9	-
Work pace	1	4	9	-
Stress	1	4	9	16

3. Results

The prevalence of MSDs and assembly activity performance among assembly plant team members were investigated using the percentages of responses, mean value, and standard deviation given on the frequency discomfort, severity discomfort, and assembly performance interference scales. Table 3 shows the frequency distribution of responses across body parts. Four body parts with MSD prevalence rates about equivalent and above 60% were included in ergonomics risk factors investigations. These body parts were the lower back (75.4%), right shoulder (61.4%), right wrist (60%) and upper back (63.2%). These four body parts involved the highest assembly process performance interference with rates of 70.1%, 50.6%, 50.6% and 54.4%, respectively.

3.1 The prevalence musculoskeletal discomforts among production team members

Table 3. The prevalence of musculoskeletal discomfort and assembly activity interference.

Body regions	Frequency Discomfort					Severity Discomfort			Interference with assembly activities			Discomfort score
	Never n (%)	1-2 times last week n (%)	3-4 times last week n (%)	Once every day n (%)	Several times every day n (%)	Slightly n (%)	Moderately n (%)	Severely n (%)	Not at all n (%)	Slightly n (%)	Substantially n (%)	Mean (SD)
Lower Back	14 (24.6)	19 (33.3)	9 (15.8)	10 (17.5)	5 (8.8)	17 (29.8)	13 (22.8)	13 (22.8)	3 (5.3)	29 (50.9)	11 (19.3)	5.16 (7.24)
Shoulder R	22 (38.6)	12 (21.1)	12 (21.1)	8 (14)	3 (5.3)	16 (28.1)	14 (24.6)	5 (8.8)	6 (10.8)	21 (37.9)	7 (12.6)	3.09 (5.53)
Lower Leg L	28 (50)	12 (21.4)	8 (14.3)	7 (12.5)	1 (1.8)	13 (23.2)	10 (17.9)	**5 (8.9)**	4 (7.1)	17 (30.4)	7 (12.5)	2.89 (5.47)
Wrist R	23 (40.4)	16 (28.1)	8 (14.0)	8 (14.0)	2 (3.5)	15 (26.3)	13 (22.8)	6 (10.5)	5 (9.0)	20 (36.1)	8 (14.4)	2.72 (4.58)
Lower Leg R	28 (48.3)	13 (22.4)	9 (15.5)	7 (12.1)	1 (1.7)	14 (24.1)	11(19.0)	**5 (8.6)**	6 (10.3)	16 (27.6)	8 (13.8)	2.70 (5.35)
Upper Back	21 (36.8)	21 (36.8)	5 (8.8)	6 (10.5)	4 (7)	15 (26.3)	13 (22.8)	8 (14.0)	5 (8.8)	27 (47.4)	4 9 (7.0)	2.59 (5.01)
Shoulder L	24 (44.4)	12 (22.2)	8 (14.8)	9 (16.7)	1 (1.9)	14 (25.9)	12 (22.2)	**4 (7.4)**	6 (11.1)	18 (33.4)	6 (11.1)	2.58 (4.81)
Wrist L	27 (47.4)	15 (26.3)	8 (14)	6 (10.5)	1 (1.8)	11 (19.3)	15 (26.3)	4 (7.0)	4 (7.0)	21 (36.8)	5 (8.8)	2.05 (3.27)
Knee L	39 (69.6)	8 (14.3)	4 (7.1)	3 (5.4)	2 (3.6)	7 (12.5)	8 (14.3)	2 (3.6)	7 (13.3)	7 (13.3)	2 (3.8)	2.02 (7.48)
Lower Arm R	34 (58.6)	12 (20.7)	5 (8.6)	5 (8.6)	2 (3.4)	12 (20.7)	8 (13.8)	4 (6.9)	4 (6.9)	15 (25.9)	5 (8.6)	1.83 (4.03)
Lower Arm L	36 (63.2)	10 (17.5)	7 (12.3)	3 (5.3)	1 (1.8)	9 (15.8)	10 (17.5)	2 (3.5)	4 (7.4)	12 (22.1)	4 (7.4)	1.46 (3.47)
Upper Arm R	27 (46.6)	22 (37.9)	4 (6.9)	4 (6.9)	1 (1.7)	17 (29.3)	11 (18.9)	3 (5.2)	6 (11.0)	20 (36.8)	3 (5.5)	1.40 (3.17)
Knee R	39 (66.1)	10 (16.9)	5 (8.5)	4 (6.8)	1 (1.7)	10 (16.9)	7 (11.86)	3 (5.08)	5 (8.9)	12 (21.4)	2 (3.6)	1.36 (4.15)
Neck	33 (55.9)	17 (28.8)	4 (6.8)	5 (8.5)	0	11 (18.7)	11 (18.7)	4 (6.8)	5 (9.2)	18 (33.1)	1 (1.8)	1.33 (3.07)
Upper Arm L	30 (51.7)	19 (32.8)	5 (8.6)	3 (5.2)	1 (1.7)	15 (25.9)	11 (19.0)	2 (3.5)	6 (10.7)	18 (32.2)	3 (5.4)	1.26 (3.00)
Thigh L	36 (64.3)	12 (21.4)	5 (8.9)	2 (3.6)	1 (1.8)	9 (16.1)	10 (17.9)	1 (1.8)	8 (13.6)	13 (22.1)	0 (0)	1.01 (2.50)
Hip/Buttocks	45 (78.9)	6 (10.5)	3 (5.3)	1 (1.8)	2 (3.5)	5 (8.8)	6 (10.6)	1 (1.8)	4 (6.0)	9 (13.6)	1 (1.5)	0.77 (3.30)
Thigh R	38 (67.9)	9 (16.1)	6 (10.7)	2 (3.6)	1 (1.8)	9 (16.1)	8 (14.3)	1 (1.8)	7 (13.6)	12 (23.2)	0 (0)	3.1 (1.9)

3.2 The exposure risk level among production team members

In accordance with the exposure levels for body regions in Figure 4, the job task assessment result reveals that 70% had very high exposure risk to worker's neck, and 60% had high exposure risk to worker's back (moving), shoulder and arm. Meanwhile out of the tasks evaluated, 80% had medium exposure risks to the wrist and hand.

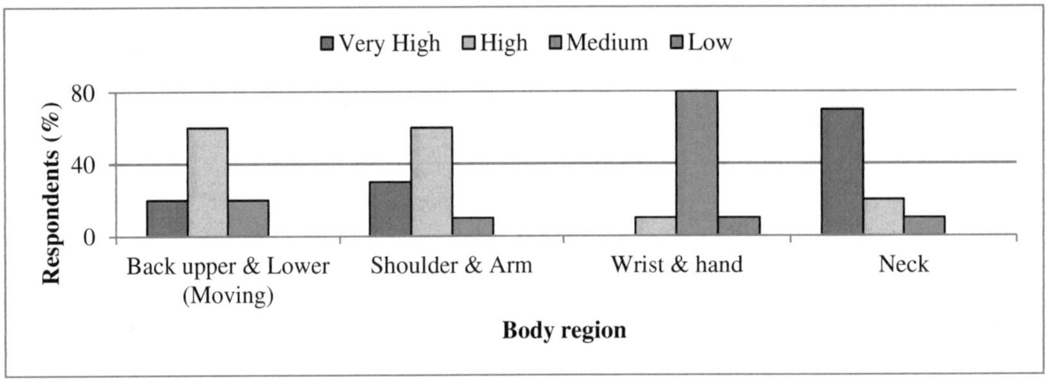

Figure 4. The exposure risk level for body regions.

3.3 Ergonomics physical and psychosocial risk factors in an automotive component assembly plant

Figure 5 is displaying the exposure risk level for ergonomics physical and psychosocial risk factors. The QEC analysis discovered that 80% of respondents have produced a high score for exposure risk to vibration and each 10% of respondents the exposure risk to vibration were low and medium. Work pace scores in 60% of the respondents were medium, in 30% of the respondents were high, and 10% of them were low. While the stress scores of respondents in 60% were medium, the scores were high in 30% and were very high in 10% of them. However, driving score is low for all workers because driving is not part of their job task.

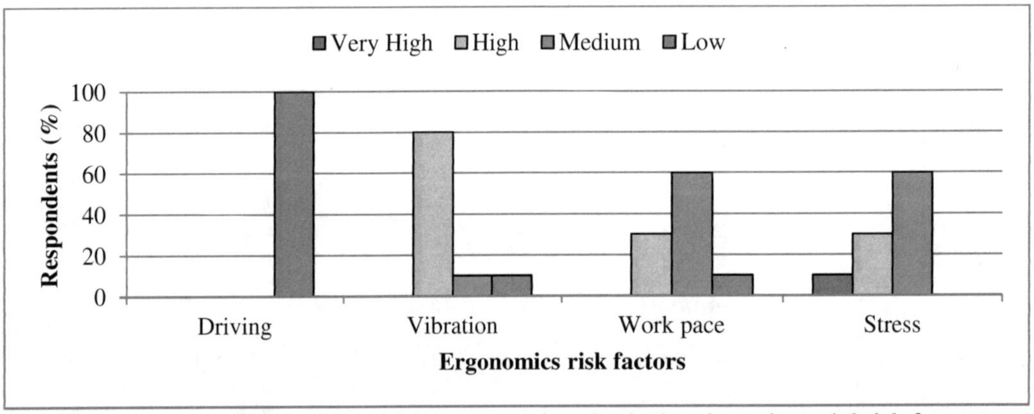

Figure 5. The exposure risk level for ergonomics physical and psychosocial risk factors.

4. Discussions

4.1 Prevalence and assembly task performance interference of MSD

The current study explored the prevalence and assembly task performance interference of musculoskeletal discomfort, among production assembly team members in the automotive component assembly plant. The study contributed additional evidence to literature about the considerable prevalence of musculoskeletal discomfort among automotive component production team members. The results of the study identified the severe musculoskeletal discomfort associated with production team members. The findings of this study are consistent with established studies conducted by Roquelaure [26], Farioli et al., [27], Yu et al., [17] which showed that musculoskeletal disorders are main causes of work disability among the employees and also can give very serious effects on employee's health and efficiency.

Production team members have involved with demanding material manual handling works such as moving, handling and lifting heavy tools like welding spot gun. It is expected that the prevalence of musculoskeletal discomfort for those production assembly team members was lower back (75.4%), upper back (63.2%), right shoulder (61.4%), and right wrist (60%). Several similar type studies have shown a variety of results which is comparable and contrast to the present study. The highest prevalence of WMSD among assembly workers in Malaysia occurs at the shoulder, feet, lower leg and lower back [28]. According to Zare et al. [29] the prevalence of musculoskeletal symptoms among truck assembly plant workers are different based on cycle time, for cycle time A were shoulders (67%), elbows (53%), and wrists (47%), and for cycle time B the prevalence of shoulder (35%), elbow (40%) and wrist (40%) symptoms respectively. In contrast, the study by Nur et al.[30] indicated that the prevalence of musculoskeletal symptoms among workers in Malaysian automotive manufacturing companies is highest for the neck (49.3%), followed by hand/wrist (48.0%), shoulder (46.7%), upper back (33.6%), lower back (21.7%), knee (15.8%), thigh/hip (14.5%), elbow 8.7%) and ankle (1.3%).

4.2 The exposure risk level among production assembly team members

This job task assessment was performed exposure risk levels for WMSDs among production assembly team members in assembly plant of an automotive component manufacturer was analysed. Working in an automotive assembly plant is known as a demanding task that requires a high level of strength and attention in order to cope with the demands of the job demands, production volume, quality of component, and variable of tasks. Most of the automotive component manufacturer's workers had significant exposure to ergonomics risk factors [31].

The results of the study indicated that about 70% of job tasks in an automotive component assembly plant had very high risks for neck posture and 60% had high risks for the back (in moving condition) and shoulder/arm postures. Current study results support the previous finding by several researchers which found that the most affected body regions for WMSDs among automotive industry's workers are the neck, shoulder, arm, and back [30,32–34]. In addition, a study conducted by de Cassia and team members [35] have found that the musculoskeletal comorbidity was high, varying between 72.2 % for lower back to 90.5 % for neck pain.

4.3 Ergonomics physical and psychosocial risk factors

The QEC analysis discovered that 80% of respondents have produced a high score for exposure risk to vibration. The high score for exposure risk to vibration among respondents may have been due to their job tasks applied manually hand tools, equipment and machines to complete assembled the components. The findings of this study are consistent with Burstrom et al., [36] who had reported that manual work involving vibrating power tools has been related to musculoskeletal disorders. Established studies had reported that manual handling of the tools and vibration were associated with increased prevalence of body pain among the workers [33,37,38]. The job task assessment analysis also exposed that 30% each of respondents have produced a high score for exposure risk to work pace and stress. This finding has support latest study conducted by Bao et al. [39] who have found that the work pace had significant effects on all biomechanical stressors and psychosocial variables. In other study conducted by Bosh et al. [40] has suggested that high work pace will cause fatigue on the workers and lead more errors, finally affect production quality. According to Thye [41] work stress in the workplace is common due to the demands of the current work environment. Authors also reported that the physical health issues and musculoskeletal disorders begun by the prolonged stress. The psychosocial risk factors like high work stress lead to the development of musculoskeletal discomforts among workers [42]. This ergonomics risk assessment study has found that the job task performed by production team members has exposed them to the physical and psychosocial risk factors.

5. Conclusions

The study explored the prevalence, severity discomfort and assembling task performance interference of MSD among production team members in an automotive component assembly plant in Malaysia. There are four body parts with MSD prevalence rates about equivalent and above 60% were included in ergonomics risk factors investigations. The prevalence of pain in the upper back, lower back, right shoulder and right wrist have been reported to be higher in comparison with pain in other parts of the body. The study results demonstrated that job tasks in automotive component assembly plant have been exposed to very high of WMSD risk at worker's neck, followed to a high of WMSDs risk at worker's back (in moving condition) and worker's shoulder/arm. This study also found that in the psychosocial factors, the exposure level is high for vibrations, work pace and stress in most of the workstations. Although the results found are constructive as a preliminary study, the inexperience of observer's assessment could lead to the misinterpretation. Moreover, the way on how inexperience assembly workers have to estimate the weight of tools and objects, and the force that existed in assembly process activities is questionable due to their limited knowledge in ergonomics. However, this job task assessment is a decent to begin for the company to actualize the participatory ergonomics program among the production employees. The main implication of the current study is that special attention should be paid to the physical and psychosocial aspects in production team members with musculoskeletal discomfort to improve their safety, health, and well-being, maintain work ability and productivity.

Acknowledgments
We thank University Malaysia Pahang (UMP) for grant support RDU160390, the Ministry of Higher Education Malaysia (MoHE) of the scholarship SLAI (Sijil latihan akademik IPTA) and Ingress Technologies Sdn Bhd for data collection.

References

1. Lawson G, Salanitri D, Waterfield B. Future directions for the development of virtual reality within an automotive manufacturer. Appl Ergon [Internet]. 2015; Available from: http://linkinghub.elsevier.com/retrieve/pii/S0003687015300260
2. Enomoto A, Yamamoto N, Suzuki T. Automatic estimation of the ergonomics parameters of assembly operations. CIRP Ann - Manuf Technol [Internet]. 2013;62(1):13–6. Available from: http://dx.doi.org/10.1016/j.cirp.2013.03.038
3. Boxall P, Macky K. High-involvement work processes, work intensification and employee well-being. Work Employ Soc. 2014 Dec;28(6):963–84.
4. Francas D, Löhndorf N, Minner S. Machine and labor flexibility in manufacturing networks. Int J Prod Econ. 2011;131(1):165–74.
5. Mossa G, Boenzi F, Digiesi S, Mummolo G, Romano VA. Productivity and ergonomic risk in human based production systems: A job-rotation scheduling model. Int J Prod Econ. 2016;171:471–7.
6. Baba Md. Deros, Dian Darina Indah Daruis, Ahmad Rasdan Ismail, Nurfarhana Abdul Sawal, Jaharah A. Ghani. Work-Related Musculoskeletal Disorders among Workers â€™ Performing Manual Material Handling Work in an Automotive Manufacturing Company Baba Md . Deros , Dian Darina Indah Daruis , Ahmad Rasdan Ismail , Nurfarhana Abdullah Sawal and Jaharah A . Ghani Depar. Am J Appl Sci. 2010;7(8):1087–92.
7. Anita AR, Yazdani A, Hayati KS, Adon MY. Association between Awkward Posture and Musculoskeletal Disorders (MSD) among Assembly Line Workers in an Automotive Industry. Malaysian J Med Heal Sci. 2014;10(1):23–8.
8. Deros BM, Daruis DDI, Basir IM. A Study on Ergonomic Awareness among Workers Performing Manual Material Handling Activities. Procedia - Soc Behav Sci [Internet]. 2015 Jul;195:1666–73. Available from: http://linkinghub.elsevier.com/retrieve/pii/S1877042815037179
9. Bao SS, Kapellusch JM, Merryweather AS, Thiese MS, Garg A, Hegmann KT, et al. Relationships between job organisational factors, biomechanical and psychosocial exposures. Ergonomics [Internet]. 2015;139(October):1–16. Available from: http://www.tandfonline.com/doi/full/10.1080/00140139.2015.1065347
10. Peppoloni L, Filippeschi a., Ruffaldi E, Avizzano C a. (WMSDs issue) A novel wearable system for the online assessment of risk for biomechanical load in repetitive efforts. Int J Ind Ergon [Internet]. 2015;1–11. Available from: http://linkinghub.elsevier.com/retrieve/pii/S0169814115300081
11. Malmberg-Ceder K, Haanpää M, Korhonen PE, Kautiainen H, Soinila S. Relationship of musculoskeletal pain and well-being at work – Does pain matter? Scand J Pain [Internet]. 2017;15:38–43. Available from: http://www.embase.com/search/results?subaction=viewrecord&from=export&id=L613686570%0Ahttp://dx.doi.org/10.1016/j.sjpain.2016.11.018%0Ahttp://sfx.aub.aau.dk/sfxaub?sid=EMBASE&issn=18778879&id=doi:10.1016%2Fj.sjpain.2016.11.018&atitle=Relationship+of+muscu

12. Nur NM, Dawal SZM, Dahari M, Sanusi J. The effects of energy expenditure rate on work productivity performance at different levels of production standard time. J Phys Ther Sci. 2015;27:2431–3.
13. Niu S. Ergonomics and occupational safety and health: An ILO perspective. Appl Ergon. 2010;41:744–53.
14. Ferguson SA, Marras WS, Lavender SA, Splittstoesser RE, Yang G. Are Workers Who Leave a Job Exposed to Similar Physical Demands as Workers Who Develop Clinically Meaningful Declines in Low-Back Function? Hum Factors J Hum Factors Ergon Soc. 2014;56(1)(1):58–72.
15. Lavender S a, Marras WS. The use of turnover rate as a passive surveillance indicator for potential low back disorders. Ergonomics. 1994;37(6)(6):971–8.
16. Allahyari T, Rangi NH, Khalkhali H, Khosravi Y. Occupational Cognitive Failures and Safety Performance in the Workplace. Int J Occup Saf Ergon. 2014 Jan 8;20(1)(1):175–80.
17. Yu W, Yu ITS, Li Z, Wang X, Sun T, Lin H, et al. Work-related injuries and musculoskeletal disorders among factory workers in a major city of China. Accid Anal Prev [Internet]. 2012;48:457–63. Available from: http://dx.doi.org/10.1016/j.aap.2012.03.001
18. Tompa E, Dolinschi R, Natale J. Economic evaluation of a participatory ergonomics intervention in a textile plant. Appl Ergon. 2013;44(3):480–7.
19. Visser S, Molen HF Van Der, Sluiter JK, Frings-dresen MHW. Guidance strategies for a participatory ergonomic intervention to increase the use of ergonomic measures of workers in construction companies : a study design of a randomised trial. BMC Musculoskelet Disord. 2014;15(132):1–11.
20. Yazdani A, Neumann WP, Imbeau D, Bigelow P, Pagell M, Theberge N, et al. How compatible are participatory ergonomics programs with occupational health and safety management systems? Scand J Work Environ Heal. 2015;41(2):111–23.
21. Hedge A. Cornell Musculoskeletal Discomfort Questionnaires (CMDQ). Cornell Univ Ergon. 1999;
22. Bulduk EÖ, Bulduk S, Süren T, Ovali F. Assessing exposure to risk factors for work-related musculoskeletal disorders using Quick Exposure Check (QEC) in taxi drivers. Int J Ind Ergon. 2014;44(6):817–20.
23. David G, Woods V, Li G, Buckle P. The development of the Quick Exposure Check (QEC) for assessing exposure to risk factors for work-related musculoskeletal disorders. Appl Ergon. 2008;39(1):57–69.
24. Occhipinti E, Colombini D. A toolkit for the analysis of biomechanical overload and prevention of WMSDs: Criteria, procedures and tool selection in a step-by-step approach. Int J Ind Ergon [Internet]. 2016;52:18–28. Available from: http://dx.doi.org/10.1016/j.ergon.2015.08.001
25. David G, Woods V, Li G, Buckle P. The development of the Quick Exposure Check (QEC) for assessing exposure to risk factors for work related musculoskeletal disorders. Appl Ergon. 2008;39(1):57–69.
26. Roquelaure Y. Promoting a Shared Representation of Workers' Activities to Improve Integrated Prevention of Work-Related Musculoskeletal Disorders. Saf Health Work [Internet]. 2016;7(2):171–4. Available from: http://dx.doi.org/10.1016/j.shaw.2016.02.001
27. Farioli A, Mattioli S, Quaglieri A, Curti S, Violante FS, Coggon D. Musculoskeletal pain in Europe : role of personal , occupational and social risk factors. Scand J Work Env Heal. 2014;40(1):36–46.
28. Aziz RA, Rebi MAT, Rani A, Rohani JM. Work-related Musculoskeletal Disorders among Assembly Workers in Malaysia. J Occup Saf Heal. 2014;2014(June):33–8.
29. Zare M, Bodin J, Cercier E, Brunet R, Roquelaure Y. Evaluation of ergonomic approach and musculoskeletal disorders in two different organizations in a truck assembly plant. Int J Ind Ergon [Internet]. 2015;50:34–42. Available from: http://linkinghub.elsevier.com/retrieve/pii/S0169814115300275
30. Nur NM, Dawal SZ, Dahari M. The Prevalence of Work Related Musculoskeletal Disorders Among Workers Performing Industrial Repetitive Tasks in the Automotive Manufacturing Companies. In: Proceedings of the 2014 International Conference on Industrial Engineering and Operations Management Bali, Indonesia, January 7 – 9, 2014. 2014. p. 1–8.
31. Mavis I, Rahman AA, Tamrin SBHM. Ergonomics Risks Reduction : In Osh Certified and Uncertifiedautomotive Parts Manufacturing Industry Workers. Middle-East J Sci Res. 2014;22(9):1272–80.
32. Zokaei M, Flahati M, Jalilian H, Faghih M, Normohammadi M, Amiry S. Assessment of musculoskeletal disorders risk using MFA method and the survey of correlation between its results and results of RULA method. Occup Med Q J. 2014;6(3):60–9.
33. Akter S, Rahman MM, Mandal S, Nahar N. Musculoskeletal symptoms and physical risk factors among automobile mechanics in Dhaka , Bangladesh. South East Asia J Public Heal. 2016;6(1):8–13.
34. Zare M, Mignot B, Sagot J, Roquelaure Y. The Pattern of Upper Arm and Back Movements for Automotive Assembly Operators: Exposure Variation Analysis of Acceleration Data. In: Advances in

Physical Ergonomics and Human Factors. Springer; 2016. p. 113–23.

35. de Cássia Pereira Fernandes R, da Silva Pataro SM, de Carvalho RB, Burdorf A. The concurrence of musculoskeletal pain and associated work-related factors: a cross sectional study. BMC Public Health. 2016;16(1):628.
36. Burstrom L, Hagberg M, Liljeind I, Lundstrom R, Nilsson T, Pettersson H, et al. A follow-up study of welders' exposure to vibration in a heavy engineering production workshop. J Low Freq Noise Vib Act Control. 2010;29(1):33–9.
37. Abaraogu UO, Ezema CI, Igwe SE, Egwuonwu AV, Okafor UC. Work-related back discomfort and associated factors among automotive maintenance mechanics in Eastern Nigeria: A cross sectional study. Work. 2016;53(4):813–23.
38. Hernández-Arellano JL, Serratos-Perez JN, Coronado PP. Ergonomic Assessment of Material Handling in CV Joint Assembly. In: Handbook of Research on Managerial Strategies for Achieving Optimal Performance in Industrial Processes. IGI Global; 2016. p. 101–15.
39. Bao SS, Kapellusch JM, Merryweather AS, Thiese MS, Garg A, Hegmann KT, et al. Relationships between job organisational factors, biomechanical and psychosocial exposures. Ergonomics [Internet]. 2016 Feb 11;59(2):179–94. Available from: https://www.tandfonline.com/doi/full/10.1080/00140139.2015.1065347
40. Bosch T, Mathiassen SE, Visser B, Looze MP de, Dieën JH van. The effect of work pace on workload, motor variability and fatigue during simulated light assembly work. Ergonomics. 2011;54(2):154–68.
41. Lee Lam Thye. Work stress and mental health status. [Internet]. News Straits Times. 2016. Available from: http://www.nst.com.my/news/2016/04/142074/work-stress-and-mental-health
42. Widanarko B, Legg S, Devereux J, Stevenson M. The combined effect of physical, psychosocial/organisational and/or environmental risk factors on the presence of work-related musculoskeletal symptoms and its consequences. Appl Ergon. 2014;45(6):1610–21.

IOP Conf. Series: Materials Science and Engineering **257** (2017) 012041 doi:10.1088/1757-899X/257/1/012041

A comparative study on performance of CBN inserts when turning steel under dry and wet conditions

Salem Abdullah Bagaber* **and Ahmad Razlan Yusoff**

Faculty of Manufacturing Engineering, Universiti Malaysia Pahang,
26600 Pekan, Pahang, Malaysia

*Corresponding email: razlan@ump.edu.my

Abstract. Cutting fluids is the most unsustainable components of machining processes, it is negatively impacting on the environmental and additional energy required. Due to its high strength and corrosion resistance, the machinability of stainless steel has attracted considerable interest. This study aims to evaluate performance of cubic boron nitride (CBN) inserts for the machining parameters includes the power consumption and surface roughness. Due to the high single cutting-edge cost of CBN, the performance of significant is importance for hard finish turning. The present work also deals with a comparative study on power consumption and surface roughness under dry and flood conditions. Turning process of the stainless steel 316 was performed. A response surface methodology based box-behnken design (BBD) was utilized for statistical analysis. The optimum process parameters are determined as the overall performance index. The comparison study has been done between dry and wet stainless-steel cut in terms of minimum value of energy and surface roughness. The result shows the stainless still can be machined under dry condition with 18.57% improvement of power consumption and acceptable quality compare to the wet cutting. The CBN tools under dry cutting stainless steel can be used to reduce the environment impacts in terms of no cutting fluid use and less energy required which is effected in machining productivity and profit.

1. Introduction

Hard finish turning is an excellent process compare to traditional grinding as a finishing process. Beside the productivity, machining process must to be attention about sustainable of product by developing a new approach with the lowest environmental impact possible. Sustainable manufacturing means concerns about energy, health safety, cost and environmental aspects. The energy is playing very big role on manufacturing sustainability. It has a negative impact on the environment. Improvement of energy efficiency one of the most the world's interests [1]. In 2010, the consumption of energy globally was 524 quadrillion Btu, and is expected to increase to 630 quadrillion Btu in the next five years [2]. Direct emission of greenhouse gases such as carbon dioxide (CO_2) generates by use of fossil fuels and released into the environment during power generation. The average emission factor for electricity was 0.5488 kg per kWh in China in 2009 [3], it is indicated that 0.5488 kg CO_2 was released when generated 1 kWh. In USA, the total CO_2 emission was 5802 million metric tons in 2008, and the industry sector was contributed by 27.4% [4]. Moreover, from economic side, the rapid increase in the price of fossil fuels and electricity. Therefore, the unit production cost will be increase. Saving energy is the concerns for most of the industrial processes. Usually, the energy consumption is not considering as machine responses and very limited research has been performed includes energy consumption [5]. Cutting fluid one of the most factors influence of environmental and workers. A common practice in machining process is mostly using cutting fluids. Although overall machinability can be improved by reducing temperature and friction at the cutting region [6], but lubrication fluids based oil considering unsustainable factor of cutting processes [7]. Addationaly, the energy consumption, maintenance and service for coolant system, and waste/disposal of cutting fluids [8], which means growing environmental issues, as well as higher manufacturing cost [9,10]. Therefore, cutting process effect should be minimized when that possible by performing dry cutting strategies [11,12] to decrease coolant impacts and costs.

Besides power consumption and cutting fluid, Cutting tool considered as one of the most factors effects in machine performance in terms of surface quality, production rate and product cost [13]. The cutting tools cost is ~4% of the total manufacturing cost [14]. For hard turning superior steel, cubic boron nitride (CBN) inserts are usually used to achieve very high accuracy and surface texture [15]. CBN tools are advanced cutting tools when machining heat resistant alloys [16], with impact resistance and good wear. its stable hot hardness, strength, toughness, wear resistance and also chemical stability at high temperatures around 1000ºC for the superior hardened steel parts [17–19]. Due to the high single cutting edge cost of CBN, the performance of significant is importance for hard finish turning. Many research [20–28] has been done in hard material machining to investigate the machine performance of variation type of cutting tool under different conditions. The stainless steel is extensively used in many applications such as aerospace, automotive, a component of contracture, chemical production, power plants, and medical instruments. Stainless steel considers as difficult to machine due to a high tool wear, work hardening, poor surface quality, low productivity, and high machining costs [29]. Machinability study and sustainable on the stainless steel still is of interest [5].

Due to importance of cutting parameters on the machining performance, the analyze impact of machining parameters helped toward reducing the power consumption and enhancement of sustaniable [30–33]. Therefore, The appropriate selection of machining parameters plays a significant role in ensuring the quality of product [31], increasing the tool life [34], reducing power consumption [35], and increasing production rate [36]. Consequently, this study proposes to evaluate the contribution of cutting parameters during machining turning of AISI 316 steel on the power consumption, surface roughness under dry and wet condition. In this research, three main cutting parameters (cutting speed, depth of cut, and feed rate) were optimized to minimize energy consumption, and surface roughness during the turning of AISI316 under dry and wet conditions using response surface methodology based box-behnken design (BBD). The comparison study has been done between dry and wet stainless steel cut in terms of minimum value of energy and surface roughness.

2. Experimental details

2.1. Work Piece Materials and Cutting inserts

The workpiece material used for this investigation is a cylindrical rod of stainless steel 316 with axial length of 80 mm and diameter of 60 mm. The composition of the stainless steel 316 tabulated in Table 1.

Table 1. Chemical composition of the AISI316.

Material	C	Si	Mn	Ni	S	Cr	Mo
Wt. %	0.078	0.337	5.81	3.68	0.29	18.28	0.049

Table 2. Cutting conditions.

Workpiece	Cutting speed	Feeds	Depth of cut	Cutting condition
AISI316 steel	(110 -140 -170) m/min	(0.1- 0.15-0.2) mm/rev	(0.8 -1.25-1.4) mm	Dry, Wet
Cutting tools	**Young's modulus**	**Tool geometry**	**Tool holder**	**Responses**
CBN, 92.4 +1 HRA	47 (10^3kgf/mm^2)	CNMG 120408, 80° diamond sharp and 0° relief angle	BCLNR2525M12	power consumed and surface roughness

The cutting tool used for this study, CBN cutting tool, the tool was designated as ISO CNMG 120408, with 80° diamond sharp and 0° relief angle (Fig.1). The details about tool holder and other cutting condition tabled in Table 2. The turning process was performed in both condition with and without cutting fluid used.

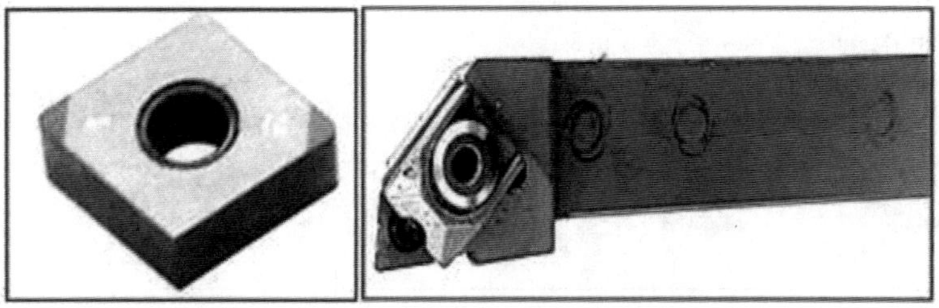

(a) (b)

Figure 1. (a) CBN inserts tools (b) Tool holder.

2.2. Method

The experiments were conducted according to the box-behnken design matrix and responses as shown in Table 3. A fresh insert was used for each experiment in order to improve the reliability of results. In present experiment, there are three (3) influences cutting parameters which are cutting speed (v_c) with the rang of (110 -140 -170 m/min), feed rate (*fr*) with the rang of (0.1- 0.15-0.2 mm/rev), and depth of cut (a_p) with the rang of (0.8 -1.25-1.4 mm) with their five (3) levels. The range each factor was selected is adapted from the Karoly handbook [37] and based on preliminary trials. Previous studies have used similar factor ranges to investigate similar responses [5,38,39]. In the present investigation, a response surface methodology (RSM) based box-behnken design was selected. A series of 15 experiments includes three center points were formulated to carry out experimental work (Table 3). The responses determined during the experiments were the power consumed during cutting stage (PW) in Wh, and average surface roughness (R_a) in μm. The turning experiments were performed using CNC lathe ROM 240 with (100 to 4000 rpm) a spindle speed range. Figure 2 present the experimental and equipment setting. Power Meter KEW6300 used to measure the power consumed, and each single data was stored during the cutting process. Surface roughness (R_a) was measured with a Mitutoyo Surftest SJ-301 portable surface roughness tester.

Table 3. Box-Behnken design matrix.

Std Order	Input variable		
	A:Vc m/min	B:fr mm/rev	C:d mm
1	110	0.1	1.25
2	170	0.1	1.25
3	110	0.2	1.25
4	170	0.2	1.25
5	110	0.15	0.8
6	170	0.15	0.8
7	110	0.15	1.7
8	170	0.15	1.7
9	140	0.1	0.8
10	140	0.2	0.8
11	140	0.1	1.7
12	140	0.2	1.7
13	140	0.15	1.25
14	140	0.15	1.25
15	140	0.15	1.25

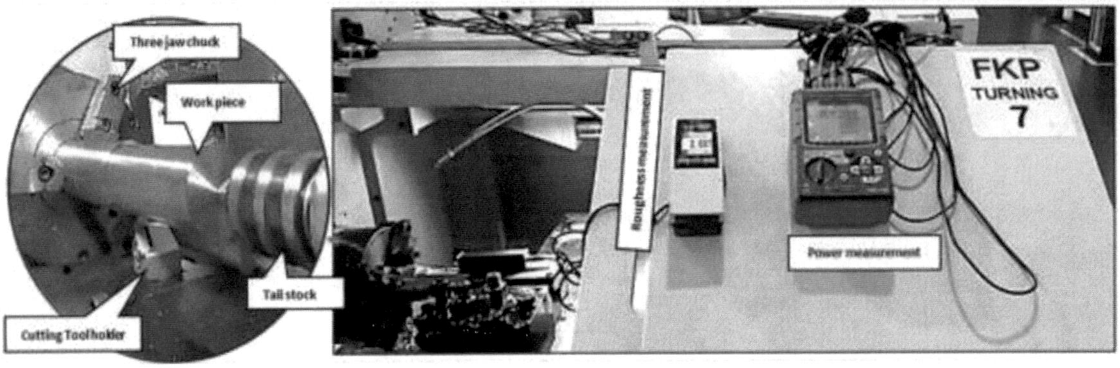

Figure 2. Experimental and instruments setting.

3. Results and discussion

Experimental results by BBD matrix under cooling and dry are presented in Table 4. The main and interaction effect plots for power consumption (PW, wh) and surface roughness average (Ra, μm) with respect to all cutting parameters: cutting speed (v_c, m/min), feed rate (*fr*, mm/rev) and depth of cut (d, mm) are presented in this section.

Table 4. Results of the experiment.

Std Order	Dry		Flood	
	PW (Wh)	Ra (μm)	PW (Wh)	Ra (μm)
1	32.6	0.478	41.5	0.415
2	35.51	0.965	43.87	0.534
3	25.65	0.64	34.31	0.498
4	30.28	1.331	39.2	0.987
5	34.11	0.491	32.92	0.613
6	37.8	0.924	45.79	0.528
7	23.5	1.37	32.14	1.29
8	22.81	1.104	31.52	1.467
9	36.2	0.845	45.6	0.677
10	35.72	1.245	44.32	1.245
11	32.8	1.53	42.04	1.53
12	23.9	1.6	32.56	1.225
13	27.8	0.72	36.09	0.814
14	29.81	0.714	38.13	0.726
15	26.5	0.741	34.98	0.711

Figure 3 shows the normal probability plot for dry condition. In this plot, the residuals are tracked along the straight line, which indicates errors are normally distributed. The residuals are randomly positioned on the positive and negative side on straight line. The residuals approximately normally distributed on each side on the line and many residuals lie between -2.5 to 2.5 for power consumption and between -0.2 to 0.2 for surface roughness. Therefore, the model accepted for validation.

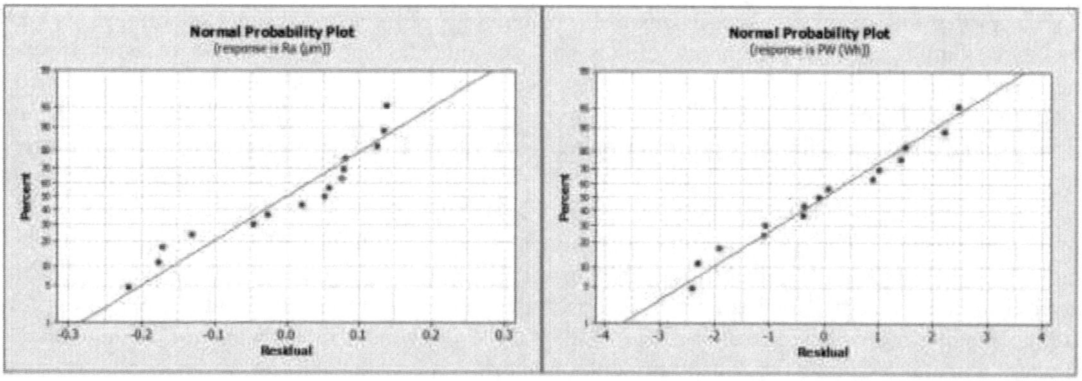

(a) (b)

Figure 3. Normal probability plot for dry condition (a) surface roughness (b) power.

Figure 4 shows the normal residual plot, which indicated that no outliers were found and deviations from the line were very minor. Therefore, the experiment provided a desirable value that satisfied the normality assumption.

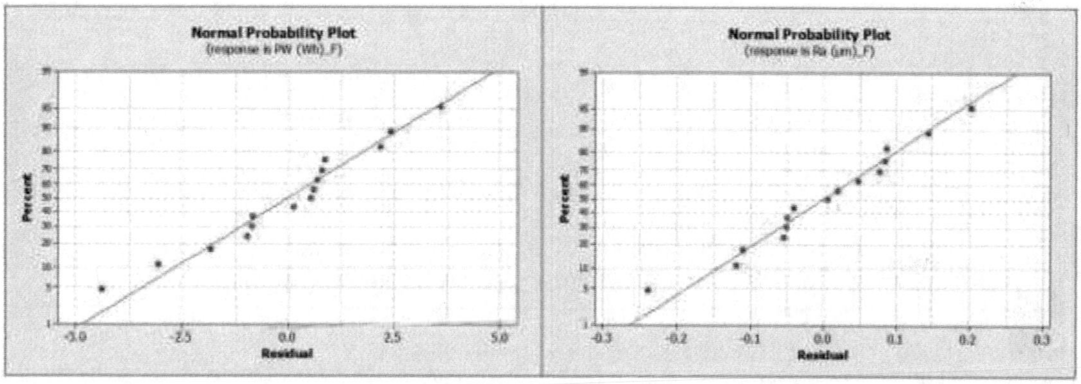

(a) (b)

Figure 4. Normal probability plot for flood condition (a) power (b) surface roughness.

Table 5 shows the regression model and R-Squared values of power consumption and surface roughness. The mathematical functions of power and surface roughness were developed using ANOVA analysis. The results for energy consumption indicated that R^2 = 87.7% and 88% for dry and wet condition, respectively. The R- Squared for surface roughness are 86.1% and 92.6% for dry and wet, respectively. Additionally, the difference between Pred R-Squared and Adj R-Squared was less than 0.2 that indicated the models is satisfactory.

Table 5. Regression model and R-Squared values.

Regression model	R-Sq.	Adj R-Sq.	Prd R-Sq.
PW (dry)= $+5.39290-0.29426*v-0.77196*f-0.21368*d$	0.8776	0.8546	0.7906
PW (wet) = $+5.79107-0.26496*v-0.85629*f-0.14931*f+0.17433*f^2$	0.8805	0.8487	0.6707
Ra (dry)= $+0.87214+0.0196*v+0.15506*f+0.068*d-0.0253*v^2+0.0228*f^2+0.0463*d^2+0.0652*v*f-0.0663*v*d$	0.8610	0.7599	0.5631
Ra (wet) = $+0.86+0.010*v+0.14*f+0.092*d-0.015*v^2+0.034*f^2+0.042*v*f-0.044*v*d+0.064*f*d$	0.9263	0.8727	0.6894

The evaluation of power consumption to machining parameters during machining are now presented. Figure 5a shows the main effects plot in indicates the mean effect of each level of parameters under dry cutting. From the plot observed that the highly-influenced factors are identified as depth of cut followed by feed rate and cutting speed. The plot for power consumption against cutting speed show a positively significant effect from low level to high level, whereas a negatively significant effect for feed rate and depth of cut. Therefore, the peak points of all independent variables as dry cutting for power consumption, 1.7mm depth of cut, feed of 0.2 mm/rev, 110 m/min cutting speed are selected. This is the optimal condition in order to minimize the power consumption to improvise the hard stainless steel turning process significantly. In Figure 5b shows the main effects plot of parameters under flood condition. It observed similar parameters influence under dry cutting with different rang of response. The high value of power consumption with use flood 42.5 Wh compare with 35 Wh under dry. The optimum points that can be achieved less power consumption are 110 m/min in cutting speed, feed rate with 0.15 mm/rev and 1.7 depth of cut.

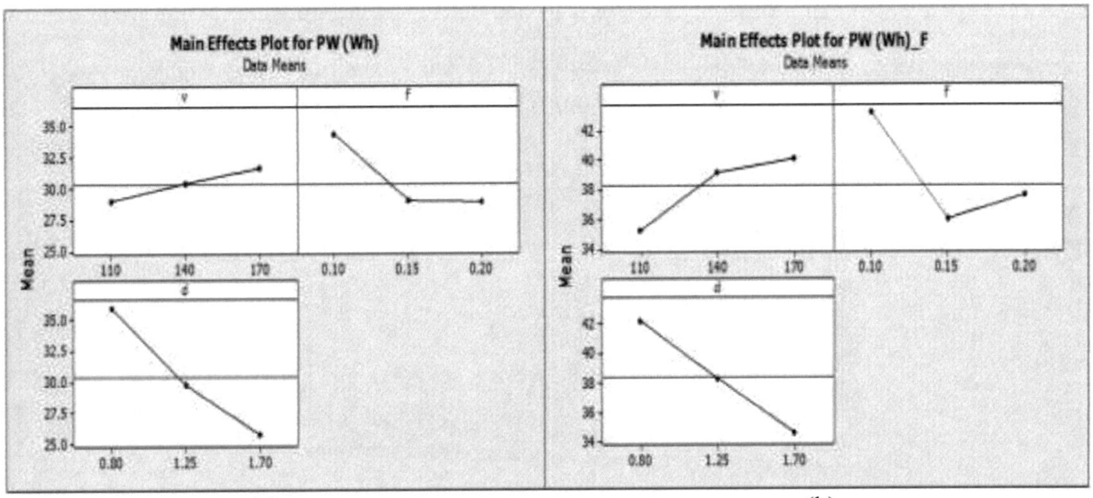

(a) (b)

Figure 5. Main plots of power consumption (PW) versus parameters under (a) dry cutting (b) flood condition.

Figure 6 (a,b) shows contour plots of power consumption for dry and flood cutting. A counter plot of feed rate Vs cutting speed as shown in Figure 6a and the depth of cut level are kept in middle level 1.5mm. For illustration, the lowest PW was reached in the region of 0.175 to 0.2 mm/rev feed rate and 110 to 132 m/min cutting speed for dry cutting (Figure 6a), whereas from 0.135 to 0.2 mm/rev for flood condition (Figure 6b).

IOP Conf. Series: Materials Science and Engineering **257** (2017) 012041 doi:10.1088/1757-899X/257/1/012041

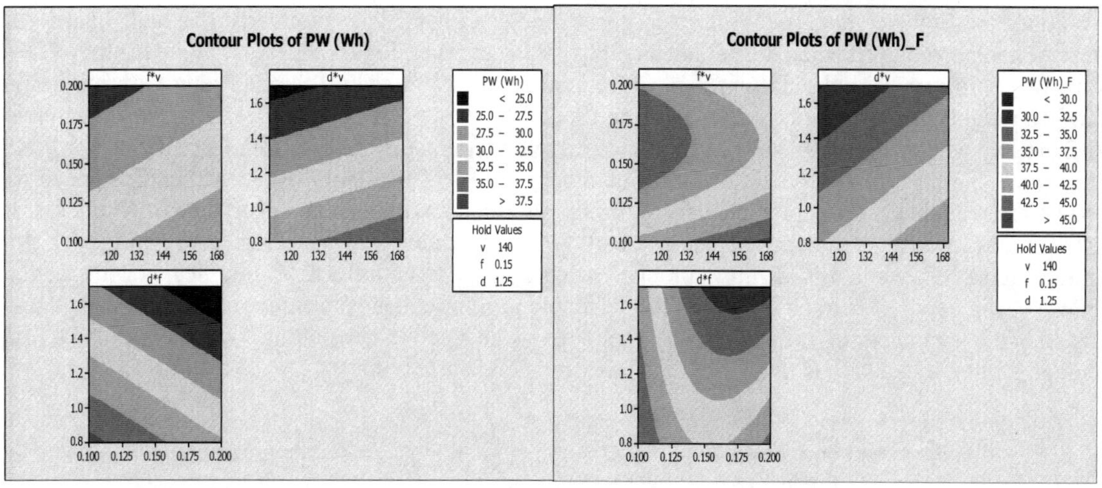

(a) (b)

Figure 6. Counter plots of power consumption (PW) versus parameters under (a) dry cutting (b) flood condition.

The contour plot of depth of cut Vs cutting speed shows the interaction effect of the depth of cut and cutting speed on the power consumption. It is seen that minimum cutting speed value and high depth of cut value facilitate to reach lower power consumption. The contour plot of depth of cut Vs feed rate shows the liner interaction effect of dry experiments (Figure 6a) and square relation under flood condition (Figure 6b) on the power consumption. The desirable region to reached minimum power consumption are (1.5-1.7 mm) depth of cut and (0.15-0.2 mm/rev) feed rate for dry cutting, and (1.6-1.7 mm) depth of cut and feed rate with (0.15-0.2 mm/rev) for flood condition.

Figure 7a shows the main plot effects parameters. It can be observed from the plot that surface roughness influenced by all factors. The depth of cut was greater influence on all responses. The plot for surface roughness against factors show a positively significant effect from low to high level, there are linear relationship between surface roughness and for both cutting parameters. The optimum points of all variables to reduce surface roughness, depth of cut with 1.7mm and 110 m/min cutting speed for both condition, and feed rate of 0.15 and 0.1 mm/rev for dry and flood, respectively.

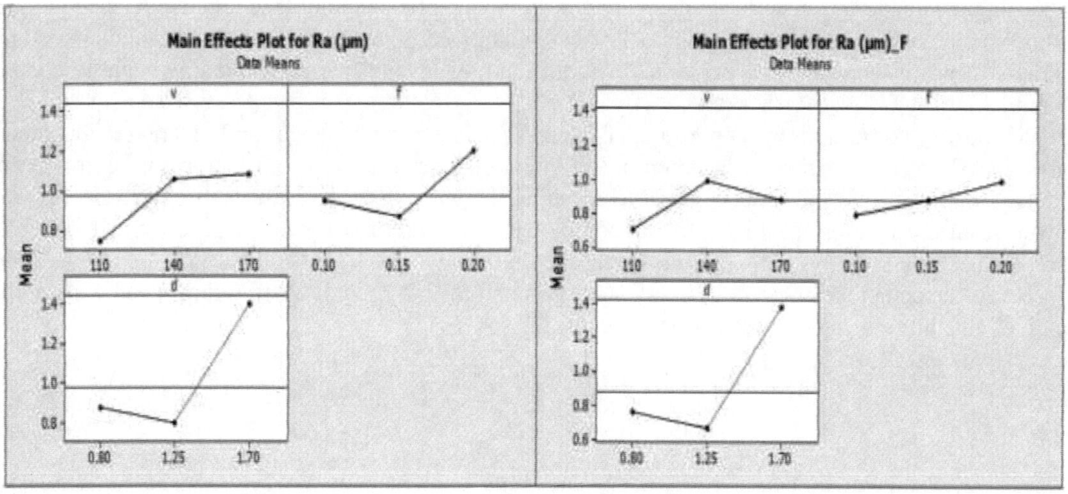

(a) (b)

Figure 7. Main plots of surface roughness (Ra) versus parameters under (a) dry cutting (b) flood condition.

Figure 8 shows contour plots of surface roughness. Generally, it is clearly the high quality of surface can be achieved when the parameters in their lower level. Figure 8a shows counter plot of feed rate Vs cutting speed with constant depth of cut value at 1.5mm. The desirable rang with minimum surface roughness for dry are 0.125 to 0.15 mm/rev of feed rate and cutting speed less than120 m/min (Figure 8a), whereas from 0.1 to 0.14 mm/rev feed and cutting speed shows small significant for flood condition (Figure 8b). The contour plot of interaction effect of the depth of cut and cutting speed on the surface roughness shows the depth of cut most significant with high range change of responses. It is seen that low value of depth of cut and cutting speed achieved lowest surface roughness for dry trials (Figure 8a), while it is slightly influence by cutting speed with flood (Figure 8b). For interaction effect of the depth of cut and feed rate, the desirable region to achieved minimum surface roughness are (0.8-1.2 μm) depth of cut and (0.5-0.15 mm/rev) feed rate for dry cutting, whereas from 0.8 to 1 mm of depth of cut and feed rate with 0.1mm/rev for flood condition.

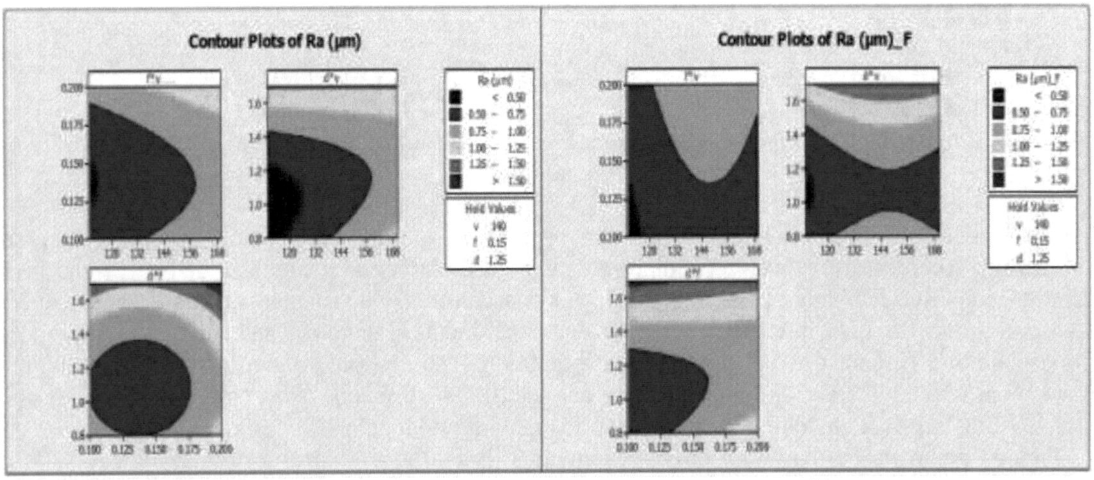

(a) (b)

Figure 8. Counter plots of surface roughness (Ra) versus parameters under (a) dry cutting (b) flood condition.

According to the statically analysis, it is found that power consumption increased with an increasing in spindle speed, whereas decreased with an increasing in both feed and depth of cut. Figure 9 shows reduce of power consumption under dry condition compare to flood during cutting stage. This increase of power consumption explained by addition power load that consumed by running coolant system.

The power consumption proportionately increase for cutting speed and decrease for others parameter (Figure 5). Increases in depth of cut and feed rate, leading to an increases the amount of removed materials and the time required to machine the workpiece decrease, subsequent decrease in power requirements for machine tool to complete the operation. Dry cutting performed better to decrease power consumption for cutting speed, feed rate and depth of cut compared to flood condition. Moreover, the compression graph (Figure 9) suggested that employing lower cutting speed with a high depth of cut and feed rate can reduce power consumption.

IOP Conf. Series: Materials Science and Engineering **257** (2017) 012041 doi:10.1088/1757-899X/257/1/012041

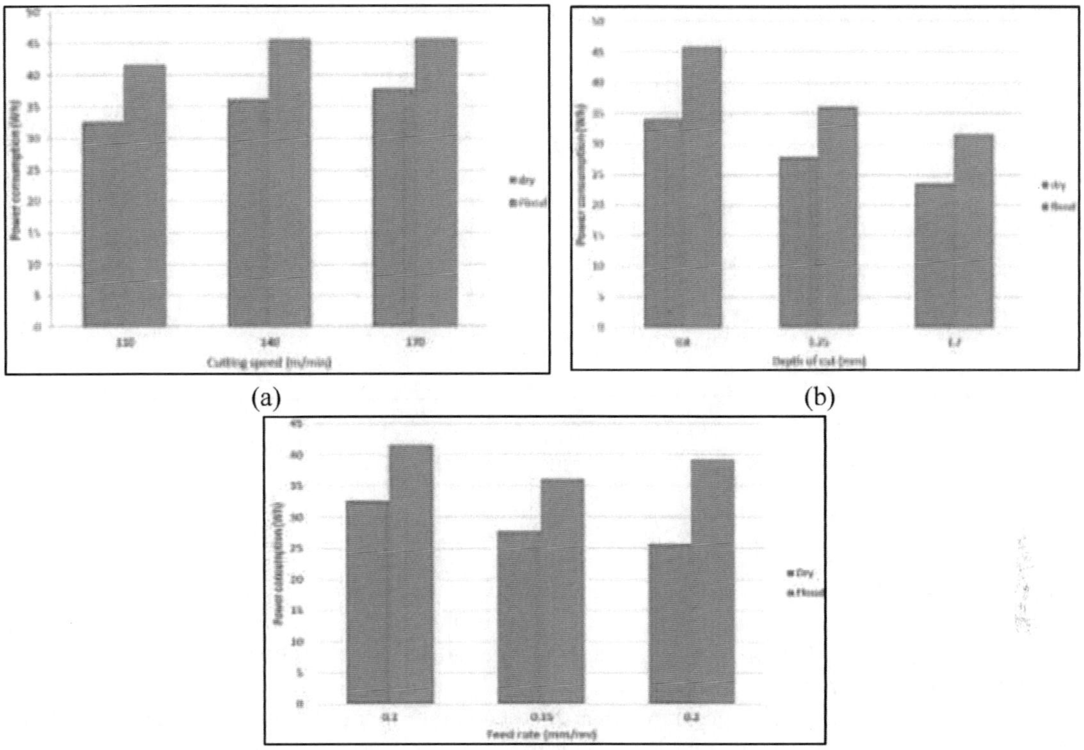

(a) (b)

(c)

Figure 9. Power consumption versus parameters (a) cutting speed (b) depth of cut (c) feed rate.

For surface roughness evaluation, surface roughness almost increases as cutting parameter increasing for both condition (Figure 10). It is shows there is linear relationship between surface roughness and three cutting parameters. From illustrate, the coolant cutting shows better than dry cutting regarding surface roughness.

At high depth of cut, a continuous chip formed which are increase of temperature in the workpiece and tool interface which is increase friction that leads to high roughness. For dry cutting, surface roughness increase when increase of speed until reach of 140 m/min than shows not effect on surface roughness. The increase of feed rate and depth of cut shows slightly decrease of surface roughness than their relation become direct proportion with high change of roughness value. Whereas the CBN cutting tool shows better under flood condition with lowest value of feed rate. In common the surface roughness increased with the higher feed rates in all experiments. The increase of speed more the middle value lead to decrease of surface roughness. All cutting process parameters showed that machining work-piece under flood better surface finish compared to dry cutting. Higher feed rate values lead to high temperature at the tool chip interface which cause adhesion and abrasion of tools. Due to these damage of tools, the surface roughness was influenced accordingly. Due to high temperature, defect occurs rapidly at dry work-piece which causes high value of roughness. The main effect plots relations established for both cutting condition are similarly.

IOP Conf. Series: Materials Science and Engineering **257** (2017) 012041 doi:10.1088/1757-899X/257/1/012041

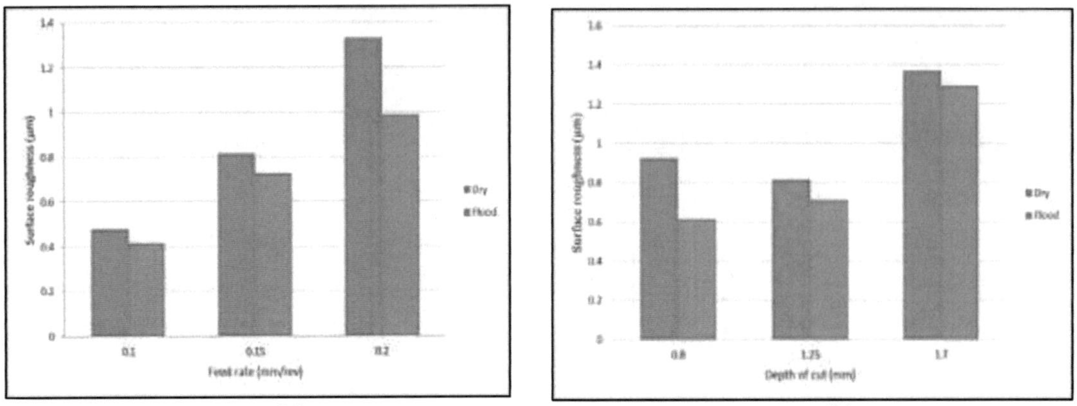

(a) (b)

Figure 10. Surface roughness (Ra) versus parameters (a) depth of cut (c) feed rate.

Thus, the optimum cutting condition was found at 110 m/min cutting speed with power consumption 28.5 Wh, 35 Wh for dry and flood, respectively, and average surface roughness 0.73 µm for dry cutting, 0.69 µm under coolant. The best surface roughness and power used value are obtained at low value cutting speed. For feed rate, the optimum value under dry condition at 0.2 mm/rev with power 28.5 Wh and at 0.15 with average surface roughness 0.85 µm. For flood, the best select for feed rate at 0.15 mm/rev with power consumption 37.8 Wh and at 0.1with 0.79µm average surface roughness. The optimum power consumption at the high depth of cut 1.7 mm with 25.5 Wh of power consumption and 0.8 µm for average surface roughness at 1.25 mm. With high value of power consumption (35 Wh) and less value of the surface roughness (0.67 µm) for wet condition. Considering both responses, it can be concluded the performance of CBN insert under dry cutting condition can be minimize the power consumption with acceptable quality. It is a good technique of environmental machining without fluids impact. The optimum parameters selected help to reduce the effect of issues associated with the dry machining such are tool life, cutting temperatures and thermal damage to the workpiece.

4. Conclusions

The performance of CBN inserts and comparison study between dry and wet cutting stainless steel on minimum value of energy and surface roughness was evaluated. The results indicated that stainless still can be machined under dry condition with low power consumption and acceptable quality when compared wet cutting. The plot graphs clearly show for both condition dry and wet that the minimum value of energy consumption of the cutting process was obtained at the lowest cutting speed value and at the high values of feed rate and depth of cut. The factor with the most significant influence on surface roughness was feed rate. Results also showed an improvement of power consumption under dry condition 18.57%, whereas for surface roughness showed coolant slightly better with 1.45% than dry cutting. The CBN tools under dry cutting can be used to reduce the environment impacts in terms of no cutting fluid use and less energy required which is affected in machining productivity and profit.

Acknowledgments

The authors would like to thank the University Malaysia Pahang for providing laboratory facilities. The

financial provided by Universiti Malaysia Pahang through GRS (20.03/13) is also acknowledged.

References

[1] Yoon H S, Kim E S, Kim M S, Lee J Y, Lee G B and Ahn S H. 2015 *Renew. Sustain. Energy* Rev. **48** 870–91

[2] EIA 2013 International Energy Outlook 2013 US Energy Inf. Administration 31
[3] NDRC 2011 Baseline Emission Factors for Regional Power Grids in China Natl. Dev. Reform Comm. 1–5
[4] EIA 2009 U.S. Carbon Dioxide Emissions from Energy Sources 2008 Flash Estimate Energy Information US Energy Inf. Adm. 1–17
[5] Nur R, Noordin M Y, Izman S and Kurniawan D 2015 *Proc. Inst. Mech. Eng. Part E J. Process Mech. Eng.* **0** 954408915624861
[6] Zheng H and Liu K 2014 *Machinability of Engineering Materials Handbook of Manufacturing Engineering and Technology* (Springer) pp 899–939
[7] Pusavec F, Krajnik P and Kopac J. 2010 *J. Clean. Prod.* **18** 174–84
[8] King N, Keranen L, Gunter K and Sutherland J. 2001 SAE Technical Paper
[9] Weinert K, Inasaki I, Sutherland J W and Wakabayashi T. 2004 *Ann. Manuf. Technol.* **53** 511–37
[10] Najiha M S, Rahman M M and Yusoff A R 2016 *Renew. Sustain. Energy Rev.* **60** 1008–31
[11] Dixit U S, Sarma D K and Davim J P 2012 *Environmentally Friendly Machining* (Boston, MA: Springer US)
[12] Schultheiss F, Zhou J, Gröntoft E and Ståhl J-E 2013 *J. Clean. Prod.* **59** 298–307
[13] Adler D P, Hii W W-S, Michalek D J and Sutherland J W 2006 *Mach. Sci. Technol.* 10 23–58
[14] Kalpakjian S and Schmid S 2006 *Manufacturing, Engineering and Technology* (Digital Designs)
[15] Galoppi G d S, Filho M S and Batalha G F 2006 *J. Mater. Process. Technol.* **179** 146–53
[16] Davoodi B and Eskandari B 2015 *J. Int. Meas. Confed.* **68** 286–94
[17] More A S, Jiang W, Brown W D and Malshe A P 2006 *J. Mater. Process. Technol.* **180** 253–62
[18] Diniz A E and de Oliveira A J 2008 *J. Mater. Process. Technol.* **195** 275–81
[19] Diniz A E, Gomes D M and Braghini A 2005 *J. Mater. Process. Technol.* **159** 240–8
[20] Ciftci I 2006 *Tribol. Int.* **39** 565–9
[21] Gokkaya H and Nalbant M 2007 *Mater. Des.* **28** 717–21
[22] Rech J 2006 *Surf. Coatings Technol.* **200** 5132–9
[23] Khrais S K and Lin Y J 2007 *Wear* **262** 64–9
[24] Bouacha K, Yallese M A, Mabrouki T and Rigal J F 2010 *Int. J. Refract. Met. Hard Mater.* **28** 349–61
[25] Bouzid L, Yallese M A, Chaoui K, Mabrouki T and Boulanouar L 2015 *Proc. Inst. Mech. Eng. Part B J. Eng. Manuf.* **229** 45–61
[26] Saini A, Dhiman S, Sharma R and Setia S 2014 *J. Mech. Sci. Technol.* **28** 2307–18
[27] Khamel S, Ouelaa N and Bouacha K 2012 *J. Mech. Sci. Technol.* **26** 3605–16
[28] Sahu S K, Mishra P C, Orra K and Sahoo A K 2015 *Proc. Inst. Mech. Eng. Part B-Journal Eng. Manuf.* **229** 251–65
[29] Wegener K, Kuster F, Weikert S, Weiss L and Stirnimann J 2016 *Procedia CIRP* **46** 512–24
[30] Bhushan R K 2013 *J. Clean. Prod.* **39** 242–54
[31] Camposeco-Negrete C 2015 *J. Clean. Prod.* **91** 109–17
[32] Campatelli G, Lorenzini L and Scippa A 2014 *J. Clean. Prod.* **66** 309–16
[33] Bagaber S A and Yusoff A R 2017 *J. Clean. Prod.* 157 30–46
[34] Bhushan R K 2013 *J. Clean. Prod.* 39 242–54
[35] Bagaber S A and Yusoff A R 2017 *AIP Conference Proceedings* **1828** 20013
[36] Lee B Y and Tarng Y S 2000 *J. Mater. Process. Technol.* **105** 61-6
[37] Korloy 2015 Korloy Cutting Tools 1–1059
[38] Philip Selvaraj D, Chandramohan P and Mohanraj M 2014 *Meas. J. Int. Meas. Confed.* **49** 205-15
[39] Berkani S, Yallese M A, Boulanouar L and Mabrouki T 2015 *Int. J. Ind. Eng. Comput.* **6** 539–52

Study on the CFD simulation of refrigerated container

Muhammad Arif Budiyanto[1*], Takeshi Shinoda[2], Nasruddin[1]

[1]Departement of Mechanical Engineering, Universitas Indonesia, Kampus Baru UI Depok, Jawa Barat, Indonesia.
[2]Departement of Marine System Engineering, Kyushu University, Motooka, Nishi-ku, Fukuoka, Japan

*Corresponding author: arif@eng.ui.ac.id

Abstract. The objective this study is to performed Computational Fluid Dynamic (CFD) simulation of refrigerated container in the container port. Refrigerated container is a thermal cargo container constructed from an insulation wall to carry kind of perishable goods. CFD simulation was carried out use cross sectional of container walls to predict surface temperatures of refrigerated container and to estimate its cooling load. The simulation model is based on the solution of the partial differential equations governing the fluid flow and heat transfer processes. The physical model of heat-transfer processes considered in this simulation are consist of solar radiation from the sun, heat conduction on the container walls, heat convection on the container surfaces and thermal radiation among the solid surfaces. The validation of simulation model was assessed uses surface temperatures at center points on each container walls obtained from the measurement experimentation in the previous study. The results shows the surface temperatures of simulation model has good agreement with the measurement data on all container walls.

1. Introduction

The measure of energy consumption in the container port is an indicator of the container port performance [1]. The breakdown of total energy consumption identify the component of energy consumption in container port and will give direction in the effort of energy saving. In the recent development of green port technology the modern container port has converted their container handlings into the electrification system, so that all of energy source comes from the power supply from the national grid. Several study of energy mapping in container port has been conducted by Tran [2], Greencranes [3], and Shinoda & Budiyanto [4]. From the statistical data refrigerated container facility contribute the most power consumption in the container port [4].

The literature of research study about refrigerated container facility were also inadequate [5]. Research study was conducted by Jolly et al [6] conclude the power consumption of refrigerated container will depend on the setting of temperature inside the cargo hold. The study conducted by Fitzgerald et al [7] assumed the mean energy consumption of refrigerated container is 2.7 kW/TEU and has values variations around 60% due to various factors. Shinoda et al [8] conducted an measurement experiment of 40 feet high cube refrigerated container consumed about 7.2 kW/h. Looking from the inadequate research study in the refrigerated container, the estimation study by CFD simulation will give opportunity for further energy analysis in container terminal operation.

In order to estimate of energy consumption of refrigeration container, the cooling load of the containers assumed as the energy to be removed from the inside cargo and equal to the power consumption of the refrigeration system. CFD simulation give the prediction of surface temperature in each container wall which is a part in the calculation of cooling load of refrigerated container. The objective this study is to performed CFD simulation of refrigerated in the container port in the transient condition. The simulation model will predict the surface temperature of container walls in various condition will the cooling load can be estimate.

2. Description of the simulation model

The heat transfer process in reefer container was simulated by means of a CFD simulation model. The PHOENICS software packages is utilized to simulate fluid flow and heat transfer [9]. In the present simulations, a cross sectional of container walls in Cartesian framework is use to simulate of the refrigerated container in storage yard. In order to reduce the computational time required by the solution of the CFD model, only middle section of reefer container is considered assuming the symmetry condition. The set of governing equations is solved by a finite control volume method (FVM) and described in the following section.

2.1 Set of governing equations

The simulation model is based on the solution of the partial differential equations governing the flow and convective heat transfer. The governing equations in the field of fluid flow based on incompressibility and heat transfer process were used of the generalized transport equations:

a. Continuity equation

$$\nabla \cdot \mathbf{v} = 0 \tag{1}$$

b. Momentum equation

$$\rho(\frac{\partial \mathbf{v}}{\partial t} + \mathbf{v} \cdot \nabla \mathbf{v}) = -\nabla p + \mu \nabla^2 \mathbf{v} + f_B \tag{2}$$

c. Energy Equation

$$\int_V \rho C_p (\frac{\partial T}{\partial t} + \mathbf{v} \cdot \nabla T) dV = \int_V k \nabla^2 T dV + \int_S (q_R + q_S) dS \tag{3}$$

Where $\boldsymbol{t}$ is the time, $\boldsymbol{V}$ is the velocity vector, $\boldsymbol{\rho}$ is the density, $\boldsymbol{p}$ is the pressure, $\boldsymbol{\mu}$ is the coefficient of viscosity, $\boldsymbol{k}$ is thermal conductivity, $\boldsymbol{C_p}$ is specific heat capacity and $\boldsymbol{T}$ is the local temperature. Natural convection was modelled using the Boussinesq approximation, which uses a constant density fluid model, but applies a local body gravitational force throughout the fluid that is a linear function of the volumetric thermal expansion coefficient $\boldsymbol{\beta}$ and of the local temperature difference. The buoyancy source is added to the momentum equation as follows [10] :

$$f_B = -\Delta \rho g = -\rho_{ref} \beta (T - T_{ref}) g \tag{4}$$

Where $\boldsymbol{\rho_{ref}}$ and $\boldsymbol{T_{ref}}$ are the density and temperature at the boundary wall condition and g is the gravitational force. Internal energy from thermal radiation ($\boldsymbol{q_R}$) is taken into account at the surface of the objective cell in FVM within the immersed solid (IMMERSOL) per unit volume as in equation:

$$q_R = \varphi_R \varepsilon_R \varepsilon \sigma (T_R^{\ 4} - T^4) \tag{5}$$

Where, $\boldsymbol{\varphi_R}$ is the radiation shape factor from a view of radiation material, $\boldsymbol{\varepsilon_R}$ is the emissivity of radiation material, ε is the emissivity of the objective solid body, $\boldsymbol{T_R}$ is the radiation temperature of the fluid and $\boldsymbol{\sigma}$ is Stefan Boltzmann constant (5.67×10^{-8}W/m^2·K^4). Particularly in equation (3), when the solid area in the model such as the container walls and road are considered, the second term of the right side of equation is neglected, and equation (3) become the equation of heat conduction except the surface of the solid body. Energy of solar radiation (q_S) was considered in this model, and the detail of q_S is described in the next section.

A Dirichlet boundary condition at the surrounding container is used to specify a uniform inlet velocity. A no-slip boundary condition for all walls is assumed for the container wall. The inlet air temperature varies with time and the walls are assumed adiabatic. In order to close the set of governing equations, we used a k–ε turbulence model.

2.2 Solar radiation and sun geometry

In the PHOENICS software packages, the energy activation from the sun utilizes SUN object to compute the intensity and direction of solar radiation heat load upon surfaces within the domain [11]. It were able to import data from weather files to obtain solar radiation at a particular date and time, in accordance to the determined also location. The solar radiation consists of the direct, diffuse and reflected radiation incident on a given surface. These radiations and the angle of incidence (θ) depends upon location on the solar geometry and also the orientation of the surface. In vertical surface, the basic parameters are defined in terms of altitude angle (γ) and azimuth angle (α) [12]. Altitude angle is described as the angle between the sun rays its projection of sun's rays onto horizontal planes. The solar azimuth angle is the angle in the horizontal plane measured from north to the horizontal projection of the sun rays. Figure 1 shows the illustration of variation of azimuth angle from sun geometry.

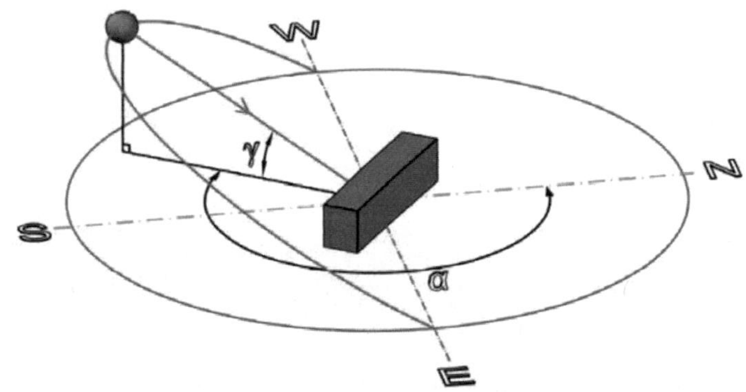

Figure 1. Sun geometry.

2.3 Geometrical model and initial condition

As mentioned in section before, the domain was assumed to be symmetric with regard to perpendicular planes through the middle section of reefer container. Middle section of container is the most representative to catch heat distribution through the surface wall. The model assumed as open system simulation which consist of four regions i.e. outside air, body of container, inside air and road concrete. Within the body container composed of three material of insulation are steel, polyurethane and aluminum [13]. The total dimension of the model is 112.3m^2 was constructed in Cartesian framework with the total gird is 59,200 cells. The time discretization was set to transient calculation in 12 hours divided into 168 time-step start from 6:00 until 18:00. The geometrical grid and illustration model shows in figure 2. The near wall uses power mesh ratio was introduced to satisfy the calculation particularly on the thin layer region.

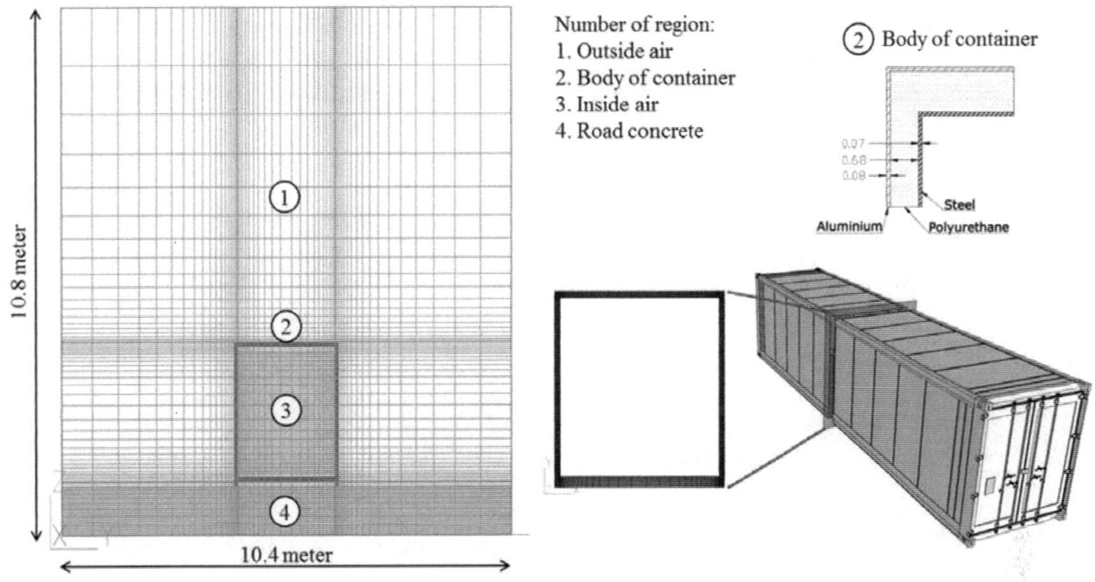

Figure 2. Grid generation and geometrical model.

In transient simulations initial values should be set for all variables to get convergence criteria. Physical domain set to be air with external pressure is 1 atmosphere and temperature is 23.39 degree Celsius. The energy activation utilize solar radiation from sun as heat source, in this case the IMMERSOL radiation model will be employed as energy equation [14]. Physical characteristic related to heat transfer was applied according to properties each material i.e. thermal conductivity and solar absorption factor. Initial value of simulation parameter and physical domain shows in table 1. Besides the energy activation, momentum flow quantities was set to complied the governing equations that had generated buoyancy effect. Wind speed was applied to the model as the momentum source. The value and direction of wind speed was imported from weather data which also contained solar radiation. For this purpose the initial value of weather data was deliberately taken from experimental measurement on summer condition [15].

Table 1. Initial condition of simulation.

Simulation Parameter	
Dimension model	10.4 x 10.8 x 1
Grid Cells	80 x 74 x 10
Velocity & Pressure	On
Energy Equation	Temperature
Gravitational Force	On
Buoyancy Model	Boussinesq
Turbulence Model	k-ε model
Domain Material	Ideal Gas Law
Time dependence	Transient
Coeff. Wall Function	Log-law
Iterations	50 steps

Properties of Air		
Density	1.22	(kg/m^3)
Velocity	0.3	(m/s)
Specific Heat	1.006	(kJ/kg.K)
Initial temperature	23.39	(°C)

Properties of Material		
	Thermal Condcutivity	
Aluminium	204	*(W/m.K)*
Polyurethane	0.003	*(W/m.K)*
Stainless steel	16	*(W/m.K)*
Road	1.13	*(W/m.K)*

2.4 Validation of simulation results

Validation of experiment and simulation model is conducted to produce an accurate results for further estimation of surface temperature on container walls. In this paper, preceding experimental result by Shinoda et al [4] is compared to the simulation results under similar condition. Figure 3 shows the schematic of measurement setup of refrigerated container. Sensor devices used in the experiment are thermocouple, power meter, velocity meter and pyranometer. Thermocouple is the temperature sensor which installed inside and outside of container surfaces. On the inside surface attached 5 sensors at the middle surfaces i.e. floor, sidewall, ceiling and the center. On the outside surface attached 15 sensors each wall surfaces at the middle point including fan and compressor surfaces. Power meter employs to measure electricity consumption during experiment. Velocity meter also employs to measure wind speed and direction were placed in around the container location. Pyranometer is to measure solar insolation on surface object, this device placed at the top position on measurement location. The measurement data from all devices is recorded every minute during the daytime on summer season in 2013.

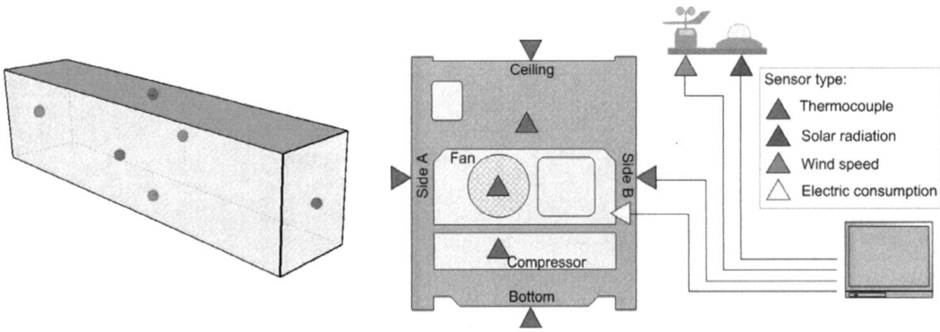

Figure 3. Schematic view of measurement set up [4].

The simulation result was validated by experimental measurement at specific points on the middle surface of container walls. The date of measurement data used in this validation is 27th August 2013. The parameter which use for validation is the temperature data in the center surface each wall i.e. ceiling, south, north, and bottom. The accuracy of the simulation was assessed by determining root mean square error ($\boldsymbol{E_{RMS}}$). The equation for $\boldsymbol{E_{RMS}}$ determination can be expressed as:

$$E_{RMS} = \sqrt{\frac{\sum_{n=1}^{n}\left(T_E - T_P\right)^2}{N}} \tag{6}$$

Where $\boldsymbol{T_P}$ is simulated temperature and $\boldsymbol{T_E}$ is measured temperature, $\boldsymbol{N}$ is the number of population. In order to ascertain the simulation result appertain in the confidence level, 95% confidence interval was predetermined for each validation to ascertain the validity of the simulation model.

3. Results and discussion

This section will discuss the thermal analysis from the CFD simulation of the refrigerated container. The CFD simulation gives evidence of the physical mechanism of the heat transfer problem and provide the parameter directions either for the improvement of its simulation model and physical condition of the reefer containers.

3.1 Validation of surface temperatures

The surface temperatures of simulation model was validated using experimental data of preceding study. Figure 4 shows the validation result in all surface of container. The simulation results were in agreement with the experimental data for the all surface, no significant differences were observed between experimental and simulated. The maximum E_{RMS} is 2.25 occurs in ceiling surface, this instance due to the instabilities of the convergence criteria at the beginning of the calculation. As seen from the validation graph, a narrow gap found in the start point began disappear after some step calculation. Second highest E_{RMS} occur in the north surface. This due to instabilities of temperature during experiment, where some indication appear in the temperature profile at certain time on all around container body. Wind speed from seaside to land side caused a turbulent velocity on the ground surface, as a consequence natural convection is occurred from the ground to vertical wall of container, this phenomenon consistent with the result conduct by Inan [16]. The lowest E_{RMS} is resulted in bottom surface of simulation condition, the reason is reasonable since the influence of wind velocity and solar radiation is very low in this surface area. From this validation, the result of surface temperature from the CFD simulation has good agreement with measurement data.

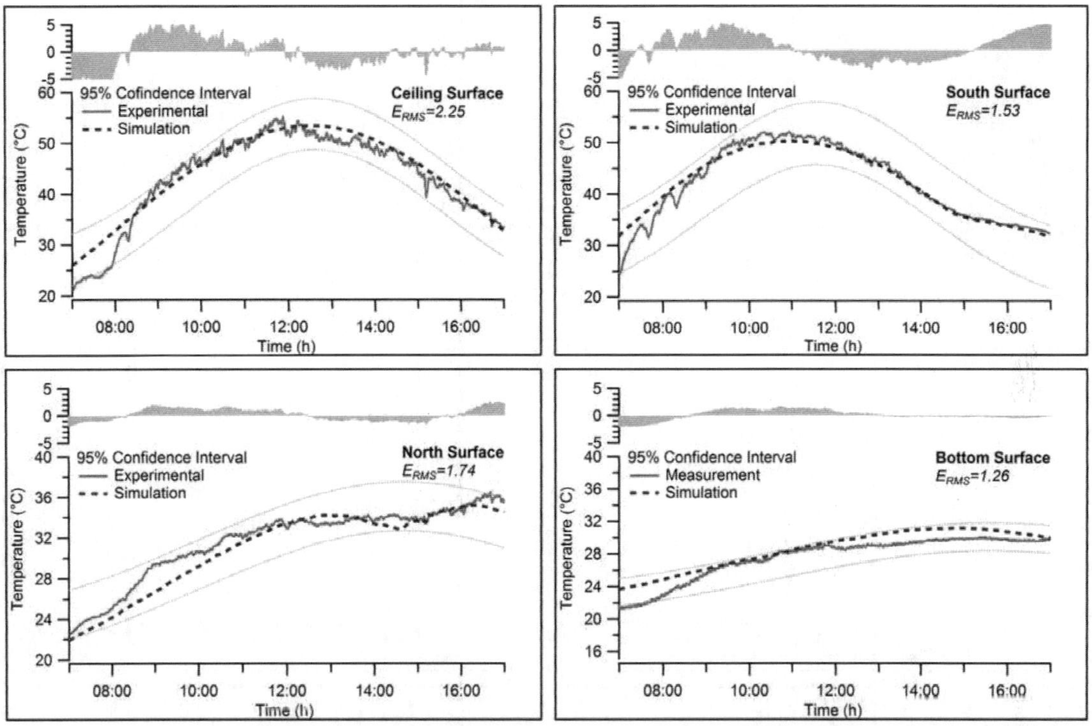

Figure 4. Validation of temperature of simulation result with experimental data.

3.2 Temperature distribution on the container walls

Figure 5 shows temperature distribution on the container walls from the morning until the afternoon. This figure shows the temperature distribution occur on the container walls of refrigerated container with maximum temperature is around 45 degree Celsius at 12:00. The temperature distribution start from the road transfer to the south container walls through the foot of the container. During the morning until the afternoon the accumulated heat move from south walls to ceiling walls and peak temperature occurs on this walls. During the afternoon until evening heat move from ceiling walls into north walls. The heat mainly invade on ceiling and south walls. This instance caused by a strong load of solar radiation at the highest position of the sun located at the south of the model, this phenomenon consistent with research was conducted by Rodriguez [18]. In normal condition heat form the sun will invade through the outer wall of container, thus increase the temperature on the outer surface exposed by solar radiation and propagates to all around container surface. Heat transfer process in numerical calculation start from the release of the heat energy from heat source. Heat source comes from the heat flux of sun and then transferred to the simulation model object via air. Thermal diffusion occurs when the flux has reached a surface of the material of the model and then, heat will be absorbed, transferred or reflected depend on the characteristic of the material [19]. Due to the turbulence model, the reflected radiation will increase the temperature of air layer in surrounding the material and caused natural convection effect. The absorbed heat will be transferred within the material, when the heat exceeds the thermal capacity per unit area, heat transfer arises inside of insulation material through thermal conduction that flows into the inner surface. Finally, thermal diffusion occurs in the inner surface between heat conduction from insulation with cooling air in the inside of container.

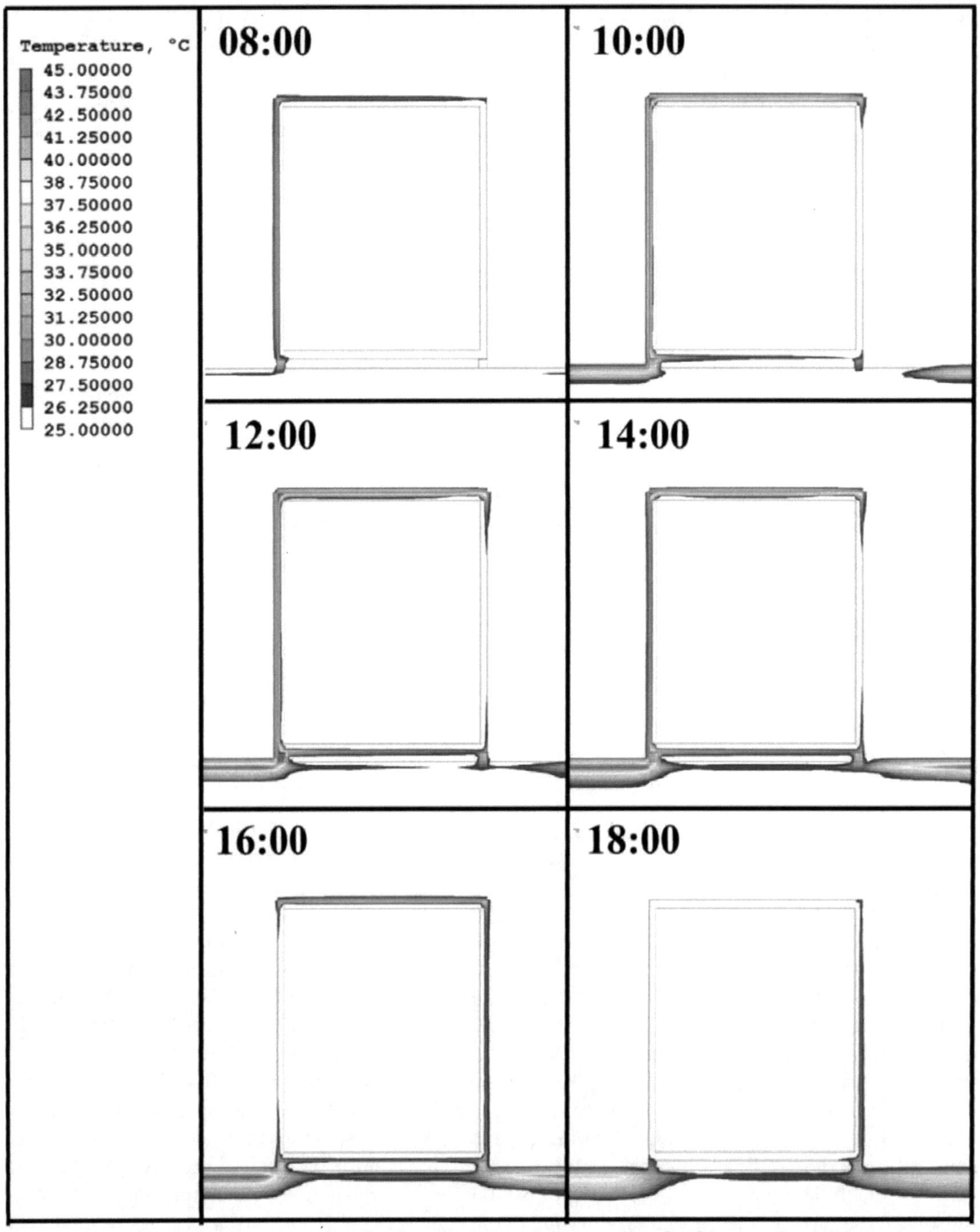

Figure 5. Temperature distribution on the container walls of refrigerated container.

3.3 Estimation of cooling load

Estimation of the cooling load of refrigerated container is defined as benchmarking of thermal performance with the following equation [20]:

$$q = K \cdot S \cdot \left(\overline{T}_o - \overline{T}_i\right) \tag{7}$$

$$q_{Total} = q_C + q_S + q_N + q_B \tag{8}$$

Here, value of q represents cooling load through the wall (*Watt*), K represents overall heat transfer coefficient of the container wall (W/m^2K), S represents area of the wall (m^2) and T represents temperature difference between inside and outside of container. Equation 8 is represent the total of heat flow from each side i.e. ceiling (*C*), south (*S*), north (*N*) and bottom side (*B*). The duration time of the calculation is from 10:00 until 16:00.

Table 2. Calculation of cooling load of refrigerated container.

Duration time		**10:00-16:00**
Azimuth Angle	*(degree)*	334.2
$\overline{T}$ Ceiling	*(°C)*	49.4
$\overline{T}$ South (Side A)	*(°C)*	43.8
$\overline{T}$ North (Side B)	*(°C)*	33.1
$\overline{T}$ Bottom	*(°C)*	29.9
Heat load	*(Watt)*	2246.7

Table 2 shows the temperature in the ceiling surface has most contribution to heat penetration, followed by south surface. From this calculation obtained the cooling load of refrigerated container is about 2.2 kW. The temperature in the wall container is the main influence of the cooling load capacity. The heat flow which appear to the inner surface is equal with balance energy which required to keep chilled condition. This condition makes cooling load increase and as consequence the energy consumption of refrigerator also increase.

3.4 Future simulation : heat island problems

Container port mostly located in the shore of urban city areas. Buildings, roads, and other infrastructure replace open land and vegetation. Surfaces that were once permeable and moist become impermeable and dry. These changes cause urban regions to become warmer than their rural surroundings, forming an "island" of higher temperatures in the landscape [21]. Looking this phenomenon the future works of this simulation model will carry out the heat island problems occurs on the container port. Figure 6 shows the illustration of simulation model in the container storage yard model. The future simulation will provide an analysis of whole phenomenon in the container port by using the current simulation model.

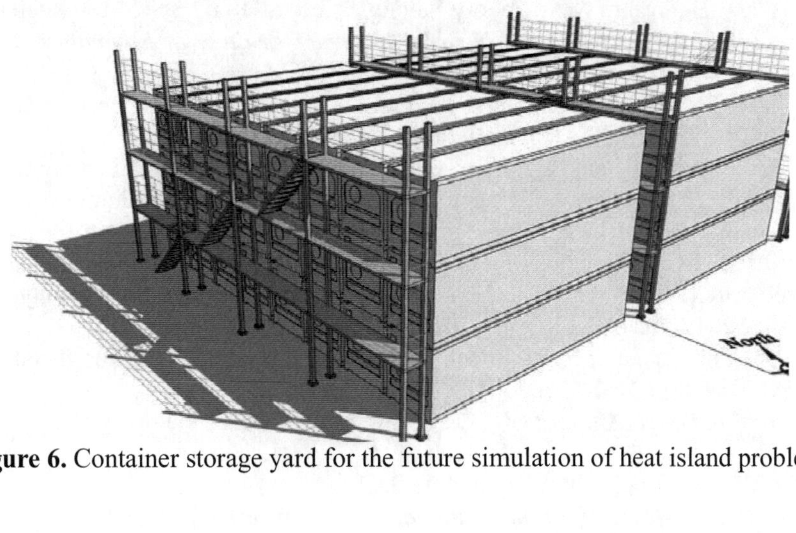

Figure 6. Container storage yard for the future simulation of heat island problems.

4. Conclusions

The CFD simulation of refrigerated in the container port was done performed use PHOENICS software packages. The simulation result of one refrigerated container in the transient condition for one day operation has been validate using measurement data carried out by Shinoda et al (2014). From the comparison result shows the simulation results has good agreement with the measurement data and obtain maximum root mean square error is 2.55 C in the ceiling surface. From the simulation obtained the average amount of energy to remove heat load inside cargo from the effect of heat penetration due to solar radiation during one day operation is 2.2 kW. Furthermore, this result give a direction for estimation of energy consumption in a container port and can be used to predict the energy consumption in other location.

Acknowledgments

Authors would like to express our gratitude to Directorate Research and Community Engagement (DRPM) Universitas Indonesia in the providing of publication funding by PITTA 2017. Author also would like to thanks Department of Marine System Engineering, Kyushu University and Hakata Port Terminal Corporation for providing required data on this study.

References

[1] UNCTAD. Review of Maritime Transport 2015. *Unctad* 204 (2015). doi:10.1017/CBO9781107415324.004
[2] Tran TK 2012 Study of Electrical Usage and Demand at the Container Terminal *Deakin University*
[3] GREENCRANES. *Report on Port Container Terminals Energy Profile: Mapping of Port Container Terminals Energy Profile*. Valencia; 2013.
[4] Shinoda T and Budiyanto M A 2014 Energy saving effect of roof shade at reefer container storage yard *Proceedings of the International Forum on Shipping, Ports and Airports (IFSPA): Sustainable Development in Shipping and Transport Logistics* pp 455-460.
[5] Wilmsmeier G. Energy consumption and efficiency : emerging challenges from reefer trade in South American container terminals. *Fal*. 2012;(329):1-9.
[6] Jolly PG, Tso CP, Wong YW, Ng SM. Simulation and measurement on the full-load performance of a refrigeration system in a shipping container. *Int J Refrig*. 2000;23(2):112-126. doi:10.1016/S0140-7007(99)00044-4.
[7] Fitzgerald WB, Howitt OJA, Smith IJ, Hume A. Energy use of integral refrigerated containers in maritime transportation. *Energy Policy*. 2011;39(4):1885-1896. doi:10.1016/j.enpol.2010.12.015.
[8] Shinoda T and Budiyanto MA. Energy Saving Effect of Roof Shade for Reefer Container in Marine Container Terminal *The Journal of Japan Institute of Navigation*. 2016;134: 103-113.
[9] CHAM. PHOENICS-VR Reference Guide. In: Documentat. Concentration, Heat and Momentum Limited (CHAM); 2010. http://www.cham.co.uk.
[10] Sanchez E, Rolando A, Sant R, Ayuso L. Influence of natural ventilation due to buoyancy and heat transfer in the energy efficiency of a double skin facade building. *Energy Sustain Dev*. 2016;33:139-148. doi:10.1016/j.esd.2016.02.002.
[11] CHAM. PHOENICS Numerics;Modelling Capabilities. Concentration Heat and Momentum Limited (CHAM) 2015. http://www.cham.co.uk/products/modellingcapabilities.php. Published 2015. Accessed July 19, 2016.
[12] Bird RE. A simple, solar spectral model for direct-normal and diffuse horizontal irradiance. *Sol Energy*. 1984;32(4):461-471. doi:10.1016/0038-092X(84)90260-3.
[13] ISO 668. *International Standard*. Vol 2013. Switzerland; 2013. www.iso.org.
[14] Spalding B. *Trends, Tricks, and Try-Ons in CFD/CHT*. Vol 45. 1st ed. Elsevier Inc.; 2013. doi:10.1016/B978-0-12-407819-2.00001-3.
[15] Port H. *Outline of Hakata Port International Container Terminal*.; 2015.
[16] İnan T, Başaran T, Ezan MA. Experimental and numerical investigation of natural convection in

a double skin facade. *Appl Therm Eng.* 2016;106:1225-1235. doi:10.1016/j.applthermaleng.2016.06.124.

[17] Liu H, Chen Z, Chen B, Xiao X, Wang X. Studies on the temperature distribution of steel plates with different paints under solar radiation. *Appl Therm Eng.* 2014;71(1):342-354. doi:10.1016/j.applthermaleng.2014.06.031.

[18] Rodríguez-Bermejo J, Barreiro P, Robla JI, Ruiz-García L. Thermal study of a transport container. *J Food Eng.* 2007;80(2):517-527. doi:10.1016/j.jfoodeng.2006.06.010.

[19] Çengel YA. *Heat Transfer: A Practical Approach 2nd Edition.* Mcgraw-Hill (Tx); 2nd edition (November 1, 2002); 2002. doi:10.1017/CBO9780511676420.004.

[20] Tanaka S, Takeda H, Iwata T, Takashi T, Michijin T. *Architectural Environment Engineering.* Inoue Shoin; 2006.

[21] Nuruzzaman MD. Urban Heat Island: Causes, Effects and Mitigation Measures - A Review. *International Journal of Environmental Monitoring and Analysis.* 2015;3(2):67-73

Monitoring the quality of welding based on welding current and ste analysis

Afidatusshimah Mazlan[1*], Hamdan Daniyal[1], Amir Izzani Mohamed[1], Mahadzir Ishak[2] and Amran Abdul Hadi[1]

[1]Faculty of Electrical & Electronic Engineering, Universiti Malaysia Pahang, 26600, Pekan, Pahang, Malaysia
[2]Faculty of Mechanical Engineering, Universiti Malaysia Pahang, 26600, Pekan, Pahang, Malaysia

*Corresponding author email: afidatusshimah@gmail.com

Abstract. Qualities of welding play an important part in industry especially in manufacturing field. Post-welding non-destructive test is one of the importance process to ensure the quality of welding but it is time consuming and costly. To reduce the chance of defects, online monitoring had been utilized by continuously sense some of welding parameters and predict welding quality. One of the parameters is welding current, which is rich of information but lack of study focus on extract them at signal analysis level. This paper presents the analysis of welding current using Short Time Energy (STE) signal processing to quantify the pattern of the current. GMAW set with carbon steel specimens are used in this experimental study with high-bandwidth and high sampling rate oscilloscope capturing the welding current. The results indicate welding current as signatures have high correlation with the welding process. Continue with STE analysis, the value below 5000 is declare as good welding, meanwhile the STE value more than 6000 is contained defect.

1. Introduction

Quality of welding is a critical role in most industries, especially in manufacturing and automotive [1], where strict procedures are in place to produce the best product with little to no flaws. Many innovations have been developed in the welding quality inspection. Post-welding non-destructive test is one of familiar process in term of inspection in quality of welding.There are several types of non-destructive test, for example visual inspection [2], radiographic inspection [3] and ultrasonic inspection [4]. Each type of testing have different process, but they have similar problems regarding to the time consumed and high cost. This is due to the post-process nature of the tests, where they can only be carried out after the welding were completed and the weld were cooled down.

In previous literatures, researchers have suggested another way to reduce welding flaws by continuously monitor the welding process via online monitoring system. Monitoring in welding is refer to several signals as the reference before summary the information of process. In past researches, radiography information analysis has been used to determine the defect while welding process is carried out [5]. On the other hand, sounds was used as reference to analyse and identify the quality control of welding [6]. The arc light spectrum also gives signal as reference in the analysis of welding process [7]. That can state the welding signal can use as one of source information of welding process. In the same

IOP Conf. Series: Materials Science and Engineering **257** (2017) 012043 doi:10.1088/1757-899X/257/1/012043

time, in this paper is purpose to monitoring the welding process as other way to detect the quality of welding.

One of common signal can be used as the signal is welding current. From the previous research, welding current was used as a reference signal because it is sensitive to any changes in the welding process. The welding current are rich of information and have own pattern to describe the welding process [8-9]. In other meaning, current signal can be utilized to determine the stability of welding process [10] and to evaluate the quality of welding [11]. The signal from a welding current can be referenced if any disturbance or spike happens when the process detect defects. Hence, it is a viable candidate of main variable in the online monitoring of welding process. Motor current signature analysis (MCSA) is one of successful research use current signature analysis to detect fault when motor running [12-14]. It is expected that method can be applied in the welding current signature analysis.

Raw welding current need to be analyzed before one can extract the welding process information inherently embedded in its shape. For example, welding current was previously processed using Artificial Neural Network (ANN) in welding control and quality estimation [11]. But the complex nature of ANN prohibits its application in real time online monitoring. Simple statistical analysis had been proven to successfully extract some of the information [15]. However, the research has tended to focus on macro level analysis, rather than on micro level analysis which hypothetically contain richer information. In this paper, we embark on an investigation of welding current analysis using Short Time Energy (STE) signal processing technique.

STE was originally intended to separate voice and unvoice segments of speech signal. It is effective in real time analysis due to its simple working principle [16]. Theoretically, it is based on short time energy calculation, short time magnitude calculation, autocorrelation of different segment of speech and zero crossing rate calculation [17]. Zero crossing rate (ZCR) provide the frequency information from signal voice[18]. When the zero crossing rate is less, the signal contain low frequency. Meanwhile in short time energy is especially use to frame the voice, unvoiced and silence signal[19]. The frame will transform to some of energy value. Until now, STE has always been applied to only speech processing application.

The objective of this paper is to predict the quality of welding based on welding signal. The welding current signal was interpret to new data information by STE analysis. In this paper, Short Time Energy is used to seperated the smooth current and disturbance current. The smooth current is created the good welding meanwhile the disturbance current made the defect in welding. The differentiate is presented in energy value. Based on the result, short time energy provides the value of energy in the welding process. Then the value is determined the condition on welding process. In this paper, the experiment of welding and analyse data using STE is presented.

2. Methodology

The workstation is carried out in Laboratory Welding at Faculty of Mechanical, Universiti Malaysia Pahang with room temperature. Figure 1 shows the workstation set up for experiment in this paper. The main of the experiment is welding machine and generate the data collection.

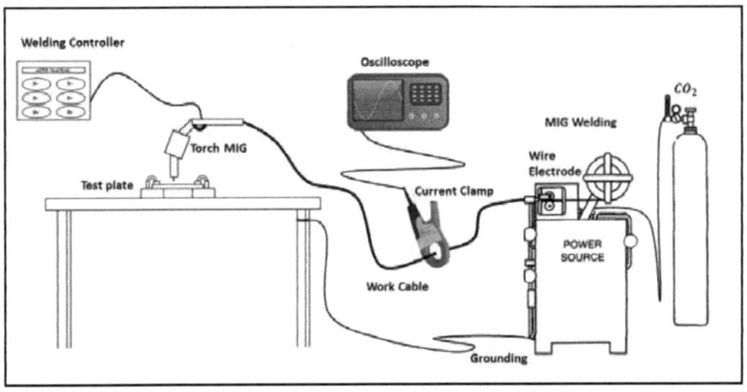

Figure 1. The workstation set up for the experiment of welding.

Present in this experiment is Gas Metal Arc Welding (GMAW) welding machine and controlled via a semi-automatic handler. Meanwhile, during the welding process, the welding current was captured on Keysight Oscilloscope DSOX3024A via Fluke i400 current probe. The semi-auto torch was 90° perpendiculars with the surface plate that was fixed to the table by clamps. The torch was fixed on a set of support with an x-axis, y-axis and z-axis adjustment. This ensures accurate control of the distance between the torch and plate.The welding parameters setting is shown in Table 1.

Table 1. List parameter in the experiment.

Parameter	Welding condition
Plate (mm)	80 × 50 carbon steel (thickness: 2.0)
Shielding gas	100% CO_2
Welding current (A)	180
Travel speed (mm/Sec)	13
Electrode type	ER70S-6 (DIA=1.2mm)

Carbon steel plate with 2.0mm thickness is used in this experiment. Each specimen uses 2 plates to weld. Carbon dioxide used as to prevent the air in welded. ER 70S-6 is an electrode type with diameter is 1.2mm. The welding power supply was operated in constant voltage mode. In this research, the current signal was captured at a very high sampling rate (200,000 samples per second) in order to preserve the current shape for further analysis.

The welding current signal was then analysed in MATLAB/Simulink using STE signal processing technique. The result of STE analysis and the welding specimen will be compared visually. The experiment repeated for three times for three different specimens with same parameter.

3. Results and discussion

After finishing the experiment, the result divide by two parts. The first part is about the welding specimen. The welding specimen was given in to Qualified Inspection Team to investigate the result of welding. In other hand, the welding current data are analysed in MATLAB/Simulink. In the MATLAB/Simulink, data are processed in Short Time Energy data.

Figure 2 shows the result of the experiment for all three specimens A the same time, from top to bottom, In the Figure 2, that consists the photos of welding specimens, the raw signals of welding current and the processed signals using STE, respectively.

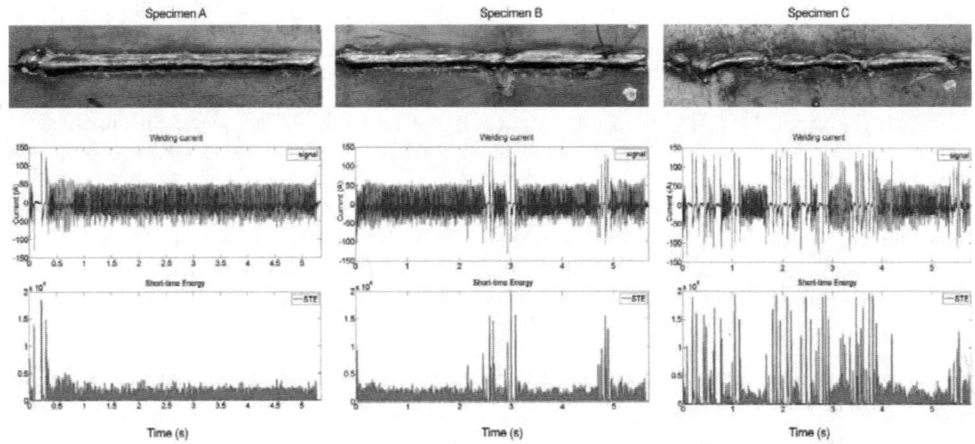

Figure 2. Welding specimens, the raw signals of welding current and the signals using STE.

From Figure 2,the specimens and the data are arranged from the best (Speciment A) on the left to the worst on the right (Specimen C). In the visual inspection report for Specimen A, on the starting welding point have the defect. It is visually apparent that the welding current "spikes" to close to 100 A. The effect of defect, cause the disturbance in welding current waveform. In the same time, the current generates maximum energy in the few second. After that, the welding is run smoothly. In this part, that is clear the current is stable and the energy in constant value. It is observed that during good welding process, the welding current waveform is in the range of ±70 A.

In the Short Time Energy (STE) signals, the contrast between good specimen and bad specimen is even clearer. In Specimen A, the welding joint looks good excepts at its starting point, where it has small defects. This defect manifests itself as a short series of spikes in welding current. Comparing this with the STE pattern, the waveform reaches more than 6000. This is obviously different from the rest of Specimen A's STE signal, where the amplitude of STE is always lower than 5000.

In visual for Specimen B, the welding contains more defects than Specimen A. The defect is captured at middle and near to end. Compare to the welding current, at the same time the current is sudden up and down. That effect on STE which is the energy is increasing in that area. Just as Specimen A, the reading of Specimen B's STE is lower than 5000 when the welding quality is good. Then, there have three defect location. Three series of spikes occurred at around 2.5 s, 3.0 s and 4.8 s, directly corresponds to the welding specimen's condition.

The worst of all three specimens is Specimen C where a lot of defects can be seen in the photo. As previously discussed, the defect can cause the disturbance on the current. That can see the unstable current waveform which define that a lot of defects is detected. At the same time, the similar pattern can be observed in the heavily spiky pattern of STE of Specimen C. In the analysis show mostly value STE is more than 6000.

It should be noted that there are several points which are the highlights of the experiment. First, the study is still an ongoing investigation where the results presented here are preliminary. Second, it is found that STE can be used to analyse welding current in the effort to differentiate between the moment of good welding joint and the moment of welding defects. In this particular experiment, welding current STE of 6000 is the threshold where any numbers below than that signifies good welding, whereas higher than that represents risks of defects. The third point is with segmented analysis such as STE, one can determine directly the moment of the defects. This is one of the great advantages of using STE instead of simple statistical analysis such as mean and variances. Simple statistical analysis evaluates the specimen as a whole rather than the process.

4. Conclusions

In this paper, the experiment is carried out to support the research. Three specimen are present with different result. The specimen from experiment is compared with welding current and STE to form the correlation between them.

As a conclusion, the welding current signal contains information of welding process, inherently embedded into its shapes and pattern. By using Short Time Energy (STE) analysis, the information of good welding and welding defects can be extracted. For further investigation, welding specimens will be processed in microstructural analysis to confirm the correlation.

Acknowledgments

The financial support provided by The Malaysian Ministry of Higher Education (MOHE) and Universiti Malaysia Pahang (UMP) is gratefully acknowledged. The authors also gratefully acknowledge the support of FRGS and internal UMP grants.

References

[1] Silva R R, Calôba L P, Siqueira M H S and Rebello JMA 2003 *American Conferences for Nondestructive Testing* 1–12.
[2] Natarajan U, Veerarajan T and Ananthan S S 2014 *Weld. J.* **93** 85–97.
[3] Kazantsev I G, Lemahieu I, Salov G I and Denys R 2002 *Signal Processing* **82** 791–801.
[4] Frank Kahmann et al 2012 *Automated ultrasonic testing of submerged-arc welded (SAW) pipes using 1 . Concepts of weld testing* 16–20.
[5] Rokhlin S I and Guu A C 1990 *Weld. J.* 83–97 1990.
[6] Saini B Y D 1998 *Weld. J.* **77** 172–179, 1998.
[7] Węglowski M S 2012 *Monitoring of Arc Welding Process Based on Arc Light Emission,* ed Radovan Kovacevic (In Tech)
[8] Song-hua J 2012 *Electronic and Mechanical Engineering and Information Technology (EMEIT), 2011 International Conference* **2** 762–766.
[9] Slania M F J and Slazak B 2014 *Archieves Metall. Mater.* **59**.
[10] Chu Y X, Hu S J, Hou W K, Wang P C and Marin S P 2004 *Weld. J.* **83** 336–343.
[11] Cho Y and Rhee S 2004 *IEEE Trans. Instrum. Meas.* **53** 330–334.
[12] Zhang J, Dhupia J S and Gajanayake C J 2014 *IEEE/ASME Int. Conf. Adv. Intell. Mechatronics* 1616-1621
[13] Lau E C C and Ngan H W 2010 *IEEE Trans. Instrum. Meas.* **59** 2683–2690.
[14] Benbouzid M E H 2000 *IEEE Trans. Ind. Electron.* 47 984–993.
[15] Murali Sachithanandam R 2014 *Development of welding signature analyzer for defect identification and quality assessment.*
[16] Yatong Z, Guizhong L and Engineering I 2002 *Signal Processing, 2002 6th International Conference* **1** 464–467.
[17] Jalil M, Butt F A and Malik A 2013 *Technological Advances in Electrical, Electronics and Computer Engineering (TAEECE), 2013 International Conference* 208–212.
[18] Nath D and Kalita S K 2014 *Business and Information Management (ICBIM), 2014 2nd International Conference* **2** 99–103.
[19] Lokhande N N, Nehe D S and Vikhe P S 2011 *International Conference in Computational Intelligence* **Iccia** 5–7.

Effect of alkaline treatment on mechanical properties of woven ramie reinforced thermoset composite

M H M Hamdan[1*], J P Siregar[1], D Bachtiar[1], M R M Rejab[1], M Samykano[1], E H Agung[2], C Tezara[3], J Jaafar[1]

[1]Faculty of Mechanical Engineering, Universiti Malaysia Pahang, 26600 Pekan, Pahang, Malaysia
[2]Mechanical Engineering Dept., Engineering Faculty, Malahayati Universit, JL.Pramuka No. 27, Kemiling, Bandar Lampung, 35153, Indonesia
[3]Department of Mechanical Engineering, Faculty of Science, Technology, Engineering and Mathematics, INTI International University 71800 Nilai, Negeri Sembilan, Malaysia;

*Corresponding author email: xhazimx8@gmail.com

Abstract. Ramie fibres are one of the strongest bast fibres with valuable potential as reinforcement in a composite. In this study, the laminated composite of ramie reinforced plain woven was prepared through hand lay-up process. Prior to moulding, the ramie reinforced woven fabric was treated by soaking the fabric in 5% NaOH solution for 1 hour at room temperature. The effects of alkaline treatment on ramie reinforced woven epoxy resin and ramie reinforced woven unsaturated polyester (UPE) resin were investigated in terms of tensile, flexural, and impact properties. Five samples of both treated ramie (TR) composite and untreated ramie (UR) composite were tested and the results were averaged for comparison with one another. Based on the results obtained, the effects of 5% (w/w) concentration of alkaline solution treatment for 1 hour at room temperature are insignificant to the tensile, flexural, and impact properties. The result of UR and TR composite was compared to the composite for pure epoxy resin and UPE resin. Based on the finding, the data obtained shown no improvement in tensile properties. However, in the case of the flexural properties are enhanced except flexural modulus of the TR-epoxy composite. Impact strength only gets better when reinforcing with the untreated ramie. On overall, the TR ramie have an admirable mechanical when reinforced with the epoxy resin than UPE resin

1. Introduction

Natural fibres are advantageous when used as a reinforcement material for the improvement of properties such as the improvement in strength, fatigues, and impact [1]. Unlike synthetic fibres, natural fibres also favourable for its characteristics such as environmentally friendly, low density, biodegradable, abundant availability, renewable, and cost effective as compared to synthetic fibres. However, several issues hindered the wide applications of natural fibres such as poor interfacial adhesion between polar hydrophilic fibre and nonpolar hydrophobic matrix and poor wetting of the fibre with the matrix polymer which resulted in a composite with weak interface [2]. The polar hydrophilic of natural fibres causes dimensional change in the composite [3].

To improve the major drawback of natural fibres as reinforcement materials, previous studies have introduced two methods of surface modification, either by physical or chemical treatment [2, 4, 5].

The main purpose of both treatments on natural fibres is to enhance the composite interface and enhance the adhesion between natural fibres. The common treatments that have been performed previously include alkaline treatment, silane treatment, sea water treatment, and hot water treatment. Among them, the alkaline treatment is frequently performed due to the low cost incurred and simplicity of the process [6]. Despite that, the alkaline treatment enhances the mechanical properties of the composites by the removing of hemicellulose and wax at the fibre surface [6]. The alkaline treatment produces a rough surface on natural fibre which consequently increases the fibre-matrix adhesion through mechanical interlocking. The alkaline treatments have been performed previously on various kind of natural fibre such as jute, kenaf, sisal, banana, hemp, banana, and ramie in the variable form of reinforcement [7-9]. The effects of treatment can either be positive or negative depending on many factors such as the type of treatments, a number of treatment agents, and the properties of natural fibres [10]. Results from the previous study mostly produced positive effects of alkaline treatment on natural fibre [5, 11, 12]. However, several adverse effects have also been reported [7, 13, 14]. The alkaline treatment of ramie woven was less discussed previously as compared to the other form of reinforcements, such as unidirectional and short directional, thus providing less information regarding this process.

Therefore, the present work focused on the effects of alkaline treatment on ramie reinforced woven epoxy resin composite. The results of tensile, flexural, and impact testing for both untreated ramie (UR) composite and treated ramie (TR) composite were compared with the data from pure/neat epoxy and polyester resin sample and are discussed in details.

2. Materials and methodology

Ramie fibre, in the form of plain woven fabric, which was purchased from Tazdiq Engineering Sdn. Bhd. Malaysia was selected as the main reinforcement material. The size of ramie fabric was 40 × 40 cm. Four plies of ramie fabric were used in each laminated composite. The matrix polymer that was selected in this study was an epoxy resin and unsaturated polyester (UPE) resin. The epoxy resin, UPE resin and hardener were purchased from Revertex (Malaysia) Sdn. Bhd. The NaOH solution was supplied from Sigma-Aldrich, United States of America.

2.1. Alkaline treatment of woven ramie

Washed. The ramie reinforced woven fabric was soaked in purified water for 6 hours to remove the impurities and dried in the oven at 60 °C for 24 hours. The simply washed fabric is referred as untreated ramie fabric (UR).

Treatment with alkaline solution. The fabric was soaked in a 5% NaOH solution for approximately 1 hour. The alkaline treatment for ramie reinforced woven fabric was performed at room temperature, followed by purification using water. Litmus paper was used to check the neutrality of the fabric. The fabric was then dried at 60°C for 48 hours. This fabric is referred as treated ramie fabric (TR).

2.2. Sample preparation of laminated composite

The composite specimens were prepared by using hand lay-up technique. At first, the realising agent was applied on the surface of the mould to easily remove the composite after curing. The epoxy resin and hardener were mixed according to the ratio of 10:1 in a beaker. The epoxy resin mixture was then stirred using a rod to ensure no air was entrapped inside the solution. Firstly, the UR fabric was placed on top of the mould and the epoxy resin mixture was poured on top of it. The hand roller and scrapper were applied to remove entrapped air bubbles. The process was repeated until the desired layering with the size of 4 was achieved. All the plies were laid up at 0° directions. After completed, the dead weight of 10 kg was applied on top of the mould and cured at room temperature for 24 hours. Upon completion, the composite was then carefully removed from the mould. The same fabrication process was repeated on another sample of UR, except by using a different resin which was UPE resin. For comparison purposes, similar fabrication process was also repeated on the sample of TR fabrics. Finally, the UR and TR laminated composite were cut by using diamond cutter into the desired dimension according to the selected standard of American Society for Testing and Materials (ASTM) based on each testing such as tensile, flexural, and impact testing.

2.3. Mechanical testing

Tensile testing. The Universal Testing Machine (UTM) model of AGI 100KN Shimadzu was used to perform the tensile testing. ASTM D-638 type I was used for preparing the sample having the dimension of 165 × 19 × 3.2 mm. The cross-head speed was set at 1 mm/min and the test was conducted at 25±20 °C with 50% humidity. Five replicated samples were used in the test to obtain a more statistical reliability

Flexural testing. The samples for flexural testing were prepared according to the ASTM D-790, having the dimension of 127 × 12.7 × 3.2 mm. The tests were carried out at an average relative humidity of 50% and at a temperature around 25±20 °C. A cross head with a constant speed of 2.5 mm/min were applied by a load on two support span with the length of 48 mm. The test was repeated for five times to acquire the average value and the standard deviation

Impact testing. The notched Charpy impact testing was performed by using the Izod impact tester. All the samples were notched and prepared according to the ASTM D-256 having the dimensions of 64 × 12.7 × 3.2 mm. The tests were performed at 25±20 °C with 50% humidity. The average of 5 tested samples was calculated for statistical reliability.

3. Result and discussion

3.1. Tensile properties

The results of tensile properties from mechanical testing are illustrated in Figure 1. Based on Figure 1, the neat epoxy resin and UPE resin have higher tensile strength. The addition of UR into epoxy resin and UPE resin did not improve the tensile strength and the tensile modulus. The tensile strength for neat epoxy resin drop from 51.37 MPa to 44.22 MPa, which about 13.9% reduction, while the neat UPE decreases from 37.26 MPa to 30.87 MPa that represents about 17.1% reduction. At the same time, the tensile strength of the neat epoxy resin reduced from 51.37 MPa to 26.56 MPa, with 48.3% reduction when reinforced with TR fabric. Furthermore, the tensile strength of the neat UPE resin also drops to 24.68 MPa with 33.7% reduction when reinforced with TR fabric. The tensile modulus also revealed a similar result as shown by the result of tensile strength. The tensile modulus of the neat epoxy resin decreased from 1.81 GPa to 1.52 GPa and 0.65 GPa when reinforced with UR and TR, respectively. Meanwhile, of the tensile modulus of the neat UPE resin drops from 1.53 GPa to 1.36 GPa and 1.09 GPa when reinforced with UR and TR, respectively.

The results obtained from tensile testing have good agreement with the previous study that performed alkaline treatment on ramie fibre. As reported by Choi et al. [4], the tensile strength and tensile modulus of the composite decreased when treated with alkaline treatment. A similar finding was reported by Goda et al. [14]which compiled the report of ramie fibre treated with 5% NaOH reinforced biodegradable polymer. It concludes that the alkaline treatment on ramie fibre causes an insignificant improvement on the tensile strength but only a slight change in the tensile modulus. In the other work, Goda et al. [13] evaluated the effects of alkaline treatment with 25% (w/w) concentrations on ramie fibre with and without applying load condition. In the case of without load applied, the tensile properties of treated ramie fibre show insignificant improvement. On the contrary, the tensile strength recorded differently in other cases. Nevertheless, both conditions recorded a reduction in the tensile modulus.

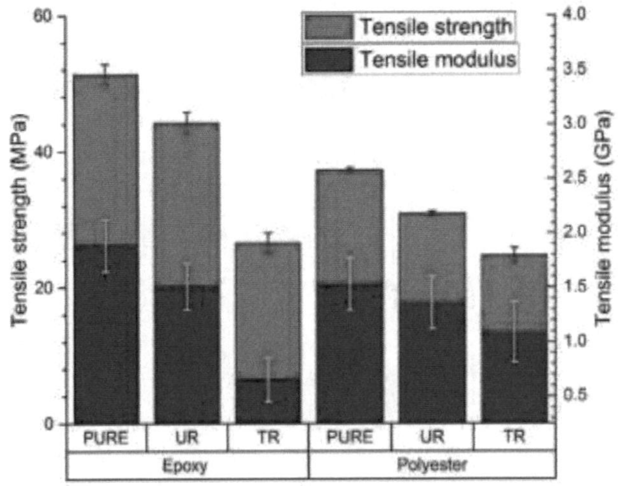

Figure 1. Tensile properties of woven ramie thermoset composite.

3.2. Flexural properties

The flexural properties of composites are shown in Figure 2. In the case of neat epoxy resin, the flexural strength increased from 45.53 MPa to 47.21 MPa and 46.05 MPa when reinforced with UR and TR, respectively. Similarly, the reinforced of UR and TR also improved the flexural modulus of neat epoxy resin from 1.15 GPa to 1.51 GPa and 1.43 GPa, respectively. As for neat UPE resin, the results obtained are different. The UR reinforced UPE improve the flexural strength of the neat UPE resin from 41.86 MPa to 44.89 MPa. However, the value for flexural strength obtained when the neat UPE resin was reinforced with TR was only 27.95 MPa. Nonetheless, both neat epoxy and neat UPE resin show significant effects on flexural modulus when reinforced with UR and TR fabrics. The UR composite is better than TR composite since it exhibited higher strength resistance. The TR composite is more brittle compared to the UR composites. Therefore, the TR composite will break first before deformation took place.

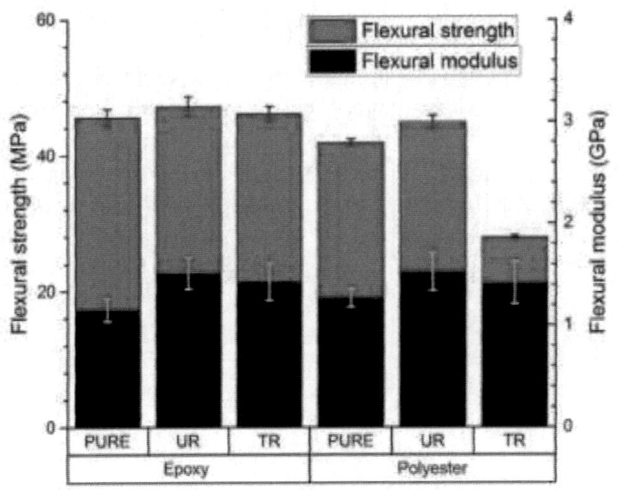

Figure 2. Flexural properties of woven ramie thermoset composite.

3.3. Impact properties
The variation of impact properties of UR and TR reinforced epoxy resin and UPE resin was presented in Figure 3. The addition of UR to the epoxy resin and UPE resin has higher influences on the impact of composites as compared to the addition of TR. The UR reinforced epoxy resin increased the impact strength of the neat epoxy resin by up to 29% and UR reinforced UPE resin enhance the impact strength of the neat UPE resin for up to 127%. However, the impact energy decreased around 32% and 40% for TR reinforced epoxy resin and TR reinforced UPE resin, respectively as compared to the neat epoxy and UPE resins. Based on the result, it can be concluded that the TR composites are more brittle compared to the UR composites. The brittle behaviour of TR composites caused by lesser fibre pull-out where the dissipation of impact energy is unable to perform. By comparing the findings with He et al. (2008)[15], the alkaline effects from the alkaline treatment of ramie fibre on the impact properties are analogous since no improvement was reported.

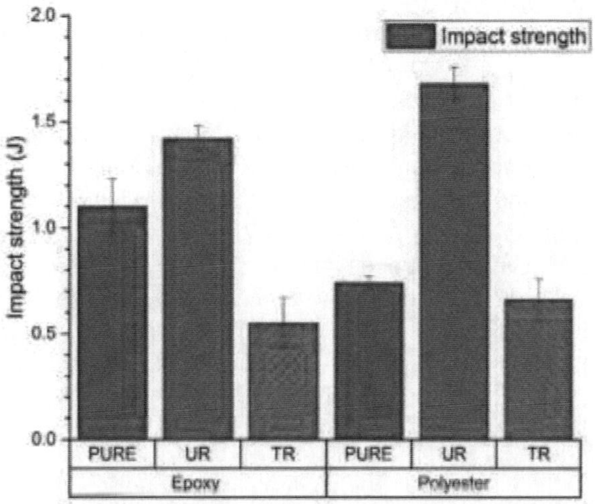

Figure 3. Impact properties of woven ramie thermoset composite.

4. Conclusions
The effects of a number of alkaline treatment on the mechanical properties of ramie reinforced woven epoxy resin and unsaturated polyester resin was explored. Based on the experimental results, the following conclusions were drawn:

- The ramie woven laminated composited treated with 5% (w/w) concentrations for 1 hour at room temperature does not improve the mechanical performance of the laminated composites for all samples. However, there are slight improvements in flexural strength and flexural modulus for the epoxy composite
- When compared to UPE composite, the epoxy composite is stiffer and brittle when reinforced with UR and TR
- The incorporation of UR composite into having significant properties than TR composite.

The scanning electron microscopic (SEM) should be performed on sample fracture to gain a better insight into the behaviour of ramie woven in the composite. Several steps should be taken into considerations to improve the development of ramie reinforced woven polymer matrix in the future. On the other hand, the immersion time and concentrations of the alkaline treatment need to be reconsidered. This is because both the duration and concentrations play a significant role in predetermining the mechanical performance of the treated woven fabric. To increase the mechanical performance of woven ramie laminated composite, it is suggested that the variable of alkali treatment such as the concentration should be defined at lower values. Through optimisation of the variable, the alkaline treatment will unleash its full potential not only to improve the interfacial adhesion strength between natural fibre and polymer matrix but at the same time eliminate the void formation and enhance the overall mechanical performance

Acknowledgments
The authors would like to be obliged to Universiti Malaysia Pahang for providing laboratory facilities and financial assistance under project no. RDU140307.

References

[1] Razera I and Frollini E 2004 *J. Appl. Polym. Sci.* **91** 1077-85
[2] John M J and Anandjiwala R D 2008 *Polym. Compos.* **29** 187-207
[3] Yu T, Jiang N and Li Y 2014 *Compos. Part A* **64** 139-46
[4] Choi H Y and Lee J S 2012 *Fibers Polym.* **13** 217-23
[5] Sever K 2010 *J. Reinf. Plast. and Comp.* **29** 1921-9
[6] Ariawan D, Salim M S, Mat Taib R, Ahmad Thirmizir M Z, Ishak M and Zainal A 2017 *Adv. Polym. Tech.*
[7] Vázquez E S R a A 2006 *FPCM8* p 5
[8] Mejía Osorio J C, Rodríguez Baracaldo R and Olaya Florez J J 2012 *Ing. Invest* **32** 83-7
[9] Hossain S I, Hasan M, Hasan M N and Hassan A 2013 *Adv. Mater. Sci. Eng.*
[10] Ticoalu A, Aravinthan T and Cardona F 2010 *Southern region engineering conference* 1-5
[11] Sudha S and Thilagavathi G 2016 *J. Text. I.* **107** 691-701
[12] Khan G M A, Terano M, Gafur M A and Alam M S 2016 *J. King Saud Univ. Sci.* 69-74
[13] Goda K, Asai T and Yamane T 2003 *Development of ramie fiber reinforced biodegradable resin matrix composites by press forming and effect of chemical treatments*
[14] Goda K, Sreekala M, Gomes A, Kaji T and Ohgi J 2006 *Compos. Part A Appl. Sci. Manuf.* **37** 2213-20
[15] He L P, Tian Y and Wang L L 2008 *Adv. Mat. Res.* 313-6

IOP Conf. Series: Materials Science and Engineering **257** (2017) 012045 doi:10.1088/1757-899X/257/1/012045

Application of response surface methodology for optimizing evacuation time in enclosed car park

Ahmad Faiz Tharima*, Md Mujibur Rahman and Mohd Zamri Yusoff

Center for Fluid Dynamics, College of Engineering, Universiti Tenaga Nasional (UNITEN), Jalan IKRAM-UNITEN, 43000 Kajang, Selangor, Malaysia

*Corresponding author email: pait.afz@gmail.com, ahmad.faiz@bomba.gov.my

Abstract. Smoke fills the car park area due to smoke back layering occurred during a fire. The presence of the beam which leads to the smoke back layering phenomena is investigated to remain smoke layer longer at the upper level with fewer occurrences of backflow. In the current study, a combination of Design of Experiment (DOE); Central Composite Design, (CCD) and statistical tools Response Surface Methodology, (RSM) were utilised to evaluate an optimal design for longer smoke residing time. The Fire Dynamic Simulator (FDS), a CFD model for the fire-driven fluid flow, was employed as a flow simulation tool. The result of six replication model produced by DOE, the error that ranged from 0.48% to 1.77% indicating that the model is reliable. It was also found that the polynomial regression result was linear with predicted R^2 of 97.64%, which was within the actual R^2 (99.45%). The effects of five control parameters such as ceiling height, beam spacing, transversal beam, extraction rate and longitudinal beam on the smoke descend time has been found to be significant. In the optimal design, the smoke remained longer at the upper level with the percentage of improvement 217.95%. The contribution of the study is the time measured in this analysis is adequate within the beam span only. Interestingly, it effects to the overall geometry with having a lengthier time of smoke to descend. The polynomial model should be used for future engineering design in an enclosed car park.

1. Introduction

A car park is constructed either at above or underground of a building. In term of fire severity, enclosed car park presents the most significant impact to the nearest property, produces more structural damage and releases larger amount of toxic gases. Due to limited air supply, the fire will be fully developed, thus generating a large amount of heat which is detrimental to the environment. In contrast, for open car park fire, the fuel-burning is localised as long as the fuel is still available. The fire might propagate to the nearest vehicles due to heat radiation or convection [1]. Smoke fills the car park area due to smoke back layering occurred during a fire. Ceiling barriers such as beam joists, miters and surface mounted light fixtures have been identified as the main components in causing ceiling jet blockage [2-3] smoke back layering, recirculation as well as stagnant smoke flow.

The effects of beam configuration on the obstruction of unconfined ceiling jet were investigated extensively by [2, 4-5]. Besides of gas temperatures, results such as smoke layer height, room pressure, oxygen concentration, flame height, smoke layer temperatures and smoke-layer depth in compartment have been reported. Johansson [6] have conducted numerical experiments to evaluate the previous correlations of ceiling jet temperature (average, maximum & maximum excess temperature) and velocities (average and maximum).

Besides that, Siang [3] has estimated the fire temperature and smoke concentration below multiple beams (19 bays and 0.54 m beam depth) by using salt-water modelling. A similar study has been conducted by Delichatsios [7]. The author has revealed the behaviours of ceiling jet flow along the unobstructed beam with no smoke leakage to the adjacent beam. However, the results were limited to square beam configuration. In contrast, Merci and Shipp [8] have devised the new research direction of car park geometry. The Fire Protection Research Foundation's Detection and Alarm Research Council has conducted a research to determine the performance of smoke detection between different ceiling structures (waffle construction) such as smooth ceiling, ceiling with deep beam and deep beam pocket [9]. Another full-scale experiment and CFD simulation in a large car park have been carried out by [10-11] in order to investigate the significances of heat release rate, extraction flow rate, opening of incoming air and transversal beam (beam which perpendicular to the longitudinal beam). Apart from affecting the smoke propagation and the flow pattern, the use of ceiling beam could affect the sprinkler activation time as well. Accordingly, the sprinkler activation times in compartments employing beamed, sloped and sloped-beamed ceilings have been measured by [12]. Apart from that, the channelling effect has been reviewed separately by [13-15]. Moreover, sidewall fire experiment was carried out by [16] in order to investigate the effect of channel width on the burning rate and the ceiling temperature distribution.

In order to ensure proper smoke propagation, mechanical and extraction fans should be used. Many studies related to mechanical fan efficiency have been carried out. Santaso et al. [17] have performed a simulation to investigate the efficiency of horizontal ventilation system. The effects of CO contribution and additional building structures have been discussed. [18-19] have utilized ventilation fans such as extraction and jet fans and investigated the temperature profiles by using the buoyant plume. A very similar smoke control approach has been adopted in an underground car park separately by [20-21]. Their studies focused on parameters such as number of jet fan, jet fan velocity, extraction rate and fire location. Similarly, Meroney et al. [22] redesigned a ventilation system in an underground military firing range tunnel due to unwanted separation, reverse flow, stagnate, dead zone and recirculation at specific point spotted in the previous design.

To the best of our knowledge, studies related to the smoke control in presence of the beam are rather limited. Correspondingly most of researchers were not investigate on how ceiling jet or smoke flow by the presence of beam allow smoke remain longer at the upper level without or fewer occurrences of backflow. For this purpose, the actual underground car park of Simulator Building at Fire and Rescue Academy of Malaysia has been chosen to examine the current problem.

2. Method

The computational fluid dynamics (CFD) simulation was performed by using FDS software, which is specialized software in modelling fire-driven fluid flow. Flow turbulence was modelled via Large Eddy Simulation. The wind effects were not taken into consideration. Figure 1 and Table 1 show the boundary conditions and the numerical settings, respectively.

Table 1. Numerical setting for the simulation.

Parameter	Description of Car Park
Geometry dimension	4m x 1.6m x 0.3m
Mesh size	0.0094m
HRRPUA	2842.7kW/m^2
Fuel	Propane (C_3H_8)
CO yield	0.005
Soot yield	0.024
Fire source area	0.11684m x 0.0762m

2.1. Boundary conditions

- The ceiling, floor and side walls were prescribed as inert boundaries.

- The surrounding environment was not modeled. The ambient temperature was simply prescribed as 28.95°C.
- The wind effects were not considered.
- Two longitudinal beams placed at the center of car park were supported by transversal beams and columns of different sizes.
- The smoke extraction rate was specified at the downstream opening which was positioned below the transversal beam depth.
- The size of fuel source area was 0.11684m x 0.0762m with height of 0.02667 m.

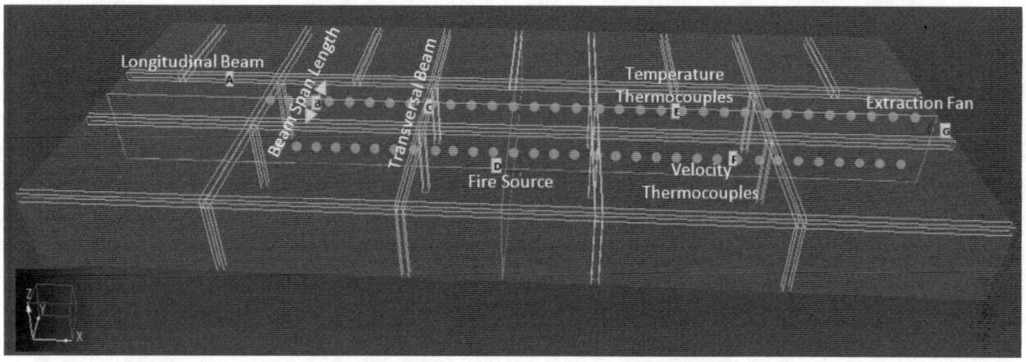

Figure 1. Boundary conditions.

The mass (1) momentum (2) and energy (3) conservation equations can be written as:

$$\frac{\partial \rho}{\partial t} + \nabla.\rho \mathrm{u} = \dot{\mathrm{m}}_{\mathrm{b}}''' \quad (1)$$

$$\frac{\partial}{\partial t}(\rho \mathrm{u}) + \nabla \cdot \rho \mathrm{uu} + \nabla p = \rho g + \mathrm{f_b} + \nabla \cdot \tau_{\mathrm{ij}} \quad (2)$$

$$\frac{\partial}{\partial t}(\rho \mathrm{h}) + \nabla \cdot \rho \mathrm{hu} = \frac{\mathrm{Dp}}{\mathrm{Dt}} + \dot{\mathrm{q}}''' - \dot{\mathrm{q}}_{\mathrm{b}}''' - \nabla \cdot \mathrm{q}''' \quad (3)$$

In these equations, ρ is density, *t* is time, **u** is velocity vector, $\dot{m}_b'''$ is net heat flux from thermal conduction and radiation, *p* is pressure, ***g*** is gravity vector, $\mathbf{f}_b$ is external force vector, τ_{ij} is viscous stress tensor, $\dot{q}'''$ is heat release rate per unit volume from a chemical reaction, $\dot{q}_b'''$ is energy transferred to the evaporating droplets, $\dot{\mathbf{q}}''$ is conductive and radiative heat flux and ε is dissipation rate.

2.2. Statistical Process

The research procedures involved are:

- obtaining the important controllable factors
- performing the Design of Experiment (DOE)
- performing CFD Simulation
- conducting the reliability test of DOE
- performing RSM and analysis of variance (ANOVA)
- Optimization

2.3. Factors Identification

In the current work, the key factors that influence the smoke descends time are ceiling height, beam span length, transversal beam depth, longitudinal beam depth and extraction fan rate. The constraints of these factors were reported in Table 2.

Table 2. Factor and response parameters.

Parameters	Name	Coded Factor	Lower	Upper
Factors	Ceiling Height (m)	X	0.3	0.442
	Beam Span Length (m)	X_1	0.213	0.57
	Transversal Beam Depth (m)	X_2	0.02	0.06
	Extraction Fan Rate (m^3/s)	X_3	0.18	0.31
	Longitudinal Beam Depth	X_4	0.02	0.061
Response	Smoke descend time (s)	Y		

2.4. Design of Experiment and Statistical Analysis

The selected DOE was Central Composite Design (CCD) because it incorporates better design points. Correspondingly, the Face Central Design was employed to obtain the 32 design points. In order to maintain a hierarchical model at each step, terms were added during the process by using the stepwise procedure. The design points and their corresponding results are reported in Table 3.

Table 3. Numerical simulation design and results

Run	Factors					Time
	X	X_1	X_2	X_3	X_4	Y
1	0.442	0.213	0.02	0.31	0.061	25.00
2	0.371	0.3915	0.04	0.245	0.02	15.30
3	0.3	0.57	0.06	0.18	0.061	9.48
4	0.3	0.57	0.06	0.31	0.02	9.27
5	0.3	0.213	0.02	0.31	0.02	10.64
6	0.371	0.3915	0.04	0.245	0.0405	12.36
7	0.371	0.3915	0.04	0.245	0.0405	12.36
8	0.3	0.57	0.02	0.18	0.02	10.52
9	0.442	0.213	0.06	0.18	0.061	22.28
10	0.371	0.3915	0.04	0.18	0.0405	12.16
11	0.371	0.213	0.04	0.245	0.0405	12.08
12	0.371	0.3915	0.04	0.245	0.061	13.40
13	0.442	0.57	0.06	0.18	0.02	25.84
14	0.3	0.3915	0.04	0.245	0.0405	7.48
15	0.371	0.3915	0.04	0.245	0.0405	12.36
16	0.371	0.3915	0.04	0.31	0.0405	13.28
17	0.442	0.213	0.02	0.18	0.02	27.64
18	0.371	0.57	0.04	0.245	0.0405	13.50
19	0.371	0.3915	0.02	0.245	0.0405	12.00
20	0.3	0.213	0.06	0.31	0.061	8.70
21	0.3	0.213	0.02	0.18	0.061	7.25
22	0.371	0.3915	0.04	0.245	0.0405	12.60
23	0.371	0.3915	0.04	0.245	0.0405	12.63
24	0.442	0.213	0.06	0.31	0.02	29.70
25	0.442	0.57	0.06	0.31	0.061	25.14
26	0.3	0.213	0.06	0.18	0.02	6.68
27	0.442	0.57	0.02	0.31	0.02	28.90
28	0.371	0.3915	0.04	0.245	0.0405	12.20
29	0.442	0.57	0.02	0.18	0.061	26.10
30	0.371	0.3915	0.06	0.245	0.0405	13.55
31	0.3	0.57	0.02	0.31	0.061	11.40
32	0.442	0.3915	0.04	0.245	0.0405	24.05

2.5. Froude Scaling

The car park model was scaled down 10 times from the actual model. As shown in Table 4, other parameters such as fire size, time, extraction flow rate, velocity, extraction fan size, fuel mass and fire source tray were scaled in accordance with the Froude scaling correlation [23-25]. However, the ceiling height was set to 3.0m in order to ensure that the flow is turbulent [26].

Table 4. List of scaling correlation for enclosed car park model. [7, 11]

Type of unit	Scaling Model
Geometric position	$X_m = X_f\left(\frac{l_m}{l_f}\right)$
Heat Release Rate, Q (kW)	$Q_m = Q_f\left(\frac{l_m}{l_f}\right)^{5/2}$
Time, *t* (s)	$t_m = t_f\left(\frac{l_m}{l_f}\right)^{1/2}$
Volume Flow (m^3/s)	$V_m = V_f\left(\frac{l_m}{l_f}\right)^{5/2}$
Velocity, u (m/s)	$u_m = u_f\left(\frac{l_m}{l_f}\right)^{1/2}$
Temperature (K)	$T_m = T_f$

3. Results and Discussion

3.1. Mesh Independence Test and Geometry Validation

The results of mesh independence test are presented in this Section. Three models employing different grid sizes such as 3.57 cm, 1.47 cm, and 0.94 cm were simulated. As reported in Table 5, the simulated maximum heat release rate is coming closer to the experimental value as the grid is refined. As observed, the error was reduced from 19% (coarse grid) to merely 4.33% (fine grid). The time histories of temperature are shown in Figure 2, showing the effectiveness of using fine grid in this model. Therefore, fine grid was employed in the subsequent simulations.

Table 5. Relative error.

Mesh	Mesh size (cm)	Number of cells	Time step	Total Time (hour)	Maximum temperature		Relative error
					Experiment	FDS	
Coarse	3.57	32,928	51766	4.4		166.33	19.01%
Moderate	1.47	471,648	166,467	107.85	205.38	189.95	7.51%
Fine	0.94	1,797,760	275,700	403.7		196.48	4.33%

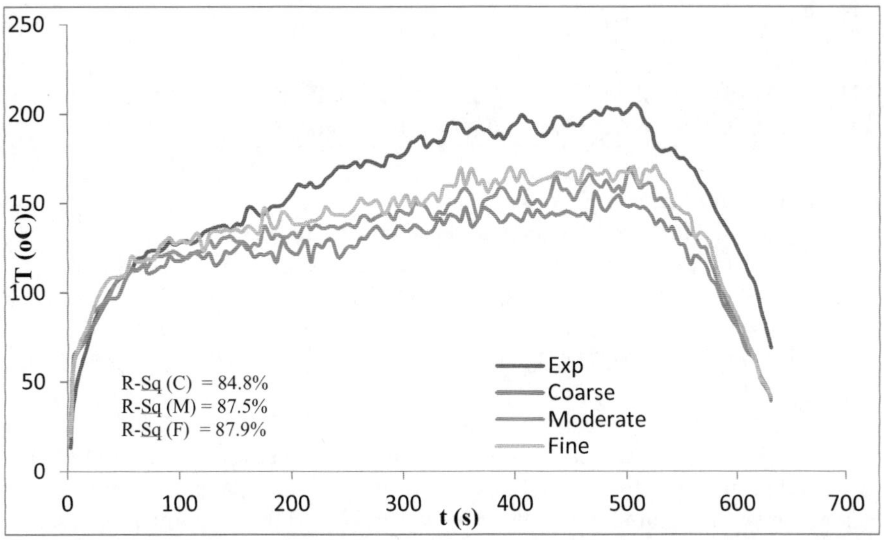

Figure 2. Temperatures at thermocouple tree A with 0.01 below the ceiling.

Apart from conducting the mesh independence test, geometric validation of the FDS model was also performed by comparing the results of an enclosed car park and a corridor-like structure as experimented by [27]. From Figure 3, it seems that HRRs of car park and corridor-like structure are comparable.

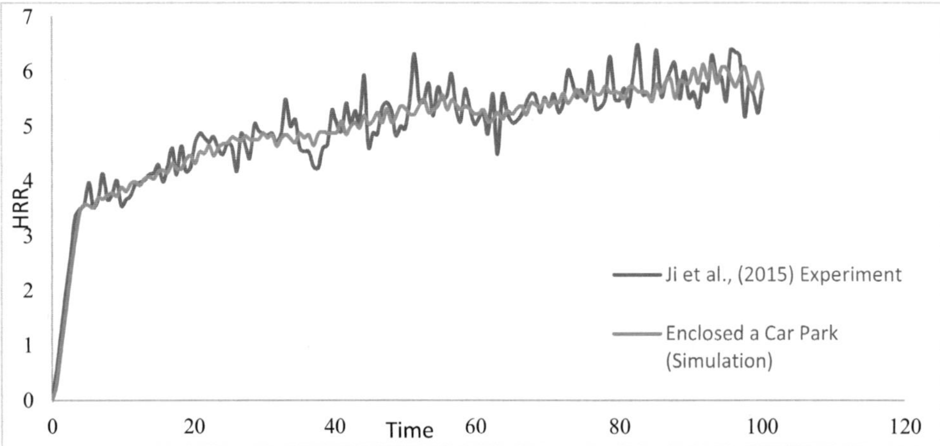

Figure 3. Validation between experiment and CFD simulation.

3.2. Reliability of Design of Experiment

The reliability test of the DOE model was performed using the replication procedure which involves six models. As shown in Table 6, the results are ranging from 0.48% to 1.77%, indicating that the experimental design for this model is reliable.

Table 6. Replication of an enclosed car park model.

Factors					Response	
Ceiling Height	Beam Span Length	Transversal Beam Depth	Extraction Rate	Longitudinal Beam Depth	Smoke Descend Time	Error
0.371	0.3915	0.04	0.245	0.0405	12.36	0.48
0.371	0.3915	0.04	0.245	0.0405	12.36	0.48
0.371	0.3915	0.04	0.245	0.0405	12.36	0.48
0.371	0.3915	0.04	0.245	0.0405	12.60	1.45
0.371	0.3915	0.04	0.245	0.0405	12.63	1.69
0.371	0.3915	0.04	0.245	0.0405	12.2	1.77

3.3. Polynomial Regression Model

Based on the CCD model, a full quadratic equation can be written as:

$$\begin{aligned} y = {} & 1.1117 - 0.27X + 0.2892X_1 - 1.372X_2 + 0.558X_3 - 3.300X_4 + 2.26X^2 + 41.74X4^2 \\ & - 0.477XX_1 + 2.108XX_2 - 0.693\,XX_3 - 2.980XX_4 - 0.475X_1X_3 \\ & + 1.093X_1X_4 + 8.72X_2X_4 \end{aligned} \quad (4)$$

From the statistical analysis, the result was linear with predicted R^2 of 97.64%, which was within the actual R^2 of 99.45%. Therefore, good agreement can be seen between actual and predicted values. Based on the P-value obtained from ANOVA, the degree of importance for main, interaction and quadratic factors can be determined (i.e. highly significant if $P < 0.001$; significant if $0.001 < P < 0.05$; insignificant if $P > 0.05$). As shown in Table 7, the P-value is less than 0.001 (with F-value 221.57), indicating that the model is highly significant and reliable. The coded coefficients for linear, interaction and quadratic variables are shown in Table 8. As seen, most variables are either highly significant or significant.

Table 7. Analysis of variance for response surface quadratic model.

Source	DF	Adj SS	Adj MS	F-Value	P-Value
Model	14	0.099424	0.007102	221.57	0.000
Linear	5	0.091557	0.018311	571.30	0.000
Height	1	0.088743	0.088743	2768.71	0.000
Span	1	0.000934	0.000934	29.15	0.000
Tranv	1	0.000404	0.000404	12.59	0.002
Extrac	1	0.001013	0.001013	31.62	0.000
Longi	1	0.000463	0.000463	14.44	0.001
Square	2	0.005729	0.002864	89.37	0.000
Height*Height	1	0.000440	0.000440	13.72	0.002
Longi*Longi	1	0.001075	0.001075	33.53	0.000
2-Way Interaction	7	0.002138	0.000305	9.53	0.000
Height*Span	1	0.000584	0.000584	18.21	0.001
Height*Tranv	1	0.000143	0.000143	4.47	0.050
Height*Extrac	1	0.000164	0.000164	5.11	0.037
Height*Longi	1	0.000301	0.000301	9.39	0.007
Span*Extrac	1	0.000485	0.000485	15.14	0.001
Span*Longi	1	0.000256	0.000256	7.99	0.012
Tranv*Longi	1	0.000204	0.000204	6.38	0.022
Error	17	0.000545	0.000032		
Lack-of-Fit	12	0.000530	0.000044	15.21	0.004
Pure Error	5	0.000015	0.000003		
Total	31	0.099969			

Table 8. Coded Coefficients for Transformed Response.

Coded Factor	Effect	Coef	SE Coef	T-Value	P-Value	Degree of Importance
X	0.14043	0.07022	0.00133	52.62	0.000	High Significant
X_1	0.01441	0.00720	0.00133	5.40	0.000	High Significant
X_2	-0.00947	-0.00474	0.00133	-3.55	0.002	High Significant
X_3	0.01501	0.00750	0.00133	5.62	0.000	High Significant
X_4	-0.01014	-0.00507	0.00133	-3.80	0.001	High Significant
X^2	0.02245	0.01122	0.00303	3.70	0.002	High Significant
X_4^2	0.03509	0.01754	0.00303	5.79	0.000	High Significant
XX_1	-0.01208	-0.00604	0.00142	-4.27	0.001	High Significant
XX_2	0.00599	0.00299	0.00142	2.11	0.050	Significant
XX_3	-0.00640	-0.00320	0.00142	-2.26	0.037	Significant
XX_4	-0.00867	-0.00434	0.00142	-3.06	0.007	High Significant
X_1X_3	-0.01101	-0.00551	0.00142	-3.89	0.001	High Significant
X_1X_4	0.00800	0.00400	0.00142	2.83	0.012	High Significant
X_2X_4	0.00715	0.00357	0.00142	2.53	0.022	High Significant

From the P-values shown in Table 8, Equation 4 should be used to relate the smoke descend time and the design parameters. Figure 4 compares the CFD results and those predicted by using Equation 4. The agreement is promising.

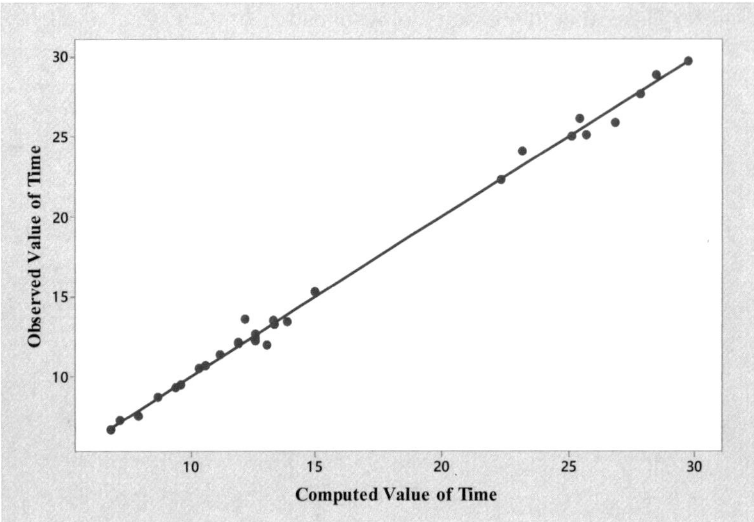

Figure 4. Scatterplot for time of smoke descending.

3.4 Optimization

The most critical things during a fire are evacuation process. The longer time for the smoke to descend, an occupant in the building can evacuate safely. To optimize, the possible factors from the literature review [2, 4-5, 7, 10, 13-14, 28-29] have been picked up and arranged as max-min parameters in Table 9. The objective function of the study is to maximize the smoke residing time at upper level for beam span that ranges between 0.4m and 0.57m. To get an ideal space, this range is applied in accordance to the preservation of Reynolds number to support the turbulent flow rule [26] With these input data, the statistical analysis has proposed an optimal design. The optimal design then simulated via FDS using the same value proposed by statistical analysis.

Table 9. Comparison between actual and optimal design.

Factors	Actual Design	Max – Min Parameters		Optimal Design	
		Lower	Upper	Statistical	FDS
Ceiling Height (m)	0.3	0.3	0.442	0.442	0.442
Beam Span (m)	0.57	0.213	0.57	0.4	0.4
Transvers (m)	0.04	0.02	0.06	0.02	0.02
Extraction (m^3/s)	0.18	0.18	0.31	0.31	0.31
Longitudinal (m)	0.04	0.02	0.061	0.02	0.02
Respond					
Smoke Descend Time (s)	9.75			30.12	31.00
Error (%)				2.93	
Improvement (%)				217.95	

Table 9 shows the design parameters of two geometrical car park designs (before and after optimization). In the optimized design, the smoke will remain longer at the upper level, which is favorable. The time predicted by Equation (4) agrees considerably well with the simulation result. It is interesting to note that the percentage of improvement is 217.95% (Figure 5). The contribution of the study is the time measured in this analysis is adequate within the beam span only. Interestingly, it effects to the overall geometry with having a lengthier time of smoke to descend as illustrated in Figure 6.

According to the parameters between actual design and optimal design as shown in Table 9, by increasing the ceiling height and extraction rate, the smoke descend time is increased. However, the effects of beam depth and beam span length on smoke descend time are not significant. It can be seen on the effect both beam depth and beam span length seems to be similar between an actual and optimal design.

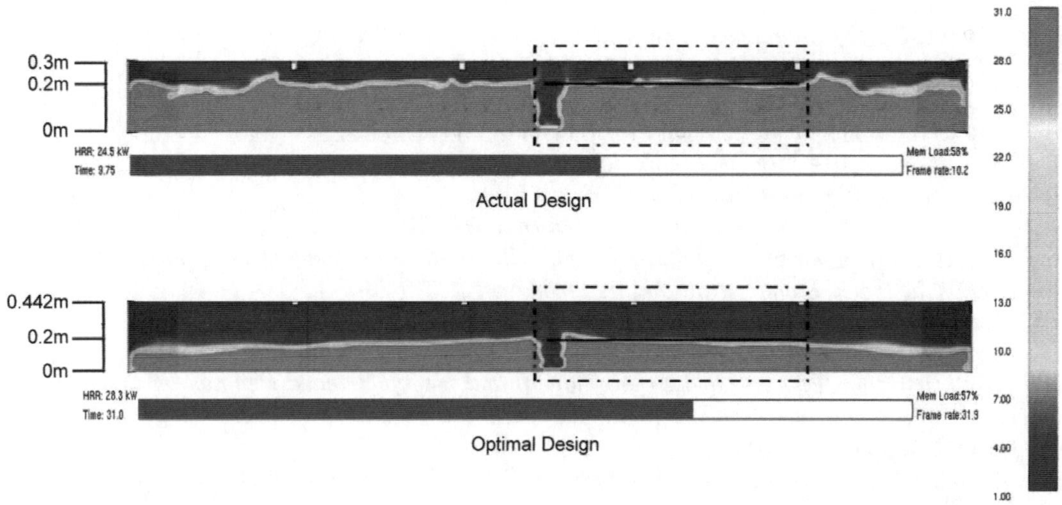

Figure 5. The different of smoke layer between actual and optimal design for enclosed car park.

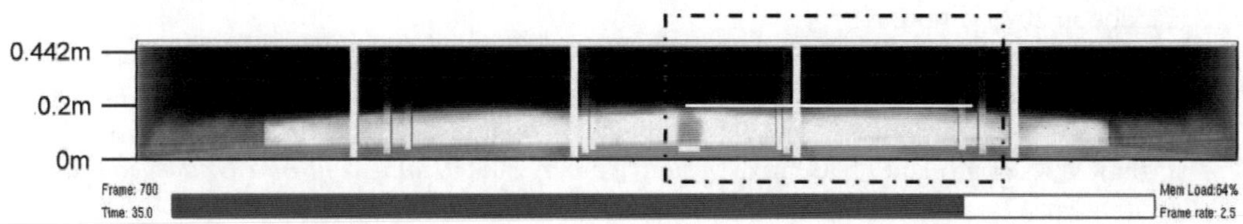

Figure 6. Overall smoke layer for optimal design of enclosed car park.

4. Conclusions

In this research, the effects of significant factors such as ceiling height, beam span length, transversal and longitudinal beam depths as well as extraction rate on the smoke descend time have been analysed by using Response Surface Methodology (RSM). The commercial software, i.e. Fire Dynamic Simulator (FDS) has been used for flow analysis. From the reliability test, it seems that RSM is able to generate a reliable model. The polynomial equation should be used to relate the smoke descend time with the design factors. In addition, these five significant factors have been optimized in order to maximize the smoke descend time, which yields 217.95% improvement as compared to the previous design. According to optimization model, the effects of beam depth and beam span length on smoke descend time are not significant but increasing the ceiling height and extraction rate was proved to be significant to the increment of smoke descend time. Therefore, engineers should focus on these five significant factors in the design stage. The research of treating temperature and smoke back layering distance as other response variables is currently underway.

Acknowledgements
This study work is supported by the scholarship provided by Malaysia Public Service Department to the first author.

References

[1] Zahirasri M 2015 *Multiple Vehicle Design Fire Scenarios in Car Parking Buildings* (University of Canterbury)
[2] Koslowski C and Motevalli V 1994 *Fire Safety Science* **4** 469-480.
[3] Siang C C 2010 *Characterizing Smoke Dispersion along Beamed Ceilings Using Salt-Water Modeling* (University of Maryland, College Park).
[4] Koslowski C C and Motevalli V 1993 *Journal of Fire Protection Engineering* **5** 97–111.
[5] Motevalli V and Yuan Z P 2008 *Fire Technology* **44** 97–112.
[6] Johansson N 2014 *Fire Science Reviews* **3** 2.
[7] Delichatsios 1981 *Combustion and Flame* **43** 1-10.
[8] Merci B and Shipp M 2013 *Fire Safety Journal* **57** 3–10.
[9] Miller C 2010 *Emergency Management: EIS International Essential Information Software Helps the World Plan for Natural and Environmental Crises.*
[10] Deckers X, Haga S and Tilley M 2013 *Fire Safety Journal* **57** 22–34.
[11] Deckers X, Haga, S, Tilley N and Merci B 2013 *Fire Safety Journal* **57** 22–34.
[12] Vettori R 2003 *Effect of a Beamed, Sloped, and Sloped Beamed Ceilings on the Activation Time of a Residential Sprinkler* (National Institute of Standards and Technology, Building and Fire Research Laboratory).
[13] Hinkley 1995 *Smoke & Heat Venting.* SFPE Handbook of Fire Protection Engineering 2nd Edition.
[14] Klote J H and Milke J A 2002 *Principles of smoke management.*
[15] Milke J A and Klote J A 2008 *Smoke Movement in Buildings* Fire Protection Fire Protection Handbook 20th Edition.
[16] Ji J, Fu Y Y, Fan C G, Gao Z H and Li K Y 2015 *Int. J. Heat Mass Transfer* **84** 562–570.
[17] Santoso M, Bey M A, Sulistyo Z and Nugroho Y S 2015 *Applied Mechanics and Materials* **758** 143–151.
[18] Hazbei M, Nematollahi O, Behnia M and Adib Z 2015 *Tunnelling and Underground Space Technology* **47**.
[19] Hu L H, Chen L F and Tang W 2014 *Int. J. Heat Mass Transfer* **79** 885–892.
[20] Lu S, Wang Y H, Zhang R F and Zhang H P 2011 *Procedia Engineering* **11** 369–378.
[21] Hu L H, Chen L F and Tang W 2014 *Int. J. Heat Mass Transfer* **79** 885–892.
[22] Meroney R N, Hill D W, Derickson R, Stroup J, Weber K and Garrett P 2015 *Journal of Wind Engineering and Industrial Aerodynamics* **136** 12–22.
[23] Ingason H and Li Y Z 2010 *Fire Safety Journal* **45** 371-384.
[24] Quintiere J G and Dillon M E 2008 *Progress in Scale Modeling* Scale Model Reconstruction *of* Fire in an Atrium p. 109–120.
[25] Tilley N, Rauwoens P, Fauconnier D and Merci B 2013 *Fire Safety Journal* **59** 160–165.
[26] Klote J H 2012 *Scaling Modeling* Handbook of Smoke Control Engineering (American Society of Heating Refrigerating and Air-Conditioning Engineers).
[27] Ji J, Fu Y Y, Fan C G, Gao Z H and Li K Y 2015 *Int. J. Heat Mass Transfer* **84** 562–570.
[28] Deckers X, Haga S, Sette B and Merci B 2013 *Fire Safety Journal* **57** 11–21.
[29] Klote J H 1984 *Fire Safety Journal* **7** 93–98.

The behavior of Aluminium Carbon/Epoxy fibre metal laminate under quasi-static loading

N K Romli[1*], M R M Rejab[1], D Bachtiar[1], J Siregar[1], M F Rani[2], W S W Harun[1], Salwani Mohd Salleh[1] and M N M Merzuki[1]

[1]Structural Materials & Degradation Focus Group, Faculty of Mechanical Engineering, Universiti Malaysia Pahang, 26600 Pekan, Pahang, Malaysia
[2]Faculty of Engineering, DRB-Hicom University of Automotive Malaysia, 26607 Pekan, Pahang, Malaysia

[*]Corresponding author email: khaleedaromli@gmail.com

Abstract. One of major concerns that related to the flight safety is impact of birds. To minimize the risks, there is need to increase the impact resistance of aircraft by developing a new material and has the good structural design of aircraft structures. The hybrid laminates are potential candidate material to be applied for the aircraft structures susceptible to bird strikes. The fibre metal laminate was fabricated by a compression moulding technique. The carbon fibre and aluminium alloy 2024-0 was laminated by using thermoset epoxy. A compression moulding technique was used for the FML fabrication. The aluminium sheet metal has been roughening by a metal sanding method which to improve the bonding between the fibre and metal layer. The main objective of this paper is to determine the failure response of the laminate under five variations of the crosshead displacement in the quasi-static loading. The FML was modelled and analysed by using Explicit solver. Based on the experimental data of the quasi-static test, the result of 1 mm/min was 11.85 kN and higher than 5, 10, 50 and 100 mm/min which because of the aluminium ductility during the impact loading response. The numerical simulations were generally in good agreement with the experimental measurements.

1. Introduction

A fibre metal laminate (FML) or also known as a hybrid laminate contains metal layers and composite materials [1-5]. The fibre metal laminate can be applied to many applications because of the fibre metal laminate can withstand to the impacts, loadings and harsh environments for a long period. The combination of the metal layers and composite fibres produce superior characteristics such as fatigue and fracture. Besides that, the substitutions of both materials are to reduce the weight of the aircraft structures and saving in oil consumption [5-11]. The Glare and Arall are members of fibre metal laminates. The fibre metal laminate started and applied in military application after the Second World War [12-23]. From the positive feedbacks of the application, the fibre metal laminate began the application widely in aircraft structures such as fuselage and wings [24, 25]. The example of the application is in the Airbus A380 due to the improvement in fatigue, resistance to corrosion and impact and weight reduced nearly 794 kg [26-29]. The other organisations such as NASA, Bombardier and EMBRAER interested with the substitution of the aluminium in a fibre metal laminate. The high strength and stiffness are a good combination and resulted in great performance in space applications [12].

An autoclave commonly used in manufacturing o fibre metal laminates or curing process at a high pressure. The curing temperature should be maintained as well. The FMLs need to be processed up to 120°C as to avoid damage on the aluminium alloy 2024-T3. The temperature will allow an excess of resin flow out and reduced its viscosity. The pressure will press and consolidate the plies then suppress the voids [23, 30-36]. The shear thickening fluid is a combination of hard metal oxide particles which suspended in the liquid polymer. The STF can lead the laminates to improve resistance from ballistic impact damage [37].

The debonding process between a sheet metal layer and composite fibre layer, delamination of the composite layers, fracture due to a plasticity behaviour of a metal are commonly happened to a fibre metal laminate structure which because of the impact of a load and depended on the layup configurations [6, 10, 11, 38-40]. The process of curing by using an autoclave will take long hours of processing and required intensive of labour cost for the whole production [9, 41].

The purpose of this paper to investigate the failure behaviour through an experiment of quasi-static impact test and predict the failure behaviour by using Abaqus/Explicit.

2. Experimental work

The behaviour of a fibre metal laminate was studied based on the aluminium alloy 2024-0 sheets and unidirectional carbon fibre prepreg. The prepreg fibre was bonded with the two sheet metal layers. The surface of the aluminium alloy was treated through a metal sanding. The grit size of sand paper is 100. The purpose of the metal sanding is for surface roughness. The surface roughness will increase the bonding strength between the sheet metal and glass fibre. The layup configuration of the FML is 2/1. The size and thickness of the FML are 150 mm x 150 mm and 2.15 – 2.52 mm. The FML was fabricated by a compression moulding technique. The glass fibre and aluminium alloy sheet metal alloy laminated by using the thermoset epoxy. A pressure was applied by placing a moderated load on it and cured in at room temperature for 24 hours. The shape and size of the indenter are hemisphere and 12.7 mm. The indenter fabricated through a machining process which is a CNC machine. The coolant was applied during the machining process as to reduce the cutting force temperature of the machining process.

Figure 1. The setup for quasi-static test.

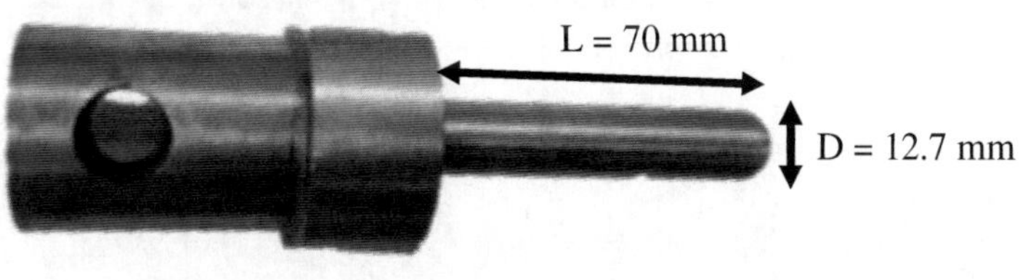

Figure 2. Hemispherical indenter.

3. Computational model

The alternative way to predict the experimental result is by using numerical analysis. Study of numerical analysis on fibre metal laminate became crucial as to know impact behaviour of it. The process of simulation could be done by adding the properties material used such as Young's modulus, Poisson's ratio, the density of the materials and so on. The simulation process required reasonable numbers of structural tests for a validation process. The previous researchers have combined two theories of failure criteria and plasticity as to simulate the degradation behaviour of the plate and study the behaviour of the nonlinear numerical model. The damage of an impact was controlled by degradation of the stiffness and fracture was because of energy releases which aimed to delamination growth. The simulation of deformation process of 3D FE model involved sensitivity of strain-rate and large changes in transverse compression. The damaged of composite plies cannot be used directly in Abaqus [42].

The Abaqus/Explicit was used to develop a numerical simulation of the FML under the low-velocity impact. The FML consisted with two different of constituent materials; 2024-0 and unidirectional prepreg carbon fibre. The failure of the FML depended on the ductility of the materials.

3.1. Modelling of the 2024-0 by using Isotropic

The aluminium alloy 2024-0 was modelled by using an isotropic elastoplasticity to predict elastic and plastic behaviour either as a rate-dependent or rate dependent model and it has a simple form.

3.1.1 Elasticity

A model of isotropic linear elasticity was generated for the elastic respond for a material in a numerical analysis. The material that performed the linear elastic behavior, the total stress defined through this an equation:

$$\sigma = D^{el}\varepsilon^{el} \tag{1}$$

Where σ the total stress, D^{el} is the fourth order of an elasticity tensor and ε^{el} is the total elastic strain. From the Equation (1) the stress-strain relationship of isotropic linear elasticity strain:

$$\begin{bmatrix}\sigma_{11}\\ \sigma_{22}\\ \sigma_{33}\\ \sigma_{12}\\ \sigma_{13}\\ \sigma_{23}\end{bmatrix} = \frac{E}{(1+\upsilon)(1-2\upsilon)}\begin{bmatrix}1-\upsilon & \upsilon & \upsilon & 0 & 0 & 0\\ \upsilon & 1-\upsilon & \upsilon & 0 & 0 & 0\\ \upsilon & \upsilon & 1-\upsilon & 0 & 0 & 0\\ 0 & 0 & 0 & \frac{(1-2\upsilon)}{2} & 0 & 0\\ 0 & 0 & 0 & 0 & \frac{(1-2\upsilon)}{2} & 0\\ 0 & 0 & 0 & 0 & 0 & \frac{(1-2\upsilon)}{2}\end{bmatrix}\begin{bmatrix}\varepsilon_{11}\\ \varepsilon_{22}\\ \varepsilon_{33}\\ \varepsilon_{12}\\ \varepsilon_{13}\\ \varepsilon_{23}\end{bmatrix} \tag{2}$$

The symbol of E is a Young's modulus and υ is a Poisson's ration. These have been defined in the equation above. The inverse relationship defined as an equation below:

$$\begin{bmatrix}\varepsilon_{11}\\ \varepsilon_{22}\\ \varepsilon_{33}\\ \gamma_{12}\\ \gamma_{13}\\ \gamma_{23}\end{bmatrix} = \begin{bmatrix}\frac{1}{E} & \frac{-\upsilon}{E} & \frac{-\upsilon}{E} & 0 & 0 & 0\\ \frac{-\upsilon}{E} & \frac{1}{E} & \frac{-\upsilon}{E} & 0 & 0 & 0\\ \frac{-\upsilon}{E} & \frac{-\upsilon}{E} & \frac{1}{E} & 0 & 0 & 0\\ 0 & 0 & 0 & \frac{1}{G} & 0 & 0\\ 0 & 0 & 0 & 0 & \frac{1}{G} & 0\\ 0 & 0 & 0 & 0 & 0 & \frac{1}{G}\end{bmatrix}\begin{bmatrix}\sigma_{11}\\ \sigma_{22}\\ \sigma_{33}\\ \sigma_{12}\\ \sigma_{13}\\ \sigma_{23}\end{bmatrix} \tag{3}$$

$$G = \frac{E}{2(1+\upsilon)} \tag{4}$$

G is a shear modulus. The elastic material in the properties of aluminium 2024-0 was used in the numerical modelling. The properties were taken from experimental results. The value of Young's modulus, $E = 70.6$ GPa and the Poisson's ratio is $\upsilon = 0.3$.

3.1.2 Yielding

An isotropic yielding is applied to a von Mises yield surface. The surface is assumed that yielding of a metal is independent where the observation was confirmed by experimental works. The metal was applied under a positive pressure stress. When the metal applied in that kind of situation which high-pressure stress, it may be inaccurate because voids can nucleate and grow in the metal.

3.1.3 Plasticity

Beyond the yield point is a permanent deformation or called as the plastic behavior of the aluminium. The plastic behavior is defined in an isotropic hardening. To describe the isotropic hardening, the yield stress, σ^0 which given in a plastic strain tabular function. The interpolation of the yield stress at any plastic strain rate can be done from the table of data. The data will remain constant when it reached to the last value that has been given in the table. Decomposition of the total increment of strain defined as below:

$$d\varepsilon = d\varepsilon^{el} + de^{pl} \tag{5}$$

For rate-dependent of material the relationship of the equivalent plastic strain rate, $\bar{\varepsilon}^{pl}$ followed by the uniaxial flow rate definition as:

$$\bar{\varepsilon}^{pl} = \varphi(q, \bar{\varepsilon}^{pl}, \theta_r) \tag{6}$$

φ known as a function, q is the von Mises equivalent stress, $\bar{\varepsilon}^{pl}$ is an equivalent of the plastic strain and θ_r is the temperature.

3.1.4 Failure Criteria

The development of damage and failure of ductile material are used in Abaqus/Explicit to model the damage behavior of the aluminium. *SHEAR FAILURE and *TENSILE FAILURE or a combination of both are two material failure models will offer in the Abaqus to account the damage and failure in a ductile metal. To calculate the failure, shear and tensile failure models can use an equivalent of the plastic strain and hydrostatic cut-off stress respectively. If the failed meshes removed from the meshes it may for both failure models.

3.1.5 Damage for Ductile Metals

There is two mechanisms for the ductile metals will be used in this study; ductile damage and shear damage models. The ductile fracture or ductile damage is an initial criterion for predicting the onset damage which due to nucleation, growth and the voids. This model will active when the following condition is satisfied:

$$\omega_D = \int \frac{d\bar{\varepsilon}^{pl}}{\bar{\varepsilon}_D^{pl}(\eta, \bar{\varepsilon}^{pl})} = 1 \tag{7}$$

ω_D is a state variable that increases monotonically with plastic deformation. The model is assumed that the equivalent plastic strain at the onset of damage, $\bar{\varepsilon}_D^{pl}$ is a function of triaxiality, $\eta = -p/q$, and the equivalent of plastic strain, $\bar{\varepsilon}^{pl}$. Note that p is the pressure stress and q is the Mises equivalent stress.

The shear damage model is a fracture which due to shear band localisation in the ductile metals. The model will active when the following condition is satisfied:

$$\omega_s = \int \frac{d\bar{\varepsilon}^{pl}}{\bar{\varepsilon}_s^{pl}\left(\theta_s, \bar{\varepsilon}^{pl}\right)} = 1 \tag{8}$$

ω_s is a state variable that increases monotonically with plastic deformation and is relative to the incremental change in equivalent plastic strain, $\bar{\varepsilon}^{pl}$. The model is assumed that the equivalent plastic strain at the onset of damage, $\bar{\varepsilon}_s^{pl}$ is a function of the shear stress ratio, $\theta_s = (q + k_s p)/\tau_{max}$ and the equivalent plastic strain rate, $\bar{\varepsilon}^{pl}$. Note that τ_{max} is the maximum shear stress k_s is a material parameter (k_s = 0.3 for the aluminium alloy) [43].

3.2 Modelling of the composite by Hashin damage criterion.

The modelling for the composite material, woven fabric carbon fibre reinforced plastic was followed based on the table below:

Table 1: Properties of woven CFRP

Property	Symbol	Value
Young's modulus in longitudinal direction	E_{11} (GPa)	48
Young's modulus in transverse direction	E_{22} (GPa)	48
Young's modulus in thickness direction	E_{33} (GPa)	1
In-plane shear modulus	G_{12} (GPa)	9
Through-thickness shear modulus	G_{13}, G_{23} (GPa)	9
In-plane Poisson's ratio	υ_{12}	0.1
Through-thickness Poisson's ratio	$\upsilon_{13}, \upsilon_{23}$	0.1
Longitudinal tensile strength	X_T (MPa)	550
Longitudinal compressive strength	X_C (MPa)	150
Transverse tensile strength	Y_T (MPa)	550
Transverse compressive strength	Y_C (MPa)	150
Transverse shear strength	S_T (MPa)	120
Longitudinal shear strength	S_L (MPa)	120

The Hashin damage criterion based on works of Hashin and Rotem and Hashin [44, 45]. This criterion is unlike other criteria like Tsai-Hill and Tsai-Wu[46, 47]. Those criteria purposed an equation to predict damage initiation. The Hashin damage presents four failure modes with four corresponding indexes (a) breakage of fibre from the tension $\left(F_f^{\ t}\right)$, (b) buckle of fibre from the compression $\left(F_f^{\ c}\right)$, (c) crack of the matrix from the tension $\left(F_m^{\ t}\right)$ and (d) crush of the matrix from the compression $\left(F_m^{\ c}\right)$. There is four equation that applied to those failure modes respectively:

$$\left(F_f^{\,t}\right)=\left(\frac{\sigma_{11}}{S_{t1}}\right)^2+\alpha\left(\frac{\sigma_{12}}{S_{12}}\right)^2\leq 1.0 and \sigma_{11}\geq 0 \tag{9}$$

$$\left(F_f^{\,c}\right)=\left(\frac{\sigma_{11}}{S_{c1}}\right)^2\leq 1.0 and \sigma_{11}<0 \tag{10}$$

$$\left(F_m^{\,t}\right)=\left(\frac{\sigma_{22}}{2S_{s23}}\right)^2+\alpha\left(\frac{\sigma_{c2}}{2S_{s23}}\right)^2\leq 1.0 and \sigma_{22}\geq 0 \tag{11}$$

$$\left(F_m^{\,c}\right)=\left(\frac{\sigma_{22}}{2S_{s23}}\right)^2+\left[\left(\frac{S_{c2}}{2S_{s23}}\right)^2-1\right]\frac{\sigma_{22}}{S_{c2}}+\left(\frac{\sigma_{12}}{S_{12}}\right)^2\leq 1.0 and \sigma_{22}<0 \tag{12}$$

σ_{11}, σ_{22} and σ_{12} are applied stresses while α is a coefficient of the shear stress σ_{12} which contributed in fibre breakage by a tension criterion. In the Abaqus, the criterion of Hashin could be implemented in a conjunction with a damage evolution law. The law based on a specification of four fracture energyvalues G_f. The values corresponded to the material degradation in each mode. The values constituent in an implementation of the Hashin-based and analysed with the damage evolution in the Abaqus. The damage variable for fibres d_f, matrix d_m and shear d_s were assessed by this expression [48]:

$$d=\frac{\delta^f{}_{eq}\left(\delta_{eq}-\delta^0{}_{eq}\right)}{\delta_{eq}\left(\delta^f{}_{eq}-\delta^0{}_{eq}\right)} \tag{13}$$

$\delta^f{}_{eq}$ is an equivalent of a displacement in an FE when the stress reached to zero in a stress-displacement while $\delta^0{}_{eq}$ is an equivalent of a displacement when the damage initiated. δ_{eq} is an equivalent of a displacement in FE for the given applied strain [48].

4. Results and Discussion

The damaged area of the laminate is depended on a crosshead displacement of the quasi-static loading. The crosshead displacement and the constituent materials, the 2024-0 aluminium alloy and carbon fibre reinforced polymer were in the present case. The prior to testing, the type of resin used and metal sanding were investigated. The thermoset epoxy was used as a matrix to laminate the carbon fibre with the aluminium alloy. The sandpaper with grit 80 was applied on the aluminium alloy as to increase the bonding between the sheet metal and matrix. From the graph, it showed that the trending of failure on the fibre metal laminate became decreased when the crosshead displacement was increased from 1 mm/min to 100 mm/min.

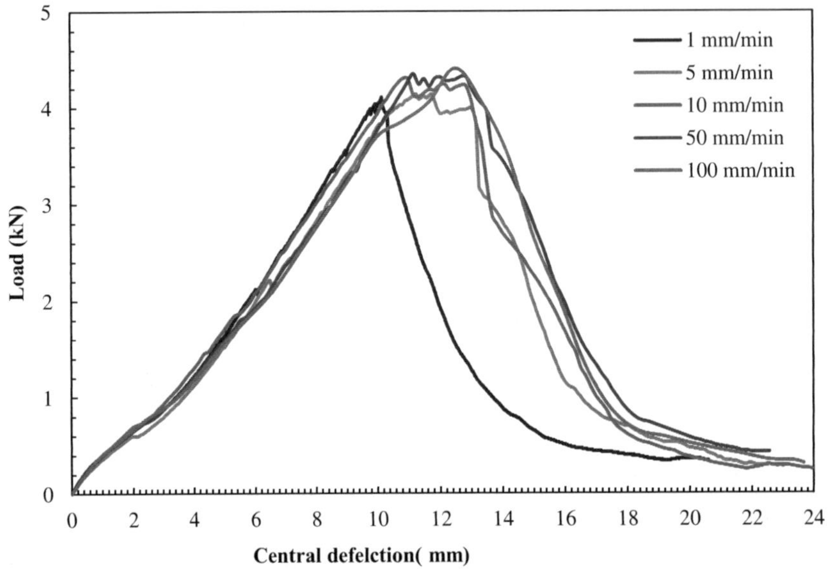

Figure 3. Graph of load against central deflection in Aluminium/CFRP FML

The failure process of the quasi-static indentation on the FML was investigated by examining the front surface and rear-surface of the impact damaged samples. The initial fracture onto the 2/1 FML by 1 mm/min crosshead displacement. The experiment was handle by using Instron 3369 Universal Testing Machine. The 1 mm/min impact shown in a Figure 4a. The damaged area has been taken through an indentation stage by the indenter at a top surface of the aluminium alloy. At the second stage, there was a localize crack on the matrix. The failure on the matrix because of the preparation of layup process. The laminate start to bend because of the maximum load that acted on it and the laminate has a maximum bending. After that, there was a fibre breakage because of the maximum bending of the loading and the stiffness of the fibre. Then, there was an indentation on the bottom ply, aluminium alloy. The perforation on the laminate happened at the third stage of failure. The passage of the indenter through the target and produce a clean hole with a diameter that similar to the indenter. From the graph, the perforation involved a local ductility.

The higher crosshead displacement resulted in a decrease of composite structure degradation. Besides that, there was a number of transverse cracks which increases the delamination of the laminate especially at the middle area of the specimen. The matrix crack also caused delamination of the laminate. When the testing was carried out with the lowest crosshead displacement of quasi-static loading it resulted in an extensive delamination at the middle part and bottom ply. When the crosshead displacement increased, the deformation on the fibre metal laminate happened at the same time. The matrix of the laminate damaged caused by impact and the connection between matrix-fibre became degradation. The epoxy resin is brittle and has low resistance to the crack propagation. The crack happened at a high transverse shearing stress which connected with the contact force. The bending crack occurred easily if the volume of the matrix is higher than the volume of the fibre.

In the experimental works, the range of crosshead displacement 1-100 mm/min involved delamination layer, failure of the fibre, plastic deformation and cracking of the aluminium layer. The energy lost

IOP Conf. Series: Materials Science and Engineering **257** (2017) 012046 doi:10.1088/1757-899X/257/1/012046

defined as $1-\eta$ will be decreasing when the impact crosshead displacement was increased. η is energy absorption factor. The lost energy would constituent the energy which used to rebound the impactor. Besides that, the energy also used as to overcome the friction from the indenter and during the impact, a sound would be produced. The sound signed for the failure of the fibre and cracking of the aluminium alloy. The deformation behavior of the top surface have been shown in the **Fig 4** respectively. The increasing of impact crosshead displacement caused the impact energy localized around the impacted area which leaded to fracture and penetration of that area for FML panels with plain weave composite fibre.

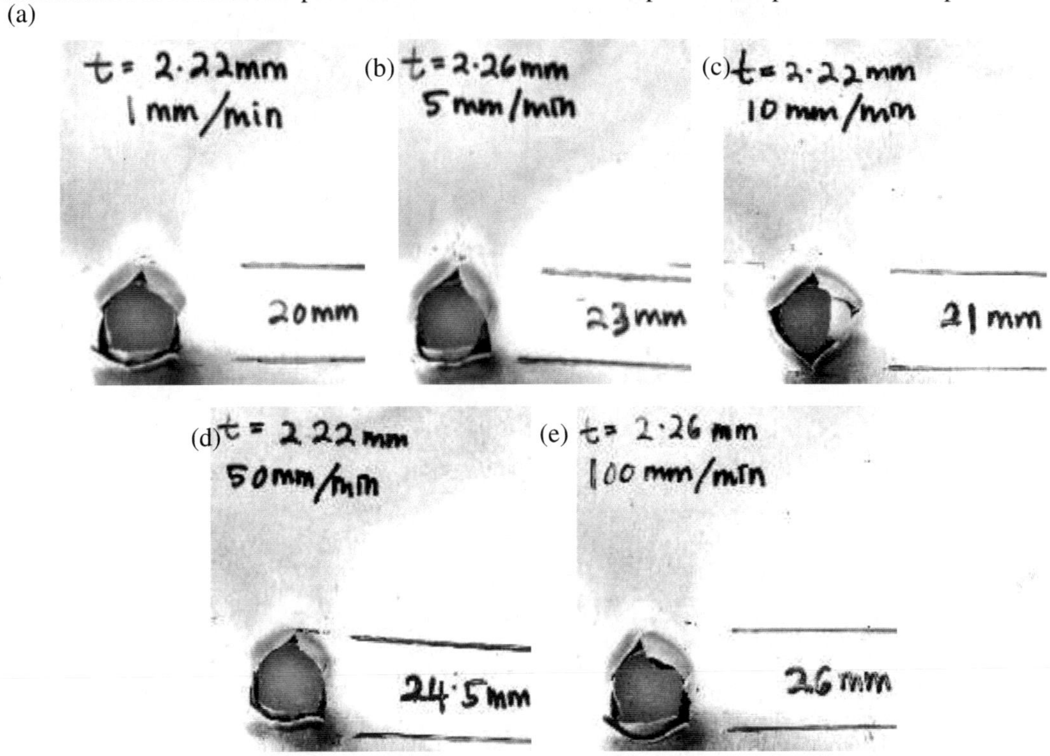

Figure 4. The visual result FML penetration based on different crosshead displacement.

The 2/1 FML model based on aluminium and carbon fibre was predicted through a finite element method; Abaqus. The numerical analysis was done as to predict the failure behavior of the FML model under the dynamic impact loading. The model was analysed by different velocities. The failure mechanism of the model was recorded. The FE model became successful when at the perforation stage of the impact event. The failure area of the FML increased because the impact velocity has been applied increasingly. The FML exhibited and appeared with little plastic deformation at the impact location. The plastic deformation was in aluminium and there was a fracture on the composite ply. The results clearly showed that the different impact velocity resulted in the flexural stiffness. The number of elements for indenter is 1345 while for the FML consisted with 4715 number of elements.

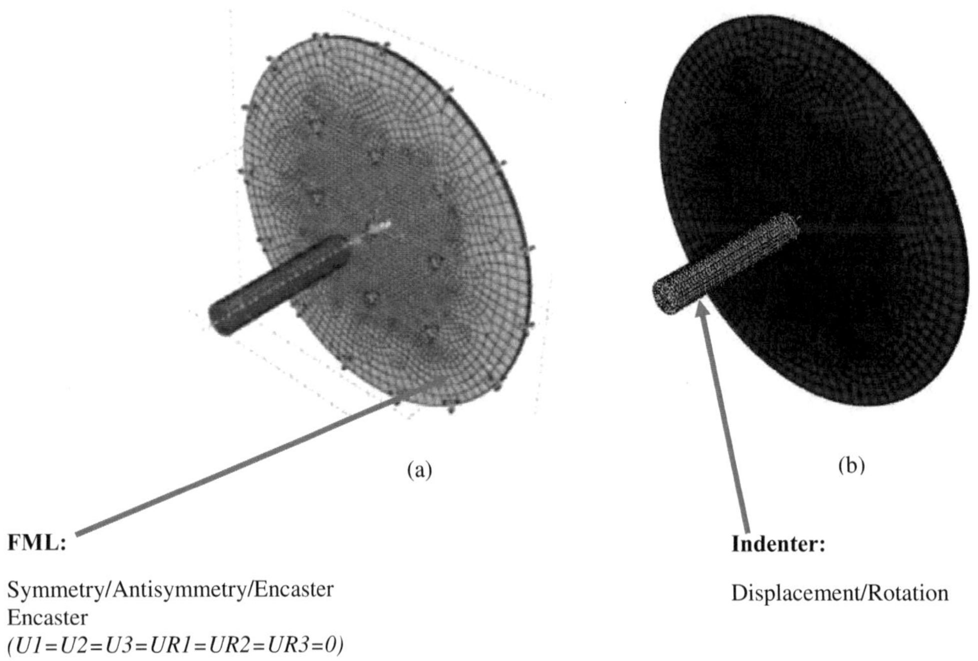

Figure 5. (a) Meshing and boundary condition of FML (b) simulation of FML.

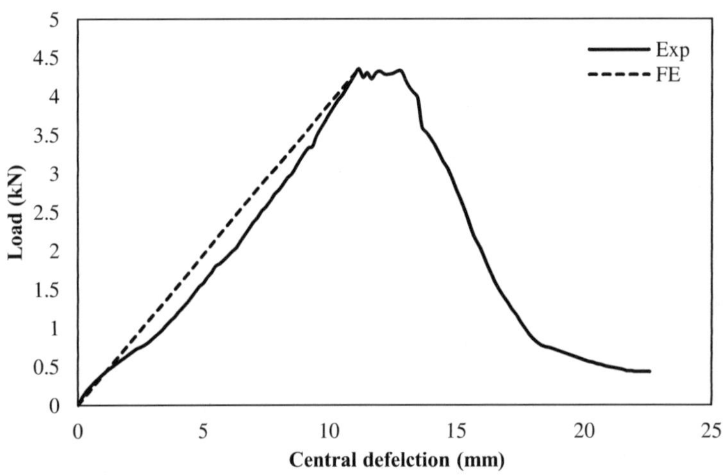

Figure 6. Validation graph between experimental and finite analysis data.

5. Conclusions

The experimental investigation was done and the finite element analysis of quasi-static test on 2/1 FMLs was presented in details. The bonding at interlayer of the FML was improved by metal sanding method. The impact area increased as the crosshead displacement was increased from 1 to 100 mm/min. The failure behaviour of the aluminium/CFRP laminate composite was modelled and analysed using finite element method, Abaqus/Explicit. As expected, the surface area from the impact loading increased when the impact velocity also has been increased. The finite element analysis was successfully predicted in load-displacement for the hybrid laminates and identifying the failure behaviour of the hybrid laminate.

The prediction offered by the numerical model was found that the validation has good agreement with the experimental data which provide an appropriate value for the initial imperfection value was ran through the analysis. The analysis was based on the impact failure response of the fibre metal laminate by the quasi-static loading. According to the fact that the numerical model was assumed as a perfect bonding between the composite material and sheet metal since did not include the effect of initial imperfection along the fibre metal laminate.

A study to investigate the failure response of fibre metal laminate based various crosshead displacement and found that when the FML indented by higher crosshead displacement, the strength of the FML would also increase. Further work will investigate the behavior of the fibre metal laminate by using Abaqus/Explicit in order to capture the crushing behavior of the laminate.

On the basis results that have been obtained through the experimental work, the behaviors of the laminates were investigated through a load-displacement curve. The high value of maximum load stipulated to high resistance of the fibre metal laminates to the quasi-static indentation. The relationship between indentation crosshead displacements and the laminates thickness have been confirmed by the experimental research. The value of the maximum load that has been recorded by the aluminium/CFRP was about 4.11 kN.

The evaluation on the damaged fibre metal laminate it is possible to compare the damaged surface areas by an experimental research. It has been found that, increasing of crosshead displacement would increase the damage surface area. A detail observation on the cross-section of the fibre metal laminates were revealed by the presence of matrix cracking and delamination between composite layers as well as delamination at metal-composite interface.

The laminate thickness was lead to degradation degree. The lower thickness of laminates would have greater deformation and lower metal layer cracking. A laminate increased with thickness was accompanied by decreasing in degree of laminate deformation and occurring delamination between the individual composite layers and delamination on metal-composite interface.

Acknowledgments
The authors are grateful to the Ministry of Higher Education and Universiti Malaysia Pahang for funding this research RDU 1603108

References
[1] Vlot A, Krull M 1997 *J. Phys. IV* **7** C3-1045-C3-50.
[2] Chai G B, Manikandan P 2014 *Composite Structures* **107** 363-81.
[3] Nakatani H, Kosaka T, Osaka K, Sawada Y 2011 *Composites Part A: Applied Science and Manufacturing* **42** 772-81.
[4] Sarlin E, Apostol M, Lindroos M, Kuokkala V-T, Vuorinen J, Lepistö T, et al. 2014 *Compos. Struct.* **108** 886-93.
[5] Zhang Z, Wang W, Rans C, Benedictus R 2016 *Procedia Structural Integrity* **2** 3361-8.
[6] Pärnänen T, Kanerva M, Sarlin E, Saarela O 2015 *Composite Structures* **119** 777-86.
[7] Sadighi M, Alderliesten R C, Benedictus R 2012 *International Journal of Impact Engineering* **49** 77-90.
[8] Bieniaś J, Jakubczak P, Surowska B, Dragan K 2015 *Archives of Civil and Mechanical Engineering* **15** 925-32.
[9] Hu Y, Zheng X, Wang D, Zhang Z, Xie Y, Yao Z 2015 *J. Mater. Process. Technol.* **226** 32-9.

[10] Su Z, Ye L 2009 *Identification of damage using Lamb waves: from fundamentals to applications* (Springer Science & Business Media).
[11] Ghadami A, Behzad M, Mirdamadi H R 2015 *Archive of Applied Mechanics* **85** 793-804.
[12] Edson Cocchieri Botelho R A S, Luiz Cláudio Pardini, Mirabel Cerqueira Rezende 2006 Materials Research 9 247-56.
[13] Rezende M C, Botelho E C 2000 *Polímeros* **10** e4-e10.
[14] Zhang P, Ruan J, Li W 2001 *Cryogenics* **41** 245-51.
[15] Botelho E C, Nogueira C L, Rezende M C 2002 *J. Appl. Polym. Sci.* **86** 3114-9.
[16] Botelho E C, Scherbakoff N, Rezende M C, Kawamoto A M, Sciamareli J 2001 Macromolecules 34 3367-75.
[17] Bhatnagar N, Ramakrishnan N, Naik N, Komanduri R 1995 I*nternational Journal of Machine Tools and Manufacture* **35** 701-16.
[18] Potter K. *Introduction to composite products: design, development and manufacture: Springer Science & Business Media*; 1996.
[19] Gutowski T G P. Advanced composites manufacturing: John Wiley & Sons; 1997.
[20] Matthews F L, Rawlings R D 1999 *Composite materials: engineering and science* (Elsevier).
[21] Pardini L C, Peres R J 1996 *Polímeros* **6** 32-42.
[22] Hou M, Ye L, Lee H, Mai Y 1998 *Composites science and technology* **58** 181-90.
[23] Hillermeier R, Seferis J 2001 *Composites Part A: Applied Science and Manufacturing* **32** 721-9.
[24] Zenkert D 1997 *Student edition.*
[25] Rejab M, Cantwell W 2013 *Composites Part B: Engineering* **47** 267-77.
[26] Chang P Y, Yeh P C, Yang J M 2008 *Fatigue Fract. Eng. Mater. Struct.* **31** 989-1003.
[27] Chang P Y, Yeh P C and Yang J M 2008 *Mater. Sci. Eng., A* **496** 273-80.
[28] Vogelesang L and Vlot A 2000 *J. Mater. Process. Technol.* **103** 1-5.
[29] Li H, Hu Y, Fu X, Zheng X, Liu H and Tao J 2016 *Composite Structures* **152** 687-92.
[30] Botelho E C, Scherbakoff N and Rezende M C 2000 *Materials Research* **3** 19-23.
[31] Botelho E C, Scherbakoff N and Rezende M C 1999 *Polímeros: Ciência e Tecnologia* **1** 59-65.
[32] Lee C L and Wei K H 2000 *J. Appl. Polym. Sci.* **77** 2149-55.
[33] Phillips R, Glauser T and Månson J A E 1997 *Polymer composites* **18** 500-8.
[34] Shim S B, Seferis J C, Eom Y S and Shim Y T 1997 *Thermochimica Acta* **291** 73-9.
[35] Wingard C D 2000 *Thermochimica Acta* **357** 293-301.
[36] Hayes B, Gilbert E and Seferis J 2000 *Composites Part A: Applied Science and Manufacturing* **31** 717-25.
[37] Lee Y S, Wetzel E D and Wagner N J 2003 *J. Mater. Sci.* **38** 2825-33.
[38] Abrate S. Impact on composite structures: Cambridge university press; 2005.
[39] Richardson M and Wisheart M 1996 *Composites Part A: Applied Science and Manufacturing* **27** 1123-31.
[40] Katnam K, Da Silva L and Young T 2013 *Progress in Aerospace Sciences* **61** 26-42.
[41] Sinmazçelik T, Avcu E, Bora M Ö and Çoban O 2011 *Mater. Des.* **32** 3671-85.
[42] Morinière F D, Alderliesten R C and Benedictus R 2014 *International Journal of Impact Engineering* **67** 27-38.
[43] Liu G, Li Q, Msekh M A and Zuo Z 2016 *Comput. Mater. Sci.* **121** 35-47.
[44] Hashin Z 1980 *Journal of applied mechanics* **47** 329-34.
[45] Hashin Z and Rotem A 1973 *J. Compos. Mater.* **7** 448-64.
[46] Tsai S W and Wu E M 1971 *J. Compos. Mater.* **5** 58-80.
[47] Tsai S W. Strength Characteristics of Composite Materials. DTIC Document; 1965.
[48] Duarte A P C, Díaz Sáez A and Silvestre N 2017 *Thin-Walled Structures* **115** 277-88.

Effect of thermal and mechanical parameter's damage numerical simulation cycling effects on defects in hot metal forming processes

Abdelouahid El Amri[1,*], Mounir el yakhloufi Haddou[1] and Abdellatif Khamlichi[2]

[1]Department of Physics, Faculty of Sciences, Tétouan 93030, Morocco
[2]Departments TITM, National School of Applied Sciences, Tetouan 93030, Morocco

[*]Corresponding author email: abdelouahid26@gmail.com

Abstract: Damage mechanisms in hot metal forming processes are accelerated by mechanical stresses arising during Thermal and mechanical properties variations, because it consists of the materials with different thermal and mechanical loadings and swelling coefficients. In this work, 3D finite element models (FEM) are developed to simulate the effect of Temperature and the stresses on the model development, using a general purpose FE software ABAQUS. Explicit dynamic analysis with coupled Temperature displacement procedure is used for a model. The purpose of this research was to study the thermomechanical damage mechanics in hot forming processes. The important process variables and the main characteristics of various hot forming processes will also be discussed.

1. Introduction

Manufacturing technologies of hot metal forming is rapidly developed in recent years. Nowadays; deep drawing, incremental hot forming, and sheet forming become very popular in hot metal forming. Ductile damage is the process of metallic material damage in conditions of monotonic loading. Evolution of the damage follows plastic straining and ends by fracture of component. Several investigations were conducted to investigate the hot forming process [1-9]. Ying et al. [1] investigate the hot forming process of 22 MnB5 steel. In metal forming processes, the plastic strain significantly outweigh the elastic strain, in many cases the rigid-plastic, or rigid-viscoplastic material behaviour is acceptable. One of the most important steps in the simulation of hot metal forming processes was the inclusion of the effects of strain-rate and temperature in material properties and the thermal coupling in forming solutions [10].The development and application of damage models was focused on hot forming process fracture. However, in recent years, several different formulations for a variety of materials and processes have been presented, such as elastic-brittle [11, 12], brittle [22, 18], creep [17, 13], fatigue [17, 18] among others. Problems of ductile damage play significant role in industry, for example in optimization of technological processes, evaluation of safety in automotive and aeronautic industry, analysis of steel civil structures. Therefore an accurate hot forming process simulation has to be thermomechanically coupled and has to incorporate flow stress and strain rate dependency on temperature [23]. Two types of material models can be distinguished. Uncoupled models are separate plastic response and ductile damage and failure. Coupled models modify plastic response in dependence on damage evolution. Process control is easier and the variety of usable hot forming process metal alloys is wider. A thermo-mechanical analysis with a three-dimensional finite element model is performed to compute stress-strain distribution. Therefore numerical simulation using the Finite Element Method (FEM), which is nowadays almost indispensable for the design of a hot metal forming process, is even more important for hot forming. As mentioned above, it has been proofed that in hot forming it is preferable to set the flange temperature higher than the punch temperature. That means that there are time dependent temperature gradients in the hot forming process metal. In additional, one finite element solver that offers adequate methods for advanced simulation of thermomechanically coupled processes is ABAQUS Software Commercial. The main aim of this study is to achieve hot forming process using a finite element analysis method. The other aim is focused on investigation into microstructure and thermal and mechanical influences on specimen and it mechanical properties. The results showed that the thermo - mechanical Manufacturing damage and microstructure of hot forming process especially Al5083-O aluminum alloy conformed to the literature.

2. The selected material

Currently, in the automotive and aeronautics industries, for instance, one has been very common to find plastic components submitted to the most diverse loading conditions. Automotive door knobs, window cranks and snap fits, which are usually manufactured with commodities, are good examples. In order to perform the experimental literature tests with the numerical, a Al5083-O aluminum metal alloy has chosen due to one reason, that can find many applications of plastic components subjected to ductile failure and manufactured with kind of material. The main mechanical and thermal properties of tested specimen are summarized in table 1. The geometry of the tested studied is shown in 'Figure 1' To obtain a thermal stress loading under homogeneous uniform temperature distribution, the steel was restrained against axial expansion by creating an interaction boundary condition at its outside edge. In order to determine thermal strain in the analysis we have need to the thermal expansion coefficient α. Poisson's coefficient does not depend on temperature and takes the constant value $\upsilon = 0.3$. A displacement boundary condition was applied at the outside extremity of the tensile specimen. The displacement was smoothly ramped up in the first portion of the test and then held constant.

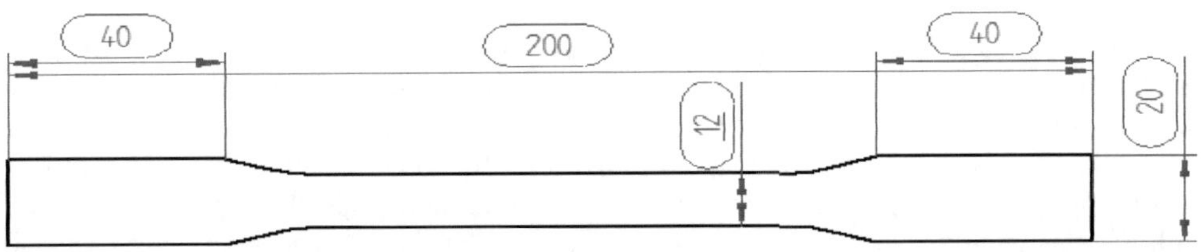

Figure1. Geometry of Al5083-O aluminum sheet metal alloy.

Table 1. Mechanical and thermal properties of the tested specimen.

Al5083-O aluminum sheet metal alloy	Values
Density	2660 kg/m^3
Young's modulus	71 GPa
Poisson's ratio	0.33
Friction coefficient	0.1
Specific heat capacity	900 J/kg.K
Thermal conductivity	117 W/mk
Thermal expansion coefficient	2.38.10^{-5} K^{-1}

3. Ductile damage model

Generally speaking, product defects in metal forming process can be addressed by employing two different categories of approaches. The first category, which is referred to as the traditional approach, consists of stress-and – strain based methods and traditional fracture mechanics. Although this type of approach has been widely used in the literature, it had some limitation in predicting fracture initiation [24-27]. The second type of approach, ductile damage criterion, is a powerful tool in prediction of material failure which has been successfully employed in different forming processes including deep drawing [28].

In this paper, for prediction of tearing of the blank, the ductile damage model proposed by Hooputra et al. [29] is employed in this study. In this model, it is assumed that the equivalent fracture strain, $\bar{\varepsilon}_f$, is a function of stress triaxiality, η, in the form of Eq. (1) [29]:

$$\bar{\varepsilon}_f = d_1 \exp(-c\eta) + d_2 \exp(c\eta) \tag{1}$$

where d_1, d_2 and c are material parameters which should be determined through experiments. Also, η is defined in Eq. (2):

$$\eta = \frac{\sigma_H}{\sigma_{eq}} \tag{2}$$

In which σ_H is the hydrostatic stress and $\sigma_{eq} = \sqrt{\frac{3}{2}\sigma_{ij}^D \sigma_{ij}^D}$ is the Misses equivalent stress.

According to ductile damage model, the damage parameter is defined as:

$$D = \int_0^{\bar{\varepsilon}_f} \frac{d\bar{\varepsilon}}{\bar{\varepsilon}_f(\eta)} \tag{3}$$

Moreover, the fracture criterion is met when D is equal to a critical value D_{cr}.

In this study, the material damage parameters were determined according to the experimental procedure proposed by Bai et al. [30]; flat-grooved tensile specimens were used to investigate the effect of stress triaxiality on fracture strain. Figure 4 shows a flat-grooved specimen from both copper and steel side views. The thickness of the specimen at the groove is represented by t and the radius of the groove is R. the groove causes stress concentration at the center of the specimen and different values of groove radii give rise to different values of stress triaxiality [30]. In order to determine the material damage parameters, one specimen with groove radii of 4mm were designed and manufactured from the produced copper/stainless steel clad sheet. The stress triaxiality at the center of the specimen is given by the modified Bridgman equation as follows [30]:

$$\eta = \frac{\sqrt{3}}{3}\left[1 + 2\ln\left(1 + \frac{t}{4R}\right)\right] \tag{4}$$

where t is the ligament thickness of the specimen and R is the groove radius. Moreover, the equivalent strain to fracture in the necking cross section of a flat-grooved specimen can be approximately determined using the logarithmic measure of strain [30]:

$$\bar{\varepsilon}_f = \frac{2}{\sqrt{3}}\ln\left(\frac{t_0}{t_f}\right) \tag{5}$$

where t_0 and t_f are the initial and fracture ligament thickness of the specimen, respectively. This formula defines the average strain through the cross section.

Tensile tests were performed on the grooved specimen and the ligament thickness of each specimen was measured after fracture. The test result are summarized in table

4. Numerical methods description

The explicit finite element calculation software, ABAQUS/ Explicit, was employed for the simulation of traction deformation and for the fracture of the Al5083-O aluminum sheet metal alloy. The size of mesh in the element model was 0.02mm×0.28mm and 0.2mm×0.12mm for the 1mm and 3mm sample, respectively. The mesh was then transitioned toward the boundaries using triangular elements and quad elements where bilinear mapping could be employed. The finite element analysis results are critical for identifying damage propagation parameters during thermal and mechanical cycles. The selected element type was an 8-node linear brick, of reduced-integration elements (C3D8R). In the finite element model, the specimen was simplified as a rigid body. The displacement load was applied on the top specimen along the 3- axis direction with the other degrees of freedom fixed. In the traction test, the Steel cylindrical rod suffered damage. The boundary element method (BEM) simplifies the meshing process and has the ability to correctly characterize the singular stress fields near the manufacturing front. The boundary conditions applied to this model were the result of a global/local approach where displacements and temperatures were taken from the results of the global model and applied to the boundaries of the local sub-model studied here. The thermal portion of the analysis should consist at least of two steps. The first step is a steady-state analysis in which the blade is brought from an initial temperature of 21.11°C to a steady operating temperature distribution and then second step simulates the cool down.

5. Results and discussion

Figure 2 shows the influence of the heat exchange factor on temperature field can be discussed by change the constant value. The total time is limited within 300s. Isotropic hardening contributes more large thermal effect than isotropic hardening when the heat exchange factor is fixed as 0.9. The more temperature change occurs with the increased the heat exchange factor at a fixed time. When the temperature increases, a large number of micro-cracks will inevitably be generated within the material and gradually expands with increasing temperature, significantly reducing the Young's modulus for macro issues, while the ability of resistance to deformation decreases, which indicates that thermal stress for high temperature effect, can cause damage to materials, namely thermal damage. In our study, we used the Young's modulus as a variable of damage, characteristic of the effect of temperature on the properties of rock damage.

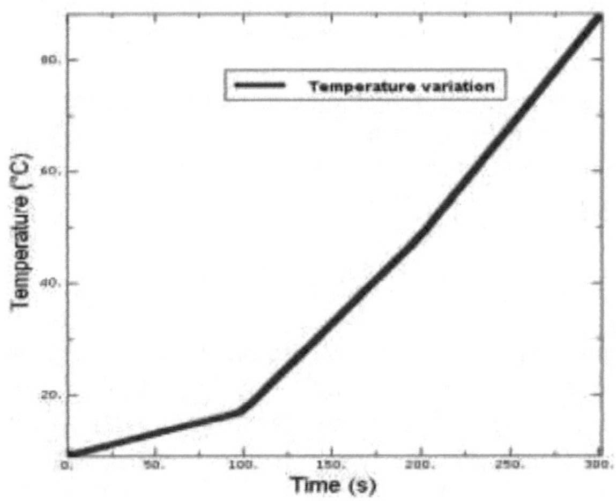

Figure 2. Temperature variation with time test.

IOP Conf. Series: Materials Science and Engineering **257** (2017) 012047 doi:10.1088/1757-899X/257/1/012047

The stress field of high-power traction with varied thermal load was analyzed. The highest stress point and its position correspond to the failure point. The simulation results shows in the Figure 3, indicate that specimen damage is always higher when the mechanical and thermal loads increases respectively. The specimen stress increases apparently when the Temperature increases with the time, the specimen damage is varied with different loads, which corresponds with the particle temperature variation of high-power traction steel under various loads. The stress at which a material exhibits a specified deviation from proportionality of stress and strain (flow stress), that point means the manufacturing.

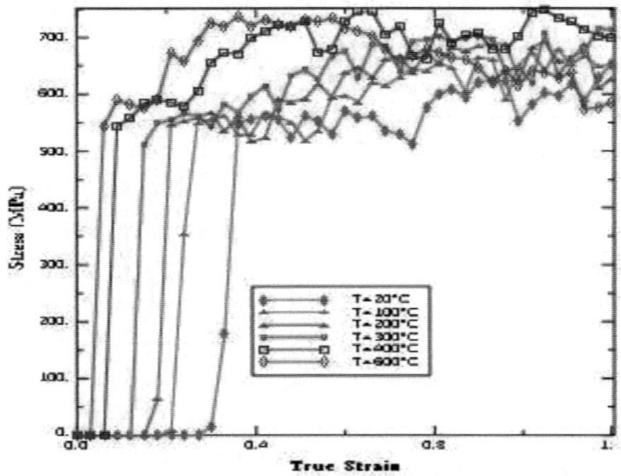

Figure 3. Stress variation with true strains at different temperature.

When the homologous temperature is relatively high Figure 4, large amounts of micro-structural of material will inevitably be produced within the material, which gradually expands as the homologous temperature rises, resulting in a significant decrease in the Young's modulus, suggesting that homologous temperature can cause failure to the steel. The basic cause of fracture strain at failure is the decrease in the Young's modulus and we can see that the modulus is a function of temperature. Therefore, we choose the Young's modulus as a damage variable to describe the effect of temperature on the mechanical properties of specimen. The damage variable becomes larger with an increase in temperature, indicating that the thermal expansion can cause a much faster rate of formation and expansion of micro-cracks within the rocks than before.

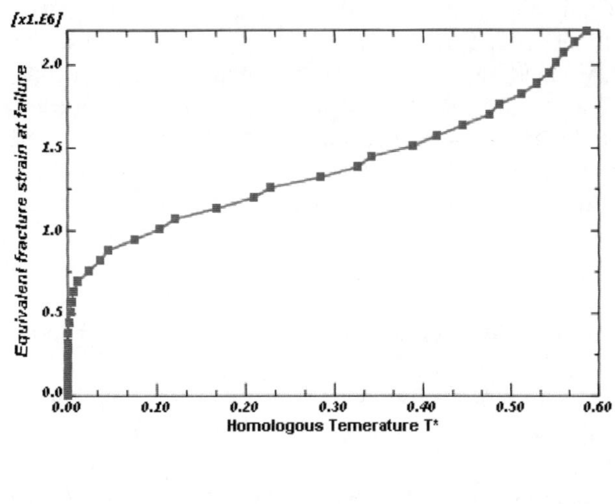

Figure 4. Evolution of equivalent strain at failure with T*.

Figure 5 shows the increase of elongation with increasing damage at different diameter and the significant drop in strength is caused by increasing over ageing, which reduces the effectiveness of the precipitates as obstacles to dislocations, as well as the increased thermal activation of dislocations, given that temperature and stress put the tests in the regime of dislocation creep. During the development of the tensile tests, one could observe three aspects. First, at the beginning of each test, due to the molecular chains orientation process, the narrower section of the specimens started becoming white and thinner, characterizing a pre-neck, prior appearing the neck, which will lead to the cold drawing process. The second aspect regards the remarkable ductility exhibited by the material, before failure. Lastly, one could observe that the fractured regions have occurred at the stretched end of the gauge section. The total elongation is 2.5 mm and the maximum damage when the displacement reached the plastic deformation.

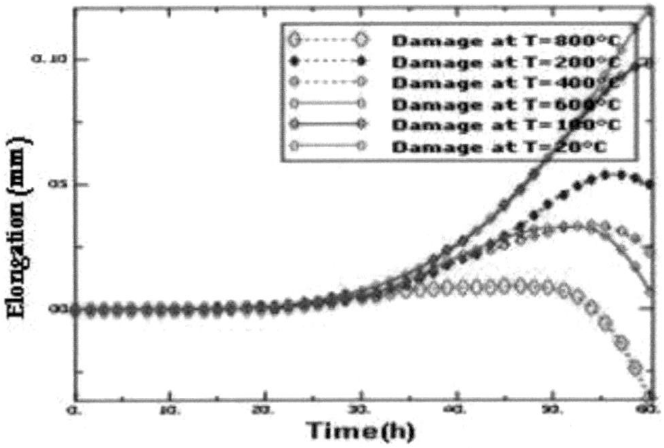

Figure 5. Damage variation at different temperature traction test.

6. Conclusions

A finite element model based on ABAQUS software was used to study the effect of thermal and mechanical parameters on Al5083-O aluminum sheet metal alloy damage. The fully coupled elastic-plastic-damage model was developed and implemented into an explicit code. It is concluded that finite element analysis (FEA) be applied to predict fracture for Al5083-O aluminum sheet metal alloy. A validation study has been performed comparing Finite Element Analysis methods based on Abaqus software of notched bar subjected to impact thermal and mechanical loadings with literature experimental results obtained in a Split Hopkinson Tension Bar. Fracture strain was calculated on the base of the specimen extension at material failure. More accurate ductile fracture description in range of calibration experiments was reached using Johnson-Cook model. Both material models successfully describe ductile damage of calibration specimens that are commonly presented in the literature.

References

[1] Ying C, Zhao-huan M, Liang Y, Xiao-dong L, Ning M and Ping H 2011 *Research International* **18** 59 – 63
[2] Yanagida A and Azushima A, 2009. *Annals Manufacturing Technology* **58** 247 – 250
[3] Merklein M, Lechler J and Stoehr T 2009 *International Journal of Material Forming* **2** 259–62
[4] Abdulhay B, Bourouga B and Dessain C 2011. *Applied Thermal Engineering* **31** 674–85
[5] Hongtu S, Ping H, Ning M, Guozhe S, Bo L and Dinglu Z 2010 *Chinese Journal of Mechanical Engineering* **23** 1–6

[6] Liu H, Bao J, Xing Z, Zhang D and Song B, Lei C 2011 *Journal of Materials Engineering and Performance* **20** 894–902
[7] Liu H, Liu W, Bao J, Xing Z, Song B and Lei C 2010 *Journal of Materials Engineering and Performance* **20** 1–10
[8] Bai-liang Z, Zhong-de S, Chao J and Wen-juan R 2010 *Proc. of Int. Conf. on Advanced Technology of Design and Manufacture*
[9] Ying X, Zhong-de S, Mi-lan Z and Chao J *Proc. of the Int. Conf. on Advanced Technology of Design and Manufacture*
[10] Boër C R, Rebelo N, Rystad H and Schröder G (1986) *Process Modelling in Metal Forming and Thermo-mechanical Treatment* (Berlin: Springer-Verlag)
[11] Murakami S 1988 *Journal of Applied Mechanics* **55** 280–6
[12] Murakami S and Kamiya K 1997 *International Journal of Mechanical Sciences* **39** 473–86
[13] Krajcinovic D and Fonseka G U 1981 *Journal of Applied Mechanics* **48** 809–15
[14] Krajcinovic D. 1983 *Journal of Applied Mechanics* **50** 355–60
[15] Simo J C and Ju J W 1987 *International Journal of Solids and Structures* **23** 821–40
[16] Simo J C and Ju J W 1987 *International Journal of Solids and Structures* **23** 841–69
[17] Chaboche J L 1988 *Journal of Applied Mechanics* **55** 59–64
[18] Chaboche J L 1988 *Journal of Applied Mechanics* **55** 65–72
[19] Lemaitre J 1984 *Nuclear Engineering and Design* **80** 233–45
[20] Krajcinovic D and Selvaraj S 1984 *Journal of Engineering Materials and Technology* **160** 809–15
[21] Saanouni K, Chaboche J L and Bathias C 1986 *Engineering Fracture Mechanics* **25** 677–91
[22] Murakami S. 1990 Role of continuum damage mechanics in fracture analysis *Inelastic solids and structures* (Swansea: Pineridge Press) p 67–80
[23] Boogaard A H 2002 *Thermally Enhanced Forming of Aluminium Sheet* (PhD. thesis) University of Twente

An overview on STEP-NC compliant controller development

M A Othman[1,2], M Minhat[1*] and Z Jamaludin[1]

[1]Faculty of Manufacturing Engineering, Universiti Teknikal Malaysia Melaka, Hang Tuah Jaya, 76100 Durian Tunggal, Melaka, Malaysia
[2]Centre of Diploma Studies, Universiti Tun Hussein Onn Malaysia, Parit Raja, 86400 Batu Pahat, Johor, Malaysia

*Corresponding author email: mohdm@utem.edu.my

Abstract. The capabilities of conventional Computer Numerical Control (CNC) machine tools as termination organiser to fabricate high-quality parts promptly, economically and precisely are undeniable. To date, most CNCs follow the programming standard of ISO 6983, also called G&M code. However, in fluctuating shop floor environment, flexibility and interoperability of current CNC system to react dynamically and adaptively are believed still limited. This outdated programming language does not explicitly relate to each other to have control of arbitrary locations other than the motion of the block-by-block. To address this limitation, new standard known as STEP-NC was developed in late 1990s and is formalized as an ISO 14649. It adds intelligence to the CNC in term of interoperability, flexibility, adaptability and openness. This paper presents an overview of the research work that have been done in developing a STEP-NC controller standard and the capabilities of STEP-NC to overcome modern manufacturing demands. Reviews stated that most existing STEP-NC controller prototypes are based on type 1 and type 2 implementation levels. There are still lack of effort being done to develop type 3 and type 4 STEP-NC compliant controller.

1. Introduction

The advanced of computing technology has heralded a significant change in Computer Numerical Control (CNC) machine tools development. CNC machine tools has come a long way, from the simple machine with controller that had no memory and driven by punched tape, NC machine tools ultimately transformed into an advanced machine tools system that embedded with the high-technology capabilities feature such as multi-axis control, multi-process manufacture, adaptive control and error compensation [1-2]. Machine tools have evolved radically, but the conventional CNC programming language has basically remained almost unaffected based on ISO 6983 (known as G&M code) standard (Figure 1). Even though, existing CNC machining systems are equipped with hundreds of sensors facilitate collection of a huge amount of data and provide opportunities for enhancing the performance of manufacturing processes, the incapability of ISO 6983 to fully utilising this opportunity is one of the reason this standard can be considered as bottleneck and obsolete for CNC enhancement. A number of problems are found with current product data of ISO 6983 and CNC architecture, which are summarized underneath [3-4].

1. **Low level programming language**. The language only focuses describe the cutter location (CL) movement and switching instruction, rather than machining tasks with respect to the part. As a result, expensive CNC machine tool is just act as executing mechanism without any intelligence.
2. **Limited information**. G&M codes controller is only provided with only elementary action and tool movement information command. The CNC part programme only contains a small portion of

the information that is available in any CAD/CAM model. This low level and incomplete data makes modification program in the shop floor is difficult.

3. **Limited control of program**. ISO 6983 is only support linear and circular movement instructions in a rigid sequential. Incapability to support the spline data makes this language unable for controlling five or more axis CNC machines.
4. **Vendor Specific code**. The limited scope covered by ISO 6983 makes each machine vendors has extended the G&M code to include their own specific functions and adaptations. As consequences, a specific post-processor is required for each CNC machine types and interchangeable of CNC programs between different controller and machine tools are restricted.

5. **Top-down programming process**. It only supports one-way information flow from design to manufacturing. If there are any modification done at shop floor level, shop floor knowledge cannot be automatically captured and directly feedback to the designer. So, companies lost their opportunities to improve their manufacturing processes using this invaluable experience.
6. **Vendor specific architecture**. This kind of architecture restricted user to access or modification the system by themselves. Manufacturing industry strictly dependent on the machine vendor to whatever machine features or requirement they needed.
7. **Limited optimisation supported**. Limited information available on the machine level due to only obtain a set of low-level and incomplete information data that make verification, simulation and optimisation are difficult, if not impossible.

Today, a new CNC challenges is not only to deal with shop floor uncertainties events such as product changeover, tool broken, job delay, machine downtime, unavailable tools and fixture, etcetera, but also fluctuating market environment. To dealing with these kinds of market situation, manufacturing companies has to increased productivity and requirements for greater product variability in order to increase or maintain their profit. Thus, to increase productivity and profitability in this globalized and competitive world, manufacturing is considered to be an integrated concept at all levels from machines tool level to production systems to an entire business level operation [5]. This will lead to the possibility of using standard data throughout the entire product process chain in the manufacturing environment and allow data to be share seamlessly with any information loss. Therefore, the next generation of CNC is required [6]: (1) To use a high level and standard data model interface for seamless integration of the entire computer-aided system (CAx) chain system; (2) To be multi-functional, intelligent and autonomous; (3) To have an open architecture with software based implementation technology.

ISO 14649, namely STEP-NC is a new standard for data exchange interface between CAD/CAM system and CNC machine and is considered as the next generation of CNC programming language for replacing ISO 6983 standard. STEP-NC compliant NC system that fully connected to manufacturing information network with legacy data are the foundation to smart NC machining system that represents the brain and knowledge repository of manufacturing system based on high bandwidth information, real time networking, and ability to be react adaptively (self-learning and flexible) [1]. Therefore, CNC manufacturing systems of the next generation are required to provide a number of new features such as high intelligence control systems for autonomy, adaptability to react with dynamically changing environment, re-configurability for agile collaboration, and sustainability for environmental benefit [7]. To achieve this aim, entire components of the CNC machining system have to benefit from the research works to discovery the better solutions and technology enhancements.

IOP Conf. Series: Materials Science and Engineering 257 (2017) 012048 doi:10.1088/1757-899X/257/1/012048

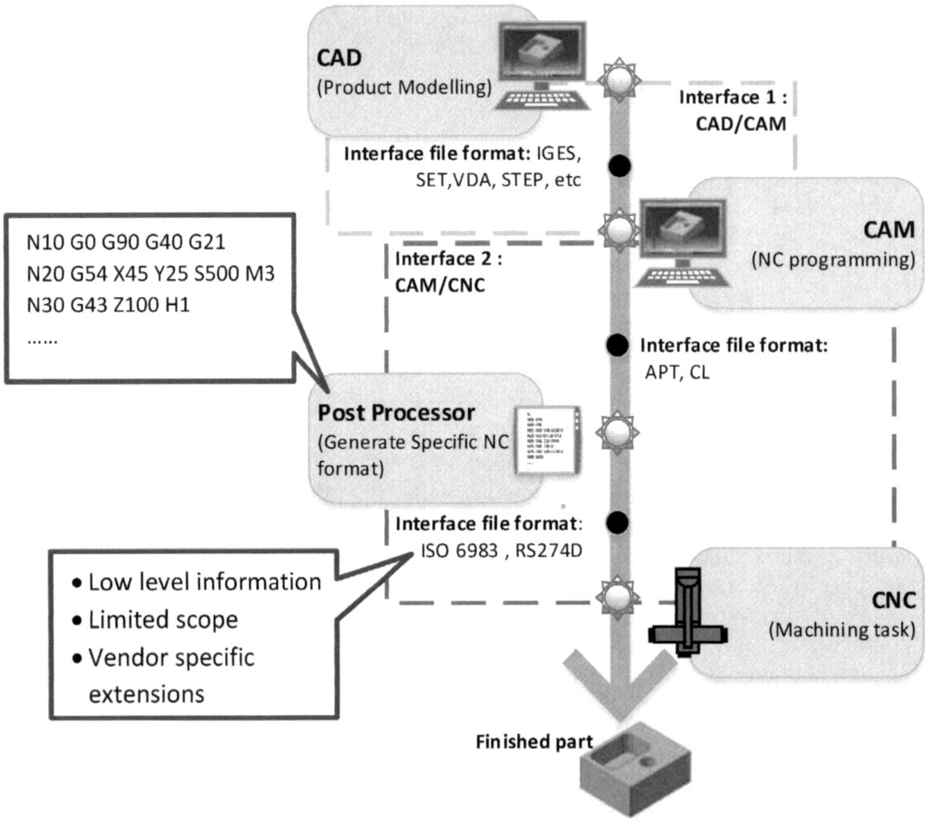

Figure 1. Current Computer Aided System (CAx) data model interface.

Up to 2016, there are many researchers having done in promoting and development of STEP-NC as a next generation of data model interface for CNC system. Based on the data gathered from the Elsevier's Scopus abstract and citation database, there are 376 journal articles/ conference papers/ review papers have been published since 2002. However, there is yet a study to be done to capture the trend of STEP-NC technology development. Hence, the objective of this paper is to review the development of new breed STEP-NC controller. This paper is organised as following: in section 2 overview of STEP and STEP-NC standard. While in section 3 implementation strategies for STEP-NC controller is discussed. Followed by section 4 where STEP-NC state of art and existing prototypes are reviewed. Finally, conclusion and future work in section 5.

2. Overview of STEP and STEP-NC standard

ISO 14649, namely STEP-NC, proposes a new vision of the CAx chain system. It provides a high level and standardized data model between CAx and CNC systems and is considered as the next generation of CNC programming languages. STEP-NC is being developed by the international community with the motivation to formalise the data model interface used at CNC machine level and remedies the shortcoming of the outdated ISO 6983 standard, which yet still dominates the control system of the most CNC machines. Unlike the current standard, STEP-NC does not defines the cutting tool movements as ISO 6983 does, but it treated high level object-oriented information and standardized information such as feature geometry, cutting tool description, operation attributes and work plan from design to CNC system (see Figure 2) [8]. Though it is possible to develop a new breed of intelligent controller NC controller that have capability to act more intelligently than ever before and with the abilities of decision making as well as control of the machine tool. This can be achieved through STEP-NC by providing generic file as a controller input and the machine-specific decisions to be made at a later stage by a CNC and its controller. As a result, part programs may be written once and used on many different types of CNC machine tool controller providing the machine has the required process capabilities [4].

Basically, STEP-NC is extended version of ISO 10303 (Standard Exchange of Product Data also known as STEP), in which both feature and model structures in STEP-NC are harmonized with ISO 10303. Unlike its other predecessors such as Drawing Transfer File (DXF), the initial Graphics Exchange Standard (IGES) and the Product Description Exchange for Standard(PDES) which primary designed to exchange geometrical data, STEP provides a neutral mechanism by specifying all the product data information throughout the life cycle of a product [4,9]. Implementation of STEP standard within CAD system to CAD system or CAD system to CAM system already solved data exchange compatibility between them. However, the programming interface between CAD/CAM and CNC systems has remained unsolved. Even though, STEP provides the unify input data for a CAM system, the output side of CAM system is still based on outdated ISO 6983. Thus, STEP-NC is introduced to overcome this ISO 6983 shortcoming issues.

Implementation of STEP and STEP-NC date model interface have bring the possibility of using single file standard data format throughout the entire product process chain in the manufacturing environment [3]. A STEP-NC file as a standard and neutral mechanism can be used on numerous machine tool controllers. As a new NC programming interface for machine level, new breed of intelligent controllers that are responsible to interpret STEP-NC program file for generate sequences of axis motions and tool functions is required. Besides that, STEP-NC also include a data model for inspection, simulation and optimisation in real time CNC system environment. Implementation of STEP-NC programming offers new possibilities for adaptive force control as a large part of the intelligence and decision making process is migrated from CAM to CNC [8]. In this system, CNC controller becomes a central element in the design to manufacturing chain system. It brings production to become more flexible and efficient. However, the migration of the way CNC is being programmed from using G&M code programming to STEP-NC demanding time and funding for changes industrial mindset.

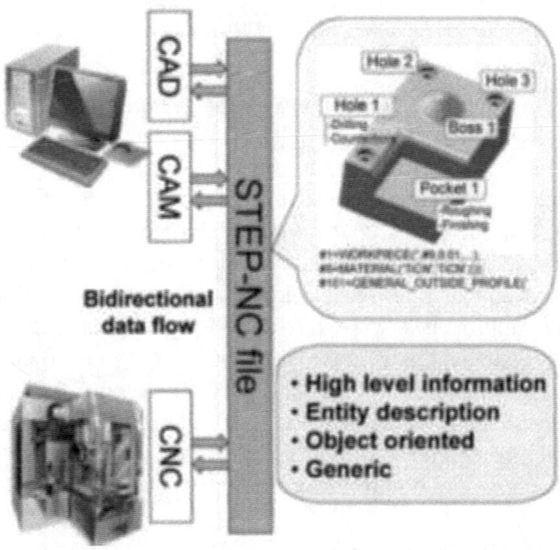

Figure 2. STEP-NC programming approach [8].

Recently, the CNC machining technology involved towards intelligently, interoperable, standardise and reconfigurable for the upcoming generation of CNC machining systems to more flexible and open. The research carried out by researcher to find the best solutions and technology improvement. Suh and Cheon in [6] emphasises a criteria have to be addressed specifically at three different requirement levels (functional level, data model level and implementation level) when developing intelligent CNC system. This requirement levels are presented graphically in the Figure 3. The first level is functional-level requirements that focuses to the capability of CNC system itself to carry out the desired activities intelligently in order to successfully achieve the machining performance and goal in shop-floor environment. Factors to be considered include autonomy, human interaction, reconfigurable, quality control, adaptive control, resource management, high-speed machining and self-learning.

The second level is data-interface level that related with the data exchange and sharing model used between CAx and CNC and data manipulation with in CNC system. The main concern is to increase data manipulation capability within the system and shortcoming of using ISO 6983 standard. This goal can be achieved by providing CNC system with intelligent product data instead of digital drawings. So the major concerns at this stage are include direct interface with CAD data by using standard data format, comprehensive data model with seamless data exchange, internet based supported, and virtual machining. The last level is implementation-level requirement. This requirement is concerns about structural and topological relationship pf the software modules and the implementation methodology. For more information on the detail description of each criteria's readers are referred to [6].

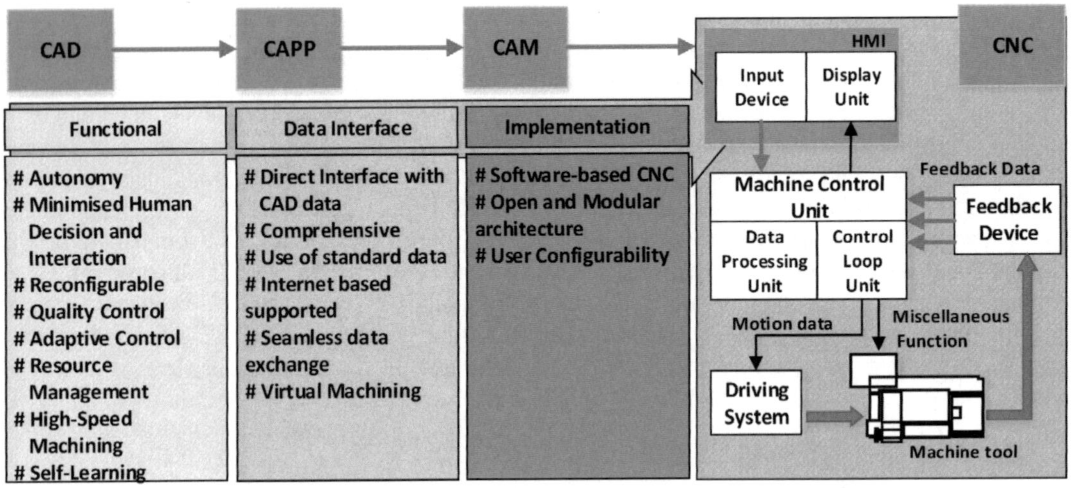

Figure 3. Requirement analysis for developing intelligent CNC system [6].

3. Implementation strategies for STEP-NC controller

STEP-NC as a new language, it is required to have a new breed of CNC controllers that capable to carrying out various intelligent tasks using this new language as an input. At the research front, multiple efforts are being made not only just to expand CNC automation and flexibility capabilities of the machine tool but also to embed intelligence and reconfigurable functionality into the machine. There are various different approaches were introduced by various scholars based on STEP-NC implementation on CNC system, those approaches can be classified into several levels of implementation in the future as present in Table 1. The lack of STEP-NC controllers has been considered as a major hindrance to popularization of STEP-NC [10]. Establishing of seamless connection, stable STEP-NC interpreter and intelligent controller are the key requirement for realise the STEP-NC concept in machine tool environment.

4. STEP-NC state of art and existing prototypes

The development of intelligent CNC systems for next generation of CNC system has been the hot topic of research. Since 2002, STEP-NC technology has been enhanced and attracted large amount of research interests among researchers, academicians, and corporates sector over the world in the race for developing technology for next generation of CNC system. Numerous technologies and various research projects have been carried out all over the world such as project OPTIMAL, project STEP-NC (EP 29708), project Super Model and Project IMS 97006 (Step-Compliant Data Interface for Numeric Controls). All these projects are being carried out with the collaboration of manufacturers (Open Mind, Dassault Cubictek, STEP Tools, Siemens, OSAI, etcetera), the user (Daimler-Chrysler,

Volvo, Wyss, Samsung, etcetera) and academic institutions from USA, Europe, Switzerland and Korea [13].

Table 1. Types of research work strategies related to STEP-NC.

Types	Strategies	Description
1	Conventional control using STEP-NC [4]; STEP-CNC with conventional control [11]; Indirect STEP-NC programming [8]	It has considered as entry level for STEP-NC by simply incorporates with legacy NC controllers. The interpreter module (act as comprehensive post processor) translated the STEP-NC program into specific machine G&M codes and then send them into CNC kernel for executes the codes. There is no intelligence at all inside this system except the capability to interpret STEP-NC program.
2	New STEP-NC enabled control [4]; STEP-CNC with new control [11]; Interpreted STEP-NC programming [8]	At this stage, machine tool motions command is directly executed based on the machining sequences and strategies defined in a STEP-NC file format. At this level, new controller architecture is required and the external data can be integrated in the tool-path generation process. Simply said, this system embedded CAM function inside the controller system. In this case, the STEP-NC controller can be employed with some intelligences, however the problems between CAM and CNC are not totally solved but hidden [12].
3	STEP-NC enabled intelligent control [4]; STEP-CNC with new intelligent control [11]; Adaptive STEP-NC programming [8]	At this point, NC controller is able to perform NC tasks intelligently and autonomously based on the online process data and optimizes machining parameters and tool-paths in real time. Intelligent function may include automatic feature recognition, automatic collision-free tool path generation (including approach and retract motion, automatic tool selection), automatic cutting condition selection, status monitoring and automatic recovery, and machining status and result feedback.
4	Collaborative STEP-NC enabled machining [4];	It has considered ultimate goal for the STEP-NC enabled machining is to support web-based, distributed and collaborative manufacturing system. This controller should be implemented with all the features inside type 3 controller plus cooperative ability via internet. The aims are parallel with e-manufacturing paradigm DA-BA-SA: Design-Anywhere, Build-Anywhere, Support-Anyways.

Most of the existing STEP-NC controllers have been reviewed by [2,8,14]. Korea STEP-NC developed by [15] consists of 5 big modules and the communication between all the modules via COBRA communication. STEP-NC part program file is generated by Shop Floor Programming system (PosSFP) module based on data input in form of STEP AP 203 or AP 224 files. Meanwhile, at Loughborough University, an Agent-Based CAM system (AB-CAM) has been developed by [16] for a generated native STEP-NC part program, and the program is able to convert into G&M code for machining execution. Xu [17] proposed a G&M code free STEP compliant machining scenario by developing a converter to convert STEP-NC files to 6K programs. Wang and colleagues in [18] developed a STEP compliant CNC system based on function blocks to translate STEP-NC data into G&M code.

Table 2. Development History of STEP-NC Controller

Authors	Strategies	Features	Limitations
Yusof and Kamran [20]	Type 2	# Can receive ISO6983 and ISO14649 as data input # Automatic tool changes and online visual video. Provides a shop floor editing facility (Working/offline) and able to generate the physical file output in .txt or .xml format.	# universal interpretation, G&M code to STEP-NC or vice versa cannot be done. # No intelligence and accurate tool path generation; # No 3D simulation # Not Accept STEP-NC P28 file # No support adaptive machining
Xiao et al. [10]	Type 2	# System able to generate and exports STEP-NC files in P21 or P28 format. #Allow shop floor adjusted #Solid 3D part viewer and tool path simulation	# No physical machining reported # Real time monitoring and on-machine inspection not included in the system # No support adaptive machining
Minhat et al [19]	Type 2	# layered structure STEP-NC controller based on IEC 61499 function blocks # Simulation and physical machining for STEP-NC Example 1&2 are performed.	# 3D modeling represents by wireframe only # No support adaptive machining
Xu [17]	Type 2	# G&M code free machining scenario # for turning process	# still use low level language (6K) # if use different NC kernel, new interpreter required # No support adaptive machining
Rauch et al. [8]	Type 1	# 3D visualization- manufacturing data, tree structure, tool path generation. # Shop floor parameter editing allowed # Collision detection simulation # On-machine inspection and real-time machining parameter optimisation	# Still using G&M code
Wang et al [18]	Type 1	# STEP-NC / G&M code mapping # Tool path generation in G&M code	# Still use G&M code for end controller # No Physical machining reported # No support adaptive machining
Newman et al. [16]	Type 1	# STEP-NC enabled controller # Tool trajectory and 3D wireframe done by commercial software	# No support adaptive machining
Suh et al. [15]	Type 2	# Develop without any commercial software #CNC generates the tool path for the cutter and touch probe (by TPG), which can be shown graphically by simulator.	# Tool path generation modules still non-real-time operation

Besides the development of interpreter and converter for STEP-NC controller, there are some researches putting their effort on closed-loop manufacturing system and system feedback in order to ensure intelligent STEP-NC controller can be realised. In [21], Brecher et al. are developed a closed-loop process chain prototype including inspection based on STEP-NC standard. Next two years, Zhao and their colleagues proposed integrated process planning system architecture for joined machining and inspection task together by provides real-time inspection feedback and On-Machine Inspection (OMI) in order to overcome relocating error during offline inspection [22]. Campos and Miguez as reported in [23] achieve vertical integration with STEP-NC in order to have a standardized process monitoring and traceability programming. The traceability is ensured at three different levels: business level, manufacturing level, and shop floor level. This allows monitoring the capitalization. Ridwan and Xu [24] define an automatic correction of cutting parameters based on the machine condition monitoring. They have developed optiSTEP-NC system which helps to perform cutting parameters optimization. With the similar work, open CNC controller named HITCNC with the ability to interpret high-level product information, inspect the part online, and monitor machining process condition in real time is proposed and developed by Hu and his friends [14]. Summaries related to STEP-NC controller development are presented in Table 2, whereas research work related to bi-direction and intelligent system is shown in Table 3.

Table 3. Research work related to enabled intelligent STEP-NC controller

Authors	Technology	Features	Limitations
Hu et al. [14]	Online Inspection	# Developed system has ability to interpret high-level product information, inspect the part online, and monitor machining process condition in real time is developed.	# Inspection involved depth of cut measurement only. # Surface roughness, features location, feature dimension not yet validate.
Danjou et al [25]	Knowledge Capitalization	# Provides guidelines to the CAM programmer in order the part can be well manufactured at 1st time.	# Involve simple shape only, freeform and complex shape parts not yet defined # Knowledge Database filling manually.
Um et al. [26]	Machine tool selection	# Automatic machine selection based on proposed machine tool data model	# Algorithm validation based on machine catalogue model (offline data) only not based on online machine resource availability. # STEP-NC machine tool data model not yet establish.
Ridwan and Xu [24]	Inspection	# Automatic correction of cutting parameters based on the machine condition monitoring # Optimisation is performed before, during or after machining operations, based on the data collected	# optimisation involve feed rate only. Overall machine performance optimisation not yet established.
Campos and Miguez [23]	Machine control monitoring	# develop a standardized process monitoring and traceability programming # CNC controllers can interpret and automatically execute programmed monitoring and traceability commands	# focusing on data traceability only, no analysis or optimisation process reported
Zhao et al. [22]	On-machine inspection	# Proposed integrated process planning system architecture for joined machining and inspection task together by provides real-time inspection feedback and On-Machine Inspection (OMI) #no data conversions are needed	# No implemented physically yet.
Brecher et al. [21]	Inspection	# Realised a closed-loop process chain with bidirectional information flow and measurement information feedback of inspection operations into the STEP-NC file.	# Conversion data still needed since CMM not support STEP-NC data yet. # No support adaptive machining

5. Conclusions and future work

In order to cope with requirements for modern CNC manufacturing environment, STEP-NC offers a new comprehensive data model remedies for the shortcomings of ISO 6983 and by incorporating various intelligent functions, which are not feasible in existing CNC control system. In order to realize the STEP-NC concept, the development of STEP-NC compliant controller is crucial. As discussed before, to date, there are multiple studies and research works have done over the world regarding developing a new breed STEP-NC compliant controller. Based on implemented strategies, the research works can be categories into four levels method (see Table 1 for detail). Considering all these evidences, it was concluded that most existing STEP-NC controller prototypes are based on type 1 and type 2 levels. Even though, the types 3 and 4 implementation level are the ultimate aims on intelligence STEP-NC controller. Up to today, there is no evidence claiming 'intelligence STEP-NC programming' controller prototype is fully developed and ready to implement on an industrial machine tool. As a Future work, the authors will develop a comprehensive framework of intelligence STEP-NC compliant controller. Feedback element, real-time control and on-machine optimisation for the CNC system also being included. The proposed system will receive generic STEP-NC file as input and machine-specific decisions including tool path generation and machining parameter setup to be made later by controller.

Acknowledgements

The authors would like to appreciate Ministry of Higher Education, Universiti Tun Hussein Onn Malaysia (UTHM), and Universiti Teknikal Malaysia Melaka (UTeM) for the financial support and facilities provided.

References

[1] Nguyen V K and Stark J 2015 *Advanced Design and Manufacturing Based on STEP* 215-232.
[2] Xu X W, Wang H, Mao J, Newman S T, Kramer T R, Proctor F M and Michaloski J L 2005 *Int. J. Prod. Res.* **43** 3703–3743.
[3] Xu X W and He Q 2004 *Robot. Comput. Integr. Manuf.* **20** 101–109.
[4] Xu X W and Newman S T 2006 *Comput. Ind.* **57** 141–152.
[5] Esmaeilian B, Behdad S and Wang B 2016 *J. Manuf. Syst.* **39** 79–100.
[6] Suh S H and Cheon S U 2002 Int. J. Adv. Manuf. Technol. **19** 727–735.
[7] Peng T, Xu X and Wang L 2014 J. Manuf. Syst. **33** 196–208.
[8] Rauch M,Laguionie R, Hascoet J Y, and Suh S H 2012 Robot. Comput. Integr. Manuf. **28** 375–384.
[9] Kemmerer S J 1999 *STEP, the grand experience* (Manufacturing Engineering Laboratory, National Institute of Standards and Technology, Gaithersburg).
[10] Xiao W, Zheng L, Huan J and Lei P 2015 *Robot. Comput. Integr. Manuf.* **31** 1–10.
[11] Suh S H, Kang S K, Chung D H, and Stroud I 2008 *Theory and design of CNC systems* (London, England: Springer-Verlag London).
[12] Zhang C, Liu R and Hu T 2006 *Int. J. Comput. Integr. Manuf.* **19** 508–515.
[13] Kržič P, Stoic A and Kopač J 2009 *Stroj. Vestnik/Journal Mech. Eng.* **55** 406–417.
[14] Hu P, Han Z, Fu H, and Han D 2016 *Int. J. Adv. Manuf. Technol.* **83** 1361–1375.
[15] Suh S h, Chung D H, Lee B E, Cho J H, Cheon S U, Hong HD and Lee H S 2002 *J. Manuf. Syst.* **21** 350–362.
[16] Newman S, Allen R, and Rosso R 2003 *Int. J. Comput. Integr. Manuf.* **16** 590–597.
[17] Xu X W 2006 *Robot. Comput. Integr. Manuf.* **22** 144–153.
[18] Wang H, Xu X, and Tedford J Des 2007 *Int. J. Prod. Res.* **45** 3809–3829.
[19] Minhat M, Vyatkin V, Xu X, Wong S and Al-Bayaa Z 2009 *Robot. Comput. Integr. Manuf.* **25** 560–569.
[20] Yusof Y and Kamran L 2016 *Int. J. Comput. Integr. Manuf.* **29** 136–148.
[21] Brecher C, Vitr M, and Wolf J 2006 *Int. J. Comput. Integr. Manuf.* **19** 570–580.
[22] Zhao F, Xu X and Xie S 2008 *Robot. Comput. Integr. Manuf.* **24** 200–216.
[23] Campos J G and Miguez L R 2011 *Comput. Ind.* **62** 311–322.
[24] Ridwan F and Xu X 2013 *Robot. Comput. Integr. Manuf.* **29** 12–20.
[25] Danjou C, Duigou J Le and Eynard B 2016 *Procedia CIRP* **41** 852–857.
[26] Um J, Suh S H and Stroud I 2016 *Int. J. Comput. Integr. Manuf.* **3052** 1–17.

A review on improving thermal-hydraulic performance of fin-and-tube heat exchangers

N. Nickolas[*], P. Moorthy, A. N. Oumer and M. Ishak

Faculty of Mechanical Engineering, Universiti Malaysia Pahang (UMP), Malaysia

*Corresponding author: nazarene40@gmail.com

Abstract. Fin-and-tube heat exchangers are one of the most common type of heat exchangers that are normally used in sectors that require small size and light weight but high heat transfer capabilities. Compact fin-and-tube heat exchangers experiences high convective thermal resistance at the air-side due to the thermo-physical properties of air. Thus, the purpose of this paper is to provide an overview of research works that are relevant to improving thermal-hydraulic performance at the air-side of fin-and-tube heat exchangers. This paper covers a variety of parameters such as tube parameters like tube arrangement, tube shapes and tube inclination angles; extended surfaces such as different shapes of fins and different parameters of vortex generators like attack angles, shapes and locations. Overall, for most modifications there was increment in heat transfer but accompanied with a pressure drop penalty. However, this varies for different combinations of parameters thus this review is to help understand how every mentioned parameters influences the thermal-hydraulic performance.

1. Introduction

Heat exchangers are devices that transfers heat energy between two fluids which are at different temperatures. This device is used in many fields such as heater and air-conditioning systems in a household or vehicle; chemical processing and as well as for power plants. There are many types of heat exchangers such as shell-and-tube heat exchanger, plate heat exchanger and compact fin-and-tube heat exchanger whereby their usage depends on the application.

Compact heat exchangers (CHEs) are normally used where weight and volume are a matter of interest. CHEs has an efficient heat transfer for a given small volume because a large surface area is obtained by attaching thin plates or fins closely to the walls separating the working fluids. In compact fin-and-tube heat exchanger, both fluids are normally in cross-flow configuration whereby they move perpendicular to each other. This produces high turbulence thus generates high heat transfer. Performance and efficiency of heat exchangers are measured through the amount of heat transfer using least area of heat transfer and pressure drop where both represents cost and power requirements of a heat exchanger respectively.

Compact fin-and-tube heat exchangers normally experiences high pressure drop which in return consumes a lot of energy. Besides, the thermal resistance experienced are the air-side convective thermal resistance, wall conductive thermal resistance and fluid-side convective thermal resistance. Due to thermo-physical property of air, the air-side convective thermal resistance is the highest thus enhancement of heat transfer should be focused on the air-side. With this in mind, the purpose of this review is to summarise experimental and numerical investigations on the thermal-hydraulic performances of fin-and-tube heat exchangers. This review is focused on different tube parameters and extended surfaces such as fins and vortex generators (VGs) for thermal-hydraulic improvements.

2. Effects of tube parameters on thermal-hydraulic performance

Tube parameters such as its arrangements, shapes and inclination angle normally influences the flow and thermal behaviour of the flowing fluid. These parameters are emphasized because they are closely related to the air-side thermal-hydraulic performance and since this part of the heat exchanger experiences the most thermal resistance, modifications to these parameters might be useful.

2.1. Tube arrangement

Tube arrays such as inline and staggered are known to produce different flow separations and fluid re-circulation. This was observed by Zhang et al. [1] where they found that the inline and staggered arrays of tubes have different performances. In the case of multiple tubes arrangement, the reduced gap between adjacent fins increased the velocity and temperature gradients resulting in a significant increase in heat transfer and pressure drop. This is more significant for staggered tubes arrangements as the gap is even smaller. This allows staggered tube arrays to distort the temperature profiles of the approaching fluid significantly. Kim et al. [2] stated that the staggered tube arrays yields higher enhancement of the heat transfer due to increment in velocity that intensifies turbulence. Their experiment showed that heat transfer coefficient of the staggered tube alignment was 10% higher than that of the inline tube arrays. Bhuiyan et al. [3] have also noted that higher heat transfer and pressured drop is observed in all flow ranges in staggered case.

2.2. Tube shape

Tube shapes influences fluid flow across it as different shapes produce different wake regions created by the boundary layer separation at the rear of tube. This can result in large pressure drop and vigorous vibrations. Besides, different shapes of tube are believed to produce swirls and vortices at varying intensity. Zhang et al. [4] conducted a numerical study on different shapes of tubes and their simulation resulted in Figure 1 where it shows that flow separation is lower for wider tubes. Smaller flow separation will reduce the pressure drop and increase heat transfer.

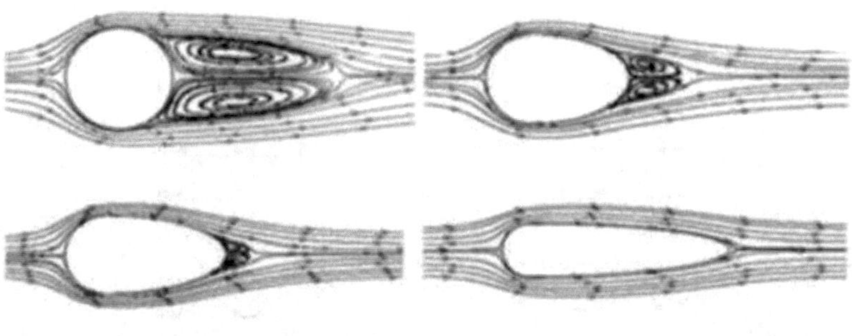

Figure 1. Comparison of streamlines produced by tubes of different widths [4].

In addition, Horvat et al. [5] conducted a study to understand the local heat transfer and fluid flow conditions in different tube shapes namely wing, elliptical and cylindrical tubes. Their numerical study further verifies the fact that different tube shapes influences flow separation. Besides, they concluded that ellipsoid tubes produces the lowest drag followed by wing tubes and cylindrical tubes. This is in line with the hypothesis that wider tubes produces smaller wake region. Furthermore, Tahseen et al. [6] mentioned that flat tube heat exchangers has smaller air-side pressure drop and improved air-side heat transfer coefficients in comparison to the round tube heat exchangers. This is due to smaller wake region for flat tubes compared to circular tube.

2.3. Tube Inclination Angle (β)

Another good way to enhance thermal performance is to induce vortex. This can be done by adjusting the inclination angles of the tubes as shown in Figure 2.

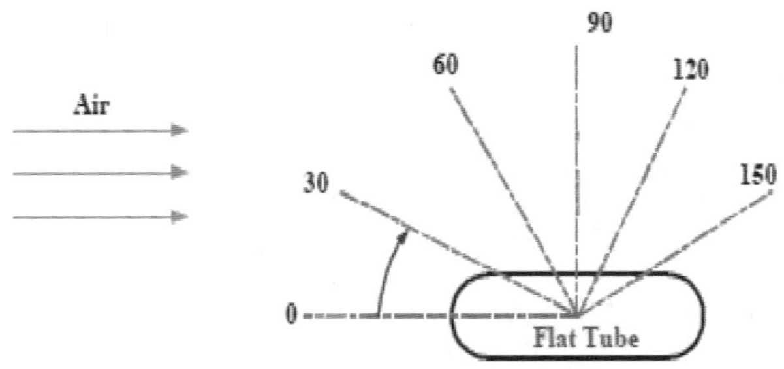

Figure 2. Different tube inclination angle with respect to air stream [7].

Badr et al. [8] visualised numerically on how inclined tubes effects thermal-hydraulic performances. They concluded that inclined tube produces counter-rotating flows in the wake region. This increases the flow velocity on the tube surface. Furthermore, they found out that increasing the tube inclination angle causes the gaps between the tubes to form a nozzle-like structure whereby the fluid flow velocity increases and this makes vortex formation more intense. Both of these effects enhances thermal performance. In addition, Ota et al. [9] claimed that increasing the tube inclination angle produces a violent transversal fluid motion that would increase heat transfer. However, this would increase the wake width so pressure drop would increase as well. Besides, Ibrahim et al. [10] concluded that as the tube inclination angle is increased from 0° to 90°, the heat transfer and pressure drop is increased because the nozzle-like flow and vortex formation increases. As the angle is further increased to 150°, the heat transfer and pressure drop reduces.

Figure 3 and Figure 4 shows the variation of normalized Nusselt number (*Nu*) and friction factor (*f*) respectively with normalized tube inclination angles. Both figures indicate that similar results were obtained by other authors [7, 11] whereby as the tube inclination angle is increased to 90° *Nu* and *f* increases and as the inclination angle is increased beyond 90°, both *Nu* and *f* reduces.

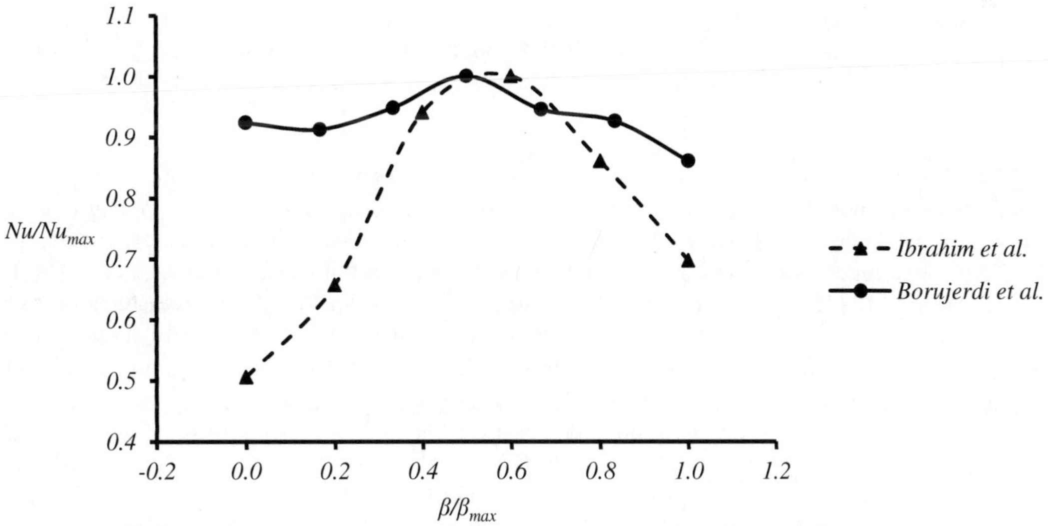

Figure 3. Variation of normalized *Nu* against normalized *β*.

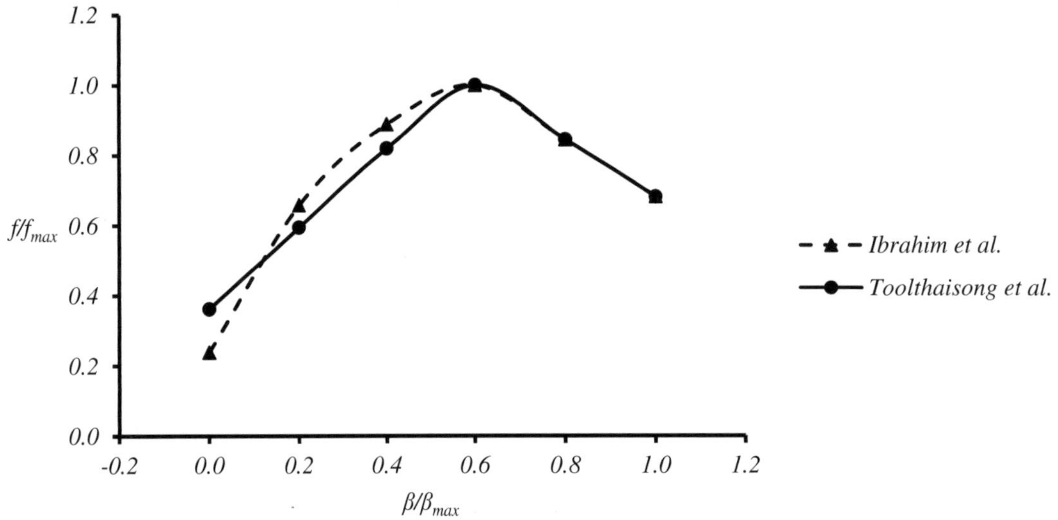

Figure 4. Variation of normalized *f* against normalized *β*.

In general, in terms of effectively distorting air flow in such a way that thermal-hydraulic performance improves, it is understood that staggered tubes produces more intense vortex and swirls whereby it increases heat transfer and pressure drop. Besides, it was also found that wider tubes produces smaller wake region. Thus, elliptical and flat tubes were identified as effective tube shapes due to their larger aspect ratio. Finally, heat transfer and pressure drop increases as the tube inclination angle is increased to 90° and further increment in the angle reduced the heat transfer and pressure drop.

3. Effect of fins on thermal-hydraulic performance

Fins are known to have increase heat transfer due to extended surface. To further enhance heat transfer, different shapes could be used to effectively distort fluid flow. However, to avoid large heat transfer area per volume, compactness was focused while ensuring efficient heat transfer and flow properties.

3.1. Plain fins

Plain fins are still popular as they are simple, durable and versatile in applications. Wang et al. [12] studied on sensible heat and friction characteristics of plate fin-and-tube heat exchangers having plain fins. They found that fin spacing has inconsistent effects on heat transfer coefficient; number of tube rows has negligible effect on friction factor; fin thickness does not affect heat transfer and friction factor. Wang et al. [13] showed that the heat transfer is independent of fin pitch for number of tube rows greater than or equal to 4 and Reynolds number (*Re*) higher than 2000. Kim et al. [2] studied heat transfer characteristics of flat plate finned-tube heat exchanger with large fin pitch. Their results show that heat transfer decreases as fin pitch decreases. Bhuiyan et al. [14] performed a 3D numerical analysis of plain fin and compared to Wang et al. [12] and the results were similar. Their result shows that heat transfer increases as fin pitch increases. Besides, heat transfer decreases with increasing longitudinal and transverse pitch of tube spacing for both inline and staggered arrangement.

3.2. Wavy fins

As researches strive for better geometry and enhanced performance, fin patterns were modified by interrupting it periodically along the stream wise direction like wavy fins. Wavy fins are vastly used in industries as they were developed to improve heat transfer where the path of the airflow is lengthen so heat transfer increases [15]. Besides, wavy fins give compactness aspect for the heat exchanger which leads to reduced space usage. Wang et al. [15] experimented on wavy fins and showed that heat transfer coefficient decreases as the number of rows increases within *Re* of 900. As the *Re* increases from 900, the heat transfer coefficient increases, and similar results were obtained by Rich et al. [16] as well.

Moreover, heat transfer coefficient is independent of wavy fin pitch and effect of tube row on friction factor is negligible. Wang et al. [17] presented empirical correlations for herringbone wavy fin on heat transfer and friction factor which implies a significant improvement on thermal-hydraulic performance compared to plain fins. Wongwises et al. [18] studied on fin pitch and number of tube rows and the results shows that wavy fin pitch has negligible effect on heat transfer meanwhile friction factor increases as fin pitch increases at *Re* greater than 2500. Pirompugd et al. [19] stated that heat transfer increases as fin pitch decreases under dehumidifying conditions. Junqi et al. [20] presented that heat transfer and flow properties decreases with increasing *Re* between 800 to 6500. Wang et al. [21] found that heat transfer decreases as number of tube rows increases and higher heat transfer is obtained for larger wavy fin pitch for number of tube rows more than two.

3.3. Rectangular offset-strip fins

Rectangular offset-strip fins provide higher degree of surface compactness, and substantial heat transfer is obtained due to the periodic starting and development of the laminar boundary layer over uninterrupted channels formed by fins and their dissipation in the fin wakes [22]. They provide better heat transfer performance due to a large amount of heat transfer area in a small volume [23] however, they are associated with increase in pressure drop due to increased friction and drag contribution. The fin offset is usually equal to the half fin spacing and uniformly arrayed in the flow direction. Suzuki et al. [24] found that thicker fins lead to increased pressure drop. Local *Nu* increases as t/S is increased where t/S is the fin thickness to hydraulic diameter ratio. They also stated that the *Nu* increases as offset length is decreased. Dong et al. [25] conducted experimental study and his results shows that heat transfer coefficient and pressure drop decreases as fin space, fin height and length increases. De Losier et al. [23] stated that thicker fins increases pressure drop so thinner fin are preferred. Lastly, it was learnt that as fin length increases, pressure drop decreases.

To summarise in terms of thermal-hydraulic performance, wavy fins are understood to have higher heat transfer coefficients compared to plain fins however, the penalty of the friction factor is even higher for wavy fins. Guo et al. [26] and Sanaye et al. [27] stated that since their staggered surfaces interrupt the flow and temperature boundary layers along the flow orientation, offset strip fins have higher heat transfer performance than plain fins. Ranganayakulu et al.'s [28] numerical investigation showed that thermal-hydraulic performance of offset fins were higher than wavy fins.

4. Effects of vortex generators (VGs) on thermal-hydraulic performance

VGs are said to be very effective in improving thermal-hydraulic performance. It mainly involves in bulk flow mixing, boundary layer development, swirl formation and flow destabilisation.

4.1. Working principle of VGs

The effects of using VGs have been documented for many years whereby majority of the research works shows that thermal-hydraulic performances were improved. In early research works such as Fiebig et al. [29], the effects of VGs on flow structure, flow losses and heat transfer were studied experimentally. They concluded that VGs can generate stable vortices whereby additional drag due to VGs are independent of *Re*. According to Ya-Ling et al. [30] when fluid flows, vortices are generated due to friction and separation on the edge of the vortex generator. This was verified numerically by He et al. [30] where Figure 5 illustrates the vector plots and streamlines generated by VGs. It can be seen that there are 3 types of vortices generated namely corner vortex, main vortex and induced vortex. They concluded that the main vortex is formed due to the flow separation and friction on the leading edge of the VGs; corner vortex is formed by the deformation of the near-wall streamlines; and the induced vortex is formed by the interaction between the main vortex and the fin surface. The combined effect of these vortices will enhance heat transfer.

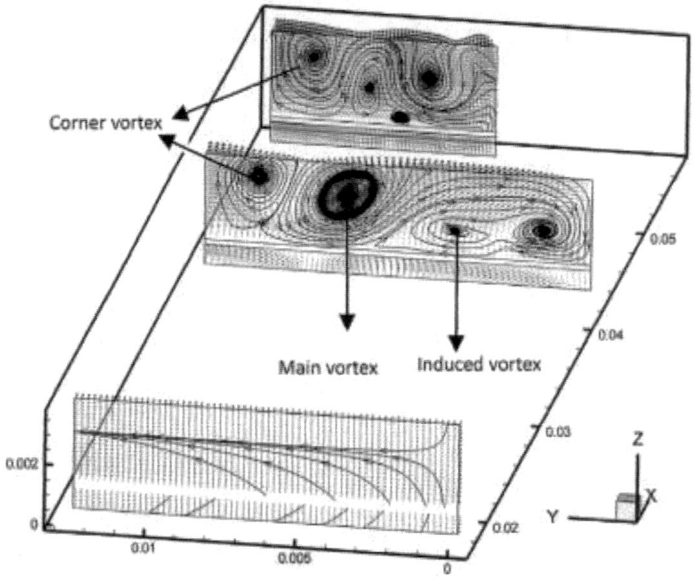

Figure 5. Vector plots and streamlines generated by VGs [30].

Besides, Fiebig et al. [31] conducted similar studies with wing type VGs for both inline and staggered tube arrays whereby there was 55-65% increase in heat transfer with 20-45% increase in friction factor for inline arrays and much lower heat transfer for staggered arrays but even lower friction factor which implies VGs with staggered tubes is more economical. Biswas et al. [32] numerically studied the flow structure with and without VGs and concluded that vortices generated by VGs in the wake region has increased the heat transfer by 240%. To further understand interactions of VGs with the boundary layer, Gentry et al. [33] conducted an experiment which resulted in an effective way of thinning the boundary layer which is by introducing vortices towards the edge of the boundary layer. This has increased the heat transfer performance to 50-60%. Wang et al. [34] conducted an experiment with the same purpose using annular and delta VGs and they concluded that fins with VGs produces more vortices compared to fins without VGs. Furthermore, Torii et al. [35] experimentally concluded that heat transfer enhancement for fin-and-tube heat exchanger with VGs are 10-30% for staggered tube arrays and 10-20% for inline tube arrays while reduction in pressure loss was 34-55% and 8-15% for staggered and inline tube arrays respectively. This is because VGs reduces form drag and wake region.

4.2. Different shapes of VGs

VGs increase heat transfer by forming vortices however this is normally accompanied with flow resistance. Hence, many research works were done to improve heat transfer with lower pressure drop and one way to accomplish this is to improve the designs of VGs. For example, in 2012, Zhou et al. [36] experimented on curved type VGs as shown in Figure 6.

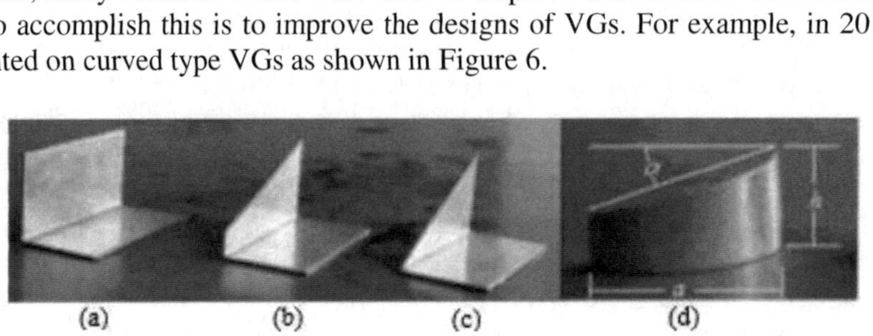

Figure 6. (a) Rectangular VG (b) Trapezoidal VG (c) Delta VG (d) Curved Trapezoidal VG [36].

They concluded that delta winglet VGs has the highest heat transfer for laminar flow while the curved trapezoidal VGs has the best thermal-hydraulic performance. They also claim that this is most likely because curved trapezoidal VGs produces a streamlined flow with low pressure drop. To further improve the thermal-hydraulic property, Zhou et al. [37] in the year 2014 experimented on different shapes of VGs such as plane and curved VGs with punched holes. They found out that for VGs with punched holes, the flow resistance was reduced. By increasing the diameter of the punched hole and positioning it at the centreline of the VG, it further improves hydraulic performances. In addition, they concluded that curved delta VG improves heat transfer better than any other VGs. This results is also similar to Song et al.'s [38] and Lin et al.'s [39] results whereby their curved delta VGs experienced heat transfer enhancement by 20.04% and 16.1% respectively. Du et al. [40] numerically investigated the influence of single and double; delta, and rectangular VGs on the thermal-hydraulic performances. They concluded that the delta winglet pairs produced the highest heat transfer enhancement at 13.5-61.2% and also at a reasonable increment in pressure drop at 5.62-46.2%.

4.3. Attack angles of VGs

One of the ways to improve thermal-hydraulic performances is to have VGs inclined at an angle to the flow as shown in Figure 7. Theoretically, as the attack angle is increased, vortices generated should be more stable and strong. Fiebig et al. [29] concluded that as the attack angle is increased from 20° to 60°, heat transfer is enhanced.

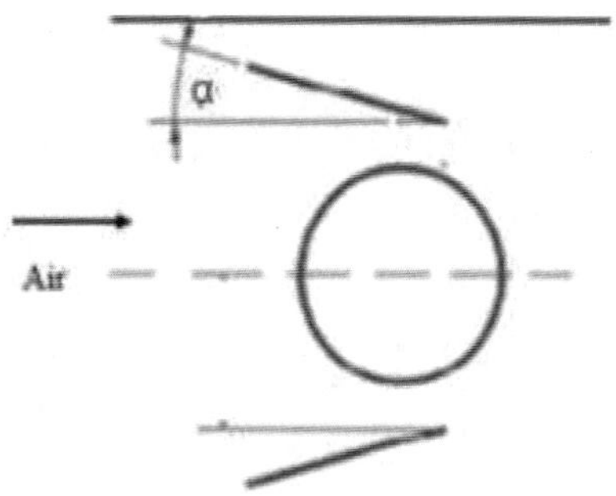

Figure 7. Placement of VGs at an angle to fluid flow [41].

Similarly, Chu et al. [42] discovered that increasing the attack angle from 15° to 30° would increase the strength of vortex generated hence increasing the heat transfer. However, they claimed that as the attack angle is increased further, the vortices may break down and reduce heat transfer. For the case of friction factor, it increases as attack angle is increased because the form drag increases with attack angle. This results is also in very much good agreement with He et al. [43] whereby their results showed that VGs at an angle of attack of 30° has better heat transfer enhancement compared to 10° and 20°. Du et al. [40] obtained a similar trend of increasing heat transfer and as well as increasing pressure drop from 10° to 25°.

Besides, Appa et al. [44] investigated the effects of attack angle numerically whereby they concluded that VGs at 45° and 30° enhances heat transfer by 38-55.68% and 18.68-46.38% respectively. They claimed that this was most likely due to reduction in wake region and as well as acceleration of fluid due to the nozzle-like region near the rear part of the tube as in Figure 7 created with the presence of VGs. On the other hand, pressure drop for VGs at 30° and 45° increased by 9.26-56.35% and 81.57-190.30% respectively. Due to very high pressure drop increment for 45°, it was concluded that VGs at an attack angle of 30° is more suitable. In a recent numerical investigation conducted by Sarangi et al. [41], their flow visualization shows that as the attack angle is increased from 5° to 20°, the wake region reduces and the flow velocity increases which is similar to Appa et al.'s findings. Besides, they observed that due to increase in flow velocity near the wall and impingement of fluid on adjacent wall, the heat transfer enhances significantly. Due to this phenomena, at VG attack angle of 20°, the heat transfer is enhanced by 60% however friction factor was increased by 100%. Figure 8 and Figure 9 shows the variation of normalized *Nu* and *f* respectively with normalized attack angles of VGs. Both figures indicate that similar results were obtained by many other authors whereby increasing attack angles of VGs will increase both *Nu* and *f* [41, 45, 46].

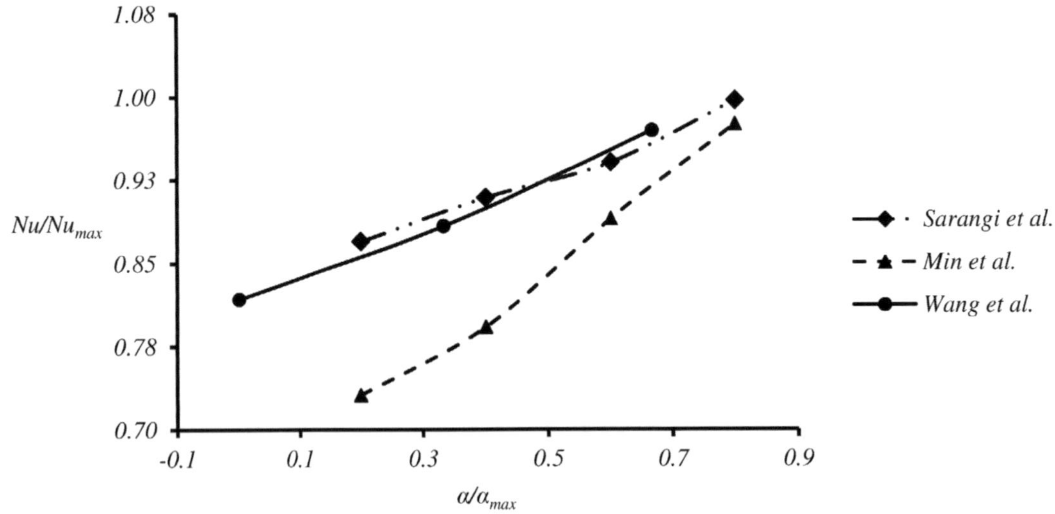

Figure 8. Variation of normalized *Nu* with normalized attack angle of VGs.

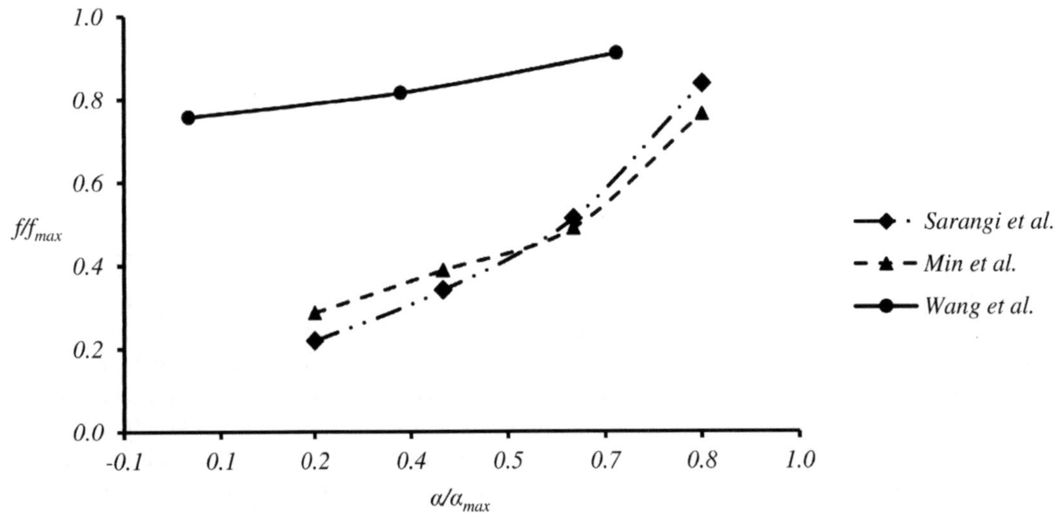

Figure 9. Variation of normalized *Nu* with normalized attack angle of VGs.

4.4. Location of VGs

The location of VGs is crucial as it plays an important role to enhance thermal-hydraulic performances by generating vortex and increasing fluid flow and both of these are influenced by the placement of VGs. According to Gentry et al. [33], by thinning the boundary layer, fluid flow will be channelled into the wake region. This can be done by bringing in free-stream fluid into the boundary layer. An efficient way of doing this is by placing the VGs upstream whereby the vortex generated will force the free-stream fluid into the boundary layer. Similarly He et al. [43] claimed that vortices generated at upstream can affect a larger downstream domain. This was in good agreement with Allison et al.'s [47] results whereby they placed VGs upstream which lead to a heat transfer and pressure drop increment of 87% and 53% respectively.

Besides, Pesteei et al. [48] experimented on different locations of VGs with respect to tube dimensions. They concluded that placing VGs slightly to the rear of the tube would enhance heat transfer by forming a nozzle-like structure so that fluid is accelerated to the wake region. Specifically, placing VGs at x = 0.5D and y = 0.5D (D = diameter of tube) will increase heat transfer and pressure drop by 46% and 18% respectively. In addition, Sinha et al. [49], suggested that placing the VGs downstream forms a narrower nozzle-like structure which might further increase the fluid velocity flowing into the wake region. Similarly, Jang et al. [50] numerically tested different locations and concluded that placing VGs further away from the tube for inline tube arrays increased heat transfer and pressure drop by 14.71-30.74% and 12.54-22.19% respectively whereas for staggered tube arrays heat transfer and pressure drop increased by 9.28-13.96% and 7.4-12.53% respectively.

In general, vortex generators functions mainly by producing vortices that enhances heat transfer. The produced vortices normally swirls into the wake region as well. Hence, improved thermal-hydraulic performance. It was understood that curved shaped VGs improved thermal-hydraulic performance very well but considering manufacturability, straight delta shaped VGs were better. Besides, increasing the attack angle to a certain value improved the performance as the vortex intensity increased and wake region reduced. In addition, some researchers concluded that placing VGs slightly to the rear end of the tube creates a nozzle-like structure that accelerates fluid into the wake region. An upstream placement should be considered as well because it helps in reducing the boundary layer thickness thus bringing the free stream fluid into the wake regions.

5. Challenges and future advancements

Further developments for fin-and-tube heat exchanger is necessary as it is a common device used in today's industry. Most of the experimental and numerical investigations has some deviations between them thus it is important to focus on reducing this gap. In addition, certain experiments were considered to be tedious so a very small range of parameters were covered. This shows that proper experimentation methods and apparatus should be devised by considering user friendliness and accuracy. The combined usage of suitable tube parameters, best fin designs and optimum configuration of VGs is an innovative idea to improve the thermal-hydraulic performance of the heat exchanger. Main challenge is to increase heat transfer capabilities with low pressure drop but there were a few combinations and configurations with good heat transfer and low pressure so there is still possibility for more improvements.

6. Conclusions

Parameters mentioned in this paper influences the performance of fin-and-tube heat exchangers by distorting the fluid flow normally by producing swirls or vortices. Besides, it was understood that for most adjustments made to the parameters to increase heat transfer, there were substantial amount of pressure drop increment as well but certain designs provided high heat transfer with low pressure drop.

In terms of tube parameters, the best tube arrangement is the staggered tube arrays. Basically, multiple tubes either inline or staggered forms a close gap between the adjacent fins creating a nozzle-like structure that accelerates fluid thus increasing both heat transfer and pressure drop. This nozzle-like effect was found to be more significant for staggered tube arrays. Besides, wider tubes like elliptical and flat tubes are said to produce small wake region thus pressure drop was low. In addition, increasing the tube inclination angle intensified the formation of vortex swirls that increased heat transfer.

Research works on variety of fins showed that it improves heat transfer by increasing the exposed area to allow more heat transfer and as well as distorting the flow to produce swirls and causing bulk fluid mixing. It was clear that standard wavy and rectangular fins provided better heat transfer but increased pressure drop as well. Offset strip fins were found to distort the temperature boundary layer effectively thus producing much better heat transfer and a much lower pressure drop.

Finally, vortex generators produces vortex that enhances heat transfer because of better fluid mixing. Considering ease of production, delta winglet VGs are said to be the best shape and increasing its attack angle improved the performance because vortex intensity increased and wake region is reduced. Besides, placing VGs carefully to produce the nozzle-like effect accelerates fluid into the wake region and helps in reducing the boundary layer thickness to bring in the free stream fluid into the wake area which increased heat transfer and reduced pressure drop.

Acknowledgement
The authors would like to thank Universiti Malaysia Pahang for providing necessary assistance for providing financial support under project number PGRS160386.

References

[1] Zhang L, Balachandar S, Tafti D and Najjar F 1997 *International Journal of Heat and Mass Transfer* **40** 2307-25
[2] Kim Y and Kim Y 2005 *International Journal of Refrigeration* **28** 851-58
[3] Bhuiyan A A, Amin M R, Naser J and Islam A 2015 *Frontiers in Heat and Mass Transfer* **6**
[4] Zhang G M, Leng X L, Zhou N X, Shi Y P and Li L M 2015 *Journal of Hydrodynamics Ser. B* **27** 76-84
[5] Horvat A, Leskovar M and Mavko B 2006 *International Journal of Heat and Mass Transfer* **49** 1027-38
[6] Tahseen T A, Ishak M and Rahman M 2015 *Renewable and Sustainable Energy Reviews* **43** 363-80
[7] Toolthaisong S and Kasayapanand N 2013 *Energy Procedia* **34** 417-29
[8] Badr H M 1998 *Heat and Mass Transfer* **34** 229-36
[9] Terukazu O, Hideya N and Yukiyasu T 1984 *International Journal of Heat and Mass Transfer* **27** 1771-79
[10] Ibrahim T A and Gomaa A 2009 *International Journal of Thermal Sciences* **48** 2148-58
[11] Nouri-Borujerdi A and Lavasani A M 2007 *International Journal of Heat and Mass Transfer* **50** 2605-11
[12] Wang C C, Chang Y J, Hsieh Y C and Lin Y T 1996 *International Journal of Refrigeration* **19** 223-30
[13] Wang C C and Chi K Y 2000 *International Journal of Heat and Mass Transfer* **43** 2681-91
[14] Bhuiyan A A, Amin M R and Islam A S 2013 *Applied Thermal Engineering* **50** 445-54
[15] Wang C, Fu W and Chang C 1997 *Experimental Thermal and Fluid Science* **14** 174-86
[16] Rich D G 1973 *ASHRAE Trans* **79** 137-45
[17] Wang C C, Hwang Y M and Lin Y T 2002 *International Journal of Refrigeration* **25** 673-80
[18] Wongwises S and Chokeman Y 2005 *Energy Conversion and Management* **46** 2216-31
[19] Pirompugd W, Wongwises S and Wang C C 2006 *International Journal of Heat and Mass Transfer* **49** 132-43
[20] Junqi D, Jiangping C, Zhijiu C, Yimin Z and Wenfeng Z 2007 *Applied Thermal Engineering* **27** 2066-73
[21] Wang C C, Liaw J S and Yang B C 2011 *International Journal of Heat and Mass Transfer* **54** 1024-9
[22] Manglik R M and Bergles A E 1995 *Experimental Thermal and Fluid Science* **10** 171-80
[23] De Losier C R, Subramanian S, Ponyavin V, Chen Y, Hechanova A E and Peterson P F 2007 *Journal of Heat Transfer* **129** 1453-8
[24] Suzuki K, Hirai E, Miyake T and Sato T 1985 *International Journal of Heat and Mass Transfer* **28** 823-36
[25] Dong J, Chen J, Chen Z and Zhou Y 2007 *Applied Thermal Engineering* **27** 306-13
[26] Guo L, Qin F, Chen J, Chen Z and Zhou Y 2007 *International Journal of Thermal Sciences* **46** 1285-96
[27] Sanaye S and Hajabdollahi H 2010 *Applied Energy* **87** 1893-902
[28] Ranganayakulu C, Luo X and Kabelac S 2016 *Applied Thermal Engineering* **111** 1588-95
[29] Fiebig M, Kallweit P, Mitra N and Tiggelbeck S 1991 *Experimental Thermal and Fluid Science* **4** 103-14
[30] He Y L, Chu P, Tao W Q, Zhang Y W and Xie T 2013 *Applied Thermal Engineering* **61** 770-83
[31] M Fiebig, A Valencia and N K Mitra 1993 *Experimental Thermal and Fluid Science* **7** 287-95

[32] Biswas G, Mitra N K and Fiebig M 1994 *International Journal of Heat and Mass Transfer* **37** 283-91

[33] Gentry M C and Jacobi A M 1997 *Experimental Thermal and Fluid Science* **14** 231-42

[34] Wang C C, Lo J, Lin Y T and Wei C S 2002 *International Journal of Heat and Mass Transfer* **45** 3803-15

[35] Torii K, Kwak K M and Nishino K 2002 *International Journal of Heat and Mass Transfer* **45** 3795-801

[36] Zhou G and Ye Q 2012 *Applied Thermal Engineering* **37** 241-48

[37] Zhou G and Feng Z 2014 *International Journal of Thermal Sciences* **78** 26-35

[38] Song K and Wang L 2016 *International Journal of Thermal Sciences* **109** 206-16

[39] Lin Z M, Liu C P, Lin M and Wang L B 2015 *Applied Thermal Engineering* **88** 198-210

[40] Du X, Feng L, Li L, Yang L and Yang Y 2014 *International Journal of Heat and Mass Transfer* **75** 368-80

[41] Sarangi S K and Mishra D P 2017 *Applied Thermal Engineering* **116** 528-40

[42] Chu P, He Y L, Lei Y G, Tian L T and Li R 2009 *Applied Thermal Engineering* **29** 859-76

[43] He Y L, Han H, Tao W Q and Zhang Y W 2012 *International Journal of Heat and Mass Transfer* **55** 5449-58

[44] Appa A and Gulhane N P 2014 *International Journal of Emerging Technologies in Computational and Applied Sciences* **3** 450-57

[45] Wang W, Bao Y and Wang Y 2015 *Applied Thermal Engineering* **86** 27-34

[46] Min C, Qi C, Wang E, Tian L and Qin Y 2012 *International Journal of Heat and Mass Transfer* **55** 7268-77

[47] Allison C and Dally B 2007 *International Journal of Heat and Mass Transfer* **50** 5065-72

[48] Pesteei S, Subbarao P and Agarwal R 2005 *Applied Thermal Engineering* **25** 1684-96

[49] Sinha A, Chattopadhyay H, Iyengar A K and Biswas G 2016 *International Journal of Heat and Mass Transfer* **101** 667-81

[50] Jang J Y, Hsu L F and Leu J S 2013 *International Journal of Heat and Mass Transfer* **67** 432-44

IOP Conf. Series: Materials Science and Engineering **257** (2017) 012050 doi:10.1088/1757-899X/257/1/012050

Condition monitoring of an electro-magnetic brake using an artificial neural network

T Gofran[1*], P Neugebauer[1] and D Schramm[2]

[1] IEEM - Institute for Energy Efficient Mobility, Karlsruhe Applied Science University, Germany
[2] University of Duisburg-Essen, Germany

*Corresponding author email: tanju-binte.gofran@hs-karlsruhe.de

Abstract. This paper presents a data-driven approach to Condition Monitoring of Electromagnetic brakes without use of additional sensors. For safe and efficient operation of electric motor a regular evaluation and replacement of the friction surface of the brake is required. One such evaluation method consists of direct or indirect sensing of the air-gap between pressure plate and magnet. A larger gap is generally indicative of worn surface(s). Traditionally this has been accomplished by the use of additional sensors - making existing systems complex, cost- sensitive and difficult to maintain. In this work a feed-forward Artificial Neural Network (ANN) is learned with the electrical data of the brake by supervised learning method to estimate the air-gap. The ANN model is optimized on the training set and validated using the test set. The experimental results of estimated air-gap with accuracy of over 95% demonstrate the validity of the proposed approach.

1. Introduction

The goal of Condition Monitoring (CM) of electric machines is to acquire a "health" indication in real-time; in order to identify possible faults and failure in advance – thus avoiding costly and unscheduled down time, upholding accurate servicing schedules [1]. An ideal CM method should be non-destructive and should depend on easily measurable parameters [2]. In contrast to sensor-based techniques, data-driven condition monitoring methods are interesting because they do not require any knowledge about the machine parameters and models; instead they only require a database of both healthy and faulty conditions of the machine for eventual feature extraction and classification [3].

Conventional methods of fault diagnosis and condition monitoring are based on well-tested mathematical models incorporating various machine faults. Generally, heavy and time-consuming computational tasks must be performed due to the complexity of the applied analytical models and the large number of required parameters [2]. The authors of [4] and [5] showed analytical methods of estimating the air-gap of a magnetic field for different electromagnetic systems. In [6] and [7] analytical methods of estimating the air-gap of a brake are shown based on the derivation of the characteristic current progression of the coil during movement of the pressure plate. A continuous method to determine the air gap using a pulsating voltage and evaluation of the resulting current is demonstrated in [8]. In industrial solutions, different sensors are also available to monitor the air-gap. One existing industrial solution is the application of eddy current sensors that can monitor the functional status of the brake [9], but it requires additional wiring. Another industrial solution is the use of recoiling micro-switch in closed contact or in open contact depending on task [10]. Unfortunately, this increases complexity in the entire construction. Adjusting the micro-switches for very low air-gap is also a difficult task. This paper presents an application of data-driven approaches to non-destructive condition monitoring of a Magnetic brake.

2. Magnetic Brakes

Magnetic brakes are an important part of electric motors requiring a regular servicing schedule. The brake considered in this paper is an electromagnetic brake with a DC coil that releases electrically and brakes using a spring force (Figure 1). The principal parts of the brake system are brake coils (acceleration coil and holding coil) comprising the brake, coil body (Magnet) with an encapsulating winding and a tap, moving pressure plate, brake springs, brake disk, brake end shield, carbon-based friction surface between pressure plate and brake carrier. The two-coil system (acceleration coil and holding coil) ensure fast braking and releasing. The acceleration coil is switched on first. The high acceleration current of the acceleration coil results in a very short response time without reaching saturation limit, and then followed by the holding coil (entire coil). The pressure plate is forced against the brake disk by the springs when the electromagnet is de-energized. When the brake coil is connected to the corresponding DC voltage, the force of the springs is overcome by the magnetic force thus bringing the pressure plate into contact with the magnet and the brake disk moves clear and the rotor can turn.

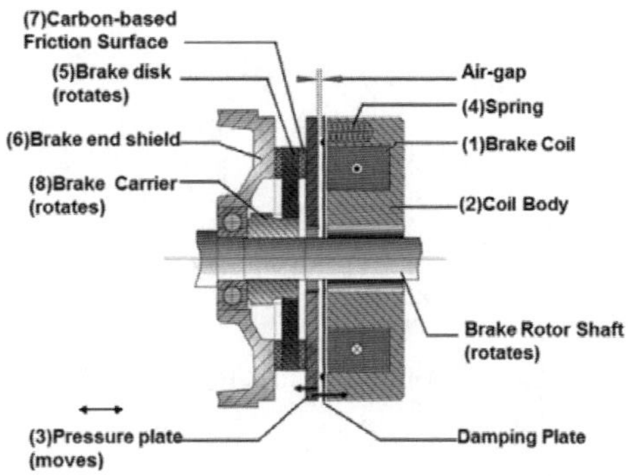

Figure 1. Principal parts of the Brake [11].

The friction caused by the sliding action eventually causes wear of carbon-based brake-discs. For high cycle rates therefore, a frequent replacement of brake-disks and/or the frictional surface is necessary. A too large air-gap between the dampening plate and stationary disk (caused by worn brake-disks) can result in a dysfunctional magnetic force; leading ultimately to a failure of the brake [11, 12].

3. Current and flux linkage analysis

The response of a brake at different air-gaps and operating conditions can be determined from the coil current (I_{coil}). Figure 2 shows one example for a coil current which contains four characteristics points representing the brake responses. The top points p1, p2 - represent the require current to move the brake and brake response time and bottom points d1, d2- represent the current and time when the brake hits the magnet (coil body). For an ideal case there should only be two points (p1 and d1), but the gap between pressure plate and coil body was adjusted manually, thus the air-gap was not equal throughout the surface which made the brake moved twice. These characteristic points are considered to estimate the air-gap using a machine learning algorithm, which will be described in section 4.3.

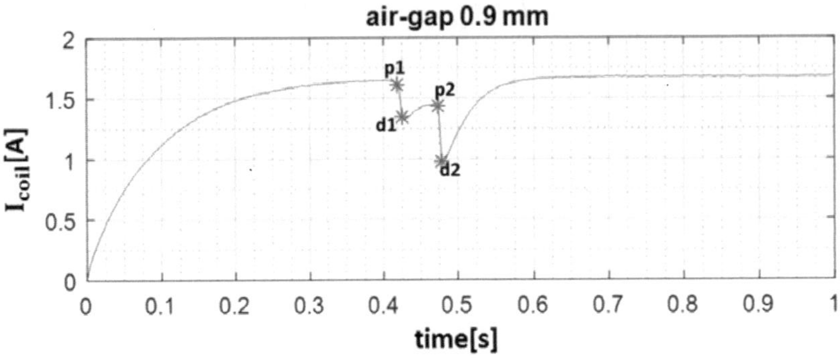

Figure 2. The four points (p1, p2, d1, d2) on current curve represent the brake's response properties.

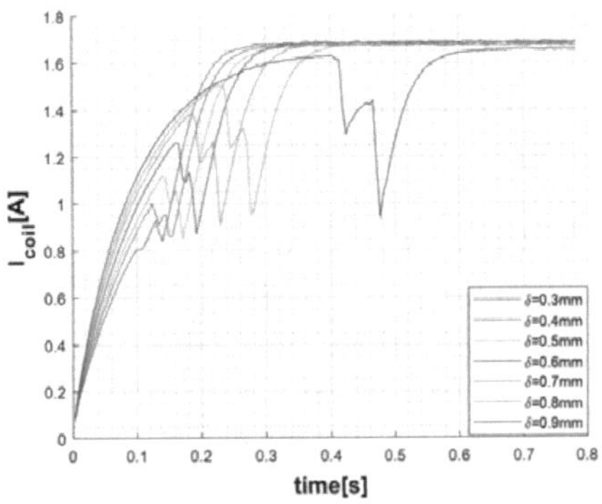

Figure 3. Coil Current (only for switch-on or brake release part) for different air-gaps.

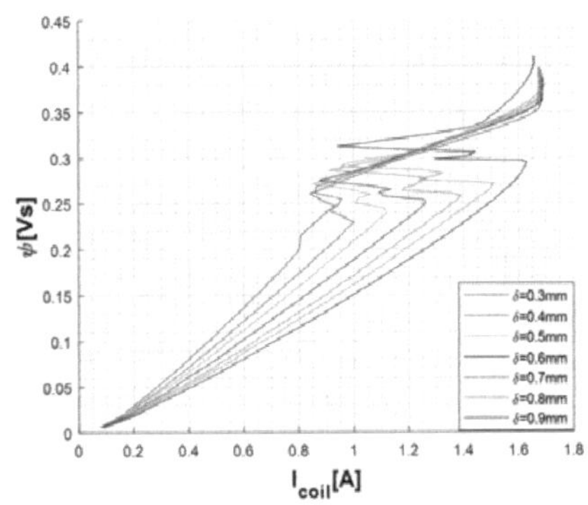

Figure 4. Flux linkage [Vs] for different air-gaps.

The measurement of magnetic properties of electromechanical systems is conventionally based on voltage, induced voltage and electromagnetic flux linkage [13]. Gadyuchko et al. in [13] presented a magnetic testing method which is considered here for electromechanical error detection. In their work, the magnetic characteristics $\Psi(\mathrm{i},\delta)$ are analyzed and utilized as testing/verification function of an electromagnet, where ψ represents magnetic flux linkage, i the current and δ the air-gap. The relationship between voltage (u) and magnetic flux linkage(ψ) is shown in Eq. (1)

$$u = iR + \frac{d\psi(i,\delta)}{dt}$$

$$u = iR + \frac{d\psi(i,\delta)}{d\delta}\frac{d\delta}{dt} + \frac{d\psi(i,\delta)}{di}\frac{di}{dt} \tag{1}$$

$$\text{where } \psi = \int (u - iR)dt \tag{2}$$

Gadyuchko et al. [13] showed that different conditions (or faults) of the electromagnetic brake i.e. short circuit between turns of the coil, broken leg of the membrane spring, wear from endurance testing with several million cycles etc. can be accurately evaluated from the $\Psi(\mathrm{i},\delta)$ character. Figure 4 shows Ψ for different δ in relation to i. This paper also considers this magnetic characteristics to be determined from electrical parameters (i,u) of the brake, in order to estimate the air-gap using a machine learning algorithm.

4. Data driven approach

Different Artificial Intelligence techniques have been widely developed in recent years in the field of machine condition monitoring. However, they are not extensively used in the industry [3]. In this work, a supervised multi-layer Artificial Neural Network (ANN) is used to determine the air-gap from the electrical parameters (i,u) measured from the brake. Figure 5 shows the different steps involved to estimate the air-gap using ANN. The steps are described in the following sub-sections and the output is discussed in section 5.

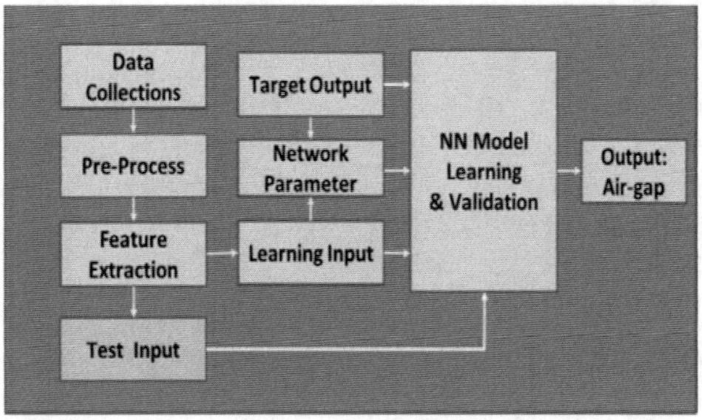

Figure 5. steps involved in the proposed data-driven approach.

4.1. Data Collection

A large number of data is recorded to learn the ANN from minimum air-gap of the brake to the maximum allowable air-gap. To achieve robust ANN solution for estimating air-gap, currents were recorded at three different operating voltages. For supervised learning purpose each recorded data was labelled for corresponding air-gap and voltage (Figure 6). The air-gap between the pressure plate and brake was adjusted by loosening or tightening the nuts as needed to achieve the necessary air-gap. Simplified diagram of data collection setup using oscilloscope as data acquisition are shown in Figure 7 and 8.

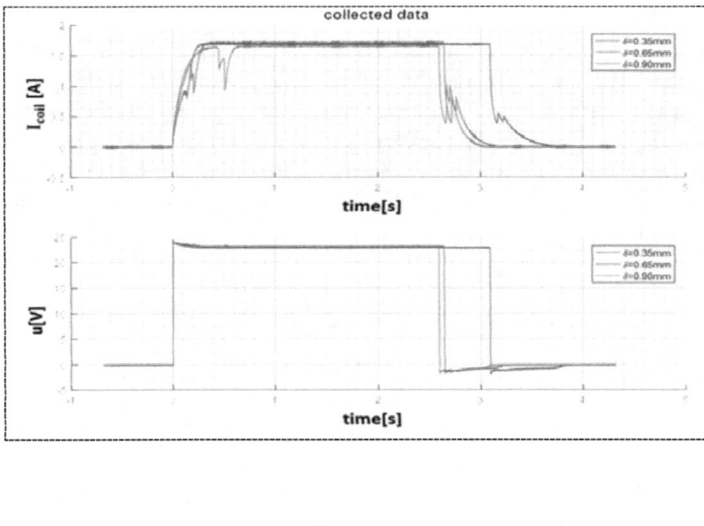

Figure 6. Example of recorded raw data for different air-gaps (sample rate 125kHz). The top figure shows coil currents during releasing and engaging of the brake and the bottom figure shows the corresponding voltages.

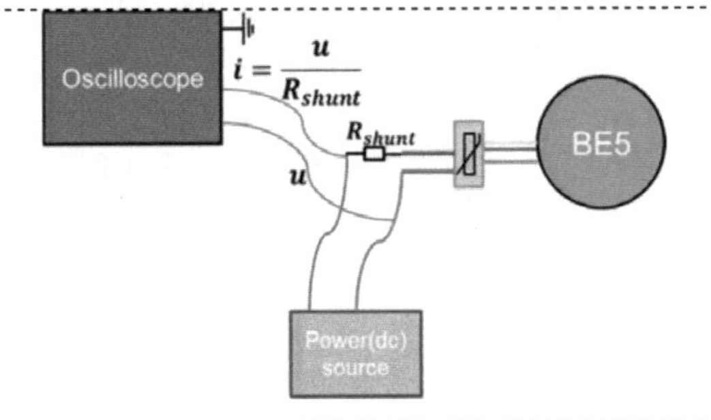

Figure 7. Simplified diagram of the experimental setup for data collection.

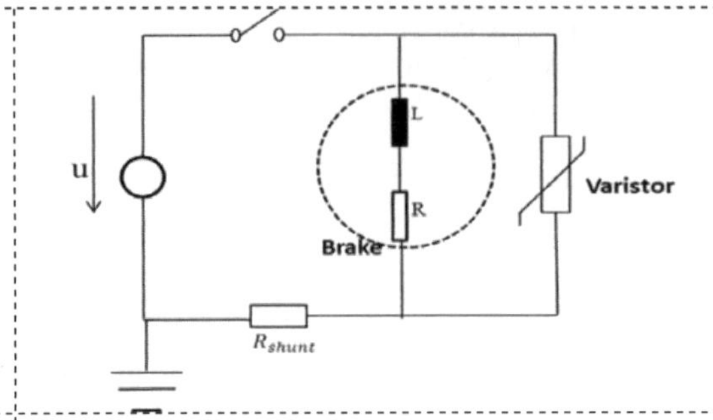

Figure 8. Simplified electric circuit of the brake's coil system with the experimental setup.

4.2. Pre-process

This step involves de-sampling and filtering of the data. Two filter algorithms are used to smooth the data to eliminate the noise within the data:

- Moving average filter:

$$y_s(i) = \frac{1}{2N+1}(y(i+N) + y(i+N-1) + \cdots + y(i-N)) \tag{3}$$

where, $y_s(i)$ is the smoothed value for the *ith* data point, N is number of neighbouring data points on either side of $y_s(i)$ and $2N+1$ is the span.

- Savitzky-Golay Filter:

$$Y_j = \sum_{i=-\frac{m-1}{2}}^{\frac{m-1}{2}} C_i y_{j+1} \quad , \frac{m+1}{2} \leq j \leq n - \frac{m-1}{2} \tag{4}$$

where, data is a set of n $\{x_j, y_j\}$ points ($j = 1, ..., n$), x is independent variable, y_j is an observed value and they are treated with a set of m convolution coefficients C_i.

4.3. Feature Extraction

A huge number of data directly to patter recognition algorithm may not be computationally feasible, require longer computational time and also too many parameters increases the complexity of network (ANN). In machine learning algorithm, the aim is to find useful features from data, where pattern recognition is hope to be easier. We calculated the characteristic points or the brake moving points described in Section 3 as the features as the learned input. We determined these points by the peak-detection algorithm based on second derivative approach (Figure 9). In one way of peak detection of signal processing is to find the downward-going zero-crossing of the first derivative of the data. But due to random noise in the experimental data zero-crossing give false peaks. Instead we calculated the second derivative of the original data and choose the maxima and minima points to locate the desired points on the original data.

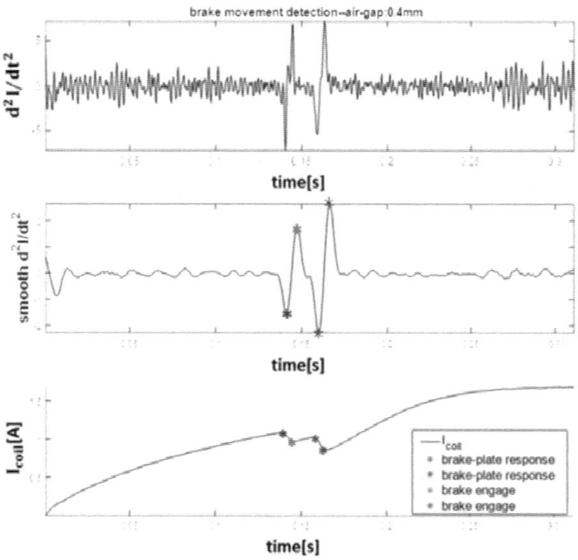

Figure 9. Example of feature extraction with second derivative approach

4.4. Learning input

Here, the input matrix used to learn the ANN is called the learning input. To achieve a robust ANN and to obtain optimize results, in this work we tested different learning conditions to train the ANN. ANN performance is compared between learning the ANN with I_{coil} data, which means no feature was extracted from the data and with the ANN learned with calculated feature data.

Different combination of input matrix depending on following parameters (*training conditions*):

- *Training dimension:* learning input (I_{coil}) contains all three or two or one voltage data set
- *Training t or t_{Imax} :* learning input (I_{coil}) contains time (t) data/not or only time (t) at maximum current (t_{Imax})

4.5. Target output

As the network was trained by supervised learning method, the corresponding target output for the input data is given as learning output matrix. Size of Target output or ANN output depends on *learning dimension* (Table 2), which means learning both air-gap and voltage or learning only air-gap.

4.6. Network Parameter

Network parameters are determined from the learning input and output matrix size.

Table 1. Contains the network parameters to determine the network size

Number of learned gap: 0.25: 0.05:0.9mm	$g = 14$
Number of measurements for each gap	$m = 10$
Number of data for each measurement or number of neurons in input layer	$N = 150(\text{with } I_{coil}) \times 2(\text{with } t) + 1(\text{with } t_{Imax})$ and 8 to $1(\text{with feature})$
Number of learned voltage	$V = 3,2,1$
Number of observation (neurons in input layer)	$g \times m$
Learning output	2 or 1
Number of hidden layer	2 or 1
Number of neurons at hidden layer-1	g
Number of neurons at hidden layer-2 (if exists)	V

4.7. ANN model

A supervised ANN is a general Function Approximator which can synthesize relationships between different variables constituting the input vectors and the output variable(s) [3]. The number of hidden layers depends on number of learning output described in Table 1. Figure 10 shows the proposed network structure of this work when considering *learning dimension* of 2.

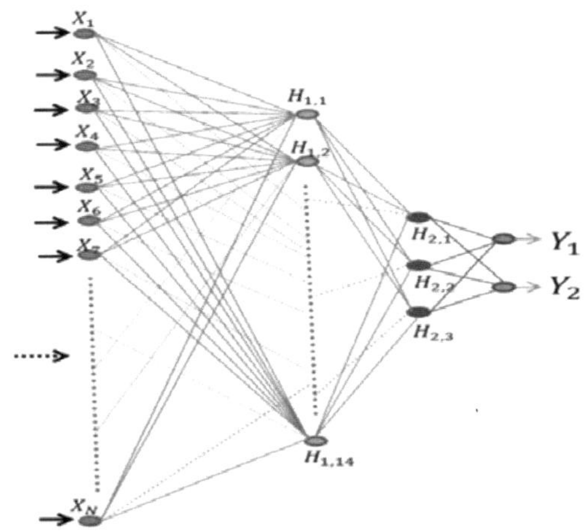

Figure 10. For learning two outputs (air-gap and voltage) the proposed structure of ANN consists two hidden layers.

5. Results and discussion

Performance of the ANN for different *training conditions* is evaluated by calculating the Mean-Square-Error (MSE). Considering *training dimension* (1 to 3), *learning dimension* (1 to 2), *Training t* (0 or 1) and *Training* t_{Imax} (0 or 1) total 18 combinations of *training conditions* were tested. Figure 11 shows the MSE (mm) of learned air-gap by ANN trained with I_{coil} for all 18 conditions. Table 2 shows comparison of different *training conditions* by comparing the learning performances of ANN trained withI_{coil}. Compared learning performances are: Number iterations required to estimated air-gap closest to target air-gap, required training time and average accuracy (%) of learned air-gap and MSE (mm) of learned air-gap. Table 2 shows also the average accuracy of estimated air-gap for Test data.

Table 2. Learning performance of ANN trained with I_{coil} for different training conditions.

Training dimension	3	3	2	2	1	1
Learning dimension	2	1	2	1	1	1
Training, *t*	0	0	0	0	0	1
Training, t_{Imax}	1	1	1	1	0	0
Num. of iteration	50	11	23	8	6	6
Training time(s)	30,185	4,96	12,303	2,517	1,807	6,133
Learning accuracy (%)	98,3	98,4	97,9	96,1	97,5	98,1
MSE (mm)	3E-04	3E-04	2E-04	1E-03	3E-04	5E-04
Estimation acc. Test data	93,1	92,1	91,3	94,0	87,1	83,1

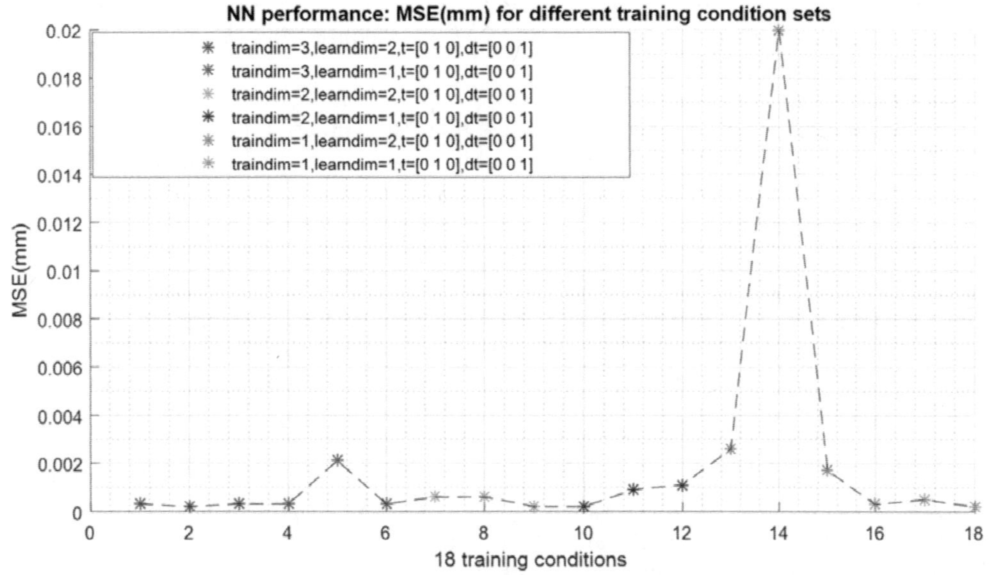

Figure 11. MSE of ANN output for different learning conditions.

The learned model was tested with newly recorded data (test data). For testing the robustness of the ANN, we recorded the data of the brake operated in different voltage source. The result shows the feature points describe best the mathematical relation between air-gap and the feature points, which the ANN can successfully learn from the training. In case of training the ANN with I_{coil}, the learning accuracy is very high but ANN does not show robustness for estimating air-gap for different voltage test data. The following figures 12 and 13 show the estimated air-gap for the test data by the leaned ANN model. The points are plotted as expected value (measured gap) versus predicted value (estimated gap). To visualize the error the heavy black line (ideal gap) is plotted in the figures showing how far or close the estimated values.

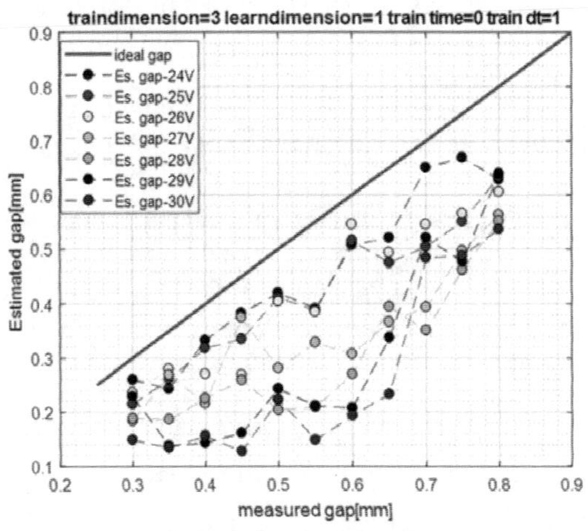

Figure 12. Estimated air-gap by ANN learned with I_{coil}(without feature extraction).

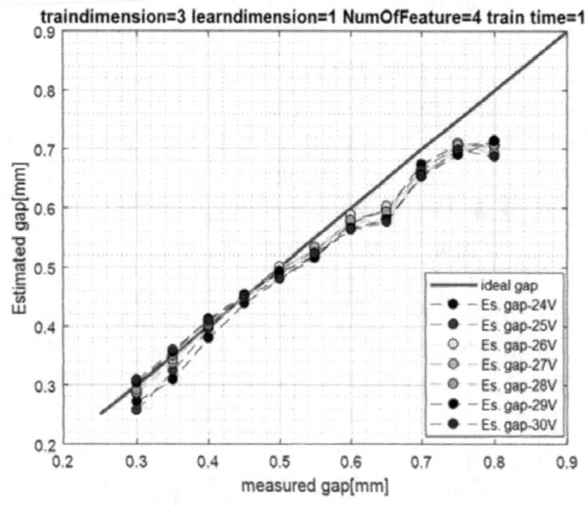

Figure 13. Estimated air-gap by ANN learned with brake response properties (with feature extraction).

The presented ANN model with coil current (I_{coil}) and feature data (p1, p2, d1, d2) as learning parameter have MSE between 1E-05 to 1E-04 mm; but output for test data of ANN learned with feature data give improved results. Figures 12 and 13 show the comparison of estimated air-gap by ANN trained with coil current and feature data or brake response properties (p1, p2, d1, d2).

6. Conclusions

Different Artificial Intelligence (AI) based machine condition monitoring is discussed in [1-3] which are generally classification of different faults. In this work we showed estimation of a physical parameter (air-gap) of electromagnetic system by using ANN regression model. By estimating the air-gap from coil current information, it is possible to determine the amount of wear of the friction surface more precisely and hence possible to make maintenance decision. The wear detection method shown in [8] also considered determining the peaks and time from coil current (similar as feature points described in section 3 and 4) to compare with stored data to generate warning message of wear. A method and device is proposed in [7] where mean of coil current is considered to determine the threshold value to monitor operating state for variety of brakes. The proposed approach in this work shows ANN learned with the feature points able to estimate air-gaps with high accuracy even the operating voltages of the test data are different than the learned data.

In conclusion, in this work, it is shown that condition monitoring of a complex electromechanical system can be performed without detailed analytical modeling or use of additional sensors. Through a discriminative learning of state of the art ANN models, the required mathematical function of condition monitoring of an electromagnetic brake through air-gap detection was successfully estimated. The work presented herein represents an interesting potential application in the area of condition monitoring for industrial usage; where handling and maintenance requirements of the machine can be substantially improved through data-driven methods.

Acknowledgement

The support and contributions of SEW-Euro drive GmbH to conduct this research work is appreciated.

References

[1] Gao R X 2003 *NATO Sci. Ser. III Comput. Syst. Sci.* **185** 167-188.
[2] Filippetti F, Franceschini G, Tassoni C and Vas P 2000 *IEEE Ind. Electron. Mag.* **47** 994-1004.
[3] Henao H, Capolino G, Fernandez-Cabanas M, Filippetti F, Bruzzese C, Strangas E, Pusca R, Estima J M, Riera-Guasp and Hedayati-Kia S 2014 *IEEE Trans. Ind. Electron.* **8** 31-42.
[4] Lubin T, Mezani S, and Rezzoug A *IEEE IEEE Trans. Magn.* **46** 1092-1099.
[5] Zarko D, Ban D and Lipo T A 2006 *IEEE IEEE Trans. Magn.* **42** 1828-1837.
[6] Chmela H 2005. EP. Patent No. 1435130 B1. (European Patent Office).
[7] Giessler and Folke Dr. 2011. DE. Patent No. 10147817C5. (Deutsches Patent-und Markenamt).
[8] Martin, John T, Reading and Berkshire GB 2015. DE. Patent No. 1011290B4. (Deutsches Patent-und Markenamt).
[9] SEW-Eurodrive. *DUE diagnostic unit option* (Accessed 2017 Jul 24) Available from: https://www.sew-urodrive.nl/products/motors/ac_motors/accessories_and_options/due/due.html
[10] SEW-Eurodrive. *Condition monitoring Systems and services for condition-based maintenance* (Accessed 2017 Jul 24) Available from: http://download.sew-eurodrive.com/download/pdf/16846028.pdf
[11] SEW-Eurodrive *SEW Brakes-Service and Maintenance* (Accessed 2017 Jul 24) Available from: http://v5.ptpilot.com/TrainingPages/resources/OnlineTraining/Maintenance AndRepairInstructions/BrakeServiceAndMaintenance.pdf
[12] SEW-Eurodrive 2006 *Drive Engineering - Practical Implementation: SEW Brake Discs / Drive Engineering – Practical Implementation* (Accessed 2017 Jul 13) Available from: http://download.sew-eurodrive.com/download/pdf/11354615.pdf.
[13] Gadyuchko, Andrey, Vladimir Kireev, and Soeren Rosenbaum 2015 *ETG/GMM Symposium Innovative Small Drives and Micro-Motor Systems Proceedings of IKMT 2015* pp. 1-6.

Integration of real-time non-surfactant emulsion fuel system on light duty lorry

Muhammad Adib Abdul Rashid*, **Ahmad Muhsin Ithnin, Wira Jazair Yahya, Nur Atiqah Ramlan, Nurul Aiyshah Mazlan, Dhani Avianto Sugeng**

Vehicle System Engineering, Malaysia-Japan International Institute of Technology, Universiti Teknologi Malaysia, Jalan Sultan Yahya Petra, 54100 Kuala Lumpur, Wilayah Persekutuan Kuala Lumpur, Malaysia

*Corresponding author email: madib52@live.utm.my

Abstract. Interest in water-in-diesel emulsion fuel (W/D) grows because of its advantages in improving fuel efficiency, reducing greenhouse emissions and retaining the quality of the lubrication oil. Recently, a device called Real-Time Non-Surfactant Emulsion Fuel System (RTES) have successfully created an emulsion without surfactant for a 5kW single-cylinder diesel engine generator. This study integrates the RTES into a light duty lorry, and the effect of the integration is investigated. The lorry was tested on a chassis dynamometer with a controlled 16.6% water ratio. The results show how fuel consumption is reduced by 7.1% compared to neat diesel. Moreover, the exhaust emission of Nitrogen Oxides (NO_x) is reduced by 52%, while as observed in other works, carbon monoxides (CO) emission also increased, in this case by 41.6%. This integration concluded to retain similar benefits and disadvantages as tested on the 5.5kW diesel generator.

1. Introduction

Documentation of ideas in water-in-diesel (W/D) emulsion fuel dated back from as early as 1931. Among its advantages are the improvement of fuel efficiency, the reduction of greenhouse gases emissions and the retainment of the quality of lubrication oil [1-4].Transportation is the main contributor of NO_x emission so integrating W/D into this sector will significantly reduce it [5].

Micro-explosion phenomena are the main reason for the improvement of the fuel efficiency and the reduction of the particulate matter (PM). It causes a "second atomization" by rapid evaporation of the water, so the diesel droplets become finer and more evenly distributed inside the combustion chamber [6-8]. Despite the benefits, emulsion fuel has not taken off in the industry because of its dependency on additives to mix the two immiscible liquids. In solving this problem, a study on a new emulsification technology was performed; it is called the Real-Time Non-Surfactant Emulsion Fuel System (RTES). This RTES was originally designed for a stationary 5kW diesel electric generator, and it has been proven to retain the similar benefits of conventional emulsion fuels [3]. It eliminates the dependency of additives by supplying freshly produced emulsion fuel to the vehicle. RTES mixes diesel and water by using three primary mechanisms; (1) water atomization, (2) high shear mixer and (3) ultrasonic transducer. Figure 1 shows the configurations of RTES.

In this research, the feasibility of RTES integrated into a 2.8L four-cylinder diesel engine vehicle was investigated. Its goal was to include it into the fuel supply system, to run the vehicle continuously without stalling, and to operate the system at different vehicle speed. Furthermore, the effect of this integration in term of the fuel consumption and exhaust emission were also studied using WVU-5 peak drive cycle on a chassis dynamometer.

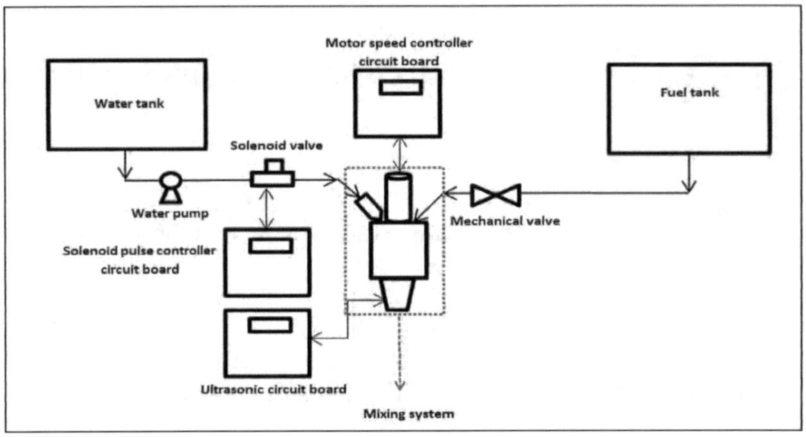

Figure 1. Real-Time Non-Surfactant Emulsion Fuel System (RTES*)* [3].

2. Experiment method

2.1. Test vehicle
In this study, a 1-ton lorry vehicle model Isuzu NHR categorised as a light-duty lorry was used as the test vehicle. The engine of this vehicle is a 2.8L 4-cylinder diesel engine. The truck was designed to fulfil Euro 2 emission regulation, and its overall specifications are given in Table 1.

Table 1. Test Vehicle Specifications.

Parameter	Specification
Model	Isuzu NHR 55E (4JB1)
Fuel injection system	Direct injection
Cylinder	4 cylinders, in-line
Cooling system	Water-cooled
Displacement (L)	2.8
Bore x stroke (mm)	93 x 102
Compression ratio	18.2:1
Fuel tank capacity (L)	75

2.2. Test fuel
Diesel and water were the only components for the emulsion fuel. Tap water was used, and its characteristics are presented in Table 2 [9]. Euro 2 diesel was used with its characteristics tabulated in Table 3 [10]. The percentage of water for emulsion utilised a controller to control the amount of water supplied. The actual average water percentage is defined with a flow rate sensor during the testing. The emulsion used in the testing is with 16.6% water percentage.

Table 2 Characteristics of tap water. [9]

Specification	Tap water
Density, g/cm³, 25 °C	1.02412
Specific conductivity, µS/cm, 25 °C	0.0532
Viscosity, milipoise, 25 °C	9.02
Vapour pressure, mm Hg, 20 °C	17.4
Isothermal compressibility, vol/atm, 0 °C	46.4 E-6
Surface tension, dyne/cm, 0 °C	72.74
Specific heat, J/g °C, 17.5 °C	3.898
Temperature of Maximum Density, °C	-3.25
Freezing point, °C	-1.91

Table 3. Malaysian Euro 2 Diesel characteristics [10].

Properties	Unit	Value
Calorific Value	MJ/kg	45.28
Cloud Point	°C	18
Density @15°C	kg/L	0.8538
Total Sulphur	mass %	0.28
Viscosity	cSt	7.6 @20°C 5.1 @40°C 4.0 @60°C
Distillation Temperature, 90% recovery	°C	367.9
Flashpoint	°C	93.0
Pour Point	°C	12
Cetane Number		54.6
Carbon	wt %	84.1
Hydrogen	wt %	12.8
Sulphur	wt %	0.2
Nitrogen	wt %	< 0.1
Oxygen	wt %	3.9
Dielectric constant		2.1

2.3. Integration of RTES

Throughout the integration of RTES, there were three main factors considered. Firstly, the non-surfactant emulsion stability. The produced emulsion fuel only lasted a short time before the water started to visually separate from diesel. Therefore, the distance of the RTES from the engine was the major factor affecting the integration design. Secondly, the space availability and RTES form factor. The engine compartment has limited space, but RTES small form factor made it possible to mount it as close possible to the engine. Thirdly, the fuel return line. The short life span of the produced emulsion prevents the return of emulsion fuel to the fuel tank. Returning the fuel to its tank was not possible because separated water from the emulsion fuel might ultimately damage the vehicle engine. Therefore, the return line was routed back to RTES. The fuel line of the vehicle with RTES is shown in Figure 2. This integration only modified the fuel line and electrical wiring without any modifications on the engine. Addressing the short stability period of emulsion fuel, after integration based on the three factors, samples of fuel from the return fuel line were withdrawn. A quick observation concluded that the emulsion was still intact with no apparent water layer. This showed that the emulsion fuel was stable enough even after the injector pump and engine.

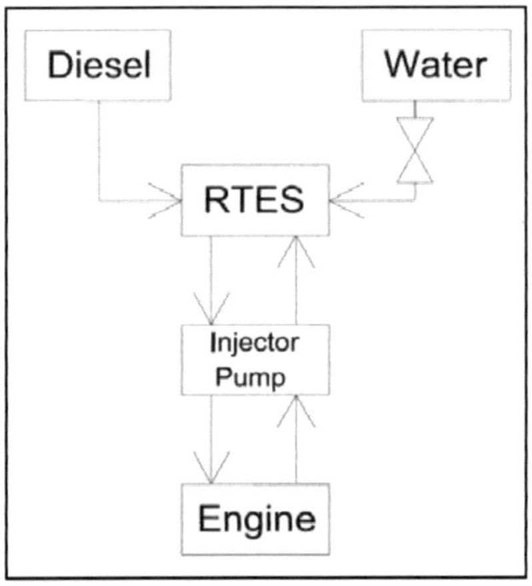

Figure 2. Vehicle fuel line with RTES.

2.4. Drive cycle

The drive cycle selected to test the vehicle followed the West Virginia University (WVU) 5-Peak drive cycle. The cycle was designed specifically for truck testing [11] with a drive cycle of 900 seconds and 8 km equivalent distance. The cycle peaks at 5 points e.g. 32.19, 40.23, 48.28, 56.33 and 64.37 km/h. Figure 3 shows the 5-peak drive cycle graph.

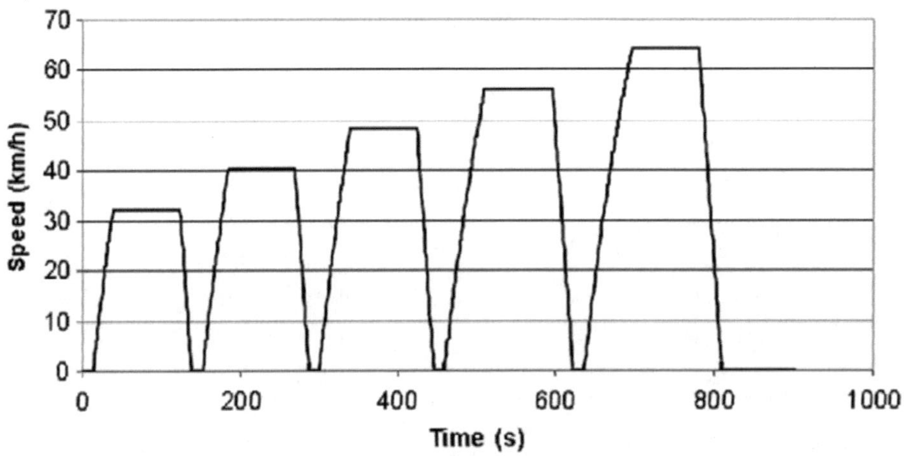

Figure 3. WVU 5-Peak drive cycle. [11]

2.5. Measurements setup

Ono Sokki FP-2240H volumetric flow detector was used to measure the fuel consumption of both fuels, and the device measured the flow located after the diesel tank. Only the diesel consumption of both fuels was taken account as the fuel consumption, thus in the case for emulsion, water was not considered. As for the exhaust emission, a Testo 350 emission analyser was used to record it. The probe of the emission analyser was placed at the tailpipe. The emission analyser measured Nitrogen Oxides (NO_x) and Carbon Monoxide (CO) in parts per million (ppm). All measuring devices simultaneously began and ended recording data with the drive cycle.

3. Results and discussion

During the testing period, RTES system supplied the non-surfactant W/D emulsion fuel to the engine, and the vehicle completed the WVU 5-peak drive cycle on a chassis dynamometer with speed ranges from 0 – 64.37 km/h without stalling or unusual behaviour.

3.1. Fuel Consumption

The result of the fuel consumption of both fuels is shown in Figure 4. The non-surfactant emulsion fuel with 16.6% of water (Emulsion 16.6) had a lower fuel consumption by 7.1% compared to neat Euro 2 diesel. This result complied with the previous study of RTES on a 5kW diesel generator which yielded a reduction of the specific fuel consumption by 3.89% [10]. The reduction of fuel consumption was mainly believed to be caused by the micro-explosion phenomenon.

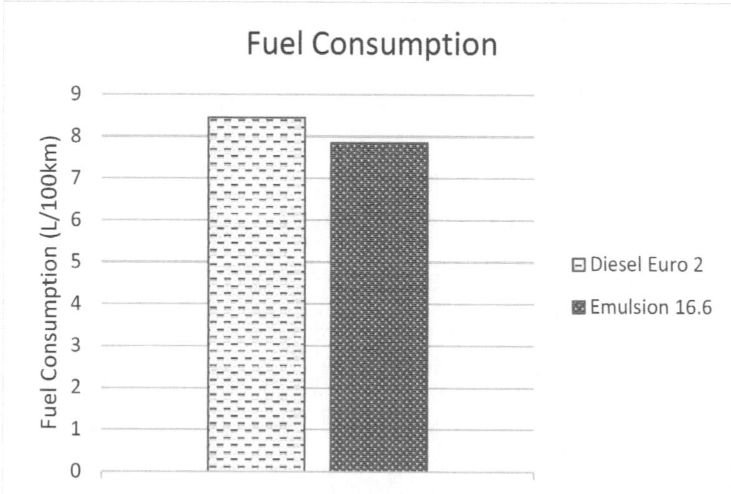

Figure 4. Fuel consumption of both fuel.

3.2. Emission Analysis

Comparison of Nitrogen Oxides (NO_x) and Carbon Monoxide (CO) of both fuels are shown in Figure 5 and Figure 6 respectively. The non-surfactant emulsion from RTES significantly reduced the NO_x emission. Under the WVU 5-peak drive cycle, 52% of average NO_x reduction for Euro 2 diesel was achieved. Bertrand D. also had a similar finding with the NO_x reduction of 50% [12]. The reduction of NO_x is primarily caused by lower peak temperature of the combustion [13]. The emulsion fuel yields lower peak temperatures because the water has the high latent heat of vaporisation that absorbs the heat inside the combustion chamber during phase transition. However, the lower combustion temperature also leads to the increase in CO emission [8]. In Figure 6, a notable increase of CO by an average of 41.6% is shown. Several other researchers also observed the similar increase in CO emission [14-15].

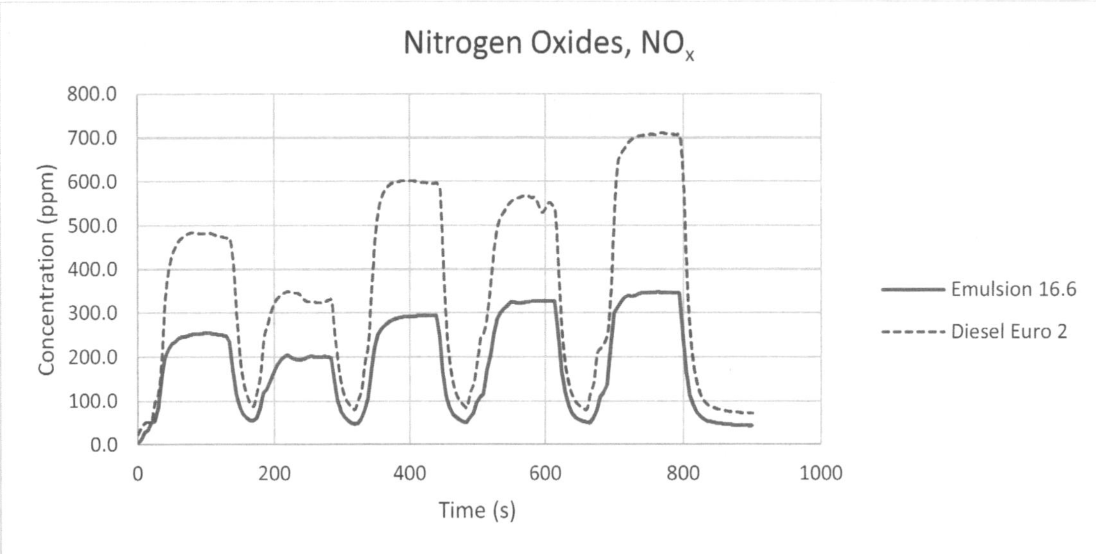

Figure 5. NO_x Emission on WVU 5-Peak Cycle.

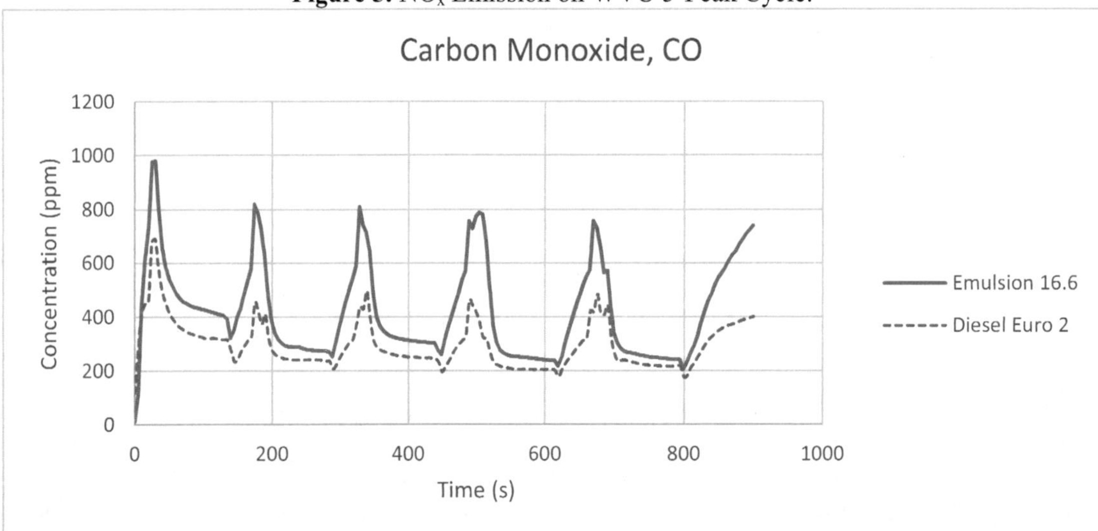

Figure 6. CO Emission on WVU 5-Peak Cycle.

4. Conclusions

The integration of RTES into a light duty lorry was successfully performed, and the concept was proven. This integration made it possible for a vehicle to take advantage of an emulsion fuel without the dependency on surfactant. This research also showed that the integration retained the same benefits of previous works on W/D of reducing the fuel consumption by 7.1% and NO_x emission by 52%. However, the significant increase of CO by 41.6% needs to be addressed. These results were similar with the previous study done on the 5.5kW diesel generator. Further research on non-surfactant emulsion fuel with innovative technology to reduce the CO emission and better integration designs for different kinds of application are recommended.

Acknowledgments

The authors would like to express their gratitude to the Malaysia-Japan International Institute of Technology (MJIIT) and Kyowa Electronics Instrument Ltd. providing the research with the needed facilities and equipment. Highest appreciation goes to the Ministry of Higher Education, Malaysia for their financial support through FRGS (R. K130000.7843.4F734).

References

[1] Joseph A Vance 1931 *Liquid fuel* US1926071A.
[2] A. K. Hasannuddin, J. Y. Wira, S. Sarah, W. M. N. Wan, S. Aqma, A. R. A. Hadi, N. Hirofumi, S. A. Aizam, M. A. B. Aiman, S. Watanabe, M. I. Ahmad, and M. A. Azrin, "Performance, emissions and lubricant oil analysis of diesel engine running on emulsion fuel," 2016.
[3] N. A. Mazlan, W. J. Yahya, A. M. Ithnin, and M. A. Ahmad, "The effect of tap water emulsified fuel on exhaust emission of single cylinder compression ignition engine."
[4] A. K. Hasannuddin, J. Y. Wira, S. Sarah, M. I. Ahmad, S. A. Aizam, M. A. B. Aiman, S. Watanabe, N. Hirofumi, and M. A. Azrin 2016 *Energy* **94** 557–568.
[5] O. V Lozhkina and V. N. Lozhkin 2016 *Transp. Res. Part D*, vol. **47** 251–264.
[6] F. Y. Hagos, A. R. A. Aziz, and I. M. Tan, "Water-in-diesel emulsion and its micro-explosion phenomenon-review," in *2011 IEEE 3rd International Conference on Communication Software and Networks*, 2011, pp. 314–318.
[7] M. E. A. Fahd, Y. Wenming, P. S. Lee, S. K. Chou, and C. R. Yap 2013 *Appl. Energy* **102** 1042–1049.
[8] A. M. Ithnin, H. Noge, H. Abdul Kadir, and W. Jazair 2014 *J. Energy Inst.* **87** 273–288.
[9] N. A. Ramlan, W. J. Yahya, A. M. Ithnin, A. K. Hasannuddin, S. A. Norazni, N. A. Mazlan, D. A. Sugeng, N. D. Bahar, and T. Koga 2016 *Energy Convers. Manag.* **130** 192–199.
[10] A. M. Ithnin 2015 *Real-Time Water-In-Diesel Emulsion Fuel Production System for Diesel Electric Generator* (Universiti Teknologi Malaysia).
[11] N. Clark, J. T. Messer, D. L. McKain, W. Wang, R. M. Bata, M. Gautam, and D. W. Lyons 1995 *Use of the West Virginia University Truck Test Cycle to Evaluate Emissions from Class 8 Trucks*.
[12] B. D. Hsu 1986 Combustion of Water-in-Diesel Emulsion in an Experimental Medium Speed Diesel Engine.
[13] A. Alahmer, J. Yamin, A. Sakhrieh, and M. A. Hamdan *Energy Convers. Manag.* **51** 1708–1713.
[14] K. A. Subramanian 2011 *Energy Convers. Manag.* **52** 849–857.
[15] A. B. Koc and M. Abdullah 2013 *Fuel Process. Technol.* **109** 70–77.

IOP Conf. Series: Materials Science and Engineering **257** (2017) 012052 doi:10.1088/1757-899X/257/1/012052

Three-dimensional finite element magnetic simulation of an innovative multi-coiled magnetorheological brake

Ubaidillah[1*], A N S Permata[1], S A Mazlan, D D D P Tjahjana[1] and P J Widodo[1]

[1]Mechanical Engineering Department, Faculty of Engineering, Universitas Sebelas Maret, Jl. Ir. Sutami 36A, Kentingan, Surakarta, 57126, Central Java, Indonesia
Corresponding email: ubaidillah@uns.ac.id
[2]Malaysia-Japan International Institute of Technology, Universiti Teknologi Malaysia, Jalan Sultan Yahya Petra, Kuala Lumpur, 54100, Malaysia

*Corresponding author email: ubaid.ubaidillah@gmail.com

Abstract. This research delivers a finite element magnetic simulation of a novel disk type multi-coil magnetorheological brake (MR brake). The MR brake axial design had more than one coil located outside of the casing. This design could simplify the maintenance process of brakes. One pair of coils was used as the representative of the entire coil in the simulation process, and it could distribute magnetic flux in all parts of the electromagnetic. The objective of this simulation was to produce magnetic flux on the surface of the disc brake rotor. The value of the MR brake magnetic flux was higher than that of the current MR brake having one coil with a larger size. The result of the simulation would be used to identify the effect of different fluids on each variation. The Magneto-rheological fluid MRF-132DG and MRF-140CG were injected in each gap as much as 0.50, 1.00, and 1.50 mm, respectively. On the simulation process, the coils were energized at 0.25, 0.50, 0.75, 1.00, 1.50, and 2.00 A, respectively. The magnetic flux produced by MRF-140CG was 336 m Tesla on the gap of 0.5 mm. The result of the simulation shows that the smaller the gap variation was, the higher the magnetic value was.

1. Introduction

Magnetorheological fluids (MRFs) have become a substitution privilege, since its rapid reaction for plenty x-by-wire applications, such as brakes, valves, dampers, clutches, and so on [1]. The novel design of a modular MR valve has been proposed by employing complicated flow path to enhance its performance [2]. MR elastomer possible utilization was investigated from three main points of view, such as its enforcement in automobile devices, prosthetic apparatus, and prospective purposes [3]. The magnetorheological brake (MR brake) is still extraordinary in the commercial markets [4].

Magneto-rheological (MR) brake is a type of MRFs application which is researchers feel strongly about. The fact is due to its various utilizations make MR brakes develop rapidly in several discipline area. It has been recently introduced by several automobile industries [5,6]. Analysis and optimal design have been investigated to enhance more capability of MR brakes, for instance, material selection, sealing, mixing scheme, and MR fluid selection [4,7]. Additionally, an MR brake which applied for prosthetic knee joint was carried out to gain its development by Gudmundsson et al. [8]. An optimal geometrical design of an MR brake considering.

It generates an interesting concern of MRFs implementation. The primary working mode of MRFs application in the MR brake is shear mode. The breaking effect is created from friction between idle and rotate a component of magnetic field vulnerable viscosity. Thus, disc type, drum type, T-shaped MR brakes, multiple disks, and multiple layers are purposed and determined to become plenty types of MR brake [9]. However, it suggests to improving the more valuable field of magnetic flux by employing the shape and numbers of the coil [10].

MR brake with drum and disk brakes are simple to manufacture. It has a cylindrical shape of the rotor and the magnetic field is performed in the radial direction. However, the results are unsatisfied enough to apply in automotive industries [11]. The T-shaped rotor brake is most accessible than other MR brake above however it contains uneasy to manufacture. This design is still rare in the literature. York et al. [12] showed an initial example used as a torque load simulator system producing a massive torque amount. Other tools requiring high torque are clutches for automatic transmission [13]. Shiao and Nguyen. [14,15] developed multi-coiled MR brake in recent years. It performed a promising brake torque due to its numbers of coils which produced much more magnetic field [14]. Its response was scrutinized beneath the free move inertial mass. The breaking time will be decreased when the boosting current exercised as the braking torque reaction [3].

Since the multi-coiled MR brake achieved an enormous amount of magnetic field, numerous researchers were desired to expand several brand new shapes. Shiao and Nguyen [14,15] built a drum brake MR brake and deliver with variations of outer and inner rotors. It managed yield resistance in the fluid, hence producing torque area for the brake [15]. Multi-coiled MR brake with disc brake performed valuable result which was firstly constructed by our group. It consists of six couples of coils and one disc brake. Due to its axisymmetric coils arrangement, it crossed much more rotor disc brake. As a result, a high value of the magnetic field was exhibited [16].

This study contributes on designing a unique type of multi-coiled MR brake for radial mode breaking system which has never been proposed before. It is that valuable for investigating the magnetic flux within the gap (breaking area) using the 3D approach of magnetic simulation. To looking after the flux density, three-dimensional of MRFs were incorporated in simulation a brief view of resulted prototype was also presented at the end of this paper.

2. Design and working principle

The coils configuration of MR brake system is displayed in Figure 1(a). Either using single or multiple coils arrangements in the earlier design is typically located inside of MR BRAKE body. Nevertheless, the suggested layout in the current MR BRAKE installs the coils on the outside of the cover. It is expected to construct more intensive magnetic flux in the radial field. The outer coils are seated on the magnetic core hubs to contact its two adjacent magnetic poles. Subsequently, the flux will pass from one pole via the magnetic core hubs, to another pole with a particular characteristic. Formerly, it will travel into the MR fluid gap, across the rotor. Finally, it will be back to the MR fluid gap and into the two adjacent poles as can be seen in Figure 1(b). By such an innovative design, the rotor disk will be entered with various coils that associated to others.

(a)

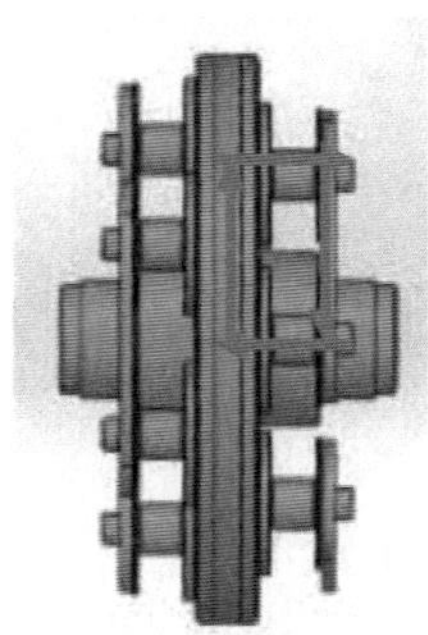

(b)

Figure 1. A model of axial multiple MR brake: (a) Isometric view, (b) Top view.

In Figure 1(b), the magnetic flux spreads over all around the disk surfaces. The result is that an appropriate shear stress will be exhibited, hence producing an area torque for the brake. From the operating concept, the head area of each pole will be triggered with an official current to stimulate the MR fluid area in the channel. Depending on the capacity of the rotor, windings (coils) and manufacture abilities, particular adjustment, and numerous of poles can be selected, from the small number, for instance, four or six poles or more. The earlier research that is integrating radially multi-coiled MR brake was founded by Shiao and Nguyen [14]. Meanwhile, the performance of 12-poles of the MR brake has been evaluated and documented in this study due to its balance between size constraints and braking performance in our future operations. The magnetic field strength in the MR fluid can be enlarged by accommodating the pole arrangement and the rotor's radial and axial dimensions. Finally, the analogous shear stress on the cylindrical surface is expanded significantly, which supports to increase the braking torque.

The components for MR brake device can be clearly seen in Figure 2 through partial cross section mode. The materials that applied in the brake have a crucial impact on the magnetic field as well as the anatomical and magnetic characteristics. Since the magnetic field simulation has a key effect on the MR brake performance, the components selection is mandatory. Considering the cost, permeability, and availability, AISI 1018 steel was preferred as the magnetic component in the MR brake. The magnetic properties of AISI 1018 are displayed in Table 1. The aluminum alloy was exploited as the non-ferromagnetic material to build the shaft and the MR BRAKE covers. The AWG-21 copper wire was employed for the coils. The polyethylene was utilized for the bobbins for the electrical insulation property.

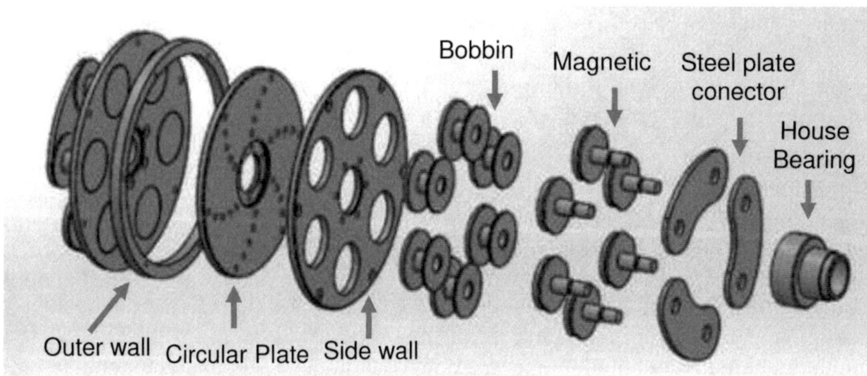

Figure 2. Exploded view of the MR brake.

Table 1. Material of MR brake.

Part	Material
Bobbin	Aluminum
Support Bobbin	Steel
House Bearing	Bronze
Bean Shaped Plate	Steel
Circular Plate	Aluminum
Hollow Circular Plate	Aluminum
Disc	Steel

3. Magnetic circuit analysis

Simulation process was used electromagnetic induction thus Maxwell equation (1) is necessary:

$$\nabla x E = \frac{\partial B}{\partial t} \tag{1}$$

where, ∇x is curl operation, E is an electric field, and B is a magnetic field. On the following equation is functioned of t (time). A simulation based on finite element model is necessary to trim the error computation of the MR brake. ANSOFT Maxwell 14 is operated to cultivate the magnetic field patterns of the expected design in a 3D environment. The electric circuit analogy of the multipoles MR brake as displayed in Figure 3 was promoted. Figure 3 also demonstrates the magnetic flux flow within the device. The magnetic device mainly consists of two coil bobbin sets, connector, and the rotor disc. The coils generate magnetic flux density to almost every part of the MRFs.

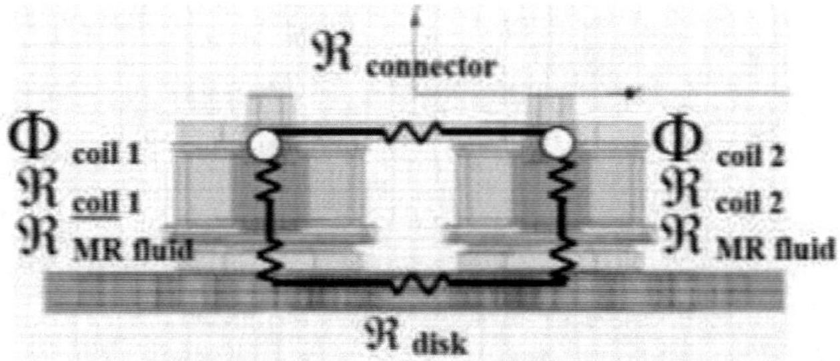

Figure 3. Magnetic circuit of a coupled coil.

One loop of differentiated Kirchhoff Laws was used to calculate magnetic circuit as portrayed in Figure 3. The reluctance of the core 1, reluctance of the core 2, reluctance from the connector, reluctance from the disc, and the reluctance of the MR fluid is personified by the symbols $\Re_{coil1}$, $\Re_{coil2}$, $\Re_{connector}$, $\Re_{disc}$, $\Re_{MR\ fluid}$, respectively. The reluctance of each section can be determined by the following equation below,

$$\Re = \frac{L}{\mu A} \tag{2}$$

where, L, μ, A represent the effective length, magnetic permeability, and effective area of the magnetic flux, correspondingly. While, the equation below displays the total magnetomotive force is the summation of the magnetomotive force produced by all of the parts in one loop. The magnetomotive force is directly proportional to the magnetic flux and reluctance as declared in below (Equation (3)),

$$\Phi_1 + \Phi_2 - \Re_{coil1} - \Re_{mrfluid} - \Re_{disc} - \Re_{mrfluid} - \Re_{coil2} - \Re_{connector} = 0 \tag{3}$$

The magnetic flux $\phi 1$ and $\phi 2$ rely on the number of coil turns and electric current pass through the coils. As a result, Equation (3) can be rewritten as the following form (Equation (4)),

$$N_1 I_1 + N_2 I_2 - \Re_{coil1} - \Re_{mrfluid} - \Re_{disc} - \Re_{coil2} - \Re_{connector} = 0 \tag{4}$$

where, N and I are numbers of wire turns in each coil and electric current passing through the coils, subsequently.

4. Simulation setup

CAD was employed to create the 3D design of proposed prototype. Then, it is converted to ANSOFT MAXWELL to perform the magnetic simulation. Nonmagnetic components were assigned as aluminum and polyethylene, while others were attributed as steel AISI 1008 for magnetic objects. Two types of MRFs i.e. MRF-132DG and MRF140CG were applied by using B-H curve from LORD technical data sheet in FIGURE. Several currents were attached to the 3D simulation, 0.25, 0.50, 0.75, 1.00, 1.50, and 2.00 A for each varied MRFs. Also, 0.50 mm, 1.00 mm, and 1.50 mm gap were operated to chase the optimum design. The measurement line was added in the middle of the gap which is located beneath of the core. For this reason, it is a proper spot to earn the magnetic flux result. Table 2 below shows the properties of each MRFs.

Table 2. MRFs Properties.

Properties	*MRF-132DG*	*MRF-140CG*
Carrier Liquid	*Hydrocarbon*	*Hydrocarbon*
Filling particle	*Carbonyl iron*	*Carbonyl iron*
***Density* g/cm^3**	3,09	3,64
Viscosity, Pa-s @ 40	0,112 ± 0,02	0,280 ± 0,070
Solids content by weight, %	80,98	85,44
***Flash point,* °C (°F)**	>150 (>302)	>150 (>302)
***Operating Temperature,* °C (°F)**	-40 to +130 (-40 to + 266)	-40 to +130 (-40 to + 266)

5. Results and discussion

The straight line in Figure 4 illustrates the position to measure the magnetic flux value. The measuring line was placed at the center position of the fluid MRFs used with 40 mm and it divided into 40 points. The position of the measuring line is located along the core diameter is pulled straight in the direction of at a central point.

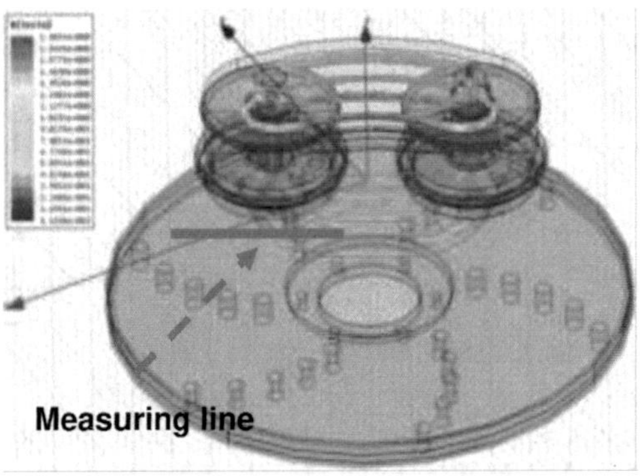

Figure 4. 3D Magnetic Flux Flow and The Measuring Line Position.

Figure 4 shows the direction of the magnetic flux generated by the coil pair. Clearly demonstrated, the distribution is done at the center of a copper coil then passes through the components that are in one loop. The highest magnetic flux is shown located at the center of the copper windings, while the lowest value was achieved when the magnetic flux passing through the disc after previously passing MRFs

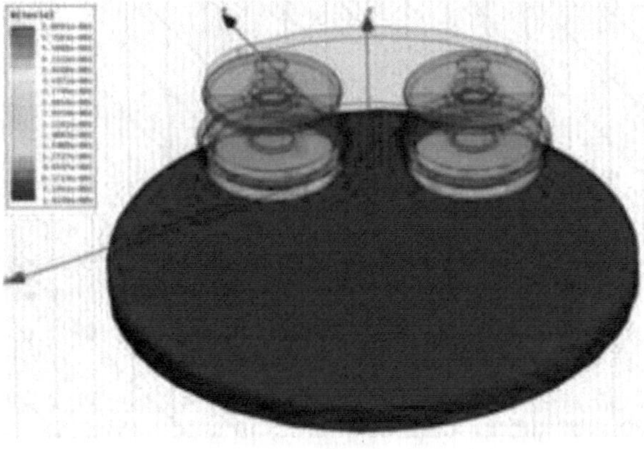

Figure 5. Axially Perspective of Magnetic Flux Flow.

IOP Conf. Series: Materials Science and Engineering **257** (2017) 012052 doi:10.1088/1757-899X/257/1/012052

Results directions of magnetic flux obtained using two-dimensional directions as shown in Figure 5. Same as depicted in Figure 4, the greatest value lies at the core of the copper windings and magnetic flux smallest value contained on the disk is used. In two dimensions is obvious leaps magnetic flux. Leaps the magnetic flux looking for the nearest magnetic material to be passed.

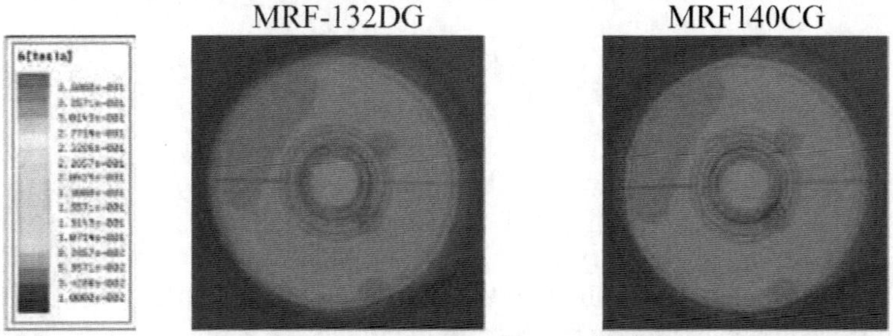

(a) Current 0.25 A

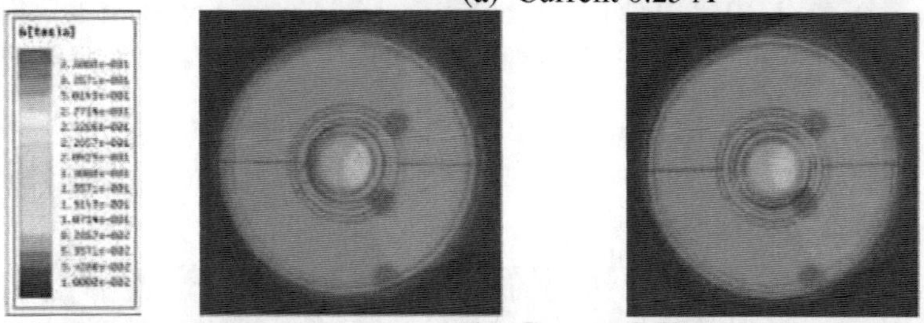

(b) Current 0.50 A

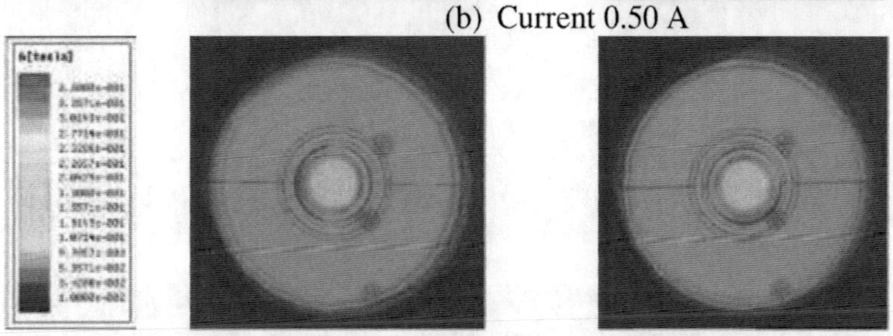

(c) Current 0.75 A

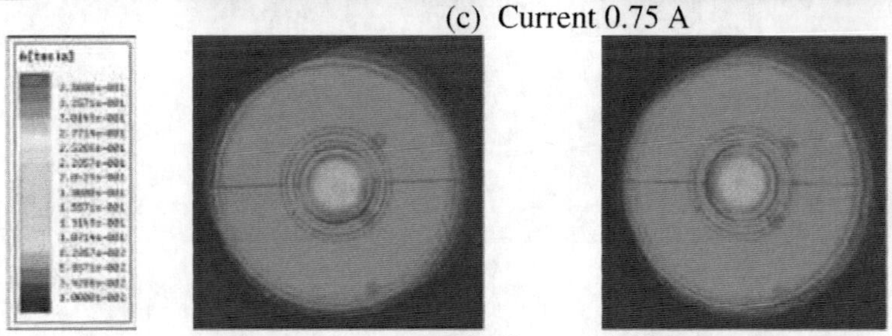

(d) Current 1.00 A

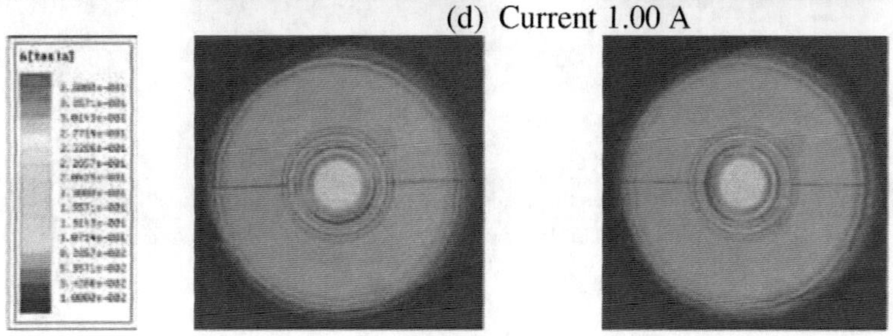

(e) Current 1.50 A

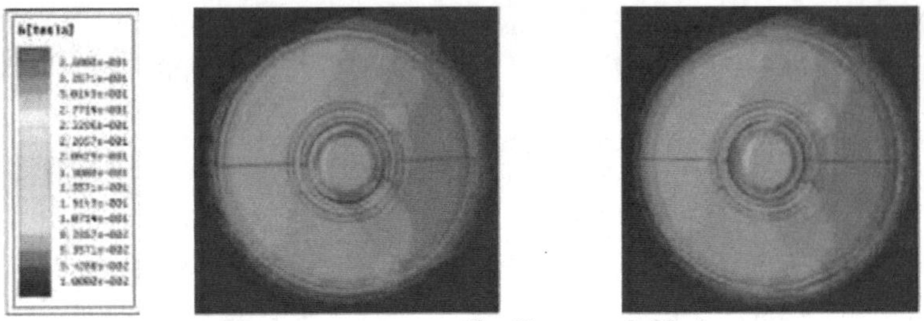

(f) Current 2.00 A

Figure 6. Flux Distribution to MRFs Variations on 0.50 mm of Gap.

Magnetic fluxes were produced by every variation of the MRFs at 0.50 mm gap is shown in Figure 6. The highest value of magnetic flux is about 339 mT was created by MRF 140 CG at 0.50 mm gap which is placed on measurement line along core diameter. That result was the best value in this simulation. Flux magnetic of MRF-140 CG is always placed higher than other MRFs. It can be caused by MRF-140 CG has greater magnetic permeability than the others. Magnetic permeability is influenced by the volume fraction of contained particle of MRFs itself. Carbonyl iron is a particle which is used to fill MRFs. The more number it contained on MRFs, the greater magnetic conductivity.

MRF-132DG MRF140CG

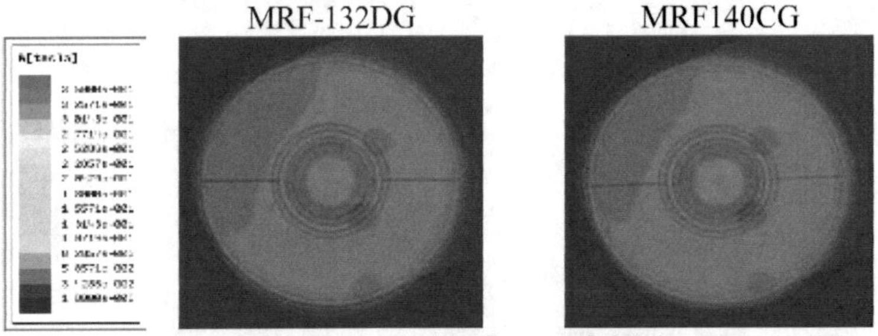

(a) Current 0.25 A

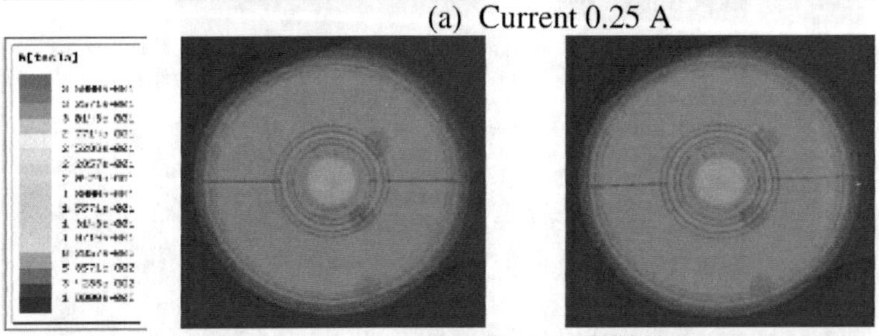

(b) Current 0.50 A

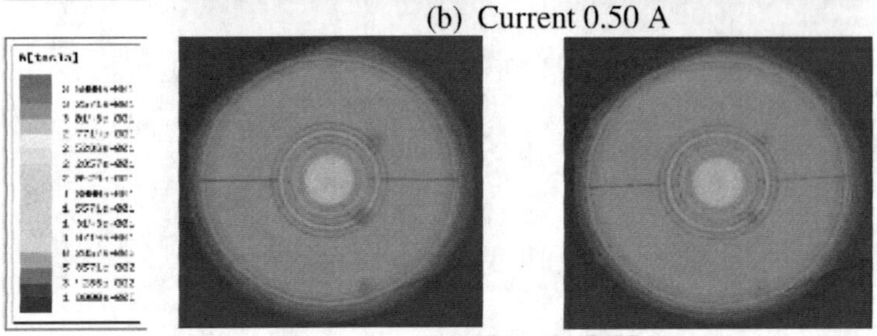

(c) Current 0.75 A

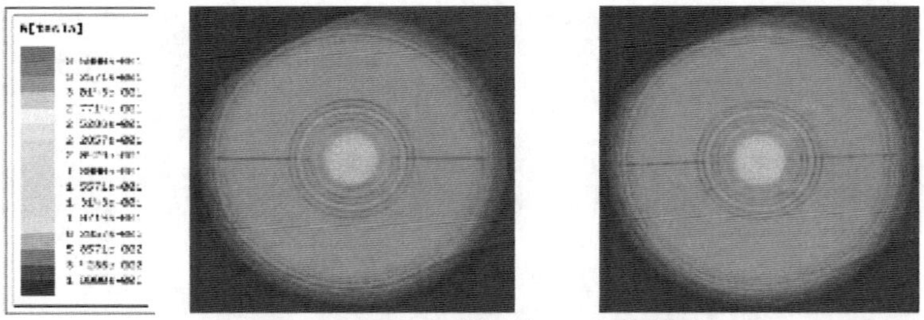

(d) Current 1.00 A

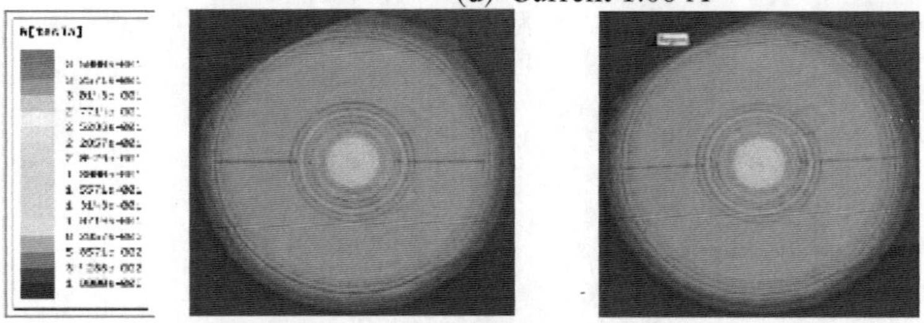

(e) Current 1.50 A

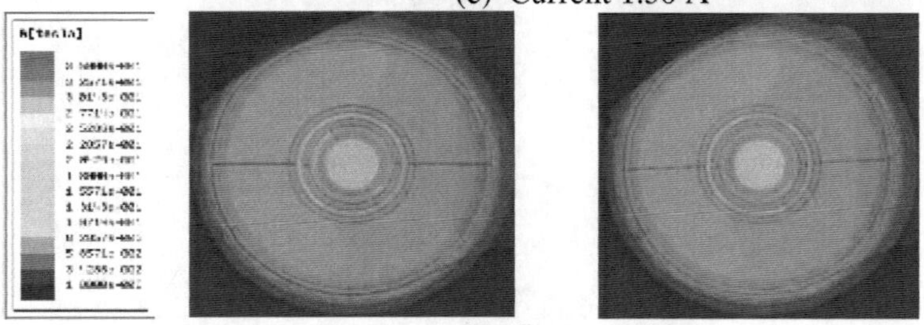

(f) Current 2.00 A

Figure 7. Flux Distribution to MRFs Variations on 1.00 mm of Gap.

Figure 7 shows magnetic flux was exhibited by gap 1.00 mm for every variation of MRFs. MRF-140 CG produced exactly 298.923 mT on measurement line of 1.0 mm gap variation which was given 2.00 A applied current. The value of the magnetic flux generated increases with the applied current. The value of the magnetic flux MRF-132DG always is lower when compared to other fluid. This is caused by the magnetic conductivity (magnetic permeability) MRF-132DG smaller than the other MRFs. The volume fraction of filler particles affects the size of the magnetic conductivity MRFs. It has the smallest volume fraction causes the magnetic flux produced the lowest.

MRF-132DG MRF140CG

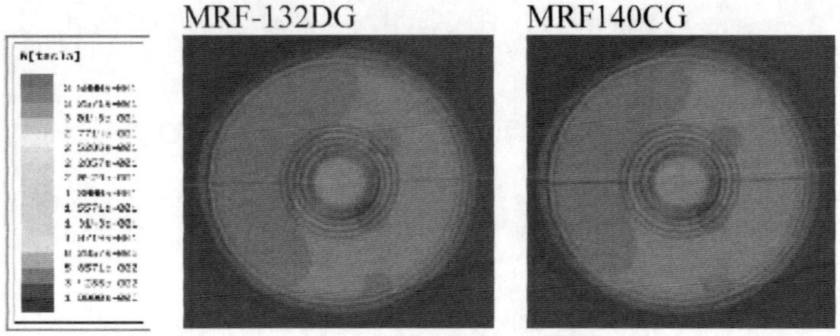

(a) Current 0.25 A

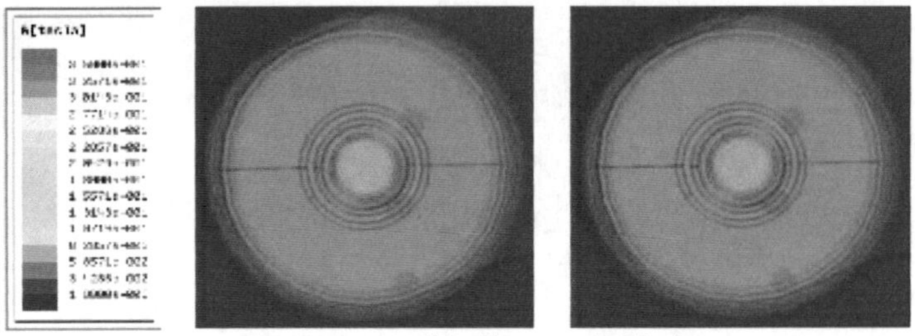

(b) Current 0.50 A

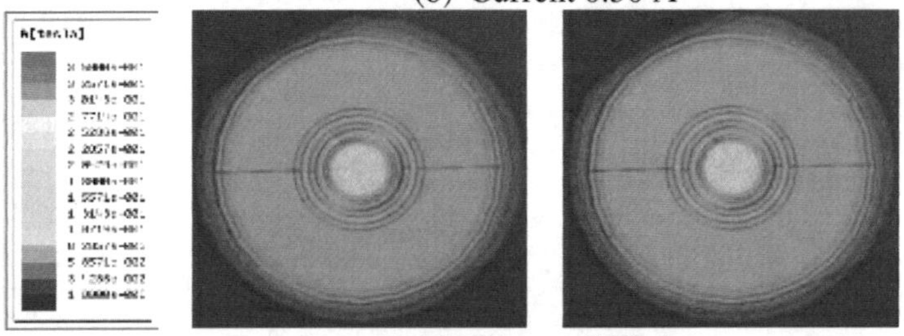

(c) Current 0.75 A

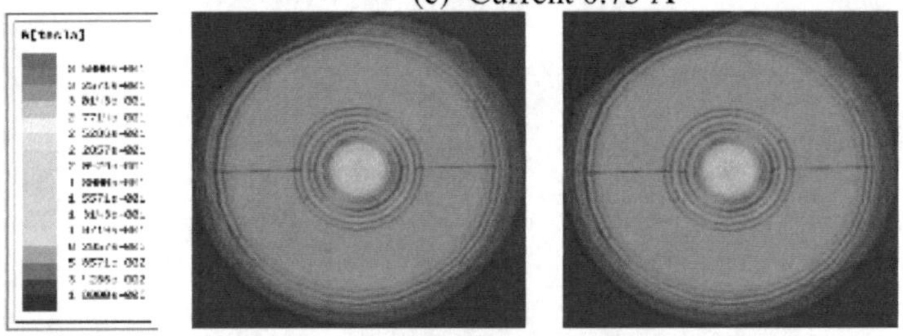

(d) Current 1.00 A

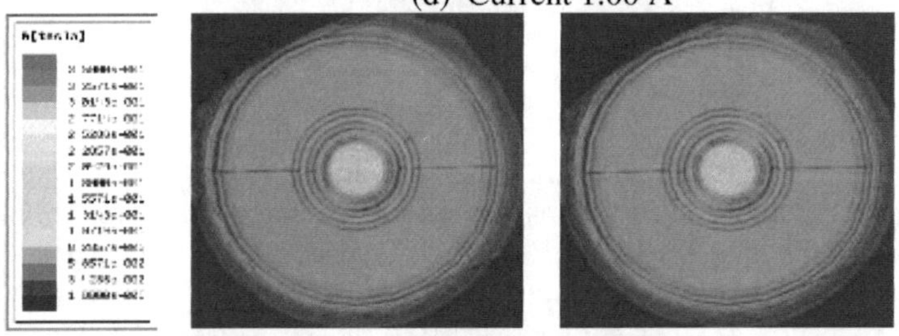

(e) Current 1.50 A

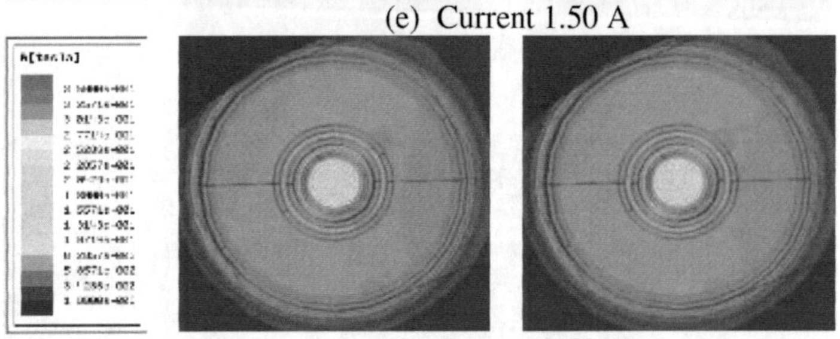

(f) Current 2.00 A

Figure 8. Flux Distribution to MRFs Variations on 1.50 mm of Gap.

The value of the magnetic flux generated by MRFs variation in the gap of 1.5 mm is shown in Figure 8. The best value magnetic flux generated by MRF-140CG amounted to 270.444 mTesla with the applied current of 2.00 A on variations in the gap of 1.5 mm along the measuring line. The increase in the provision resulted in an increased flow of magnetic flux generated value. The value of the magnetic flux MRF-140CG is always higher when compared to other fluid. Magnetic conductivity (magnetic permeability) MRF-140CG greater than MRFs are other causes magnetic flux value is always higher. The value of the magnetic flux is influenced by the volume fraction of particles in the MRFs. Carbonyl iron is used as filler particles in MRFs. The more the iron particles used in MRFs, the greater the magnetic conductivity and magnetic flux generated.

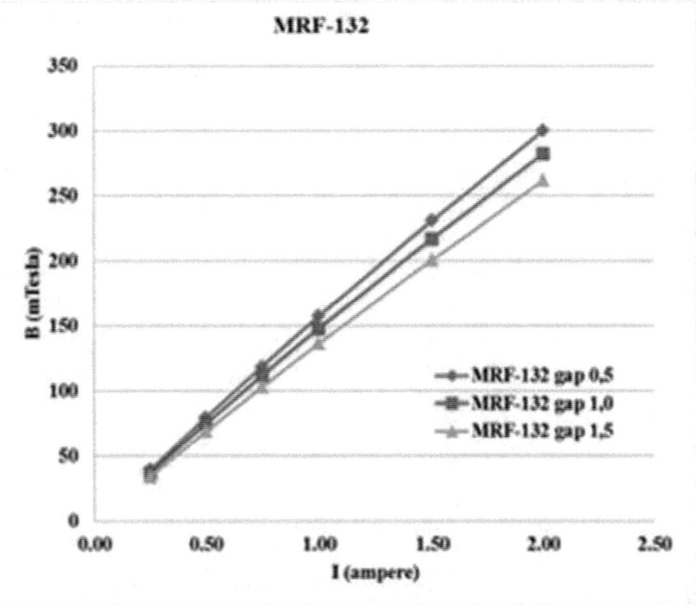

(a) Magnetic flux density at each gap MRF-132-DG

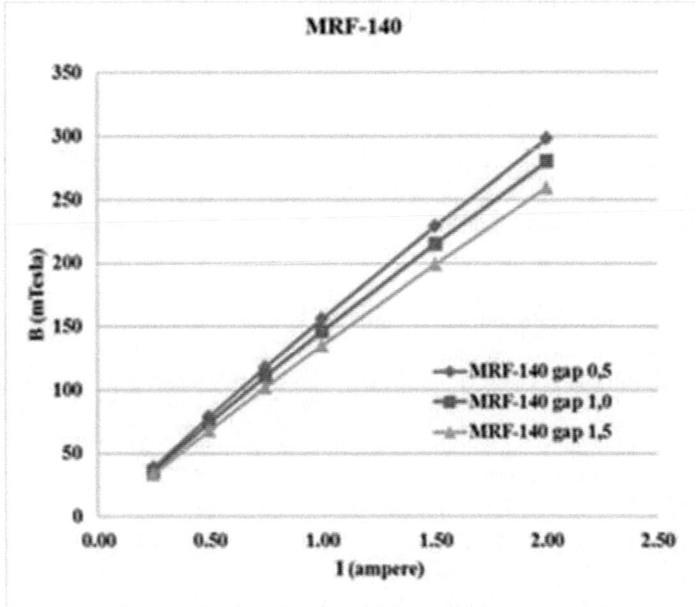

(b) Magnetic flux density at each gap MRF-140CG

Figure 9. Magnetic Flux Density Versus Applied Current.

Current distribution result graphs were delivered has the same pattern for all variations of MRFs, MRF-132DG, and MRF140CG. It distinguishes on every gap variations are shown in Figure 9. Comparison of current to the magnetic flux of MRF-132DG for any variations gap which was used is shown in Figure 9(a). It is apparent that it produces a linear graph. The magnetic flux density for all of three variations at 0.25 A was given tends to equal at 40 mT. However, the differences seen at the time line on 0.75 A. The distinguished of the magnetic flux started to occur although not too significant at 99-110 mT. The highest value of the magnetic flux was exhibited by 0.5 mm at 2.00 A was located at 280 mT. The larger the gap is used, the smaller the magnetic flux generated.

Figure 9(b) shows magnetic flux density MRF-132DG which was delivered at each variation. It provided 300 mT at 0.5 mm which lies in the middle of the MRFs along the core diameter. The comparison of current to the magnetic flux of 0.5, 1.0, and 1.5 mm for MRFs MRF-140CG is clearly displayed in Figure 9(c). The graph results in a linear curve. The greater the current supplied to the coil, the greater the magnetic flux developed. The smaller gap is used, the greater the magnetic flux actualized.

Magnetic flux increased as long as the current added. The gaps and MRFs were stimulated the increasing of magnetic flux. The larger the gap was used, the smaller magnetic flux would be obtained. MRF-140CG performed the biggest magnetic flux due to a magnetic conductivity. Volume fraction was affecting the magnetic conductivity of MRFs. Magnetic flux earned by MRF-132 and MRF-140 was slightly different since volume fraction of both MRFs is not much contrast. The magnetic flux total at whole cores in the simulation is 2.4 T. The magnetic flux is greater than the study by Shiao and Nguyen [14], which was mounted 2.3 T. The magnetic flux in the simulation is enough for application on static bicycles.

6. Conclusions

This paper has addressed several variations of magnetostatic simulation from an innovative multi-coiled disk type MR brake. Its procedure approach and dimensional are the same from a previous study. The results display that some modification can exhibit acceptable magnetic flux value which validates the performance of proposed device. Magnetic permeability placed MRF-140CG became the best MRFs for magnetic flux distribution. The smallest gap produces the most sophisticated amount. Therefore, it can be used for next study on the MR brake system for medical device purposes.

Acknowledgements

Authors highly appreciate the financial support from Universitas Sebelas Maret through Mandatory research grant 2017.

References

[1] Karakoc K, Park E J and Suleman A 2008 *Mechatronics* **18** 434-447.
[2] Ichwan B, Mazlan S A, Imaduddin F, Ubaidillah, Koga T and Idris M 2016 *Smart Mater. Struct.* **13**.
[3] Ubaidillah, Mazlan S A, Sutrisno J and Zamzuri H 2014 *Applied Mechanics Materials* **663** 695-699.
[4] Ubaidillah, Imaduddin F, Nizam M and Mazlan S A 2015 *Int. J. Electr. Eng. Informatics* **7** 308–22
[5] Han Y M, Noh K W, Lee Y S and Choi S B 2010 *Smart Mater. Struct.* **19** 075016.
[6] Park E J, da Luz L F, Suleman A 2008 *Comput. Struct.* **86** 207–16.
[7] Li W and Du H 2003 *Int. J. Adv. Manuf. Tech.* **21** 508–515.
[8] Gudmundsson K H, Jonsdottir F and Thorsteinsson F 2010 *Smart Mater. Struct.* **19** 035023.
[9] Avraam M T 2009 *MR-fluid Brake Design and Its Application to a Portable Muscular Rehabilitation Device* (Universite Libre de Bruxelles)
[10] Rossa C, Jaegy A, Lozada J, Micaelli A, ASME Trans. Mechatronics, 1, 2014
[11] Huang J, Zhang J Q, Yang Y and Wei Y Q 2002 *J. Mater. Process. Technol.* **129** 559-562.
[12] York T M, Gilmore C D and Libertiny T G 1997 US Patent. 5 598 908.
[13] Gopalswamy S, Linzell S M and Jones G L 1999 US Patent. 5 896 965.
[14] Shiao Y and Nguyen Q 2013 *Smart Mater. Struct* **22** 1-13.
[15] Shiao Y, Nguyen Q, Procedia Eng. Elsevier B.V., pp. 12–23, 2014
[16] Ubaidillah, Permata A N S, Wibowo A, Budiana E P, Yahya I and Mazlan S A 2016 *AIP Conference Proceedings* 1717 4943480.

IOP Conf. Series: Materials Science and Engineering **257** (2017) 012053 doi:10.1088/1757-899X/257/1/012053

An overview of thixoforming process

N H Husain, A H Ahmad* and M M Rashidi

Faculty of Mechanical Engineering, Universiti Malaysia Pahang, 26600, Pekan, Malaysia.

Corresponding author email: asnul@ump.edu.my

Abstract. Thixoforming is a forming process which exploits metal rheological behaviour during solidus and liquidus range temperature. Many research works in thixoforming are currently focusing on the raw material used to produce superior mechanical properties and excellent formability components, especially in automotive industries. Furthermore, the thixoforming process also produced less casting defect component such as macrosegration, shrinkage and porosity. These advantages are sufficient to attract more exploration works of thixoforming operation. However, the weakness of this process such as high production cost due to leftover billet which cannot be recycled, encourage researcher works to overcome thixoforming limitations by using various methods. The thixoforming methods that widely used are thixocasting, thixoforging, thixorolling, thixoextrusion and thixomoulding. Each method provides varieties of final product characteristics; hence offer the extensive possibility of component invention. On the other hand, new thixoforming method leads to exploration research such as microstructure evolution, heating and pouring temperature, die temperature, mechanical properties, viscosity and final product quality. This review paper presents findings in the rheological material behaviour of thixoforming, advantages and disadvantanges of thixoforming, parameters affecting the thixoforming operation, morphology of thixoforming and various methods which have been used in this research area.

1. Introduction

Semisolid metallic alloys were discovered by Spencer in 1972 at MIT while investigated viscosity of Sn-15%Pb and found out during the experiment, semisolid state without stirring produce dendritic while the continuously stirred material was spheroidal [1]. Since then, extensive research regarding semisolid metal processing (SSMP) leads to various processes development. SSMP were applied in the industry due to the final product has less defect such as porosity, shrinkage, gas entrapment and macrosegration [2]. SSMP can be categorized into two routes, namely rheocasting and thixoforming [3]. Rheocasting is a forming process start from liquid alloy, directly introduced into a die without any intermediate solidification step. The semisolid slurry in this route produced from an entirely liquid regular alloy. While thixoforming is a route consists of reheating and forming process. Thixoforming steps are feedstock billet with globular microstructure specially prepared, cut to length, reheated into solid-liquid temperature to achieve suitable solid fraction, then final product formed by various methods [4].

In the early years of SSMP research, scientists mainly focus on thixotopic material development because it was essential in both routes. The material in thixotropic condition was described as the ability of material to flow when shear force was given and it will thicken again when shear force was released [5]. Thixotropic condition happened due to globular microstructure in the material. Methods to gain fine globular microstructure such as Magnetohydrodynamic stirring, Strain Induced Melt

Activated (SIMA), Mechanical stirring, Cooling slope and Direct Thermal Method are widely used either for scientific research or industry application [6].

Thixoforming particularly needs feedstock billet which prepared by special method because solidification step is essential compared to rheocasting. Thixoforming also described as near net shape forming due to billet reheated into solid-liquid temperature and partially melted non-dendritic slug into a metal die [7].

In industry application, thixoforming was considered high-cost manufacturing because of raw material was not supplied widely [4]. However, this method offers broader ranges of design option which attract people in this area improves it continuously [8].

This paper focused on thixoforming and divided into four sections. The first section highlights the rheological material behaviour, the second part emphases on advantages and disadvantages of thixoforming, the third section focuses on parameters affecting the thixoforming operation, the fourth division explained morphology of thixoforming and final section presents various methods have been used in this process.

2. Rheological material behaviour

Rheology is a term to describe deformation and flow behaviour of material. In SSMP, non-dendritic microstructures play the main role of semisolid state behaviour. Globular microstructure which gains from continuous stirring during cooling exhibits lower viscosity compared to a material with dendritic microstructures. Spheroid solid surround by the liquid matrix, is a requirement for semisolid processing and thixotropic condition. The thixotropic condition is explained as material decreases in viscosity if shear but stand still again if allowed [9].

Material or alloy in the semisolid state also referred as slurry, which can be divided into two categories: a liquid-like and a solid-like. Liquid-like named for slurry behaves like liquid under external shear due to dispersed solid particles. Meanwhile, solid-like slurry exhibits solid behaviour because of containing interconnected solid phase [6]. In thixoforming process, solid-like slurry is involved. Solid-like slurry with thixotropic condition has the ability to flow in laminar, which uniform flow during forming is possible.

Viscosity is the main parameter for rheology in semisolid processing. Viscosity indicates capabilities of semisolid material in filling mould, flow behaviour and required force for deformation [10]. The viscosity in semisolid metal was influenced by solid fraction, particle size, and particle distribution. Solid fraction was defined as the density of solid in the semisolid slurry. Solid component was generated by partial solidification of slurry [11]. For thixoforming processing, effective solid fraction is between 0.5 and 0.6. If the solid fraction is below 0.5, the slug cannot support its weight and consider as too soft. Whereas, if solid fraction is over 0.6, the slug will not be able to fill the die because too stiff to flow [12].

3. Advantages and disadvantages

Thixoforming process like any other operation, hold its own benefits and limitation. As for advantages, thixoformed parts reported have higher quality compared to conventional produce. Better quality products of thixoforming were gain due to globular microstructure. Spheroidal microstructure provides flow ability in billet during forming. Less shrinkage, gas entrapment and porosity parts were created from slug flow in laminar. Other than that, mechanical properties of final product were enhanced by fine and uniform microstructure. Due to fine and uniform structure, thixoforming operation able to create wide range of component shape with different wall thickness, sharp edged and radii edge [4-6, 8, 9, 11-15].

However, limitations of thixoforming process may cause less response in industry. The most well-known shortcoming of thixoforming was special requirement for feedstock billet may increase production cost. Besides, scrap from forming process cannot be recycled. Furthermore, to ensure final product quality, knowledgeable supervise were needed for each step. Moreover, crucial points in thixoforming operation were viscosity and solid fraction. Both were dependent on temperature control, therefore, high technical personnel were needed for accurate temperature control. Even though SSMP has taken places in research light for 40 years, continuous research in thixoforming is necessary to support development of new components [4-6, 11, 16].

Identifying leads and shortcomings of thixoforming will lead to deeper investigation, thus improve thixoforming in many aspect. Recently, researchers focus on thixoforming as changed from development of technique produce feedstock billet into improvement of cost saving thixoforming operation.

4. Parameters affecting thixoforming process

Since the expansion of thixoforming research takes place, many research appoint parameter affecting thixoforming based on critical criteria for thixoforming to be successful.

4.1. Solid fraction

Viscosity of material during thixoforming highly influence by solid fraction. In SSMP, suitable solid fraction usually in range of 30% - 70% [17, 18]. Low solid fraction indicates high viscosity, thus slurry flowing ability during forming will be lower.

Ahmad et al. conduct an experiment to find effect of cooling rates on thermal profiles aluminium alloy 7075. They pour molten metal into chamber with Kaowool blanket then, thermocouples were inserted at the centre and near cubicle wall. Temperature differences were plotted against time to determine dendritic coherency point. Through the experiment, they found that fraction solid were affected by cooling rates. From graph plotted, the different value of fraction solid for slow cooling rates and high cooling rates were at 0.7 fractions solid. This investigation exhibits the effect of cooling rate on solid temperature and liquid temperature which is crucial during reheat in thixoforming process [17].

Earlier than Ahmad et al., Nafisi et al. investigate fraction solid during processing temperature with the same set up. Melts were prepared in silicon carbide crucible which later transferred into cooling stand and K-type thermocouple inserted at two locations. In this experiment, they processed data using computer aided cooling curve analysis (CA-CCA), computational thermodynamic and quantitative metallography. They concluded inefficient during quenching methods lead to more primary α-Al formation as the main reason for fraction solid differences in the experiment [19].

4.2. Particle size

Thru understanding of SSMP, smaller or finer particle size of material is expected to provide smooth filling movement and less viscose. However, through various experimental, particle sizes are much affected by temperature such as pouring temperature, tool temperature and cooling rate.

El-Mahallawi et al. investigated effects of pouring temperature and water cooling on semisolid microstructure of A319 aluminium cast alloy. This experiment started with A319 melted in electric furnace and allowed to cool down to specified pouring temperature before continuously poured on cooling slope. For each temperature, pouring was carried out with and without water cooling. They found that ingots with water cooling show lower bulk porosity and increasing pouring temperature slightly reducing bulk porosity. Other than that, they discovered ingot with water cooling exhibits lower shape factor and with increasing pouring temperature accumulate shape factor of α-Al grains. Furthermore, they also found that water cooled ingot shown lower average grain size and with increasing pouring temperature accumulate average size of primary grains [20].

5. Morphology of thixoforming

In investigation of thixoforming, microstructures were frequently used to analyse relationship between process and mechanical properties. Based on experiment by Tavakoli et al., cooling slope method and thixoforming were proven improved mechanical properties of A380 alloy, such as tensile strength, elongation and hardness [21].

Lee et al. has produced 6061 wrought aluminium alloy feedstock billet from thermo-mechanical method and thixoextruded it into automotive component. The research was concentrated on the possibility of thixoformed 6061 into steering knuckles. The grain size distribution of the steering knuckles from this research is presented in figure 1. This research finding showed that 6061 was able to be produced by thixofroming operation which ductility and strength were significantly increased after T6 heat treatment [22].

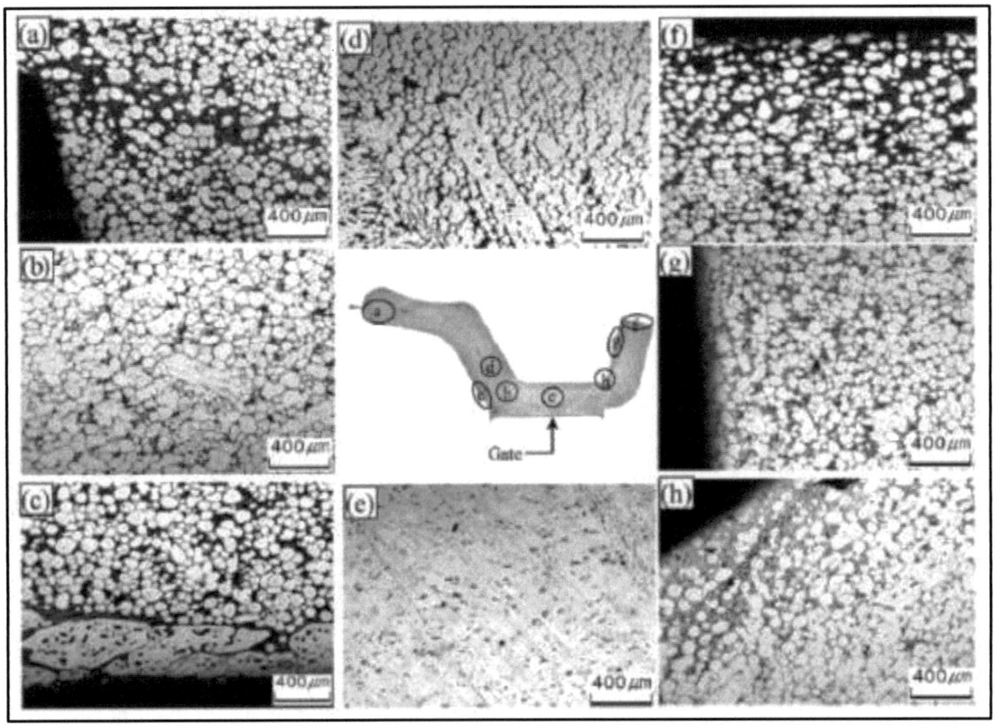

Figure 1. Microstructures of steering knuckles which shown fraction solid distribution at different area [22].

6. Various forming method

In the early years of SSMP, forming method mainly same as conventional, the difference is SSMP using globular microstructure feedstock billet. For thixoforming, after billet is reheated into semisolid temperature range, forming shortly takes place. Following are few of thixoforming methods.

6.1. Thixoforging

Thixoforging is a process similar to closed-die forging. Billet is placed between open die halves and pressed to form final product. In this process, billet with lower liquid fraction is significantly used. This method able to produce product with variable wall thickness, sharp radii and sharp edge thickness changed [16].

Jung et al. conducted an experiment to determine reheating process for cast and wrought aluminium alloy for thixoforging. They achieve optimal reheating by controlling holding time and relevant temperature for each step using induction heating. From this experiment, they find that suitable temperature for each step in thixofoging will lead to high quality product. Since fluidity of billet during forming have high influenced on final product, reheating holding time were take account. If the reheating holding time were too long, risk for grain coarsening will increase too [23].

Meanwhile, Cho et al. has done experiment on mechanical properties of thixoforging product. This experiment, researcher observing filling behaviour during forming, investigates defects on product and final product mechanical properties. They used A356 and Al2024 (wrought) with different die temperature to find die temperature filling limitation. From the research, results shown that ultimate tensile strength and yield strength were increase at higher applied pressure. Other than that, highest hardness degree was shown at the center of area. This happened due to edge of the area have higher liquid distribution during forming [24].

As research of thixoforging using aluminium alloy expanding tremendously, Wang et al. investigated microstructural evolution and mechanical properties of 9Cr18 steel after thixoforging. Through this research, they found out optimal heat treatment for 9Cr18 steel thixoforging is at 550° C for 2 hour. Strong hardening at the inner area was observed and compressive strength of 4680 MPa were achieve with compressive strain of 53.3% [25].

6.2. Thixocasting
In thixocasting, billet is shot into a closed-die by piston to form and usually used billet with liquid fraction in range of 40% - 60%. During forming in thixocasting, laminar flow of slug is critical aspect in ensure high quality product [6]. Shown in figure 2 is difference process for thixoforging and thixocasting method.

In order to understand filling or flow in thixocasting, scientists develop software to calculate and have better picture of product forming using thixocasting method. Modigell et al. investigate tin lead alloy (Sn-15%Pb) rheology by changing shear rate under isothermal conditions. Data from experimentation later compared to model develop by using commercial CFD software. Results from this experiment exhibit filling velocity give high impact on product quality [26].

Another experiment on rheology behaviour for thixocasting is conducted by Yang et al.. Aluminium alloys (A356) were thixocasted and compared to model. From their calculation, it is proven rheological behaviour of semisolid metal during thixocasting is pseudoplasticity of non-Newtonian. Their calculation has open to better understanding of filling flow during thixocasting and provide improvement in producing higher quality product using thixocasting method [27].

While, Kim et al. used thixocasting method to produce possible high strength Mg-Cu-Y alloy. Magnesium alloy is an attractive material for aeronautic transportation application industry because of its light weight properties and excellent damping capacity [28, 29]. However, magnesium alloy were limitedly found in industry is cause by its poor corrosion resistance. Thru thixocasting, $Mg_{90}\,Cu_5\,Y_5$ alloy exhibits high tensile strength compared to conventional magnesium. Researchers explained that globular magnesium as primary particles were surrounded by liquid quenched (partly amorphized) with later transform into hard amorphous. This microstructure has improved Mg-Cu-Y alloy strength[29].

Other than using slug, researchers also extend their interest towards powder thixocasting. Chen et al. investigation shown powder thixocasted Al-Si powder provide good formability with satisfied strength and uniform distribution of Si particles [30].

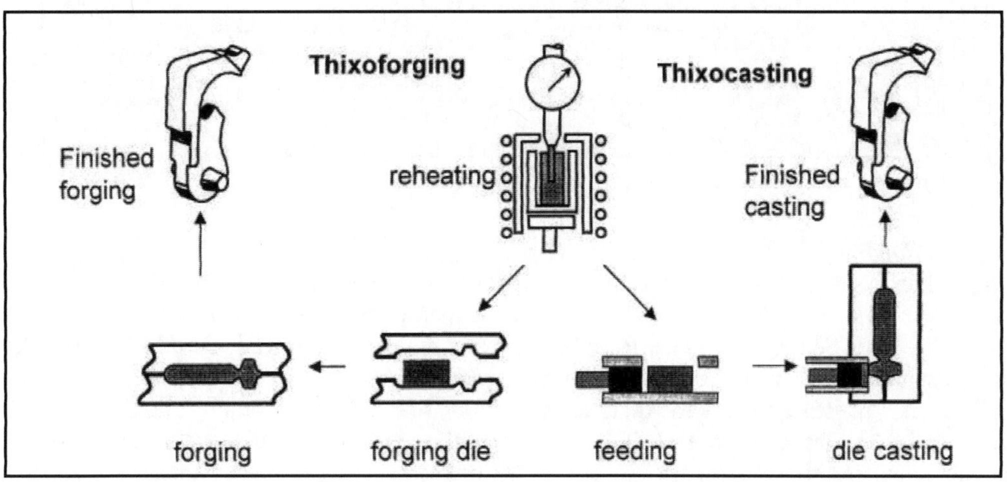

Figure 2. Illustration of thixoforging and thixocasting process[6].

6.3. Thixomoulding
Thixomoulding is a process to produce near net shaped component from magnesium alloy. This process took place in a single integrated machine as shown in figure 3. Magnesium alloy chips in

range size of 2-5 mm feed into machine and heated until partial melted. Partial melted alloys transform into semisolid slurry. Huge and high speed screw later injects slurry into mould for final product forming. As magnesium alloy would oxidize when heated, an argon atmosphere was maintained at in-feed. Thixomoulding is an effective process of thixoforming because involved only one step process[6, 31, 32]. However, thixomoulding only available for magnesium alloy production because other alloy such as aluminium alloy would likely attack the barrel and screw while in semisolid state[5].

To evaluate capabilities of thixomoulding system in processing magnesium alloy, Czerwinski operate an experiment using a Husky's prototype system named TXM500-M70. The researcher used mechanically chipped magnesium alloy as material. In this experiment, major operating parameters are cylinder barrel temperature, screw rotation speed, injection velocity, injection pressure and mould temperature. Thru this experiment, Czerwinski conclude that temperature along injection mould is the key factor in controlling process. While, solid fraction and injection velocity were influenced flow behaviour of magnesium alloy slurry during mould filling [33].

Furthermore, thixomoulding is seen as potential near-net shape method which could provide effective energy usage and less pollution. Thus, Patel et al. conduct an experiment focusing on evaluating thixomoulded AZ91D and AM60B due to its flexibility in fabrication, high strength-to-weight ratio besides increasing usage in automotive industry. from this experiment, for fracture characteristic, AZ91D exhibited cleavage like fracture with secondary crack while AM60B shown dimple like features due to formation and coalescence microvoids. Ultimate tensile strength for both were shown equivalent within experimental scatter. AZ91D alloys displayed higher yield strength and lower ductility compared to AM60B. Characteristics of both alloys revealed in this experiment were lead to deeper understanding of their production through thixomoulding method [34].

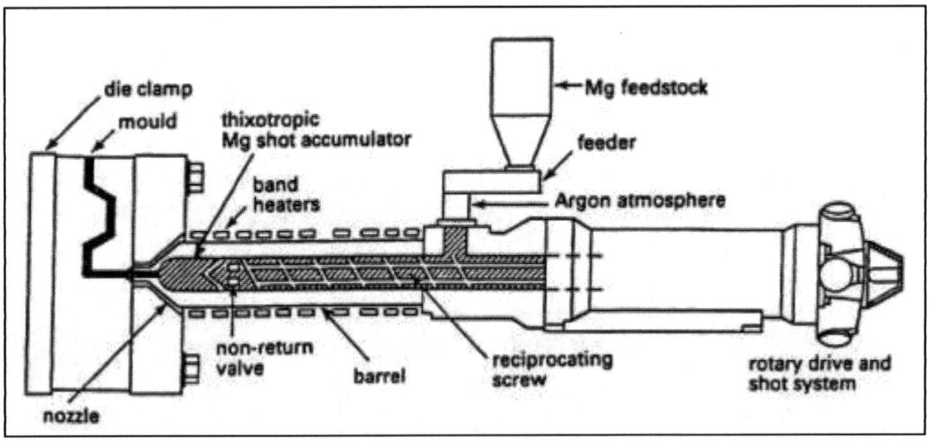

Figure 3. Diagram of a thixomoulding machine [6].

6.4. Thixorolling

Rolling is a process where materials were undergoes two rollers which produce flat like metal [35]. In thixorolling, slug or reheated billet with globular microstructure were flowed through two roller and produce alloy strip. Semisolid Powder Rolling (SSPR) is said possible for industrial application and its product possible to be used in aerospace application [36].

Liu et al. investigated microstructural evolution of 7050 aluminium alloy during SSPR. Researchers used gas atomized 7050 aluminium alloy powder which, reheated to semisolid temperature and inserted into inert atmosphere with different period. They observe that the most suitable liquid fraction for SSPR is 45 – 65 %. Alloy strips can be produce with nearly full density and high micro-hardness. They highlight three important mechanisms in this technique which are filling and flowing of liquid, solidification by rolling and recrystallization [37].

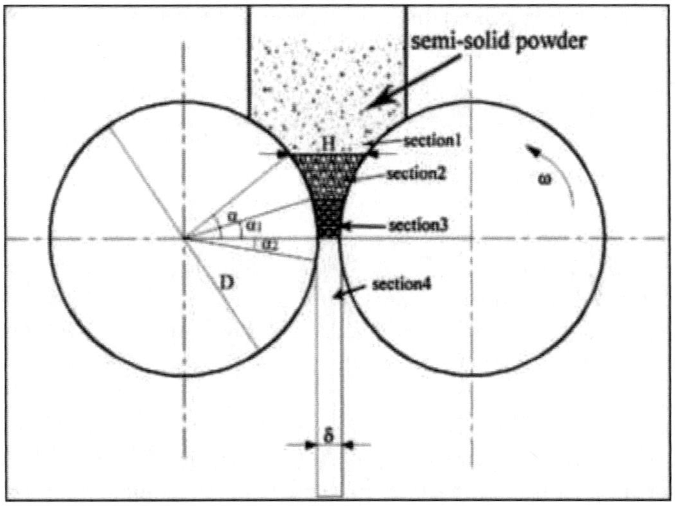

Figure 4. Schematic picture of semisolid powder rolling [37].

6.5. Thixoextrusion

In thixoextrusion, reheated billet with globular microstructure were squeezed into an already closed die [38-40]. Like any other thixoforming method, thixoextrusion offer several advantages compared to conventional extrusion such as higher material fluidity, lower pressure and longer tools life. Thixoextrusion also offer production of hard-to-form aluminium alloys for automotive industry [38].

Rovira et al. has done experiment thixoforming of Al-Cu alloys by using thixoforging and thixoextrusion. They compared thixoforming method with conventional processing, proved thixoextrusion produce final product with fine and equi-axial structure besides lower energy use [41].

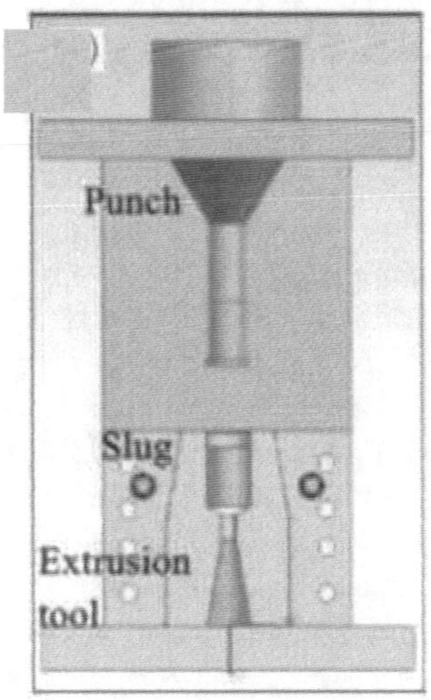

Figure 5. Example outline of extrusion tool by Forn et al. [38].

7. Conclusions

This overview shows thixoforming process for component forming by using semisolid principle. These methods were applied in several industries worldwide such as automotive and aerospace. This review appoints that effective route for thixoforming development is understanding rheology behaviour of material during process. Likewise, highlighted identified benefit and limitation of thixoforming gives deeper comprehension of this operation. Although thixoforming operation was applied in industry, continuous study on material aspect for this process still needed as an effort to accommodate future industry requirement. Nevertheless, thixoforming is still providing wider possibility production of high quality components compared to conventional.

Acknowledgements
The authors would like to be obliged to Universiti Malaysia Pahang for providing laboratory facilities and financial assistance under project no. RDU160311.

References
[1] Spencer D, Mehrabian R and Flemings MC 1972 Rheological behavior of Sn-15 pct Pb in the crystallization range *Metallurgical and Materials Transactions* B **3** 1925-32
[2] Kumar SD, Mandal A and Chakraborty M 2014 Cooling Slope Casting Process of Semi-solid Aluminum Alloys: A Review *International Journal of Engineering Research and Technology: IJERT*
[3] Aziz AM, Omar MZ, Sajuri Z and Salleh MS 2016 Microstructural morphology of rheocast A319 aluminium alloy *Advances in Mechanical Engineering* **8**1687814016649354
[4] Rosso M 2012 Thixocasting and rheocasting technologies, improvements going on *Journal of Achievement in Materials and Manufacturing Engineering* **54** 9
[5] Atkinson HV 2005 Modelling the semisolid processing of metallic alloys *Progress in materials science* **50** 341-412
[6] Fan Z 2002 Semisolid metal processing *International materials reviews* **47** 49-85
[7] Mohammed M, Omar M, Salleh M, Alhawari K and Abdelgnei M 2014 An overview of semi solid metal processing *Australian Journal of Basic and Applied Sciences* **8** 369-73
[8] Hirt G, Cremer R, Witulski T and Tinius H-C 1997 Lightweight near net shape components produced by thixoforming *Materials & design* **18** 315-21
[9] Chayong S, Atkinson H and Kapranos P 2005 Thixoforming 7075 aluminium alloys *Materials Science and Engineering* A **390** 3-12
[10] Lashkari O and Ghomashchi R 2007 The implication of rheology in semi-solid metal processes An Overview *Journal of materials processing technology* **182** 229-40
[11] Kiuchi M and Kopp R 2002 Mushy/semi-solid metal forming technology–Present and Future *CIRP Annals-Manufacturing Technology* **51** 653-70
[12] Salleh MS, Omar MZ, Syarif J and Mohammed MN 2013 An overview of semisolid processing of aluminium alloys *ISRN Materials Science 2013*
[13] Birol Y 2007 A357 thixoforming feedstock produced by cooling slope casting *Journal of Materials Processing Technology* **186** 94-101
[14] Hirt G, Khizhnyakova L, Baadjou R, Knauf F and Kopp R 2009 Semi-Solid Forming of Aluminium and Steel–Introduction and Overview *Thixoforming: semi-solid metal processing* 1-27
[15] Koke J and Modigell M 2003 Flow behaviour of semi-solid metal alloys *Journal of Non Newtonian Fluid Mechanics* **112** 141-60
[16] Kopp R, Neudenberger D and Winning G 2001 Different concepts of thixoforging and experiments for rheological data *Journal of Materials Processing Technology* **111** 48-52
[17] Ahmad A, Naher S and Brabazon D 2014 Effects of Cooling Rates on Thermal Profiles and Microstructure of Aluminium 7075 *International Journal of Automotive and Mechanical Engineering* **9** 1685
[18] Vaneetveld G, Rassili A, Pierret J-C and Lecomte-Beckers J 2010 Conception of tooling adapted to thixoforging of high solid fraction hot-crack-sensitive aluminium alloys *Transactions of Nonferrous Metals Society of China* **20** 1712-8
[19] Nafisi S, Emadi D and Ghomashchi R 2009 Semi solid metal processing: The fraction solid dilemma *Materials Science and Engineering: A* **507** 87-92
[20] El-Mahallawi IS-E, Mahmoud TS, Gaafer AM and Mahmoud FH 2015 Effect of pouring temperature and water cooling on the thixotropic semi-solid microstructure of A319 aluminium cast alloy *Materials Research* **18** 170-6
[21] Tavakoli MM, Shabestari SG and Ghanbari M 2014 Effect of thixoforming on

morphological changes in iron-bearing intermetallics and mechanical properties of Al–Si–Cu alloys *International Journal of Materials Research* **105** 1210-7

[22] Lee S-Y and Oh S-I 2002 Thixoforming characteristics of thermo-mechanically treated AA 6061 alloy for suspension parts of electric vehicles *Journal of materials processing technology* **130** 587-93

[23] Jung H and Kang C 2000 Reheating process of cast and wrought aluminum alloys for thixoforging and their globularization mechanism *Journal of Materials Processing Technology* **104** 244-53

[24] Cho W and Kang C 2000 Mechanical properties and their microstructure evaluation in the thixoforming process of semi-solid aluminum alloys *Journal of Materials Processing Technology* **105** 269-77

[25] Wang Y, Song R and Li Y 2017 Microstructural evolution and mechanical properties of 9Cr18 steel after thixoforging and heat treatment *Materials Characterization* **127** 64-72

[26] Modigell M and Koke J 2001 Rheological modelling on semi-solid metal alloys and simulation of thixocasting processes *Journal of Materials Processing Technology* **111** 53-8

[27] Yang X, Jing Y and Liu J 2002 The rheological behavior for thixocasting of semi-solid aluminum alloy (A356) *Journal of materials processing technology* **130** 569-73

[28] Hirai K, Somekawa H, Takigawa Y and Higashi K 2007 Superplastic forging with dynamic recrystallization of Mg–Al–Zn alloys cast by thixo-molding *Scripta materialia* **56** 237-40

[29] Kim J-M, Shin K, Kim K-T and Jung W-J 2003 Microstructure and mechanical properties of a thixocast Mg–Cu–Y alloy *Scripta materialia* **49** 687-91

[30] Chen C, Yang C and Chao C 2005 A novel method for net-shape forming of hypereutectic Al Si alloys by thixocasting with powder preforms *Journal of Materials Processing Technology* **167** 103-9

[31] Mansoor B, Mukherjee S and Ghosh A 2009 Microstructure and porosity in thixomolded Mg alloys and minimizing adverse effects on formability *Materials Science and Engineering:* A **512** 10-8

[32] Yang L-j, Wei Y-h and Hou L-f 2010 Microstructure evolution of thixomolding AZ91D magnesium alloy during heat treatment *Journal of materials science* **45** 3626-34

[33] Czerwinski F 2003 Assessing capabilities of Thixomolding system in semisolid processing of magnesium alloys *International Journal of Cast Metals Research* **16** 389-96

[34] Patel H, Chen D, Bhole S and Sadayappan K 2010 Microstructure and tensile properties of thixomolded magnesium alloys *Journal of Alloys and Compounds* **496** 140-8

[35] McHugh KM, Delplanque J-P, Johnson S, Lavernia E, Zhou Y and Lin Y2004 Spray rolling aluminum alloy strip *Materials Science and Engineering:* A **383** 96-106

[36] Li F-x, Liu Y-z and Yi J-h 2014 Microstructural evolution of gas atomized Al–Zn–Mg–Cu–Zr powders during semi-solid rolling process *Transactions of Nonferrous Metals Society of China* **24** 2475-81

[37] Liu Y, Luo X and Li Z 2014 Microstructure evolution during semi-solid powder rolling and post-treatment of 7050 aluminum alloy strips *Journal of Materials Processing Technology* **214** 165-74

[38] Forn A, Vaneetveld G, Pierret J-C, Menargues S, Baile M, Campillo M and Rassili A 2010 Thixoextrusion of A357 aluminium alloy *Transactions of Nonferrous Metals Society of China* **20** s1005-s9

[39] Gu G, Pesci R, Langlois L, Becker E and Bigot R 2015 Microstructure investigation and flow behavior during thixoextrusion of M2 steel grade *Journal of Materials Processing Technology* **216** 178-87

[40] Kapranos P, Nakamura R, Bertoli E, Pola A, Azpilgain Z and Hurtado I 2008 Thixo-extrusion of 5182 aluminium alloy *Solid State Phenomena: Trans Tech Publ* 115-20

[41] Rovira M, Lancini B and Robert M 1999 Thixo-forming of Al–Cu alloys *Journal of Materials Processing Technology* **92** 42-9

The numerical assessment of motion strategies for integrated linear motor during starting of a free-piston engine generator

M Razali Hanipah[1,3*] and Akhtar Razul Razali[2,3]

[1]Automotive Engineering Research Group (AERG), Faculty of Mechanical Engineering,
[2]Manufacturing Focus Group (MFG), Faculty of Mechanical Engineering,
[3]Automotive Engineering Centre (AEC),
Universiti Malaysia Pahang, 26600, Pekan, Pahang, Malaysia.

*Corresponding author email: mohdrazali@ump.edu.my

Abstract. Free-piston engine generator (FPEG) provides a novel method for electrical power generation in hybrid electric vehicle applications with scarcely reported prototype development and testing. This paper is looking into the motion control strategy for motoring the FPEG during starting. There are two motion profiles investigated namely, trapezoidal velocity and S-curve velocity. Both motion profiles were investigated numerically and the results have shown that the S-curve motion can only achieve 80% of the stroke when operated at the proposed motoring speed of 10Hz.

1. Introduction

A free-piston engine (FPE) operates on dynamic balance of forces, which produces linear reciprocation motion. In the crankshaft engine, the piston position is consistent and can be represented by a kinematic relationship between crank radius, connecting rod length and crank angle. On the contrary, an FPE has non-fixed piston stop positions (top-dead-centre (TDC) and bottom-dead-centre (BDC)) and its piston motion is not governed by any mechanical component. Further, due to the absence of the crank-slider mechanism, the fundamental principle of operation of this engine requires a new approach.

A popular configuration of the FPE is the free-piston engine generator (FPEG) which has attracted commercial interest in recent years (for review, see e.g. Hanipah, Mikalsen and Roskilly [1]). However, the key step towards commercialisation is overcoming the hardware and software challenges [2]. One of the key challenges is the piston motion which posed numerous issues such as start-ability, misfiring and unstable operation [3]. This paper outlines the piston motion characterisation approach using numerical investigations.

2. Prior research

The typical configuration of a free-piston engine is a single piston configuration. Primarily, for cyclic operation to be possible, a free-piston engine requires a bounce device to ensure the piston returns to initial top-dead-centre position for the next engine cycle.

Free-piston engine is said to be dynamically constrained as opposed to kinematically constrained crankshaft engine [4]. A dynamics constraint means the piston stop positions (TDC and BDC) is not consistent and its motion profile is not governed by any mechanical component as in crankshaft engine. Further, due to the absence of the crank-slider mechanism, the fundamental principles of operation of this engine require new approach.

The most popular configuration of a free-piston engine generator (FPEG) is a dual-piston type configuration allows each cylinder to act as bounce chamber [5, 6] for cyclic operation to be possible. The dual-piston type is shown Figure 1, where two pistons are connected by a shaft integrated with permanent magnet assembly as a single moving mass called a translator. The translator oscillates as a result of alternate combustion event occurring in the opposing combustion chambers.

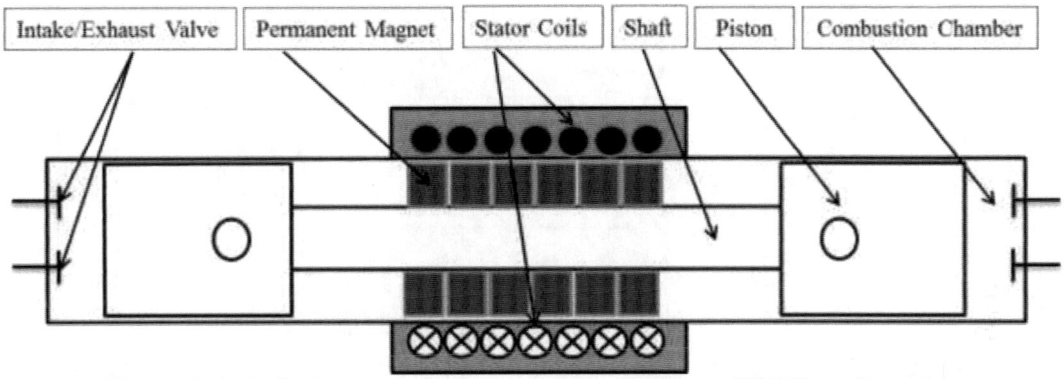

Figure 1. A dual-piston type free-piston engine generator (FPEG) configuration.

However, as oppose to conventional crankshaft engine, which can be cranked several times before it starts by using starter motor to drive the flywheel, the starting of an FPEG is arguably the key problem which can be addressed electrically [7-9]. Thus, any development of such engine must consider its starting scheme and strategy to ensure sufficient linear motor capability to overcome compression force at sufficient starting speed [6]. Further, the omission of flywheel creates a critical issue since the engine must obtain its energy from each stroke. Thus, researchers have suggested the integrated linear motor to provide assistance during misfire for continuous operation [6, 10].

In terms of engine cycle, FPEG must operate in two-stroke cycle although a complex and bulky four-stroke cycle version was theoretically shown to be possible [11]. The two-stroke version is simpler and thus, is widely adopted since the combustion occurs at every stroke to provide kinetic energy require for reciprocation thereby increasing its power output. Nevertheless, reported dual-piston configuration FPEG running on two-stroke seems to be plagued with scavenging issues [12, 13]. Some researchers eliminate this issue by using compressed air boost intake [5] and air-assisted direct injection [14].

The unique feature of free-piston engine generator (FPEG) lies in the piston dynamics. A single linear motion is converting the chemical energy from combustion process into kinetic energy of the moving mass and finally into electrical energy through the integrated generator. The electrical energy can then be stored in an energy storage system to be utilized for various applications such as for series hybrid electric vehicles and electrical generators.

Various researchers have applied Newton second law [10, 15-25] to model the dynamics of free-piston engine. It is limited to single piston and dual piston configuration since the opposed piston requires synchronizer mechanism, which introduce kinematic constraint to the pistons, which should be treated similar to crankshaft engine.

The principles of the numerical modelling of this paper is based on the dynamics model as established in [26] which is presented in the next section.

3. Piston motion numerical modelling

The piston motion profiles are the key aspect in controlling the energy input during starting of the free-piston engine generator (FPEG). This section establishes the dynamics model of the FPEG, which is based on the free-body diagram shown in Figure 2. The main forces comprise of:

- The in-cylinder pressure forces acting on both cylinders, F_{p_1} and F_{p_2}
- The frictional forces due to contact surfaces on the moving part of the engine, F_f
- The net force or inertial load of the moving mass, $m\ddot{x}$,(where $\ddot{x}$ is the acceleration of the piston in the direction of motion and m moving mass).
- The electrical force acting on the permanent magnet assembly of the generator, F_{cog}
- The motoring force during starting, F_{mot}

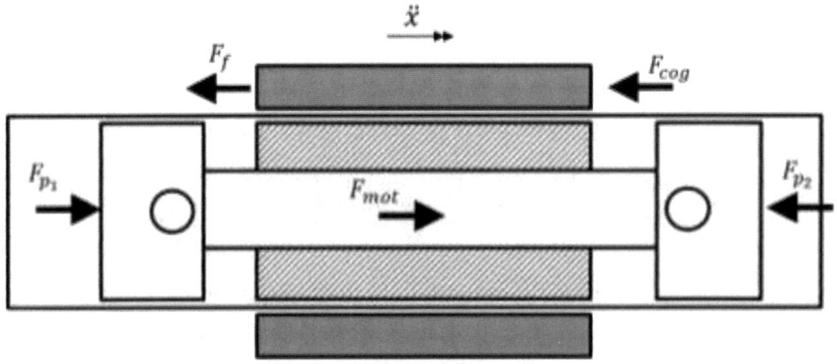

Figure 2. Free-body diagram of the dual piston free-piston engine generator dynamics model.

The dynamics equation can be obtained by applying Newton second law in the direction of acceleration as shown in Equation 1.

$$\sum F = F_{mot} + F_p - F_f - F_{cog} = m\ddot{x} \quad (1)$$

F_p is the resulting in-cylinder pressure forces acting on the translator can be expressed in the Equation 2:

$$F_p = F_{p_1} - F_{p_2} = (p_1 - p_2) \times \frac{\pi B^2}{4} \quad (2)$$

Where, p_1 and p_2 is the in-cylinder pressure in cylinder 1 and cylinder 2 respectively and B is the cylinder bore diameter.

Generally, there are three modes of free-piston engine generator operation as shown in Table 1. In motoring mode, which occurred during starting, the motoring force is the most dominant force. When combustion occurs, the combustion force $(F_p)_c$ becomes the most dominant. During this mode, motoring forces shall be gradually decreased until the sustainable reciprocation is achieved.

Table 1. Summary of the three main modes for a free-piston engine generator and the corresponding general equation of motion.

Mode	Equation of motion
Motoring	$\sum F = F_{mot} + (F_p)_m - F_f - F_{cog}$
Combustion	$\sum F = F_{mot} + (F_p)_c - F_f - F_{cog}$
Generating	$\sum F = (F_p)_c - F_f - F_{cog}$

The model developed in this paper is based on the design specifications presented in Table 2. The design has variable compression ratio capability of up to 20:1 but the maximum stroke is limited to 38mm for sufficient clearance between the cylinder head and the piston at TDC. The FPEG is expected to produce electrical power of 5kW at 50Hz by running in two-stroke cycle operation at a minimum of 2.8kW per-cylinder produced from combustion.

Table 2. The free-piston engine generator (FPEG) specifications.

Parameter	Value
Bore [mm]	50
Minimum/Maximum stroke [mm]	33/38
Minimum/Maximum Geometric compression ratio [-]	9.5:1 to 20:1
Moving mass [kg]	6.5
Minimum engine power/cylinder [kW/cyl] @50Hz	2.8
Maximum electrical power output [kW] @50Hz	5.0
Motoring speed [Hz]	10
Idling speed [Hz]	20
Generating speed [Hz]	50

There are two types motion profiles investigated in the numerical modelling:

•Trapezoidal velocity profile

•S-curve velocity profile.

The trapezoidal velocity profile is simpler and widely employed while the S-curve velocity profile, although complex, leads to smoother vibration-less motion. Figure 3 shows the modelled motion based on trapezoidal velocity profile. The translator undergoes constant acceleration at the beginning of the motion, zero acceleration when it reaches maximum velocity and steadily decelerated when it reaches the targeted stroke. The main problem with trapezoidal motion profile is the bumpy motion resulting from the infinite jerk characteristic to produce constant acceleration and deceleration values.

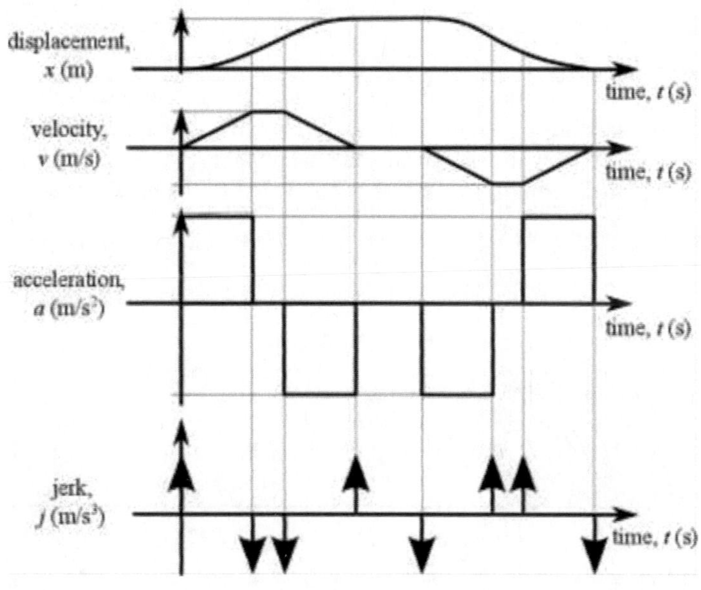

Figure 3. Trapezoidal velocity profiles with resulting acceleration and displacement profiles. The jerk is the derivative of acceleration.

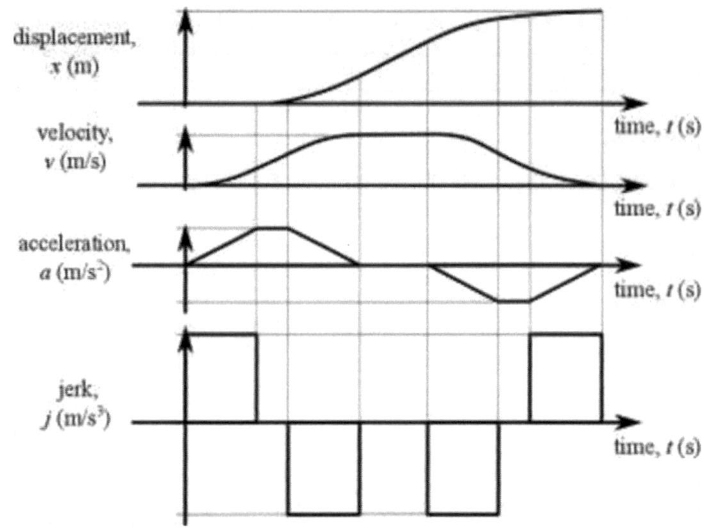

Figure 4. S-curve velocity profile with resulting acceleration and displacement profiles showing finite jerk characteristics.

Figure 4 shows the S-curve velocity profile with the resulting acceleration, and position profiles. In this profile, the jerk now has finite values due to the S-shaped velocity profiles. Both motion profiles are investigated in numerical model develop using Matlab and the results are discussed in the following section.

4. Results and discussion

The main aim of the paper is to investigate the piston motion resulting from two motion profiles commonly used for motion control presented earlier namely, trapezoidal and S-curve velocity profiles. Figure 5 shows a typical crankshaft piston velocity against piston position which produced higher piston velocity around TDC as compare to lower piston velocity for FPEG operating at similar engine cyclic speed.

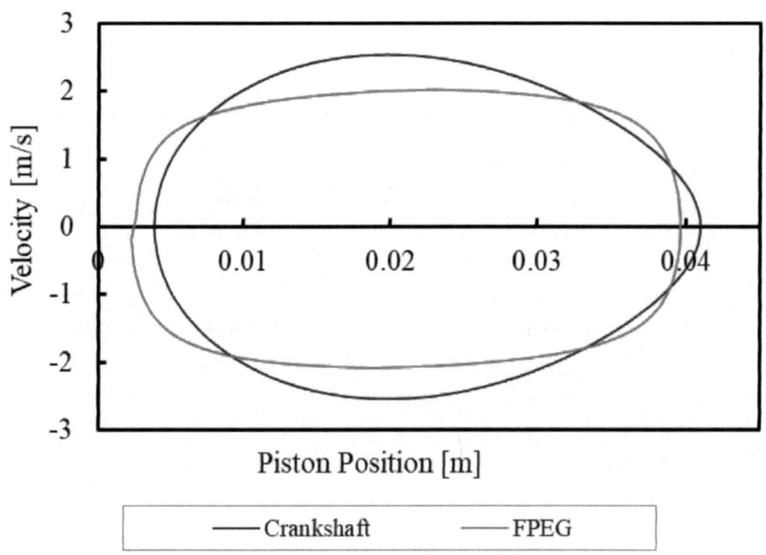

Figure 5. The typical velocity against piston position curves for crankshaft IC engine and free-piston engine generator (FPEG) used as baseline [26].

Figure 6 shows the piston velocity and acceleration at 10Hz motoring speed and 33mm stroke produced using trapezoidal velocity motion. There is no constant speed region since the piston is accelerated for 25ms and decelerated quickly so that the targeted stroke can be achieved. The typical trapezoidal velocity profile is observed with the maximum velocity occurs at mid-stroke, coincidental with the deceleration timing. This profile produced discontinuities for the jerk positions as presented in Figure 3 earlier.

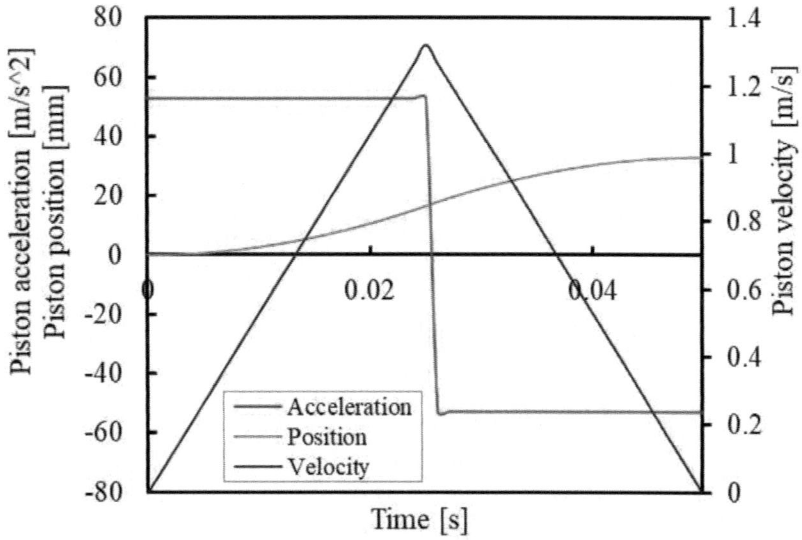

Figure 6. The acceleration and velocity produced at 10Hz motoring speed for the starting using trapezoidal velocity profile and its resulting piston position.

Figure 7 shows the resulting piston position using trapezoidal velocity profile at minimum stroke length of 33mm and maximum stroke of 38mm. Both targeted stroke positions are achieved within the cyclic duration at 10Hz. This is the main reason why this method is preferred widely for motion control applications.

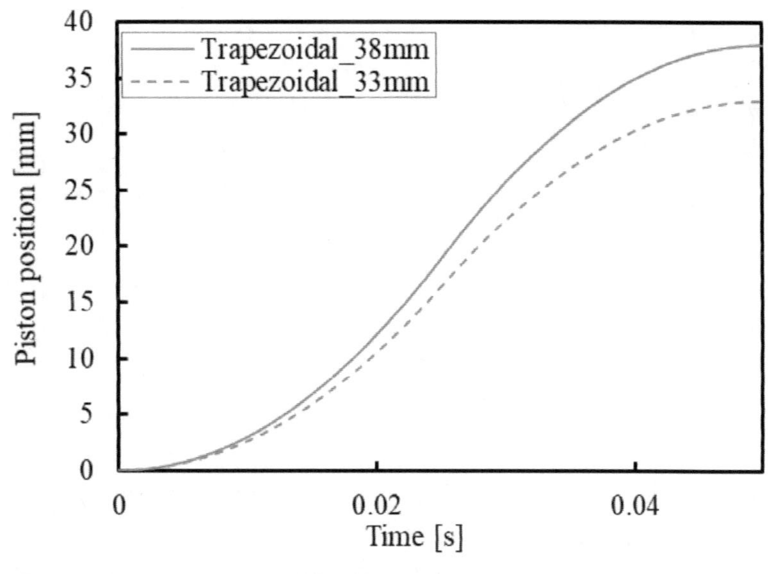

Figure 7. The piston position resulting from trapezoidal motion simulated at minimum and maximum stroke length.

Figure 8 shows the piston velocity and acceleration at 10Hz motoring speed and 33mm stroke produced using S-curve velocity motion. This motion produced favourable smooth acceleration transition for dual-piston type free-piston engine generator in order to reduce the cyclic vibration. However, the final position is 6mm shorter than the targeted stroke, which needs further tuning before this profile can be implemented.

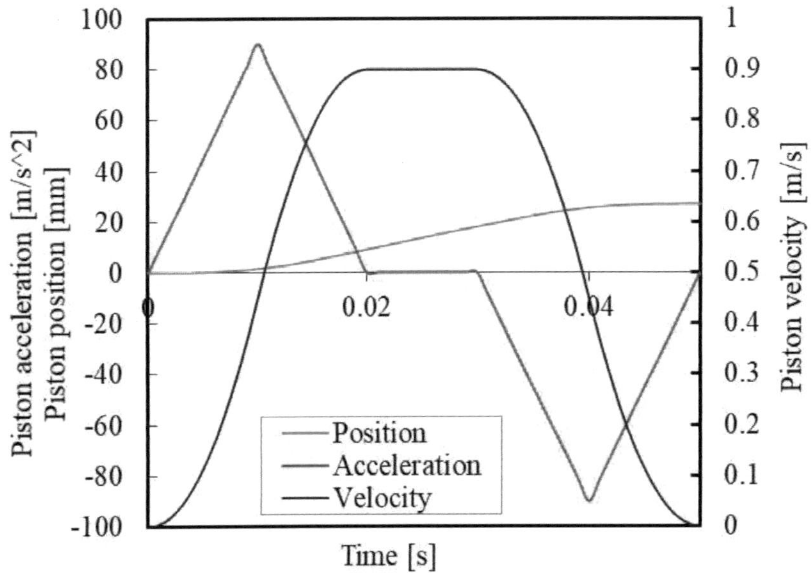

Figure 8. The acceleration and velocity produced at 10Hz motoring speed for the starting using S-curve velocity profile and its resulting piston position.

Figure 9 illustrates the severity of this situation where the stroke reduction is about 18% when S-curve motion profile is used. It can be observed that the piston motion has slower ascend during the first and last 10ms of the motion due to the S-shaped velocity profile which has contributed to the shorter stroke.

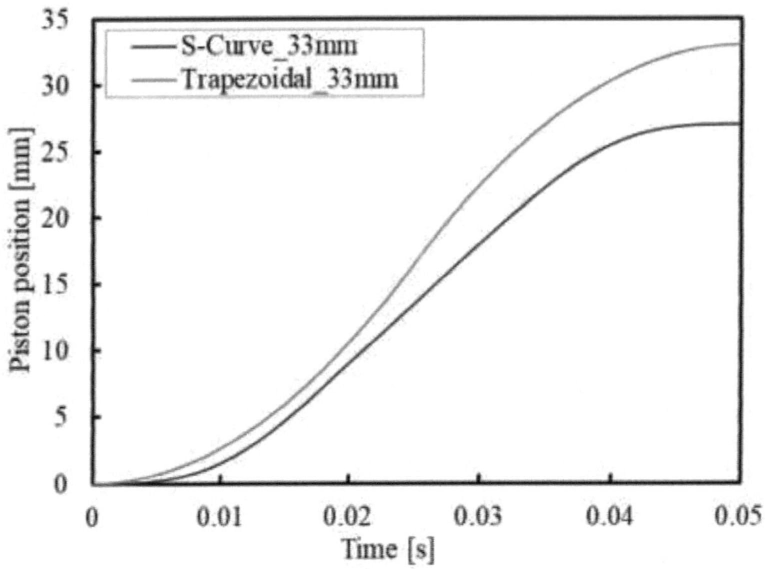

Figure 9. The piston position comparison between trapezoidal and S-curve motion simulated at 33mm.

5. Conclusions

In this paper, a numerical method is used to investigate two types of motion profiles, trapezoidal velocity and S-curve velocity for starting of free-piston engine generator. The trapezoidal motion profile is able to move the piston to the targeted stroke of 33mm and 38mm. However, the final position produced via S-curve motion profile is 18% shorter due to the slower ascend and descend during the first and final 20% of the cyclic duration. Further tuning of the model is necessary before the S-curve motion can be implemented in the free-piston engine generator starting strategy.

Acknowledgments

The author would like to thank Universiti Malaysia Pahang (www.ump.edu.my) for providing laboratory facilities and financial assistance under project no. RDU160384.

References

[1] Hanipah M R, Mikalsen R and Roskilly A P 2015 *Appl. Therm. Eng.* **75** 493-503

[2] Kock F, Heron A, Rinderknecht F and Friedrich H 2013 *MTZ Worldw* **74** 38-43

[3] Jia B, Mikalsen R, Smallbone A, Zuo Z, Feng H and Roskilly A P 2016 *Applied Energy* **179** 1166-75

[4] Aichlmayr H T 2002 *Design Considerations, Modeling and Analysis of Micro-Homogeneous Charge Compression Ignition Combustion Free-Piston Engines.*: UNIVERSITY OF MINNESOTA)

[5] Nandkumar S 1998 *Two-Stroke Linear Engine.* In: *Department of Mechanical and Aerospace Engineering,* (Morgantown, West Virginia: West Virginia University)

[6] Saiful Azrin M Z 2007 *Modeling, Simulation and Implementation of Rectangular Commutation for Starting of Free-Piston Linear Generator.* In: *Electrical and Electronic Engineering Department*: Universiti Teknologi PETRONAS)

[7] Arof H, Eid A M and Nor. K M 2004 *Australasian Universities Power Engineering Conference (AUPEC 2004)*

[8] Zulkifli S A, Karsiti M N and Aziz A R A 2008 Rectangular current commutation and open-loop control for starting of a free-piston linear engine-generator. In: *Power and Energy Conference, 2008. PECon 2008. IEEE 2nd International,* pp 1086-91

[9] Zulkifli S A, Karsiti M N and A-Aziz A R 2009 Investigation of linear generator starting modes by mechanical resonance and rectangular current commutation. In: *Electric Machines and Drives Conference, 2009. IEMDC '09. IEEE International,* pp 425-33

[10] Něměček P, Šindelka M and Vysoký O 2003 *IFAC Symposium on Advances in Automotive Control.*

[11] Hunt B R, Lipsman R L and Rosenberg J M 2001 *A guide to MATLAB for Beginners and Experienced Users*: Cambridge University Press)

[12] Goldsborough S S 2002 *Optimizing the Scavenging System for High Efficiency and Low Emissions: A Computational Approach.* In: *Department of Mechanical Engineering,* (Fort Collins, Colorado: Colorado State University)

[13] Mao J L, Zuo Z X, Li W and Feng H H 2011 *Applied Energy* **88** 1140-52

[14] Něměček P and Vysoký O 2006 *Proceedings of the 6th Asian Control Conference* **VOL.1**

[15] Atkinson C M, Petreanu S, Clark N N, Atkinson R J, McDaniel T I, Nandkumar S and Famouri P 1999 *SAE Internationl: Hybrid Vehicle Engines and Fuel Technology.* **1999-01-0921**

[16] Goldsborough S S and Blarigan P V 1999 *SAE International* **1999-01-0619**

[17] Houdyschell D 2000 A Diesel Two-Stroke Linear Engine. In: *Department of Mechanical and Aerospace Engineering*: West Virginia University) p 64

[18] Johansen T A, Egeland O, Johannessen E A and Kvamsdal. R 2001 *Proc. American Control Conference,*

[19] Fredriksson J and Denbratt I 2004 *SAE International* **2004-01-1871**

[20] Johansen T A, Egeland O, Johannessen E A and Kvamsdal. R 2003

[21] Shoukry E F 2003 *Numerical Simulation for Parametric Study of A Two-Stroke Compression Ignition Direct Injection Linear Engine*. In: *Department of Mechanical and Aerospace Engineering,* (Morgantown, West Virginia: West Virginia University)
[22] Hansson J, Leksell M, Carlsson F and Sadarangani. C 2005
[23] H.Xia, Y.Pang and M.Grimble 2006 *Hybrid Modelling and Control of a Free-Piston Energy Converter*. In: *International Conference on Control Applications,* (Munich, Germany
[24] Deutsch P and Vysoký O 2007 *MECCA Journal of Middle European Construction and Design of Cars* **VOL.5**
[25] Mikalsen R and Roskilly A P 2008 *Appl. Therm. Eng.* **28** 589-600
[26] Hanipah M R 2015 Development of a spark ignition free-piston engine generator. In: *School of Mechanical and Systems Engineering,* (Newcastle upon Tyne, United Kingdom: Newcastle University)

IOP Conf. Series: Materials Science and Engineering **257** (2017) 012055 doi:10.1088/1757-899X/257/1/012055

Cogging force investigation of a free piston permanent magnet linear generator

I I Abdalla[1], E Z Zainal A.[2], N A Ramlan[2], Firmansyah[2], A R A Aziz[2]* and M R Heikal[2]

[1]Department of Electrical and Electronic Engineering, Universiti Teknologi PETRONAS, 32610 Bandar Seri Iskandar, Perak, Malaysia
[2]Centre for Automotive Research and Electric Mobility, Universiti Teknologi PETRONAS 32610 Bandar Seri Iskandar, Perak, Malaysia

*Corresponding author email: rashid@utp.edu.my

Abstract. Better performance and higher efficiency of the vehicles can be achieved by using free piston engine, in which the piston is connected directly to the linear generator and waiving of any mechanical means. The free piston engine has the ability to overcome or reduce many of the challenges, such as the carbon dioxide (CO_2) emission and fossil fuel consumption. The cogging force produces undesired vibration and acoustic noise in the generator. However, the cogging force must be minimized as much as possible, in order to have a high performance. This paper studies the effects of ferromagnetic materials on the cogging force of the permanent magnet linear generator (PMLG) to be used in a free piston engine using nonlinear finite-element analysis (FEA) under ANSYS Maxwell. The comparisons have been established for the cogging force of the PMLG under various translator velocities and three different ferromagnetic materials for the stator core, namely, Silicon Steel laminations, Mild Steel and Somaloy. It has been shown that the PMLG with a stator core made of Somaloy has a lower cogging force among them. Furthermore, the induced voltage of the PMLG at different accelerations has been studied. It is found that the PMLG with Mild Steel and Somaloy, respectively give larger induced voltage. Moreover, as the translator speed increase the induced voltage increased.

1. Introduction

As the fossil fuels and many of energy sources are contributing to CO_2 emission and other pollutants in both land and sea-based transport, many countries are promoting the development process of clean energy [1][2]. However, within the automotive industry, there are many kinds of research have been conducted to reduce the fossil fuel consumption which leads to high cost and environmental problems. The hybrid electric vehicle is one of the recently studied solutions [3, 4].

The configuration of the conventional internal combustion engines powering the hybrid electric vehicles generally used mechanical transmission means such as crank mechanism, which restricts the motion of the piston. Moreover, the major part of the total friction losses occurring in the conventional combustion engine because of the crank mechanism [5, 6]. Besides, it limits the range of the compression ratio of the engine. Hence, better performance and high efficiency of the conventional engine can be achieved by eliminating the crank mechanism. This can easily be realized by using free piston engine, in which the piston reciprocates linearly with PMLG without the need of any transmission means [7-9].

The free-piston engine converter composed of a permanent magnet linear generator coupled to a free-piston engine as illustrated in Figure 1, the major parts have been indicated. Recently, this technology is being a major of concern of many researchers worldwide. Basically, the flexibility and easy controllability as well as the high efficiency of electrical machines, make them an interesting concept [10, 11]. The growing interest of the automotive industry in the technology of electric hybrid vehicles is a driving force behind the interest in free-piston engine generators. The single piston and dual piston of free-piston engine generator designs have been reported [9, 12]. Moreover, the control can be well done by implementing appropriate power electronics control.

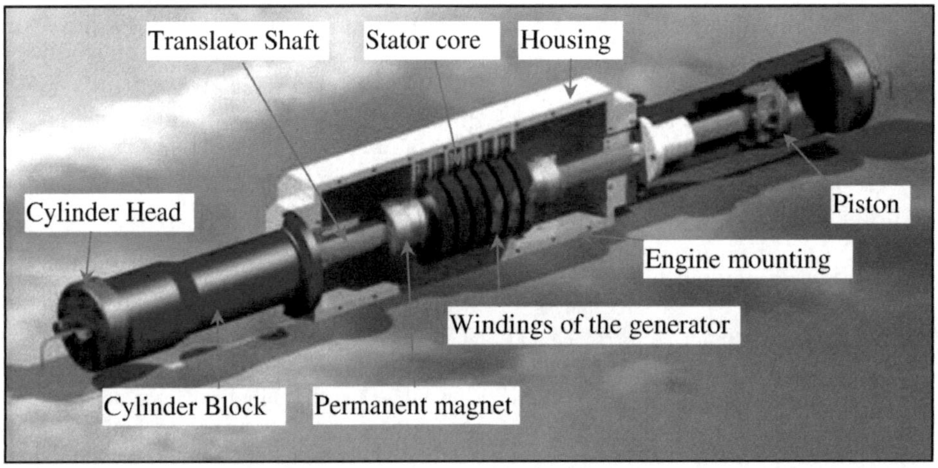

Figure 1. Free piston permanent magnet linear generator engine.

Mainly, there are four different approaches for producing a linear energy conversion. The first approach is to use the electrostatic properties. Thus, a maximum force density of around 16 N/m^2 can be obtained. The second approach is to produce a linear energy conversion by an electromagnetic way. The third and fourth approaches are based on the mechanical friction that uses the piezoelectric or magnetostrictive properties to interact with the translator [13, 14].

The developments based on the PMLG are very likable owing to efficient electromagnetic performance, despite, suffers from the cogging force. This force produces due to the attraction between ferromagnetic core and magnetic with zero current in the winding of the machine [15, 16]. The periodic waveform of the cogging force is depending on the relative position of the translator. When the excitation current assigned to the winding of the machine, the cogging force will be added to a thrust force. The cogging force makes a ripple in thrust force. The ripple resulted by the cogging force will deteriorate the position control and precise speed in many applications. The low-speed applications are more suffering from such ripple; moreover, it produces undesirable acoustic noises and vibrations. Thus, at the design stage must be minimized [15]. Numerous of techniques have been used to reduce the cogging force in permanent magnet machines, but most of these techniques contribute to the reducing of the main electromagnetic performance. Alternatively, air-cored PMLGs are preferred in terms of unavailability of cogging force, lightweight and simple. Nonetheless, they have limitations in electromagnetic performance and power generation [17-19].

An accurate and fast calculation of the magnetic field distributions are necessary for many of electromagnetic machines, they can provide more efficient design and execution of such machines, subsequently, higher performance of the machine can be obtained [20, 21]. However, numerous modeling methods exist for prediction and analysis the electromagnetic behavior of the electric machines. These methods vary from simple and accurate to complicated and time-consuming models [22]. The finite element analysis (FEA) is offering many features, thus, it is widely used for the modeling and simulation of the electrical machines. However, this technique empowers us to perform a complicated analysis of electrical machines in a minimum estimated time. This paper presents the cogging force investigation of a PMLG utilizing three different ferromagnetic materials for the stator core by using FEA under ANSOFT Maxwell software.

2. Research Method

The FEA has been used to compute the magnetic field along the cross-section of the proposed generator. By the fact that, with the rare-earth magnet materials and Halbach array configuration, the high magnetic field will be achieved. especially, with the. Moreover, the quasi-Halbach magnetization technique has been selected in this study because it has many advantages over the conventional PM array [23, 24].

The finite element two-dimensional (2-D) and three-dimensional (3-D) models adopted from ANSYS Maxwell simulation software for the proposed PMLG is shown in Figure 2. The PMLG with single phase and long translator. The stator contains 6 coils and 6 slots. The translator is made of the neodymium-iron-boron (NdFeB) permanent magnet with quasi-Halbach magnetized magnets. Quasi-Halbach magnetization provides higher air gap magnetic field distribution [25, 26]. The FEA is carried out for the PMLG with the three different ferromagnetic materials. The axisymmetrical coordinate system and vector orientation for magnets have been adopted in the software.

The FE mesh affects the FEA calculation, especially in terms of time and accuracy of the computation. The automatic mesh allowed for faster simulation and shorter execution time than a fine mesh. Nevertheless, the computation accuracy is low because the number of degrees of freedom is low [27]. However, the fine mesh has been utilized for this study.

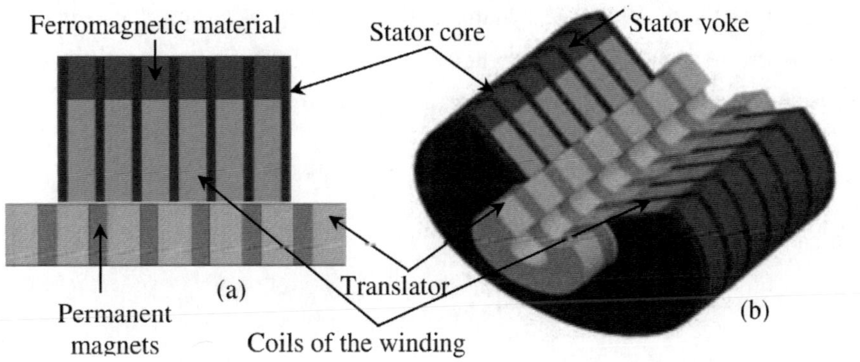

Figure 2. Configuration of the proposed PMLG. (a) two-dimensional (b) three-dimensional.

The magnetic field analysis is confined to two regions, namely the air region with a permeability of μ_0, and the magnetic region with a permeability of $\mu_0\mu_r$. μ_r is the relative recoil permeability and for rare-earth PMs is close to unity. Therefore, the magnetic flux density, B related to the magnetic field intensity, H in airspace region and magnetic region, respectively, can be expressed as [28-32]

$$B = \mu_0 H \tag{1}$$

$$B = \mu_0\mu_r H + \mu_0 M \tag{2}$$

The magnetization, M of the linear machine in the cylindrical coordinate system can be expressed as [30, 32-34]

$$M = M_r e_r + M_z e_z \tag{3}$$

The magnetization distribution was expandable into Fourier series, with M_r and M_z expressed as a function of z as in (4) and (5), respectively [16, 32].

$$M_r = \sum_{n=1,2,...}^{\infty} M_{rn} \cos m_n z \tag{4}$$

$$M_z = \sum_{n=1,2,...}^{\infty} M_{zn} \sin m_n z \tag{5}$$

where M_r and M_z denoted the components of M in the radially and axially directions, respectively, and $m_n = 2\pi n / T_{lp}$.

When the flux waveform of the open circuit is known, it can be used to calculate the induced voltage of the machine. Hence, with a time-varying magnetic flux, $\phi(t)$, the induced voltage can be calculated as [35, 36]:

$$e_c = -\frac{d\phi(t)}{dt} = -\frac{dB}{dt} A \tag{6}$$

where B is the magnetic flux density and A is the area that is occupying B. On the other hand, the thrust force for a given machine's current can be calculated from the electromagnetic power related to the translator speed, v_t. Therefore, F_T is quantified as in (7) [32]:

$$F_T = \frac{e_c i_a}{v_t} = K_E(z_d) i_a = K_T(z_d) i_a \tag{7}$$

Based on Alembert's equation, the dynamics of the system governing the armature movement of the proposed generator along the z-axis when the mass, m_m, is moving at speed; V_t with a damping coefficient, b, and spring elasticity, k, can be expressed as in (8) [37, 38]:

$$m_m \frac{dv_t}{dt} + bv_t + k\int v_t dt = K_T(z_d) i_a \tag{8}$$

By substituting the value of F_T from (7), equation (8) can be rewritten as in (9):

$$m_m \frac{dv_t}{dt} + bv_t + k\int v_t dt = F_T \tag{9}$$

When the linear velocity is related to the displacement and time, the velocity of the translator can be expressed as in (10) [39]:

$$v_t = \frac{dz_d}{dt} \tag{10}$$

By taking the integration of (9) and substituting the value of v_t, it results in:

$$m_m \frac{d^2 z_d}{dt^2} + b\frac{dz_d}{dt} + kz_d = F_T \tag{11}$$

where z_d, $d^2 z_d / dt^2$, K_T, i_a and dz_d / dt are the displacement of the translator, linear acceleration, thrust force constant, coil current and velocity of the translator, respectively.

3. Results and Analysis

In this study, the machine is running at no-load, hence the winding current is zero. The permeability into the magnets and the coils without demagnetization of the magnet is μ_0. The magnetic properties of silicon steel lamination, mild steel and Somaloy have been identified by their B-H curves and other quantities. The design specification and main dimensions of the proposed PMLG are tabulated in Table 1.

Table 1. Design specification and main dimensions of PMLG.

Parameter	Value	Unit
Magnet thickness	29.50	mm
Mechanical air gap	1.00	mm
Magnetic remanence	1.14	Tesla
Stroke	45.00	mm
The total length	221.00	mm

The quasi-Halbach provided the magnetization for the translator and created the flux lines in the round or the trapezoidal closed loop pattern. The conducted analysis and comparisons were based on various velocities and three different ferromagnetic materials, namely Silicon Steel laminations, Mild Steel and Somaloy. Further discussions on the outcomes are explained as below. The cogging force in the PMLG leads to oscillations of the generator speed and therefore output voltage and power fluctuations. The cogging force corresponds to the force due to the shape of the teeth and the permanent magnets when the current in the coil of the machine is zero. The cogging force evaluation is very sensitive to the mesh and its value is small as compared to the full load force. However, the preferred and accurate method to compute the cogging force is the use of the transient solver with motion; because the mesh will remain unchanged at all positions of the translator. Basically, the stator is fixed and the translator will move with steps. Thus, only the magnetic field from the magnets existed and then the effect of the slot will present. As the translator of the PMLG moved forward and backward, this effect was computed by using the FEA. Figure 3 shows the comparison of the cogging force resulted at the translator velocity of 1.0 m/s and using three different ferromagnetic materials for the stator core. The result is fluctuating between the positive and negative value as can be clearly observed. Furthermore, it can be observed that the average cogging forces have been obtained for the three ferromagnetic materials, namely Silicon steel, Mild Steel and Somaloy, respectively, are 121.5238 N, 55.0707 N and 5.2115 N. It can be concluded that in this range of speed Somaloy is preferred for the stator core of the proposed generator.

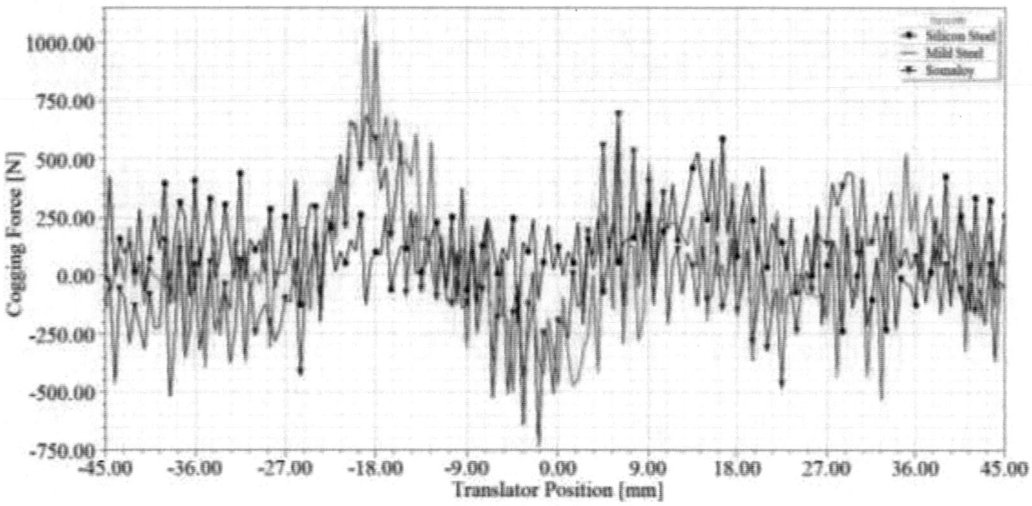

Figure 3. Comparison of the cogging force of the generator under three ferromagnetic materials when the excitation current is zero and 45.0 mm displacement of the translator each side and a linear velocity of 1.0 m/s.

Figure 4 shows the comparison of the cogging force of the generator under the three different materials at translator velocity of 3.0 m/s. It can be observed that the generator with a stator core made of Somaloy has a lower cogging force among the three materials. Furthermore, it can conclude that in this range of speed the Somaloy is preferred for the stator of the proposed generator.

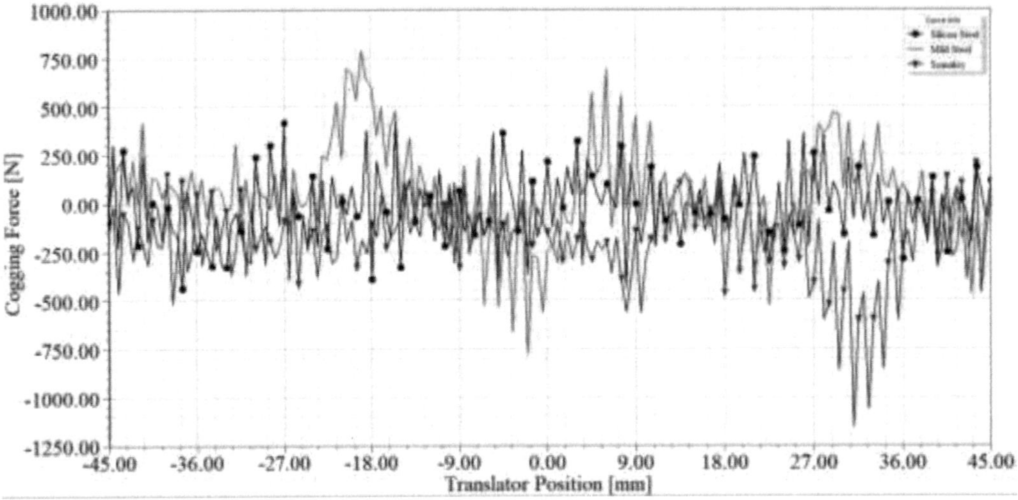

Figure 4. Comparison of the tangential electromagnetic force of the generator when the excitation current is zero and 45.0 mm displacement of the translator as well as a linear velocity of 3.0 m/s.

Figure 5 shows the comparison of the cogging force of the proposed generator that using three different materials for the stator core at translator velocity 5.0 m/s. It can be observed that the generator with a stator core made of silicon steel laminations has higher average cogging force, whereas the generator with Somaloy stator core has lower average cogging force.

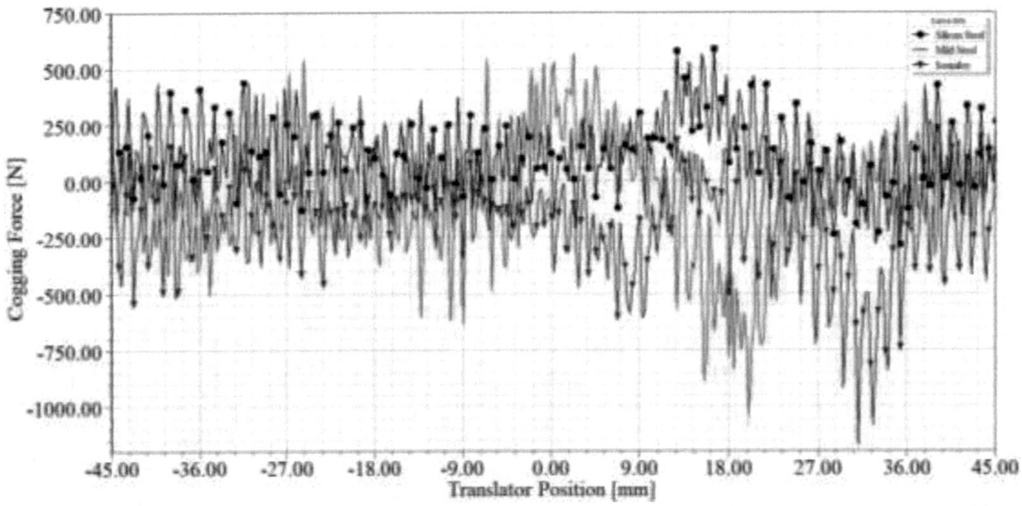

Figure 5. Comparison of the tangential electromagnetic force of the generator when the excitation current is zero and 45.0 mm displacement of the translator, under a linear velocity of 5.0 m/s.

Figure 6 shows the comparison of the cogging force of the linear generator for the three different ferromagnetic materials for the stator core at translator velocity 7.0 m/s. It can be seen that the generator with a stator core made of mild steel laminations has higher average cogging force, 122.0425 N, whereas the generator with Somaloy stator core has a lower average cogging force, 183.5258 N.

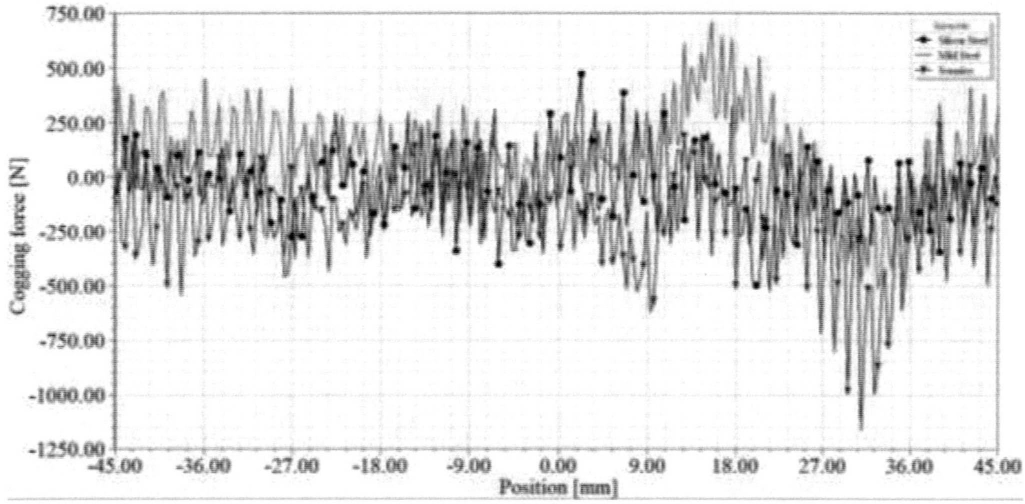

Figure 6. Comparison of the tangential electromagnetic force when the excitation current is zero and 45.0 mm displacement of the translator on both sides under a linear velocity of 7.0 m/s.

Because of inherent lower magnetic permeability, would Somaloy components be chosen to compete on magnetic performance at either adequate level or to be better than steel lamination assemblies. The improvement of the densities of Somaloy components correspondingly leads to increment of the magnetic induction and permeability as well. The cogging force, in general, has a positive value and is an undesirable force, but the negative cogging force will add to the current-generated force (main force), mainly from the perspective of engineering education [40].

Figure 7 shows the comparison of the average cogging force of the PMLG at various accelerations of the translator and three different ferromagnetic materials for the stator core. It can be observed that there is an optimal point for the operation of the PMLG at which the generator run with minimum cogging force, such as 6.0 m/s for the Somaloy, 5.0 m/s for mild steel and 7.0 m/s for laminated silicon steel.

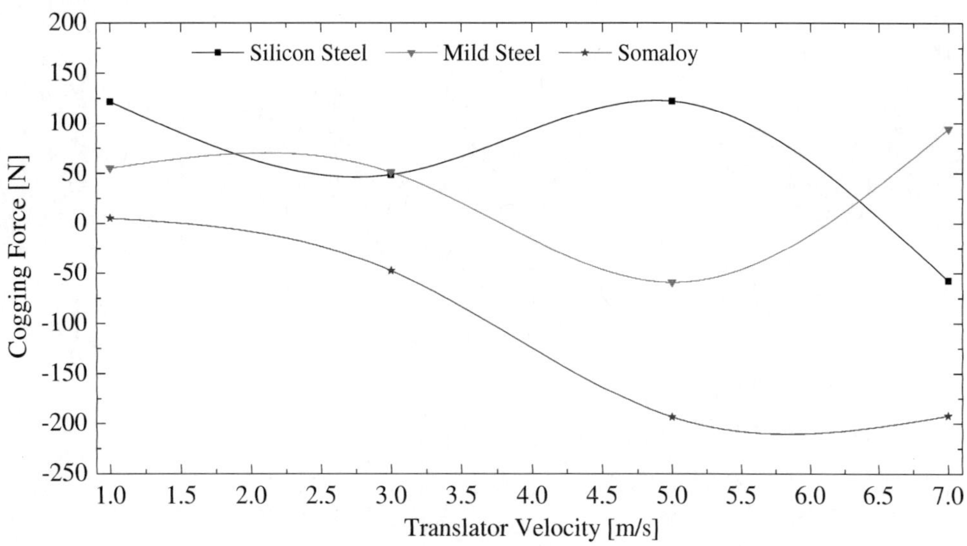

Figure 7. Comparison of average cogging force of PMLG under three different ferromagnetic materials and different translator velocities.

By moving the translator of the PMLG, the induced voltage is computed using the FEA. Figure 8 shows a comparison of the induced voltage waveforms of PMLG using three different ferromagnetic materials for the stator core at a constant translator velocity of 1.0 m/s. It can be seen that the PMLG with the mild steel and Somaloy gives an induced voltage of 13.0 V and 11.92 V, respectively, while the PMLG with silicon steel gives an induced voltage of 5.88 V. This means that the generator with mild steel and Somaloy can generate larger electrical power.

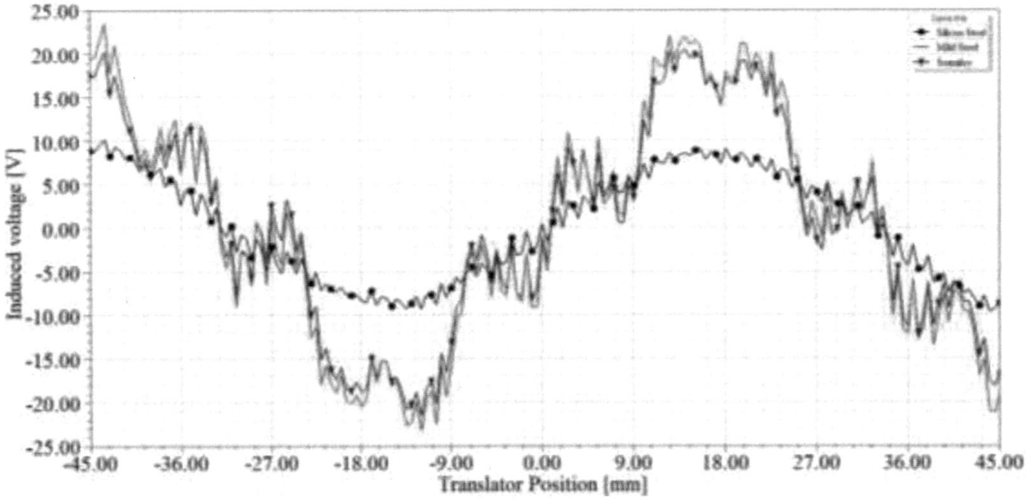

Figure 8. Comparison of the induced voltage among the three ferromagnetic materials for the stator core of the PMLG at the linear speed of 1.0 m/s.

Figure 9 shows a comparison of the induced voltage waveforms of PMLG using three different ferromagnetic materials for the stator core at a constant translator velocity of 3.0 m/s. It can be seen that the PMLG with the mild steel and Somaloy gives an induced voltage of 39.47 V and 36.27 V, respectively, while the PMLG with silicon steel gives an induced voltage of 18.03 V.

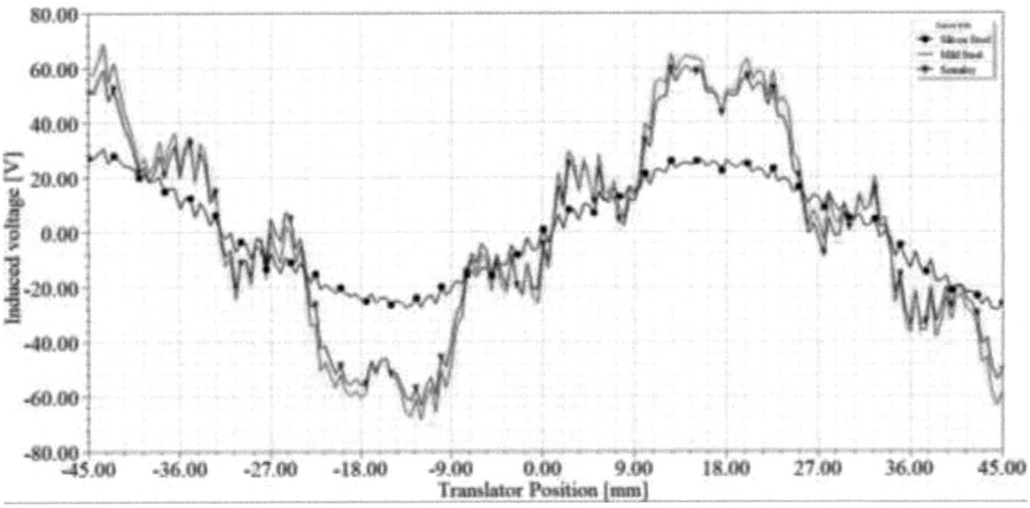

Figure 9. Comparison of the induced voltage among the three ferromagnetic materials for the stator core of the PMLG at the linear speed of 3.0 m/s.

Figure 10 and Figure 11 shows the comparison of the induced voltage waveforms of PMLG using three different ferromagnetic materials for the stator core at constant translator velocities of 5.0 m/s and 7.0 m/s, respectively. It can be seen that the PMLG with the mild steel and Somaloy at a speed of 5.0 m/s gives an induced voltage of 64.5 V and 59.15 V, respectively, while the PMLG with silicon steel gives an induced voltage of 29.35 V. Meanwhile, the PMLG with the mild steel and Somaloy at a speed of 7.0 m/s gives an induced voltage of 93.6 V and 85.76 V, respectively, while the PMLG with silicon steel gives an induced voltage of 42.71 V.

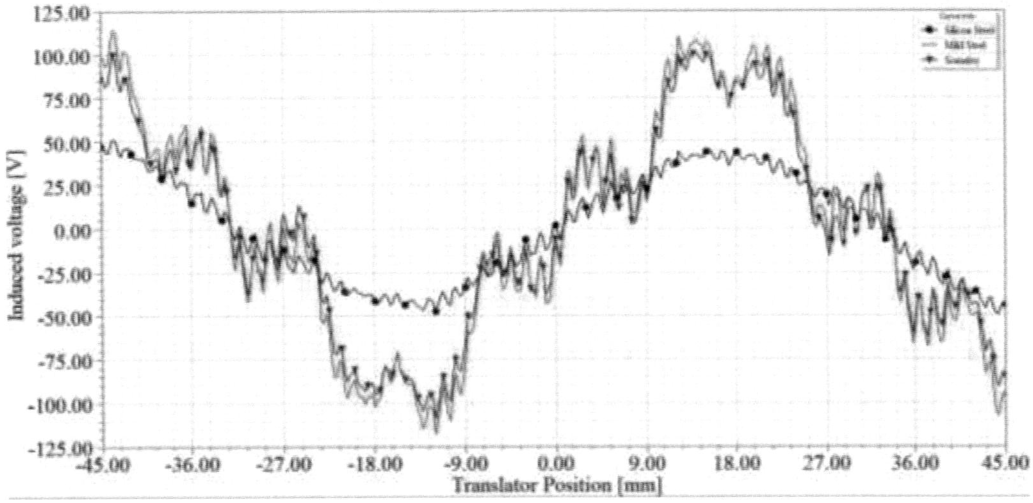

Figure 10. Comparison of the induced voltage among the three ferromagnetic materials for the stator core of the PMLG at the linear speed of 5.0 m/s.

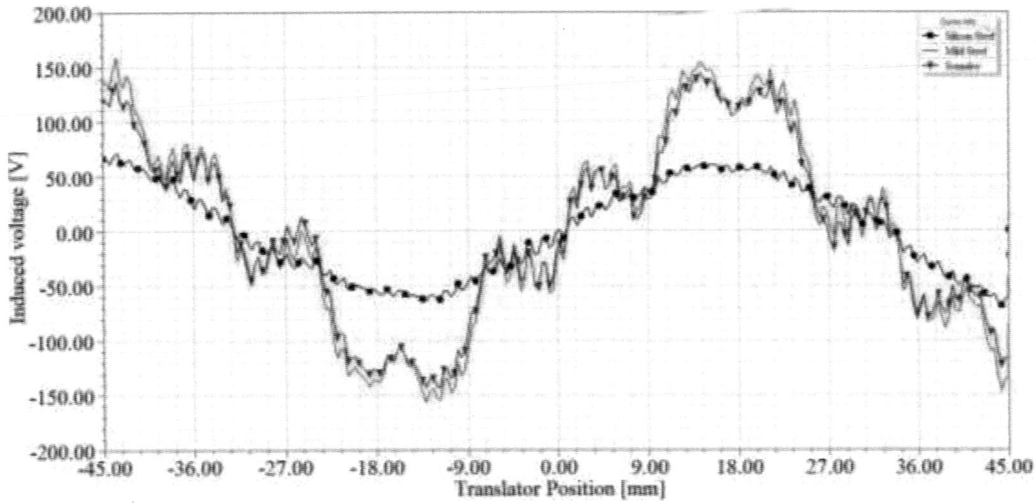

Figure 11. Comparison of the induced voltage among the three ferromagnetic materials for the stator core of the PMLG at the linear speed of 7.0 m/s.

Figure 12 shows the comparison of the RMS induced voltage of PMLG at different translator velocities and three different ferromagnetic materials for the stator core. It can be seen that the PMLG with mild steel has the highest induced voltage. The PMLG with Somaloy has second highest induced voltage, whilst the PMLG with silicon steel laminations has a lowest induced voltage. Furthermore, it can be observed that there is a direct proportion between the translator speed and the induced voltage.

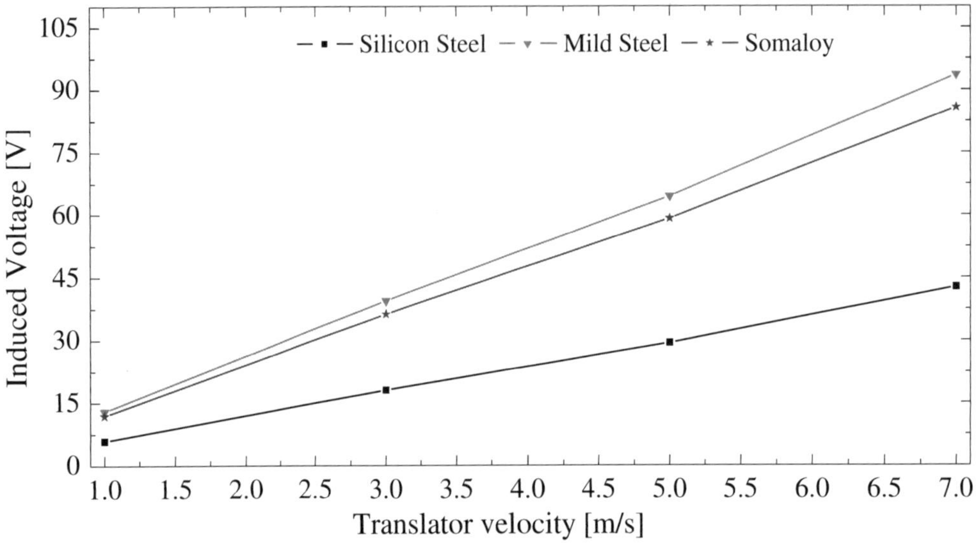

Figure 12. Comparison of the RMS induced voltage of PMLG at different translator velocities.

4. Conclusions

This paper investigated the influence of ferromagnetic materials of the stator core on the amount of induced cogging force in a free piston permanent magnet linear generator (FP-PMLG). The PMLG with laminated silicon steel, mild steel, and Somaloy is analyzed using FEA. The cogging force investigation is carried out for PMLG at various translator velocities along with main dimensions and specifications that have been given in Table 1. It is found that the properties of the material have a significant influence on the cogging force produced. Also, the velocity of the translator influences the cogging force dramatically. From the cogging force comparisons of the PMLG among the three different ferromagnetic materials, namely silicon steel laminations, mild steel and Somaloy, it has been found that Somaloy material for stator core is preferred for the less cogging force. However, when the translator velocity greater than 3.0 m/s and less than 7.0 m/s, the mild steel lamination for the stator core is preferred. Whereas, if the proposed PMLG with a velocity higher than 6.0 m/s, it is preferred to use silicon steel lamination in terms of minimum cogging force. Meanwhile, the comparison of the induced voltage showed that there is a direct proportion between the induced voltage and velocity of the translator. The highest, second highest and lower induced voltage has been obtained for the silicon steel laminations, mild steel, and Somaloy, respectively. Furthermore, with the check for the fabrication availability, it is found that is difficult to fabricate such generator with silicon steel laminations because it is difficult to assemble and de-assemble the stator core.

Acknowledgments

Authors gratefully would like to thank Yayasan Universiti Teknologi PETRONAS (UTP) (0153AA-A92) and the Petroleum Research Fund (0153AB-A34) for funding this research work.

References

[1] Si J, Feng H, Su P and Zhang L 2014 *The Scientific World Journal* 2014.
[2] HOEVEN M V D 2013 *International Energy Agency: highlights. França: International Energy Agency.*
[3] Maloberti O, Figueredo R, Marchand C, Choua Y, Condamin D, Kobylanski L and Bommé E 2014 *IEEE Trans. Magn.* ***50*** 1-11.
[4] Wang J and Baker N J 2016 *International Conference on Power Electronics, Machines and Drives (PEMD 2016)* 6-6.

[5] Zainal A E Z B 2013 *International Journal of Automotive Engineering* **4** 47-53.
[6] Hanipah M R, Mikalsen R and Roskilly A 2015 *Appl. Therm. Eng.* **75** 493-503.
[7] Guo C, Feng H, Jia B, Zuo Z, Guo Y and Roskilly T 2017 *Energy Convers. Manage.* **131** 32-43.
[8] Hung N B and Lim O 2016 *Applied Energy* **178** 78-97.
[9] Jia B, Mikalsen R, Smallbone A, Zuo Z, Feng H and Roskilly A P 2016 *Appl. Energy.* **179** 1166-1175.
[10] Abidin E Z Z, Ibrahim A A, Aziz A R A and Zulkifli S A 2012 *J. Appl. Sci.* **12** 2592.
[11] Ibrahim A A, Aziz A R A, Abidin Z, Ezrann Z and Zulkifli S A 2011 *J. Appl. Sci.* 1-6.
[12] Mikalsen R and Roskilly A 2007 *Appl. Therm. Eng 27* 2339-2352.
[13] Chevailler S *Comparative study and selection criteria of linear motors*.
[14] Ezrann Z A, Abdulwehab A I, Rashid A A A, Saiful A and Zulkifli 2015 *J. Mech. Eng. Sci.* **8** 1393-1400.
[15] Youn S W, Lee J J, Yoon H S and Koh C S 2008 *IEEE Trans. Magn.* **44** 1785-1790.
[16] Abdalla I I, Ibrahim T and Nor N B M 2016 *Int. J. Elect. Power* **77** 263-270.
[17] Gargov N and Zobaa A 2012 *IET Renew. Power Gen.* **6** 171-176.
[18] Yuan C, Feng H and He Y 2017 *Fuel* **188** 390-400.
[19] Sui Y, Zheng P, Yu B, Cheng L and Liu Z 2015 *IEEE Trans. Magn.* **51** 1-4.
[20] Lin Z, Wang J and Howe D 2007 *Electric Machines & Drives Conference, 2007 IEMDC'07 IEEE International* **1** 370-375.
[21] I. I. Abdalla, T. Ibrahim, and N. M. Nor 2013 *Power Engineering and Optimization Conference (PEOCO), 2013 IEEE 7th International* 29-34.
[22] Janssen J L G, Paulides J J H and Lomonova E A 2010 *COMPEL-The international journal for computation and mathematics in electrical and electronic engineering* **29** 984-993.
[23] Han Q 2004 A*nalysis and modeling of the EDS maglev system based on the Halbach permanent magnet array* (Department of Electrical and Computer Engineering University of Central Florida Orlando, Florida, Orlando, Florida).
[24] Gieras J F, Wang R J and Kamper M J 2008 *Axial flux permanent magnet brushless machines vol. 1* (Springer).
[25] Abdalla I I, Ibrahim T B and Nor N M 2016 *Appl. Comput. Electrom.* **31**.
[26] Abdalla I I, Ibrahim T and Nor N B M 2017 *9th International Conference on Robotic, Vision, Signal Processing and Power Applications* 777-789.
[27] Silwal B 2012 *Computation of eddy currents in a solid rotor induction machine with 2-D and 3-D FEM*.
[28] Wang J, Howe D and Lin Z 2008 *IET Electr. Power App.* **2** 193-200.
[29] Wang J, Jewell G W and Howe D 1999 *IEEE Trans. Magn.* **35** 1986-2000.
[30] Meessen K J, Gysen B, Paulides J and Lomonova E A 2008 *IEEE Trans. Magn.* **44** 4305-4308.
[31] Tsai N C and Chiang C W 2010 *Mechatronics* **20** 596-603.
[32] Wang J, Howe D and Lin Z 2007 *Industry Applications Conference, 2007. 42nd IAS Annual Meeting. Conference Record of the 2007 IEEE* 311-318.
[33] Zhu Z and Howe D 2001 *IET Electr. Power App* **148** 299-308.
[34] Abdalla I I, Ibrahim T B and Nor N M 2015 *Appl. Mech. Mater.* **793** 274-279.
[35] Feng H, Song Y, Zuo Z, Shang J, Wang Y and Roskilly A P 2015 *Energies* **8** 765-785.
[36] Dall'Ora L 2014 *Analysis and Design of a Linear Tubular Electric Machine for Free-piston Stirling Micro-cogeneration Systems*.
[37] Tomczuk B and Sobol M 2005 *IEEE Trans. Magn.* **41** 2362-2367.
[38] Chen J, Liao Y, Zhang C and Sun P 2014 *Electrical Machines and Systems (ICEMS), 2014 17th International Conference* 2211-2215.
[39] Hasanien H M 2011 *IEEE Trans. Ind. Electron.* **58** 4048-4056.
[40] Kikuchi T and Kenjo T 1998 *IEEE Trans. Edu.* **41** 16.

Risk assessment in infrastructure in educational institution: A study in Malaysia

Ahmad Rasdan Ismail[1], Noor Adilah Hamzah[1*], Nor Kamilah Makhtar[2], Khairul Azhar Mat Daud[1], Nik Zulkarnaen Khidzir[1], Nurul Husna Che Hassan[1] and Muhamad Arifpin Mansor[3]

[1]Faculty of Creative Technology and Heritage, University Malaysia Kelantan, Locked Bag 01, 16300 Bachok, Kelantan. Malaysia.
[2]Institut Pendidikan Guru Campus Kota Bharu, Jalan Maktab, Pengkalan Chepa, 16109 Kota Bharu Kelantan, Malaysia.
[3]Faculty of Engineering Technology, University Malaysia Pahang, Lebuhraya Tun Razak, 26300 Gambang, Malaysia.

*Corresponding author email: nooradilahhamzah@gmail.com

Abstract. This particular study was conducted to assess the hazard exposure in education institution and to highlight the possible risk level available. The assessment utilised is Hazard Identification, Risk Assessment and Risk Control (HIRARC). There was a 2008's form in order to determine the risk level of the hazard. There were over 111 of education institutions were selected around Malaysia to perform this assessment. Area chosen for each institution was office, playing field, canteen, classroom, toilet and drainage. By referring HIRARC Guideline 2008, the determination of risk rank is measure based on the formula likelihood multiply severity and the rank need to refer from risk matrix standard. There are several hazard have be found and shows the high, medium and low of risk level. The higher level of risk was discussed in the study which is hazard found in playing field and hazard in office. There several hazard that need to be control by education management to avoid increase of case accident in Education Sector, Malaysia. As conclusion, the exposure hazard among the staff and educators is high and further action and control are needed. Further study need to explore the best recommendation for control measure of the hazard exposed by education institution.

1. Introduction

An education institution is considered as a place of work as stated in the Occupational Safety and Health Act (OSHA) 1994, "place of work" means premises where persons work or premises used for the storage of plant and substance [4]. In Occupational Safety and Health Act (ACT 514), stated in section 1(2), OSHA 1994 should apply throughout Malaysia to the industries specified in the First Schedule [20].There were 10 sector in Malaysia. Based on Article 132 Federal constitution, Education Institution was under Public Services and Statutory Authorities [24]. Besides, Education Institution is divided into preschool education, primary education, secondary education, post-secondary education and tertiary education [25]. Therefore, this study only covered result from schools due the increased of accident cases reported.

In Malaysia, the disclosure of occupational safety and health aspects in school, especially among teachers and parents is very low compared to another sector. This aspect has been underestimated because it is considered can delay a job and wasted. However, since 2005, local news reported the news of the accident involving the school and students. Statistics accidents have been recorded daily based on reports by local newspaper. From January 2016 to May 2017, there were 54 accidents have been reported in schools. Here is a summarisation of statistics based on the accident happens at school that have been recorded from newspaper (**Table 1**). Besides NIOSH Chairman, Tan Sri Le Lam Thye pointed out that, safety of school children should not be taken lightly in view of recent incidents where a few cases of primary school pupils were killed after being hit by cars or heavy vehicles while crossing the road in front of the school [27].

Table 1. Statistic of accident occur in school 2016.

No	Hazards	No of cases
1	Epidemic tuberculosis	1
2	Hit by goalpost	1
3	Sexual harassment	1
4	Fall from high	3
5	Biological- snake bite	3
6	Structural failure	3
7	Fire	7
8	Food poisoning	9
9	Accidents due negligence	11
10	Mercury spill	15
	Total	**54**

*sources: collection from newspaper

School safety means students and staff should feel themselves free in physically, psychologically and emotionally manners [7]. The basic reason why schools exist is because of the students and their education. Therefore, school safety provides a social and physical environment that effect appropriate behavior of the students. The physical environment includes the way in which the building and the school's routines are managed to prevent problems [1]. Young people are more vulnerable to accidents when entering any of the workplace, compared to older people [8]. Therefore, the community of school need to conduct safety program to avoid from unexpected situation occur in the workplace [2].

Schools become one of the largest categories of work place because children, young people, teachers and other staff spend a lot of their daily time in schools where variety of risks and hazards can occur including physical, psychical and social that may negatively affect their health and wellbeing [2]. Meanwhile, the school area is not only for teachers, management and staffs, even parents also are people who should be protected because they send and co-opted their children in school area. According to Japan Sport Council, a largest set of school injury data in Japan, approximately one million injuries occur in elementary, junior high and high schools each year and this remains relatively stable year by year. School accidents causing injuries have not been as widely investigated as those occurring at home or in road traffic accidents [3]. This is due to the general belief that children are relatively protected in school because of the safe environment and supervision.

Thus, the probability of the risk and hazard occurred in school is high. The risks and hazards that occur may negatively affect the health and behaviors of the students. In order to prevent danger situations in schools, safety condition in school are need to be considered in all of aspects which is the school physical environment and also the social environment [2]. Today, safety has become a central issue which has been growing in the social context that increasing the concern for prevention of danger situations [5]. Certain safety conditions are being explored in entirely life scopes which are related to quality and welfare. One of the scopes for ensuring safety is from the educational activities in school. Children, young people, teachers and other staff spend most of their time in school. In Turkey, previous study reported that the period of students spend their daily in school is about 180 days of a year and 6 hours [6].

Under OSHA 1994 [20] stated that the responsibility of the employer to ensure and provide a safe workplace for their workers. School management also is able to set the priority in improving their school area that will benefit to the educators, staff, students and visitors. The success of the system and safety of workplace can be achieved with the full commitment of all levels of the organization. Ideas, cooperation and commitment of employees in the workplace can help the education institution to reduce accidents number. Perceptions and opinions of the staff, educators and students are important to improve the level of risk at the workplace [22-23]. Therefore, management must take corrective action in order to ensure hazards in controls and appropriate measures should be taken. Simultaneously, assessment should be carried out to identify and improve level of risk in education institution in order to reduce accidents. By doing this, the management of school will be able to identify which area in school that can be considered as safer zone environment [21].

Risk assessment conducted to make sure that no one gets hurt or becomes ill. Accidents and ill health can worst lives and affect an output or product of process. Therefore, carrying out risk assessment, and provide corrective action will prevent workers from accident happens [19]. The main objective of this paper is to highlight the findings on risk assessment and identification of hazard exposed to staff and educators in education institution. Further measurement the level of risk was conducted by using risk matrix to highlight the higher risk found in education institution.

2. Methodology

Data collection was collected by walkthrough observation and calculated by using Hazard Identification, Risk Assessment, and Risk Control form. Educational institution made up from universities, colleges, secondary schools and primary schools. Therefore, collection of data was collected over 111 of schools randomly in Kelantan, a state located in the East Coast of Peninsular Malaysia which 53 primary school and 58 secondary school. The observation area included 6 areas available in schools which are office, playing field, classroom, toilet, canteen and drainage. To determine the risk level, the formula 'risk=likelihood x severity' was utilised. By referring to the Guidelines of HIRARC 2008, Table C, page 12, the risk level were obtained through multiplying the severity and likelihood. The relative value can be used to prioritize necessary action that can be taken in order to improve the effectiveness the management of workplace hazards available in education institutions. The higher risk level was discussed further.

3. Results and Discussions

The finding tabulation in Table 2 which the data analyzed by using formula, rating of severity multiply to likelihood rating. There are 5 groups of severity rating; catastrophic-5, fatal-4, serious-3, minor-2, negligible-1 and likelihood rating; most likely-5, possible-4, conceivable-3, remote-2 and inconceivable-1. Ranking hazard represent by number. First ranking (1) indicates that hazard in playing field was the highest compare to another area. Office or teacher's room indicates second ranking which differ one point to risk at playing field. Canteen (3), classroom (4) and toilet (5) were shows medium level of risk followed by the lowest (6) result shows to drainage.

Table 2. The result of hazard identification and risk analysis at selected schools area.

No	Area	Possible hazards	Consequences	L	S	LXS	Risk ranking
1.	Playing field 	- Hit by goalpost - Habitat of venomous animals; snake	- Body fractured - Hit by goalpost - Snakebite - Death	4	4	16	High (1)
2.	Office/ Teacher's room 	- Ergonomic hazard - Psychological hazards - Eyestrain and glares - Slip, trip and falls hazards - Wire tangled	- Muscle stress - Mental Stress - Workload demand - Body injured - Falls	5	3	15	High (2)
3.	Canteen 	- broken pipe - leakage - slippery floor - ergonomic - ventilation	- minor injury - body fractured - muscle stress	4	3	12	Medium (3)
4.	Classroom 	- Slip, trip and falls hazards - Broken chair and desk - Heat - Glare - Sharp object: scissors, stapler	- Body injured and fracture - Not focus and cannot concentrate - Facility damaged	5	2	10	Medium (4)
5.	Toilet 	- slippery floor - line pipe leaks in ceiling - air and smell not retained - toilet clog	- slip, trip and falls - mouldy ceiling - smelly - cannot use - waste and dirty	5	1	5	Medium (5)
6.	Drainage 	- Slippery floor - Object falls into drain - Hand trapped at drain covered	- Falls - Minor injury - Body fractured	4	1	4	Low (6)

*Source: DOSH Guidelines 2008

According to the Table 2, hazard at the playing field is the highest compared with other areas. Playing field or sport field is an outdoor sports ground for various sports. Local press had reported the death of a student who hit by the goal post during playing football [13]. During walkthrough observation conducted, there are two types of goalposts was used; planted in the ground (Figure 1b) and mobile goalposts (Figure

1a). However, most of the goalposts were found in poor condition which is rusted and broken (Figure 1c). During observation, some of playing field was placed behind the bushes which may attend by venomous animals; snake. According the verbal communication with person-in-charge, state that their workers found snake on the tree. Besides, local press reported there were 3 cases were recorded students bitten by snake [17-18].

Besides, the risk level in office is high but lower than risk level at playing field. Among the hazards that are found in the office is ergonomic. Employees have their own workstation but not safe to use. In terms of ergonomics concept and Guidelines on Occupational Safety and Health in Office, sitting shall have legroom to ensure worker's leg can be moveable. However, the room had been misused illustrate in Figure 3, by placing books and papers [14-16]. Ergonomics in workplace is important in providing an appropriate and comfortable working area [14]. Space of legroom important to avoid leg muscles stress. Another hazard that found in school office was photocopiers and laser printer. Normally the modern photocopiers and laser printer, Figure 4, fitted with an ozone filter do not present any hazard to health, provided they are properly maintained. It is recommended that photocopiers are not placed on or in close proximity to the personal workstations of office workers because of possible discomfort from the heat, light and noise generated during the photocopying process [14]. The exit of ozone may cause to fatigue and sleepy in short term effect.

(a) (b) (c)

Figure 1. The condition of goalposts: (a) mobile goalpost, (b) Planted goalpost, (c) rusted goalposts.

(a) (b)

Figure 2. Playing field behind the bushes (a) Fenced invaded by animals, (b) Field archery near bushes.

(a) (b) (c)

Figure 3. Condition under the desk. a) A pile of paper under the desk, b) Things storage under the desk, c) No space for legroom.

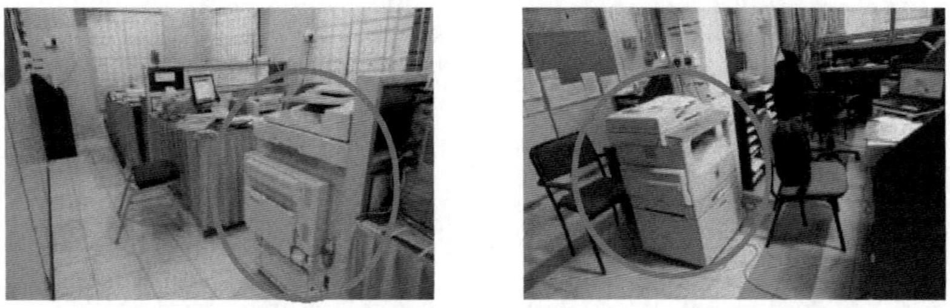

(a) (b)

Figure 4. Photocopiers and laser printer was placed behind the workstation area. a) Photocopiers and workers in the same place, b) Photocopiers behind the workstation.

4. Conclusions

In conclusion, the risk of hazards found was high and potential to cause harm and accident to employer, employees, students and visitors. Based on the findings of the study, there are some unsafe and risky and deadly situations which were rusted goalposts and school field near the bushes or forests. Compared to the reports of accident in school, accidents hit by goalposts and bitten by snakes were the highest cases after the road accident cases. As a conclusion, the result of the study shows that the cause of accident was caused by a rotten and decayed goalpost and falls on students. Besides, school fields near the forest or bushes infested with poisonous animals unnoticed. Therefore, schools management need to monitor every single of risky area of the school by fencing the area with safety tapes or placing warning signs as a warning, so that students were cautious and aware of risky areas.

Acknowledgements

The authors would like to extend gratitude and to Ministry of Higher Education in providing grant for this Knowledge Transfer Programme (R/KTP/A02.00/01163A/001/2015/000269). Special thanks goes to Jabatan Pendidikan Negeri, Pejabat Pendidikan Daerah, school headmaster, school principal and teachers for willingness to participate in this study.

References

[1] Dwyer K and Osher D 2004 *Safeguarding our children: An action guide* (American Institute for Research) 154-161.

[2] Vicario A D 2012 *Social and Behavioral Sciences* 3324-3328.

[3] Scheps SB D E 1987 *Pediatrics* **79** 69-75.

[4] NSKC *Promoting Child Safety to Prevent Unintentional Injury Education Institution Injury Fact Sheet Washington DC* 2004 (Accessed 2015 March 10) Available from: http://www.safekids.org.
[5] Anna Diaz Vicario 2012 *Social and Behavioral Sciences* **46** 3324 – 3328.
[6] Erkan Tabancalia and Talha Bektas 2009 *Procedia Social and Behavioral Sciences 1* 281–284
[7] Tabancalia E and Bektas T 2009*Procedia Social and Behavioral Sciences 1* 281-284.
[8] Balany J, Adesina A, Kearny G, and Richards S 2014 *Assessment of Occupational Health and Safety Hazard Exposure among Working College Students*. 114-124.
[9] Sha Jaya M 1996 *The development of occupational health in the oil industry* (Occuptional Safety and Health in Malaysia. Institut Keselamatan dan Kesihatan Pekerjaan Negara) 43-54.
[10] Natarajan 1996 *Occupational safety and health in the mining industry*. (Occuptional Safety and Health in Malaysia. Institut Keselamatan dan Kesihatan Pekerjaan Negara) 80-94.
[11] Rozario M 1996 *Construction safety*. (Occupational Safety and Health in Malaysia. Institut Keselamatan dan Kesihatan Pekerjaan Negara) 88-94.
[12] Corn J K 1992 *Response to Occuptional Health Hazard, A Historical Perspective*. (Van Nostrand Reinhold, New York) 124.
[13] Ramli Ibrahim 2017 *The goal is rusty and decayed* (Accessed 10th April 2017) Available from: www.hmetro.com.my/node/151281
[14] Department of Occupational Safety and Health 1996 *Guidelines on Occupational Safety and Health in The Office*.
[15] Department of Occupational Safety and Health 2002 *Guidelines on Occupational Safety and Health on Seating Work*.
[16] Health and Safety Executive 1992 *Workplace health, safety and welfare*.
[17] Alias Abd Rani 2016 *The other students were stabbed by a snake* (Accessed 20th April 2016) Available from: www.hmetro.com.my/node/132117
[18] Raja Noraina R R 2016 *The students are again stabbed by snakes* (Accessed 25th April 2016) Available: www.hmetro.my/node/133280
[19] Health and Safety Authority 2017 *Healthy, Safe and Productive Lives* (Retrieved 2017 May 17) Available from: www.hsa.ie/eng/Topics
[20] Department of Occupational safety and Health 2012. *Occupational Safety and Health ACT 1994 (ACT 514) Regulations and Orders* (ILBS) 16-19.
[21] Kadir A, Shaharuddin A, Kadaruddin A, Lukman Z M, Azahan A, Azmi A, Razman M R and Samsu A M 2011 *International Business Management* **5** 3-18.
[22] Kadir A, Shaharuddin A, Kadaruddin A, Azahan A, Lukman Z M, Azmi A, Samsu A M and Razman M R 2010 *Journal of Applied Sciences* **5** 165-171.
[23] Kadir A, Shaharuddin A, Jamaluddin M J, Razman M R, Kadaruddin A, Azahan A and Azhar J 2012 *The Social Science* **7** 424-430.
[24] Malaysia Government 2010. *Article 132 Federal Constitution* (The Commissioner Law Revision, Malaysia) pp 32, 228
[25] *Education in Malaysia* (Retrieved 2017 May 18 2017) Available from: http://en.wikipedia.org/wiki/Education_in_Malaysia.
[26] Mutiara A M 2012. *Analysis on Quantitative Chemical Practicum Practice Activity*. (University of Indonesia)
[27] *Niosh introduces 'OSH in School' programme to make schools safe place for work, study. Sibu, Sabah.* Borneo Post Online. 8 March 2017.

IOP Conf. Series: Materials Science and Engineering **257** (2017) 012057 doi:10.1088/1757-899X/257/1/012057

Injection characteristics study of high-pressure direct injector for Compressed Natural Gas (CNG) using experimental and analytical method

Z Taha[1], MF Abdul Rahim [2,3*] and R Mamat[3]

[1,2]Innovative Manufacturing and Mechatronics and Sports (IMAMS) Laboratory, Faculty of Manufacturing Engineering, Universiti Malaysia Pahang, Pahang, Malaysia.
[3]Faculty of Mechanical Engineering, Universiti Malaysia Pahang, Pahang, Malaysia

*Corresponding author: mfadzil@ump.edu.my

Abstract. The injection characteristics of direct injector affect the mixture formation and combustion processes. In addition, the injector is converted from gasoline operation for CNG application. Thus measurement of CNG direct injector mass flow rate was done by independently tested a single injector on a test bench. The first case investigated the effect of CNG injection pressure and the second case evaluate the effect of pulse-width of injection duration. An analytical model was also developed to predict the mass flow rate of the injector. The injector was operated in a choked condition in both the experiments and simulation studies. In case 1, it was shown that mass flow rate through the injector is affected by injection pressure linearly. Based on the tested injection pressure of 20 bar to 60 bar, the resultant mass flow rate are in the range of 0.4 g/s to 1.2 g/s which are met with theoretical flow rate required by the engine. However, in Case 2, it was demonstrated that the average mass flow rate at short injection durations is lower than recorded in Case 1. At injection pressure of 50 bar, the average mass flow rate for Case 2 and Case 1 are 0.7 g/s and 1.1 g/s respectively. Also, the measured mass flow rate at short injection duration showing a fluctuating data in the range of 0.2 g/s – 1.3 g/s without any noticeable trends. The injector model able to predict the trend of the mass flow rate at different injection pressure but unable to track the fluctuating trend at short injection duration.

1. Introduction

Emission produced by internal combustion engine running on conventional fossil fuels are of major concern nowadays and have become major research topic. Vehicle manufacturers are imposed by more stringent emission regulation year by year. One of the most established approaches is converting conventional fossil fuels to cleaner alternatives. The most widely accepted alternative to gasoline and diesel is compressed natural gas (CNG). Significant advantages of CNG are large proven reserves, cleaner combustion and competitive market price [1] CNG also has a higher RON number compared to gasoline and diesel hence promoting better knock resistance [2] and higher hydrogen to carbon ratio [3] therefore making it suitable as a transportation fuel. Previous studies have concluded that the power loss of CNG engine can be compensated by the use of direct injection CNG fuel system [4].

Based on previous studies, it is obvious that the application of CNG is mostly restricted to the framework of compression ignition engine. It is understood that the compression ignition engine is preferable due to their higher compression ratio which can exploit the higher knock resistance of CNG fuel. This will lead to a higher engine thermal efficiency. As an overview, in the framework of compression ignition engine, the studies on the jet structure have been performed experimentally and numerically by [5], [6], and [7]. Other studies focus on the injection system [8] and [9], combustion development and engine emission levels [10]. Chiodi et al. [11] and Muhammad et al. [12] investigated the methane injection spray structure using Laser Induced Fluorescent (LIF) technology. The axial and radial penetration length of methane issued from multi-hole injector were measured. Chiodi et al. study was based on a multi-hole injector design whereas Muhammad et al. used an innovative spark-plug fuel injector design. Chan et al. investigated the distribution of natural gas in a partially stratified charge engine using a modified injector [13].

The previous study had also investigated the transient injection of helium and hydrogen through multi-hole gaseous injectors using Schlieren visualisation [14]. Chitsaz and Hajialimohammadi investigated the jet structure of helium injected through different nozzle diameters experimentally [15] and [16]. Both had present a correlation of tip penetration for both near and far field regions of the jet. The study on tip penetration of gaseous jets was also conducted by [17], [18], [19], and [20]. They reported different correlations for tip penetration of gaseous jets. Chitsaz et al proposed a semi-analytical solution for the transient start of a under-expanded jet using numerical Laplace transform.[20]. The injection system for the compression ignition engine has been studied in detail by [21], [6], [7], [8] and [9]. However, the major difference between previous studies and natural gas spark ignition engine is the pressure ratios of the injection and the jet structure. Also, the injection pressure of CNG in compression ignition engines is higher than those based on spark ignition engines [21] and [6] by almost two times.

The CNG application in the context of the spark ignition engine is mostly based on the multi-point port injection approach. Comparison of carburetted gasoline, port injected gasoline and direct injected gasoline has been performed by Kalam and Masjuki [4]. It is important to note that the results obtained showed that the maximum power of CNG-DI is only 5% lower than the performance of gasoline port injection. It is expected that by performing strategic optimization on the CNG-DI engine, the resulted performance can be at least at the same par with gasoline port injection engine. Commercialized CNG direct injector is hardly found in the market. Thus a commercial GDI injector is mostly used to inject CNG. The changes of fuel properties from liquid to gases affect the injector characteristic benchmarked by the manufacturer. This is the utmost important reason on why the injector characterization for the natural gas engine is highly desirable.

Conventional fuel injectors are mostly used solenoid drive unit. The electrical energy is converted to mechanical energy in term of pintle displacement via electromagnetic principle. The movement of the pintle defines the opening and closing of the nozzle which in fact controlled the fuel flow through the nozzle. The pintle movement and corresponding time scale are used to define the temporal characteristics of an injector. The temporal characteristics of an injector include injection duration, rate shape, opening delay time and closing delay time. The temporal characteristics of the injector are controlled by the ECU based on sensors data and predefined optimal setting stored in the base maps of the ECU. Sensors data are such as engine rotational speed, manifold absolute pressure, engine operating temperature, and air to fuel ratio are used by the ECU to select the corresponding optimal values for injector parameters. The ECU will send a control signal to the fuel injector power driver which instantly provide proper excitation to the solenoid to generate magnetic force. The induced magnetic force will pull the pintle to produce nozzle flow area.

The linear displacement of injector's pintle is an important variable for injector control. However, the pintle's linear displacement is affected by numbers of associated parameters which physically interacted with the pintle or armature body. The displacement occurs only when the magnetic force overcomes the opposing forces. The opposing forces are composed of inertia force, contact friction force, gas pressure force and spring compression force. Therefore, the physics involved in fuel injector operation is an interaction of electromagnetic, mechanical dynamic and fluid dynamics. High-performance injector required that pintle dynamics should have the following characteristics: (1) quick response, (2) repeatability, and (3) least power consumption [22]. Short opening delay and short closing delay are characteristics of a quick response injector. As the engine cycle completed on a scale of milliseconds, these are very crucial parameters for engine control nowadays. The fuel injection process must be completed efficiently in a very small interval at each engine cycle. The opening and closing delay affect the minimum total time for injection duration and injection timing. Actual mass injected in a very short injection duration might not resemble theoretical values. The pintle response is also a decisive factor for multiple injections strategy. By using such approach, multiple injections must be performed in each cycle. The pintle needed to open and close at a higher frequency. Repeatable fuel injection allows a lessen fuel injection quantity and injection rate per shot but the total mass injected at each cycle is expected to be greater than a single injection approach. Moreover, it is anticipated that a less shot-to-shot variations injection quantity is produced.

Injector current affects the response of the pintle. In general, current is load dependent. The higher the resistance to pintle movement (gas pressure, friction, spring forces), then higher current shall be withdrawn from the power supply to overcome the load. Higher current may as well ensure the repeatability of the injection by creating more than enough magnetic force to hold the pintle at its desired position. Insufficient current supply may cause the pintle partially pulled and vibrated during the opening and closing of the injector. Increasing the current ramp rate theoretically will reduce the opening delay and closing delay. However, in practical, the current increment rate is influenced by the inductance of the solenoid. Also, the higher current would create more heat in the solenoid, which increases the temperature of the surrounding magnetic material. The increased heat of the magnetic material will deteriorate the magnetic material properties. Among the distressed magnetic properties are lowered saturation limit and increased hysteresis. The deterioration of these properties will distract the pintle response time. Other factors affecting the pintle response time are hydraulic hysteresis, magnetization hysteresis and saturation of the magnetic material.

Earlier discussions have focused mainly on the experimental characterization of injectors. The studies of direct injector were also explored using modelling approaches. The modelling of injectors had extended from the injector physical modelling until complete injection system modelling. The direct injectors were mainly modelled using an analytical approach. The extent of models can be categorized into two major areas which are (1) injector dynamics modelling and (2) power driver design and optimization. The numerical models usually include or partially include coupled multi-physics problem [23-25]. Major components of a model are an electromagnetic model, the mechanical/ dynamic model, and the dynamic flow model. A model coupled one-dimensional hydraulics with a two-dimensional (2D) axisymmetric electromagnetic finite element code has been developed by Digesu to calculate the magnetic force regarding exciting current and working gap [26]. He validated simulated results of needle lift, pressures (pipes, control volume, rail and accumulative volume) and flow rate with experimental data. He had demonstrated the effect of pipe design on the needle lift profile due to the hydraulic wave. Ficarella had evaluated the instability phenomena in a common rail injection system using numerical modelling. He had also used the model for predicting injection characteristics [27]. He concluded that the control valve residual motion had caused re-opening of the valve after a pilot injection and was considered as the most challenging part of the modelling.

Coppo's had developed a model of a common-rail injector with a combined pipe, fluid, mechanical and electromagnetic model [24, 25]. He experimentally measured exciting current signals as the input to the model which included magnetic saturation effects. Good agreement was observed between calculated and experimental profiles of rail pressure, needle lift, and injection quantity. Hu model excluded the used of electromagnetic sub-model. Instead, a measured magnetic force profile is used as the input to the model's calculation. His model was able to correlate the pressure profile with the measured one [28]. Ando developed a one-dimensional injector model which coupled the electromagnetic model with simplified rigid body and fluid models [29]. Ando used the model to study the electromagnetic effects including magnetization time lag.

The power driver is the interim unit between injectors and ECU which manage the power supplied to the injector. The power driver is essential to drive the injector operation at their optimum and designed condition. Tsai had developed a power driver for high-pressure gasoline direct injectors (GDI). The injector is a peak and holds injector, and the proposed power driver by Tsai was a three-stage power driver. The first two stages of the driver used a power Metal-Oxide-Semiconductor Field-

Effect Transistor (MOSFET) to switch between power supplies. While the 3rd stage used pulse-width-modulation (PWM) control. The results showed that the pintle closed faster with the 3rd stage PWM control than with a direct voltage control method [30]. Tsai had also performed a parametric study to investigate the effect of voltage, current, frequency and fuel pressure on injection quantity.

From the literature review, it was found out that the previous studies on direct fuel injector were very crucial, but the characterization study was mostly centred on the application of conventional fuel such as gasoline and diesel. The characterization of a direct fuel injector for natural gas is rarely found. The characteristics of direct injection CNG affects the overall engine performance. Therefore it is very crucial to investigate the injection characteristics, fuel metering, design and optimization of such system. Hence, this study is conducted to analyze the fuel metering of direct injector used for CNG fueling. This is to confirm the sufficiency of the fuel supplied to the engine requirement throughout the engine operating regime. The work of Erfan et al. [31], and Chitsaz et al. [15] are among the significant work concentrated on the direct fuel injector characterization for CNG. However, their work only used experimental approach. In this study, the characterization was also utilized modelling approach. The injector dynamic-analytical model is vital for detail cause-effect relationship studies and able to be coupled with analytical engine models for the development of a fully predictive engine model. The method of the modelling is using electromagnetic model coupled with mechanical and fluid flow model.

2. Experimental apparatus

The physical layout of the baseline injector is shown by figure 1(a). The injector is a Bosch HDEV 1.2 injector for gasoline direct injection (GDI) engine. General specification of the injector and flow characteristics for gasoline fuel operation is provided in table 1. The maximum allowable fuel pressure for the injector is 200 bar. Calibration of the injector with gasoline fuel at 100 bar produced an estimated volumetric flow rate of 30 cm3/min or 0.36 g/s (by taking gasoline density as 719.7 kg/m^3). In this study, the fuel pressure was regulated in between 20 bar to 60 bar. The mass flow rate for different pressure and injection duration are to be determined. Conversion of the injector for running on CNG is studied in this paper using an experimental and modelling approach. The main injector's components consist of an armature, return spring, fine fuel filter, solenoid driver, and nozzle tip. The arrangement of components is shown in figure 1(b). Based on the injector construction, the injector armature is pulled backwards as the solenoid is energized causing the nozzle pintle lift up and the nozzle flow area opened. As the solenoid is de-energized, the return spring pushes the pintle back to its seat, closing the nozzle flow area.

The experimental setup is shown schematically in figure 2. This setup comprises of a CNG supply tank, fuel pressure regulator, pressure gauge, test injector, injector micro controller, and a control computer. The microcontroller act as a pulse generator, provides square wave signals with adjustable frequency and amplitude to a PWM driver who drive the injector in different modes. A digital scale is used to record the instantaneous cylinder mass tanks after some injection is completed. The mass changes is used to determine the mass flow rate of the injector. The time factor is calculated by the total length of the "on" signals produced by the microcontroller. In the experiment, a constant 12 V supply is provided through the PWM driver. The supply current is limited to 8 Amp to replicate the saturated current limit used in the actual electronic control unit (ECU) of the prototype vehicle.

Table 1. General injector specification, model HDEV 1.2.

Attributes	Values
Mechanical specifications	
Allowable maximum pressure (bar)	200
Volume flow rate (gasoline fuel/cm3/min) at 100 bar	30
Weight (g)	78
Length (mm)	85
Electrical specifications	
Resistance (Ohm)	0.9 @ 1.5
Voltage (Volt)	90 V
Allowable peak current (Amp)	20 A
Operating Condition (Gasoline fuel)	
Fuel Input	Axial (top feed)
Operating Temperatures (^oC)	30-120
Permissible Fuel Temperatures (^oC)	<80

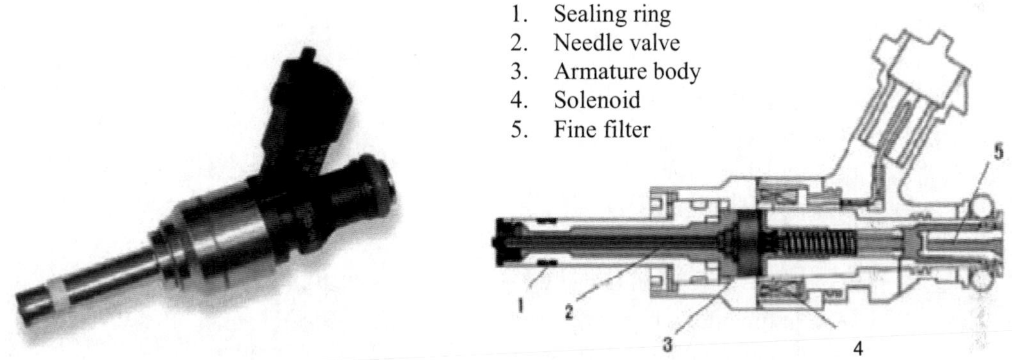

(a) (b)

Figure 1(a). Photo of injector physical appearance and (b) Injector cut-view of Bosch single-hole gasoline direct injector [30].

3. Experimental procedures

The varied parameters in this study are the injection pressure and the injection duration. In Case 1, the injection pressure is varied from 20-60 bar with an increment of 20 bar. At each pressure setting, the CNG fuel is injected at a 25 ms duration from 0 to 3000 injection count with an increment of 100 counts per step. The fuel is injected into ambient condition concerning the actual HPDI-CNG system which is utilized in a vehicle prototype. The ambient condition is assumed to be the downstream pressure because the direct injection of gas in the prototype vehicle is injected at the valve opening event. In Case 2, the injection duration is varied in the range of 2 to 26 ms. At each duration setup, the injection count is set to 500 counts.

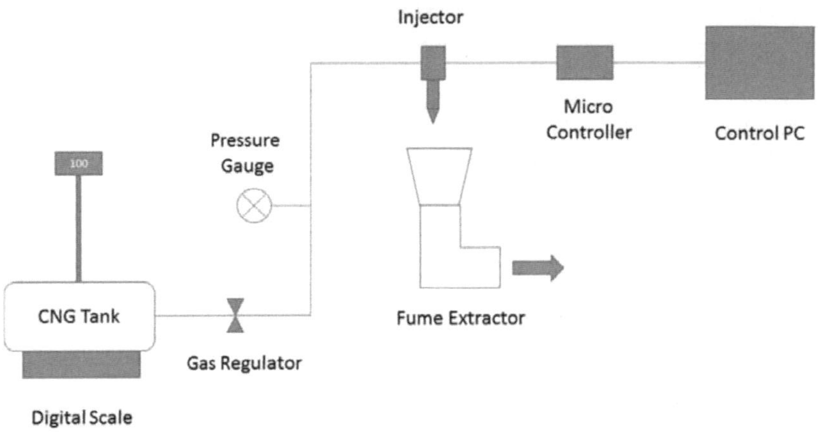

Figure 2. Schematic diagram of experimental setup.

Table 2. Experimental conditions.

Parameters (units)	Values
Number of nozzles	Single hole injector
Spray Angle (Degree)	11
Injected gas	Compressed natural gas
CNG molar mass kgkmol)	18.12
Chamber gas	Nitrogen
Chamber temperature (K)	295.5
Ambient temperature (K)	295.5
Ambient pressure (bar)	1
Injection pressure (bar) (Case 1)	20,30,40,50,60
Injection time duration (ms)	25
Varied time duration (ms) (Case 2)	2-26 (interval of 2)

4. Injector modelling approach

The modelling section is elaborated in four different section (1) The electromagnetic model, (2) the mechanical model and (3) the flow sub models. These models are combined and developed in Matlab/Simulink environment. The Matlab ODE solver is selected as the baseline solver in the study.

4.1 The electromagnetic model

The electromagnetic is based on the work of Schimpf [32]. The solenoid driver presented by Schimpf is a derivation of electromagnetic force regarding coil diameter, coil length, wire gauge, supply voltage, packing density, and the number of turns. This model eliminates the current term which in most cases are difficult to solve. Detail discussion and explanation of the model can be found in Schimpf. The electromagnetic force is given by the following expressions:

$$F_{\mathrm{mag}}=\frac{-V^2\mu_r\mu_o}{8\pi\gamma^2 l^2}\left(\frac{r_o}{r_a}\right)^2\alpha e^{-\frac{\alpha}{l}x} \qquad (1)$$

or

$$F_{\mathrm{mag}}=\frac{-V^2\mu_r\mu_o}{8\pi\gamma^2 l^2}W_f\alpha e^{-\frac{\alpha}{l}x} \qquad (2)$$

where

V : Supply voltage
μ_r : Relative magnetic permeability of armature/pintle material
μ_o : Air gaps magnetic permeability
r_o : Inner radius of coil cross section
r_a : Average radius of coil cross section
γ : Ratio of coil material resistivity to coil wire cross section area
W_f : Winding factor, equal to square of r_o/r_a ratio
l : Length of coil body
α : Ratio of inductance to relative permeability of armature material
x : Instantaneous position of armature

To utilize the model, one needed to have detail information of the coil material. For example, the type of coil material, coil material resistivity and coil wire cross-section area. In general, the larger the coil wire cross section area, the larger the generated force. In the study, the selected coil wire is estimated to be based on AWG 43 which has a diameter of 0.07874 mm and copper type.

4.2 The mechanical model

The mechanical system of the direct injector is represented by a mass-spring-damper system. In the initial state, it was assumed that the pintle sits on the valve seats. In this initial state, a total of five forces are acting on the pintle. The gas pressure force, contact friction force, gravitational force, initial spring force, and finally normal reaction force. Figure 3 presents the free body diagram of the pintle in the study. The gas force is due to the CNG fuel pressure, the contact friction force is due to the contact between the pintle and surface of the valve seats, and the gravitational force is due to the mass of the pintle. The initial spring force is due to the compression of the spring at the initial state. Additional spring force will be generated as the pintle is pulled by the solenoids. The normal reaction force defines the existence of the lower (and upper) stopper of valve seats. This force is represented by the virtual spring, and damper unit which is equal and opposite direction of all other forces when the pintle rest or hit bottom and upper stopper.

The pintle act as a plunger which open and closes the nozzle flow area. During the opening state of the injector, the pintle will overcome all the resistant forces by withdrawing required current from the power supplies. The relationship between the all the forces is described by the mathematical expressions of the pintle's equation of motion which is given by equation 3. Based on the equation, the displacement of the pintle can be obtained by a twice integration of the acceleration. The mass considered in the equation is only the mass of moving the rigid body of the pintle.

$$m \cdot \ddot{x} = F_{\text{sol}} + F_{\text{spring}} + F_{\text{contact friction}} + F_{\text{upper wall}} + F_{\text{bottom wall}} + F_{\text{pressure}} \tag{3}$$

The spring compression force F_{spring} Is defined as the sum of initial compression force and the force due to the additional compression during the solenoid activation. The equation for the spring compression force is given by the following expression.

$$F_{\text{spring}}(\mathrm{x}) = F_0 + K_{\text{spring}} \cdot x \tag{4}$$

where,

F_0 : the initial compression force, N
K_{spring} : spring constant, N/m

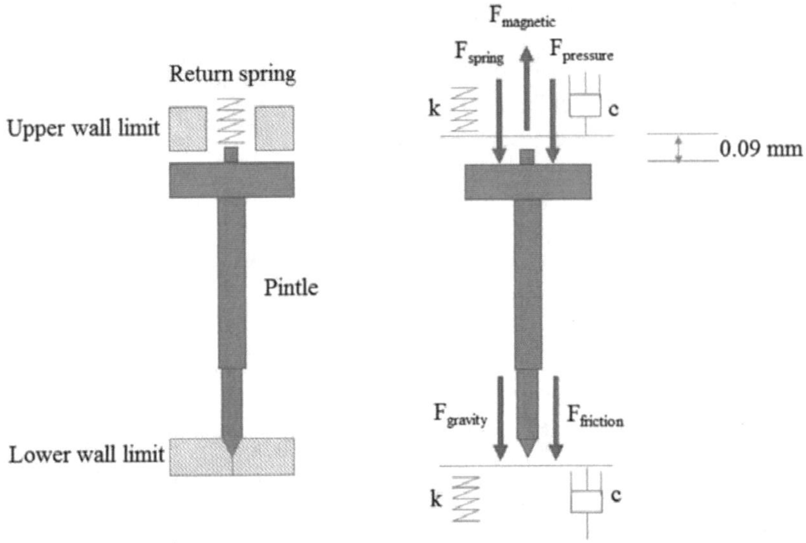

Figure 3. Pintle free body diagram.

The initial contact friction force is given a constant value of 13.8 N at its rest position (at the upper stator or lower stator). The value will reduce to zero as the position of the pintle surpassed a threshold distance of 0.001 mm. The mathematical form of the contact friction force is given by equation 5.

$$F_{\text{initial contact friction}}(x)=\begin{cases}\text{-13.8N}, & x\leq\text{1e-3mm}\\ 0, & x>\text{1e-3mm}\end{cases} \tag{5}$$

Based on the findings of Zhang et al. [22] the displacement of the pintle is limited by the existing of the bottom and upper stator. The maximum displacement allowable for the pintle is estimated to be 0.09 mm upwardly. Farther than that, a barrier is enforced by imposing a stiff spring and damper. This is to simulate a condition where the pintle hits the stator. The reaction force imposed on the pintle at the upper wall is expressed mathematically by equation 6.

$$F_{\text{upper wall}}(x)=\begin{cases}-K*(x\text{-0.09mm})-c*\dot{x}, & x<0\\ 0, & x>0\end{cases} \tag{6}$$

As the pintle moves back to the initial position and hit the lower stator, another barrier is imposed to simulate the reaction force from the bottom stator. The mathematical expression for the lower stator reaction force is given by equation 7.

$$F_{\text{bottom wall}}(\mathrm{x})=\begin{cases}-K*x-c*\dot{x}, & x<0\\ 0, & x>0\end{cases} \tag{7}$$

where,

K : the wall stiffness, N/m
C : the damping coefficient, N·s/m

4.3 The flow model

The flow model is based on one-dimensional compressible flow equation. The model considers choking and non-choke flow situations which are determined by the critical pressure ratio. Choked flow occurs when the ratio P_1/P_2 exceeds the critical pressure ratioP_c, which is given by the following equation.

$$P_c=\left(\frac{\gamma+1}{2}\right)^{\frac{\gamma}{(\gamma-1)}} \tag{8}$$

The mass flow rate through the orifice for non-choked and choked flow conditions are given by equation 9 and 10 respectively. The only varying parameter in the equation 9 is the effective area of the nozzle which is calculated as a function of pintle displacement.

$$\dot{m}=K_{nv}AP_1\sqrt{\frac{2M}{RT}\left(\frac{\gamma}{\gamma-1}\right)\left[\left(\frac{P_2}{P_1}\right)^{\frac{2}{\gamma}}-\left(\frac{P_2}{P_1}\right)^{\frac{(\gamma+1)}{\gamma}}\right]} \tag{9}$$

or

$$\dot{m}= K_{nv}AP_1\sqrt{\frac{\gamma M}{RT}\left(\frac{2\gamma}{\gamma-1}\right)^{\frac{(\gamma+1)}{(\gamma-1)}}} \tag{10}$$

Where,

K_{nv}	: the discharge flow coefficients
A	: the orifice area (m^2)
P_1	: the H_2 upstream pressure (Pa)
P_2	: the in-cylinder pressure (Pa)
T	: the H_2 upstream temperature (K)

The crucial part of the flow model is the definition of the nozzle effective flow area. Figure 4 presents a schematic definition of the nozzle effective flow area for the pulled-in injector. The effective flow area is defined as a surface area of a truncated cone. The effective flow area is formulated based on the work of Antunes [33].

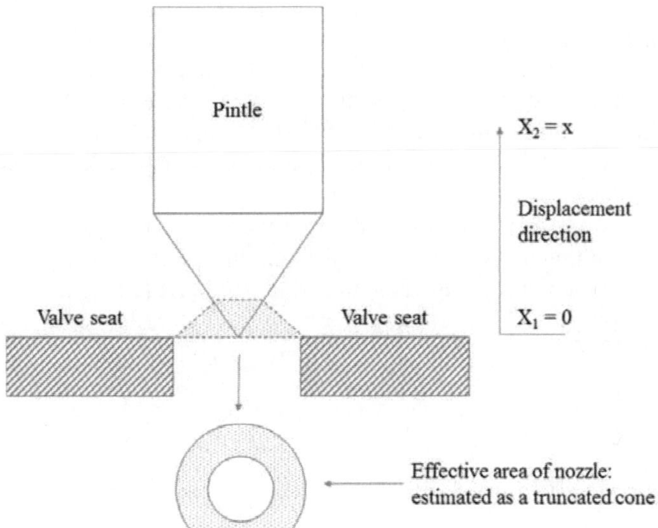

Figure 4. Definitions of effective flow area for nozzle flow calculation.
Source: Antunes [33].

In mathematical form, the effective area for the nozzle flow, as illustrated in Figure 4, can be calculated by using equation 11.

$$A_o=\pi\left[\left(R\sqrt{R^2+(R\sin\varphi)}\right)-(R-X\sin\varphi\sin\varphi)\sqrt{(R-X\sin\varphi\sin\varphi)^2+(R\tan\varphi-X\sin^2\varphi)^2}\right] \quad (11)$$

Where,

R	: the nozzle hole radius (m)
θ	: the needle tip angle (rad)
ϕ	: is equal to $\theta/2$

The maximum area of the nozzle is given by:

$$A_{o\ Max}=\pi R^2 \quad (12)$$

5. Simulation Setup

5.1 Model Input

The simulation was carried out to replicate the two cases from the experiment. A significant amount of inputs are required to ensure the model can predict as close as possible to the measured data. Table 3 listed all the general and specific inputs required by the model.

The selected injection pressure of 50 bar is equivalent to the injection pressure used in actual prototype vehicle of CNG direct injection. The varied values are used to study the effect of injection pressure on the injected mass quantity. The electro-magnetic inputs, as well as the mechanical inputs parameter of the injector, are mostly based on Zhang et al. [22]. Figure 5 presents the layout of the simulation model developed in Matlab/Simulink. The pulse generator produced square wave signal which represents the output of the PWM driver.

6. Results and discussion

The injected mass flow rate is assumed to be equal to the rate of change of mass from the CNG storage tank. In case 1, the injector frequency and injection duration are 8 Hz and 25 ms respectively. Based on Erfan et al. [31], the injection duration should be less than 50% of the time between pulses. Therefore, in the study, the injection duration cannot exceed 50 ms at the frequency of 8 Hz. The results of figure 6(a) and 6(b) shows the example of CNG storage tank mass for injection pressure of 20 bar and 60 bar plotted against the injection time. The value of mass reduction can be approximated by a linear equation for all cases. The slope of the plot is representative of the injector mass flow rate for different injection pressure. It is noticeable that the mass flow rate is increased as the injection pressure increased. The resulting mass flow rate at different injection pressure are summarized in table 4. The trend of the results obtained is comparable to earlier results of [31].

Table 3. Simulation model input.

Attributes	Values
General Input Parameter (unit)	
Actuator radius (m)	0.005
Gas pressure (bar)	20, 30,40,50, 60
Nozzle Diameter(m)	0.00068
Spring Constant (N/m)	12140 N/m
Electro-magnetic Input Parameters (unit)	
Length of air gap (m)	0.00009
Magnetic Permeability of Air (H/m)	1.256×10^{-6}
Magnetic Permeability of Steel (H/m)	3290
Magnetic circuit length (m)	~0.001
Number of Turns	160
Coil Resistance (Ohm)	0.9
Resistance (ohm)	1.5 (Zhang)
Inductance (mH)	1.9 mH @ 1kHz 3.9 mH @0.12 kHz
Voltage (Volt)	90
Peak Current (Amps)	20
Mechanical Input Parameters (unit)	
Static Spring Force (N)	40
Spring mass (kg)	0.001
Actuator Mass (kg)	0.003
Actuator Damping Constant (Nm/s2)	14.97 N.s/m
Overall Weight (kg)	0.078
Flow Input Parameters (unit)	
Gas specific heat ratio	1.32
Gas valve flow co-efficient	0.65
Universal gas constant (J/kg.K)	8314
Gas Molecular Mass (kg/kmol)	16.04
Gas Supply Temperature (K)	300

Table 4. Case 1: Effect of injection pressure results summary.

Injection Pressure (bar)	Mass flow rate (g/s)
20	0.4
30	0.4
40	0.7
50	1.1
60	1.2

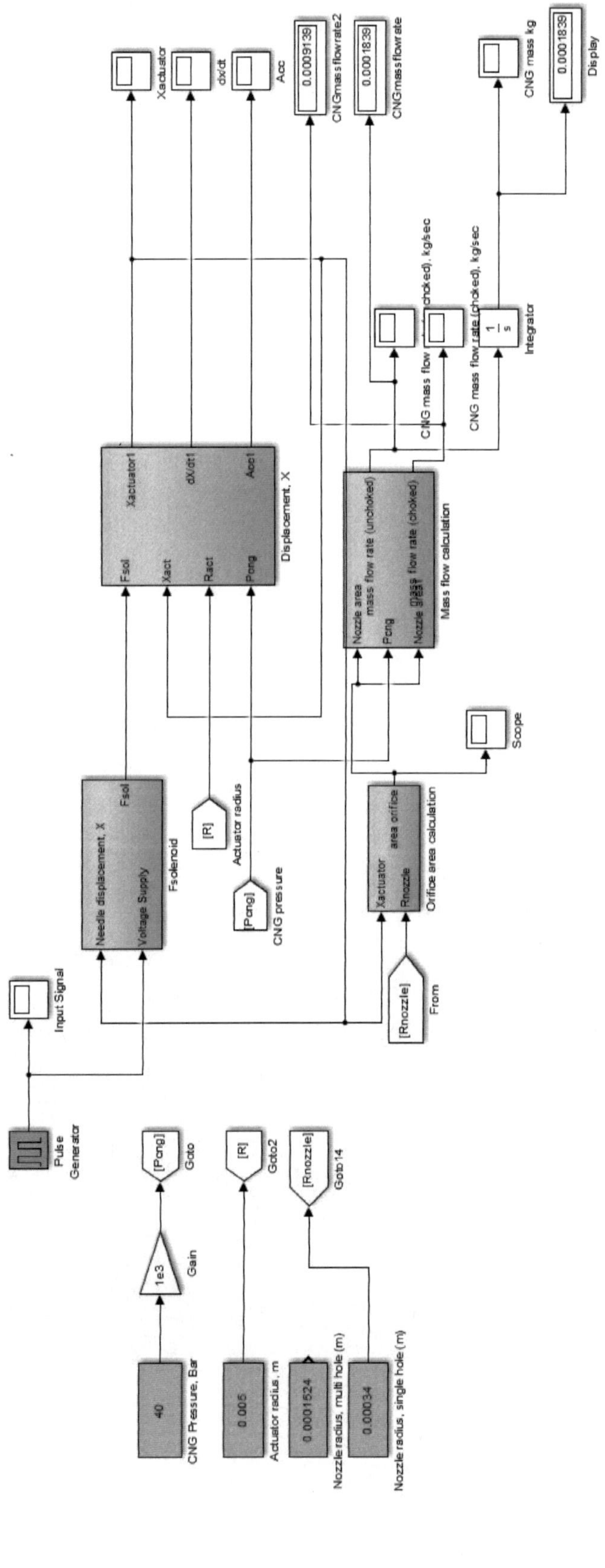

Figure 5. Matlab/Simulink model layout.

As a guideline of fuel requirement for the injector utilization in prototype CNG-DI vehicle, the mass flow rate which is needed for a conventional 1.6-liter engine have been calculated. Under stoichiometric chemical reaction for methane by assuming methane as a main composition of CNG, we have a stoichiometric air to fuel ratio for the engine of about 0.058 (or 17.24 air to fuel ratio). The required mass of fuel is proportional to the mass of air. By assuming that the engine displacement is 1.6 litre and the average volumetric efficiency of the engine is about 90%, the mass of air for stoichiometric combustion is about 0.36 g. Therefore, the required mass of fuel for stoichiometric combustion is about 0.02g. However, these requirement is also affected by the engine speed. Table 5 summarised the fuel requirement as a function of engine speed for stoichiometric combustion.

Table 5. Fuel requirement for different engine speed by assuming the fuel can be injected 300 crank angle in each cycle.

Engine speed (rpm)	Required mass flow rate (g/s)
1000	0.2
2000	0.4
3000	0.6
4000	0.8
5000	1.0
6000	1.2

Based on table 4 and table 5, the maximum requirement of fuel flow rate for a 1.6-liter engine at 6000 rpm is about 1.2 g/s while the minimum requirement is about 0.2 g/s at 1000 rpm (idle speed). By considering the effect of injection pressure, it is noteworthy that the best injection pressure which suits the engine requirement at all engine speed is 60 bar. This is the condition if the engine theoretically needs to be operated in a stoichiometric combustion. If the engine is operated at a lower injection pressure, the utmost significant effect on the engine performance is the engine will operate at a leaner mixture which will result in critical power drop, especially at high engine speed.

For Case 2, the injector is operated by varying the injection duration from 2 to 26 ms at the injection pressure of 50 bar and delay between pulses is retained at 100 ms. Therefore, as the injection duration is increased, the frequency of the injector is decreased. In this context, the frequency is defined as the number of injection per second. Based on figure 7, it is noticeable that the total mass injected will increase linearly with increased injection duration. The increased of injection duration has provided a longer delivery time for the injector. This has enabled higher amount of mass flowing through the nozzle. Figure 8 shows that the average mass flow rate of the injector is fluctuating with increased injection duration. Based on the case setup, the expected mass flow rate is about 1.1 g/s as in Case 1(p = 50 bar). However, the measured results in Case 2 possess a fluctuating trend of this variable. The maximum and minimum mass flow rates are 1.33 g/s and 0.2 g/s respectively.

The highly fluctuated mass flow rates measured at short injection duration. This indicates that, the length of injection duration is crucial in direct injection process. As the injection duration increased, the degree of fluctuation is reduced as demonstrated by Case 1. But if the injection duration is reduced as in Case 2, the consistency of the mass delivered by the injector is deteriorated. Detail study of the inconsistency of injector mass flow rate during short injection duration has never been presented previously. The results suggested that at short injection duration, the resultant mixture air to fuel ratio will fluctuate as well. This may suggest unstable combustion process due to mixture ratio inconsistency.

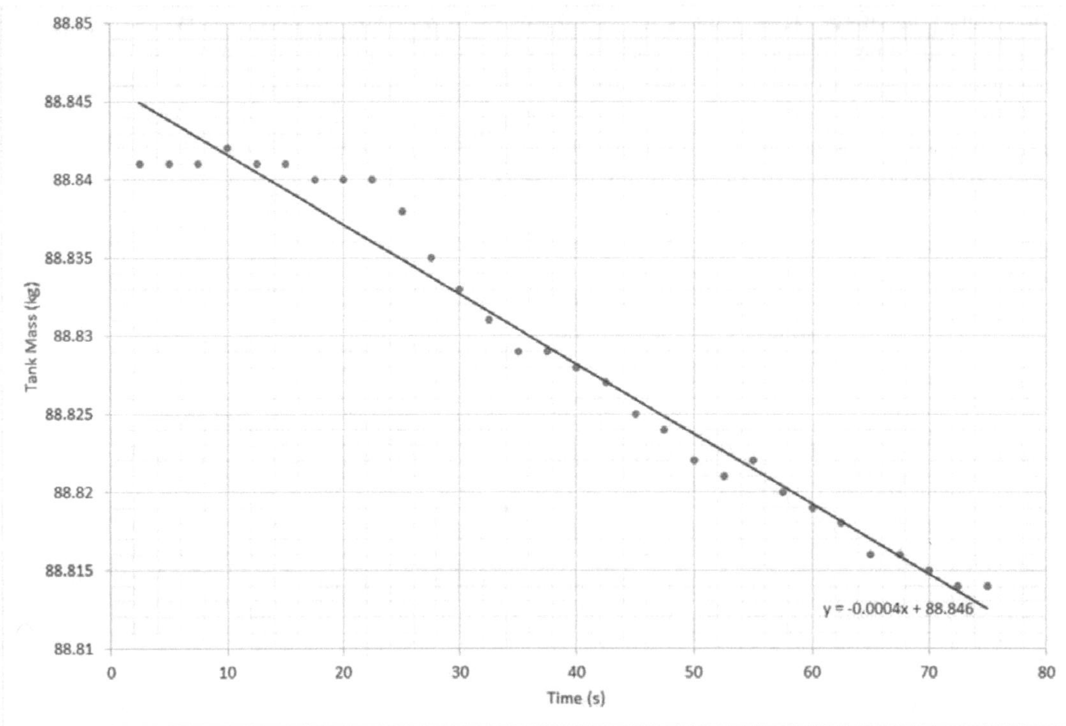

(a)

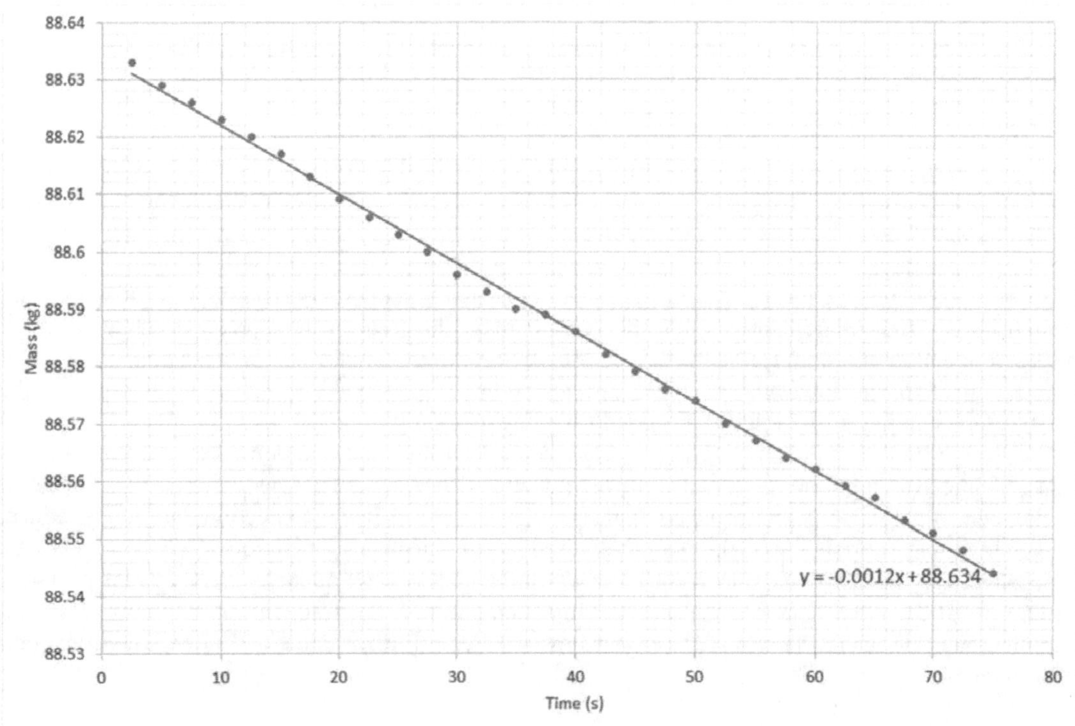

(b)

Figure 6. Instantaneous CNG storage tank mass for (a) injection pressure = 20 bar, and (b) Injection pressure = 60 bar.

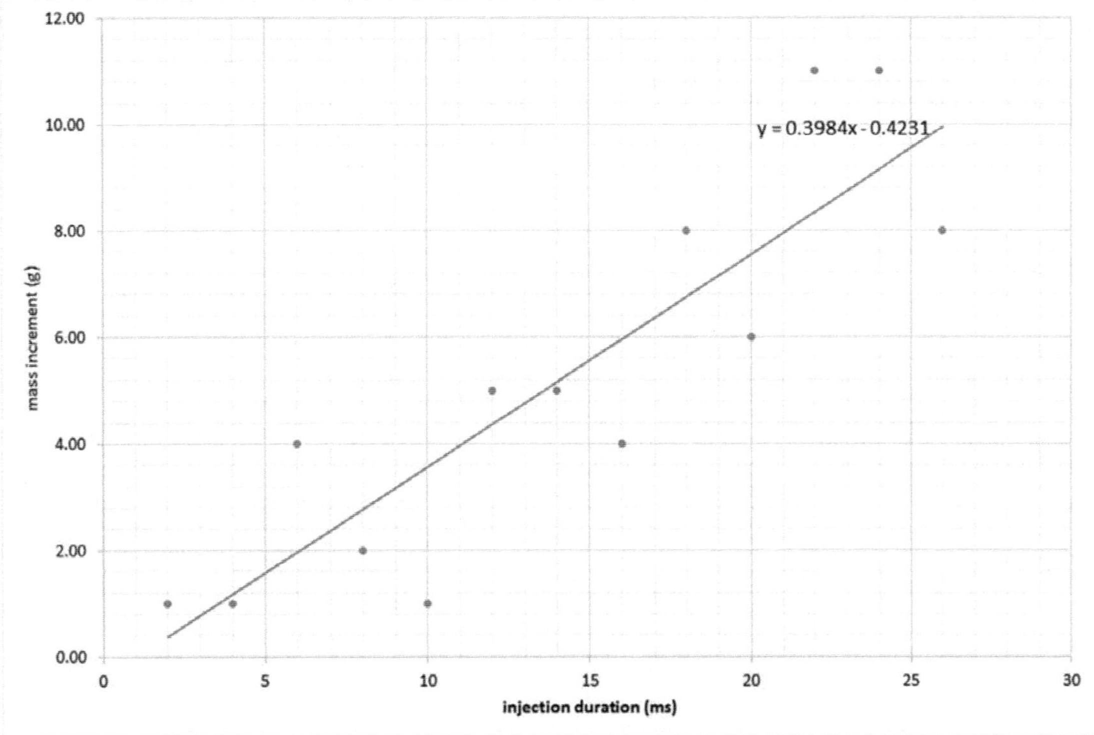

Figure 7. Mass increment at increased injection duration for Case 2.

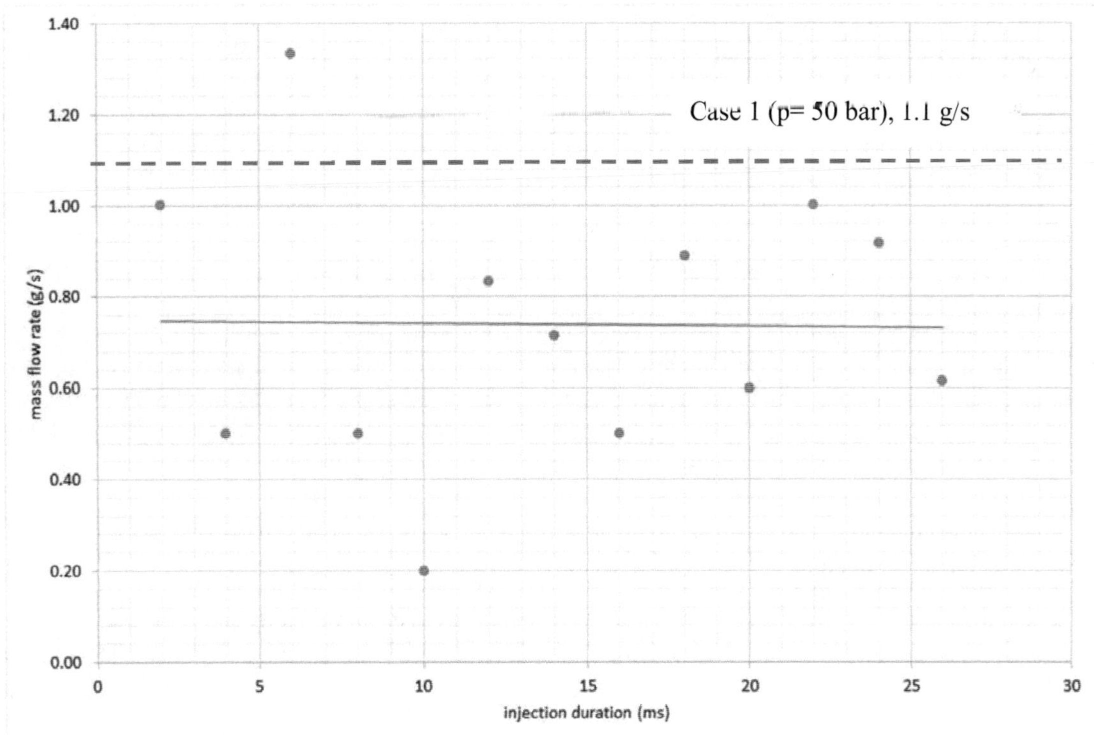

Figure 8. Mass flow rate at increased injection duration for Case 2.

The results also suggest that the frequency of the injector may affect the mass flow rate of the injector. Table 6 presents the cycle time and frequency used in Case 2. The degree of fluctuation of mass flow rates are reduced as the frequency of the injector is lowered.

Table 6. Cycle time and frequency used in Case 2.

Duration (ms)	cycle time (s)	frequency (Hz)
2	0.102	9.804
4	0.104	9.615
6	0.106	9.434
8	0.108	9.259
10	0.11	9.091
12	0.112	8.929
14	0.114	8.772
16	0.116	8.621
18	0.118	8.475
20	0.12	8.333
22	0.122	8.197
24	0.124	8.065
26	0.126	7.937

Figure 9 presents comparison of measured and simulated results for mass flow rate at different injection pressure of Case 1. The simulated mass flow rate is higher than the measured data. However, both predicts the same trend where the mass flow rate is linearly increased as the injection pressure is increased. Figure 10 presents comparison of measured and simulated mass increment for different injection duration of Case 2. The simulated mass increment is higher than the measured mass. Both the simulation and experiment predicted linear increment of mass as the injection duration increased. The rate of mass increment is higher for the simulation model compared to the experiment. Figure 11 presents comparison of measured and simulated mass flow rate at different injection duration of Case 2. Both the simulation model and experiment possess a constant trend of mass flow rate and independent of injection duration. However, the simulated mass flow rate is higher than measured results. The simulated model predict a mass flow rate of 1.06 g/s whereas the mass flow rate plot for Case 2 is about 0.75 g/s.

Based on the simulated results, it is concluded that the simulation is confirmed to be capable to predict the trend of the injector mass flow rate at different injection pressure and different injection duration. But the injector model predicted higher values of mass flow rate compared to measured results in both cases. This discrepancy is suspected due to certain properties of the model which are adapted from previous studies which may not fully suitable for the injector in current study. Detail parametric study is needed to clarify the effect of each injector parameters towards injector's mass flow rate. In addition, the theoretical calculation of the injector unable to track the fluctuating trend of mass flow rate at each injection duration. This will required detail analysis of the losses inside the injector solenoid driver and its effect on the injector opening and closing time. It is possible that at short injection duration, the opening and closing of the injector are not perfectly done due to extremely short period of cyclic operation. The simulation model is only simulate an idealized process of fuel injection.

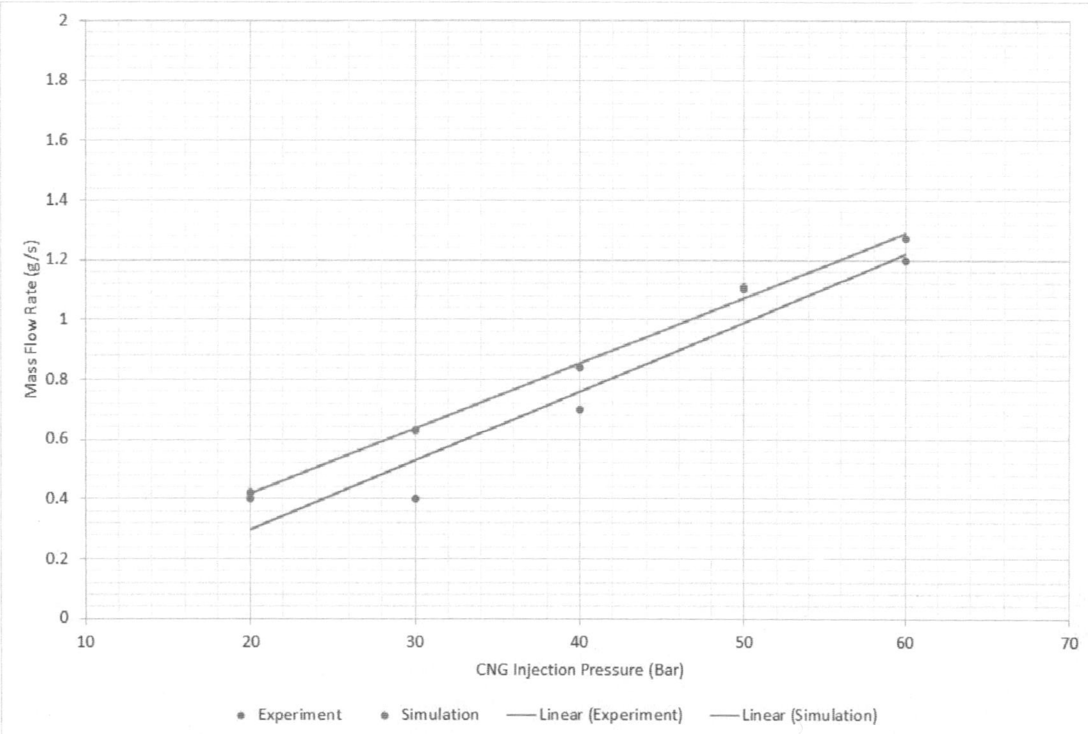

Figure 9. Comparison of simulated and measured mass flow rate for different gas injection pressure for Case 1.

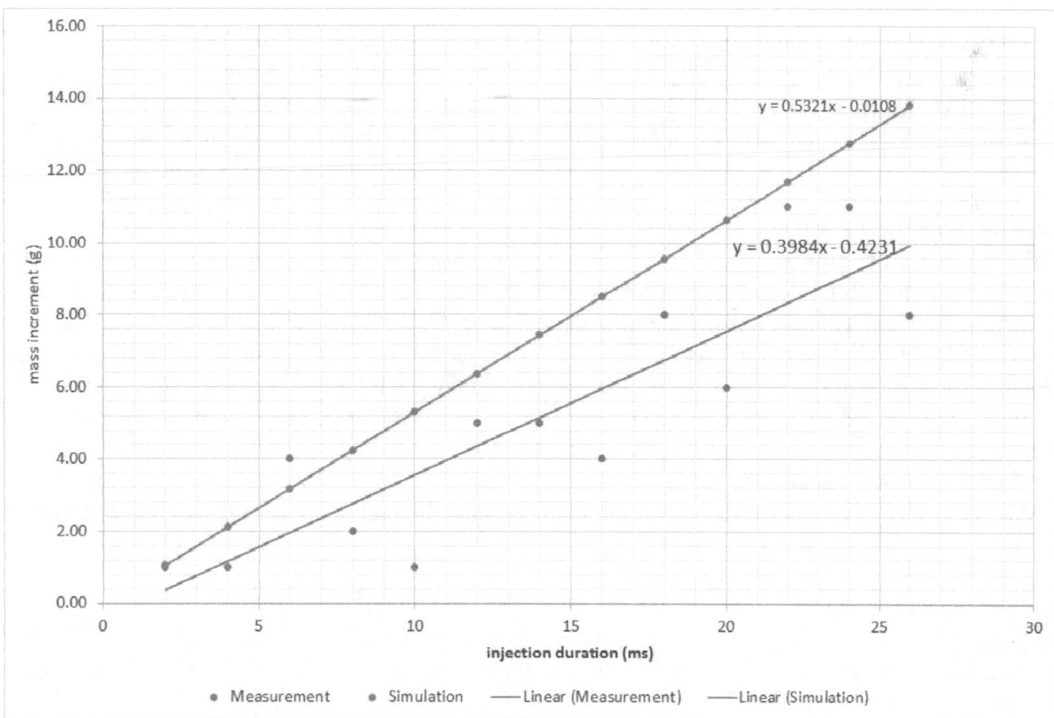

Figure 10. Comparison of measured and simulated mass increment for different injection duration for Case 2.

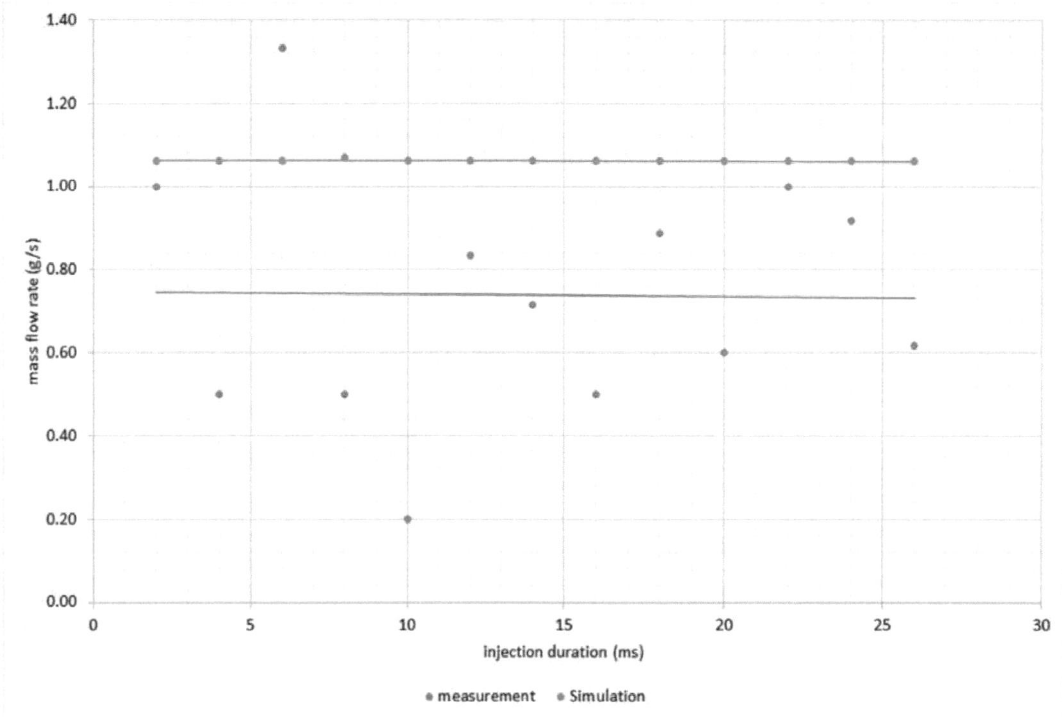

Figure 11. Comparison of measured and simulated of mass flow rate for different injection duration for Case 2.

7. Conclusions

Experimental work has been carried out to describe the jet characteristics of CNG injection for a single-hole direct injector for different injection pressure and injection duration. An analytical model of the injector was also developed to predict the injection mass flow rate. The quantitative results of the injection experiment are useful to understand the characteristics of the injector at different injection pressure. The mass flow rate through the injector is affected by injection pressure linearly as shown in table 4. The mass flow rate is only dependent on upstream pressure because the injector nozzle is operated in chocked condition. It was also shown by Case 2 that average mass flow rate at short injection duration is lower than recorded data in Case 1 of 50 bar injection pressure eventhough the range of the resultant mass flow rate is comparable to each other. The injection pressure of 50 bar is used as the baseline case for Case 2. The deviation is due to the fluctuating mass flow rate data produced for Case 2. We believed that that the fluctuating values of the mass flow rate at short injection duration is due to the internal losses of the injector. The internal losses are hydraulic and electrical losses as well as the hysteresis (magnetic and hydraulic) of the electromagnetic material. Based on simulation results, the analytical model of the injector are able to predict the trend of the injector mass flow rate at different injection pressure and different injection duration. However, the fluctuated trend of experimental data in Case 2 were unable to be predicted by the model.

Acknowledgments

The present research activity has been carried out as part of the CNG Direct Injection Project of the Universiti Malaysia Pahang. The authors would like to thank the technical staff of Automotive Laboratory of Automotive Engineering Centre (AEC) of Universiti Malaysia Pahang, members of Innovative Manufacturing and Mechatronics (IMAMS) laboratory, Faculty of Manufacturing Engineering, Universiti Malaysia Pahang, Malaysia. Moreover, finally to Universiti Malaysia Pahang for providing the funding for the project under short grant scheme.

References

[1] Pourkhesalian A M, Shamekhi, A H, and Salimi, F 2010 Alternative fuel and gasoline in an SI engine: A comparative study of performance and emissions characteristics *J. Fuel* **89** 5 p 1056-1063

[2] Cho H M and He, B Q 2007 Spark ignition natural gas engines-A review *J. Energy Conversion and Management* **48** 2 p 608-618

[3] Sen A K, Zheng, J, and Huang, Z 2011 Dynamics of cycle-to-cycle variations in a natural gas direct-injection spark-ignition engine *J. Applied Energy* **88** 7 p 2324-2334

[4] Kalam M A and Masjuki, H H 2011 An experimental investigation of high performance natural gas engine with direct injection *J. Energy* **36** 5 p 3563-3571

[5] Ouellette P, Mtui, P L, and Hill, P G 1998 Numerical Simulations of Directly Injected Natural Gas and Pilot Diesel Fuel in a Two-Stroke Compression Ignition Engine *J. Sae Technical Paper Series* 724

[6] Banapurmath N R, Tewari, P G, and Hosmath, R S 2008 Experimental investigations of a four-stroke single cylinder direct injection diesel engine operated on dual fuel mode with producer gas as inducted fuel and Honge oil and its methyl ester (HOME) as injected fuels *J. Renewable Energy* **33** 9 p 2007-2018

[7] Carlucci A P, de Risi, A, Laforgia, D, and Naccarato, F 2008 Experimental investigation and combustion analysis of a direct injection dual-fuel diesel-natural gas engine *J. Energy* **33** 2 p 256-263

[8] Gebert N, Beck, R, Barkhimer, H, and Wong, A K. Wells, Development of pilot fuel injection system for CNG engine, Diesel Engine *J.* **2012** 1996 SRC - GoogleScholar p 12-04

[9] Hodgins K B, Hill, P G, Ouellette, P, and Hung, P 1996 Directly Injected Natural Gas Fueling of Diesel Engines *J. Sae Technical Paper Series*

[10] Papagiannakis R G, Rakopoulos, C D, Hountalas, D T, and Rakopoulos, D C 2010 Emission characteristics of high speed, dual fuel, compression ignition engine operating in a wide range of natural gas/diesel fuel proportions *J. Fuel* **89** 7 p 1397-1406

[11] Chiodi M, Berner, H-J, and Bargende, M 2006 Investigation on different Injection Strategies in a Direct Injected Turbocharged CNG-Engine *J. SAE Technical Paper*

[12] Mohamad T, Harrison, M, Jermy, M, and How, H 2010 The structure of the high-pressure gas jet from a spark plug fuel injector for direct fuel injection *J. Journal of Visualization* **13** 2 p 121-131

[13] Chan E C, Evans, R L, Davy, M H, and Cordiner, S 2007 Pre-ignition characterization of partially-stratified natural gas injection *J. SAE Technical Paper*

[14] Petersen B R, 2006 Transient high-pressure hydrogen jet measurements [Ph.D. thesis]. University of Wisconsin

[15] Chitsaz I, Saidi, M H, Mozafari, A A, and Hajialimohammadi, A 2013 Experimental and numerical investigation on the jet characteristics of spark ignition direct injection gaseous injector *J.* **105** p 8-16

[16] Hajialimohammadi A, Honnery, D, Abdullah, A, and Mirsalim, M A 2013 Time resolved characteristics of gaseous jet injected by a group-hole nozzle *J.* **113** p 497-505

[17] Hill P G and Ouellette, P 1999 Transient Turbulent Gaseous Fuel Jets for Diesel Engines *J. Journal of Fluids Engineering* **121** 1 p 93-101

[18] Abraham J 1996/09/01 1996 Entrapment Characteristics Of Transient Gas Jets *J. Numerical Heat Transfer, Part A: Applications* **30** 4 p 347-364

[19] Ouellette P and Hill, P G 1999 Turbulent Transient Gas Injections *J. Journal of Fluids Engineering* **122** 4 p 743-752

[20] Chitsaz I, Saidi, M H, and Mozafari, A A 2011 Semi Analytical Solution to Transient Start of Weakly Underexpanded Turbulent Jet *J. Journal of Fluids Engineering* **133** 9 p 091204-091204-8

[21] Ouellette P, 1996 Direct injection of natural gas for diesel engine fueling [Ph.D. thesis]. University of British Columbia

[22] Zhang X, Palazzolo, A, Kweon, C-B, Thomas, E, Tucker, R, and Kascak, A 2014 Direct Fuel Injector Power Drive System Optimization *J. SAE Int. J. Engines* **7** 3
[23] Digesu P, Ficarella, A, Laforgia, D, Bruni, G, and Ricco, M 1994 Diesel Electro-injector:A Numerical Simulation Code *J. SAE Technical Paper*
[24] Coppo M, Dongiovanni, C, and Negri, C 2004 Numerical analysis and experimental investigation of a common rail-type diesel injector *J.* **126** 4 p 874-885
[25] Coppo M and Dongiovanni, C 2007 Experimental validation of a common-rail injector model in the whole operation field *J.* **129** 2 p 596-608
[26] Digesu P, Ficarella, A, Laforgia, D, Bruni, G, and Ricco, M 1994 Diesel Electro-injector:A Numerical Simulation Code *J. SAE Technical Paper*
[27] Ficarella A and Laforgia, D 1998 Experimental and Numerical Investigation on Cavitating Flows in Diesel Injection Systems *J. Meccanica* **33** 4 p 407-425
[28] Hu Q, Wu, S F, Stottler, S, and Raghupathi, R 2001 Modelling of Dynamic Responses of an Automotive Fuel Rail System, Part I: Injector *J. Journal of Sound and Vibration* **245** 5 p 801-814
[29] Ando R, Koizumi, M, and Ishikawa, T 2001 Development of a simulation method for dynamic characteristics of fuel injector *J. Magnetics, IEEE Transactions on* **37** 5 p 3715-3718
[30] Tsai W C and Yu, P C 2011 Design of the Electrical Drive for the High-Pressure GDI Injector in a 500cc Motorbike Engine *J. International Journal of Engineering and Industries* **2** 1 p 70-83
[31] Erfan I, Chitsaz, I, Ziabasharhagh, M, Hajialimohammadi, A, and Fleck, B 2015 Injection characteristics of gaseous jet injected by a single-hole nozzle direct injector *J. Fuel* **160** p 24-34
[32] Schimpf P H 2013 A Detailed Explanation of Solenoid Force *J. Int. J. on Recent Trends in Engineering and Technology* **8** 2
[33] Antunes J M G, 2010 The Use Of Hydrogen As A Fuel For Compression Ignition Engines Phd School of Marine Science and Technology Newcastle University England.

Sintering of FeCuAl powder compacts formed through die compaction process at elevated temperature

M M Rahman[1*], M.A. Ismail, H.Y. Rahman and Z. Hasan

[1]Department of Mechanical Engineering, Universiti Tenaga Nasional,
Putrajaya Campus, Jalan IKRAM-UNITEN, 43000 Kajang, Selangor, Malaysia.

*Corresponding author email: mujibur@uniten.edu.my

Abstract. This paper presents the outcomes of an experimental investigation regarding the sintering characteristics of FeCuAl powder compacts formed through uniaxial die compaction process at elevated temperature. A lab-scale uniaxial die compaction rig was designed and fabricated in order to shape powder mass at elevated temperature. Iron powder ASC 100.29 (92 wt%) was blended mechanically together with other elemental powders, i.e., copper (7.5 wt%), and aluminum (0.5 wt%) at a rotation of 30 rpm for 30 min. The prepared feedstock was shaped at 200°C through simultaneous upward and downward axial loadings of 325 MPa and the product generated through this process is known as green compacts. The defect-free green compacts were subsequently sintered in argon gas fired furnace at constant temperature, i.e., 800°C at three different sintering rates, i.e., 5, 10 and 15°C/min for three different holding times, i.e., 30, 60 and 90 min, respectively. The final products were then characterized for their physical, electrical, and mechanical properties and their microstructures were evaluated. The results revealed that the sample sintered for a shorter period of time, i.e., 30 min at a moderate heating/cooling rate, i.e., 10°C/min obtained the better final characteristics, i.e., higher relative density, lower volumetric expansion, lower electrical resistivity, higher strength and more uniform microstructure.

1. Introduction

In the area of materials processing technology, there are several approaches to create or design new materials where powder metallurgy has been selected by most of the manufacturers since this method has been already recognized and used in worldwide with numerous significant advantages compared to others. Powder metallurgy is a process where a powder or powder mass is compacted by pressing it in order to form into a desired shape [1-2]. Powder metallurgy is divided into (i) isostatic pressing, (ii) injection moulding, and (iii) powder compaction. Isostatic pressing can be conducted either at room temperature, i.e., cold isostatic pressing or at elevated temperature, i.e., hot isostatic pressing. Powder compaction is also can be conducted at room temperature, i.e., cold compaction or at elevated temperature but below the recrystallization temperature of the main powder constituent called warm compaction [3-4].

There are many reasons why powder metallurgy has been selected in manufacturing industries compared to others such as (i) shorter production time, i.e., it does not require lengthy process compared to other manufacturing methods, (ii) cheaper production cost or more economical, (iii) no extra scrap losses, (iv) energy savings, and (v) suitable for large volume of production. Near-net-shape products or components are very important in manufacturing industries hence powder metallurgy is given as the top position compared to other manufacturing methods in manufacturing final products or components with absolute near-net-shape [5-7].

The combination of iron, copper, and aluminum are selected to overcome their individual material weaknesses such as iron corrodes easily and faces strength to weight ratio problem, copper and aluminum are expensive and very soft type of materials. Composition of elemental powder is very important because it has major effect on properties. Each elemental powder has different properties and to combine with other types of powder, there are several significant factors need to be considered, i.e., their weaknesses, advantages, characteristics, quantities and qualities along with their applications in daily life [8].

Powder compaction is the production of any component from powder material by pressing it to form into desired shape. The product after compaction process is known as a green compact, which has enough strength to be handled to the next process. The strength of green compact, which is mechanical strength is known as green strength. Green strength is formed from interaction of mechanical interlocking that occurred on the surfaces of particle and at the same time, it is contributed by plastic deformation during process of pressing. Powder compaction can either be conducted at room temperature or at elevated temperature but below the recrystallization temperature of the main powder constituent. The first type of compaction is termed as cold compaction process whereas the second type is termed as warm compaction process. The warm compaction process was introduced in late nineties by realizing the fact that temperature rise softens the metal powder hence higher green density could be achieved if a powder mass is formed at above room temperature [9-10].

Sintering is the last step in a full cycle of powder compaction process, which is a thermally activated material transport in a powder compact in order to reduce surface contact among particles by growing process, shrinkage of pore volume and change of pore geometry. Sintering is known as a process of powder particles bonding by molecular or atomic attraction to form a coherent body and at the same time due to the heat application, the strength of the powder mass becomes higher and resulting some changes in densification and recrystallization by material transport. The application of heat for this treatment is set up below melting point of main powder constituent, which is around 60% - 80% for a certain duration of time in order to improve inter-particle bonding. Sintering is governed by several significant variables, i.e., temperature, time, and rate. These variables can be controlled in order to increase density of a product with better microstructure and less porosity [11-12].

Sintering can be divided into solid state sintering and liquid state sintering. Both types of sintering are possible to occur at the same time based on the melting point of individual material. Diffusion acts an important role during sintering, which is governed by surface diffusion, volume or grain boundary diffusion. Density is a significant factor for many applications, which can be improved through solid state sintering, liquid state sintering or combination of both. Products with increased density could simultaneously improve the other properties, i.e., fatigue, ductility, strength, and modulus of elasticity. The transport mechanisms that commonly involved in solid state sintering are volume diffusion, surface diffusion, grain-boundary diffusion, atoms evaporated on surfaces, and viscous or plastic flow. Liquid state sintering is related to formation of liquid phase during sintering due to lower melting point of element inside green compacts. Liquid state sintering is commonly practised to sinter alloys and composites that melt below temperature range used for sintering. Due to this solubility process, the attraction occurs among the grains during the wetting process of liquid to solid in order to unite the grains. Solid-liquid solubility relations between liquid material and solid material are very significant in this process in order to create coherent bonding among the particles [13-15].

A lot of research activities are reported on the warm metal powder compaction by using a single metallic material or by adding powder type lubricants or additives to enhance the characteristics of final products. Since sintering is a key step to manufacture new products or components through warm powder compaction method and for combination of iron (Fe), copper (Cu) and aluminium (Al), there is still a lot of research gaps need to be explored in order to find out the most suitable variables and moreover. Furthermore, FeCuAl is not commercially available in the market because it is a new kind of alloy. Therefore, the objective of this paper is to study the effects of sintering schedule to the final properties of FeCuAl powder mass formed at elevated temperature.

2. Materials and Methods

Four consecutive steps are involved in this experiment, i.e., (i) feedstock preparation, (ii) green compact generation, (iii) sintering in argon gas fired environment, and (iv) sample characterization. Iron powder ASC 100.29 with particle size range of 20 - 180 μm (92 wt%) was used as the main powder constituent whereas copper (7.5 wt%) and aluminum (0.5 wt%) were used as other elemental powders. The main powder and elemental powder were mixed mechanically for 30 min at a rotation of 30 rpm. The mixed powder mass was filled into the die cavity located at the warm compaction rig (Figure 1(a)) and rectangular shape green compacts (Figure 1(b) were generated by compacting the powder mass at 200°C through simultaneous upward and downward axial loadings of 325 MPa.

(a)

(b)

Figure 1. (a). T-15 compaction rig with heating system.
(b). FeCuAl green compact.

The defect-free green compacts were subsequently sintered in a custom made argon fired furnace (Model: HT3-1400-SIC, S/N: LT007) at a constant sintering temperature of 800°C at three different sintering rates, i.e., 5, 10 and 15°C/min for 30, 60 and 90 min, respectively. The sintering furnace was customized in order to ensure the ceramic tube can be mounted as it was 150 mm long, 50 mm OD, and 40 mm ID. The final products were characterized for their physical, electrical, and mechanical properties and their microstructures were evaluated. The relative density of the products was calculated from the dimension and mass of the final products. The dimensional measurement was conducted by using a digital Vernier calliper (Brand: Mitutoyo, Model: CD-6"CS, S/N: 04171546, Accuracy: +/- 0.001 in). The electrical resistivity was measured by using a digital multimeter (Brand: Fluke, Model: Fluke 115, S/N: 28341190ws) whereas the flexure stress was measured through three point bending test machine (Brand: Instron, Model: Instron 3365, S/N: SAA61569) following the standard (ASTM E290-09). The image of the fractured surface of each sample was captured through scanning electron microscopy (Brand: JEOL, Model: JSM- 6010PLUS/LA).

3. Results and Discussion

Relative densities of samples sintered at different schedules are depicted in Figure 2. It can clearly be observed in Figure 2 that the highest relative density was obtained by the sample sintered for 90 min at a rate of 5°C/min whereas the lowest relative density was obtained by the sample sintered for 30 min at the same rate. Sintering for longer time permits the perfect necking of particles hence the size of the product becomes smaller. Neck growth also strengthens the product.

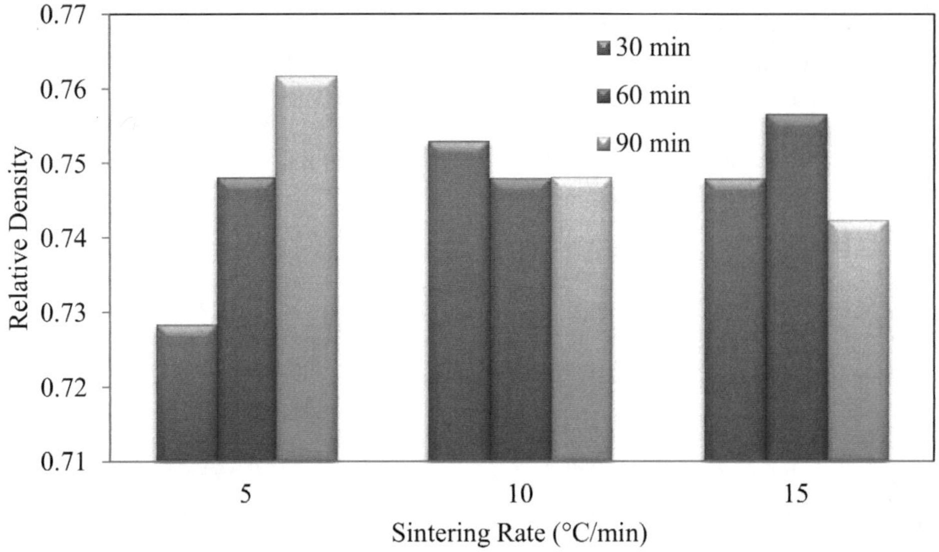

Figure 2. Relative density of sintered products.

Figure 3 shows the dimensional change of the samples sintered at different schedule. It is clear from the figure that every sample experienced swelling during sintering. The highest swelling was occurred at the sample sintered for 30 min at a rate of 10ºC/min whereas the lowest volumetric expansion was occurred at the sample sintered for 90 min at a rate of 5ºC/min. Swelling during sintering occurs mainly due to some factors, i.e., size of the particles, elemental composition, density of green compact, sintering temperature, and heating rate during the sintering processes. The results are also in line with relative density, i.e., higher volumetric expansion during sintering means lower relative density after sintering.

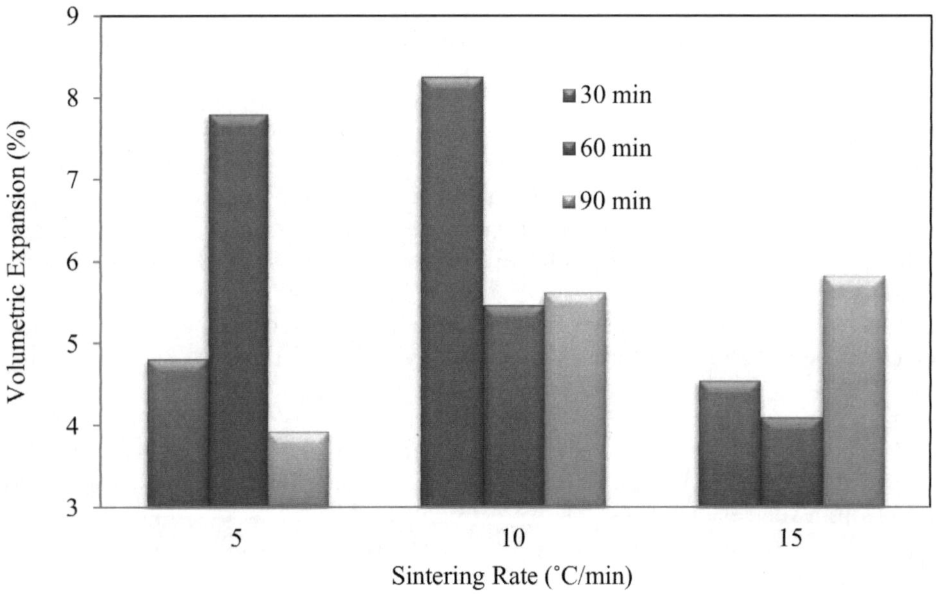

Figure 3. Volumetric expansion (percentage) of samples during sintering.

The electrical resistivity of the samples sintered at different schedule is depicted in Figure 4. Overall, the electrical resistivity values are found to be different at each sintering schedule. Electrical resistivity is the ability of a material to conduct electrical current. Since all the samples have the same composition, the electrical resistivity values were used as indicator of the integrity of the sample. Lower electrical resistivity value means the sample is able to flow electrical current easily means the possibility of the existence of interconnected pores or cracks inside the samples are smaller. Generally, the electrical resistivity was found to be higher at the samples sintered for longer period of time. Electrical resistivity was also found to be higher at the samples sintered at a higher sintering rate. Prolong exposure of the sample at argon gas fired environment might cause the formation of micro-cracks inside the sample which might lower down the electrical conductivity means increased the electrical resistivity.

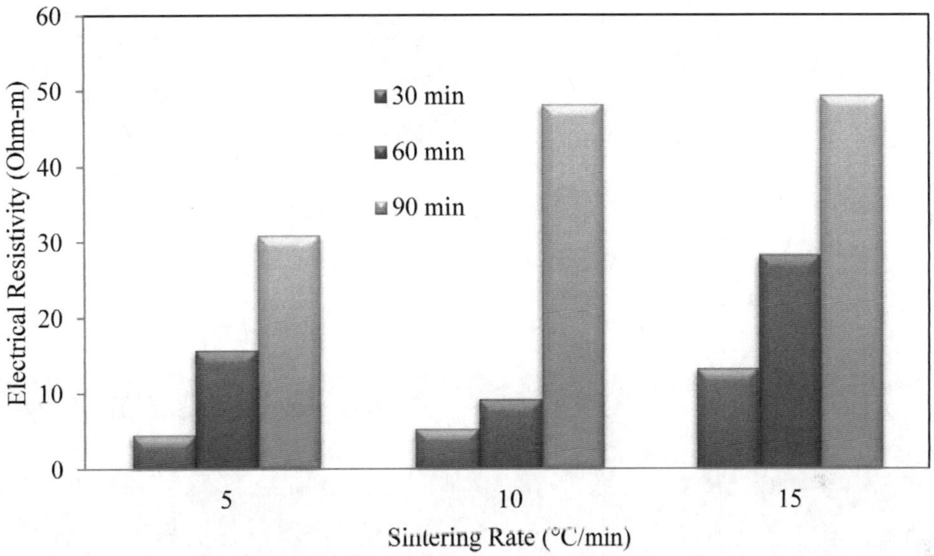

Figure 4. Electrical resistivity of the sintered products.

Flexure stress or three point bending strength of the sintered samples are presented in Figure 5. It is clear that higher bending strength was obtained by the sample sintered for 90 min at rate of 5°C/min. Generally, the material particles are affected during the sintering process. The particles of the material started to react on each other as the transport mechanism begins in order to fill the pores. Since the melting point of aluminum is less than the sintering temperature, liquid state sintering by aluminum helped to improve inter-particle bonding among particles as molten aluminum would fill the pores and functioned to attract iron and copper particles to close the gaps/pores among them. Effectiveness of sintering schedule to the flexure stress has been identified where most of the samples that have lower flexure stress were sintered for 30 min compared to others. Therefore, a holding time of 30 min is found to be not enough to particles movement and rearrangement of the grains to develop higher strength of the final products.

The SEM images of fractured surface of samples sintered at different schedule are shown in Figures 6-8. The microstructures are found to be in line with the strength where the sintered samples of higher flexure stress have good inter-particle bonding and low interconnected pores. The microstructure of samples sintered at a rate of 10°C/min shown uniform grain rearrangement compared to the samples sintered at 5°C/min and 15°C/min.

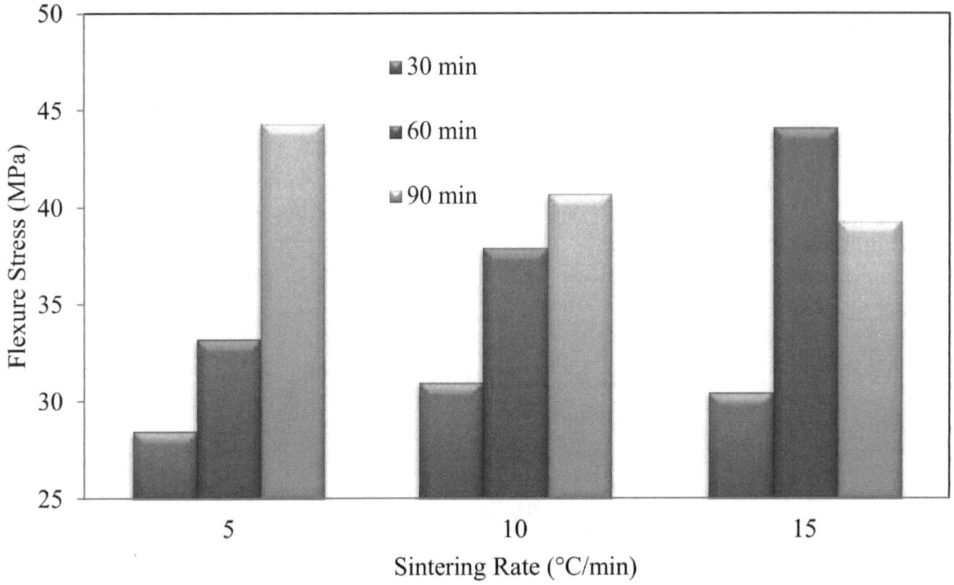

Figure 5. Flexure stress of sintered products.

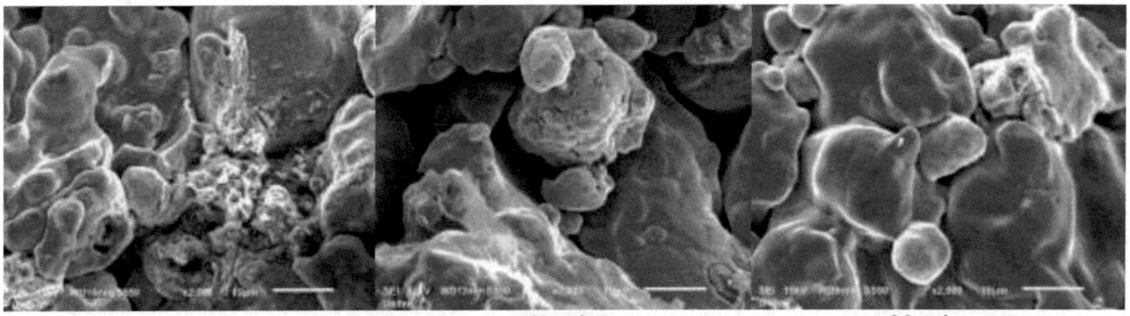

30 min 60 min 90 min

Figure 6. Microstructure of products sintered at a rate of 5°C/min.

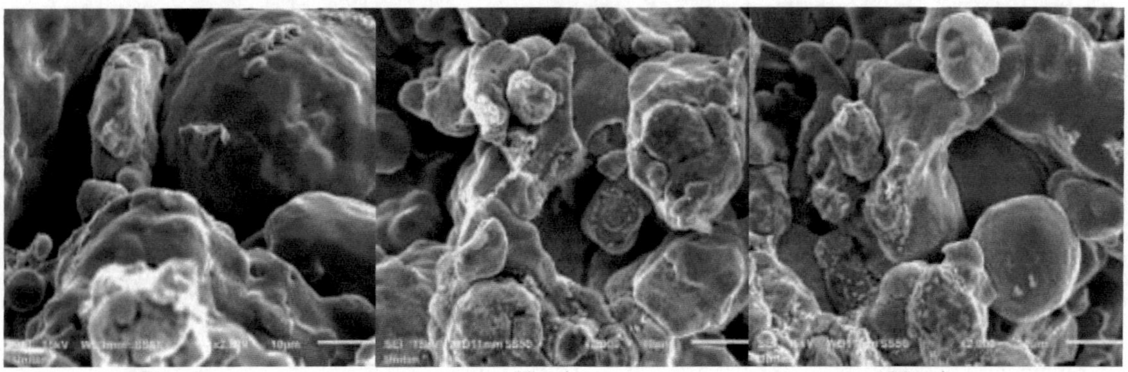

30 min 60 min 90 min

Figure 7. Microstructure of samples sintered at a rate of 10°C/min.

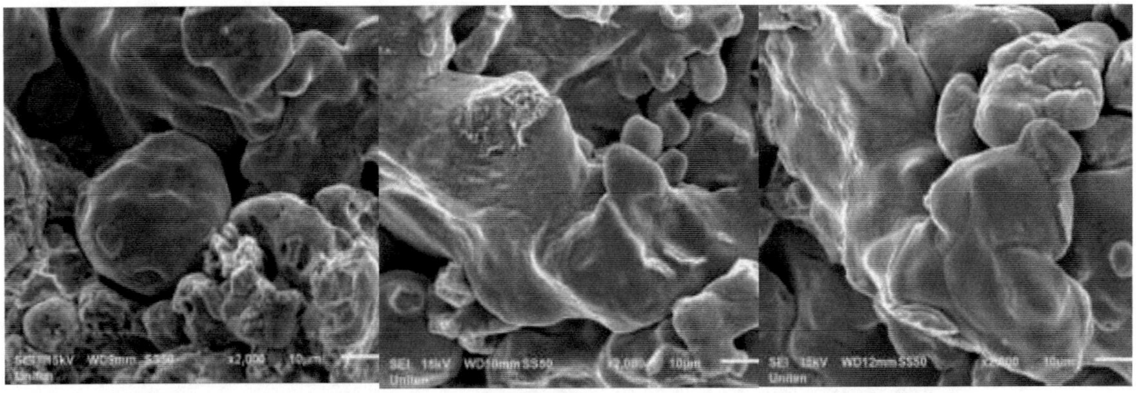

30 min 60 min 90 min

Figure 8. Microstructure of samples sintered at a rate of 15°C/min.

4. Conclusions

Sintering schedule was found to give impacts to the final properties of the products. Higher density final products of FeCuAl experienced lower volumetric expansion when sintered for 90 min at slow rate, i.e., 5°C/min. However, prolong sintering time caused the formation of micro cracks inside the products, which caused the lower electrical conductivity. However, prolong sintering time also caused the proper necking among particles, which is manifested by the high flexure stress. Sintered products having more uniform microstructure were obtained by sintering the green compacts for 90 min at a slow heating/cooling rate, i.e., 5°C/min.

Acknowledgements
This research is financially supported by Ministry of Higher Education (MOHE) Malaysia through the research project 20140117 FRGS.

References

[1] Ramakrishnan P 1983 Indian Journal of History of Science **18(1)** 109-114.
[2] Angelo PC and Subramanian R 2008 *Powder Metallurgy: Science, Technology and Applications* (New Delhi: PHI Learning Pvt. Ltd.) pp 140-197.
[3] Shima S and Shimada M 1991 *Isostatic Pressing Technology and Applications* Elsevier Applied Science ed M Koizumi and M Nishihara (England: Elsevier Science Publishers Ltd.) pp 7-25.
[4] Rahman MM Wardi MAS Nor SSM 2013 *International Journal of Automotive and Mechanical Engineering* **8** 1132-1139.
[5] Rahman MM Talib AAA 2015 *Mater. Today Communications* **4** 42-49.
[6] Sharma D Chandra K Misra PS 2011 *Mater. Des.* **32** 3198-3204.
[7] Rahman MM Nor SSM Rahman HY 2011 *Mater. Des.* **32** 447-452.
[8] Sun J Zhao J Li Z Ni X Zhou Y Li A 2017 *Ceramics International* **43** 2686-2696.
[9] Sonsino CM Ratz IR 2004 *Powder Metall.* **47** 352-357.
[10] Rahman MM Ariffin AK Nor SSM 2009 *Applied Mathematical Modelling* **33** 4031-4048.
[11] Almomami MA Shatnawi AM Alrashdan MK 2014 *Adv. Mater. Res.* **1064** 32-37.
[12] Narasimhan KS 2001 *Mater. Chem. Phys.* **67** 56-65.
[13] Martin CL Schneider LCR Olmos L Bouvard D 2006 *Scripta Mater.* **55** 425-428.
[14] Liang F Liu J 2017 *Ceram. Int.* **43** 5856-5861.
[15] Chen CL Ma SH 2017 *J. Alloys Compd.* **711** 488-494.

IOP Conf. Series: Materials Science and Engineering **257** (2017) 012059 doi:10.1088/1757-899X/257/1/012059

Finite element normal mode analysis of resistance welding jointed of dissimilar plate hat structure

N A Nazri [1] and M S M Sani [1,2*]

[1]Advanced Structural Integrity and Vibration Research (ASIVR), Faculty of Mechanical Engineering, Universiti Malaysia Pahang,
26600 Pekan, Pahang, Malaysia
[2]Automotive Engineering Centre (AEC), Universiti Malaysia Pahang,
26600 Pekan, Pahang, Malaysia.

Corresponding author email: mshahrir@ump.edu.my

Abstract. Structural joints offer connection between structural element (beam, plate, solid etc.) in order to build a whole assembled structure. The complex behaviour of connecting elements plays a valuable role in characteristics of dynamic such as natural frequencies and mode shapes. In automotive structures, the trustworthiness arrangement of the structure extremely depends on joints. In this paper, top hat structure is modelled and designed with spot welding joint using dissimilar materials which is mild steel 1010 and stainless steel 304, using finite element software. Different types of connector elements such as rigid body element (RBE2), welding joint element (CWELD), and bar element (CBAR) are applied to represent real connection between two dissimilar plates. Normal mode analysis is simulated with different types of joining element in order to determine modal properties. Natural frequencies using RBE2, CBAR and CWELD are compared to equivalent rigid body method. Connection that gives the lowest percentage error among these three will be selected as the most reliable joining for resistance spot weld. From the analysis, it is shown that CWELD is better compared to others in term of weld joining among dissimilar plate materials. It is expected that joint modelling of finite element plays significant role in structural dynamics.

1. Introduction

Lightweight based materials experience high demand in the application of automotive industries due to environment reason and social pressure [1]. Properties of joints for lightweight automotive parts are greatly influenced the stiffness and fatigue behaviour of the structure model [2]. Connections joining are crucial parts of complex structure and play a significant function in the assembled structure's behaviour in the aspect of flexibility and damping. There are numerous kinds of connection used in the engineering structure currently, such as bolts, screw, rivet and clinching which provides good advantages in term of is assembly ability, simple surface preparation, simple expectation of joint failure, ease in the inspection process, easy handling and machining, short assembly and joining times [3]. However, joints can be sensitive to temperature and moisture besides may enhance more weight to the structure [4]. Predicting rigid connections between substructures in modelling by neglecting the joints effect may lead to the different characteristics than actual physical structure. Modal analysis can be defined as an approach of characterizing process for dynamic properties of an elastic structure by identifying its modes of vibration and become a major technique to determine dynamic characteristics of engineering structures and its components [1, 5]. Normally, each mode has a specific natural frequency, mode shapes and damping factor which can be identified from practically any point on the structure either by numerical or experimental methods [6, 7]. In order to design a structure model to be applied in real life, it is essential to execute a materials suitability and structural analysis first so that the result from analysis can predict the effectiveness of material, methods and properties that going to be used practically [8]. In this paper, modal analysis which focusing on numerical study of top hat structure is going to be explores. A major concern in analyzing practical mechanical structures is to reliably identify their dynamic characteristics, i.e., their natural frequencies and vibration mode shapes. These vibration characteristics are needed in order to achieve effective design and control of the vibrations of structural components [9]. The main purpose of this paper is to investigate the most suitable joining method for the analysis of top hat structure which resembles spot welding connection.

2. Plate Hat Modelling

Nowadays, current automotive industry focusing on reducing fuel consumption and cost deduction in the producing vehicle [8]. One of the methods to achieve these goals is by choosing the right joining method between materials. Usually, complete set of automotive body-in-white (BIW) is made up from several combinations of substructures which also contribute to the stiffness and strength of the car body. In this paper, one of the substructures which is top hat is designed to resemble the automotive part and can be seen in Figure 1.

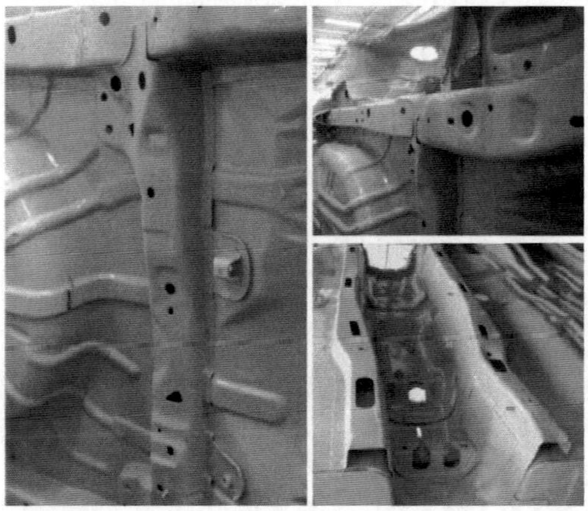

Figure 1. Substructure of automotive parts. [10]

Top hat structures are designed using MSC Nastran Patran with specified dimension as shown in Figure 2. In geometry section for modelling purpose, the easiness of stainless steel to be bend allow the shaping of this material to be at the top while bottom surface is made up of mild steel.

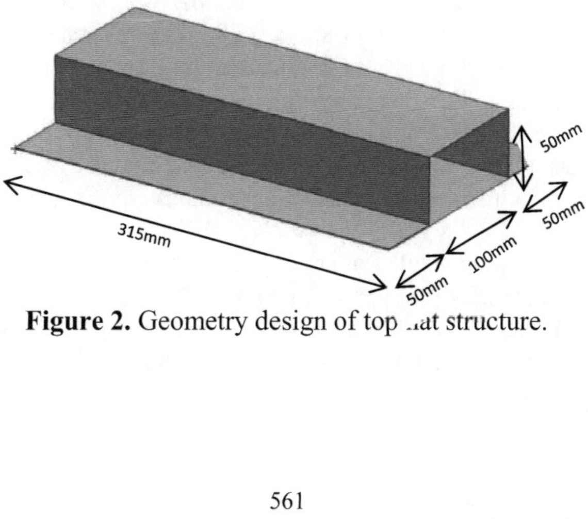

Figure 2. Geometry design of top ..at structure.

After design stage is completed, top hat structure is meshed equally and there are 2050 elements involved. Choosing the right meshing size is very crucial because it allow the correct simulation results while deducted the model's complexity as much as possible [11, 12]. Conversely, tiny mesh will also upturn the calculation period needed to finish the simulation [13]. The size of element is picked and confirmed after conducting convergence test where the element size is lessened gradually until the value of natural frequency is converge [12]. Furthermore, based on FEA theory, the smaller the element size which is due to the fine mesh, will increase the accuracy of the result even though the computing time is rather higher than that one with bigger element size [14, 15]. Figure 3 shows top hat structure after meshing process.

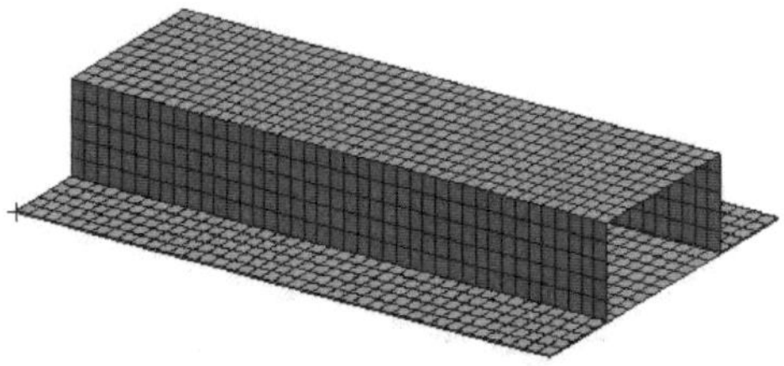

Figure 3. Meshed top hat structure.

Mild steel and stainless-steel properties such as Young's modulus, density and poisson ratio are inserted in finite element calculation. Table 1 shows the properties of those materials.

Table 1. Nominal values of materials

Properties	Mild Steel 304	Stainless steel 1010
Poisson Ratio	0.30	0.27
Young Modulus	210 GPa	200 GPa
Density	7870 kg/m^3	8000 kg/m^3

3. Finite Element Analysis (FEA)

As the behavior of bolted joints plays an imperative function in the dynamic characteristics of structures, the necessity for emerging precise predictive models of the joints is very demanding [16]. The purpose of numerical modelling can be categorized into three different main parts which are analysis, prediction and design [7]. That is the reason why development of finite element has been introduced in order to ease process of analyzing. Finite element analysis can be defined as a computer simulation technique for modeling and analyzing the effect of the part or model built. This tool is very crucial to rectify the structure failure before manufacturing and test is carried on [17]. It is definitely vital to certify that the prevailing finite element model that was created during design process is a reliable model and able to provide accurate prediction of structural behaviour and performance before the genuine structure experience mass production in manufacturing area [18]. FE stands as the most suitable tool for numerical modeling in structural engineering as it is having ability of handling complex structural geometry, large complex assemblies of structural components and can build many different types of analysis [19]. This method of analysis becomes one of top crucial analysis in structural dynamics analysis. This method of analysis becomes one of top crucial analysis in structural dynamics analysis.

In order to get the most accurate result from the analysis, there are numerous existing factors which need to take an observation such as the type of element used, number of elements used, accuracy of geometric modeling, simulation of actual constraints, loading and boundary conditions [20]. The finite element analysis of a structure entails a modeling of fastener joints which connecting either composite parts, metallic parts, or amalgamation of these two materials [21] . Analysis is extends by choosing required type of joints to connect those two material declared earlier. Three joining methods are chosen which are RBE2, CBAR and CWELD which have dissimilar properties to each other.

3.1. Equivalence

Equivalency in model is conducted frequently in order to remove any coincides nodes and elements in a structure besides to keep the rigid connectivity in joining dissimilar plates together as explained in Figure 4. For a structure with a multiple surfaces, applying equivalence in analysis is unavoidable in order to connect meshes [22]. Besides, equivalence is needed in order to abolish the redundant nodes generated by meshing each surface independently [23, 24]. In this simulation, there are 168 nodes were eliminates due to the redundancy in the model and highlighted point of nodes can be seen in Figure 5 below.

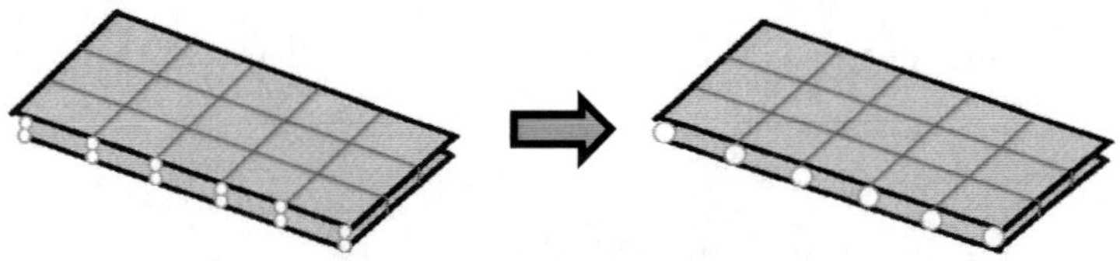

Figure 4. Equivalence.

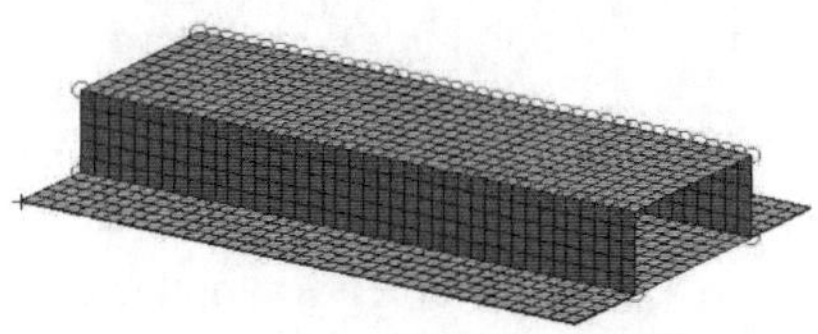

Figure 5. Eliminated nodes.

3.2. RBE2

RBE is an acronym stands for Rigid Body Element and RBE2 is a kind of formulation which consists of one independent node and one or more dependent node as shown in Figure 6, that does not have mass contribution [20, 25]. Rigid body using independent degree of freedom (DOF) at one grid and dependent DOF at an random number of grids called RBE2 [22]. These elements rigidly weld multiple grids to other grid. The independent and dependent term at the required nodes and joint must be considered. All dependent nodes of RBE-2 are coupled with independent nodes located at the specified location which resemble the spot weld points with total of 10 points each side in translation x-axis (Ux), translation y-axis (Uy), translation z-axis (Uz), rotational x-axis (Rx), rotational y-axis (Ry) and rotational z-axis (Rz) directions. Figure 7 shows the RBE2 connector of top hat structure, with total 20 RBE2.

IOP Conf. Series: Materials Science and Engineering **257** (2017) 012059 doi:10.1088/1757-899X/257/1/012059

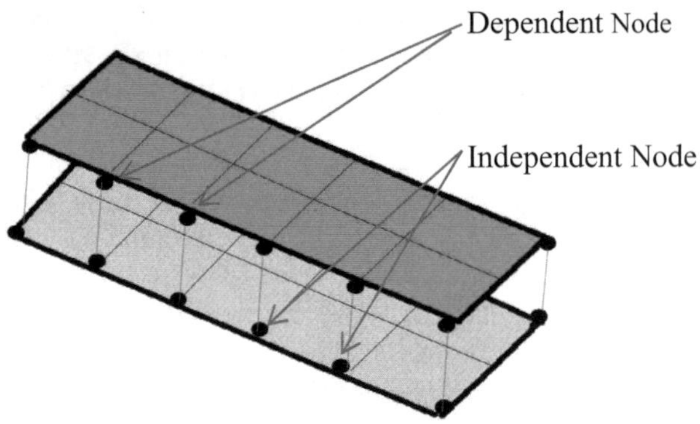

Figure 6. RBE2 with dependent and independent joints.

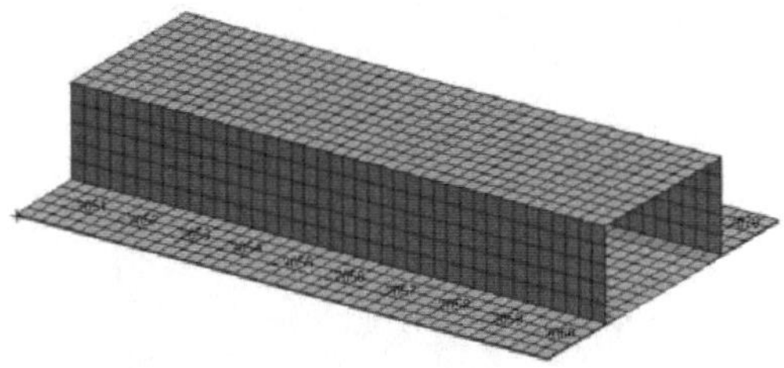

Figure 7. RBE2 element in top hat structure

3.3. CBAR

CBAR element is one example of joining provides in MSC Patran that able to withstand various kind of loads such as tension, compression and also bending forces. In the simulation, bar element is declared as a form of properties for curve [26]. Bar is responsible to joined two plates together at node location as displayed in Figure 8 below. The diameter for bar is 0.001m and its geometrical properties which is Young Modulus and Poisson ratio had been set to 1000 GPa and 0.27 respectively. The reason why Young Modulus of CBAR is declared much more higher than material is to represent the rigidity of welded joint. Figure 9 shows total of 20 CBAR elements top hat structure.

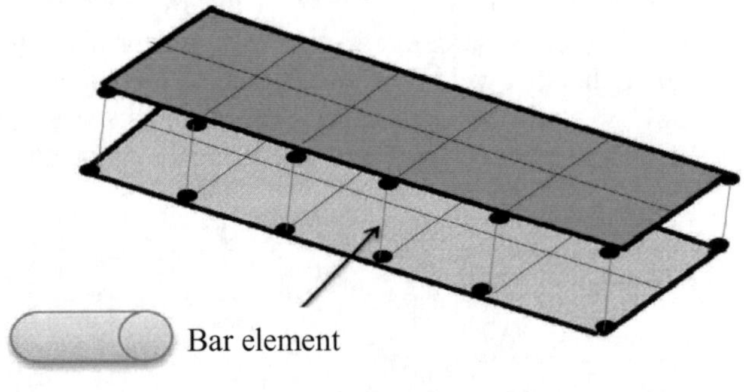

Figure 8. CBAR.

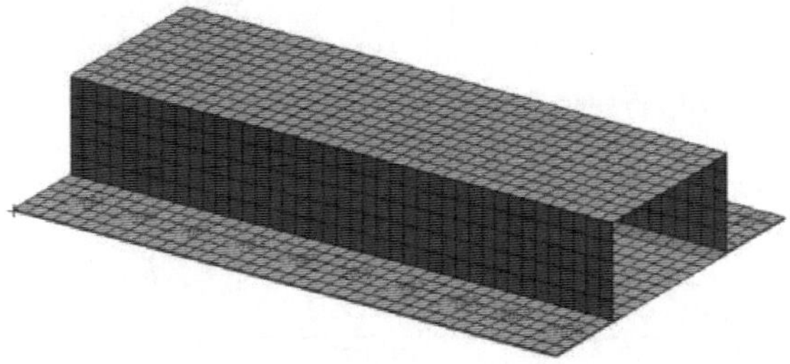

Figure 9. CBAR element in top hat structure.

3.4. CWELD

Weld connector, CWELD is a type of joining in which is extensively picked and utilized in modelling spot welds especially for Resistance Spot Welding process and finite element analysis due to goodness in flexibility in the type of mesh and ability to be update the element in model updating. Among three different connection types which is point-to-point, patch-to-point, and patch-to-patch connection [27, 28], the most versatile and frequently chosen in structural dynamic is the third one as displayed in Figure 10 below. In this paper, the connection is made up by connecting patch from upper element with corresponds lower element by choosing specified connecting node. The property of CWELD element is defined in welding properties (PWELD) entry. The property parameters are the material identification number and the diameter D of the spot weld. The diameter of spot weld is set to 0.001m with range length to diameter ratio between 0.2 to 5. If the diameter is larger than the surface patch, the spot weld element may underestimate the stiffness of the connection. Figure 11 shows how CWELD is modelled in top hat structure.

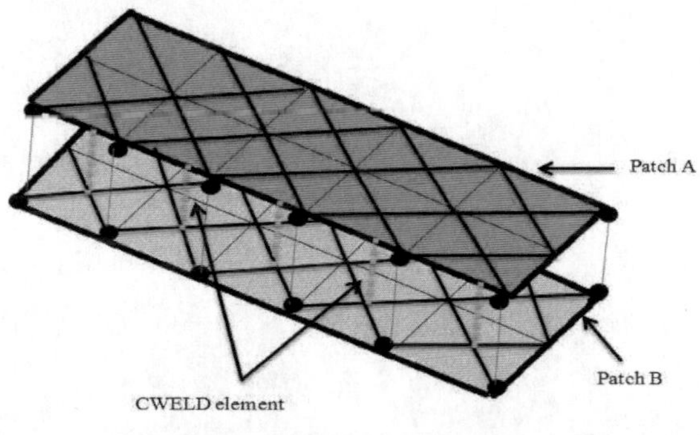

Figure 10. CWELD element.

Figure 11. CWELD element in top hat structure.

4. Results from FEA

After conducting the joining process, the result is analysed and they are differ to each other which prove that joining method strongly affect the dynamic behaviour of structure. Natural frequencies

IOP Conf. Series: Materials Science and Engineering **257** (2017) 012059 doi:10.1088/1757-899X/257/1/012059

which can be defined as a frequency at which a system tends to oscillate in the absence of any driving or damping force are acquired from simulation along with mode shape which is an illustration of a structure at certain frequency. From the simulation, there are bending, torsion and both in the structure for each mode shapes. Tabulated results can be seen in Table 2 and Table 3 below.

Table 2. Natural frequencies of top hat structures.

Mode	Natural Frequency (Hz)			
	Equivalence	RBE2	CBAR	CWELD
1	175.58	177.87	175.68	175.60
2	253.14	253.15	253.14	253.14
3	278.14	279.88	278.26	278.19
4	338.05	342.29	338.72	338.21
5	408.98	409.08	409.00	408.99

Table 3. Mode shapes of top hat structures.

Mode	Mode Shape			
	Equivalence	RBE2	CBAR	CWELD
1	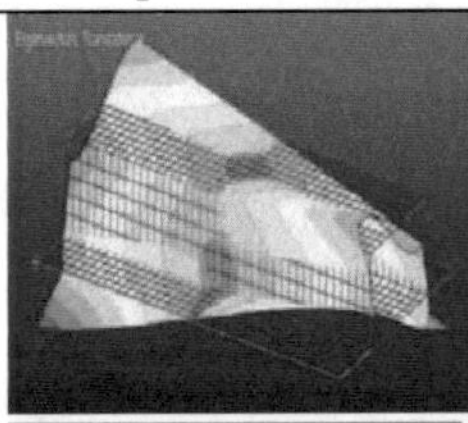	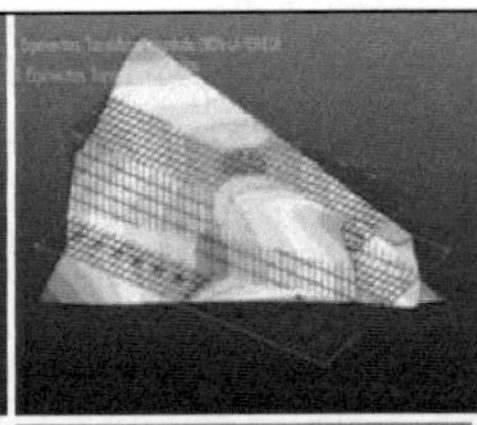	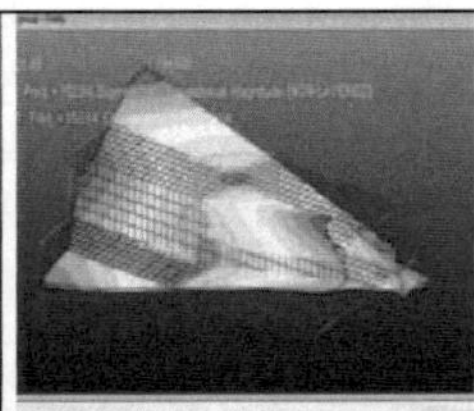	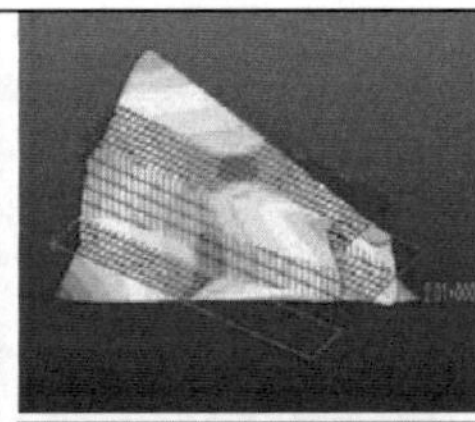
2	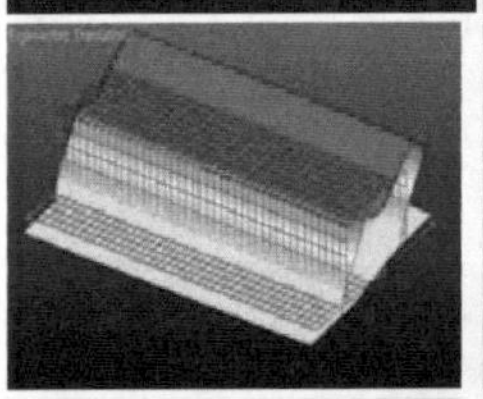	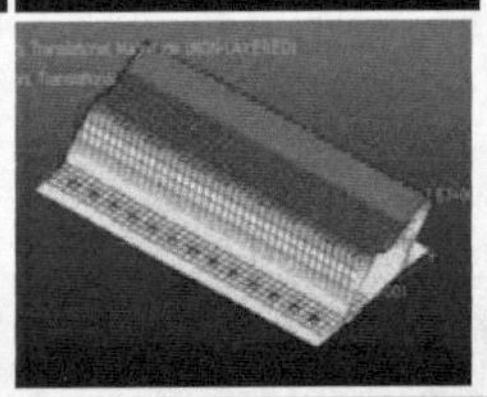	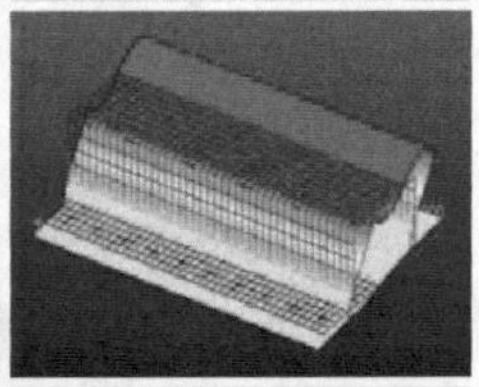	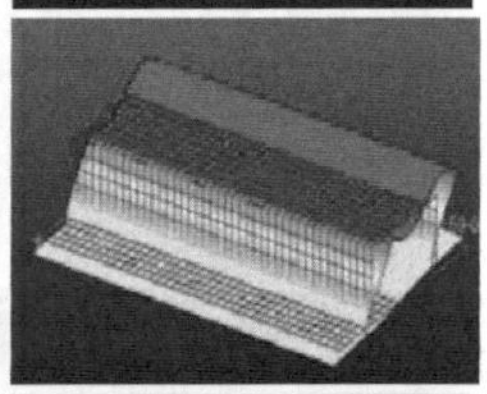
3	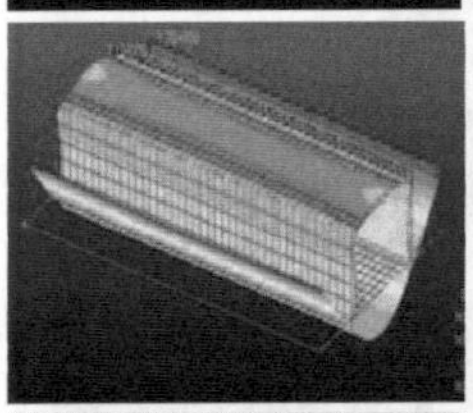	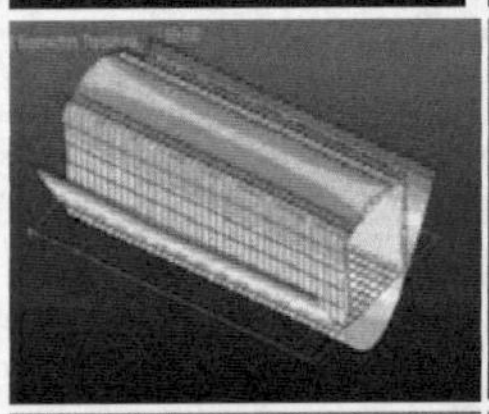	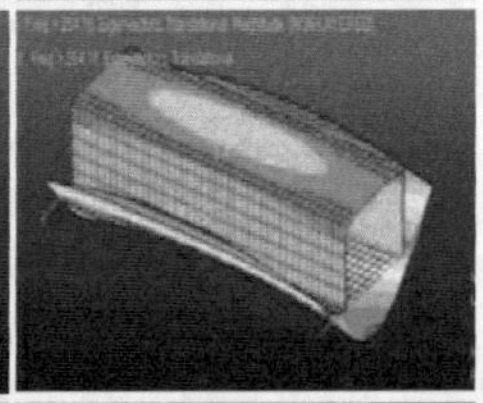	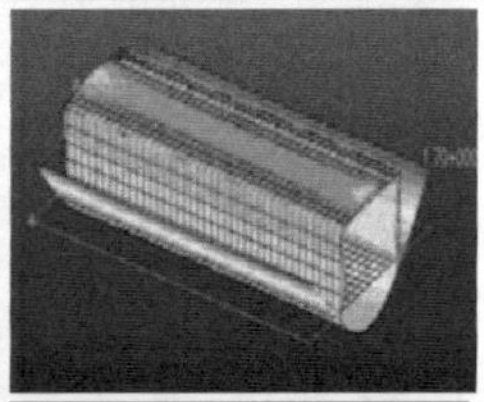
4	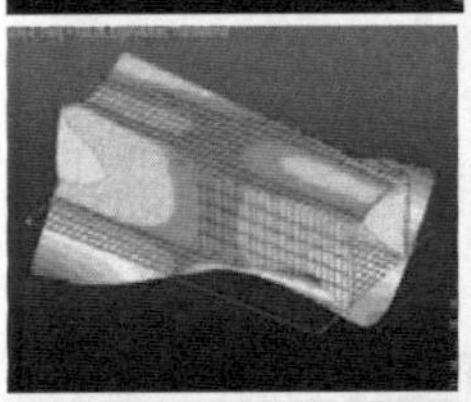	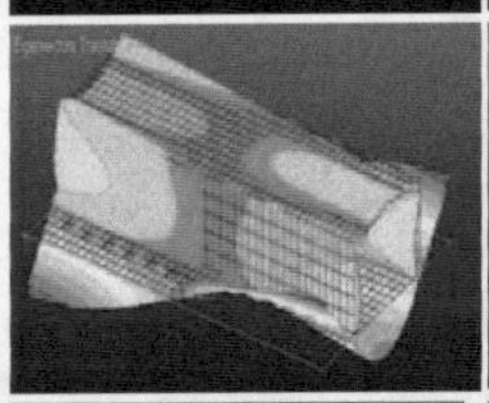	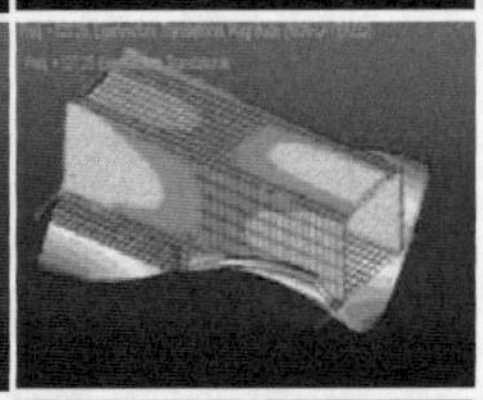	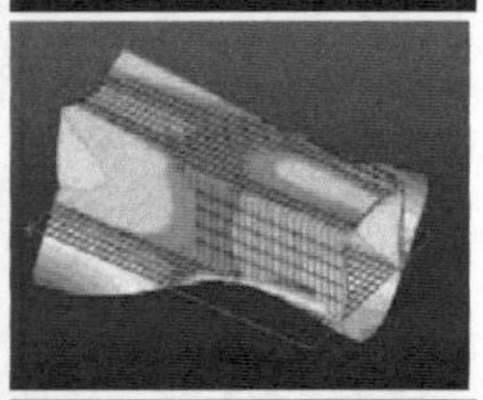
5	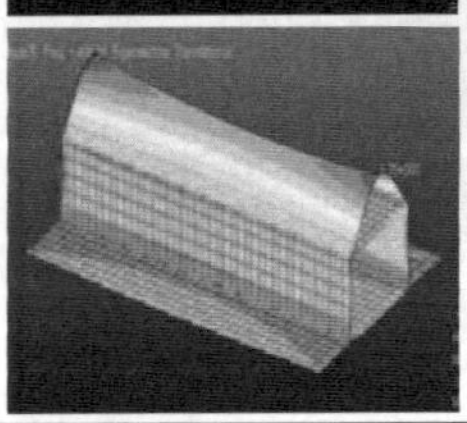	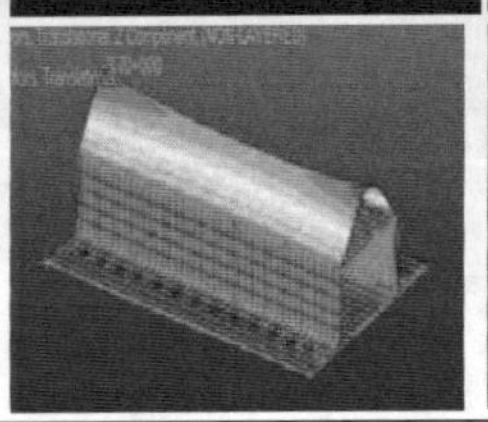	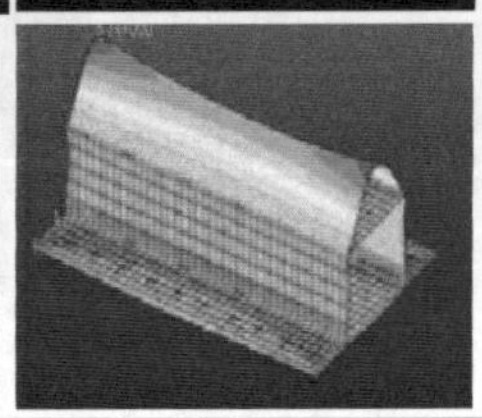	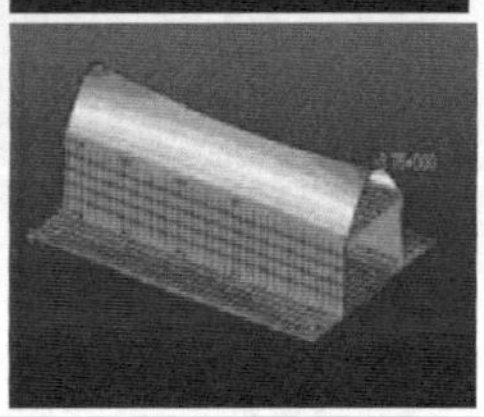

5. Discussion

Regarding joining strategy in this study, all connection types can be modelled when compatible mesh was applied to the structure [22]. Three types of connecting elements are compared with a purpose to select the most reliable to resemble rigid joined between material which is assumed as equivalence rigid body connection. Errors might be as a result of the improperly declaring materials properties such as Young's Modulus, density or Poisson ratio. Based on comparison of natural frequencies with different joint structure in Table 4, it is clearly seen that CWELD offered the lowest percentage of errors as it is closely resembling rigid body connection with almost similar mode shapes. In this paper, the diameter is set to be less than surface patch to prevent underestimation of connection stiffness. Result gained indicates that CWELD joining almost nearly reached rigid body connection with only 0.016% errors while RBE2 and CBAR are failed to gives lower percentage errors.

Table 4. Comparison of natural frequencies with different joint structure.

Mode	Equivalence	RBE2	Error (%)	CBAR	Error (%)	CWELD	Error (%)
1	175.58	177.87	1.300	175.68	0.057	175.60	0.011
2	253.14	253.15	0.004	253.14	0.000	253.14	0.000
3	278.14	279.88	0.626	278.26	0.043	278.19	0.018
4	338.05	342.29	1.254	338.72	0.198	338.21	0.047
5	408.98	409.08	0.024	409.00	0.005	408.99	0.002
Total Average Error			0.642		0.061		0.016

6. Conclusions

This effort has examined some of the FE models developed to represent spot welds normally produced by RSW in the past. In between all these models, CWELD is chosen to model the spot welds. Generally, the aim of this study is to investigate some methods on modelling joined in order to estimate the dynamic behaviour. Three types of connection modelling which is RBE2, CBAR and CWELD are chosen and trustworthiness of these three joining were compared as referring rigid body connection which become as benchmark in calculating errors. RBE2 is less preferable for the future study due to the disability in model updating. In a nutshell, for joining resistance spot welding in dissimilar material plates, CWELD joining is the most reliable to be nominated due to its most exactly prediction of mode shapes and less total average errors in natural frequencies for all modes.

Acknowledgments

The author would like to greatly acknowledge the support by focus group of Advanced Structural Integrity of Vibration Research (ASIVR) and Universiti Malaysia Pahang for providing all the facilities and equipment used for this project.

References

[1] Hassin N, Yunus M, Rani M A and Sulaiman M 2016 *Journal of Engineering and Applied Sciences* **100** 1451-5
[2] Dannbauer H, Gaier C and Hofwimmer K 2005 Fatigue analysis of welding seams and spot joints in automotive structures. SAE Technical Paper)
[3] Blaga L, dos Santos J F, Bancila R and Amancio-Filho S T 2015 *Construction and Building Materials* **80** 167-79
[4] Schneider J and Radzilowski R 2014 *JOM* **66** 2123-9
[5] Sharma J K and Parashar S K 2016 *IJIRSE* **2** 128-36

[6] Sani M S M, Abdullah N A Z, Zahari S N, Siregar J P and Rahman M M 2016 Finite element model updating of natural fibre reinforced composite structure in structural dynamics. In: *MATEC Web of Conferences*: EDP Sciences) pp 1-4
[7] Zahari S N, Zakaria A A R and Sani M S M 2015
[8] Kiani M, Shiozaki H and Motoyama K 2015 *International Journal of Vehicle Design* **67** 219-36
[9] Sani M S M, Nazri N A, Zahari S N, Abdullah N A Z and Priyandoko G 2016 Dynamic Study of Bicycle Frame Structure. In: *IOP Conference Series: Materials Science and Engineering*: IOP Publishing) p 012009
[10] Husain N A, Haddad H K and Ouyang H 2010 *International Journal of Vehicle Noise and Vibration* **6** 163-75
[11] Sukumar N, Chopp D L, Moës N and Belytschko T 2001 *Computer Methods in Applied Mechanics and Engineering* **190** 6183-200
[12] Liu Y and Glass G 2013 *British Journal of Engineering and Technology, Vols* **1** 13-28
[13] Chan W M, Chew H I, Lee H P and Cheok B T 2004 *Journal of Materials Processing Technology* **148** 15-24
[14] Kennedy F, Colin F, Floquet A and Glovsky R 2014 *Developments in Numerical and Experimental Methods Applied to Tribology* 138-50
[15] More S T and Bindu R 2015 *Int J Eng Sci Innov Technol* **4** 181-5
[16] Husain N A, Khodaparast H H, Snaylam A, James S, Dearden G and Ouyang H 2010 *Proceedings of the Institution of Mechanical Engineers, Part C: Journal of Mechanical Engineering Science* **224** 851-61
[17] Sani M S M, Arbain M, Noor M, Ming G, Zohari M, Nizwan C and Mon T 2009 Stress Analysis and Modal Transient Response of Car Chassis. In: *International Conference on Advance Mechanical Engineering (ICAME09),* pp 22-5
[18] Abdullah N A Z, Sani M S M, Rahman M M and Zaman I 2017 Correlation of numerical and experimental analysis for dynamic behaviour of a body-in-white (BIW) structure. In: *MATEC Web of Conferences*: EDP Sciences) p 01020
[19] Zienkiewicz O C and Taylor R L 2000 *The finite element method: solid mechanics* vol 2: Butterworth-heinemann)
[20] Ahmed S and Ahmed M 2014 *Applied Mechanics and Materials* **629** 82
[21] Rutman A, Boshers C, Pearce L and Parady J 2009 Fastener modeling for joining composite parts. In: *Americas Virtual Product Development Conference,* pp 1-28
[22] Zahari S N, Sani M S M, Husain N A, Ishak M and Zaman I 2016 *Jurnal Teknologi (Sciences and Engineering)* **78** 57-65
[23] Han Q P and Peng B 2014 Research of the FEM Mesh Generation Based on Patran. In: *Applied Mechanics and Materials*: Trans Tech Publ) pp 731-5
[24] Li W and Robinson J 2016 Automated Generation of Finite-Element Meshes for Aircraft Conceptual Design. In: *16th AIAA Aviation Technology, Integration, and Operations Conference,* p 3287
[25] Chen H and Qi P 2004 Training Tutorial and Examples for MSC. Patran/Marc. Science Press, Beijing, China)
[26] Zahari S N and Sani M S M 2017
[27] Fang J, Hoff C, Holman B, Mueller F and Wallerstein D 2000 Weld modeling with MSC/Nastran. In: *Second MSC Worldwide Automotive User Conference, Dearborn, MI, USA*: Citeseer)
[28] Abdullah N A Z, Sani M S M, Husain N A, Rahman M M and Zaman I 2017 *International Journal of Automotive and Mechanical Engineering* **14** 3887-97

IOP Conf. Series: Materials Science and Engineering **257** (2017) 012060 doi:10.1088/1757-899X/257/1/012060

Performance of hybrid nano-micro reinforced mg metal matrix composites brake calliper: simulation approach

N Fatchurrohman* and S T Chia

Faculty of Manufacturing Engineering, Universiti Malaysia Pahang, Kuala Pahang, 26600 Pekan, Pahang, Malaysia

Corresponding author email: fatchurrohman@ump.edu.my

Abstract. Most commercial vehicles use brake calliper made of grey cast iron (GCI) which possesses heavy weight. This contributes to the total weight of the vehicle which can lead to higher fuel consumption. Another major problem is GCI calliper tends to deflect during clamping action, known as "bending of bridge". This will result in extended pedal travel. Magnesium metal matrix composites (Mg-MMC) has a potential application in the automotive industry since it having a lower density, higher strength and very good modulus of elasticity as compared to GCI. This paper proposed initial development of hybrid Mg-MMC brake calliper. This was achieved by analyzing the performance of hybrid nano-micro reinforced Mg-MMC and comparing with the conventional GCI brake calliper. It was performed using simulation in ANSYS, a finite element analysis (FEA) software. The results show that hybrid Mg-MMC has better performance in terms of reduction the weight of the brake calliper, reduction in total deformation/deflection and better ability to withstand equivalent elastic strain.

1. Introduction

Brake calliper, acts as a U-shaped housing that wraps around the brake rotor and is mostly made of grey cast iron (GCI). The current problem of existing GCI brake calliper is it possesses heavy weight which contributes to higher fuel consumption as has been highlighted by [1]. Brake calliper (floating) has one or more pistons and only located on the inboard side of the calliper. Calliper body is mounted on a pin to give cylindrical support and allow moving linearly. During braking, hydraulic pressure forces the piston inwards, pushing the pads against the revolving brake disc. The calliper creates a reaction force cause the calliper slide over the pin leading to clamp action on rotor. Figure 1 presents an illustration of a floating brake calliper.

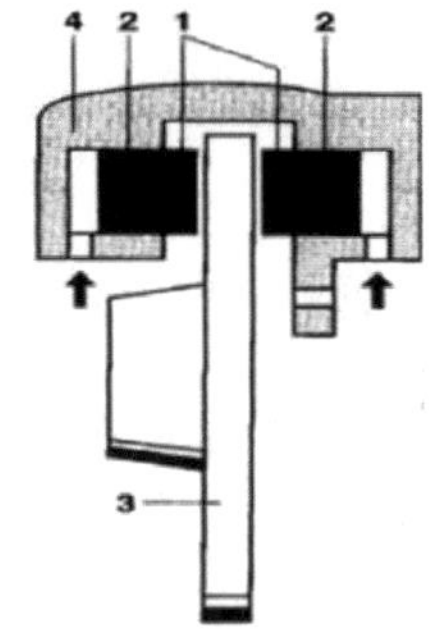

1 Brake pads, 2 Piston, 3 Brake disc, 4 Floating calliper, 5 Support

Figure 1. An illustration of a floating brake calliper [2].

The clamping force results in frictional force and generates heat may be transferred to the calliper body through the brake pads. Single piston is preferable due to light weight, less leakage points and also performs least uneven pad wear as compared to more pistons [3]. High structural stiffness could prevent uneven wear of brake pads and large deformations of brake calliper results in short pedal travel, improving comfort, driving feelings and safety [4]. Another problem of GCI brake calliper is deflection due to the hydraulic pressure acting on the piston and calliper housing, which is also known as "bending of bridge". This problem will result in extended pedal travel since additional fluid volume is required to compensate for deflections. Therefore, minimizing the mass and maximizing the stiffness should be primary design considerations.

Kumar et al. [5] have highlighted that the usage of magnesium (Mg) alloys has considerably increased in automotive sector. They have mentioned that the potential of Mg alloy as substitution to aluminium alloys and iron alloys. Mg-Al-Zn alloys are widely used in many applications due to its both high strength and ductility. The addition of nano particles reinforced metal matrix composites are being studied in recent years, due to their promising properties are suitable for the used in functional and structural applications [6]. According to Macke et al. [7], advanced nano and micro reinforced metal matrix composites can effectively reduce mass, improve reliability and efficiency.

Hybrid nano-micro reinforced Mg-MMC has a potential application in the automotive industry due to their significant improved properties including high specific strength and stiffness, temperature resistance, low thermal expansion coefficient and light weight. This appear to offer more advantages over GCI in brake calliper application. Nguyen et al. [8] have studied the microstructure and mechanical behaviour of Mg alloy AZ31 hybridized with nano sized alumina to improve the ductility and micro-sized copper particulates to enhance the microstructural characteristics, hardness and strength of AZ31 alloy which synthesized through the technique of disintegrated melt deposition. The primary goal of this paper is to perform simulation and to compare hybrid nano-micro reinforced Mg-MMC material with conventional GCI for brake calliper application.

2. Methodology

To perform the analysis finite element analysis (FEA) is applied using ANSYS simulation software. Three stages are involved (Figure 2); pre-processing stage include which type of analysis, material properties, select fine element size of mesh analysis, loads and boundary condition are defined. The processing stage where the desired result is computed and solved; and the results are interpreted during post-processing stage. The material properties used is presented in table 1, both for GCI and hybrid nano-micro Mg-MMC (AZ31-14.0vol% SiCmicro-1.0vol% SiCnano) respectively.

Table 1. Material properties of GCI and hybrid nano-micro Mg-MMC.

Properties	GCI	Hybrid nano-micro Mg-MMC
Density (kg/m^3)	7200	1995
Young's modulus, (GPa)	100	103
Poisson's ratio	0.28	0.27
Tensile yield strength, (MPa)	276	300
Tensile ultimate strength, (MPa)	250	380

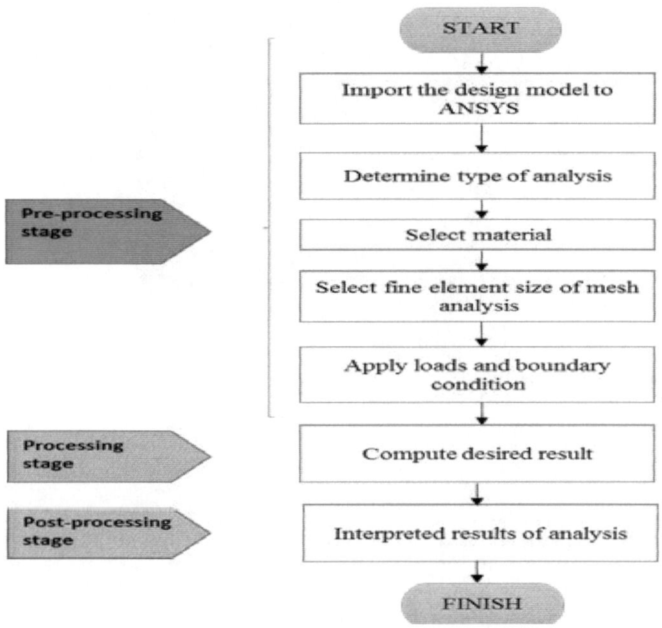

Figure 2. Flowchart for FEA analysis.

To improve the mesh quality, additional control be added to the default mesh before solving. Mesh parameters such as element quality, skewness and orthogonal quality were considered. An average element mesh quality of 0.8 and above is considered acceptable. In order to achieve high mesh quality, different meshing techniques were used. There are three advanced size functions can be employed: proximity, curvature and fixed. Both proximity and curvature were turned on for this model in order to have a much better mesh along the curve regions and varying cross sections. "Patch conforming" mesher under "tetrahedrons" were the most suitable as they capture the curvatures more accurately as compared to other method such as multizone, hexagonal dominant and sweep. Standard skewness and orthogonal quality mesh metrics spectrums were shown in Figure 3.

The brake calliper model was meshed consisting of 457326 nodes and 302619 elements as shown in Figure 4. Through the suitable meshing method, mesh parameter like average element quality achieved 0.82, average skewness quality has reached 0.26, and average orthogonal quality have achieved 0.85 shown in Figure 5-7. These elements confirm the high meshing quality used in this work.

Skewness mesh metrics spectrum

Excellent	Very good	Good	Acceptable	Bad	Unacceptable
0-0.25	0.25-0.50	0.50-0.80	0.80-0.94	0.95-0.97	0.98-1.00

Orthogonal Quality mesh metrics spectrum

Unacceptable	Bad	Acceptable	Good	Very good	Excellent
0-0.001	0.001-0.14	0.15-0.20	0.20-0.69	0.70-0.95	0.95-1.00

Figure 3. Skewness and orthogonal quality mesh metrics spectrums.

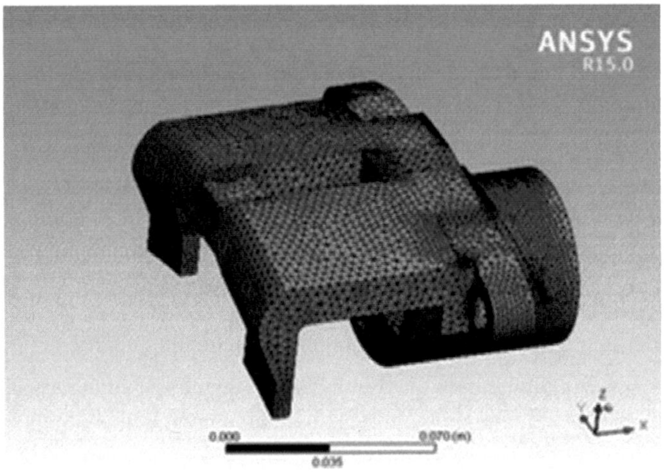

Figure 4. Refined mesh original brake calliper model.

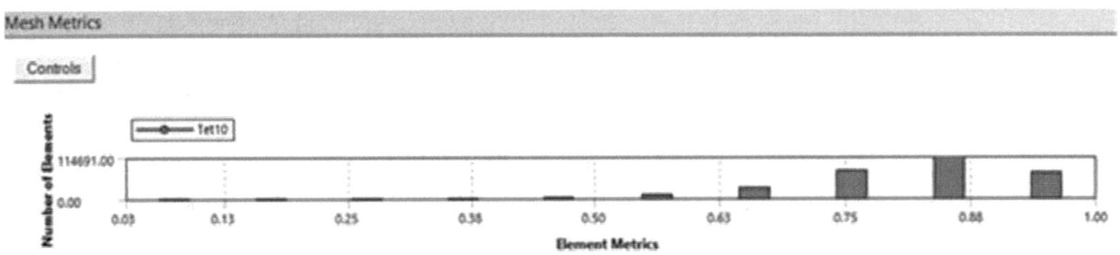

Figure 5. Mesh average element quality.

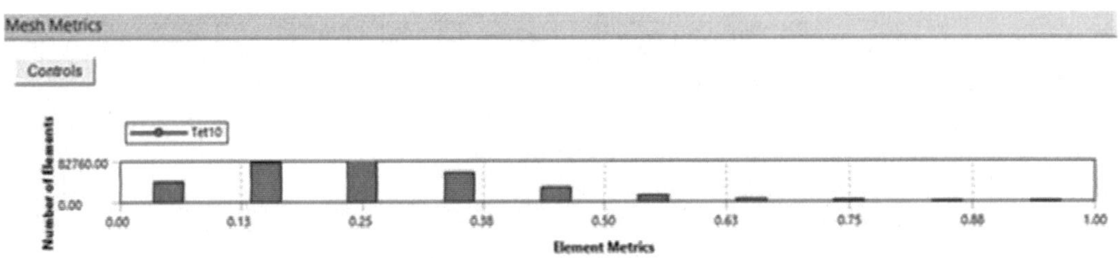

Figure 6. Mesh average skewness quality.

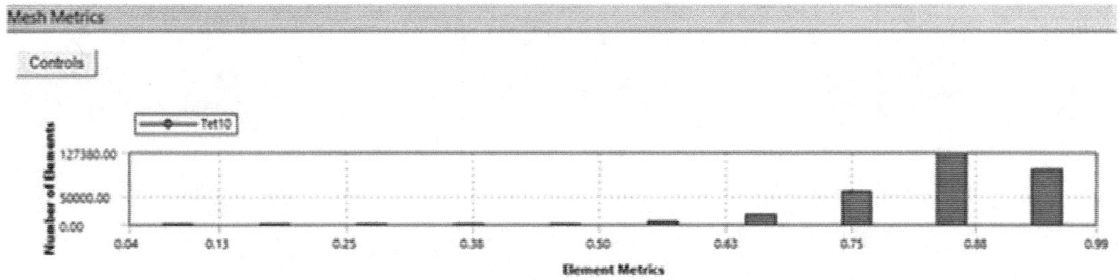

Figure 7. Mesh average orthogonal quality.

Loads and boundary condition were set according to the brake pipeline pressure (fluid pressure). The load was set at the extreme condition, which is the panic braking situation. In this condition the brake fluid pressure was set at 7 MPa [9]. This pressure was transferred onto the calliper housing. Then it was converted to force at the inner face of the cylinder and reaction force on finger area [4]. Both loads have the same values because they are equal and opposite force caused due to fluid pressure.

3. Results and discussion

This section will discuss the results of the present FEA analysis. The results for GCI calliper are presented in Figure 8, while for hybrid Mg-MMC are shown in Figure 9. It is revealed that from the analysis brake calliper made of hybrid Mg-MMC have smaller total deformation distribution and maximum equivalent elastic strains compared to GCI. The values were reduced by 2.76% and 2.93% respectively. This can be achieved because hybrid Mg-MMC was stiffer than GCI since it possesses larger Young's modulus than GCI. Through the study of literature review, deflection of brake callipers, also known as "bending of bridge" was highlighted as the priority problem that affected the performance of brake calliper. This problem can be improved by bridge design features, as Sergent et al. [4] stated that bridge design features are the most important in maintaining structural stiffness. The FEA analysis results are summarised in table 2. The table indicates that hybrid Mg-MMC has the desired results having low maximum equivalent von misses stress, small total deformation and low equivalent elastic strain. From the present model and materials properties comparisons, hybrid Mg-MMC brake calliper had reduced the weight almost 72.29% to the GCI calliper.

The comparison of maximum equivalent von misses stress on the stress distribution of hybrid Mg-MMC brake calliper exceed the GCI calliper by 0.89%. This is the area where improvement can be made. In fact, larger von misses stress implies that the material is close to the yield point. When the stress exceeded the yield point, the material can be suffered some level of permanent distortion which mean it does not return to its original shape. Von misses stress means the stresses which act to distort the shape of the part. Engineers will typically try to design such that the peak stresses as low as possible to reduce the distortion of material [10]. Related to this some works have been done by one of the author to modify the original hybrid Mg-MMC brake calliper with few alternative designs. This is targeted to improve the performance of hybrid Mg-MMC brake calliper.

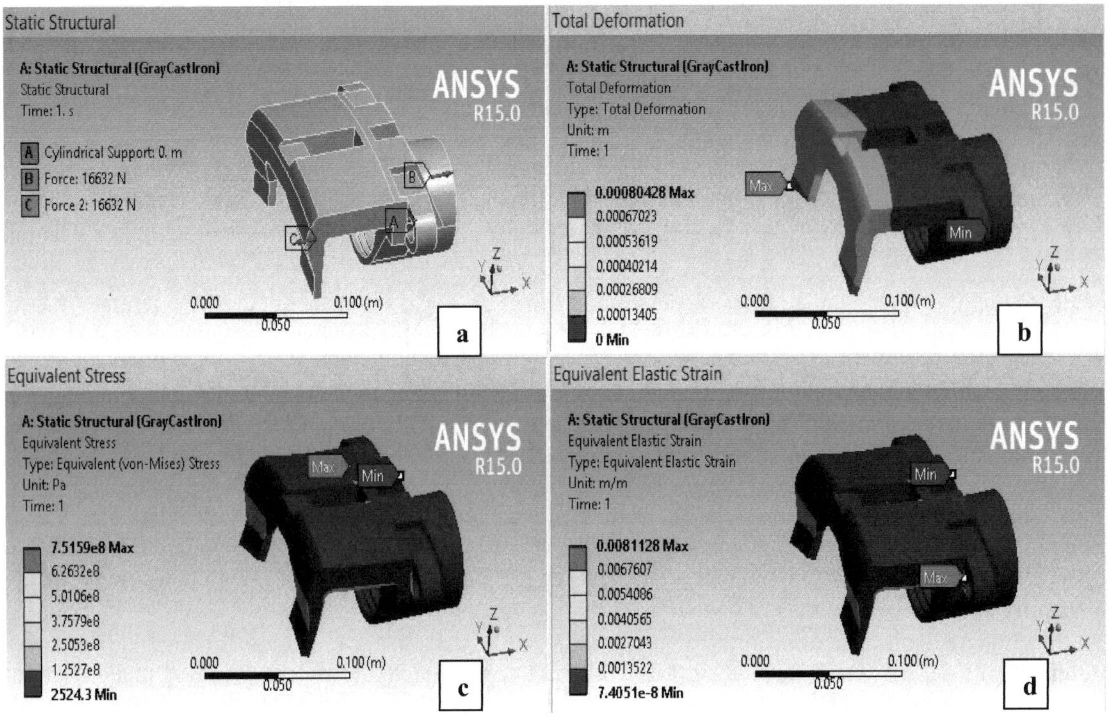

Figure 8. ANSYS analysis results GCI calliper (a) structural loadings (b) total deformation (c) equivalent stress (d) equivalent stress.

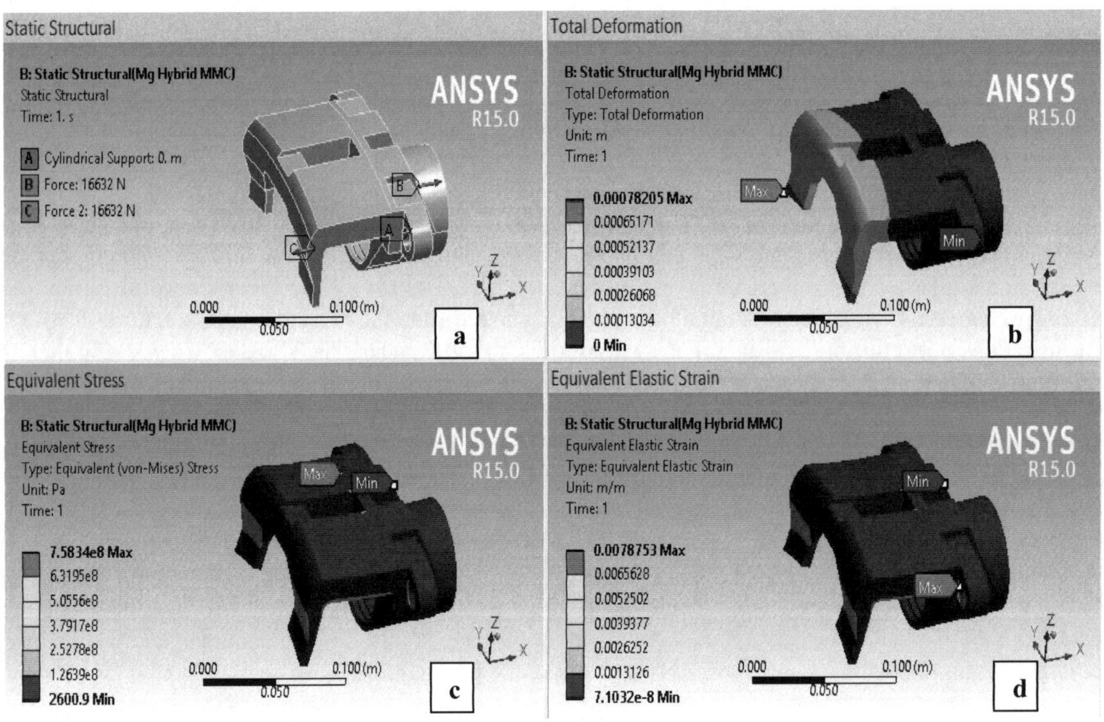

Figure 9. ANSYS analysis results Mg-MMC calliper (a) structural loadings (b) total deformation (c) equivalent stress (d) equivalent stress.

Table 2. Summarised FEA results of GCI and hybrid Mg-MMC brake callipers.

Properties	GCI	Hybrid nano-micro Mg-MMC	Percentage of improvement (%)
Total mass, kg	2.17250	0.60196	72.29
Max. Total Deformation, m	0.00080428	0.00078205	2.76
Max. Equivalent Stress, Pa	7.5159e8	7.5834e8	- 0.89
Max. Equivalent Elastic Strain, m/m	0.0081128	0.0078753	2.93

4. Conclusions

From this paper, it can be concluded that:

- The material of a brake calliper body must be rigid to allow less deflection and should be light to reduce the final weight.
- The proposed hybrid Mg-MMC, AZ31-14.0SiCmicro-1.0SiCnano can be a good alternative to replace GCI.
- The new material has potentially met the present demands for high performance brake calliper based on the extreme panic braking situation.
- From the results, the hybrid Mg-MMC calliper compared to the GCI has reduced the weight of the brake calliper, reduced in total deformation/deflection and better ability to withstand equivalent elastic strain.

Acknowledgements

The authors gratefully acknowledge Universiti Malaysia Pahang for the financial support through Internal Research Grant RDU150371.

References

[1] Pishdad A R 2012 *Advanced design of brake calipers* PhD Thesis (Polytechnic University of Milan Italy)
[2] Wagh N P 2005 *Design and analysis of modular caliper assembly* PhD Thesis (Shivaji University India)
[3] Ballo F, Gobbi M, Mastinu G and Previati G 2015 Race Motorcycle Smart Wheel SAE World Congress and Exhibition Detroit Michigan USA paper no. 15SS-0277
[4] Sergent N, Tirovic M and Voveris J 2014 *Eng. Optim.* **46** 1520-1537
[5] Kumar D S, Sasanka C T, Ravindra K and Suman K N S. 2015 *Am. J. Mater. Sci. Technol.* **4** 12-30
[6] Casati R and Vedani M 2014 *Met.* **4** 65-83
[7] Macke A, Schultz B F and Rohatgi P 2012 *Adv. Mater Processes* **170** 19-23
[8] Nguyen Q, Tun K, Chan J, Kwok R, Kuma J M, Phung T and Gupta M 2012 *Mater. Sci. Technol.* **28** 227-233
[9] Rajaram L S, and Sudharsan S 2005 Optimization of calliper housing using FEM. Symposium on International Automotive Technology 695-700
[10] http://www.learnengineering.org/2012/12/what-is-von-mises-stress.html accessed 10/8/2017

Biocomposites from (*Anadara granosa*) shells waste for bone material applications

S H Saharudin[1], J H Shariffuddin[1,2*] and N I A A Nordin[1]

[1]Faculty of Chemical and Natural Resources Engineering, Universiti Malaysia Pahang, Lebuhraya Tun Razak 26300 Gambang, Kuantan, Pahang, Malaysia.
[2]Centre of Excellence for Advanced Research in Fluid Flow, Universiti Malaysia Pahang, Lebuhraya Tun Razak 26300 Gambang, Kuantan, Pahang, Malaysia.

Corresponding author email: junhaslinda@ump.edu.my

Abstract. The overall objective for this study is to determine the effect of biopolymer types and ratio on producing biocomposites derived from cockle (*Anadara granosa*) shells waste. In this study, two types of biopolymer were used i.e. sodium alginate and carboxymethyl cellulose. These biocomposites are meant for bone material applications. The processes involved in producing the biocomposites were pre-treatment of the cockle shells, formation of $CaCO_3$ in aragonite form and finally the synthesis of biocomposites. All samples have undergone physicochemical and mechanical analyses to determine their crystallinity, purity, functional group, surface morphology, elemental compounds and compressive strength using X-ray diffraction (XRD), Fourier Transform Infrared Spectroscopy (FTIR), Scanning Electron Microscope (SEM), Electron Dispersive X-ray Analysis (EDX) and Universal testing machine (UTM). The $CaCO_3$-biocomposites have been prepared using different ratio of biopolymer and different types of biopolymers. The physicochemical properties of the biocomposites are not affected by the ratio and types of biopolymers. Yet, the ratio and types of biopolymers have influenced the mechanical properties of biocomposites. The biocomposite with lesser amount of biopolymer has denser structure and greater mechanical strength. Comparison between types of polymers revealed that biocomposite with carboxymethyl cellulose has greater mechanical strength compared to biocomposite with sodium alginate.

1. Introduction

Shellfish aquaculture mainly cockle (*Anadara granosa*) is one of the most important industries worldwide and generates abundant of shell wastes. In Malaysia, marine shells are mostly used in handicrafts industry to produce decorative stuffs like key chains. Shell wastes are mostly dumped in landfills without any post-treatment due to high cost of disposal procedure [1,2]. The abandoned shell wastes often cause unpleasant view to the landfills and give out nauseating smell.

In order to minimise the negative impacts on the environment, researchers around the world are studying on the transformation of cockle shell wastes into valuable products. For examples, it has been used as a potential aggregate replacement in concrete [3–5], as catalysts [6,7], drug delivery and bone materials [8,9]. In fact, seashells consist of high composition of calcium carbonate ($CaCO_3$) of more than 95% by weight [10–12]. $CaCO_3$ especially in aragonite polymorph crystal is suitable in bone tissue engineering as it is biocompatible, dense in nature, and can replace bone structure [13–15].

However, application of $CaCO_3$ alone as bone material has limitation due to inadequate in mechanical properties. Therefore, combining $CaCO_3$ with other polymer such as sodium alginate, cellulose and polyvinyl alcohol could improve its mechanical properties while maintaining its biological properties [16–18]. The combination of $CaCO_3$ with natural sources polymer is known as biocomposite.

As to date, a few study on the formation of biocomposite derived from waste materials were done using cockle shell wastes [9,19], *Paphia undulates* shell wastes [20], and fish scale wastes [21]. Hence, this study is investigating the feasibility of producing $CaCO_3$-biocomposite derived from cockle shell wastes that suitable as bone material. This current study is focusing on the effect of biopolymer ratio which is sodium alginate and effect of biopolymer types which are sodium alginate and carboxymethyl cellulose on physicochemical and mechanical properties of $CaCO_3$-biocomposites.

2. Materials and Methods

2.1. Materials

The cockle shells used in this study were obtained from local market in Kuantan, Pahang. Sodium alginate was purchased from R&M (Malaysia) and carboxymethyl cellulose was purchased from Sigma Aldrich (USA). Deionized water (from Millipore Elix 5, USA) was used in the entire studies.

2.2. Pre-treatment of shell and synthesis of $CaCO_3$

The seashells were boiled in water for 15-20 minutes to remove the organic matters that attached on them. Then, the shells were dried in the oven at the temperature of 110 °C for 2 hours. The dried shells were then crushed by using grinder (Retsch ZM200, Germany) into $CaCO_3$ powder and sieved over a set of clean test sieve (Retsch AS200, Germany) with sizes between 63 and 71 μm.

2.3. Synthesis of $CaCO_3$-biocomposites

The biocomposites were synthesised according to the method of [22,23] with slight modifications. Five percent of sodium alginate was dissolved in deionised water and hydrated overnight. The sodium alginate solution was mixed with $CaCO_3$ powder in mass ratios of 1:0.3 and 1:1. The ceramic/sodium alginate mixture was kneaded using flat end spatula into a homogenous paste and subsequently moulded into silicone mould. Moulded green bodies were incubated at 60 °C for 24 h to remove moisture and water content. The experiment was repeated using different biopolymer which is carboxymethyl cellulose. Table 1 shows biocomposites that were produced in this study.

Table 1. Different types of samples produced in this study.

No	Type of Sample	Abbreviation
1	Raw shell powder	$CaCO_3$
2	Calcium carbonate-sodium alginate biocomposite with ratio 1:0.3	CA3
3	Calcium carbonate-sodium alginate biocomposite with ratio 1:1	CA10
4	Calcium carbonate-carboxymethyl cellulose biocomposite with ratio 1:0.3	CC3

2.4. Physicochemical and mechanical characterization of $CaCO_3$-biocomposites

The physicochemical analyses of biocomposites were done using X-ray diffraction (XRD), Fourier transform infrared (FTIR) and scanning electron microscope (SEM) along with energy dispersive X-ray (EDX). The XRD analysis was carried out using X-ray powder diffractometer (Rigaku MiniFlex II, Japan) in order to assess the phases, crystallinity and purity of the biocomposites. Scans were carried out in the range of 2θ range from 20°-80° with step size 0.02° at 1s step time. FTIR (Thermo Fisher Nicolet iS5, USA) with diamond attenuated total reflectance (ATR) system over a range of 400 to 4000 cm^{-1} was used to identify functional groups in the biocomposites. The morphology and elemental analysis of the biocomposites were observed using SEM along with EDX (Hitachi TM3030

Plus, Japan). The samples were sputter-coated with platinum to prevent charging during analysis process. The compressive strength was determined by crushing a biocomposite between two flat platens using a computer controlled universal testing machine with a ramp rate of 5 mm/min. Three identical specimens for each sample group were used in the compressive testing [24,25].

3. Results and discussion

3.1. Effect of $CaCO_3$ to sodium alginate ratio on properties of $CaCO_3$-biocomposite

Figure 1 shows the XRD patterns of sodium alginate powder, $CaCO_3$ powder, CA3 biocomposite and CA10 biocomposite. It is confirmed that CA3 and CA10 possessed aragonite polymorph when compared to PDF Card (No: 00-041-1475). Although both biocomposites contain sodium alginate, the peak that resembles sodium alginate did not show on the XRD patterns. It may due to very limited concentration of sodium alginate and it did not shown in XRD patterns. Furthermore, the XRD might not detect the sodium alginate due to the detection limit of XRD [26] and amorphous nature of sodium alginate [27]. Thus, it did not disturb the crystallinity and purity of $CaCO_3$. This finding is slightly different with [9] and [28] where the sodium alginate concentration in both studies are higher and there were sodium alginate peaks appeared on its XRD patterns.

Figure 2 illustrated the FTIR spectrum of sodium alginate, $CaCO_3$, CA3 biocomposite and CA10 biocomposite. It is observed that the bands of both biocomposites (Figure 2(c) and 2(d)) are identical to $CaCO_3$ band as indicated by carbonate (CO_3^{2-}) group bands appeared at 1790, 1490, 1170, 861, and 716 cm^{-1}. The spectra of both biocomposites did not contain any band that related to sodium alginate. This may due to the concentration of sodium alginate in the biocomposites are too low and the FTIR could not detect its presence. At the meantime, the IR spectrum of sodium alginate in Figure 2(a) shows a wide band at 3,430 cm^{-1}, which related to the characteristic absorption of hydroxyl groups. Whereas, bands that appeared at 1,615 and 1,418 cm^{-1} were attributed to the asymmetric and symmetric stretching vibrations of carboxyl groups (COO–), respectively. A band around 1,029 cm^{-1} (C–O–C stretching) was observed corresponding to its saccharide structure. The similar sodium alginate band is also observed in study that has been done by [29] and [9].

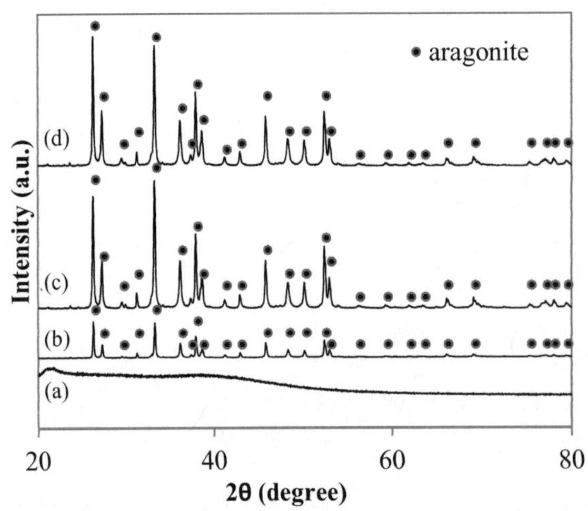

Figure 1. XRD patterns of (a) sodium alginate, (b) $CaCO_3$, (c) CA3 biocomposite and (d) CA10 biocomposite.

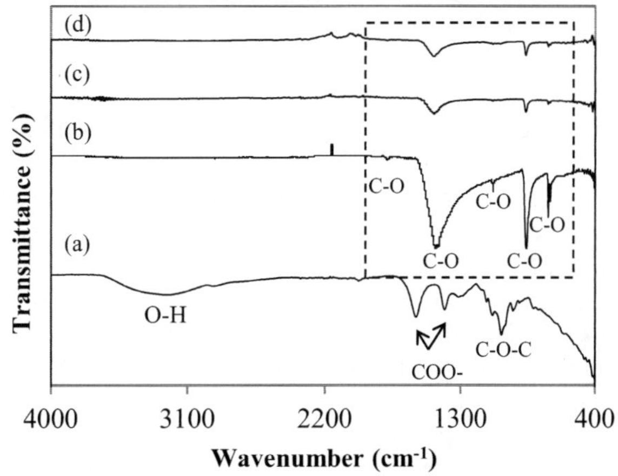

Figure 2. FTIR spectra of (a) sodium alginate, (b) $CaCO_3$, (c) CA3 biocomposite and (d) CA10 biocomposites.

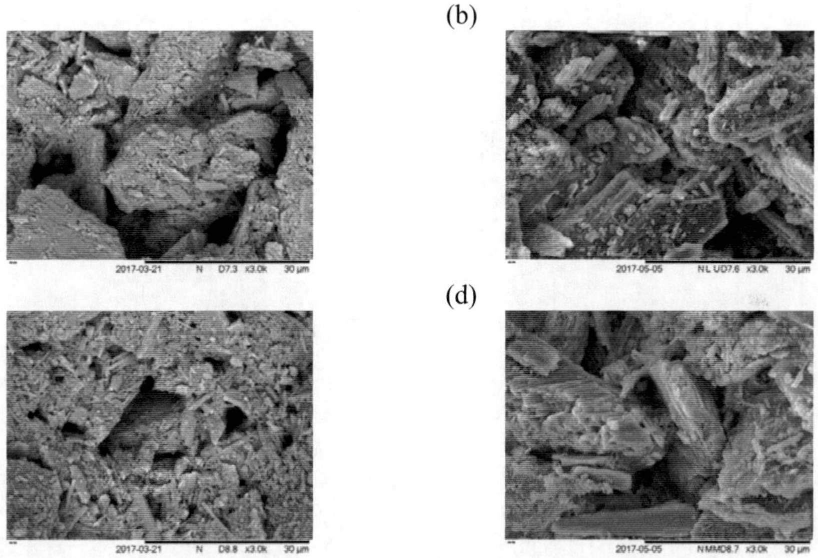

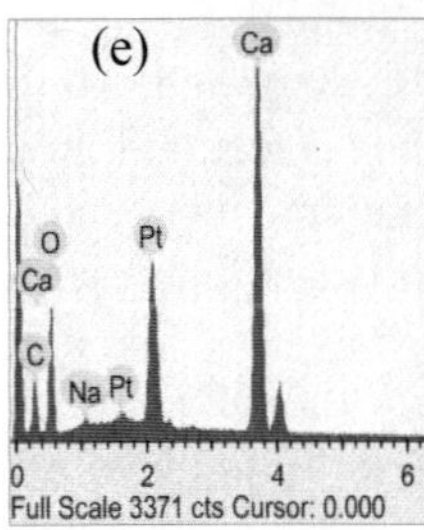

Element	Atomic %
C	22.05 ± 2.1
O	54.12 ± 2.56
Na	0.48 ± 0.1
Ca	20.58 ± 4.7
Pt	2.78 ± 0.2

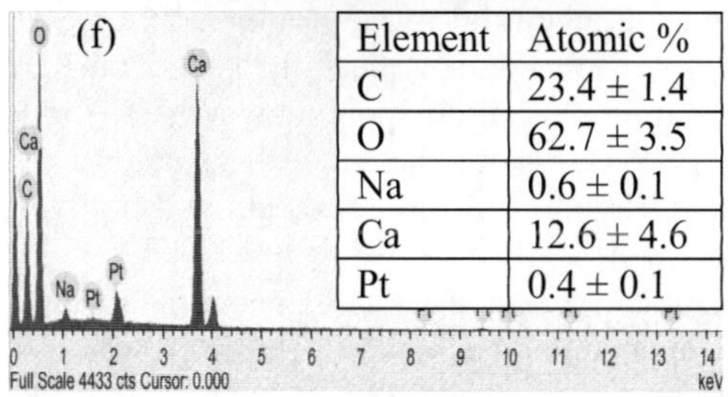

Element	Atomic %
C	23.4 ± 1.4
O	62.7 ± 3.5
Na	0.6 ± 0.1
Ca	12.6 ± 4.6
Pt	0.4 ± 0.1

Figure 3. SEM micrographs and EDX analyses of CA3 (a - surface, c – cross section and e - EDX analysis) and CA10 (b - surface, d - cross section and f - EDX analysis) at 3000× magnification.

Figure 3 displays the SEM observation and EDX analysis of CA3 and CA10. It can be seen that the $CaCO_3$ in both biocomposites were agglomerated. This may due to the binding capability of sodium alginate [30,31] thus stacking the powders together. Although the powder particles have been clumped, it still can be observed the lamellar structure of $CaCO_3$. This finding is correlated with [32] and [33] where $CaCO_3$ possessed lamellar structure.

Meanwhile, EDX analyses of both biocomposites show sodium (Na) element peak existed. The Na element proves that sodium alginate is presence in the biocomposites. This similar results was reported by [34]. The difference between CA3 and CA10 is the percentage of Na in the biocomposite. CA10 (Figure 3(f)) has higher percentage of Na compared to CA3 (Figure 3(e)) due to higher concentration of sodium alginate in CA10. From Figure 3, it can be observed that CA3 biocomposite has denser structure compared to CA10. This finding is proven by [22] where the density of biocomposite is increased when concentration of biopolymer decreased.

The compressive strength of CA3 and CA10 biocomposites were shown in Figure 4. The CA3 biocomposite which has lesser sodium alginate ratio exhibit higher compressive strength (243.3 MPa) compared to CA10 biocomposite (93.3 MPa). Increase in the concentration of biopolymer which in this study is sodium alginate has decreased the density of the biocomposite. This finding was supported by [22] and [25] where the compressive strength is inversely proportional with the amount of polymer. This result is correlated with SEM results in Figure 3 where the higher concentrations of sodium alginate in biocomposite make its morphology less dense. Kokubo et al. [35] reported, the compressive strength of human cortical bone is around 100-230 MPa. Thus, only CA3 biocomposite has passed the bone material properties.

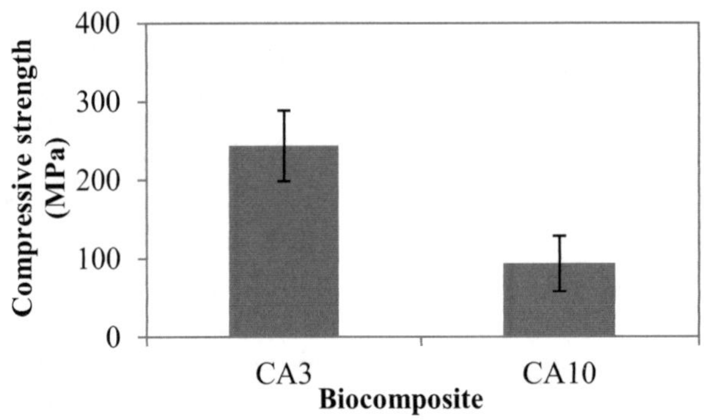

Figure 4. Compressive strength of (a) CA3 biocomposite and (b) CA10 biocomposite.

3.2. Effect of polymer types on properties of $CaCO_3$-biocomposites

Figure 5 shows the XRD patterns of carboxymethyl cellulose, sodium alginate, $CaCO_3$, CA3 biocomposite and CC3 biocomposite. The biocomposites, CA3 and CC3 both displayed aragonite crystal peaks by comparing with $CaCO_3$ pattern (Figure 5(c)). The CA3 biocomposite contained sodium alginate while CC3 contained carboxymethyl cellulose. However, there is no peak which resembles both biopolymers appeared on CA3 and CC3 patterns. This may due to the low concentration of both biopolymers in the biocomposites. Furthermore, XRD has detection limit around 5-10% [26]. Besides that, the XRD pattern of carboxymethyl cellulose (Figure 5(a)) and sodium alginate (Figure 5(b)) are broad and did not have sharp peak. This indicates that both are amorphous biopolymer [36].

Figure 6 shows the FTIR spectrum of carboxymethyl cellulose, sodium alginate, $CaCO_3$, CA3 biocomposite and CC3 biocomposite. It was observed that the bands appeared in both biocomposites spectrum are identical to $CaCO_3$ band as indicated by the CO_3^{2-} bond at 1790, 1480, 1130, 854 and 714 cm^{-1}. By comparing both biocomposites with carboxymethyl cellulose (Figure 6(a)) and sodium alginate (Figure 6(b)) spectrum, it can be seen that there is no band in the biocomposites (Figure 6(d) and 6(e)) that identical with the biopolymers. This may due to the detection limit of FTIR that could not detect the presence of carboxymethyl cellulose and sodium alginate in very low concentrations

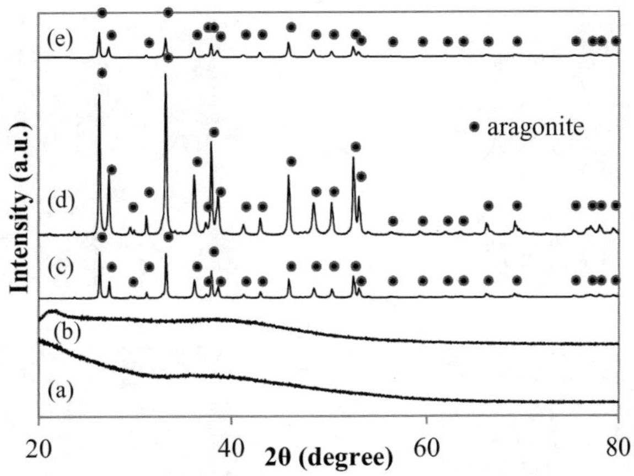

Figure 5. XRD patterns of (a) carboxymethyl cellulose, (b) sodium alginate, (c) $CaCO_3$, (d) CA3 biocomposite, (e) CC3 biocomposite.

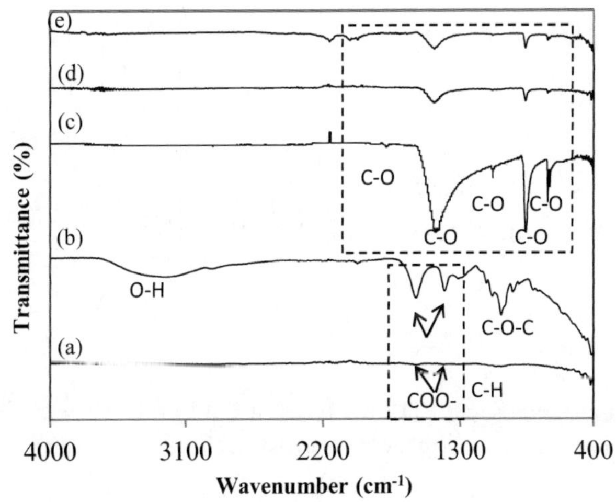

Figure 6. FTIR spectrum of (a) carboxymethyl cellulose, (b) sodium alginate, (c) $CaCO_3$, (d) CA3 biocomposite, (e) CC3 biocomposite.

Figure 7 shows the SEM observation and EDX analysis of CA3 and CC3 at 3000× magnification. The morphologies of CA3 and CC3 biocomposites are almost similar in term of density. This may due to the similar concentration of biopolymer used in both biocomposites. Meanwhile, EDX analyses show Na element peaks exist in both biocomposites. This Na element confirms that sodium alginate and carboxymethyl cellulose are presence in the CA3 and CC3, respectively. A study done by [34] stated that Na is representing sodium alginate in biocomposite as Na is one of the elements in sodium alginate. Meanwhile, [37] reported that Na in his study representing carboxymethyl cellulose as Na is also one of the elements in carboxymethyl cellulose. Both biopolymers are identified with Na element due to their chemical formula. Sodium alginate has chemical formula $C_6H_9NaO_7$ while carboxymethyl cellulose has chemical formula $C_8H_{15}NaO_8$. The atomic percentage of Na is similar due to same concentration of sodium alginate and carboxymethyl cellulose presence in both bio composites.

IOP Conf. Series: Materials Science and Engineering **257** (2017) 012061 doi:10.1088/1757-899X/257/1/012061

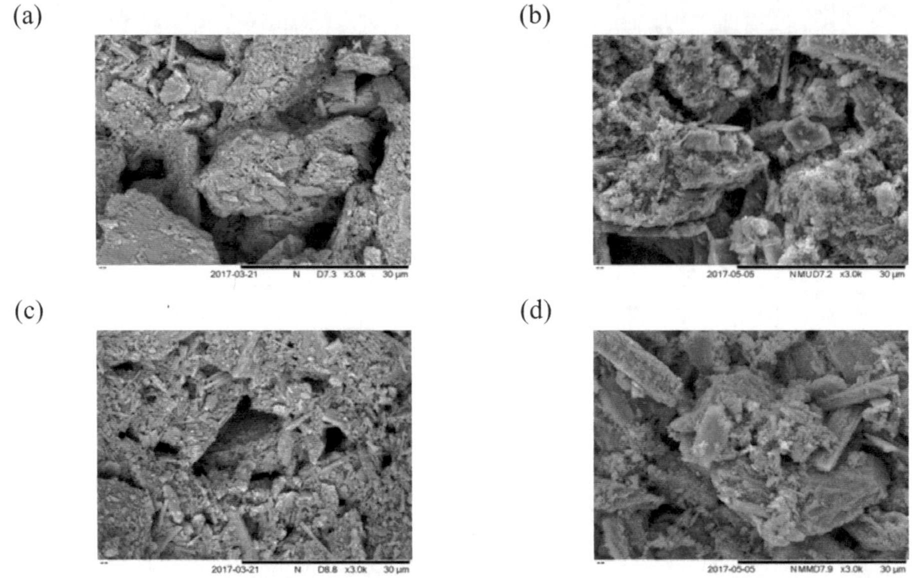

(e)

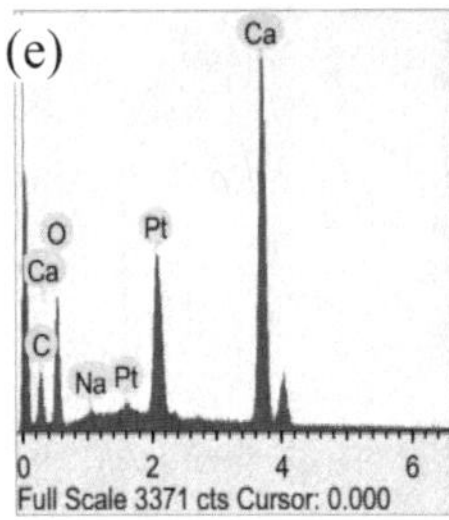

Element	Atomic %
C	22.05 ± 2.1
O	54.12 ± 2.56
Na	0.48 ± 0.1
Ca	20.58 ± 4.7
Pt	2.78 ± 0.2

(f)

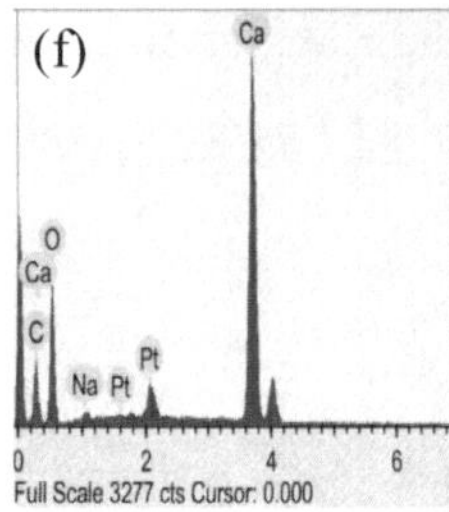

Element	Atomic %
C	21.62 ± 2.1
O	62.4 ± 1.2
Na	0.47 ± 0.1
Ca	15.1 ± 2.8
Pt	0.4 ± 0.1

Figure 7. SEM micrographs and EDX analyses of CA3 (a - surface, c – cross section and e - EDX analysis) and CC3 (b - surface, d - cross section and f - EDX analysis) at 3000× magnification.

Figure 8 shows the comparison of compressive strength between CA3 and CC3. Both biocomposites consist of similar concentration of biopolymers. CA3 contained sodium alginate while CC3 contained carboxymethyl cellulose. However, the result shows CC3 exhibited higher compressive strength which is 343.3 MPa while CA3 possessed 243.3 MPa. According to [38], carboxymethyl cellulose acts as a skeleton in biocomposite because of the bridging effect of the molecules which contributes to improvement of the compressive strength of the biocomposite. Thus, biocomposite containing carboxymethyl cellulose has higher compressive strength compared to biocomposite with sodium alginate.

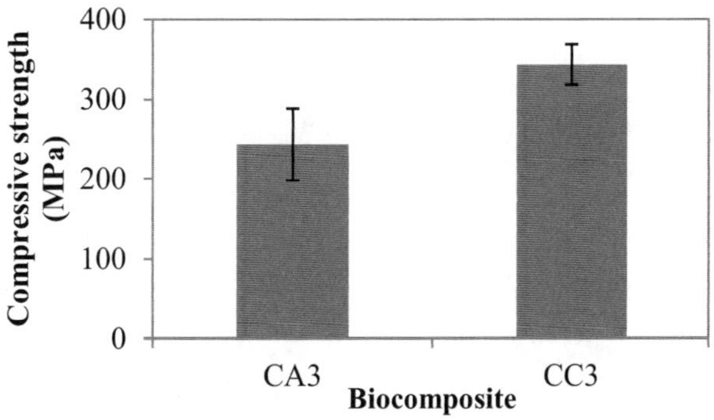

Figure 8. Compressive strength of (a) CA3 biocomposite and (b) CC3 biocomposite

4. Conclusions

$CaCO_3$-biocomposites have been synthesised using cockle shell wastes. Physicochemical analyses (XRD, FTIR, SEM/EDX) have proven that $CaCO_3$-biocomposites contained aragonite crystal. The $CaCO_3$-sodium alginate biocomposites show that the sodium alginate ratio in the system has affected the surface and cross section morphology and mechanical characteristics. Lower the amount of sodium alginate made the structure of biocomposite denser thus increased the compressive strength. Therefore, CA3 is better compared to CA10 in term of mechanical characteristic. Comparison between $CaCO_3$-sodium alginate (CA3) and $CaCO_3$-carboxymethyl cellulose (CC3) revealed that different types of biopolymers give out almost similar morphology but different in mechanical characteristics. The CC3 has greater mechanical strength compared to CA3.

Acknowledgement

The authors acknowledge Universiti Malaysia Pahang for the financial support through grant (RDU150394) and the Doctorate Research Scheme (DRS) for student Siti Hajar Saharudin.

References

[1] Hangun-Balkir Y 2016 Green biodiesel synthesis using waste shells as sustainable catalysts with *Camelina sativa* Oil *J. Chem.* **2016** 1–10

[2] Hayes M, Carney B, Slater J and Bruck W 2008 Mining marine shellfish wastes for bioactive molecules: Chitin and chitosan - Part A: extraction methods *Biotechnol. J.* **3** 871–7

[3] Muthusamy K and Sabri NA 2012 Cockle Shell: A potential partial coarse aggregate replacement in concrete *Int. J. Sci. Environ. Technol.* **1(4)** 260–7

[4] Motamedi S, Shamshirband S, Hashim R, Petkovic D and Roy C 2015 Estimating unconfined compressive strength of cockle shell – cement – sand mixtures using soft computing methodologies *Eng. Struct.* **98** 49–58

[5] Othman NH, Badorul Hisham, Abu Bakar Mashitah MD and Johari MAM 2013 Cockle shell ash replacement for cement and filler in concrete *Malaysian J. Civ. Eng.* **25** 201–11

[6] Boey PL, Maniam GP, Hamid SA and Ali DMH 2011 Utilization of waste cockle shell (*Anadara granosa*) in biodiesel production from palm olein: Optimization using response surface methodology *Fuel* **90** 2353–8

[7] Boey PL, Maniam GP, Hamid SA and Ali DMH 2011 Crab and cockle shells as catalysts for the preparation of methyl esters from low free fatty acid chicken fat *J. Am. Oil. Chem. Soc.* **88** 283–8

[8] Saidykhan L, Bakar MZBA, Rukayadi Y, Kura AU and Latifah SY 2016 Development of nanoantibiotic delivery system using cockle shell-derived aragonite nanoparticles for treatment of osteomyelitis *Int. J. Nanomedicine* **11** 661–73

[9] Bharatham BH, Abu Bakar MZ, Perimal EK, Yusof LM and Hamid M 2014 Development and characterization of novel porous 3D alginate-cockle shell powder nanobiocomposite bone scaffold *Biomed. Res. Int.* **2014** 1–11

[10] Mohamed M, Yusup S and Maitra S 2012 Decomposition study of calcium carbonate in cockle shell *J. Eng. Sci. Technol.* **7** 1–10

[11] Barros MC, Bello PM, Bao M and Torrado JJ 2009 From waste to commodity: transforming shells into high purity calcium carbonate *J. Clean. Prod.* **17** 400–7

[12] Nakatani N, Takamori H, Takeda K and Sakugawa H 2009 Transesterification of soybean oil using combusted oyster shell waste as a catalyst *Bioresour. Technol.* **100** 1510–3

[13] Islam KN, Ali ME, Bakar MZBA, Loqman MY, Islam A, Islam MS, et al. 2013 A novel catalytic method for the synthesis of spherical aragonite nanoparticles from cockle shells *Powder Technol.* **246** 434–40

[14] Islam KN, Bakar MZBA, Ali ME, Hussein MZ Bin, Noordin MM, Loqman MY, et al. 2013 A novel method for the synthesis of calcium carbonate (aragonite) nanoparticles from cockle shells *Powder Technol.* **235** 70–5

[15] Natali I, Tempesti P, Carretti E, Potenza M, Sansoni S, Baglioni P, et al. 2014 Aragonite crystals grown on bones by reaction of CO2 with nanostructured Ca(OH)2 in the presence of collagen. Implications in archaeology and paleontology *Langmuir* **30** 660–8

[16] Chen PY and Lai JT 2013 Mechanical analysis of biocomposite materials from bacterial cellulose and hydroxyapatite *J. Med. Bioeng.* **2(4)** 228–31

[17] Kamalaldin NA, Yahya BH and Nurazreena A 2016 Cell evaluation on alginate/hydroxyapatite block for biomedical application. *Procedia Chem.* **19** 297–303

[18] Mollazadeh S, Javadpour J and Khavandi A 2007 Biomimetic synthesis and mechanical properties of hydroxyapatite/poly (vinyl alcohol) nanocomposites *Adv. Appl. Ceram.* **106** 165–70

[19] Mahmood S, Zakaria M, Razak I, Yusof L, Jaji A, Isa T, et al. 2017 Preparation and characterization of cockle shell aragonite nanocomposite porous 3D scaffolds for bone repair *Biochem. Biophys. Reports* **10** 237–51

[20] Siriprom W, Witit-Anun N, Choeysuppaket A and Ratana T 2016 Characterization of cellulose/calcium carbonate biocomposite film *Key Eng. Mater.* **675-676** 209–12

[21] Belouafa S, Bennamara A and Abourriche A 2017 Low-cost processing technology for the synthesis of a biocomposite for biomedical applications : A preliminary study *J. Mater. Environ. Sci.* **8** 825–30

[22] McNamara SL, Rnjak-Kovacina J, Schmidt DF, Lo TJ and Kaplan DL 2014 Silk as a biocohesive sacrificial binder in the fabrication of hydroxyapatite load bearing scaffolds *Biomaterials* **35** 6941–53

[23] Neira IS, Kolen'ko YV, Lebedev OI, Van Tendeloo G, Gupta HS, Matsushita N, et al. 2009 Rational synthesis of a nanocrystalline calcium phosphate cement exhibiting rapid conversion to hydroxyapatite *Mater. Sci. Eng. C* **29** 2124–32

[24] Uswatta SP, Okeke IU and Jayasuriya AC 2016 Injectable porous nanohydroxyapatite/chitosan/tripolyphosphate scaffolds with improved compressive strength for bone regeneration *Mater. Sci. Eng. C* **69** 505–12

[25] Roohani-Esfahani SI, Nouri-Khorasani S, Lu ZF, Appleyard RC and Zreiqat H 2011 Effects of bioactive glass nanoparticles on the mechanical and biological behavior of composite coated scaffolds *Acta Biomater.* **7** 1307–18

[26] Cullity BD 1956 *Elements of X-ray Diffraction* Addison-Wesley Publishing Company (Addison-Wesley metallurgy series)

[27] Xiong G, Luo H, Zuo G, Ren K and Wan Y 2015 Novel porous graphene oxide and hydroxyapatite nanosheets-reinforced sodium alginate hybrid nanocomposites for medical applications *Mater. Charact.* **107** 419–25

[28] Ionita M, Pandele MA and Iovu H 2013 Sodium alginate/graphene oxide composite films with enhanced thermal and mechanical properties *Carbohydr. Polym.* **94** 339–44

[29] Wang QS, Zhu XN, Jiang HL, Wang GF and Cui YL 2015 Protective effects of alginate-chitosan microspheres loaded with alkaloids from *Coptis chinensis* Franch. and *Evodia rutaecarpa* (Juss.) Benth. (Zuojin Pill) against ethanol-induced acute gastric mucosal injury in rats *Drug Des. Devel. Ther.* **9** 6151–65

[30] Li J, Zhao Y, Wang N, Ding Y and Guan L 2012 Enhanced performance of a MnO2–graphene sheet cathode for lithium ion batteries using sodium alginate as a binder *J. Mater. Chem.* **22** 13002–4

[31] Bigoni F, De Giorgio F, Soavi F and Arbizzani C 2017 Sodium alginate: A water processable binder in high-voltage cathode formulations *J. Electrochem. Soc.* **164** A6171–7

[32] Arma LH, Saitoh A, Ishibashi Y, Asahi T, Sueoka Y, Sakakibara M, et al. 2014 Red fluorescence lamellae in calcitic prismatic layer of *Pinctada vulgaris* shell (Mollusc, bivalvia) *Opt. Mater. Express.* **4(9)** 1813–23

[33] Almagro Í, Drzymala P, Rodríguez-Navarro AB, Sainz-Díaz CI, Willinger MG, Bonarski J, et al. 2016 Crystallography and textural aspects of crossed lamellar layers in Arcidae (bivalvia, mollusca) shells *Key Eng. Mater.* **672** 60–70

[34] Daemi H, Barikani M and Barmar M 2013 Novel Compatible Compositions Based on aqueous anionic polyurethane-urea dispersions and Sodium Alginate *Carbohydr. Polym.* **92** 490–6

[35] Kokubo T, Kim HM and Kawashita M 2003 Novel bioactive materials with different mechanical properties *Biomaterials* **24** 2161–75

[36] Gaihre B and Jayasuriya AC 2016 Fabrication and characterization of carboxymethyl cellulose novel microparticles for bone tissue engineering *Mater. Sci. Eng. C* **69** 733–43

[37] Wang J, Lin X, Luo X and Long Y 2014 A sorbent of carboxymethyl cellulose loaded with zirconium for the removal of fluoride from aqueous solution *Chem. Eng. J.* **252** 415–22

[38] Liu H, Zhao Y, Peng C, Song S and López–Valdivieso A 2016 Improvement of compressive strength of lime mortar with carboxymethyl cellulose *J. Mater. Sci.* **51** 9279–86

CFD analysis of temperature imbalance in superheater/reheater region of tangentially coal-fired boiler

A F Zainudin*, H Hasini and S S A Fadhil

Fluid Dynamics & Risk Reduction Research Group, Institute of Sustainable Engineering, Universiti Tenaga Nasional, Jalan IKRAM-UNITEN, 43000 Kajang, Selangor, Malaysia

*Corresponding author: abdullahfarhan2@gmail.com

Abstract. This paper presents a CFD analysis of the flow, velocity and temperature distribution in a 700 MW tangentially coal-fired boiler operating in Malaysia. The main objective of the analysis is to gain insights on the occurrences in the boiler so as to understand the inherent steam temperature imbalance problem. The results show that the root cause of the problem comes from the residual swirl in the horizontal pass. The deflection of the residual swirl due to the sudden reduction and expansion of the flow cross-sectional area causes velocity deviation between the left and right side of the boiler. This consequently results in flue gas temperature imbalance which has often caused tube leaks in the superheater/reheater region. Therefore, eliminating the residual swirl or restraining it from being diverted might help to alleviate the problem.

1. Introduction

Despite the current demand for cleaner and environment-friendly energy, fossil fuel remains to be the major source of energy in the world. The U.S. Energy Information Administration (EIA) reported that in 2012, coal continues to be the most widely used source for electricity generation, covering 40% of the world's total electricity generation [1]. This proves the relevance of coal as the main source for electricity generation thus studies are continuously being carried out to find methods to overcome its downsides, especially the environmental-related ones. For example, techniques such as over fire air (OFA), low NO_x burner and flue gas recirculation are designed to abate the NO_x emission problem.

For most coal-fired power plants, tangential firing boiler is used, thanks to its advantages of providing complete combustion and uniform heat distribution. However, there is an inherent problem associated with the tangential firing boiler which is steam temperature imbalance in the superheater/reheater region. Excessively higher steam temperature on one side of the boiler frequently causes superheater or reheater tube leaks which interrupt the boiler operation, affecting plant's reliability.

There have been a number of studies on the temperature imbalance problem, most of which are with the aid of Computational Fluid Dynamics (CFD) technique [2-4]. When modelling the boiler, some literatures include the platen superheater/reheater in the upper part of the boiler. These platen superheaters/reheaters are most commonly modelled as porous media to account for the effect of the plates on the flow and pressure drop [5-7]. There are also researchers who modelled the plates as solid domain [8, 9] and some even model their boilers without the plates [10]. In Malaysia particularly, to the authors' knowledge, there has been no published CFD study on this problem that includes the platen superheaters/reheaters in the boiler model. All the available literatures model 'empty' boiler and disregard the presence of platen superheater/reheater. Each of these available methods of modelling the boiler has its advantages and drawbacks, and the relevancy and validity of its application depend primarily on the objective of the study.

The objective of this study is to analyze the flow, velocity and temperature distribution in one of the coal-fired power plant boilers in Malaysia, in order to understand the temperature imbalance occurrence better. The platen superheaters/reheaters are modelled and treated as porous media. With the knowledge and findings gained from this study, feasible mitigation scheme could be figured out to alleviate or eliminate the temperature imbalance occurrence.

2. Boiler modelling

2.1. Geometry

The boiler studied is of tangential firing type and has generating capacity of 700 MW. The cross section of the furnace has 18.4 width × 19.5 depth dimension, and is 72.58 high. It comprises several sections namely hopper, windbox, furnace, nose, horizontal pass and rear pass, as depicted in Figure 1. In the horizontal pass, there are 2 groups of 6 plates low temperature superheater (LTS), 2 groups of 22 plates high temperature superheater (HTS), and a group of 46 plates high temperature reheater (HTR). In the windbox section, there are 7 elevations of pulverized coal (PC) burners, 13 elevations of secondary air (SA) nozzles, and 4 elevations of fuel oil (FO) burners. At normal operating condition, only 6 PC burners and 12 SA nozzles are operated while the remaining are set on standby. For the purpose of analysis, 7 horizontal planes are created, of which 3 are in the furnace section (F1, F2, F3), 1 at the furnace exit (FE), and 3 in the horizontal pass (HP1, HP2, HP3), while 2 vertical planes are created upstream the second group of HTS (X1) and just before the HTR (X2), respectively. These planes are illustrated in Figure 2.

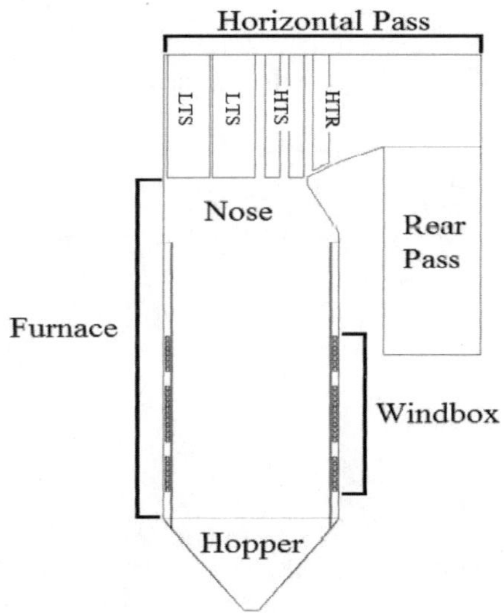

Figure 1. Sections in the boiler.

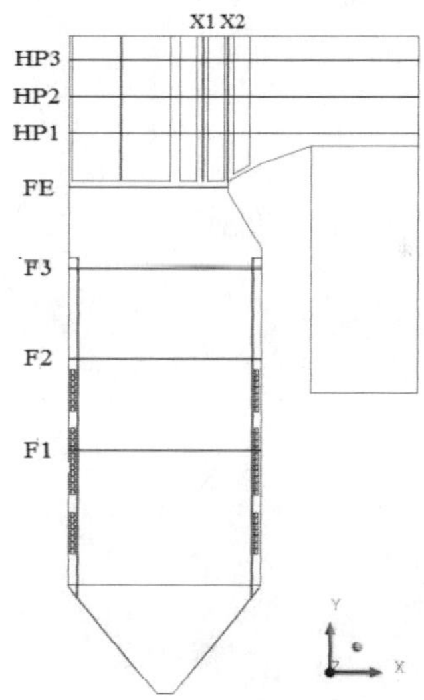

Figure 2. Locations in the boiler used for analysis.

2.2. Mesh

The mesh employed for this study consists primarily of hexahedral meshes, except for the nose and hopper sections which consist of tetrahedral meshes. This is shown in Figure 3. To reduce the effect of pseudo-diffusion, the mesh of the horizontal cross section in the furnace is made to follow the dominant swirling direction of the flow in the furnace, as displayed in Figure 4. The overall quality of the mesh is deemed very good, based on the values of equiangle skewness and orthogonal quality. The average and maximum values of equiangle skewness are 0.08 and 0.82, respectively, while the average and minimum values of orthogonal quality 0.965 and 0.265, respectively. Mesh independence test is performed for three simulations which respective total number of mesh are 1500620, 1923879, and

2100856. Due to complexity of the geometry, the mesh in the furnace section only is refined. Figure 5 shows the plots of temperature along the boiler height for the three simulations. Based on the plots, the solution can be considered mesh-independent and the mesh with 1500620 cells is employed for further analysis to save computational time.

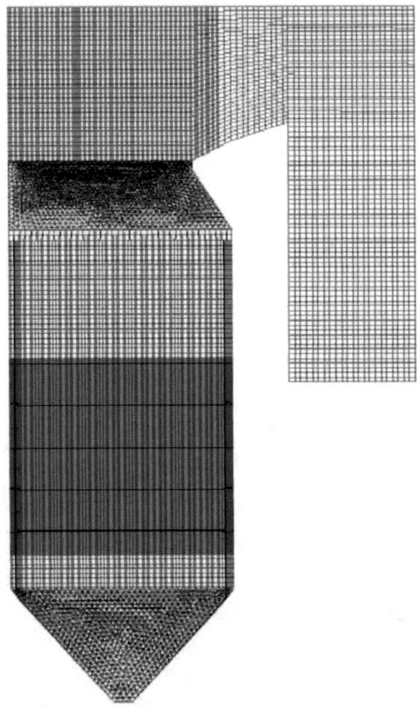

Figure 3. Mesh of the whole domain.

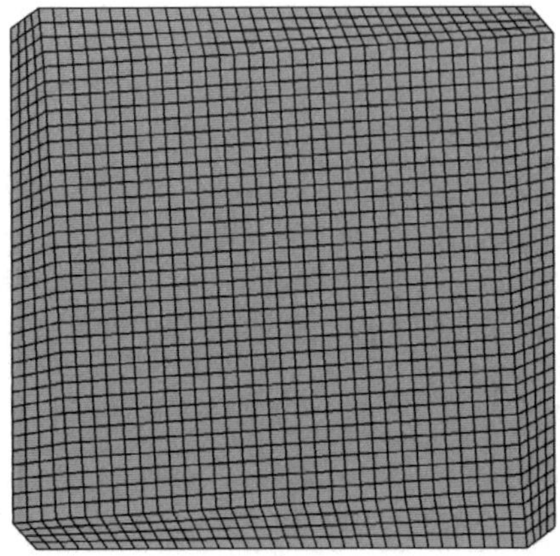

Figure 4. Mesh of horizontal cross section in the furnace.

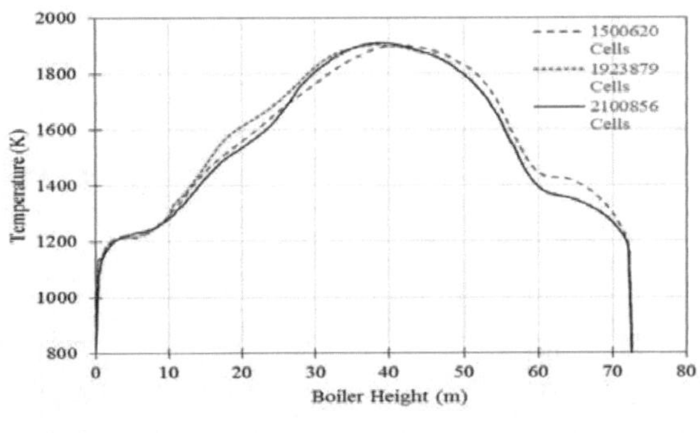

Figure 5. Result of mesh independence test.

2.3. Boundary conditions

The LTS, HTS and HTR in the horizontal pass are treated as porous media to consider the effect of the plates or tube bundles on the flow and pressure drop. The porosity of the plates is set to 0.2 which is determined based on the ratio of total area of gaps in a plate to the total area of the plate. The inertial resistance in x- and y-direction is set to 0.35 while in the z-direction is set to 50 to consider anisotropic influence of the plates on the flow. These values are assumed as in [11] due to the absence of experimental or on-site data. The LTS, HTS and HTR are also treated as heat sinks, to include heat transfer from the flue gas to the plates. Heat transfer to the water walls is considered by fixing its temperature at 800 K.

Adaro sub-bituminous coal is used for the study. The proximate and ultimate analysis of the coal are given in Table 1. The high heating value (HHV) for the coal is 20.64 MJ/kg. Coal and air are injected into the furnace from the corners, forming anti-clockwise swirling flow at the center. The firing angles are illustrated in Figure 6. The mass flow rate and temperature of PA, SA, and PC at the inlets are summarized in Table 2. Excess air of 20% is supplied into the furnace, which corresponds to air-fuel ratio (AFR) of 11.28.

Table 1. Proximate and ultimate analysis of Adaro coal.

Proximate Analysis (% as-received)	
Volatile Matters	35.6
Fixed Carbon	33.8
Moisture	27.8
Ash	2.8
Ultimate Analysis (%)	
Carbon	73.7
Hydrogen	5.17
Oxygen	20.09
Nitrogen	0.89
Sulphur	0.15

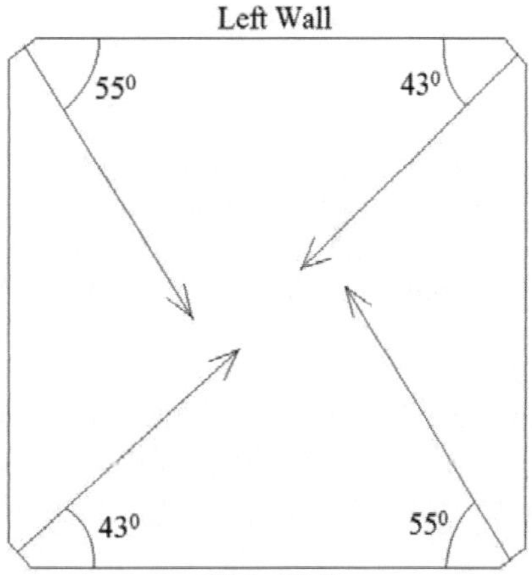

Figure 6. Firing angles of coal and air.

Table 2. Mass flow rates and temperatures of PA, SA and PC at the inlets.

	Primary air (PA)	Secondary air (SA)	Pulverized coal (PC)
Mass flow rate at one inlet (kg/s)	14.143	13.152	3.585
Total mass flow rate (kg/s)	339.432	631.296	86.04
Temperature (K)	630	623	350

2.4. CFD model

The numerical simulations for this study are carried out using CFD commercial code – FLUENT. Turbulence is modelled using realizable k-epsilon model as it has been proven to provide the best performance of all the k-epsilon models [12]. Turbulent intensity, I of 10% [13] and hydraulic diameter of 0.54 are set at the inlets. Radiation is the most dominant mode of heat transfer in a flow involving combustion. P1 gray radiation model is used to model radiation heat transfer as it is relatively simpler than other radiation models and works reasonably well when the optical thickness is large. The emissivity of the walls is set to 0.7 while weighted-sum-of-gray-gases model (WSGGM) is used to calculate the emissivity of the flue gas.

Coal combustion generally involves three sub-processes which are devolatilization, volatiles combustion and char combustion. Devolatilization starts when the coal temperature reaches

IOP Conf. Series: Materials Science and Engineering **257** (2017) 012062 doi:10.1088/1757-899X/257/1/012062

vaporization temperature, which is set to 350 K in this case. During this process, mineral matters or volatiles are released from the coal particles. The rate at which this process occurs is determined by two-competing-rates model [14], which suggests that two kinetic rates control the devolatilization process over two different temperature ranges. The combustion of the volatile matters is modelled using non-premixed combustion model, under the assumption of chemical equilibrium. Two-mixture-fraction approach is employed, in which the thermochemical state of the fluid is related to two conserved scalar mixture fractions. Instantaneous species mass fractions, density and temperature of the fluid are derived from the calculated mixture fractions. Turbulence-chemistry interaction is modelled using beta probability density function (PDF). Kinetics/diffusion limited model is used to model char combustion. In this model, the diffusion rate coefficient and kinetic rate are weighted to obtain a char combustion rate.

The pulverized coal particles trajectories are tracked using Lagrangian approach. The Rosin-Rammler diameter distribution method is used for the coal particles size distribution. The minimum, maximum and mean diameters of coal particles in each stream are 1.7×10^{-5} m, 12.7×10^{-5} m, and 5.5×10^{-5} m, respectively. There are 10 different diameters in each stream with spread diameter of 1.3. Turbulence effect on the particles trajectories is considered using stochastic tracking technique (random walk model) with 10 number of tries.

3. Results and discussion

3.1. Model validation

The solution is validated based on the flue gas temperature at the furnace exit (FE) and the general trend of this parameter along the furnace height. The general design value for flue gas temperature at furnace exit is approximately 1600 K [11], while the predicted value is 1621 K. Therefore the solution can be deemed acceptable with only about 1% error.

3.2. Flow and velocity distribution

As coal and air are injected into the furnace, intense swirling flow in anti-clockwise direction is observed in the windbox. Figure 7 shows the velocity profiles on several representative elevations (F1, F2, F3 and FE).

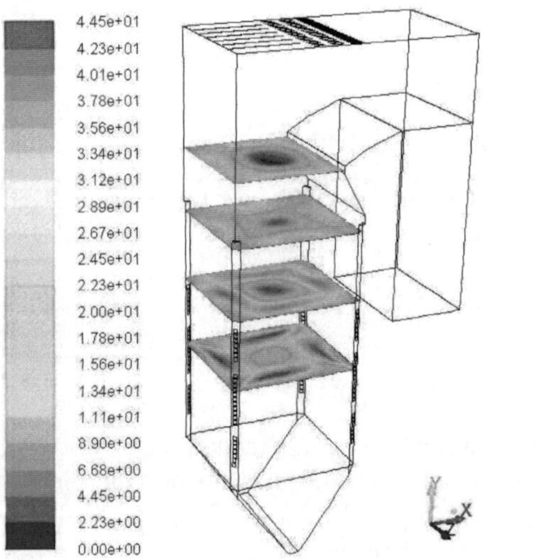

Figure 7. Velocity contours on several elevations (F1, F2, F3 and FE).

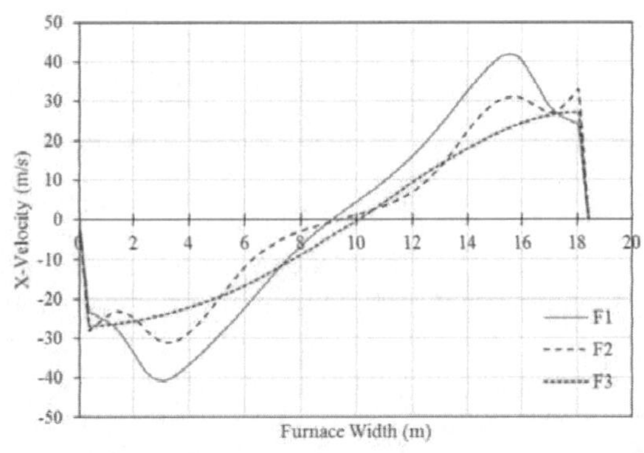

Figure 8. X-velocity distribution along furnace width.

The swirl shape can be seen at all elevations, preserved even until the furnace exit (FE). This is well known as residual swirl which is believed in some previous studies to be the cause of temperature imbalance problem. It can also be noticed that the magnitude of velocity decreases as the flue gas flows upwards due to energy loss from heat transfer to the walls. As the elevation increases, higher velocity spots are found to be closer to the walls. This suggests that as the height increases, the swirl becomes weaker but its diameter is enlarged. The x-velocity distribution along the furnace width on F1, F2 and F3 is plotted as in Figure 8. From the graph, it is evident that the x-velocity decreases and higher x-velocity is observed nearer to the walls as the height increases. In the furnace, the swirl remains uninterrupted at the center along the height. As it passes through FE, the swirl seems to be a little shifted towards the rear wall due to the significant reduction of cross-sectional area.

Figure 9 (a) and 9 (b) display the velocity vectors near the right wall and left wall, respectively. On the right side, the flue gas flows towards the rear wall and enters the horizontal pass directly from the lower part. On the left side, the flue gas initially flows towards the front wall and later reverses into the horizontal pass from the upper part. Such flow patterns result from the anti-clockwise swirling flow in the furnace. From the observation of the streamline in the horizontal pass, it is found that after passing the furnace exit (FE), the swirl is shifted towards the left wall as it flows upwards, which may be due to the abrupt expansion of the cross section. A large portion of high-speed flue gas flowing on the left side in the furnace turns to flow through the middle plates, leaving slow-moving flue gas flowing near the left wall. As the flue gas flows into the rear pass after passing through the middle plates, it curves back towards the left wall, forming a small swirl. This can be observed clearly from the velocity vectors on planes HP1, HP2 and HP3 in Figure 10 (a) – 10 (c).

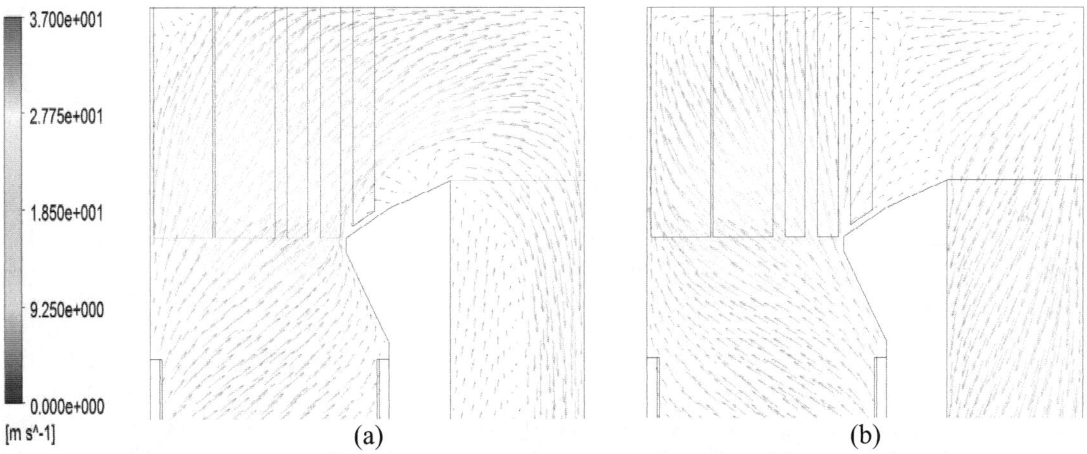

Figure 9. Velocity vectors on plane (a) near right wall and (b) near left wall.

In these figures, the swirl is seen to bias towards the left wall as the height increases and subsides as it reaches HP3. This explains the trend of velocity plot along the furnace width in Figure 10 (d) and 10 (e) which show the velocity on the left side declines as the elevation increases. The velocity in the middle is higher in the upper part (HP2 and HP3) compared to the lower part (HP1). This justifies the earlier observation that a larger portion of flue gas flowing near the left side in the furnace curves inwards and enters the horizontal pass through the upper part of the middle plates. Before the second group of HTS (plane X1), the velocity deviation between the right and left sides exists but is not really serious. However, before flowing through the HTR (plane X2), velocity deviation is very large between the right and left sides. Looking at the velocity vectors in Figure 10 (a) – 10 (c), the swirl shifts towards the left and rear wall and happens to settle and subside at the left side of HTR. Based on all these findings, it can be inferred that the residual swirl in the horizontal pass causes flow imbalance thus is the key factor of the velocity deviation in this region.

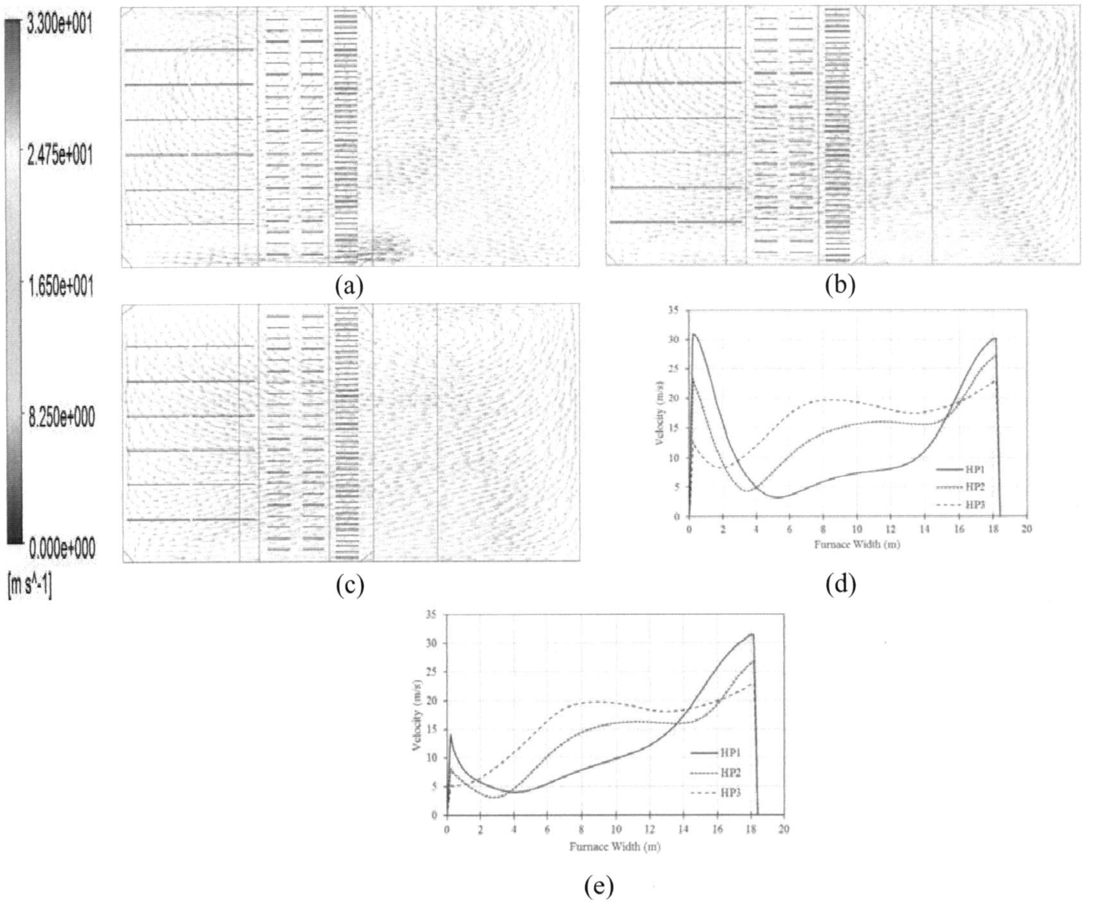

Figure 10. Velocity vectors on (a) HP1 (b) HP2 (c) HP3 and plot of velocity along furnace width on plane (d) X1 and (e) X2.

3.3. Temperature distribution

The trend of temperature along the furnace height can be referred from Figure 5. Basically, the temperature starts to increase rapidly in the windbox section where the combustion starts. After the windbox, the temperature begins to drop as heat energy is being transferred to the water walls and the platen heat exchangers. As the objective of this study is on the temperature imbalance problem occurring in the horizontal pass, the analysis and discussion of temperature distribution will focus only on the horizontal pass section.

Figure 11 shows the temperature contour on HP1, HP2, and HP3. Generally, the temperature is seen higher on the right side, except at the LTS area which is relatively balanced between the right and left sides. Low temperature spot can be obviously noticed at the location of the residual swirl. This evidently implies that the residual swirl is the main factor responsible for the temperature imbalance occurrence. The left side of the HTR is the most affected by the low temperature spot compared to other platen heat exchangers. To have a clearer view, temperature contours on plane X1 and X2 are observed, as depicted in Figure 12 (a) and 12 (c). Figure 12 (b) and 12 (d) are their corresponding temperature plots along the boiler width. On both planes, temperature imbalance is evident between the right and left sides. On plane X1, the deviation is less severe (between 200-300 K) compared to that on plane X2 which ranges between 400-550 K. Notably higher temperature is observed at the lower right part of plane X2. The temperature deviation on plane X1 is about the same irrespective of the elevation while on plane X2 larger temperature deviation occurs at the middle elevation (HP2).

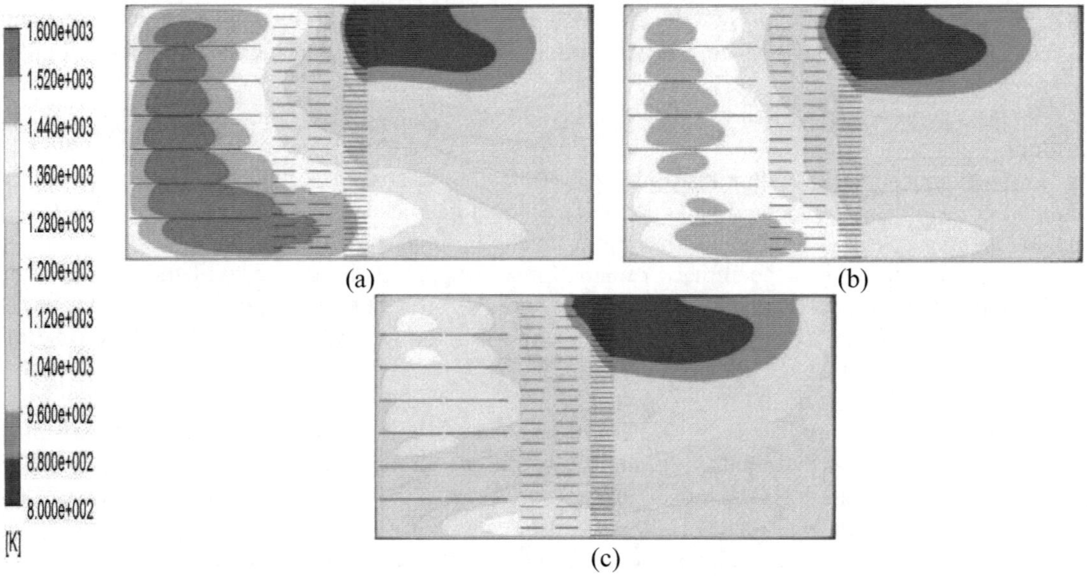

Figure 11. Temperature contour on plane (a) HP1 (b) HP2 (c) HP3.

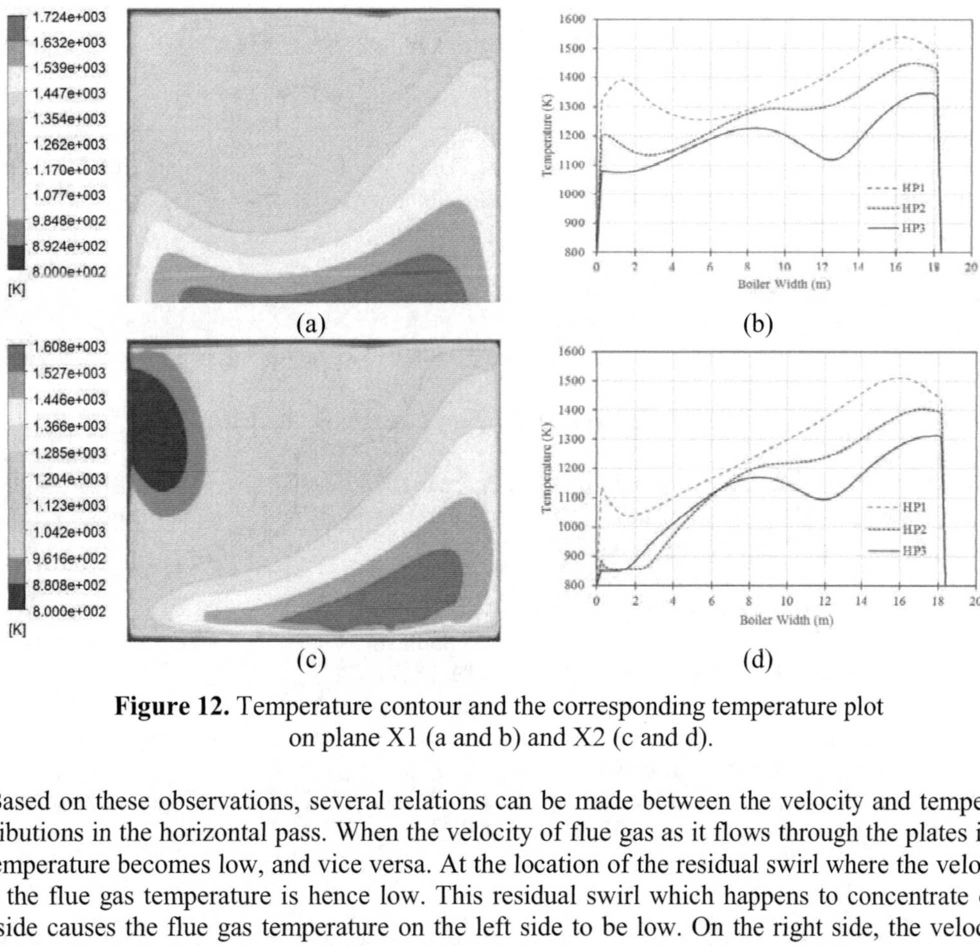

Figure 12. Temperature contour and the corresponding temperature plot on plane X1 (a and b) and X2 (c and d).

Based on these observations, several relations can be made between the velocity and temperature distributions in the horizontal pass. When the velocity of flue gas as it flows through the plates is low, its temperature becomes low, and vice versa. At the location of the residual swirl where the velocity is low, the flue gas temperature is hence low. This residual swirl which happens to concentrate on the left side causes the flue gas temperature on the left side to be low. On the right side, the velocity of flue gas is high thus its temperature is high. The more the velocity deviation, the larger the temperature imbalance between the right and left sides. This is because when the flue gas flows through the plates with low velocity, its residence time increases allowing more heat to be transferred to the plates which consequently lowering its temperature. High velocity flue gas flowing on the right side has shorter residence time hence less heat is transferred to the plates resulting in only small drop of its temperature.

4. Conclusions

The flow, velocity and temperature distribution in the 700 MW coal-fired tangential firing boiler has been studied using CFD numerical simulation. The model has been well validated and proven to be mesh-independent. Several conclusions that can be made from the study include:

a) The swirl formed in the furnace remains uninterrupted at the center until the furnace exit where it is shifted towards the rear wall due to the significant reduction of cross-sectional area. After the furnace exit, it is further shifted towards the left wall due to sudden increase of cross-sectional area and eventually subsides. A new swirl is formed in the rear pass after the flue gas flows pass the HTR.
b) The shift of the residual swirl to the left side causes the flue gas from the left side to curve inwards and flows through the middle plates from the upper part.
c) The residual swirl is the factor that causes low-velocity flue gas on the left side. This causes velocity deviation between the right and left sides.
d) Temperature imbalance is the consequence of velocity deviation due to the difference of flue gas residence time between the plates and amount of heat transferred to the platen heat exchangers.
e) Since the root cause of the temperature imbalance is the residual swirl, the solution would be to eliminate the residual swirl in the horizontal pass region or to prevent it from shifting and remain at the center of the furnace cross section.

Acknowledgement
The authors would like to thank the Ministry of Education Malaysia through the Fundamental Research Grant Scheme 2015 (20150211FRGS) for the financial support.

References

[1] U.S. Energy Information Administration 2016 *International Energy Outlook 2016*
[2] Liu Y, Fan W and Li Y 2016 *Appl. Energy* **177** 323–34
[3] Park H Y, Baek S H, Kim H H, Kim Y J, Kim T H, Lim H S and Kang D S 2016 *Fuel* **166** 509–16
[4] Tian D, Zhong L, Tan P, Ma L, Fang Q, Zhang C, Zhang D and Chen G 2015 *Fuel Process. Technol.* **138** 616-28
[5] Yin C, Caillat S, Harion J-L, Baudoin B and Perez E 2002 *Fuel* **81** 997–1006
[6] He B, Chen M, Yu Q, Liu S, Fan L, Sun S, Xu J and Pan W P 2004 *Comput. Fluids* **33** 1201–23
[7] Diez L I, Cortes C and Pallares J *Fuel* **87** 1259–69
[8] Zhou Y, Zhang M, Xu T and Hui S 2009 *Energy and Fuels* **23** 5375–82
[9] Baek S H, Park H Y and Ko S H 2014 *Fuel* **128** 62–70
[10] Ahmad A, Hasan H, Kannan M, Noor N A W M and Lim M T 2015 *J. Mech. Autom.* **5** 12–18
[11] Yin C, Rosendahl L and Condra T J 2003 *Fuel* **82** 1127–37
[12] *ANSYS FLUENT Theory Guide* 2011 (Canonsburg: ANSYS, Inc)
[13] Cengel Y A and Cimbala J M 2014 *Fluid Mechanics: Fundamentals and Applications 3rd Ed. in SI Units* (McGraw-Hill) p 904
[14] Kobayashi H, Howard J B and Sarofim A F 1976 Coal devolatilization at high temperatures *Proc. 16th Symp. (Int.) on Combustion* p 411–25

IOP Conf. Series: Materials Science and Engineering **257** (2017) 012063 doi:10.1088/1757-899X/257/1/012063

Phosphorus removal by electric arc furnace steel slag adsorption

J W Lim[1,2*], K F Lee[1], Thomas S.Y. Chong[2], L C Abdullah[2], M A Razak[2], C Tezara[1]

[1] Department of Mechanical Engineering, Faculty of Engineering & Quantity Surveying, INTI International University, Nilai, Negeri Sembilan, Malaysia.
[2] Department of Chemical and Environmental Engineering, Faculty of Engineering, UPM Serdang, Universiti Putra Malaysia.

*Corresponding author: jinwei.lim@newinti.edu.my

Abstract. As to overcome the eutrophication in lakes and reservoirs which is resulted from excessive input of phosphorus due to rapid urbanization or uncontrolled agricultural activities, Electric Arc Furnace steel slag (EAFS), a steelmaking by-product, in which the disposal of this industrial waste considered economically unfavourable yet it's physical and chemical properties exhibits high potential to be great P adsorbent. The objective of this study was to identify most suitable mathematical model in description of adsorption by using traditional batch experiment and to investigate the effect on Phosphorus removal efficiency and Phosphorus removal capacity by EAFS adsorption through variation of parameters such as pH, size of slag and initial concentration of Phosphorus. Result demonstrated that, Langmuir is suitable in describing Phosphorus removal mechanisms with the Maximum Adsorption Capacity, Q_m of 0.166 mg/g and Langmuir Constant, K_L of 0.03519 L/mg. As for effect studies, smaller size of adsorbent shows higher percentage (up to 37.8%) of Phosphorus removal compared to the larger size. Besides that, the experiment indicated a more acidic environment is favourable for Phosphorus removal and the amount of Phosphorus adsorbed at pH 3.0 was the highest. In addition, the adsorption capacity increases steadily as the initial Phosphorus concentration increases but it remained steady at 100mg P/L. Eventually, this study serves as better understanding on preliminary studies of P removal mechanisms by EAFS.

1. Introduction

Based on United Nations Environment Programme (UNEP), lakes could be defined as "natural, standing, freshwater or saline enclosed water body surrounded by land found on the Earth's continental land masses." It is playing an important role to the aquatic ecosystem by providing habitat for aquatic livings, serving as huge storage basins for municipal and industrial water supply, flood mitigation or power generation station. Furthermore, it could be one of the tourisms spot or water based recreational sites by implementation commercial or sports activities accordingly [1]. However, based on a recent research with regards to water quality of lakes in Malaysia, although most of the lakes studies classified under Class II which is appropriate for water supply with treatment system and served for recreational purposes but the studies indicated that most of the lakes and reservoirs studied consist of high total phosphorus (TP) concentration which exceeding 0.1 mg/L or could be classified as hypereutrophic that lead to algae bloom such as Aman and Ayer Keloh lakes. Eventually, this will affect human uses for drinking, aesthetic and recreation values [2]. Besides that, another research indicated 62% of lakes were eutrophic out of 90 lakes and reservoir evaluated in Malaysia by using Carlson's Trophic State Index (TSI), which is developed from TP and chlorophyll relationship [3]. Hence, it reflected the severity of eutrophication phenomenon in Malaysia, constant follow-up, enforcement of laws and regulations or any appropriate solutions should be implement as soon as possible to prevent status of lakes gone severely.

Eutrophication can be defined as the summation of the impacts of the uncontrolled growth of planktons causing imbalanced primary and secondary productivity as well as faster rate of succession from existence to higher serial stage as resulted from excessive of nutrient (phosphorus & nitrogen) in water bodies through runoffs, leaching and sewage disposal. In fact, it is a slow natural process but contribution of human activities such as rapid industrialization, discharge of wastewater, wetland drainage, deforestation or heavy development agricultural activities on nutrient input will greatly altered the process [4]. As for the process eutrophication, it started with a body of water contained excess amount of phosphorus from sewage and watershed runoff that exceed the limiting factors of growth, eventually it accelerated the growth of plant biomass under photosynthesis which called algal bloom based on equation (1) below. It showed nitrogen and Phosphorus are the dominant control factors for the propagation of algae, particularly Phosphorus [5, 6].

$$106\,CO_2 + 16\,NO_3{}^- + HPO_4{}^{2-} + 122\,H_2O + 18H^+ \xrightarrow{energy+microelemnt} C_{106}H_{263}O_{110}N_{16}P\ (bioplasm\ of\ algae) + 138O_2 \quad (1)$$

Consequently, aquatic plants that live underneath of algae bloom unable to conduct photosynthesis due to blockage of sunlight. Then, decomposition by bacteria take into place once bottom of the lake accumulated by dead biomass plant. As the decomposition required oxygen, the concentration of dissolved oxygen in lake decreased which affected the lake aquatic ecosystem. Then, body of water filled with plant through formation of marsh, these phenomenon is called eutrophication. Eutrophication creates negative impacts to environment and community, either physically, chemically or biologically. The general negative impacts of eutrophication on lakes and reservoirs are listed in table 1 below. The most significant effects of eutrophication as it can interrupt the natural equilibrium of aquatic ecosystem by declining functions of water ecosystem. Furthermore, eutrophication will decrease transparency of water and deteriorate the water quality. In addition, decomposition of dead plants underneath of algae bloom will lead to lack of dissolved oxygen level in water which endanger the species or aquatic animals. Toxins could be release from algae which is poisonous to seashell and fish causing shortage of drinking water supply since the toxin is poisonous to human as well [6].

In order to tackle and combat this global phenomenon, various approaches either in biological, chemical or physical technologies such as development of constructed wetlands (CW) were implemented for phosphorus or nitrogen removal but these methods had several drawbacks and disadvantages particularly. Since 1980s, different researches have been conducted to investigate the potential of steel slag in Phosphorus removal from wastewater [7]. Therefore, the utilization of steel slag (an industrial by-product in the steelmaking process) in phosphorus adsorption seems to be a potential alternative in tackling eutrophication. Various researches indicated that Phosphorus removal efficiency was highly depended on initial P concentration, pH, size of adsorbent, dosage of adsorbent, agitation mode and temperature. In addition, different combinations of these variables will affect P removal efficiency in their own ways [8-10]. However, only several parameters will be selected to be investigated while others remained as controlled variables in this paper.

2. Materials and methods

2.1. Materials

The samples of EAFS tested in this study were collected from Southern Steel Berhad, a steel manufacturing company located in Penang, Malaysia. The samples were crushed and sieved into different size stated on table 2 below. Sieved EAFS were washed with distilled water to remove impurities. The, EAFS were dried in oven at 110°C for 24 hrs and stored in heavy duty plastic bags until used [11].

Table 1. Classifications on size of EAFS.

Classification of size	Size (mm)
A	1.18 – 2.36
B	2.36 – 3.35
C	3.35 – 4.75
D	4.75 – 5.00
E	5.00 – 9.50

2.2. Phosphorus batch adsorption experiments

Table 2. List of controlled parameters in batch experiment [10].

Controlled parameters	Value
Mass of EAFS (g)	5
Volume of P Solution (mL)	100
Slag to Solution Ratio (g/mL)	0.05
Speed of Agitation (rpm)	175
Time of contact – on shaker (min)	90
Time of contact – after shaking (h)	168
pH (Original)	5.02-6.05
Temperature (°C)	22-25 (Room Temp.)

Based on Table 2 above, 100 mL of 6 different initial concentrations which ranged between 5 – 100 mg KH_2PO_4/L of Phosphorus solution with its original pH was mixed with 5 g of EAFS in size C in order to achieve ration of slag to solution of 0.05 g/mL. Then, the adsorption studies were conducted on an orbital shaker at 175 rpm for 90 minutes under room temperature between 22 – 25°C. All flasks were tightly covered to avoid contamination. Mixture was let to rest for 168 hours after agitation to achieve equilibrium. Then, 50 mL of mixture was then filtered with 0.45µm membrane filter with their pH adjusted to 2 and stored in 4°C chiller until used. This experiment was conducted in duplicate. Based on Table 3 above, the effects on variation of initial Phosphorus concentration (5 – 100 mg KH_2PO_4/L) on efficiency of P removal by EAFS was investigated. The series of phosphorus solution will be sent absorbance test using spectrophotometer. With similar conditions and procedures above, mixture of ration 0.05 g/mL which consist of Phosphorus solution with initial concentration of 30 mg KH_2PO_4/L and 5g of EAFS in size C, the effects of pH value on P removal efficiency is investigated. The pH value was moderated by using 1M Sodium Hydroxide (NaOH) and 1M Hydrochloric Acid (HCl). The pH values of P solution were adjusted from 3.0 – 12.0 in this experiment. This experiment was conducted in duplicate. With similar conditions and procedures above, mixture of ratio 0.05 g/mL which consists of Phosphorus solution with initial concentration of 30 mg KH_2PO_4/L and 5g of different size EAFS, the impact of different sizes of EAFS was investigated. This experiment was conducted in duplicate.

2.3. Removal efficiency and adsorption capacity

Removal efficiency or removal percentage (RP) of P by EAFS adsorbent can be calculated by using equation (2) below,

$$RP = \frac{C_o - C_e}{C_0} \; x \; 100\% \tag{2}$$

where C_o and C_e are initial and final concentration of phosphorus (mg/L), respectively; the amount of Phosphorus adsorbed by EAFS during equilibrium (q_e) can be obtained by using equation (3) below,

$$q_e = \frac{C_o - C_e}{m} \, x \, V \quad (3)$$

where q_e represents amount of Phosphorus adsorbed by EAFS during equilibrium (mg P/g) , m indicates the mass of EAFS used (g) and V represents as volume of solution (L).

2.4. Langmuir adsorption isotherm

It assumed single layer adsorption on the outer surface of adsorbent with adsorption can only occur at a fixed number of definite localized sites which all the sites are equivalent and identical. Furthermore, lateral interaction and steric hindrance between adsorbed molecules and adjacent sites were not existed [12, 13].

$$\frac{C_e}{q_e} = \frac{1}{Q_m K_L} + \frac{C_e}{Q_m} \quad (4)$$

where Q_m, C_e and K_L represents maximum adsorption capacity (mg/g), equilibrium concentration (mg/L) and Langmuir constant (L/mg) respectively. The Langmuir constant can be obtained by slope and intercept from linear graph of $\frac{C_e}{q_e}$ versus C_e.

In order to determine the favourability of Langmuir isotherm on adsorption, R_L, a dimensionless constant known as separation factor or equilibrium parameter shown as equation (5) can be used.

$$R_L = \frac{1}{1 + K_L C_0} \quad (5)$$

where C_0 indicated as initial concentration of Phosphorus.

2.5. Freundlich adsorption isotherm

It is a well-known model that used to describe non-ideal adsorption of multilayer adsorption over heterogeneous surface with assumption of stronger binding sites will be occupied first. It also stated the degree of site occupation increased, the binding strength decreased [13].

$$ln\, q_e = ln\, K_F + \frac{1}{n} ln\, C_e \quad (6)$$

where K_F represents Freundlich constant (mg/g), n indicates adsorption intensity and C_e defined as the equilibrium concentration of adsorbate (mg/L). Value of K_F and n can be determined through linear graph of $\ln q_e$ versus $\ln C_e$.

3. Results and discussion

3.1. Characterization of adsorbent

The specific surface area and pore size distribution of steel slag were determined from BET analysis method by using using N_2 adsorption/desorption at 77.3 K. The data were tabulated in Table 3 below.

Table 3. Surface data of EAFS samples.

Parameters	Value
BET surface area, m^2/g	5.5264
Pore Volume, cm^3/g	0.09610
Average Pore Diameter, Å	713.7362

As for the surface images of the EAFS (before & after Phosphorus adsorption), it was captured by SEM test as shown in Figure 1 and 2 below. Since size particle, diameter of pore and specific surface area of adsorbent are the main parameters that affecting the adsorption capacity of adsorbent [14]. Figure 1 showed that existence of pore on the surface of steel slag hence confirmed the porosity of the steel slag. Based on the observation obtained from SEM shown in Figure 2, a fine distributed crystalline layer was identified on the surface of EAFS which believed the crystal is Hydroxylapatite (HAP) [15].

Figure 1. Surface image of EAFS before phosphorus adsorption (7.3mm x 300)

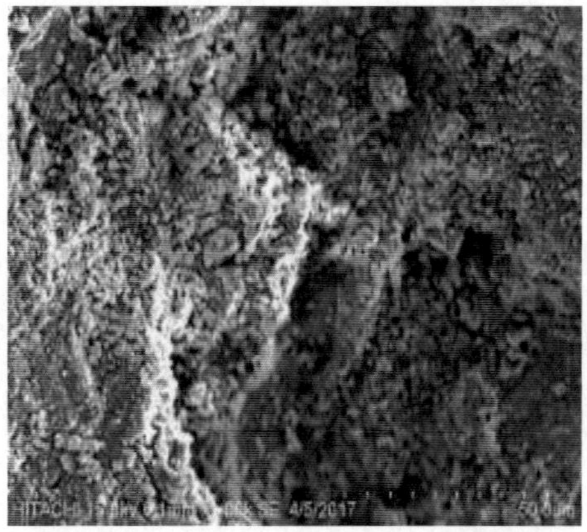

Figure 2. Surface image of EAFS after phosphorus adsorption (6.1mm x 1.00k)

EDX analyses revealed the difference of chemical components of EAFS (before and after Phosphorus adsorption) as shown in table 4 below. Before Phosphorus adsorption, the main elements existed on surface of EAFS are Ca & Fe (Metals) followed by Si (Metalloids) in their oxides form which is CaO, Fe_2O_3 and SiO_2 and these results agreed with the findings of other study [16]. However, due to inconsistency and high variation of the chemical composition in EAFS, the proportion of certain elements might be different, but the main highlight of this comparison is on the existence of the phosphorus, it confirmed the adsorption of Phosphorus by EAFS.

Table 4. Weight proportion of element in EAFS before and after adsorption.

	Before adsorption			After adsorption		
Element	Weight (%)	Formula	Compound (%)	Weight (%)	Formula	Compound (%)
C	6.00	C	6.00	*6.56*	*C*	*6.56*
O	37.77S			30.72S		
F	4.64	F	4.64	3.18	F	3.18
Mg	2.95	MgO	4.89	1.91	MgO	3.16
Al	5.08	Al_2O_3	9.60	4.14	Al_2O_3	7.82
Si	10.01	SiO_2	21.41	6.57	SiO_2	16.22
P	0		0	***2.80***	**P**	***2.80***
Ca	16.64	CaO	29.93	16.37	CaO	22.90
Fe	15.39	Fe_2O_3	22.01	*24.69*	*Fe_2O_3*	*35.30*
Co	1.51	Co	1.51	*2.05*	*Co*	*2.05*
Total	100		100	100		100

Based on Figure 3 below, XRD test revealed the EAFS was made up by several constituents including Ca_3O_5Si, Periclase (MgO), Gehlenite ($Al_{1.91}Ca\ Fe_{0.02}Mg_{0.05}O_7Si_{1.02}$), Tricalcium Oxynitride Silicate ($Ca_{2.894}N_{1.76}O_{4.24}Si_2$) and Franklinite ($Fe_{1.96}\ Mn_{0.36}O_4Zn_{0.68}$). Therefore, it concludes Ca is the dominant element in the EAFS, by ignoring the variation of Fe proportion in table 5, the decrement of certain element was noticeable. For instance, the decrease in weight of CaO might due to its dissolution throughout the adsorption mechanisms. Eventually this finding is support the proposed Phosphorus removal mechanisms by one of the research [15]. The dissolution process can be explained in the equations 7 and 8 below.

$$CaO + H_2O \rightarrow Ca^{2+} + 2OH^- \quad (7)$$

Then Ca^{2+} will react with PO_4^{3-} to form HAP, [22]

$$10\ Ca^{2+} + 6PO_4{}^{3-} + 2OH^- \leftrightarrow Ca_{10}(PO_4) * 6(OH)_2 \downarrow \quad (8)$$

In addition, the decrement of weight in Al_2O_3 also can be identified in table 5. Therefore, the available Al^{3+} ions might react with PO_4^{3-} to form $AlPO_4$ shown as equation (9) below [17].

$$Al^{3+} + H_nPO_4{}^{3-n} \leftrightarrow Al(PO_4) + nH^+ \quad (9)$$

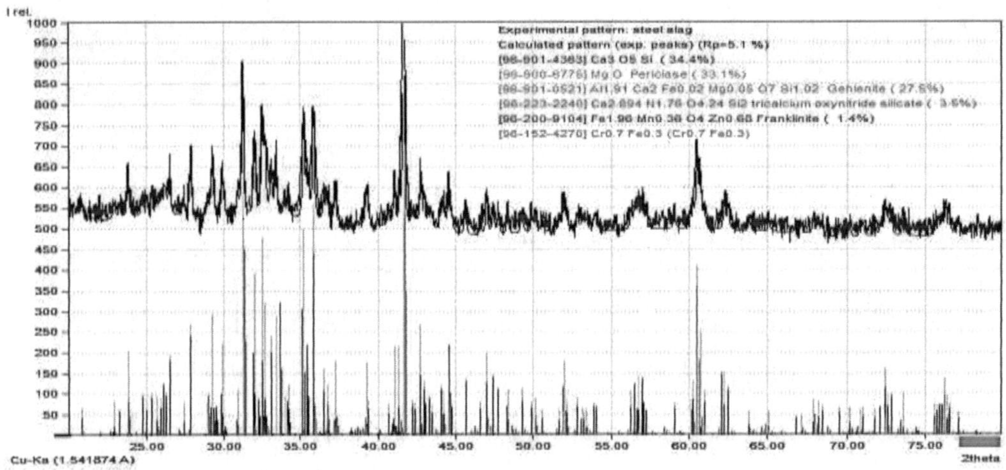

Figure 3. X-ray diffraction pattern of EAFS (before phosphorus adsorption).

3.2. Adsorption isotherm

In this paper, the equilibrium data were analysed with 2 well-established isotherm mathematical models; Langmuir and Freundlich. The adsorption data have been fitted to liner form of these mathematical model as shown as Figures 4 and 5 below. As for the Langmuir and Freundlich constant, it showed as Table 5 below.

Table 5. Parameters obtained from Langmuir and Fruendlich Isotherm Plot.

Langmuir		Freundlich	
Q_m, mg/g	0.166	K_F, mg/g	0.0118
K_L, L/mg	0.03519	N	1.8311
R^2	0.9603	R^2	0.9469

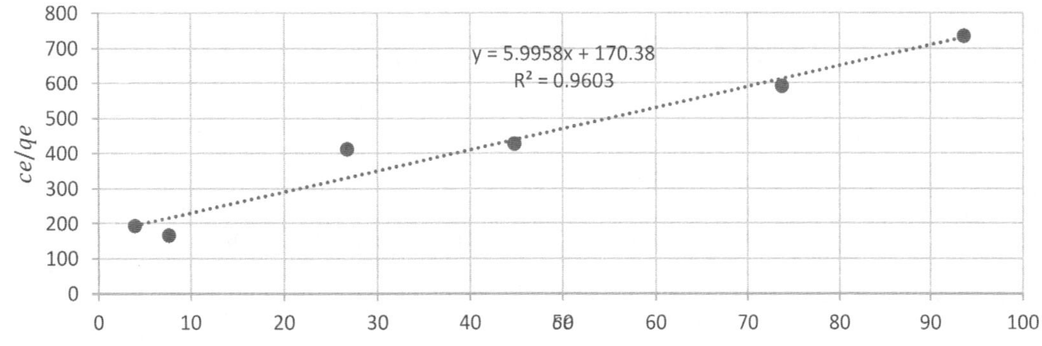

Figure 4. Langmuir plot for phosphorus adsorption onto EAFS.

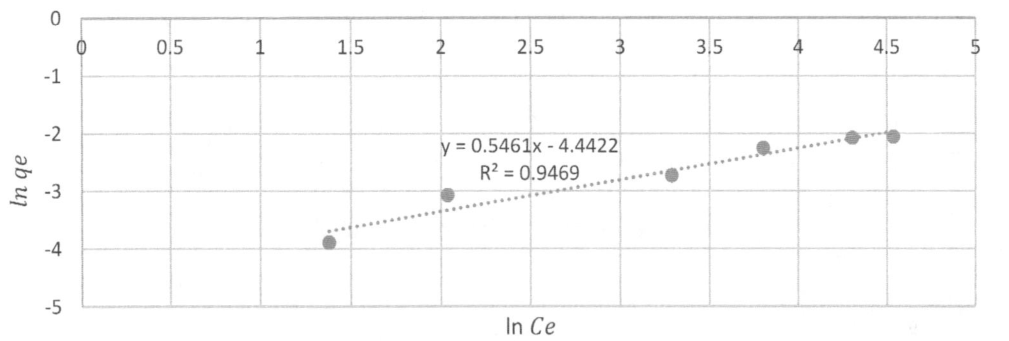

Figure 5. Freundlich plot for phosphorus adsorption onto EAFS.

Based on the coefficient of determination (R^2), the values for Langmuir model and Freundlich are 0.9603 and 0.9469 respectively which showed Langmuir model is more suitable in description of adsorption process compared to Freundlich model. As mention on previous section, R_L value will identify the favourable level of isotherm for adsorbent. Based on Figure 6 below, it showed the relationship between R_L value and initial Phosphorus concentration. By referring to Table 6 below, it indicated the range of R_L with respective favourability, as the R_L value ranged between 0.85 – 0.22 (fall under$0 < R_L < 1$ category), it further affirmed the suitability of Langmuir model in description of P removal by EAFS.

Table 6. R_L value and level of favourability.

Value	Indicator
$0 < R_L < 1$	Favourable
$R_L > 1$	Unfavourable
$R_L = 1$	Linear
$R_L = 0$	Irreversible

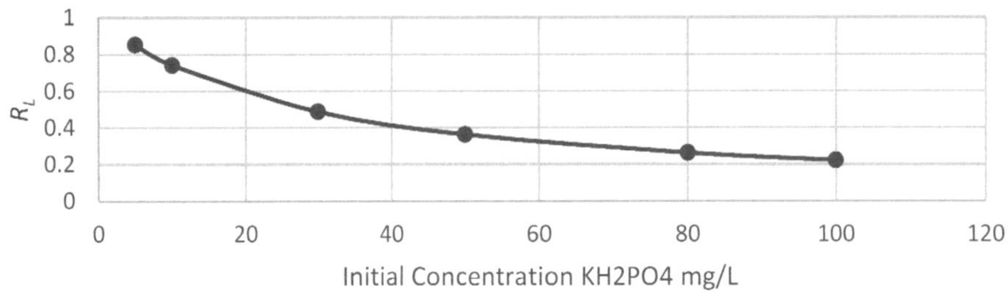

Figure 6. R_L Value versus initial concentration of KH_2PO_4.

3.3. Effect of various conditions on phosphorus removal efficiency

3.3.1. Effect of slag size. Figure 7 below showed the effect of size of slag with the phosphorus removal efficiency of EAFS. It indicated the P-removal percentage increased as the size of slag decreased with a fixed mass of slag (5g) in the experiment despite the rebound on Phosphorus removal percentage on the size between 4.75- 5.00 mm. The rebound result might be caused by large range of difference between size D and E, 4.75-5.00 and 5.00-9.50 respectively due to limited availability of sieve.

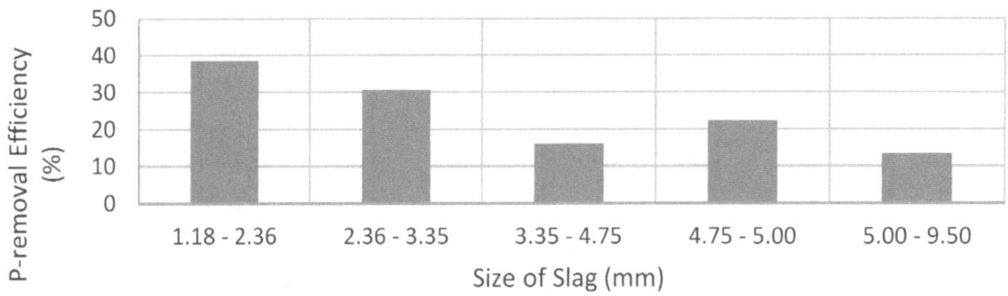

Figure 7. Effect of slag size on phosphorus removal (Experimental conditions: Initial phosphorus concentration=30mg P/L; solid to solution ratio=0.05 mg/ml; agitation speed = 175rpm; original pH=5.02-6.05; temperature = 22-25°C)

According to the BET results on pore size and pore area earlier which confirm the porosity of the EAFS. Hence, smaller size of slag has a larger surface area resulting higher adsorption on phosphorus and its removal percentage. These results agreed with the finding in other study [9]. Since the diffusion path decreased as the size of slag decreased, eventually it increased the effectiveness of the penetration of adsorbate as well as then depth of penetration in adsorbent particle resulting higher rate of adsorption. In addition, mass transfer limitation and internal diffusional resistance will decrease as the size of slag decrease which further increased the rate of adsorption of P [18].

3.3.2 Effects of pH. The effect of pH of solution (Initial pH) on phosphorus removal efficiency has been demonstrated in Figure 8 below. As the initial pH increase by becoming a higher basicity solution, the P removal efficiency decreased exponentially. Besides that, the results demonstrated the amount of P adsorbed at pH 3.0 was the greatest which indicated the adsorption of P by EAFS is effective under acidic environment. Although these findings are likely support with several studies but in terms of the optimum value of pH, the value is not similar where their studies indicated pH 5 - 5.5 consisted of highest P-removal efficiency [8, 9]. Due to the variation of chemical composition in the EAFS hence the optimum pH for highest P-removal efficiency is subjected to the chemical composition.

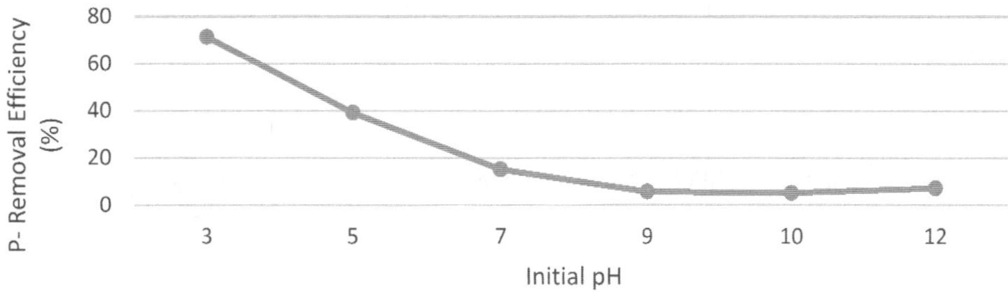

Figure 8. Effect of initial pH on P-removal (Experimental conditions: Initial P-concentration=30mg P/L; solid to solution ratio=0.05 mg/ml; size of slag: 3.35-4.75mm; agitation speed = 175rpm; temperature = 22-25°C)

Based on the result on chemical composition of EAFS through EDX, most of the metal existed in their respective oxides such as CaO, Fe_2O_3 and SiO_2 which indicated the large existence of oxygen atoms and hydroxyl groups on the surface of slag. According to a recent research, the acid or basic properties and the reactivity of these surface is highly depending on the hydroxyl group on the surface [19]. In a higher pH or a more basic condition, negative charges (OH^-) may generated on the surface of slag thus would repulse the negative charged phosphate ion (PO_4^{3-}) from being adsorbed.

3.3.3. Effect of Initial Phosphate Concentration. Based on Figure 9 below, it showed the relationship between the initial phosphate concentration with the Adsorption capacity of the EAFS, the adsorption capacity is increased steadily as the P-concentration increase because there was more P existed and available for sorption in solution per unit mass of slag. However, it was tending to be flatten at 100 mg P/L because for a fixed mass of adsorbent (5g in this study), the total available adsorption sites is constant hence it will only absorb same amount of phosphate. Therefore, it led constant amount of percentage removal at 100 mg P/L [20].

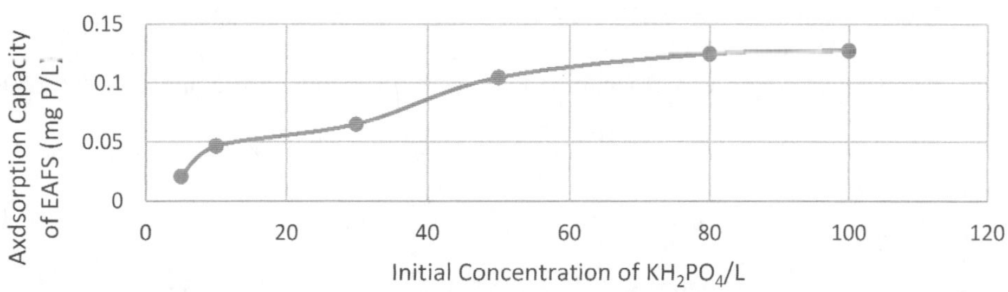

Figure 9. Effect of initial phosphate concentration on P-removal (Experimental conditions: solid to solution ratio=0.05 mg/ml; size of slag: 3.35-4.75mm; agitation speed = 175rpm; original pH=5.02-6.05; temperature = 22-25°C)

4. Conclusions

This paper serves as the purpose of preliminary studies on phosphorus removal by EAFS. It showed the Langmuir model is more suitable in description of phosphorus removal by EAFS in compared to Freundlich model with the maximum adsorption capacity, Q_m of 0.166 mg/g and Langmuir constant, K_L of 0.03519 L/mg. Furthermore, this study clearly indicated the importance of size of slag, pH and initial concentration of phosphorus on effectiveness of its removal by EAFS. Based on the experiment, small size of adsorbent shows higher percentage (up to 37.8 %) of Phosphorus removal compared to the larger size. Besides that, the experiment showed a more acidic environment is favourable for phosphorus removal and the amount of Phosphorus adsorbed at pH 3.0 was the highest. In addition, the adsorption capacity increases steadily as the initial Phosphorus concentration increase but it remains nearly constant at 100mg P/L due to fixed adsorption site at given mass throughout experiment.

Acknowledgment
The authors would like to gratefully acknowledge Ministry of Higher Education (MOHE), Universiti Putra Malaysia (UPM) and INTI International University for the financial support of this work.

References

[1] Sharip Z and Zakaria S 2008 *Lakes and Reservoir in Malaysia: Management and Research Challenges* (Seri Kembangan, Malaysia)
[2] Sharip Z, Zaki A T, Shapai M, Sauratmaan S and Shaaban A J 2014 *Lakes and Reservoirs: Research and Management* **19** 130-41
[3] Sharip Z and Zulkifi Y 2007 *Overview: The Status of Lakes Eutrophication in Malaysia* (Putrajaya)
[4] Khan F A and Ansari A A 2005 *The Botanical Review* **71** 449-82
[5] Mainstone C P and Parr W 2002 *The Science of the Total Environment* **282-283** 25-47
[6] Yang X E, Wu X, Hao H L and He Z L 2008 *Journal of Zhejiang University SCIENCE B* **9** 197-209
[7] Lim J, Chew L, Choong T, Tezara C and Yazdi M H 2016 *IOP Conference Series: Earth and Environmental Science* p 36
[8] He Z, Mahmood Q, Liu D, Yang X, Islam E and Jibing X 2008 *Journal of Hazardous* **152** 211-5
[9] Okochi N C and Mc Martin D W 2012 *Water* **4** 321-44
[10] Cucarella V and Renman G 2009 *J. Environ. Qual.* **38** 381-92
[11] Worch P D E 2012 *Adsorption Technology in Water Treatment: Fundamentals, Processes, and Modeling* (Dresden: Walter de Gruyter)
[12] Vijayaraghavan K, Kandasamy P, Padmesh T and Velan M 2006 *Journal of Hazardous Materials* **133** 304-8
[13] Foo K and Hameed B 2009 *Chemical Engineering Journal* **156** 2-10
[14] Wang F Y, Wang H and Ma J W 2010 *Journal of Hazardous Materials* **177** 300-306
[15] Barca C, Ge´rente C, Meyer D, Chazarenc F and Andre`s Y 2012 *Water Research* **46** 2376-84
[16] Drizo A, Comeau Y, Forget C and Chapius R P 2002 *Environmenral Science and Technology* **36** 4642-8
[17] Leentech B V 2016 *Phosphorus removal from wastewater* (Retrieved on 30th Sept., 2016) Available from: http://www.lenntech.com/phosphorous-removal.htm.
[18] Yusuf M, Luqman C A, Mohammed M A and Shitu A 2013 *Research Journal of Chemical Sciences* **3** 29-37
[19] Al-Degs Y, Kharaisheh M, Allen S and Ahmad M 2000 *Wat. Res.* **34** 927-35
[20] Ozacar M 2003 *Chemosphere* **51** 321-7

IOP Conf. Series: Materials Science and Engineering **257** (2017) 012064 doi:10.1088/1757-899X/257/1/012064

Development of references of anomalies detection on P91 material using Self-Magnetic Leakage Field (SMLF) technique

Shuib Husin*, **Ahmad Afiq Pauzi, Salmi Mohd Yunus, Mohd Hafiz Abdul Ghafar, and Saiful Adilin Sekari**[1]

Material Engineering Group, TNB Research, Malaysia

*Corresponding author: shuib.husin@tnb.com.my

Abstract. This technical paper demonstrates the successful of the application of self-magnetic leakage field (SMLF) technique in detecting anomalies in weldment of a thick P91 materials joint (1 inch thickness). Boiler components such as boiler tubes, stub boiler at penthouse and energy piping such as hot reheat pipe (HRP) and H-balance energy piping to turbine are made of P91 material. P91 is ferromagnetic material, therefore the technique of self-magnetic leakage field (SMLF) is applicable for P91 in detecting anomalies within material (internal defects). The technique is categorized under non-destructive technique (NDT). It is the second passive method after acoustic emission (AE), at which the information on structures radiation (magnetic field and energy waves) is used. The measured magnetic leakage field of a product or component is a magnetic leakage field occurring on the component's surface in the zone of dislocation stable slipbands under the influence of operational (in-service) or residual stresses or in zones of maximum inhomogeneity of metal structure in new products or components. Inter-granular and trans-granular cracks, inclusion, void, cavity and corrosion are considered types of inhomogeneity and discontinuity in material where obviously the output of magnetic leakage field will be shown when using this technique. The technique does not required surface preparation for the component to be inspected. This technique is contact-type inspection, which means the sensor has to touch or in-contact to the component's surface during inspection. The results of application of SMLF technique on the developed P91 reference blocks have demonstrated that the technique is practical to be used for anomaly inspection and detection as well as identification of anomalies' location. The evaluation of this passive self-magnetic leakage field (SMLF) technique has been verified by other conventional non-destructive tests (NDTs) on the reference blocks where simulated defects/anomalies have been developed inside at the weldment. The results from the inspection test showed that the signatures of magnetic leakage field gradient distribution prove that the peak is found on the location of defect/anomaly in the reference block. It is in agreement with the evidence of anomaly that seen in the radiography test film (RT).

1. Introduction

A passive self-magnetic leakage field (SMLF) technique is a technique that use natural magnetization and the after displayed results as the magnetic memory of metal to actual strains and structural changes in products and equipment metal. It is quick detection of the stress concentration zones (SCZs) based on 100% inspection of equipment and structures. The SCZ is defined as the zone with high stress-strained state inhomogeneity [1]. The SCZs are the main sources of equipment damages. Degradation of mechanical properties, corrosion, fatigue, creeping are developed in the SCZs most intensively [2]. Hence, detection of the SCZs is very important for early diagnostics of fatigue damage and evaluation of equipment and structure lifetime. SMLF technique is non-destructive testing (NDT) method based on analysis of self-magnetic leakage fields (SMLF) distribution on component's surfaces for determination of stress concentration zones (SCZs), imperfections, and heterogeneity in microstructure of the material and in welded joints [3]. This method is deemed as passive technique because it uses natural magnetization and no any artificial magnetization used. Its equipment is very simple and portable (hand-held equipment). It has fluxgate transducers that have to in-contact with surface of component during inspection. Fluxgate transducers are usually equipped with two coils: excitation coil and pick-off coil (see Figure 1). These coils usually have a core (wireshaped, toroid, oval, etc.). The basic property of fluxgate sensor is that the sensor core is periodically saturated by the excitation current and that its output is on the second harmonic (or higher even harmonic) of the excitation frequency. Harmonic (sinusoidal or pulsed) voltage will apply on exciting coil. In case of external field absence similar harmonic (sinusoidal or pulsed) voltage will be measured on measurement coil. The second harmonic (doubled frequency) will appear when flux-gate subjected by external field. Voltage of the second harmonic will be filtered out and measured.

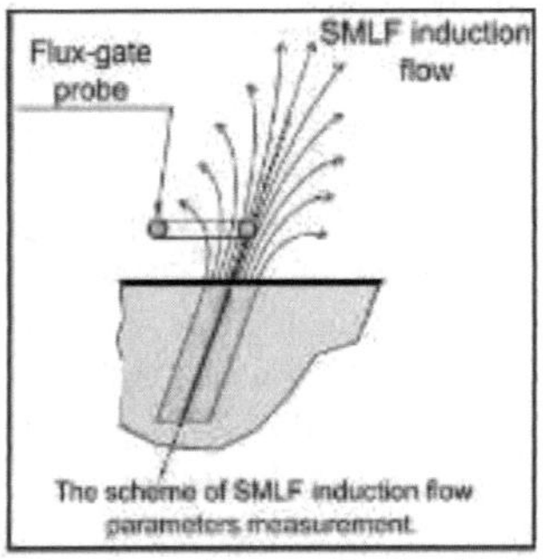

Figure 1. Schematic of working principal of SMLF technique.

This method is the only passive method among the magnetic NDT methods. The method allows detecting SCZs and locations of defects by using the pattern of SMLF and its gradient distribution. The basic task of testing by using this method is assessment of test object stressed state with SCZ detection at simultaneous determination of defects locations. The advantages of this method are: small-sized instruments are used for the testing (e.g. for boiler tube inspection, it enables the scanning device/sensor to reach at the inspection spot which has limited space for accessibility of the normal sensors (e.g. in between of stubs), not required any special test surface preparation, not required special magnetization, the method allows carrying out quick testing of components made of carbon, ferritic, pearlitic and austenitic steel grades and last but not least, the technique allows carrying out quick quality testing.

The output from the measurement that fundamentally from SMLF is actually considered as after-effect result which occurs as residual magnetization in components and welded joints formed in the course of their fabrication and due to irreversible change of the local magnetization state of components in zones of stress concentration and damage under working [3] (see Figure 2). SMLF method has been demonstrated its capability to assess material conditions or properties such as residual stress and SCZs measurement which both can be used to assess material integrity or strength. Therefore, it is a basis to be used as after-effect metal magnetic memory measurement and assessment for material or component. Figure 2 shows physical fundamental or condition based of stressed material that associated with "metal magnetic memory (MMM)"; it is about the measurement of magnetic field scattering that by structural and mechanical heterogeneities at metal natural magnetization. Figure 2 explains that a very specific feature of the method, in which MMM method is distinguished from other NDT methods, is that it detects concentration of stresses on defects. Based on this basis, an equipment called Metal Magnetic Memory (MMM) technology which attributed by SMLF technique has been developed by Energodiagnostika, Russia. Bin et al, (2012) used MMM technology for detection of fatigue fracture on diesel engine crankshaft [4]. They found that MMM technology can be applied to ferromagnetic components such as diesel engine crankshaft in daily routine to dynamic monitor crack failure of crankshaft. Significantly, they found that MMM technology is capable to monitor and give indication for early warn on early fatigue and this can be as a tool to provide solutions to prevent serious accidents.

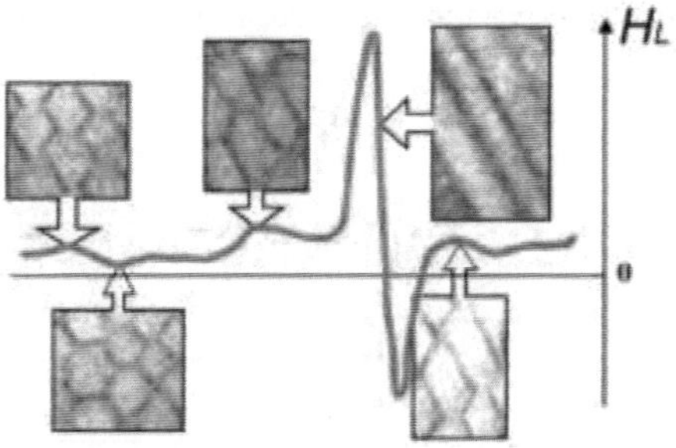

Figure 2. Interrelation between SMLF distribution and magnetic domain structure

SMLF is a self-magnetic leakage field occurring on the component's surface in the zones of stable lines of dislocations under operational or residual stresses or in the zones of maximal heterogeneity of metal structure [3]. The technique has unique opportunity to carry out integral assessment of the weld's actual state taking into account structural inhomogeneity, distribution of residual stresses and welding defects. Wenjian et al. [5] has proved the effective of MMM testing for early damage diagnosis of ferromagnetic materials. They found that magnetic domain structure's characteristic changes with the change of external loads. It was demonstrated that the magnetic flux leakage signal of specimen surface increases with the external loads increases. This shows the technique is reliable, effective and has basis for further study of the microscopic mechanism of ferromagnetic materials magnetic memory testing.

The comparison of measurement parameters of magnetic fields; Magnetic field intensity (H) and Magnetic Field Gradient distribution (dH/dx) has been done by Wang et al. [6]. In their experimental work, the Hp(y) value and Hp(y) gradient dHp(y)/dx of the middle point of the sample were measured using MMM testing instrument. They found that from the Hp(y) value plot, Hp(y) value was around zero with the applied tensile stress. While, dHp(y)/dx gradient has a good linear relationship with applied tensile stress. However, dHp(y)/dx value showed arbitrary change with the applied stress.

Jiancheng et al. [7] has used MMM method where SMLF technique was applied to detect stress concentration and defining critical stress state that leading to structural damage. They used the technique to evaluate the stress concentration quantitatively due to ambiguous physical mechanism. They demonstrated that the technique has potential to detect early defects in ferromagnetic materials. In their experiment they used mild steel (Q235) defective specimens in demagnetization state were loaded in tension up to visible necking, with magnetic memory signals measurement made at increasing stress levels. They found that in the plastic stage, an abnormal wave occurred in the stress concentration zone and its height value was sensitive to plastic deformation levels. Different magnetic signal characteristics in the plastic-plastic region indicate that the magnetic memory method can identify macro-yielding and early damage. This detection and indication is of profound significance for ensuring safe operation of equipment in service. Their findings show that the Hy signals show different variations in elastic and plastic regions due to the influence of either applied stresses or different deformation levels. They concluded that the abnormal wave in Hy signal observed in the plastic stage can detect the critical state of the macroscopic yield point.

Dubov [8] has used MMM method to investigate of the actual stress states in the hoisting vessel bearing rods as an alternative for other methods within the scope of diagnostics and reliability prediction of hoisting vessel structure during operation. This method were used for precise detection of the location of stress concentration zones that representing the main sources of fatigue damages development, determination of the maximum fatigue stress state of the bearing rods relative to the mechanical properties of the material. This is including of detection of the existing cracks and estimation of the assumed non-failure operation time (lifetime) of the bearing rods. Dubov [8] concluded that MMM method can be used for early diagnostics of fatigue damages (cracks) of bearing rod. This demonstrates that SMLF technique that principally used in MMM method is capable for determination for the investigated rod of the actual fatigue stress state and estimation of the assumed non-failure operation time of bearing rods.

The MMM method is a non-destructive testing method with a great potential and it can be adopted for many applications in inspection and material assessment. However, many of its basic issues, such as the problems of its application procedures or the criteria of the evaluation of its indications still remain unsolved [9]. This is a justification for research of using this technique as a tool for material integrity assessment. In most presented application, Roskosz et al. [9] commented that the specialists using the MMM method usually, through their own experience, work out the appropriate way of conducting tests and defining the evaluation criteria. Therefore, this technical paper is meant for providing references on the SMLF signatures or output obtained based on the known defects types, parameters and location that specifically developed in the sample blocks.

2. Methodology

The aim of the project is to verify the application of SMLF with the known type of simulated defects in weldment, size and location. The simulated defects in the weldment on a thick P91 material blocks (thickness 1 inch) were developed for the project of verification SMLF technique on MMM method in detecting defects within material. The blocks are called "reference blocks". 3 types of defect were developed for this purpose; voids, cavity and crack. Basically, the simulated defects developed in the material P91 have difference length sizes, but the height is the same, e.g crack height is 2.5 mm (see Figure 3). The summary of simulated defects in the reference block is shown in Table 1. The defects were checked and verified with other NDT techniques; Ulrasonics test (UT), Dye Penetrant test (DP), Magnetic Particle test (MP, Phase Array Ultrasonic Test (PAUT) and Radiography test (RT). The measurement or inspection results using SMLF technique was compared with the abovementioned techniques.

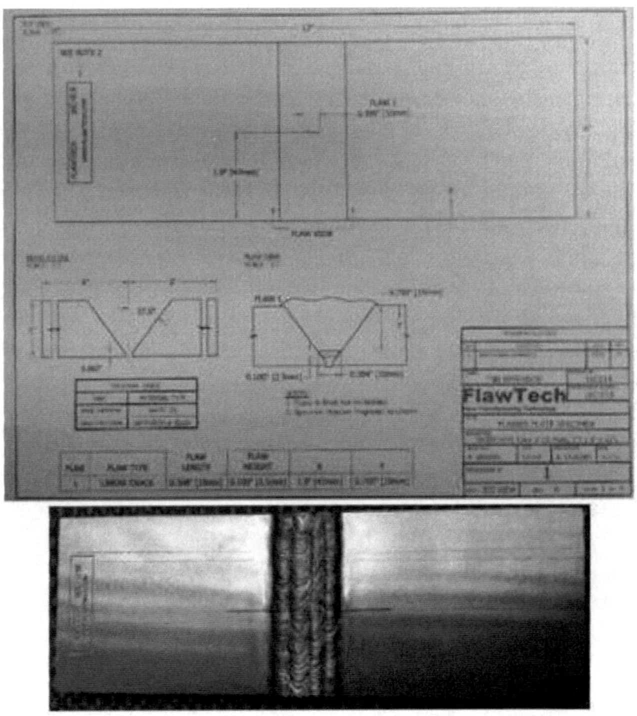

Figure 3. Simulated defect of linear crack 10mm in weldment as depicted in blue print in above drawing, X position of defect location is 47 mm, depth of defect from surface is 19 mm.

Table 1. Summary of simulated defects manufactured and conventional NDTs tested by Flawtech.

Thickness: 1 inch

Anomaly Type	Flaw parameter	Definition
Porosity	Porosity 1) Cluster 10mm porosity, Depth 1/4 depth (6mm) 2) Cluster 20mm porosity, Depth 1/2 depth (12mm) 3) Cluster 30mm porosity, Depth 3/4 depth (19mm)	• Porosity is gas pores found in the solidified weld bead. • Pores may vary in size and are generally distributed in random manner
Cavity	Cavity 1) 10mm cavity, Depth 1/4 depth (6mm) 2) 20mm cavity, Depth 1/2 depth (13mm) 3) 30mm cavity, Depth 3/4 depth (19mm)	• Hole(s) in weld • discontinuity formed by gas entrapment during solidification
Crack	Linear Crack 1) 10mm crack, Depth 1/4 depth (19mm) 2) 20mm crack, Depth 1/2 depth (13mm) 3) 30mm crack, Depth 3/4 depth (6mm)	• Also called as discontinuity. • Discontinuity is defined as an interruption of the typical structure of a material

In evaluating the accuracy of SMF technique in detecting anomalies in weldment, a pipe T-joint reference has been fabricated which contains of 3 cracks at difference locations as can be seen in Figure 4. The thickness of the pipe is 1 inch.

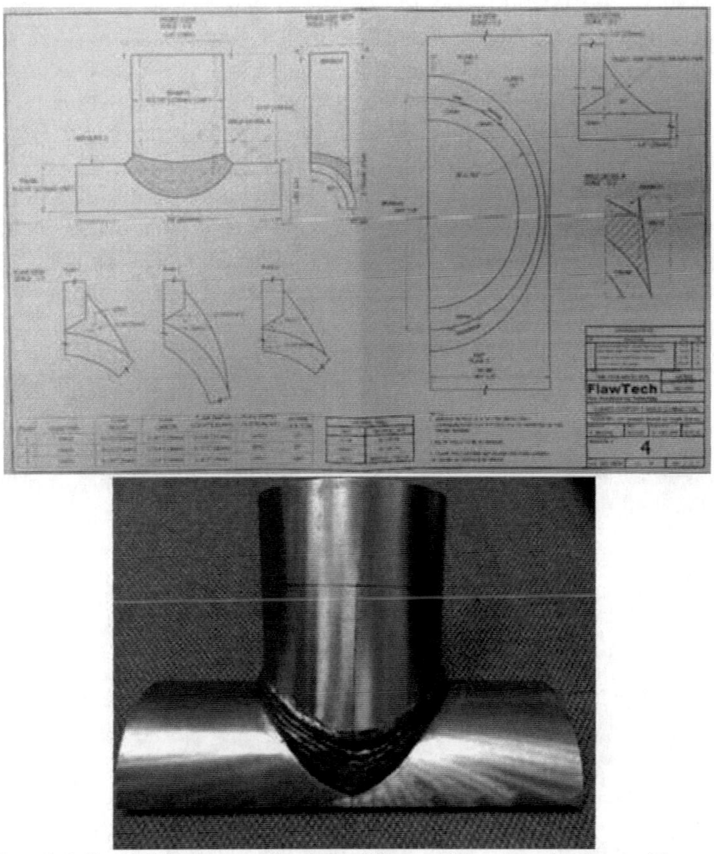

Figure 4. Simulated defect of linear cracks at 3 different locations in the weldment of T-joint P91 pipes as shown in the product drawing in above picture.

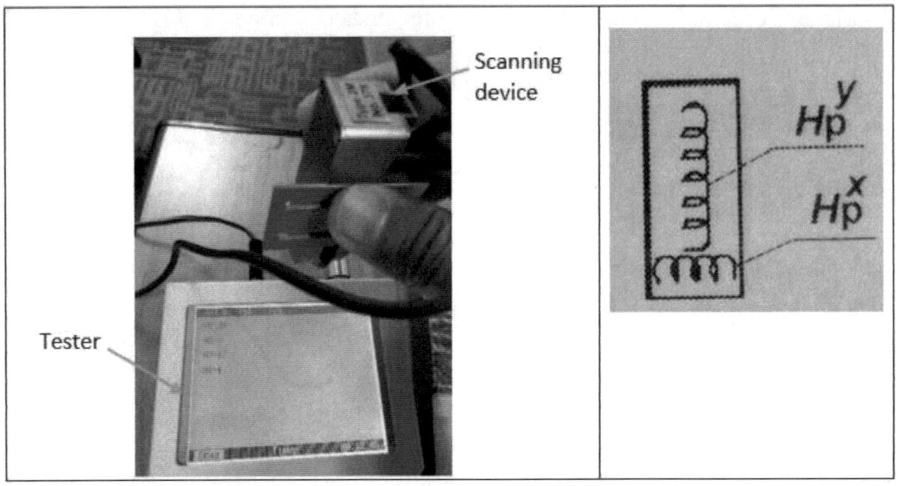

Figure 5. Calibration of scanning device (sensor) and its functionality checking: vertical down sensor's position (Left). Flux-gate transducers in two-component sensor (Right).

For SMLF measurement, tester equipment and scanning device (sensor) type TSC-3M-12 (Energodiagnostika) and 2M were used. Figure 5 shows the equipment is positioned during sensor calibration and functional checking process. The reading of H2 and H4 (even sensor numbers) are normal components (Hpy) of SMLF gradient, while the reading at odd sensor numbers (H1 and H3) are the tangential components of SMLF (see Figure 5-Right). Scanning device type 2M has 2 sensors. Sensor 1 has tangential and normal components of SMLF gradient H1 and H2 respectively. While sensor 2, the tangential and normal component of SMLF gradient are indicated by H3 and H4. During the calibration process, as shown in Figure 6, the readings of normal component of SMLF gradient at vertical down sensor position for sensor 1 and 2 show the reading difference for H2 (sensor 1) and H4 (sensor 2) is 2 (A/M)/mm. Similarly, the reading difference of H1 (sensor 1) and H3 (sensor 2) shows close reading, 2 (A/M)/mm. These show the scanning devices (sensors 1 and 2) are in good condition. For this verification and assessment work of the application of SMLF technique on a thick component/sample, the reading of normal component of SMLF on the surface of component/sample was taken into consideration for assessing defect. Figure 7 show a base-line reading of SMLF at the surface of reference block on the position of non-defect (good condition). The average of reading recorded for tangential component of SMLF gradient is about 138 (A/M)/mm.

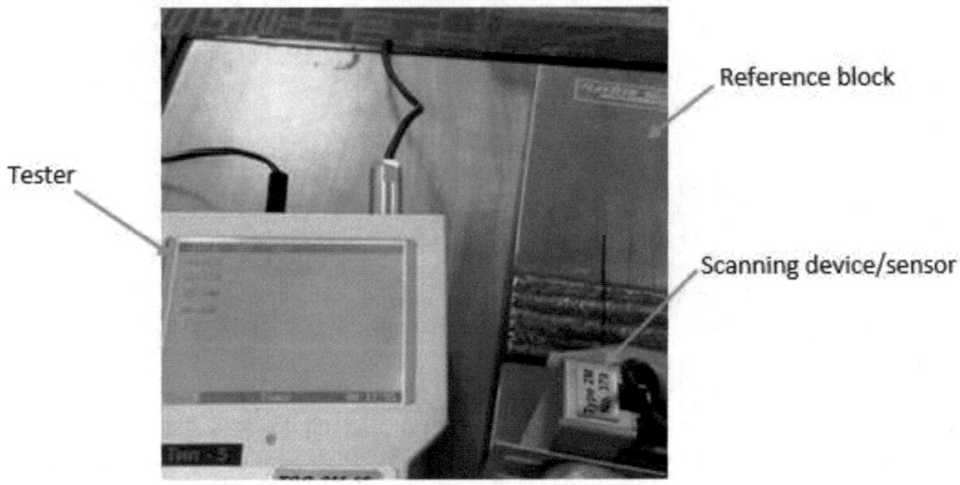

Figure 6. Scanning device (sensor) in-touching with P91 reference block's surface at non-defect location.

Figure 7. Scanning direction of the reference block's weldment.

In this verification project on the SMLF technique, the developed reference blocks have defects with various depths from surface, as can be seen the summary of defect types and parameters in Table 1. The capability of the SMLF technique was studied for any significant different results for various depths of defect in the blocks. Commonly, the threshold used in determining defect or critical condition of defect in weldment is 10 (A/M/mm). In this work, if the output (signature) of SMLF is over the threshold, hence, we confirmed that the defect is identified at the location where the scanning device is positioned on the surface of component/reference block during the inspection. The scanning device of the system has length decoder (roller), therefore the length of the sensor travels is shown on the output/result and the defect location which based on the peak over the threshold can be determined accurately. Human error for this inspection is about ±3 mm. The measurement of normal component of SMLF gradient performed on the surface where the known anomaly type and location will be discussed in details on results section.

3. Results and discussion

Magnetic particle (MP) test is unable to detect subsurface anomaly. Similarly, DP test only applicable for surface defect detection. Internal defect can be clearly shown by RT, however RT has limitation where its equipment is expensive, it needs high level of skill and it has obvious limitation on the weldment features such as fillet pipe joint. In addition, accessibility for the RT equipment to be used at limited space such as in between of stubs in power plants is an issues and limitation of RT.

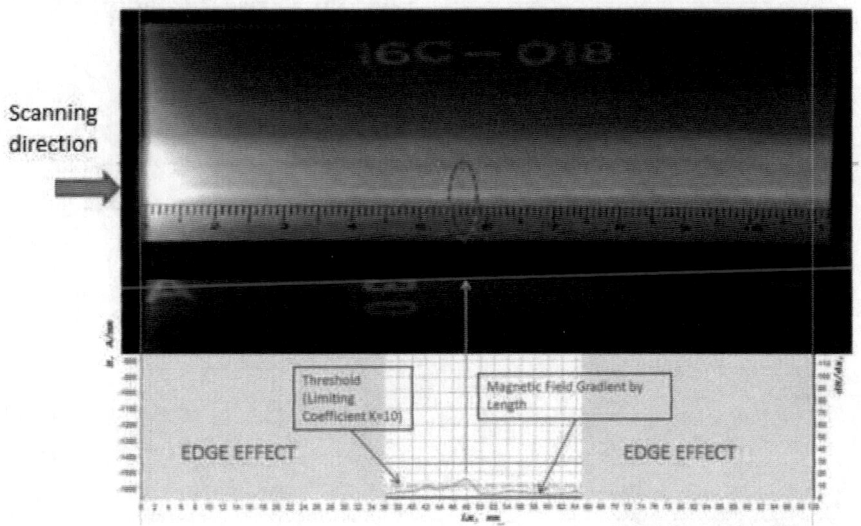

Figure 8. Comparison of Radiography Test (RT) and SMLF results on the sample block #018, linear-crack 10 mm at depth 19 mm from surface: showing the location of defect is found in agreement for both technique RT and SMLF.

Figure 8 shows the agreement of results between RT and SMLF technique. It shows that SMLF technique detected the anomaly exactly at the location as indicated by RT. The scale in the Figure 8 shows the defect is located at 47mm from the edge (see Figure 8-top), The signature of normal component of SMLF gradient is above the threshold 10 (A/M)/mm. It is noted that as can be seen in Figure 8, SMLF output has edge effect. The edge effect is attributed by cutting process at the edge of the block during fabrication process. The secondary process such as cutting has caused the residual stress and this stress remains in the material. It is found that the signature of SMLF gradient is in "U" shape. The higher SMLF gradient at the beginning and end of the measurement is due to the edge effect where the scanning device was started and ended at the edges of the reference block. It is demonstrated that the edge effect length is about 32mm from the edge. However, the peak that over the threshold at the middle/centre of output length is attributed to the known defect at the centre (about 47 mm from the edge-datum).

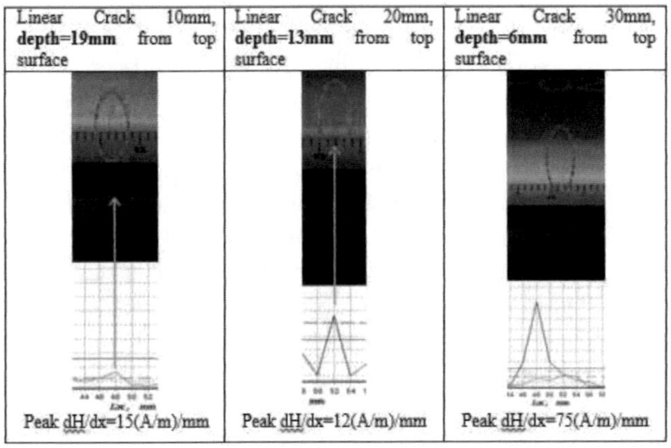

Figure 9. Linear crack's depth and SMLF output defect is found in agreement for both technique RT and SMLF.

Table 2. Reference on SMLF signature (curve and spike).

Type of Anomaly	Spikes Shape	Remarks
• Porosity 	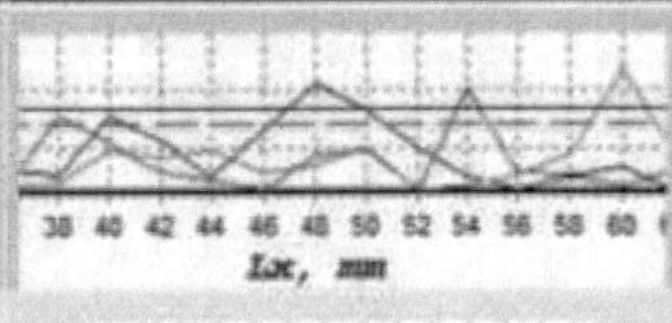	• Many spikes with close legs • At random/cluster porosity, every Scanning Device (SD) gives spikes. 
• Cavity 	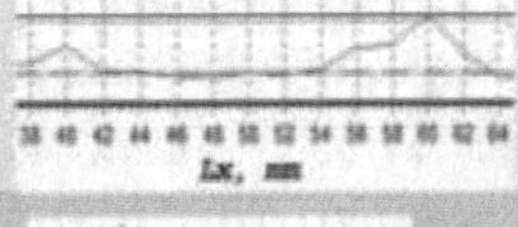	• Cavity is synonym with "hole", it shows spike with wide opening "floating" leg
• Linear (transverse) Crack 	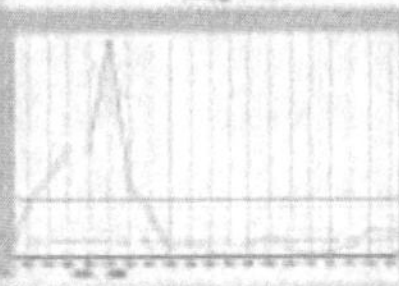	• Single sharp spike
• Planar Crack 	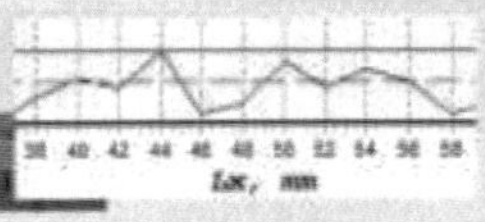	• Spike and wide opening leg (right) above threshold.

The summary of SMLF gradient results and its signature/pattern distribution according to the type of defects is shown in Table 2. The result shows obvious shape different with the different types of anomaly. Significantly the output shows single sharp peak with very high amplitude when the scanning device passed the crack compared to porosity and cavity. The gradient of SMLF on normal component (Hpy) (in this study, refer to 1 sensor), the output of SMLF for porosity defect shows many spikes with much closed legs, while for cavity defect it shows single spike with wide opening "floating" leg. Planar-crack can be differentiate with linear crack, it was demonstrated that planar crack has the output of SMLF signature with single spike above the threshold with the characteristic of wide opening leg floating for certain length before it falls back to the floor/axis (Table 2-planar crack). It is coincident with the sensor moves along the crack, which is the reason the signature has wide opening leg (right) before it falls back to the floor/axis when the scanning device has over the defect length.

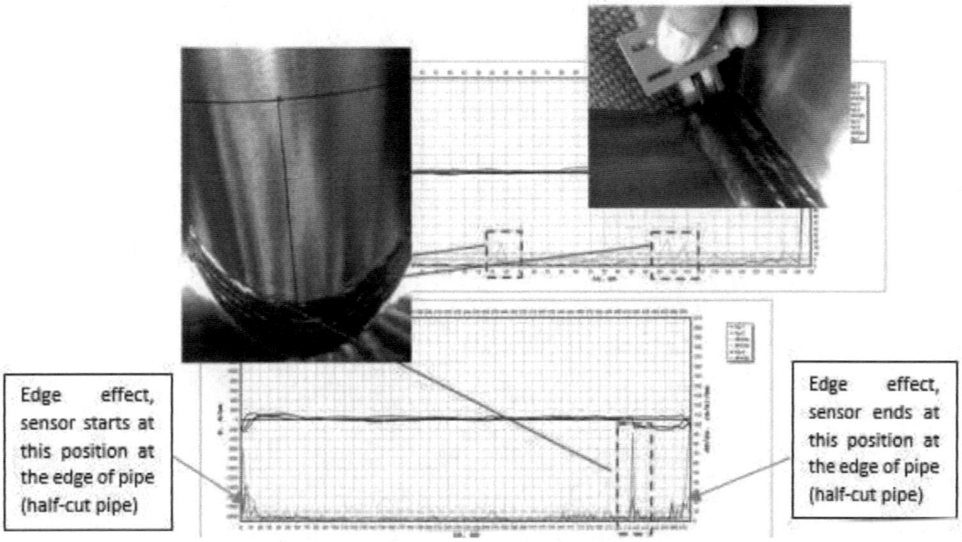

Figure 10. Defect (crack) detection and location identification using SMLF technique and T-joint pipe welding.

Refer to the blue print of graphical drawing as can be seen in Figure 4 (top), the results as shown in Figure 10 demonstrates that SMLF technique is capable of scanning, detecting and identifying the type of crack. The results of SMLF scanning obviously detects 3 significant peaks at different locations. The results of SMLF inspection is in agreement with the blue print in the drawing. The output show very sharp spike with very high amplitude, thus easily the NDT technician can define the type of anomaly inspected is linear crack. This can be confirmed and referred the signature pattern to the Table 2; very sharp peak with high amplitude over the threshold, then the defect type must a linear (transverse) crack. Figure 9 demonstrates that depth of defect from the surface does not contribute to significant different of amplitude of the peak. Obviously for crack type of anomaly, as can be seen in Figure 9, the shape where sharp single spike is similar for the 3 cracks that developed in the reference P91 pipe-joint (see the output in Figure 10).

4. Conclusions

The main purpose of the application of this technique is the determination of stress concentration zones (SCZs) and simultaneously identification of defect in ferromagnetic materials. The validation works on the developed reference blocks have demonstrated that SMLF technique can detect internal defects in ferromagnetic material P91. It also demonstrated that SMLF technique can detect flaws/defects in a thick ferromagnetic material. The SMLF technique can identify the location of defect with the accuracy ±3 mm. Based on the signatures of SMLF output of its gradient, the type of anomaly within the material can be determined. Comparatively. The SMLF technique is practical, save cost, faster and can be done via automation with special design of scanning device, compared with other NDTs such as Dye Penetrant, Magnetic Particle, Ultrasonic and Radiography tests.

References

[1] Dubov A and Kolokolnikov S 2008 *Proc.of the IIW Int. Conf. Safety and Reliability of Welded Components in Energy and Processing Industry* (Graz, Austria) p 709–14
[2] Dubov A and Kolokolnikov S 2010 *Welding in the world* **54** 9/10
[3] Dubov A and Kolokolnikov S 2008 *17th World Conf. on Nondestructive Testing* (Shanghai, China) p 25-8
[4] Bin L L Feng Z H and Di S 2012 *Information Technology Journal II* **4** 516-9
[5] Wenjian R, Jinli S, Xi C, Zhen W and Jilin R 2013 *Journal of Mechanical Engineering* **43** 2
[6] Wang P, Zhu S, Tian G Y, Wang H and Wang X 2008 *17th World Conf. on Destructive Testing* p 25-28
[7] Jiancheng L, Minqiang X, Jianwei L and Jiazhong Z 2010 *Chinese Journal of Mech Eng* **23** 4
[8] Dubov A, Kawka A and Juraszek 2010 *Proceedings of ECNDT* (2017) Available at http://www.ndt.net/article/ecndt2010/reports/1_14_03.pdf
[9] Roskosz M 2010 *Proc.of the 15th Int. Workshop on the Electromagnetic Non-Destructive Evaluation ENDE* (Szczecin, Poland)

Coefficient of friction and wear rate effects of different composite nanolubricant concentrations on Aluminium 2024 plate

N N M Zawawi[1*], **W H Azmi**[1], **A A M Redhwan**[1,2] **and M Z Sharif**[1]

[1]Faculty of Mechanical Engineering, Universiti Malaysia Pahang, 26600 Pekan, Pahang, Malaysia
[2]Faculty of Manufacturing Engineering Technology, TATI University College, 24000 Kemaman Terengganu, Malaysia

[*]Corresponding author: naal30@gmail.com

Abstract. Wear of sliding parts and operational machine consistency enhancement can be avoided with good lubrication. Lubrication reduce wear between two contacting and sliding surfaces and decrease the frictional power losses in compressor. The coefficient of friction and wear rate effects study were carried out to measure the friction and anti-wear abilities of Al_2O_3-SiO_2 composite nanolubricants a new type of compressor lubricant to enhanced the compressor performances. The tribology test rig employing reciprocating test conditions to replicate a piston ring contact in the compressor was used to measure the coefficient of friction and wear rate. Coefficient of friction and wear rate effects of different Al_2O_3-SiO_2/PAG composite nanolubricants of Aluminium 2024 plate for 10-kg load at different speed were investigated. Al_2O_3 and SiO_2 nanoparticles were dispersed in the Polyalkylene Glycol (PAG 46) lubricant using two-steps method of preparation. The result shows that the coefficient friction and wear rate of composite nanolubricants decreased compared to pure lubricant. The maximum reduction achievement for friction of coefficient and wear rate by Al_2O_3-SiO_2 composite nanolubricants by 4.78% and 12.96% with 0.06% volume concentration. Therefore, 0.06% volume concentration is selected as the most enhanced composite nanolubricants with effective coefficient of friction and wear rate reduction compared to other volume concentrations. Thus, it is recommended to be used as the compressor lubrication to enhanced compressor performances.

1. Introduction

For mechanical machines, lubrication is important to avoid wear of sliding parts and operational machine consistency enhancement. Failure due to surface wear at the sliding boundary is the commonly problem arise for mechanical systems with sliding contacts [1]. Lubrication reduce wear between two contacting and sliding surfaces and decrease the frictional power losses in compressor and avoid rust and corrosion [2]. Frequent usage of the systems without proper lubrication cause increment in wear rate and coefficient of performance (COP) on the system. Tribological behaviour mainly refer to coefficient of friction (COF) and wear rate plays an important role to identify the tribological performance of nanolubricant additives. Encouragement to this problem, different properties lubrications are studied to be utilised in many applications especially in transportation, heavy industries, energy generation and refrigeration system.

From literature, a few scientists focused on in enhancing lubricant properties for the energy efficiency increment. Recent development in increasing efficiency of lubricant is by dispersing nanoparticle in lubricant base and most well known as nanolubricant. Tribological performance can be further improve when nanolubricant are used as additives in automotive engines [2] through protective film formation and a rolling effects friction surfaces [3, 4]. Nanolubricants has benefits in reducing the friction of coefficient and wear rate [5]. Nanoparticles dispersed in mineral oils has effective effects in reducing wear and enhancing load carrying capacity [6]. Friction controlled through nanolubricants usage reduced wear [2]. Thus, this behaviour helps in reducing maintenance cost and improve the compressor performance.

Tribological properties of piston ring in a compressor using Al_2O_3-SiO_2/PAG composite nanoparticles as nanolubricant additives have been studied in this recent work. The coefficient of friction and wear rate effects study were carried out to measure the friction and anti-wear abilities of Al_2O_3-SiO_2 composite nanolubricants. The properties were investigated using three different speeds with a 10 kg load. Optimum composite nanolubricant volume concentration is suggested for further investigations.

2. Materials

In preparation of composite nanolubricants, different metal oxides nanoparticles namely Al_2O_3 and SiO_2 are used. Al_2O_3 (99.8% purity) with 13 nm in size and SiO_2 (99.5% purity) with 30 nm in size. These nanoparticles are procured from Sigma-Aldrich. The properties of the nanoparticles are given in Table 1. The characterizations of these nanoparticles are obtained by the field emission scanning electron microscopy (FESEM) technique. The images of FESEM are shown in Figure 1. PAG 46 lubricant properties at the atmospheric pressure is given in Table 2.

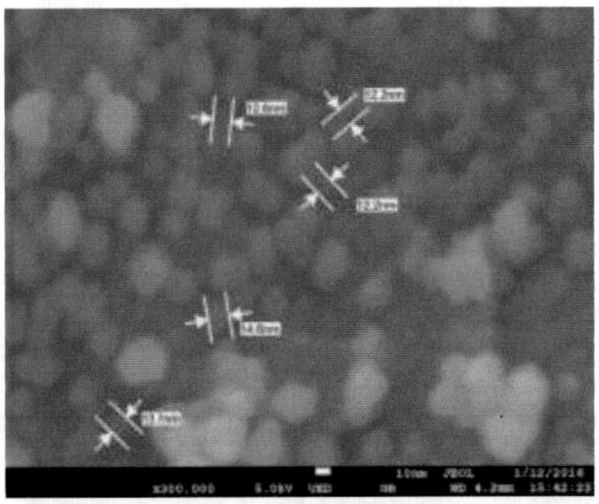

(a) Al_2O_3

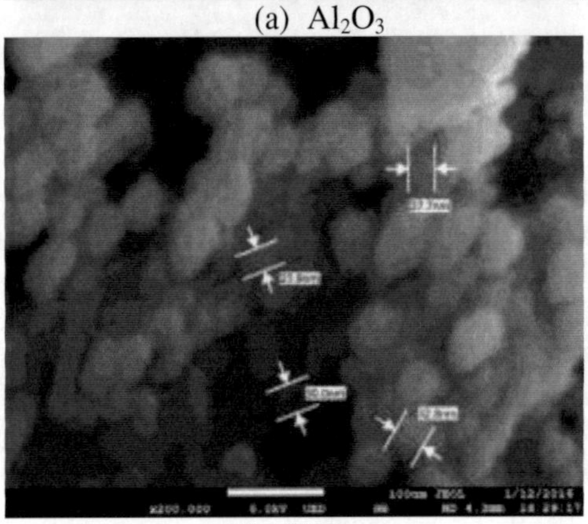

(b) SiO_2

Figure 1. FESEM analysis nanoparticles [7].

Table 1. Properties Al_2O_3 and SiO_2 nanoparticles [8-10].

Properties	**Al_2O_3**	**SiO_2**
Molecular mass, g mol^{-1}	101.96	60.08
Average Particle diameter, nm	13	30
Density, kg m^{-3}	4000	2220
Thermal Conductivity, W m^{-1} k^{-1}	40	1.4
Specific heat, J kg^{-1} K^{-1}	773	745

Table 2. Properties of polyalkylene glycol (PAG 46) [11, 12].

Properties	**PAG 46**
Density, g.cm^{-3}@ 20.0°C	0.9954
Flash Point, °C	174
Kinematic viscosity, cSt @ 40°C	41.4-50.6
Pour Point, °C	-51

A Sanden SD 7H10 piston type compressor is used in the investigation [13]. Prior the tribology test, Positive Material Identification (PMI) method is done on the compressor cylinder wall to identify the wall's material. Further examination on hardness and surface roughness are done consequently. Table 3 shows the result of the investigations. Aluminium Al 2024 was used as sample plate in tribology investigation. Table 4 shows the properties of Al 2024.

Table 3. Compressor Cylinder Wall Properties.

Properties	Outcome
Material	Al 2024
Hardness (Rockwell B)	74.52
Surface roughness, Ra, (µm)	0.121

Table 4. Properties of Aluminium 2024 [14].

Properties	**Aluminum 2024**
Ultimate Tensile Strength, psi	70,000
Yield Strength, psi	50,000
Brinell Hardness	120
Rockwell Hardness	B75

3. Methodology

3.1. Composite nanolubricant characterization

The two-step method was used in the preparation of composite nanolubricant for 0.02 to 0.1% volume concentration. During first part of preparation, all nanolubricants used in the experiment; Al_2O_3/PAG nanolubricant and SiO_2/PAG nanolubricants were prepared separately. The volume concentrations of the composite nanolubricant were calculated using equation (1) [7, 9, 15].

$$\phi = \frac{m_p/\rho_p}{m_p/\rho_p + m_l/\rho_l} \times 100 \quad (1)$$

The composite nanolubricants were then mixed and homogenized together by a 50:50 mixture ratio for each volume concentration for two hours. The ultrasonic bath was used for stabilization and for the reduction of agglomerates size of the composite nanolubricant. Nanoparticles are unable to spread homogeneously without sonication in the base fluid [16]. Colloidal stability of the composite nanolubricant dispersions was then measured for each volume concentration using UV-Vis Spectrophotometer (with wavelength accuracy of ±1 nm) within wavelength range 190 to 1800 nm at room temperature. This step is important to measure stability of composite nanolubricants as it is possible that agglomeration of nanoparticles would occur causing damage to the sliding surfaces [2]. Dispersion stability of the composite nanolubricant was visually observed after preparation and after 14 days.

3.2. Test rig description

The tribology test rig bench was designed to imitate the lubricated reciprocating sliding motion of the piston ring interface of compressor. Figure 3 illustrates the piston ring assembly of designated tribology test rig bench. The test was conducted in according to ASTM G181-11 [17] standard. Coefficient of friction performance and specific wear rate for composite nanolubricant were evaluated using test rig. Tribology test was operated under a reciprocating movement which similar to the piston head operating in real condition and contact occurred when the attached piston ring around the piston head had a sliding contact with the cylinder wall. Material that were used for the ring segment is cast iron which is the common material for the piston ring. The main structural features of the test rig are shown in Figure 2.

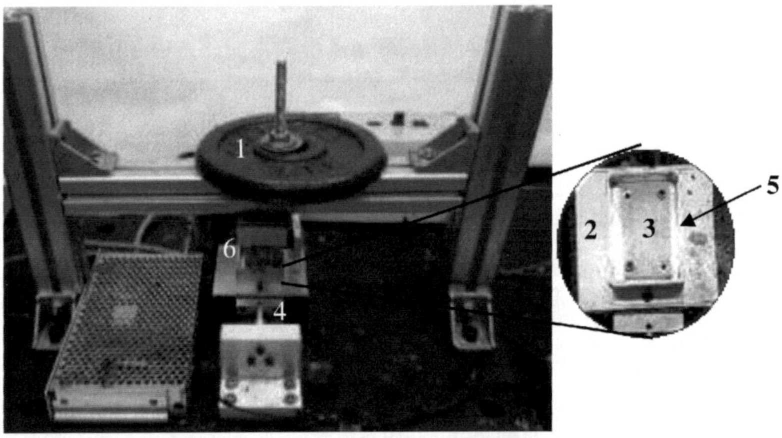

Figure 2. Tribology test rig bench. The main component of the test rig; 1- load, 2-holder, 3- specimen, 4- force sensor, 5- specimen bath, 6- ring segment.

Load was applied to the test rig by placing the load on the bearing lever above the specimen plate. A 10 kg load was chosen and the experiments were conducted for 5 minutes for each specimen with different speed ranging from 200 to 300 rpm. To generate larger friction in compressor, low speed intervals (200 rpm to 300 rpm) were used during testing [18]. A small amount of 7 ml composite nanolubricants was supplied between ring segment and specimen (in the specimen bath). The coefficient of friction was then measured and recorded using ARDUINO Software. Wear rate (k_i) was then calculated by calculating using the following equation (2):

$$k_i = \frac{V}{F_n S} \tag{2}$$

where V is volume loss, and can be determine using equations (3) and (4) meanwhile F_n is the applied load on ring and S is the sliding distance.

$$V = \frac{1}{\rho}(\Delta W) \quad (3)$$

$$\Delta W = W_1 - W_2 \quad (4)$$

The tests were carried out at least three times under the same conditions in an effort to replicate experimental results then the average results were taken to minimize measurement error. Before tests were done, specimen surface polishing was done to minimize the surface roughness of the sample plates and to avoid unnecessary test rig vibration during test. Furthermore, surface polishing is important to imitate the real roughness of cylinder wall and piston head in the real AAC compressor system. Tribology properties such as friction of coefficient and the wear rate reflect the tribological performance of Al_2O_3-SiO_2 composite nanolubricant.

4. Results and discussion

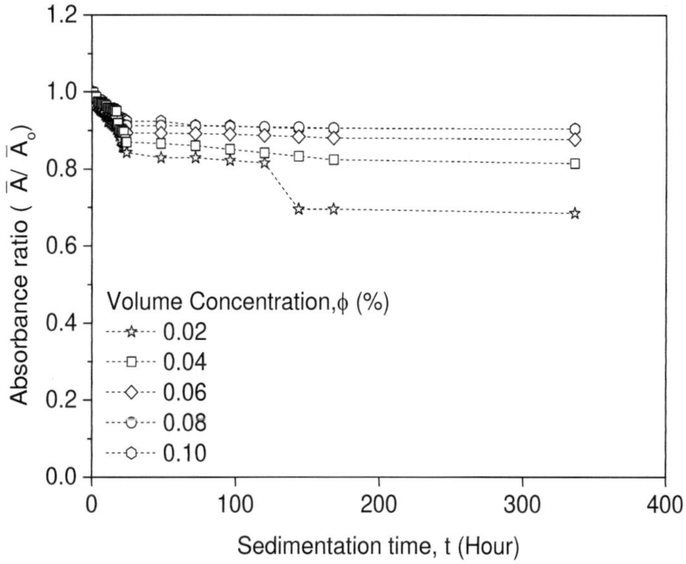

Figure 3. Absorbance ratio versus sedimentation time.

Figure 3 shows the absorbance ratio against sedimentation time for different volume concentration. The highest absorbance ratio represents the most stable nanolubricant. Absorbance ratio is observed to be more than 80% for all volume concentrations with exception of 0.02%. The 0.02% volume concentration absorbance ratio dropped to less than 80% and then maintained. Hence, it can be concluded Al_2O_3-SiO_2/PAG composite nanolubricant for all concentrations in this work were stable over a time period of 400 h sedimentation time. The dispersion stability of the composite nanolubricant was observed visually for all samples with conditions after preparation and fourteen days later. It was observed that no sedimentations occurred in the samples as shown in Figure 4. It should be noted that no surfactant was used in this preparation.

IOP Conf. Series: Materials Science and Engineering **257** (2017) 012065 doi:10.1088/1757-899X/257/1/012065

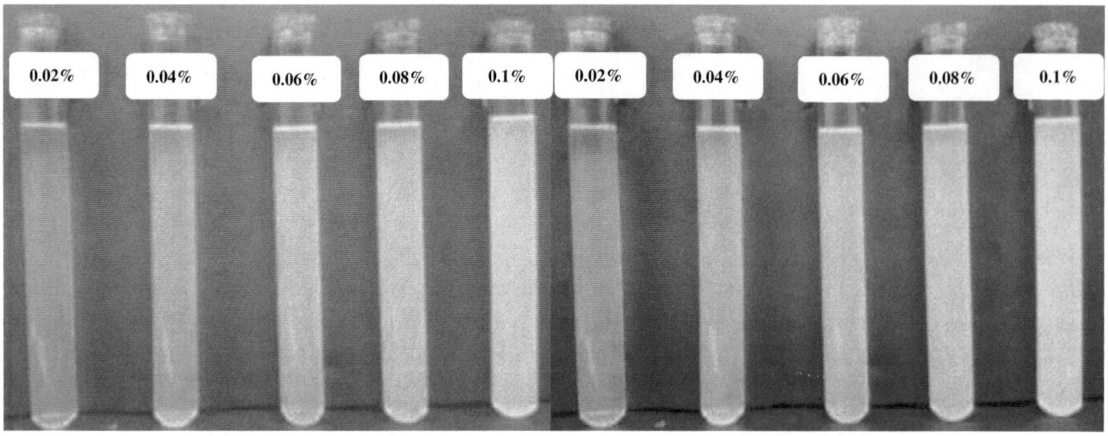

(a) After preparation (b) After 14 days

Figure 4. Composite nanolubricant (a) after preparation (b) after 14 days.

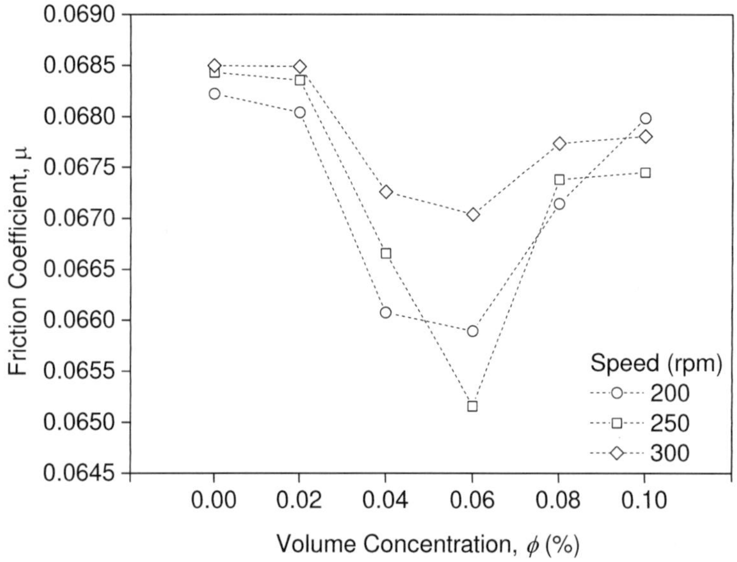

(a) Coefficient of friction vs volume concentration for 10 kg load.

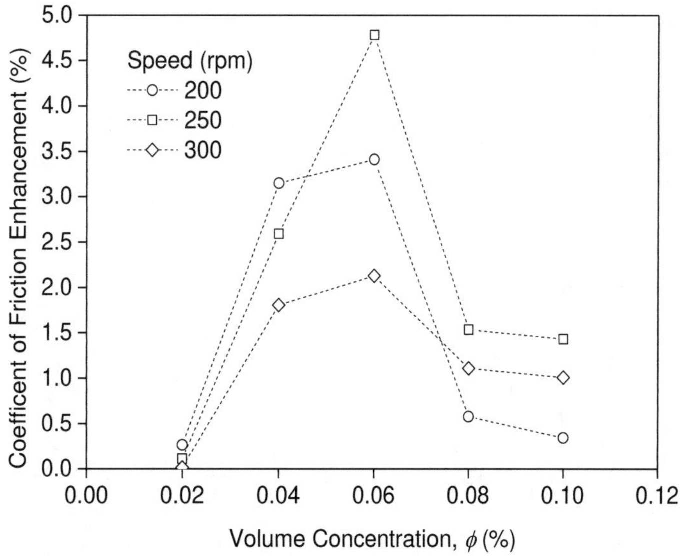

(b) Coefficient of friction enhancement.

Figure 5. Coefficient of fiction.

Figure 5(a) illustrates the effect of composite nanoparticles concentration on coefficient of friction for different speeds and a contact load of 10 kg. All volume concentrations of Al_2O_3-SiO_2 composite nanolubricants showed a decrement pattern in the coefficient of friction. Based on the obtained results, it was observed that the composite nanolubricants with 0.06% volume concentration of nanoparticles was the best sample out all nanolubricants investigated. All enhancement of coefficient of friction for all volume concentrations of Al_2O_3-SiO_2/PAG composite nanolubricant are shown in Figure 5(b). From Figure 5(b), the maximum reduction achievement for friction of coefficient by Al_2O_3-SiO_2/PAG composite nanolubricants by 4.78% with 0.06% volume concentration. Meanwhile, the lowest is recorded at 0.01% for 0.02% volume concentration. Based on the obtained results, 0.06% volume concentration is the most optimum concentration composite nanolubricant for reducing coefficient of friction. The friction coefficient was reduced due to nanoparticles mechanism [2]. This mechanism involves converting sliding into rolling friction. Beside that, tribo-films formation on the sliding surfaces enhanced the compressor efficiency.

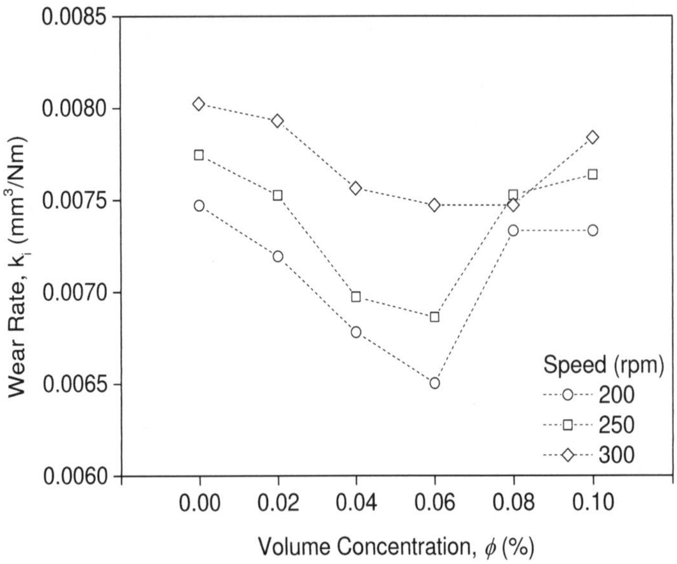

(a) Wear rate vs volume concentration for 10 kg load.

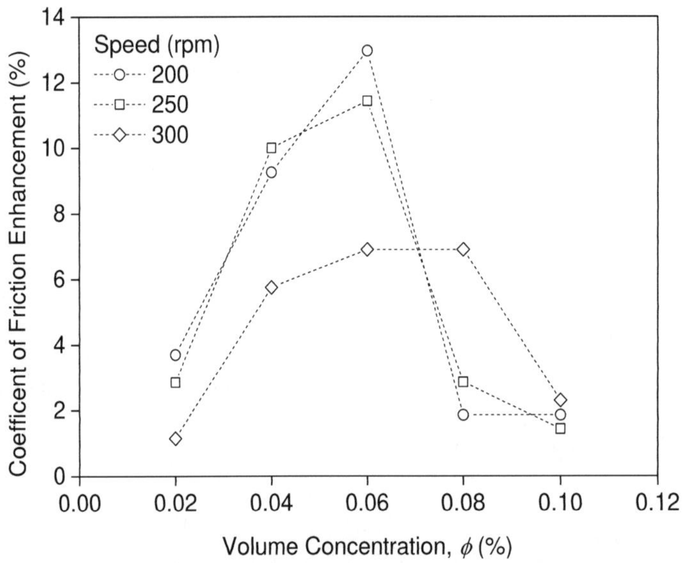

(b) Wear rate enhancement.

Figure 6. Wear rate.

The effect of composite nanoparticles concentration on wear rate is illustrated in Figure 6(a) for different speeds and a contact load of 10-kg. All volume concentrations of Al_2O_3-SiO_2 composite nanolubricants show a decrement pattern in the coefficient of friction. Figure 6(b) shows the enhancement of wear rate for all volume concentrations of Al_2O_3-SiO_2/PAG composite nanolubricant. From Figure 6(b), the maximum reduction achievement for wear rate by Al_2O_3-SiO_2 composite nanolubricants by 12.96% with 0.06% volume concentration. Meanwhile, the lowest is recorded at 1.15% for 0.02% volume concentration. Based on the obtained results, 0.06% volume concentration is the most optimum concentration composite nanolubricant for reducing wear rate. Ali et al [2] noted that optimum concentration of nanoparticle is a crucial parameter to investigate tribological performance as can negatively effect the performance because of the excess of nanoparticles.

5. Conclusions

Coefficient of friction and wear rate of composite nanolubricant were determined at 0.02 to 0.1% and various speed from 200 to 300 rpm with a 10 kg load. Al_2O_3-SiO_2/PAG composite nanolubricant with 0.06% volume concentration recorded the highest COF reduction of 4.78%. while, the maximum wear rate reduction of 12.96% recorded with the same 0.06% volume concentration sample. The present composite nanolubricant was intended to be tested in automotive air conditioning systems that use compressors and lubricated by polyalkylene glycol (PAG). Therefore, it is recommended to use the Al_2O_3-SiO_2/PAG composite nanolubricants with volume concentrations of 0.06% for application in refrigeration systems as it yields higher enhancement in coefficient of friction and wear rate which were important in tribological performance. Further investigations on the tribological performance using the Al_2O_3-SiO_2/PAG composite nanolubricants are required to extend the present work.

Acknowledgment

The authors are grateful to the Universiti Malaysia Pahang (UMP) and Automotive Engineering Centre (AEC) for financial supports given under PGRS170381 and RDU1603110.

References

[1] Williams J and Le H 2006 *J. Appl. Phys. D: Appl. Phys.* **39** R201
[2] Ali M K A, Xianjun H, Mai L, Qingping C, Turkson R F and Bicheng C 2016 *Tribol. Int.* **103** 540-54
[3] Ali M K A, Xianjun H, Turkson R F, Peng Z and Chen X 2016 *RSC Adv.* **6** 77913-24
[4] Dai W, Kheireddin B, Gao H and Liang H. 2016 *Tribol. Int.* **102** 88-98
[5] Bi S, Guo K, Liu Z and Wu J 2011 *Energy Convers. Manage.* **52** 733-7
[6] Zhang Z, Xue Q and Zhang J 1997 *Wear* **209** 8-12
[7] Sharif M Z, Azmi W H, Redhwan A A M, Mamat R and Yusof T 2017 *Int. J. Refrig.* **75** 204-216
[8] Sharif M Z, Azmi W H, Redhwan A A M and Mamat R. 2016 *Int. J. Refrig.* **70** 93-102
[9] Sharif M Z, Azmi W H, Redhwan A A M, Zawawi N N M 2017 *MATEC Web of Conferences:EDP Sciences* **90** 01049
[10] Redhwan A A M, Azmi W H, Sharif M Z and Zawawi N N M 2017 *MATEC Web of Conferences:EDP Sciences* **90** 01051
[11] Brown W L 1993 *CRC Handbook of Lubrication and Tribology* vol 3, ed R Booser (CRC Press) p 253-67.
[12] Dow 2013 Material Safety Data Sheet *Ucon Refrigerant Lubricant 213*
[13] Redhwan A A M, Azmi W H, Sharif M Z and Hagos F Y 2017 *MATEC Web of Conferences: EDP Sciences* **90** 01050
[14] ASTM B209-14 2014 *Standard Specification for Aluminum and Aluminum-Alloy Sheet and Plate* (West Conshohocken, PA: ASTM International)
[15] Zawawi N N M, Azmi W H, Redhwan A A M, Sharif M Z and Sharma K V 2017 *Int. J. Refrig.* **80** 1-10
[16] ASTM G181-11 *Standard Test Method for Conducting Friction Tests of Piston Ring and Cylinder Liner Materials Under Lubricated Conditions* 2011 (West Conshohocken, PA: ASTM International)
[17] Cesur İ, Ayhan V, Parlak A, Savaş Ö and Aydin Z 2014 *Adv. Mech. Eng.* **6** 503212
[18] Mahbubul I M, Shahrul I M, Khaleduzzaman S S, Saidur R, Amalina M A and Turgut A 2015 *Int. J. Heat Mass Trans.* **88** 73-81

Heat transfer and friction factor of composite TiO_2–SiO_2 nanofluids in water-ethylene glycol (60:40) mixture

M F Nabil [1*], W H Azmi [1,2], K A Hamid [1] and R Mamat [1,2]

[1] Faculty of Mechanical Engineering, Universiti Malaysia Pahang, 26600 Pekan, Pahang, Malaysia
[2] Automotive Engineering Centre, Universiti Malaysia Pahang, 26600 Pekan, Pahang, Malaysia

[*]Corresponding author: nabilfikri0112@gmail.com

Abstract. The need for high performance of heat transfer has been evaluated by finding different ways to enhance heat transfer rate in fluid. One of the methods is the combination of two or more nanoparticles and it is known as hybrid/composite nanofluids which can give better performance of heat transfer. Thus, the present study focused on combination of Titanium oxide (TiO_2) and Silicon oxide (SiO_2) nanoparticles dispersed in 60:40 volume ratio of water and ethylene glycol mixture as the base fluid. The TiO_2-SiO_2 hybrid nanofluids are prepared using two-step method for different concentration of 2.0%, 2.5% and 3.0%. The experimental determination of heat transfer coefficients are conducted in the Reynolds numbers range from 2000 to 10000 at a bulk temperature of 30°C. The experiments are undertaken for constant heat flux in a circular tube. The Nusselt number of composite TiO_2-SiO_2 nanofluids is observed to be higher than the base fluid. The finding on heat transfer coefficient shows that 3.0% volume concentration is the highest enhancement with 45.9% compared with base fluid. While at concentration 2.0% and 2.5%, the enhancement recorded were 29.4% and 33.2%, respectively. The friction factor of nanofluids shows a decreased with the increasing of Reynolds numbers. However, the friction factor slightly increased with the increased of concentration.

1. Introduction

In recent decades, the researchers have been carried out on the development of convective heat transfer enhancement techniques. Previously, the used of conventional fluids such as water, ethylene glycol and oil as an additives is one of the techniques. However, the conventional fluids have relatively low thermal conductivity compared with solid particles. The addition of micron sized particles as an additive in a base fluid enhanced the heat transfer. Some disadvantages with micron-sized particles are large particles sized, its caused some problems such as clogging, settle down rapidly, erosion of the heat transfer device and pressure drop increases rapidly [1]. Therefore, nanofluids was introduced by Choi et al. [2] with size dimensions of less than 100 nm in a liquid.

According to Hatwar et al. [3], nanofluids have potential to be used for wide applications in industries such as microelectronics, transportations and manufacturing because of several factors which are (i) higher thermal conductivity and the larger surface area enhanced the heat transfer rate, (ii) pressure drop is minimum because of small size particles, (iii) most effective in rapid heating and cooling system [3]. A study was done by Azmi et al. [4] for the heat transfer coefficient and friction factor of single TiO_2 and SiO_2 water based nanofluids. To observe the heat transfer performance, the experimental study is conducted at working temperature of 30°C for volume concentration up to 3.0%.

The findings from the study showed that the TiO_2 nanofluids enhanced 26% in heat transfer coefficient at 1.0% concentration, while 33% enhancement for SiO_2 nanofluids at 3.0% concentration.

Recently, the researchers gave a comprehensive review about the combination of nanoparticles which are known as hybrid nanofluids [5-8]. Hybrid nanofluids is the combination of two or more dissimilar nanoparticles in the base fluid [9-12]. The aim of synthesizing hybrid nanofluids is to improve the properties and heat transfer of single materials. Suresh et al. [7] studied the effect of hybrid Al_2O_3-Cu water based nanofluids in heat transfer. The convective heat transfer increases with increasing of Reynolds numbers with 13.56% enhancement compared with pure water at volume concentration 0.1%. For the friction factor of hybrid nanofluids, it slightly higher compared with Al_2O_3/water single nanofluids. However, the friction factor correlations are in good agreement with the experimental data. Therefore, hybrid nanofluids have potential in improving the performance of heat transfer.

Since there are very limited information on convective heat transfer for hybrid nanofluids, the present study is conducted experimentally using hybrid TiO_2-SiO_2 nanofluids dispersed in 60:40 (water/EG) volume ratio for various concentration with bulk temperature of 30°C.

2. Methodology

2.1. Composite nanofluids preparation

The nanomaterials of TiO_2 and SiO_2 nanofluids were procured from US Research Nanomaterials, Inc. are used in the present study. The water/EG based nanofluids at volume ratio of 60:40 contained TiO_2 (40 wt.%) and SiO_2 (25 wt.%) nanoparticles with an average diameter of 50 nm and 22 nm, respectively. Equation (1) is used to convert the weight concentration to volume concentration. The single nanofluids are prepared separately at concentration of 3.0%. Both TiO_2 and SiO_2 nanofluids are then mixed together to produce composite nanofluids by mixing process for an hour. The composite nanofluids is then subjected to ultrasonic hemogenization for 2 hours. This to ensure the solution are stable. For record, the samples were found to be stable for more than a month. The dilution from 3.0% concentration to lower concentration 2.5% and 2.0% are prepared by adding the base fluid water/EG into the nanofluids using equation (2).

$$\phi = \frac{\omega \rho_{bf}}{\frac{\omega}{100}\rho_{bf} + \rho_p\left(1 - \frac{\omega}{100}\right)} \tag{1}$$

$$\Delta V = (V_2 - V_1) = V_1\left(\frac{\phi_1}{\phi_2} - 1\right) \tag{2}$$

where ϕ is volume concentration in %; ω is weight concentration in %; ρ_{bf} is density of base fluid in kg/m^3; ρ_p is density of nanoparticles in kg/m^3; ΔV is additional volume in mL; V_1 is initial volume in mL; V_2 is final volume in mL; ϕ_1 is initial volume concentration in %; and ϕ_2 final volume concentration in %.

2.2. Thermo-physical properties

The density and specific heat of composite nanofluids for different concentrations are obtained from the solid-liquid mixture relation as shown in equations (3) and (4), respectively.

$$\rho_{hnf} = (1-\varphi)\rho_{bf} + (0.5\times\varphi)\rho_{p_1} + (0.5\times\varphi)\rho_{p_2} \tag{3}$$

$$C_{hnf} = \frac{(1-\varphi)\rho_{bf}C_{bf} + (0.5\times\varphi)\rho_{p_1}C_{p_1} + (0.5\times\varphi)\rho_{p_2}C_{p_2}}{\rho_{hnf}} \tag{4}$$

where ρ_{hnf} is density of hybrid nanofluids in kg/m^3; φ is volume fraction; ρ_{bf} is density of base fluid in kg/m^3; ρ_p is density of nanoparticles in kg/m^3; C_{hnf} is specific heat of hybrid nanofluids in J/kg.K; C_{bf} is specific heat of base fluid in J/kg.K; and C_p is specific heat of nanoparticles in J/kg.K.

The dynamic viscosity and thermal conductivity of TiO_2-SiO_2 nanofluids were measured for each concentration using Brookfield LVDV-III Ultra Rheometer and KD2 Pro Thermal Properties Analyzer. The thermo-physical properties for the TiO_2-SiO_2 nanofluids in water/EG mixture based are presented in Table 1.

Table 1. Thermo-physical properties of TiO_2-SiO_2 nanofluids at temperature 30°C.

Volume concentration, ϕ (%)	**Density, ρ (kg m^{-3})**	**Viscosity, μ (kg m^{-1}s^{-1})**	**Thermal conductivity, k (W m^{-1}K^{-1})**	**Specific heat, C (J kg^{-1}K^{-1})**
2.0	1098.78	3.06	0.450	3338.12
2.5	1109.63	3.09	0.456	3299.15
3.0	1120.48	3.12	0.467	3260.94

2.3. Experimental setup

As shown in Figure 1, the force convection experimental setup is integrated with a collecting tank, pump, flow rate meter, test section, differential pressure transducer, voltage regulator, chiller and the data logger. The 1.5 m length of copper tube as a test section (ID = 16 mm and OD = 19 mm) was insulated with fibre glass insulators. Six K-type thermocouples are fixed at different location on the body of tube wall (surface temperature) whereas the other two of thermocouples is placed at the inlet and outlet of test section to measure the temperature of working fluid (bulk temperature). A constant 7955 W/m^2 heat flux is supplied by the heater regulator, while the chiller is provided in the system to cool and maintain the bulk fluid temperature of 30°C. Thermocouple and pressure transducer are connected to the data logger (data acquisition system). The flow meter is used to measure the fluid flow rate in LPM by controlling the bypass regulator. The experimental value of Nusselt number for water/EG mixture base fluid is compared with single-phase liquid relation by Dittus-Boelter [13] as in equation (5) for the validation of experimental setup.

$$Nu = 0.023Re^{0.8}Pr^{0.4} \tag{5}$$

where Nu is Nusselt number; Re is Reynolds numbers; Pr is Pradtl number.

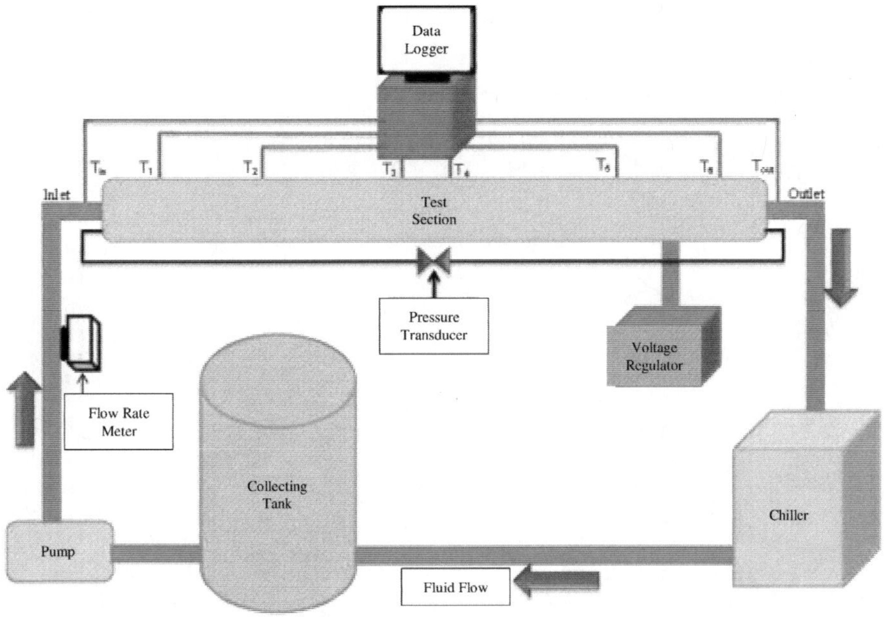

Figure 1. Schematic diagram of test rig.

3. Results and discussion

The force convection heat transfer experiment is started with water/EG base fluid and the data was compared to Dittus-Boelter [13] equation as shown in Figure 2. The experiment is conducted at working temperature 30°C with volume concentration of 3.0%, 2.5% and 2.0%. Figure 3 shows the variation of Nusselt number against Reynolds numbers ($2000 < Re < 10000$). The variation of graph shows that, the fluid that contain large amount of particles (high concentration) in base fluid solutions has higher Nusselt number. The same pattern is founded by Chiam et al. [14] which indicated that the thermal conductivity depend on volume concentration and affect the Nusselt number. The highest enhancement of thermal conductivity could be achieved on high volume concentration. At this condition, the Nusselt number is efficient for the convection process.

The distribution of heat transfer coefficient versus Reynolds number for the TiO_2-SiO_2 nanofluids in mixture water/EG is shown in Figure 4. From the figure, it shows that the heat transfer coefficient increases with increasing of Reynolds numbers and volume concentration. The maximum enhancement of heat transfer coefficient is achieved at volume concentration of 3.0% with 45.9% compare to the base fluid. While, at concentrations 2.0% and 2.5%, the heat transfer coefficient enhanced to 29.4% and 33.2%, respectively. The average enhancement percentage of heat transfer coefficient for each concentration is shown in Figure 5 and Table 2, respectively. This shows that the addition of small amount of the combination of TiO_2-SiO_2 nanoparticles to the mixture base of water/EG improved the heat transfer performance. Previous studies also encounter similar findings [4, 15-17]. However, in their studies the single nanofluids were used such as Al_2O_3, TiO_2, and SiO_2.

Table 2. Heat transfer coefficient enhancement of TiO_2-SiO_2 nanofluids at temperature 30°C.

Volume concentration, ϕ (%)	Average enhancement, (%)
2.0	29.4
2.5	33.2
3.0	45.9

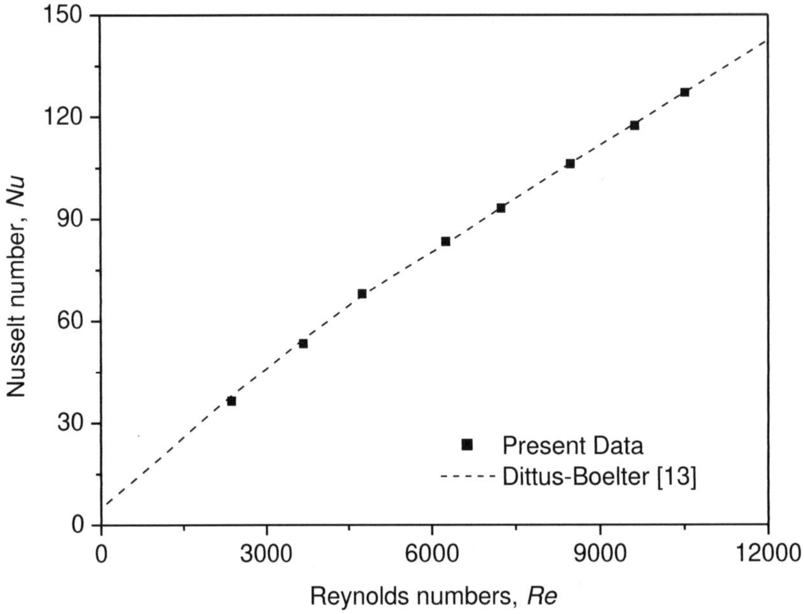

Figure 2. Validation of Nusselt number base fluid with Dittus-Boelter [13] relation.

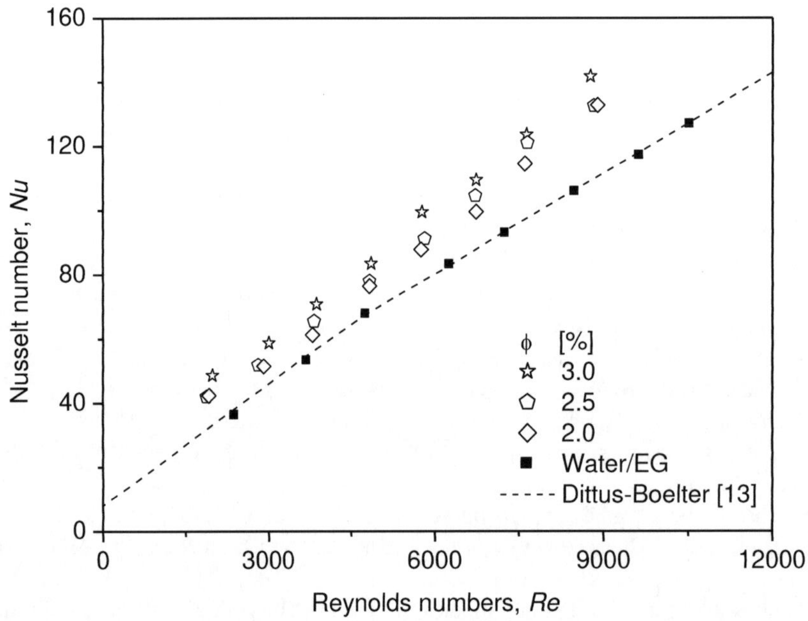

Figure 3. Variation of Nusselt number against Reynold numbers for concentration $2.0 < \phi < 3.0\%$.

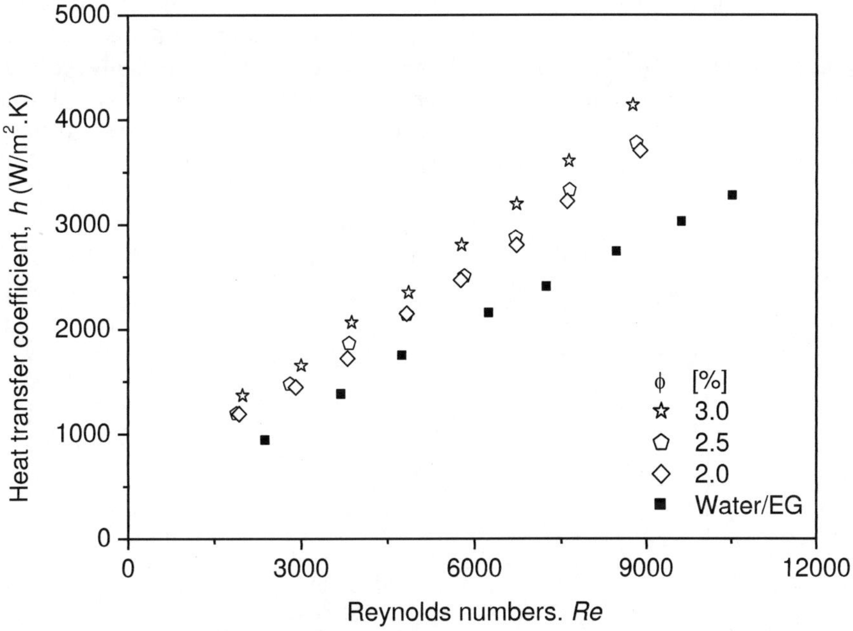

Figure 4. Distribution of heat transfer coefficient versus Reynolds numbers for concentration $2.0 < \phi < 3.0\%$.

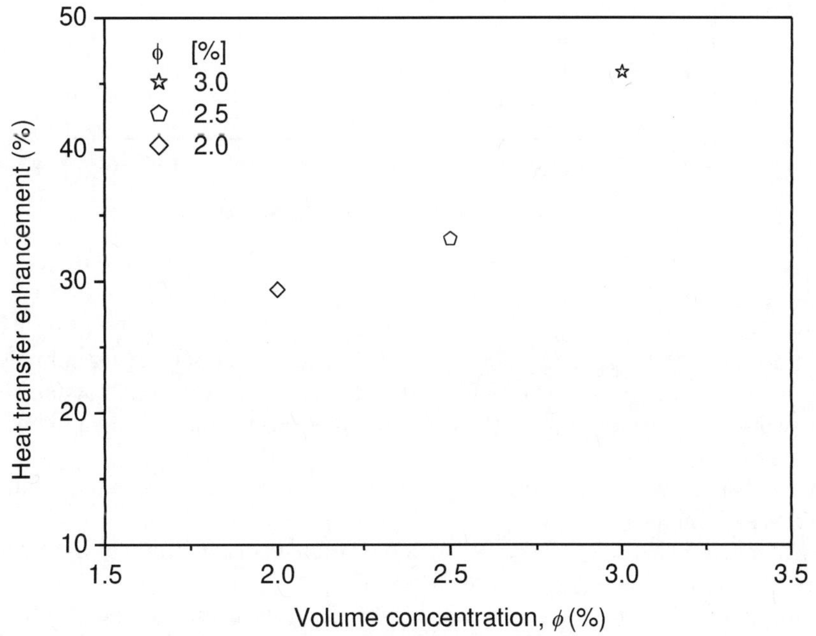

Figure 5. Enhancement of heat transfer coefficient for concentration $2.0 < \phi < 3.0\%$ at 30°C.

The present data of friction factor for composite TiO_2-SiO_2 nanofluids in mixture water/EG base fluid at different volume concentrations are shown in Figure 6. The figure shows that, the experimental friction factor is decreasing with the increment of Reynolds numbers. The friction factor of base fluid followed closely to the Blasius [18] line and decreased exponentially with the increasing of Reynolds numbers. The friction factor of hybrid nanofluids at all concentration increased slightly at about 7 to 8 % and follow the base fluid trend. However, the friction factor are higher at lower Reynolds numbers and then its decrease more closer to the Blasius [18] line.

$$f_B = \frac{0.3164}{Re^{0.25}} \tag{6}$$

where f_B is Blasius friction factor; *Re* is Reynolds numbers.

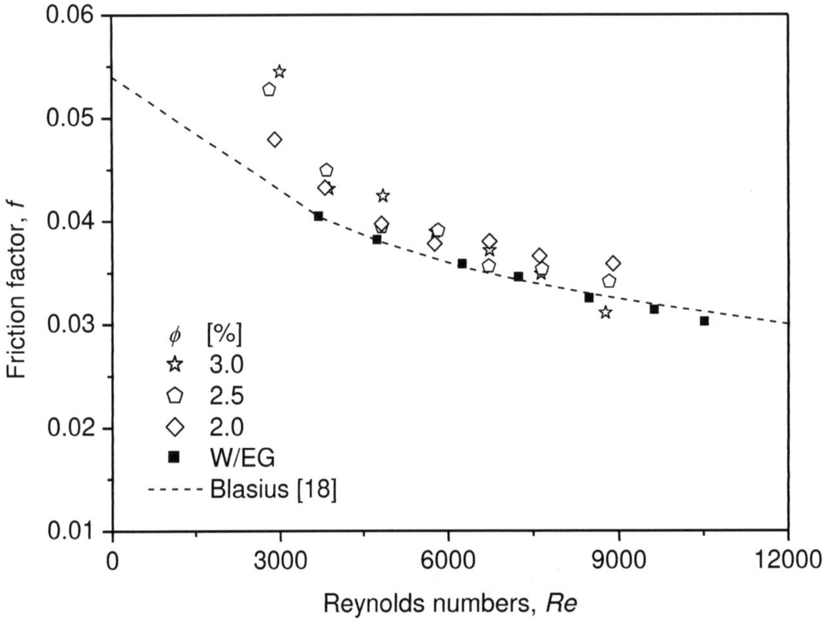

Figure 6. Variation of friction factor against Reynolds numbers for concentration $2.0 < \phi < 3.0\%$.

4. Conclusions

The present study is focused on the augmentation of convective heat transfer for composite TiO_2-SiO_2 nanoparticles in mixture of water/EG base fluid. The force convection heat transfer experiment was performed at working temperature of 30°C. From the experimental observations, the following conclusions are made:

a. The fluid that contain large amount of particles (high concentration) in base fluid solutions has higher Nusselt number.
b. The heat transfer coefficient increases with the volume concentration of hybrid nanofluids and enhanced more than 29% compared with base fluid.
c. The maximum enhancement of convective heat transfer is found at 3.0% volume concentration by 45.9%, then followed by volume concentration of 2.5% and 2.0% with the enhancement of 33.2% and 29.4%, respectively.
d. The friction factor of composite TiO_2-SiO_2 nanoparticles in the water/EG base fluid is decreased with increasing of Reynold numbers slightly increased for concentration 2.0-3.0%.

Acknowledgment

The authors are grateful to the University Malaysia Pahang (UMP) for financial support. This work was supported by the UMP AEC RDU under grant No RDU 160395 and RDU 170375.

References

[1] Das S K, Choi S U S and Patel H E 2006 *Heat Trans. Eng.* **27** 3-19

[2] Choi S U S 1995 *"Enhancing thermal conductivity of fluids with nanoparticles in developments and applications of non-newtonian flows,"* Ed D A Singer and HP Wang (New York: American Society of Mechanical Engineers) p 99–105

[3] Hatwar S and Kriplani V M 2014 *Int. J. Adv. Res. Sci. Eng.* **3** 175-83

[4] Azmi W H, Sharma K V, Sarma P K, Mamat R and Najafi G 2014 *Int. Commun. Heat Mass* **59** 30-8

[5] Sarkar J, Ghosh P and Adil A 2015 *Renew. Sust. Energ. Rev.* **43** 164-77

[6] Hemmat Esfe M, Wongwises S, Naderi A, Asadi A, Safaei M R and Rostamian H 2015 *Int. Commun. Heat Mass* **66** 100-4

[7] Suresh S, Venkitaraj K P, Selvakumar P and Chandrasekar M 2012 *Exp. Therm. Fluid Sci.* **38** 54-60

[8] Moghadassi A, Ghomi E and Parvizian F 2015 *Int. J. Therm. Sci.* **92** 50-7

[9] Sundar L S, Sharma K V, Singh M K and Sousa A C M 2017 *Renew. Sust. Energ. Rev* **68** 185-98

[10] Minea A A 2017 *Int. J. Heat Mass Transf.* **104** 852-60

[11] Jana S, Salehi-Khojin A and Zhong W H 2007 *Thermochim. Acta* **462** 45-55

[12] Hormozi F, ZareNezhad B and Allahyar H R 2016 *Int. Commun. Heat Mass* **78** 271-6

[13] Dittus F W and Boelter L M K 1930 *Heat transfer in automobile radiators of the tubular type* (University of California Publications on Engineering)

[14] Chiam H W, Azmi W H, Usri N A, Mamat R and Adam N M 2017 *Exp. Therm. Fluid Sci.* **81** 420-9

[15] Abdul Hamid K, Azmi W H, Mamat R, Usri N A and Najafi G 2015 *J. Mech. Eng. Sci.* **8** 1367-75

[16] Usri N A, Azmi W H, Mamat R, Hamid K A and Najafi G 2015 *Energy Procedia* **79** 403-8

[17] Abdolbaqi M K, Mamat R, Sidik N A C, Azmi W H and Selvakumar P 2017 *Int. J. Heat Mass Transf.* **108** 1026-35

[18] Blasius H 1913 *Mitteilungen über Forschungsarbeiten auf dem Gebiete des Ingenieurwesens* (Springer) p 1-41

IOP Conf. Series: Materials Science and Engineering **257** (2017) 012067 doi:10.1088/1757-899X/257/1/012067

Improved thermal conductivity of TiO_2-SiO_2 hybrid nanofluid in ethylene glycol and water mixture

K A Hamid[1*], **W H Azmi**[1,2], **M F Nabil**[1] **and R Mamat**[1,2]

[1] Faculty of Mechanical Engineering, Universiti Malaysia Pahang, 26600 Pekan, Pahang, Malaysia
[2] Automotive Engineering Centre, Universiti Malaysia Pahang, 26600 Pekan, Pahang, Malaysia

*Corresponding author: khamisah0301@gmail.com

Abstract. The need to study hybrid nanofluid properties such as thermal conductivity has increased recently in order to provide better understanding on nanofluid thermal properties and behaviour. Due to its ability to improve heat transfer compared to conventional heat transfer fluids, nanofluids as a new coolant fluid are widely investigated. This paper presents the thermal conductivity of TiO_2-SiO_2 nanoparticles dispersed in ethylene glycol (EG)-water. The TiO_2-SiO_2 hybrid nanofluids is measured for its thermal conductivity using KD2 Pro Thermal Properties Analyzer for concentration ranging from 0.5% to 3.0% and temperature of 30, 50 and 70°C. The results show that the increasing in concentration and temperature lead to enhancement in thermal conductivity at range of concentration studied. The maximum enhancement is found to be 22.1% at concentration 3.0% and temperature 70°C. A new equation is proposed based on the experiment data and found to be in good agreement where the average deviation (AD), standard deviation (SD) and maximum deviation (MD) are 1.67%, 1.66% and 5.13%, respectively.

1. Introduction

The nanofluids are proven to be one of the suitable heat transfer fluid as its thermal properties are improved compared to conventional fluid such as water, ethylene glycol (EG) and oil based fluid [1, 2]. Recently, hybrid nanofluid gain attention among researcher due to its augmentation in thermal properties compared to single nanofluid. Hybrid nanofluid can be defined as combination of two or more different types of nanoparticles dispersed in a base fluid [3, 4].

A lot of studies conducted on investigation of themal conductivity where several factors mention affected the properties. Han et al. [5] studied on hybrid carbon nanotubes (CNT) attached to alumina/iron oxide sphere and dispersed in poly-alpha-olefin. The measurement was taken for temperature range from 10 to 90°C. They obtained an enhancement of 21% for volume fraction of 0.2%. The diffusive heat conduction and CNT's thermal conductivity contribute to the enhancement. A study by Suresh et al. [6] used Al_2O_3–Cu/water hybrid nanofluids with volume concentrations from 0.1% to 2%. The two nanoparticles Al_2O_3 and Cu were mixed at weight ratio of 90:10. At 2% concentration, the maximum enhancement of thermal conductivity observed was 12.11% compared to single Al_2O_3-water nanofluid. The study also obtained a significant increase in thermal conductivity ratio with the increasing in concentration where formation of larger particle-free regions in the liquid offer greater thermal resistances by the agglomerated particles. For that reason, the thermal conductivity of Al_2O_3–Cu/water hybrid nanofluids is improved. Another study by Madhesh and Kalaiselvam [7] used Cu-TiO_2 in water based as a coolant. The thermal conductivity of the coolant is improved resulting in enhancement of conductive heat transfer at about 48.4% for concentration 0.7%. The surface functionalized and highly crystalline nature of the hybrid nanofluid is said to be the factor that contributed to the enhancement of thermal conductivity. Baghbanzadeh et al. [8] investigated thermal conductivity of ssilica (SiO_2) nanosphere/multiwall carbon nanotube (MWCNT). Two ratios of SiO_2 to MWCNT were used in the investigation which were 80:20 and 50:50. The enhancement in thermal conductivity were 16.7% for SiO_2:MWCNT at ratio 50:50, and 12.3% at ratio 80:20. The structure of the CNT that have high aspect ratio and its high thermal conductivity influence the thermal conductivity. The hybrid nanofluid with more percentage of CNT showed higher thermal conductivity.

From the literature survey, there were very few investigations concern on hybrid nanofluid mixed at ratio other than 50:50. Hence, the purpose of the present study is to investigate the thermal conductivity of TiO_2-SiO_2 hybrid nanofluid at ratio 40:60 (vol.%), dispersed in EG/water (40:60, vol.%). The measurement of thermal conductivity is carried out at three temperatures 30, 50 and 70°C for nanofluid concentration ranging from 0.5% to 3.0%.

2. Materials and solutions

Two types of nanoparticles used in the present study to prepare hybrid nanofluid are TiO_2 and SiO_2, procured from US Research Nanomaterials, Inc. (USA) with weight concentration of 40 wt.% and 25 wt.%, respectively. The average size of TiO_2 and SiO_2 are 50 nm and 30 nm, respectively. A solution of EG and water is mixed at volume ratio of 40:60, used as the base fluid. Properties of nanoparticles TiO_2 and SiO_2, and EG are presented in Table 1 and Table 2.

Table 1. Properties of TiO_2 and SiO_2 nanoparticles [9-12].

Property	TiO_2	SiO_2
Average Particle diameter, nm	50	30
Density, kg m^{-3}	4230	2220
Thermal Conductivity, W m^{-1} k^{-1}	8.4	1.4
Specific heat, J kg^{-1} K^{-1}	692	745

Table 2. Properties of ethylene glycol (EG) [13].

Property	EG
Formula	$C_2H_6O_2$
Molecular mass, $gmol^{-1}$	62.07
Freezing point, °C	-12.7
Boiling point, °C at 101.3 kPa	198
Viscosity, mPas at 20°C	20.9
Density, kgm^{-3} at 20°C	1113
Specific heat, $Jkg^{-1}K^{-1}$ at 20°C	2347

3. Methodology

This section elaborates on the preparation of TiO_2-SiO_2 hybrid nanofluid samples at six volume concentrations. The measurement of the thermal conductivity is also included covering the measurement technique and the description of the instrument related.

3.1. Preparation of hybrid nanofluid

The TiO_2-SiO_2 hybrid nanofluid is prepared by mixing the single nanofluids of TiO_2 and SiO_2, undergo mixing process and sonication. For both single nanofluid TiO_2 and SiO_2, the weight concentration is converted to volume concentration using equation (1) [14, 15] where ϕ, ω, and ρ are volume concentration, weight concentration and density, respectively. The subscript p and bf correspond to nanoparticles and base fluid. Equation (2) [14, 16] is used to dilute the volume concentration at higher concentration to lower concentration by adding the EG/water base fluid. ΔV is the volume of the base fluid required to dilute the nanofluid at higher concentration (ϕ_1) to the lower concentration (ϕ_2), prior to V_1 is known. TiO_2 nanofluid is then mixed with SiO_2 nanofluid at volume ratio 40:60 with magnetic stirrer for 15 minutes; and subjected to sonication with ultrasonic bath for 2 hours. Six samples with 100 mL volume each is prepared at concentration 0.5%, 1.0%, 1.5%, 2.0%, 2.5% and 3.0%. The stability of the sample is determined by Transmission Electron Microscopy (TEM) observation as shown in Figure 1. The TiO_2 and SiO_2 nanoparticles are well dispersed in the base fluid.

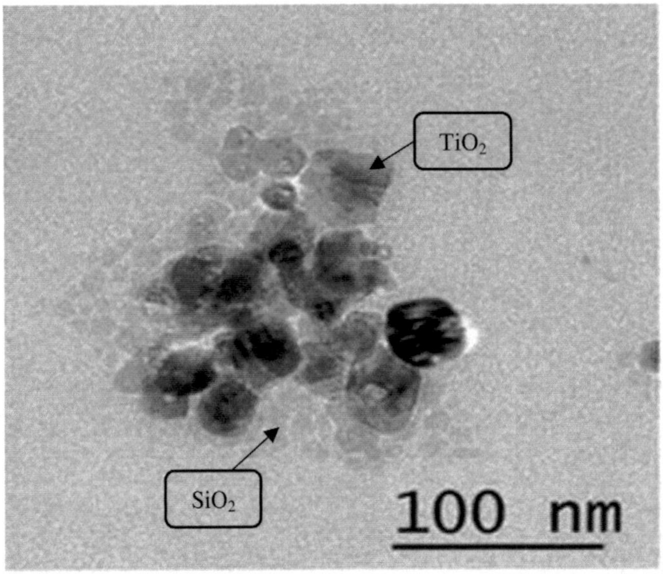

Figure 1. TEM image (100 nm scales) for TiO_2-SiO_2 hybrid nanofluids at ϕ =1.0%.

$$\phi = \frac{\omega \rho_{bf}}{\left(1 - \frac{\omega}{100}\right)\rho_p + \frac{\omega}{100}\rho_{bf}} \tag{1}$$

$$\Delta V = V_2 - V_1 = V_1\left(\frac{\phi_1}{\phi_2} - 1\right) \tag{2}$$

3.2. Measurement of thermal conductivity

The measurement of thermal conductivity is conducted using KD2 Pro Thermal Properties Analyzer. The device measures the thermal conductivity by evaluating the time and temperature response of the sudden electric signals. The device is equipped with handheld controller and needle sensor KS-1. The range of thermal conductivity measured is 0.02 to 2.00 W/m.K with an accuracy of ±0.01%. The glycerine solution is provided for calibration/verification of the sensor, where the thermal conductivity for glycerin at 25°C is 0.285 W/m.K (±5%). A water bath is used to control the temperature of the sample along the measurement. A nanofluid sample with 70 mL is inserted in a bottle, immersed in the water bath until the temperature reach the desired temperature (30, 50 and 70°C). The arrangement of the measurement is shown in Figure 2. The interval of each measurement is set at 15 minutes. A minimum of five readings were taken and the mean value was considered in the analysis for accuracy within 5%. Previous study by Abdolbaqi et al. [17] and Zakaria et al. [18] also used the same device.

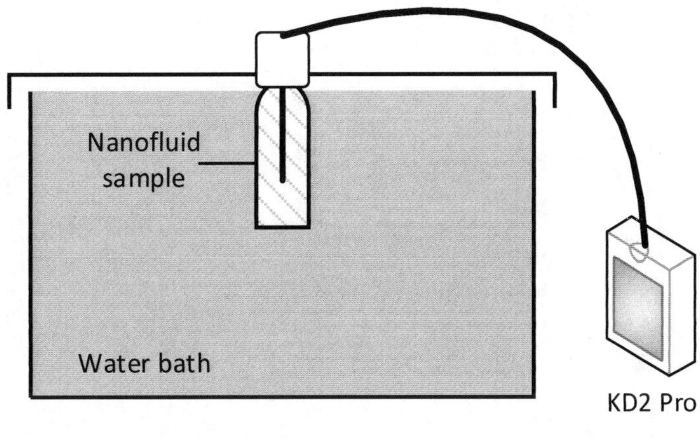

Figure 2. Measurement of thermal conductivity with KD2 Pro and water bath.

4. Results and discussion

Figure 3 shows the effect of concentration to the thermal conductivity of TiO_2-SiO_2 hybrid nanofluid for three temperatures; 30, 50 and 70°C. The thermal conductivity at all concentrations of the nanofluids are higher than the base fluid (ϕ =0%). As the concentration increased, the thermal conductivity of the nanofluid is also increased. The maximum thermal conductivity is occurred at highest concentration 3.0%. The range of enhancement compared to base fluid are from 3.9-17.4%, 5.8-18.9%, and 8.7-22.1%, respectively for temperature at 30, 50 and 70°C. The effect of concentration is noticeable due to the addition of the nanoparticles. The rate of collision between the nanoparticles is increased hence affect the thermal conductivity of the nanofluid [19].

The effect of temperature to the thermal conductivity of TiO_2-SiO_2 nanofluid is shown in Figure 4. The figure clearly shown that the thermal conductivity is increased with the increasing of the temperature. It can be observed that all concentrations at temperature measured have higher thermal conductivity than the base fluid, which follow the trend in Figure 3. The temperature mainly affected the thermal conductivity because of the Brownian motion which is strong function of temperature [20]. When the temperature in the nanofluid is raised, the contribution to micro-convection is greater resulting to the enhancement in effective conductivity [21]. The similar behaviour is also found in previous studies by Usri et al. [22] and Megatif et al. [23].

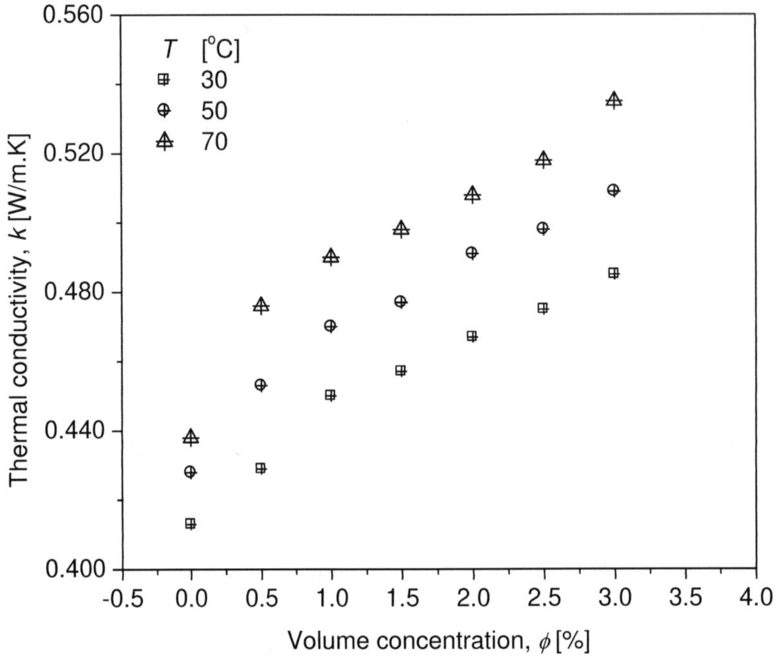

Figure 3. Effect of concentration to thermal conductivity.

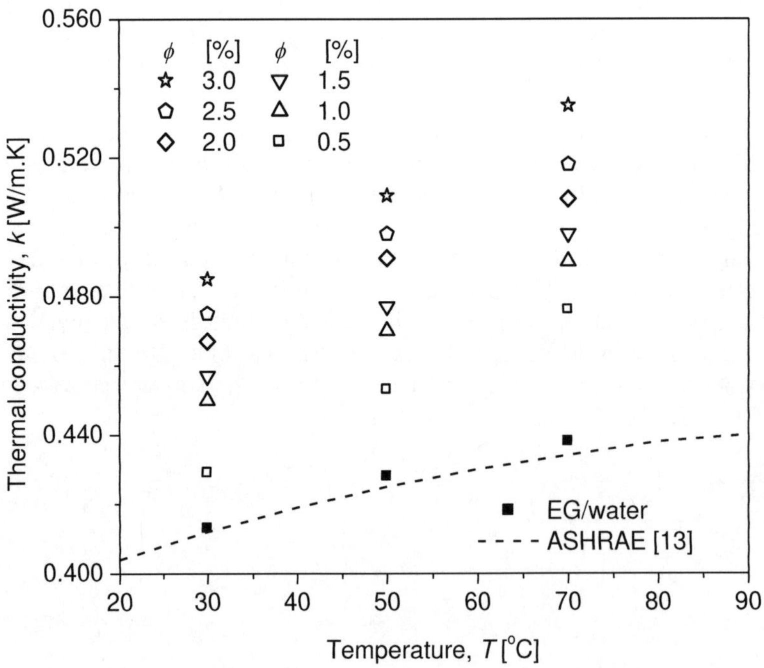

Figure 4. Effect of temperature to thermal conductivity.

Equation (3) is proposed based from experimental data where k_r, k_{nf} and k_{bf} represent the ratio of thermal conductivity, thermal conductivity of nanofluid and thermal conductivity of base fluid; while ϕ and T are volume concentration and temperature. This equation valid for range of concentration $0 \le \phi \le 3.0\%$ and range of temperature $30 \le T \le 70°C$. The average deviation (AD) and standard deviation (SD) are 1.67% and 1.66%. Maximum deviation (MD) is found to be 5.13%. The tabulation of thermal conductivity from the experiment and from equation (3) is presented in Figure 5. The figure shows that the experiment data is in good agreement with the equation proposed.

$$k_r = \frac{k_{nf}}{k_{bf}} = \left(1 + \frac{\phi}{100}\right)^{5.25} \left(1 + \frac{T}{70}\right)^{0.076} \quad (3)$$

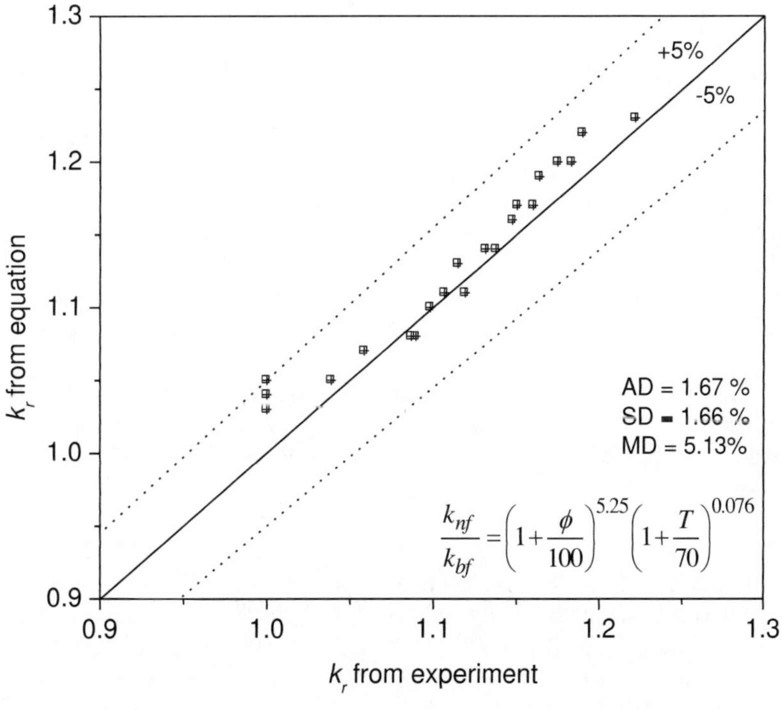

Figure 5. Comparison between experimental thermal conductivity and Equation (3).

5. Conclusions

The TiO_2-SiO_2 hybrid nanofluid is prepared with single nanofluid TiO_2 and SiO_2, mixed and sonicated until the stable solution formed. TiO_2 and SiO_2 is mixed at ratio of 40:60 (vol.%) and dispersed in base fluid of EG/water at ratio 40:60 (vol.%); for volume concentration of 0.5% to 3.0% The measurement of thermal conductivity is conducted at 30, 50 and 70°C. From the findings discussed in the previous section, in can be conclude that the thermal conductivity of TiO_2-SiO_2 hybrid nanofluid dispersed in EG/water is affected by the concentration and temperature. The thermal conductivity of the hybrid nanofluid is enhanced up to 22.1%, observed at concentration 3.0% and temperature 70°C. The proposed equation as in Equation (3) valid for volume concentration range of $0 \le \phi \le 3.0\%$ and temperature range of $30 \le T \le 70°C$.

Acknowledgment
The authors are grateful to the Universiti Malaysia Pahang (UMP) and Automotive Engineering Centre (AEC) for financial supports given under RDU160395, PGRS170375 and RDU170375.

References
[1] Azmi W H, Sharma K V, Mamat R, Najafi G and Mohamad M S 2016 *Renew. Sust. Energ. Rev.* **53** 1046-58
[2] Devendiran D K and Amirtham V A 2016 *Renew. Sust. Energ. Rev.* **60** 21-40
[3] Sundar L S, Sharma K V, Singh M K and Sousa A C M 2017 *Renew. Sust. Energ. Rev.* **68**, Part 1 185-98
[4] Leong K Y, Ku Ahmad K Z, Ong H C, Ghazali M J and Baharum A 2017 *Renew. Sust. Energ. Rev.* **75** 868-78
[5] Han Z H, Yang B, Kim S H and Zachariah M R 2007 *Nanotechnology* **18** 105701
[6] Suresh S, Venkitaraj K P, Selvakumar P and Chandrasekar M 2011 *Colloids Surf.* A **388** 41-8
[7] Madhesh D and Kalaiselvam S 2014 *Procedia Eng.* **97** 1667-75
[8] Baghbanzadeh M, Rashidi A, Rashtchian D, Lotfi R and Amrollahi A 2012 *Thermochim. Acta* **549** 87-94
[9] He Y, Jin Y, Chen H, Ding Y, Cang D and Lu H 2007 *Int. J. Heat Mass Transf.* **50** 2272-81
[10] Heris S Z, Etemad S G and Nasr Esfahany M 2006 *Int. Commun. Heat Mass* **33** 529-35
[11] Vajjha R S, Das D K and Kulkarni D P 2010 *Int. J. Heat Mass Transf.* **53** 4607-18
[12] Pak B C and Cho Y I 1998 *Exp. Heat Transfer* **11** 151-70
[13] ASHRAE 2009 *ASHRAE Handbook-Fundamentals* SI Edition, ed M S Owen and H E Kennedy (Atlanta: American Society of Heating, Refrigerating and Air-Conditioning Engineers, Inc.) p 31.4
[14] Hamid K A, Azmi W H, Mamat R and Sharma K V 2016 *Int. Commun. Heat Mass* **73** 16-24
[15] Azmi W H, Sharma K V, Sarma P K, Mamat R, Anuar S and Dharma Rao V 2013 *Exp. Therm. Fluid Sci.* **51** 103-11
[16] Azmi W H, Sharma K V, Sarma P K, Mamat R and Najafi G 2014 *Int. Commun. Heat Mass* **59** 30-8
[17] Abdolbaqi M K, Azmi W H, Mamat R, Sharma K V and Najafi G 2016 *Appl. Therm. Eng.* **102** 932-41
[18] Zakaria I, Azmi W H, Mamat A M I, Mamat R, Saidur R, Abu Talib S F and Mohamed W A N W 2016 *Int. J. Hydrogen Energ.* **41** 5096-112
[19] Charab A A, Movahedirad S and Norouzbeigi R 2017 *Appl. Therm. Eng.* **119** 42-51
[20] Akilu S, Sharma K V, Baheta A T and Mamat R 2016 *Renew. Sust. Energ. Rev.* **66** 654-78
[21] Murshed S M S, Leong K C and Yang C 2008 *Int. J. Therm. Sci.* **47** 560-8
[22] Usri N A, Azmi W H, Mamat R, Hamid K A and Najafi G 2015 *Energy Procedia* **79** 397-402
[23] Megatif L, Ghozatloo A, Arimi A and Shariati-Niasar M 2016 *Exp. Heat Transfer* **29** 124-38

Investigation on wear characteristic of biopolymer gear

Wafiuddin Bin Md Ghazali[1], Daing Mohamad Nafiz Bin Daing Idris[1*], Azizul Helmi Bin Sofian[2], Mohamad Firdaus bin Basrawi[1] and Thamir Khalil Ibrahim[1]

[1]Faculty of Mechanical Engineering, Universiti Malaysia Pahang, 26600, Pekan, Pahang, Malaysia
[2]Faculty of Chemical and Natural Resources Engineering, Universiti Malaysia Pahang, 26300, Gambang, Kuantan, Pahang, Malaysia

Corresponding author email: daingnafiz@ump.edu.my

Abstract. Polymer is widely used in many mechanical components such as gear. With the world going to a more green and sustainable environment, polymers which are bio based are being recognized as a replacement for conventional polymers based on fossil fuel. The use of biopolymer in mechanical components especially gear have not been fully explored yet. This research focuses on biopolymer for spur gear and whether the conventional method to investigate wear characteristic is applicable. The spur gears are produced by injection moulding and tested on several speeds using a custom test equipment. The wear formation such as tooth fracture, tooth deformation, debris and weight loss was observed on the biopolymer spur gear. It was noted that the biopolymer gear wear mechanism was similar with other type of polymer spur gears. It also undergoes stages of wear which are; running in, linear and rapid. It can be said that the wear mechanism of biopolymer spur gear is comparable to fossil fuel based polymer spur gear, thus it can be considered to replace polymer gears in suitable applications.

1. Introduction

Polymer has been chosen to replace steel in various application such as gearing due to its mechanical advantages and economical value. Its ability to operate with a little lubrication, low noise operation and light weight material are some of the advantages [1-6].

The abundance of polymer from the non-renewable source which is fossil fuel have raised problems such as disposing and recyclability. This brings about the emergence of biopolymer made from natural resource and more environmentally friendly. The usage of biopolymer in the gearing section has not yet gained traction which is what this paper target to overcome. Despite the same function, polymer gear does not behave as same as its counterpart, the steel gears even in the designing stage. They even have specific standards which are found in [7-9]. For polymer and engineering polymer gears, several types of research have been conducted which focuses on gear life [10], the effect of temperature[11-13] and the failure methods [14-17].

However, research on biopolymer gear is not yet available during the time of research. This research focuses on identifying the wear mechanism of biopolymer gear and whether it is comparable to fossil fuel based polymer gear. In this research, the following were used to identify wear mechanism and the failure of biopolymer gear; tooth fracture, tooth deformation, debris and weight loss. This method was also used by [12, 16, 18-20] in their research to identify failure of polymer gear.

IOP Conf. Series: Materials Science and Engineering **257** (2017) 012068 doi:10.1088/1757-899X/257/1/012068

2. Methodology

The gears were designed in SolidWorks and followed the geometry shown in Table 1. The gears were then produced using injection moulding process. The parameters of injection moulding process are seen in **Table 2**. The test was done on a dedicated test rig seen in Figure 1.

Table 1. Gear geometry.

Name	Symbol	Pinion	Gear
Module	m	$m = \frac{d_1}{z_1} = \frac{d_2}{z_2}$ $m = 2$	
Tooth	z	$z_1 = \frac{d_1}{m}$ $z_1 = 25$	$z_2 = \frac{d_2}{m}$ $z_2 = 30$
Pressure angle	a	$20°$	
Face width		$15mm$	
Reference pitch circle diameter	d	$d_1 = z_1 m$ $d_1 = 25 \times 2 = 50mm$	$d_2 = z_2 m$ $d_2 = 30 \times 2 = 60mm$
Tooth tip circle diameter	d_a	$d_{a1} = d_1 + 2h_a$ $d_{a1} = 50 + 2(2) = 54mm$	$d_{a2} = d_2 + 2h_a$ $d_{a2} = 60 + 2(2) = 64mm$
Root circle diameter	d_f	$d_{f1} = d_1 - 2h_f$ $d_{f1} = 50 - 2(2.5) = 45mm$	$d_{f2} = d_2 - 2h_f$ $d_{f2} = 60 - 2(2.5) = 55mm$
Addendum	h_a	$h_a = m$ $h_a = 2$	
Dedendum	h_f	$h_f = h_a + c \geq 1.25 \times m$ $h_f = 2 + 0.5 = 2.5$	
Tip and root clearance	c	$c = h_f - h_a \geq 0.25 \times m$ $c = 2.5 - 2 = 0.5$	
Tooth depth	h	$h = h_a + h_f \geq 2.25 \times m$ $h = 2 + 2.5 = 4.5mm$	
Centre distance	a	$a = \frac{d_1 \pm d_2}{2} = m\left(\frac{z_1 \pm z_2}{2}\right)$ $a = \frac{50 \pm 60}{2} = 2\left(\frac{25 \pm 30}{2}\right) = 55mm$	
Backlash	b_c	$b_c = 2\Delta C \tan 20$ $b_c = 2(55.05 - 55) \tan 20 = 0.36 \times 10^{-3}$meter	

Table 2. Injection Moulding Parameter.

Parameter	Value
Screw Temperature, °C	160
Dosage Stroke, mm	25, 35
Injection Speed, mm/s	75
Injection Pressure, Bar	210
Cooling time, $second$	110
Back Pressure, Bar	50

Figure 1. Test rig for polymer gear.

The gear pair were run on different RPM to create the effect of different speed. The gears were run to a certain amount of revolutions. The conditions are 500, 1000, and 1500 RPM. Each pair undergoes the test for 2 hours and a non-load and 10N load condition. The standard practice to test gear fatigue is at 1×10^6 however this result is for preliminary and to determine whether the conventional method to investigate wear characteristic in polymer gear is applicable for biopolymer gear.

The surrounding area of testing was then investigated to determine the presence of debris. Then the gears are inspected visually to detect fracture and crack formation on the tooth region, including tip and root.

The microstructure before and after the test was then studied using a microscope. The weight reduction of gears is also recorded using a digital scale.

3. Results and discussion

3.1. Weight Loss

Table 3 and 4 shows the weight loss and its rate. Figure 2 and 3 shows the graph of weight loss. The steeper the gradient indicates that the gear loses weight much faster. The slower RPM have a much steep curve compared to the higher speed. This can be explained by the effect of the gear pump as seen in [13, 16, 21-23]. When meshing, the space between the gears or backlash creates an air pocket which helps to improve the heat distribution. A much faster gear pair will distribute more heat to the surrounding, helping it to regulate the temperature. If too much heat is produced, the gear may undergo thermal deformation. The teeth may become soft, and flakes occurs which is explained in 3.3.

The weight loss is contributed by the debris formed during the test. The chipping of tooth also influences the weight reduction. The weight loss can be grouped into three stages; running in wear, where the wear occurs for a short time but high, followed by linear wear, where the wear is small but progressively increase and finally rapid wear, where the wear causes deformation leading to gear failure.

Table 3. Weight Loss of Non-Load Gear Pair.

Pair No.	Biopolymer Gear	Weight Before → After	% of Weight Loss	Gradient
500 RPM without load for 60,000 revs				
Pair 1	Driver	24.389 → 24.290	0.4059	-1.65×10^{-6}
	Driven	37.3370 → 36.340	2.6702	-1.66×10^{-5}
1000 RPM without load for 120,000 revs				
Pair 2	Driver	24.399 → 24.348	0.2090	-4.25×10^{-7}
	Driven	37.455 → 37.345	0.2945	-9.16×10^{-7}
1500 RPM without load for 180,000 revs				
Pair 3	Driver	24.390 → 24.370	0.0820	-1.11×10^{-7}
	Driven	37.342 → 36.380	2.5761	-4.00×10^{-6}

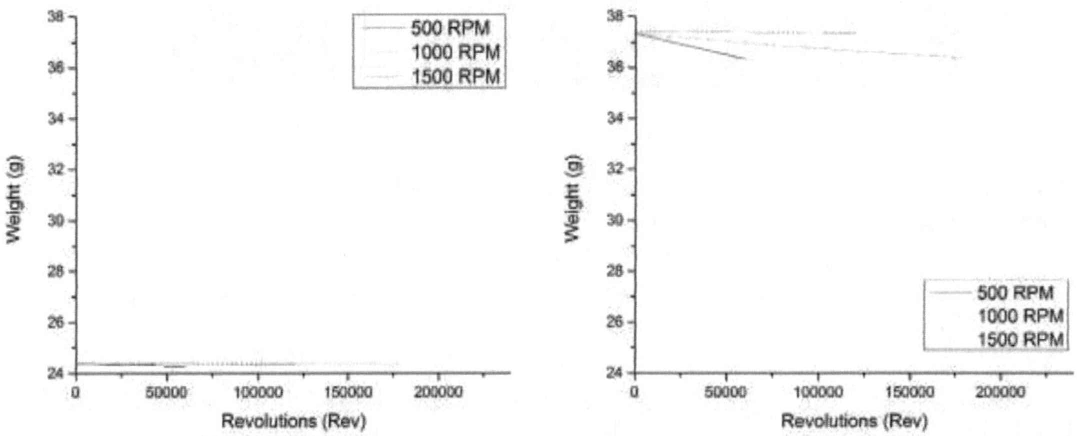

a) Driver gear without load b) Driven gear without load

Figure 2. Weight loss of non-load gear pair.

The percentage of weight loss for driven gear is much higher compared to the driver gear as it carries the load. Even without a load attached at the shaft, the shaft itself is also a load as power is needed to turn it.

The slower RPM have a higher percentage of weight loss, this is due to more debris produced. A slower RPM reduces the heat transfer rate, therefore contributing to the amount of debris produced. The gradient for gear Pair 2 does not follow the trend of other pairs. This can be caused by the defect of product during production process or misalignment of test equipment. Since the data sampling is not enough the result was taken, however for future works more pair will be tested for each parameter giving more valid data to be selected.

Table 4. Weight Loss of 10N Load Gear Pair.

Pair No.	Biopolymer Gear	Weight Before → After	% of Weight Loss	Gradient
500 RPM with load for 60,000 revs				
Pair 4	Driver	24.581 → 24.563	0.0732	-3.00×10^{-7}
	Driven	37.582 → 37.538	0.1170	-7.33×10^{-7}
1000 RPM with load for 120,000 revs				
Pair 5	Driver	24.487 → 24.477	0.0408	-8.33×10^{-8}
	Driven	37.132 → 37.067	0.1750	-5.41×10^{-7}
1500 RPM with load for 180,000 revs				
Pair 6	Driver	24.466 → 24.458	0.0326	-3.33×10^{-8}
	Driven	37.120 → 37.067	0.1427	-2.20×10^{-7}

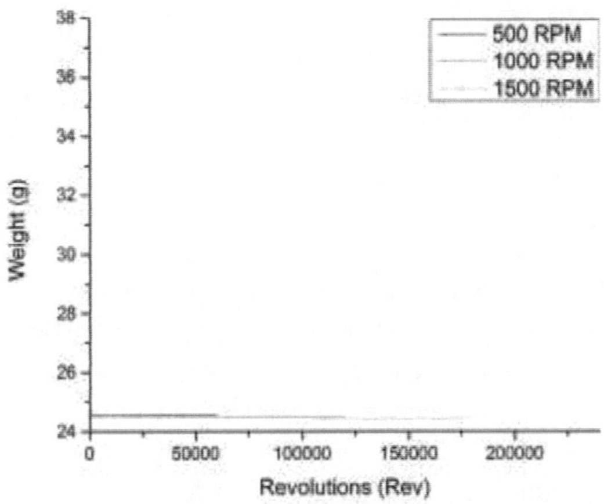

a) Driver gear with load

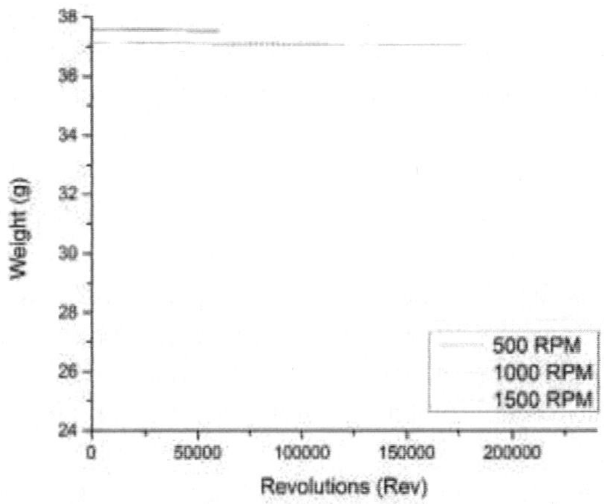

b) Driven gear with load

Figure 3. Weight loss of 10N load gear pair.

Figure 4. Backlash checking.

3.2. Tooth fracture and deformation

The gears were checked for any fracture and deformation after completing their cycles. It was found that only several pairs undergo tooth fracture and deformation. Figure 5 shows tooth fracture occurring on the biopolymer gear.

a) Fracture occurring at the pitch circle at driven gear at running condition of 1500 RPM without load

b) Fracture occurring at the tooth root at driven gear at running condition of 1000 RPM with 10N load

Figure 5. Tooth fracture on driven gear Pair 3 (a) and Pair 5 (b).

3.3. Debris formation

Debris and flake formation on the surrounding are of the gear test is due to the gear disintegrating caused by the temperature rise. The flakes mostly come from the meshing of gear pair where the tooth tip of driver gear is digs out the softened material on the tooth of the driven gear which is also found in [14].

These flakes influence the weight reduction of the gear. When the scraping is severe, it may lead to the teeth deformation and finally fracture of the said teeth. These can be reduced by implementing several steps. First is by ensuring the gear profile is accurate. This can be done by having an accuracy test on the gears. Second is checking the tooth contact. The contact can be checked by applying a gear marking compound dedicated for testing the contact of the gear. The backlash of the gears also plays and important role in reducing the temperature and tooth damage. This can be checked using a dial gauge on the meshed gear at the teeth pitch region, root, and tip. The backlash checking can be seen in Figure 4.

By increasing the speed and adding load, the heat produced is increased, which leads to more debris formation. The debris formed can be seen in Figure 6.

a) Debris formed around gear at running condition of 1000 RPM without load

b) Debris formed around gear at running condition of 1000 RPM with 10N load

Figure 6. Debris formed around gear Pair 2 (a), Pair 5 (b).

3.4. Microstructure condition

The surface was studied before and after testing to get a better understanding of the surface condition resulting from the test. The main area of interest is the pitch line, where the contact of teeth is at the maximum.

From the resulting images, it was seen that the surface becomes smooth due to the contact. The materials in contact experience higher temperature and softening may occur in that region. This leads to the smooth region on the pitch section.

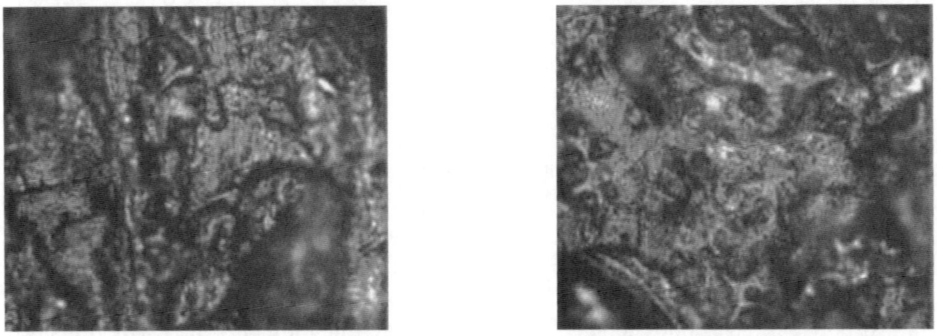

a) Surface condition of driver gear b) Surface condition of driven gear

Figure 7. Microstructure condition of gear tooth surface before test.

a) Surface condition of driver gear with running condition of 1000 RPM with 10N load
b) Surface condition of driven gear with running condition of 1000 RPM with 10N load
c) Surface condition of driver gear with running condition of 1500 RPM with 10N load
d) Surface condition of driven gear with running condition of 1500 RPM with 10N load

Figure 8. Microstructure condition of gear tooth surface after test.

Surface seen in Figure 8 a) is much smoother compared to b) which is caused by the wear is higher in b). Since the RPM is lower, less heat is released to the surrounding during the test, causing the material to soften. This makes the surface mush easier to wear compared to surface in Figure 8 b).

4. Conclusions

During this experiment, the wear characteristic of biopolymer gear is investigated. Methods used in this experiment to characterize wear was widely used in the polymer gear research field. Weight loss was used by [11, 14, 19], tooth fracture and deformation as a failure was studied by [15, 24], [16, 18, 25]used debris formation to study failure and gear microstructure was studied by [26, 27]. From those research, it can be concluded that the methods used to determine wear for polymer gear are also viable for biopolymer gear.

The wear formed on the biopolymer gear can be identified by visual examination on the gear tooth for crack and deformation. The debris formed also indicates that wear is occurring on the gear pair. The weight loss of gears is also an indication of wear. By using a microscope, the microstructure image can be investigated to prove the wear is occurring on the gear pair.

It can be noted that the methods above can be used to determine wear of biopolymer gear. The gears undergo three stages of wear; running in, linear and final rapid wear. The wear debris increase as the gear approaches final wear period.

Weight loss is due to gear tooth contact which causes friction and heat. Gear will undergo moisture loss and debris formation leading to the weight loss. Running temperature affects the material as at a certain temperature; the material will start to degrade and causing the gear to fail. For future improvements, more load value will be used and the frequency of weight recording will be increased, and the running temperature of the gears will be observed.

From these results, the biopolymer gears can be considered as an alternative to be used in application where polymer gears are used. By using biopolymer gears, the application can be more sustainable and environmental friendly.

Acknowledgements

This research work has been done under the support of Universiti Malaysia Pahang under grant RDU150386. The authors gratefully acknowledge the support given throughout this project.

References

[1] L. Snyder. (2010) At the "PEEK" of the Polymer Food Chain. *Gear Technology*. Available: http://www.geartechnology.com/articles/0610/At_the_PEEK_of_the_Polymer_Food_Chain
[2] J. L. Elmquist. (2014) Deciding When to Go Plastic. *Gear Technology*. 46-47.
[3] R. Walter. (2004) Engineering Principles for Plastic Gears. *Gear Solutions*. Available: http://www.gearsolutions.com/article/detail/5621/engineering-principles-for-plastic-gears
[4] A. Ulrich. (2012) Exploiting the Potential of Plastic Gears. *Gear Solutions*. Available: http://www.gearsolutions.com/article/detail/6255/exploiting-the-potential-of-plastic-gears
[5] M. Fletcher and Z. Smith. (2012, Improving Appliances with Plastic Gears. 5. Available: http://www.thomasnet.com/white-papers/abstract/101817/improving-appliances-with-plastic-gears.html#null
[6] M. Jaster. (2002) The Efficiency Experts: Bradley University And Winzeler Gear Collabrate On Urban Vehicle Project. *Gear Technology*. Available: http://www.geartechnology.com/articles/0910/The_Efficiency_Experts
[7] A. G. M. Association, "ANSI/AGMA 1106-A97, Tooth Proportions for Plastic Gears," ed. 1500 King Street, Suite 201, Alexandria, Virginia, 22314: American Gear Manufacturers Association, p. 47.
[8] B. S. Institution, "BS 6168. Specification for non-metallic spur gears," ed. London, 1987.
[9] V. D. Ingenieure, "VDI 2545:1981-01, Zahnräder aus thermoplastischen Kunststoffen," ed.
[10] A. Bravo, D. Koffi, L. Toubal, and F. Erchiqui, "Life and damage mode modeling applied to plastic gears," *Engineering Failure Analysis,* vol. 58, Part 1, pp. 113-133, 12// 2015.
[11] C. H. Kim, "Durability improvement method for plastic spur gears," *Tribology International,* vol. 39, pp. 1454-1461, 11// 2006.
[12] A. J. Mertens and S. Senthilvelan, "Effect of mating metal gear surface texture on the polymer gear surface temperature," *Materials Today-Proceedings,* vol. 2, pp. 1763-1769, 2015.

[13] C. J. Hooke, K. Mao, D. Walton, A. R. Breeds, and S. N. Kukureka, "Measurement and Prediction of the Surface Temperature in Polymer Gears and Its Relationship to Gear Wear," *Journal of Tribology,* vol. 115, pp. 119-124, 1993.

[14] S. Senthilvelan and R. Gnanamoorthy, "Damage Mechanisms in Injection Molded Unreinforced, Glass and Carbon Reinforced Nylon 66 Spur Gears," *Applied Composite Materials,* vol. 11, pp. 377-397, 2004.

[15] S. Senthilvelan and R. Gnanamoorthy, "Effect of rotational speed on the performance of unreinforced and glass fiber reinforced Nylon 6 spur gears," *Materials & Design,* vol. 28, pp. 765-772, // 2007.

[16] K. Mao, W. Li, C. J. Hooke, and D. Walton, "Friction and wear behaviour of acetal and nylon gears," *Wear,* vol. 267, pp. 639-645, 6/15/ 2009.

[17] H. Düzcükoğlu, "PA 66 spur gear durability improvement with tooth width modification," *Materials & Design,* vol. 30, pp. 1060-1067, 4// 2009.

[18] W. Li, A. Wood, R. Weidig, and K. Mao, "An investigation on the wear behaviour of dissimilar polymer gear engagements," *Wear,* vol. 271, pp. 2176-2183, Jul 29 2011.

[19] N. A. Wright and S. N. Kukureka, "Wear testing and measurement techniques for polymer composite gears," *Wear,* vol. 251, pp. 1567-1578, 10// 2001.

[20] S. Senthilvelan and R. Gnanamoorthy, "Effect of gear tooth fillet radius on the performance of injection molded Nylon 6/6 gears," *Materials & Design,* vol. 27, pp. 632-639, // 2006.

[21] K. Mao, "A numerical method for polymer composite gear flash temperature prediction," *Wear,* vol. 262, pp. 1321-1329, 5/10/ 2007.

[22] K. Mao, "A new approach for polymer composite gear design," *Wear,* vol. 262, pp. 432-441, 2/4/ 2007.

[23] H. Düzcükoğlu, "Study on development of polyamide gears for improvement of load-carrying capacity," *Tribology International,* vol. 42, pp. 1146-1153, 8// 2009.

[24] H. İmrek, "Performance improvement method for Nylon 6 spur gears," *Tribology International,* vol. 42, pp. 503-510, 3// 2009.

[25] K. Mao, P. Langlois, Z. Hu, K. Alharbi, X. Xu, M. Milson, *et al.*, "The wear and thermal mechanical contact behaviour of machine cut polymer gears," *Wear,* vol. 332–333, pp. 822-826, 5// 2015.

[26] H. Düzcükoğlu, R. Yakut, and E. Uysal, "The Use of Cooling Holes to Decrease the Amount of Thermal Damage on a Plastic Gear Tooth," *Journal of Failure Analysis and Prevention,* vol. 10, pp. 545-555, 2010.

[27] K. D. Dearn, T. J. Hoskins, D. G. Petrov, S. C. Reynolds, and R. Banks, "Applications of dry film lubricants for polymer gears," *Wear,* vol. 298–299, pp. 99-108, 2/15/ 2013.

The challenges of lean manufacturing implementation in kitting assembly

A F H Fansuri[1*], A N M Rose[1], N M Z Nik Mohamed[1] and H Ahmad[2]

[1] Manufacturing Focus Group, Faculty of Mechanical Engineering, Universiti Malaysia Pahang, 26600, Pekan, Pahang, Malaysia
[2] TATI University College, 24000 Kemaman, Terengganu, Malaysia

[*]Corresponding author: anithfanz@yahoo.com

Abstract. Literature studies shows that lean manufacturing goes way back with the original founder Eli Whitney in year 1799. The main purpose of lean manufacturing is to identify and eliminate waste in production. The application of lean manufacturing can be carried out in any industrial processes with regards to the understanding of lean principles, theories and practices. Kitting is one of the important aspects in a successful production. The continuous supply of materials from store to production has to be systematic and able to achieve lean standard for it to be successful. The objective of this paper is to review the implementation of lean manufacturing in kitting assembly. Previous papers show that, the implementation of lean manufacturing in kitting assembly may be beneficial to the organization such as reduce in space occupancy, part shortages, lead time and manpower. Based on previous research, some industries may tend to change between kitting and line stocking which are due to lack of understanding when implementing kitting and causes longer lead time and materials overflow in store. With a proper understanding on what to kit, where to kit, how to kit, why to kit and who kits the material with a standardised process flow may ensure the success of kitting.

1. Introduction

Lean manufacturing has been recognised as the centre spotlight of the current global industry to implement it as to achieve the higher level of leanness in today's competitive industries. Lean system is focusing on the demands of the customers and the lean philosophies should be flexible enough to meet the needs of customers [1].

Most of the companies have implemented lean manufacturing principles such as just-in-time (JIT) production which are deemed to be an ideal solution [2]. In the ever changing of the world economy, the manufacturing industries must be able to produce products or services with shorter lead times and therefore lean manufacturing has been a well-accepted philosophy by many organisations [3]. Lean manufacturing is also focused on the elimination of waste and non -value added activities [4].

Through lean manufacturing, the concept of a proper material feeding flow will determine the ability of the organisation to meet customer's demand. In this paper, the authors will review the implementation of lean manufacturing in kitting assembly. In assembly lines, parts are needed to be restored without any interruption, and the parts are needed to be replenished along the assembly line based on production plan [5].

Kitting is a process of materials feeding to the assembly station. Parts are sorted and transferred to assembly station in pre-sorted kits where one assembly product parts is in each kit [6]. The initial reasons of implementing kitting in a production assembly line is usually involves parallelised assembly system, product with various part numbers, quality of the assembly and products with high volume [7]. With kitting, parts can be kitted in advance (temporarily stored in store) before it can be delivered to the production assembly line. Therefore, with kitting, no inventories are kept at the production floor [8] .

2. Literature review

Kitting is one of the important processes of material feeding in a manufacturing company. However, the implementation itself needs a thorough understanding in lean principles before it can be carried out. It is believed that through lean manufacturing practices, the company could eliminate wastes [8]. Based on previous research, kitting is implemented successfully in many organisation [9]. However, it is also causes major problems when the implementation is not based on the production requirement such as level of work in progress, space allocation in production and walking distance for operator [10]. Due to increased production volume and demands, kitting system is considered the best option for material feeding instead of line stocking [11]. In addition, the critical success factors for a successful lean manufacturing implementation must be considered especially the top management commitment [12].

3. Kitting assembly

With the growing demand from customer, line stocking is deemed to be unsuitable for material feeding to production assembly. There are factors need to be considered by the organisation such as floor space, work-in-progress (WIP) and material security before implementing kitting process. Many companies are unsure about when a kitting system should be implemented and what is the method fits to their environment. Many researches developed kitting methods, but still these theories are difficult to be applied in production assembly [13]. Besides that, there are also few highlighted issues on the confusion of the implementation of kitting assembly due to lack of understanding in kitting principle on when and where each type of material feeding system such as (line stocking or kitting) should be used [14]. The advantages and disadvantages of kitting are listed in Table 1. It shows that the company will gain benefits on parts management and production flexibility including quality control. As a result, the production can be run smoothly and successfully towards production target.

Table 1. Advantages and disadvantages of kitting assembly [11].

Advantages	Disadvantages
Saves production space	Increase storage if kit is done in advance
Reduce operators walking and searching time.	Time consuming on making kit
Better control of WIP at assembly line since the components are stored in storage area	Part shortage from store may cause operator to kits short thus reducing the overall efficiency.
Increase flexibility at assembly line as product changeover can be done with ease.	Defect components during assembly causes additional actions and time to provide the spare piece.
Increase in product quality due to possible quality checks earlier before material is transacted to production.	Poor kitting design in picking process may cause injuries and unmotivated personnel.

Based on advantages and disadvantages of kitting, it may be implemented successfully if the companies have taken into consideration on the following factors [15]:

i. Products quality and line assembly support.
   According to Hanson and Brolin [15] research, kitting may improve product quality since the operator at the assembly line does not have to concern what are the parts to be assemble and what material to pick. It also provides a learning support when the kitted material is arranged in a manner which reflects the assembly operation.

ii. Man-hour consumption.
Kitting can reduce man-hour consumption as the operator does not need to spend their time to search for parts. Besides that, without kitting production with high variety and high volume of products requires more space to store parts in production line. In addition, with increased number of part numbers, the walking distance and time for the assembly operator is also increased. Thus, through kitting, the kitted material is easily accessible by the assembly operator.

iii. The flexibility in product changeover.
Through kitting it will create more flexibility in changing one product to another, or product of different colour shades as the parts are not stationed at the assembly area. Besides that, racks at the assembly station causes hindrances to the quantity of parts which can be presented at the workstation. Thus, kitting is more flexible in this area as kitting can handle high product volume, varieties and can be kitted even with uncertain demand.

iv. Inventory volumes and space requirement.
Even when kitting can save space at the assembly station however, kitting needs more space for kit preparation before transact to assembly station. Product with high variations may requires multiple storage location in line stocking. Therefore, kitting could be advantages as it only requires a kitting space.

To overcome the kitting disadvantages as stated before, research by [5] proposed an economic model in order to identify the cost breakeven points based on measurable parts features to determine which material feeding policy is more convenient and suitable [13].

4. Material flow in kitting assembly

Before kitting process, component parts were placed in racks at the assembly station. The parts are supplied directly by continuous supply from store to assembly line, hence consumed a lot of production space. As for kitting, the kit are done in kit preparation area and transferred to assembly line which reduces material occupancy in assembly line as shown in Figure 1 [15].

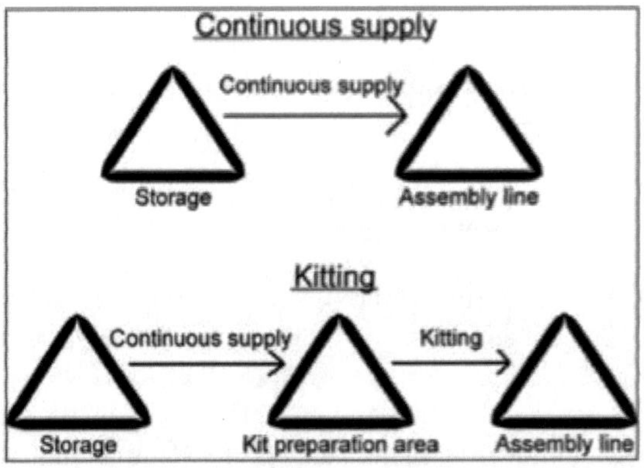

Figure 1. Schematic diagram of material flow [15].

Kitting is calling for components to be prepared and gathered before it could be sent to the assembly line. The components are retrieved from its storage location by picking operations and then it will be placed in a containers or trays or trolleys to hold all the kitted parts. Once the operation is completed, the kit is then transferred to assembly line depending on production sequence [16].

Kitting can be done in the central pick store or a decentralised area which are close to assembly workstations. Kitting may be done by either the assembly operators or operators that specialised in kitting assembly only. The materials that need to be kitted can be moved towards operators (part-to-picker) or the operators has to move to the picking materials location (picker-to-part). Supermarket style of picking is one of the example of picker-to-part scenario [17].

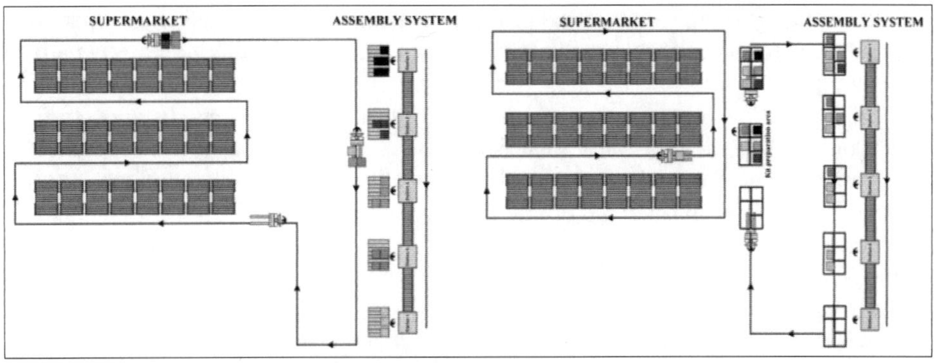

Figure 2. Line stocking *(left)* and kitting picker-to-part (*right)* [16].

Figure 2 shows the material feeding choices between line stocking and kitting assembly. The illustration shows material flow in kitting assembly. The materials needed for production is able to be determined before production assembly thus reduce the risk of part shortages. Kitting requires operators to pick the items in warehouse (picker-to-part) thus increases material handling [16].

5. Lean kitting

Kitting is considered a waste due to it needs component racks, Kanban signals, and operators to pick materials from warehouse. However, the elimination of these activities may influence more wastes to the organisation such as inflexible, parts mixed up and etc. Thus, it may cause disruption to production process and reduce line efficiency. In year 1997, Toyota has implemented a kitting process call Set Part Supply (SPS) in their production assembly [18].

The implementation of SPS is to contribute in the reduction of wastes known as *Muda* (waste), *Muri* (Overburden) and *Mura* (Unevenness). The SPS is known by using *minomi* concept. This concept used no containers for supplying parts to the production line assembly as in Figure 3 [19]. At Toyota, the minomi concept is used after stamping process of coiled steel in which the parts are stacked together based on their shapes which can reduce motion waste, scarps and scratches.

Figure 3. Examples of Minomi Concept [24].

There are few factors need to be considered in implementing a lean-kitting assembly [18] such as:

- Waste related to machine downtime which causes in invalid kitting has to be eliminated.
- Kitting assembly has to be done right for the first time.
- Any waste found in kitting assembly has to be eliminated.

The kitting crew will receive list of materials to be kitted from the store department and then, the parts will be delivered by using pallets to the production area. Therefore, production operators do need to walk to pick required materials at decided area. As a result, this method could give advantages to the Toyota such as;

- The elimination of non-value-added activities such as walking distance from assembly station to flow racks to pick materials. This could increase operator's concentration in installing parts. Thus, increase the value-added time for assembly operator.
- Reduce space occupancy in production area and creates clean work areas with better control.
- Provides leaner and easier training for assembly operator as the job scope for the operators is narrowed down.

However, based on Carlsson and Hensvold [11] research, in order to achieve lean manufacturing system, a pulling of material flow has to be implemented. By implementing the kitting system as continuous supply, production area will be able to reduce the usage of bins and provide more spacious on floor area. The level of understanding and correct kitting implementation are the crucial factors in contributing for a leaner production in assemblies with high variations [11].

6. Conclusions

Based on previous research, most researchers agreed that the implementation of kitting is the best way for material feeding in order to achieve a status of lean manufacturing. Factors such as inventory, floor space occupancy and operator walking distances have to be reduced and therefore, lean kitting assembly is considered the suitable method to be implemented in production area which has high variation and volume compared to line stocking [10]. In addition, kitting assembly can also be affected by three of these factors such as the sharing of components, uncertain demands and also lead time [20].

Kitting may be beneficial if the implementation is done correctly. Common problem in kitting or material feeding is when the synchronisation in material purchasing and supply chains is not in equal. It might cause part shortage thus halted production assembly. Besides that, non-value added activities in kitting have to be identified and a proper kitting procedure has to be implemented [21].

There is still not much research in the implementation of lean manufacturing principles in the kitting production assembly line. Therefore, many authors from previous research suggested further or continuation of research in:

- Identifying and analysing the factors that influenced the material feeding policy and where to implement the Kanban and kitting assembly system [22].
- To identify the appropriate measure to in analysing work exposures levels during materials picking [23].

As been discussed earlier there are many hidden wastes in the production line including kitting process, the authors would like to further investigate and analyse the possible causes and suggestion for better kitting methodology.

Reference

[1] Brown C B, T R Collins and McCombs E L 2006 *Engineering Management Journal* **18** 3-14
[2] Lyonnet B and Toscano R 2014 *Production Planning & Control* **25** 346-54
[3] Panwar A 2015 *Production Planning & Control* **26** 564-87
[4] Green J C, Lee and Kozman T A 2010 International Journal of Production Research **48** 2975-93
[5] Caputo A C, Pelagagge P M and Salini P 2015 *IFAC-PapersOnLine* **48** 338-44
[6] Hanson R and Medbo L 2012 *International Journal of Production Research* **50** 1115-25
[7] Brynzér H and Johansson M I 1995 *International Journal of Production Economics* **41** 115-25
[8] Rose A N M, Nik Mohamed N M Z and Ahmad H 2017 *MATEC Web of Conferences* **87**
[9] Bozer Y A and McGinnis L F 1992 *International Journal of Production Economics* **28** 1-19
[10] Ramnath B V, Elanchezhian C and Kesavan R 2010 Suitability assessment of lean kitting assembly through fuzzy based simulation model

[11] Carlsson O and Hensvold B 2008 *BCP-E* **11**
[12] Rose A N M, Deros B M and Ab.Rahman M N 2014. *Research Journal of Applied Sciences, Engineering and Technology* **8** 1191-200
[13] Caputo A C, Pelagagge P M and Salini P 2016 *IFAC-PapersOnLine* **49** 185-90
[14] Hua S Y and Johnson D J 2010 *International Journal of Production Research* **48** 779-800
[15] Hanson R and Brolin A 2013 *International Journal of Production Research* **51** 979-92
[16] Faccio M 2014 *The International Journal of Advanced Manufacturing Technology* **72** 543-60
[17] Limère V 2012 *International Journal of Production Research* **50** 4046-60
[18] Vujosevic R 2012 *Lean Kitting: A Case Study* (Austin: Departamento de Ingeniería Mecánica, Universidad de Texas)
[19] Jainury S M 2013 *Research Journal of Applied Sciences, Engineering and Technology* **6** 3682-5
[20] Choobineh F and Mohebbi E 2004 *Production Planning & Control* **15** 63-70
[21] Barker R C 2001 *Production Planning & Control* **12** 408-17
[22] Faccio M, Gamberi M and Persona A 2013 *International Journal of Production Research* **51** 2997-3017
[23] Christmansson, M., L. Medbo, G. Å. Hansson, K. Ohlsson, J. Unge Byström, T. Möller and M. Forsman 2002 *International Journal of Industrial Ergonomics* **30** 49-65
[24] Makhmudov Iskandar 2012 *Degree Project in Production Engineering Management* p 495

The use of transmission line modelling to test the effectiveness of I-kaz as autonomous selection of intrinsic mode function

Hanafi M Yusop[1,*], M F Ghazali [1], M F M Yusof [1], M A PiRemli [1], B Karollah[2], Rusman[2]

[1]Advanced Structural Integrity and Vibration Research (ASIVR, Faculty of Mechanical Engineering, Universiti Malaysia Pahang, Pekan, Pahang, Malaysia
[2]STIES, Jl. Prada Utama, Peurada, Syiah Kuala, Kota Banda Aceh, Aceh 24415, Indonesia

*Corresponding author: m.hanafiyusop@yahoo.com.my

Abstract. Pressure transient signal occurred due to sudden changes in fluid propagation filled in pipelines system, which is caused by rapid pressure and flow fluctuation in a system, such as closing and opening valve rapidly. The application of Hilbert-Huang Transform (HHT) as the method to analyse the pressure transient signal utilised in this research. However, this method has the difficulty in selecting the suitable IMF for the further post-processing, which is Hilbert Transform (HT). This paper proposed the implementation of Integrated Kurtosis-based Algorithm for z-filter Technique (I-kaz) to kurtosis ratio (I-kaz-Kurtosis) for that allows automatic selection of intrinsic mode function (IMF) that's should be used. This work demonstrated the synthetic pressure transient signal generates using transmission line modelling (TLM) in order to test the effectiveness of I-kaz as the autonomous selection of intrinsic mode function (IMF). A straight fluid network was designed using TLM fixing with higher resistance at some point act as a leak and connecting to the pipe feature (junction, pipefitting or blockage). The analysis results using I-kaz-kurtosis ratio revealed that the method can be utilised as an automatic selection of intrinsic mode function (IMF) although the noise level ratio of the signal is lower. I-kaz-kurtosis ratio is recommended and advised to be implemented as automatic selection of intrinsic mode function (IMF) through HHT analysis.

1. Introduction

Water play an important role in the contribution to the economic growth of a country. In the recent decade, water is one of the major issues addressed in research worldwide due to the increase in water pollution and non-revenue water (NRW). Water Economy Network (WEN), reported that water scarcity rapidly enhancing the number one global resource concern. Consequently, with the increase in the number of accessible clean and fresh water, the water network will increase thus the possibility of the increment of water issues is higher. The first global survey conducted by Lai [1] reported that's water loss for most country falling into the range of 20%-30% and for Malaysia, the number is 43% of NRW. Non-revenue water (NRW) is when the amount of water reach to the consumer is lesser than the amount of water that has been produced. The losses are due to real losses such as pipeline network leakage or the apparent loss such as through theft or bypass of water network (the water reach to the consumer not metered) and metering inaccuracies. Nowadays, minimising of NRW during transportation process through pipeline network is very essential. The problem of water leaks in pipeline distribution system is the main cause of Non-revenue water (NRW). Lately, water authorities concern of leak in water distribution system, then provide some necessary incentives for investment in leak detection technology and leak reduction strategies [2]. The main cause of leaks is due to the aging of pipelines, corrosion, erosion, excessive pressure of water resulting from operational error and water hammer generate by rapid opening and closing valve.

There are two types of leak detection technique which are an external and internal method. External method is a way to inspect pipe condition externally such as by visually inspecting the pipe condition and the internal method inspects the pipe condition, internally such as by mathematical computation and signal processing [3]. Signal processing is a famous method among the researchers due to its ability to detect and localise any disturbance (leak) and blockage in the pipeline system. Pressure Transient analysis is a recently developed method and attracted researchers due to cost effectiveness. Theoretically, transient phenomena are generated due to sudden changes of fluids propagation filled in the pipeline's system caused by rapid pressure and flow fluctuation in the systems such as rapid closing and opening valve. These phenomena generate pressure wave propagation which travels with the speed of sound along the pipeline's system then the wave characteristic represents the pipe information and condition from various boundaries such as pipe feature, junction outlet and the presence of the leak. Signal analysis is a way to extract the characteristics, identity, and information of pressure transient signal captured along the pipeline distribution system. A variety of pressure transient analysis method has been practiced among the researchers in order to extract information and characteristic of pressure transient signal such as cepstrum analysis [3], cross-correlation [4], orthogonal wavelets transform (OWT) [3], instantaneous frequency analysis [5], and inverse transient analysis [6].

The possibility to analyse the phenomenon in both time and frequency domain has been satisfied by implementation of time-frequency signal processing techniques such as Hilbert-Huang Transform (HHT). HHT analysis was proposed since the ability of this method to detect and capture transient phenomena occur in the non-stationary signal. In recent years, HHT has been more widely used as time-frequency analysis as signal processing method [7, 8]. HHT was proposed as newly developed and powerful method to analyse non-linear and non-stationary signal [9].

However, this method has the difficulty in selecting the suitable IMF for the next data post-processing method which is Hilbert Transform (HT). Since HHT through Empirical Mode Decomposition (EMD) will decompose the signal into the different monocomponent and symmetric component by the sifting process. The monocomponent signal called as intrinsic mode function (IMF) [10]. The main drawback of this approach is the necessity to know a priori the frequencies and level of original signal that's should be analysed [11]. The HHT analysis, in this case, is not completely automatic since need the interaction of the skilled user to select which IMF level was suitable for further analysed. Therefore, automatic selection for better IMF needed to overcome this problem. The mathematical computational technique and statistical value analysis were applied by the researcher to increase the degree of automation then eliminate the interaction of the skilled user to select the relevant IMF.

There have been relatively few recent studies and statistical approach on measurable quality for automatic selecting relevant and appropriate intrinsic mode function (IMF). Maji et al. [12] proposed variance and standard deviation as statistical value to differentiate each level of the decomposition of intrinsic mode function (IMF). The research was done using electrocardiogram (ECG) signal in order to detect Artificial Fibrillation (AF) rhythm and normal rhythm. Ricci and Pennacchi [11] demonstrated Merit Index as an indicator that allows automatic selection of IMF level. The research utilises rotary element (gear) signal to test the efficiency of merit index in selecting of relevant and appropriate IMF. The method works as a measure of periodicity degree and absolute skewness of each IMF level. In monitoring rotating machines, Kedadouche et al. [13] introduced computation of correlation coefficient value for each level of IMF as quantifiable quality in selecting appropriate IMF level. In the research vibration signal was captured from rolling element (bearing), then the signal extracted using a hybrid method of Minimum Entropy Deconvolution (MED), Empirical Mode Decomposition (EMD) and Teager Energy Operator (TEO). The selected IMF is the IMF which presents the higher value of correlation coefficient compare to the original signal. De Souza et al. [14] present an Energy-Based approach through mutual information (MI) coefficient as a method of selection relevant IMF. The research was done by applying the synthetic signal with embedded in white noise and real-world signal. Mutual Information (MI) also utilised as a method to select appropriate IMF in biomedical signal processing done by Ricci and Pennacchi [11].

This research proposed Integrated Kurtosis Algorithm for Z-notch filter hybrid with higher order statistical method (Kurtosis) as the automatic selection of intrinsic mode function (IMF). The Ikaz has chosen as since the approach of this method was adaptive in general and detects very well any changes and uncertainties of the measured signal [15]. Unlike the existing statistical analysis such as variance, standard deviation, and kurtosis, I-Kaz method was capable of indicating both amplitude and frequency difference by simultaneously obtaining the I-Kaz representation and I-Kaz coefficient, Zσ. Details experimental result was discussed by Nuawi et al. [15] and Rizal et al. [16] using vibrational signal acquire from rotating machinery part (Bearing) and detect tool wear during turning process respectively. The research conducted by Nuawi et al. [15], the I-kaz was compared to variance and the results are, variance parameter was unable to detect both amplitude and frequency changes in the non-stationary signal. Thus, the Ikaz method was reliable especially for monitoring purpose where the observation on the changes of the signal amplitude and frequency were commonly required.

The effectiveness and reliability of the Ikaz-kurtosis ratio as the automatic selection of IMF was done using artificial pressure transient signal and random signal generated using Matlab and the results revealed IMF contain maximum value of Ikaz-kurtosis ratio coefficient was relevant and appropriate to be further analysed. The research also was done by comparing the effectiveness of kurtosis compared to I-kaz as the automatic selection method for better IMF. The aim of this paper focused on the implementation of Ikaz-kurtosis ratio coefficient as the automatic selection of intrinsic mode function using synthetic pressure transient response generate using transmission line modelling (TLM). TLM is a technique, computationally very efficient, for studying a wide range of wave and diffusion phenomena. The white Gaussian noise was added into each of the signal response acting as the background noise in order to make the signal response was similar to the real one. The result for synthetic signal response with added background noise generates using TLM presented the IMF contain highest value of Ikaz-kurtosis coefficient was suitable and appropriate to be further analysed. Furthermore, the Ikaz-kurtosis ratio was proven suitable as an automatic selection method for better and relevant IMF and highly proposed to test the effectiveness using real pressure transient response. Therefore, the degree of automation for Hilbert-Huang Transform (HHT) was improvised to detect pipe leakage in live water distribution system using pressure transient signal.

2. Integrated based Kurtosis algorithm for z-filter technique to Kurtosis ratio

2.1. Integrated based Kurtosis algorithm for z-filter technique

Integrated Kurtosis-based algorithm for z-filter technique (Ikaz) is the method developed based on the concept of data scattering about its centroid. The sampling frequency of the raw signal was chosen as 2.56 referring to Nyquist number [15]. Most of the researcher in signal analysis and processing are comfortable with the number. To avoid the aliasing effect, the maximum frequency span will be of equation 1.

$$F_{\max} = \frac{fs}{2.56} \tag{1}$$

Ikaz decomposes the time domain signal into three level of the frequency range, which are x-axis represent as low frequency (LF) with a range of 0 – 0.25 of f_{max}, followed by y-axis represent as high frequency (HF) with a range of 0.25 – 0.5 of f_{max}. Finally, z-axis represents as very high frequency (VF) with range 0.5 of f_{max}. The 0.25 f_{max} and the 0.5 f_{max} was selected as low and high-frequency range limit respectively done with referring to the 2nd order of the Daubechies concept in signal decomposition process [17]. Referring to the kurtosis, I-kaz method contributes of three-dimensional graphical representation of the measured signal frequency distribution. The variance, σ^2 of each frequency band which is σL^2 represent as a low-frequency band, σH^2 represent as a high-frequency band and σV^2 represent as a very-high-frequency band which calculated as in equation 2,3 and 4 to measure scattering of data distribution.

$$\sigma L^2 = \frac{\sum_{i=1}^{N}(x_i - \mu_L)^2}{n} \tag{2}$$

$$\sigma H^2 = \frac{\sum_{i=1}^{N}(x_i - \mu_H)^2}{n} \tag{3}$$

$$\sigma V^2 = \frac{\sum_{i=1}^{N}(x_i - \mu_V)^2}{n} \tag{4}$$

The I-Kaz coefficient, Ƶσ can be simplified in term of variance, σ as in equation 5 [16].

$$Ƶ\sigma = \sqrt{(\sigma L^2)^2 + (\sigma H^2)^2 + (\sigma V^2)^2} \tag{5}$$

2.2. *Kurtosis*

Kurtosis describes as a measure of spikiness and hence a good indicator of peak analysis for spikes detection in a non-stationary signal component such as pressure transient. Kurtosis is expressed as;

$$\text{Kurtosis}(x) = \frac{E\{(x-\mu)^4\}}{\sigma^4} \tag{6}$$

where μ and σ represent the mean and standard deviation of time series signal respectively. The E illustrate the expectation operation. The kurtosis demonstrates the spikiness and peakedness of probability distribution associated to the instantaneous amplitudes of the time-series analysis [18]. Therefore, the I-kaz-kurtosis ratio expressed as equation 7.

$$ZK\sigma = \frac{\left(\sqrt{(\sigma L^2)^2 + (\sigma H^2)^2 + (\sigma V^2)^2}\right)(\sigma^4)}{\left(E\{(x-\mu)^4\}\right)} \tag{7}$$

2.3. *Algorithm*

Automatic selection of intrinsic mode function (IMF) based on pressure transient signal

1. Perform EMD on a synthetic pressure transient response generate from TLM
2. EMD decomposed the signal into so-called level of Intrinsic Mode Function
3. Calculate the Integrated Kurtosis Algorithm for a z-filter technique to kurtosis ratio (Ikaz-Kurtosis) coefficient using Eq. 7.
4. Identified the IMF that corresponds to the largest Ikaz-kurtosis ratio coefficient.
5. Perform Hilbert Transform (HT) and Hilbert Spectrum (HS) for IMF level that contains the largest coefficient of Ikaz-Kurtosis ratio.

3. Methodology

3.1. *Simulated pressure transient signal*

In this research, two simulated pipeline models were constructed with different pipe length, water pressure and a number of leaks fitted on the pipeline model. The first model (Figure 1(a)) simulated was a pipe with the leak and without a leak. The model consists of Medium High-Density Polyethylene (MDPE) pipe, which is divided into 3 section by the valve and junction (Figure 1(a)). The total distance of pipe for the first model is 70 meter with water pressure by 1 bar. The second model (Figure 1(b)) consist of MDPE pipe , which is divided into 4 section by valve and 2 junctions. For the second model, the number of the leak is 2 with total distance and water pressure acting on the pipe is 73 meter and 2 bars respectively. For simulated signal without a leak, the resistance of the junction is set to be 3000 meanwhile for the simulated signal with the leak, the resistance at the junction set to be 30,000,000. Both models has pipe diameter with 60 mm respectively.

The design of the model was made as simple as possible in order to validate the identification approach. Opening and closing the valve (Figure 1) cause water hammer pulse which then propagates through the system. The system was applied time-varying valve as detailed describe by Beck et al. [19]. The software allows the user to set the geometry, length, and diameter of the pipes. By inserting the higher value of resistivity at the junction in order to simulate the leak feature, then some of the pressure wave generated by the valve will be reflected. By knowing the distance of the reflection point from the end of the pipe and the speed of sound that pressure waves travel through the medium, the location of the leak and pipe outlet can be determined.

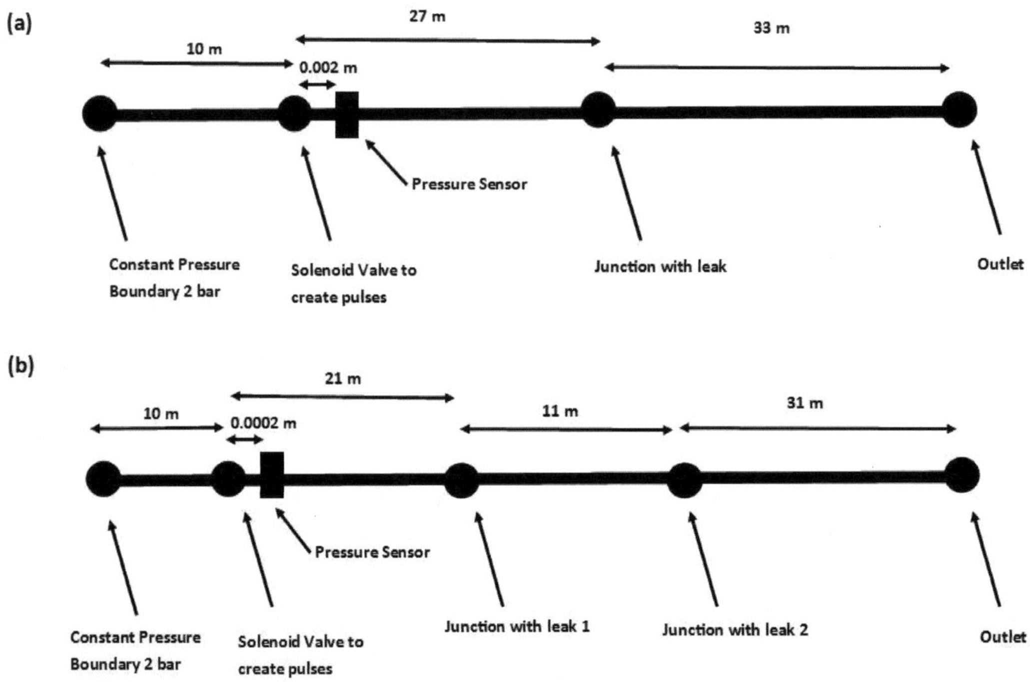

Figure 1. Pipe network model. (a) First model (b) Second model

In this research, water was utilized as simulated fluid as the medium flow in both pipeline model. It was assumed that water has Standard Temperature and Pressure (STP). The speed of sound is equal to 493 m/s since the material and pipe diameter apply on the model is MDPE and 60 mm respectively. The distance between the valve and the pressure sensor was set to 0.001 meters (Figure 1). This simulates the capture of a time history of signal response of pressure using a pressure sensor located very close to the valve. The software also allows the user to choose any number of flow and pressure data as the output from the system.

3.2. Added white Gaussian noise

The white Gaussian noise then added to each of signal response generates from TLM software in order to make the signal very close to the real pressure transient signal. The value of Signal to Noise Ratio (SNR) is a key parameter when deciding to generate white noise signal response. If the added noise is too weak to bring the changes of extrema of the original signal. Conversely, if the added noise is too strong, the EMD/EEMD method will derive meaningless results which are mainly controlled by the added noise and scarcely associated with the original signal [20]. According to Zhang et al. [21], it appropriates for EMD/EEMD method to set SNR between the original signal and added noise is approximately 37 dB. Therefore, the value of SNR is set to be 37 dB. As presented in Figure 2 below, the graph shows the signal response from TLM with added noise. The red line indicated the original response from TLM without noise and the light blue line indicated the signal response with added noise. The signal response from TLM with added noise then going to analyse using Hilbert-Huang Transform (HHT) and Integrated Kurtosis Algorithm for Z-notch filter technique (IKAZ) as the automatic selection of intrinsic mode function (IMF). The results of the analysis will show the characteristic of the pipeline that has indicated the pipe feature (leak, blockage and pipefitting).

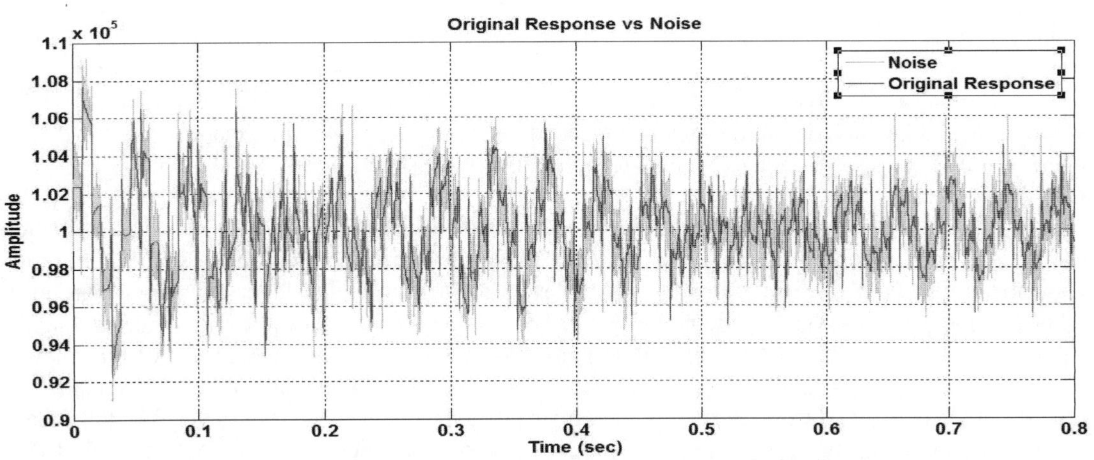

Figure 2. Signal response with added white Gaussian noise.

4. Results and discussion

Figure 3 illustrate the signal response with added noise for the first pipe network model. Figure 3 (a) and (b) show the no leak and with leak response respectively. Meanwhile Figure 4 present the signal response from second pipe network model with added noise. Presented in Figure 4 (a) and (b) are the signal response of no leak and leak respectively. By using Hilbert-Huang Transformed that's contained of Empirical Mode Decomposition (EMD) and Hilbert Transform (HT), the instantaneous frequency and phase angle of the original signal was obtained.

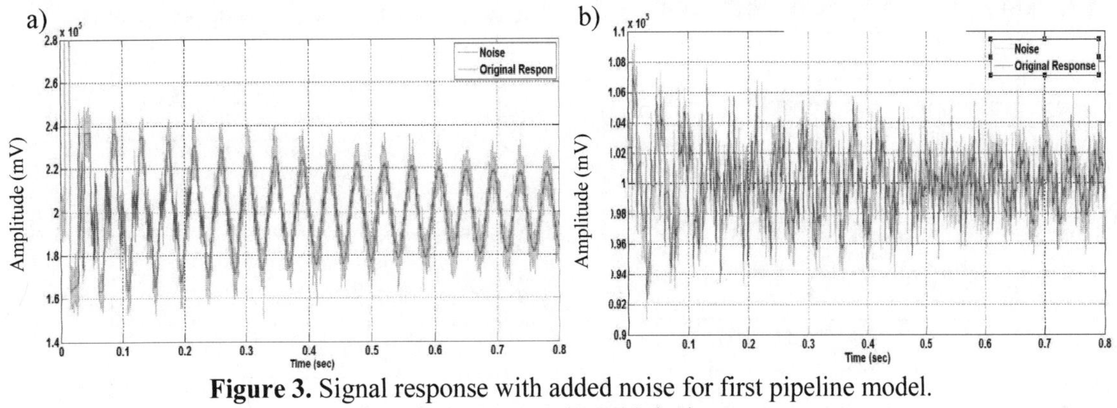

Figure 3. Signal response with added noise for first pipeline model.
a) No leak response b) With leak response

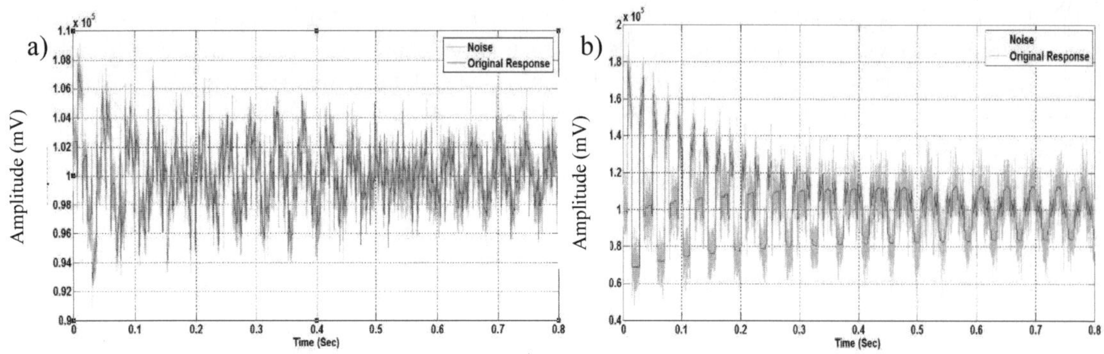

Figure 4. Signal response with added noise for second pipeline model.
a) No leak response b) With leak response

The step of analysis began with the signal response was decompose into Intrinsic Mode Function (EMD). EMD decompose the signal response into 13 levels of intrinsic mode function (IMF). Figure 5 shows the amplitude versus time for the first 12 levels of IMF. The first level of IMF is a group of higher frequency signals which is the noise signal. The last level was a reserve for the lower frequency signals. The first and second level of IMF was avoided for further analysis because of these levels contain noise frequency signals. Meanwhile, IMF level 7 and the residue contain basic respond of the network. All these IMF were, therefore, discarded. The rest which is IMF level 3-6 have been recombined to produce a signal without noise [22]. Therefore, in order to identify which level of IMF is suitable for going through the final step of Hilbert-Huang Transform (HHT) which is Hilbert Transform (HT), Integrated Kurtosis Algorithm for Z-filter technique to kurtosis ratio (Ikaz-Kurtosis) was implemented.Since EMD decomposed the signal response into 13 levels of IMF, the Ikaz-Kurtosis ratio was utilised to computed the coefficient of each level.

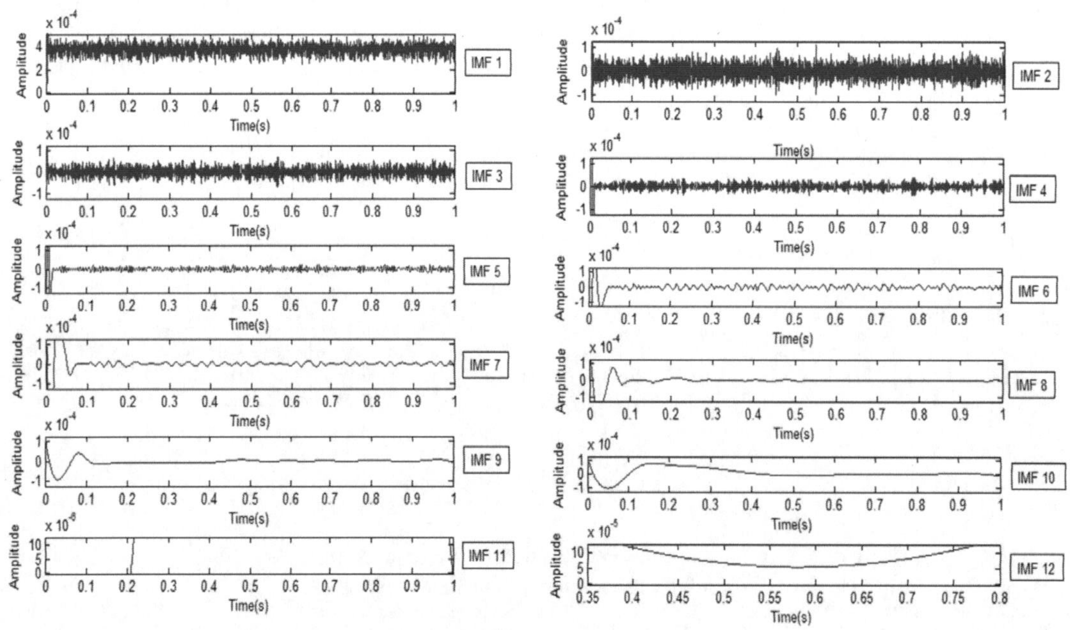

Figure 5. Intrinsic Mode Function (IMF) from level 1-12.

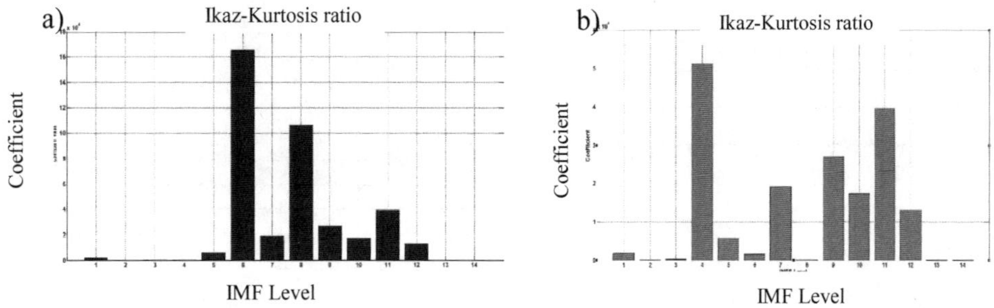

Figure 6. Ikaz-Kurtosis ratio coefficient for first pipeline model. a) No leak signal b) With leak signal

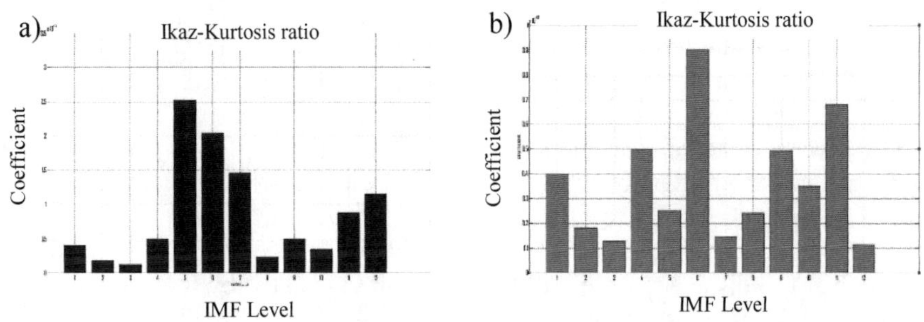

Figure 7. Ikaz-Kurtosis ratio coefficient for second pipeline model.
a) No leak signal b) With leak signal

Figure 6 (a,b) and 7 (a,b) represent the Ikaz-Kurtosis ratio coefficient histogram for the first and second pipe network module for the Without leak and with leak response respectively. The X-axis show the IMF level. Meanwhile, Y-axis show the value of Ikaz-Kurtosis ratio coefficient. From the observation, the maximum and highest value of Ikaz-Kurtosis coefficient were the most suitable IMF level for going through the post-processing and the final step of Hilbert-Huang Transform (HHT) analysis which is Hilbert Transform (HT). This is because the IMF level which contains the highest value of Ikaz-Kurtosis ratio consist of the IMF selection criterion which is clear and narrow spikes. The lower the value of Ikaz-Kurtosis ratio, the obscure of clear and narrow spikes which are the main selecting criterion of IMF.In addition, the IMF level that contains the highest value of Ikaz-Kurtosis ratio coefficient was reserved on level range of 3-6. As explained above, level 3-6 have been recombined to produce a signal without noise [22].Therefore, the level is relevant and appropriate to further analysed. As illustrated in Figure 6 (a) the IMF level 6 contained the highest Ikaz-Kurtosis ratio meanwhile 6 (b) reserved on IMF level 4. The Figure 7 (a) , show IMF level 5 was contain the highest coefficient followed by IMF level 6 for Figure 7 (b).

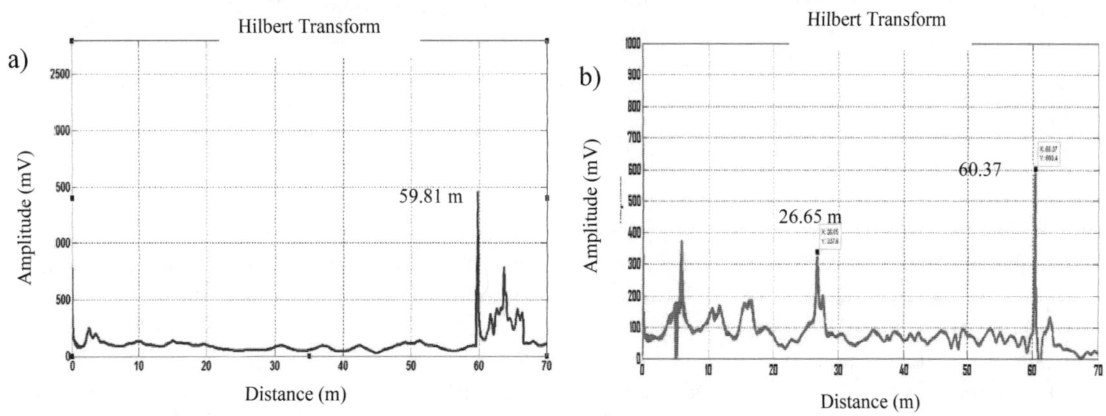

Figure 8. Instantaneous characteristic of HT analysis for first pipe network model.
a) Signal without leak b) Signal with leak

IOP Conf. Series: Materials Science and Engineering **257** (2017) 012070 doi:10.1088/1757-899X/257/1/012070

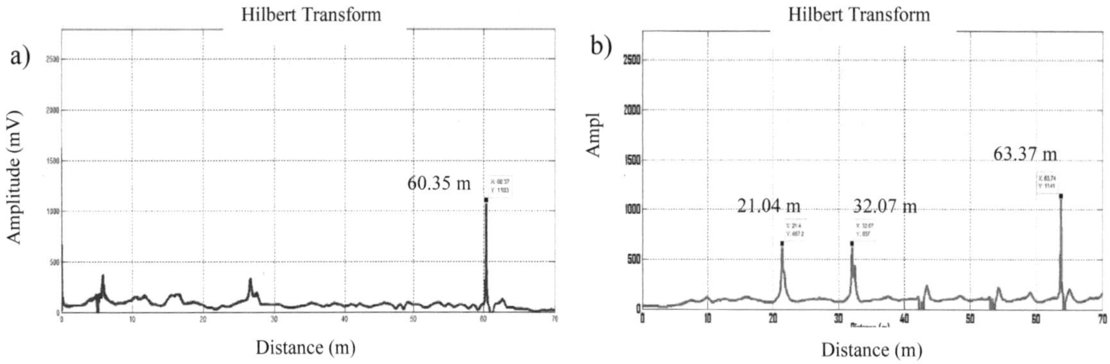

Figure 9. Instantaneous characteristic of HT analysis for first pipe network model.
a) Signal without leak b) Signal with leak

Figure 8 and 9 show the instantaneous characteristic of the signal for first and second pipe network model Meanwhile a and b represent for a signal without leak and with the leak. The instantaneous frequency clearly highlights the presence of a reflection. The peak of the analysed signal matched up with the time taken for the wave to travel along the pipe network to the reflection point and return to the measurement point. The distance of the reflection point is calculated by multiplying the time delay corresponding to the peak by the speed of sound in the pipe network (a=493 m/s) and halving this value to account for the return journey. As shown in Figure 8 (a) and (b), the presence of reflection corresponding to outlet signature is clearly seen by 59.81 m and 60.37 m respectively. Meanwhile, the reflection corresponds of leak signature seen in Figure 8 (b) was measured by 26.65 m. Figure 9 (a) and (b) illustrated the reflection with the signature of the outlet was measured by 60.35 m and 63.37 m. The reflection with the signature of leak 1 and leak 2 was measured by 21.04 m and 32.07 respectively as presented in Figure 9 (b). There are some errors when compared the experimental positioned and simulated positioned. The comparison between simulated and experimental position is tabulated in Table 1.

Table 1. Comparison between simulated and experimental position.

Water pressure	Signal response	Pipe feature	Simulated position (m)	IMF contain maximum Ikaz-Kurtosis ratio coefficient	Experimental position (m)	Error %
First Pipeline Model	Leak Data	Leak	27	5	26.65	0.3
		Outlet	60		59.81	1.2
	No Leak Data	Outlet	60	4	60.37	0.6
2nd Pipeline Model	Leak Data	Leak 1	21	5	21.04	0.2
		Leak 2	32		32.07	0.2
		Outlet	63		63.74	0.5
	No Leak Data	Outlet	63	6	60.35	4.4

As presented in Table 1, for first pipe network model contain leak signature with water pressure of 1 bar, the position of leak shows 0.3% error and position of outlet signature show 1.2% error. Meanwhile for no leak data presented 0.6% error for outlet signature position. Besides that, second pipe network model contains two leak signature with water pressure of 2 bars, the position of signature leak 1 and 2 shows 0.2% for both error and position of outlet signature show 0.5% error. Meanwhile for no leak signal response presented 4.4% error for outlet signature position. From the final part of HHT analysis, it is clearly shown that the position of pipe feature that's is leak and pipe outlet appear at the same position compared to each original position and simulated positioned. It also proves HHT analysis able to detect and positioning the transient event occurred in the non-stationary pressure transient signal.

5. Conclusions
This paper discussed the self-decision method for IMF selection through Hibert Huang Transform (HHT) analysis. The result proves the I-kaz-kurtosis ratio is suitable and advisable self-decision method for IMF selection to implement in Hilbert-Huang Transform (HHT) analysis. The development of automated self-decision of IMF through HHT has been built and statistically analysed using I-kaz-kurtosis ratio. Therefore, this method was proposed and advised to be implemented. By efficiently utilizing I-kaz-kurtosis ratio, the issue of IMF selection was overcome. Therefore degree of automation for Hilbert Huang Transform (HHT) was improvised to detect pipe leakage in live water distribution system using pressure transient signal..

Acknowledgement
University Malaysia Pahang fully supports the facilities and resources for this research. The authors would like to acknowledge the support from the University Malaysia Pahang internal grant RDU170386.

References
[1] Lai C 1991 *Water Supply* **9** 4
[2] Colombo A F, Lee P and Karney B W 2009 *Journal of Hydro-environment Research* **2** 212-27
[3] Taghvaei M, Beck S and Staszewski W 2006 *Measurement Science and Technology* **17** 367
[4] Liou C P 1998 *Journal of Fluids Engineering* **120** 833-8
[5] Ghazali M F 2012 *Leak detection using instantaneous frequency analysis* (Thesis) University of Sheffield
[6] Stephens M L, Simpson A R, Lambert M F, Vítkovský J and Nixon J 2002 *South Australian Regional Conf.*
[7] Huang N E, Shen Z, Long S R, Wu M C, Shih H H, Zheng Q, Yen N C, Tung C C and Liu H H 1998 *Proc. of the Royal Society of London A: Mathematical, Physical and Engineering Sciences* (The Royal Society) p 903-95
[8] Shi C X and Luo Q F 2003 *Acta Seismologica Sinica* **16** 422-9
[9] Bin G, Gao J, Li X and Dhillon B 2012 *Mechanical Systems and Signal Processing* **27** 696-711
[10] Manjula M and Sarma A 2012 *Energy Procedia* **14** 1156-62
[11] Ricci R and Pennacchi P 2011 *Mechanical Systems and Signal Processing* **25** 821-38
[12] Maji U, Mitra M and Pal S 2013 *Procedia Technology* **10** 45-52
[13] Kedadouche M, Thomas M and Tahan A 2014 *Advances in Acoustics and Vibration* 2014
[14] De Souza D B, Chanussot J and Favre A C 2014 2014 *IEEE Int. Conf. on Acoustics, Speech and Signal Processing* (ICASSP): IEEE) p 325-9
[15] Nuawi M Z, Nor M J M, Jamaludin N, Abdullah S, Lamin F and Nizwan C 2008 *Journal of Applied Sciences* **8** 1541-7
[16] Rizal M, Ghani J A, Nuawi M Z and Haron C H C 2013 *Jurnal Teknologi* **66**
[17] Daubechies I 1992 *Ten lectures on wavelets* vol 61 (SIAM)
[18] Wang Y, Xiang J, Markert R and Liang M 2016 *Mechanical Systems and Signal Processing* **66** 679-98
[19] Beck S, Haider H and Boucher R 1995 *Journal of Mechanical Engineering Science* **209** 419-27
[20] Lin J 2012 *International Journal of Computer Science* **9** 194-9
[21] Zhang J, Yan R, Gao R X and Feng Z 2010 *Mechanical Systems and Signal Processing* **24** 2104-23
[22] Ghazali M, Beck S, Shucksmith J, Boxall J and Staszewski W 2012 *Mechanical Systems and Signal Processing* **29** 187-200

IOP Conf. Series: Materials Science and Engineering **257** (2017) 012071 doi:10.1088/1757-899X/257/1/012071

Prevention of crack in stretch flanging process using hot stamping technique

Y Mohd Syafiq, Z Hamedon*, Wan Azila Aziz and Ahmad Razlan Yusoff

Faculty of Manufacturing Engineering, Universiti Malaysia Pahang,
26600 Pekan, Pahang Darul Makmur, Malaysia

*Corresponding author: zamzuri@ump.edu.my

Abstract. Demand for enhancing of passenger safety as well as weight reduction of automobiles has increased the use of high strength steel sheets. As a sheet metal is a lightweight having high strength is suitable for producing automotive parts such as white body panel. The stretch flanging of the high strength steel sheet is a problem due to high springback and easy to crack. This study uses three methods to stretch flange the sheets; using lubricants, shear-edge polishing and hot stamping. The effectiveness of these methods will be measured by comparing the flange length of each methods can achieved. For stretch flange with lubricant and polished sheared edge, the flange length failed to achieve the target 15 mm while hot stamping improved the formability of the sheet and preventing the occurrence of the springback and crack. Hot stamping not only improved formability of the sheet but also transformed the microstructure into martensite thus improve the hardness and the strength of the sheet after been quenched with the dies.

1. Introduction

Environmental sustainability in transportation especially passenger vehicle is important and crucial issue and the main dispute agenda are linked to have a better materials and efficient energy [1]. For passenger vehicles, the main contribution which can be consider is to reduce fuel consumption and exhaust emission levels is related to the weight reduction thus increased the application of high strength steel sheets. High strength steel sheet is one of the commonly used material that having high tensile strength about 490 MPa to 790 MPa. The highest grades are known to be used, 1180 MPa and 980 MPa class ultra-high strength steels for cold stamping was applied in structural parts. Table 1 shows the types of steel sheets with their respective tensile strength and strength-to-specific gravity ratio.

Table 1. Metals together with their tensile strength and strength-to-specific gravity ratio [2].

Sheet materials	Tensile strength (MPa)	Strength-to-specific gravity ratio (MPa)
Ultra-high strength steel sheet	980 - 1470	126 - 188
High strength steel	490 -790	63 - 101
Mild steel SPCC	340	44

Besides, 980 MPa grade hot rolled ultra-high strength steel sheets is used in stamping of chassis although it is difficult to adopt high strength steels due to safety, durability and rigidity issues [3]. For exposed panel, 590 MPa grade high strength steel sheets are applied although the exposed parts need a very fine surface quality. In automotive industry, specifically, a high strength steel sheet can produce a white body part to reduce the weight of body panels [4]. As the weight of the body panels decreases, the fuel consumption also is decrease. Takahashi mentioned an automotive industry has been trying to reduce the fuel consumption of vehicle with development of program; for example, are Partnership for a New Generation of Vehicles (PNGV) of United States; which are development of 80 miles per gallon car, and the development of 100 km per 3 litter cars in Europe [5]. However, the problems encounter as it's produced a large effect of springback and the crack formation towards the material during the stamping operation either bending or flanging process. These problems will affect the dimensional accuracy and appearance of the manufactured parts. Besides, this will contribute to assembly difficulties [6]. Hamedon et al. used stopper too prevent the fracture at the flip edge of the hemming [7]. Mori et al. discussed the method to improve the stretch flange ability by smoothing of sheared edge [8]. The in-situ deformation of the sheet during stretch flanging was observed using borescope was used by Hamedon et al. to understand actual phenomenon occurred [9]. The quality of the sheared edge influenced the limiting expansion ratio. By improving the limiting expansion ratio, the stretch flange ability is increased. Kim and Park discussed that the formability is affected by the parameters considerations during the stamping thus suggesting that improvement of the formability can be done by using ball tools with certain size is used with a small feed rate and a little friction. This is due to the plane-anisotropy; the formability is differed base on the direction of the tool movements [10]. Kusumi et al. stated that it is possible to predict fracture of steel sheet by studying the thickness reduction and mentioned that hot stamping permits the high strength steel sheets to be formed with good shape fix ability [11]. Maeno et al. improved the formability of titanium using resistance hot stamping [12]. This is the effect of the martensitic transformation during quenching after forming.

The objective of this study is to obtain the best method for stretch flanging of the high strength steel without fracture. In this research, the flange length is targeted to achieve 15 mm for each method that are going to be examined.

2. Experimental setup

The experimental setup is shown in the Figure 1 below. The blank size is 170 x 100 x 2 mm. The die set are as shown in the figure. High strength steel sheet is hold by the blank holder and the flange length are measured. The blank is sheared into the dimension first before sheared at the edge to produce V-shape edge for the stretch flanging process.

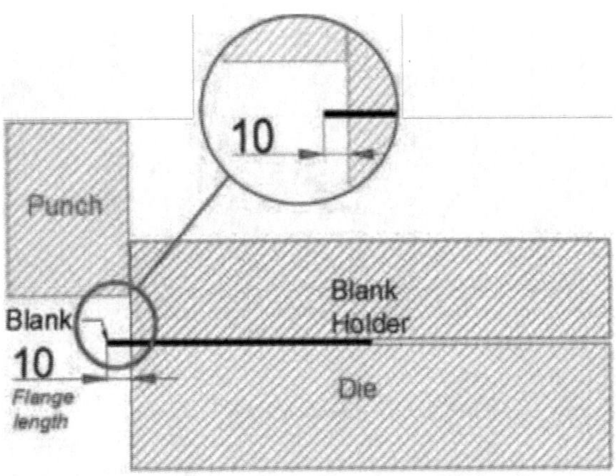

Figure 1. The experimental setup for stretch flanging and the flange length measurement.

Figure 2 show the stretch flanging operation. The corner edge will have a large tensile stress during the stretch flanging process.

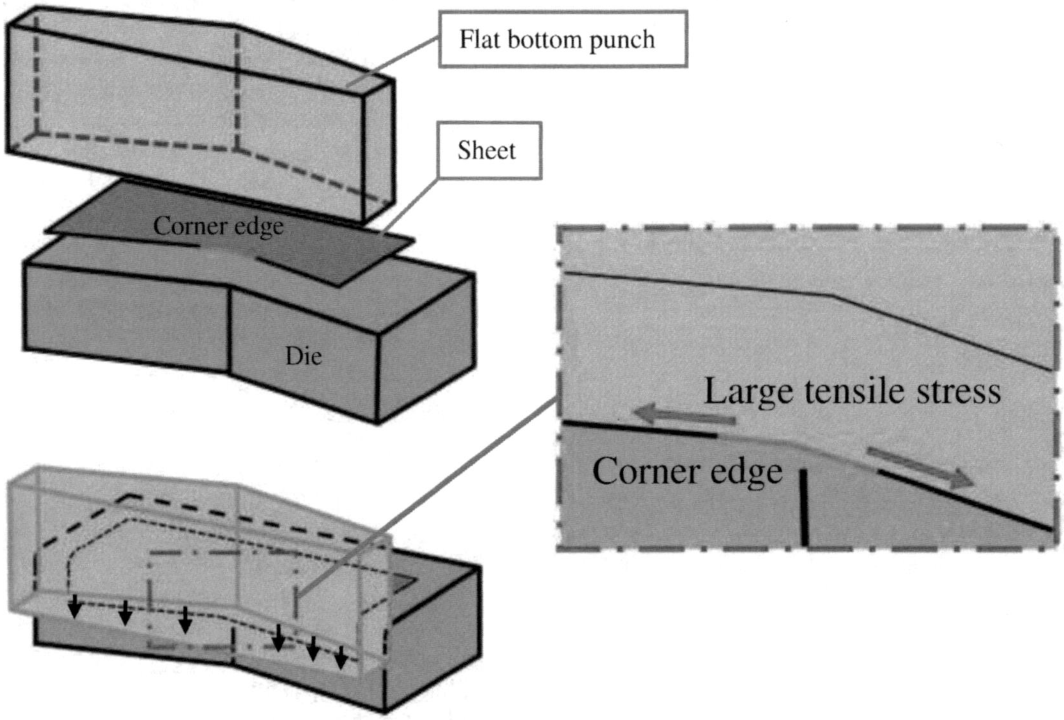

Figure 2. Stretch flanging process with flat bottoming punch.

3. Preliminary experiment

Specimens are produced from the process of EDM wire cutting. The wire speed is set at 50 mm/min and the spindle feed at 5 rev/min. The finished product is shown in figure 3(a) with the dimension 100 x 10 x 2 mm. The cutting direction with consider the orientation is shown in figure 3(b). Tensile test is conducted for a specimen under a room temperature, heated at 700°C, 800°C and 900°C. Figure 3(c) shows the physical changes after the specimen is extended. The result from the tensile test are shown in Figure 4.

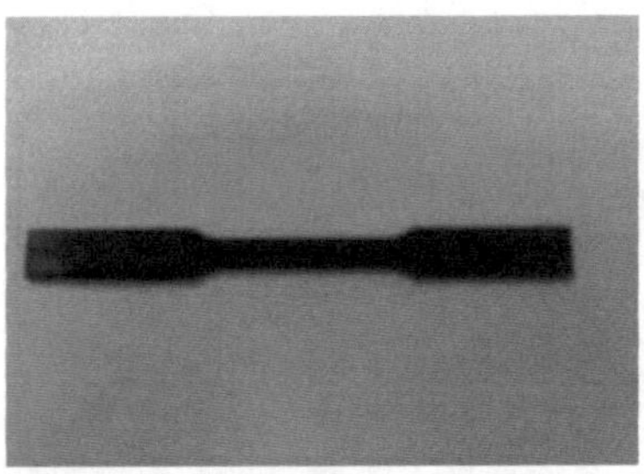

Figure 3(a). The finished specimen after wire cutting process.

Figure 3(b). The cutting direction with considering different orientation.

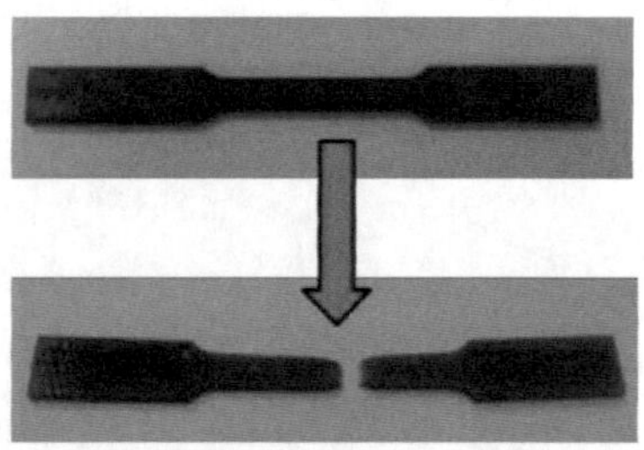

Figure 3(c). The specimen before and after tensile test

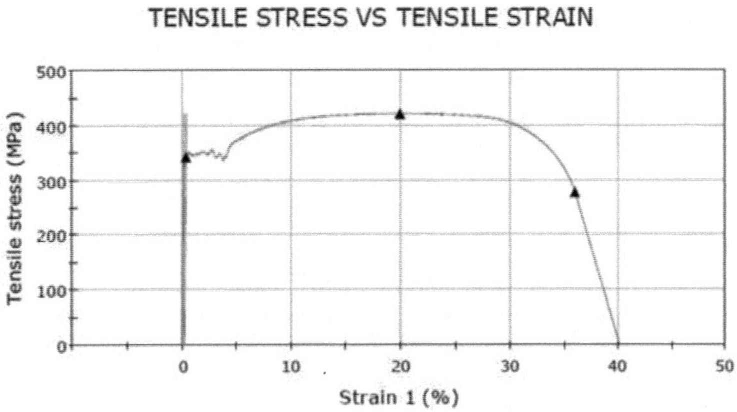

Figure 4. Stress strain graph for 0° orientation. The specimen is heated at 700°C.

From this test, the results show the mechanical properties obtained from the test for a different orientation heated under different temperature. Stress strain graph acquired shows the pattern of the graph before and after the materials undergo plastic deformation. By referring to the graph, the value of tensile stress at maximum load is 420 MPa whereas yield stress is about 342 MPa. The specimen starts to deform and fracture at 278 MPa.

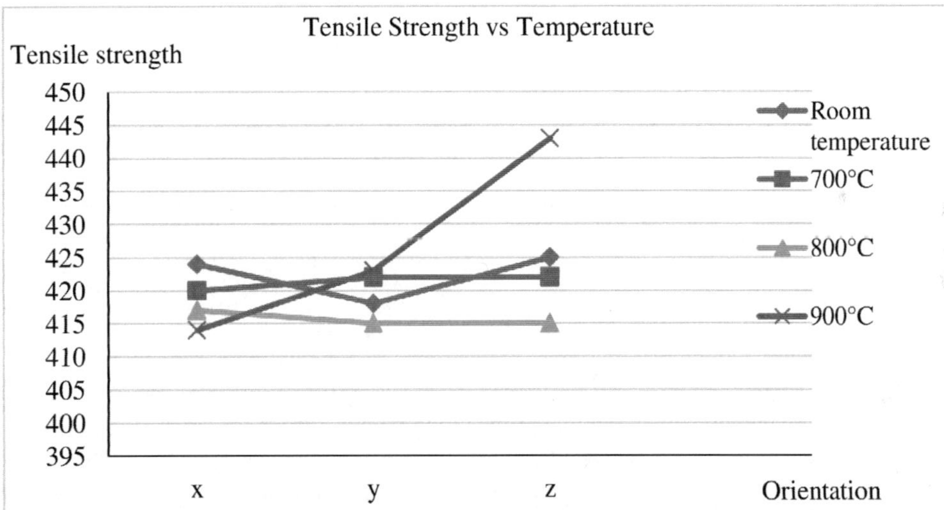

Figure 5. Graph of tensile strength against temperature.

For the x-orientation, tensile strength decreases the temperature increases. Differ from y and z-orientation; they achieved a maximum tensile strength at 900°C. For z-orientation, it gives the highest value of tensile stress compare to the others (refer Figure 5). However, the higher the strength of the material, manufacturing will be more challenging and difficult. Based on this results, the process proceeds with x-orientation, the commonly used direction when perform a stamping with high temperature ranging 700°C to 800°C when heated. This is because the material was at the best condition so that the stamping process can easily conducted and fracture may be prevented.

Generally, good hardness means that the material is high resistant to scratching and wear. It is commonly used because they are quick and convenient for assessing material properties. Hardness Vickers test is applied during this research. The surface of the material was grind and polished to have flat and smooth surface. The mould is prepared for the specimen as shown in Figure 6. Then, diamond is indented along the surface. The hardness values appear on the PC screen after the indented area is calculate. Based on the graph, the distribution of hardness values under different temperature do not shows consistency in plotted graph. The values range between 120HV to 180HV. The high strength steel sheets achieved the highest hardness when heated with 800°C.

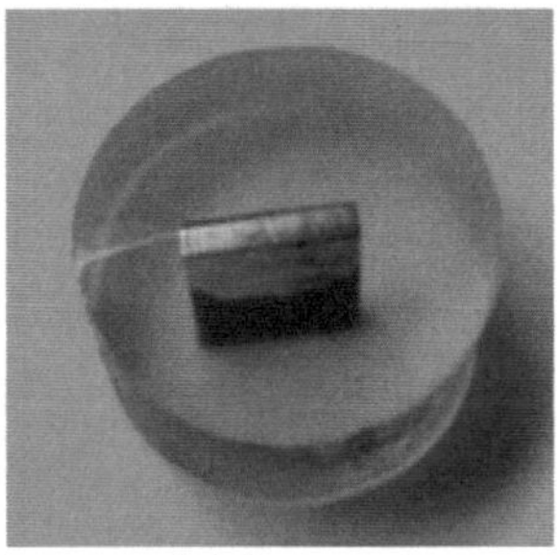

Figure 6. After polished and mounted. The specimens are prepared for hardness test.

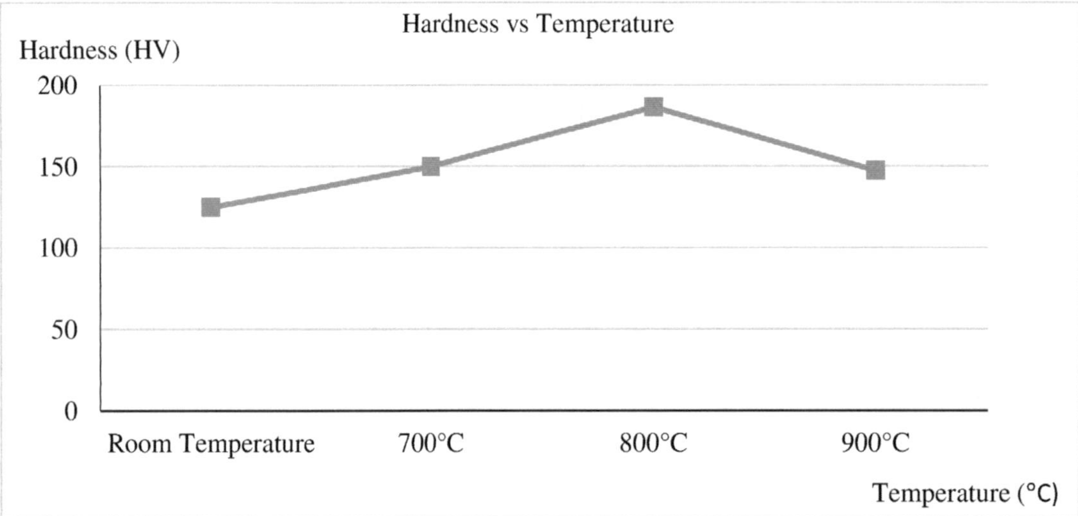

Figure 7. Graph of hardness versus temperature.

4. Results and discussion

Result obtained from the stamping process are shown in Figure 8. For the first attempt, the sheet is tested at room temperature. As can be seen, the fracture formed at the flanged area. This is due to the unheated material has a high tensile strength compared to heated material. Therefore, the material is hard to form and crack is highly occurring in this case.

Figure 8. Specimen after stamping operation. At room temperature, the high strength steel sheets fractured as shown in the circle.

The next test is with the same condition; the sheet does not undergo heating process but the sheet is flip to another side to identify whether there is a crack formation when stamping is performed. The result remains the same as the fracture still occurs on the material. It is due to the burr formation after cutting the material into V-shape.

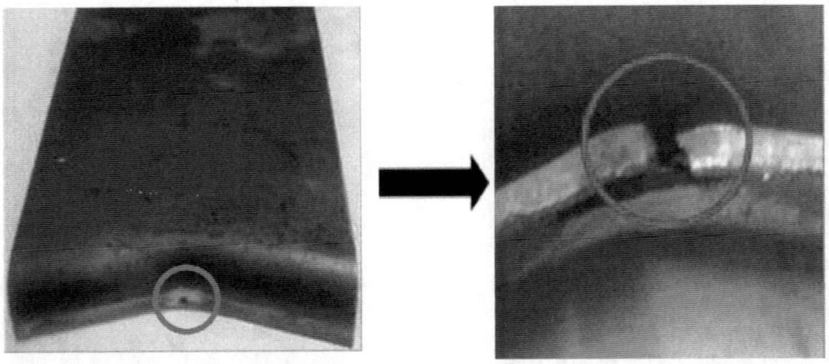

Figure 9. Specimen after stamping operation. In reverse position, where the high strength steel sheets are flipped to other side; fracture occurs as shown in the circle.

Another steel sheets are applied with light oil which is coolant oil around the surface cutting area. However, the problem still occurs as there is no effect although the oil is applied to the surface. Oil usually acts as the cooling reagent and heat can be reduces during stamping operation.

Figure 10. Specimen after stamping operation with applying light oil (coolant oil) to the surface. Fracture occurs as shown in the circle.

The next high strength steel sheets are applying with heavy oil which is grease. But the crack still occurs around the flange part. It shows that there is no effect although the material is applied with oil.

Figure 11. Specimen after stamping operation with applying heavy oil (grease) to the surface. Fracture occurs as shown in the circle.

Another sheet is polished with sand papers to remove the burr. Burr is the main problems as more stress will be concentrated at the flange area. The effect of excessive stress to the material are the fracture may occur as shown in Figure 12.

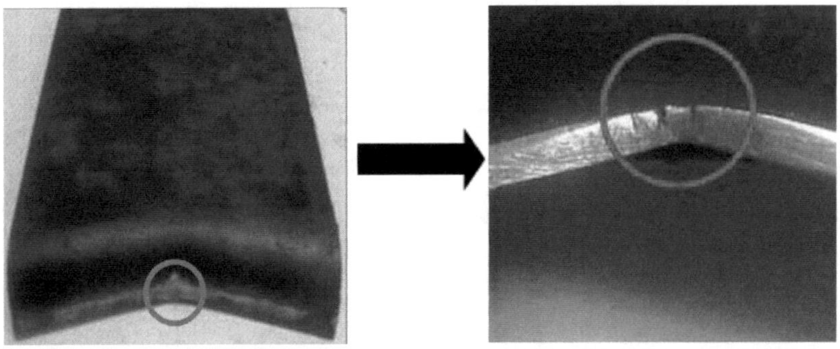

Figure 12. Specimen after stamping operation. The surface was polished to remove the burr. Fractures still occur as shown in the circle.

The remaining high strength steel sheets are heated under different temperature which are 700°C, 725°C, 750°C and 800°C. heating process assist the mechanical structure of the sheets so that the fracture can be prevented. As shown in Figure 13, for all temperature used; there is no fracture form but only contribute to the thickness distribution along the edge.

Figure 13. Specimen that are heated. The flange part is good (no fracture)

Thickness distribution of the flanged part shows a slightly different for the different temperature. Figure 14 shows the distribution sheets thickness that has been measured. The measured point is 2 mm from the centre line of the flanged sheets and the distances from point to point are fixed to 2 mm.

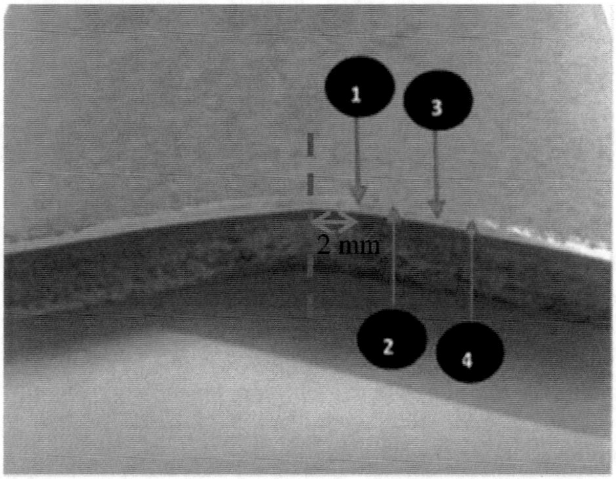

Figure 14. Specimen that are heated are only showing different in thickness distribution at the numbered region.

From the graph; at point 1, thinning occurred where the thickness was reduced for all temperature tested. At point 2 also thinning are occurred except for the sheets that are heated with 800°C where the thickness remains at 2 mm. Point 3 and 4 are where thickening occurred and the highest value is 2.15 mm, for sheets that are heated at 800°C. Based on the figure, the lowest thickness distribution is for the sheets with heating temperature 800°C where the thinning at point 1 are by 0.2 mm and the thickening recorded is 0.15 mm. Thickness reduction will contribute to fracture of the sheets. Therefore, the smaller reduction is better for prevention of fracture in high strength steel sheet.

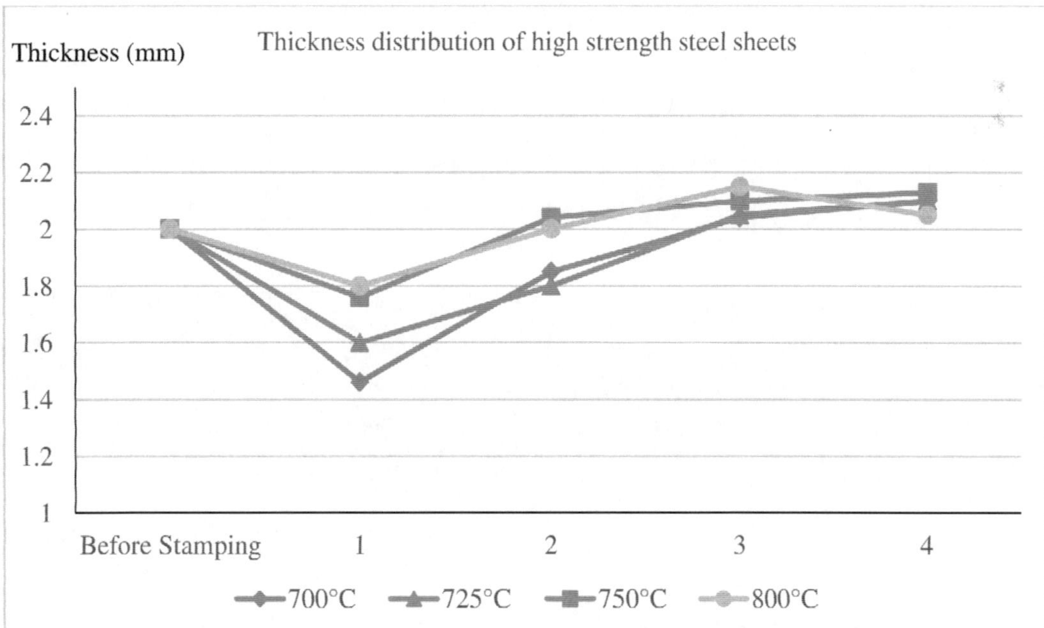

Figure 15. thickness distribution of high strength steel sheets.

5. Conclusions

It is found that among the factor that contributed to the fracture is the formation of burr and the crack at the edge of the sheet initiate the fracture thus polishing the sheet edge is one of the best solutions to this problem and it will be more successful if the material is polished before heated. Study also showed that applying oil does not contributed the prevention of fracture to the steel sheets. However,

IOP Conf. Series: Materials Science and Engineering **257** (2017) 012071 doi:10.1088/1757-899X/257/1/012071

through heating up the sheet before flanging had increase the formability of the high strength steel. Fracture or crack are prevented by hot stamping. The length of the flange is increased by using this method. Heating process helps to reduce the strength of the materials so that the forming process will easily performed. Lastly, the result obtained will be much better when a stamping is done.

References

[1] Ingarao G, Di Lorenzo R and Micari F 2011 *J. Cleaner Production* **19** 337-47
[2] Mori K, Maki S and Tanaka Y 2005 *CIRP Annals – Manufac.Tech.* **54** 209-12
[3] Hasnulhadi J, Mori K and Abe Y 2014 *Procedia Eng* **81** 843-48
[4] Hamedon Z, Abe Y and Mori K 2001 *Materials. Sci. and Eng.* **114** 01
[5] Takahashi M, Uenishi A and Kuriyama Y 2000 *Nippon Steel Technical Report* 81.0
[6] Gan W and Wagoner R H 2004 *Inter. J. Mech. sciences* **46** 1097-113
[7] Hamedon Z, Mori K and Abe Y 2014 *Procedia Eng.* **81** 2074-79
[8] Mori K, Abe Y and Suzui Y 2010 *J. Materials Processing Tech.* **210** 653-59
[9] Hamedon Z, Mori K and Abe Y 2014 *J. Materials Processing Tech.* **214** 945-50
[10] Kim Y H and Park J J 2002 *J. Materials Processing Tech.* **130–131** 42–6
[11] Kusumi K, Nomura N and Maki J 2013 *Nippon Steel Technical Report* no. 103
[12] Maeno T, Mori K and Hamedon Z 2012 *Steel Research Inter.* **SPL. ISSUE** 287-90

IOP Conf. Series: Materials Science and Engineering **257** (2017) 012072 doi:10.1088/1757-899X/257/1/012072

Weld bead profile of laser welding dissimilar joints stainless steel

Ghusoon R Mohammed[1,2*], M Ishak[1], S N Aqida[1] and Hassan A Abdulhadi[1,2]

[1]Faculty of Mechanical Engineering, University Malaysia Pahang, Pekan 26600, Pahang, Malaysia
[2]Middle Technical University, Baghdad, Iraq

*Corresponding author email: ghusoon_ridha@yahoo.com

Abstract: During the process of laser welding, the material consecutively melts and solidifies by a laser beam with a peak high power. Several parameters such as the laser energy, pulse frequency, pulse duration, welding power and welding speed govern the mode of the welding process. The aim of this paper is to investigate the effect of peak power, incident angle, and welding speed on the weld bead geometry. The first investigation in this context was conducted using 2205-316L stainless steel plates through the varying of the welding speed from 1.3 mm/s to 2.1 mm/s. The second investigation was conducted by varying the peak power from 1100 W to 1500 W. From the results of the experiments, the welding speed and laser power had a significant effect on the geometry of the weld bead, and the variation in the diameter of the bead pulse-size. Due to the decrease in the heat input, welding speed affected penetration depth more than bead width, and a narrow width of heat affected zone was achieved ranging from 0.2 to 0.5 mm. Conclusively, weld bead geometry dimensions increase as a function of peak power; at over 1350 W peak power, the dimensions lie within 30 μm.

1. Introduction

In the critical industrial technologies, stainless steel is a vital material owing to its fine and excellent mechanical characteristics. Recently, many methods have been adopted by the researchers to improve further on the surface or total performance of steels to make it as fit as possible against any forms of severe conditions [1-3]. In addition, when joining stainless steel plates, it is necessary that most requires are achieved. One of the most common ways of joining materials is welding. Over the years, the welding of dissimilar metals has attracted much investigation interests; their increasing interest been justified by their economic and technical potentials. The idea of joining dissimilar metal has made it possible to achieve flexible product designs through an efficient use of each component material, (maximally utilizing the good specific attributes of individual material). The welding of dissimilar metals or alloys confers flexibility to the design but this often results in problems that have a negative impact on the weld performance [4].

There is a chance of failure of dissimilar welds which were performed on alloys with a wide thermal expansion coefficients variation either at the fusion zone (FZ), solidification, or during the service life of the product as a result of thermal fatigue [5, 6]. The joining of two metals with wide variation in their melting temperature or thermal conductivities is challenging because of the possibility of the material melting at different temperatures [7].

To control these problems, there is a need for a proper choice of the welding process. One of the most common methods of metal joining is the fusion welding. There have been continuous efforts to adopt these methods when joining dissimilar metals. Based on the temperature-pressure combination of the individual metals, several welding processes such as arc welding, gas welding, solid-state welding, resistance welding, thermo-chemical welding and high energy beam welding have been developed. These welding processes produce different weld bead profiles and angular distortions to the weld pieces, as governed by the inherent characteristics of the process and the processing parameters.

Out of the whole welding processes, the high-energy beam welding process which uses a laser is more beneficial in terms of faster cooling rate, localized heating, less heat affected zone, and ease of access to the weld seam through fiber delivery, facilitating the welding of intricate geometries [8]. It often results in minimized distortion of the workpieces and the surface profile of the weld bead in controlled and ambient environments [9-11].

The adoption of heating sources with high-power density such as laser beam or electron beam can minimize some of these issues, owing to the greater monitoring of the shape and dimensions of the molten region and beam alignment; the high-power density allows for the immediate fusion of metals even when the melting temperatures vary. Many studies have reported the effect of fusion welding of materials that are characterized by different thermal, optical, electrochemical, and mechanical properties [12-14].

A careful selection of process parameters and requirements determines the successes of laser welding. A pre-knowledge of the expected temperature rise of the material is required when heating sheet metals with a laser beam. The increase in temperature can affect the geometry, velocity, workpiece hardness, gas flow rate, and microstructure of the weld piece. Many studies have investigated and estimated the geometry, hardness, heat affected zone, and microstructure of the weld beads.

The power of the laser beam and the welding speed have been reported as the major parameters that affect the penetration depth and bead width [15, 16]. The most significant parameter effect can lead to a weld with acceptable quality and efficiency. Lee et al have characterized the depth of the higher fusion zone to the width ratio, the cooling rate and the amount of porosity during welding using laser beam by varying the welding power, speed and focal position [17]. Similarly, the impact of different inert gases on the appearance of the weld line, the fusion zone dimensions, microhardness, solute evaporation, post weld tensile properties, and the porosity distribution has also been investigated. The increasing of the heat input, the area of the fusion zone, and the ratio of the dimensions of the weld area was shown to have an increasing trend. Paleocrassas [18] studied and reported the maximum welding speed as between 2 and 3 mm/s for a maximum penetration of 1.02 mm through the defining of the effluence and weld energy per weld length, considering the power, focal length and speed. The effect of the angle of incidence on the penetration depth, bead width and length have been investigated by Liao and Yu [19]. The study found reported an increase in the penetration depth and bead width and decreases in bead length when the incident angle was increased.

The width of the weld and penetration depth are the weld bead characteristics lengths that determine the cross section of the weld and with them, the strength or the shearing force of the weld joint can be determined [20-22]. Due to this, it is crucial to attain the right weld length characteristics. The fabricated weld geometric parameters must satisfy the requirements of the design and remain constant through the entire seam to achieve the performance desired from the weld joint in service. Therefore, in this paper the effect of welding speed, laser incident angle and the peak power on the weld bead dimensions (weld width and depth) and of the fusion zone welding will be presented.

2. Experimental details

The laser welding was studied in this report using a pulse wave fiber laser with an output power of 2 kW (Figure 1). A refractive optics based on focusing head with a focal length of 150 mm was used to direct the laser beam on the substrate. To avoid back reflection, the focusing head was tilted by 10° to the vertical axis. The surface of the workpieces was shielded with argon gas at the flow rate of 20 L/min. The austenitic stainless steel type (AISI316L) and duplex (AISI2205) which are some of the most common weld materials was used in this study. The materials used and their chemical composition and mechanical properties are shown in Tables 1 and 2.

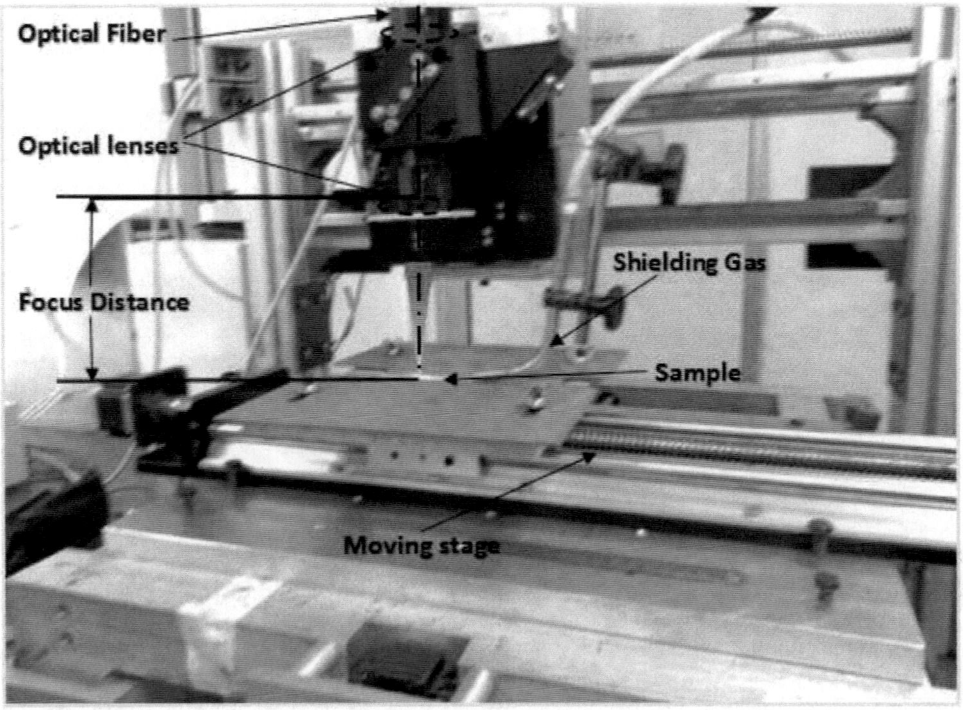

Figure 1. The set up for the fiber laser welding.

Table 1. The chemical composition of the base metal AISI 316L (wt. %).

Elements	C	Cr	Ni	Mn	Si	P	S	Mo	Nb
Weight (%)	0.03	16.5	9.6	1.15	0.48	0.003	0.005	2.4	0.002

Table 2. The chemical composition of AISI 2205 (wt. %).

Elements	C	Cr	Ni	Mn	Si	P	S	Mo	Nb
Weight (%)	0.03	22.07	4.8	1.15	0.53	0.003	0.005	3.65	0.002

To identify the process parameters for the laser welding, the laser power at 2 kW butt weld experiments were performed at different welding power and speeds; the plates were transversely cut towards the path of the laser beam propagation, and the cross section of the beam was examined under an optical microscope. The achieved aspect ratio (weld depth to width) of the weld bead cross-section was used for the selection of the optimum weld parameters based on the depth of penetration and minimal weld bead width. After this, the samples were prepared with the dimension of size 100 mm x 75 mm and thickness 2 mm, having square edges. The sketch of weld bead dimensions and the prepared samples is shown in Figure 2 and Figure 3, respectively.

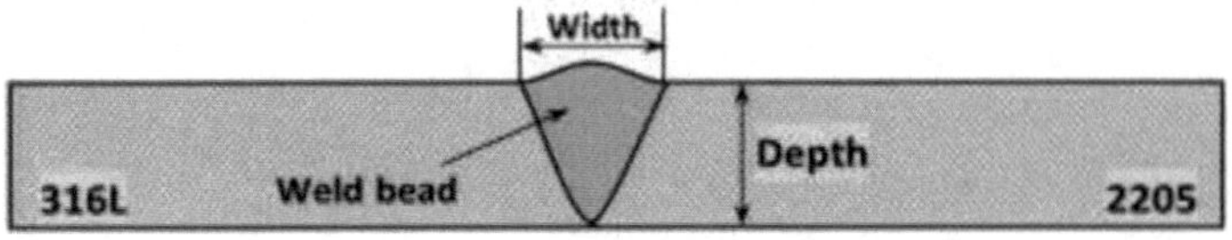

Figure 2. Sketch show the weld bead dimensions.

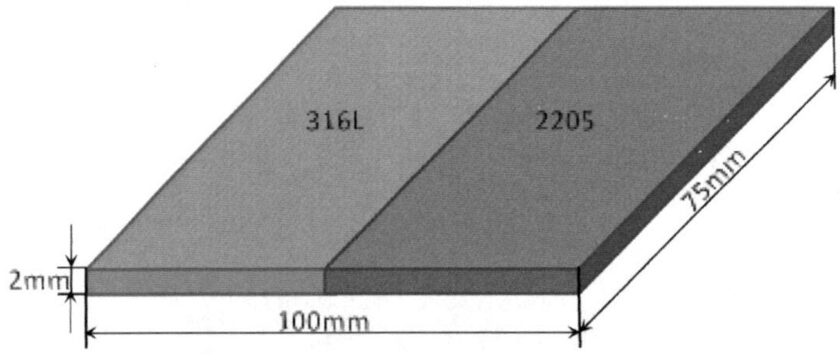

Figure 3. The sketch of the prepared samples.

The experiments for the welding were performed to investigate the effect of the welding speed, incident angle and peak power on geometry of the weld bead by varying the speed from 1.3 mm/s to 2.1 mm/s, the peak power from 1100 W to 1500 W, and the incident angle from 10° to 20°. The weld samples were transversely cut and the cross-sectional surfaces prepared for metallographic investigation. Different grades of polishing papers were used to polish the weld joints before etching with modified Kalling's regent to observe for the geometry of the weld bead. The prepared samples for metallographic investigation are shown in Figure 4 (a) to Figure 4(e), and Figure 5 (a) to Figure 5 (e). An optical microscope and image analyzer were used to measure the geometry of the weld bead parameters. Figure 6 shows the different zones of the 316L-2205 weld joints.

The heat affected zones of the weld samples were examined and found not to be in a straight line along the direction of the weld. This was because the base materials were non-homogeneous, as well as the power fluctuation and machine process capabilities.

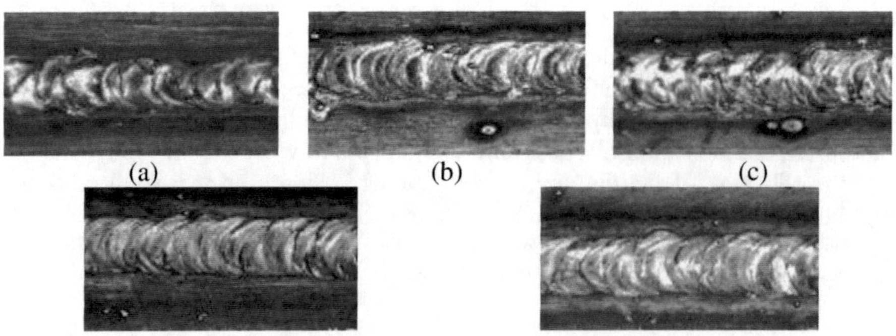

(a) (b) (c)

(d) (e)

Figure 4. Top views of welded joints at welding speed. (a) 1.3 mm/s (b) 1.5 mm/s(c) 1.7 mm/s (d) 1.9 mm/s (e) 2.1 mm/s

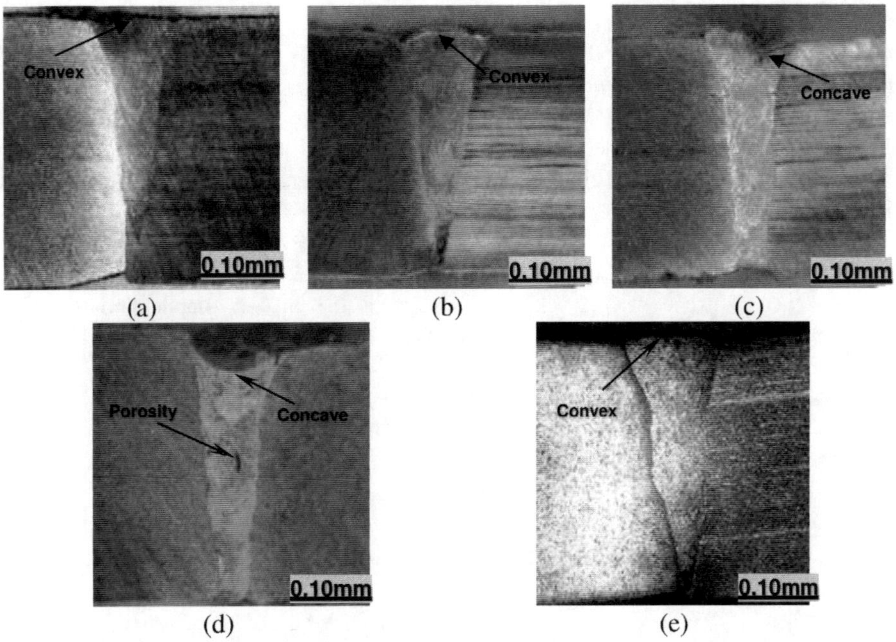

Figure 5. Transverse cross section of welded joints at welding speed. (a) 1.3 mm/s (b) 1.5 mm/s (c) 1.7 mm/s(d) 1.9 mm/s (e) 2.1 mm/s

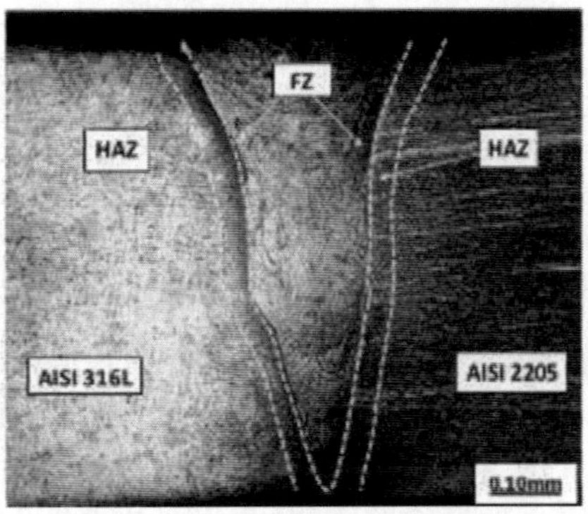

Figure 6. Micrograph of the different zones of the dissimilar welded joint.

Table 3. Parameters of fiber laser welding process.

Parameters	Values				
Peak power (W)	1100	1250	1350	1450	1500
Speed (mm/s)	1.3	1.5	1.7	1.9	2.1
Pulse width (ms)	3	4	5	6	6
Pulse repetition rate (Hz)	9	11	11	12	12
Incident angle	10°	12°	15°	18°	20°
Beam diameter (mm)	0.37	0.37	0.37	0.37	0.37
Laser energy density (W/mm^2)	10.4	11.8	12.7	13.7	14.1

3. Results and discussion

3.1. Effect of welding speed on weld bead profile
The surface changes of the welded pool were convex or concave in shape during the fusion welding. Few studies have investigated the mechanisms that are responsible for the shape of weld bead profile [23, 24]. In Figure 7, it was shown that the bead width and penetration depth depends on the weld speed as they decrease with increase in the welding speed.

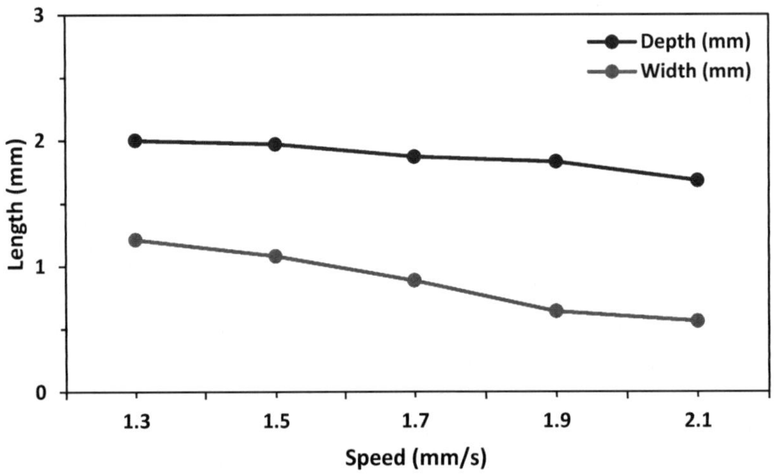

Figure 7. Effect of welding speed on bead width and depth of penetration of the welded joints.

The results suggested that the penetration depth depends more on the welding speed than the bead width due to the decrease in the heat input as well as less time of interaction between the source of the laser beam and the weld material. The weldment was found to be V-shaped, indicating the occurrence of the laser weld in keyhole mode [25, 26]. A narrow width of HAZ was achieved ranging from 0.2 to 0.5 mm and the width decreased as the welding speed increased.

The porosity of the welds was observed at some isolated locations, especially at the tip of the keyhole while welding at a lower speed due to the trapped gas [27] as shown in Figure 5d. At a higher speed, no porosity was observed. These observations agreed with the earlier reports [28, 29]. It was noted that the penetration depth is always greater than bead width in all consecutive spots due to the variation in the incident angle of the laser beam and the high heat input [19, 30].

3.2. Effect of peak power on weld bead profile
The dimensions of the weld pool were measured using image analyser. The geometry photographs of the weld pool created by the supply of pulse laser power input of 1100 W, 1250 W, 1350 W, and 1500 W with incident angle 20° are shown in Figure 8. The photographic inspection revealed an increase in the power input, leading to an increase in the dimensions of the weld bead and promoting spattering. In welding, spattering results in the reduction of the weld joint strength [31].

Figure 8 showed the bead width and penetration depth to depend on the peak power. As these values indicates, an increase in the heat input proportionately increases the bead width and penetration depth of the joints. For the HAZ area, a similar trend was followed with each of the associated joints. A similar trend has been reported by Yan [32] and Jana [33] while investigating TIG welded 304 SS and SMAW welded duplex SS, respectively. They found that an increase in the heat input increases the fusion zone and HAZ area. The size of the bead width and bead depth increased over a peak power range of 1100 -1500 W, and beyond 1350 W, the fluctuation in the weld bead geometry dimensions lied within 30 μm.

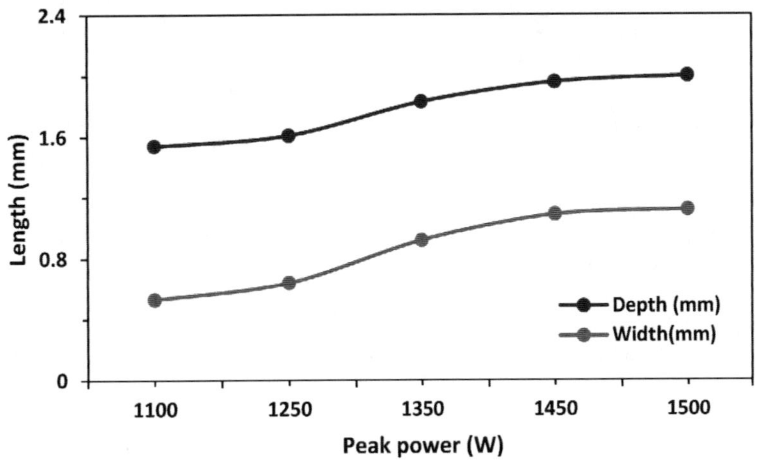

Figure 8. Effect of peak power on bead width and depth of penetration of the welded joints.

3.3. Effect of incident angle on weld bead profile

Figure 9 showed the depth of penetration to increase with increases in the incident angle at the laser energy range of 1100 - 1500W. The bead width, however, decreased with increase in the angle of incidence. Figure 9 also showed that the penetration depth and bead width are relatively short when the laser peak power (P.P) was 1100W.

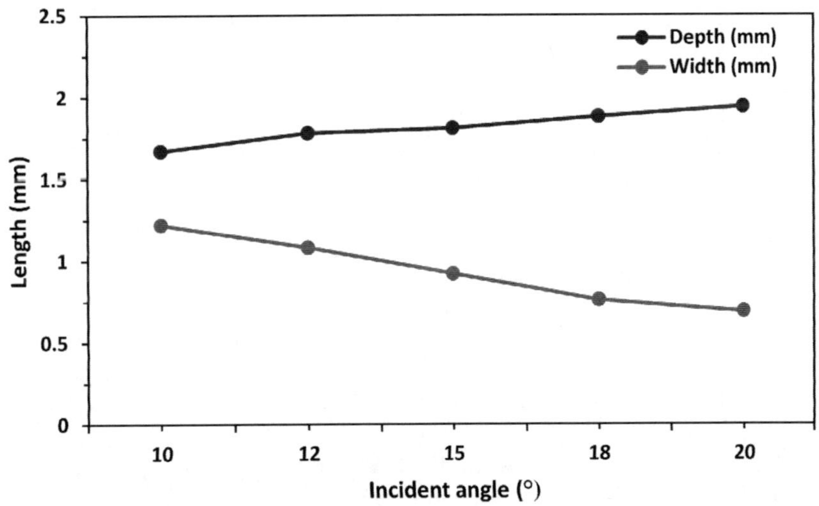

Figure 9. Effect of incident angle on bead width and depth of penetration of the welded joints.

The top view of the associated cross-sectional welded spot is presented in Figure 10 (a and b). The small depth of penetration and bead length were due to the low power density of the laser 967 W/mm^2 (<10^3 W/mm^2). When irradiated at this low density, the welding process is mainly conductive in nature [34, 35]. The keyhole welding which leaves a larger welding spot can occur only at a power density of more than 2×10^3 W/mm^2 since the laser power is not enough to cause boiling and hence, generate a keyhole [36]. The influenced zone in the specimen was basically caused by the conductive heat transfer, and the depth of penetration is limited (Figure 10b).

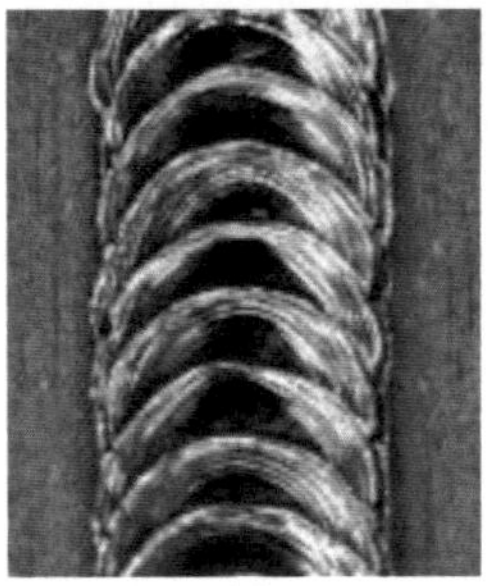

(a)

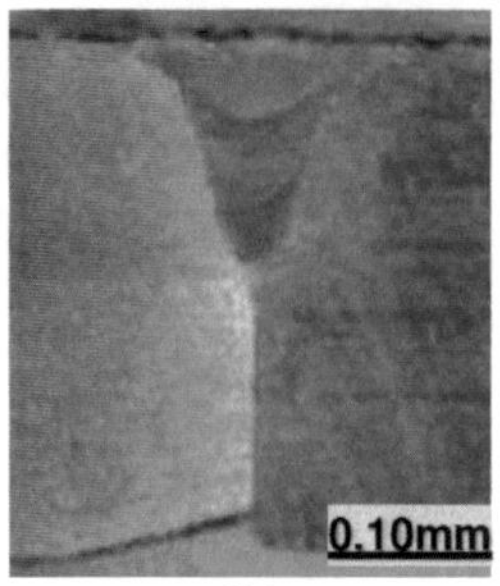

(b)

Figure 10. (a) Top views of seam laser welded and (b) Metallographic of a cross-sectional specimen. Laser weld parameter: laser incident angle=10◦ and laser peak power =1100 W

4. Conclusions

From the experimental outcomes of the fiber laser welding process to joint dissimilar stainless steel plates, the following conclusions may be concluded:

1. There was a decrease in the fusion zone depth and width with increasing welding speed.
2. The width of the zone affected by heat is very narrow, ranging from 0.2 to 0.5 mm and decreases with increasing welding speed.
3. There was no solidification crack in the fusion zone, though porosity was seen occasionally.
4. The elevation in the laser energy promote the characteristic lengths of the penetration depth (D), and the bead width (W) as the laser beam incident angle increased.
5. The characteristic length of the bead width and weld penetration depth at the same laser beam incident angle increases with an increase in the laser energy.
6. The bead width variation was found to be very minimum and not significantly affected by the beam power, welding speed, and beam angle.
7. The characteristic lengths of the weld bead and the shape depend not only on the welding speed but also on the peak power and the laser beam angle of incidence.
8. A proper use of parameters during fibre laser welding in the industries can enhance the quality of weld joints and prolong the durability of finished products.

Acknowledgements

The authors wish to acknowledge the financial support provided by the Universiti Malaysia Pahang (UMP)-FKM under grant no. RDU150378 and GRS 1503107.

References

[1] Zhang K M and Zou J X 2012 *Thin Solid Films* **526** 28-33.
[2] Liu R L and Yan MFF 2010 Surf. Coat. Technol. 204 2251-6.
[3] Wickström L and Hinds G, Turnbull A 2015 *Corrosion* **71** 1036-47.
[4] Shah L H and Ishak M 2014 *Mater. Manuf. Process.* **29** 928-33.
[5] Lee H Y, Lee S H, Kim J B and Lee L H 2007 *Int. J. Fatigue* **29** 1868-79.
[6] Paventhan R, Lakshminarayanan P R and Balasubramanian V 2011 *Mater. Des.***32** 1888-94.
[7] Velu M and Bhat S 2015 *Mater. Des.* **67** 244-60.
[8] Mohammed G R, Ishak M, Aqida S N and Abdulhadi H A 2017 *MATEC Web Conf.* **90** 01024.
[9] Nosyrev N A 2016 *Int. Conf Laser Opt.* (LO) S1-30-S1-.
[10] Katayama A, Kawahito Y and Mizutani M 2012 *Physics Procedia* **39** 8-16.
[11] Cui C, Cui X, Ren X, Liu T, Hu J and Wang Y 2013 *Mater. Des.* **49** 761-5.
[12] Tomashchuk I, Sallamand P, Belyavina N and Pilloz M 2013 *Mater. Sci. Eng. A-Struct. Mater.* **585** 114-22.
[13] Devendranath R K, Singh A, Raghuvanshi S, Bajpai B, Solanki T, Arivarasu, M 2015 J. Manuf. Process **19** 212-32.
[14] Romoli L and Rashed C A A 2015 Int. J. Adv. Manuf. Technol. **81** 563-76.

[15] Balasubramanian K R, Buvanashekaran G, and Sankaranarayanasamy K 2010 CIRP *J. Manuf. Sci. Technol.* **3** 80-4.
[16] Mohammed G R, Ishak M, Aqida S N and Abdulhadi H A 2017 *Metals* **7** 39.
[17] Lee M F, Huang J C, and Ho N J 1996 *J. Mater. Sci.* **31** 1455-68.
[18] Paleocrassas A G 2009 North Carolina State University.
[19] Liao Y C and Yu M H 2007 *J. Mater. Process. Technol.* **190** 102-8.
[20] Khan M A A, Romoli L, Fiaschi M, Dini G and Sarri F 2011 *Opt. Laser Technol.* **43** 158-72.
[21] Khan M A A, Romoli L, Fiaschi M, Sarri F and Dini G 2010 *J. Mater. Process. Technol.* **210** 1340-53.
[22] Khan M A A, Romoli L, Ishak R, Fiaschi M, Dini G and De Sanctis M 2012 *Opt. Laser Technol.* **44** 1611-9.
[23] Thomy C, Seefeld T and Vollertsen F 2008 *Weld World* **5** 9-18.
[24] Zhou J, Tsai H L and Wang P C 2005 *J. Heat Transf.-Trans. ASME* **128** 680-90.
[25] Fabbro R 2010 J. Phys. D-Appl. Phys. **43** 445501.
[26] Khan M A A, Romoli , Fiaschi , Dini G and Sarri F 2012 *J. Mater. Process. Technol.* **212** 856-67.
[27] Sales A M, Westin E M and Jarvis B L 2017 *Weld World*
[28] Katayama S, KawahitoY and Mizutani M 2010 *Physics Procedia* **5** 9-17.
[29] Kawahito Y, Matsumoto N, Abe Y, Katayama S 2013 *Weld Int.* **27** 129-135.
[30] Manonmani K, Murugan N, Buvanasekaran G 2007 *Int. J. Adv. Manuf. Technol.* **32** 1125-1133.
[31] Vasantharaja P, Vasudevan M and Maduraimuthu V 2017 *Trans. Indian Inst. Metals.*
[32] Yan J, Gao M and Zeng X 2010 *Opt. Lasers Eng.* **48** 512-517.
[33] Jana S 1992 *J. Mater. Process. Technol.* **33** 247-61.
[34] Curcio F, Daurelio G, Minutolo F M C and Caiazzo F 2006 *J. Mater. Process. Technol.* **175** 83-9.
[35] Triantafyllidis D, Schmidt M J J and Li L 2003 *J. Mater. Process. Technol.* **138** 102-8
[36] Li S, Chen G, Zhang Y, Zhang M, Zhou Y and Deng H 2014 *Laser Phys.* **24** 106003.

IOP Conf. Series: Materials Science and Engineering **257** (2017) 012073 doi:10.1088/1757-899X/257/1/012073

Control of non-linear actuator of artificial muscles for the use in low-cost robotics prosthetics limbs

Nurul Anis Atikah[4*], Leong Yeng Weng[1, 4], Adzly Anuar[2, 4], Chau Chien Fat[1, 4], Khairul Salleh Mohamed Sahari[2, 4] and Izham Zainal Abidin [3, 4]

[1]Department of Electrical and Electronics Engineering, College of Engineering, Universiti Tenaga Nasional, Jalan IKRAM-UNITEN, 43000 Kajang, Selangor, Malaysia.
[2]Department of Mechanical Engineering, College of Engineering, Universiti Tenaga Nasional, Jalan IKRAM-UNITEN, 43000 Kajang, Selangor, Malaysia.
[3]Department of Electrical Power Engineering, College of Engineering, Universiti Tenaga Nasional, Jalan IKRAM-UNITEN, 43000 Kajang, Selangor, Malaysia.
[4]CAMARO Research Group, Universiti Tenaga Nasional, Jalan IKRAM-UNITEN, 43000 Kajang, Selangor, Malaysia.

[*]Corresponding author: anisatikah17@gmail.com

Abstract. Currently, the methods of actuating robotic-based prosthetic limbs are moving away from bulky actuators to more fluid materials such as artificial muscles. The main disadvantages of these artificial muscles are their high cost of manufacturing, low-force generation, cumbersome and complex controls. A recent discovery into using super coiled polymer (SCP) proved to have low manufacturing costs, high force generation, compact and simple controls. Nevertheless, the non-linear controls still exists due to the nature of heat-based actuation, which is hysteresis. This makes position control difficult. Using electrically conductive devices allows for very quick heating, but not quick cooling. This research tries to solve the problem by using peltier devices, which can effectively heat and cool the SCP, hence giving way to a more precise control. The peltier device does not actively introduce more energy to a volume of space, which the coiled heating does; instead, it acts as a heat pump. Experiments were conducted to test the feasibility of using peltier as an actuating method on different diameters of nylon fishing strings. Based on these experiments, the performance characteristics of the strings were plotted, which could be used to control the actuation of the string efficiently in the future.

1. Introduction

The efficiency of prosthetic limbs today is limited by the actuator technology such as the motor. The motor being a staple of electric to mechanical transducer increases the mass of the prosthetic limbs while at the same time requiring additional electrical energy in the form of bulky batteries to move it. This results in a limited mobility time for users attached with prosthetic limbs [1]. Hence, researcher around the world are finding ways to create a relatively low-cost, lightweight and sufficiently powerful actuator to replace the current actuator technology which will significantly benefit the users. Typical users of the prosthesis are amputees. Young amputees in particular require their prosthetic limbs to be refit once or twice every year as they continue to grow. [2] As such, costs to replace them are expensive and research has shown usage of SCP [2], can reduce the costs significantly.

Natural muscles have many desirable properties such as high power-to-weight ratios, inherent compliance and damping, fast actuation and high dynamic range [2]. Unfortunately, current robotics actuators are unable to attain similar properties especially in a slender muscle-like form factor. A recent discovery showed that artificial muscles made from nylon fishing strings are strong, tough and cheap to manufacture. In addition to that, capable of actively contracting and expanding with much larger force than real muscles[3]. This artificial muscle can be also called lightweight actuators since it does not need any supplementary component and have been proven to produce significant mechanical power when thermally cycled. The aesthetic appearance of the limb could also be more human like with smoothed and rounded contours [2]. Compared to carbon nanotubes constructed artificial muscles, the cost of artificial muscles using nylon fishing string are much lower and furthermore is able to spin rotor at 100,000 revolutions per minutes or more [3]–[5].

Researchers utilizing nylon fishing string mostly employ passive cooling and active heating, to generate fast and large force in contraction but slow stretching during the cooling phase. In order to make to process more efficient, active cooling should be employed. Based on recent literature search, active cooling were only done by scientists at UT Dallas's Alan G. MacDiarmid Nano-Tech Institute teamed with scientist from universities in Australia, South Korea, Canada, Turkey and China [4].
We have also embarked on a research project to explore the feasibility of using Peltier system that can provide compact active cooling and heating as an alternative control method to the artificial muscle actuation of SCP created. It was expected that the contractions and expansions of the artificial muscle would behave in a non-linear manner, and thus requiring experimental study to model this characteristics. Initial experiments were conducted to evaluate the performance of different diameters strings and under similar conditions.

What if there is a simpler and cheaper solution to produce artificial muscles that one day can be used for prosthetic purposes, the first discovery reported that nylon is strong, tough and low-cost nylon strings can actively contract and expand with much greater force than muscles. This new discovery on artificial muscles enabled new opportunities in medical devices among others areas that are just beginning to be explored.[4] Our own skeletal muscle, have similarities with other mammals, which can contract 20% in length and can lift a gallon of milk (3.78 L) [4]. Nylon was found to be capable of contracting in a similar fashion but able to lift a hundred times more load. This means that nylon strings can provide an alternative to conventional motors. It also opens up the possibility of more natural, simple, low-cost and compact devices. Nylon thread can contract by almost 2% in length when heated. It also increased in diameter by 5%. Surprisingly, both of these effects can be combined to make a 20% or more increase in length by producing a coil [4].

Active actuators can be driven using heating and cooling. Transforming a nylon fishing string into artificial muscle is relatively easy. The nylon fishing string need to be twisted until it forms a coil. Activation of the strings can be achieved by any means, provided it can create high temperature such as hot air, hot water, or electric current. For example, using a hot air gun will be sufficient to see the actuators react [3].An activation cycle can be produced by placing the SCP within a tube that allows hot or cool water to be alternately pumped through. A thin metal coating can be used with nylon strings where current can be applied to generate high temperature. Maximum displacement takes place when SCP is heated to just below the melting point. To attain a full displacement, a nylon based SCP needs temperature of above 180°C [3].

2. Peltier Module

Peltier module is the key component in the proposed control circuit. A thermoelectric (TE) module or Peltier module is a semiconductor based electronic component. It can function as a small heat pump and mainly used in high end refrigerators and other cooling devices. A Peltier module acts as a heat pump when current is flowed through its terminals, where one of the sides will start to heat up and vice-versa for the other side.

IOP Conf. Series: Materials Science and Engineering **257** (2017) 012073 doi:10.1088/1757-899X/257/1/012073

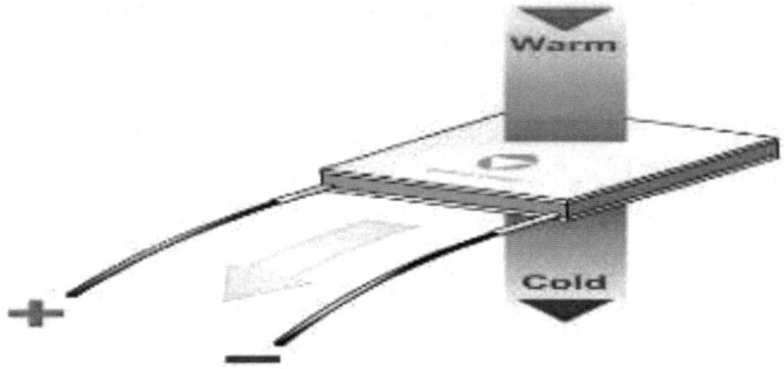

Figure 1. Peltier module[6]

A practical Peltier module, as shown in figure 1, consists of two or more elements of n and p-type doped semiconductor material. Both material are connected electrically in series and thermally parallel. These Peltier modules are mounted between two ceramics substances. The ceramics hold all structure together mechanically and electrically. These substances were insulated with individual elements from one another and from external mounting surfaces. Peltier module has variety ranges in size approximately 2.5 – 50 mm (0.1 to 2.0 inches) square and 2.5 – 5 mm (0.1 to 0.2 inches) in height. Peltier modules available in the market come in different shapes, different substrate materials, and different metallization patterns and different in mounting options [6]. Peltier module is commonly marked with some technical data presented in the form of TEC1-12709 (figure 2). [6]

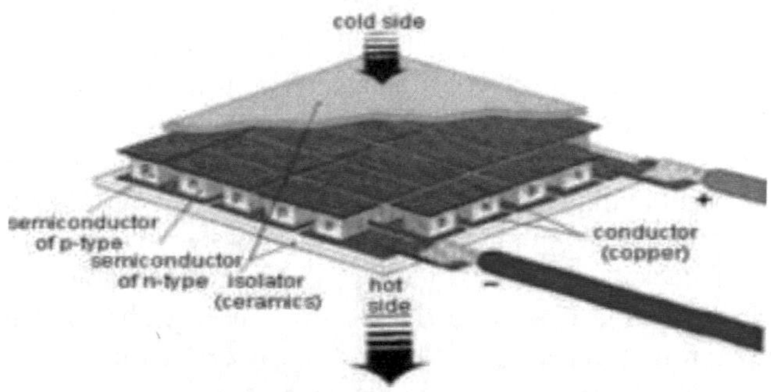

Figure 2. TEC1-12709 construction[6]

During the experiment, hot sides of the peltier module was the base in which the SCP was placed upon while immersed in thermal paste to improve heat conducting from the peltier devices to SCP.

3. Insulated Gate Bipolar Transistor (IGBT)

The Insulated Gate Bipolar Transistor (IGBT) is another important component used in these control circuit. It is a minority-carrier device with large bipolar current-carrying capability and high input impedance. Most designers consider IGBT as a device with MOS input characteristic and bipolar output characteristic that is a voltage-controlled bipolar device. IGBT also can be used in switched-mode power supply (SMPS) and other high switching repetition rates of power circuits. In this paper, the device was used to control the power supply to the Peltier module. The IGBT was connected to the peltier module which was used in this experiment.

IOP Conf. Series: Materials Science and Engineering **257** (2017) 012073 doi:10.1088/1757-899X/257/1/012073

4. Methodology

In this experiment, the nylon fishing strings were twisted using a portable drill that was attached to one end of the line while the other was attached to a 50g mass to create tension while the fishing string was being twisted. After the string was constructed to an SCP, it was stretched across a peltier module attached to a Perspex base. This base can be made up any other materials that can withstand the tension of the SCP during testing.

A Peltier system was constructed by using a 60W Peltier and IGBT, which acted as soft switch for the Peltier. The circuit was connected to two power supplies which were set at 11.2V and 7.3V respectively. The 7.3 volts was supply to turn on the IGBT while the 11.2V was used to energize the peltier. To simplify the circuit, the author's used the functionality of the power supply to control the maximum current which is to be supplied to the individual circuits. The peltier was allowed to sink 3.27 A while the turn on voltage was given a nominal current rating of 0.05A. The circuit diagram is as shown in figure 3. A thermal paste was used to ensure the temperature was evenly distributed from the peltier's hot surface to the SCP. Thermal paste was applied onto the SCP which made it a thermal interface between the peltier module and the string.

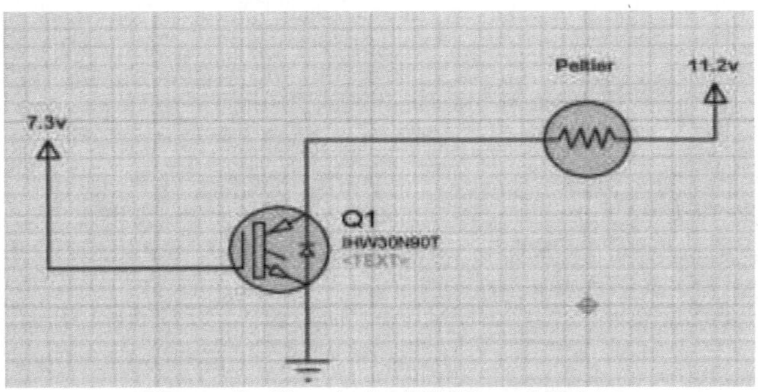

Figure 3. Circuit for IGBT and Peltier

A data acquisition device was connected a load cell rated at 3kg which was then attached to the end of the SCP to collect data during the experiment. The data acquisition device used in this experiment was the Agilent U2351A. In this research, the strings were selected from the same brand, but of varying diameter. Even thermal distribution is important to ensure different thickness of strings reacted within 10 seconds. The experimental set up to test the feasibility of using peltier module to actuate the SCP is shown as in figure 4. Data in this experiment was collected for less than 10 seconds each time and having a cool down period of 10 seconds before the next test was carried out. The forces in this experiment were collected via the load cell in voltage units. The strings with different diameters were tested in this experiment.

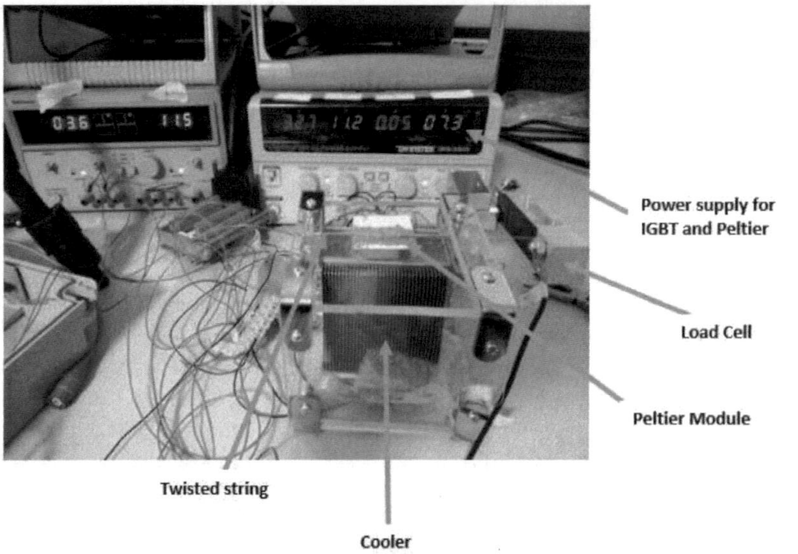

Figure 4. Experiment apparatus setup

5. Results and Discussion

Figure 5 shows the activation point (dotted lines) for the Peltier and the moment for which each of the differently sized strings reacted (solid lines) after the Peltier was turned on. On average, all of the strings took 20 -25 milliseconds to reach their respective maximum contraction force before they started to unravel. The bright red line representing string thickness of 0.55 mm had the shortest reaction time which may due to its small diameter compared to the 0.65 mm and 1.00 mm diameter string. It is thought that small diameter SPC allowed for even and faster heat transmission.

Figure 6 and figure 7 show the reaction times of strings with diameter of 0.45mm and 0.55mm respectively. As shown, smaller diameter strings reacted faster than larger diameter strings. This may due to the uneven distribution of thermal paste along the strings on the Peltier module. The uneven paste distribution could have contributed to uneven heating of the string, which may have led to slower reaction times. The 0.45 mm string had lower reaction times, this may be due to the time needed to heat up the string evenly was shorter compared to the thicker string. From the graph, equations were developed from the respective curves. These graphs were created using polynomial regression functions in Microsoft Excel. The 'y' represents the voltage in volts and the 'x' represents the time in samples from the sampling rate of 10ksa/s. The turn on voltage can be calculated at any given time in samples with the provided equation for specific curve.

Based on the points plotted on the graph, the equation for each line was produced. These equations could be used to describe the characteristics for different diameter strings. In Figure 6 and 7, the force data of each experiment were extract from the times the peltier was activated; to the times, the twisted string achieved maximum force in voltage units. We are particularly interested in this region is firstly because this is the region which effectively generates the maximum force without losing its mechanical properties such as elasticity.

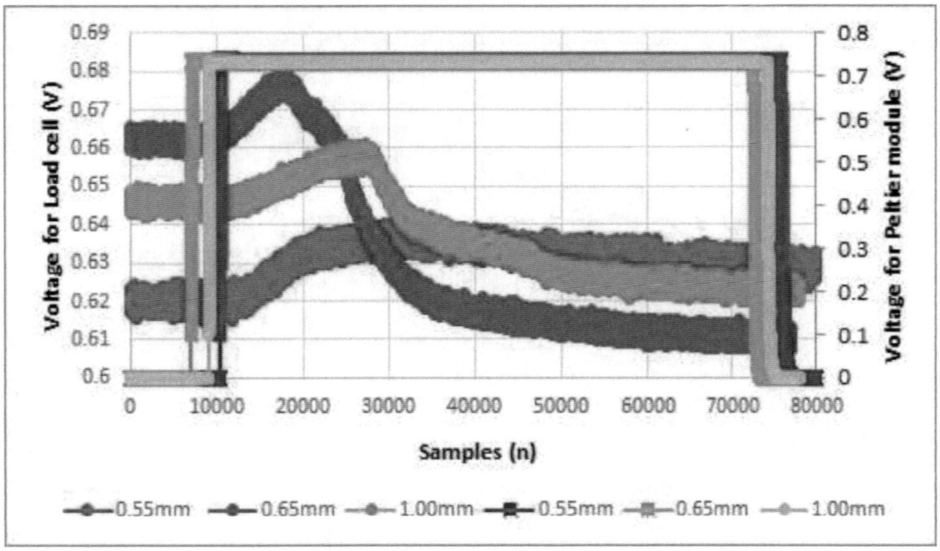

Figure 5. Activation point and characteristics of different diameter strings.

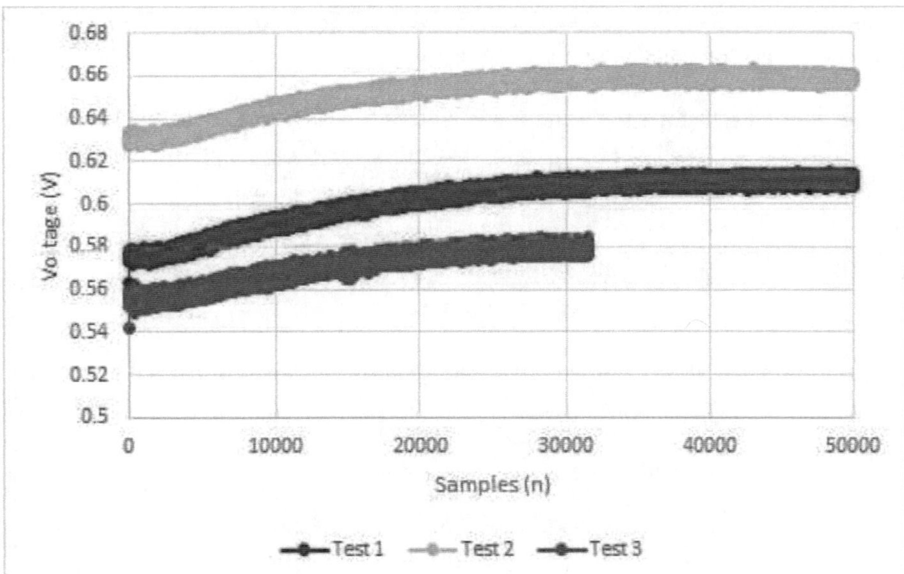

Figure 6. Reaction time for 0.45mm diameter string

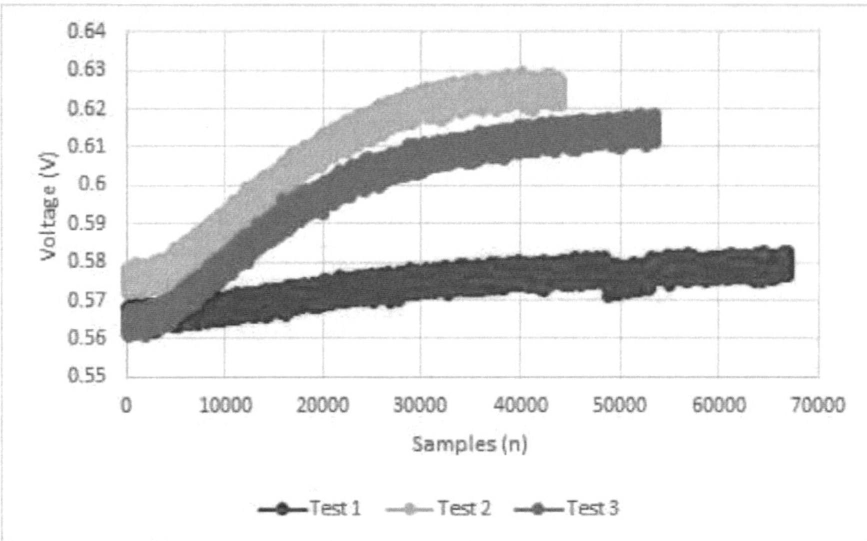

Figure 7. Reaction time for 0.55mm diameter string

In this particular case of figure 6 and 7, it can be seen that the polynomial fits yield equations with low residues of 0.9 and above. This indicates that the base polynomial fit is able to correctly characterize the strings force generation properties in voltage with reference to time.

Equation 1 Equations representing each line for each test of the 0.45 mm string

$$y = 3E\text{-}16x^3 - 4E\text{-}11x^2 + 2E\text{-}06x + 0.5725 \quad (1a)$$
$$R^2 = 0.9887$$
$$y = 3E\text{-}16x^3 - 4E\text{-}11x^2 + 2E\text{-}06x + 0.6275 \quad (1b)$$
$$R^2 = 0.9836$$
$$y = -7E\text{-}16x^3 + 7E\text{-}12x^2 + 1E\text{-}06x + 0.5538 \quad (1c)$$
$$R^2 = 0.9666$$

Equation 2 Equations representing each line for each test of the 0.55 mm string

$$y = 7E\text{-}18x^3 - 4E\text{-}12x^2 + 4E\text{-}07x + 0.5644 \quad (2a)$$
$$R^2 = 0.9233$$
$$y = -6E\text{-}16x^3 + 1E\text{-}11x^2 + 2E\text{-}06x + 0.5717 \quad (2b)$$
$$R^2 = 0.992$$
$$y = 9E\text{-}17x^3 - 3E\text{-}11x^2 + 2E\text{-}06x + 0.5589 \quad (2c)$$
$$R^2 = 0.9909$$

Each equation were the result of one specific diameter string being tested multiple times. Based on the residues, equation 1a and 1b, which have the highest score of reside also exhibit similar coefficients for the equation. The only difference between equation 1a and 1b is the constant, where the difference was only 0.055. As for equation 2a, 2b and 2c, the consistency was not as good as that of equation 1a, 1b and 1c. The residue value varied greatly between the three equations. Despite the equation 2b having the largest residue, the authors think the equation 2c is more likely be the actual characterisation equation for the string of 0.55mm diameter, first being that the coefficient are closer to that of equation 1a and 1b, and also the residue is the second best. Further studies need to be done, but as of now, it shows that the response of the string can be characterised experimentally.

6. Conclusions

The experiment successfully archived the objective of proving that the actuation of SCP can be produced with peltier module. Even so, it was inferred that the string would behave better if it were to be woven or bundled up so that the string can withstand the heating that were supplied by the Peltier module and also increasing the force threshold for the strings. A compact solution to allow for even heating and cooling should be found as the current solution using thermal paste does not perform well. Although other researchers have done the same experiment and resulted with only a 3ms of reaction time, but due to the different usage of material in this experiment, the data collected can be added as a resource to be further researched on. The result looks promising for further improvement and can be further expanded to ensure the artificial muscle made from nylon string works well for the young amputees.

Acknowledgments

This research is funded under UNITEN Internal Research Grant no: J510050638. The authors would like to thank those who had directly and indirectly provided insights which greatly assisted this research.

References

[1] B. Tondu, "A dynamic model of the McKibben artificial muscle contraction," *IEEE/ASME Int. Conf. Adv. Intell. Mechatronics*, p. 64, 1997.

[2] M. C. Yip and G. Niemeyer, "High-Performance Robotic Muscles from Conductive Nylon Sewing Thread," *Proc. IEEE Int. Conf. Robot. Autom.*, pp. 2313–2318, 2015.

[3] C. S. Haines, N. Li, G. M. Spinks, A. E. Aliev, J. Di, and R. H. Baughman, "New twist on artificial muscles.," *Proc. Natl. Acad. Sci. U. S. A.*, vol. 113, no. 42, p. 701-708, 2016.

[4] J. D. W. Madden and S. Kianzad, "Twisted Lines : Artificial muscle and advanced instruments can be formed from nylon threads and fabric," *IEEE Pulse*, vol. 6, no. 1, pp. 32–35, 2015.

[5] U. of Hanyang, "Hybrid carbon nanotube yarn muscle," no. May, pp. 2013–2014, 2014.

[6] A. N. Nikam and J. A. Hole, "A Review on use of Peltier Effects," vol. 2, no. 2, pp. 6–12, 2014.

[7] Data Sheet, "Keysight U2300A Series USB Multifunction Data Acquisition Devices."

IOP Conf. Series: Materials Science and Engineering **257** (2017) 012074 doi:10.1088/1757-899X/257/1/012074

Potential utilization of biodiesel as alternative fuel for compression ignition engine in Malaysia

M A Wahab[1,*] M I N Ma'arof[1], I N Ahmad[2] and H Husain[2]

[1]Department of Mechanical Engineering, Faculty of Engineering and Quantity Surveying, INTI International University, Persiaran Perdana BBN, Putra Nilai, 71800 Nilai, Negeri Sembilan, Malaysia
[2]Faculty of Mechanical Engineering, Universiti Teknologi Mara 40450 Shah Alam, Selangor, Malaysia

*Corresponding author: i12000703@student.newinti.edu.my

Abstract. Biodiesel is a type of fuel which is derived from various sources of vegetable plants and waste fuels. Today, numerous biodiesels have been engineered to be at par or even better in term of performance in comparison to pure diesel. Therefore, biodiesel has shown a promising sign as one of the best candidate in overcoming total dependency on pure diesel. This paper gives review on various tests and experiments conducted on biodiesel in order to highlight the potentials given by this particular fuel. In addition, providing the supporting evidences to further endorse for a mass usage of biodiesel in Malaysia – simultaneously, driving the country to become a potential global biodiesel producer in the near future. The reviewed studies were obtained mainly via indexed journals and online libraries. Conclusively, every test and study for every blend of biodiesel had shown consistent positive results in regards to performance and in overcoming emission related issues. Thus, providing the evidence that biodiesel is highly reliable. Malaysia as a semi-agricultural nation could take the advantage in becoming one of the leading global biodiesel producers. Nevertheless, this will requires total cooperation of every concerned government bodies and authorities.

1. Introduction

Global energy crisis along on both socioeconomic and environmental aspects on continuous use of petroleum derived fuels had given rise on demands for alternative fuels [1-3]. This is proven by the European Union (EU) promotion on the use of biodiesel fuels for transportation in 2003. This particular promotion has given an increment from 2%-5.75% from the year 2005-2010 [2-4]. Biodiesel is a fuel derived from natural, renewable sources, such as from new and used vegetable oils, animal fats, and recycled restaurant grease [5, 6]. This is achieved via the transesterification process to bring the viscosity and density level of the produced biodiesel fuels to be close to that of diesel [7-9]. In order to be called biodiesel, the fuel must meet the strict quality specifications of ASTM D 6751 [7]. Additionally, biodiesel is found to be diesel engine friendly [8, 3]. The aim of this study is to give review on the unique advantages in utilizing biodiesel as an alternative source of fuel and also to highlight on the extraordinary potentials of biodiesel for mass application in Malaysia in the near future.

2. Literature Review

2.1. Performance of pure biodiesel and biodiesel blends

Literature has discussed on various positive outcomes with the use of biodiesel. The advantages of biodiesel fuels are summarized as follow [1, 8, 26]:

a) Biodiesel fuels show excellent lubricity characteristics, though, most are sulphur free. These fuels were found by Alam et al. [10] and Chuepeng et al. [11] to have the capability to increase the overall engine performance and reduce particulate matter emissions.
b) Biodiesel fuels are safe to be stored and handled since they have a flash and fire point higher than 150^0C.
c) High oxygen percentage (10%) in biodiesel fuels "significantly contributes" to complete fuel combustion.
d) The cetane number (CN) for biodiesel fuels is almost similar or higher than that of diesel, though still dependent on the source. Higher CN would lead to lower ignition delay [12].

Biodiesel could be utilized either as a pure form (100% biodiesel) or blended with petroleum-based diesel fuel (biodiesel blend) "designated BXX, where XX represents the volume percentage of biodiesel fuel in the blend" (biodiesel.org). The commonly tested types of biodiesel are the 10% (B10), 20% (B20) and 100% (B100). The summary of the tests from the previous studies is presented as in Table 1 [4, 10, 11, 13-20, 30].

Table 1. Summary of tests conducted on biodiesel.

Test	Result	Percentage
Particulate matters (PM)	Significance decrement in percentage of emission	0.25±0.13%
Carbon monoxides (CO)		0.277±0.041%
Hydrocarbon (HC)		0.49±0.05%
Smoke emissions		-
Nitrogen oxides (NOx)	Increment in percentage of emission	0.186±0.022% for every 1% of biodiesel in the blend
Specific fuel consumption	Increment	0.116±0.009% (95% confidence interval) for every 1% of biodiesel
Engine relative energy conversion efficiency	Increment	Approximately 3-5% of the petrodiesel fraction

As a result, from the existence of several blends of biodiesel; Suryanarayanan et al. [21] had conducted a study to seek for the optimum blend of biodiesel "for which the best performance and/or lowest emissions are obtained". They concluded that the selection of blend must be made dependent on the goal or objective concerning the aforementioned parameters (SFC, NOx, and HC emission), as well as other factors such as cost. The summary of the comparison test from previous studies [1, 4, 8, 12-14, 19, 31, 32] among pure biodiesel (B100) and other blends of biodiesel is as shown in Table 2.

In order to totally exploit the true potential of biodiesel, the CI engine has to be specifically calibrated. Kim et al. [22] reported that "there exists optimum fuel injection pressure that has minimum droplet size when the ambient gas pressure is constant". Basavaraja et al. [3] concluded that the optimum injection pressure for operation of biodiesel is at 20 MPa. At 20 MPa calibration, exhaust temperature was reduced, maximum brake thermal efficiency obtained and combustion emissions being smoke, CO, HC, NOx and UBHC were lowered. For this operating calibration, they also recommended the usage of a B20 biodiesel (off vegetable oil ester) since the study regarded that particular blend to be the best blend in term of overall performance as oppose to other blends.

Table 2. Summary of comparison test of B100 and B20 biodiesel blends with other biodiesel blends and diesel.

Biodiesel Blends	Results summary
B100	The use of pure biodiesel (B100) in comparison to other biodiesel blends is not recommended due to having a lower brake thermal efficiency, mainly as a result from having a high viscosity level and inferior peak power of as oppose to diesel. B100 also shows a higher level of smoke, CO, HC and combustion duration, yet, relatively lower peak power NO emission and heat release rate (decrease of 9%-12%) in the maximum value of BMEP (brake mean effective pressure) and a minimal increase of fuel conversion efficiency were observed. This suggested that a pure biodiesel is not the best fuel to be used for compression ignition engines.
B20	B20 shows higher exhaust temperature due to higher mixing and diffusion of controlled combustion fractions. Also, the specific fuel consumption is higher resulting from a lower heating value and the high percentage of oxygen atom leads to late burning of heavier components of biodiesel in the expansion stroke, thus, leading to the increment of the mass average temperature and the lowering the emission of HC (by 9% lower), CO (by 3% lower), PM and smoke (by 10% lower) at 95% level of confidence. Soluable organic fraction (SOF) and sulphates were observed to have been reduced with the use of B20. Maximum engine power and torque of B20 was also observed to be at par with pure diesel blends which do not really give much disadvantages in regards to engine power, fuel consumption and NOx emission (at low injection pressure), although, minimal drop in power is almost negligible (approximately 3% and 8%).

Verhaeven et al. [4] specifically highlighted on the advantages of low blends biodiesel fuel over pure biodiesel and mineral diesel fuel. Firstly, there is no need for engine adjustments (also noted by Malhotra et al. [1], Basavaraja et al. [3], Bushnoor et al. [15] in regards to biodiesel in general). Secondly is there is no effect on fuel consumption and vehicle power and potential of emission reduction. Next is that there is better lubrication of the fuel injection system and a comparable or somewhat higher energy efficiency. Finally, there is improvement of the biodegradability of the fuel [4]. Patterson et al. [9] showed that different sources of biodiesel could give different outcome of combustion output. However, Malhotra et al. [1] specifically concluded that various biodiesels (but with B20 blends) were able to meet the diesel BIS specification. Hence, this indicated that the B20 is the most optimum blend of biodiesel which could be incorporated at a much larger scale. Though Patterson et al. [9] had carried out a study in comparing three different biodiesel fuels (rape, soy and waste oil) this study suggested for a higher variant of biodiesel fuels to be tested simultaneously in the future.

2.2. Issues with NOx emission: impact on human and the environment

Although, biodiesel have shown such a promising characteristic for a substitute from total dependent on diesel, it should be noted that almost all aforementioned studies had highlighted on the increase of NOx emission as a result from biodiesel fuel combustion. This however, will give negative impact to human and the environment.

From the observation made by this study, the main trade-off in controlling smoke emission is the increase of NOx emission [4, 10, 11, 13-20] and excessive emission of NOx that could bring catastrophic impact to both human as well as to the environment such as chronic bronchitis and pulmonary edema [24, 27, 28]. Nevertheless, continuous studies – for example, Zhang et al. [12] and Patterson et al. [9] have suggested that a specific paper that discussed on reviewing past and current studies in achieving both the control of NOx and smoke should be presented in the near future since it is a wide topic.

3. Malaysia as potential user and global producer of biodiesel

Malaysia as a semi-agricultural country has both the high potential for mass utilization of biodiesel fuel and as global producer of the fuel. Southeast Asia has a variety of sources for biofuel for instances, palm oil and coconut oil, and Malaysia is already the world's leading producer for palm oil [26]. Verhaeven et al. [4] had outlined several "conditions" which need to be considered prior to introducing biodiesel for public usage and "a number of these are already being looked into on a European level":

1. There has to be guarantee of product quality, that is, the produced biodiesel should have fulfilled certain product specifications, hence, being safe for usage.
2. The availability of raw materials either naturally or through imports
3. Industrial and government support and co-operation

Hence, it is clear that the "conditions" denoted by Verhaeven et al. [4] require the involvement of every level of individuals in the country to make full introduction and application of biodiesel fuel in the country. A report released by USDA suggests that the consumption of B10 will reach by 770 million litters in 2017 and by 2020, a 15% blend in expected for a roll-out. Hence, over the next few years, the higher blending rates combined with the increasing size of the biodiesel pool, will result in the expansion of biodiesel usage [29]. This paper truly believes that Malaysia could definitely do so, given the time and opportunity.

4. Conclusions

Conclusively, every test and study had shown consistent positive results for every blend of biodiesel in regards to performance and in overcoming emission related issues. Thus, providing the evidence that biodiesel is highly reliable. Malaysia as a semi-agricultural nation could take the advantage in becoming one of the leading countries in the world's supplier of biofuel. Nevertheless, this will require total cooperation of every concerned government bodies and authorities.

References

[1] Malhotra R K, Jain P, Chopra A and Maheshwari M 2005 *SAE Technical Paper Series 2005* **26** 032
[2] Payri F, Macián V, Arrègle J, Tormos B and Martínez J L 2005 *SAE Technical Paper Series 2005* **01** 2205
[3] Basavaraja T, Reddy R P and Swamy V 2005 *SAE Technical Paper Series 2005* **26** 030
[4] Verhaeven E, Pelkmans L, Govaerts L, Lamers R and Theunissen F 2005 *SAE Technical Paper Series* **01** 2201
[5] U.S. Department of Energy *Alternative and Advanced Fuels* (Retrieved on 14th June, 2012) Available from: http://www.afdc.energy.gov/afdc/fuels/biodiesel.html
[6] Yahoo Education *Biodiesel* (Retrieved on 14th June, 2012) Available from: http://education.yahoo.com/reference/encyclopedia/entry/biodiesel
[7] biodiesel.org *Biodiesel* (Retrieved on 26th December, 2012) Available from: http://www.biodiesel.org/what-is-biodiesel/biodiesel-basics
[8] Pradeep V and Sharma R P 2005 *SAE Technical Paper Series* 2005 **26** 353
[9] Patterson J, Hassan M G, Clarke A, Shama G, Hellgardt K and Chen R 2006 *SAE Technical Paper Series 2006* **01** 0234
[10] Alam M, Song J, Acharya R, Boehman A and Miller K 2004 *SAE Technical Paper Series 2004* **01** 3024
[11] Chuepeng S, Tsolakis A, Theinnoi K, Xu H M, Wyszynski M L and Qiao J 2007 *SAE Technical Paper Series 2007* **01** 0072
[12] Zhang X, Li G G L, Wu Z, Hu Z and Deng J 2008 *SAE Technical Paper Series 2008(a)* **01** 1832
[13] Grimaldi C N, Postrioti L and Battistoni M 2002 *SAE Technical Paper Series 2002* **01** 0865
[14] Xiaoming L, Yunshan G, Sijin W and Xiukun H 2005 *SAE Technical Paper Series 2005* **01** 2199
[15] Bhusnoor S S, Babu M K G and Subrahmanyam J P 2007 *SAE Technical Paper Series 2007* **01** 0613

[16] Nigro F E B, Trielli M A and da Costa C M 2007 *SAE Technical Paper Series 2007* **01** 2635
[17] Kong S C and Kimber A 2008 *SAE Technical Paper Series 2008* **01** 1389
[18] Zhang X, Li H W L, Wu Z, Hu Z and Zhao H 2008 *SAE Technical Paper Series 2008(b)* **01** 1834
[19] Bielaczyc P and Szczotka A 2008 *SAE Technical Paper Series 2008* **01** 1398
[20] Reksowardojo K, Lubis I H, Wishnu Manggala S A., Brodjonegoro T P, Soerawidjaja T H, Arismunandar W, Dung N N and Ogawa H 2007 *SAE Technical Paper Series 2007* **01** 2025
[21] Suryanarayanan S, Janakiraman V M, Sekar J, Rao G L N and Sampath S 2006 *SAE Technical Paper Series 2006* **01** 3534
[22] Kim G Hyun, Goto S and Ehara R 2001 *SAE Technical Paper Series 2001* **01** 3636
[23] Shudo T, Hiraga K and Ogawa H 2007 *SAE Technical Paper Series 2007* **01** 0622
[24] Saini K *Effect of environmental pollutants on human health* (Retrieved on 14th June, 2012) Available from: http://www.environmentabout.com/723/effect-of-environmental-pollutants-on-human-health
[25] Clover C 2007 *Malaysia defends palm oil production* (Retrieved on 14th June, 2012) Available from: http://www.telegraph.co.uk/earth/earthnews/3296977/Malaysia-defends-palm-oil-production.html
[26] Demirbas A 2007 *Energy Policy* **35** 4661-670
[27] Davidson C I, Phalen R F and Solomon P 2005 *Aerosol Science and Technology* **39** 737-49
[28] Kagawa J 2002 *Toxicology* **181-182** 349-53
[29] Wahab A G 2016 *USDA Foreign Agriculture Service: Bio Fuels Annual, Malaysia* p 7-9
[30] Murillo S and Miguez J L 2007 *Fuel* **86** 1765-71
[31] Buyukkaya E 2010 *Fuel* **89** 3099-105
[32] Gumus M and Kasifoglu M 2009 *Biomass and* **34** 134-9

IOP Conf. Series: Materials Science and Engineering **257** (2017) 012075 doi:10.1088/1757-899X/257/1/012075

Dynamic analysis of composite beam with piezoelectric layers under thermo-mechanical load

A Toudehdehghan[1,2,*], M M Rahman[2] and Farrukh Nagi[2]

[1]Department of Mechanical Engineering, INTI International University, Jalan BBN 12/1, Bandar Baru Nilai, 71800 Nilai, Negeri Sembilan
[2]Department of Mechanical Engineering, Universiti Tenaga Nasional Malaysia, km7, Jalan Kajang-Puchong, 43009 Kajang, Selangor, Malaysia

*Corresponding author: abdolreza.toudehdehghan@newinti.edu.my

Abstract. In this paper, the control of composite beam vibrations with sensor and actuator connected layers is considered with consideration of the effect of thermal environment. The coupling relation between electrical field and mechanical deformation with uncoupled thermal impact are used. The mathematical model of shear deformation (Timoshenko's theory) has been applied and basic equations for piezoelectric sensors and actuators have been proposed. The equation of motion for the beam structure is obtained by the Hamilton principle and analyzed by finite element method. The control algorithm is based on proportional velocity control. Hence, the purpose of this article is to investigate the direct and inverse effects of piezoelectric on control of simply supported beam vibration under uniform temperature.

1. Introduction

Temperature variation is one of the effective or sometimes a major factor in the destruction of engineering structures. The structure members, for example, in reactor shells, spatial structures, turbines and other components of the mechanism, are prone to breakage due to buckling, deflection with high ranges or additional stresses due to thermal loads or mechanical-thermal composite loads. The development of composite materials with piezoelectric components has great potential in terms of safety and structural control [1-4]. Initially, piezoelectric phenomenon is used as a collection of mechanical energy from vibrations [5]. Due to their direct and reciprocal effects, these new materials, with their inherent and adaptive properties, can demonstrate better mechanical and thermal properties and, in fact, eliminate problems and problems caused by thermal loads or mechanical thermal loads. Given that these materials are lightweight, compact, simple, solid and also small in size, a large number of them can be used in the structure without significantly increasing the structure's mass. Piezoelectricity was first discovered by the Curie Brothers in 1880. They found that by applying pressure to a quartz crystal, an electric field is produced in the crystal [6]. The piezoelectricity theory, along with the thermal field, is called the theory of thermo-piezoelectric and has been worked out in limited articles [7, 8]. The governing equations for thermo-piezoelectric were first extracted by Mindlin [9]. Using the general theory of Hamilton, Nowacki obtained the general principle of thermo-piezoelectric [10]. Tauchert has been instrumental in controlling the force vibrations caused by very rapid thermal or mechanical arousal [11]. Saravanos and Lee have performed thermo-piezoelectric analysis of composite beam using layerwise theory [12]. Sunar and Rao have done research on the use of piezoelectric materials in places where temperature changes are significant [13].

Here, the two sides pin beam with piezoelectric layers attached to the upper and lower surfaces is considered, and the effect of piezoelectric materials on the behavior of the beam in a dynamical state is investigated. A finite element model using the Gelarkin method has been used based on the Timoshenko beam theory to analyze the beam performance, under the influence of distributed force on the upper beam and thermal load, using the Newmark method.

2. Extracting governing equations

The displacement field components in Timoshenko's beam are considered in the form below [14]:

$$u(x,z,t) = z\varphi_x(x,t) \tag{1}$$

$$w(x,z,t) = w_0(x,t) \tag{2}$$

Thus, the axial strain components and the shear strain are

$$\varepsilon_{xx} = z\frac{\partial \varphi_x}{\partial x} \tag{3}$$

$$\varepsilon_{xz} = \varphi_x + \frac{\partial w_0}{\partial x} \tag{4}$$

2.1. Thermo-piezoelectric equations

$$\{\sigma\} = [c]\{\varepsilon\} - [e]^T\{E\} - [\lambda]\theta \tag{5}$$

$$\{D\} = [e]\{\varepsilon\} + [\varepsilon_P]\{E\} - [P]\theta \tag{6}$$

The equation (5) is called the inverted thermo-piezoelectric effect, and equation (6) is called direct thermos-piezoelectric effect [12, 13]. It is assumed that the layer of the actuator with thickness h_p, voltage (V^e) is applied only in the direction of the thickness of the actuator layer. As a result, the electric field vector is defined as follows [13]:

$$\{E\} = \begin{bmatrix} 0 & 0 & \dfrac{1}{h_P} \end{bmatrix}^T V^e \tag{7}$$

2.2. Governing equations of beam

To derive the governing equations for moving the beam, we use the principle of virtual displacement defined as:

$$\delta\int_{t_0}^{t_1}(T - U + W)dt = 0 \tag{8}$$

where $\delta T, \delta U$ and δW are virtual kinetic energy, virtual strain energy and virtual work respectively, which are done by external forces.

Using the displacement field of the beam (Eqs. (1) and (2)) and strain equations (Eqs. (3) and (4)), and using the dynamic model of the principle of virtual displacement, It can be obtained the governing equations of motion beam, taking into account Timoshenko's theory, according to the following:

$$\frac{\partial Q_x}{\partial x}+q=I_0\frac{\partial^2 w_0}{\partial t^2}$$
$$\frac{\partial M_{xx}}{\partial x}-Q_x=I_2\frac{\partial^2 \varphi_x}{\partial t^2} \tag{9}$$

where

$$M_{xx}=b\int_{-h/2}^{h/2}\sigma_{xx}zdz \tag{10}$$

$$Q_x=bk_s\int_{-h/2}^{h/2}\sigma_{xz}dz \tag{11}$$

$$\begin{Bmatrix} I_0 \\ I_2 \end{Bmatrix}=b\int_{-h/2}^{h/2}\rho\begin{Bmatrix} 1 \\ z^2 \end{Bmatrix}\rho dz \tag{12}$$

Parameters M_{xx}, Q_s, (I_0, I_2), b, ρ and (k_s) are respectively result moment, shear force, mass moments inertia, width of beam, density and shear correction coefficients. Thus, The shear correction factor is considered $k_s = 5/6$ [14].

2.3. One dimensional thermal loading

A double-sided pin beam with a thickness of h, where boundary conditions ($z=-h/2$ and $z=h/2$) are constant. So that θ_b and θ_t are represented respectively temperature of the below and top of the beam. Thermal energy is not produced inside the beam.

The temperature distribution in thermos-piezoelectric composite beam is

$$\Delta\theta=\frac{\theta_b+\theta_t}{2}+z\frac{\theta_b-\theta_t}{h} \tag{13}$$

Thus, thermal stress is obtained from equation (13). The mechanical stress is acquired with substituting equations ((3) and (4)) into equation (5). Substituting the thermal and mechanical stresses into equations ((10) , (11)), and then substituting along with equation (12) into equation (9), the governing equations motion of thermo-piezoelectric of beam will be obtained.

2.4. Sensor and actuator equations

2.4.1. Sensor equation. A layer is used as a sensor layer in a distributed thermo-piezoelectric composite system. Based on the theory of direct piezoelectric effects in a distributed system, it can obtain the effects of the strain of the beam and the field of heat. To avoid error, the distributed thermo-piezoelectric sensor layer has a very small thickness compared to the total thickness of the beam. Electric charge using Gaussian theory is as follows [15]:

$$Q_s^k=\frac{1}{2}\left[\int_{z=z_{k+1}}D_z dz+\int_{z=z_k}D_z dz\right] \tag{14}$$

The relation of electric field with the difference of electrical potential is as follows [16, 17]:

$$E=\frac{V_s}{h_{ps}} \tag{15}$$

where h_{ps} is thickness of sensor layer.

By putting the equation (6) in equation (14), the electric field is obtained and then by placing in the equation (15), the difference in the electrical potential of the sensor layer is obtained.

2.4.2. Actuator equation. In the direct proportional control system, the output voltage equation of the sensor with the applied voltage of the operator is as follows:

$$V_a = GV_s \tag{16}$$

By placing the equation (16) into equation (15), the electric field of actuator is obtained. Then, the result put into equation (6).

3. Finite element

Here, by finite element method, the Timoshenko beam equations with the aid of the Galerkin method, are obtained mass, hardness and force matrices.

3.1. Integral form

By means of applying the Gelerkin method which equation (9) are multiply by weight functions (M_1, M_2). Then with integration relative to a length of an element is

$$\int_0^L \left[\frac{\partial Q_x}{\partial x} + q - I_0 \frac{\partial^2 w_0}{\partial t^2} \right] (-M_1)\, dx = 0$$
$$\int_0^L \left[\frac{\partial M_{xx}}{\partial x} - Q_x - I_2 \frac{\partial^2 \varphi_x}{\partial t^2} \right] (-M_2)\, dx = 0 \tag{17}$$

3.2. Model of finite element

The equation (17) are the integral form of the equations of the beam. Then with approximation φ_x and w_0 according to the integral form of φ_x and w_0 are relative to x. Therefore, these approximation are approximated throughout the length of element:

$$\varphi_x \approx \varphi_x^e = \sum_{j=1}^{m} \varphi_j^e n_j^e \tag{18}$$
$$w_0 \approx w_0^e = \sum_{j=1}^{n} w_j^e m_j^e$$

where, φ_j^e and w_j^e are at j node. The weight functions also are defined as follows:

$$M_1 = m_j$$
$$M_2 = n_j \tag{19}$$

m_j^e and n_j^e are obtained by choosing the type of element and its form that is considered linear. Replace the approximation equation (equation (19)) into the integral forms (equation (17)). Thus, the results are obtained in form of mass, stiffness and force matrices:

$$
\begin{aligned}
K_{ij}^{11} &= \int_0^L GA\frac{dm_i}{dx}\frac{dm_j}{dx}dx & K_{ij}^{12} &= \int_0^L GA\frac{dm_i}{dx}n_j dx \\
K_{ij}^{21} &= \int_0^L GA\frac{dm_j}{dx}n_i dx & K_{ij}^{22} &= \int_0^L \left(EI\frac{dn_i}{dx}\frac{dn_j}{dx} + GAn_i n_j \right) dx \\
m_{ij}^{11} &= \int_0^L I_0 m_i m_j dx & m_{ij}^{22} &= \int_0^L I_2 n_i n_j dx \\
F_i^1 &= Q_{2i-1} + \int_0^L m_i q_m dx & F_i^2 &= Q_{2i}
\end{aligned}
\tag{20}
$$

3.3. Shape function

The functions w and φ are considered first-order linear:

$$
\begin{aligned}
& w = m_1 w_1 + m_2 w_2 , \qquad \varphi = n_1\varphi_1 + n_2\varphi_2 \\
& m_1 = n_1 = 1 - \frac{x}{L_e}, \qquad m_2 = n_2 = \frac{x}{L_e}
\end{aligned}
\tag{21}
$$

By placing equation (21) in equation (20), the matrix form of the equation (20) are obtained. As a result, the general form of the finite element of uncoupled dynamical thermo-piezoelectric equations in closed-circuit mode in the direct proportional control are as follows:

$$
\begin{aligned}
& (Mx)_1 + (Kx)_1 = (F_{mec})_1 + (F_{th})_1 + (F_{boundary})_1 \\
& (Mx)_2 + (Kx)_2 = (F_p)_2 + (F_{th})_2 + (F_{boundary})_2
\end{aligned}
\tag{22}
$$

4. Results and discussion

The double-headed joint (simply supported) is made of graphite-epoxy which a combination of two symmetrical angular layers (angle-ply) $[45/-45]_s$ with a total thickness of beam (35 mm), a width of beam (50 mm) and a length of composite beam (350 mm) in initial temperature 298.15° K. Piezo-electric layers on the sides of the beam are made of piezoceramic (PZT) with thickness of 0.8 mm. The properties of the material are in Table 1.

Table 1. Material properties [12].

Property	Graphite/epoxy	Piezoceramic (PZT) APC 840
Elastic stiffness, E_{11} (Gpa)	39	68
Elastic stiffness, E_{22} (Gpa)	8.6	68
Density (Kg/m^3)	2100	7600
Shear modulus G_{12} (Gpa)	3.8	26.2
Major poisson's ratio v_{12}	0.28	0.3
Major poisson's ratio v_{12}	0.06	0.3
Piezo-electric Constant, e_{31} (m/V)	0	$-125\text{x}10^{-12}$
Therma Expansion Coefficient λ_{11} (1/^{0}C)	$7\text{x}10^{-6}$	$3.8\text{x}10^{-6}$
Pyroelectric constant, P_3 (C/m^2 ^{0}C)	0	$-0.25\text{x}10^{-3}$
Electric permitivity, ε_{33} (N/V^2)	0	$11.06\text{x}10^{-9}$

IOP Conf. Series: Materials Science and Engineering **257** (2017) 012075 doi:10.1088/1757-899X/257/1/012075

4.1. Dynamic analysis

In dynamic analysis, a uniform impact force $(F = 400\exp(-1000t))$, applied on the upper surface of the beam is used to examine the vibration of the beam.To solve, we use the Newark method with linear coefficients $\beta = 0.25, \gamma = 0.5$ .The upper piezoelectric layer and the lower layer are considered as a sensor and actuator respectively.The decrease in the range of vibrations, depends on the feedback control gain.

The following graph shows the thermal and mechanical distribution effect on the vibrating system. As the coefficient of damping (gain) increases, the damping force increases and the vibration velocity decreases. Heat term acts as an external force (temperature on top and bottom surface respectively are ($\theta_t = 373\ K$, $\theta_b = 298.15\ K$), it increases the range of vibrations. In Figure 1 free vibration of the middle of simply supported beam with thermal force effect and without damping.

The boundary condition of simply supported beam as

$$x = 0, a \quad ; \quad w_0 = w = \varphi_x = M_{xx} = 0$$

The cross section of side boundary are considered to be insulated thermally.

In Figure 2, with the application of feed back control, the electrical gain is increased, with regard to the effect of temperature rise, the amplitude of the vibration of the composite beam is reduced earlier and more.

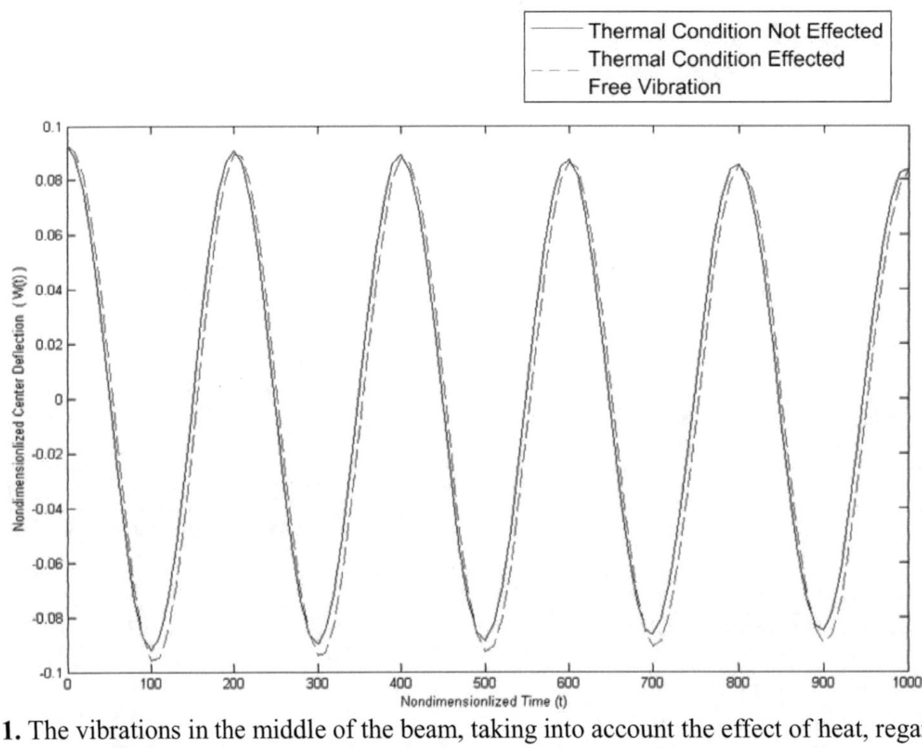

Figure 1. The vibrations in the middle of the beam, taking into account the effect of heat, regardless of the distributed effect of the sensor layer and piezoelectric actuator layer.

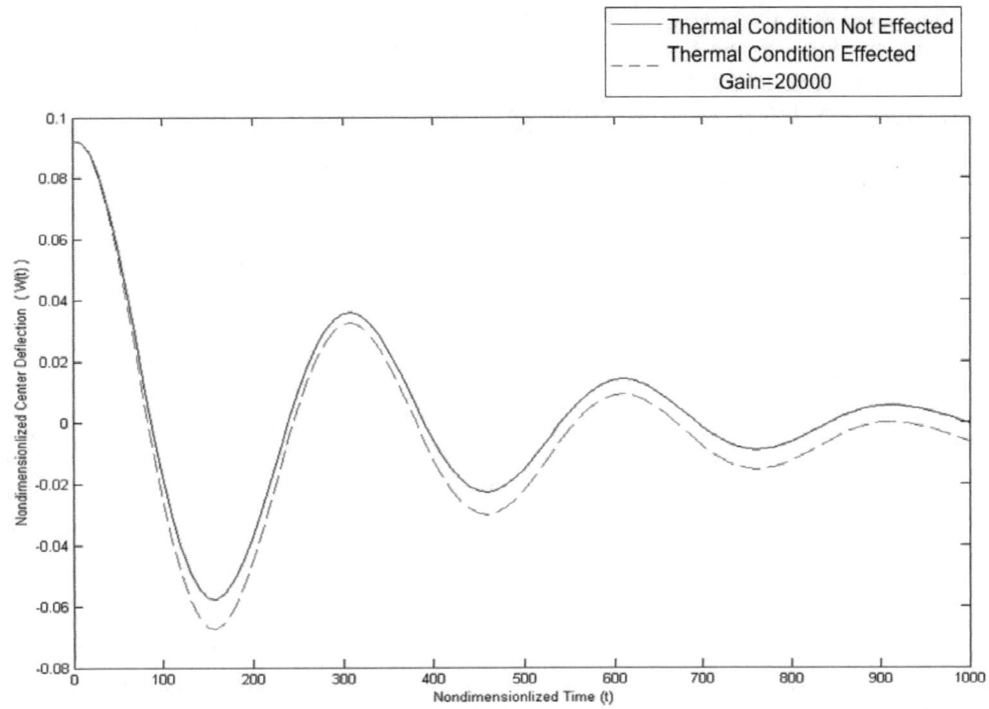

Figure 2. Damping the vibrations in the middle of the beam due to the direct distributed proportional control effect in terms of time, taking into account the effect of temperature rise.

5. Conclusions

Based on the Timoshenko beam model and the principle of virtual work, the governing equations were extracted. For a linear thermal distribution, the deflection of the beam and the effect of the voltage were checked for a decrease.Also, in dynamic analysis, a simple feedback control algorithm called Direct proportional Controlling, which couples the direct and reverse effects of piezoelectric couplings, and the effect of increasing the gain rate on vibration reduction, was investigated. A special thermal distribution that dynamically increases the amplitude of the vibration. It can be concluded that piezoelectric materials have a relatively high efficiency for controlling the shape of structures in environments that are characterized by significant temperature variations.

References

[1] Raghavan A and Cesnik C E 2005 *Smart Materials and Structures* **14** 1448
[2] Saravanos D A and Heyliger P R 1995 *Journal of Intelligent Material Systems and Structures* **6** 350-63
[3] Mancini S, Tumino G and Gaudenzi P 2006 *Journal of Intelligent Material Systems and Structures* **17** 577-85
[4] Han B, Wang Y, Dong S, Zhang L, Ding S and Yu X 2015 *Journal of Intelligent Material Systems and Structures* **26** 1303-45
[5] Bowen C, Kim H, Weaver P and Dunn S 2014 *Energy & Environmental Science* **7** 25-44
[6] Qin Q 2012 *Advanced Mechanics of Piezoelectricity* (Springer Science & Business Media)
[7] Aouadi M 2006 *International Journal of Solids and Structures* **43** 6347-58
[8] Chattopadhyay A, Li J and Gu H 1999 *AIAA Journal* **37** 1633-38
[9] Mindlin R 1974 *International Journal of Solids and Structures* **10** 625-37
[10] Nowacki W 1985 *Mathematical models of phenomenological piezoelectricity* vol. 1 (Banach Center Publications) p 593-607
[11] Tauchert T, Adali S and Verijenko V 2001 *4th Int. Cong. on Thermal Stresses* (Cracow) p 8-11
[12] Lee H J and Saravanos D D 1996 *AIAA Journal* **34** 1231-37

[13] Sunar M and Rao S S 1997 *AIAA Journal* **35** 534-9
[14] Wang C, Reddy J N and Lee K 2000 *Shear deformable beams and plates: Relationships with classical solutions* (Elsevier)
[15] Lam K, Peng X, Liu G and Reddy J 1997 *Smart Materials and Structures* **6** 583
[16] Shegokar N L and Lal A 2014 *Meccanica*, **49** 1039-68
[17] Liew K, Yang J and Kitipornchai S 2003 *International Journal of Solids and Structures* **40** 3869-92

A significant effect on flow analysis & simulation study of improve design hydraulic pump

M N Harith*, R A Bakar , D Ramasamy, Ma Quanjin

Automotive Engineering Research Group (AERG), Faculty of Mechanical Engineering, Universiti Malaysia Pahang, Malaysia

*Corresponding author: najmiharith@gmail.com

Abstract. Hydraulic ram pump is device that can lift water to the higher position without use of external energy. Several ways have been tested to decrease the water waste rate to maximize efficiency. The purpose of this research work is to develop a new design of hydraulic pump and conduct simulation study to justify improvement of the design. Major addition on making threaded waste valve system in order to control the opening and closing of the valve. At the same time reduce time taken for creating enough momentum and hammering effect. The results revealed that with the advancement of the proposed well improve designs, the water losses at waste valve reducing about 20-30% compare to existing design across the mass flow rate of 0.10 kg/s.The min/max velocity and pressure along the pipeline also has been increase for both open and closed condition.

1. Introduction

A hydraulic ram or hydram is a pump that utilize momentum energy from a constantly falling quantity of water to pump some of it to an elevation much higher than the original level at the source. No other external energy is required as long as there is constant supply of water, the pump will remain to work continuously and automatically. This makes it a highly useful mechanism for pumping water where there is a large source of available water, such as a river, channel, dam or any other form of reservoir. Thus, the hydraulic ram pump is completely self-sustaining and green in nature.It is well known that hydraulic pump are far more environment friendly to pump water compared to conventional electrical or fossil fuel based energy pump. In addition the waste water from the hydram can easily be cycle back into the reservoir, thus effectively creating zero wastage of water. Most of the conventional pump technology are limited by the high cost maintenance and electrical usage cause it impossible to be develop in rural areas. Furthermore, these conventional system require close monitoring and need special expertise to handle it. Thus, specialized characteristics such as free energy usage, low expertise and cost of maintenance, long life span, relative pressure stability and recyclable waste have turned researchers to make use of the hydraulic pump as alternative option pump in the future. Research of hydraulic ram pump has a long history. Begin in 1797 with unsuccessful theoretical attempts to explain the hammering action by J. M. Montgolfier until the end of the 19th century, when the water hammer was first explained in detail.Over the past decade, most research in hydram has emphasized on design optimization and self-built valves for high water supply.

In an analysis of experimentation and analysis involved in the performance optimization of a hydraulic ram pump, Prasanna et al. [1] found that are multiple parameters that affect the performance of a waste valve such as shape of the valve, size of the valve, size of the bush/valve seat, length of stroke and weight of the valve, whether spring operated or weighted, and orientation of the valve. These discovery are supported by the findings of Li J and Yang K [2] which showed that efficiency of the novel hydraulic ram pump reaches 70% by adding adjustable local resistance components in parallel in the delivery valve. With this technology, an adjustable and high-head experimental platform has been developed and experiments for the newly developed hydraulic ram pump were conducted on the platform. The adjustability and stability of the platform were verified. On the other hand, interestingly Harish Kumar [3] suggesting deeper research especially in design and manufacturing considerations to get the best design with higher efficiency. Every aspects of involved components being studied including dimensions, type of wielding preferred and material selection. Besides that, one study by E.J Schiller [4] demonstrate derivation of computerized theoretical model for hydram design, prediction of performance and surveying suitable sited. These result were contradicted by the experiments of author who considered modification of design only to improve overall hydram performance.

In particular, no investigation has been made on opening and closing condition of improve design of hydraulic pump using fluent and computational fluid dynamics (CFD).Therefore the aim of this work is to study significant effect on flow analysis and simulation study of improve design hydraulic pump. For this purpose, the opening and closing effect of valve on pump performance being investigate in order to reduce water losses at the waste valve and increase the pumping rate. To the author knowledge.no findings on CFD analysis of opening and closing condition of hydraulic pump regardless new or old design.

The contribution of this study is obvious as the resulting outcomes can be capitalized as guidelines for modelling and fabrication of more efficient hydraulic pump. Both modelling and actual fabrication can be compare and further analysis to obtain more details and improvement.

1.1. Model Designing

By using Solidwork 2017, roughly 1000 mm length and 1283 mm height hydraulic pump was designated as shown in isometric, top and bottom figure below. Total assembled of about more than 30 components.

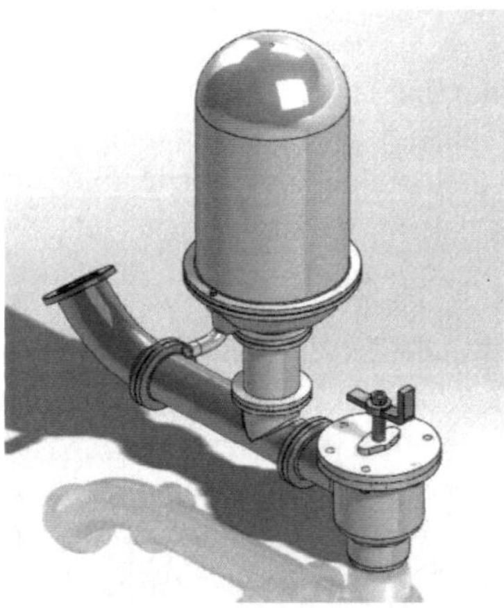

Figure 1. Isometric View

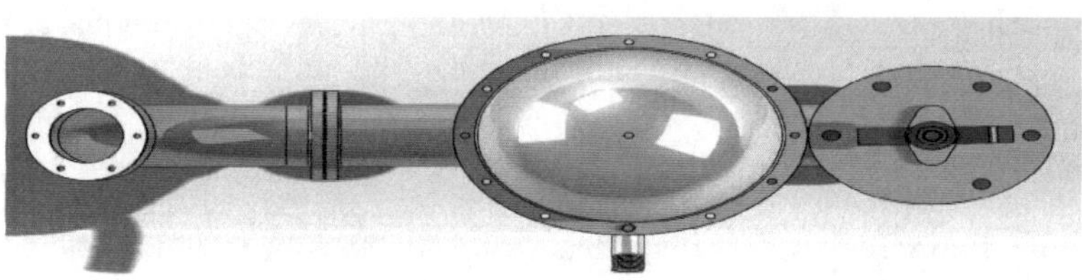

Figure 2. Top View

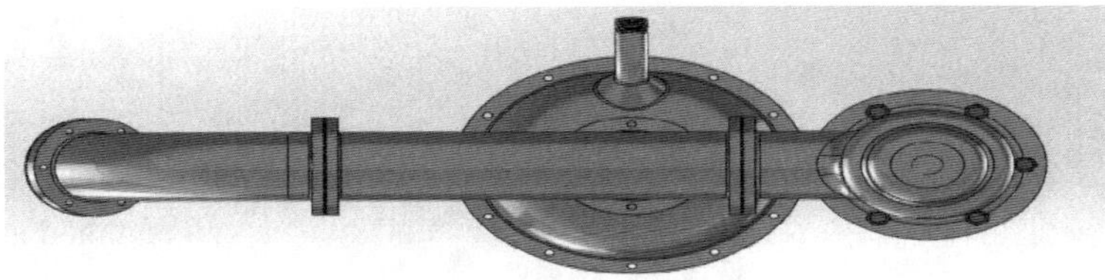

Figure 3. Bottom View

1.1.1 Improve Design of Delivery Valve. For delivery valve design as shown in figure 4, it being created with adjustable threaded nut to enable control of waste valve opening and closing clearance dimension. Meaning this design is more dynamics because it enable time-control water volume from source to pressure chamber. Differ from existing design that just rely on valve weightage and spring loaded that were supported by Poonam et al. [5]. Next installation of spring will aid the distribution of force equally around the waste valve. Hence maintain well the clearance gap that being set based on initial requirement.

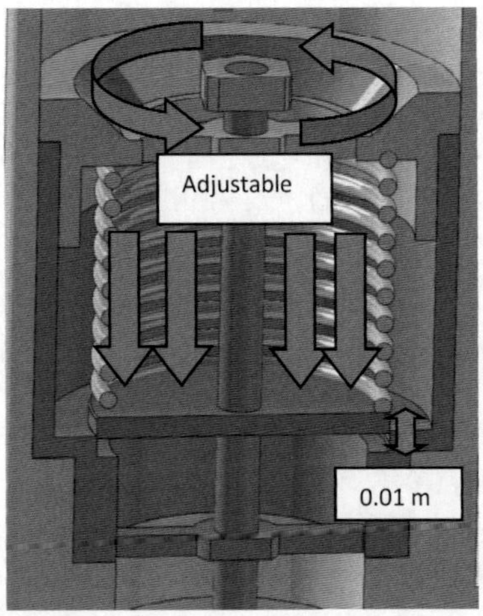

Figure 4. Well Improve Delivery Valve Design

1.1.2 Improve Design of Waste Valve. Gap at the outlet being reduced to prevent big volume of water overflow. Compare with existing design that allow so much of water out just to wait enough water momentum to shut the valve close, so that hammering effect can be created as being described by Young and his co-workers [6]. Furthermore, this design also similar to delivery valve by which we can limit/control the water flow by adjustable threaded nut mechanism.

IOP Conf. Series: Materials Science and Engineering **257** (2017) 012076 doi:10.1088/1757-899X/257/1/012076

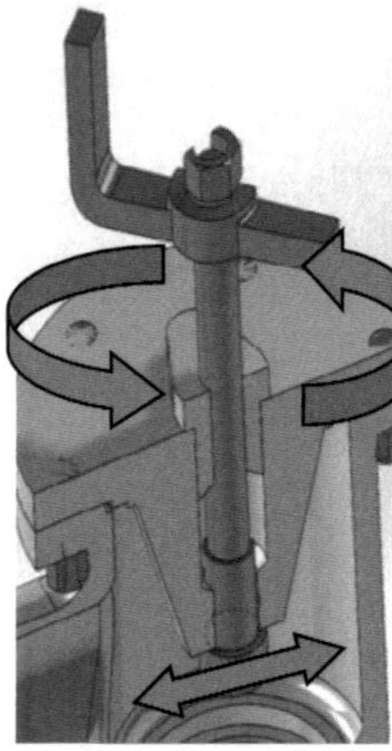

Figure 5. Small Gap and Adjustable Clearance Waste Valve

1.2 Meshing

For meshing and analysis projection, Fluent and CFD Ansys software being applied in this study as being shown by figure 6.Tetrahedrons type mesh is used for the assembly meshing method as they are easy to create automatically while the generation of other type elements is very difficult for complex shapes such as those being designated. This method supported by the finding of S.K Choundary et al. [7]. About 145977 elements and 28131 nodes successfully created for analysis preparation.

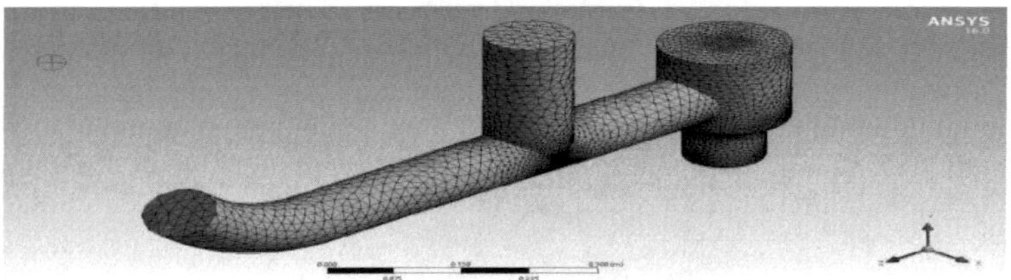

Figure 6. Meshing Result

1.3 Boundary Condition

Table 1. Boundary Condition Settings

Velocity Inlet	**Inlet Intake**
	Velocity Inlet or Mass Flow Inlet
	Values Can Be Added For Different Head
Outlet	Mass Outlet
Wall	Rest of Parts Geometry/Boundary Condition

2. Analysis for Existing Design

The dynamic meshing or transient analysis approach is used to treat the closing and opening of waste/delivery valve over time in the computational area. In other words, the mesh generation approach was used to treat the moving waste/delivery valve as a moving solid body in the computational domain. Valve moves upward and downward position considering open and closed condition similar to Reza Fatahi et. al findings [8]. Standard initialization was used with relative to cell zone over 1000 iterations. Figures above shown velocity vectors, pressure vectors for open and closed condition.

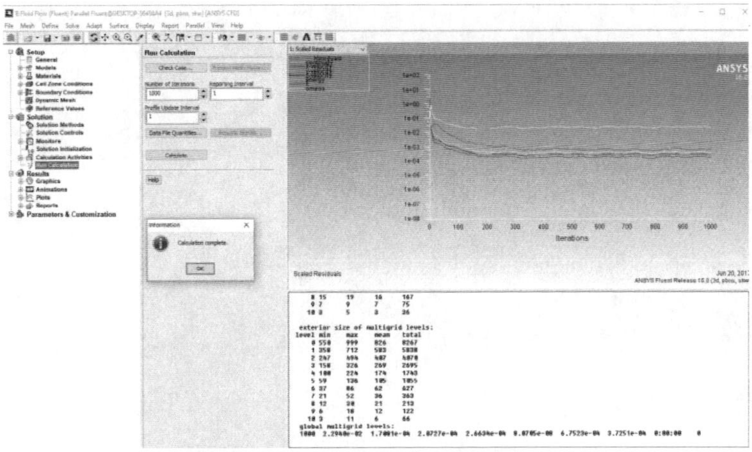

Figure 7. Solution Calculation Result

2.1 Result Contour at Intake Flow Rate 0.10 kg/h in Closed Condition

Figure 8. Velocity Contour in X-axis

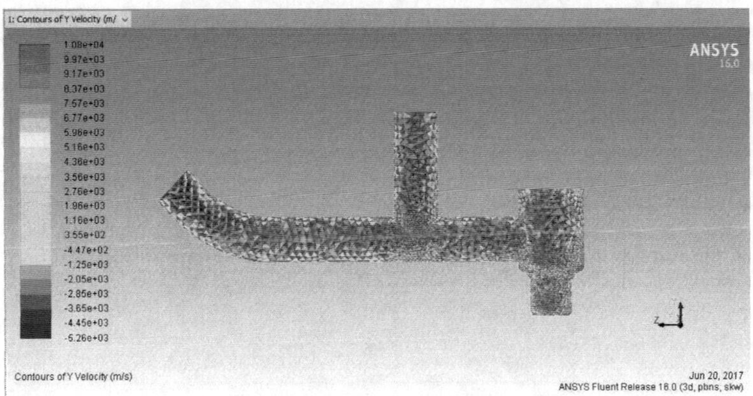

Figure 9. Velocity Contour in Y-axis

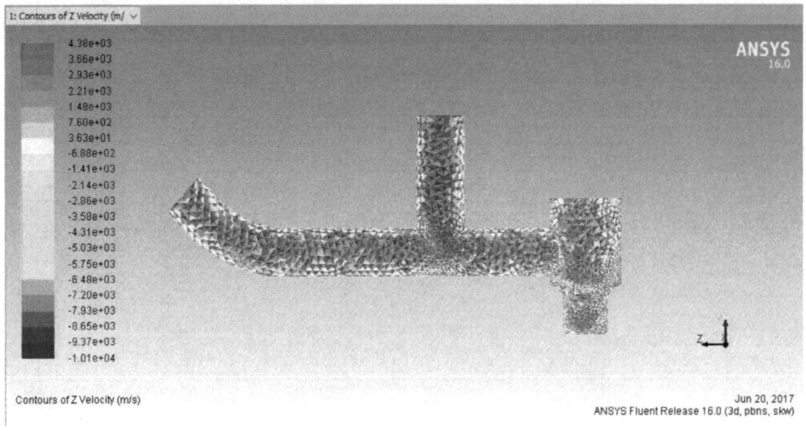

Figure 10. Velocity Contour in Z-axis

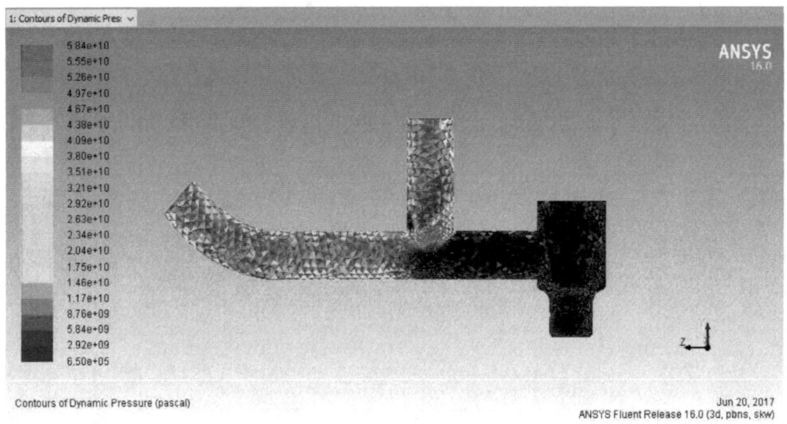

Figure 11. Dynamic Pressure Contour

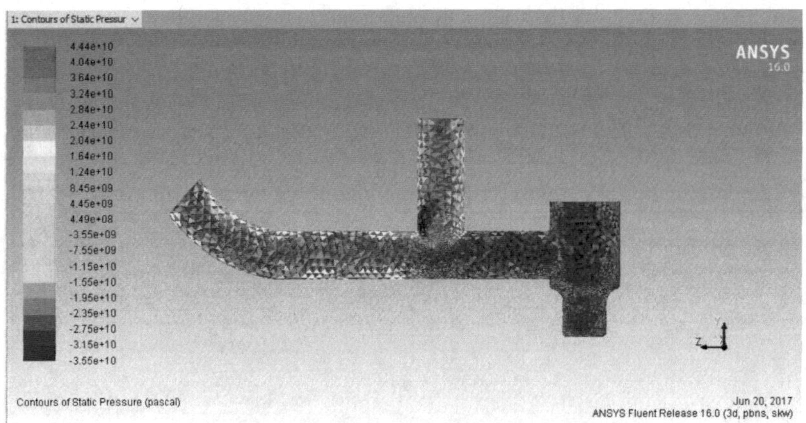

Figure 12. Static Pressure Contour

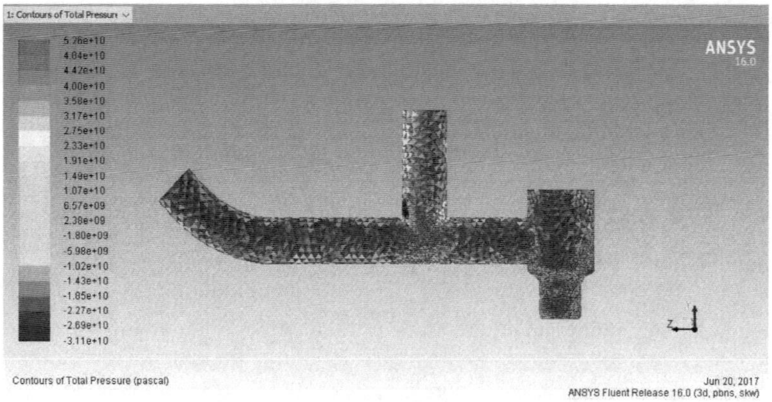

Figure 13. Total Pressure Contour

*2.1.1 Mass Flow Rate for Existing Design (Closed Condition).*The output generated for this simulation is 0.050kg/s.About 50% losses created (based on the 0.100 kg/s mass flow rate input). Hence the losses created is far better than previous ram pump design conducted by Matthias Inthachot and his co-workers [9] which is between 60-70%.

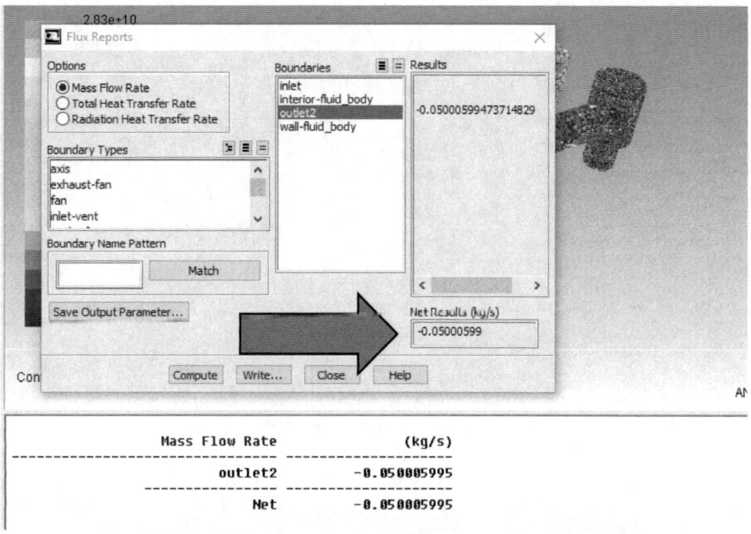

Figure 14. Mass Flow Rate Result Simulation

IOP Conf. Series: Materials Science and Engineering **257** (2017) 012076 doi:10.1088/1757-899X/257/1/012076

2.2 Result Contour at Intake Flow Rate 0.10 kg/h in Open Condition

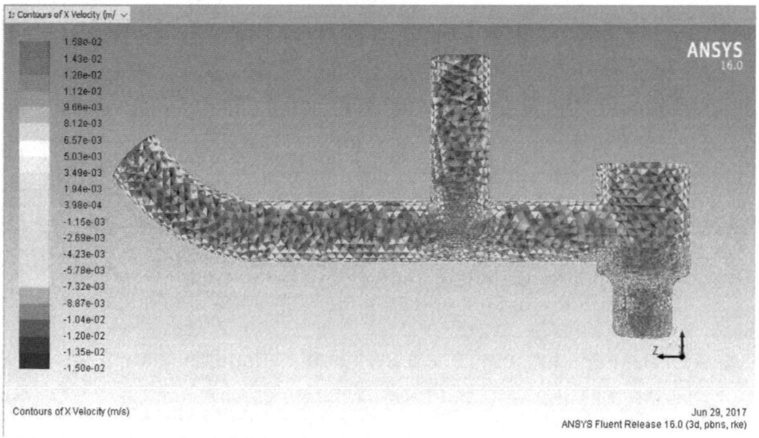

Figure 15. Velocity Contour in X-axis

Figure 16. Velocity Contour in Y-axis

Figure 17. Velocity Contour in Z-axis

IOP Conf. Series: Materials Science and Engineering **257** (2017) 012076 doi:10.1088/1757-899X/257/1/012076

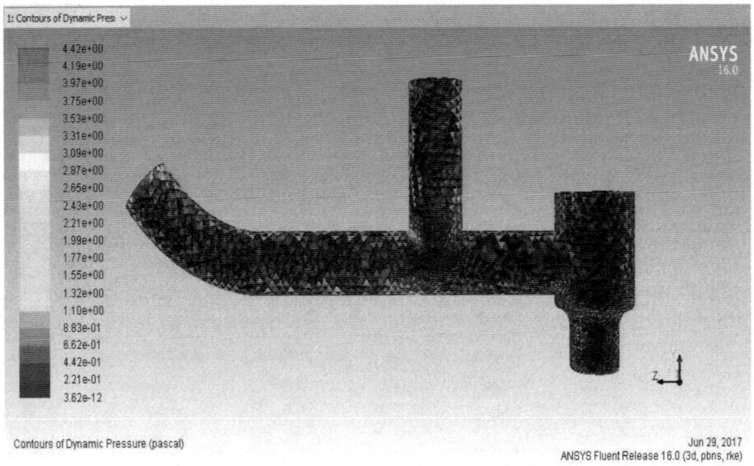

Figure 18. Dynamic Pressure Contour

Figure 19. Static Pressure Contour

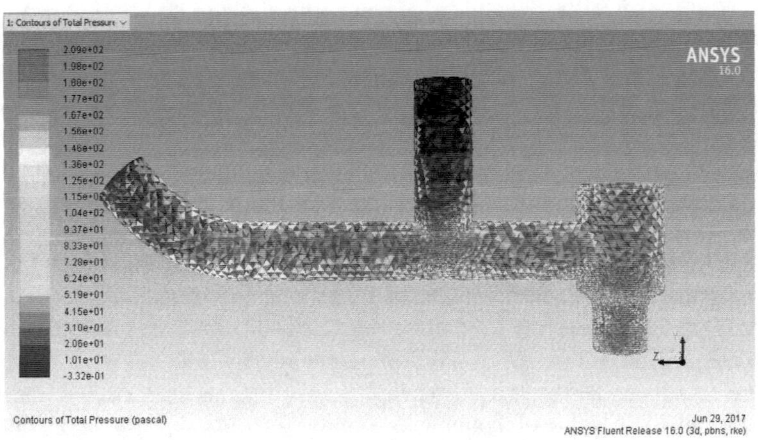

Figure 20. Total Pressure Contour

2.3 Mass Flow Rate for Existing Design (Open Condition)

The output generated for this simulation is 0.003kg/s.About 97% losses created (based on the 0.100 kg/s mass flow rate input). Theoretically accepted as during this process, lot of water wasted to the surroundings due to waste valve opening.

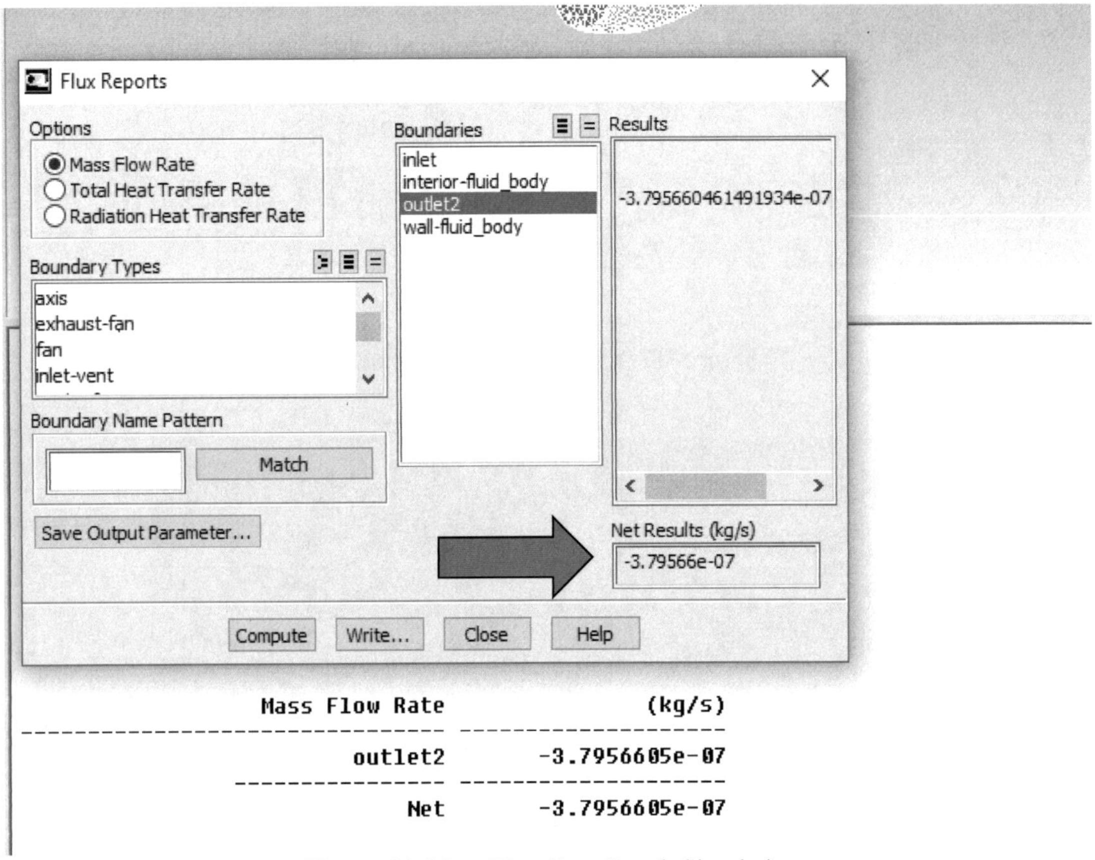

Figure 21. Mass Flow Rate Result Simulation

3. Discussion

The discussion of the results begin with an explanation of two conditions mainly open and closed condition. During open condition the waste valve remain open in 10mm (refer figure 22) diameter gap and water starts to flow from the source and escapes through the waste valve. The finding of velocity and pressure contour provides evidence of force flow distribution that accelerates under the effect of the supply head about mass flow of 0.10 km/h till enough velocity is attained in the drive pipe to pump water out of delivery valve. Our finding revealed that for both open and close condition along 3 axis (X,Y and Z axis) velocity analysis as shown in figure 8, figure 9, figure 10, figure 15, figure 16 and figure 17 will generated different results and indicate area of improvement that need to be done. These result support previous research by GK Awari et al [10] into this brain area which links open and closed condition to the significant velocity effect.

The most striking result to emerge from these analysis is the total, dynamics and pressure forces that further analyse the pressure illustrated in 3 different condition within the same pump (Refer figure 11, figure 12, figure 13, figure 18, figure 19 and figure 20). For instance, experiment in open condition recorded 4.42kPa and 15.15 value of dynamics and static pressure respectively. Contradict with closed condition that generated 128.00kPa and 97.8, a tremendous difference that justify the operation of alternating valve in open and close condition.

Next for a good hydram design, the valve closure must be rapid or instantaneous as being experimented by Nambiar and his friends [11]. Thus for this analysis we set 0.20 seconds for the waste valve to close (refer figure 22). 4 sets of data as shown in figure 21 were simulated to demonstrate the transient analysis for the waste and delivery valve. This ways is somewhat surprising given that other research do not consider this parameter in their analysis.

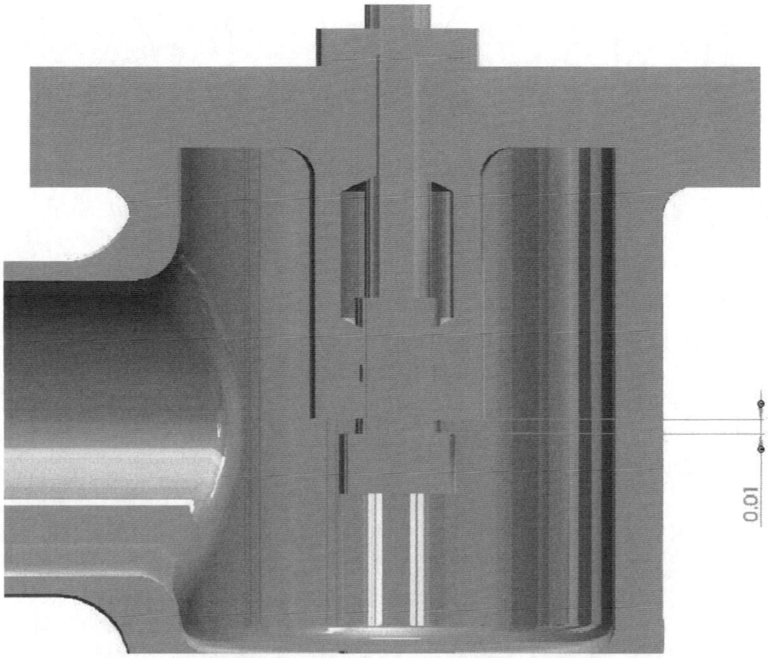

Figure 22. Outlet Gap for Waste Valve

The sudden closure creates a high pressure in the hydram and on the delivery valve that is in excess of the static delivery pressure. What is surprising is that 15.44kPa (open condition) and 115.89kPa (close condition) total pressure force were recorded, far better than Gan Shu San el at findings [12]. The waste valve is forced open again and pumping takes place till the velocity becomes zero and pumping stops as the force value drop drastically. After that, delivery valve then closes. Resulted the pressure near the waste valve is much higher than the static supply pressure and the flow is reversed towards the supply source. This action is also termed as recoil. The recoil action creates a vacuum condition in the hydram caused temporarily forcing a small amount of air to be sucked into the hydram through the air valve. The pressure on the underside of the waste valve is also reduced and together with the effect of its own weight, the waste valve opens automatically. The water in the drive pipe returns to the static supply pressure as before and the next repeating cycle begins.

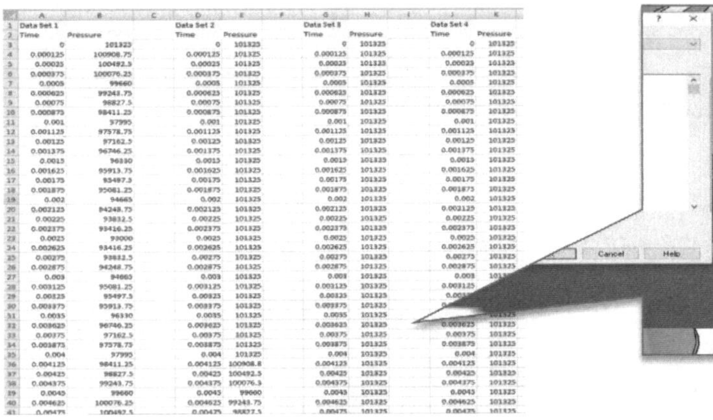

Figure 23. Data for Transient Flow Analysis

Overall based on analysis, this improve design especially on the waste and delivery valve manage to reduce losses generated in order to create hammering effect by establish control mechanism in closing the waste valve. The single most striking observation to emerge from the data comparison as tabulated in Table 2 is the reducing of overall waste percentage for both open and close condition compare to previous Sakhare et at research [13].

Table 2. Summary of Analysis Data for Both Condition

Condition	Parameter Vectors	Min	Max	Mass Flow Rate	Waste Percentage %
Open	Velocity X-axis(m/s)	0.20	0.21		
	Velocity Y-axis(m/s)	0.82	1.26		
	Velocity Z-axis(m/s)	0.47	1.27		
	Dynamics Pressure(kPa)	2.22x10^-5	4.42	0.003kg/s	97%
	Static Pressure	0	15.15		
	Total Pressure(kPa)	1.22	15.44		
Close	Velocity X-axis	49.21	57.04		
	Velocity Y-axis	46.68	58.97		
	Velocity Z-axis	55.14	89.97		
	Dynamics Pressure(kPa)	0.96	128.00	0.050kg/s	50%
	Static Pressure	78.19	97.80		
	Total Pressure(kPa)	68.50	115.86		

4. Conclusions

This present study was designed to determine the effect of improved design to significant effect on flow analysis and simulation study. This study set out to evaluate how effective the new design towards improving the overall pump performance. These experiment confirmed that the by adding control mechanism to the newly design component delivery and waste valve have enhance about 20% more efficiency than current design. The main finding can be summarized as follow:

I. A hydraulic ram pump create a lot of water waste at the waste and delivery valve. 90% from the incoming water will be left as waste, just 10% will be pump out to the greater heights. Additional of threaded control valve for both valve will help to control the water losses created.

II. The projection of velocity and pressure contour will help to understand of water flow profile within designated pump. These will help to analyze and discover area of improvement for better pump performance.

More information on the new design hydraulic ram in actual model would help us to establish a greater degree of accuracy on this matter. If the debate is to be moved forward, a better understanding of comparison between simulated and actual fabricated modelling need to be developed.

Acknowledgments
The authors would like to thank the Faculty of Mechanical Engineering, University Malaysia Pahang for providing laboratory facilities and advice support.

References

[1] Nambiar, P., Shetty, A., Thatte, A., Lonkar, S., & Jokhi, V. (2015). Hydraulic Ram Pump: Maximizing Efficiency. International Conference On Technologies For Sustainable Development (ICTSD) (pp.1–4).
[2] Li, J., Yang, K., Guo, X., & Guo, Y. (2015). Adjustable And High-Head Model Experiment On The Hydraulic Ram Pump. Shuili Xuebao/Journal Of Hydraulic Engineering, 46(1), 102–108.
[3] Harish. Kumar, Temesgen,Tedesse Beyene,Guttu Ofgaa,M.D Kaso,Shemellis,P. N. Rao,Teressa Negassa. (2016).Pollution Free Design And Manufacturing Of Hydraulic Ram Pump For Villages Hill Areas. International Journal of Mechanical Engineering Research and Technology, 2(2).
[4] Schiller, E. J., & Kahangire, P. (1984). Analysis And Computerized Model Of The Automatic Hydraulic Ram Pump. Can. J. Civ. Eng., 11, 743–750
[5] Poonam Diwan,Aman Patel,Lavish Sahu (2016). Design And Fabrication Of Hydraulic Ram With Methods Of Improving Efficiency. International Journal Of Current Engineering And Scientific Research (IJCESR), 3(4)
[6] Young, B. W. (1998). Generic Design Of Ram Pumps. Proceedings Of The Institution Of Mechanical Engineers Journal Of Power And Energy, 212(2), 117–124.
[7] Piyush B. Shende,D.K Choundary,A.P.Ninawe. (2015). Analysis And Enhancement of Hydraulic Ram Pump Using Computational Fluid Dynamics (CFD). International Journal For Innovative Research In Science And Technology.2(3)
[8] Reza Fatahi-Alkouhi,Babak Lashkar-Ara,Alireza Keramat, (2015). Determine The Efficiency Of Hydraulic Ram Pumps. IAHR World Congress. 202(4),249-256
[9] Inthachot, M., Saehaeng, S., Max, J. F. J., Müller, J., & Spreer, W. (2015). Hydraulic Ram Pumps for Irrigation in Northern Thailand. *Agriculture and Agricultural Science Procedia*, *5*, 107–114.
[10] Awari, G. K., Ardhapurkar, P. M., Wakde, D. G., & Bhuyar, L. B. (2004). Performance analysis of air-lift pump design. Proceedings of the Institution of Mechanical Engineers, Part C: Journal of Mechanical Engineering Science, 218(10), 1155–1161.
[11] Nambiar, P., Shetty, A., Thatte, A., Lonkar, S., & Jokhi, V. (2015). Hydraulic ram pump: Maximizing efficiency. In Proceedings - International Conference on Technologies for Sustainable Development, ICTSD 2015.
[12] Gan Shu San, & Gunawan Santoso. (2002). Studi Karakteristik Volume Tabung Udara dan Beban Katup Limbah Terhadap Efisiensi Pompa Hydraulic Ram. Jurnal Teknik Mesin, 4(2), 81–87.
[13] Sakhare mayor Mahadev . (2016) International Journal Of Trend In Research And Development (IJTRD),3(3).

A study of the stabilities, microstructures and fuel characteristics of tri-fuel (diesel-biodiesel-ethanol) using various fuel preparation methods

K H Lee[1,2], M Mukhtar N A[1,2], Ftwi Yohaness Hagos[1,2] and M M Noor[1,2]

[1]Faculty of Mechanical Engineering, Universiti Malaysia Pahang, 26600 Pekan, Pahang, Malaysia
[2]Automotive Engineering Centre, Universiti Malaysia Pahang, 26600 Pekan, Pahang, Malaysia

[*]Corresponding author: ftwi@ump.edu.my

Abstract. In this study, the work was carried out to investigate the effects of ethanol proportions on the stabilities and physicochemical characteristics of tri-fuel (Diesel-Biodiesel-Ethanol). For the first time, tri-fuel emulsions and blended were compared side by side. The experiment was done with composition having 5%, 10%, 15%, 20% and 25 % of ethanol with fixed 10% of biodiesel from palm oil origin on a volume basis into diesel. The results indicated that the phase stabilities of the emulsified fuels were higher compared to the blended fuels. In addition, tri-fuel composition with higher proportion of ethanol were found unstable with high tendency to form layer separation. It was found that tri-fuel emulsion with 5% ethanol content (D85B10E5) was of the best in stability with little separation. Furthermore, tri-fuel with lowest ethanol proportion indicated convincing physicochemical characteristics compared to others. Physicochemical characteristics of tri-fuel blending yield almost similar results to tri-fuel emulsion but degrading as more proportion ethanol content added. Emulsion category had cloudy look but on temporarily basis. Under the microscope, tri-fuel emulsion and blending droplet were similar for its active moving about micro-bubble but distinct in term of detection of collision, average disperse micro-bubble size, the spread and organization of the microstructure.

1. Introduction

Owing to their excellent drivability and fuel economy, diesel engine has been utilized in most of the sector including transportation and agricultural industry [1]. Meanwhile, the level of air pollution and greenhouse gases emission are now a wide concern. Indirectly, diesel engine contributed via incomplete combustion that release polluted hazardous gases, including carbon monoxide (CO), carbon dioxide (CO2), nitrogen dioxide (NO2), unburned hydrocarbon (HC) and particulate matter (PM). In the same way, this is a serious risk to the environment and could affect human health eventually. Thus, the research and development of alternative fuels for transportation are preferable niche area [2]. In addition, studies on the applicability of clean energy technologies continue to receive attention [3]. Mixing biodiesel, alcohol and diesel are one of the latest attempt as part of renewable alternative energy sources solution with environmentally friendly characteristics in contrast to fossil fuels [2, 4]. Tri-fuel new lineup includes the possibility of secondary atomization attribute capability [5]. The work has started but is at preliminary stage. Good news is that biodiesel and ethanol are under oxygenated fuel categories and many believe that oxygenated fuel into diesel could improve the combustion efficiency [6-8]. Also, the interest in tri-fuel category is mainly driven by the potential reduction of exhausted CI engine gas emission namely carbon monoxides, hydrocarbon and particulate matter [9]. Besides, air-fuel mixture formation could be improved by the influence mechanism of ethanol [10]. And the best part is, no minor or major engine modification needed [11].

Unfortunately, the solubility of diesel with ethanol would be the problem considering ethanol immiscibility nature and water content [1, 12]. It was reported that ethanol is not completely miscible in the composition and only with 5% of ethanol demonstrate complete miscibility in diesel [11]. Fortunately, biodiesel could be the solution to the phase separation problem of diesel-ethanol [13]. Beside economically viable [14], biodiesel could increase the solubility of ethanol in diesel [15]. Compatibly and suggested in the literature, phase separation of diesel-ethanol could be prevented by emulsifier [16]. It was not until recently, Hagos et al. [15] initiated the concern towards the distinction between tri-fuel emulsions and blending in the review article last year. In other word, to date, little research has been done to compare tri-fuel under blending and emulsions category side by side. For commercialization, it is imperative to ensure the fuel stability to be consistent. The fuel sit in a tank for extended period is likely give chance for the gravity to do the work. Destabilization of the fuel during operation or storage may cause engine failure [17]. Few minutes operation of variable fuels due to the phase separation in the engine could do the damage and may trigger costly engine repair. In addition, stability study is necessary to determine if additional mixing or additive is required besides the main component in the composition [18]. Bottom line, understanding the physicochemical properties of tri-fuel is essential as basis to rationalize the formulation potential. The propensity to downgrade the properties with time due to phase separation is not something to be taken for granted. Potential degradation may appear as the alteration of cetane number, viscosity and density. As an example, consequently, density degradation alone with the effect on the injection timing could promote engine deficiency and unwanted emission [18].

The main objective of this work was to evaluate diesel-biodiesel-ethanol as tri-fuel and compared blended against emulsions category with biodiesel constant volume percentage from the stability and physicochemical characteristics viewpoint. Concurrently, microstructures of both tri-fuel categories were immediately observed upon prepared and compared.

2. Design of experiment

Diesel, biodiesel and ethanol were the main ingredient to form tri-fuel composition. The study was carried out in three stages. Firstly, to acknowledge the effect of ethanol on the stability and physicochemical characteristics of tri-fuel blended. Next, the effect of ethanol on the stability and physicochemical characteristics on tri-fuel emulsions type. Finally, under the microscope, the comparison between the two-mixing strategy; droplet of blended versus emulsions were outlined and discussed.

3. Preparation of tri-fuel blends and emulsions

Diesel, ethanol and biodiesel from palm oil origin were obtained from nearby convenient supplier. Tri-fuel emulsions were compounded using Hielscher Ultrasonic Processor UP400S Emulsifier with 0.7% cycle and 40% amplitude maintaining the temperature between 30°C to 40°C for 5 minutes. Approximately 100 ml on a beaker per samples were prepared. Temperature was monitored throughout the exercise with infrared thermometer. For blending, magnetic stirring has been used for 30 minutes. Again, the temperature was monitored consistently the entire time. The use of magnetic stirrer meant no physical contact of the sample with the blade. Stirring hotplate brand FAVORITE with variable control on temperature and magnetic stirrer speed was the selected apparatus to do the job. The increase temperature can be controlled from 90°C to 525°C. The embedded heating mechanism were switched off in this case. The intensity level of magnetic stirring can be adjusted from 1 to 9. For the experiment, the level 5 speed was utilized. According to the model specification, maximum speed of the stirrer could reach up to 1000 rpm. For the sake of argument, it is fair to approximate level 5 speed stirring speed to represent range of output between 500 to 600 rpm. For blended tri-fuels, regardless of percentage ethanol content, it was found that the temperature of the composition remains under room temperature the entire time. Ethanol content were varied from 5%, 10% ,15%, 20% and 25% with biodiesel remain 10%. The experiments began with the preparation of the composition. The volume of each fuel was measured and mixed according to the decided proportion. 10 samples of tri-fuel were tested for the stability and fuel characteristics. The proportions were varied as shown in Table 1.

Table 1. Proportion of fuel for each of the sample

Sample	Diesel (%)	Biodiesel (%)	Ethanol (%)	Purpose
1	85	10	5	Blending
2	80	10	10	Blending
3	75	10	15	Blending
4	70	10	20	Blending
5	65	10	25	Blending
6	85	10	5	Emulsification
7	80	10	10	Emulsification
8	75	10	15	Emulsification
9	70	10	20	Emulsification
10	65	10	25	Emulsification

The stability behavior tests were conducted at room temperature for approximately one month. All the data on the stability behavior was done by visual inspection of the blends physical appearance every day for 35 days. Subsequently, there were four types of testing conducted mainly density, calorific value, viscosity and flash point for each of the tri-fuel emulsions following diesel standard test. Besides that, the microstructure of the tri-fuels was also observed to differentiate between blended and emulsion categories.

4. Experimental procedure

4.1. Stability behavior

The stability test method in this experiment is widely known as gravitational stability [17, 19]. The stability of the fuels was observed for 1 month under room temperature. The blended fuels and emulsified fuels are being filled into 10 test tubes respectively as shown in Figure 1. The sample can be differentiated by comparing the layer of separation during the stability process. Photos were taken every 2 hours on the first day. For the following 7 days, the photos were set to be taken once a day. Subsequently, the photos were taken continuously each week for 4 weeks.

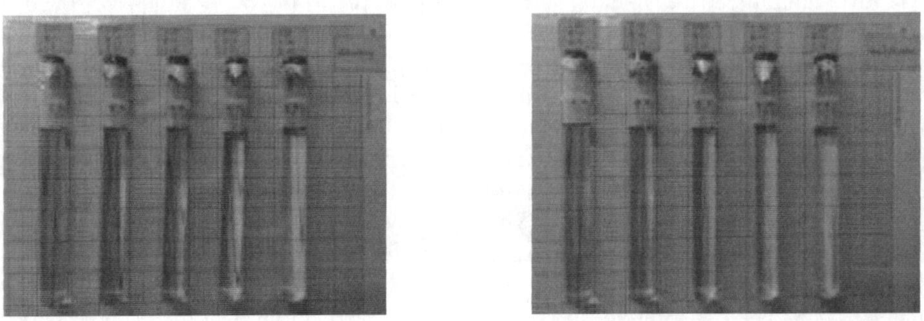

(a) (b)

Figure 1. The samples which are undergo blending process (a) and emulsification process (b).

4.2. Microstructure

Digital microscope brand Dolomite Calestron was used to observe the microstructure of the samples. Microstructure of the tri-fuel blends and emulsion fuels were observed to differentiate between blended fuels and emulsified fuels. Immediate inspections (within the same day of the fuel preparation) were done to spot the significant different between blending and emulsions. Each of the tri-fuels were taken 3 times for convincing result. 40x objective lens was used and the total magnification of the images was up to 400x. On top of that, all the images were additionally analyzed using image processing software called Image J (automatic particle sizing).

4.3. Physicochemical properties – density, calorific value, viscosity and flash point

The density reading was not obtained directly from hydrometer instrumentation but was calculated from separated volume and mass measurement. The density was obtained applying weight measurement technique. Weight measurements machine was required to measure the weights of the fuels and calibrated beforehand. The volumes of the fuels were measured by using a 10 ml syringe. To be specific, 10 ml of fuel is taken by using a syringe. Then, every 2 ml of fuel is weighted and the weight is recorded. The experiment was conducted at room temperature and repeated for three times. An average reading was taken before the densities of tri-fuel blended and tri-fuel emulsified were calculated with the equations below:

$$\rho = \frac{m}{V} \tag{1}$$

where m = mass of the sample
V = volume of the sample
ρ = density

Research indicate that comparable calorific value of blending fuel to diesel meant almost similar thermodynamic parameters for engine can be expected [5]. The calorific value of the tri-fuels was measured with Bomb Calorimeter and Model 6772 Calorimetric Thermometer in accordance to diesel international standard procedure ASTM-D240. The dynamic viscosity was measured using Brookfield Viscometer Model DV-III ULTRA Programmable Rheometer. The reading should be accepted within the equivalent percentage of torque, ranging between 10% to 100% for any combination of spindle rotation. Kinematic viscosity was calculated by the equations below:

$$\upsilon = \frac{\mu}{\rho} \tag{2}$$

where v = kinematic viscosity
ρ = density
μ = dynamic viscosity

The measurement was in accordance to diesel international standard procedure ASTM-D7042. Tri-fuel flashpoint was evaluated using Koehler model K16591, rapid flash point tester. Flash point is widely known as the lowest temperature of the fuel when it starts to burn or ignite. Slightly differ than autoignition reading as flash point is the lowest temperature to ignite without the influence of ignition source. In this experiment, there were five samples which contain different ethanol proportion tested. The measurement was in accordance to diesel international standard procedure ASTM D93.

IOP Conf. Series: Materials Science and Engineering **257** (2017) 012077 doi:10.1088/1757-899X/257/1/012077

5. Results and discussion

5.1. Stability behaviour

The stabilities of all the samples were observed for more than a month at room temperature. Center of attention was on any detected separation layer in centimeter with observation conducted every 2 hours on the first day. Subsequently, once a day till day seven. Onward, observation was taken every once a week till the fifth week. In term of appearance, under tri-fuel emulsions categories, it was prominently translucent but that was only last for a short while. The cloudy looking was noticeable only applied to emulsions category and not on blending type. Best bet the cause of this could be because the intensive disruption effect of ultrasonic emulsifier to the composition. As for blending mixtures, it appeared transparent all along from the beginning. Overall, all categories presented itself with clear look eventually. Table 2 shows the separated layer of tri-fuel blend and emulsion for 35 days via visual observation. The blended fuels were observed more profound to encounter phase separation compared to emulsified fuels. From Figure 2, D65B10E25 with the highest ethanol content, separated layer of blended fuel was observed higher than the emulsified category. Whereas in Figure 3, the emulsified fuel which contained the lowest ethanol content (D85B10E5) was found to be the best in stability with no separated layer detected.

Table 2. Separated layer of tri-fuel blend and emulsion within one month.

	Separated layer (cm)									
	Blended fuels					Emulsified fuels				
Duration	D85 B10 E5	D80 B10 E10	D75 B10 E15	D70 B10 E20	D65 B10 E25	D85 B10 E5	D80 B10 E10	D75 B10 E15	D70 B10 E20	D65 B10 E25
2 hours	0	0	0	0	0	0	0	0	0	0
4 hours	0	0	0	0	0	0	0	0	0	0
6 hours	0	0	0	0	0.3	0	0	0	0	0.2
8 hours	0	0	0	0	0.3	0	0	0	0	0.2
10 hours	0	0	0	0	0.4	0	0	0	0	0.2
12 hours	0	0	0	0	0.4	0	0	0	0	0.2
2 days	0	0	0	0	0.5	0	0	0	0.4	0.4
3 days	0	0	0	0	0.5	0	0	0	0.4	0.4
4 days	0	0	0	0	0.6	0	0	0	0.4	0.5
5 days	0	0	0	0	0.7	0	0	0	0.4	0.5
6 days	0	0	0	0	0.7	0	0	0.2	0.5	0.6
7 days	0	0	0	0.2	0.8	0	0	0.2	0.6	0.7
14 days	0	0	0.2	0.4	0.8	0	0.2	0.3	0.6	0.8
21 days	0	0.2	0.4	0.5	1	0	0.2	0.3	0.7	0.8
28days	0.1	0.3	0.4	0.6	1.1	0	0.2	0.3	0.7	0.8
35 days	0.2	0.4	0.5	0.7	1.2	0	0.2	0.3	0.7	0.8

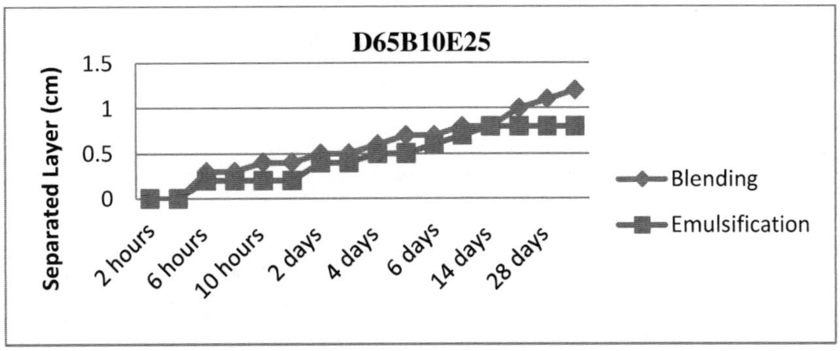

Figure 2. Comparison of tri-fuel (D65B10E25) blend and emulsion.

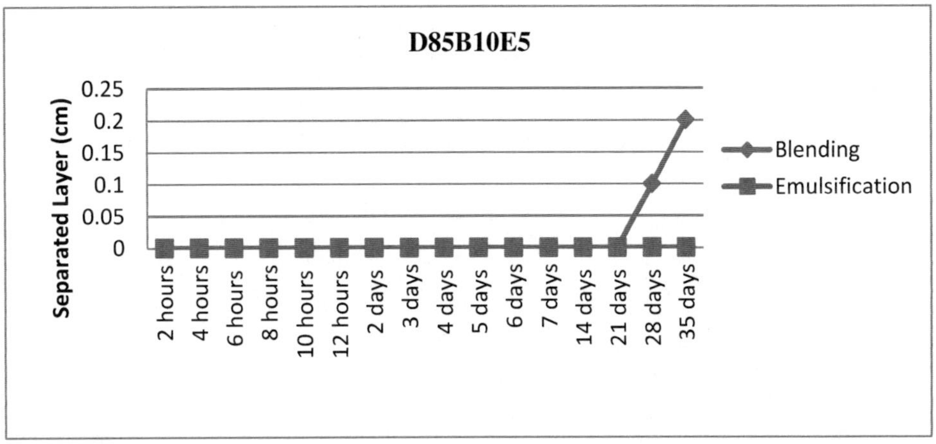

Figure 3. Comparison of tri-fuel (D85B10E5) blend and emulsion.

The effect of ethanol proportion on phase stability for tri-fuel blends and emulsions were studied at room temperature. Figure 4 and 5 depicted the effect of ethanol proportion on the stability of tri-fuel blends and emulsions. More ethanol content in the composition depicted faster phase separation. Nevertheless, notice that, ethanol content with the exact same proportion or more than biodiesel had cause phase separation. Tri-fuel emulsions with 2:1 ratio between biodiesel and ethanol was found effective with none phase separation detected for more than a month. As for the blending, phase separation on day 21, likely the interpretation was that during mixing process, the addition of biodiesel did not really perform its trade in the mixture. It was understood in the theory that the miscibility was possible reflecting biodiesel acted as an amphiphile (surface active agent) which contained nonpolar tails and polar heads [20]. The basis was that, the polar head in a biodiesel molecule oriented itself to the ethanol and the nonpolar tail oriented to the diesel [3]. Thus, figure 4 and 5 are tide back to the previous study reigniting the debate on whether can biodiesel really act as surfactant or just a source for string of setback. However, the skepticism on the basic entitlement that biodiesel did not act as surfactant was not however relevant in this case. It was important to acknowledge under D85B10E5 both blended and emulsions category, biodiesel was seen did contributed to the delay of phase separation and to one extend preventing the phase but not at all cost. The trend revealed that constant volume of biodiesel into each of the composition as per the experiment design could not keep up with the increment of higher ethanol content. Thus, argument on constant volume of biodiesel with weak miscible propensity tolerance in the tri-fuel mixture cannot stands in one row with the previous dispute by Kwanchareon et al. [1] on biodiesel inability to accommodate solubility.

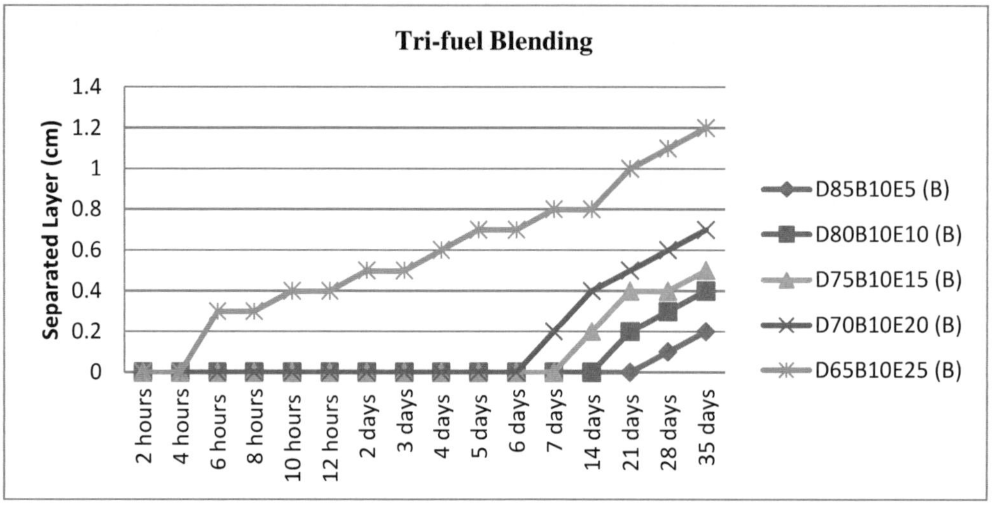

Figure 4. The effect of ethanol proportion on the stability in tri-fuel blend.

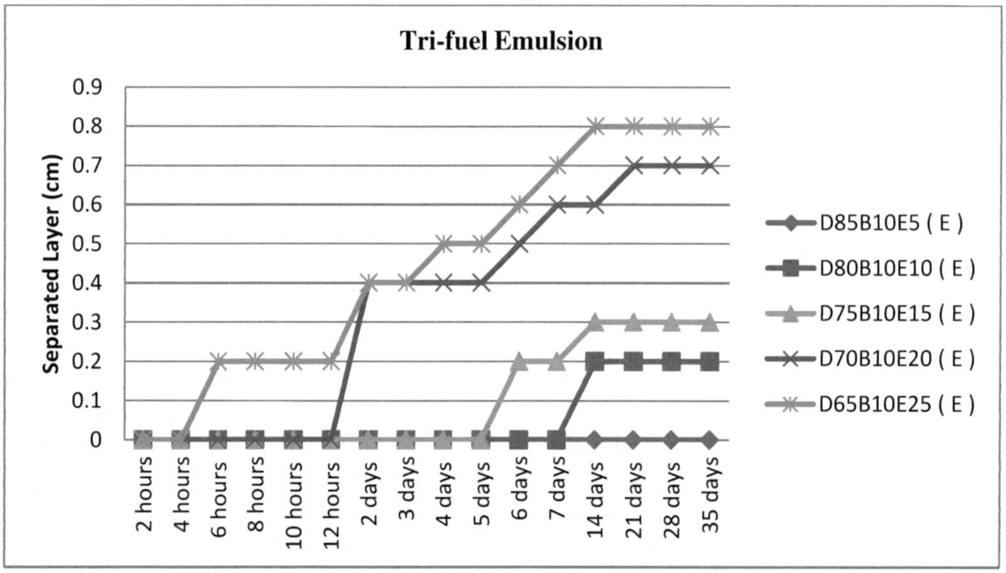

Figure 5. The effect of ethanol proportion on the stability in tri-fuel emulsion.

Furthermore, something in common between the two figures and quite noticeable was that the pattern indicated the delay of the separation phase per different mixing strategy. What it suggests was that, underestimating mixing procedure and the assumption that improper mixing strategy carry no weight has been significantly proven wrong here. Unexpectedly, half an hour blending approach beat 5 minutes emulsification technique. 5 days variant on D70B10E20 between blending and emulsification approach remark the substantial distinction upon the cost for miscibility. The reversing news was unprecedented but recall back on the mentioned momentarily translucent look upon 5 minutes emulsification process, the bafflement was somehow cured and self-explained. The temporarily cloudy phenomenon on the composition noticed under emulsification category deserved to be given a second look seriously in the future to get better understanding. One idea is to increase emulsification time or perhaps be more creative in seeking for better results. After all, the quick 5 minutes emulsification time was proven inadequate to substantiate desire outcome. Hence, another suggestion would be to try on a mix mood approach. The rational to the urge on being creative to get the solution is based on the revealed stability period of the blended category yield better than emulsions plus the desire microstructure characteristics that will be explained in the next section. Anyhow, as expected, both mixing categories with the highest ethanol contain were the most unstable composition and easily prone to separate. The highest separated layer could be seen on the sample with the highest percentage of ethanol (D65B10E25). Meanwhile, the lowest separated layer could be seen on the sample with the lowest percentage of ethanol (D85B10E5). When the ethanol contents increased in the tri-fuel blends and emulsions, the instability were seen more noticeable and formed separated layer in a brief period. Although the scenario was a sign of failure, lesson to be learned here is the basic to be grabbed.

5.2. Microstructure

The microscopic images of tri-fuels were used to be further analyzed by using software called Image J as shown in Figure 6. The particle size of the blended fuels and emulsified fuels are depicted in Table 3.

Table 3. The particle size of the blended fuels and emulsified fuels.

Slice	Count	Total area (μm^2)	Ave size (μm)	Area (%)	Mean grey value
D75B10E15 (B)	1183	1045041	883.382	21.261	255
D75B10E15 (E)	5751	802083	139.468	16.318	255
D80B10E10 (B)	1608	968467	602.28	19.704	255
D80B10E10 (E)	6455	716168	110.948	14.57	255
D85B10E5 (B)	2114	941134	445.191	19.147	255
D85B10E5 (E)	8863	588155	66.361	11.966	255

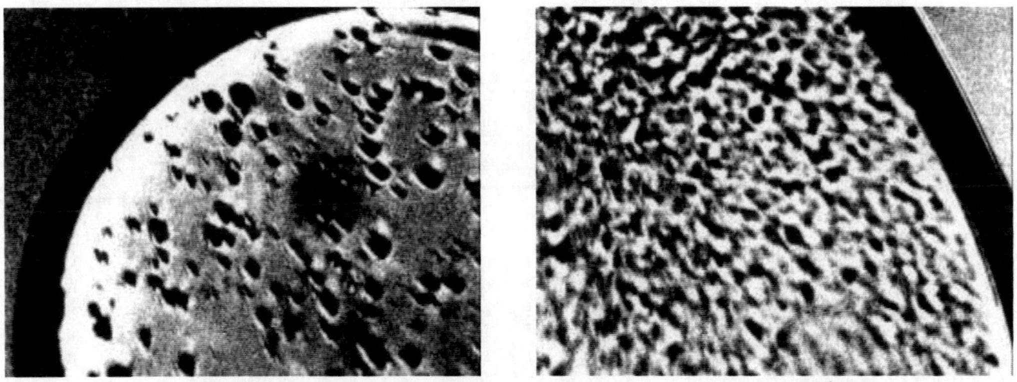

(a) (b)

Figure 6. Microscopic image of (D75B10E15) tri-fuel blend (a) and emulsion (b).

Under the microscope, the significant difference between the two mixing categories were apparent. Emulsions spreading micro bubble dispersant were found incontestable compared to blending category. 30 minutes stirring period, as suspected, tri-fuel blending category appeared with non-uniform size micro-bubble. Furthermore, there were many active micro-bubbles detected in the tri-fuel emulsion droplet. The blended fuels show higher average particle size than the emulsified fuels. In addition, it was observed that the micro-bubbles were constantly moving and active. Somehow the movement was more syncronized compared to blending category. The question rises quickly whether this was due to the mixing strategy or due to other ciscumstances such as unconcious normal handling quivering, surrounding wind interaction or liquid density energy. Again, it not a bad idea to explore the reason behind this behavior in the future.

It was convincingly demonstrated that effective meant of dispersing through emulsification approach were needed to obtain evenly distributed micro size bubble in the droplet as in Figure 7 and not on figure 8. It was also observed that tri-fuel emulsions appeared with particle size smaller than the blended fuels. Agglomeration behavior detected in blending category could be the root cause of this. From Figure 9, the agglomeration, attraction and interaction between the active micro bubble in the droplet was detected microscopically. As mentioned earlier, random movement of the micro-bubble under the tri-fuel emulsion droplet were more organized and so non-collide or sticking into one another was detected as found in the blending droplet. Whether this was due to the effectiveness of emulsification or not remain unanswered. The undesired collision was observed on blending category but not on tri-fuel emulsions as shown in Figure 10. When this distinction came, it is important to go back checking on the quality of the composition mixing procedure seriously before conducting any engine test in the future. Last, but not least, it is worth point taking that blending category demonstrate ineffective dispersing via just magnetic stirring mechanism.

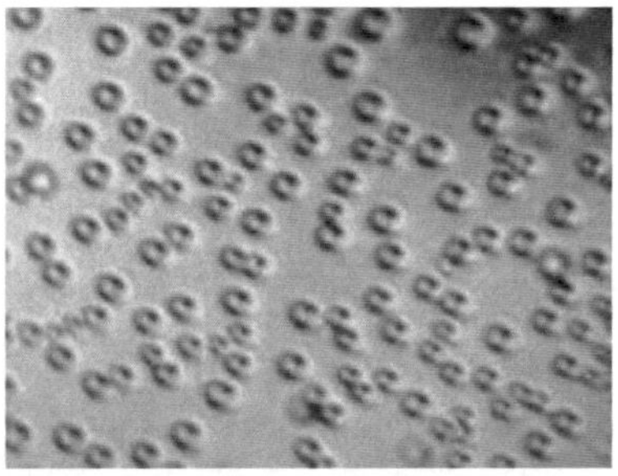

Figure 7. D65B10E25 emulsion category.

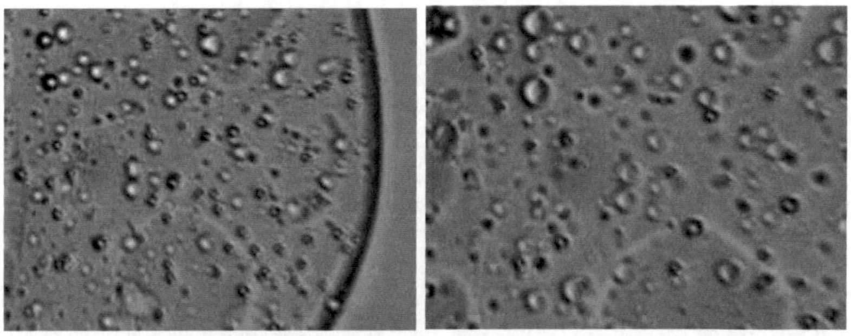

Figure 8: D80B10E10 blending category.

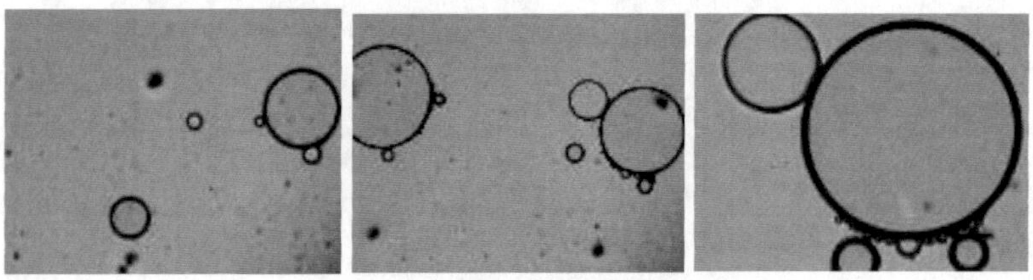

Figure 9. D75B10E15 blending category with bridging interaction between particle.

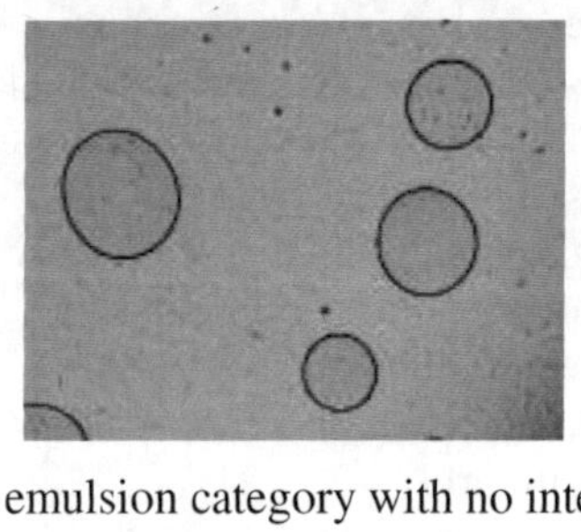

Figure 10. D75B10E15 emulsion category with no interaction between particle

As mentioned in the previous section, it is reasonable to pause and discuss what not many recognize, the merit of droplet with uniform, active and evently disburst of micro-bubble. What embedded inside the narrative appearance of the microscopic image may or may not signal the contribution of the composition to the combustion behavior. But it is wise to acknowledge the droplet microstructure as a tool to unlock, connect and compensate the limited knowledge on tri-fuel secondary atomization attribute with microstructure identity of each of tri-fuel composition setting. Adapting knowledge from the study of micro-explosion phenomenon on water in diesel, the evently distributed disperst of the liquid in micro-scale would be a desired appearance that should tip the scales in favor of the tri-fuel study relative to secondary atomization as subject matter. No research so far testified that the quality of tri-fuel mixing prepared in favor or otherwise to the micro-explosion phenomenon theory. The reason this is important to note is that through this, the appreciation on the signature of secondary atomization can be appreciated and the phenomenon can be predicted based on the identity of the droplet microstructure uniqeness. It is not enough to assume or just swallow that now. The idea is to take this argument seriously as the essence to explore the composition and proceed with further discovery. Moreover, this sentiment was not just synthesized from the outcome of the experiment but also extracted and integrated from previous work related to puffing and micro-explosion phenomenon study of tri-fuel.

5.3. Physicochemical properties

5.3.1. Density

Table 4 shows the density of tri-fuel blend and emulsion at different ratios of diesel, biodiesel and ethanol. Plotted on graft, Figure 11 depicted the variation of density reading for all categories. It can be observed that the density of the blended fuels and emulsified fuels decreased with an increasing of the percentage of ethanol content. Density reading of blended fuels were found almost like emulsified fuels in view of both had the same proportion. Perhaps, mixing strategy has no significant influence on the density of the composition. Measured separately as individual component, biodiesel showed the highest density reading followed by diesel. Ethanol has lower density and as such lowered the density of the mixture. Tri-fuel which contained 5% of ethanol content (D85B10E5) showed highest density value while tri-fuel which contained 25% of ethanol content (D65B10E25) showed the lowest density value. Biodiesel clearly compensated the drop trend caused by ethanol. Comparing the density drop trend of all proportion and category to diesel as the base guideline, 20% ethanol in the composition can be considered too much with unacceptable density reading. Nonetheless, it is worth to note that higher density reading means higher energy content could be expected.

Table 4. Density of tri-fuel blend and emulsion.

Sample fuel	Density (g/ml)	
	Blending	Emulsification
D85B10E5	0.8505	0.8347
D80B10E10	0.838	0.8273
D75B10E15	0.8298	0.8153
D70B10E20	0.7859	0.7995
D65B10E25	0.7753	0.7854

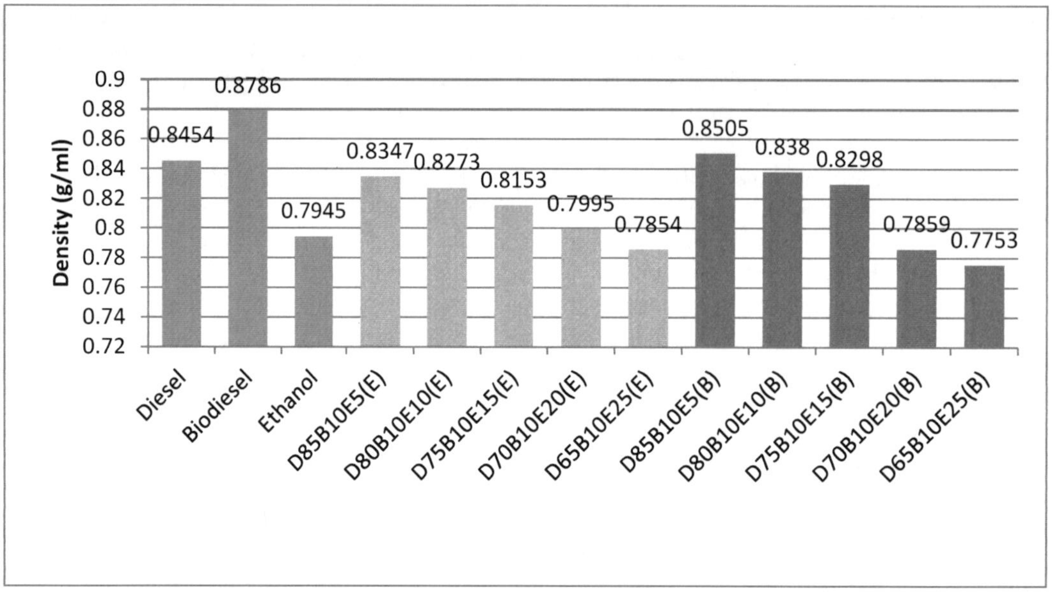

Figure 11. The variation of density of tri-fuel blends and emulsions.

5.3.2. Calorific value

Table 5 shows the calorific value of tri-fuel blends and emulsions at different ratios of diesel, ethanol but constant biodiesel proportion. Plotted on graph as shown in Figure 12, the calorific value of the blended fuels and emulsified fuels decreased with an increasing of the percentage of ethanol content. Side by side plotted on graft, all the calorific values of the blended fuels were found almost indistinguishable to the emulsified fuels considering both had the same proportion of tri-fuel. The variation however was not significant and small. In other word, emulsification or blending, mixing strategy did not tailored the calorific value as seen on the density reading. On the other hand, the calorific value of the mixture reduced with the increasing of ethanol content. This is seemingly familiar case in bi-fuel study with biodiesel into diesel. As more biodiesel into diesel resultant in reduction of calorific value [21]. The tri-fuel which contained highest percentage of ethanol (D65B10E25) had the lowest calorific value. Because fundamentally, lower calorific value means lower heat of combustion, subsequently, lower heating value of the tri-fuel may affect the power output of the engine. The highest calorific value could be seen at the tri-fuel blend which contained 5% of ethanol (D85B10E5) and it was almost the same as the calorific value of pure diesel. Thus, it makes sense to include oxygenation fuel with the purpose of obtaining better combustion via realizing more oxygen is added. Nevertheless, to gamble by adding too much, the impact will be on deterioration of calorific value which in turn could also disturb the combustion efficiency. It is therefore advisable to stay right in between the two extreme values. The argument on calorific value decrement pattern versus increasing trend of oxygenated content due to ethanol and biodiesel in tri-fuel however have not much received attention in the literature. To sum up, it is crucial to recognize the balance in between the conflicting decision that one must go through when dealing with tri-fuel because of these two key physicochemical properties.

Table 5. Calorific value of tri-fuel blend and emulsion.

Sample fuel	Calorific value (MJ/kg)	
	Blending	Emulsification
D85B10E5	48.1567	47.7769
D80B10E10	45.901	45.7008
D75B10E15	44.7689	44.4648
D70B10E20	41.9735	42.6654

D65B10E25	41.929	41.0202

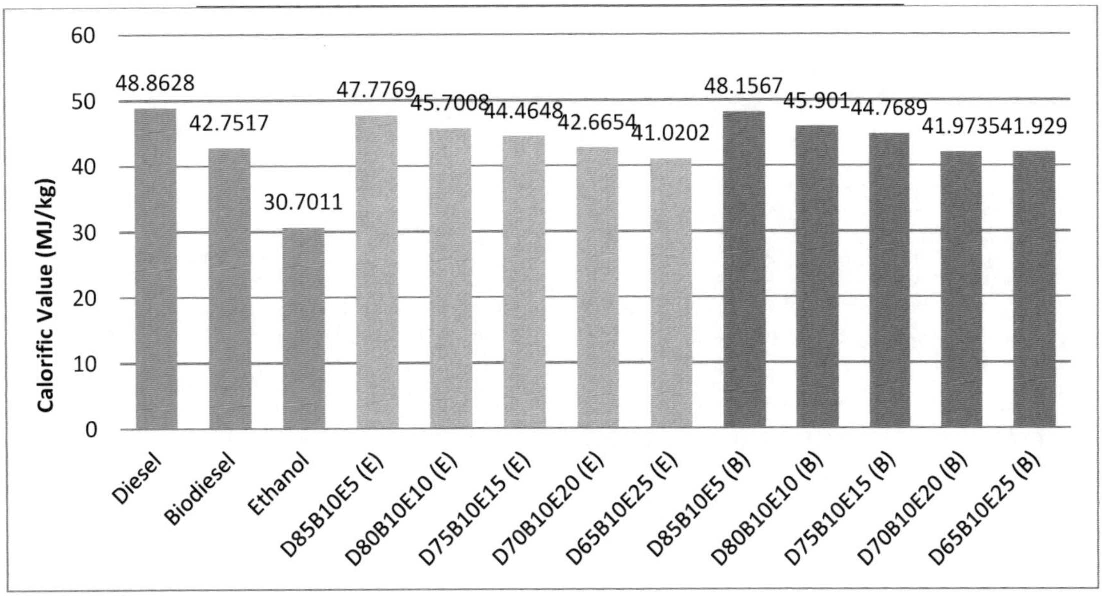

Figure 12. The variation of calorific value of tri-fuel blends and emulsions.

5.3.3. Kinematic viscosity

Figure 13 illustrates the kinematic viscosities of the blended fuels were almost on the same track as emulsified fuels perhaps because both had the identical component proportion. Again, it was safe to say that mixing strategy has no significant impact on the kinematic viscosity of the composition. However, it is widely accepted that the increase in temperature affected the molecular structure of the tri-fuel. Theoretically, as the temperature increased, the molecules started to get excited. Because of the overwhelm inner molecules gross motion, bond of the molecules got loosen. Hence, the tri-fuel decrease it viscosity. Figure 14 and 15 show the effect of ethanol proportions on kinematic viscosities in tri-fuel at various temperatures. Overall, the kinematic viscosities of the tri-fuel decreased with the increase of the ethanol contents. Nonetheless, the kinematic viscosities reading obtained can be considered compatible with diesel grade No 2-D with normal range between 1.9 to 4.1. It is reasonable to discuss the constraint subject to closer to the peak and closer to the lower level of acceptable viscosity. Too high viscosity would cause high pumping load requirement. In addition, the limitation will be on flow characteristics of the fuel in the nozzle and spray. Furthermore, micro-explosion attribute is almost impossible with high viscosity fuel. On the other hand, tendency towards too low viscosity level would be the dilution. Possible leaking into the cylindrical wall of combustion chamber and lubrication issue are the two-main problem.

Table 6. Kinematic viscosity of tri-fuel blend and emulsion at various temperatures.

Sample fuel	Kinematic viscosity					
	30°C (Cst)		40°C (Cst)		50°C (Cst)	
	Blending	Emulsification	Blending	Emulsification	Blending	Emulsification
D85B10E5	4.3483	4.4306	3.4973	3.5635	2.8189	2.8723
D80B10E10	4.0286	4.0807	3.2648	3.307	2.4953	2.5276
D75B10E15	3.6678	3.733	3.0998	3.155	2.4346	2.4779
D70B10E20	3.6773	3.6148	3.0883	3.0358	2.4673	2.4254
D65B10E25	3.5363	3.4908	2.9287	2.891	2.2844	2.255

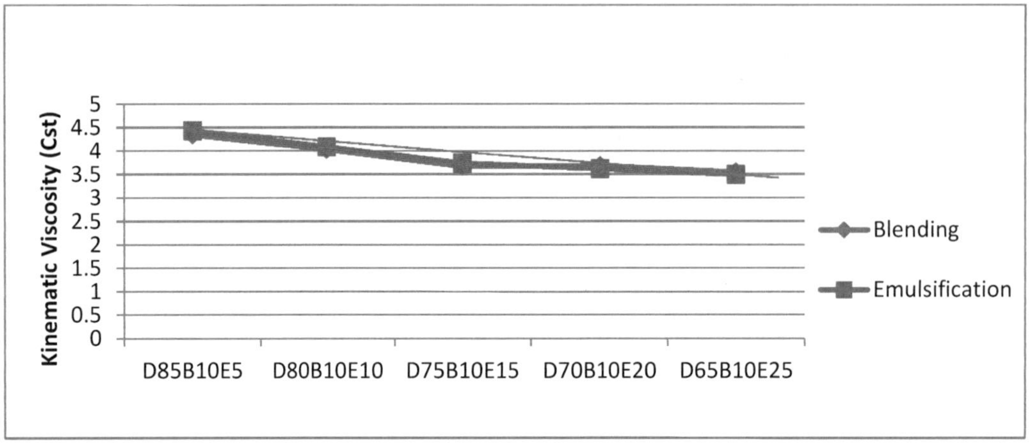

Figure 13. The variation of kinematic viscosity of tri-fuel blend and emulsion at 30°C.

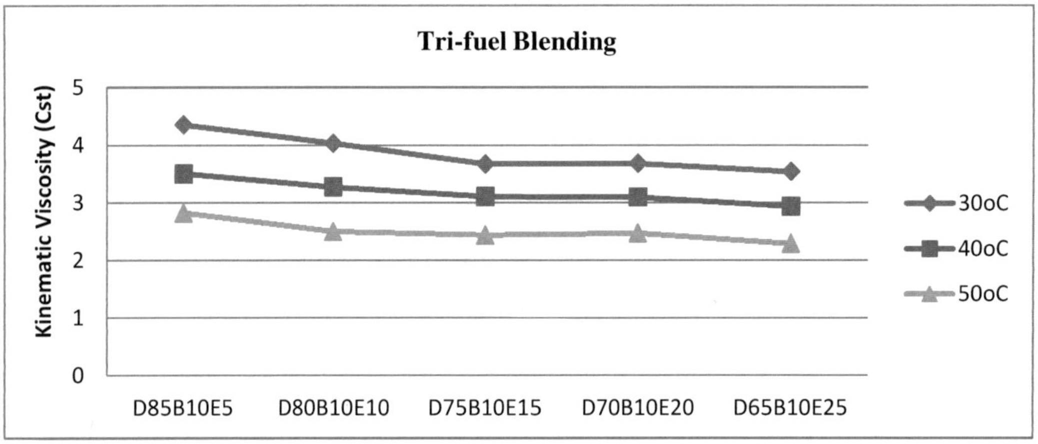

Figure 14. The effect of ethanol proportion on kinematic viscosity in tri-fuel blend at various temperatures.

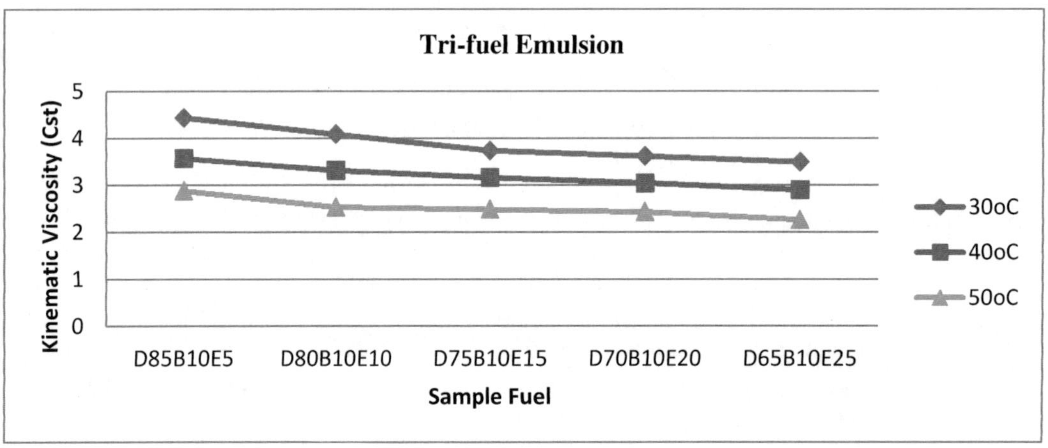

Figure 15. The effect of ethanol proportion on kinematic viscosity in tri-fuel emulsion at various temperatures.

5.3.4. Flash Point of tri-fuel Blends and emulsions

The concern on flash point limit is the handling and safety issue. Because the composition involved three different type of fuels, flash point reading presented in range is not a surprise. Dynamic behavior of the flash point is seen relatively closer to each other, approximately 5°C. Not to be confused with the temperature of autoignition, similar or higher reading of flash point from conventional diesel is desirable. The ethanol contents in the mixtures affected the flash points of the mixtures. Table 7 shows that the flash points of the tri-fuels gradual decreased with the increase of ethanol proportions. Tri-fuel blends and emulsions, both category had the same values of flash points. In other word, no impact of blending or emulsion, either way, the study confirms the flash point reading was identical. Instead, the impact is coming from the type and level of content of the composition. From the previous study, tri-fuel which contained higher flash point were relatively safe for transportation and distribution in current fuel systems [22]. Eyes on the acceptable limit of at least 60oC or above, D70B10E20 and D65B10E25 can be regarded as highly flammable composition for both category and should be avoided.

Table 7. Flash point of tri-fuel blend and emulsion.

Sample fuel	Flash point (°C)
D85B10E5	110°C – 115°C
D80B10E10	85°C – 90°C
D75B10E15	65°C – 70°C
D70B10E20	50°C – 55°C
D65B10E25	45°C – 50°C

6. Conclusions and recommendations

Comparative study of tri-fuel between blending strategy and emulsions under one roof had never been done before. The pattern on period taken for phase separation between different proportion likely been driven by way the compositions were prepared beforehand. Temporarily translucent alike detected on tri-fuel emulsions category should have been given the fair attention. Calorific value and density decline trend as more ethanol content was a complementary outcome. In the sense that degradation in-term of calorific value because of ethanol content could be complemented by the decreasing trend of density. This was so because as a result of density decrease, wider spread of fuel spray in combustion chamber and better atomization behavior could be expected. The emulsified fuel with lower ethanol content was the most stable and had better fuel characterization compared to others. Thus, the emulsion of 85% diesel, 10% biodiesel and 5% ethanol were suggested to be the optimum choice as an alternative fuel. Tri-fuel emulsions and blending had support from science, but the level of acceptance on the stability with time remain a questionable drawback. More experiment needs to be conducted with diverse ways of preparing the emulsions and blending. Perhaps, additive could be included in the future direction on tri-fuel blending or emulsions comparative study. Nonetheless, it was established that, blending or emulsification could not directly tailor the reading of density, flashpoint, viscosity or calorific value but different proportion could. What's clear was that, different mixing strategy could influence the stability against time and phase separation could potentially promote variable reading of physicochemical properties. Tri-fuel emulsification identity was on the more uniform dispersed micro-bubbled compared to blending. On the other hand, non-uniform in size, collision and interaction between micro-bubble were the signature of blending category under the microscope. Both categories with active consistent interrupted behavior. It is believed that the motion is essential in delaying the phase separation and encourage miscibility. To maintain the desire disruption influence, in the storage it is wiser to consider stabilizer mechanism by maintaining mild stirring force in the storage tank. Perhaps, In the future, to sustain evenly distributed droplet size, further study to explore the potential of stabilizer additive on top of biodiesel into tri-fuel should be proposed. The bottom line is that, the microstructure study may give an insight for future researchers to address the lack of continuity found between the study of tri-fuel with different category, micro-explosion phenomenon and engine combustion outcome. This study is only small portion of the equation and more needs to be done. Perhaps, next, surface tension study could be done with the same comparison proportion and mixing approach. Finally, another future work recommendation would be to compare tri-fuel emulsions and blending microstructure subsequent to stream under common rail injection mechanism.

Acknowledgement
The authors would like to acknowledge the financial support by Universiti Malaysia Pahang under the internal grant RDU 150124 and RDU 160152.

References

[1] Kwanchareon P, Luengnaruemitchai A and Jai-In S 2007 *Fuel* **86** 1053-61
[2] Celikten İ 2011 *Gazi University Journal of Science* **24** 341-5
[3] Fernando S and Hanna M 2004 *Energy & Fuels* **18** 1695-1703
[4] Cheenkachorn K and Fungtammasan B 2010 *Energy Sources, Part A: Recovery, Utilization, and Environmental Effects* **32** 637-44
[5] Avulapati M M, Ganippa L C, Xia J and Megaritis A 2016 *Fuel* **166** 59-66
[6] Yilmaz N 2012 *Energy* **40** 210-3
[7] Guarieiro L L N, de Souza A F, Torres E A and de Andrade J B 2009 *Atmospheric Environment* **43** 2754-61
[8] Alptekin E 2017 *Energy* **119** 44-52
[9] Mofijur M, Rasul M G, Hyde J, Azad A K, Mamat R and Bhuiya M M K 2016 *Renewable and Sustainable Energy Reviews* **53** 265-78
[10] Han K, Yang B, Zhao C, Fu G, Ma X and Song G 2016 *Experimental Thermal and Fluid Science* **70** 381-8
[11] Pidol L, Lecointe B, Starck L and Jeuland N 2012 *Fuel* **93** 329-38
[12] Pidol L, Lecointe B and Jeuland N 2009 *SAE Technical Paper* **01** 1807
[13] Park S H, Kim S H and Lee C S 2009 *Energy & Fuels* **23** 5228-35
[14] Gunatilake H, Pohit S and Sugiyarto G 2011 *Economy-Wide Impacts of Biodiesel Production and Use in India: A Computable General Equilibrium Model Assessment* Available from: http://hdl.handle.net/11540/1411
[15] Hagos F Y, Ali O M, Mamat R and Abdullah A A 2017 *Renewable and Sustainable Energy Reviews* **75** 1281-94
[16] He B Q, Shuai S J, Wang J X and He H 2003 *Atmospheric Environment* **37** 4965-71
[17] Reham S S, Masjuki H H, Kalam M A, Shancita I, Fattah I R and Ruhul A M 2015 *Renewable and Sustainable Energy Reviews* **52** 1566-79
[18] Torres-Jimenez E, Jerman M S, Gregorc A, Lisec I, Dorado M P and Kegl B 2011 *Fuel* **90** 795-802
[19] Nesterenko A, Drelich A, Lu H, Clausse D and Pezron I 2014 *Colloids and Surfaces A: Physicochemical and Engineering Aspects* **457** 49-57
[20] Park S H, Kim S H and Lee C S 2009 *Energy & Fuels* **23** 5228-35
[21] Churchill G V and Srinivasan C A 2017 *International Research Journal of Engineering and Technology* **4** 1523-9
[22] Kansal H K, Singh S and Kumar P 2005 *Journal of Materials Processing Technology* **169** 427–36

IOP Conf. Series: Materials Science and Engineering **257** (2017) 012078 doi:10.1088/1757-899X/257/1/012078

Experimental study of the vehicle dynamics behavior during lane changing in different speeds

P.M. Heerwan[1], S.M. Asyraf [2], A.N. Efistein[2], C.H. Seah[2], J.M. Zikri[2], J.N. Syawahieda[2]
[1]Automotive Engineering Research Group (AERG), Faculty of Mechanical Engineering,
[2]Faculty of Mechanical Engineering,
Universiti Malaysia Pahang, 26600 Pekan, Pahang, Malaysia

*Corresponding author: mheerwan@ump.edu.my

Abstract. During lane changing, the speed of the vehicle is related to the stability of the vehicle. If the driver changes the lane at a high speed, the vehicle will lose its stability and it can increase the possibility of an accident. In this study, the experiment has been developed to analyse how the speed of the vehicle can affect the vehicle dynamics behavior. To achieve this objective, the UMP Test Car which employed with global positioning system (GPS), steering torque and angle sensor, displacement sensor and gyro sensor is used in the experiment. The experiment is run at the UMP test track and the track has 2 lanes which can allows the vehicle to change the position from the left to the right. In the experiment, when the GPS monitor shows 30 km/h, the driver will maintain the speed and start to turn the steering just after the test car reaches to the first skittle. Then, the driver will turn again the steering when the test car reaches to the second skittle. This method is repeated two times and the same methods is used for the speed 50 km/h. The data from the sensors is recorded in the Dewetron software and the graph is plotted. From the experimental results, the steering angle, steering torque, yaw rate and displacement for the speed 30 km/h is smaller than 50 km/h. It means that during lane changing, the speed 30 km/h is more stable compared with 50 km/h.

1. Introduction

In general, ground vehicle can be divided into two categories; track vehicle such as train and untrack vehicle such as cars [1]. For untrack vehicle, it can move in various direction on the ground according to the steering angle given from the driver. The steering angle is not just change the direction of the vehicle, it also can affect the stability of the vehicle[2]. To analyse the stability of the vehicle, the vehicle dynamics equation of motion is used and the vehicle is considered to have six degree of freedom; vertical motion in the z-direction, lateral motion in the y-direction, longitudinal motion in the x-direction, rolling motion in the x-axis, pitching motion around the y-axis, and yawing motion around the z-axis [3]. By analysing of these motion, the control system to improve the stability of the vehicle can be developed.

The other aspects need to be considered during lane changing is the road traffic. Xiang Li et al have studied the optimal traffic control during lane changing. The effect of the lane changing to the traffic efficiency, safety and environmental impact are considered. Their findings provide guidance for the traffic management and important for the researcher to achieve high traffic efficiency [4].

Impact of in-vehicle navigation information on lane-change behavior has been studied by Meingping Yun et al. The experiment is done in the simulator and the road is in the urban expressway. Six indicators which are merging gap, position, delay, steering angle, deceleration, and safe distance of lane changing are analysed to determine the lane changing characteristics. In the experiment, the traffic flow density is take into account and the result shows that the proposed navigation had positive impact under medium to high traffic density [5].

A dynamic automated lane changing has been developed by Yugong Luo et al. In their study, the vehicle-to-vehicle communication is used to develop the automated lane changing. The trajectory planning in their model will be the reference trajectory because the trajectory planning satisfying the safety, ride comfort and road traffic [6]. The lane changing of the autonomous vehicle also has been studied by Maksat Atagoziyev et al. They have developed an algorithm for the vehicle to change the lane in a short period of time. In the algorithm, they have considered the trajectory of all vehicles and also considered the traffic safety. Their algorithm have been applied in the various situation to ensure the functionality [7].

Lane changes prediction based on adaptive fuzzy neural network (AFNN) has been developed by Jinjun Tang et al. By using the information from the vehicle sensors, the AFNN can predict the steering angle. The experiment is developed in the simulator and the experimental result shows that the proposed method is able to follow the desired steering angle [8].

In term of vehicle stability during lane changing, the yaw rate of the vehicle can be an indicator to measure either the vehicle is in a stable condition or not. Over the past years, there are many studies on yaw moment control. M.Heerwan et al have developed the direct yaw moment control using tire steer angle to improve the steer performance in the disturbance condition. The control strategy that they have used in their model is optimal control. In the analysis, the wind disturbance is added at the lateral axis and the result shows that by using direct yaw moment control, it can improve the steer performance of the vehicle [9].

To improve the stability of the vehicle, the vehicle must have the optimum yaw rate and this optimum yaw rate can be achieved if the control system of the vehicle can know the error between the desired and actual yaw rate. Dong-Hyung et al have developed an adaptive direct yaw-moment control method for electric vehicle based on identification yaw rate model. The identification yaw-moment control can minimize the yaw rate errors and finally the vehicle will follow the desired vehicle dynamic model [10]. Based on the advantages of the in-wheel motor at the electric vehicle (EV), which is electric motor can be controlled independently, Bingtao et al have developed MPC-based yaw stability control in in-wheel motored EV via active front steering and motor torque distribution. From their study, it is approved that yaw rate and vehicle stability can be improved by controlling both the steer angle and individual torque at each tire [11].

Although many studies on vehicle stability during lane changing, the analysis of the vehicle dynamics behavior at the different speed is very crucial because it can be a reference for the researcher to develop their control system. The objective of this study is to analyse the vehicle dynamics behavior during lane changing at the different speed. UMP test car that equipped with the high technology sensors is used as the experimental vehicle model. The track that has been used in the experiment is inside the UMP. Considering the vehicle is changing the lane in the UMP road, two types of the speed which is 30 km/h and 50 km/h are used in the experiment. The experimental result shows that the steering torque, steering angle, yaw rate and displacement for the 30 km/h is smaller than 50 km/h. From this results, it is understood that at a higher speed, the vehicle exhibits more violent behavior.

2. Experimental Vehicle Model

Figure 1 shows the UMP Test Car that has been used in the experimental and the specification of this vehicle is shown in Table 1.

Figure 1. Test car UMP.

Table 1. Specification of the experimental vehicle model.

Drive type	Front wheel drive
Gearbox	4-speed automatic
Front brake system	Ventilated Disc
Rear brake system	Drum
Tire Size	195/60/R15
Length	447.8 cm
Width	172.5 cm
Height	143.8 cm
Front/rear track	147.6/147.1 cm
Wheelbase	260.1 cm
Unladed weight	1195 kg

The steering angle and torque sensors, gyro sensor and displacement sensor are shown in the Fig.2,3 and 4. The function of the steering angle and torque sensors are to measure the steering angle and torque when the driver turn the steering wheel. If the value of the steering angle and torque are high, means that the driver turn the steering wheel very fast. During lane changing, the load will transfer from the one position to another position. This will affect the displacement at the upper arm and lower arm. In this experiment, the displacement sensor is used to measure the displacement from the upper arm to the lower arm. Lane changing also can affect the yaw rate of the vehicle. The gyro sensor that located at the center of the vehicle is used to measure the yaw rate of the vehicle. If the value of yaw rate is high, means that the changes of the angle per second is very fast and it can contribute to the instability of the vehicle.

Figure 2. Steering Angle and torque sensors.

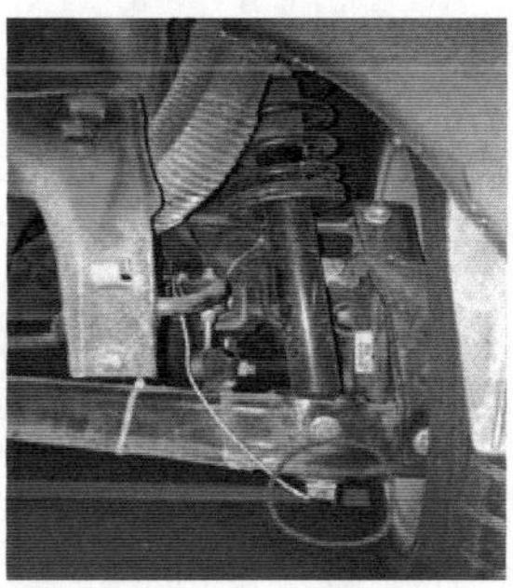

Figure 3. Displacement Sensor.

Figure 4. Gyro sensor.

3. Experimental Method

The experiment has been conducted in the UMP and the route for the experiment is shown in Fig 2. In the experiment, the driver will change the lane from the left side to the right side and then turn back to the left side as shown in the Fig.6. Before start the experiment, all the sensors are calibrated in the Dewetron software. Dewetron software is the data acquisition setup in the experimental vehicle model. After the sensors are calibrated, the driver will start the engine and drive the vehicle to the experiment area. When the GPS monitor shows 30 km/h and the vehicle approach to the first skittle, the driver will turn the steering to the right side. Then, when the vehicle is approach to the second skittle, the driver will turn the steering to the left side. All the data from the sensors are recorded in the Dewetron software and the results are plotted in the graph. This method is repeated for the speed 50 km/h and each experiment is repeated three times.

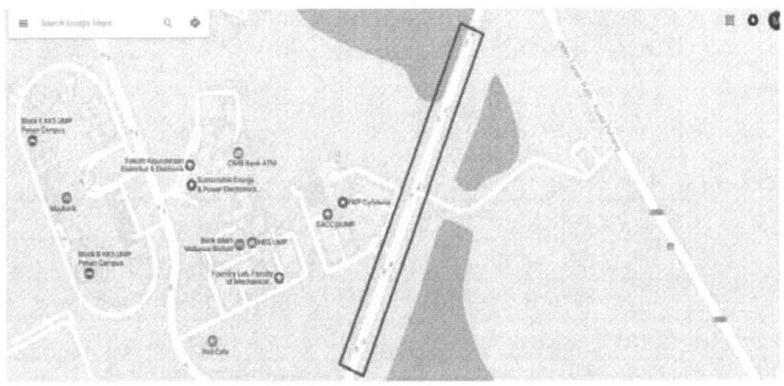

Figure 5. Route for the experiment.

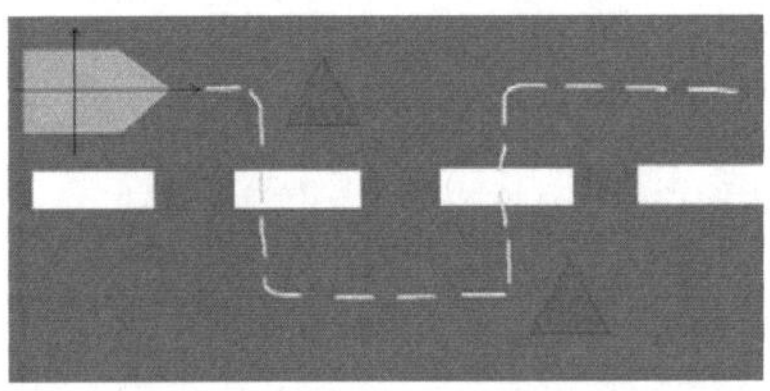

Figure 6. Lane changing of the vehicle in the experiment.

4. Results and Discussion

4.1 Relation between steering angles with the vehicle speed

Figure 7 shows the graph of the comparison of effect on steer angle with the different vehicle speed. As seen in the graphs above the patterns of the data are quite similar because the test vehicle is moving with the same pattern which is changing lanes from left – right – left. The comparison is using the data of the 30km/h and 50km/h which are the lowest and highest speed respectively. From the data it can observed that the graphs are similar but the base value of steer angle is different. 50km/h has the smallest base while 30km/h has the largest base. The change in steer angle in 50km/h is more abrupt than the 30km/h as seen in the graph. Thus, we can conclude that, the faster the speed of vehicle, the faster the time it takes to lane change or to navigate around road obstacles.

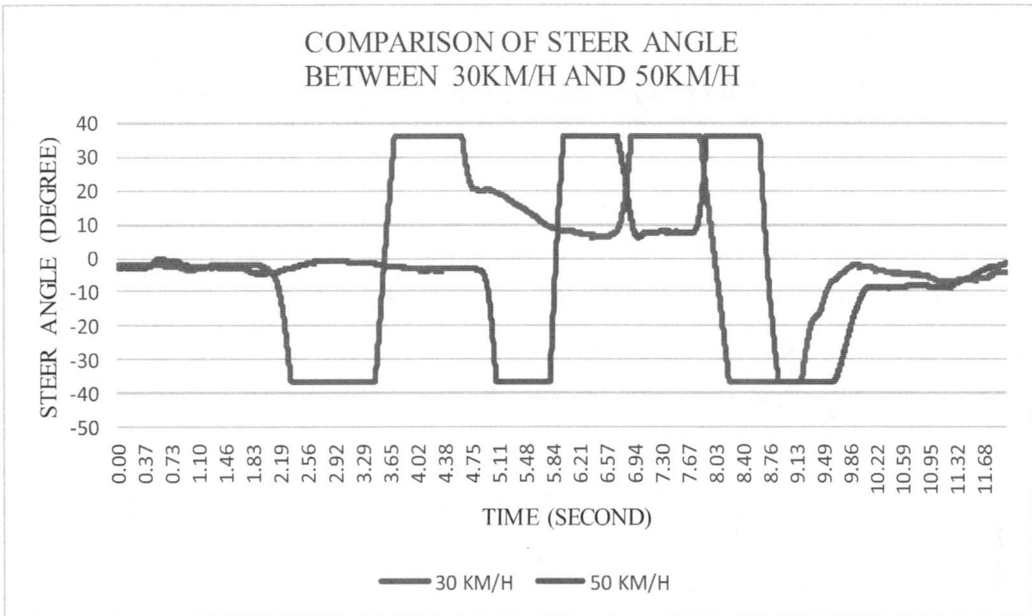

Figure 7. Steer Angle at 30km/h and 50 km/h.

4.2 Relation between steering torques with the vehicle speed

Figure 8 shows the graph of the comparison of steering torque with different vehicle speed. From this data, it can be seen that the pattern is quite similar between those two speeds. Even the peaks of the graphs are more or less the same value. The steering torque at the different speed is similar because at a high speed, a small torque at the steering wheel can increase the steering angle immediately. In contradict, at the small speed, the driver need to push the steering wheel harder to steer the steering wheel.

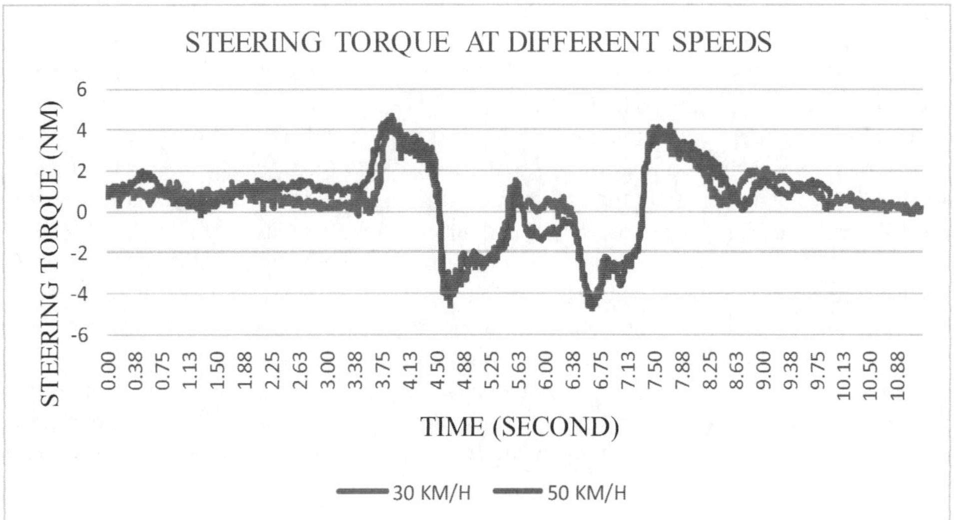

Figure 8. Steering Torque at 30 km/h and 50 km/h.

4.3 Relation between yaw angles with the vehicle speed

Figure 9 shows the graph of comparison on the effect to yaw angle at different speed of the vehicle. Based on the Fig.7 and Fig.8, the data curves are relatively similar in shape and pattern. This similarity are due to the fact that the test vehicle is moving at the same pattern which is changing lane from left – right – left. From the data of 30km/h, and 50km/h we can observe that the graphs are similar in shape but the peak amplitude of yaw angle is different. At 50km/h has the highest peak with the value of 45.34° while 30km/h has the lowest peak with the value of 39.02°. A faster moving vehicle will have to turn more quickly than a slower moving vehicle in order to clear the same distance when lane

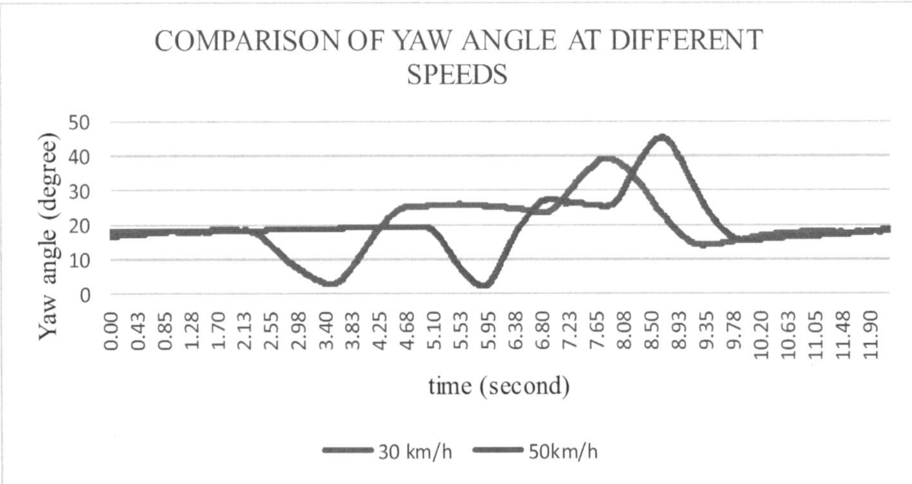

Figure 9. Yaw angle at 30 km/h and 50 km/h.

Changing or avoiding obstacles on the road. A faster moving car carry bigger inertia, and will produce larger yaw angle (γ). Thus, we can determine that, the faster the speed of vehicle, the bigger the yaw angle produced.

4.4 Relation between displacements with the vehicle speed

Figures 10 and 11 shows the displacement sensor data for 30km/h and 50km/h respectively. When the graphs indicate a positive peak value, it means that the specific sensor is going in a positive vertical direction. Vice versa, if negative value is recorded this implies that the sensor is going negative vertical displacement. From Fig 10 and Fig 11, the higher speed records a greater peak value compared to the slower speed. This can be attributed to the fact that the car will tend to roll more at higher speed when making a lane change or turning.

In a turn, the car's tires are generating mechanical grip which is pulling the car inside the turn (centripetal force), the reaction to this force (i.e. the centrifugal force) pushes the car outside the turn. This centrifugal force is acting about the car's center of mass. Since the center of mass is always above the point of the car touching the ground, there is a turning force produced (the rolling moment) about the ground contact point. This becomes the pivot point. Higher the center of mass, more the rolling moment. This turning force causes the car to roll. This forces is resisted by the suspension.

Body roll is not desirable. Rolling causes the weight of the car to shift to the outside of the turn. Thus there is less downward force pushing down on the inside wheels. Thus the inside wheels produce less grip. This affects the car's handling and can cause the car to understeer. If the rolling moment is too high (too much speed or too tight a turn) then the whole car can roll-over and crash. This is because the rolling moment exceeded the resisting force of the suspension springs of the outer wheels.

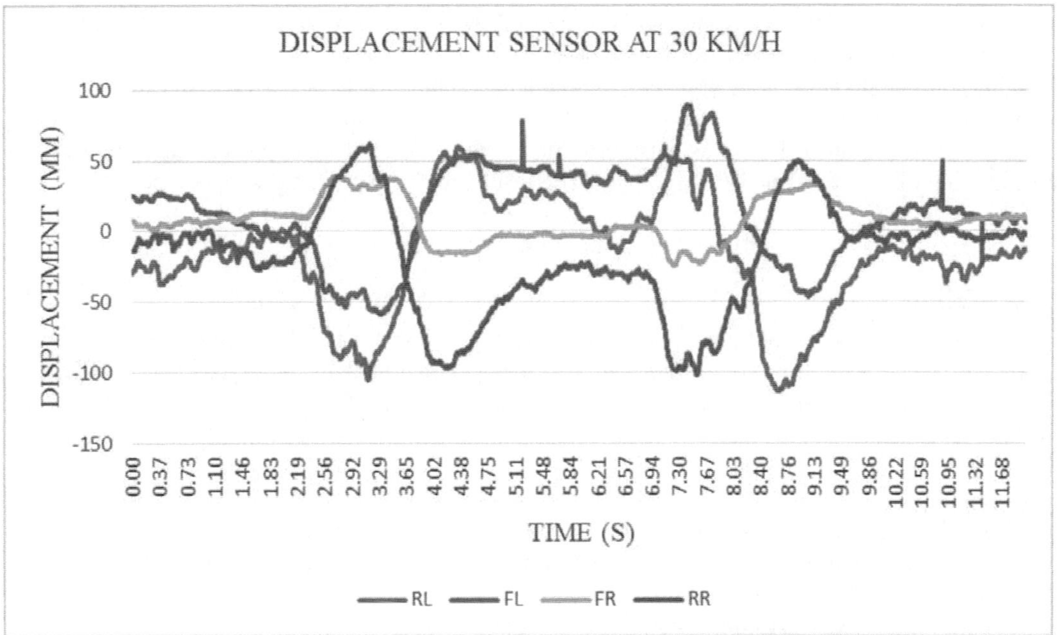

Figure 10. Displacement at each tire at 30 km/h.

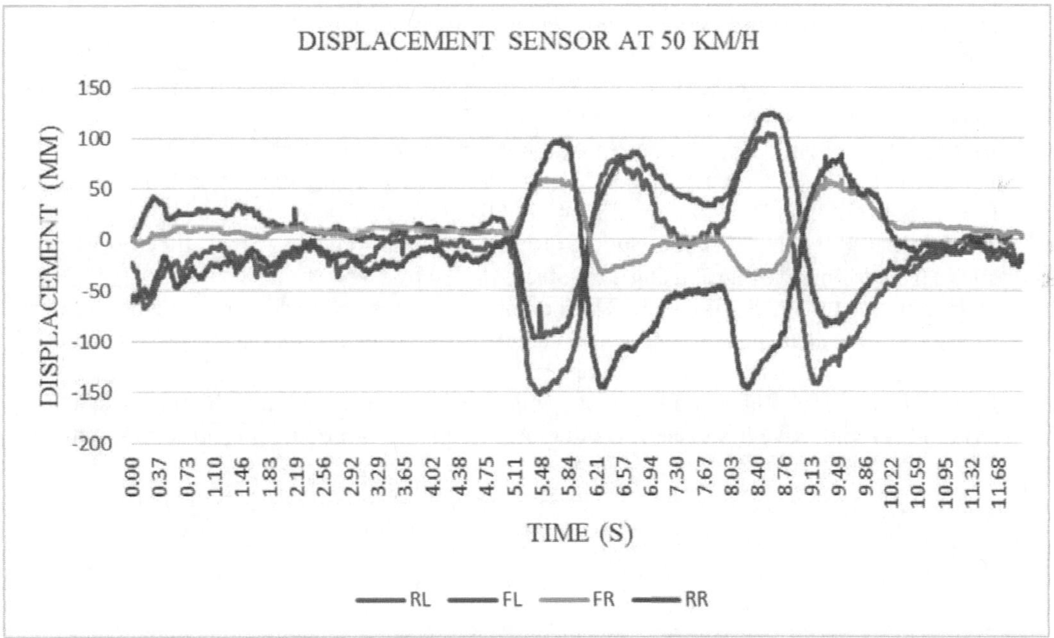

Figure 11. Displacement at each tire at 50 km/h.

5. Conclusions

From the experiment, the vehicle behavior during lane changing in the different vehicle speed is obtained. A vehicle moving with faster speed will produce higher yaw angle when lane changing. While if the vehicle is moving at a slower speed, the steering angle base will be big when changing the lane. If the vehicle is moving with a higher speed, a more abrupt change of steer angle can be seen from the graph resulting in a smaller base area of steer angle.

It can be concluded that if changing lane at a faster speed is quite hazardous as the yaw produced is significantly higher. The abrupt change in steer angle also can be dangerous as the tires might experience slipping which will result in a loss of control in a car. For a front wheel drive such as the test vehicle, the vehicle might suffer from understeer.

Changing lane at slower speed, will have a smaller yaw angle and larger base of steer angle. A smaller yaw angle can prevent the vehicle from spinning or rolling too much and goes out of control. A larger base of steer angle can give us more gripping on tire as the tire changes direction gradually in the same amount of time it takes to change lane. The results in the experiment support the objective to analyse the vehicle dynamics behavior during lane changing at different speed. From the results, researcher can consider the stability factor in order to develop the lane changing control system.

Acknowledgements

The authors would like to be obliged to University Malaysia Pahang for the financial support under RDU160312. The authors also would like to thanks Faculty of Mechanical Engineering (FKM) and Automotive Engineering Center (AEC) University Malaysia Pahang for providing Test Car with the high technology sensors.

References

[1] M. Abe, *Vehicle handling dynamics: theory and application*. 1994.
[2] O. Mokhiamar and S. Amine, "Lateral motion control of skid steering vehicles using full drive-by-wire system," *Alexandria Eng. J.*, 2016.
[3] R. Rajamani, *Vehicle Dynamics and Control*. 2006.
[4] X. Li and J. Sun, "Studies of vehicle lane-changing dynamics and its effect on traffic efficiency , safety and environmental impact," vol. 467, pp. 41–58, 2017.
[5] M. Yun, J. Zhao, J. Zhao, X. Weng, and X. Yang, "Impact of in-vehicle navigation information on lane-change behavior in urban expressway diverge segments," vol. 106, no. May, pp. 53–66, 2017.
[6] Y. Luo, Y. Xiang, K. Cao, and K. Li, "A dynamic automated lane change maneuver based on vehicle-to-vehicle communication," vol. 62, pp. 87–102, 2016.
[7] Maksat Atagoziyev, "Lane Change Scheduling for Autonomous Vehicle," 2016.
[8] J. Tang, F. Liu, W. Zhang, R. Ke, and Y. Zou, "Lane-changes pre diction base d on adaptive fuzzy neural network," vol. 91, pp. 452–463, 2018.
[9] M. Heerwan, B. Peeie, H. Ogino, and Y. Oshinoya, "Skid Control of Small Electric Vehicles (Direct Yaw Moment Control using Tire Steer Angle) by," vol. XXXIX, pp. 73–80, 2014.
[10] D. H. Kim, C. J. Kim, S. H. Kim, J. Y. Choi, and C. S. Han, "Development of adaptive direct yaw-moment control method for electric vehicle based on identification of yaw-rate model," *IEEE Intell. Veh. Symp. Proc.*, no. Iv, pp. 1098–1103, 2011.
[11] B. Ren, H. Chen, H. Zhao, and L. Yuan, "MPC-based yaw stability control in in-wheel-motored EV via active front steering and motor torque distribution," *Mechatronics*, vol. 38, pp. 103–114, 2016.

IOP Conf. Series: Materials Science and Engineering **257** (2017) 012079 doi:10.1088/1757-899X/257/1/012079

Study of alcohol fuel of butanol and ethanol effect on the compression ignition (CI) engine performance, combustion and emission characteristic

M A Aziz, A F Yusop*, M H Mat Yasin, M A Hamidi, A Alias, H Hussin and S Hamri

Faculty of Mechanical Engineering, Universiti Malaysia Pahang, 26600, Pekan, Pahang, Malaysia.

*corresponding author: fitriy@ump.edu.my

Abstract. Diesel engine which is one of the larger contributors to total consumption for petroleum is an attractive power unit used widely in many fields. However, diesel engines are among the main contributors to air pollutions for the large amount of emissions, such as CO, CO_2 and NOx lead to an adverse effect on human health. Many researches have been done to find alternative fuels that are clean and efficient. Biodiesel is preferred as an alternative source for diesel engine which produces lower emission of pollutants. This study has focused on the evaluation of diesel and alcohol-diesel fuel properties and also the performance, combustion and exhaust emission from diesel engine fuelled with diesel and alcohol. Butanol and ethanol is blend with diesel fuel at 1:9 ratio. There are three test fuel that is tested which Diesel (100% diesel), D90BU10 (10% Butanol and 90% diesel) and D90E10 (10% Ethanol and 90% diesel). The comparison between diesel and alcohol-diesel blend has been made in terms of fuel properties characterization, engine performance such as brake power (BP) and brake specific fuel consumption (BSFC) also the in cylinder maximum pressure characteristic. Thus, exhaust gas emission of CO, CO_2, NOx and O_2 emission also has been observed at constant load of 50% but in different operating engine speed (1100 rpm, 1400 rpm, 1700 rpm, 2000 rpm and 2300 rpm). The results show the addition of 10% of each butanol and ethanol to diesel fuel had decreased the fuel density about 0.3% to 0.5% compared to mineral diesel. In addition, viscosity and energy content are also decrease. The addition of 10% butanol had improved the fuel cetane number however the ethanol blends react differently. In term of engine performance, as the engine speed increased, BP output also increase respectively. Hence, the alcohol blends fuel generates lower BP compared to diesel, plus BSFC for all test fuel shows decreasing trend at low and medium speed, however increased gradually at higher engine speed. Thus, D90BU10 had higher BSFC compared to mineral diesel and D90E10. In general, the addition of alcohol blend in diesel fuel had increase the BSFC. In term of in cylinder pressure, as the engine speed is increased, the crank angle noted to move away from TDC for all test fuel. The maximum cylinder pressure increased at low and medium speed, but decrease in higher engine speed. The addition of 10% of butanol and ethanol in the mineral diesel decreased the maximum cylinder pressure. Meanwhile, O_2 emission of D90E10 is higher compared to D90BU10 due to higher oxygen content found in ethanol. The CO_2 emission of D90BU10 recorded higher compared to mineral diesel due to the high oxygen contents in the alcohol. CO emission of alcohol blend on the other hand had lower emission at higher engine speed compared to mineral diesel. As engine speed is increased, NOx emission of mineral diesel and D90E10 had decreased gradually. However, D90BU10 had increased of NOx emission at lower to medium engine speed, than gradually decreased at higher engine speed.

1. Introduction

The major air pollution sources are mostly coming transportation, fuel combustion by stationary sources, industrial processes, oil and gas industries solid wastes and many more. In recent decades, the number of vehicles on the roads has steadily increased. The exhaust fumes such as CO, CO_2, NOx and others from this rapidly expanding fleet of vehicles is one of a principal cause of the worsening air quality nowadays.

Research on adverse health effects of air pollution recorded and led to several changes in practices and regulations since the great smog of 1952 in London. Thus, many of the scientist and policy makers work together to solve the problems in different angles of view. The investigation included the epidemiological and the toxicological effect of airborne pollutants. Although the pollution problems are alarming, but the needs of ICE unit in industries, transportation and in other field still getting high demand.

Diesel engine is an example of attractive power unit used widely in many fields due to its great advantages over gasoline engines such as lower fuel consumption, lower carbon monoxide emissions, better torque characteristics and higher reliability [1, 2]. Hence, diesel engine being the main contributors to air pollutions for the large amount of emissions, especially particulates and nitrogen oxides (NOx), diesel engine without a doubt still getting its demand over the years. Therefore, controlling these emissions is one of the most important aspects of modern air quality management.

Moreover, a few studies related comparative engine performance and emission characteristic of different techniques involved in diesel engine as dual-fuel engine operation had being studied by Abedin et al. [3] and reported to have decrease in PM emission in different mode. Besides, Huang et al. reported that by using diesel blends with pine oil under low temperature combustion, the morphology and reactivity of soot emitted was decrease gradually by the engine load condition and fuel composition [4].

Table 1 illustrates how the European Union emission standards for heavy-duty diesel engines have tightened since EURO I, which came into force in 1992. Diesel engines are associated with combustion noise, engine vibration, and the problem of nitrogen oxides (NOx) - particulate matter (PM) trade-off emissions. Researchers have made lots of effort to reduce toxic and greenhouse gases emitted from these engines.

Several advanced technologies for clean diesel engines have been introduced and been categorized into four strategies such as fuel and fuel additive, fuel injection systems including in-cylinder technology, lubricant oil development and exhaust gas after-treatment devices.

All these approaches have been developed into more technology advanced levels since internal combustion (IC) engines are invented with many researchers and engine manufacturers have greatly involved in improving the technology behind the diesel engine [5-7].

Table 1. Past, current and proposed future for European emission standards [8].

Type of emission		Diesel			
		EURO4	EURO 5a	EURO 5b	EURO 6
		Jan.2005	Sept. 2009	Sept. 2011	Sept.2014
THC	mg/km	-	-	-	-
NMHC	mg/km	-	-	-	-
HC + NOx		300	230	230	170
NOx	mg/km	250	180	180	80
CO	mg/km	500	500	500	500
PM	mg/km	25	5	4.5	4.5
PN#	#/km	-	-	6.0E+11	6.0E+11

With the internal combustion engine are still leading the needs of industries. The usage of fossil fuel is still the number one fuel for this type on engine. In recent years, the non-renewable energy source is decreasing in amount, becoming more and more limited and surely will be finish in approximately less than 100 years from now. However, thanks to the advancement of technologies, human now days able to create new form of renewable energy sources, that can be replaced in a short period of time. For example, bio fuel is one of the natural alternative renewable energy sources which come out by using ethanol from naturally grown plant. Biodiesel is an example of bio fuel that is eco-friendly fuel made from natural, renewable sources.

Previous studies have been proving that the biodiesel can decrease the emission compare to diesel. However, contain emission such as NOx cannot be reduced. It was reported that by present of water, the NOx can be reduced. Thus, this study is to know the performance, combustion and emission of alcohol blend fuel as alternative fuel. Biodiesel is a pure natural, renewable fuel suitable to be used in any situation where basic petroleum diesel is currently used. Biodiesel is made by combining alcohol with vegetable oil or recycled cooking greases. The physical properties of biodiesel are similar to petroleum diesel, plus it's a cleaner-burning alternative. Biodiesel is biodegradable, nontoxic, and essentially free of sulphur and aromatics. Diesel fuel will lead to the high emission of nitrogen compounds and particulate matter. The emission of these chemicals will lead to acid rain, smog and poor health conditions. However, according to the research, biodiesel will increase the nitric oxide and nitrogen oxide. Nitrogen oxide very dangerous to human and environment. So, by using the biodiesel, the target to decrease the emission cannot be achieve.

Biodiesel with alcohol additives is the combination of biodiesel and alcohol. To investigate whether it practically or not, the experiment on engine testing is carry out to investigate emission properties and what is the effect to the engine performance. This paper will introduce the diesel mixed with alcohol fuel (butanol and ethanol) as a fuel for the compression ignition engine with mineral diesel set as a baseline fuel.

2. Methodology and experimental setup

2.1. Engine setup and specification

Fuel engine testing was conducted at Automotive Engine Centre (AEC) laboratory located in University Malaysia Pahang. Experiment and tests were conducted at Faculty of Mechanical Engineering laboratory located at University Malaysia Pahang. The naturally aspirated YANMAR TF120 single cylinder, horizontal diesel engine with water cooling system, a 15kW eddy current dynamometer equipped with dump load, two fuel beakers which can accommodate up to 1 liter, a Meriam Laminar Flow Element (LFE) for air intake system and TFX data acquisition system as well as a KANE Autoplus 5-2 gas analyser is used for this research. Thus, Figure 1 shows the overall project test rig schematic and Table 2 shows the specification of the engine.
Figure 1: Schematic diagram of the test rig.

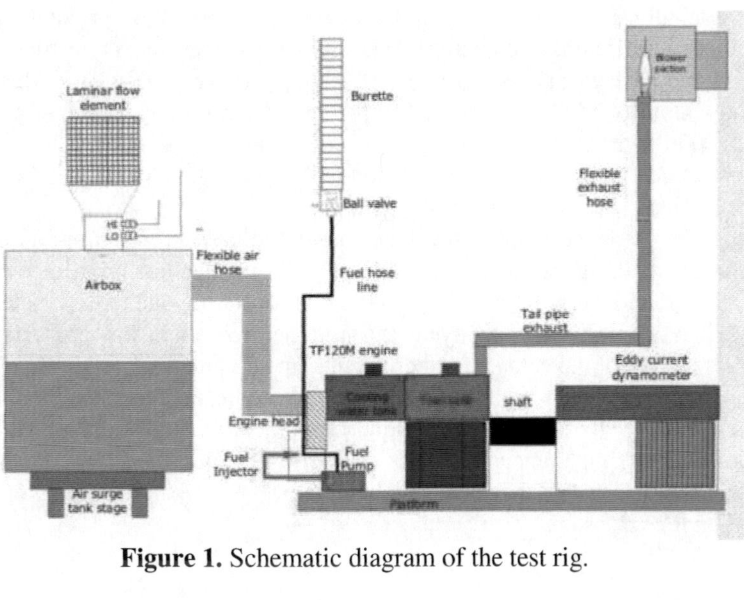

Figure 1. Schematic diagram of the test rig.

Table 2. Engine specification.

Description	Specification
Engine model	YANMAR TF120
Engine type	Horizontal, diesel 4 stroke cycle
Combustion system	Direct injection
Number of cylinders	1
Bore x Stroke (mm)	92 x 96
Displacement (L)	0.638
Dimensions (mm)	Length: 776 Height: 621 Width: 379.5
Injection timing	17° BTDC
Fuel injection pump	Bosch injection
Pressure (kg/cm2)	200
Compression ratio	17.7
Continuous output (HP)	10.5 HP at 2400 RPM
Rated output (HP)	12 HP at 2400 RPM cooling
System	Water cooled (radiator type) cooling water
Capacity (L)	2.3
Dry weight (kg)	102
Connecting rod length (mm)	149.5

2.2. Test fuel preparation and measurement

For experiment testing, three different type of fuel are going to be experimented which is diesel, D90BU10, and D90E10. Diesel fuel will have 100% diesel, while D90BU10 will contain 10% of butanol, 90% of diesel fuel and for D90E10 contains 10% of ethanol and 90% of diesel fuel. Sample of diesel fuel is mixed with alcohol (butanol and ethanol) and mixed at low stirring rate using electric magnetic. For 15 minutes, the mixture is blended continuously, at room temperature.

In order to measure the viscosity of sample made, The Standard Test Method which is specified for Kinematic Viscosity of Transparent and Opaque Liquids is ASTM Standard D445-88. This test method is frequently used to measure the kinematic viscosity of liquid petroleum products. Based on ASTM Standard, the measurement is repeated twice with the tolerance between those two results is 0.02 [9]. Kinematic viscosity values were determined with Cannon-Fenske viscometers [10]. The manufacturer of the Cannon-Fenske type viscometer tubes supplied calibration constants at 40 and 100°C. These constants were generally different by approximately 0.5%, probably due to dimensional changes in the tubes at different temperatures. Density can be measured referring to standard ASTM D1298 using Portable Density/Specific Gravity Meter model DA-130N from industry. The sample is measured at temperature of 15°C. This is a microprocessor controlled system with an LED display. It has a range of 0.0000 to 2.0000 g/cm^3 with an accuracy of +/- 0.001 g/cm^3. In order to measure the heating value of blended fuel, standard that needed to follow is ASTM D4809. Calorimeter system is used when there is a heat leak during the calometric pre-period

The test procedure of cetane number has been subject to criteria based on the difference between the cetane numbers test configuration and operating condition. The test procedure to determine cetane number is complicated and might involves considerable uncertainty due to experimental error. The test needs engine while testing. One of the ways of solving the problem of this properties determination is to develop models to predict the cetane number when some parameters are known. The predicated CN of biodiesel is comparable to that of the actual CN of the biodiesel, and it has been concluded that the CN of biodiesel can be predicted based on thermal properties. In determining cetane number, ignition quality tester is used referring standard of ASTM D613. The measured Cetane number and physical properties of the biofuels were analysed to obtain regression equation and to rank the physical properties based on R2 values.

2.3. Operating condition
The tests were carried out at a steady-state testing condition. It was conducted with 3 different test fuels (diesel fuel, D90BU10 and D90E10) at 5 different engine speeds which were 1100 rpm, 1400 rpm, 1700 rpm, 2000 rpm and 2300 rpm. The load applied on the engine was kept constant at 50%.

3. Results and discussion

3.1. Analysis of fuel properties
Summarized fuel properties are illustrated in Table 3.

Table 3. Fuel properties summary.

Properties	Testing standard	Types of fuel				
		Reference fuel			Test fuel	
		Ethanol	**Butanol**	**Diesel**	**D90BU10**	**D90E10**
Formula		C2H5OH	CH3CH(OH)CH2CH3 / C4H9OH	C12H24		
Viscosity (mm2/s) (cSt)	D 445	1.2	4.21	5.144	3.576	3.674
Density at 25°C(g/m^3)		0.79	0.81	0.8264	0.8236	0.8226
Cetane number	D 613	8	25	47.8	51	43.82
Energy content (MJ/kg)		26.8	32.5	44.8	43.57	43

3.1.1. Fuel density. Figure 2 (a) below shows the test fuel density for Diesel, D90BU10 and D90E10. From the result obtained, test fuel density ranging from 0.8226 9 g/cm³ to 0.8264 g/cm³. Baseline diesel fuel 100% density is 0.8264 g/cm³. The addition of 10% of butanol and ethanol both have different effect on the test fuel density. The addition of 10% butanol to diesel fuel had resulted in decreased of fuel density approximately 0.34% of fuel density. In addition to that, the addition of 10% ethanol had resulted in decreased of fuel density about 0.5%.

3.1.2. Fuel viscosity. Figure 2 (b) shows the test fuel viscosity for diesel fuel, D90BU10 and D90E10 respectively. The viscosities of fuels fall between the ranges of 3.576 to 5.144 mm²/s, where diesel fuel have the highest fuel viscosity compare to the other two test fuel. The addition of alcohol to diesel shows similar result where the test fuel viscosity is decreased. The addition of 10% of butanol and ethanol to diesel had decreased the fuel viscosity approximately 30%. The decrease of viscosity of fuel will decrease the fuel resistivity to the flow of liquid which is due to the internal friction in the moving fluid.

3.1.3. Energy content. Figure 2 (c) show the energy content for Diesel fuel 100%, D90BU10 and D90E10 respectively. Diesel fuel have the highest energy content at 44.8 MJ/Kg while D90BU10 and D90E10 had 43.57 MJ/Kg and 43.0 MJ/Kg of energy content respectively. From the graph we can see that the addition of alcohol to the diesel fuel will decrease the test fuel energy content.

3.1.4. Cetane number. The cetane number is one of the most commonly cited indicators of diesel fuel quality. It measures the readiness of the fuel to auto ignite when injected into the engine. It is generally dependent on the composition of the fuel and can impact the engine's start ability, noise level, and exhaust emissions [11]. Higher cetane number means that the fuel auto ignites more easily in the engine cylinder. Figure 2 (d) shows the cetane number for Diesel 100%, D90BU10 and D90E10 respectively. The higher the cetane number, the easier for the fuel to ignite the engine. Diesel fuel had lower cetane number compared to D90BU10. However, D90E10 recorded to have lower cetane number compared to diesel fuel 100%.

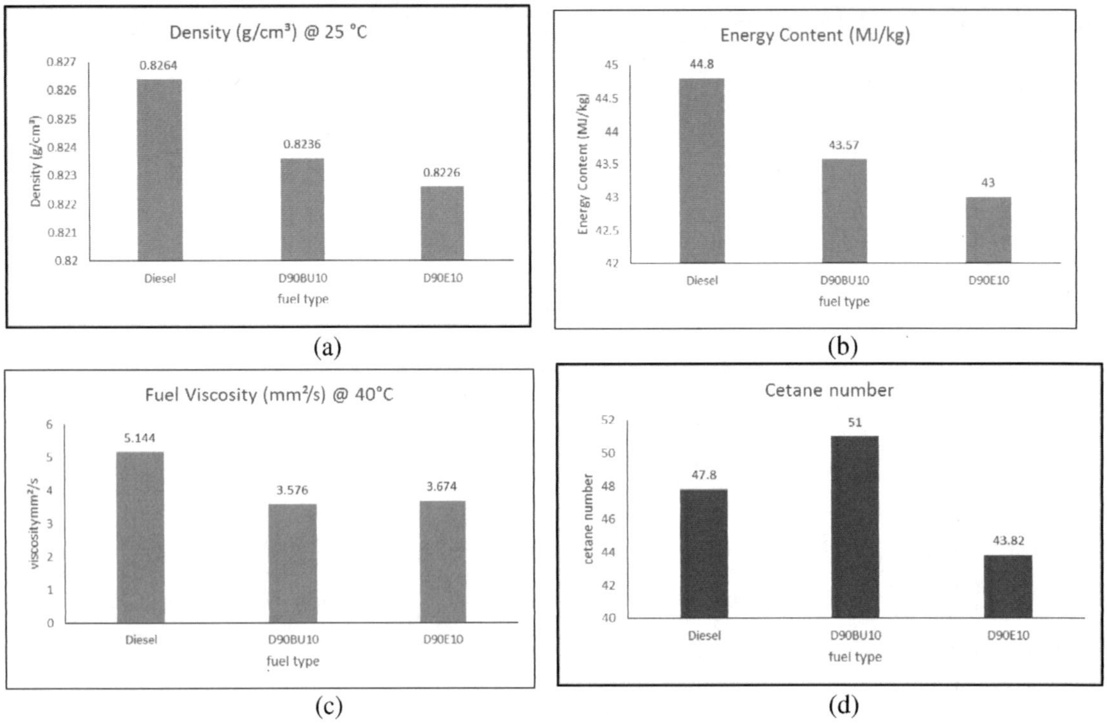

Figure 2. (a) Density, (b) Energy content, (c) Fuel viscosity and (d) Cetane number

3.2. Analysis of engine performance

3.2.1. Brake Power (BP). Figure 3 shows the comparison of engine brake power (BP) output of Diesel, D90BU10 and B90E10 at difference engine speeds. Engine brake power indicates the amount of work of one engine does per unit of time. The output curves are plotted in Fig A with engine speed ranging from 1000 RPM to 2400 RPM.

Some researchers in their respective studies also found similar trend for BP versus RPM [12, 13]. Diesel BP output found to be higher at 1100 RPM and 2300 RPM compared to other two dual fuels. Thus, diesel BP output recorded higher compared to D90E10 at 1700 RPM and 2000 RPM. However, Diesel fuel BP output are lower, compared to D90BU10 at 1400 RPM to 2000 RPM. The maximum BP output recorded at 2000 RPM of 3.5kW for D90BU10, which is 11% higher compared to diesel fuel BP output.

Diesel, D90BU10 and D90E10 have approximately similar density at range of 0.8226 to 0.8286 g/cm3, however due to alcohol blends, the fuel viscosity in dual fuel blend mode is lower compared to diesel fuel. Moreover, lower fuel viscosities lead to greater injector leakage reducing maximum fuel delivery and ultimately reducing the engine power output [14].

Referring from the previous studies of Tuccar et al., the engine brake power output reduced with the butanol addition in the fuel. The decreased in brake power value of the engine was due to the lower energy content of butanol additive compared to the diesel fuel. In addition, oxygen content of the fuel blends also lead to decrease in brake torque [15].

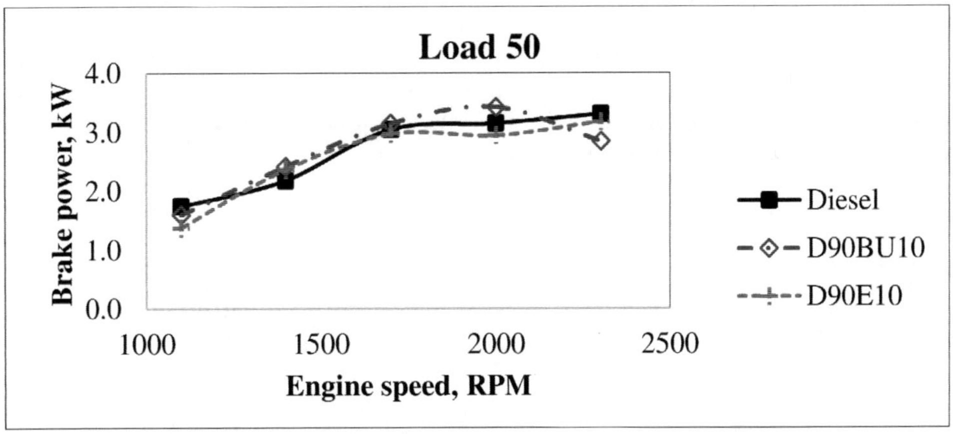

Figure 3. Brake Power versus Engine Speed at 50% load.

3.2.2. Brake Specific Fuel Consumption (BSFC). Brake specific fuel consumption (BSFC) is the amount of engine fuel consume in a certain period to produce one kilowatt of brake power. The brake specific fuel consumption is influenced by physiochemical properties of fuel such as the fuel viscosity, density, cetane number and heating value. Generally, the specific fuel consumption of the alcohol blended fuel is more because of the lower heating value of alcohol.

The analysis result of BSFC in this study can be found in Figure 4. From the graph, shows the BSFC of diesel fuel, D90BU10 and D90E10 at 50% engine load versus engine speed ranging from 1100 RPM to 2300 RPM. BSFC for all test fuel initially shows the decreasing trend from 1100 RPM to 1700 RPM. However gradually increase until 2300 RPM. The increase of BSFC at high speed is due to the increased of friction. At lower speeds. BSFC increase due to increased time for heat losses from the gas too engine cylinder and pistons wall [16].

D90BU10 have higher BSFC output compared to diesel and D90E10 at all engine speed. At 1400 RPM to 2100 RPM, D90BU10 recorded 15.92%, 5.83%, 6.94% and 5.16% higher BSFC output compared to diesel fuel. Thus, D90BU10 recorded highest BSFC output at 2300 RPM with 26.87% higher compared to diesel fuel. While D90E10 recorded lowest BSFC at 1400 RPM with 22.04% lower compared to diesel fuel BSFC. This is incline with the previous studies where the addition of butanol and ethanol in the diesel fuel increases the engine fuel consumption for alcohol and diesel fuel blends due to the lower calorific value [17]. Meaning that, more fuel is needed during combustion stage in order to generate same amount of energy when using pure diesel fuel.

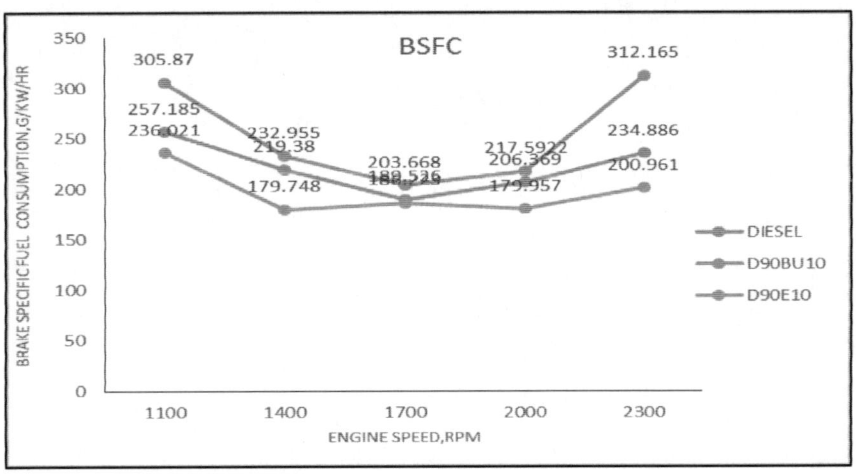

Figure 4. BSFC versus Engine Speed at 50% load.

IOP Conf. Series: Materials Science and Engineering **257** (2017) 012079 doi:10.1088/1757-899X/257/1/012079

3.3. Analysis of engine combustion

3.3.1. In cylinder pressure. In cylinder pressure data must be analysed to understand the engine combustion process and events occurring in the combustion chamber. The relationship between the in cylinder pressure and crank angle indicates the performance of the engine. Few factors that will affect peak pressure for example are, compression ratio, types of fuel used, air fuel mixture ratio and engine type. Figure 5 (a) to (e) shows the crank angle versus the peak pressure at different engine speed ranging from 1100 RPM to 2300 RPM for Diesel, D90BU10 and D90E10.

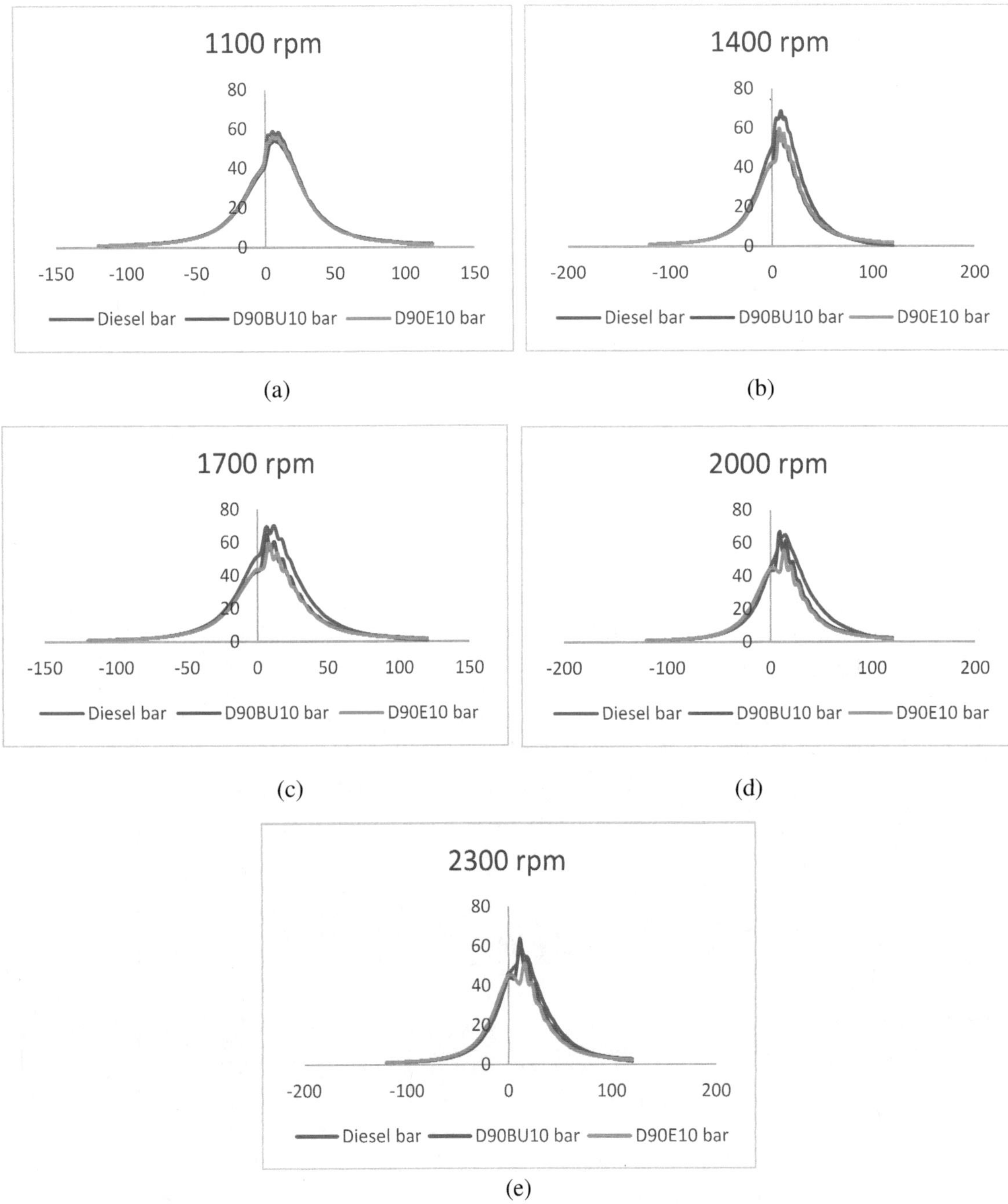

Figure 5. In Cylinder Pressure versus Engine Speed at 50% load.

At 1100 RPM, diesel fuel recorded peak pressure 58.8 bar at 5° ATDC where D90BU10 and D90E10 recorded 54.22 bar at 7° ATDC and 56.27 bar at 5° ATDC respectively. As the engine speed increase, the peak pressure for diesel fuel increase to 68.55 bar at 9° ATDC and 70.09 bar at 12° ATDC respectively for 1400 RPM and 1700 RPM. While for D90BU10, the peak pressure increases to 57.95 bar at 5° ATDC and 64.65 bar at 6° ATDC. D90E10 on the other hand recorded 59.64 bar and 59.34 bar peak pressure at 7° and 8° ATDC respectively for 1400 and 1700 RPM.
At 2000 RPM, diesel fuel and D90E10 in cylinder pressure start to decrease to 64.75 at 14° ATDC and 55.38 bar at 13° ATDC and continue to decrease at 2300 RPM where diesel recorded 59.76 bar at 12° ATDC and D90E10 recorded 51.16 bar at 16° ATDC. Meanwhile, D90BU10 continue to increase in pressure at 2000 RPM where it recorded 66.63 bar at 9° ATDC but the pressure drop to 63.66 bar at 11° ATDC at 2300 RPM.

Datta and Mandal in their reports found that the maximum pressure is noted after TDC and the addition of ethanol and methanol moves the maximum pressure away from the TDC [17]. Compare to the result in this study, the addition of butanol and ethanol not showing similar trends. In average, diesel fuel in cylinder pressure recorded to be in higher number compared to D90BU10 and D90E10. In cylinder pressure of diesel fuel greater than the alcohol mixture fuel is due to the prolong ignition delay of alcohol mixture fuel that causing the combustion period become shorter [18].

3.4. Analysis of engine gas emission

3.4.1. Oxygen (O_2). Figure 6 shows the oxygen, O_2 emission of diesel fuel, D90BU10 and D90E10 at 50% engine load with engine speed ranging from 1100RPM to 2300RPM. Diesel fuel recorded highest O2 emission at 2000 RPM with 17.46% emission. While both D90BU10 and D90E10 recorded highest O_2 emission at engine speed 1100 RPM with 17.28% and 17.40% emission. D90BU10 recorded a decreasing trend as engine speed increase from 1100 RPM to 2300 RPM. D90E10 found to have a higher O_2 emission compared to D90BU10 at all engine speed due to the higher oxygen content found in D90E10 compared to D90BU10.

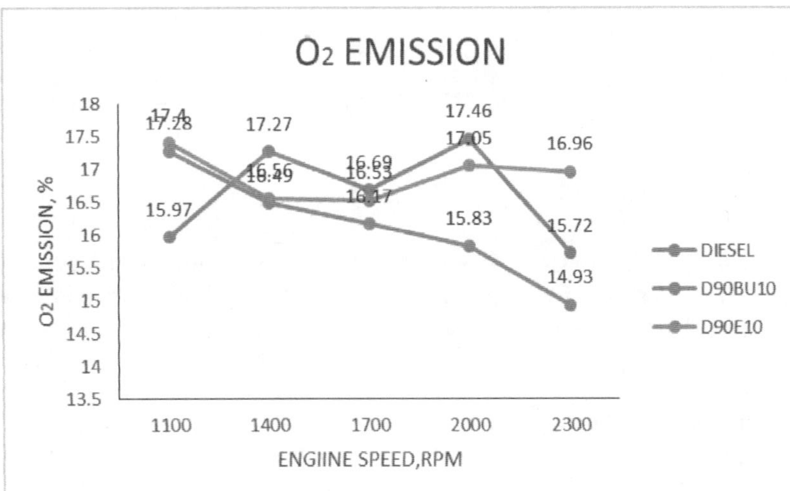

Figure 6. Emission of Oxygen at 50% load.

3.4.2. Carbon Dioxide (CO_2). Carbon dioxide (CO_2) is an ordinary end combustion product of each fuel that contains carbon (biomass, wood, coal, and its variations, oil and petroleum) When enough oxygen is available, hydroxyl radical OH is one of the principal oxiding agents that converts CO into CO_2 [12]. Plus, it is a well-known fact that complete combustion inside the engine combustion chamber will increased the CO_2 emissions. Although there is no possibility of perfect or complete combustion occurs, there may be a near complete or near perfect combustion happen. Hence, that's depends on the engine operating condition and the nature of fuel used to run the engine [17].

Figure 7 shows the CO_2 emissions of diesel fuel, D90BU10 and D90E10 at engine speed ranging from 1100 RPM to 2300 RPM. CO_2 emission from D90BU10 shows an increasing trend and recorded highest emission at 2300 RPM. CO_2 emission of diesel fuel at 1100 RPM is 26% and 30% higher compared to D90BU10 and D90E10 and also recorded highest emission at 2300 RPM. CO_2 emission for D90E10 shows no significant changes as engine speed increased. Higher CO_2 emission in alcohol blend fuel could be due to their high oxygen contents which may improve the combustion quality, thus increase the CO_2 emission [19].

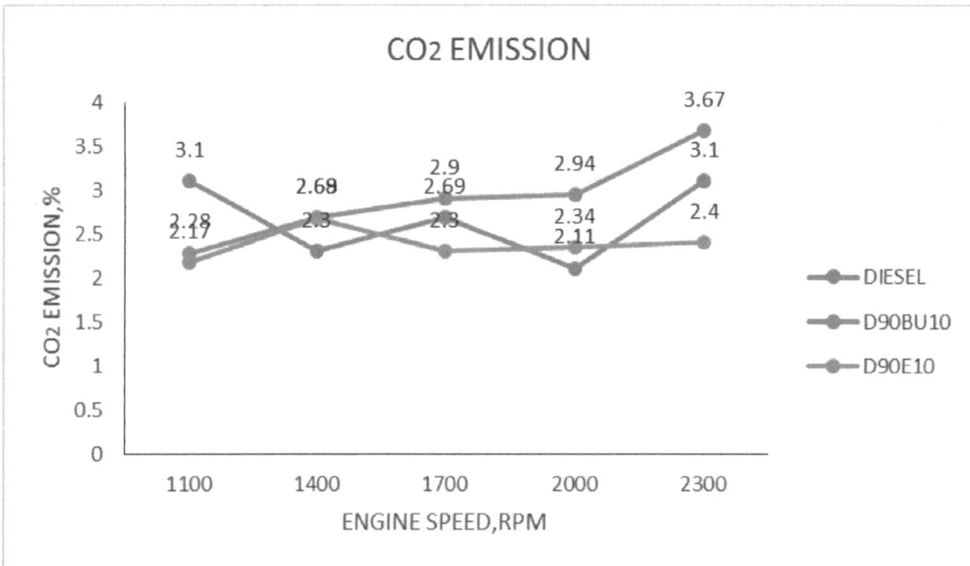

Figure 7. Emission of Carbon Dioxide at 50% load.

3.4.3. Carbon Monoxide (CO). Carbon monoxide, (CO) can be produced into the engine cylinder when combustion occurred with insufficient air flow and low flame temperature. However, formation of CO generally occurs during rich air and fuel mixture as a result of O_2 inaccessibility to totally oxidize all CO proportion in the fuel [12]. Figure 8 shows the engine CO emissions versus the engine speed at 50% load.

As reported by Imdadul et al., in comparative study of alcohol blend in term of diesel engine performance and emissions. The CO emission decreased as engine speed increased, thus with increase of alcohol blends in the fuel the CO emission also decreased. The lower density of alcohols blends compared to diesel making the alcohol blends fuel evaporates easily in the combustion chamber. Plus, the length of splash entrance become smaller and this will help blending process and at the same time decrease the CO emission. Furthermore, alcohol content in the blend fuel provide more oxygen and that will lower the stoichiometric air/fuel ration of alcohols, thus lowering CO emission [12]. Same findings also reported by Sanli et al. [19].

In this study, no significant trend can be observed. Referring to the graph of CO emission in Figure 8, diesel fuel recorded highest CO emission at 2300 RPM with 28.6% higher compared to D90BU10 and D90E10. D90BU10 and D90E10 both recorded highest CO emission at 1700 RPM with 20% and 33.3% higher compared to diesel fuel CO emissions.

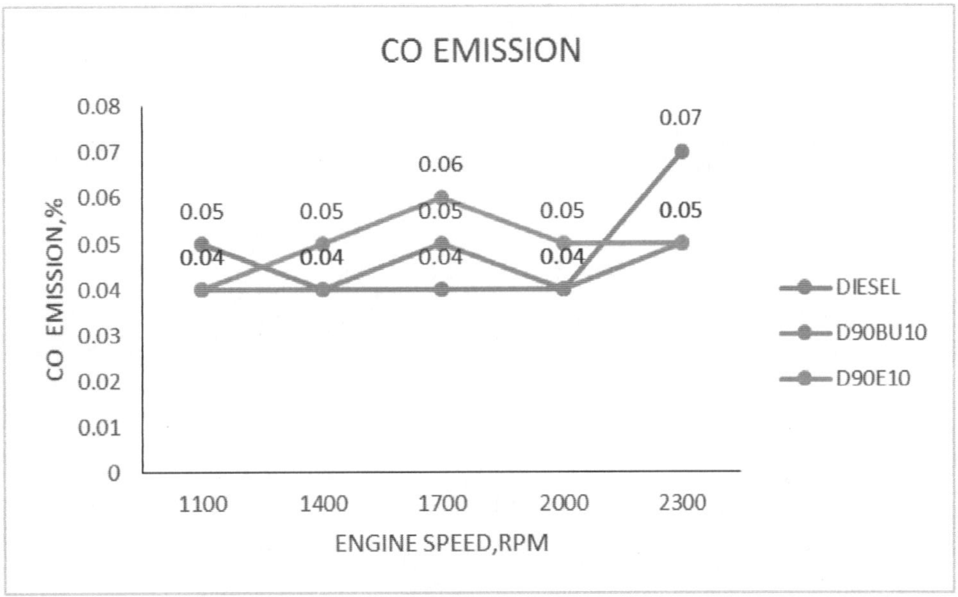

Figure 8. Carbon Monoxide Emission at 50% load.

3.4.4. Oxide of Nitrogen (NOx). NOx emission is a function of residence time related to oxygen concentration and also related to combustion chamber temperature. NOx formation rates increase with increase on combustion temperatures and oxygen concentrations [20]. Figure 9 shows the NOx emission for diesel fuel, D90BU10 and D90E10 with respect of engine speed at 50% engine load in speed ranging from 1100 RPM to 2300 RPM.

As reported in many research, as engine speed increase, and the alcohol mixture increase in the diesel fuel. The emission of NOx emission will increase respectively [4, 12, 15, 21, 22]. More oxygen content exists in the alcohol blends which attributed to the amount of NOx formation. Plus, higher atomic weight species combust in delayed combustion stages due to the lower atomization at the beginning of controlled combustion [23]. Atmanli reported that NOx emission increased as engine speed increased. However, Atmanli [23] reported that NOx emission decreases in the presence of higher alcohol in diesel or biodiesel blends. Similar finding also reported by Tuccar et al. [15] where NOx emission is lower in high engine speed and in butanol blends fuel compared to diesel fuel.

NOx emission for the test diesel fuel, D90BU10 and D90E10 react slightly different. Diesel fuel show a decrease trend of NOx emission, however at 1700 RPM the NOx emission is slightly high compared to the emission at 1400 RPM. Where, for D90BU10 the NOx emission increase from 1100 RPM to 1700 RPM, however starting to decrease from 1700 RPM to 2300 RPM. D90E10 on the other hand, shows decreasing trend for both graph. The decreasing trend of NOx emission may due to the lower calorific value and higher heat of evaporation of alcohol which leading to higher temperature during premixed part of combustion [15].

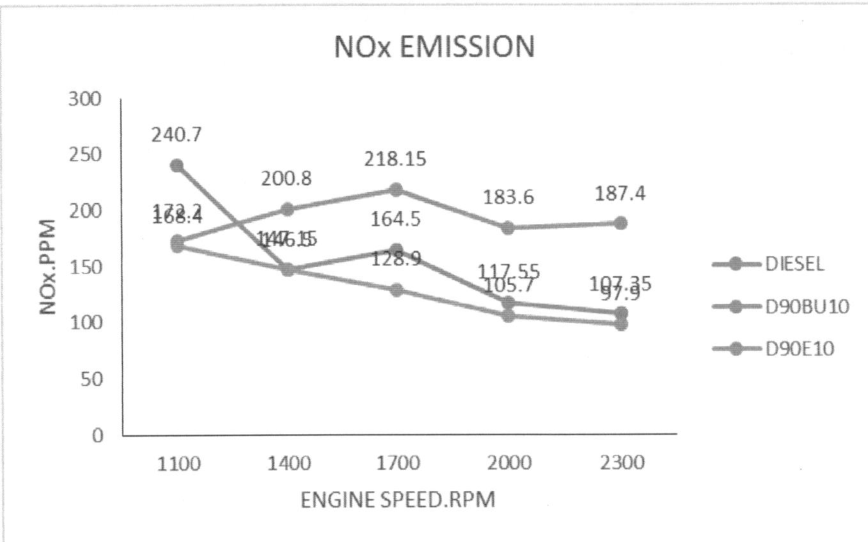

Figure 9. NOx Emission at 50% load.

4. Conclusions

Engine brake power indicated the amount of work of one engine does per unit of time. The engine BP output is increased for all test fuel as the speed is increased. Here, we can conclude that the brake power output depends extensively on engine speed. In addition of that, at 10% of alcohol mixture, the density of alcohol and diesel blends and mineral diesel does not have a significant effect on the BP. Hence, the difference in viscosity of test fuel bring effect on the BP output at higher engine speed as the diesel generated highest BP compared to the other two test fuel. The lower fuel viscosities in both alcohol and diesel blends mode leads to the greater injector leaking, thus resulting reduce in maximum fuel delivery and ultimately reducing the engine power output. On top of that, the lower energy content of dual-fuel compared to diesel fuel may resulted in decreased of engine BP value.

Brake specific fuel consumption is the amount of engine fuel consume in a certain period to produce one kilowatt of brake power. Generally, engine BSFC will increased as the engine speed increased, and addition of alcohol mixture in the test fuel. Although the BSFC in this study have a decreasing trend at lower engine speed, it gradually increased at higher engine speed. The increase of BSFC was due to the lower heating values of dual fuel compared with mineral diesel. Most studies found that the diesel-alcohol mixture have a lower heating values and could cause significant increases in BSFC.

It can be concluded from the overall results that at lower engine speed, in cylinder maximum pressure at lower engine speed increase as the engine speed is increased for all test fuel. However, there is a decreased for the maximum cylinder pressure at the higher engine speed. Thus, mineral diesel maximum cylinder pressure was higher compared to the alcohol and diesel blend fuel. However, D90BU10 produces closest maximum cylinder pressure compared to mineral diesel. From previous researched had found that the maximum cylinder pressure is noted after TDC and the addition of ethanol mixes the maximum cylinder pressure away from TDC.

Dual fuel recorded a decreasing trend of O_2 emission, especially for D90BU10. For D90E10, O_2 emission decreased at lower rpm however gradually increase at higher engine rpm. Ethanol fuel mixture found to have higher O_2 content compared to mineral diesel and butanol fuel mixture. It is due to higher O_2 content in ethanol compared to the butanol. CO_2 is an ordinary exhaust gas of each fuel that contains carbon. Should there is enough O_2 available during combustion stage, hydroxyl radical OH will react and converts CO into CO_2. CO_2 emission from D90BU10 and D90E10 are increased as engine speed is increased. Higher CO_2 emission in alcohol and diesel fuel was due to their high O_2 contents which improve combustion quality, thus increase the CO_2 emission. On the other hand, no significant trend can be observed for CO emission. Looking at the higher side rpm, diesel generates more CO compare to the alcohol and diesel blend fuels and it can be conclude that addition of alcohol to mineral diesel will lower the CO emission. In line with most of the previous studies, NOx emission for diesel and D90E10 test fuel showing a decreasing NOx emission as engine speed is increased. However, for D90BU10, NOx emission was higher at low or medium engine speed, then gradually decreased at higher engine speed. The addition of 10% ethanol to mineral fuel can reduce NOx emission.

Acknowledgments
The authors would also like to acknowledge the Universiti Malaysia Pahang for providing financial support through RDU 1603126 and RDU 160309 which made this study possible.

References

[1] Heywood J B 1988 *Internal Combustion Engine Fundamentals* (New York: Mcgraw-Hill)
[2] Stone R 1999 *Introduction to Internal Combustion Engines*
[3] Abedin M J, Imran A, Masjuki H H, Kalam M A, Shahir S A and Varman M 2016 *Renewable and Sustainable Energy Reviews* **60** 306-16
[4] H Huang, Teng W, Liu Q, Zhou C, Wang Q and Wang X 2016 *Energy Conversion and Management* **128** 317-26
[5] Ali O M, Mamat R and Faizal C K M 2013 *Journal of Renewable and Sustainable Energy* **5** 012701
[6] Miyaura T, Morikawa A, Ito Y, Ishizuka K and Tsuiki T 2013 *SAE Technical Paper*
[7] Rakopoulos D C, Rakopoulos C D, Giakoumis E G, Dimaratos A M and Kakaras E C 2013 *Journal of Energy Engineering*
[8] Piock W, Hoffmann G, Berndorfer A, Salemi P and Fusshoeller B 2011 *SAE Paper* **01** 1212
[9] Tat M and Van Gerpen J 1999 *Journal of the American Oil Chemists' Society* **76** 1511-13
[10] Knothe G and Steidley K R 2005 Fuel 84 1059-65
[11] Van Gerpen J 1996 *Proc. Third Liquid Fuel Conference: Liquid Fuel and Industrial Products from Renewable Resources* p 197-206
[12] Imdadul H K, Masjuki H H, Kalam M A, Zulkifli N W M, Alabdulkarem A and Kamruzzaman M 2016 *Fuel* **179** 281-8
[13] Heydari-Maleney K, Taghizadeh-Alisaraei A, Ghobadian B and Abbaszadeh-Mayvan A 2017 *Fuel* **196** 110-23
[14] Campos-Fernández J, Arnal J M, Gómez J and Dorado M P 2012 *Applied Energy* **95** 267-75
[15] Tüccar G, Özgür T and Aydın K 2014 *Fuel* **132** 47-52
[16] Tan Y H, Abdullah M O, Nolasco-Hipolito C, Zauzi N S A and Abdullah G W 2017 *Energy Conversion and Management* **132** 54-64
[17] Datta A and Mandal B K 2016 *Applied Thermal Engineering* **98** 670-82
[18] Lee S and Kim T Y 2015 *Fuel* **162** 65-73
[19] Sanli H, Canakci M, Alptekin E, Turkcan A and Ozsezen A N 2015 *Fuel* **159** 179-87
[20] Ong H C, Masjuki H, Mahlia T, Silitonga A, Chong W and Yusaf T 2014 *Energy* **69** 427-45
[21] Tutak W, Lukács K, Szwaja S and Bereczky Á 2015 *Fuel* **154** 196-206
[22] Yilmaz N, Vigil F M, Benalil K, Davis S M and Calva A 2014 *Fuel* **135** 46-50
[23] Atmanli A 2016 *Fuel* **176** 209-215

Development of Fibre Bragg grating (FBG) dynamic pressure transducer with diminutive voltage inconsistency

A M Aizzuddin[1], Z M Hafizi[1,*], L V Kee[1], E Vorathin[1] and K S Lim[2]

[1]Faculty of Mechanical Engineering, Universiti Malaysia Pahang, Pekan, Pahang, Malaysia; muhammad.aizzuddin.a@gmail.com
[2]Photonics Research Centre, Faculty of Science, University of Malaya, 50603 Kuala Lumpur, Malaysia

[*]Corresponding author: hafizi@ump.edu.my

Abstract. In this work, a diaphragm-type Fibre Bragg grating (FBG) dynamic pressure transducer was designed and developed. Pressure transducer in this study consists of elastic thin metal diaphragm which acts as a primary sensing element; and integrated with an FBG sensor for the secondary sensing element. In the common match filter interrogation system, converting optical signal to voltage is the challenging due to voltage reading inconsistency, which would cause variation in pressure reading. New alternative arrangements of matched filter interrogation system are used to address the issue. For automated pressure measurement, the optical signal from FBG was converted into voltage by using the proposed arrangement of matched filter interrogation system. Additional FBG was added as reference sensor which installed in the system for detection the change of broadband light source. Reduction of voltage inconsistency was achieved by normalizing the voltage reading from sensor with the voltage reading from reference FBG. The result shows that the FBG pressure transducer is proven to be suitable for pressure measurement of gas or liquid with an average error of 5.348%. Furthermore, the FBG sensor has a good linearity with a linear correlation coefficient of 97.29% in pressure measurement.

1. Introduction

Recently, increasing interest in worldwide industries towards the application of Fiber Bragg grating (FBG) in sensing technology has brought to the rapid development and deployment of optical sensors especially in the monitoring of temperature and strain. FBG has a number of advantages, such as relatively small size and long life span. Besides, it is also inexpensive to manufacture, lightweight, multiplexing ability, self-referencing with a linear response, ease of installation, durability and immune to electromagnetic interference (EMI) [1-4]. Moreover, a pressure quasi-distributed measure can be realized by multiplexing in one single optical fiber to provide multiple FBG sensing elements without the need of installing a huge amount of strain gauge [5]. Sensing process in extremely harsh environments such as explosive gas exposure setting, high-temperature combustion chamber, and environment which contains high electromagnetic interference is feasible by using FBG sensors due to its passive nature. Although naked FBG sensor is fragile, an approach to encapsulate the FBG sensor could solve the problem by adding carbon fiber for reinforcement and solidified by epoxy resin with encapsulation technique [6]. The encapsulation technology protects the fiber from the severe environment in mounting processing without influencing the transmission of the strain applied to the FBG [7, 8].

In recent years, various types of the FBG-based sensor [9, 10] has been developed. Vengel Rao Pachava et al. [11] used the longitudinal FBG deformation principle by the transverse deflection of a diaphragm which induced an axially stretched-strain along the length of the FBG thereby creating a red shift of Bragg wavelength with the increased pressure. However, according to Frantisek Urban et al. [12], a longitudinal FBG's deformation will result in bigger optical signal wavelength displacement (FBG central frequency moves from 10 to 30 nm). They also emphasized that, by using the method of pressing the FBG laterally to obtain an ellipsoidal fiber cross-section shape only, will generate a maximum of 300pm spectrum peak spread.

In 2008, Zhang et al. [13] has reported their new FBG pressure sensor based on a flat diaphragm with enhanced responsibility by using a single FBG and an L-shaped lever with a curve of Archimedes spiral. It achieved ultrahigh sensitivity, 244 pm/kPa and reduced temperature sensitivity of 2.8 pm/°C. The L-shaped lever is made up of quartz glass and laser machine. It requires high precision and the sensor design will be complicated compared to conventional diaphragm transducer. In the research works [13, 14] temperature was compensated because a temperature cross-sensitivity can lead to inaccurate measurement result [15, 16]. Zhao et al. [17] proposed a solution to compensate the temperature by adding a reference FBG. Sengupta et al. [18] solved this problem by utilized two FBG in different diameter fibers and the combination of FBG with the Fabry-Perot cavity is reported by Lin et al. [19].

A conventional pressure sensor based on strain gauge, vibration wire, mechanics, and etc. [20, 21] are unable to adapt to the harsh environments and also impossible to realize the online monitoring for long distance. Therefore, the FBG based pressure sensors become important in recent researches [22, 23]. Several studies have proposed improvement to the pressure measurement sensitivity, such as embedding FBG in the polymer, soldering metal-coated FBGs on a free elastic cylinder, and attaching the FBG to a diaphragm [5, 24].

In this research work, an FBG pressure transducer was designed and applied for pressure measurement by attaching FBG on the diaphragm. However, there are a lot of challenges to make the FBG based pressure transducer possible for a robust pressure measurement. For example, the issue of optic signal conversion to the voltage signal by using photodetector (PD); which shows the significant inconsistencies in output voltage [25]. This has led to huge variation in desired readings. To overcome the problem, a new arrangement of optical components has been proposed. In this configuration, additional FBG was added to the system which acts as reference FBG in order to eliminate the variation of voltage. The proposed pressure transducer, works under constant room temperature and the temperature cross-sensitivity is not being considered.

2. Working principles of FBG

FBG sensors are produced by creating periodic variations in the refractive index of the core of an optical fiber by using a high energy optical source and a phase mask [26]. Figure 1 shows the internal structure of an FBG sensor. Reflection of certain wavelength occurs when the light is traveling at the Bragg wavelength, which is a grating feature and appears missing in the transmission spectrum also shown in the diagram below when the grating structure only allows wavelengths of light [27].

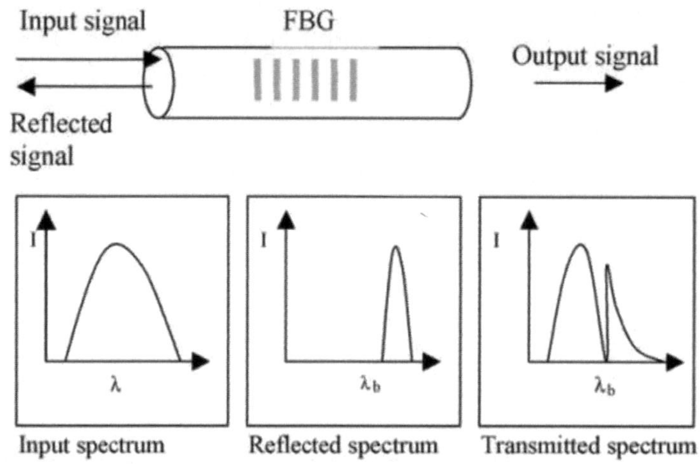

Figure 1. Transmission and reflection spectra from an FBG [28].

Bragg wavelength is a narrowband spectral output or a peak reflected wavelength from the FBG sensor after being illuminated by broadband light source and the light interacts with the grating of an FBG. The detection of local strain from a deformation is done through the variation of grating period and the reflected wavelength via the Bragg equation. According to the Bragg condition, the Bragg wavelength can be expressed in a well-known formula [29]:

$$\lambda_B = n_{eff}\,\Lambda \tag{1}$$

where λ_B is Bragg grating wavelength, Λ is grating periodic spacing, n_{eff} is the effective reflective index of the fiber core. This formula is established by Nobel Laureate Sir William Lawrence Bragg in 1915 about diffraction of X-Ray from crystals and he expressed it into a simple mathematical formula.

The Bragg wavelength is sensitive to physical changes in the grating due to strain and temperature [30]. Thermal expansion in the FBG causes the effective refractive index and the spacing of the gratings to change simultaneously and produced a wavelength shift. This fractional Bragg wavelength shift for change in temperature ΔT can be written as [31]:

$$\frac{\Delta\lambda_B}{\lambda_o} = (\alpha + \xi)\,\Delta T \tag{2}$$

where α is thermal coefficient, ξ is the thermo-optic coefficient. Therefore, the strain effect on an optical fiber under a constant temperature, the Bragg wavelength shift $\Delta\lambda_B$ can be expressed in the form [32]:

$$\frac{\Delta\lambda_B}{\lambda_o} = (1 - p_e)\,\varepsilon \tag{3}$$

where $\Delta\lambda_B$ wavelength difference compared to original Bragg wavelength, λ_o. p_e is the gage factor or the effective photo-elastic constant and ε is the strain applied to the optical fiber. The linear change of the strain or pressure results in variation of FBG central wavelength.

$$p_e = \frac{n^2{}_{eff}}{2}\,[p_{12} - v(p_{11} + p_{12})] \tag{4}$$

where p_{12} are the silica photoelastic tensor components and v is the Poisson ratio. For an FBG central wavelength of 1550nm, typical strain sensitivity is approximately 1.2 pm/microstrain [33].

3. FBG interrogation system

There are two interrogation methods that are available for high-frequency vibration signals with FBG and suitable for pressure measurement; the edge filter detection and power detection methods [34]. In the edge filtering technique, an optical filter, which is also an FBG, is used for filtering. A broadband light source is supplied to FBG sensor through a three port optical circulator. Conversion between the intensity of light and voltage signal by a photodetector (PD) after the reflected light passed through the filter for signal acquisition [35]. The interrogation principle can be briefly explained in Figure 2. Bragg wavelength reflected from the FBG sensor which in tension is longer than original Bragg wavelength and shorter as the sensor is subjected to a compression.

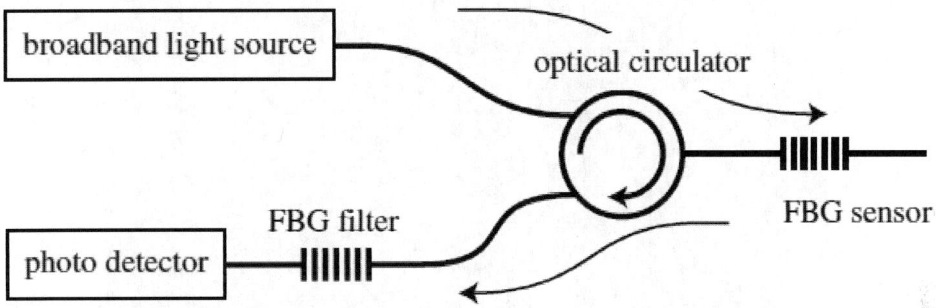

Figure 2. Matched edge filter arrangement for dynamic sensing by Tsuda [35].

Figure 3 show the schematic diagram of the movement of light in the circulator-based system. The matched edge filter interrogation method used in this experiment makes cheap interrogation method for FBG possible without sacrificing the accuracy measurement of the output voltage. The market product solid state FBG interrogator is expensive. Interrogation of FBG has enabling the optical signal to be converted to an electrical signal; therefore, the transducer can be easily connected to a data acquisition unit or an electronic controller, such as Programmable Logic Controller (PLC).

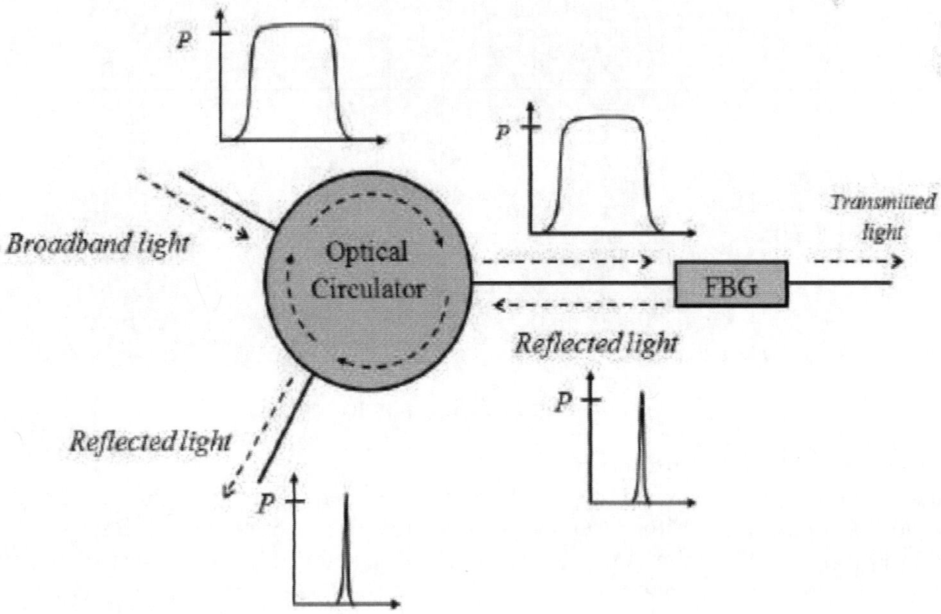

Figure 3. Schematic diagram of the moving light in the circulator-based system by Zohari [36].

IOP Conf. Series: Materials Science and Engineering **257** (2017) 012080 doi:10.1088/1757-899X/257/1/012080

4. Experimentation

An FBG was attached to the center of the diaphragm with 3-second glue as shown in Figure 4. The FBG pressure transducer was connected to the pipe leak detection test rig while an air compressor was used to supply pressure as shown in Figure 5.

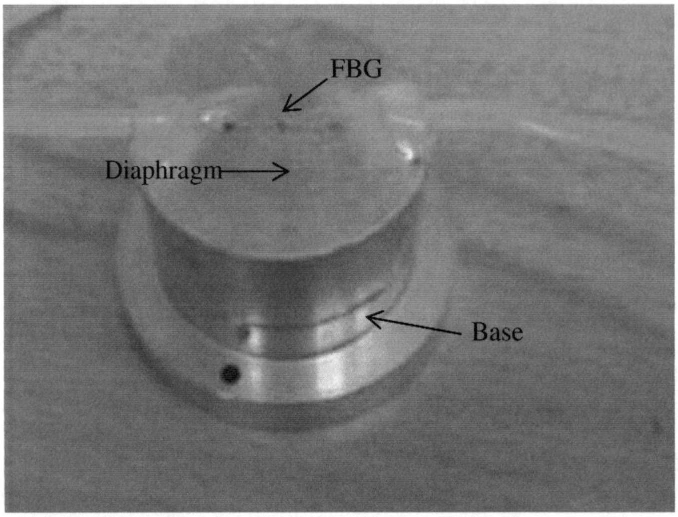

Figure 4. FBG pressure transducer.

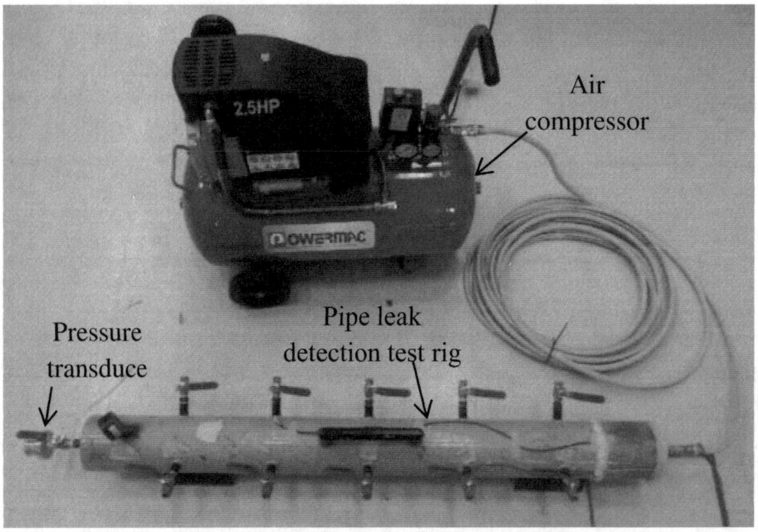

Figure 5. Set up of test rig.

4.1. Bragg wavelength against pressure

The pressure transducer was mounted to the interrogation system as shown in Figure 6; to determine the reflected Bragg wavelength for FBG sensor (set A) and FBG filter (set B). Bayspec (FBGA-F-1525-1565-FP) Optical Spectrum Analyzer (OSA) was used to display the reflected spectrum. The sensitivity of the FBG on pressure sensing was obtained from the relationship between pressure and wavelength shift in FBG sensor. A pressure increment was implemented gradually to get the strain in the FBG sensor through the deformation of the diaphragm.

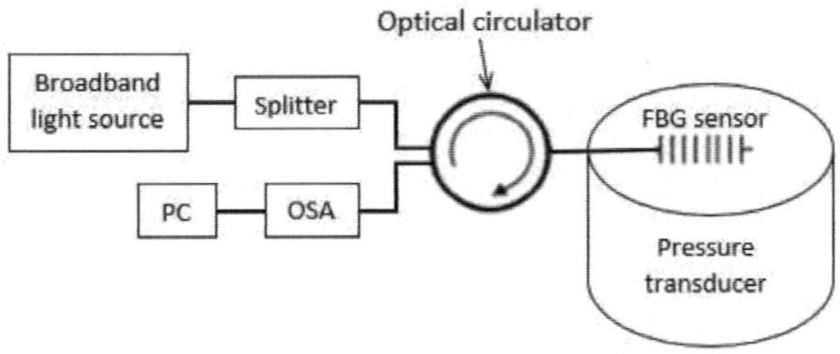

Set A

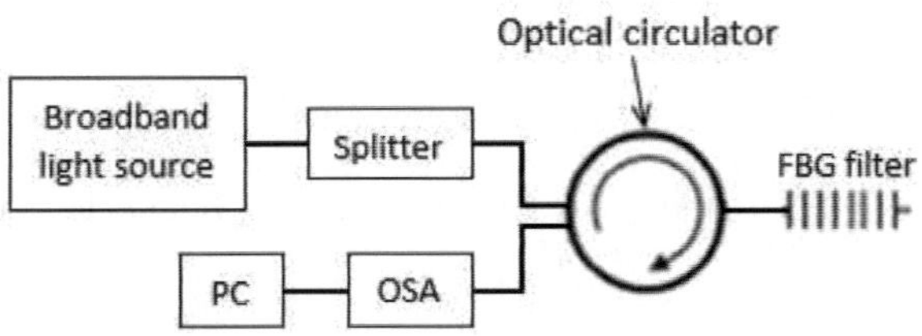

Set B

Figure 6. Configuration of interrogation system of set A (top) and set B (bottom).

Observation through OSA shows that there was significant wavelength shift after increasing pressure scheme was applied to the test rig. Once the FBG sensor detected the deformation of the diaphragm caused by the pressure exerted, the grating period changes to a longer or shorter wavelength, therefore causing a shift in Bragg wavelength of the FBG. It can be observed from Figure 7 that the variation of the Bragg wavelength under the increased pressure is linear with pressure applied; and exhibited no considerable variation in the peak power levels. Figure 8 shows the variation of the wavelength of FBG versus pressure. The variation pattern is linear and the slope is 0.278; and the fitting linear correlation coefficient reached 97.29%.

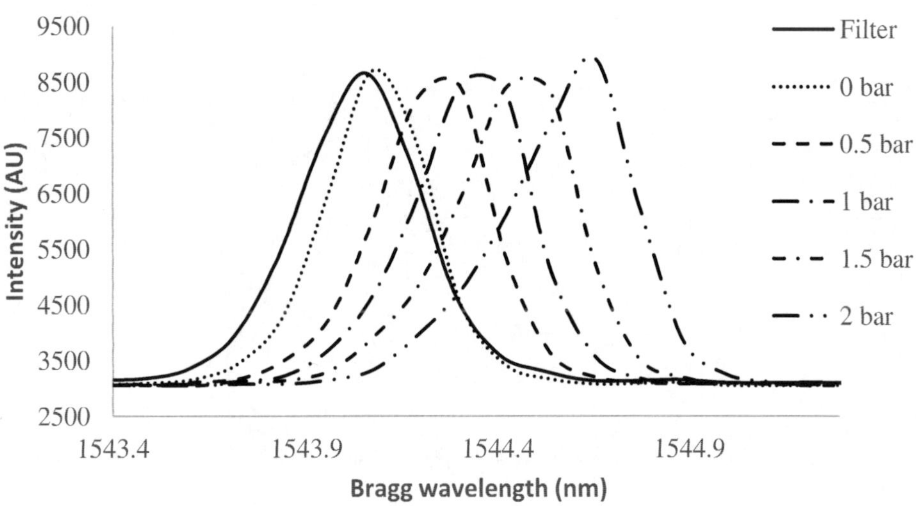

Figure 7. The spectra of Bragg wavelength shift of the sensing FBG at different applied pressure values 0 until 2 bar.

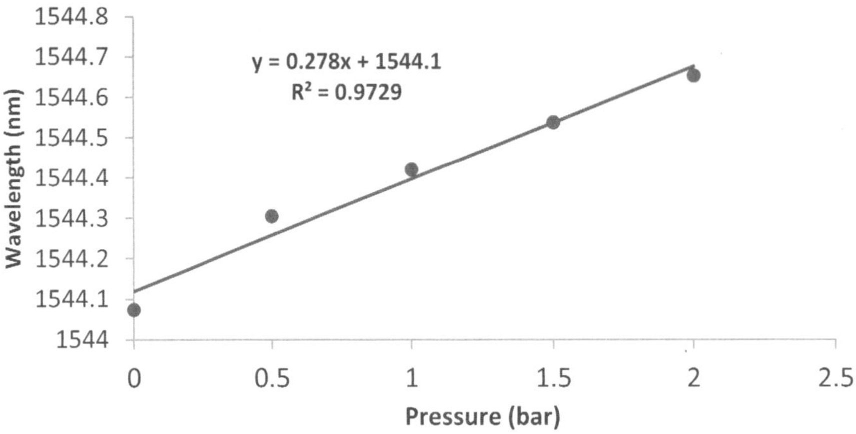

Figure 8. Variation of wavelength of FBG versus pressure.

4.2. Pressure against voltage
For this study, an improved interrogation system was used. Two FBG were used, whereby one of the FBG acts as a sensor and the other as a reference. Figure 9 shows the interrogation system proposed.

Principally, the broadband signal from an (ASE) source is launched into the splitter to emit light through two optical circulators; and to both of the FBGs. The reflected wavelength from FBG sensor will pass through the FBG filter, whereby the matched wavelengths were filtered. Here, the remaining wavelengths were transformed into intensity shift; and converted to voltage readings by photodetector CH1. At the same time, the reflected wavelength from FBG reference will also be converted to the voltage readings by photodetector CH2 as a voltage reference. Note that, FBG reference only detect any intensity variations from the broadband light source.

The normalizing the voltage from FBG sensor with the voltage from FBG reference was done in order to reduce the voltage inconsistency. Voltage reading obtained from different applied pressures setting ranging from 0 to 2 bar within steps of 0.2 bar was recorded. The experiment was repeated for six times to calibrate the pressure transducer before analysis was conducted. All the measurements were monitored utilizing MATLAB real-time graphical user interface (GUI) monitoring system.

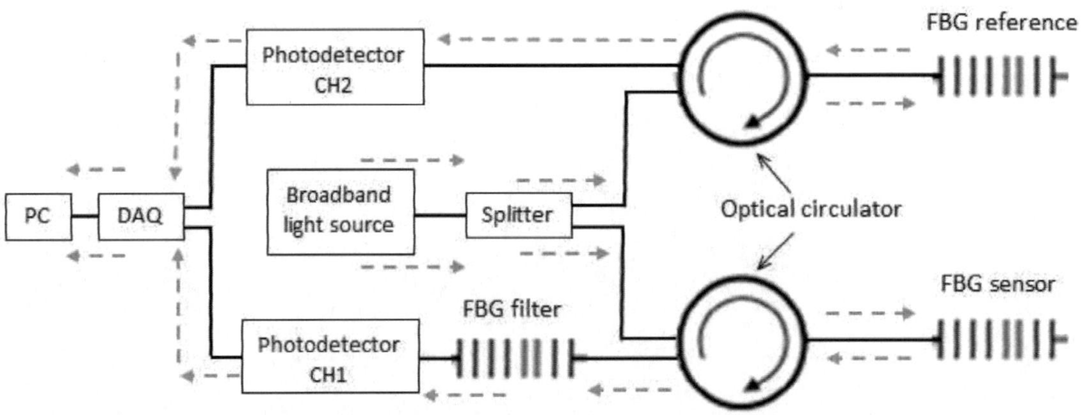

Figure 9. New interrogation system.

The FBG dynamic pressure transducer was connected to the interrogation system as shown in Figure 9 which contains a photodetector to convert the intensity into the voltage. The conversion optical into an electrical signal is aimed in order for the FBG dynamic pressure transducer to be easily connected to an electronic controller or signal processing in software such as DASYLAB and MATLAB via NI (National Instruments) devices. Figure 10 shows the graph of the calibration process. A linear relation was calculated from the averaging of the calibration data and was used in GUI for converting the voltage to the pressure. The linear equation of the calibration process as shown in Figure 11.

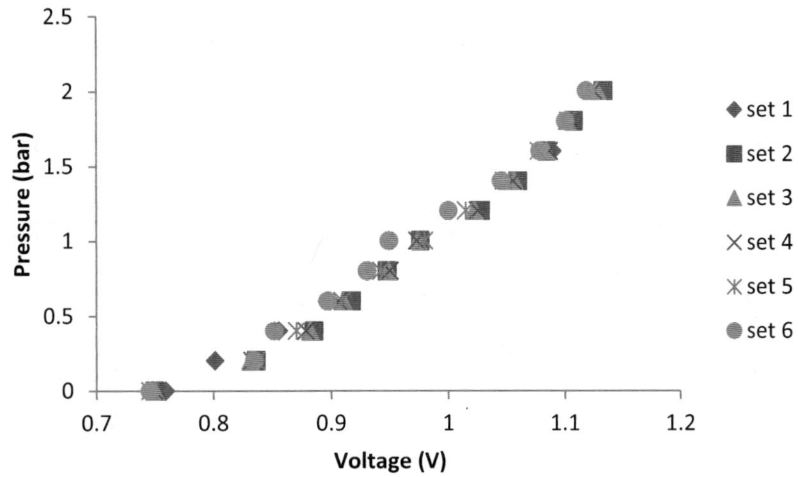

Figure 10. Calibration process.

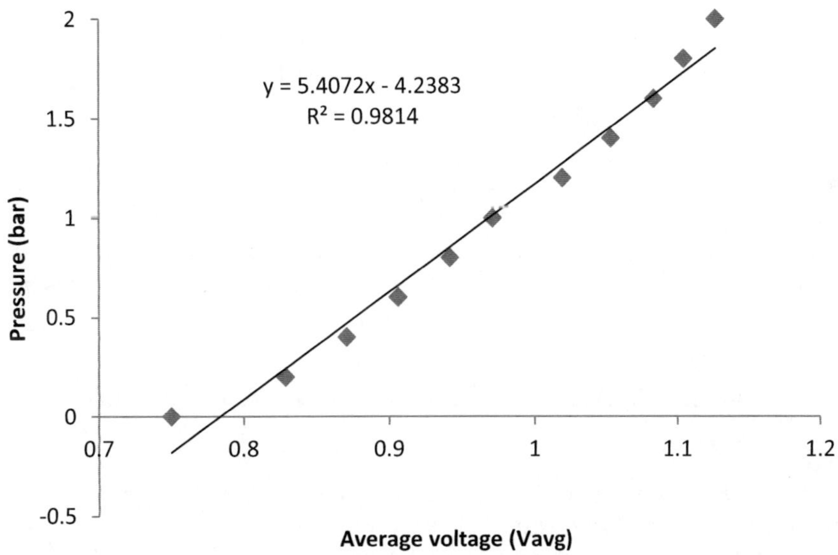

Figure 11. Relationship between pressure and voltage.

4.3. Data validation

A new set of experiment is carried out to find the error of pressure between pressure gauge readings and pressure from FBG pressure transducer readings. 5 samples were recorded as shown in Table 1 below and the errors between two output readings were calculated. Figure 12 shows the data validation between pressure gauge readings and FBG pressure transducer readings. An average error for 5 pressure measurement is found to be 5.348%. From the result, we can conclude that the FBG dynamic pressure transducer has good accuracy and high reliability in measuring pressure.

Table 1. Summarized data of comparison between pressure gauge and FBG pressure transducer.

Sample	Pressure gauge readings (bar)	FBG pressure transducer reading (bar)	Error %
1	0.4	0.350589	12.350
2	0.8	0.764673	4.420
3	1.2	1.14443	4.630
4	1.6	1.5626	2.340
5	2	1.94005	3.000
		Average error	**5.348**

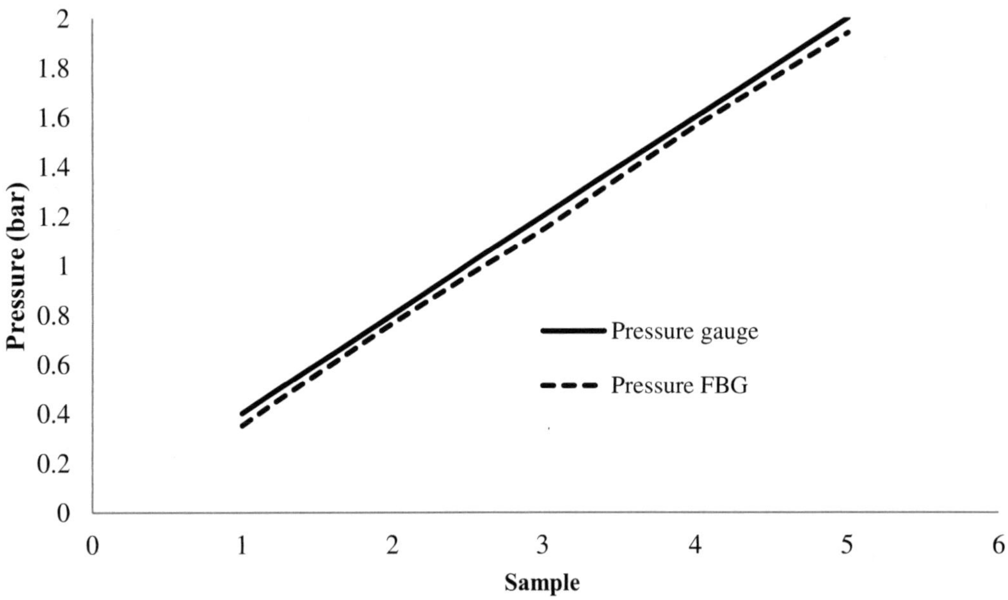

Figure 12. Data validation between pressure gauge and FBG.

5. Conclusions

In this study, new interrogation system was successfully applied to reduce inconsistency voltage. The FBG sensor has a good linearity with a linear correlation coefficient of 97.29% in pressure measurement. From the pressure measurement test between pressure gauge readings and FBG pressure transducer, an average error is found to be 5.348%. Note that, temperature compensation of the pressure transducer has not been taken into consideration as the operating temperature was fixed at room temperature.

Acknowledgements

The authors thank the Photonics Research Centre at the University of Malaya, Malaysia for providing some research equipment; and Faculty of Mechanical Engineering and UMP Research & Innovation Department of University Malaysia Pahang, Malaysia for financial support to carry out this present work, under RDU170388.

References

[1] Liang M F, Fang X Q, Wu G, Xue G Z and Li H W 2017 *International Journal for Light and Electron Optics* **145** 503-12
[2] Gu Y F, Zhao Y, Lv R Q and Yang Y 2016 *IEEE Photonics Technology Letters* **28** 2569-72
[3] Zhao Y, Hu H F, Bi D J and Yang Y 2016 *Optics and Lasers in Engineering* **82** 122-6
[4] Dai Y, Li P, Liu Y, Asundi A and Leng J 2014 *Optics and Lasers in Engineering* **59** 19-24
[5] Huang J, Zhou Z, Wen X and Zhang D 2013 Measurement **46** 1041-6
[6] Liu L, Chen W, Zhang P, Wu J and Liu H 2012 *Opto-Electronic Engineering* **39** 57-61
[7] Shu Y, Chen W, Zhang P, Wu J, Liu H and Zhang W 2013 *12th International Conference of the Slovenian Society for Non-Destructive Testing* (Portorož, Slovenia: Application of Contemporary Non-Destructive Testing in Engineering)
[8] Mamidi V R, Kamineni S, Ravinuthala L S P, Thumu V and Pachava V R 2014 *Fiber Integr. Opt.* **33** 325-35
[9] Liu L, Zhang H, Zhao Q, Liu Y and Li F 2007 *Optical Fiber Technology* **13** 78-80
[10] Wang Y, Wang M and Huang X 2010 *Measurement Science and Technology* **21** 065304
[11] Pachava V R, Kamineni S, Madhuvarasu S S and Putha K 2014 *Journal of Optics* **43** 117-21
[12] Urban F, Kadlec J, Vlach R and Kuchta R 2010 *Sensors* **10** 11212-25
[13] Zhang W, Li F and Liu Y 2009 *Optical Engineering* **48** 024402-4
[14] Huang J, Zhou Z, Wen X and Zhang D 2013 *Measurement* **46** 1041-6
[15] Hsu Y, Wang L, Liu W F and Chiang Y 2006 *IEEE Photonics Technology Letters* **18** 874-6
[16] Wang H L, Song J, Feng D Q and Wu H C 2011 *Optics and Precision Engineering* **3** 010
[17] Zhao Y, Yu C and Liao Y 2004 *Optics & Laser Technology* **36** 39-42
[18] Sengupta D, Sai Shankar M, Saidi Reddy P, Sai Prasad R and Srimannarayana K 2012 *Microwave and Optical Technology Letters* **54** 1679-83
[19] Lin Y and Wang W 2009 *Symp. Photonics and Optoelectronics* (IEEE) p 1-3
[20] Lan C, Zhou Z, He J and Ou J 2008 *19th International Conference on Optical Fibre Sensors* (International Society for Optics and Photonics) p 700463--4
[21] Wang X X and Wu C Y 2012 *Proc. Advanced Materials Research* (Trans Tech Publ) p 160-3
[22] López-Higuera J M, Cobo L R, Incera A Q and Cobo A 2011 *Journal of Lightwave Technology* **29** 587-608
[23] Pinet É 2011 *21st International Conference on Optical Fibre Sensors* (International Society for Optics and Photonics) p 775304-4
[24] Ahmad H, Harun S, Chong W, Zulkifli M, Thant M, Yusof Z and Poopalan P 2008 *Microwave and Optical Technology Letters* **50** 60-1
[25] Allwood G, Wild G, Lubansky A and Hinckley S 2015 *Optical Fiber Technology* **25** 25-32
[26] Meltz G, Morey W W and Glenn W 1989 *Optics Letters* **14** 823-5
[27] Mihailov S J 2012 *Sensors* **12** 1898-918
[28] Majumder M, Gangopadhyay T K, Chakraborty A K, Dasgupta K and Bhattacharya D K 2008 *Sensors and Actuators A: Physical* **147** 150-64
[29] Rao Y J 1997 *Measurement Science and Technology* **8** 355
[30] Lawson N J, Correia R, James S W, Partridge M, Staines S E, Gautrey J E, Garry K, Holt J C and Tatam R P 2016 *Measurement Science and Technology* **27** 104001
[31] Wang T, Yuan Z, Gong Y, Wu Y, Rao Y, Wei L, Guo P, Wang J and Wan F 2013 *Photonic Sensors* **3** 267
[32] Wang Y, Wang M and Huang X 2010 *Measurement Science and Technology* **21** 065304
[33] Rao Y J 1998 *Fibre Bragg Grating Sensors: Principles and Applications* vol 2 (London: Chapman and Hall)
[34] Wild G and Hinckley S 2010 *Proc. 20th International Congress on Acoustics* p 23-7
[35] Tsuda H 2006 *Composites Science and Technology* **66** 676-83
[36] Zohari M H B 2014 *The applications of near infra-red Fibre Bragg grating sensors for wave propagation based structural health monitoring of thin laminated composite plates* (Thesis) University of Southern Queensland

Geometrical parameters influence on the stiffness of steel sandwich plates with web-core

X.X. Jiang[1,2,3*], H. Ji1, M.R.M. Rejab[2], S. Zhang1, M. Ishak[2], L. Zhu[3]

[1]School of Mechanical Engineering, Ningxia University, 750021 Yinchuan, China
[2]Faculty of Mechanical Engineering, Universiti Malaysia Pahang, 26600 Pekan, Pahang, Malaysia
[3]State Key Laboratory of Advanced Processing and Recycling of Non-ferrous Metals, Lanzhou University of Technology, 730050, Lanzhou, China

*Corresponding e-mail: jiangxx@nxu.edu.cn

Abstract. Laser welded sandwich plates with web-core have found their position in the marine and land vehicles. To implement such design accurately, knowledge about the influence of geometrical parameters on the stiffness is necessary. In this paper, the over-hanging three points bending tests were performed on the laser welded web-core steel sandwich plate under quasi-static conditions, together with the finite element (FE) simulations. The following parameters were analysed: the thickness of the face plate, the spacing of the two core plates and the height of the core plates. The agreement between experimental measurements and FE results was considered to be good. It is shown that changes of these parameters can contribute to increase or decrease of the stiffness of web-core sandwich plate, but the height of the core plate has no effect on the shear stiffness as the spacing of the two core plates is known. There are linear, exponential and polynomial fitted relationships between the geometrical and the stiffness of the web-core sandwich plates.

1. Introduction

The demand for faster and lighter marine and land transports has increased the need for more efficient structures. All-metal sandwich plates offer an option that can fulfill these requirements. By using sandwich structures, it is possible to obtain high stiffness and strength to weight ratio, i.e., the sandwich structures were found to be 30-50% lighter than traditional stiffened plates in marine structures [1, 2]. Contemporary interest in steel sandwich plates has been awakened by the developments in laser welding technology which enables the efficient production of these plates. Elevated pre-fabrication accuracy of the components, high welding speed and the possibility to connect internal stiffeners with the face sheets from outside has led to a wide application of laser welding in the construction of steel sandwich plates [3, 4].

The web-core sandwich panel has perhaps the simplest core topology of all steel sandwich plates, as showed in figure 1. The laser-welded web-core steel sandwich panel is a set of steel plates assembled by utilizing a state-welded T-joint. The joint connects a vertical core plate called the web plate to the horizontal face plates that are located far from the mid-plane of the cross-section of the panel. The so-called sandwich effect is achieved by the coupled response of the connected web and face plates[5]. The design and production aspects of web-core panels in the shipbuilding industry have been explored by many authors [6-8] and in civil engineering [9-11]. The breadthwise penetration of the laser beam determines the thickness of the weld, which is lower than half of the thickness of the web plate. Consequently, the joint has two crack-like notches on each side of the weld. This makes the joint considerably less stiff than an equivalent fillet-welded T-joint. So numerous researchers are interesting in the stiffness of web-core sandwich plate. Ref. [12-15] have given us a good suggestion that we should consider the influence of weld width and notches of the T-joint on the stiffness laser welded web-core sandwich plate in the process of design and optimization. In addition to the effects of T-joint, the structural parameters are the primary factors affecting the stiffness of the web-core sandwich panel.

This paper presents a typical three-point bending testing on web-core steel sandwich plate to analyze its stiffness and to validate with finite element model. The FEM was performed to analyze the effect of the geometrical parameters (the thickness of the face plate t_f, the height of the core plate and the spacing of the two core plates) on the stiffness of web-core sandwich plate. Total of 17 finite element models were established by the simulation software Abaqus 6.13, and the bending and shear stiffness were calculated respectively according to the over-hanging three-points bending testing method.

2. Experimental and finite element model

2.1 Web-core plate geometrical characteristics and dimensions

The sandwich plate with the web-core as shown in figure 1, consists of two flat plates – faces of thickness t_f and the core plate thickness t_w. The dimensions of the plate are as shown in figure 1. L is the length of the sandwich plate, B is the width of the sandwich plate, H is the height of the sandwich plate, h_c is the height of the core-plate, and s is the distance between two adjacent core plates. The base metal used in this paper is high strength steel SPFH780. Its yield strength is 631MPa. The face and core plates are connected by laser welding, as shown in figure 2. The welding power is 12 kW, and the speed is 1000 mm/min. The shielding gas is carbon dioxide.

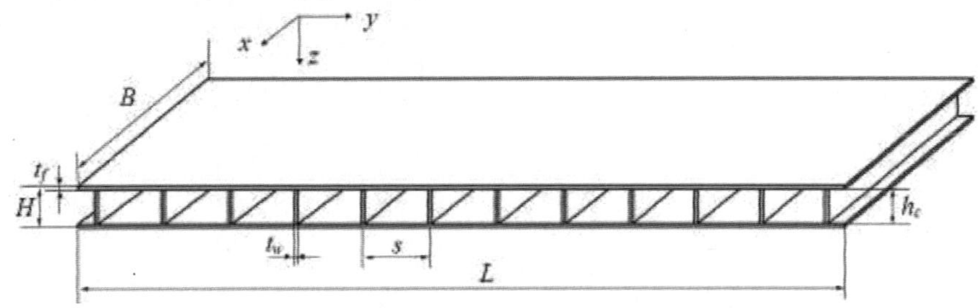

Figure 1. Geometrical characteristics of web-core sandwich plate

Figure 2. The laser welding is manufacturing the web-core sandwich plate

Table 1 is the dimensions of sandwich plates with different geometrical parameters. For all the web-core sandwich plates, they have the same length and width, 1200 mm x 500 mm. The other geometrical parameters will be changed. Among, the variable is the thickness of the face-plates for the 1st plate to the 4th plate, and the variable is the height of the core-plate h_c and the distance between two adjacent core-plates s for the sandwich plate from the 4th to the 16th.

Table 1. Dimensions of sandwich plates with different geometrical parameters

Parameters	L/mm	B/mm	H/mm	h_c/mm	s/mm	t_f/mm	t_w/mm
$Test_{1,2,3}$	1200	500	60	50	100	2.5	5
1st	1200	500	60	50	100	1.5	5
2nd	1200	500	60	50	100	2.5	5
3rd	1200	500	60	50	100	3	5
4th	1200	500	60	50	100	4	5
5th	1200	500	60	50	100	5	5
5th	1200	500	40	35	55	2.5	5
6th	1200	500	40	35	70	2.5	5
7th	1200	500	40	35	85	2.5	5
8th	1200	500	40	35	100	2.5	5
9th	1200	500	50	40	55	2.5	5
10th	1200	500	50	40	70	2.5	5
11th	1200	500	50	40	85	2.5	5
12th	1200	500	50	40	100	2.5	5
13th	1200	500	60	50	55	2.5	5
14th	1200	500	60	50	70	2.5	5
15th	1200	500	60	50	85	2.5	5
17th	1200	500	60	50	100	2.5	5

The thickness of the face-plate is 2.5 mm for the sandwich plate from the 4th to the 16th. As a result, there are twelve web-core sandwich plates to research more on their stiffness. Besides 3 testing specimens, which are marked as $Test_1$, $Test_2$ and $Test_3$, the others are FE models. Among them, there are 5 web-core sandwich plates will be used to study the relationship between the face plates's thickness and the stiffness of the sandwich plates. There are twelve web-core sandwich plates will be used to study the relationship between the geometrical parameters of core plates and the stiffness of the new plate. Three sets of data for the height of the core plates h_c, and the one height of the core plates h_c corresponds 4 sets of data of the spacing of the two core plates s.

2.2 Testing method and equipment

2.2.1 Over-hanging three points bending testing. Generally, to measure the bending stiffness and shear stiffness of one web-core sandwich plate, it should be conducted in two types of testing which are three points bending testing and four points bending testing for the same web-core sandwich plate. This method is not only increasing the number of testing, but also increases the error of the measurement data. It is noteworthy that a method called over-hanging three points bending testing, as showed in figure 3, can be used to determine the bending stiffness together with shear stiffness.

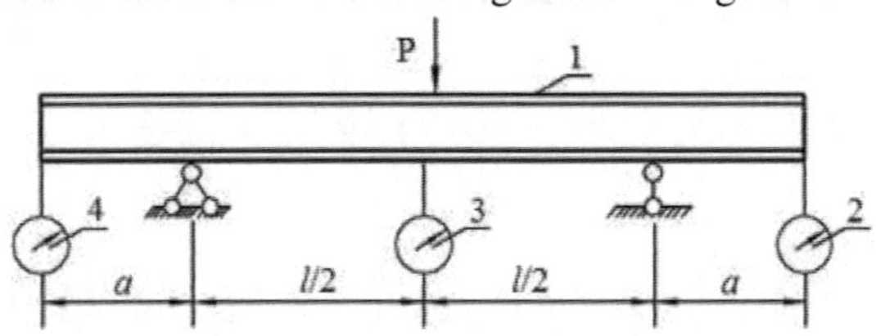

*where 1 is a testing specimen; 2,4 is the left and right over-hanging displacement sensor, respectively; 3 is the middle displacement sensor; P is loading sensor; l is span, in m; a is the distance between over-hanging displacement sensor and the middle displacement sensor.

Figure 3. Over-hanging three points bending testing

In the elastic segment, according the loading which is measured by the loading sensor, the left deflection which is measured by the left over-hanging deformation sensor, the middle deflection which is measured by the left middle deformation sensor and the right deflection which is measured by the right over-hanging deformation sensor, the bending stiffness and the shear stiffness are determined. The formula as follows:

$$D_y = \frac{\Delta P \cdot l^3}{48 \cdot B \cdot \omega_b} \tag{1}$$

$$D_{Qy} = \frac{\Delta P \cdot l}{4 \cdot B \cdot (\omega - \omega_b)} \tag{2}$$

*where D_y is bending stiffness, in N.m; D_{Qy} is shear stiffness, N/m; ΔP is load increment, in N; l is span, in m; B is the width of specimen, in m; ω is the increment of mid-span deflection, which is total deflection, in m; ω_b is the increment of left or right over-hanging deflection, which is the bending deflection, and $\omega_b = (\omega_l + \omega_r)/2$, in m; ω_s is the shear deflection, and $\omega_s = \omega - \omega_b$, in m.

2.2.2 Equipment. According to ASTM C393-00 Standard test method for flexural properties of sandwich structures and GB/T 1456-2005 Performance test methods for sandwich structures, a special test setup which is suitable for simply support, as be shown in figure 3, was developed to measure the bending and shear stiffness of web-core sandwich plates . The experiments were carried on a 3150 kN four-column hydraulic machine, as be shown in figure 4. Here, we can see a sandwich plate is during three points bending testing. The span is 936 mm, and the loading speed is 5 mm/min.

Figure 4. Four - column hydraulic press machine and a sandwich plate during testing

2.3 Finite element modeling

The Abaqus 6.13 standard solver was employed to analyze the stiffness of sandwich plates with various structural dimensions. Figure 5 shows one of the 2D models. In the modeling of sandwich plate, the welds are all at the middle of each core plate, and the root gap is zero. In this model, the elements used are CPE8R. In general the aspect ratio in rectangular elements is kept close to 1:2 and the angle in quadrangular between 45°-135° to avoid problems in accuracy of solution. After the model has been verified by the experimental results, some modification of variables of the thickness of the panel and other geometric parameters to study the stiffness of the web-core sandwich plate. The geometric parameters are studied in this paper has been shown in table 1.

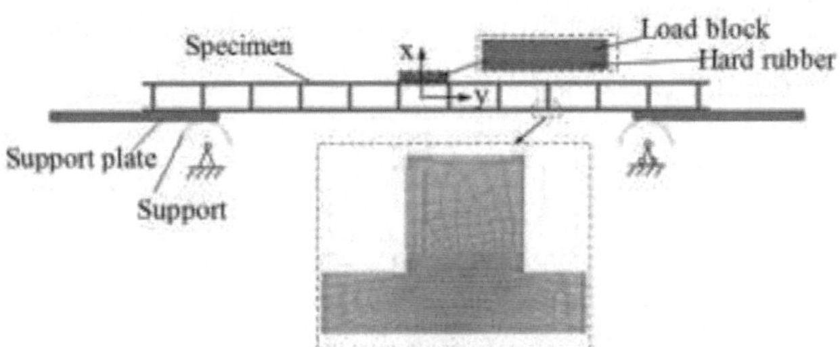

Figure 5. 2D FEM model

3. Results and discussion

3.1 The results of experimental and FEM

In order to ensure that the deformation of the sandwich plate is in the elastic range, the load is lower than 20 kN. Each specimen was tested for three times, and figure 6 shows the typical load-deflection curve of the sandwich plate. It is found that the load-left deflection and load-right deflection curve are almost completely coincident for each plot, which indicates: (1) the testing plate has excellent welding quality. If one welding has defects, there is no doubt that the two curves will be deviated, which can be got a strong support from my previous research. (2) the experimental installation is accurate, especially for the displacement and load sensor installation location. If the location of the two over-hanging displacement is asymmetric about the displacement sensor in the middle, where assume that the position of the middle sensor is accurate, then the result is that the curve does not coincide. (3) the experimental result has a good repeatability and reliability. So according to the equations (1) and (2), it is calculated that the bending stiffness is $2.92\times10^5\,\mathrm{N\cdot m}$, and the shear stiffness is $8.12\times10^5\,\mathrm{N/m}$. Certainly, the stiffness is an average value which is calculated by 9 sets data from 3 specimens.

Figure 6(b) shows the load-deflection curve of the experimental method and the FE method. Obviously, the simulation results agree well with the experimental values. Because the expensive manufacture costing, the experimental testing and analysis costing, finite element analysis method is really at a best choice to analyze the sandwich plates with different geometric parameters. This is not simply to solve the costing problems, but to modify the model is very convenient. After the model has been validated and improved, it is a perfect method to calculate the stiffness of the sandwich plate which we are stimulating, as the following contents.

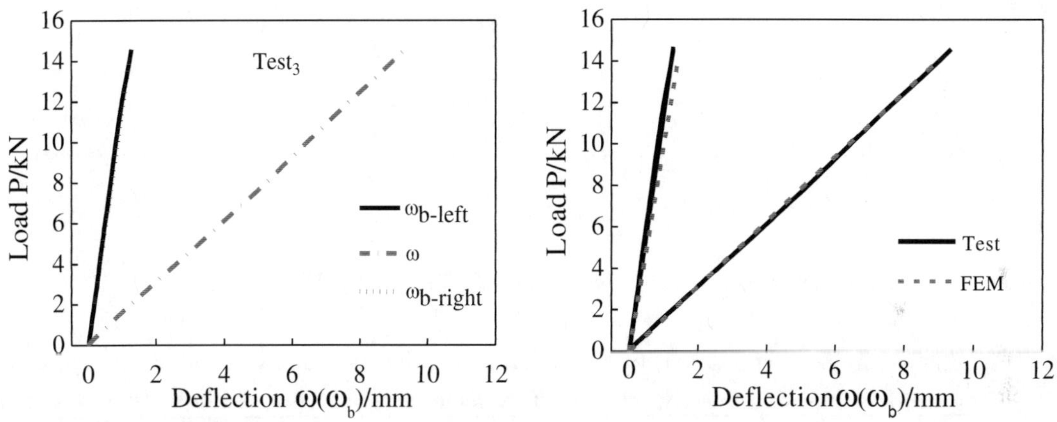

(a) Load-deflection curve of testing (b) Comparation of Testing and FEM

Figure 6. Load-deflection curves for different face plate thickness

3.2 The influence of the thickness of face plates on stiffness

Figure 7 shows the curves of load-deflection from the finite simulation results. As the thickness of the face plate increases, we can make a conclusion that the slope of the curve is changed, and the stiffness value will be changed too. The bending stiffness be calculated according to the formula (1), and the shear stiffness be calculated according to the formula (2). The stiffness be shown in figure 8, and there are 5 discrete data points in each figure. As the thickness of the faceplate increasing, the bending stiffness and shear stiffness are increased, though the form of increasing is different.

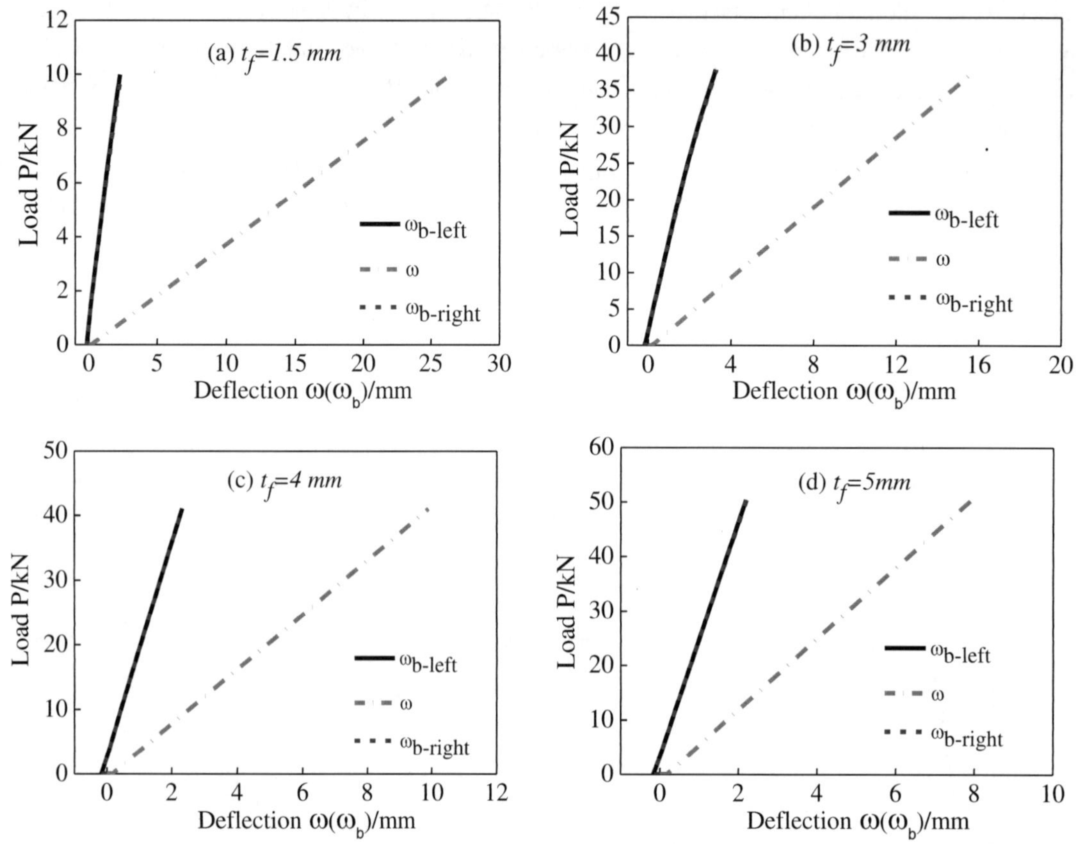

Figure 7. Load-deflection curves for different face plate thickness

Figure 8 is also the relationship between the face-plate thickness and stiffness. From figure 8(a), the bending stiffness increased linearly, when the face-plate increasing from 1mm to 5mm. Using the Origin software tool, the best fitting equation is:

$$D_{\mathrm{y}} = 0.17 \cdot t_f - 0.1 \tag{3}$$

From figure 8(b), the shear stiffness increased non-linearly, and the fitting equation is:

$$D_{\mathrm{Qy}} = 1.12 \cdot \mathrm{e}^{0.34 \cdot t_f} - 1.67 \tag{4}$$

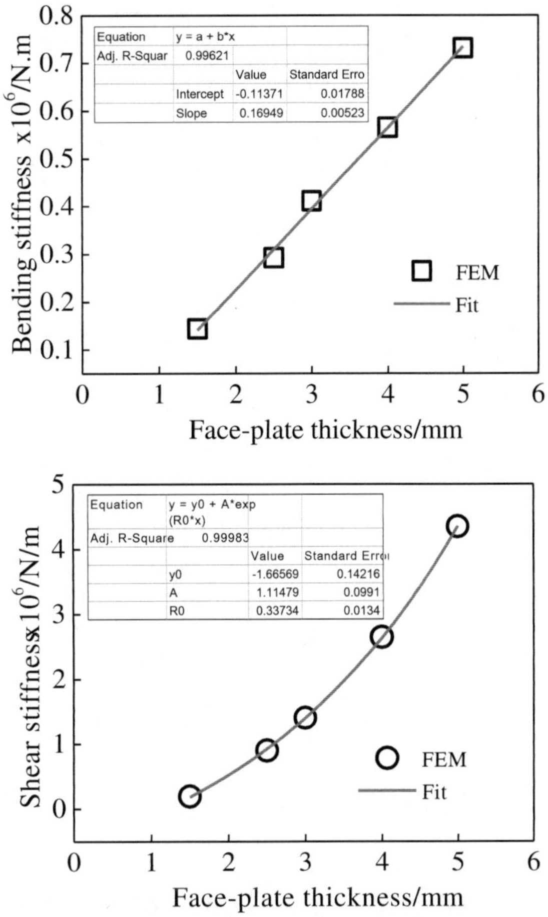

(a) Face-plate thickness and bending stiffness (b) Face-plate thickness and bending stiffness

Figure 8. The relationship between face-plate thickness and stiffness

3.3 The influence of core plates's geometrical parameters on stiffness

According to the data of table1, there are twelve finite element models were conducted to study the stiffness of web-core sandwich plates which have different core plate's geometrical parameters. As we know that the height of core plate h_c is 35 mm, 45 mm and 55 mm, respectively. Also the spacing of the two core plates s is 55 mm, 70 mm, 85 mm, and 100 mm, respectively. Figure 8 shows the load-deflection curves for web-core sandwich plate (h_c=35mm) with different s. Because the other figures are similar with these figures, not all are shown in this paper. Certainly, the bending stiffness for each web-core sandwich plate is calculated by equation (1) and the shear stiffness is calculated by equation (2). The values are shown in figure 10.

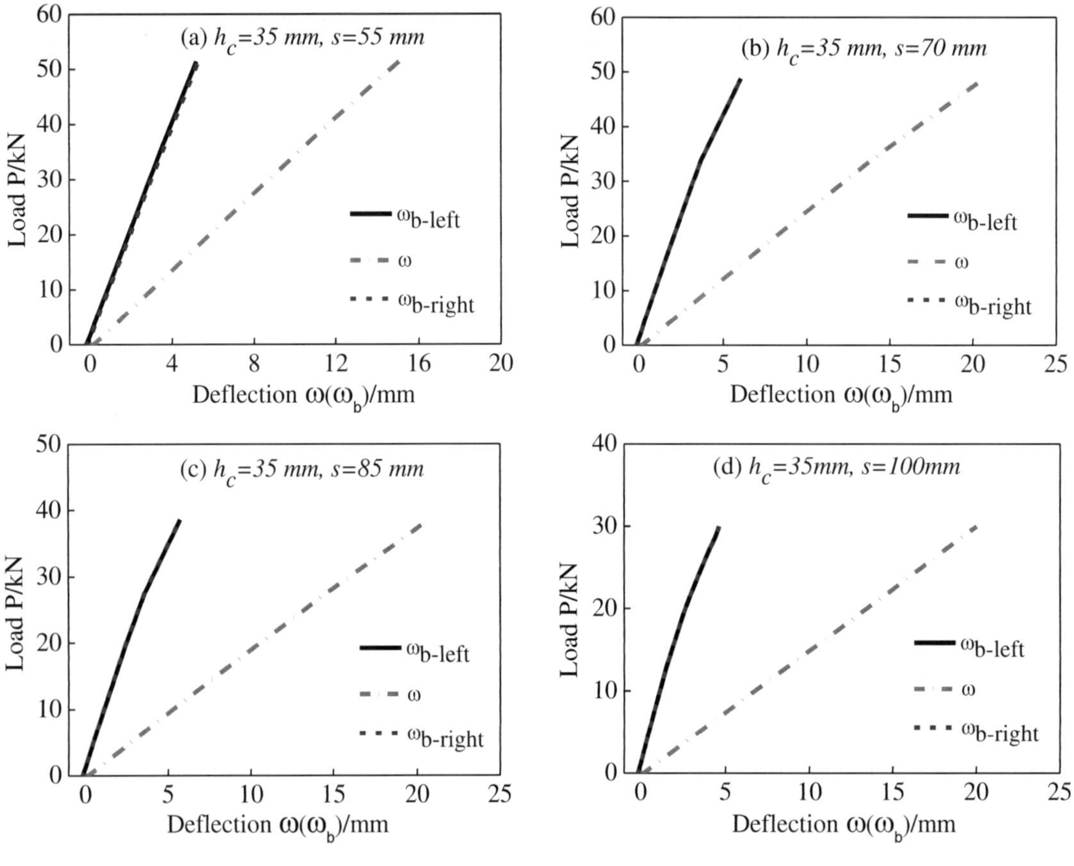

Figure 9. Load-deflection curves for web-core sandwich plates with different core plate geometrical parameters

Figure 10(a) shows the bending stiffness of web-core sandwich with different height of the core plate *hc* and the spacing of the two core plates *s*. Analysis of the results as follows:

(1) For the web-core sandwich plate with the same core plate's geometrical parameter *s*, the higher of the core plates, the greater the bending stiffness, which except the bending stiffness of web-core sandwich (*hc*=55 mm, *s*=100 mm). Obviously, its stiffness is lower than the bending stiffness of web-core sandwich (*hc*=45 mm, *s*=100 mm).

(2) For the web-core sandwich plate with the same core plate's geometrical parameter *hc*, the larger spacing of the two core plates *s*, the smaller of the bending stiffness, and the reduction is proportional. The higher of the core plate, the more the bending stiffness decreased as the height of the core plate increasing. This phenomenon is due to the fact that the higher of the core plate, the greater the spacing of the two core plates, the local deformation of the face plate could reduce the total stiffness of the web-core sandwich plate. The future research works will provide a more comprehensive explanation.

Figure 10(b) shows the shear stiffness of web-core sandwich with different height of the core plate *hc* and the spacing of the two core plates *s*.

(1) For the web-core sandwich plate with the same core plate's geometrical parameter *s*, all of the web-core sandwich plates almost have the same shear stiffness. In other words, the

height of the core plates hasn't any influence on the total shear stiffness of the web-core sandwich plates.

(2) For the web-core sandwich plate with the same core plate's geometrical parameter *hc*, the larger spacing of two core plates *s*, the smaller of the shear stiffness. The reduction is not proportional. The larger the spacing of the two core plates, the smaller the shear stiffness decreased. The future research works will also give a perfect answer.

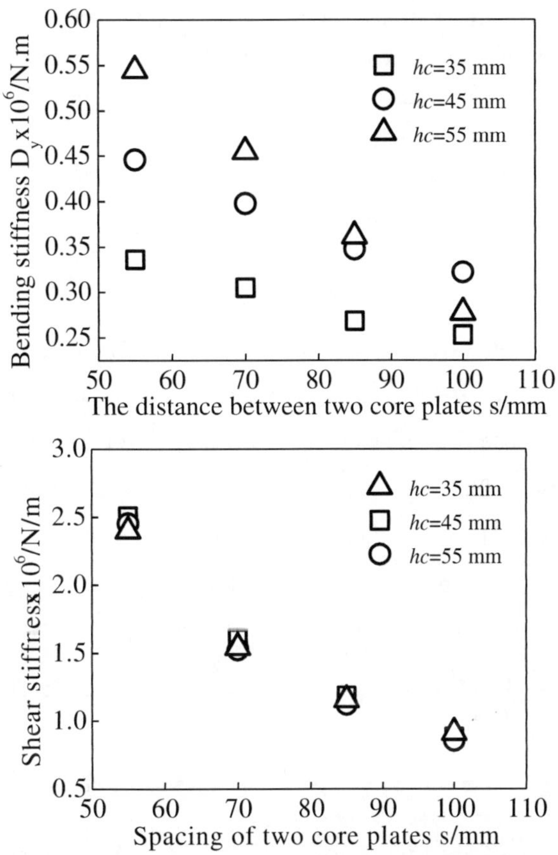

(a) Core plate's parameters - bending stiffness (b) Core plate's parameters - shear stiffness

Figure 10. (a) Bending and (b) Shear stiffnesses of web-core sandwich with different height of the core plate *hc* and the spacing of the two core plates *s*

4. Conclusions

Based on the experimental investigation, and validation with the finite element models to numerically analyze the stiffness of the web-core sandwich plate with various geometrical parameters, such as thickness of the face plates t_f, the height the core plates h_c and the spacing of the two core plates *s*. The following conclusions can be drawn:

(1) Over-hanging three points bending testing were conducted to measure the bending and shear stiffness of the web-core sandwich plate(t_f=2.5 mm, h_c=55 mm, s=100 mm). As the result of the bending stiffness is $2.92\times10^5\,N\cdot m$, and the shear stiffness is $8.12\times10^5\,N/m$. The FEM results show great agreement with these testing values.

(2) As the thickness of the faceplate increasing, the bending stiffness and shear stiffness are increased, though the form of increasing is different. The fitting eqautions (3) and (4) expresses the influence of the face-plate thickness on the bending and shear stiffness, respectively.

(3) The geometrical parameters core plates a significant effect on the stiffness of the web-core sandwich plates. If the spacing of two core plates s is not changing, the higher of the core plates, the greater the bending stiffness, which except the bending stiffness of web-core sandwich (h_c=55 mm, s=100 mm). If the height of the core plates h_c is not changing, the larger of the spacing of the two core plates s, the smaller of the bending stiffness, and the reduction is proportional. The higher of the core plate, the more the bending stiffness decreased as the height of the core plate increasing.

(4) If the spacing of two core plates s is not changing, the height of the core plates h_c has no any influence on the total shear stiffness of the web-core sandwich plates. If the sandwich plate has the same height of the core plates h_c, the larger spacing of two core plates s, the smaller of the shear stiffness, and the reduction is not proportional.

Acknowledgement

The authors are grateful to the Ningxia Hui Autonomous Region (Nature Science Foundation: No. NZ16049). Ningxia University (www.nxu.edu.cn) and Ningxia High School Scientific Research Project(Excellent Young Teacher Training Fund Project: No. NGY2017016) for funding this project. The authors are also thankful to the Ministry of Education, Malaysia (No. FRGS/1/2014/TK01/UMP/02/3) and Universiti Malaysia Pahang (www.ump.edu.my) through research grant UIC171505 for supporting this international research collaboration between Malaysia and China.

References

[1] Kujala, P. and Klanac, A. 2005. Steel sandwich panels in marine applications. *Brodogradnja*, **56**(4), 305-314.

[2] Allen, H. G. 2013. Analysis and design of structural sandwich panels: the Commonwealth and international library: Structures and Solid Body Mechanics Division. Elsevier.

[3] Kozak, J. 2009. Selected problems on application of steel sandwich panels to marine structures. *Polish Maritime Research*, 9-15.

[4] Romanoff, J. and Varsta, P. 2007. Bending response of web-core sandwich plates. *Composite Structures*, **81**(2), 292-302.

[5] Frank, D., Romanoff, J. and Remes, H. 2013. Fatigue strength assessment of laser stake-welded web-core steel sandwich panels. *Fatigue & Fracture of Engineering Materials & Structures*, **36**(8), 724-737.

[6] Klanac, A., and Kujala, P. 2004. Optimal design of steel sandwich panel applications in ships. In The Proceedings of The Ninth International Symposium on Practical Design of Ships and Other Floating Structures, Lübeck-Travemunde (pp. 907-914).

[7] Metschkow, B. 2006. Sandwich panels in shipbuilding. *Polish Maritime Research*, 5-8.

[8] Romanoff, J., Naar, H., and Varsta, P. 2011. Interaction between web-core sandwich deck and hull

girder of passenger ship. *Mar Syst Ocean Technol*, **6**(1), 39-45.

[9] Pyszko, R. 2006. Strength assessment of a version of joint of sandwich panels, *Polish Maritime Research*, Special Issue 2006/S1, pp. 17-20.

[10] Caccese, V., and Yorulmaz, S. 2009. Laser Welded Steel Sandwich Panel Bridge Deck Development: Finite Element Analysis and Stake Weld Strength Tests. *Finite Element Method.*

[11] Briscoe, C. R., Mantell, S. C., Davidson, J. H., and Okazaki, T. 2011. Design procedure for web core sandwich panels for residential roofs. Journal of Sandwich Structures & Materials, **13**(1), 23-58.

[12] Jiang, X. X., Zhu, L., Qiao, J. S., Wu, Y. X., Li, Z. G., and Chen, J. H. 2014. Bending properties of laser welded web-core steel sandwich plates.. *Advanced Materials Research*, **936**, 1451-1455.

[13] Jiang, X. X., Fei, S. C., Zhang, S., Ji, H. and Zhu, L. (2017). Failure Analysis of the Laser-Welded Web-Core Steel Sandwich Panel with Narrow Weld Width T-Joints. *Applied Mechanics & Materials,* **863**.

[14] Romanoff, J., Remes, H., Socha, G., Jutila, M and Varsta, P. 2007. The stiffness of laser stake welded T-joints in web-core sandwich structures. *Thin-Walled Structures*, **45**(4), 453-462.

[15] Romanoff, J. 2014. Optimization of web-core steel sandwich decks at concept design stage using envelope surface for stress assessment. *Engineering Structures*, **66**, 1-9.

IOP Conf. Series: Materials Science and Engineering **257** (2017) 012082 doi:10.1088/1757-899X/257/1/012082

Tri-fuel (diesel-biodiesel-ethanol) emulsion characterization, stability and the corrosion effect

M H Low[1,2], M Mukhtar N A[1,2], Ftwi Yohaness Hagos[1,2,*] and M M Noor[1,2]

[1]Faculty of Mechanical Engineering, Universiti Malaysia Pahang, 26600 Pekan, Pahang, Malaysia
[2]Automotive Engineering Centre, Universiti Malaysia Pahang, 26600 Pekan, Pahang, Malaysia

*Corresponding author: ftwi@ump.edu.my

Abstract. This paper presents the result of experimenting emulsified tri-fuel in term of stability, physico-chemical properties and corrosion effect on three common metals. The results were interpreted in terms of the impact of five minutes emulsification approach. Tri-fuel emulsions were varied in proportion ratio consist of biodiesel; 0%, 5%, 10%, and ethanol; 5%, 10%, 15%. Fuel characterization includes density, calorific value, flash point, and kinematic viscosity. Flash point of tri-fuel emulsion came with range catalog. Calorific value of tri-fuel emulsion appeared in declining pattern as more ethanol and biodiesel were added. Biodiesel promoted flow resistance while ethanol with opposite effect. 15% ethanol content in tri-fuel emulsion separated faster than 10% ethanol content but ethanol content with 5% yield no phase separation at all. Close cap under static immersion with various ratio of tri-fuel emulsions for over a month, corrosiveness attack was detected via weight loss technique on aluminum, stainless steel and mild steel.

1. Introduction

Diesel engine has been widely used in various industrial sectors due to its durability, better fuel economy and greater fuel to power conversion efficiency as compared to gasoline engine [1]. It is still not the time to give up on diesel engine especially when the demand is still strong and even manifested with market outlook forecasted 9% steady growth of diesel engine demand globally [2]. Furthermore, widely known that diesel engine could support high compression ratio much higher than race car with petrol engine.

Despite the advantages, people with concern over environmental and health have been casting blame and continuously pointing finger to the issue of diesel engine incomplete combustion as part of the serious source of air pollution due to exhaustive hazardous gases emission, including Nitrogen Oxide (NOx), Carbon Monoxide (CO), Carbon Dioxide (CO2) and Sulphur Oxide (SOx). It is disturbing to comprehend that engine emission affects human health on a large scale [3]. Research also shown that widespread especially in compactly populated urban area, regular exposure to diesel exhaust could in due course bring harm to human health [4,5]. Furthermore, diesel engine emissions contain many mutagens, carcinogens and toxic substances such as NOx, SOx, CO and CO2. Moreover, lung cancer has become a major health risk in animal and human research. Studies on railroad workers and truck drivers occupationally exposed to diesel engine emission show that these workers have been found tend to have increasing in getting lung cancer risk with increasing years of work in the exposed job [6].

Bearing in mind the statement that poor air quality mainly caused by the emission of carbon monoxide and carbon dioxide [7], diesel engine emission somewhat partly responsible for the global ecological disturbance. According to the US National Oceanic and Atmospheric Administration (NOAA), it was announced that 2015 was the warmest year in the record, where the global surface temperature was increased by 0.29°F compared to last year and this is the largest margin since records began in 1880s. According to the State of the Climate in 2015, the highest contribution for the annual combined temperature was long term warming and strong El-Nino since records. This is well related to global warming, which causing ice arctic melting, extreme weather events, changing seasons and disruptions to food supplies. Intergovernmental Panel on Climate Change (IPCC) has claimed that the major causes of global sea level rise are thermal expansion of the oceans and the loss of land-based ice due to increased melting. Meanwhile, total dependency on diesel fuel as the only reliable sole source of combustion to the compression ignition engine call for emerging popular biofuel such as biodiesel [8] and bioethanol [9]. Nonrenewable source couple with fear of shortage supply on fossil fuel appeared to be the strong motive to proceed with the research on developing blending solution with conventional diesel fuel.

All the challenges outlined so far offered entree to innovative approach, presenting an opportunity for emulsion fuel technology combo with three fuel composition categories to play key role in automotive. Optimistically, with tri-fuel emulsion, the hope is to become an effective solution as alternative fuel and in time to earn the claim for better fuel efficiency and lesser harmful engine emissions. If not hundred percent. at least, partial replacement with multicomponent biofuel components combined with diesel would be a motivating option. Since bioethanol and diesel got immiscibility issue, biodiesel influence with innovative strategic mixing approach hopefully could be useful. Hagos et al. (2016) signified the possible distinction between blending and emulsion on tri-fuel [10]. Thus, instead of casual mixing, this study experimented emulsification approach on tri-fuel with multiple proportion. As a practical matter, tri-fuels under emulsion category may have some special qualities to contest with tri-fuel blending category [11] on secondary atomization trait.

Apart from that, because corrosiveness was mentioned as one of the concern with blended tri-fuel by Zöldy in 2006 and 2011 [12, 13], the study to investigate corrosiveness on mild steel, stainless steel and aluminum by weight loss technique subsequent to the static immersion with various proportion of tri-fuel emulsions was not yet revealed. Number of studies that dogged on corrosion behavior relative to biodiesel have been found in the literatures to date [14-18]. Other example, corrosion of copper in blended tri-fuel category [19] has also been found reported. But corrosiveness effect on common metals such as mild steel, stainless steel and aluminum under tri-fuel emulsion categories according to the best of author knowledge, was never been explored. Hence, it was aimed to fill this gap partially by detecting any weight loss due to corrosiveness consequent to the immersion on various tri-fuel emulsified proportions and compared to one another. At the same time, stability test for different composition ratios was conducted. Last but not least, the study also took some initiative to obtain primary properties of the tri-fuels composition with varies proportion such as density, viscosity, flash point and calorific value. The objective was to provide some rationalization to the influence of tri-fuel components on its fundamental properties.

2. Experimental set up

2.1. Fuel preparation

Conventional diesel, ethanol and biodiesel from palm oil origin were attained from close convenient provider. The amount of biodiesel was varied from 0% to 10%, while ethanol was varied from 5% to 15% as can be seen in Table 1. The tri-fuels were emulsified using Hielscher Ultrasonic Processor UP400S Emulsifier as shown in Figure 1 for 5 minutes with 0.7% cycle and 40% amplitude setting. The temperature of the tri-fuel was observed consistently and not to exceed 50°C using an infrared thermometer.

Table 1. Tri-fuel matrix ratio.

No	Diesel (%)	Biodiesel (%)	Ethanol (%)
1	95	0	5
2	90	5	5
3	85	5	10
4	80	5	15
5	85	10	5
6	80	10	10
7	75	10	15

2.2. Fuel properties

Density of the tri-fuel sample was calculated by measuring the mass and volume in equation (1):

$$\rho = \frac{m}{v} \quad (1)$$

where ρ is the density of the tri-fuel in g/ml, m is the mass of the tri-fuel in g, v is the volume of the tri-fuel in ml.

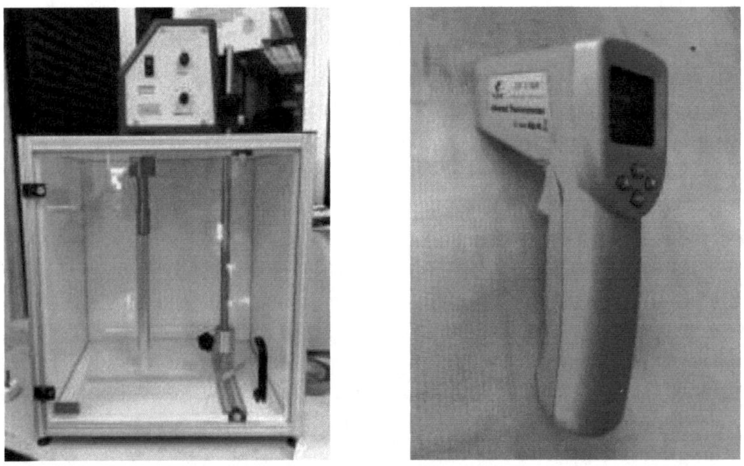

(a) (b)

Figure 1. Experimental setup. (a) Hielscher Ultrasonic Processor UP400S Emulsifier (b) Infrared thermometer.

Calorific value of tri-fuel emulsions was determined using oxygen bomb calorimeter and Model 6772 Calorimetric Thermometer observing the standard ASTM-D240. Dynamic viscosity was determined using Brookfield Viscometer Model DV-III ULTRA Programmable Rheometer observing the standard ASTM-D7042. The dynamic viscosity of the tri-fuel was then converted to kinematic viscosity by dividing the density of the tri-fuel sample. Flash point was determined using Koehler K16591 Rapid Flash Point Tester observing standard ASTM D93.

2.3. Fuel stability

To study the stability effect of tri-fuel emulsions, all the samples were monitored by visual observation for any phase separation to occur. 20 ml of all tri-fuel emulsions samples are filled into test tube and observed for 35 days period. The test tubes were examined every 2 hours for the first day, then every day for the first week and every week until 35 days. Result were plotted for analysis.

2.4. Corrosion detection

The fuels used were diesel, biodiesel and ethanol while the metal piece materials used for corrosion testing were mild steel, aluminum and stainless steel. Firstly, all the specimens undergone general procedure of specimen preparation for metallographic examination such as cleaning and surface polishing to obtain scratch free condition. The weight of each metal pieces was measured before immersion. Mild steel, aluminum and stainless steel undergone static immersion into each tri-fuel sample for the period of 2 months and 2 weeks under close cap condition. After immersion, each of the specimen was dried, cleaned, and weighed again.

3. Results and discussion

3.1. Properties

Table 2 shows the density, calorific value, flash point and kinematic viscosity for all tri-fuel samples. Overall, in view of the result obtained, the increase of ethanol content in tri-fuel decrease the density, calorific value, flash point, and kinematic viscosity. The result in this experiment is in agreement to the literatures reviewed due to the characteristics of ethanol [20-22]. In addition, the results are comparable with blending category by Rajesh et al. [23] that studied higher percentage by volume of ethanol and biodiesel into tri-fuel blended category.

Table 2. Physical characteristics of tri-fuel emulsions.

Tri-fuel	Density (kg/m^3)	Calorific value (MJ/kg)	Flash point
D95E5	838.6	48.6673	86°C - 90°C
D90B5E5	833.1	48.4884	78°C - 80°C
D85B5E10	818.0	47.5606	80°C - 85°C
D80B5E15	814.9	46.5904	85°C - 90°C
D85B10E5	834.7	47.7769	110°C - 115°C
D80B10E10	827.3	45.7008	85°C - 90°C
D75B10E15	815.3	44.4648	65°C - 70°C
Tri-fuel	**Kinematic viscosity at 30°C (mm^2/s)**	**Kinematic viscosity at 40°C (mm^2/s)**	**Kinematic viscosity at 50°C (mm^2/s)**
D95 E5	4.2323	3.5639	2.8236
D90 B5 E5	4.2872	3.5957	2.8760
D85 B5 E10	4.0213	3.4340	2.7313
D80 B5 E15	3.8164	3.1851	2.5965
D85 B10 E5	4.2946	3.6147	2.9099
D80 B10 E10	4.0758	3.4112	2.6606
D75 B10 E15	3.8001	3.2671	2.5967

3.2. Calorific value

Calorific value of tri-fuel emulsions was found apparently degrading as can be seen in Figure 2. The degradation is inevitable considering biodiesel is known with slightly lower calorific value as compared to diesel. Moreover, ethanol carries substantial lower level of calorific value compared to the two fuels individually. Thus, this possibly explain the degradation. Meanwhile, D95E5 and D90B5E5 calorific value were not much different from PD100 as the base fuel. D75B10E15 is the composition with the lowest calorific value among all, 44.4648 MJ/kg. Despite of the calorific value decreases, the level is still within the range of 40 MJ/kg to 49 MJ/kg for the used in diesel engine as stress out in the previous literature [24].

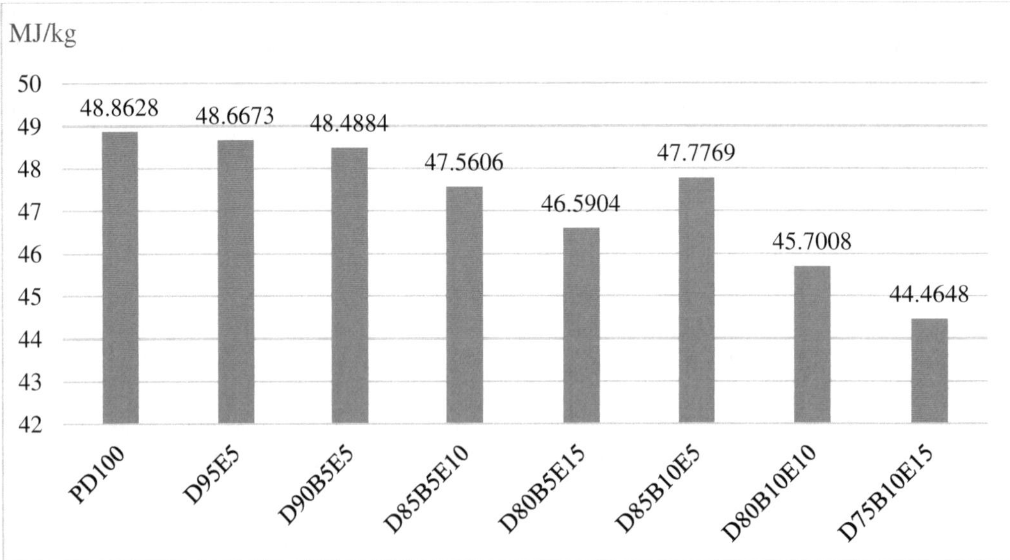

Figure 2. Calorific value of tri-fuel emulsions

3.3. Kinematic viscosity and density

Keep in mind that higher viscosity means weaker fuel injection as the fuel are more viscous and adherent [24]. Figure 3 shows the kinematic viscosity of all the composition at 3 different temperatures. But, it is important to look deeper by carefully track and appreciate the source of influence to the differential value between the ups and down of viscosity level with either 5% ethanol or 5% biodiesel increment gap. Hence, side by side variant assessments were done as in Table 3 for kinematic viscosity and table 4 for density. It is well known individually that density of ethanol is lesser than diesel and much lower than biodiesel. Biodiesel in fact carry higher density than diesel. The highest variant detected between D90B5E5 and D85B5E10 with 15.1 kg/m^3 density drop as 5% of ethanol was added. From viscosity reading, this is the second biggest drop among all with 0.2659 mm^2/s at 30 °C. The highest drop detected on the viscosity reading was between D80B10E10 and D75B10E15 with 5% ethanol addition for 0.3615 mm^2/s changed. Again, comparing in term of density, the smallest variant detected between D80B5E15 and D75B10E15 with 5% biodiesel add on with influence only 0.4 kg/m^3 density increment. D95E5 which by right under bi-fuel emulsion category was compared with tri-fuel emulsion with the presence of 5% biodiesel (D90B5E5), yield density drops of 5.5 kg/m^3. Furthermore, increasing 5% biodiesel by comparing D85B5E10 and D80B10E10 yield increase in density of 9.3 kg/m^3. Meanwhile, second highest density drop (12 kg/m^3) can be seen on D80B10E10 versus D75B10E15 with 5% ethanol influence. While biodiesel increase the density level, ethanol compelled on shifting the reading the other way around. Nevertheless, back to back trait between ethanol and biodiesel can be noticed in the viscosity reading between D85B5E10 and D80B10E10 at different temperatures. At 30°C, viscosity increased by 0.0545 but at 40°C and 50°C, viscosity drop for 0.0228 mm^2/s and 0.0707 mm^2/s respectively. The rest of the viscosity reading appeared regular as expected.

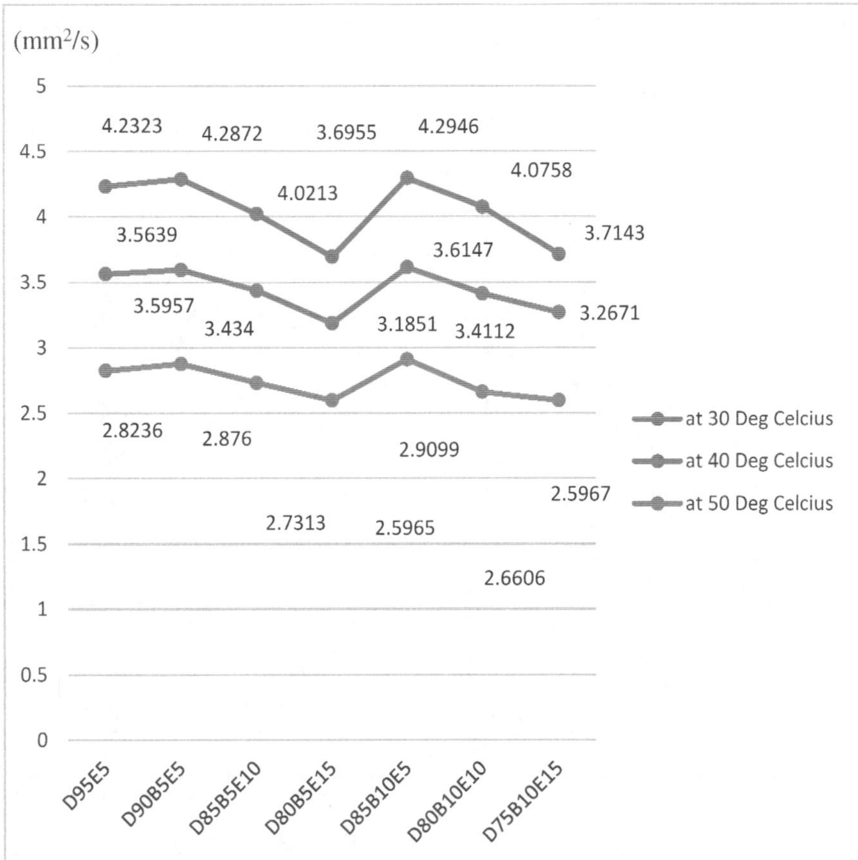

Figure 3. Kinematic viscosity of tri-fuel emulsions at 3 different temperature.

3.4. Flash point

As expected, Table 2 also suggests that ethanol content has significant effect on the flash point of tri-fuel emulsions. Flash point of all proportion were observed appeared with range and with increasing trend parallel to the increasing level of ethanol content. As compared to the well-known conventional diesel flash point base with over 52°C, the range for all the samples tested are under acceptable practical level. The lowest range detected for D75B10E15 with the range from 65°C to 70°C. Providentially, it is well known that flash point has no significant effect on engine performance [25] but the issue lies on safety concern with regards to handling, producing and storing.

3.5. Emulsification stability

Tri-fuel emulsions appeared to be transparent and was observed with noticeable cloudy look occurred upon emulsified but only last temporarily and not on all samples. Sometimes, the samples turned to cloudy appearance the next day with no consistency detected. Perhaps this is some grey areas that require intense observation and further investigation under microscope is needed. The range of appearance detected probably subject to droplet size and difference in refractive indices between the phases. Perhaps, optical microscope can be used in the future to determine the size and distribution upon physical appearance variation. The stability of tri-fuel emulsions of different proportion was studied for 35 days as shown in Table 5 and Figure 4. The study indicates that as ethanol content increase, the tri-fuel undergone phase separation faster, especially for tri-fuel D75B10E15 and D80B5E15. Tri-fuel with ethanol content of 5% did not experience phase separation until the end of observation of 35 days. As indicated, composition with highest ethanol content 15% initiated as early as before it reaches 24 hours. As evident from the figure, five minutes preparation tri-fuel emulsions were seeming to be inadequate to stabilize the composition against the effect of gravitational separation. Perhaps, considering the use of surfactant addition would not be a bad idea after all as demonstrated in the recent study by Tan et al. [26] on engine performance and emissions test with tri-fuel emulsions. In other word, the result in this experiment is comparable to Tan et al. [26].

Table 3. Analysis of biodiesel and ethanol influence on kinematic viscosity.

Temp at (°C)	D95E5	D90B5E5	Influence	Viscosity	Variant
30	4.2323 mm^2/s	4.2872	Biodiesel	Increment	0.0549
40	3.5639 mm^2/s	3.5957	Biodiesel	Increment	0.0318
50	2.8236 mm^2/s	2.876	Biodiesel	Increment	0.0524
Temp at (°C)	D90B5E5	D85B5E10	Influence	Viscosity	Variant
30	4.2872 mm^2/s	4.0213	Ethanol	Decrement	-0.2659
40	3.5957 mm^2/s	3.434	Ethanol	Decrement	-0.1617
50	2.876 mm^2/s	2.7313	Ethanol	Decrement	-0.1447
Temp at (°C)	D85B5E10	D80B10E10	Influence	Viscosity	Variant
30	4.0213 mm^2/s	4.0758	Biodiesel	Increment	0.0545
40	3.434 mm^2/s	3.4112	Biodiesel	Decrement	-0.0228
50	2.7313 mm^2/s	2.6606	Biodiesel	Decrement	-0.0707
Temp at (°C)	D80B5E15	D75B10E15	Influence	Viscosity	Variant
30	3.6955 mm^2/s	3.7143	Biodiesel	Increment	0.0188
40	3.1851 mm^2/s	3.2671	Biodiesel	Increment	0.082
50	2.5965 mm^2/s	2.5967	Biodiesel	Increment	0.0002
Temp at (°C)	D80B10E10	D75B10E15	Influence	Viscosity	Variant
30	4.0758 mm^2/s	3.7143	Ethanol	Decrement	-0.3615
40	3.4112 mm^2/s	3.2671	Ethanol	Decrement	-0.1441
50	2.6606 mm^2/s	2.5967	Ethanol	Decrement	-0.0639

Table 4. Analysis of biodiesel and ethanol influence on density.

Density	D95E5	D90B5E5	Variant	Add Biodiesel
	838.6 kg/m^3	833.1 kg/m^3	-5.5	
Density	D90B5E5	D85B5E10	Variant	Add ethanol
	833.1 kg/m^3	818 kg/m^3	-15.1	
Density	D85B5E10	D80B10E10	Variant	Add biodiesel
	818 kg/m^3	827.3 kg/m^3	9.3	
Density	D80B5E15	D75B10E15	Variant	Add biodiesel
	814.9 kg/m^3	815.3 kg/m^3	0.4	
Density	D80B10E10	D75B10E15	Variant	Add ethanol
	827.3 kg/m^3	815.3 kg/m^3	-12	

Table 5. Phase separation of all tri-fuel samples for 35 days.

	Phase separation (cm)							
Fuel types	**2 hrs**	**4 hrs**	**6 hrs**	**8 hrs**	**10 hrs**	**12 hrs**	**2 days**	**3 days**
D95 E5	0	0	0	0	0	0	0	0
D90 B5 E5	0	0	0	0	0	0	0	0
D85 B5 E10	0	0	0	0	0	0	0	0
D80 B5 E15	0	0	0	0	0	0	0.10	0.16
D85 B10 E5	0	0	0	0	0	0	0	0
D80 B10 E10	0	0	0	0	0	0	0	0
D75 B10 E15	0	0	0	0	0	0	0.14	0.14
	Phase separation (cm)							
	4 days	**5 days**	**6 days**	**7 days**	**14 days**	**21 days**	**28 days**	**35 days**
D95 E5	0	0	0	0	0	0	0.20	0.30
D90 B5 E5	0	0	0	0	0	0	0	0
D85 B5 E10	0	0	0	0	0	0.20	0.30	0.40
D80 B5 E15	0.16	0.16	0.20	0.30	0.60	0.70	0.80	0.80
D85 B10 E5	0	0	0	0	0	0	0	0
D80 B10 E10	0	0	0.08	0.16	0.34	0.40	0.40	0.40
D75 B10 E15	0.18	0.18	0.30	0.40	0.60	0.60	0.80	0.80

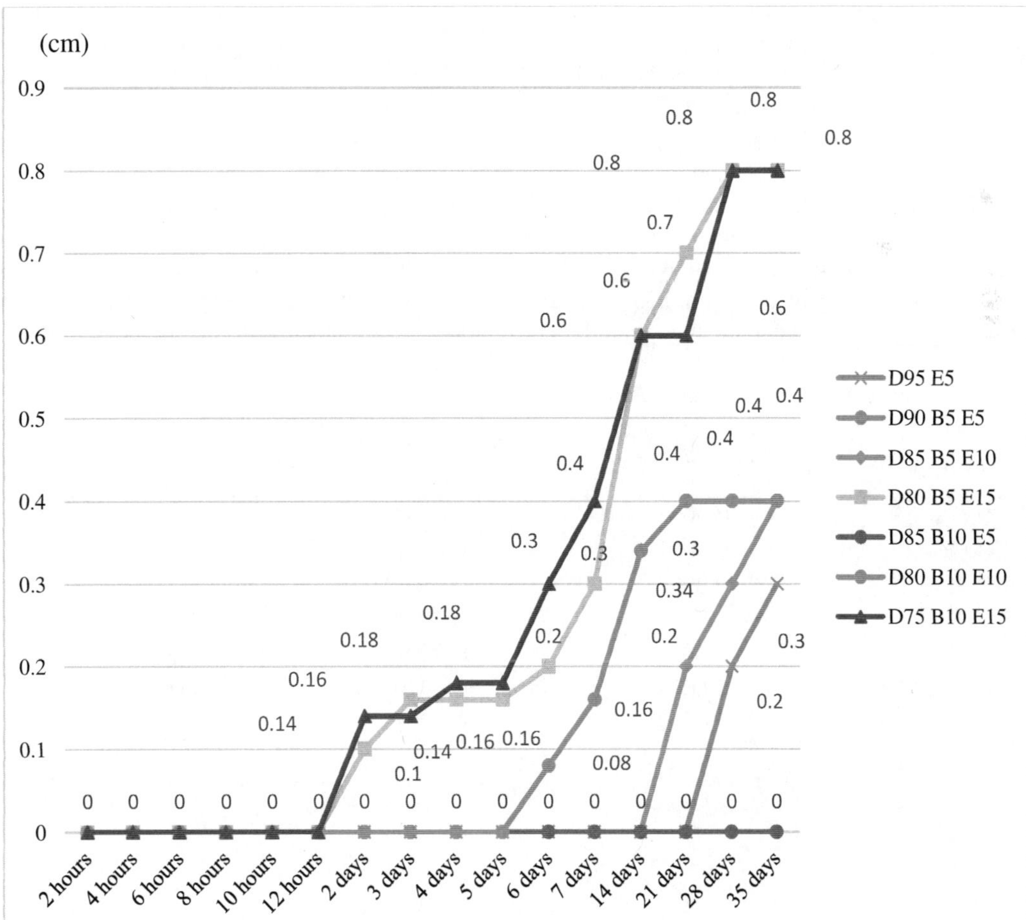

Figure 4. Stability of tri-fuel emulsions.

3.6. Corrosion detection

In this section, it is important to stress that upon corrosion, decreasing in weight should be expected due to metal loss because of the unwanted oxidation activity. Chromium, nickel content in the stainless steel supposed to act as corrosion resistance. Back then, Torres et al. [27] and Hansen et al. [28] outlined ethanol corrosion under 3 categories caused by water content cause, ionic impurities and polarity of ethanol molecule. Less corrosion attack because of low acid content in biodiesel also mentioned in the recent study by Gautam [29] but the fact that weight loss was detected considering brief period of 2 months under close cap room temperature, the plot suggested that corrosion could happened for all three types of metal immersed in tri-fuel emulsions. The data however is not sufficiently enough for deeper comparison for addressing the question on which different proportion ratio contribute more or less to the corrosiveness of the three metals. It was suspected that corrosion mechanism in stainless steel was due to the contact with tri-fuel emulsions D95B0E5 and this could explain with the 0.0125 g weight loss. Further work is required to verify all the finding in this section. Table 6 shows there is significant weight loss on aluminum and stainless steel for D95E5 and this is the highest weight loss among all the tri-fuel samples. The weight loss for most of the metal pieces for all tri-fuel samples between 0.0001 g and 0.0002 g could be neglected and consider none corrosiveness attack detected. The weight losses were too small to consider due to the short immersion period 2 months and 2 weeks for close cap condition. The unnoticed but present of contaminant could also play a role in the detected weight loss. Therefore, it is wise to consider reading above 0.0002 g. It is believed that weight loss suspected of corrosion attack on all the samples mostly under crevice corrosion group. The geometry of the surface of the sample with limited oxygen content because of the emersion plus close cap condition fit into the classification criteria. Further work is required to validate this. Detection of weight loss was recognized on aluminum emersion with B80B10E10 and D90B5E5 while stainless steel emersion with D80B5E15 and D95B0E5. Weight loss of mild steel was noticeable with the emersion in D85B5E10 as depicted in Figure 5. The limitation of the study however that the detection is not by stages within the time frame. Longer period is needed for more significant weight loss on the metal piece to be traced. Hence it is not possible to track which of the composition contribute to the corrosiveness strike first. It would also be useful to obtain a more quantitative understanding of the influence of the component to the corrosiveness of metals that subject to tri-fuel blending immersion. Comparison of weight loss of the complete five standard known types of stainless steel will be also the recommended future study. Finally, in the future, surface analysis is suggested with Scanning Electron Microscope and Energy Dispersive X-Ray Spectroscopy for detail element detection and identify corrosion initiation hot spot on metal pieces.

Table 6. Weight loss for several types of metal piece for all tri-fuel samples.

Fuel type	Weight loss (g)		
	Mild steel	Aluminium	Stainless steel
D95 B0 E5	0.0001	0.0049	0.0125
D90 B5 E5	0.0001	0.0021	0.0002
D85 B5 E10	0.0024	0.0001	0.0000
D80 B5 E15	0.0001	0.0001	0.0038
D85 B10 E5	0.0001	0.0001	0.0000
D80 B10 E10	0.0001	0.0040	0.0001
D75 B10 E15	0.0002	0.0001	0.0002

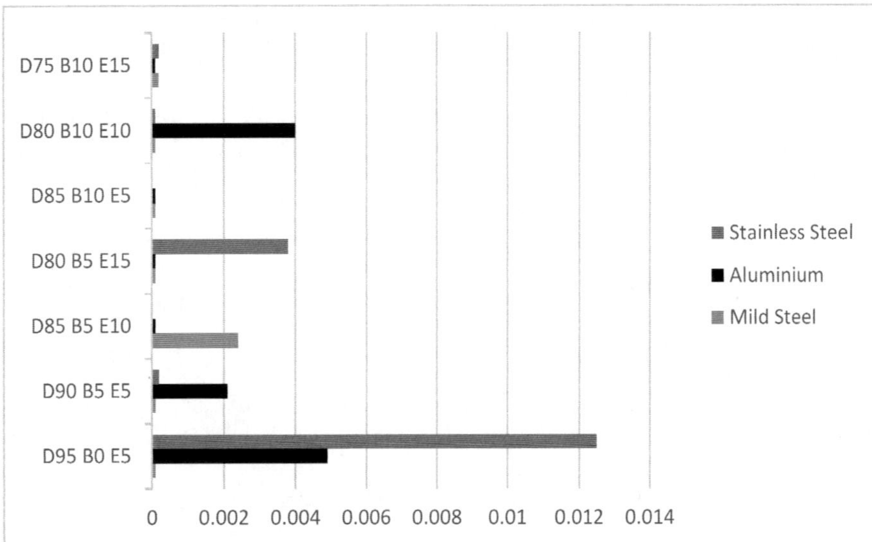

Figure 5. Weight loss detection.

4. Conclusions

Partial key properties, stability and corrosion effect of seven various tri-fuel emulsions ratios were determined in this paper. The highest density reading was the one with the highest biodiesel content and lowest ethanol content. Furthermore, the calorific value degraded as the biodiesel and ethanol content level is amplified but comparatively still under acceptable limit. Flash point of tri-fuel emulsion came with range classification and decreases as more ethanol is added. In other word, more ethanol meant lower flash point. The level however is comparable with diesel standard limit and worth noting for safe handling. Meanwhile, ethanol content pulled the level of kinematic viscosity down whereas biodiesel influence on the opposite direction. Meanwhile, information about the kinetic of cloudy appearance of tri-fuel emulsions mentioned ought for future research consideration to obtain better insight. Tri-fuel emulsion with higher ethanol content could undergo phase separation faster. In summary, 15% ethanol content in tri-fuel emulsion separate faster than 10% ethanol content but ethanol content of 5% yield satisfactory result with no phase separation up to 35 days experimentation period. Weight losses were detected on all three metals under close cap static immersion with tri-fuel emulsion which an indicator to the corrosive initiation. The section on corrosiveness effect can be improved further in the future with comparison between immersion under close and open cap. Perhaps, it is more effective to conduct similar future work with frequent weight loss inspection on weekly basis to know which metal corrode first among all. Furthermore, extended time frame is highly recommended in addition to metallographic study upon significant weight loss detection.

Acknowledgement

The authors would like to acknowledge the financial support by Universiti Malaysia Pahang under the internal grant RDU 150124.

References

[1] Sonar D, Soni S L, Sharma D, Srivastava A and Goyal R 2015 *Clean Technologies and Environmental Policy* **17** 1499

[2] Technavio's Research 2016 *Global Diesel Market 2016-2020* Available from: https://www.technavio.com/report/global-engineering-tools-diesel-engine-market

[3] Steiner S, Bisig C, Petri-Fink A and Rothen-Rutishauser B 2016 *Archives of Toxicology* **90** 1541.

[4] Brook R D, Rajagopalan S, Pope C A, Brook J R, Bhatnagar A, Diez-Roux A V and Peters A 2010 *Circulation* **121** 2331-78

IOP Conf. Series: Materials Science and Engineering **257** (2017) 012082 doi:10.1088/1757-899X/257/1/012082

[5] Adenuga A A, Wright M E and Atkinson D B 2016 *Chemosphere* **147** 396-403

[6] Garshick E, Laden F, Hart J E, Rosner B, Smith T J, Dockery D W and Speizer F E 2004 *Environmental Health Perspectives* **112** 1539

[7] Rinkesh 2009 *Causes and Effects of Particular Matter* Available from: http://www.conserve-energy-future.com/causes-and-effects-of-particulate-matter.php

[8] Shahid E M and Jamal Y 2011 *Pakistan Journal of Engineering and Applied Sciences* **9** 68-75

[9] Ballesteros R, Guillén-Flores J and Barba J 2015 *Fuel* **157** 191-201

[10] Hagos F Y, Ali O M, Mamat R and Abdullah A A 2017 *Renewable and Sustainable Energy Reviews* **75** 1281-94

[11] Avulapati M M, Ganippa L C, Xia J and and Megaritis A 2016 *Fuel* **166** 59-66

[12] Zöldy M 2006 FISITA 2006 (Yokohama, Japan) p 2006.10.22

[13] Zöldy M 2011 *Transport* **26** 303–9

[14] Hu E, Xu Y, Hu X, Pan L and Jiang S 2012 *Renewable Energy* **37** 371-78

[15] Cursaru D L, Brănoiu G, Ramadan I and Miculescu F 2014 *Industrial Crops and Products* **54** 149-58

[16] Norouzi S, Eslami F, Wyszynski M L and Tsolakis A 2012 *Fuel Processing Technology* **104** 204-10

[17] Fazal M A, Haseeb A S M A and Masjuki H H 2011 *International Journal on Advanced Science, Engineering and Information Technology* **1** 505-10

[18] Chew K V, Haseeb A S M A, Masjuki H H, Fazal M A and Gupta M 2013 *Energy* **57** 478-83

[19] Thangavelu S K, Piraiarasi C, Ahmed A S and Ani F N 2015 *Advanced Materials Research* **1098** 44-50

[20] Abdullah A Z, Salamatinia B, Mootabadi H and Bhatia S 2009 *Energy Policy* **37** 5440-48

[21] Barabas I, Todoruţ A and Băldean D 2010 *Fuel* **89** 3827-32

[22] Hulwan D B and Joshi S V 2011 *Applied Energy* **88** 5042-55

[23] Rajesh S, Kulkarni B, Kumarappa S and Shanmukhappa S 2014 *Int. J. Res. Eng. Technol.* **2** 181-88

[24] Kwanchareo P, Luengnaruemitchai A and Jai-In S 2007 *Fuel* **86** 1053-61

[25] Song C 2000 *Chemistry of diesel fuels* (CRC Press)

[26] Tan Y H, Abdullah M O, Nolasco-Hipolito C, Zauzi N S A and Abdullah G W 2017 *Energy Conversion and Management* **132** 54-64

[27] Torres-Jimenez E, Jerman M S, Gregorc A, Lisec I, Dorado M P and Kegl B 2011 *Fuel* **90** 795-802

[28] Hansen A C, Zhang Q and Lyne P W 2005 *Bioresource Technology* **96** 277-85

[29] Gautam R and Kumar N 2016 *Energy Sources, Part A: Recovery, Utilization, and Environmental Effects* **38** 3464-69

IOP Conf. Series: Materials Science and Engineering **257** (2017) 012083 doi:10.1088/1757-899X/257/1/012083

Noise and vibration analysis for automotive radiator cooling fan

N F D Razak[1], M S M Sani[1,2*], W H Azmi[1,2] and B Zhang[3]

[1]Advanced Structural Integrity and Vibration Research (ASIVR), Faculty of Mechanical Engineering, Universiti Malaysia Pahang,26600 Pekan, Pahang, Malaysia
[2]Automotive Engineering Centre (AEC), Universiti Malaysia Pahang, 26600 Pekan, Pahang, Malaysia.
[3]School of Mechanical Engineering, Ningxia University, 750021 Yinchuan, China

[*]Corresponding author: mshahrir@ump.edu.my

Abstract. This paper aims to analyse the noise and vibration of the automotive radiator specifically focused on its cooling fan for different fan conditions and different coolants used namely Ethylene Glycol (EG) water-based and Titanium Oxide (TiO_2) nanofluid. Noise source identification is carried out by utilizing the sound intensity mapping method while an accelerometer is used to measure the vibration results. Both of these experiments are conducted when the fan was both in static and working conditions. The maximum cooling fan speed for the working fan detected by a tachometer for EG water-based is 1990 rpm while TiO_2 nanofluid is 2030 rpm. The difference in speed is due to the different physical properties such viscosity of each coolant has where TiO_2 nanofluid has lower viscosity than EG water-based. The maximum sound power level produced by EG water-based is 53.73 dB while TiO_2 nanofluid is 101.94 dB. Meanwhile, the vibration frequencies of EG water-based are higher than TiO_2 nanofluid. The noise level increases with the cooling fan speed but decreases with the vibration frequency. Apart from studying the noise and vibration of the automotive radiator, this research also analysed the potential application using nanofluid due to its great properties according to its major use in the heat transfer enhancement. As a conclusion, nanofluid as a radiator coolant could improve heat transfer rate, and could also reduce the presence of vibration in the automotive cooling system.

1. Introduction

A typical automotive engine cooling system consists of a radiator, cooling fan, water pump, coolant reservoir, thermostat, heater core and necessary plumbing for both radiator and heater core. This study had focused on car radiators as heat exchangers that are used diversely in the automotive industry for cooling internal combustion engines. Radiator cooling fan is a device that works to pull the air through a radiator to regulate the engine temperature. However, this operating cooling fan has been known as a major contributor to noise in a vehicle other than the engine. It became more crucial to figure out a solution in reducing this vulnerability according to the unwanted bad effects that gives distraction to the community specifically to the passengers. In consequence to this risk, it is absolutely necessary to conduct a detailed experimental study on noise analysis for automotive radiator cooling fans hence the automotive cooling system. Sound is produced by the vibrating body which the vibrations that propagate through the medium to the listeners and it also such a common part of everyday life [1].

Sound has affected humans in a positive and negative way. In a positive way, sound can be defined as pleasant and relaxing. But the opposite, the unpleasant and unwanted sound is called noise which occurs in high decibel (dB) frequency [2]. When a source produces sound power (P), it will create a certain sound intensity (I) at a distance away from the source [3]. Noise measuring devices are typically used a sensor such microphones to receive noise signals from the source [4]. Noise source identification (NSI) techniques are used to optimise the noise emission from a wide range of products including vehicles, household goods and wind turbines [5]. Sound waves which consist of a pure tone are only characterised by amplitude, wavelength, frequency and period [6]. Sound transmits at numerous speeds and this movement gradually spreads to adjoining air particles that assist far from the source and depends on the medium [7]. Frequency of sound is the number of pressure difference per second and is measured in hertz (Hz) [8]. Sound source emits power and these results in a sound pressure. Sound power is the cause while sound pressure is the effect [9]. Hearing damage may be caused by high sound pressure [10].

The measure of the total sound energy produced is called sound power, where sound source produces per unit time and free of its surroundings [11]. Sound intensity describes the rate of energy flowing through a unit area. In the SI system, the unit area is 1 m^2. There will be energy flow in certain directions but not in others because sound intensity gives a measure of direction [12]. Noise is judged to be unpleasant, loud or disruptive to hearing and according to a physics standpoint, noise is indistinguishable from sound, as both are vibrations through a medium, such as air or water [13, 14].

Tones or changes in sound level are the noise features that will make human listen and take notice [15]. Noise can be produced by many sources such as a man's vocal cord, a running engine, a vibrating loudspeaker diaphragm, an operating machine tool and others [16]. The dimensions of a noise source are small compared to the distance to the listener, it is called a point source and sound energy spreads out spherically. So the sound pressure level is the same for all points at the same distance from the source, and decreases by 6 dB per doubling of distance [6]. If a noise source is narrow in one direction and long in the other compared to the distance to the listener, it is called a line source and the sound level spreads out cylindrically. This means the sound pressure level is the same at all points at the same distance from the line, and decreases by 3 dB per doubling of distance [17].

Objective measurements of sound levels are an indispensable part of any environmental noise protection program. The environmental noise levels vary greatly. Noise is often impulsive or contains pure tones [18]. A sound field is a region where there is sound. It is classified according to the manner and the environment in which the sound waves travel [19]. Sound intensity measurement offers several ways of doing this which have considerable advantages over older techniques. Contour plots give a more detailed picture of the sound field generated by a source [20]. Sound intensity measurements normal to the surface are made from a number of equally spaced points on the surface. Same measurement can be used to calculate the sound power over the grid [19, 21]. Noise source identification by using sound intensity mapping was investigated in diesel engines for different engine speed [22]. An experiment carried out utilizing eddy current dynamometer by sound intensity mapping method for the engine speed range of 1500 to 2000 rpm. The noise source of flywheel cover and manifold are presented. It can be observed that the critical cut-off frequency is 1250 Hz. The sound level powers are within 86.42 dB to 96.34 dB in the range of engine speed. The noise level increases with engine speed.

In this paper, noise analysis is focused on the sound intensity mapping method where a grid is directly setup from Hand-held analyser and an accelerometer is used for vibration analysis. The main purpose of this project is to identify the location of maximum noise level and to measure the vibration result from Frequency Response Function (FRF) obtained using different types of coolants which are Ethylene Glycol (EG) water-based (60% water, 40% EG) and Titanium Oxide, TiO_2 nanofluid. Other than studyof noise and vibration of an automotive radiator, this research also analysed the potential application of using nanofluid due to its many great properties according to its major use in heat transfer enhancement.

2. Noise, vibration and heat transfer enhancement relations in automotive cooling system

Many applications such as aircraft engines, motorcycles and other application especially in the automobile field use radiators as heat exchangers for internal combustion cooling devices. Coolant is a liquid that circulates through the engine block to absorb the heat from the operating engine, keeping the engine cool and then flowing into the radiator after reaching high temperatures around 98 °C to lose the heat to the atmosphere. The coolant will return to the engine when its temperature is dropped to ambient temperature and this process will repeat continuously. Engine coolant is usually water-based, but some may also be oil. It is common to employ a water pump to force the engine coolant to circulate, and also for an axial fan to force air through the radiator.

There are two main types of radiator fans which are mechanical and electrical and this research investigation is focusing on mechanical fans. Mechanical fans are belt driven which it's attached to a pulley that is connected to the crankshaft via a drive belt. The fan pulley will rotate when the crank shaft rotates then the fan will push air through the automobile radiator. Therefore, when the engine operates at high speed, it will contribute to high fan speed. An automobile travels at diverse ranges of speeds. The more power is needed by the engine to generate speed because it increases when it travels faster; hence the cooling process needs to be enhanced. Thus, the velocity of air striking the radiator becomes a crucial parameter during the cooling phenomenon through the fins. Although, different velocity will affect the noise level [23].

It is difficult for circular fans to cover the whole surface area of the radiator that is rectangular in size. It generates lower velocity area at the corners which then resulted in less heat transfer. Therefore, the lower velocity zones will induce the lower noise level zones [24]. The use of nanofluid based coolant and its effect on cooling capacity was found that nanofluid has a higher thermal conductivity than base coolant which is 50:50 ratio between water and ethylene glycol [25]. It is theorized that different thermal conductivity will result in different rates of heat transfer hence affecting the velocity of the cooling fans. The various parameters of coolants including its mass flow rate, inlet and outlet coolant temperature, viscosity and other properties are varied [26]. The equations for friction factor and Nusselt number to relate it with the enhancement of heat transfer for both laminar and turbulent flow conditions have already been derived [27]. The effects of momentum and heat transfer induced effects in assessing the relations between heat transfer and friction factor [28]. The noise of cooling fans can be summarized that many parameters in an automotive cooling system will affect the noise level because when engines run at high values of rpm, the speed of the vehicle will be increased; the heat generated in the parts of the engine will also increase drastically. Hence, at a higher speed the cooling process should also be effective in order to dissipate the heat to the atmosphere. This study does not only is looking for a solution for the enhancement of the heat transfer rate, but also how to improve in terms of noise and vibration of the car radiator.

3. Experimental measurement method

The experimental methodology starts from the preparation of EG water-based and TiO_2 nanofluid coolant in an advance automotive fluid lab, followed by the preparation of acoustic device and setting up the radiator cooling fans. Next, this is followed by setting up the calibration of intensity mapping prop and vibration analysis method. Then, the noise and vibration measurement of radiator cooling fans is conducted in the alternative energy lab. The total sound power is computed and the vibration result is measured.

3.1. Noise analysis contribution by sound intensity

The noise analysis of the car radiator cooling fan in this study was set up properly in the alternative energy laboratory before the experiment is conducted. The hand-held analyser is placed 1 m away from the front side of the radiator according to the standards of noise measurement method as shown in Figure 1. The devices and instruments used in this noise investigation are Bruel & Kjaer intensity probe, a laptop, a pair of microphones and a hand-held analyser. Before the experiment of noise source identification is conducted, a rectangular grid was directly set up by taking a photo of the car radiator using the hand-held analyser, where a 4x4 grid system is plotted to define its surface. Then, the sound intensity technique was carried out from the front side of the radiator as shown in Figure 2.

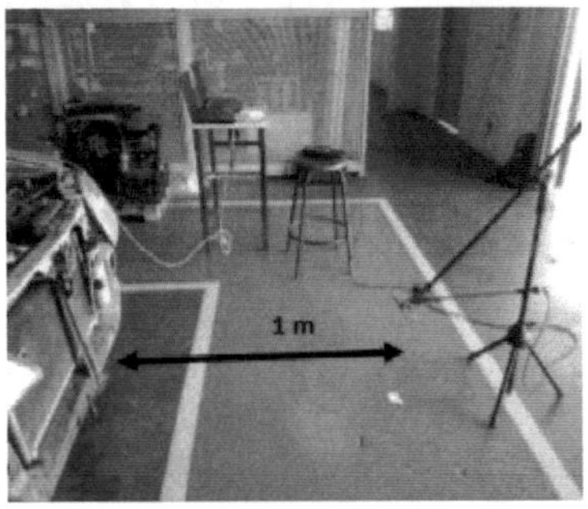

Figure 1. Experimental set up.

Figure 2. Noise and vibration measurement.

The data obtained is displayed on the screen of the hand-held analyser and is interpreted by using post processing software, PULSE LabShop Version 16.1.0. The noisier part of the radiator is determined by observing the colour its contour mapping.

3.2. Vibration device setup

There are numerous tools and devices used to perform this case study and each tool and device has its own functions and specifications. A uniaxial accelerometer sensor shown in Figure 3 is used to measure automobile acceleration hence measuring the vibrations on cars. Figure 4 shows another device that is used in this analysis which is the NI Acoustic and Vibration Data Logger to measure the data produce from the accelerometer sensor in order to run the data acquisition using ME'ScopeVES software.

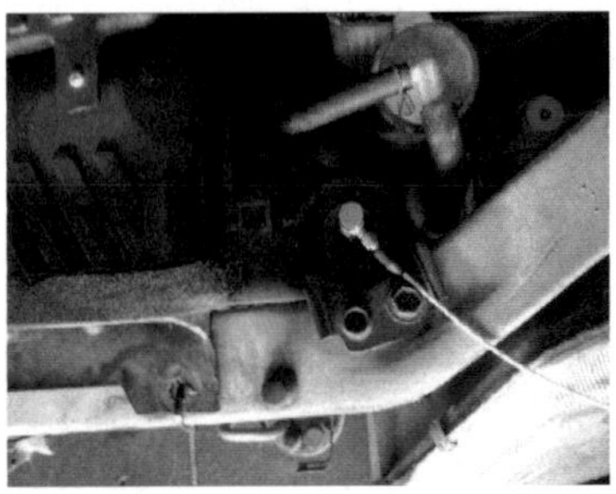

Figure 3. Uniaxial accelerometer.

Figure 4. NI Acoustic and vibration data logger.

4. Results and discussion

4.1. Sound intensity mapping results

Intensity mapping is the contour that explains a more detailed picture of sound field generated by a noise source hence giving a quick overview of the sources in the selected frequency range. The noise measurements are performed in an alternative energy laboratory with less than 20 to 30 dB background noise and are already filtered by the hand-held analyser itself. Contour plot and the sound power level results when the cooling fan is in static condition for EG water-based coolant is shown in Figure 5 and Table 1. Based on the measured results presented, the highest sound power level produced during the experiment is 50.83 dB and is located at point 16. The green colour area where point 16 is located indicates the radiator part that produced the highest noise level while the blue colour area indicates the lowest noise level which at point 3 with 42.61 dB.

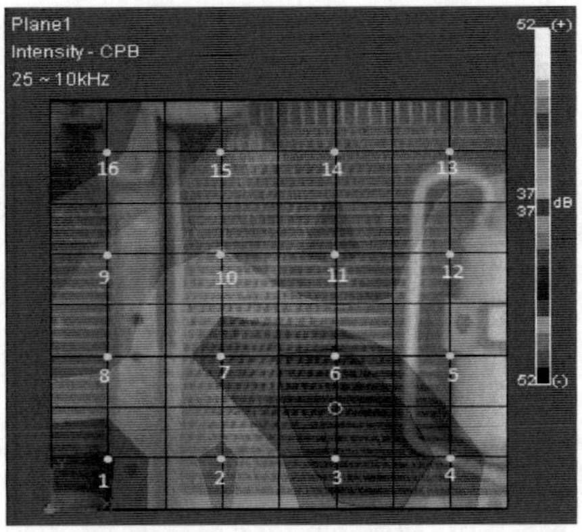

Figure 5. Sound intensity mapping when the cooling fan is in static condition for EG water-based coolant.

Table 1. Sound power level for sound intensity mapping when the cooling fan is in static condition for EG water-based coolant.

Point	Sound power level (dB)	Point	Sound power level (dB)	Point	Sound power level (dB)	Point	Sound power level (dB)
1	44.22	5	47.27	9	44.00	13	48.23
2	48.06	6	43.88	10	49.62	14	48.24
3	42.61	7	45.13	11	47.38	15	48.69
4	45.08	8	47.97	12	48.69	16	50.83

Next, the contour plot of the sound intensity mapping when the cooling fan is working at the speed range from 1970 to 1990 rpm is shown in Figure 6 and Table 2 shows another sound power level data. Based on the measured results presented, the highest sound power level produced during the experiment is 53.73 dB and located at point 12 which is represented by the green colour. The lowest noise level located at point 11 with 34.89 dB and there is no colour present there because the sound power level value is below than 40 dB.

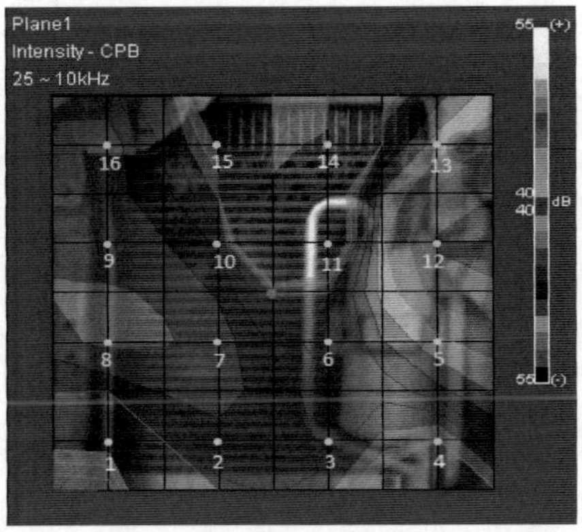

Figure 6. Sound intensity mapping when the cooling fan is in working condition for EG water-based coolant.

Table 2. Sound power level for sound intensity mapping when the cooling fan is in working condition for EG water-based coolant.

Point	Sound power level (dB)	Point	Sound power level (dB)	Point	Sound power level (dB)	Point	Sound power level (dB)
1	38.62	5	50.23	9	48.18	13	49.73
2	47.05	6	43.38	10	45.69	14	38.97
3	45.65	7	48.36	11	34.89	15	39.66
4	41.37	8	49.38	12	53.73	16	48.11

The contour plot of the sound intensity mapping when the cooling fan is in static condition for TiO_2 nanofluid is shown in Figure 7 and another data is shown in Table 3. Based on the measured results presented, the highest sound power level produced during the experiment is 56.48 dB and located at point 13 which is represented by the green colour. The blue colour area where point 8 is located indicates the lowest noise level of the car radiator with 45.46 dB.

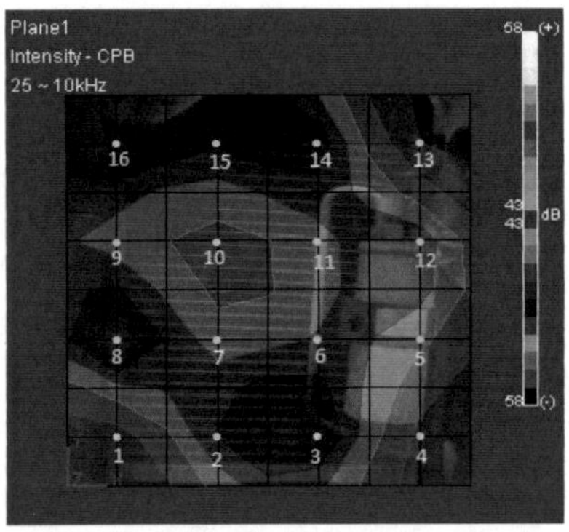

Figure 7. Sound intensity mapping when the cooling fan is in static condition for TiO_2 nanofluid coolant.

Table 3. Sound power level for sound intensity mapping when the cooling fan is in static condition for TiO_2 nanofluid coolant.

Point	Sound power level (dB)	Point	Sound power level (dB)	Point	Sound power level (dB)	Point	Sound power level (dB)
1	55.82	5	53.12	9	52.75	13	56.48
2	50.42	6	51.85	10	54.14	14	49.42
3	48.50	7	52.92	11	52.62	15	50.22
4	55.42	8	45.46	12	50.98	16	47.61

The contour plot of the sound intensity mapping when the cooling fan is working at a speed range from 2000 to 2030 rpm is shown in Figure 8 and another data is presented in Table 4. Based on the measured results presented, the highest sound power level produced during the experiment is 101.94 dB and located at point 4 represented by the dark green colour. The lowest noise level is located at point 15 with 44.59 dB and there is no colour present because the sound power level value is below than 87 dB.

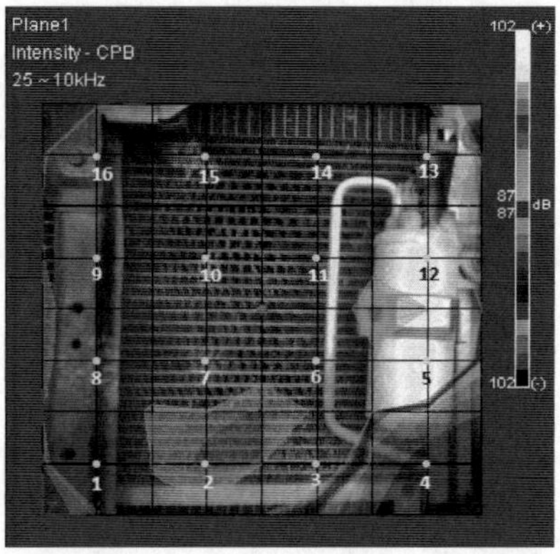

Figure 8. Sound intensity mapping when the cooling fan in working condition for TiO_2 nanofluid coolant.

Table 4. Sound power level for sound intensity mapping when the cooling fan in working condition for TiO_2 nanofluid coolant.

Point	Sound power level (dB)	Point	Sound power level (dB)	Point	Sound power level (dB)	Point	Sound power level (dB)
1	70.13	5	50.66	9	58.79	13	70.94
2	92.16	6	53.27	10	58.28	14	49.34
3	76.43	7	69.44	11	51.55	15	44.59
4	101.94	8	80.09	12	81.79	16	56.24

Many possibilities that contributed to the car radiator noises such as where the location is nearest to the engine compartment, as well as the belt and pulley of the car cooling system. All the belts are driven by an engine pulley which is connected to the automotive engine. Noise is produced when this belt and pulley system starts to work. A screeching sound is normally heard from the belt and pulley noise and this will happen when the pulley attached to an engine accessory that is starting to freeze up or caused by the drive belt is loosened. Therefore, these noises induced to the highest sound power level.

4.2. Vibration analysis (FRF results)

The data results were taken from the Me'scopeVES software using the curve fitting method. Table 5 shows the FRF result of the radiator when the fan is in static condition and Table 6 shows the results when the fan is in working condition for EG water-based coolant. Most of the peaks range for the static fan is below 20 Hz and the highest peak occurred at 13.3 Hz of its vibration frequency. For the working fan, most of the peaks range is below than 100 Hz and the highest peak occurred at the vibration frequency of 30.2 Hz when the fan is running at maximum speed of 1990 rpm.

Table 5. Operating frequency when the fan is in static condition for EG water-based coolant.

Mode	Frequency (Hz)	Mode	Frequency (Hz)
1	3.71	6	9.84
2	5.31	7	12.3
3	6.03	8	13.3
4	7.09	9	14.5
5	9.39	10	15.4

Table 6. Operating frequency when the fan is in working condition for EG water-based coolant.

Mode	Frequency (Hz)	Mode	Frequency (Hz)
1	29.6	6	60.5
2	30.2	7	61.3
3	33.4	8	63.6
4	58.7	9	66.8
5	59.9	10	93.3

Table 7 shows the FRF result of the radiator when the fan is in static condition and Table 8 shows the results when the fan is in working condition for TiO_2 nanofluid coolant. Most of the peaks range for static fan is below 20 Hz and the highest peak occurred at 4.56 Hz of its vibration frequency. For the working fan, most of the peaks range is below than 100 Hz and the highest peak occurred at the vibration frequency of 4.62 Hz when the fan is running at maximum speed of 2030 rpm.

Table 7. Operating frequency when the fan is in static condition for TiO_2 nanofluid coolant.

Mode	Frequency (Hz)	Mode	Frequency (Hz)
1	2.65	6	9.17
2	3.27	7	10.4
3	4.56	8	11.0
4	5.27	9	12.6
5	8.33	10	13.5

Table 8. Operating frequency when the fan is in working condition for TiO_2 nanofluid coolant.

Mode	Frequency (Hz)	Mode	Frequency (Hz)
1	3.45	6	9.21
2	4.62	7	9.94
3	5.70	8	10.7
4	8.13	9	11.1
5	8.74	10	59.7

5. Conclusions

Different coolants are used in the automotive cooling system which will result in different engine speeds due to the different properties of the coolants used such as viscosity and thermal conductivity hence influencing the radiator cooling fan speed. When using EG water based coolant, the maximum speed of the radiator fan in working condition detected by tachometer is 1990 rpm while 2030 rpm when using TiO_2 nanofluid. The sound power level for TiO_2 nanofluid is higher than EG water-based coolant for both fans in static and working conditions. Meanwhile, the vibration result shows TiO_2 nanofluid has a lower vibration frequency than EG water-based coolant for both fans in static and working conditions. The maximum sound power level produced by EG water-based is 53.73 dB while TiO_2 nanofluid is 101.94 dB. Meanwhile, the vibration frequencies of EG water-based are higher than TiO_2 nanofluid. Therefore, the noise level increases with the engine speed but decrease the vibration frequency. The location of the maximum noise level is identified and mostly occurred nearest to the belting and engine system. Overall, results of this study shows that the use of nanofluid as a coolant is not just for improving the enhancement of the heat transfer rate, but also can improve the presence of vibration in automotive cooling system. However, the improvement in noise reduction still need to be studied and improved due to high noise level produced when using nanofluid.

Acknowledgments
The author would like to greatly acknowledge the support by focus group of Advanced Structural Integrity of Vibration Research (ASIVR) and Universiti Malaysia Pahang for providing all the facilities and equipment used for this project.

References

[1] .Bruel KC 1984 *Measuring Sound* (Naerum, Denmark: DK)
[2] Kang J 2006 *Urban Sound Environment* (CRC Press)
[3] Beranek L L and Ver IL 1992 Noise and vibration control engineering-Principles and applications (John Wiley & Sons, Inc.) p 814
[4] ISVR KK, ISVR DT 2007 *Noise Source Identification Using Microphone Arrays*
[5] Ginn K B and Haddad K 2012 *Société Française d'Acoustique* (Nantes, France: Acoustics)
[6] Hansen C H and Sehrndt C 2001 *Occupational Exposure to Noise: Evaluation, Prevention and Control* (Geneva: World Health Organization)
[7] Malcoci I 2014 *Applied Mechanics and Materials* **657** 584-88
[8] Vaillancourt V, Nélisse H, Laroche C, Giguère C, Boutin J and Laferrière P 2013 *Noise and Health* **15** 420-36
[9] Baken R J and Orlikoff R F 2000 *Clinical measurement of speech and voice* (Diego, CA: Thomson Learning)
[10] Fahy F 2002 *Sound intensity* (CRC Press)
[11] Sani M S M, Zaman I, Rahman M 2015 *Journal of Advanced Research in Fluid Mechanics* **9** 28-33
[12] Davy J L 2010 *The Journal of the Acoustical Society of America* **127** 1168-9
[13] Allard J and Atalla N 2009 *Propagation of sound in porous media: modelling sound absorbing materials 2e* (John Wiley & Sons)
[14] Cogan R and Escot P 1976 *Sonic design: The nature of sound and music* (Prentice Hall)
[15] Broch J T 1980 *Mechanical vibration and shock measurements* (Brüel & Kjær)
[16] Long M 2014 *Architectural Acoustics* 2nd ed. (Boston: Academic Press) p 495-528
[17] Ţârulescu S 2013 *Annals of the Oradea University* 401-4
[18] Bruel K 2000 *Signal Analysis, Signal Processingand Noise Reduction Measures*
[19] Bruel K 1993 *Sound Intensity* (Naerum, Denmark)
[20] Washburn K B, Frazer T and Kunio J 2005 *SAE Technical Paper* 0148-7191
[21] Tandon N 1996 *Symp. Emerging Trends in Vibration and Noise Engineering* (New Delhi) 171-85
[22] Sani M S M, Rahman M M, Baharom M Z and Zaman I 2015 *International Journal of Automotive and Mechanical Engineering* **12** 2820-8
[23] Mounika P, Sharma R K and Kishore P 2016 *International Journal on Recent Technologies in Mechanical and Electrical Engineering* **3** 35-8
[24] Chavan D K and Tasgaonkar G S 2014 *International Journal of Modern Engineering Research* **1** 225-8
[25] Leong K, Saidur R, Kazi S and Mamun A 2010 *Applied Thermal Engineering* **30** 2685-92
[26] Yadav J and Singh B R 2011 *S-JPSET* **2** 47-56
[27] Sarma P K, Kishore P S, Rao V D and Subrahmanyam T 2005 *International Journal of Thermal Sciences* **44** 393-8
[28] Sarma P, Chada K, Rao V D, Kishore P, Subrahmanyam T and Bergles A 2010 *International Journal of Heat and Mass Transfer* **53** 1237-42

IOP Conf. Series: Materials Science and Engineering **257** (2017) 012084 doi:10.1088/1757-899X/257/1/012084

Effect of oxygenate additive on diesel engine fuel consumption and emissions operating with biodiesel-diesel blend at idling conditions

H M Mahmudul[1]*, F Y Hagos[1], R Mamat[1], M M Noor[1], I M Yusri[1]

[1]Automotive Engineering Research Group, Faculty of Mechanical Engineering, Universiti Malaysia Pahang, 26600 Pekan, Pahang, Malaysia

*Corresponding author: hassan.mahmudul62@yahoo.com

Abstract. Biodiesel is promising alternative fuel to run the automotive engine but idling is the main problem to run the vehicles in a big city. Vehicles running with idling condition cause higher fuel supply and higher emission level due to being having fuel residues in the exhaust. The purpose of this study is to evaluate the impact of alcohol additive on fuel consumption and emissions parameters under idling conditions when a multicylinder diesel engine operates with the diesel-biodiesel blend. The study found that using 5% butanol as an additive with B5 (5% Palm biodiesel + 95% diesel) blends fuel lowers brake specific fuel consumption and CO emissions by 38% and 20% respectively. But the addition of butanol increases NO_x and CO_2 emissions. Based on the result it can be said that 5% butanol can be used in a diesel engine with B5 without any engine modifications to tackle the idling problem.

1. Introduction

Transportation system has a great importance for the social and economic development of any country. The rising issue for transportation sector is the energy, which is mainly fulfilled by gasoline and diesel fuel. Globally 1.1% in average energy consumption is increased in the transportation sector every year [1]. The transportation sector accounts for the largest share (63%) of the total growth in world consumption of petroleum and other liquid fuels from 2010 to 2040 [2]. There are about 22% global GHG (greenhouse gas) emissions come from the transportation sector due to the increasing demand for vehicles. The rapid growth of the number of vehicle industry in the world has resulted in an increase of exhaust emissions to the environment. Vehicular emissions such as particulate matter (PM), hydrocarbon (HC), carbon dioxide (CO_2), carbon monoxide (CO) and nitrogen oxides (NO_x) are hugely responsible for the air quality deterioration [3]. The International Energy Agency (IEA) forecasts that the emissions of CO_2 from transport sector will increase by 92% between 1990 and 2020 [4]. Also, it is estimated that 8.6 billion metric tons of CO_2will be released to the atmosphere from 2020 to 2035. It is very urgent to find out alternative fuels for transportation sector as this sector is emitting higher GHG emission and contributing to the rapid growth of global oil demand. Recently, attention has been drawn to develop cleaner alternative fuels from renewable sources to reduce the harmful emissions to air and to reduce the dependency on the petro-diesel fuel [5]. Researchers are using biodiesel in diesel engines with or without additives to solve the energy and environment issue. The advantages of biodiesel as compared to diesel fuel are such as the negligible sulphur and aromatic content. Moreover, higher flash point and lubricity added some advantages[6]. The majority of the researchers agrees that biodiesel reduces the emissions like particulate matter (PM), unburnt total hydrocarbons (THC) and carbon monoxide (CO) emissions than diesel.[7] Kumar et al. [8] applied methanol in Jatropha oil and ethanol in animal fat to a diesel engine. The results showed that drastic reduction in smoke, NO_x, HC and CO emissions were observed with the emulsion as compared to neat fat and neat diesel at high power outputs. Guarieiro et al. [9] used various blends of diesel and ethanol with soybean oil and biodiesel, castor oil and biodiesel, where 7% to 15% by volume ethanol was added. The author reported that improvement of combustion efficiencies of diesel fuel can be achieved by the adding oxygenated fuels such as ethanol, biodiesel and vegetable oils, due to having a complete combustion. But, there was no substantial difference in case of CO emission. Alcohols such as ethanol, methanol, pentanol, and butanol are being used as an additive in a diesel engine to overcome the cold flow problems and mitigate the NOx emission. Also, alcohols are low cost oxygenate additive having high oxygen content (up to 34% by weight) which can significantly reduce further PM emission. Among the alcohols, butanol is better additive for internal combustion engine combustion compared to ethanol and methanol. Butanol offers higher cetane number, heating value, lower vaporization heat and better miscibility with biodiesel fuel than ethanol and methanol. Also, butanol is a better choice to be used as an additive with biodiesel-diesel blends as it lowers CO, HC, and soot emission than other alcohols [10].

When the engine is run with biodiesel and their blends with diesel fuel it affects the engine performance and emissions of a diesel engine.so many researchers are already done using diesel-biodiesel blends. Currently, this is the major problem truck industry is currently facing. Though biodiesel is a very effective solutions energy source for the engine, high idling is a critical problem for the transport industry. The idling operational condition is defined as a condition in which the engine runs at low load and at low-rated speed. Engine running at idling condition results in higher fuel consumption [11]. During this condition, the engine cannot work at peak operating temperature. This leads to incomplete combustion and emissions level increase due to having fuel residues in the exhaust. Many studies have been conducted to evaluate engine performance and emission using palm biodiesel blends in a diesel engine at idling condition [12, 13] but very few have been conducted using a small portion of alcohol as an additive with palm biodiesel-diesel blend at idling condition. Thus, the aim of this study is to evaluate the effect of a small proportion of butanol as a fuel additive in the B5 blend on engine fuel consumption and exhaust emissions at idling condition. In this study, 5% butanol has been chosen as an additive with 5% palm oil biodiesel and 95% diesel fuel because butanol is less corrosive and energy content of butanol is higher than ethanol and methanol.

1.1. Engine setup and specification

In this study B5, biodiesel has been used as a reference fuel which is suggested by Malaysian government [11]. Biodiesel-diesel blended fuels were collected from the local market in Malaysia. A small proportion butanol was added as an additive which was also acquired from local markets. Table 1 shows the properties of biodiesel, diesel and butanol. An ISUZU (4JJ1) multi-cylinder vehicular engine was used to perform the performance and emission test. The engine was run at two idling conditions at 1000 rpm and 1500 rpm. Figure 1 shows the schematic diagram of engine test bed. Table 2 shows the details specification of the engine. In order to acquire accurate and repeatable engine test data for diesel engine performance and emission characteristics, the engine test bed was instrumented with speed sensors, pressure transducers, fuel flow meters, in-line torque meter and a gas analyser. The emissions of the engine were measured by Kane gas analyser (Kane auto 4-3).

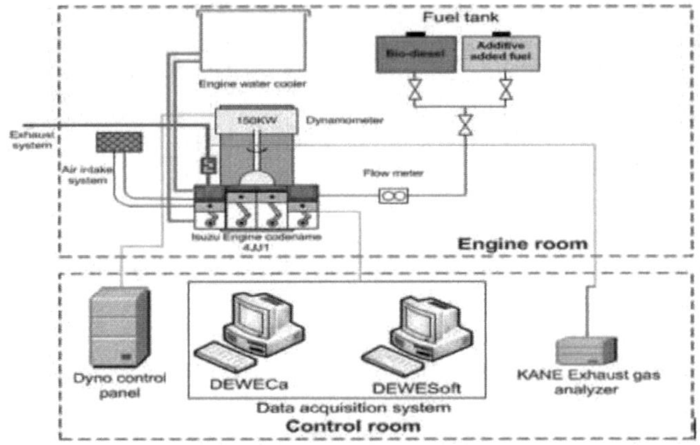

Figure 1. Project Test Rig Schematic

Table 1. Engine Specification

Engine parameters	**Value**
Model	ISUZU 4JJ1 (Turbocharge)
Type	4-cylinder, water cooled, vertical in-line
Bore (mm)	95.4
Stroke (mm)	104.9
Displacement (L)	3 L
Number of cylinders	4 in-line
Compression ratio	17.5
Connecting rod length (mm)	150.0
Piston pin offset (mm)	1.0
Rated output at 2500 rpm	61.0±3% kW
Maximum torque at 1800 rpm	280±5% Nm
Injection system	Direct injection common rail

2. Result and Discussion

2.1. Analysis of Engine Performance

i. Engine Torque

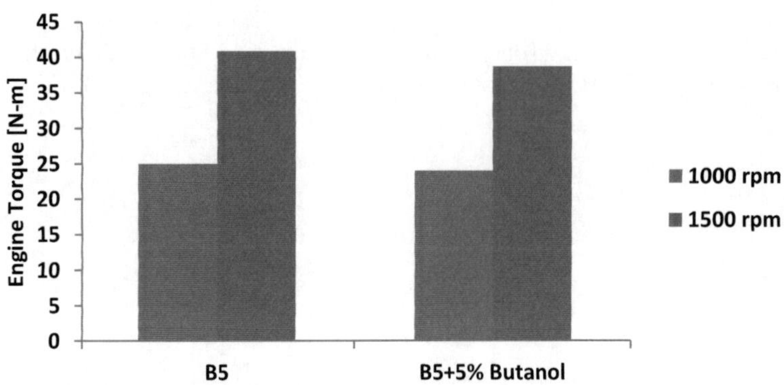

Figure 2. Engine torque at idling condition for B5 and B5+5% butanol.

Figure 2 shows the engine torque for B5 and B5 with 5% butanol additive (B5+5% butanol) at idling conditions for 1000 and 1500 rpm. It can be seen that at both idling condition torque for B5 was 25 N-m and 40.8 N-m but after adding 5% of butanol with B5, the engine torque was found to be 24.03 N-m and 38.63 N-m. So the B5 fuel increased engine torque by 3.88% and 5.31% than B5+5% butanol fuel at 1000 rpm and 1500 rpm, respectively. The reason of increasing torque with that of B5 fuel is due to the higher calorific value as compared to B5+5% butanol [5, 14].

ii. Brake Power

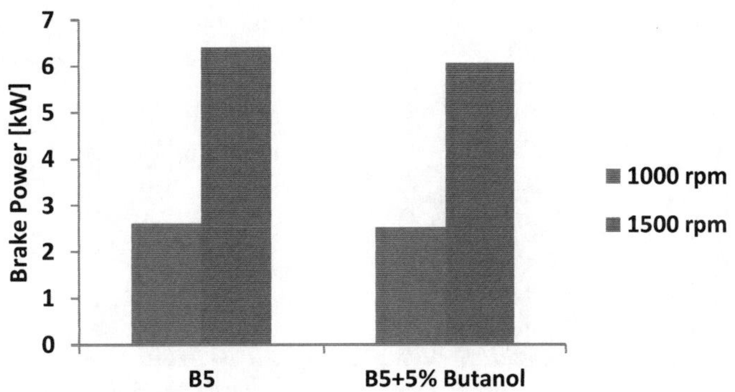

Figure 3. Brake power at idling condition for B5 and B5+5% butanol.

Figure 3 shows the brake power for B5 and B5 with 5% butanol additive at idling conditions. It can be seen that at both idling conditions brake power for B5 was 2.62 and 6.41 kW but after adding 5% of butanol with B5, the brake power found 2.52 and 6.07 kW. It is clear that B5 fuel increased brake power at both idling conditions by 3.81 and 5.30% respectively. The reason of increasing power with B5 fuel is due to the higher calorific value of as compared to B5+5% butanol [5, 14].

iii. Fuel Consumption

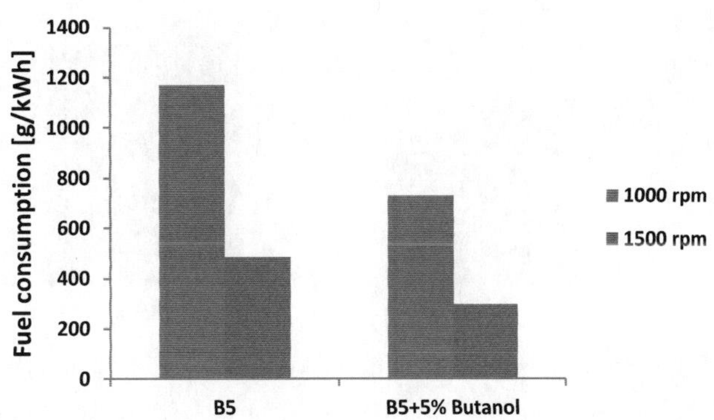

Figure 4. Fuel consumption at idling condition for B5 and B5+5% butanol.

Figure 4 shows the fuel consumption for B5 and B5 with 5% butanol additive at idling conditions. It is clear from the graph that at both idling conditions, fuel consumption for B5 was 1169.44 g/kWh and 485.50 g/kWh but after adding 5% of butanol with B5, the fuel consumption was found to be

485.50 and 297.65 g/kWh. It is clear that B5+5% butanol fuel decreased fuel consumption by 38% and 39% at both 1000 and 1500 rpm, respectively. The reason of reducing fuel consumption by the additive added fuel can be attributed to the lower density and viscosity of butanol additive [10]. Fuel is delivered to the engine on a volumetric basis per stroke; thus, lower quantities of additive added fuels are fed into the engine due to the lower viscosity [15].

2.2. Analysis of Engine Gas Emission

i. Carbon mono oxide (CO)

Carbon monoxide is formed during the combustion process with rich air-fuel mixtures regions and when there is insufficient oxygen to fully burn all the carbon in the fuel.

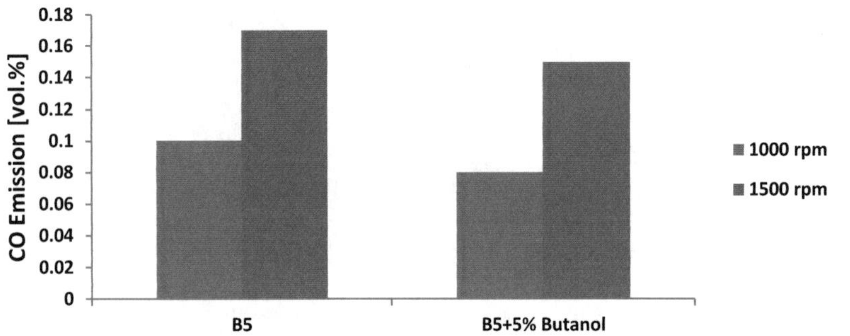

Figure 5. CO emission at idling condition for B5 and B5+5% butanol.

Figure 5 shows the CO emissions at different idling conditions for B5 and B5+5% butanol. The CO emission for B5 and B5+5% butanol is 0.1 and 0.08 vol% respectively at idle speed. The lowest CO emissions are found for B5 with 5% at idling conditions and B5+5% butanol reduces CO emissions by 20%. The reason can be explained by the 21.6% higher oxygen contents of butanol additive which reduced the viscosity and helps to proper mixing and spray formation to complete the combustion[13].

ii. Nitrogen Oxide (NOx)

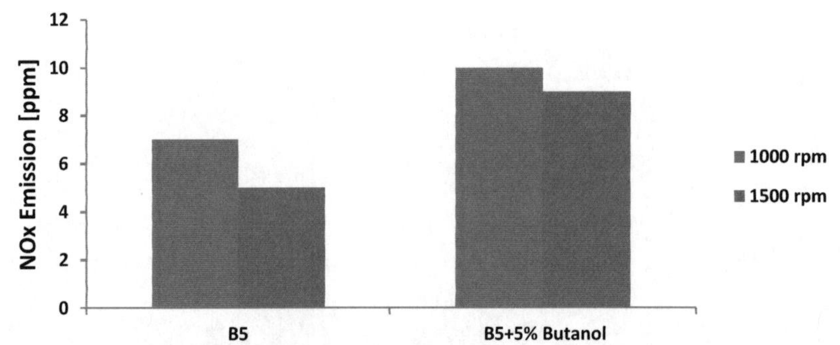

Figure 6.. NO_x emission at idling condition for B5 and B5+5% Butanol.

NO_x emission is strongly related to the higher peak combustion temperature. Figure 6 shows the NO_x emissions at different idling conditions for B5 and B5+5% Butanol at idling conditions. The NO_x emission for B5 and B5+5% butanol is 7 and 10 ppm respectively. Additive added fuel produces 43% NO_x emission that diesel-biodiesel blend. This can be attributed to the lean air/fuel ratio, as butanol contains 21.6% more oxygen in its molecular structure. Many researchers reported that oxygenated fuel blends help to complete combustion which causes higher combustion temperature. Thus, NO_x emission increased [13, 14].

iii. Carbon-di-oxide (CO_2)

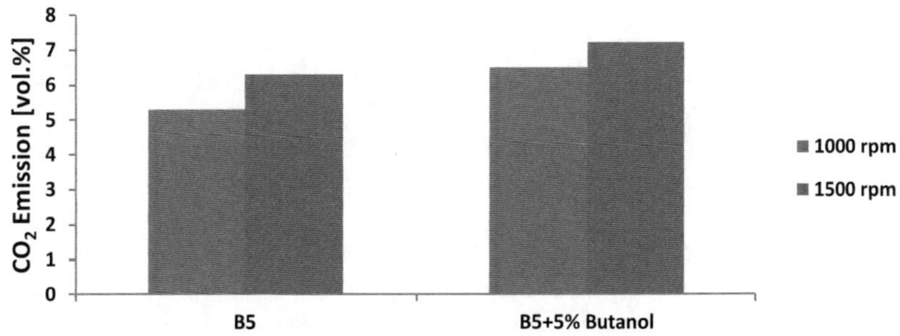

Figure 7. CO_2 emission at idling condition for B5 and B5+5% Butanol

When the carbon completely burns with oxygen then CO_2 is produced. The variation of CO_2 emission at idling condition is shown in figure 7. The CO_2 emission for B5 and B5+5% butanol is 5.3 and 6.5 vol. % respectively. Additive added fuel increased the CO_2 emissions by 23% than B5 fuel. This could be attributed to the higher oxygen contents in butanol additive fuel which helps to burn the fuel completely in diesel engines. The production of carbon dioxide from biofuel can be neglected as it is absorbed by the crops, thus helps to maintain CO_2 levels constant [4].

3. Conclusions

An experimental investigation has been carried out to find the fuel consumption and emissions characteristics of a diesel engine at idle conditions. Based on the findings it can be concluded that, during idling condition, the use of B5+5% butanol fuel makes a significant improvement in fuel consumption and CO emissions. Fuel consumption for B5+5% butanol is 38%lower than B5 fuel. The butanol additive added fuel reduces CO emissions by 20% but increases NO_x and CO_2 emissions by 43% and 23% respectively. Because butanol contains 21.6% more oxygen in its molecular structure which helps to complete combustion. Finally, it can be recommended that 5% butanol with B5 blend can slightly improve engines performance and emission during idling thus it can be used to run diesel engines.

Acknowledgement

Authors wishing to acknowledge assistance or encouragement from colleagues, special work by technical staff and the financial support (RDU 160304) from University Malaysia Pahang.

References

[1] Newell, R.G. and S. Iler, *The global energy outlook.* 2013, National Bureau of Economic Research.

[2] Mahmudul, H., et al., *Production, characterization and performance of biodiesel as an alternative fuel in diesel engines–A review.* Renewable and Sustainable Energy Reviews, 2017. **72**: p. 497-509.

[3] Liaquat, A., et al., *Potential emissions reduction in road transport sector using biofuel in developing countries.* Atmospheric Environment, 2010. **44**(32): p. 3869-3877.

[4] Gorham, R., *Air pollution from ground transportation: an assessment of causes, strategies and tactics, and proposed actions for the international community.* 2002: Tournesol Conseils.

IOP Conf. Series: Materials Science and Engineering **257** (2017) 012084 doi:10.1088/1757-899X/257/1/012084

[5] Mofijur, M., et al., *Comparative evaluation of performance and emission characteristics of Moringa oleifera and Palm oil based biodiesel in a diesel engine.* Industrial Crops and Products, 2014. **53**: p. 78-84.

[6] Mahmudul, H., et al. *Impact of oxygenated additives to diesel-biodiesel blends in the context of performance and emissions characteristics of a CI engine.* in *IOP Conference Series: Materials Science and Engineering*. 2016. IOP Publishing.

[7] Mahmudul, H., et al. *Experimental investigation of the impact of using alcohol-biodiesel-diesel blending fuel on combustion of single cylinder CI engine.* in *IOP Conference Series: Materials Science and Engineering*. 2016. IOP Publishing.

[8] Kumar, M.S., A. Ramesh, and B. Nagalingam, *An experimental comparison of methods to use methanol and Jatropha oil in a compression ignition engine.* biomass and bioenergy, 2003. **25**(3): p. 309-318.

[9] Guarieiro, L.L.N., et al., *Assessment of the use of oxygenated fuels on emissions and performance of a diesel engine.* Microchemical Journal, 2014. **117**: p. 94-99.

[10] Mofijur, M., M. Rasul, and N. Hassan. *Effect of butanol additive on the performance and emission of Australian macadamia biodiesel fuel in a diesel engine.* in *Sustainable and Renewable Energy Engineering (ICSREE), 2017 2nd International Conference*. 2017. IEEE.

[11] Rahman, S.A., et al., *Effect of idling on fuel consumption and emissions of a diesel engine fueled by Jatropha biodiesel blends.* Journal of cleaner production, 2014. **69**: p. 208-215.

[12] Rahman, S.A., et al., *Production of palm and Calophyllum inophyllum based biodiesel and investigation of blend performance and exhaust emission in an unmodified diesel engine at high idling conditions.* Energy Conversion and Management, 2013. **76**: p. 362-367.

[13] Rahman, S.A., et al., *Assessing idling effects on a compression ignition engine fueled with Jatropha and Palm biodiesel blends.* Renewable Energy, 2014. **68**: p. 644-650.

[14] Xue, J., T.E. Grift, and A.C. Hansen, *Effect of biodiesel on engine performances and emissions.* Renewable and sustainable energy reviews, 2011. **15**(2): p. 1098-1116.

[15] Rao, P., *Experimental investigations on the influence of properties of jatropha biodiesel on performance, combustion, and emission characteristics of a DI-CI engine.* World academy of science, engineering and technology, 2011. **75**: p. 855-868.

IOP Conf. Series: Materials Science and Engineering **257** (2017) 012085 doi:10.1088/1757-899X/257/1/012085

Investigation of pouring temperature and holding time for semisolid metal feedstock production

N. A. Razak[1,2*], A. H. Ahmad[1,2] and M. M. Rashidi[1,2]

[1]Manufacturing Focus Group (MFG), Faculty of Mechanical Engineering, Universiti Malaysia Pahang, 26600 Pekan, Pahang, Malaysia
[2]Faculty of Mechanical Engineering, Universiti Malaysia Pahang, 26600 Pekan, Pahang, Malaysia

*Corresponding author: azhani@ump.edu.my

Abstract. Semisolid metal (SSM) processing, as a kind of new technology that exploits forming of alloys between solidus and liquidus temperatures, has attracted great attention from investigators for its thixotropic behaviour as well as having advantages in reducing porosity, macrosegregation, and forming forces during shaping process. Various techniques are employed to produce feedstock with fine globular microstructures, and direct thermal method is one of them. In this paper, the effect from different pouring temperatures and holding times using a direct thermal method on microstructure and hardness of aluminium alloy 6061 is presented. Molten aluminium alloy 6061 was poured into a cylindrical copper mould and cooled down to the semisolid temperature before being quenched in water at room temperature. The effect of different pouring temperatures of 660 °C, 680 °C, 700 °C, and holding time of 20 s, and 60 s on the microstructure of aluminium alloy 6061 were investigated. From the micrographs, it was found that the most globular structures were achieved at processing parameters of 660 °C pouring temperature and 60 s holding time. The highest density and hardness of the samples were found at the same processing parameters. It can be concluded that the most spheroidal microstructure, the highest density, and the hardness were recorded at lower pouring temperature and longer holding time.

1. Introduction

Semisolid metal (SSM) processing is an attractive and advanced technology to produce lower cost and higher quality of engineering parts [1]. It occurs between liquidus and solidus temperature range, where within this range, fluidity of the molten metal changes significantly. SSM processing enables the production of a spheroidal microstructure instead of a dendritic microstructure as normally achieved by conventional processing technique. The thixotropic effect of the alloys produced by this method allows them to flow when shear force is given, but thicken if shear force is released [2]. SSM processing has the ability to produce near-net-shape products with high mechanical properties [2–5]. Therefore, this process is often desired for producing products with complex shape geometries. Furthermore, it has the advantages of prolonged die life due to less thermal shock, and low shrinkage porosity defect as compared to other conventional processes [3,6]. Gas entrapment is eliminated due to laminar cavity fill during SSM process. As a result, a high integrity product which has a uniform and fine microstructure is produced. Due to the potential advantages of this novel technology, research has been accelerating along with the adoption of the technology by industry since the 1990s.

SSM processing mainly comprises of three important stages; (a) preparation of globular feedstock billet, (b) reheating to the semisolid condition, and (c) forming process [7,8]. The key towards SSM processing is by having a non-dendritic or spheroidal feedstock billet. In recent years, many different processing routes have been developed to produce spheroidal microstructures, such as liquid metal routes, solid state routes and combination methods [9,10]. Among all these methods, the direct thermal method (DTM) is a simple route that can develop non-dendritic or spheroidal microstructure with reduced equipment and processing costs [8,11–13]. In this method, the liquid alloy is poured into a cylindrical copper mould to produce spheroidal microstructures. High thermal conductivity of copper mould is the reason that enables rapid cooling to a low superheat liquid alloy. This action creates numerous nucleation sites within the molten alloys during solidification. Heat matching between the mould and the molten alloy results in a pseudo isothermal hold within the solidification range [12]. As a result, spheroidal microstructure is formed during the holding time. Then, quenching in room temperature water is performed to freeze the formation of spheroidal microstructure.

Aluminium alloys are widely used in mechanical construction and automotive industries. This stems from the fact that aluminium is a lighter structural metal in use as its density is 2.7 g/cm^3 as compared to 4.51 g/cm^3 for titanium alloys and 7.2 g/cm^3 for steels [14,15]. Most of aluminium alloys products are made from conventional casting process. According to Kenney et. al. [16], the energy needed to heat aluminium alloys for semisolid processing is 35% less than that required to heat the same aluminium alloys for casting process. Plenty of studies utilizing SSM processing were conducted on aluminium silicon alloy [1,12,17–22]. It is due to the alloy has better flow ability, but fared poorly in mechanical properties as compared to wrought aluminium alloys. However, recent trend has seen that several research studies are moving towards wrought aluminium alloys [8,23–26] by taking advantage of the higher mechanical properties. Ahmad et. al. [8] had conducted a study on wrought aluminium alloy 7075 to investigate the relationship between pouring temperature and holding time of DTM by using aluminium 7075 in order to produce a spheroidal microstructure for the SSM feedstock. They found that fraction solid was strongly dependent on temperature whilst combination of pouring temperature and holding time of 665 °C and 60 s, respectively, produced the best spheroidal primary phase microstructures. Furthermore, higher pouring temperature led to production of higher amount of secondary phase, which in turn result in greater fluidity of the semisolid material. In addition, structures with greater amount of secondary phase were found to have the lowest strength and hardness properties.

Based on the literature, aluminium silicon alloys were the most popular choices of material for SSM processing, as mentioned previously. Limited information about material behaviour, especially on wrought aluminium 6061 as a raw material in DTM was the motivation for this experimental works. The aim of this present work was to investigate the relationship between pouring temperature and holding time of DTM by using aluminium alloy 6061 to produce a spheroidal microstructure for the SSM feedstock.

2. Experimental procedure

Commercial aluminium alloy 6061 was used for this experimental work with the chemical compositions determined by using Optical Emission Spectrometer and is presented in table 1.

Table 1. Chemical composition of aluminium alloy 6061.

Element (wt. %)	Al	Si	Fe	Cu	Mn	Mg	Zn	Cr	Ni	Ti
As-received sample	97.400	1.00	0.290	0.030	0.530	0.570	0.009	0.011	0.019	0.020

2.1. Direct thermal method experiment

1 kg of aluminium alloy 6061 ingots was placed in a graphite crucible and heated to a temperature of 715 °C by using the resistance heated box furnace. The melt was poured into a cylindrical copper mould with 1 mm wall thickness, 25 mm in diameter, and 90 mm in height once the desired temperature of the melt was obtained. The temperature of the molten metal was measured by using a

K-type thermocouple that was attached to a data logger. The K-type thermocouple was used because it has good sensitivity and can measure the temperature up to 1350 °C. The different pouring temperatures were set at 660 °C, 680 °C, and 700 °C. Then, the molten metal was held for 20 s, 40 s, or 60 s respectively after pouring it into the mould. It was later quenched into room temperature water as soon as the holding time for each copper mould was achieved. Another billet with pouring temperature of 700 °C was allowed to solidify without quenching in order to achieve microstructure for the billet with the normal solidification condition. Schematic diagram for DTM apparatus is presented in figure 1 as follows:

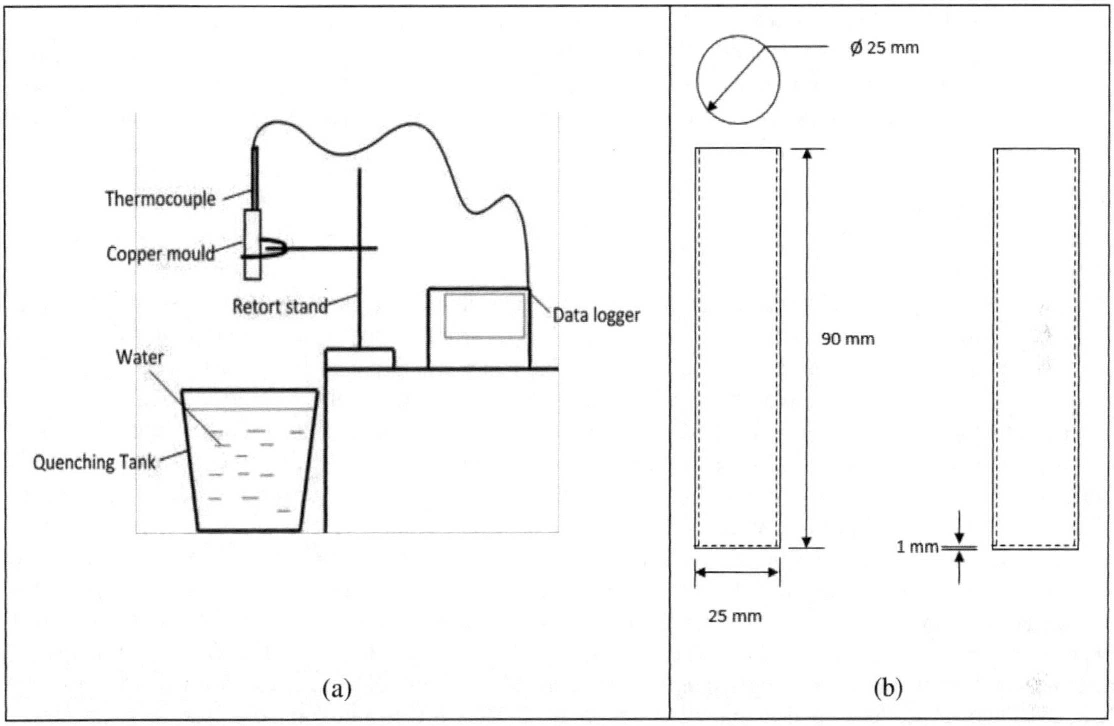

Figure 1. Schematic diagram for the DTM used in the experimental work with (a) experimental set-up and (b) dimension for the copper mould.

2.2. Density test
Feedstock billets from the DTM process were subjected to a density test. This test was performed to determine the porosity levels for each feedstock billets. The feedstock billets were first weighted using a mass balance. Next, the feedstock billets were submerged into water. The water readings before and after the feedstock billets being submerged into water were recorded. The density of feedstock billets was then calculated by Archimedes' principle as presented in equation (1) as follows:

$$\rho_f = \frac{W_f \times \rho_w}{W_w} \quad (1)$$

where ρ_f is the density of feedstock billet in g/cm^3, W_f is the weight of the feedstock billet measure by a mass balance in g, and W_w is the weight of feedstock billets fully submerged in water with the reading in ml. Whilst, ρ_w is the density of the water with the value of 1 g/cm^3.

2.3. Metallography sample preparation

Samples from the DTM process were cut in a rectangular shape and approximately 2 mm in thickness by using a sectioning machine. Then, all samples were mounted in Bakelite to prepare them for grinding and polishing stages. The samples were then ground using 240, 600, 800, and 1200 grit paper, followed by polishing using 6, 3, and finally 1-micron diamond paste. Next, the samples were etched by using Keller's reagent. Optical microscope was then used to view the microstructure and a specific software was used to capture the microstructure images.

2.4. Vickers hardness test

Room temperature hardness was performed with a Vickers hardness instrument. Samples were mounted in Bakelite and subsequently lightly polished to ensure a flat surface, whilst minimizing loss of material. At least 3 measurements were made for each sample and ensuring different segments were analysed. Vickers hardness values were calculated by taking the average of the two diagonal lengths for the indent.

3. Results and discussion

3.1 Density test

Average densities and porosity levels of the samples were determined by Archimedes' principle. Based on results in table 2, there are samples that have large standard deviations than others. This shows that the results between the specimens are not consistent. This might be caused by impurities which contaminate the aluminium alloy 6061 during the DTM experiment. It might be due to graphite, which was placed along the wall of the mould for easier removal of the samples. Other than that, another explanation which contribute to the differences of results might be due to porosity that happen within the samples. Porosity normally can occur when the material undergoes shrinkage during solidification process. A cavity of air within the specimen will cause the billet to have lower weight to volume ratio which will result in the specimen having less mass while maintaining volume and ultimately results in having lower density than expected. One of the densities obtained from this experimental works was slightly higher than the density found in literature for aluminium alloy 6061 which normally at 2.70 g/cm^3 [27]. The porosity level in the samples was determined by comparing the density obtained from the experimental work and the literature. The average density of the sample which obtained slightly higher than 2.70 g/cm^3 indicated lower porosity and sample with lower average density showed higher porosity content. The results in table 2 is presented in a form of graph as shown in figure 2.

Table 2. Average density and porosity level of aluminium alloy 6061 samples at different pouring temperature and holding time determined by Archimedes' principle.

Pouring temperature (°C) / Holding time (s)	Average density (g/cm^3)	Porosity level
660 / 20	2.70 ± 0.09	High
680 / 20	2.68 ± 0.05	Moderate
700 / 20	2.63 ± 0.03	Low
660 / 60	2.74 ± 0.04	Highest
680 / 60	2.49 ± 0.11	Lowest
700 / 60	2.69 ± 0.01	Moderate
700 / NQ[a]	2.61± 0.03	Low

[a] No quenching

According to figure 2, the specimen with the highest density which is 2.74 g/cm^3, is achieved by the combination of pouring temperature and holding time of 660 °C and 60 s, respectively. This finding suggests that the specimen has the lowest porosity. Combination of pouring temperature 680 °C with holding time 60 s produces the lowest density of sample that is 2.49 g/cm^3 which indicated that it has the highest porosity content. The sudden drop of density might be due to porosity which occur during solidification. The standard casting technique sample (700 °C of pouring temperature and without quenching) has a density of 2.61 g/cm^3, which is lower than the density of sample which is subjected to 660 °C pouring temperature with 60 s holding time. This finding might suggest that the mechanical properties of the thixoformed billet is better than the one which is formed by the standard casting technique, as increased in density would lead to increase of strength and hardness.

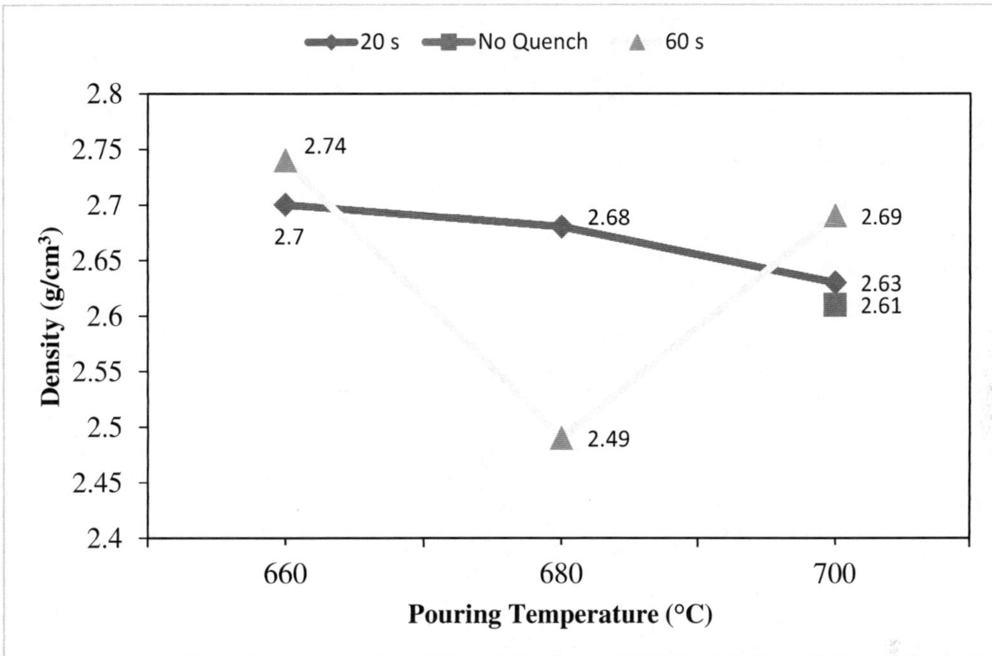

Figure 2. Density of the samples at different pouring temperature and holding time.

3.2 Characterization analysis

The microstructure for pouring temperature of 660 °C, 680 °C, and 700 °C are presented in figure 3. It was observed that the microstructure for pouring temperature of 660 °C was more spheroidal compare with the microstructure for pouring temperature of 680 °C and 700 °C. This finding shows that lower pouring temperature leads to the formation of more spheroidal microstructure and uniform in shape. According to Ahmad et. al. [8], lower pouring temperature retarded the formation of microstructure and was considered the most significant factor in DTM. This in turn, determined the successful of the process. Furthermore, cooling rate is also affected by the pouring temperature. Higher cooling rate is achieved when lower pouring temperature is used, as less superheat needs to be extracted from the sample. On the other hand, higher pouring temperature can cause reduction in cooling rate. The samples need longer time to extract the heat from the above to below liquidus temperature. In this study, it was obvious that the low pouring temperature of 660 °C produced more spherical microstructure compared with the pouring temperature of 680 °C and 700 °C, as the low pouring temperature helps to increase the cooling rate. This result is in agreement with the findings by Ahmad et. al. [8].

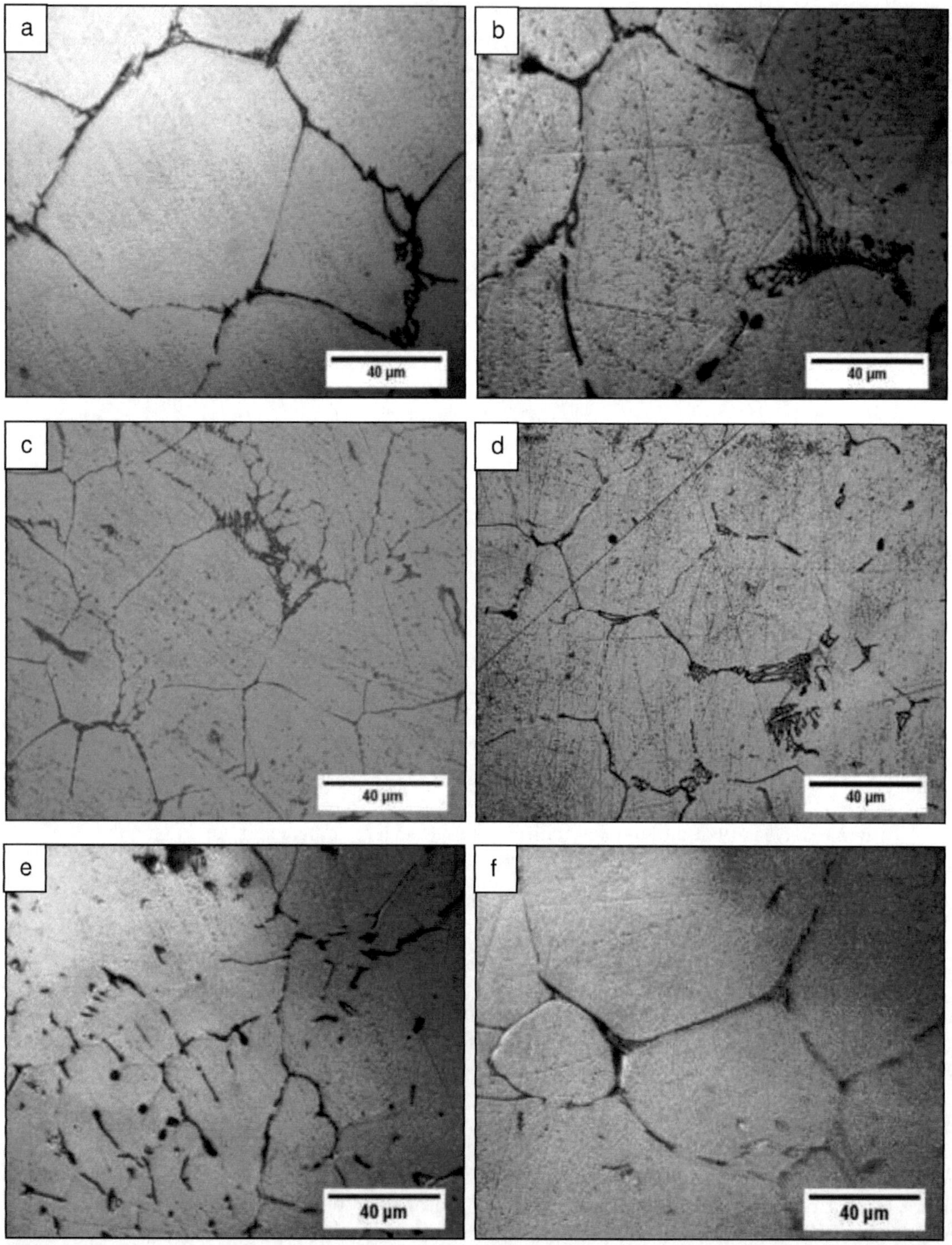

Figure 3. Microstructures for different pouring temperatures and holding times with (a) pouring temperature 660 °C and holding time 20 s, (b) pouring temperature 660 °C and holding time 60 s, (c) pouring temperature 680 °C and holding time 20 s, (d) pouring temperature 680 °C and holding time 60 s, (e) pouring temperature 700 °C and holding time 20 s, and (f) pouring temperature 700 °C and holding time 60 s.

3.3 Vickers hardness test

Figure 4 shows the graph of samples hardness at different pouring temperature and holding time. From this data, the highest hardness is 62.1 HV which experienced by the combination of 660 °C pouring temperature with 60 s holding time. Whilst sample subjected to 700 °C pouring temperature with 60 s holding time has the lowest hardness of 48.4 HV. Based on the results, it can be seen that sample with the highest hardness corresponds to the sample with the highest density in section 3.2, where both of these attributes occur at 660 °C pouring temperature with 60 s holding time. This finding is in agreement with the previous work by Ahmad [28], where higher density led to increase in strength and hardness.

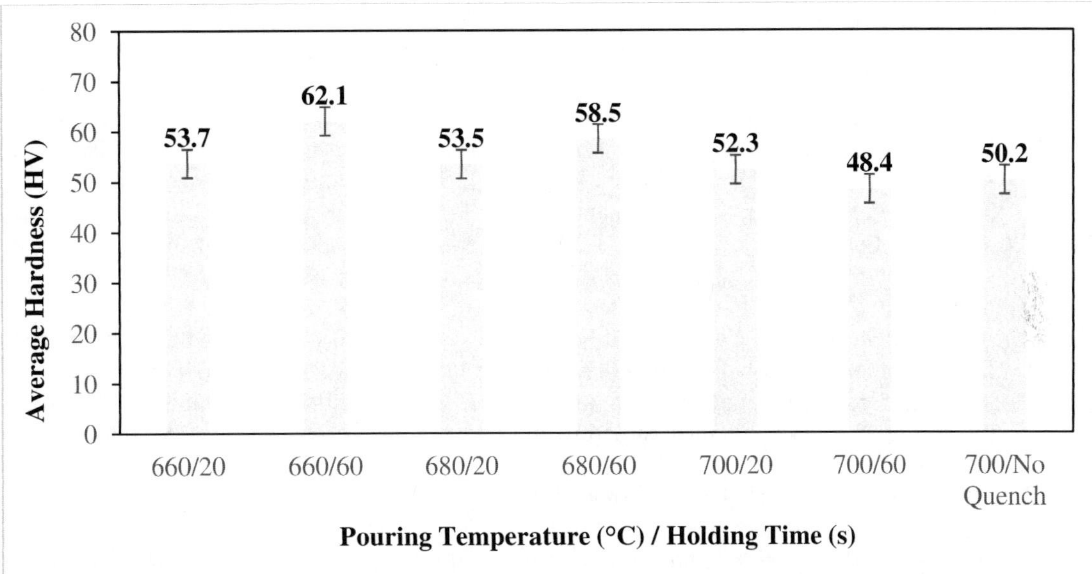

Figure 4. Hardness of the samples at different pouring temperature and holding time.

4. Conclusions

The effect of pouring temperature and holding time to the evolution of the microstructure, density, and hardness of aluminium alloy 6061 were successfully investigated. The results obtained from the experimental works have shown that the combination of lower pouring temperature and longer holding time produced a globular microstructure. Lower pouring temperature has led to higher cooling rate, which in turn results in more spheroidal microstructure. The same processing parameters have been observed to produce the highest density and hardness within the material. It can be concluded that the combination of pouring temperature at 660 °C and the holding time of 60 s, produced a spherical microstructure feature, highest density, and hardness for the 6061 feedstock billets.

Acknowledgements

The authors would like to be obliged to University Malaysia Pahang for the financial support under RDU1603125.

References

[1] Kolahdooz A, Nourouzi S, Bakhshi Jooybari M and Hosseinipour S J 2016 Experimental investigation of the effect of temperature in semisolid casting using cooling slope method *Proc. Inst. Mech. Eng. Part E J. Process Mech. Eng.* **230** 1–10

[2] Atkinson H V. 2005 Modelling the semisolid processing of metallic alloys *Prog. Mater. Sci.* **50** 341–412

[3] Kirkwood D H 1994 Semi-solid metal processing *Int. Mater. Rev.* **47** 173–89
[4] Flemings M C 1991 Behavior of metal alloys in the semisolid state *Metall. Trans. A* **22A** 957–81
[5] Fan Z 2002 Semisolid metal processing *Int. Mater. Rev.* **47** 1–37
[6] McLelland A R A, Henderson N G, Atkinson H V and Kirkwood D H 1997 Anomalous rheological behaviour of semi-solid alloy slurries at low shear rates *Mater. Sci. Eng. A* **232** 110–8
[7] Salleh M S, Omar M Z, Syarif J and Mohammed M N 2013 An overview of semisolid processing of aluminium alloys *ISRN Mater. Sci.* **2013** 1–9
[8] Ahmad A H, Naher S and Brabazon D 2014 Direct thermal method of aluminium 7075 *Adv. Mater. Res.* **939** 400–8
[9] Atkinson H V. 2010 Semisolid processing of metallic materials *Mater. Sci. Technol.* **26** 1401–13
[10] Ahmad A H, Naher S, Aqida S N and Brabazon D 2014 Routes to Spheroidal Starting Material for Semisolid Metal Processing *Comprehensive Materials Processing* vol 5, ed S Hashmi, C J Van Tyne, G F Batalha, S T Button, D Cameron, J A McGeough, E Bayraktar, B S Yilbas, N Bassim, S H Masood, M Rahman and G Krauss (Oxford, UK: Elsevier, Ltd.) pp 135–148
[11] Brabazon D, Browne D J and Carr A J 2003 Experimental investigation of the transient and steady state rheological behaviour of Al-Si alloys in the mushy state *Mater. Sci. Eng. A* **356** 69–80
[12] Browne D J, Hussey M J, Carr A J and Brabazon D 2003 Direct thermal method: new process for development of globular alloy microstructure *Int. J. CAST Met. Res.* **16** 418–26
[13] Brabazon D, Browne D J and Carr A J 2002 Mechanical stir casting of aluminium alloys from the mushy state: Process, microstructure and mechanical properties *Mater. Sci. Eng. A* **326** 370–81
[14] Polmear I J 2006 *Light Alloys, From Traditional Alloys to Nanocrystals* (Oxford, UK: Butterworth-Heinemann)
[15] Mallick P K 2010 Overview *Materials, Design and Manufacturing for Lightweight Vehicles* ed P K Mallick (Cambridge, UK:Woodhead Publishing Limited) pp 1–32
[16] Kenney M P, Courtois R D, Evans G M, Farrior C P, Koch A A and Young K P 1998 Semi solid Metal Casting and Forging *Metals Handbook* (Materials Park, Ohio, USA: ASM International) pp 327–38
[17] Yang X R, Mao W M and Pei S 2007 Preparation of semisolid A356 alloy feedstock cast through vertical pipe *Mater. Sci. Technol.* **23** 1049–53
[18] Ning Z L, Wang H and Sun J F 2010 Deformation behavior of semisolid A356 alloy prepared by low temperature pouring *Mater. Manuf. Process.* **25** 648–53
[19] Jarfors E W 2004 Melting and coarsening of A356 during preheating for semisolid forming *Int. J. Cast Met. Res.* **17** 229–37
[20] El-Mahallawi I and Shash Y 2010 Influence of nanodispersions on strength–ductility properties of semisolid cast A356 Al alloy *Mater. Sci. Technol.* **26** 1226–31
[21] Forn A, Vaneetveld G, Pierret J C, Menargues S, Baile M T, Campillo M and Rassili A 2010 Thixoextrusion of A357 aluminium alloy *Trans. Nonferrous Met. Soc. China (English Ed.* **20** s1005–9
[22] Birol Y 2007 A357 thixoforming feedstock produced by cooling slope casting *J. Mater. Process. Technol.* **186** 94–101
[23] Ahmad A H, Naher S and Brabazon D 2014 Injection Tests and Effect on microstructure and properties of aluminium 7075 direct thermal method feedstock billets *Key Eng. Mater.* **611–612** 1637–44
[24] Ahmad A H, Naher S and Brabazon D 2015 Mechanical properties of thixoformed 7075 feedstock produced via the direct thermal method *Key Eng. Mater.* **651–653** 1569–74
[25] Lee S Y and Oh S Il 2002 Thixoforming characteristics of thermo-mechanically treated AA 6061 alloy for suspension parts of electric vehicles *J. Mater. Process. Technol.* **130** 587–93

[26] Zhang X Z, Chen T J, Chen Y S, Wang Y J and Qin H 2016 Effects of solution treatment on microstructure and mechanical properties of powder thixoforming 6061 aluminum alloy *Mater. Sci. Eng. A* **662** 214–26
[27] ASM International 1990 *ASM Metals Handbook, Properties and Selection:Nonferrous Alloys and Special-Purpose Materials* vol 2, ed ASM International (Materials Park, OH: ASM International)
[28] Ahmad A H 2015 *Effect of Temporal Thermal Field on Quality of Semi-Solid Metal Formed Components* (Dublin City University)

IOP Conf. Series: Materials Science and Engineering **257** (2017) 012086 doi:10.1088/1757-899X/257/1/012086

High temperature oxidation in boiler environment of chromized steel

F F Alia[1], T Kurniawan[1*], Y P Asmara[1], M H B Ani[2] and A B D Nandiyanto[3]

[1]Structural and Material Degradation Focus Group, Faculty of Mechanical Engineering, University Malaysia Pahang, 26600 Pekan, Pahang, Malaysia.
[2]Department of Manufacturing and Materials, Faculty of Engineering, International Islamic University Malaysia, PO Box 10, 50728 Kuala Lumpur, Malaysia.
[3]Department of Chemistry, Universitas Pendidikan Indonesia, Jl. Dr. Setiabudhi No. 229, Bandung 40154, Indonesia.

*Corresponding author: tedikurniawan@ump.edu.my

Abstract. The demand for increasing efficiency has led to the development and construction of higher operating temperature power plant. This condition may lead to more severe thickness losses in boiler tubes due to excessive corrosion process. Hence, the research to improve the corrosion resistance of the current operated material is needed so that it can be applied for higher temperature application. In this research, the effect of chromizing process on the oxidation behaviour of T91 steel was investigated under steam condition. In order to deposit chromium, mixture of chromium (Cr) powder as master alloy, halide salt (NH_4Cl) powder as activator and alumina (Al_2O_3) powder as inert filler were inserted into alumina retort together with the steel sample and heated inside furnace at 1050°C for ten hours under argon gas environment. Furthermore, for the oxidation process, steels were exposed at 700°C at different oxidation time (6h-24h) under steam condition. From FESEM/EDX analysis, it was found that oxidation rate of pack cemented steel was lower than the un-packed steel. These results show that Cr from chromizing process was able to become reservoir for the formation of Cr_2O_3 in high temperature steam oxidation, and its existence can be used for a longer oxidation time.

1. Introduction

The demand to increase the efficiency of the development advance power plant such as supercritical (SC) and ultra-supercritical (USC) is becoming an important issue. This is because by increasing the efficiency of the power plant, it would reduce the environmental impact caused by the greenhouse gasses emission [1]. The easiest way to increase the efficiency of the power plant is to increase their operating temperature as stated in Carnot cycle [2, 3]. Currently, the maximum operated temperature is about 650°C, and targeted operating temperature for future power plant is above 700°C [1, 4]. However, by increasing the temperature of the power plant, the operating conditions of the plant component will become more severe for oxidation resistance of the material [5-6]. Therefore, further study to improve corrosion resistant of the current operated material is needed so that it can be applied for higher temperature application.

In the superheaters and reheater of boiler in power plant, the inside part is exposed to high temperature and high pressure steam (steam side) and the outer part is exposed to combustion gases (fire side). These steam condition gives the higher rate of oxide growth and produced the corrosion product. Many report clarified in their research the oxidation in steam condition is more severe than in dry condition [7-9]. The expected protective chromia (Cr_2O_3) that will be formed on the surface will be retarded in steam condition due to the growth of duplex scale, which consists of magnetite as the outer scale and Fe-Cr spinel as the inner scale. Galerie [9] in their study stated 15-18% Cr ferritic steels in environment containing water vapour at 800°C-1000°C, the failure for the formation of protective Cr_2O_3 will cause the iron oxide to form, and the initial oxide $(Fe, Cr)_3O_4$ shows the growth with oxidation time. The mass gain on the surface of the tube will cause the thickness loss and may lower the thermal conductivity, lead to premature creep failure [10]. Furthermore, the exfoliated scale may accumulate inside the bends tube and cause blockage to the tube. If the oxide particles enter the steam turbine, erosion blades and nozzles may occur.

The aim of the present work is to improve the oxidation behavior of the steel in steam condition. The surface of steel is enriched by chromium in the form of diffusion coating by the chromizing process (pack cementation process). The result for steam exposure test was done by comparing the weight gain for oxidation after chromizing and also the steel that does not undergo the chromizing process. The same process has been conducted for the P11 steel [11]. In this work, the test was carried out with pure water vapor at ambient pressure and 700°C with different exposure time.

2. Experimental procedure

2.1. Chromizing process

The ferritic-martensitic steel T91 was used as a substrate material to apply the diffusion coatings. The chemical composition of this material is summarized in Table 1. The samples were cut into the required dimension 10 x 10 x 4mm^3. The material then was prepared and polished their surface finishing up to 2000 grit finish using SiC abrasive papers and washed with acetone in an ultrasonic bath to remove its surface from rust and grease.

Table 1. Nominal composition for T91 steel.

Alloy	Composition wt.%
C	0.11
Mn	0.47
Si	0.30
Nb	0.08
Mo	0.96
V	0.23
N	0.05
Cr	9.26
Fe	Balance

The chromizing process was assembled as illustrated in the Figure 1. For chromizing, there are four component that will make the chromizing process complete, they are substrate to be coated, master alloy, activators, and filler. For this chromizing process, the master alloy used was chromium (Cr) powder, the activator is NH_4Cl powder, and Al_2O_3 powder as filler. The pack powders for chromizing process were set up for 20wt.% Cr; 20wt.% NH_4Cl; balance, Al_2O_3. Powder pack mixture was homogenized by using a porcelain mortar for one hour. Samples and powder then embedded in the alumina crucible with lid glued by alumina cement. The crucible then was put into a horizontal tube furnace and loaded with argon gas. The exposure time for chromizing process was 10 hours at temperature 1050°C with constant flow rate 150 ml/min. The samples were allowed to cool naturally in the crucible at room temperature. Lastly, the samples were taken out from the crucible and the surfaces were cleaned again by using ultrasonic cleaning. The cross sectional image of the chromized steel was investigated by using FESEM/EDX.

IOP Conf. Series: Materials Science and Engineering **257** (2017) 012086 doi:10.1088/1757-899X/257/1/012086

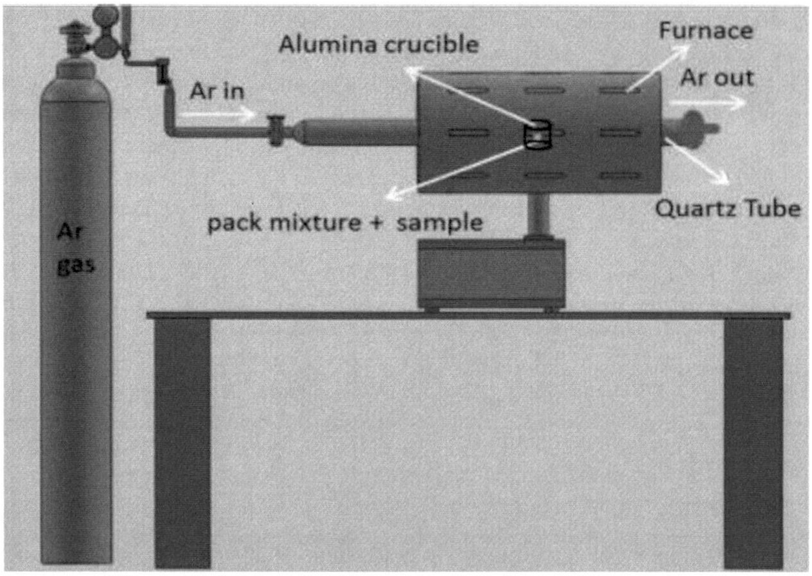

Figure 1. Schematic illustration for chromizing process.

2.2. Oxidation tests in steam condition

The schematic illustration for experimental setup which was used in this experiment is illustrated in Figure 2. The temperature for steam oxidation tests was 700ºC with constant flow rate at 150 ml/min of high purity argon was used as a carrier gas. The distilled water was set at a constant value at the mantle heater. The argon gas was carried in the glass grit to make sure a good contact between argon and water. For this experiment, the exposure time for oxidation is exposed to 6h, 12h and 24h for chromized steel and non-chromized steel. The samples were weighing before and after oxidation to check their weight change.

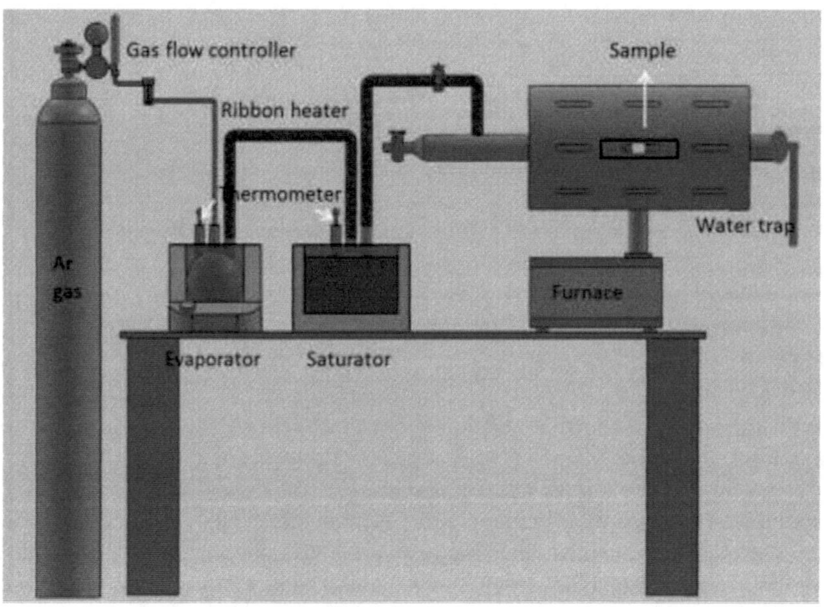

Figure 2. Schematic illustration for oxidation tests in steam condition.

3. Results and discussion

Figure 3(a) and 3(b) shows the image of as polished steel, chromized steel respectively, while Figure 3(c) and 3(d) shows the steel exposed to steam condition for chromized and non-chromized steel in 12h exposure. It can be seen that in Figure 3(b) shows the changes happened when Cr diffused in the steel. The colour changes were observed as soon as the steel taken out from the crucible. This happened because of the reaction of content in pack mixture during chromizing process. The exposed steel in steam condition for 12h in Figure 3(d) shows a very bad condition due to the formation of iron oxide which is coincide with previous work stated that steel that exposed in water vapour existence will cause severe exfoliation and spallation to the steel. However, the chromized steel in Figure 3(c) does not formed iron oxide when exposed in the same condition. This results show the formation of iron oxide can be reduced by using the suitable parameters for chromizing process.

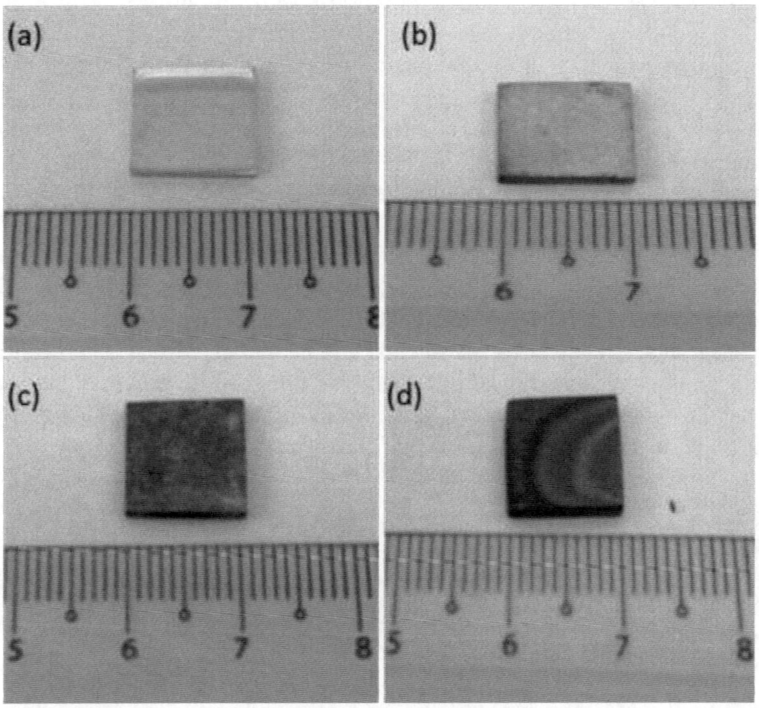

Figure 3. The image for (a) as polished steel; (b) the steel after chromizing process; (c) chromized steel after 12h oxidation; (d) non-chromized steel after exposed to 12h oxidation.

This result was strengthened by surface morphologies investigated in FESEM/EDX in Figure 4. The results show that the content of Cr increase whiles the content of Fe decrease in 10h chromizing time at 1050°C. This suggests the Cr content in the steel prevents the diffusion of Fe into the coating from steel substrate. In order to determine the thickness of obtained coating, the back scattered electron (BSE) image of the cross section were measured as shown in Figure 4. It is found that chromized layer was diffuse into the steel until 15 μm. The continuous and homogenous coating are composed of Cr-rich layer and diffusion rich layer which both determine the effective of the chromizing process. One of the parameters to increase the thickness of Cr-rich layer is the diffusion time. The relatively short diffusion time cause the Cr has less time to diffuse inside the coating layer. The using of 10h exposure time is suitable for the formation of higher and stable Cr-rich layer.

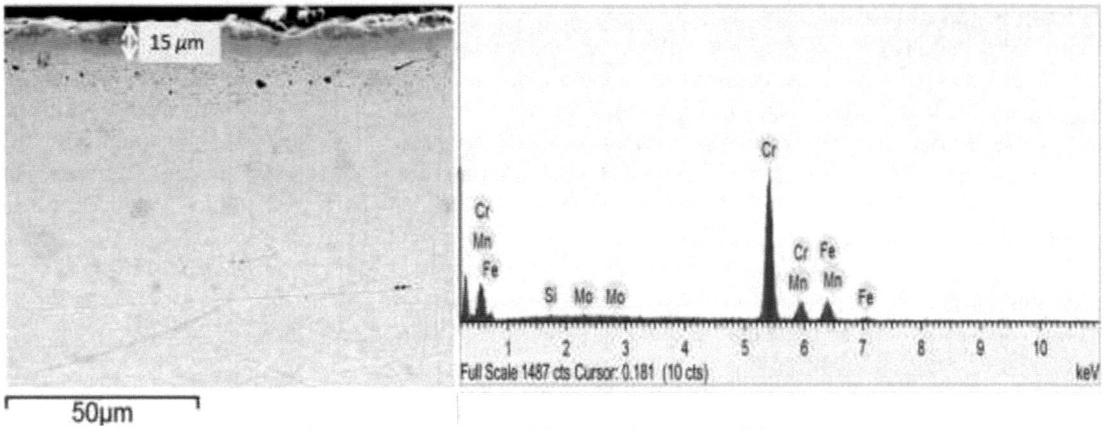

Figure 4. BSE cross sectional image after chromizing process.

The increase in the mass gain of the non-chromized steel after exposed to steam condition lead to the test of chromized steel in steam condition. The morphology of the cross section image is shown in Figure 5. the results exposed in 6h does not have much changes from the chromized steel. The exposure time for 6h is too short for the oxidation to occur. Hence, the image in Figure 5 shows the BSE image for steel exposed in steam condition for 12h. From the cross sectional image, we can see the Cr-rich layer was formed on top of the steel. By using the spot analysis in EDX analysis, the element depth of the diffusion was measured as 20 μm. After the exposure, there are small amount of Fe was present in the layer. There was interdiffusion zone between inner layer, outer layer and steel substrate as can be seen in the image. The interdiffusion zone is not visible on the EDX graph, but can recognized on the Cr and Fe concentration measured by EDX spot analysis. Cr atoms that diffused across the Cr-rich layer enabled the growth of the interdiffusion zone.

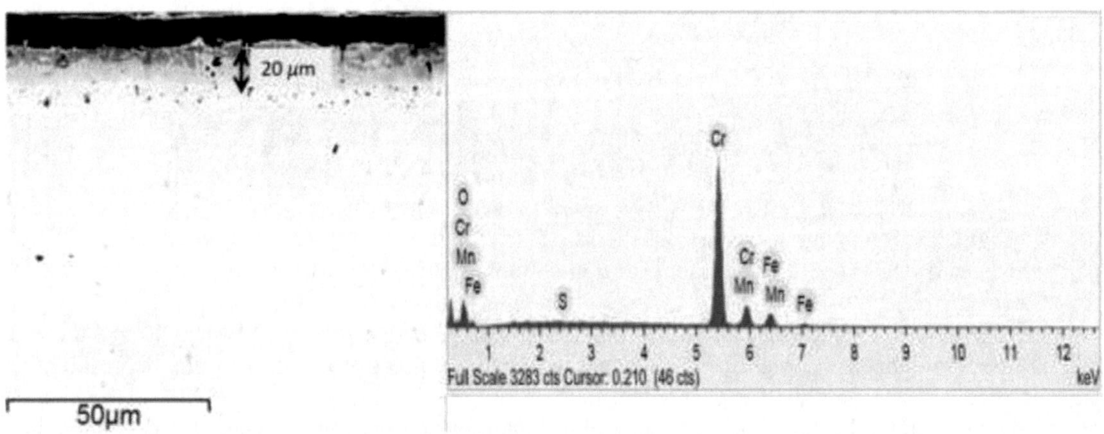

Figure 5. BSE cross sectional image after exposed to 12h in steam condition.

The diffusion depth for chromized steel exposed in 24h also show the same behaviour with 12h exposure time. From the results, authors suggested to extend the time exposure in future work. The image for 24h exposure can be seen in Figure 6. In this condition, the wt.% of Cr is higher compare to Cr wt.% in steel exposed in 12h steam condition with value 84.8wt.% and 79.9wt.% respectively.

IOP Conf. Series: Materials Science and Engineering **257** (2017) 012086 doi:10.1088/1757-899X/257/1/012086

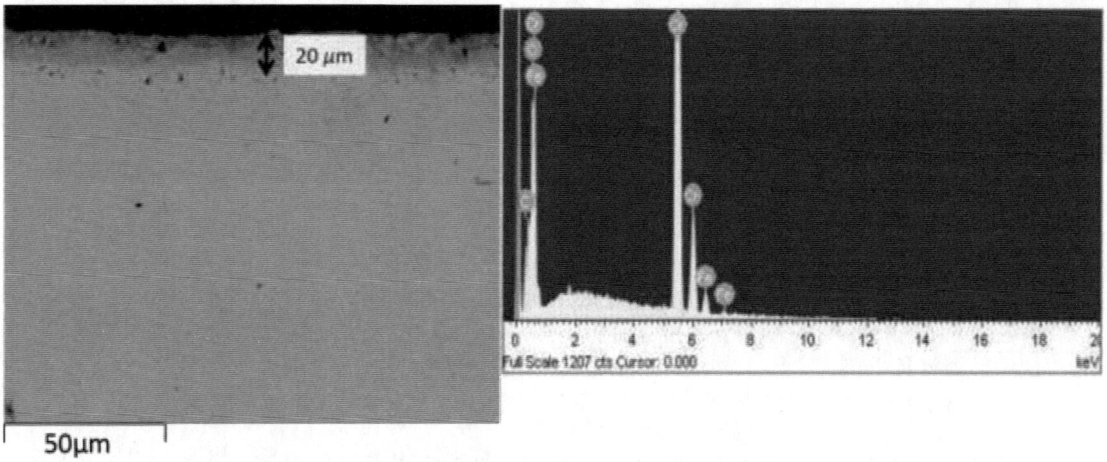

Figure 6. BSE cross sectional image after exposed to 24h in steam condition.

The graph of mass change versus time for after oxidation time was plotted in Figure 7. It showed that lower mass gains were observed for the chromized steels when it compared to the non-chromized steels. It is assumed that for the non-chromized steel, the chromium content was not enough to form continuous protective Cr_2O_3 layer when it is exposed to high temperature steam. The difference for weight after exposure was divided with total area of the sample to obtain the total mass gain of the sample. The difference of mass gain between chromized steel and non-chromized steel suggests that the chromized steel able to slow down the formation of iron oxide when oxidation occur. Meanwhile, for the chromized steel, the Cr reservoir can maintain the formation of Cr_2O_3, so that when the oxidation process occur, the formation iron oxide in the steel can be prevented.

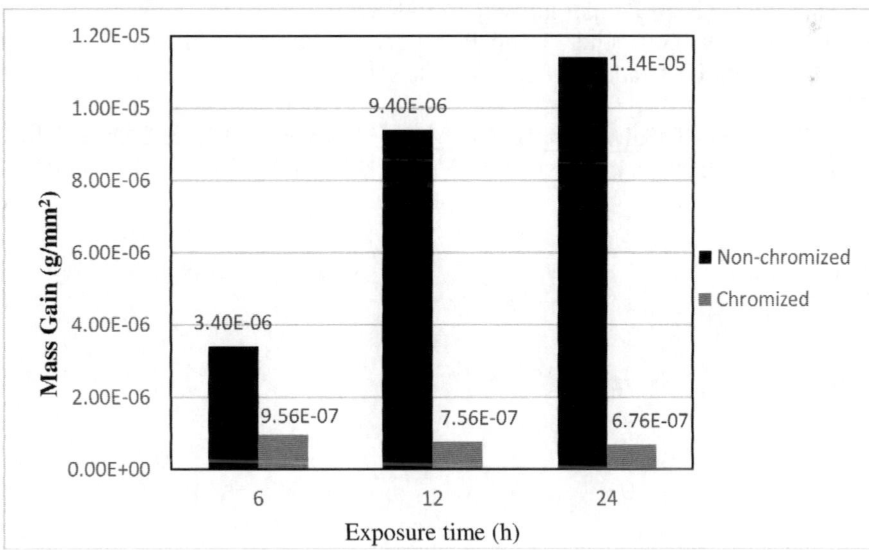

Figure 7. Graph for mass gain versus exposure time.

4. Conclusions

Chromizing process at 1050°C for 10 hours has successfully diffuse Cr and formed Cr-rich layer in the surface of T91 steel. The diffused Cr able to become chrome reservoir for the formation of protective Cr_2O_3 layer during high temperature steam oxidation at 700°C and reduce corrosion rate of the steel.

Acknowledgments
This research is supported by Fundametal Research Gran Scheme from Ministry of Higher Education Malaysia under RDU 140110. One of the author (F.A.B.F) is grateful for the PGRS UMP grant under PGRS160310.

References

[1] Viswanathan, R and Bakker W 2001 Materials for ultrasupercritical coal power plants—Boiler materials: Part 1. *Journal of Materials Engineering and Performance* **10**(1) 81-95.
[2] Cengel, Y A, Boles M A and Kanoğlu M 2014 Thermodynamics: an engineering approach 8th Ed. McGraw-Hill New York.
[3] Kurniawan, T, Fauzi FAB and Asmara YP 2016 High-temperature Oxidation of Fe-Cr Steels in Steam Condition–A Review *Indonesian Journal of Science and Technology* **1**(1) 107-114.
[4] Viswanathan R, Henry J F, Tanzosh J, Stanko G, Shingledecker J, Vitalis B and Purgert R 2005 U.S. program on materials technology for ultra-supercritical coal power plants *Journal of Materials Engineering and Performance* **14** 281-292.
[5] Essuman E, Meier G H, Zurek J, Hansel M and Quadakkers W J 2008 The Effect of Water Vapor on Selective Oxidation of Fe–Cr Alloys. *Oxidation of Metals* **69** 143-162.
[6] Nakai M, Nagai K, Murata Y, Morinaga M, Matsuda S and Kanno M 2005 Correlation of high-temperature steam oxidation with hydrogen dissolution in pure iron and ternary high-chromium ferritic steel *ISIJ International* **45**(7) 1066-72.
[7] Aguero A, Gonzalez V, Gutierrez M and Muelas R 2013 Oxidation under pure steam: Cr based protective oxides and coating *Surface and Coatings Technology* **237** 30-38.
[8] Kurniawan T, Ueda M, Kawamura K and Maruyama T 2013 Phase Stability of Iron Oxides on Palladium-Iron Alloy at Elevated Temperatures and Its Application to High Temperature Oxidation *Materials Transaction* **54**(09) 1829-37
[9] Galerie A, Henry S, Wouters Y, Mermoux M, Petit J -P and Antoni L 2005 Mechanisms of chromia scale failure during the course of 15–18Cr ferritic stainless steel oxidation in water vapour *Materials at High Temperatures* **22** 105-112.
[10] Bates B L, Zhang Y, Dryepondt S and Pint B A 2014 Creep behavior of pack cementation aluminide coatings on Grade 91 ferritic–martensitic alloy. *Surface and Coatings Technology* **240** 32-39.
[11] Fauzi F A, Kurniawan T, Salwani M S, Bin Y S and Harun W S W 2016 Chromium Enrichment on P11 Ferritic Steel by Pack Cementation *MATEC Web of Conferences* **74** 00036.

IOP Conf. Series: Materials Science and Engineering **257** (2017) 012087 doi:10.1088/1757-899X/257/1/012087

Morphology evaluation of ZrO_2 dip coating on mild steel and its corrosion performance in NaOH solution

M A Anwar[1,] T Kurniawan[1], Y P Asmara[1], W S W Harun[2], A N Oumar[3], A B D Nandyanto[4]

[1] Structural Materials and Degradation Focus Group, Faculty of Mechanical Engineering, Universiti Malaysia Pahang, 26600 Pekan,Pahang, Malaysia
[2] Human Engineering Group, Faculty of Mechanical Engineering, Universiti Malaysia Pahang, 26600 Pekan,Pahang, Malaysia
[3] Energy Sustainability Focus Group, Faculty of Mechanical Engineering, Universiti Malaysia Pahang, 26600 Pekan,Pahang, Malaysia
[4] Chemistry Department, Universitas Pendidikan Indonesia, Jl. Dr. Setiabudi No. 229, Bandung 40154, Indonesia

*Corresponding author: tedikurniawan@ump.edu.my

Abstract. In this work, the morphology of ZrO_2 thin film from dip coating process on mild steel has been investigated. Mild steel was dip-coated on solution made of zirconium butoxide as a precursor, ethanol as solvent, acetylacetone as chelating agent and water for hydrolysis. Number of dipping was adjusted at 3, 5 and 7 times. The dipped sample then annealed at 350°C for two hours by adjusting the heating rate at 1°C/min respectively. The optical microscope showed that micro-cracks were observed on the surface of the coating with its concentration reduced as dipping sequence increased. The XRD result showed that annealing process can produce polycrystalline tetragonal-ZrO_2. Meanwhile, SEM image showed that the thicknesses of the ZrO_2 coatings were in between 400-600 nm. The corrosion resistance of uncoated and coated substrates was studied by polarization test through potentio-dynamic polarization curve at 1mV/s immersed in with 3.5% NaCl. The coating efficiency was improved as the number of layer dip coated increased, which showed improvement in corrosion protection.

1. Introduction

Mild steel has been used in many engineering fields such as tooling, construction and transportation. It has an important role as it shows characteristics of high mechanical strength and ductility. However, mild steel do exhibit disadvantages, such as chemically unstable and corrosion happened on it easily. This deterioration may effect on premature and sudden failure with the risk of maintenance cost, unscheduled downtime, and lifetime. Several methods have been used for to protect mild steel from corrosion, such as coatings, cathodic protection, and inhibition [1-4].

Ceramic based coating was known as type of coating, which not only can improve corrosion resistance of metals, but also it can increase wear and erosion resistance and also provide thermal insulation [5,6]. Zirconium oxide is a type of ceramics materials that used largely for its excellent properties such as high hardness and good chemical stability and innertness [7-10]. It also known as ceramic material with thermalexpansion coefficient close to steel, which can reduce cracking during thermal treatment [11]. Furthermore, sol-gel technique was chosen to deposit ceramic coatings as it provide several advantages, such as produce good homogeneity coating, can accommodate complex shape and low annealing temperature [12,13].

In this work, ZrO_2 coating was prepared using zirconium butoxide as a precursor. The coating was applied on mild steel substrate through dip-coating method at different dipping sequence followed by annealing at 350°C. The structure of the coating was characterized by optical microscope, XRD, and SEM/EDX. The corrosion test for the coated substrates were carried out to find out their performance as protection against corrosion using potentiodynamic polarization in 3.5% NaCl. It is expected that ZrO_2 sol-gel dip coating can improve the corrosion performance of mild steel.

2. Experimental Procedure

2.1 Sample and solution preparation

Figure 1 shows the mild steel sample used in this experiment. The steel was mechanically cut into half-rounded shape using electrical discharge machining (EDM) with thickness of 5 mm and diameter of 16 mm. For dip-coating purpose, 1 mm hole was drilled on the top of the sample. After that, the sample was polished by emery paper until 800 grit and ultrasonically cleaned by acetone.

Figure 1. Mild Steel Substrate.

For sol-gel solution, zirconium (IV) butoxide, ethanol, acetylaceton, and distilled water were prepared at fixed molar ratio of: 1.25:20:1:3.2. Xirconium (IV) butoxide is a precurosor, ethanol act as solvent, acetylacetone act as chelating agent, and distilled water for hydrolysis. First, zirconium (IV) butoxide and ethanol were mixed and stirred at room temperature using magnetic stirrer for 30 minutes. Then acetylacetone was added to the mixture continued to stir for another 30 minutes. Distilled water then added to continue the for another 30 minutes.

2.2 Sol-gel dip coating process

The sol-gel dip coating process was conducted on PTL-MM01 Dip Coater. The substrates were dip coated for 3, 5 and 7 number of coatings. The parameters of the dip coater was set as followed; travel distance of 35 mm and immersion speed of 20mm/s at room temperature. The substrates were fully immersed in zirconium oxide solution for 1 minute after which air dried for another 2 minutes. These steps were repeated according to the number of coatings of each substrate. Once the deposition of the coating was completed, the substrates were thermally treated. During this phase, excess organic compounds found on the surface of coated substrates will evaporate. The substrate then annealed at 350°C for two hours by adjusting the heating rate at 1°C/min.

The morphology and thickness of coated substrates were examined using optical microscope and SEM (back scattered emission gun), while the chemical composition was studied using EDX. The SEM was equipped with a detector for energy dispersive X-ray analysis (EDX) and conducted at 15 kV.]The X-ray diffraction (XRD) Rigaku Miniflex using detector CuKα amp radiation at 30 kV and 15 mA. The diffractogram was obtained at scanning range 20 -70 degrees at scan step of 0.02 degree.

2.3 Corrosion testing

The rates of corrosion of the coated substrates were tested using electrochemical corrosion test. Electrochemical corrosion behaviours of ZrO_2 films were investigated using potentiostat to determine the corrosion protection efficiency of ZrO_2 coating in 3.5 % NaCl solution. Three electrodes were used to setup the electrochemical reaction, namely; platinum foil as a counter electrode, saturated calomel electrode (SCE) as a reference e electrode and substrate as working electrode. At a scan rate of 1mV/s, the experiment was carried out to reach the required equilibrium. The scanning range for voltage was set between -1.0 to 1.5 volt. The open circuit potential (OCP) was measured for 2 minutes to confirm the condition equilibrium prior to obtaining the potentiodynamic polarization curve.

3. Results and Discussion

The surface morphology and quality of the resulting ZrO_2 coatings was investigated using an optical microscope. Minor cracks were observed on the surface of the substrates under the microscope, as shown in Figure 2. More cracks were observed in the area close to the hole of the sample, meanwhile, less visible cracks were observed in the substrate with 7 times coatings compared to 3 times coating. These cracks may came from contamination of the solution due to environmental impurities and undeveloped dip coating technique which lacks control and consistency [14, 15].

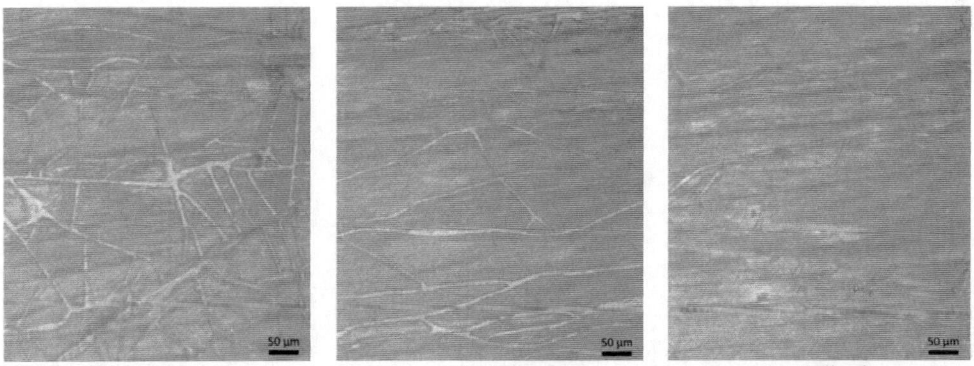

(a) 3 dipping (b) 5 dipping (c) 7 dipping

Figure 2. Optical microscopic images of the coated substrates at different number of dipping.

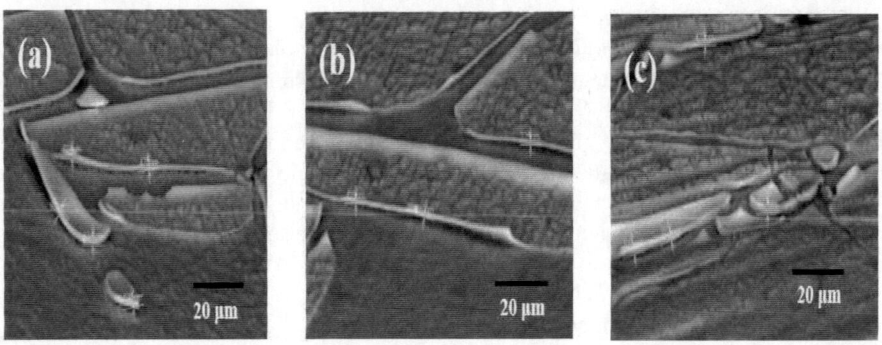

Figure 3. Thickness of ZrO_2 from different dipping time: (a) 3 dipping, (b) 5 dipping, and (c) 7 dipping.

Figure 3 shows the SEM images of ZrO_2 thin films from different dipping time. Cracks were clearly observed from all dipped sample, as has been shown in optical microscope. Thicker ZrO_2 was clearly shown by 5 and 7 dipping, as compared to 3 dipping. The thicknesses of coating were measured at several points and the average thickness of the coating was shown in Figure 4. From the graph, it was understood that the thickness was almost proportional to the number of coatings. Somehow, the change from 5 dipping to 7 dipping was not significant. The thickness for each dipping time are: 3 dip = 409±35 nm, 5 dip = 571±107 nm, and 7 dip = 623±18 nm. The EDX results shows the compositions of the element in all coated substrates consist of Fe, O and Zr. Fe was came from the mild steel substrate, since the ZrO_2 films was very thin, so electron can penetrate to the substrate area.

The characterization of thin film was continued by analysing ZrO_2 powder which were obtain by heating the sol-gel solution at 350°C at rate of 1°C/min for 2 hours. This work was conducted since ZrO_2 films was too thin so it will be difficult to get the XRD peak for the coating materials. From the XRD result in Figure 5, it was found that the annealing of ZrO_2 was successfully transformed amorphous ZrO2 into polycrystalline tetragonal ZrO_2 (t-ZrO_2). This result was in a good agreement with the work done by Soo et.al [16].

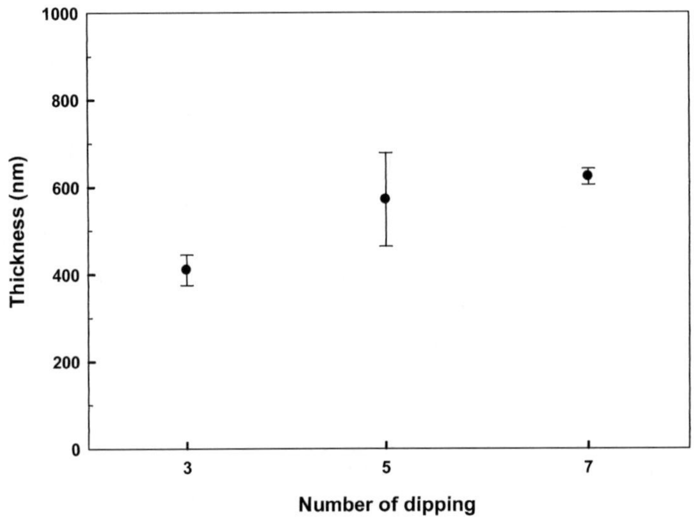

Figure 4. Average thickness of ZrO2 thin film at different number of dipping.

Figure 6 shows a comparison of potentiodynamic polarization curve between uncoated substrate and coated substrates with different number of coating. It can be obverse that all the substrate shows a plateau state for the cathodic reaction, this might be limiting current controlled by O_2 diffusion in the NaCl solutions.

The corrosion current density I_{corr} decreases. Normally E_{corr} can be used to estimate the tendency of the substrates. Based on the data summarized the corrosion current density decreases while polarization resistance increases as the number of coating increases. Based on the discussion can it can be conclude that corrosion resistance of zirconium oxide coating was gradually improved with increasing of number of coatings.

The substrate with 7 number of coating shows excellent corrosion resistance. This is because the quality of the films has an important effect on their corrosion behavior. Structural defects such as major and minor cracks permits corrosion on substrates so that the corrosion was far more serious than the substrates with minimal cracks.

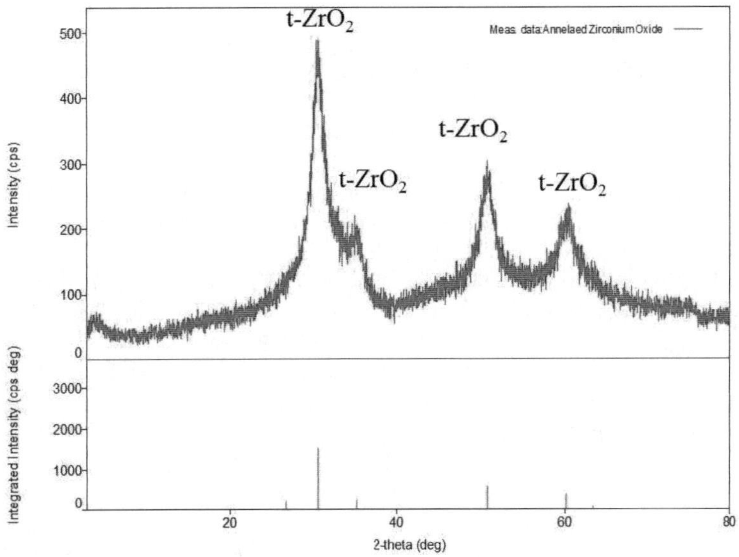

Figure 5. XRD analysis of ZrO_2 anneal at 350 °C with heating rate of 1 °C/min for 2 hours.

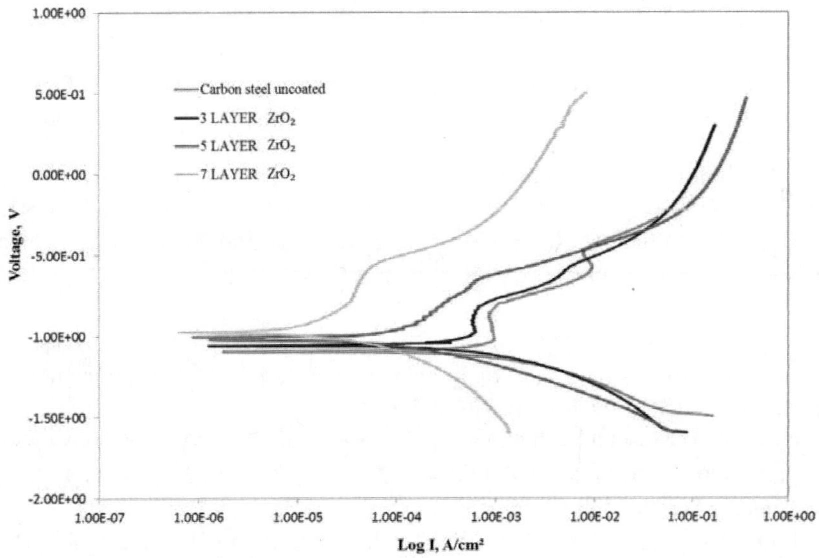

Figure 6. Potentiodynamic polarization curve of coated and uncoated substrates.

4. Conclusions

The morphology of ZrO_2 thin film produced by sol-gel dip coating techniques on mild steel has been studied. The performance of the coating on corrosion protection NaOH has also been investigated. It was found that number of dipping has a strong influence on the crack formation and film thickness. Higher dipping number produce thicker film with lower crack concentration. This properties effect on corrosion performance of the mild steel. The steel coated with higher dipping number produce better corrosion protection. These results confirmed that zirconium oxide coating can be applied as a barrier for the corrosion protection of mild steel.

Acknowledgements

This research is supported by Internal Research Grant Scheme from University Malaysia Pahang under grant number RDU1403112

References

1. Ghaziof S and Gao W 2014 Electrodeposition of single gamma phased Zn–Ni alloy coatings from additive-free acidic bath *Appl. Surf. Sci.* **311** 635–642.
2. Zhang Y, Shao Y, Liu X, Shi J, Wang Y, Meng G, Zeng X and Yang Y 2017 A study on corrosion protection of different polyaniline coatings for mild steel *Progress in Organic Coatings,* **111** 240-247.
3. Sepideh P, Mohammad R V, Alimorad R and Mohammad R B 2017 Exploring corrosion protection properties of solvent based epoxy-graphene oxide nanocomposite coatings on mild steel *Corrosion Science,* **115** 78-92.
4. Kai Y, Jian R, Jingwei F, Yin Z, Huayu Z, Liang W, Jinxing N, Shunyan T, Fang S and Chuanxian D 2017 Excellent wear resistance of plasma-sprayed amorphous Al_2O_3–$Y_3Al_5O_{12}$ ceramic coating *Surface and Coatings Technology* **309** 959-968
5. Hesam, Rezvani S, Mehdi S, Hossein E and Mansoor T F 2017 The effect of APS parameter on the microstructural, mechanical and corrosion properties of plasma sprayed Ni-Ti-Al intermetallic coatings *Surface and Coatings Technology.* **27** 1319–1323.
6. Adraider Y, Pang Y X, Nabhani F, Hodgson S N, Sharp M C and Al-Waidh A 2013 Fabrication of zirconium oxide coatings on stainless steel by a combined laser/sol-gel technique. *Ceramics International*, **39** (8) 9665-70.
7. Chen S, Xiang J, Huang J and Zhao X 2015 Microstructures and properties of double-ceramic-layer thermal barrier coatings of La2(Zr0.7Ce0.3)2O7/8YSZ made by atmospheric plasma spraying *Appl. Surf. Sci.* **340** 173–181.
8. Chang F, Zhou K, Tong X, Xu L, Zhang X and Liu M 2014 Microstructure and thermal shock resistance of the peg-nail structured TBCs treated by selective laser modification *Appl. Surf. Sci.* **317** 598–606.
9. Hany M, Abd E L, Mai M.and Khalaf 2015 Corrosion resistance of ZrO_2–TiO_2 nanocomposite multilayer thin films coated on carbon steel in hydrochloric acid solution *Materials Characterization*, **108** 29-41.
10. Luciano L 2017 Pitting and Crevice Corrosion *Engineering Tools for Corrosion* **4** 61-80.
11. El-Hadad A A, Barranco V, Samaniego A, Llorente A, García-Galván F R, Jiménez-Morales A, Galván J C and S Feliu Jr. 2014 Influence of substrate composition on corrosion protection of sol–gel thin films on magnesium alloys in 0.6 M NaCl aqueous solution *Progress in Organic Coatings*, **77** 1642-52.
12. Balaji J and Sethuraman M G 2017 Chitosan-doped-hybrid/TiO2 nanocomposite based sol-gel coating for the corrosion resistance of aluminum metal in 3.5% NaCl medium *International Journal of Biological Macromolecules* **104** 1730-39.
13. Babhu Vignesh R, Balaji J and Sethuraman M G 2017 Surface modification, characterization and corrosion protection of 1,3-diphenylthiourea doped sol-gel coating on aluminium. *Progress in Organic Coatings* **111** 112-123
14. Mohd Yusoff, M F, Abdul Kadir M R, Iqbal N, Hassan M A and Hussain R 2014 Dip coating of poly (caprolactone)/hydroxyapatite composite coating on Ti6Al4V for enhanced corrosion protection *Surface and Coatings Technology*, **245** 102-107
15. Online V A, Diaz L, Garc F R, Llorente I, Jim A, Galv J C and Jr F 2015 RSC Advances Effect of heat treatment of magnesium alloy substrates on corrosion resistance of a hybrid organic – inorganic sol–gel film *RSCAdv* **5** 105735-46
16. Soo M T, Prastomo N, Matsuda A, Kawamura G, Muto H, Noor A F M and Cheong K Y 2012 Elaboration and characterization of sol-gel derived ZrO 2 thin films treated with hot water. *Applied Surface Science,* **258** 5250-5258

Mapping of unknown industrial plant using ROS-based navigation mobile robot

G. Priyandoko[1], T. Y. Ming[1], M. S. H. Achmad[2]

[1]Faculty of Mechanical Engineering, University Malaysia Pahang (UMP), Pekan 26600 Pahang, Malaysia
[2]Universitas Teknologi Yogyakarta, 55285 Sleman, Yogyakarta, Indonesia

*Corresponding author: gigih@ump.edu.my

Abstract. This research examines how humans work with teleoperated unmanned mobile robot inspection in industrial plant area resulting 2D/3D map for further critical evaluation. This experiment focuses on two parts, the way human-robot doing remote interactions using robust method and the way robot perceives the environment surround as a 2D/3D perspective map. ROS (robot operating system) as a tool was utilized in the development and implementation during the research which comes up with robust data communication method in the form of messages and topics. RGBD SLAM performs the visual mapping function to construct 2D/3D map using Kinect sensor. The results showed that the mobile robot-based teleoperated system are successful to extend human perspective in term of remote surveillance in large area of industrial plant. It was concluded that the proposed work is robust solution for large mapping within an unknown construction building.

1. Introduction

Inspection robots enable operators to plan outages more precisely and efficient and thereby reduce downtime. This increases the availability of facilities and makes them more profitable, as well as boosting the safety of people and the environment. Eliminate human entry in confined spaces and perform the inspection from a safe & remote location. It can save the life of workers and allows the operator to access areas that could not be accessed by human, get more coverage, inspect behind baffles and obstacles.

While performing inspection, mapping is an important step when a mobile robot wants to explore the unknown environment [1-3], and the robot should be able to move and find the pathways to avoid obstacles and achieve the goal. Mobile robot navigation technology is now increasingly sophisticated, with the technology of 2D/3D mapping [4]. Many studies have been done on the implementation of 2D/3D SLAM (simultaneous localization and mapping) to determine the goal position respect to the current position of the robot in an area that has not been recognized previously [5]. The differential mobile robots are commonly used among researcher due to the ease of kinematic run in a narrow area, as in the halls of industrial plant.

ROS is an open source framework which has the capability to handle all software layers from low-level up to high-level layers. Because of that, everyone all over the world can build and share ROS stack and package for certain purposes. High demand on ROS stack from researchers was increasing exponentially the number of ROS stack production year to year. Therefore, ROS provides better methods which were can be chosen based on our own desires and purposes. It is also the product of trade-offs and prioritization made during its design cycle [6].

IOP Conf. Series: Materials Science and Engineering **257** (2017) 012088 doi:10.1088/1757-899X/257/1/012088

The contribution of this work is an industrial inspection solution which is able to describe the environment in the entire industrial plant in the form of large 2D/3D map that can be used for evaluation regard to safety and maintenance issues instead of using captured images or videos. The mobile robot was able to simultaneously localize and build a real-time map more than 2 km^2 of our robotic laboratory areas less than one hour. The purpose of this experiment was not merely to demonstrate implementation of differential mobile robot for industrial inspection, but also describes the methods how to develop mobile robot platforms run on robust operating system, that is ROS.

2. Differential Mobile Robot

In this study, the differential mobile robot (DMR) is utilized, which commonly used by researchers in addition to the Ackermann model. The DMR has two independent wheel drive system that can be controlled independently and it has a turning radius more varied than the Ackermann, even capable of rotating exactly on the centre of gravity (cog) of the robot body. So the DMR has the manoeuvrability to avoid obstacles better than the Ackermann model in term of narrow space.

Figure 1(a) shows the kinematic model of the real robot shown in figure 1(b) which is accommodates two input, linear velocity (v) and angular velocity (ω) toward the target position. Basically, the DMR model adopts unicycle kinematic model shown in Equation (1-3), as follows:

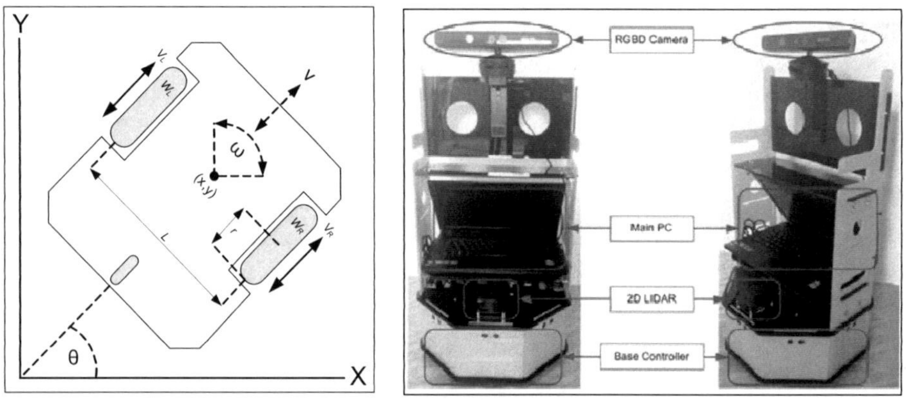

(a) (b)

Figure 1. The DMR model used in this study. (a) The kinematic model. (b) The real model.

$$\dot{x} = v \cos\theta \tag{1}$$

$$\dot{y} = v \sin\theta \tag{2}$$

$$\dot{\theta} = \omega \tag{3}$$

($\dot{\boldsymbol{x}}$,$\dot{\boldsymbol{y}}$) and ($\dot{\boldsymbol{\theta}}$) are dynamics position and orientation respectively in discrete time. While in the DMR kinematic model, **v** and **ω** are obtained from the variations in the speed of the right wheel ($\boldsymbol{v_R}$) and the left wheel ($\boldsymbol{v_L}$) simultaneously as shown in Equation (4-6), as follows:

$$\dot{x} = {}^{r}/_{2}\,(v_R + v_L) \cos\theta \tag{4}$$

$$\dot{y} = {}^{r}/_{2}\,(v_R + v_L) \sin\theta \tag{5}$$

$$\dot{\theta} = {}^{r}/_{L}\,(v_R - v_L) \tag{6}$$

Where, $\boldsymbol{r}$ denotes the wheel radius, and $\boldsymbol{L}$ denotes the distance between right wheel and left wheel. Thus, between Equation (1-3) and (4-6) are become equal, $\boldsymbol{v_R}$ and $\boldsymbol{v_L}$ can be calculated accordingly as shown in Equation (7-8), as follows:

$$v_R = (2v + \omega L)/2r \tag{7}$$

$$v_L = (2v - \omega L)/2r \tag{8}$$

Based on Eq. (1-8) then the base controller of the robot can translate the twist message from main controller contains linear and angular velocity as set points into speed of wheels for both of the right and the left side as shown in figure 2.

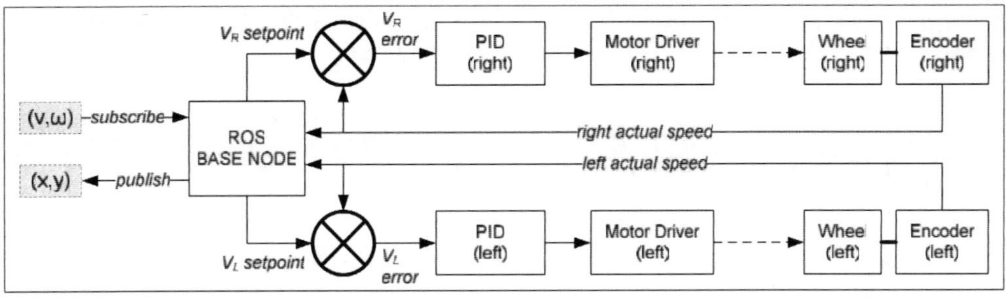

Figure 2. The DMR base controller system connected to the ROS environment.

Figure 2 describes the diagram of base controller system as ROS node. The base controller node subscribes ***v*** and ***ω*** in the form of geometry_msgs/Twist as ROS standard message from the other ROS node as remote operator (human). Simultaneously, nav_msgs/Odometry is published by the base controller, which is consist of *x,y,z* (=0) obtained from the encoders. The odometry message from the base controller can be used for pose estimation together with laser and visual odometry for better result. ROS geometry_msgs/Twist is translated into vR and vL based on Equation (7-8) as a set points for two PID controllers. Afterward, PID controllers maintain linear velocity and angular velocity of the mobile robot toward the target position as desired.

3. Robot Operating System

Robot Operating System (ROS) is an open source software framework primarily based on UNIX platform for operating robots. It is widely conducted for robotics research in the last decade and an obvious overview of ROS has been presented by [6-7]. ROS has three levels of concepts: the file system level, the Computation Graph level, and the Community level. ROS provides the service user would expect from an operating system, including hardware layer abstraction, low-level device control, implementation of commonly-used functionality, message-passing between processes, and package management. It also provides tools and libraries for obtaining, building, writing, and running code across multiple platforms. The primary goal of ROS is to support code reuse in robotics research and development. ROS is a distributed framework of processes (nodes) that enables executables to be individually designed and flexible at runtime. These processes can be grouped into Stacks and Packages, which can be easily shared and distributed. ROS also supports a federated system of code Repositories that enable collaboration to be distributed as well. Figure 3 shows the basic concept of ROS with distributed system capabilities.

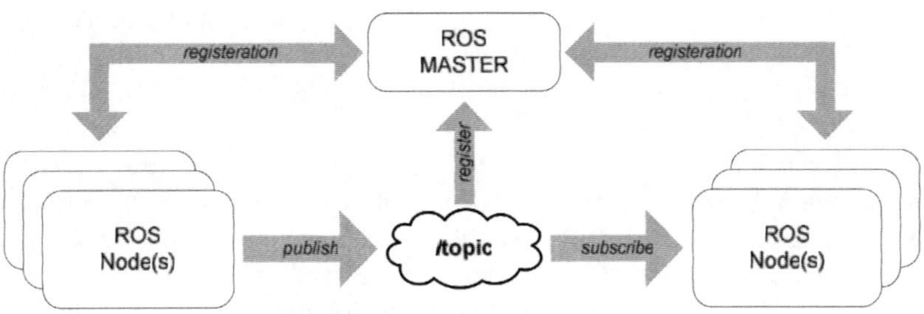

Figure 3. ROS basic concept.

ROS Master served as a major control the entire distribution of messages within the network. Each node will register the identity to the ROS Master include the IP address and the name of the topic will be published or subscribed. When ROS nodes subscribe to a particular topic, ROS nodes will contact ROS master to get the IP address of other ROS nodes which publish the associated topics. The name of topic is unique, so ROS nodes should not publish the topic with the same name in the same ROS network. Relationships between ROS nodes not merely publish and subscribe, but also client-server relationship. ROS node can ask the service (request) to the other nodes to do something and then get the results (response) without processing it themselves. It indicates the advantages of ROS system associated with the concept of distribution of information, so as to facilitate users to build a robotic system that has a complicated mission with simple solution.

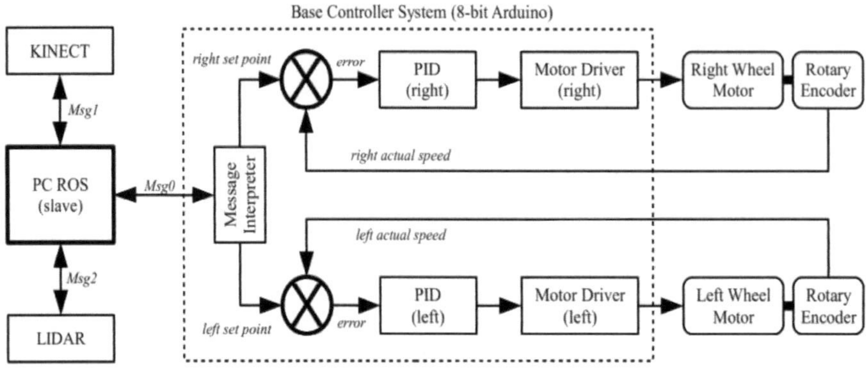

Figure 4. ROS-DMR ecosystems.

In figure 4, PC ROS-slave receives messages from Kinect and LIDAR and transmits a signal to the base controller. The Kinect sensor produces RGB-D image data, laser sensor generates 2D depth scanning data, and the base controller receives two types of input velocity, 3-axis linear velocity (***vx, vy, vz***) and 3-axis angular velocity (***ωx, ωy, ωz***). The base controller receives velocity messages were interpreted in two correspondence values of speed (left wheel & right wheel) based on robot kinematics model as seen in equations (1-2). Set points for two PID controllers are Vleftwheel and Vrightwheel which correspondence with targeted linear velocity and angular velocity given by a PC ROS-slave. Afterward, PID controllers maintain linear velocity and angular velocity of the mobile robot toward the target position. Figure 3 shows the real mobile robot which is completely designed based on model and component configuration as seen in figures 2 and 3.

4. Results and Discussion

The purpose of this study is to develop an interface between the human as an an operator with the mobile robot to get information about 2D/3D map in an unknown area. The distributed system of ROS ecosystems makes it easy for people to control the robot remotely and get a lot of information at the same time.

Figure 5 shows the reconstructed 2D map which is compared with ground truth. Apparently, the reconstructed 2D map is sync with ground truth map. Ground truth map does not cover the entire obstacle, just walls, drawers, cabinets, partition board are used as a comparison. The light gray section shows the occupied area, while at the dark gray indicates the area that have not occupied during exploration.

Figure 6 shows the occupied map from explorations in our experiment and visualizes using RVIZ (GUI). A 3D map obtained from Kinect sensor using Octomap was exactly located on the top of 2D map, shows that localization conducted by Hector SLAM was well performed. RVIZ runs on ROS platform which is provides attractive visual interaction facilities to accommodate human-robot interaction impressively. All of the topics from any kind of messages which belongs to nodes are could be obtained by RVIZ. As seen in figure 5 operator can use RGB image from Kinect sensor, kind of nodes contain the topics to be visualized, and 2D/3D map to be analysed as the main purpose in term of human-robot interaction that has been done. The numbering index on the 3D map shown in figure 6 illustrates each robot position when taking a picture using the RGB camera. This information is used to associate a 3D perception with 2D RGB image, and it is enable people to understand the situation around the robot either 2D or 3D.

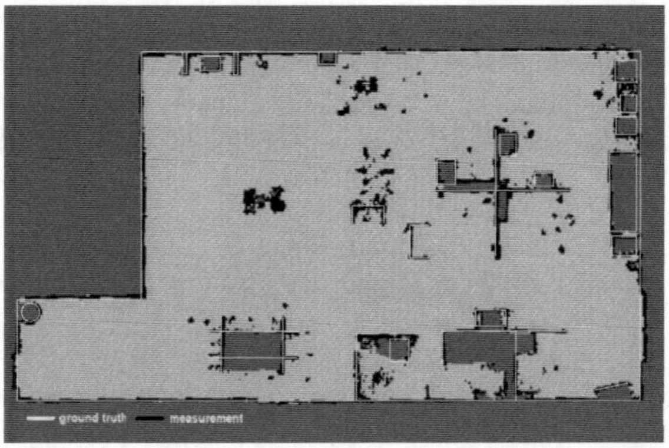

Figure 5. Constructed 2D map compared with ground truth

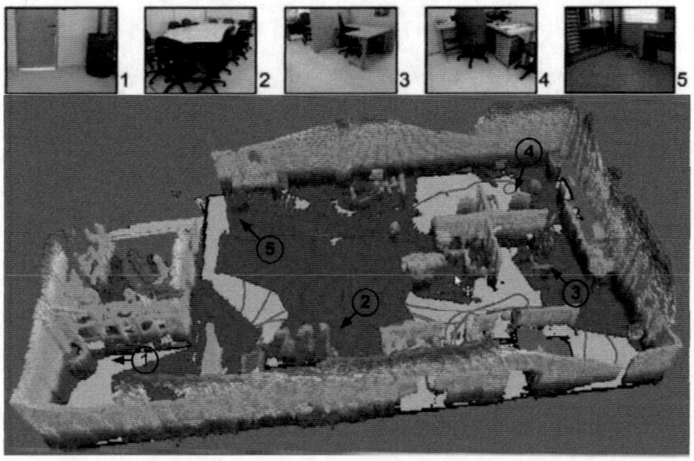

Figure 6. Visualize 2D & 3D occupied map using RVIZ

5. Conclusions

In this experiment, the odometry function becomes a big issue in term of accuracy and precision. Visual odometry which comes up with uncertainty error depends on image features detection and tracking is relying on how SURF algorithm deals with the texture-less object and the acceleration of the camera movement relative to the object. Texture-less area and high acceleration of camera movement produces high error in odometry function. The translation and rotational speed of the robot are the challenges in order to perform the detail inspection resulting low drift 3D map. However, this works has successfully demonstrated the methods how to apply mobile robot on ROS platform in order to performs the exploration and generate the 3D map which can be applied in the remote inspection of industrial plant.

Acknowledgments

The authors would like to thank the Faculty of Mechanical Engineering, University Malaysia Pahang for providing laboratory facilities and financial support. This project is supported by Universiti Malaysia Pahang under Research Grant Scheme (RDU140358).

References

[1] Chen C and Yinhang C 2008 Research on Map Building by Mobile Robots *Proceedings - 2nd International Symposium on Intelligent Information Technology Application (IITA)* **2** 673–677
[2] Wolf D. F. and Sukhatme G. S.,2008 Semantic mapping using mobile robots *IEEE Trans. Robot* **24**(2) 245–258
[3] Rusu R B 2010 Semantic 3D Object Maps for Everyday Manipulation in Human Living

Environments *KI - Kunstl. Intelligenz* **24** 345–348
[4] Zlot R and Bosse M 2014 Efficient large-scale 3D mobile mapping and surface reconstruction of an underground mine *Springer Tracts Adv. Robot* **92** 479–494
[5] Cortes B B, Sole X C and Salvi J 2011 Indoor SLAM using a range-augmented omnidirectional vision *Proceedings of the IEEE International Conference on Computer Vision* 280–287
[6] Quigley M, Conley K, Gerkey B, FAust J, Foote T, Leibs J, Berger E, Wheeler R and Mg A 2009 ROS: an open-source Robot Operating System *IEEE International Conference on Robotics and Automation*
[7] Cousins S 2010 ROS on the PR2 *IEEE Robot. Autom. Mag* **17** 23–25
[8] Labbe M and Michaud F 2014 Online Global Loop Closure Detection for Large-Scale Multi-Session Graph-Based SLAM *IEEE/RSJ International Conference on Intelligent Robots and Systems (IROS)* 2661–2666
[9] Labbe M and Michaud F 2013 Appearance-Based Loop Closure Detection *IEEE Trans. Robot* **29**(3) 734–745
[10] Grisetti G, Grzonka S, Stachniss C, Pfaff P and Burgard W 2007 Efficient Estimation of Accurate Maximum Likelihood Maps in 3D *IEEE/RSJ International Conference on Intelligent Robots and Systems* 3472–3478

Development of green vapour corrosion inhibitor

Y P Asmara*, **Suraj V, J P Siregar, T Kurniawan, D Bachtiar, N M Z N Mohamed**

Faculty of Mechanical Engineering, Universiti Malaysia Pahang 26600 Pekan, Pahang, Malaysia

*Corresponding author: ypanca@ump.edu.my

Abstract. Corrosion control using inhibitor is an effective method to protect carbon steel from corrosion. Due to environmental toxicity of chemical inorganic corrosion inhibitors (synthetic), green inhibitors are potentially to develop. In atmospheric conditions, green vapour corrosion inhibitors are the best solutions to replace the uses of inorganic corrosion inhibitors. This research used chemical acid extraction from the key lime (citrus aurantiifolia) leaves and seeds. They are used as the main ingredients to produce this effective green corrosion inhibitor. The experiments investigated effects of corrosion inhibition on corrosion rate of low carbon steel in 3% NaCl solution using both fog salt chamber and electrochemical cell. Using salt fog chamber to represent atmospheric conditions, and corrosion rates are evaluated visually and calculated using weight loss methods. Corrosion rate on electrochemical cell were calculated using linear polarization resistance (LPR) methods. All of the experiments were set in natural conditions at pH 7. Using weight loss for three days exposure time, the efficiency of the inhibitor reached 82.39%.

1. Introduction

Metallic structures are prone to corrode in atmospheric conditions, especially in marine environments. They are exposed to corrosive elements such as water vapour, and rain which is contained NaCl particles. Corrosion will occur quickly for the metallic structures made of carbon steel. There will be severe forms of localized corrosion, erosion corrosion and uniform corrosion. These forms of corrosion occur when water is absorbed and collected in the local area of the structures. To prolong the service life, the protection system should be applied to provide a barrier of corrosive elements penetration. Currently, inhibitor is the preferred protection choice from corrosion damages [1-4]. Corrosion inhibitor which are widely used in industries are commonly in liquid forms. In the forms of liquid, inhibitor cannot effectively protect carbon steel corrosion because liquid inhibitor cannot reach dry location. Thus, it needs inhibitor which can evaporate called volatile corrosion inhibitor (VCI) to protect steel corrosion in dry location. There are various kind of evaporating corrosion protection inhibitors had been develop since 1950's in the industrial of metal fields. VCI has been widely used for the protection of metal surfaces in a container. VCI often combined with other corrosion protection medium in a packaging of machined parts, joining fasteners and more. The wrapping of the packaging usually will be coated to protect the object inside it. The usage of VCI inhibitors mostly applied in confined spaces to protect the product for temporary only [5]. The application of VCI is to protect stored tools or metal made product inside bags, boxes, storage during transportation or protecting a newly arrived metal product. Effectiveness of corrosion inhibitors also depend on quantity. Low inhibitor quantity cannot protect effectively, hence promote corrosion process. While excessive quantity of inhibitor will affect production cost. The effectiveness of an inhibitor is shown by its impact on reducing corrosion rate. Recent researchers have used various types of inhibitor such as organic inhibitor or un-organic depending on environmental conditions [6].

The Key Lime Plant The Key lime (Citrus aurantiifolia) is a citrus hybrid of Citrus micrantha and Citrus nedica with a spherical shaped fruit with the diameter from 2.5cm to 5cm (1–2 in). The colour of the fruit is yellow when ripe but usually picked green commercially due to it have stronger sour taste. In addition, it is smaller and seedier, with a higher acidity, a stronger aroma, and a thinner rind, than the Persian lime (Citrus latifolia). In Malaysia it is commonly known as 'Limau Nipis' which is very popular amongst Malaysia's foods and drinks. On the other hand, key lime also famous for its health benefits than only giving taste to foods. For instance, it is used to treat nausea, headache, fainting, malaria, and sore throat. Many studies had been carried out regarding to the ability or function of the key lime plan food and health benefits. For example the studies that had been carried out based on the benefits of key lime plant are in the field of antioxidants, where the juice and peel do demonstrate antioxidant properties. In the field of corrosion inhibition, acid extract from both leaves and seed show impressive properties that can reduce the corrosion rate of metal especially in mild steel [7-10].

2. Research methodology

2.1. Electrochemical Set-up
A typical schematic three-electrode set-up used in all electrochemical experiments is shown in figure 1. A low carbon steel was used as the working electrode. Glass cell was fitted with graphite electrodes as auxiliary electrode and Ag/AgCl as a reference electrode.

2.2. Specimen Preparation
The working electrodes were carbon steels. The cylindrical specimens have diameter of 12 mm and 10 mm thickness which is used to electrochemical test. Weight loss and salt fog chamber test were used samples as shown in figure 1. Before immersion, the specimen surfaces were polished successively with 240, 400 and 600 grit SiC paper, rinsed with methanol and degreased using acetone. The experiments were repeated at least twice in order to ensure reasonable reproducibility.

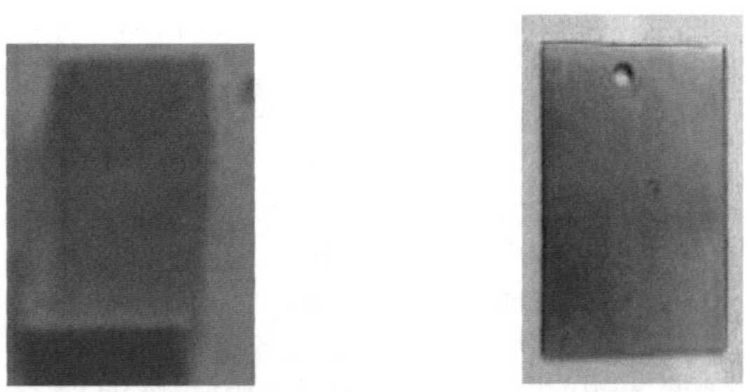

(a) (b)

Figure 1. Sample for (a) electrochemical test 1 x 3 cm and (b) sample for salt fog chamber tests.

2.3. Green inhibitor preparation
Key Lime Leaves and Seeds were used of inhibitors. The key lime leaves are obtained free from the fruit or groceries sellers in the market. Big amount of leaves are managed to be collected which is approximately to 1kg of key lime leaves. Meanwhile, the key lime seeds are obtained from the Pekan. Both the leaves and the seeds have almost the same concentration of acid compared to the key lime fruit itself. 500 gr of both of the leaves and the seeds are let to be dried under the sun to reduce the moisture content for three to four days until it turns into light brownish colour and the leaves becomes brittle. For the extraction of the plant materials, 25 gr of the powdered dry leaves and seeds poured

into 500mL of 1M HCL in a beaker. The powder concentration in the mixture is 5% of (wt/vol) in the HCl solution.

2.4. Cell Solutions
The experiments were performed in stagnant solutions condition. Schematic electrochemical test was presented in figure 2. The test pressure was 1 bar, the glass cell was filled with 1 liter of distilled water and 3% wt NaCl which was stirred with magnetic stirrer. Temperature was set using a hot plate. The Linear Polarization Resistance (LPR) technique was used to measure the corrosion rate. The procedure is similar to ASTM Experimental test G 5-94 [13].

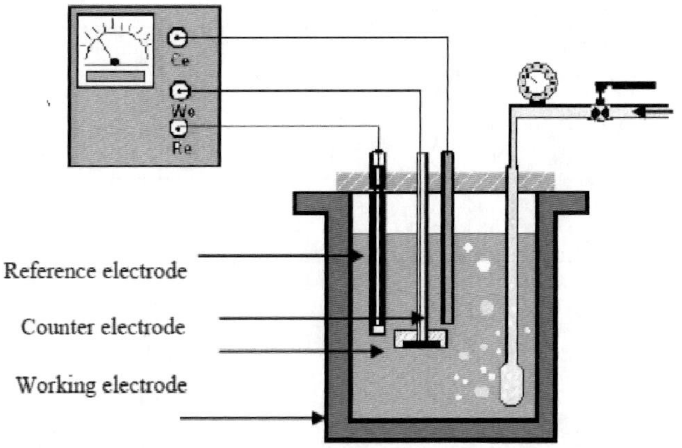

Figure 2. Schematic of corrosion test cell.

2.5. Salt Chamber Experiment
Since this project is mainly focus in producing vapour corrosion inhibitor, thus this salt chamber experiment plays an important role in providing a suitable condition to test the inhibitor solution in vapour form and it is carried out by using the ECO Instrument Salt Chamber machine as presented in figure 3. This salt chamber is set into 35°C as the control temperature in the chamber, spray time is set to five minutes and the overall process time is set to one hour. Before starting, it is a must to sand and polish the mild steel metal plate to get rid of oxidants or other residuals on the surface. The testing equipment consists of a closed testing chamber, where a salt solution mainly is atomized by means of a nozzle. This produces a corrosive environment in the chamber so that parts exposed in it are attacked under this severe corroding atmosphere.

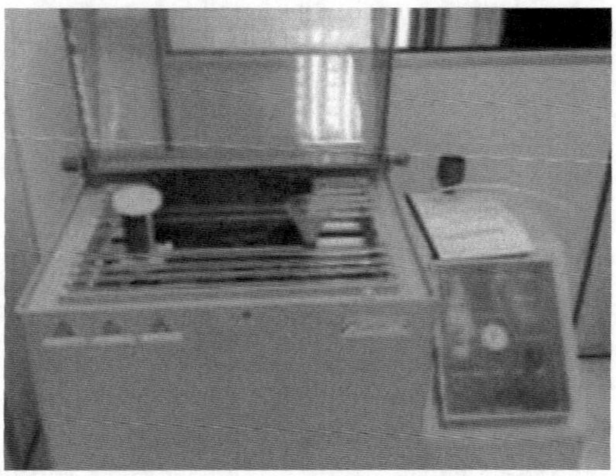

Figure 3. Salt fog chamber for the corrosion resistance testing, ASTM B117.

3. Results and Discussion

3.1. Potentiodynamic Polarization Measurement.

The measurement is carried out with 500 mL and 20% (v/v) of NaCl solution or known as Blank solution. Then it is followed by the adding and increasing the amount of the inhibitor from 5mL (1% v/v) and 7mL (1.4% v/v) in the NaCl solution. Every measurement is obtained by the plotting of Tafel graph is presented in table 1. As calculated, the key lime inhibitor's efficiency increases as the concentration increases in the NaCl solution.

Table 1. The I_{corr} values from the Tafel plot

Concentration, v/v %	Icorr , A/cm^2	Efficiency
0 (Blank)	0.00007	-
1 (5mL)	0.00006	14.57
1.4 (7mL)	0.00003	57.14

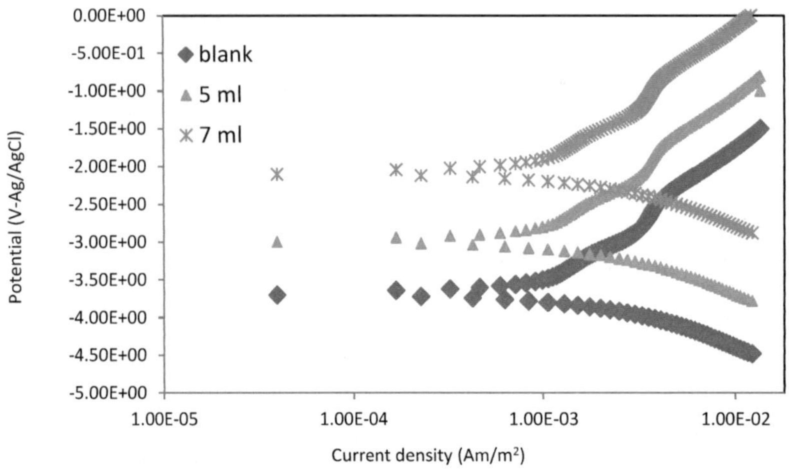

Figure 4. Polarisation graph of green corrosion inhibitor on carbon steel.

From the polarisation graph (figure 4), it is shown that green corrosion inhibitor can improve corrosion potential. Using 7 ml of inhibitor, potential corrosion of carbon steel on NaCl solution increased by 2V. It means that tendency of corrosion to start will reduced. Green corrosion inhibitors also may give the protection for both anodic and cathodic process [12, 13]. They work as a mixed corrosion inhibitor and change polarity of the samples. Inhibitors form a passivation films which inhibits the anodic metal dissolution reaction and causes largest anodic shift of the corrosion potential. Besides, it is important that the concentration of the inhibitor should be high enough. This is because, if may affect the formation of the film if the concentration is low and the metal surface might not be fully covered with the anodic corrosion inhibitor and localized corrosion may take place [14].

3.2. Weight Loss

Based on the figure 5, container I salt solution had turned from colourless to brownish-red colour. This shows that corrosion of the mild steel does occurs in this container. The cause of the colour change of the NaCl solution is due to the metal oxides that are formed during the corrosion process. The mild steel turn in to dark silver colour which shows that, the surfaces do faces oxidation in the three weeks period time. The salt formation is in smaller size like grains and sticks onto the surface of the cuboid together with the metal oxides. On the other side container II, the solutions remains the same greenish colour as the beginning of the experiment. The metal surface remains shiny as the initial condition and

IOP Conf. Series: Materials Science and Engineering **257** (2017) 012089 doi:10.1088/1757-899X/257/1/012089

there is no presence of metal oxides on the surfaces too compared to the sample from container I. By using weight loss method, the average of corrosion rate of samples I was 2.8775 mm/year is higher compared to the corrosion rate of samples II which us just 0.5066 mm/year. This shows that samples I corrodes faster than samples II during the presence of salt. Thus, the green inhibitor from the key lime reduces or slows down the corrosion rate of mild steel in during the presence of salt. The inhibitor efficiency of this project in this experiment is 82.39% as calculated.

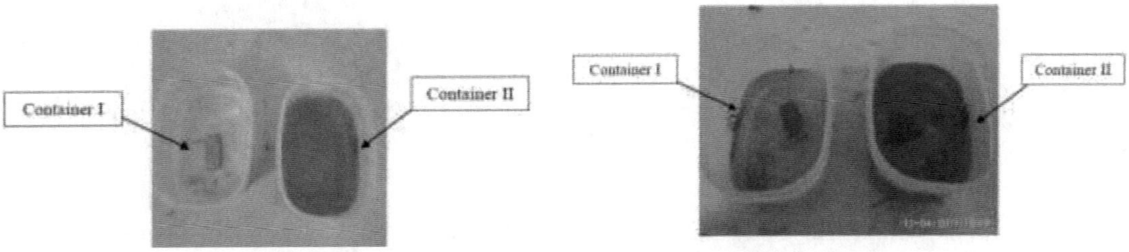

(a) (b)

Figure 5. Comparison samples after three weeks without inhibitor (container I) and with inhibitor (container II) after three days experiments (b).

Table 2. The weighing of samples after 3weeks immersion time.

Weighing Term	Samples I (gr)	Samples II (gr)
Term 1	10.1134	10.3681
Term 2	10.1137	10.3682
Term 3	10.1133	10.3684

3.3. Salt Fog Chamber Experiment

For the first test, the sanded and washed metal plate that is left for one hour in the salt chamber without the presence of the inhibitor. The metal plate surface shows the corrosion precipitation to be more vigorous compared to the experiment with the presence of the inhibitor. Visually proven that the inhibitor does show some inhabitancy to fight against corrosion that occurs on the mild steel plate surface there is not much difference (figure 6). VCI usually have low or significant vapour pressure which includes the effective of the inhibiting properties. Besides that VCI also provides protection against moisture in the air through its vaporization and cover fully the surface of the metal by forming a thin film of protective layer. In addition, it also neutralize the carbon dioxide which is acidic and shifts the Ph of the metal surface towards less corrosive and acidic value. In other words, when the VCI come into contact with the metal surface it condensed and hydrolysed by the moisture for the liberation of the protective ions. If the vapour pressure is too low, another volatile substance such as water vapour or urea can be used as carrier of the VCI. The volatility can also be enhanced by the use of host guest systems where the VCI are incorporated to a porous substrate. Then, the VCIs are transferred to the metal surface by diffusion through the gas room and are adsorbed directly on the metal or dissolved in a surface moisture layer and reduce the corrosion rate markedly [15].

Without the presence of inhibitor

With the presence of inhibitor

Figure 6. Salt fog chamber test results for 1 hour of green vapour corrosion inhibitor.

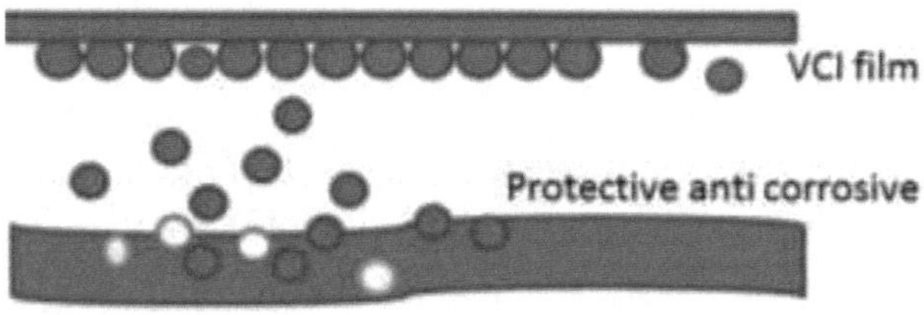

Figure 7. Model formation of VCI film on metal surface.

The mechanism of a corrosion inhibitor acts by undergoing chemisorption process, in which it is a chemically adsorb by the metal surface through the combination of the inhibitor ions and the metallic ions on the metal surface (figure 6). This will form a thin layer of protective film as the corrosion protection of the metal surface. Moreover, the inhibitor also causes the formation of oxide protection film of the base metal. The inhibitor undergoes a reaction together with the corrosive component which is potential in the aqueous media and a complex product is formed [16].

4. Conclusions

It is proven that application of this vapour green anticorrosion using key lime is reliable to reduce corrosion process. It can be applied both on dry and wet conditions. The highest efficiency was obtained in wet conditions which was 82 %. In dry conditions, efficiency inhibitor seemed not contributed significantly. Mechanism of inhibitor to protect carbon steel may be related to chemisorption process, in which it is a chemically adsorb by the metal surface through the combination of the inhibitor ions and the metallic ions on the metal surface to form film formation.

Acknowledgments

The authors are thankful to Universiti Malaysia Pahang for providing grant and facilities for the research with grant number RDU150372.

References

[1] Asmara, Y. P., Ismail, M., 2011, Study Combinations Effects of HAc in H2S/CO2 Corrosion, Journal of Applied Sciences, Volume 11, Page No.: 1821-1826.

[2] Asmara Y P, Juliawati, J. Jaafar, K. Azuar, J.P. Siregar, Kurniawan T, 2014, Effects Pre-strain of Carbon Steel on Stress-Strain Diagram in CO_2 Environment with the Presence of H_2S

[3] Bastidas, J. M. and E. M. Mora (2013). "A Laboratory Study of Mild Steel Vapour Phase Corrosion and Its Inhibition by Dicyclohexylamine Nitrite." 97: 57-65.

[4] Daniyan, A. A., et al. (2011). "Effect of Palm Oil as Corrosion Inhibitor on Ductile Iron and Mild Steel." Engineering Materials Development Institute, Akure, Nigeria 12: 45-53.

[5] Dariva, C. G. and A. F. Galio (2014). "Corrosion Inhibitors – Principles, Mechanisms and Applications." Developments in Corrosion Protection: 365-379.

[6] Finšgar, M. and J. Jackson (2014). "Application of corrosion inhibitors for steels in acidic media for the oil and gas industry: A review." Corrosion Science: 17–41.

[7] Goyal, G., et al. (2008). "Effect of superficially applied ZrO2 inhibitor on the high temperature corrosion performance of some Fe-, Co- and Ni-base superalloys." Applied Surface Science(254): 6653–6661.

[8] Nor Asma, R.B.A., Asmara, Y. P., Ismail, M., 2011, *Study on the Effect of Surface Finish on Corrosion of Carbon Steel in CO_2 Environment,* Journal of Applied Sciences, Volume 11, Page No.2053-2057.

[9] Azhani, Asmara, Y. P., Khairuazlan, M., 2014, Influences of H_2SO_4 and NaCl concentrations on stress corrosion cracking of AISI 304 stainless steel, Advanced Materials Research Vol. 893, pp. 410-414.

[10] M.A. Quraishib, et al. (2010). "Green approach to corrosion inhibition of mild steel in hydrochloric acid and sulphuric acid solutions by the extract of Murraya koenigii leaves." Materials Chemistry and Physics(122): 114–122.

[11] Pandian Bothi Raja and Mathur Gopalakrishnan Sethuraman (2007). "Natural products as corrosion inhibitor for metals in corrosive media — A review." Materials Letters(62): 113 - 116.

[12] Singh, A. and M. A. Quraishib (2015). "Acidizing Corrosion Inhibitors: A Review." Environ. Sci.: 224-235.

[13] ASTM G 5-94 - Standard Reference Test Method for Making Potentiostatic and Potentiodynamic Polarization Measurements.

[14] International Journal of Material Science Innovations (IJMSI), v: 2, 3 pp.52-58.

[15] Wormwell, F. and A. D. Mercer (1952). "Sodium Benzoate And Other Metal Benzoates As Solutions Corrosion-Inhibitors In Water And In Aqueous." CORROSION-INHIBITORS(2): 150 - 160.

[16] Zhang, D.-q., et al. (2006). "Volatile corrosion inhibitor film formation on carbon steel surface and its inhibition effect on the atmospheric corrosion of carbon steel." Applied Surface Science(253): 343–1348.

IOP Conf. Series: Materials Science and Engineering **257** (2017) 012090 doi:10.1088/1757-899X/257/1/012090

Application of response surface methodology method in designing corrosion inhibitor

Y P Asmara[1], Athirah[1], J P Siregar[1], T Kurniawan[1*] and D Bachtiar[1]

[1]Structural Materials and Degradation Focus Group, Faculty of Mechanical Engineering, Universiti Malaysia Pahang 26600 Pekan, Pahang, Malaysia

*Corresponding author email: tedikurniawan@ump.edu.my

Abstract. In oil and gas pipelines and offshore structure, inhibitors have been considered to be the first choice to reduce corrosion rate. There are many corrosion inhibitor compositions available in the market. To produce the best corrosion inhibitor requires many experimental data which is not efficient. These experiments used response surface methodology (RSM) to select corrosion inhibitor compositions. The experiments investigated effects of corrosion inhibition on corrosion rate of low carbon steel in 3% NaCl solution with different concentrations of selected main inhibitor compositions which are ethyl acetate (EA), ethylene glycol (EG) and sodium benzoate (SB). Corrosion rate were calculated using linear polarization resistance (LPR). All of the experiments were set in natural conditions at pH 7. MINITAB® version 15 was used for data analysis. It is shown that a quadratic model is a representative model can predict best corrosion inhibitor composition comprehensibly.

1. Introduction

Corrosion is the destructive attack of materials through reaction with its environment. Corrosion is an important and critical issue faced in various industrial sectors [1-3]. The offshore structures, which always exposed to NaCl solutions, rain, condensation, and water, constantly exposed to severe corrosion. One method for controlling the metal losses that occurs due to corrosion in pipelines is to use a corrosion inhibitor.

Corrosion inhibitors are chemical substances or mixtures which are added in low concentrations to an aggressive environment for reducing corrosion rate of a material [4]. Typically they work by adsorbing themselves on the metal surface, and form film on it to protect the surfaces of metal. Among various methods to prevent the destruction or degradation of metal surface, corrosion inhibitor is the best methods of corrosion protection due to its low cost and practical method.

According to National Association of Corrosion Engineers (NACE) [5-7], inhibitors can be classified into anodic, cathodic or mixed depends on the intervention by the reaction of corrosion inhibitor preferentially in the anodic or cathodic sites or engage in both sites. They are widely used to minimize or reduce the rate of corrosion of metal structures.

The common inhibitors that were being used are nitrates, benzoates, phosphates, and chromates. Since it is almost impossible to prevent corrosion, it is more apparent that controlling the rate of corrosion can be the most economical solution. Hence, this study is set to design a novel corrosion inhibitor of selected chemicals which are ethyl acetate, ethylene glycol, and sodium benzoate for corrosion protection on offshore structure that will be useful for oil and gas industry in future. The main objective of this research is to design a novel inhibitor with optimum corrosion performance that can be used as a corrosion protection in oil and gas environments. (RSM) methods have become a feasible tool to simplify empirical corrosion models [8-10].

2. Research methodology

2.1. Modelling work

Experimental design for corrosion using central composite design (CCD) is an accepted experimental approach in RSM. It enables the construction of a second order (quadratic) model without the need to perform a full factorial design experiment. Using response information, the optimum data between factors can be developed, and model improvements can be achieved. The application of RSM with a CCD technique suggests a design model to calculate the response variables at specific environmental values [11,12].

Table 1. Relationship between natural and coded variable levels of dependent variables used in RSM to study effects of components to select inhibitor.

Level	Code	Concentration of Inhibitor		
		Ethyl Acetate (A) (ppm)	Ethylene Glycol (B) (%)	Sodium Benzoate (C) (ppm)
Axial point	1.7	500	50	500
High	1	400	40	400
Centre	0	300	30	300
Low	-1	200	20	200
Axial point	-1.7	100	10	100

2.2. Electrochemical Set-up

A typical schematic three-electrode set-up used in all electrochemical experiments is shown in Figure 1. A low carbon steel was used as the working electrode. Glass cell was fitted with graphite electrodes as auxiliary electrode and Ag/AgCl as a reference electrode [13].

2.3. Specimen Preparation

The working electrodes were carbon steels. The cylindrical specimens have diameter of 12 mm and 10 mm thickness. Before immersion, the specimen surfaces were polished successively with 240, 400 and 600 grit SiC paper, rinsed with methanol and degreased using acetone. The experiments were repeated at least twice in order to ensure reasonable reproducibility.

2.4. Cell Solutions

The experiments were performed both in stagnant solutions condition. The test pressure was 1 bar, the glass cell was filled with 1 liter of distilled water and 3% wt NaCl which was stirred with magnetic stirrer. Temperature was set using a hot plate. The Linear Polarization Resistance (LPR) technique was used to measure the corrosion rate. The procedure is similar to ASTM Experimental test G 5-94 [14].

Table 2. Central-composite experimental design of the independent variables with the observed values for the response, (Y_i).

CODE			NATURAL		
A	B	C	A (ppm)	B (%)	C (ppm)
0	-1.7	0	300	10	300
0	0	0	300	30	300
0	0	-1.7	300	30	100
-1	-1	1	200	20	400
1	1	-1	400	40	200
-1.7	0	0	100	30	300
1.7	0	0	500	30	300
0	0	0	300	30	300
1	-1	-1	400	20	200
0	0	0	300	30	300
0	0	0	300	30	300
1	-1	1	400	20	400
-1	1	1	200	40	400
0	1.7	0	300	50	300
0	0	1.7	300	30	500
-1	-1	-1	200	20	200
-1	1	-1	200	40	200
0	0	0	300	30	300
1	1	1	400	40	400
0	0	0	300	30	300

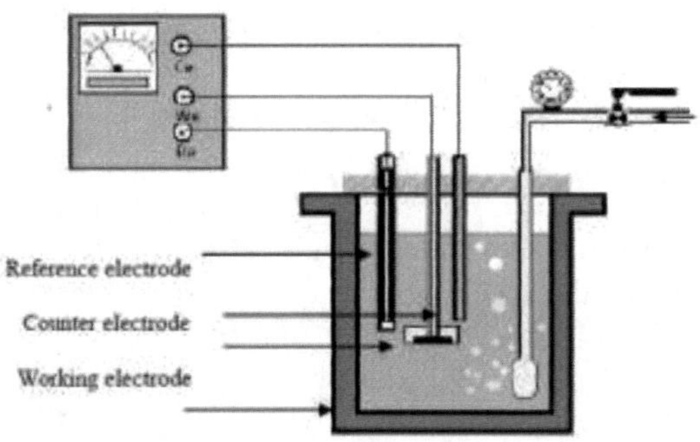

Figure 1. Schematic of corrosion test cell.

3. Results and discussion

3.1. Fitting corrosion predictions model at pH 5.5

Corrosion predictions model at pH 7 was fitted using second-order polynomial pattern. The model proposed for the response is given below. The model was calculated using Minitab software program [15].

$$\textit{Corrosion Rate} = 0.122 - 0.000220\,EA - 0.00141\,EG - 0.000156\,SB + 0.000017\,EG^2 + 0.000001\,EA*EG + 0.000001\,EG*SB \quad (1)$$

Where:
EA: Ethyl Acetate
EG: Ethylene Glycol
SB: Sodium Benzoate

For linear relationship, EA shows the most effect in reducing the corrosion rate followed by SB and EG respectively. On the other hand, for square relationship, SB*SB shows the lowest value of corrosion rate, then followed by EG*EG and EA*EA respectively. For two way interaction, EG*SB indicates the most effect in reduce the corrosion rate, and then followed by EA*EG and EA*EA.

3.2. Analysis and interpretation of response surface design

In order to examine how the factors influence the response, the factorial plots were done by plotting the data means or also known as fitted means. Figure 2 below shows the main effects plot for corrosion rate.

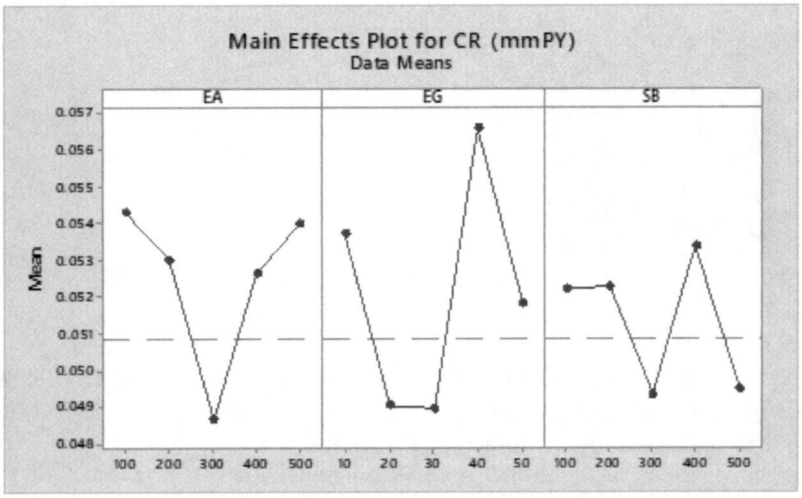

Figure 2. Main effects plot for corrosion rate.

Figure 2 shows the main effects plot of corrosion rate for low carbon steel in NaCl with different concentration of inhibitors. For ethyl acetate, when the concentration decrease from 100 to 300 ppm, the corrosion rate decreases until it reaches a minimum value of 0.0485 mmPY, then increased with increasing concentrations from 300 to 500 ppm. As for ethylene glycol, the minimum value of corrosion rate is at 30 % in water with 0.0489 mmPY and the maximum value is at 40 % in water with 0.0565 mmPY. For sodium benzoate, the minimum corrosion rate is at 300 ppm with 0.0492 mmPY and the maximum is at 400 ppm with 0.0531 mmPY.

The contour plots with 2D contour lines were done by plotting the relationship between two continuous predictors and a fitted response. They were produced for all pairs of continuous variables in separate panels on the same graph for corrosion rate and Icorr. Due to synergistic effect of these mixed inhibitor, the simultaneous combined effects of the concentration with different types of inhibitor on the corrosion rate and Icorr are shown in Figure 3.

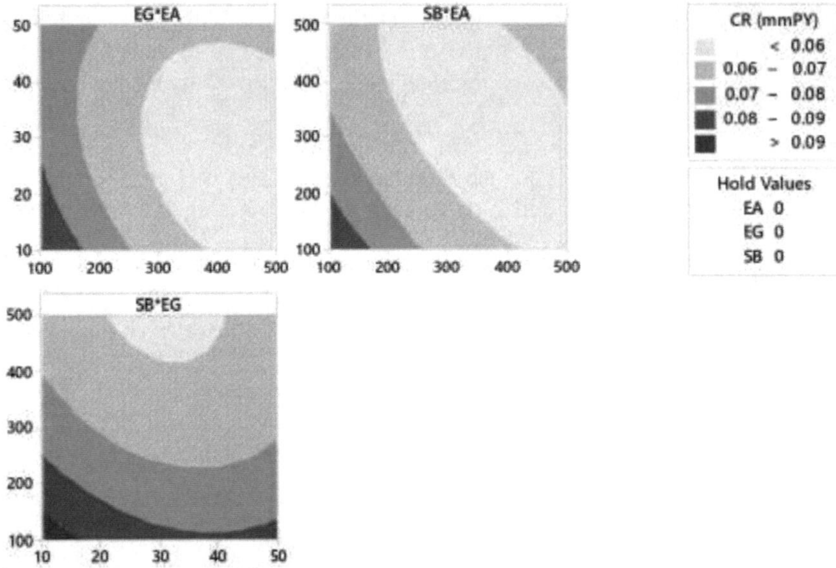

Figure 3. Contour plots of corrosion rate.

As illustrated in Figure 3, the contour plot of ethyl acetate and ethylene glycol indicates that the greatest corrosion rate is about 100 ppm of ethyl acetate with around 20 to 30 % in water of ethylene glycol, and the area appears at the bottom left of the plot. The smallest corrosion rate is when the concentration of ethyl acetate is about 480 ppm with about 20 to 30 % in water of ethylene glycol which appears at the centre of the plot. Besides, from the contour plot of ethyl acetate and sodium benzoate, the greatest corrosion rate takes place with 100 ppm for both sodium benzoate and ethyl acetate, which appears at the lower left of the plot. The smallest corrosion rate for these chemical is when the concentration of both of them around 300 to 400 ppm which located at the centre of the plot. Next, from the contour plot of sodium benzoate and ethylene glycol, it shows that the greatest corrosion rate is at 100 ppm of sodium benzoate and around 10 % in water of ethylene glycol where it appears at the lower left of the plot, while the smallest corrosion rate is when the concentration of sodium benzoate is around 420 to 500 ppm, with around 30 % of ethylene glycol in water.

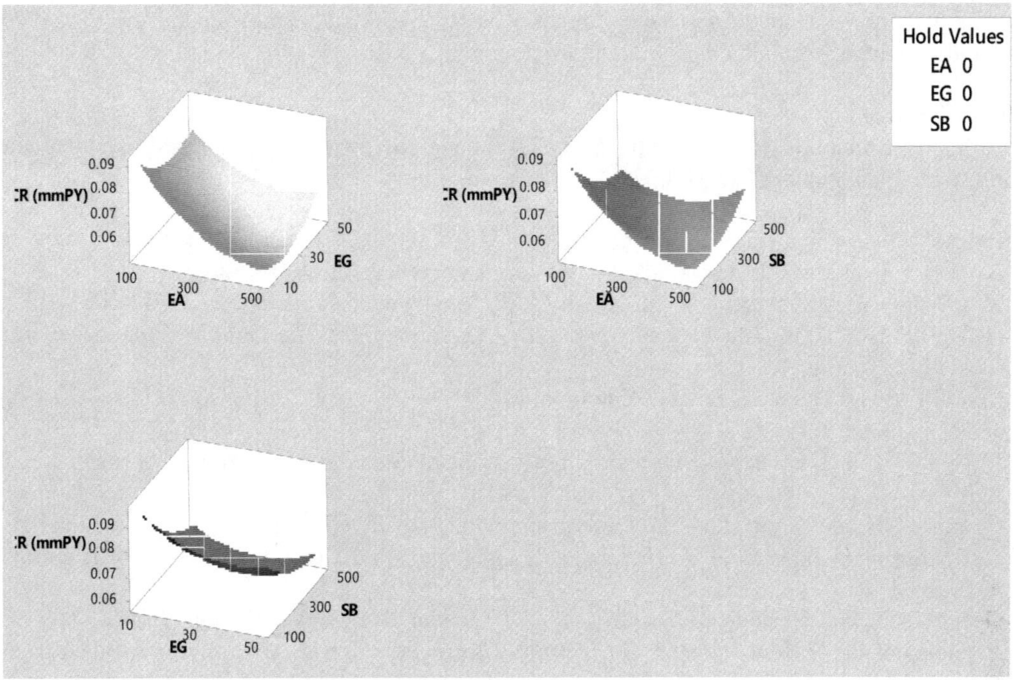

Figure 4. Surface plots of corrosion rate [15].

Figure 4 shows the results of three dimensional view that provides a clear vision of corrosion rate. From the figure, it is clearly shown that the area represent for the lowest corrosion rate of the surface plot of ethyl acetate and ethylene glycol is the largest among the three surface plots, then it is followed by the surface plot of ethyl acetate and sodium glycol, and the last one is the surface plot of ethylene glycol and sodium benzoate. In the other hand, the surface plot of ethylene glycol and sodium benzoate has the greatest area of the highest value of corrosion rate. Therefore, it can be conclude that the combination of ethyl acetate and ethylene glycol is the most efficient, then followed by the combination of ethyl acetate and sodium benzoate, then the last one is ethylene glycol and sodium benzoate.

In this project, the occurrence of the results may cause by the synergistic effect when the three inhibitors are mixed together. After reviewing a number of journals, there might be some possible situations to occur during the experiment. Sodium benzoate may affect the solubility of ethyl acetate and ethylene glycol in NaCl. When different concentration of sodium benzoate were used, the solubility of other corrosion inhibitors will be different. In addition, the results also can be affected by the combination of ethyl acetate and ethylene glycol since their combination will increase the toxicity. Besides, when ethylene glycol mixed with ethyl acetate, acetic acid in ethylene glycol would be extracted out. Hence, the most synergistic effect of the mixed inhibitor is ethyl acetate and ethylene glycol among the three pairs of inhibitors.

4. Conclusions

From the experiments that have been carried out, it indicates the lowest corrosion rate happened with 300 ppm ethyl acetate, 30 % in water of ethylene glycol, and 300 ppm sodium benzoate whose value is 0.0439 mmPY. The combination of ethyl acetate and sodium benzoate has a medium area of the lowest corrosion rate, where the combination of ethylene glycol and sodium benzoate implies the smallest area of the lowest corrosion rate. From this results, it indicates that the best combination or synergistic effect of the chemicals are ethyl acetate and ethylene glycol. Furthermore, It can be concluded that the selected model adequately represent the data for all responses obtained. Second order polynomial regression model is adequate to represent the data for all responses obtained. Using RSM, further analyses regarding individual and interaction effects between the variables can be study more comprehensive. Corrosion product formation is an important parameter in defining corrosion model and predicting corrosion rates in systems containing NaCl.

Acknowledgments
The authors are thankful to Universiti Malaysia Pahang for providing grant and facilities for the research with grand number RDU150372.

References

[1] Asmara Y P and Ismail M 2011 *Journal of Applied Sciences* 11 1821-1826.
[2] Nor Asma R B A, Asmara Y P and Ismail M *2011 Journal of Applied Sciences* **11** 2053-2057.
[3] Hedges B L M 1999 *The Role of Acetate in CO_2 Corrosion: The Double* Whammy Corrosion (NACE International).
[4] Box G E P and Draper N R 1987 Empirical Model Building and Response Surfaces (John Wiley and Sons, New York).
[5] Devore P 2005 *The Exploration and Analysis Data* Statistic 5th edition (Thomson Learning, Belmont, USA).
[6] George K S and Nesic 2004 *Electrochemical Investigation and Modeling of Carbon Dioxide Corrosion of Carbon Steel in The Presence of Acetic Acid* Corrosion (NACE International) p. 4379.
[7] George K S and Nesic 2007 *Investigation of Carbon Dioxide Corrosion of Mild Steel in The Presence of Acetic Acid— Part 1: Basic Mechanisms* Corrosion (NACE International) p. 4379.
[8] Gulbrandsen E 2007 *Acetic Acid and Carbon Dioxide Corrosion of Carbon Steel Covered with Iron Carbonate* Corrosion (NACE International) p. 7322.
[9] Sutjipto A G E, Asmara Y P and Jusoh M A 2017 *Procedia Engineering* **17** 88-92.
[10] Azhani, Asmara Y P and Khairuazlan M 2014 *Advanced Materials Research* **893** 410-414.
[11] Nesic S 2007 *Corrosion Science* **49** 4308 – 4338.
[12] Asmara Y P and Ismail M 2012 Corrosion Engineering, Science and Technology 47 10-18.
[13] Asmara Y P, Ismail M F, Chui L G and Halimi J 2016 *IOP Conference Series: Materials Science and Engineering* **114** 012128.
[14] ASTM G 5-94 - Standard Reference Test Method for Making Potentiostatic and Potentiodynamic Polarization Measurements.
[15] Minitab Inc. 2006. Minitab Statistical Software, Release 15 for Windows, State College, Pennsylvania.

IOP Conf. Series: Materials Science and Engineering **257** (2017) 012091 doi:10.1088/1757-899X/257/1/012091

Pipe leak diagnostic using high frequency piezoelectric pressure sensor and automatic selection of intrinsic mode function

Hanafi M Yusop[1*], M F Ghazali [1], M.F.M.Yusof [1], M.A Pi Remli [1] and M H Kamarulzaman[2]

[1]Advanced Structural Integrity and Vibration Research (ASIVR, Faculty of Mechanical Engineering, Universiti Malaysia Pahang, Pekan, Pahang, Malaysia;
[2] PERMATA Pintar National Gifted Center, National University of Malaysia, Malaysia

*Corresponding author email: m.hanafiyusop@yahoo.com.my

Abstract. In a recent study, the analysis of pressure transient signals could be seen as an accurate and low-cost method for leak and feature detection in water distribution systems. Transient phenomena occurs due to sudden changes in the fluid's propagation in pipelines system caused by rapid pressure and flow fluctuation due to events such as closing and opening valves rapidly or through pump failure. In this paper, the feasibility of the Hilbert-Huang transform (HHT) method/technique in analysing the pressure transient signals in presented and discussed. HHT is a way to decompose a signal into intrinsic mode functions (IMF). However, the advantage of HHT is its difficulty in selecting the suitable IMF for the next data post-processing method which is Hilbert Transform (HT). This paper reveals that utilizing the application of an integrated kurtosis-based algorithm for a z-filter technique (I-Kaz) to kurtosis ratio (I-Kaz-Kurtosis) allows/contributes to/leads to automatic selection of the IMF that should be used. This technique is demonstrated on a 57.90-meter medium high-density polyethylene (MDPE) pipe installed with a single artificial leak. The analysis results using the I-Kaz-kurtosis ratio revealed/confirmed that the method can be used as an automatic selection of the IMF although the noise level ratio of the signal is low. Therefore, the I-Kaz-kurtosis ratio method is recommended as a means to implement an automatic selection technique of the IMF for HHT analysis.

1. Introduction

Water plays an important role in the contribution of the economic growth of a country. In the recent decade, water is one of the major issues addressed in research worldwide due to the increase in water pollution and non-revenue water (NRW). Water Economy Network (WEN), reported that water scarcity is rapidly becoming the number one global resource concern. Consequently, with the increase in the number/demand of accessible clean and fresh water, the water network will increase thus the possibility of the increment of water issues is higher. The first global survey conducted by [1] reported that water loss for most country falls into the range of 20%-30% and for Malaysia, the figure is 43% of NRW. NRW refers to the circumstance in which the amount of water reaches to the consumer is lesser than the amount of water that has been produced. The losses are due to real losses such as pipeline network leakage or the apparent loss such as through theft or bypass of water network (the water reaches to the consumer not metered) and metering inaccuracies. Nowadays, minimizing of NRW during transportation process through pipeline network is very essential. The problem of water leaks in pipeline distribution system is the main cause of NRW. Lately, water authorities' concern of leak in water distribution system is growing, and thus provide some necessary incentives for investment in leak detection technology and leak reduction strategies [2].The main cause of leaks is due to the aging of pipelines, corrosion, erosion, excessive pressure of water resulting from operational error and water hammer generated by rapid opening and closing of valves. In the recent year, water authorities voiced out such concern in reducing the NRW. Pressure Transient analysis is a recently developed technique/method for leak detection and has attracted researchers to conduct more studies due to its cost effectiveness.

The possibility to analyse the pressure transient phenomenon in both time and frequency domain has been satisfied by the implementation of time-frequency signal processing techniques such as Hilbert-Huang Transform (HHT). HHT analysis was proposed due to the ability of this method to detect and capture the occurrence of transient phenomena through the non-stationary signal. In recent years, HHT has been more widely used as time-frequency analysis as signal processing method [3] ,[4]. HHT was proposed as newly developed and powerful method to analyze non-linear and non-stationary signal [5].

However, this method has the difficulty in selecting the suitable IMF for the next data post-processing method which is Hilbert Transform (HT). Through the Empirical Mode Decomposition (EMD), HHT will decompose the signals into the different monocomponents and symmetric components by the sifting process. The monocomponent signal is called intrinsic mode function (IMF) ([6]).The main drawback of this approach is the necessity to know a priori of the frequencies and level of original signal that should be analysed [7]. The HHT analysis in this case is not completely automatic because it requires the interaction of skilled user to select which IMF level is suitable for further analysis. Therefore, automatic selection for better IMF is needed to overcome this problem. The mathematical computational technique and statistical value analysis were applied by the researcher to increase the degree of automation, and then eliminates the interaction of skilled user to select the relevant IMF.

There have been relatively few recent studies and statistical approach on measurable quality for automatic selection of relevant and appropriate intrinsic mode function (IMF). [8] proposed variance and standard deviation as statistical value to differentiate each level of the decomposition of intrinsic mode function (IMF). The research was done using electrocardiogram (ECG) signal in order to detect Artificial Fibrillation (AF) rhythm and normal rhythm. [7] demonstrated Merit Index as an indicator that allows automatic selection of IMF level.The research/study utilizes rotary element (gear) signal to test the efficiency of merit index in the selection of relevant and appropriate IMF. The method works as a measure of periodicity degree and absolute skewness of each IMF level. In monitoring rotating machines, [9] introduced computation of correlation coefficient value for each level of IMF as quantifiable quality in selecting appropriate IMF level. In the research/study, vibration signal is captured from rolling element (bearing), then the signal is extracted using a hybrid method of Minimum Entropy Deconvolution (MED), Empirical Mode Decomposition (EMD) and Teager Energy Operator (TEO). The selected IMF is the IMF which presents the higher value of correlation coefficient compared to the original signal. [10] presents an Energy-Based approach through mutual information (MI) coefficient as a method of selection relevant IMF.The research/study was done by applying the synthetic signal embedded in white noise and real world signal. Mutual Information (MI) is also utilized as a method to select appropriate IMF in biomedical signal processing conducted by [7].

This paper proposed Integrated Kurtosis Algorithm for Z-notch filter hybrid with higher order statistical method (Kurtosis) as the automatic selection of intrinsic mode function (IMF). The I-Kaz was chosen since the approach of this method was adapted in general [leak detection techniques] and detects very well any changes and uncertainties of the measured signals [11]. Unlike the existing statistical analysis such as variance, standard deviation , and kurtosis, I-Kaz method was capable of indicating both amplitude and frequency difference by simultaneously obtaining the I-Kaz representation and I-Kaz coefficient, $Z\sigma$. Detailed experimental result was discussed by [11] and [12] using vibrational signal acquired from rotating machinery part (Bearing) and detected tool wear during turning process respectively. In a study conducted by Nuawi 2014 et al, the I-Kaz was compared to variance and the results revealed that the variance parameter was unable to detect both amplitude and frequency changes in the non-stationary signal. Thus, the I-Kaz method was reliable especially for monitoring purposes where the observation on the changes of the signal amplitude and frequency were commonly required ([11]).

The effectiveness and reliability of the I-Kaz-kurtosis ratio as the automatic selection of IMF is shown/demonstrated using artificial pressure transient signal and random signal generated using Matlab, and the results revealed that IMF with maximum value of I-Kaz-kurtosis ratio coefficient is relevant and appropriate for further analysis. The research also comprises the comparison of the effectiveness of kurtosis with IKaz as the automatic selection method for better IMF. This paper focuses on the implementation of I-Kaz-kurtosis ratio coefficient as the automatic selection means/technique of intrinsic mode function using real pressure transient signal captured along the 57.9 meter medium density polyethylene pipe with single artificial leak simulator attached within 37.8 meter distance away from the point of analysis. The result for real pressure transient signal showed that the IMF contains highest value of I-Kaz-kurtosis coefficient, which is suitable and appropriate for further analysis. Furthermore, the I-Kaz-kurtosis ratio is proven suitable as automatic selection method for better and relevant IMF. Therefore, the degree of automation for Hilbert Huang Transform (HHT) was enhanced in detecting pipe leakage in live water distribution system using pressure transient signal.

2. Integrated based kurtosis algorithm for z-filter technique to kurtosis ratio

2.1. *Integrated Based Kurtosis Algorithm for z-filter technique*

Integrated Kurtosis-based Algorithm for z-filter technique (I-Kaz) is the method developed based on the concept of data scattering about its centroid. The sampling frequency of the raw signal was chosen as 2.56 referring to Nyquist number ([11]). Most of the researchers in signal analysis and processing are comfortable with the number. To avoid the aliasing effect, the maximum frequency span will be of equation 1.

$$F_{max} = \frac{fs}{2.56} \tag{1}$$

I-Kaz decomposes the time domain signal into three levels of the frequency range, which are x-axis represent as low frequency (LF) with a range of 0 – 0.25 of f_{max}, followed by y-axis represent as high frequency (HF) with a range of 0.25 – 0.5 of f_{max}, and finally, z-axis represents as very high frequency (VF) with range 0.5 of f_{max}. The 0.25 f_{max} and the 0.5 f_{max} were selected as low and high-frequency range limits respectively done by referring to the 2nd order of the Daubechies concept in signal decomposition process [13]. Referring to the kurtosis, I-kaz method contributes to three-dimensional graphical representations of the measured signal frequency distribution. The variance, σ^2 of each frequency band which is σL^2 represent as a low-frequency band, σH^2 represent as a high-frequency band and σV^2 represents as a very-high-frequency band which is calculated as in equation 2, 3 and 4 to measure the scattering of data distribution.

$$\sigma L^2 = \frac{\sum_{i=1}^{N}(x_i - \mu_L)^2}{n} \tag{2}$$

$$\sigma H^2 = \frac{\sum_{i=1}^{N}(x_i - \mu_H)^2}{n} \tag{3}$$

$$\sigma V^2 = \frac{\sum_{i=1}^{N}(x_i - \mu_V)^2}{n} \tag{4}$$

The I-Kaz coefficient, Ƶσ can be simplified in term of variance, σ as in equation 5 [12].

$$Z\sigma = \sqrt{(\sigma L^2)^2 + (\sigma H^2)^2 + (\sigma V^2)^2} \quad (5)$$

2.2. Kurtosis

Kurtosis describes the measure of spikiness and hence a good indicator of peak analysis for spikes detection in a non-stationary signal component such as pressure transient. Kurtosis is expressed as;

$$\text{Kurtosis}(x) = \frac{E\{(x-\mu)^4\}}{\sigma^4} \quad (6)$$

where μ and σ represent the mean and standard deviation of time series signal respectively. The E illustrates the expectation operation. The kurtosis demonstrates the spikiness and peakedness of probability distribution associated to the instantaneous amplitudes of the time-series analysis [14]. Therefore, the I-kaz-kurtosis ratio is expressed as equation 7.

$$ZK\sigma = \frac{(\sqrt{(\sigma L^2)^2 + (\sigma H^2)^2 + (\sigma V^2)^2})(\sigma^4)}{(E\{(x-\mu)^4\})} \quad (7)$$

2.3 Algorithm

Automatic Selection of Intrinsic Mode Function (IMF) Based on Pressure Transient Signal

1. Perform EMD on a synthetic pressure transient response generate from TLM
2. EMD decomposed the signal into so-called level of Intrinsic Mode Function
3. Calculate the Integrated Kurtosis Algorithm for a z-filter technique to kurtosis ratio (I-Kaz-Kurtosis) coefficient using Eq. 7.
4. Identified the IMF that corresponds to the largest I-Kaz-kurtosis ratio coefficient.
5. Perform Hilbert Transform (HT) and Hilbert Spectrum (HS) for IMF level that contains the largest coefficient of I-Kaz-Kurtosis ratio.

3. Piezoelectric Pressure Sensor

Piezoelectric pressure sensor was designed and suited to measure dynamic pressure. It is typically not suited for static pressure measurements. Sensor with special capabilities are needed to measure dynamic pressure including turbulence, blast, ballistic, and engine combustion. High frequency piezoelectric pressure sensor has the capabilities to measure fast response of pressure change, ruggedness, high stiffness, extended range, and has the ability to measure quasi static pressures. This type of pressure sensor has the capabilities to capture signal with continuous operating system even for underwater, 'dirty' experiment and also field test application for a long range cable. Basically, this transducer is ideal for all dynamic pressure applications for a temperature range from -196 to +135 ^{0}C. With the advantages of detecting small pressure changes at high pressure levels, this sensor is one of the most practically used ones in real applications. This type of pressure transducer is practically used in the application that includes hydraulic and pneumatic systems, blast and ballistic, very repeatable, wide dynamic range, fast rise time, and also fluid borne raise.

Figure 1. Piezoelectric pressure sensor.

Figure 1 illustrates the piezoelectric pressure sensor that was utilized in this experiment. Basically, the sensor has the ability in detecting fast frequency response in the pipeline system. The signal and frequency response from the pipeline system represent the pipeline condition. The reflection signal acquired in the pipeline system basically represents the response from the pipe features such as pipefitting, blockage, leak and strangest precipitates in the pipelines system. The excitation frequency, vibration from test rig, water pump frequency and other noise provides some effect to the signal acquired by the piezoelectric sensor. In this research Hilbert Huang Transform (HHT) acts as the Signal analysis method, and then was utilized in extracting the noisy signal in order to get the real reflection signal that represents the pipeline information.

4. Methodology

The experimental test was regulated in a circular loop pipeline system with 57.9-meter long. The network was constructed using Medium High-Density Polyethylene pipes (MDPE) with an outer diameter 60 millimeters, 55 millimetres of internal diameter and 2.6 millimeters mean thickness. An artificial leak was installed at a point of 37.8 meters away from the point of analysis in the pipeline system. The outlet of the pipe is kept connected to a free surface tank where the water from the pipe ends is discharged. This is to prevent sudden expansion phenomenon of the pressure waves, and minimizing negative pressure wave, as it will affect the collected data from the transducer. The speed of sound calibrated from the test rig is 524.3 m/s.

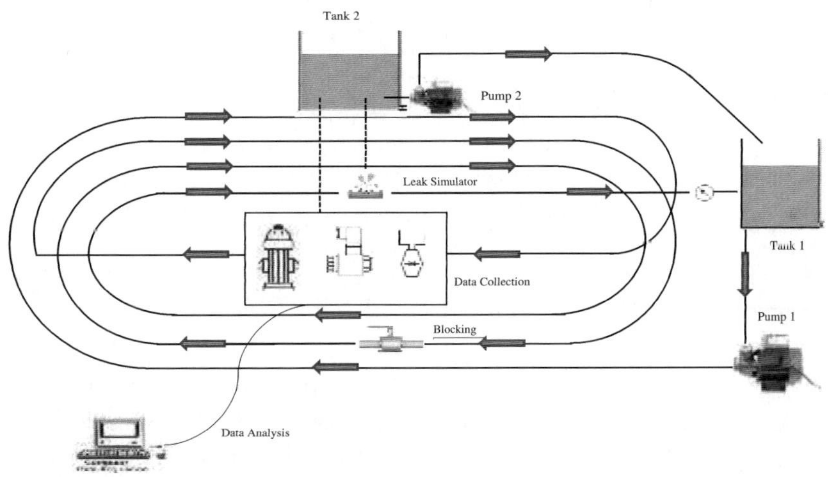

Figure 2. Schematic diagram of test rig design for laboratory and experimental test.

The schematic diagram (Figure 2) shows the piezoelectric pressure sensor and solenoid valves installed 10 meters away from the electric pump to avoid too much noise during the collection of the data. This is due to wave characteristic as turbulent flow when the water exits from pump outlet, the flow is approaching in laminar or steady state condition due to the distance traveled. This phenomenon is generated by friction from wall pipe due to water flow and damping. This research mainly utilized pressure transient respond as the method of detection and location in the pipeline system. The idea is proposed from the usage of pressure transient flow of water in the pipeline network. Theoretically, pressure transient responds occur when there is a sudden change in the flow of water by closing or opening of the valves in the pipeline network. As the phenomena happened inside the pipeline network, wave propagation of water will be created along the pipeline network, and also the characteristic of wave propagation will be used as a medium to detect leak and pipe features along the pipeline network. The wave inside the pipeline network will propagate with different signal depending on the size of the leak, distance of leak, type of pipe features and distance of pipe feature.

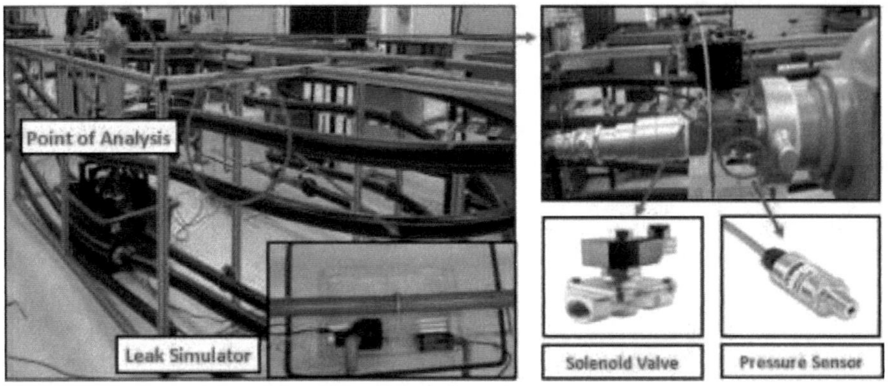

Figure 3. Experimental test rig in laboratory scale.

Figure 3 shows the experimental test rig installed at the Fluid Mechanic Laboratory in University Malaysia Pahang Malaysia. The fire hydrant cap was designed and fabricated to fix the solenoid valve and pressure sensor as well as to make the system robust for a field test. Both components play an important role in each task as solenoid valve is used to generate the water hammer phenomenon and piezoelectric pressure sensor is used to acquire the reflection signal. In a few cases, the system will shut down by itself due to sudden opening or closing of valve and malfunction of the pump. This will generate the occurrence of "water hammer" in the pipeline system. This phenomenon took in all of the pressure pipe systems, frequently causes vibration and destruction on pipeline system [15]. To generate this phenomenon, solenoid valve (Figure 3) was utilized as a tool to create water hammer along the pipeline system.

5. Results and Discussion

The data of the experiment acquired from the transducers using Matlab represent the pipeline system behavior and characteristics. The signal acquired retrieves the characteristics of the whole pipeline system itself. The pressure transient waves propagated along the pipeline system in both directions, away from the burst origin through the speed of sound in water distribution system. In this paper, Hilbert-Huang Transform (HHT) which contains Empirical Mode Decomposition (EMD) as data pre-processing method and Hilbert Transform (HT) as data post-processing method were employed as the tools to analyze the pressure transient signal.

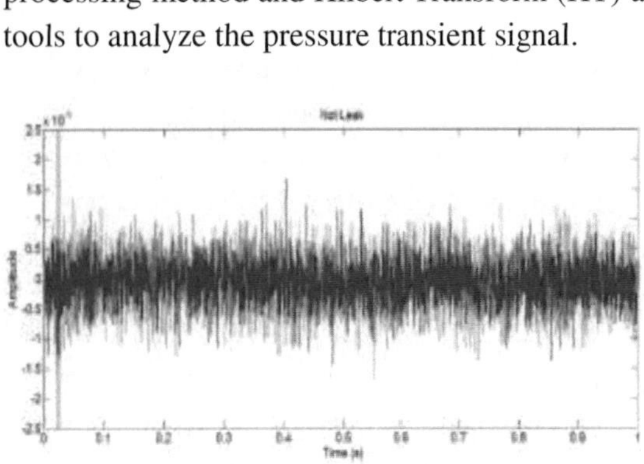

Figure 4. Signal Response for No Leak Data.

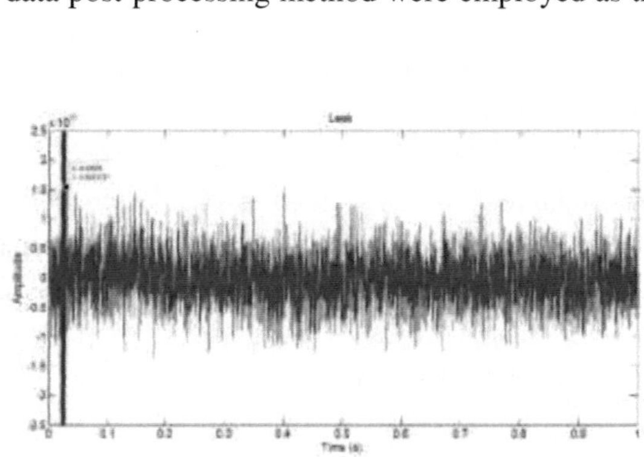

Figure 5. Signal Response for Leak Data.

Figure 4 represents the response signal for no leak data while Figure 5 represents the signal response for the leak data. The first maximum amplitude on the signal response presented in both figure 4 and 5 represents the opening and closing of the solenoid valve that produces pressure transient which is transmitted and reflected around the pipe system. The original response was recorded at three different levels of pressure: 1 bar and 2 bar. The signals captured using a different level of pressure demonstrated different results. Higher pressure creates higher frequency of the signal and this causes the higher amplitude of the signal.

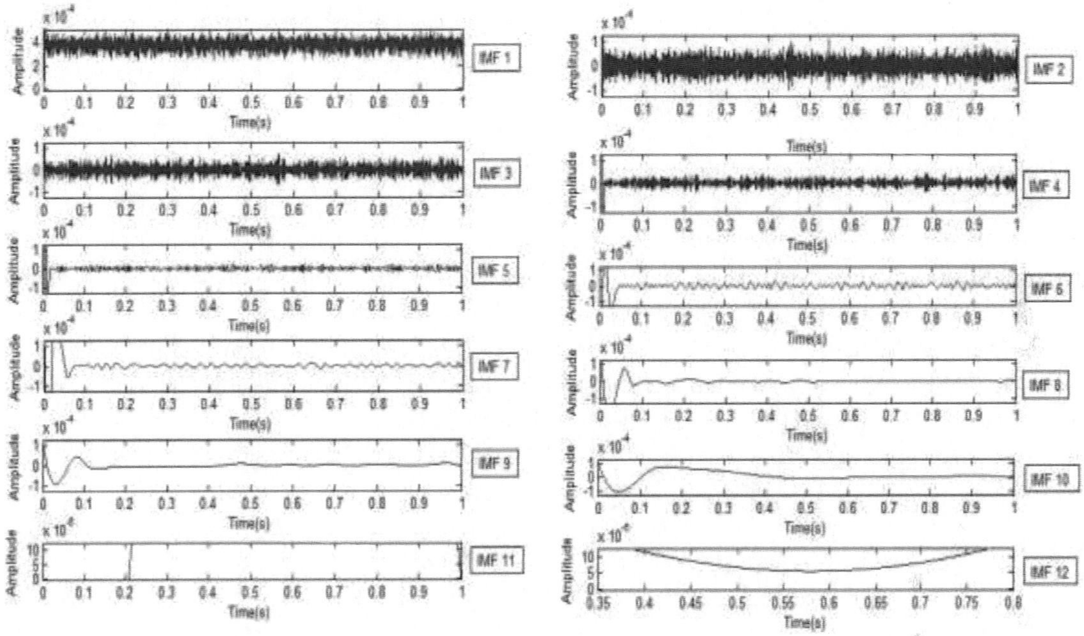

Figure 6. IMF level 1-12 for no leak data.

EMD decomposes the signal response acquired from the test rig into level 13 of intrinsic mode function (IMF). Figure 6 shows amplitude versus time for the first level 12 of IMF. The first level of IMF is a group of higher frequency signal which is noise signal. The last level was a reserve for the lower frequency signal. Commonly, the first and second levels of IMF are not considered for further analysis because these levels contained frequency signal noise. Meanwhile, IMF level 7 and the rests contain basic response of the network. All these IMF were, therefore, discarded. The rest which is IMF level 3-6 have been recombined to produce a signal without noise [16]. Therefore, to identify which level of IMF that is suitable for the final step of Hilbert-Huang Transform (HHT) which is Hilbert Transform (HT), Integrated Kurtosis Algorithm for Z-filter technique to kurtosis ratio (I-Kaz-Kurtosis) was employed. Since EMD decomposes the signal response into level 13 of IMF, the I-Kaz-Kurtosis ratio was utilized to compute the coefficient for each level. Therefore each level of IMF contains differents value of I-Kaz-Kurtosis ratio coefficient.

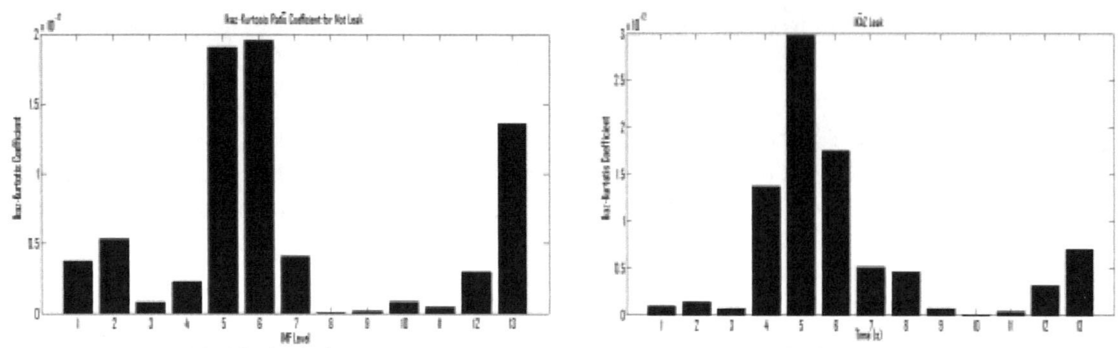

(a) No leak data (b) Leak data

Figure 7. I-Kaz-Kurtosis ratio coefficient

Figure 7 (a) and 7 (b) represent the I-Kaz-Kurtosis ratio coefficient histogram.The X-axis shows the IMF level meanwhile, Y-axis shows the value of I-Kaz-Kurtosis ratio coefficient. From the observation, the maximum and highest value of I-Kaz-Kurtosis coefficient were suitable IMF level to be considered for the post-processing and the final step of Hilbert-Huang Transform (HHT) analysis which is Hilbert Transform (HT). This is because the IMF level which contains the highest value of I-Kaz-Kurtosis ratio consists of the IMF selection criterion which is clear and narrow spikes. The lower the value of I-Kaz-Kurtosis ratio, the more obscure of the IMF selection criterion. The final results of Hilbert-Huang Transform (HHT) are presented in Figure 8 and 9.

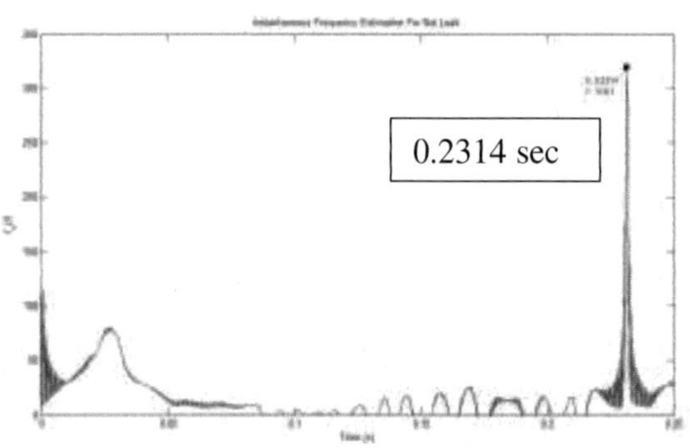

Figure 8. Instantaneous frequency estimation for no leak data

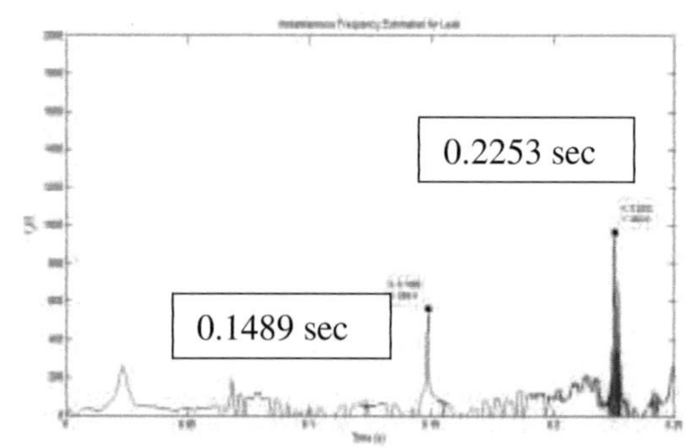

Figure 9. Instantaneous frequency estimation for leak data

Figure 8 and 9 illustrates the final results of Hilbert-Huang Transform (HHT) which is Hilbert Transform (HT) for IMF level that contains the highest value of I-Kaz-Kurtosis ratio coefficient (I-Kaz-Kurtosis). The graph presented in figure 8 and 9 is the instantaneous frequency estimation for no leak and leak data. During this phase, nonstationary signal, in which frequency value changes at any moments, is more useful to characterize a signal in terms of its instantaneous frequency. Instantaneous frequency describes the frequency that locally fits the signal [17]. From the observation, clear and obvious transient can be seen from the final part of HHT analysis. The clear spikes and transient are more clearly seen after Hilbert Transform (HT) take charge in this phase. Figure 8 shows clear spikes at the position of 0.234 seconds compared to the theoretical calculation of the spikes generated by the outlet of the pipeline system. Meanwhile, in Figure 9, obvious/salient spikes were present at the position of 0.1489 seconds and 0.2253 seconds compared to theoretical calculation of the spikes generated by leak and pipe outlet. Therefore, it proves the I-Kaz-kurtosis ratio work properly as self-decision method for IMF.

Table 1. Comparison of position spikes between measured position and experimental position.

Water Pressure	Signal Response	Pipe Feature	Measured Position (m)	Measured Position (sec)	IMF contain Maximum I-Kaz-Kurtosis ratio Coefficient	Experimental Position (m)	Experimental Position (sec)	Error %
1 bar	Leak Data	Leak	37.8	0.144	6	39.03	0.149	3.0
		Outlet	57.9	0.221		59.06	0.225	2.0
	No Leak Data	Outlet	57.9	0.221	5	61.34	0.234	5.9
2 bar	Leak Data	Leak	37.8	0.144	4	38.03	0.138	3.2
		Outlet	57.9	0.221		58.06	0.204	2.2
	No Leak Data	Outlet	57.9	0.221	5	61.01	0.231	6.1

Table 1 illustrates the comparison of position spikes between theoretical calculation and experimental position. For leak data, the position of leak shows 3% error and position of outlet shows 2% error. Meanwhile, no leak data present 5.9% error for outlet position. The final part of HHT analysis clearly shows the position of pipe feature that is leak and pipe outlet appears at the same position compared to each the original position and theoretical calculation. It also proves that HHT analysis was able to detect and positioning the transient event occurred in the non-stationary pressure transient signal.

6. Conclusions

This paper discusses self-decision method for IMF selection through Hibert Huang Transform (HHT) analysis by using high frequency piezoelectric pressure sensor. The result proves that the I-Kaz-kurtosis ratio is suitable and advisable as a self-decision method for IMF selection to implement in Hilbert-Huang Transform (HHT) analysis. The development of automated self-decision of IMF through HHT has been built and statistically analyzed using I-Kaz-kurtosis ratio. Therefore, this method was proposed and advised to be implemented. By efficiently utilizing I-Kaz-kurtosis ratio, and the real-time monitoring system for live water distribution system, it can be used to manage the issue of IMF selection; therefore online monitoring system of leak detection can be developed. The results also prove the ability of piezoelectric pressure sensor in detecting fast response frequency change in the pipeline system.

Acknowledgements

University Malaysia Pahang fully supports the facilities and resources for this research. The authors would like to acknowledge the support from the University Malaysia Pahang internal grant RDU170386 and Ministry of Higher Education(MOHE), Malaysia grant RDU170121.

References

[1] Lai C 1991 *Water Supply* **9** 4
[2] Colombo A F, Lee P and Karney B W 2009 **2** 212-27
[3] Huang N E, Shen Z, Long S R, Wu M C, Shih H H, Zheng Q, Yen N-C, Tung C C and Liu H H 1998 *Proceedings of the Royal Society of London A: Mathematical, Physical and*

IOP Conf. Series: Materials Science and Engineering **257** (2017) 012091 doi:10.1088/1757-899X/257/1/012091

Engineering Sciences: The Royal Society) pp 903-95
[4] Shi C-x and Luo Q-f 2003 *Acta Seismologica Sinica* **16** 422-9
[5] Bin G, Gao J, Li X and Dhillon B 2012 *Mechanical Systems and Signal Processing* **27** 696-711
[6] Manjula M and Sarma A 2012 *Energy Procedia* **14** 1156-62
[7] Ricci R and Pennacchi P 2011 *Mechanical Systems and Signal Processing* **25** 821-38
[8] Maji U, Mitra M and Pal S 2013 *Procedia Technology* **10** 45-52
[9] Kedadouche M, Thomas M and Tahan A 2014 *Advances in Acoustics and Vibration* **2014**
[10] de Souza D B, Chanussot J and Favre A-C 2014 *IEEE International Conference on Acoustics, Speech and Signal Processing (ICASSP)*: IEEE) pp 325-9
[11] Nuawi M Z, Nor M J M, Jamaludin N, Abdullah S, Lamin F and Nizwan C 2008 *Journal of applied sciences* **8** 1541-7
[12] Rizal M, Ghani J A, Nuawi M Z and Haron C H C 2013 *Jurnal Teknologi* **66**
[13] Daubechies I 1992 *Ten lectures on wavelets* vol 61: SIAM)
[14] Wang Y, Xiang J, Markert R and Liang M 2016 *Mechanical Systems and Signal Processing* **66** 679-98
[15] Amin M, Ghazali M, PiRemli M, Hamat A and Adnan N 2015 *IOP Conference Series: Materials Science and Engineering*: IOP Publishing pp 012007
[16] Ghazali M, Beck S, Shucksmith J, Boxall J and Staszewski W 2012 C *Mechanical Systems and Signal Processing* **29** 187-200
[17] Ghazali, M.F., Staszewski, W.W.J., Shucksmith, J.D., Boxall, J.B., Beck, S.B.M. 2011 *Structural Health Monitoring* **10 (4)**, pp. 351-360